D0894428

Japanese/English English/Japanese Glossary of Scientific and Technical Terms

和英・英和

科学工学用語辞典

Japanese/English English/Japanese Glossary of Scientific and Technical Terms

和英・英和
科学工学用語辞典

Louise Watanabe Tung
渡 辺 久 子

A Wiley-Interscience Publication
JOHN WILEY & SONS, INC.
New York / Chichester / Brisbane / Toronto / Singapore

This text is printed on acid-free paper.

Published by John Wiley & Sons, Inc.

Library of Congress Cataloging in Publication Data:

Tung, Louise Watanabe.
Japanese/English English/Japanese glossary of scientific and technical terms
Louise Watanabe Tung.
p. cm.
ISBN 0-471-57463-5
1. Science—Dictionaries 2. Engineering—Dictionaries—
Japanese. 3. Science—Dictionaries. 4. Engineering—Dictionaries.
5. Japanese language—Dictionaries—English. 6. English language—
Dictionaries—Japanese. I. Title.
Q123.T844 1993
503—dc20 92-28867

Printed in the United States of America

10 9 8 7 6 5 4 3 2

Contents

Preface

The years since 1946, and especially the 1960s and 1980s, have seen phenomenal growth in science and technology in Japan. Incredibly, some 500,000 patent applications are filed annually there now. Nearly twenty percent of all patents granted in the United States has been issued to Japanese companies for several years. As a translator of patent and technical papers from Japanese into English, I have enjoyed the enlightening experience of watching the evolution of new ideas in many competitive fields such as polymer science, textiles, ceramics, optics, electronics, and biotechnology.

As I translated these papers regularly, I felt the need for good Japanese and English bilingual dictionaries in the fields of science and technology. Although excellent dictionaries have been and are being published one after another in Japan, they are not easily obtained outside the country; and unfortunately, most accommodate the user who already knows Japanese.

The language is not readily mastered, and pronunciation of long words is tricky. New nouns and verbs are coined all the time, but many such words are not found even in the newest dictionaries. The translator must rely on his or her trained sense of meaning and on the language aids which are available nearby to perform his or her work. Few Japanese and English bilingual science dictionaries are found in Western bookstores.

More and more foreign words are entering our vocabulary these days as television brings distant scenes into the home and boardroom. New achievements of technology all over the world are introduced on the screen from their sources of origin. Competitiveness is fostered at the workplace. Schools are upgrading their curricula.

The main languages of science now are English, Japanese, and German. Many American colleges are recommending Japanese or German for doctoral candidates of science as their foreign language. That there is a need for quick-access general science glossaries among these leading languages should be obvious to anyone who does research internationally with other members of the scientific community.

The pressure to understand each other's terminology can be very urgent, such as when international infringement of patents is being questioned. Banks of translators must work around the clock for weeks, even months, on both sides to prepare for trials. Definitions must be found. Correct character combinations must be used for the terms. Nearly illegible handwriting must be deciphered, and nuances of meaning must be accounted for. Dictionaries play an indispensable role in the heated atmosphere of litigation. The more information aids there are, the better we can understand, and ultimately cooperate with, each other.

I have compiled this bilingual glossary for the benefit of my fellow translators and interpreters, scientists and engineers, patent process experts, media researchers, teachers and students of language, and others who look up English as well as Japanese terms. It is based on my experience with Japanese texts and on my own early compilations of difficult-to-find terms which were begun many years ago.

Japanese patents sometimes contain extraordinary expressions. Verb phrases of mechanical engineering such as "hang axially," "attach pivotally," and "slide contacting against," which are compound Japanese terms ("juku'go"), never enter the mainstream of language and are not found in dictionaries.

They are jargon of limited use, but are encountered rather often in certain types of text. There has been such an explosive growth of scientific knowledge that these days one barely knows which way to turn for definitions of words. I have built on my rudimentary lists so that the glossary would be more extensive and useful, but have not intended it to be exhaustive.

The entries I have chosen are terms expected to be looked up by students of science and technology in their dealings with Japanese matters and by the Japanese researcher in English matters. This volume should be useful as a reliable quick-reference tool in many fields. It should help in pronouncing Japanese terms since I have segmented longer strings. It will dispel questions regarding the meaning of homonyms since the *kanji* (Chinese character) and *kana* (Japanese alphabet) forms are incorporated into the entries.

This feat, the inclusion of the *kanji/kana* forms into the line of information, was accomplished by the use of the computer. With the aid of my husband, I converted a Chinese word processor into a Japanese one for this work. It was a time-consuming and rather tedious task since at least one Japanese pronunciation had to be assigned, to be consistently accessible, to each of thousands of Chinese characters. The results were rewarding indeed. Next, I set about constructing uniquely Japanese *kanji* (termed "koku'ji") and Japanese contemporary *kanji*, which are not used in Chinese. The great range of *kanji* which is found in this compilation was provided by the Chinese word processor. By superimposing English text upon a page of *kanji/kana* text to obtain correct matching, camera-ready copies were prepared. This technique enabled me at last to dream of a dictionary such as this.

Another dividend arising from the use of a Chinese word processor is the fact that I have been able to give not only the contemporary Japanese forms of *kanji* but the traditional forms as well. The traditional *kanji* were used in Japan until about 1946, and they appear in literature published before then, but many are still used elsewhere in Asia. Correlation between the new and old forms is shown in this glossary. The contemporary Japanese form always precedes the traditional form when both are given.

The subject fields which are covered number 120. In round numbers there are 3,000 chemical terms, 1,600 mathematical terms, 3,700 physics terms, 900 computer and logic terms, 3,200 engineering terms, and 3,700 biology terms. Entries beginning with the words "law of" span two-and-a-half pages. Other fields which may be of interest to special groups of users are patent processes, textiles and clothing, meteorology, and food.

As a departure from the usual, I have provided some information on constituent elements—called *radicals*—of Chinese characters (pages 635–636). A few non-standard radicals were wrought using the computer. The reader might be amused to find that there is a radical which denotes a "short-tailed bird," and one for a "long-tailed bird"; this besides several standing simply for "bird." The radicals for moon and flesh are the same, for some obscure reason. The list gives samples and is not exhaustive.

The spelling of the English terms has been checked in current general and specialized science and technology dictionaries and in Webster's dictionaries. The *kanji* and *kana* forms too have been checked in the most authoritative current Japanese dictionaries, including Izuru Shinmura's *Kōjien*, 4th edition, 1991; *Daijirin*, complied by Akira Matsumura, 1989; and *Gakken Kanwa Daijiten*, compiled by A. Tōdō, 1990.

Although it has been a laborious task, taking five years in concerted preparation, I have enjoyed the challenges encountered while compiling this volume. I trust that the user of this work will find it easy to look up words and understand the makeup of terms. I hope that it will save time and will encourage translation of literature in more extensive subject areas than are available now, toward broadening of mutual horizons of knowledge in science and technology between Japan and the English-speaking lands of the world.

ACKNOWLEDGMENTS

Many friends have shared their expertise with me in the compiling of this volume. Their invaluable knowledge and insight were essential factors leading to its completion. Mr. Tim Li discussed his experiences with Chinese word processing with my husband and me. Mr. Richard Forbes sent volumes and pages of literature related to the work. Mr. James Gambrell gave me valuable advice and suggestions over several times on how to go about realizing my hope for publication. Mr. William Marutani of Philadelphia, my beloved brother Warren's friend, guided me to Mr. Gambrell. Dr. Arata and Mrs. Fumiyo Sugimura airmailed a copy of the latest edition of *Kōjien* from Tokyo to help me define the words. Without their care and assistance, this quality of work would not have been achieved. I thank them all, deeply.

My husband, Lloyd Tung, has been my mentor throughout the work. He trained me to the computer. It was he who designed and printed the pages, programmed macros, shot all the trouble I kept generating, and stayed up late to assuage my concerns and analyze the latest antics of the laser printer. He studied the Chinese word processor and set the stage for me for its conversion into a Japanese one. Besides my lasting gratitude, I owe him this very book.

Our sons have energized me and have been a source of determination to see this dictionary through to finish. Glen's ideas and suggestions, and the many programs he sent, have added depth and quality to the tasks. Roger was my inspiration to begin compiling the dictionary, though he may not know it. Here, my sons, is the fruit of our endeavors; yes, yours, too.

To Ruth and Philip Chang, thank you for your unflagging support through the many years of work spent on this volume.

HISAKO WATANABE
(*Louise Watanabe Tung*)

An Explanation of the Style of the Glossary

ROMANIZATION

The Hepburn (James Curtis Hepburn) romanization system is used for this glossary. The Hepburn system is known in Japan as the "Hebon" system. It is felt that this system is the most natural for English speakers.

The following chart gives the Hepburn spelling of Japanese *kana* combinations, contrasted with the other, kunrei, romanization where the two differ.

Hepburn	Kunrei	Hepburn	Kunrei	Hepburn	Kunrei
shi	si	shu	syu	ja	zya
chi	ti	sho	syo	ju	zyu
tsu	tu	cha	tya	jo	zyo
ji	zi	chu	tyu		
sha	sya	cho	tyo		

Thus, "tea," the Japanese national beverage, is represented as "cha" by the Hepburn system and "tya" by the kunrei system. "Tea ceremony room" is "cha-shitsu" by the former, contrasted with "tya-situ" by the latter representation.

ALPHABETIZATION

Alphabetization is letter-by-letter, with word spacing, hyphens, and apostrophes disregarded in the sequencing.

SELECTION OF THE TERMS

The fields covered are diverse, with the interpretation that technology touches not only on engineering, architecture, mining, and so forth, but also on clothing, furniture, and food. Many names of birds, fish, animals, flowers, trees, minerals and gems, and so forth, are included in light of today's interest in ecology. The Japanese-to-English section also contains many local-color words related to food, clothing, and transportation, and this should be helpful to persons who visit Japan or who have not lived there for long.

It is a feature of this glossary that few of the terms which are usually phoneticized in Japanese are included. Stress has been laid on defining terms which are in the Japanese vernacular (as distinct from imported words) and in giving the *kanji* (Chinese character, singular and plural form are the same spelling) and *kana* (Japanese letter, singular and plural form are the same) forms correctly. Much work has been done to verify the entries in scholarly reference sources as to the spelling, pronunciation, plural (sometimes singular) form, preferred characters, attendant *kana* (called "okuri-gana"), and so on.

HOW THE ENTRIES ARE LAID OUT

An effort has been made to contain all information pertaining to a given entry in a single line in the belief that this format aids rapid access to information.

From Japanese into English

The Japanese term in romanized phonetics is given first; this is followed by the *kanji/kana* forms in the central column; and lastly the English definitions are given in the rightmost column. The English definition is followed by the part of speech, whether noun, adjective, adverb, etc., when pertinent, and the field with which the term is associated is given last in brackets.

From English into Japanese

The term in English is given in the leftmost column, with the part of speech and field category following. The *kanji/kana* forms are given in the center column. The romanized Japanese phonetic rendition is given in the rightmost column.

ORDER OF THE ENTRIES

When two or more entries are identical, the words are given according to the alphabetical order of the first of the fields in which they are used. For example, "nephelometer [meteor(ology)]: unryō-kei," appears before, "nephelometer [opt(ics)]: hidaku-kei." Thus, the meteorological nephelometer precedes the optical nephelometer (page 894).

When the terms and the fields are the same for two entries, the words are listed according to the alphabetical order of the romanized Japanese. For example, "neve [geogr(aphy)]: funjō-hyō'setsu" precedes "neve [geogr]: mannen'setsu-gen" (page 896).

When the same term is used in several fields, the fields are given in alphabetical order. Thus, we have, "degree [astron(omy)] [geogr] [geom(etry)] [thermo(dynamics)]: do," (page 682) and, "node [astron] [electr(onics)] [eng(ineering)]: setten" (page 898). Fields associated with specific terms take precedence over terms with no association. Thus, "identification [bio(logy)] [ch(emistry)]: dōtei," which is associated with biology and chemistry, appears before "identification: dō'itsu-sei," and "identification: kaku'nin," which are general terms with no special association (page 802).

DITTOING AS A FEATURE OF THE LAYOUT

English into Japanese

The ditto mark to indicate repetition of the preceding term is a visual mechanism to alert the user that there are many definitions for a given same word. One can see quickly that there are, for example, five definitions listed for the word "inclination" and that one definition is as an engineering term, two are as geology and math terms, three are as math terms, two are as sci-tech terms, and one is as a psychology term (page 806). It can be seen on page 740 that the term, "field," is listed eleven times as "hatake, ta, tanbo, tai, ran," etc., and a glance shows that in this glossary the term used in the areas of agriculture, algebra, computers, ecology, electricity, geography, industrial engineering, patents, and physics. One can select the appropriate character(s) and find their pronunciation.

Japanese into English

In this section too, ditto marks serve a special function. Homonyms abound in the language. It is important to find the correct *kanji/kana* for homonyms since differences in meaning among homonyms can be great. Dittos in the Japanese section are used only for identical words using the same characters. When the characters are not identical, a ditto is not used. An example is "hen'i," with many definitions. The reader is referred to page 128.

It can be seen there that the first "hen'i" is written with a different set of characters than the succeeding three, which are the same in form. Then, four more terms, "hen'i," follow and each is written differently. Here, too, the dittos are a visual aid to detecting sameness, but in this section they also serve to alert one that other terms nearby are pronounced the same, but to watch out as they are written differently and might have quite different meanings.

PRONUNCIATION AND ACCENTING

To aid pronunciation of romanized Japanese words, hyphens and apostrophes have been used. Basically, the hyphen divides major components of an entry, while an apostrophe divides minor components.

Many Japanese terms are translations of English (or other non-Japanese) terms, and because of this, there is often a one-to-one correspondence between segments of terms in the two languages. It is helpful for the translator to be shown this correspondence, especially when the Japanese term is a long string of characters.

With this borne in mind, other factors too have been considered. Many first and last *kanji* of a term are in common to other terms and have weight which deserves notice. More often, these *kanji* arrive at the end of a term. Examples of such ending *kanji* are those which denote machinery and instruments (-ki), constellations of stars (-za), acid (-san), medicine (-yaku and -zai), method (-hō), system (-kei), and illustration (-zu). These *kanji* are suffixes in Japanese, but are not suffixes in English; that is, there is no corresponding English suffix which means machine or constellation. Examples of significant beginning *kanji* are specific (hi-), high (kō-), opposite (gyaku-), opto (kō-) and half (han-). These are prefixes in Japanese, but are not necessarily so in English. These *kanji* benefit by being set apart in some fashion. It helps in finding the correspondence with the English word when a hyphen or apostrophe is used to break down the Japanese.

Long strings of letters in a foreign language are difficult to pronounce. The logical way to manage a long string is to segment it between the *kanji* which are its building blocks. This segmenting must be done with an awareness of the natural clustering of the constituent *kanji*. Since consciously and unconsciously, Japanese is spoken with emphasis on characters, it turns out that when division points between *kanji* clusters are indicated, it aids correct pronunciation. The hyphen and apostrophe locations are where slight hesitation in speech occurs.

The hyphen shows major divisions. Thus, "emery cloth" in English is "nuno-yasuri" in Japanese. Although the order is reversed, it is clear that "nuno (cloth)" is a major component, and that "yasuri (file)" too is a major component. "Echo check" is "hankyō-kensa," and there is a one-to-one correspondence, not reversed in this case. Heavy-weight ending *kanji* too are set apart, with either the hyphen or the apostrophe. Thus, "accounting machine" is "kaikei-ki." "Specification" is "meisai'sho." An apostrophe is used, instead of a hyphen, in the latter case since the English term is a single word.

Certain suffix *kanji* have always been assigned an apostrophe just preceding them. Examples are "'sho," meaning "document," "'sha" meaning "person," and "'butsu," meaning "thing." The Japanese "'en" is a suffix which corresponds to the English "-itis," meaning inflammation. Even though at times the English counterpart actually states document, person, or thing as an independent word, the majority rule was applied in these cases since they are very integral to the term. On the other hand, "-tai," meaning "body, object, or entity," has been assigned an apostrophe or hyphen depending on the English term or on other considerations.

Some *kanji* are pronounced according to Chinese phonetics (as the Japanese use them) as well as Japanese. The Chinese sounds tend to be briefer, while the Japanese sounds are often drawn out. This is because Chinese is a monosyllabic language, while Japanese is a multisyllabic one. For example, the *kanji* for "wine" is "shu" in Chinese phonetics, with one syllable, while it is "sake" in Japanese, with two syllables. In compounded terms denoting specific wines, the Chinese reading is usually preceded by an apostrophe, as "'shu," while the Japanese is preceded by a hyphen, as "-sake" or "-zake." An example is "ume-zake" and "ume'shu," translations of the term, "apricot brandy." This distinction is unique to this compilation.

At the discretion of the reader who is acquainted with Japanese, these devices can be omitted, altered, or even reversed. There is no hard and fast rule. One Japanese word meaning "closet" is "mono'oki." The rendition "mono-oki" can also be used since both "mono" and "oki" are Japanese phonetics, not Chinese; the former form has been favored here since the English is a single word. Either the apostrophe or the hyphen must be used for this word since otherwise it will become "monooki," and will be confusing to pronounce or even interpreted to read "monōki." The *kanji* counterpart cannot be imagined from "monōki." Too many apostrophes in a segment is not advised; three or four is usually the limit.

It must be cautioned that certain words must contain the apostrophe or the meaning will change. "Tani" means "valley," while "tan'i" means "unit." "Kani" is a crab, but "kan'i" means "spreadsheet." "Sen'i," meaning "fiber," must contain the apostrophe, for otherwise the word will be mispronounced. The same applies to "bo'on," which means "vowel." "Boon," without the apostrophe, can be pronounced wrongly, as "buun." The location of the apostrophe can also be sensitive. "Ka'nen" means "twisting" in one form and "flammable" in another. "Kan'en" means "hepatitis."

A term may be hyphenated onto a new line of text at the hyphen. It is less desirable to divide a term onto a new line at an apostrophe; at the discretion of the writer, a long term may be so divided. Further explanation of the hyphen will be given in the section titled "Word Division."

PUNCTUATION

Periods

Periods are rarely used in the glossary. Periods are used in "q.v." to denote "which see" and for abbreviations, as in "in." to denote "inch," "sq. ft." to denote "square foot," "ex." to denote "example," and, "Mrs."

Commas

Commas are used to set apart definitions of a similar or related nature, to mean "or," as in "cabin (of aircraft, seacraft)" and "axle (shaft, axis)." Commas are used to set apart descriptions of botanical types, e.g., "almond (tree, flower)." Commas are also used to distinguish between plural forms, e.g., "cacti, cactus, -es."

Colons

Colons are used to amplify meaning of a subordinate nature, e.g., "minimal: smallest" and "minimal: least." A colon is also used after the word "example," as in "bird (example: hawk)."

Semicolons

Semicolons are used to separate definitions of a mutually distinct nature, e.g., "damp; humid; moist" and "energy; spirit; essence." They are also used to separate scientific names and generic names, e.g., "Ctenophora; comb jellies."

In the *kanji* and *kana* column, semicolons are used between different forms of representation, as in the current *kanji*, the traditional *kanji*, and *kana*.

Parentheses

Parentheses are used to elaborate explanations of terms, e.g., "autopilot (of a ship)" and "balance (a pair of scales)." They are used to categorize a term, e.g., "melon (fruit)." They are used to set apart a part of a definition which may not necessarily have to be used, e.g., "optical (photo)micrograph," "storm(y) petrel," and "cross(wise) direction." They are used to give the plural, or sometimes singular, form of a term, e.g., "phylum (pl phyla)"; or to set apart "-s" or "-ing" at the end of a word.

Brackets

Brackets are used to enclose subject categories, e.g., [*meteor*], which denotes meteorology, and [*ocean*], which denotes [*oceanography*].

Braces

Braces are used to enclose parts of speech, such as {*vb*} for verb, {*adj*} for adjective, and {*conj*} for conjunction.

ITALICS

Subject fields and parts of speech are italicized.

BOLDFACED WORDS

Only those words at the beginning of each entry which are alphabetized in the glossary are boldfaced. Thus, words which are not boldfaced serve to further explain the boldfaced part but do not figure in the alphabetization.

SPELLING

American spelling is used. "Gage" is preferred over "gauge."

CAPITALIZATION

Proper nouns in English are always capitalized, but proper nouns in Japanese are sometimes not capitalized. Names of Japanese cities, Japanese eras, and Japanese geographical locations are capitalized, but names of constellations are not.

COMPOUND WORDS

Words are compounded differently in different source literature. Several references were checked to arriving at the most favored compounding, among them McGraw Hill's *Dictionary of Scientific and Technical Terms*, 4th edition, 1989, Webster's *Ninth Collegiate Dictionary*, 1985, Kenkyusha's *New English-Japanese Dictionary*, 5th edition, 1980, and Kenkyusha's *New Japanese-English Dictionary*, 4th edition, 1974.

WORD DIVISION

Hyphenation

Hyphenation within Japanese words is used maximally at the first or an early introduction in both the Japanese-into-English glossary and English-into-Japanese glossary in order to aid pronunciation. When the same word appears repeatedly in a cluster, fewer hyphens (and apostrophes) may be used. By the same token, when a term seems to be long and complicated to pronounce, sometimes an easier way of pronouncing it will be found by tracing back to more basic forms of the word.

Hyphens are also used for prefix terms, as in "counter-: hantai no," and for suffix terms, as in "-shi: magazine." Hyphenation in romanized Japanese is an arbitrary practice; an attempt has been made to be as logical, reasonable, and helpful as possible.

When laying out the entry lines in the body of the glossary, words were hyphenated into the next line as little as possible. A special attempt has been made to keep information within a single line of text.

KANJI AND KANA

Kanji for daily use, known as "tōyō-kanji," is given first, and this is sometimes followed by the older form, which is found in traditional literature. When it was felt helpful to give the *kana* form, the *kana* rendition has been given last. Since the romanized form is the key to pronunciation, *kana* forms have been given only as an extra aid. *Kata-kana* (the script form) is usually used for terms derived from foreign words, e.g., "aluminum," "diode," and "phenolphthalein," often for the names of denizens of the natural world, and for terms originating in sounds, e.g., "pika-pika," meaning "glitter." *Hira-gana* (the cursive form) is used in other cases of *kana* use.

HOMONYMS IN KANJI

Homonyms in *kanji* are listed in ascending order of complexity of the strokes used in writing the *kanji*. Among the several Kan'wa dictionaries referred to establish the *Kanji* strokes are Ueda's *Daijiten*, 1963, and Tōdō's *Gakken Kanwa Daijiten*, 1990.

OKURI-GANA

Okura-gana, namely, the *kana* added to *kanji* for declension, is important when writing Japanese. Preferred forms of *okuri-gana* have been verified by consulting Shinmura's *Kōjien*, 4th ed., 1991, and Matsumura's *Daijirin*, 1988.

PRONUNCIATION

A macron over a vowel denotes a long vowel. Thus, ā is pronounced "aa" as in "baa"; ē is pronounced "eh" as in "Behring"; "ī" is pronounced "ie" as in "shield"; "ō" is pronounced "o" as in "robot"; and "ū" is pronounced "u" as in "cruise." "Ky" followed by a long u is pronounced "kew." It is accepted practice to use the caret, ˆ, instead of the macron. The macron is the traditional form for Japanese.

NAMES OF CHEMICALS AND MINERALS

The names of chemical substances are usually not qualified, whether organic or inorganic. In the main, names of chemicals which are expressed in Japanese text using *kanji* have been included, and terms imported into Japanese and phoneticized in *kana* have not been included. Elements are a notable exception to this rule, as are chemicals which are of special importance.

Names of minerals may be found to be accented somewhat differently from the other terms. A unity was found for this group of terms by the system used in this work.

GRAMMATICAL FORMS: PARTS OF SPEECH

Nouns are by far the most numerous form of word in this glossary. Nouns are looked up the most often in work dealing with science and technology. Therefore, the noun form is given first when a term can be given in several grammatical forms. The identification {*n*}, denoting "noun," is given only when it is ambiguous as to whether a term is a noun. Words ending in "-ing" are usually qualified as to whether they are nouns. Adjectives {*adj*}, adverbs {*adv*}, and other forms follow nouns in alphabetical order of the part of speech. Verbs may be identified as intransitive {*vb i*} or transitive {*vb t*} when this distinction is necessary. Otherwise a verb is merely identified as {*vb*}.

PLURAL FORMS

Plural forms of English words are given in parentheses only when they do not follow the standard rule of addition of the suffix "-s" or "-es."

Abbreviations of the Subject Fields

acous	acoustics	音響学
adhes	adhesives	粘着物工学
aero-eng	aerospace engineering	航空宇宙工学
aerosp	aerospace; aircraft	航空宇宙；航空機
agr	agriculture	農業
alg	algebra	代数学
anat	anatomy	解剖学
an-ch	analytical chemistry	分析化学
an-geom	analytic geometry	解析幾何学
arch	architecture; buildings	建築学；建物
archeo	archeology	考古学
arith	arithmetic	算術
astron	astronomy	天文学
astrophys	astrophysics	天体物理学
atom-phys	atomic physics	原子物理学
bio	biology	生物学
bioch	biochemistry	生化学
bot	botany	植物学
calc	calculus	微分積分学
cer	ceramics	窯業
ch-eng	chemical engineering	化学工学
ch	chemistry	化学
civ-eng	civil engineering	土木工学
cl	clothing; headgear; footwear	衣類；被り物；履き物
climat	climatology	気候学
comm	communication	通信情報伝達
comput	computer science	計算機科学
constel	constellation	星座
cont-sys	control systems	制御系学
cook	cookery; food; drinks	料理；食物；飲み物
crys	crystallography	結晶学
cyt	cytology	細胞学
dent	dentistry	歯科医学
ecol	ecology	生態学
econ	economics; finance; business	経済学；財務；営業
elec	electricity	電気学
elecmg	electromagnetism	電磁気学
electr	electronics	電子工学

embryo	embryology	発生学
eng	engineering; design engineering	工学；設計工学
eng-acous	engineering acoustics	工学音響学
fl-mech	fluid mechanics	流体力学
food-eng	food engineering	食品工学
for	forestry	林学
furn	furniture; furnishings	家具；調度
gen	genetics	遺伝学
geoch	geochemistry	地球化学
geod	geodesy	測地学
geogr	geography	地理学
geol	geology	地質学；地学
geom	geometry	幾何学
geophys	geophysics	地球物理学
gram	grammar	文法
graph	graphics	図学
hist	histology	組織学
horol	horology	測時学
hyd	hydrology	陸水学
immun	immunology	免疫学
ind-eng	industrial engineering	生産管理工学
inorg-ch	inorganic chemistry	無機化学
i-zoo	invertebrate zoology	無脊椎動物学
lap	lapidary	宝石細工術
ling	liguistics	言語学
map	mapping	地図作成
mat	materials science	材料工学
math	mathematics	数学
mech	mechanics	力学
mech-eng	mechanical engineering	機械工学
med	medicine	医学
met	metallurgy	冶金学
meteor	meteorology	気象学
microbio	microbiology	微生物学
mil	military	軍事
min-eng	mining engineering	鉱山工学
miner	mineralogy	鉱物学
music	musicology	音学
mycol	mycology	菌学
nav-arch	naval architecture	造船学
navig	navigation	航法

nuc-phys	nuclear physics	核物理学
nucleo	nucleonics	原子核工学
ocean	oceanography	海洋学
opt	optics	光学
ord	ordnance	軍需品
org-ch	organic chemistry	有機化学
paleon	paleontology	古生物学
paper	paper technology	紙工学
part-phys	particle physics	素粒子物理学
pat	patent process	特許手順
path	pathology	病理学
p-ch	physical chemistry	物理化学
petr	petrology	岩石学
petrol	petroleum technology	石油工学
pharm	pharmacology	薬理学
photo	photography	写真術
phys	physics	物理学
physio	physiology	生理学
plas	plastics	合成樹脂化学
poly-ch	polymer chemistry	高分子化学
pr	printing; publications	印刷術；刊行物
prob	probability	確率
psy	psychology	心理学
quant-mech	quantum mechanics	量子力学
rail	railroad	鉄道
rela	relativity	相対性原理
rub	rubber technology	ゴム工学
sci-t	science/technology; basic terms	基本科学工学用語
sol-st	solid state physics	固体物理学
spect	spectroscopy	分光学
stat	statistics	統計学
stat-mech	statistical mechanics	統計力学
steel	steel technolgy	鋼工学
syst	systematics	分類学
tex	textiles	織物工学
thermo	thermodynamics	熱力学
traffic	traffic engineering	交通工学
trans	transportation; travel	輸送；旅行
trig	trigonometry	三角法
v-zoo	vertebrate zoology	脊椎動物
zoo	zoology	動物学

Other Abbreviations

abbrev	abbreviation; abbreviated	省略形
adj	adjective; adjective phrase	形容詞；形容詞句
adv	adverb; adverb phrase	副詞；副詞句
conj	conjunction	接続詞
n	noun; noun phrase	名詞；名詞句
pl	plural	複数
prep	preposition	前置詞
pron	pronoun	代名詞
q.v.	quod vide: which see	その項を見よ
sing	singular	単数
vb	verb; verb phrase	動詞；動詞句
vb i	intransitive verb	自動詞
vb t	transitive verb	他動詞

PART I

Japanese-into-English Section

A

a-	亜；亞	sub-; near- *{prefix}*
abara(-bone)	肋(骨)	rib (bone) *{n} [anat]*
abi	阿比；あび	loon (bird) *[v-zoo]*
abu	虻；あぶ	gadfly; horsefly (insect) *[i-zoo]*
abumi	鐙	stirrup (riding gear) *[trans]*
abumi-bone; tō-kotsu	鐙骨	stirrup; stapes (ear bone) *[anat]*
abunai	危い	dangerous; hazardous *{adj}*
abura (yu)	油；あぶら；ゆ	oil *{n} [geol]*
abura (shi)	脂；あぶら；し	fat *{n} [bioch]*
"	脂	sevum *[v-zoo]*
aburāge; abura'age	油揚げ；油煬げ	fried bean curd *[cook]*
abura-ana	油穴	oil hole *[eng]*
abura-dame	油溜め	oil sump *[eng]*
abura-e	油絵；油繪	oil painting *[graph]*
abura-enogu	油絵具	oil paints; oil colors *[mat]*
abura-hensei-jushi	油変性樹脂	oil-modified resin *[poly-ch]*
abura-kasu	油粕；油糟	oil cake *[agr]*
abura'ko	油蚕；あぶらこ	translucent silkworm *[i-zoo]*
abura-kuda	油管	oil pipe *[eng]*
abura-mushi	油虫	aphid (insect) *[i-zoo]*
abura'na	油菜	cole; rape (vegetable) *[bot]*
abura-sashi	油差し	oil feeder; oil can *[eng]*
abura-seijō-ki	油清浄器	oil cleaner; oil purifier *[eng]*
abura-shōhi'ryō	油消費量	oil consumption *[econ]*
abura-shōji	油障子	oil-paper sash *[agr]*
abura-to'ishi	油砥石	oilstone *[mat]*
abura-tsuno-zame	油角鮫	spiny dogfish; Squalus acanthias (fish) *[v-zoo]*
abura-yaki'ire	油焼入れ	oil quenching *{n} [met]*
abura-yaki'ire-kō	油焼入鋼	oil-hardened steel *[met]*
aburi'yaki-kigu	焙り焼き器具	roaster *[eng]*
aburu	焙る；あぶる	roast *{vb t} [cook]*
a'chishi	亜致死	sublethal *[gen] [med]*
adorenarin	アドレナリン	adrenaline *[bioch]*
ae'mono	和え物	sauce-dressed dishes *[cook]*
a'en	亜鉛	zinc (element)
aen-guro	亜鉛黒	zinc black *[ch]*
aen-ka	亜鉛華	zinc white; Chinese white *[ch]*
aen-kō; aen-ki	亜鉛黄	zinc yellow *[mat]*

aen'kujaku'seki	亜鉛孔雀石	rosasite *[miner]*
aen-matsu	亜鉛末	zinc dust *[met]*
aen-mekki-kōban	亜鉛めつき鋼板	galvanized sheet iron *[met]*
aen-mekki-kōsen	亜鉛めつき鋼線	galvanized steel wire *[met]*
aen-san	亜鉛酸	zincic acid
aen'san-en	亜鉛酸塩	zincate *[ch]*
a'enso-san	亜塩素酸	chlorous acid
a'enso'san-en	亜塩素酸塩	chlorite *[ch]*
a'enso'san-sōda	亜塩素酸ソーダ	sodium chlorite
aen-ten	亜鉛点	zinc point *[ch]*
afure	溢れ；あふれ	overflow *{n} [civ-eng] [comput]*
afure-hyō'shiki	溢れ標識	overflow indicator *[comput]*
a'gata	亜型	subtype *{n} [microbio]*
age-abura	揚げ油	frying oil *[cook] [mat]*
age-chi	揚げ地	unloading port *[geogr]*
age'ha-chō	揚羽蝶	swallowtail (butterfly) *[i-zoo]*
age'mono	揚げ物	fried dishes *[cook]*
ageru	上げる；揚げる	raise; lift *{vb t} [eng]*
ageru	揚げる	fry *{vb t} [cook]*
ageru	上げる	give *{vb t}*
ago	顎	jaw *[anat]*
ago'saki	顎先	chin *[anat]*
a'gun	亜群	subgroup *[math]*
ahen	阿片	opium *[pharm]*
ahiru	家鴨；あひる	duck (bird, domesticated) *[v-zoo]*
a'hisan	亜砒酸	arsenious acid
a'hisan-en	亜砒酸塩	arsenite *[ch]*
ahō-dori	信天翁；阿房鳥；= あほうどり	albatross (bird) *[v-zoo]*
a'hokkyoku no	亜北極の	subarctic *{adj} [geogr]*
ai	藍	indigo *[org-ch]*
aida	間	between *{adv} [math]*
"	間	space; opening; distance *{n}*
"	間	time; interim; a while *{n}*
ai-dama	藍玉	indigo (leaf) ball *[org-ch]*
aigamo	間鴨	crossbreed of wild and domestic ducks *[v-zoo]*
aigo	藍子	rabbit fish; Siganus fuscescens (fish) *[v-zoo]*
ai-iro	藍色	indigo blue (color) *[opt] [org-ch]*
ai-kagi	合い鍵	duplicate key *[eng]*
ai'kotoba	合言葉	password *[comm] [comput]*

aima	合間	interlude; interval; interstice *{n}*
aimai(-sei)	曖昧(-性)	ambiguity; vagueness *{n}*
ainame	鮎並; あいなめ	rock trout; Hexagrammos otakii (fish) *[v-zoo]*
ainsutainyūmu	アインスタイニウム	einsteinium (element)
airo	隘路; あいろ	bottleneck *[traffic]*
aita	開いた	open *{adj}*
aite(-kata)	相手(方)	opponent; opposite party; adversary
ai-zome	藍染め	indigo dyeing *[tex]*
aji	味	flavor *{n} [cook] [physio]*
aji	鯵; あじ	horse mackerel; saurel (fish) *[v-zoo]*
ajisai	紫陽花; あじさい	hydrangea (flower) *[bot]*
aji'sashi	鯵刺; あじさし	tern; Sterna hirundo (bird) *[v-zoo]*
aji'tsuke	味付け	flavoring; seasoning *{n} [cook]*
ajiwau koto	味わうこと	tasting *{n} [physio]*
a'jūryoku	亜重力	subgravity *[mech]*
aka	赤	red (color) *{n} [opt]*
a'ka	亜科	subfamily *[bio] [syst]*
a'kabu	亜株	substrain *[microbio]*
aka-budō'shu	赤葡萄酒	red wine *[cook]*
aka-ei	赤鱏; あかえい	stingray; Dasyatis akajei (fish) *[v-zoo]*
aka-fūkin-chō	赤風琴鳥	scarlet tanager (bird) *[v-zoo]*
aka-gaeru	赤蛙	ranid; true frog (amphibian) *[v-zoo]*
aka-gai	赤貝	ark shell (shellfish) *[i-zoo]*
aka'ge-zaru	赤毛猿	rhesus monkey; Macaca mulatta (animal) *[v-zoo]*
aka-hata	赤旗	red flag *[comm]*
akai	赤い	red (color) *{adj} [opt]*
a'kai	亜界	subkingdom *[bio] [syst]*
a'kai'chōseki	亜灰長石	bytownite *[miner]*
aka-ie-ka	赤家蚊	house mosquito; Culex pipens pallens (insect) *[i-zoo]*
aka-kabu	赤蕪	beet; red turnip (vegetable) *[bot]*
aka-mi	赤身	red meat *[cook]*
akane	茜; あかね	madder (dye) *[bot] [tex]*
akane-iro	茜色	crimson; madder red (color) *[opt]*
a'kantai	亜寒帯; 亞寒帯	subpolar zone *[geogr]*
a'kantai-rin	亜寒帯林	taiga *{n} [ecol]*
akari	明かり	light; lamp *{n} [opt]*
akarui	明るい	bright; light *{adj} [opt]*
akaruku naru	明るくなる	become light; become bright *{vb i}*
akaruku suru	明るくする	make light(er); brighten *{vb t}*

akaru-sa	明るさ	brightness; lightness *{n} [opt]*
aka-sabi	赤錆	red rust *[met]*
aka-sango	赤珊瑚	red coral *[i-zoo]*
aka-shio	赤潮	red tide *[bio]*
akatsuki	暁; 曉	dawn; daybreak *[astron]*
aka-umi'game	赤海亀	loggerhead (turtle) *[v-zoo]*
akaza	藜; あかざ	wild spinach; goosefoot; pigweed; lamb's quarters (vegetable) *[bot]*
a'kei	亜系	substrain *[microbio]*
akeru	開ける	open; undo; unlock; untie *{vb t}*
aki	秋; あき	autumn; Fall (season) *[astron]*
Akita-inu; Akita'ken	秋田犬	Akita dog (animal) *[v-zoo]*
akkan-setchaku-zai	圧感接着剤	pressure-sensitive adhesive *[mat]*
akka-ritsu	圧下率; 壓下率	draft *{n} [steel]*
akka-ryoku	圧下力	rolling force *[steel]*
akkon	圧痕	impression *[geol]*
"	圧痕	press mark *{n}*
a'kō	亜綱	subclass *[bio] [syst]*
a'kōzan-tai	亜高山帯	subalpine zone *[geogr]*
a'kōzan-tai no	亜高山帯の	alpestrine; subalpine *{adj} [ecol]*
aku	灰汁; あく	lye; ash; harshness (of taste)
akubi	欠伸; あくび	yawn *{n} [physio]*
akubi o suru	欠伸をする	yawn *{vb} [physio]*
aku'chi	悪地; 惡地	badlands *[geogr]*
akuchinyūmu	アクチニウム	actinium (element)
akufu	握斧; あくふ	handaxe *[eng] [paleon]*
aku'nuki	灰汁抜き	removal of harsh taste *[cook]*
aku-ryoku	握力	grip strength *[physio]*
aku'shitsu no	悪質の	bad-quality *{adj}*
aku-tenkō	悪天候	adverse weather *[meteor]*
ama	亜麻; あま	flax *[bot]*
ama-ashi	雨足; あまあし	wisp of rain *{n} [meteor]*
amacha	甘茶	hydrangea, big-leaved *[bot]*
ama'do	雨戸	storm window *[arch]*
ama-gaeru	雨蛙	tree frog; Hyla arborea (amphibian) *[v-zoo]*
ama'gasa-hebi	雨傘蛇	krait (reptile) *[v-zoo]*
ama'gu	雨具	rain gear *[cl]*
amai	甘い	sweet *{adj} [physio]*
amai yori	甘い撚り	soft twist *[tex]*
ama'mi	甘味	sweet flavor; sweetness *[cook]*
amani; amanin	亜麻仁; あまに(ん)	flaxseed; linseed *[bot]*

amani(n)-yu	亜麻仁油	linseed oil *[mat]*
ama-no-gawa	天の川	Milky Way *[astron]*
ama-nuno	亜麻布	linen *[tex]*
amari	余り; 餘り	remainder *[math]*
amarugamu	アマルガム	amalgam *[met]*
ama-sa	甘さ	sweetness *{n} [cook] [physio]*
ama-tsubame	雨燕	white-rumped swift (bird) *[v-zoo]*
ama-zake	甘酒	sweet fermented rice drink *[cook]*
ame	雨	rain *{n} [meteor]*
ame	飴	starch-saccharified jelly *[cook]*
ame-agari	雨上り	after a rain *{adv} [meteor]*
ame'dama	飴玉	toffee (candy) *[cook]*
ame-furashi	雨降らし	sea hare; Aplysia kurodai *[i-zoo]*
ame-osae	雨押え	flashing *{n} [arch]*
Amerika	アメリカ	America *[geogr]*
amerishūmu	アメリシウム	americium (element)
ame'yoke-ita	雨避け板	flashing *{n} [arch]*
ami	網	web *{n} [bio]*
"	網	network *[comput]*
"	網	net *{n} [tex]*
ami	糠蝦; あみ	mysid; opossum shrimp *[i-zoo]*
amihan	網版	halftone *{n} [pr]*
ami-heimen	網平面	net plane *[crys]*
amijō-jūgō'tai	網状重合体	network polymer *[poly-ch]*
amijō-kōbunshi	網状高分子	network polymer *[poly-ch]*
amijō-kōshi	網状格子	mesh grid *[elec]*
amijō-kōzō	網状構造	network structure *[met]*
amijō-kōzō no	網状構造の	reticulate *{adj} [bio] [geol]*
ami'me	網目	network *[poly-ch]*
"	網目	mesh; stitch *{n} [tex]*
ami'me-kōzō	網目構造	network structure *[met] [poly-ch]*
ami'me-sa	網目鎖	network chain *[poly-ch]*
ami-mono	編み物	knitted work *[tex]*
amin	アミン	amine *[org-ch]*
amino-san	アミノ酸	amino acid *[bioch]*
ami-yaki	網焼き	wire-net grilling *{n} [food-eng]*
a'moku	亜目	suborder *{n} [bio] [syst]*
a'mon	亜門	subdivision *{n} [bot] [syst]*
"	亜門	subphylum (pl -phyla) *[zoo] [syst]*
an	餡; あん	bean jam *[cook]*
ana	穴; 孔	hole *{n}*
anago	穴子; 海鰻; あなご	conger; sea eel (fish) *[v-zoo]*

ana-guma	穴熊	badger (animal) *[v-zoo]*
ana'guri'gu	穴刳り具	reamer *[eng]*
ana'hori-game	穴掘り亀	gopher tortoise (reptile) *[v-zoo]*
ana'hori-ki	穴掘(り)機	earth drill *[civ-eng]*
anchimon	アンチモン	antimony (element)
an-denryū	暗電流	dark current *[electr]*
andon	行灯	lantern; lamp stand *[furn]*
an'ei	暗影	black spot *[electr]*
a'nettai	亜熱帯	subtropical zone *[geogr]*
angō	暗号; あんごう	cipher; cryptogram *[comm]*
angō-hō	暗号法; 暗號法	cryptography *[comm]*
angō'ka suru	暗号化する	encrypt; encipher *{vb}* *[comm]*
an-hannō	暗反応; 暗反應	dark reaction *[bioch]* *[ch]*
anji suru	暗示する	imply; suggest *{vb}* *[psy]*
an-junnō	暗順応	dark adaptation *[opt]*
anki	暗記	memorization *[psy]*
anki suru	暗記する	memorize *{vb}* *[psy]*
ankō	鮟鱇; あんこう	anglerfish; frogfish; sea toad; Lophius litulon *[v-zoo]*
ankoku-sei'un	暗黒星雲	dark nebula *[astron]*
anma	按摩; あんま	massage *{n}* *[med]*
anmoku'ri ni	暗默裡に	implicitly; tacitly *{adv}*
anmoku-shiki'betsu	暗默識別	implicit identification *[psy]*
annai-ban	案内板	guide plate *[mech-eng]*
annai-bō	案内棒	guide bar *[mech-eng]*
annai-bōenkyō	案内望遠鏡	guiding telescope *[astron]*
annai-chūshin	案内中心	guiding center *[elecmg]*
annai-guruma	案内車	guide pulley *[eng]*
annai-jiku'uke	案内軸受	guide bearing *[mech-eng]*
annai-men	案内面	guideway *[eng]*
annai suru	案内する	guide; conduct *{vb t}*
anpō'butsu	罨法物	compress; poultice *{n}* *[pharm]*
anpō-zai	罨法剤	fomentation *{n}* *[pharm]*
anryū	暗流	dark current *[electr]*
ansen	暗線	dark line *[spect]*
an'shimen'dōkō	安四面銅鉱	tetrahedrite *[miner]*
an-shitsu	暗室; あんしつ	darkroom *[photo]*
an'shiya-zō	暗視野像	dark-field image *[opt]*
an-shoku	暗色	dark color *[opt]*
ansok(u)kaku	安息角	angle of repose *[mech]*
ansok(u)kō'ju	安息香樹	benzoin *[bot]*
ansok(u)kō-san	安息香酸	benzoic acid

ansok(u)kō'san-en	安息香酸塩	benzoate *[ch]*
an'sōon	暗騒音	background noise *[acous]*
antai	暗帯；暗帶	dark lane *[astron]*
antei	安定	stablility; steadiness; balance *{n}*
antei-dōi'tai	安定同位体	stable isotope *[nuc-phys]*
antei'do-teisū	安定度定数	stability constant *[ch]*
antei'ka suru	安定化する	stabilize *{vb t}*
antei'ka-zai	安定化剤	stabilizer *[ch] [poly-ch]*
antei-ki	安定器	ballast *[elec]*
antei na	安定な	stable *{adj}*
antei-sei	安定性	stability *[electr]*
"	安定性	robustness *[math]*
antei-teikō	安定抵抗	ballast *[elec]*
antei-zai	安定剤	stabilizer *[mat]*
anten	鞍点	saddle point *[math]*
anten-hō	鞍点法	steepest descent, method of *[math]*
anzan	暗算	mental calculation *[math]*
anzan'gan	安山岩	andesite *[miner]*
anzen	安全	safety *{n}*
anzen-ben	安全弁；安全瓣	safety valve *[eng]*
anzen-beruto	安全ベルト	safety belt *[eng]*
anzen-chitai	安全地帯	safety zone *[civ-eng]*
anzen-megane	安全眼鏡	safety glasses *[ind-eng] [opt]*
anzen-ritsu	安全率	safety factor; margin of safety *[mech-eng]*
anzu	杏；あんず	apricot (fruit, tree) *[bot]*
ao	青	blue (color) *{n} [opt]*
ao-bō	青棒	green rouge *[poly-ch]*
ao-gai	青貝	limpet (shellfish); mother of pearl; Notoacmea schrenckii *[i-zoo]*
aoi	葵；あおい	hollyhock; mallow (flower) *[bot]*
aoi	青い	unripe; green *{adj} [bot]*
"	青い	blue (color) *{adj} [opt]*
ao-kabi	青黴	green mold; penicillum (pl -lia) *[mycol]*
ao'kusa-sa	青臭さ	grassy smell; unripe smell *[physio]*
ao-mame	青豆	green peas (vegetable) *[bot]*
aomi-zuke	青味付	bluing (for laundry) *{n} [mat]*
ao'muke ni naru	仰向けになる	turn to face upward *{vb i}*
ao'muke ni suru	仰向けにする	make supine; face upward *{vb t}*
a'onsoku	亜音速	subsonic speed *[mech] [phys]*
aoru	煽る	fan; agitate *{vb t}*
aosa	石蓴；あおさ	sea lettuce; laver *[bot]*
aosa-moku	石蓴目；あおさ目	Ulvales *[bot]*

ao-shashin	青写真	blueprint; cyanotype *[arch]*
ao'tsuzura-fuji	青葛藤	Cocculus trilobus (herb) *[bot]*
ao-ume	青梅; 青梅	unripe plum *[bot]*
ao-umi'game	青海亀; 青海蠵	green turtle *[v-zoo]*
ao-yaki	青焼; 青焼	blueprint *[graph]*
ao-zame	青鮫; あおざめ	bonito shark; mako; Isurus glaucus (fish) *[v-zoo]*
appaku-kan	圧迫感	sense of pressure *{n} [physio]*
appun-tai	圧粉体	green compact *[mat]*
appun-tesshin	圧粉鉄心	pressed powder core *[met]*
ara'biki	荒引き; あらびき	roughing *{n} [mech-eng]*
araha-giri	荒歯切り	gear roughing *[mech-eng]*
araha'giri-kakō	荒歯切加工	rough gear cutting *[mech-eng]*
arai-guma	洗い熊	raccoon (animal) *[v-zoo]*
arame	荒布; あらめ	Eisenia bicyclis (seaweed) *[bot]*
ara-neri	粗練り	mixing *{n} [rub]*
arare	霰; あられ	hail; snow pellets; graupel *{n} [meteor]*
arare-ishi	霰石	aragonite *[miner]*
ara-sa	荒さ	coarseness; roughness *{n} [eng]*
arashi	嵐	storm *{n} [meteor]*
arasou	争う; 爭う	contest; contend; argue *{vb}*
arau	洗う	wash; rinse *{vb t}*
arawasu	現わす	manifest; reveal *{vb}*
arayuru	あらゆる	all; every *{adj}*
a'rekisei-tan	亜瀝青炭	subbituminous coal *[geol]*
a'ren	亜連; あれん	subtribe *[bio] [syst]*
arerugī	アレルギー	allergy *[med]*
ari	蟻; あり	ant (insect) *[i-zoo]*
ari'kui	食蟻獣; ありくい	anteater (animal) *[v-zoo]*
arimaki	蟻巻; 蟻巻	aphid (insect) *[i-zoo]*
a'rin-san	亜燐酸	phosphorous acid
a'rinsan-en	亜燐酸塩	phosphite *[ch]*
a'rinsan-namari	亜燐酸鉛	lead phosphite
ariru-arukōru	アリルアルコール	allyl alcohol
arīru-ki	アリール基	aryl group *[org-ch]*
Arita-yaki	有田焼	Arita porcelain *[cer]*
ari-tsugi	蟻継ぎ	dovetail joint *[eng]*
aru	在る	is; exist(s) *{vb i}*
arudehido	アルデヒド	aldehyde *[org-ch]*
arufā-sen	アルファー線	alpha rays *[nucleo]*
arugon	アルゴン	argon (element)
a'rui-kei	亜類型; 亞類型	subtype *[syst]*

arukari	アルカリ	alkali *[ch]*
arukari'sei no	アルカリ性の	alkaline *{adj}* *[ch]*
arukiru	アルキル	alkyl *[org-ch]*
arukōru	アルコール	alcohol *[ch]*
aruku	歩く	walk *{vb}*
aruminyūmu	アルミニウム	aluminum (element)
a'ryūsan	亜硫酸	sulfurous acid
a'ryūsan-aen	亜硫酸亜鉛	zinc sulfite
a'ryūsan-en	亜硫酸塩	sulfite *[ch]*
a'ryūsan-gasu	亜硫酸ガス	sulfur dioxide gas; sulfurous acid gas
a'ryūsan-namari	亜硫酸鉛	lead sulfite
asa	麻	hemp; linen; ramie *[tex]*
asa	朝	morning *{n}* *[astron]*
asa-gao	朝顔	morning glory (flower) *[bot]*
asa-gohan	朝御飯	breakfast *{n}* *[cook]*
asai	浅い; 淺い	shallow *{adj}*
a'saibō'sei no	亜細胞性の	subcellular *{adj}* *{n}* *[bio]*
a'saiteki no	亜最適の	suboptimal *{adj}* *{n}*
asaku suru	浅くする	make shallow *{vb t}*
asami-sen	浅海線	shallow-water cable *[elec]*
asami-yu	麻実油	hempseed oil *[mat]*
a-san	亜酸	-ous acid *[ch]*
a'sanka'butsu	亜酸化物	suboxide *[ch]*
a'sanka-chisso	亜酸化窒素	nitrous oxide
a'sanka-dō	亜酸化銅	copper suboxide; cuprous oxide
asa no mi	麻の実	hempseed *[bot]* *[cook]*
asa-nuno	麻布	hemp cloth *[tex]*
asari	浅蜊; あさり	short-necked clam (shellfish) *[i-zoo]*
asa-sa	浅さ; 淺さ	shallowness *[ocean]*
asatte	明後日; あさって	day after tomorrow *{n}*
ase	汗	perspiration; sweat *{n}* *[physio]*
asebi	馬酔木; あせび	andromeda *[bot]*
ase'dome	汗止め	antiperspirant *{n}* *[pharm]*
ase'kaki	汗かき	perspiring; sweating *{n}* *[physio]*
a'sen-yaku	阿仙薬	catechu; Japanese earth *[bot]* *[pharm]*
a'senzō	亜潜像	latent subimage *[graph]*
aseru	焦る	be impatient; be in a hurry *{vb}*
aseru	褪せる	fade *{vb}* *[tex]*
ashi	足; 脚	foot (pl feet); leg *[anat]*
"	足; 脚	lower-part radical of a Chinese character *[pr]*
ashi	葦	reed *[bot]*

ashi-ato	足跡	footprints; spoor *{n}*
ashi'ba	足場	footing; scaffold(ing) *{n} [civ-eng]*
ashiba-izon-sei	足場依存性	anchorage dependence *[microbio]*
ashi'daka-kumo	足高蜘蛛	daddy longlegs (spider) *[i-zoo]*
ashi-kubi	足首	ankle *[anat]*
ashi no kō	足の甲	instep (of foot) *[anat]*
ashi no ura	足の裏	sole (of foot) *[anat]*
ashita	明日；あした	tomorrow *{n}*
ashi-yubi	足指	toe *{n} [anat]*
a'shōsan	亜硝酸	nitrous acid
a'shōsan-en	亜硝酸塩	nitrite *[ch]*
a'shu	亜種	subspecies *[bio] [syst]*
a'shu-ika	亜種以下	infrasubspecies *[bio]*
a'shūso-san	亜臭素酸	bromous acid
a'shūso'san-en	亜臭素酸塩	bromite *[ch]*
asobi-ha'guruma	遊び歯車	idler gear *[mech-eng]*
assaku	圧搾	expression; pressing *{n} [cook]*
"	圧搾	compression *[mech]*
assaku-ki	圧搾器；壓搾器	press *{n} [mech-eng]*
assaku-yu	圧搾油	pressed oil *[mat]*
asshi	圧子	penetrator *[eng]*
asshin-ki	圧伸器	compandor *[electr]*
asshuku	圧縮	compression; compaction *[comput] [mech]*
asshuku-eikyū-hizumi	圧縮永久歪	compression (permanent) set *[poly-ch]*
asshuku-fuku'gen-ritsu	圧縮復元率	compression recovery rate *[poly-ch]*
asshuku-hakai-tsuyo-sa	圧縮破壊強さ	compression breaking strength *[mech]*
asshuku-hi	圧縮比	compression ratio *[poly-ch]*
asshuku-inshi	圧縮因子	compressibility factor *[thermo]*
asshuku-ki	圧縮機	compressor *[comput] [mech-eng]*
asshuku-kōtei	圧縮行程	compression stroke *[mech-eng]*
asshuku-kūki	圧縮空気	compressed air *[mech]*
asshuku-ō'ryoku	圧縮応力	compressive stress *[mech]*
asshuku-ritsu	圧縮率	compressibility *[mech]*
asshuku-seikei	圧縮成形	compression molding *[poly-ch]*
asshuku-setsu'zoku	圧縮接続	compressive joint *[eng]*
asshuku-tsuyo-sa	圧縮強さ	compressive strength *[poly-ch]*
asshuku-zai	圧縮材	compression member *[eng]*
asshuku-zanryū-ō'ryoku	圧縮殘留応力	compressional residual stress *[mech]*
āsu	アース	earth(ing); ground(ing) *{n} [elec]*
asutachin	アスタチン	astatine (element)
atai; chi	値；あたい；ち	value *{n} [math]*
a'tairiku	亜大陸	subcontinent *[geogr]*

atama	頭	head *{n} [anat]*
atama-kazari	頭飾り	tiara *[lap]*
atama-uchi	頭打ち	reaching the top; topping out *{n}*
a'tan	亜炭	lignite *[geol]*
atarashī	新しい	new *{adj}*
atatakai	暖かい; 温かい	warm *{adj}*
atatakai ame	暖い雨	warm rain *[meteor]*
atatameru	暖める; 温める	warm: make warm; heat *{vb t}*
atchaku suru	圧着する	press on; press together *{vb}*
ate-ban	当て盤	dolly *[eng]*
ate'na	宛名	address *{n} [comm]*
ate'saki	宛先	destination *[comput]*
ato	址	ruins *{n} [archeo]*
ato; ushiro	後	behind (location) *{adv}*
ato	痕	impression *[geol] [graph]*
ato	跡	trace; track *{n} [sci-t]*
ato-aji	後味	aftertaste *[cook] [physio]*
ato-bunrui	後分類	post-categorization; post-sorting *{n}*
ato'gaki	後書き	trailer *{n} [comput]*
ato'ire-saki'dashi-hyō	後入れ先出し表	pushdown list *[comput]*
ato-jūgō	後重合	postpolymerization *[poly-ch]*
ato-jukusei	後熟成	afterripening *{n} [photo]*
ato-kakō	後加工	afterprocessing *{n} [mech-eng]*
ato-karyū	後加硫	aftervulcanization (cure) *[rub]*
ato-kōka	後硬化	postcure; postcuring *{n} [plas]*
ato-moe	後燃え	afterburning *{n} [aero-eng]*
ato-rokō	後露光	postexposure *[photo]*
ato-seikei	後成形	postforming *{n} [poly-ch]*
ato-shori	後処理	aftertreatment *[ind-eng]*
ato-shūshuku	後収縮; 後收縮	aftershrinkage *[poly-ch]*
atsu'bai-tai	圧媒体	pressure medium *[mat]*
atsuden-hasshin-ki	圧電発振器	piezoelectric oscillator *[electr]*
atsu-denki	圧電気; 壓電氣	piezoelectricity *[elec]*
atsuden-kōka	圧電効果	piezoelectric effect *[sol-st]*
atsuden-kyōshin'shi	圧電共振子	piezoelectric resonator *[electr]*
atsuden-ritsu	圧電率	piezoelectric coefficient; piezoelectric modulus *[sol-st]*
atsuden'sei-kōbunshi	圧電性高分子	piezoelectric polymer *[poly-ch]*
atsuden'sei-seramikkusu	圧電性セラミックス	piezoelectric ceramics *[cer]*
atsuden-shindō'shi	圧電振動子	piezoelectric vibrator *[sol-st]*
atsu'en	圧延	rolling *{n} [steel]*

atsu'en-jiki-ihō'sei	圧延磁気異方性	roll magnetic anisotropy *[elecmg]*
atsu'en-ki	圧延機	roll mill *[mech-eng]*
atsu'en-kōzai	圧延鋼材	rolled steel *[met]*
atsu'en-ryoku	圧延力	rolling force *[met]*
atsu'en-zai	圧延剤	sheeting agent *[rub]*
atsui	厚い	thick *{adj} [mech]*
atsui	熱い	hot *{adj} [phys] [physio]*
atsui denshi	熱い電子	hot electron *[electr]*
atsu-kōban	厚鋼板	plate; steel plate *{n} [mat]*
atsu'maki-dōtai	厚巻導体	thick-film conductor *[elec]*
atsu'maki-haisen	厚巻配線	thick-film wiring *{n} [elec]*
atsumeru	集める	collect; gather; assemble *{vb t}*
atsu'mi	厚み	thickness *[math]*
atsumi-kei	厚み計	thickness gage *[eng]*
atsu'mitsu	圧密	consolidation *[civ-eng] [geol]*
atsu'ro-ki	圧沪器; 壓濾器	filter press *[eng]*
atsu'ryoku	圧力; 壓力	pressure *{n} [mech]*
atsu'ryoku-kei	圧力計	manometer; pressure gage *[eng]*
atsu'ryoku-kōbai	圧力勾配	pressure gradient *[fl-mech]*
atsu'ryoku-teikō	圧力抵抗	pressure resistance *[fl-mech]*
atsu-sa	厚さ	thickness *[math]*
atsu-sa	暑さ	hot weather; warmth; heat *{n}*
atsu-sa	熱さ	hotness *{n} [meteor] [physio]*
atsusa-kei	厚さ計	thickness gage *[eng]*
atsu'teikō-kōka	圧抵抗効果	piezoresistance effect *[elec]*
attō'teki na	圧倒的な	overwhelming *{adj}*
awa	泡; 泡	bubble; foam *{n} [ch] [phys] [poly-ch]*
"	泡	froth *{n} [fl-mech]*
awa	粟	millet (cereal, grain) *[bot]*
awa-bako	泡箱	bubble chamber *[nucleo]*
awabi	鮑; あわび	abalone (shellfish) *[i-zoo]*
awa'dachi	泡立ち	foaming *{n} [ch-eng]*
awa'date-ki	泡立て器	egg beater; whisk *[eng] [cook]*
awa'date-zai	泡立て剤	foaming agent *[poly-ch]*
awa'keshi	泡消し	defoaming *{n} [ch-eng]*
awase-garasu	合せガラス	laminated glass *[mat]*
awaseru	合わせる	unite; fit together; add; match *{vb t}*
awa'tachi	泡立ち	effervescence *[ch]*
ayamachi	過ち	fault; mistake; misstep; error *{n}*
ayamari	誤り	error; fault *[comput]*
ayamari-bangō	誤り番号	wrong number *[comm]*
ayamari-han'i	誤り範囲	error range *[comput]*

ayamari-hyō'shiki	誤り標識	error indicator *[comput]*
ayamari-jōtai	誤り状態	error condition *[comput]*
ayamari-kaiseki	誤り解析	error analysis *[comput] [math]*
ayamari-kaifuku	誤り回復	error recovery *[comput]*
ayamari-seigyo	誤り制御	error control *[comput]*
ayatsuru	操る	manipulate; maneuver *{vb t}*
ayu	鮎	ayu; Plecoglossus altivelis (fish) *[v-zoo]*
ayumi	歩み	pace; stepping; walking *{n}*
ayumi-ita	歩み板	scaffold board *[civ-eng]*
aza	字; あざ	Minor Section (of village) *[geogr]*
aza	痣; あざ	birthmark *[med]*
azami	薊; あざみ	thistle *[bot]*
azami'uma	薊馬; あざみうま	thrip (insect) *[i-zoo]*
azarashi	海豹; あざらし	seal (marine mammal) *{n} [v-zoo]*
a'zoku	亜族	subtribe *[bio] [syst]*
a'zoku	亜属	subgenus (pl -genera) *[bio] [syst]*
azuki	小豆	adzuki bean (legume) *[bot]*
azuki-zō'mushi	小豆象虫	adzuki bean weevil *[i-zoo]*

B

ba	場	field *{n} [phys]*
ba'ai	場合	case; instance *{n}*
bagu	馬具	harness *{n} [trans]*
bai	倍	double *{n} [math]*
"	倍	times as much *[math]*
baichi	培地	culture medium; medium *[microbio]*
bai-den'atsu-seiryū-kan	倍電圧整流管	voltage doubler tube *[electr]*
bai'en	煤煙；ばいえん	sooty smoke *[eng]*
bai'hōsen	陪法線	binormal *{n} [math]*
baikai	媒介	parameter *[comput]*
baikin	黴菌	bacteria (sing bacterium) *[microbio]*
(bai)kin no	(黴)菌の	bacterial *{adj} [microbio]*
bai no	倍の	double *{adj} [math]*
bai'on	倍音	harmonic(s); harmonic sound; overtone *[acous]*
bai-ritsu	倍率；ばいりつ	multiplying factor; multiplying ratio; scale factor *[eng] [math]*
"	倍率	magnification *[opt]*
bai'ritsu-ki	倍率器	multiplier *[elec]*
baisen	媒染；ばいせん	mordanting *{n} [ch]*
baisen-senryō	媒染染料	mordant dye; mordant color *[opt]*
baisen-zai	媒染剤	mordant *[ch]*
bai'shitsu	媒質	medium *{n} [ch-eng] [phys]*
baishō	焙焼；焙燒	roasting *{n} [met]*
baishū-ki	倍周器	frequency multiplier *[electr]*
baisū	倍数；倍數	multiple *[math]*
baisū-hirei no hōsoku	倍数比例の法則	multiple proportions, law of *[ch]*
baisū-sei	倍数性	polyploidy; ploidy *[gen]*
baitai	媒体；媒體	medium *{n} [ch-eng] [comput] [phys]*
baito	バイト	byte *[comput]*
bai'u-ki; tsuyu-ki	梅雨期	rainy season *[meteor]*
baiyaku	売薬；賣藥	patent medicine *{n} [pharm]*
baiyō	培養；ばいよう	culture; cultivation; incubation *[microbio]*
baiyō-ki	培養基	culture medium *[microbio]*
baiyō suru	培養する	cultivate *{vb} [microbio]*
bakkaku	麦角；麥角	ergot *[mycol] [org-ch]*
bakki	曝気	aeration *[eng]*
bakki-eki	曝気液	aeration solution *[ch]*

bakkin	罰金	penalty; fine *{n} [legal]*
bakki-sō	曝気槽	aeration tank *[ch]*
baku	貘；ばく	tapir (animal) *[v-zoo]*
bakudan	爆弾	bomb *[ord]*
baku'fu	瀑布	cataract; waterfall *[hyd]*
bakufū	爆風	blast *{n} [phys]*
baku'ga	麦芽；麥芽	barley; germ wheat; malt *[bot] [cook]*
bakuga-jū	麦芽汁	malt juice *[cook]*
bakuga'tō	麦芽糖	maltose; malt sugar *[bioch]*
bakugeki-ki	爆撃機	bomber *[aero-eng]*
bakugō	爆轟	detonation *[ch]*
baku'ha	爆破	blasting *[eng]*
baku'hatsu	爆発	explosion *[ch]*
bakuhatsu'sei-kaku-bunretsu	爆発性核分裂	explosive fission *[nuc-phys]*
baku'jū	麦汁	wort *[cook]*
bakumei-ki	爆鳴気	detonating gas *[eng]*
bakumei suru	爆鳴する	fulminate *{vb} [ch]*
baku'nen	爆燃	deflagration; detonation *[ch]*
baku'on	爆音	detonating sound *[acous]*
bakurai	爆雷	depth charge *[ord]*
bakuro	暴露；曝露	exposure *[graph] [opt]*
bākuryūmu	バークリウム	berkelium (element)
baku'shuku-gōsei	爆縮合成	implosion synthesis *[ch]*
baku'tō	爆筒	firecracker *[eng]*
baku'yaku	爆薬	explosives; high explosives *[ord]*
ban	晩	nighttime *[astron]*
ban	番；ばん	number *{n} [comput] [math]*
"	番	guard *{n}*
ban	盤；ばん	panel *[comput]*
ban	礬；ばん	vitriol *[inorg-ch]*
banajūmu	バナジウム	vanadium (element)
bancha	番茶	tea (inferior grade) *[bot] [cook]*
banchi	番地	address *{n} [comput]*
bando	礬土	alum *[inorg-ch]*
bane	発条；發條；ばね	spring *{n} [mech-eng]*
bane-bakari	発条秤；ばね秤	spring balance *[eng]*
bane-kō	発条鋼；ばね鋼	spring steel *[met]*
bane-teisū	ばね定数	spring constant *[mech-eng]*
bangō	番号；番號	number *[comput] [math]*
bangō-kigō	番号記号	number sign *[comput]*
bangō-tō	番号灯	licence plate light *[elec]*
bangō-tsūwa	番号通話	station-to-station call *[comm]*

banjō-kesshō	板状結晶	plate crystal *[crys]*
bankoku no	万国の；萬國の	universal; of all nations *{adj}*
Banko-yaki	万古焼	Banko ware *[cer]*
bannō-bunryū-ki	万能分流器	universal shunt *[elec]*
bannō-shiken-ki	万能試検機	universal testing machine *[eng]*
ba no tsuyo-sa	場の強さ	intensity of field; field strength *[phys]*
bansan	晩餐；ばんさん	dinner *[cook]*
bansan'kai	晩餐会	banquet *[cook]*
bansō-kō	絆創膏	adhesive plaster (or tape) *[pharm]*
ban-te	番手	yarn (count) number *[tex]*
ba'nyō-san	馬尿酸	hippuric acid
ba'nyōsan-en	馬尿酸塩	hippurate *[ch]*
ban'yū-in'ryoku-teisū	万有引力定数	universal gravitation constant *[phys]*
bara	薔薇；ばら	rose (flower) *[bot]*
baraita-shi	バライタ紙	baryta paper *[paper]*
bara'kiseki	薔薇輝石	rhodonite *[miner]*
bara'tsuki	ばらつき	dispersion *[math]*
"	ばらつき	scatter *{n}*
bareisho	馬鈴薯	potato (pl -toes) (vegetable) *[bot]*
ba'riki	馬力；ばりき	horsepower *[mech]*
baryūmu	バリウム	barium (element)
basho	場所	place; location *{n}* *[geogr]*
bashō	芭蕉	plantain *[bot]*
bashō-kajiki	芭蕉梶木；芭蕉旗魚	sailfish *[v-zoo]*
bassen	抜染	discharge printing *{n}* *[graph]*
basshi-kyōdo	抜糸強度	stitch extraction strength *[tex]*
bassui	抜粋	abstract; excerpt; extract *{n}* *[pr]*
bata'tsuki	ばたつき	flutter *{n}* *[eng]*
batei-gata; batei-kei	馬蹄形	horseshoe shape *{n}*
batei'kei-ji'shaku	馬蹄形磁石	horseshoe magnet *[elecmg]*
batō-sei'un	馬頭星雲	horsehead nebula *[astron]*
batsu'mō	抜毛	depilation *[eng]*
batta	飛蝗；蝗；ばった	short-horned grasshopper; locust (insect) *[i-zoo]*
batten	罰点	demerit mark; black mark
beibaku	米麦；米麥	rice and wheat; grains *[bot]*
Beikoku	米国；米國	America *[geogr]*
beikoku	米穀	rice *[bot]*
Beikoku no	米国の	American *{adj}* *[geogr]*
beki	冪；べき	power *{n}* *[math]*
beki-kyūsū	冪級数	power series *[math]*
beki-shisū	冪指数	exponent *[math]*

bekko	別個	separate *(n)*
bekkō	鼈甲；べっこう	tortoiseshell *[v-zoo]*
bekutoru	ベクトル	vector *[math] [phys]*
ben	弁；瓣；べん	valve *[mech-eng]*
ben-annai	弁案内	valve guide *[mech-eng]*
ben-bane	弁ばね	valve spring *[mech-eng]*
ben'betsu	弁別；辨別	discrimination *[comm]*
"	弁別	distinction *[psy]*
benbetsu-ki	弁別器；辨別器	discriminator *[electr]*
benchi; tako	胼胝；べんち；たこ	callus *[med]*
bengo	弁護；辯護	defense *[legal]*
bengo'shi	弁護士	attorney-at-law; lawyer *[legal]*
beni'bana	紅花	safflower (herb) *[bot]*
beni-shōga	紅生姜	ginger, red pickled (herb) *[cook]*
beni-tengu-take	紅天狗茸	fly agaric; Amanita muscaria (poison mushroom, fly killer) *[bot]*
beniya-ita	ベニヤ板	plywood *[mat]*
beni-zuru	紅鶴	flamingo (pl -gos, -goes) (bird) *[v-zoo]*
benjo	便所	rest room *[arch]*
benkei'sō	弁慶草；辨慶草	sedum; orpine *[bot]*
benki	便器	toilet bowl; commode *[arch]*
benmō	鞭毛；べんもう	flagellum (pl flagella) *[bio]*
benmō'chū-rui	鞭毛虫類	Mastigophora *[i-zoo]*
benmō'kin-rui	鞭毛菌類	Mastigomycotina *[mycol]*
benri'shi	弁理士；辯理士	patent attorney *[pat]*
benron	弁論；辯論	argument *[legal]*
ben-sayō	弁作用；瓣作用	valve action *[mech-eng]*
bentō	弁当；辨當	lunch, boxed or picnic *[cook]*
ben-za	弁座；瓣座	valve seat *[eng]*
bera	遍羅；べら	wrasse; seawife (fish) *[v-zoo]*
beriryūmu	ベリリウム	beryllium (element)
bessatsu	別冊；別册	separate edition; supplement *[pr]*
besshi	別紙	separate sheet *[pat]*
betabeta suru	べたべたする	be sticky *(vb i)*
beta-tsuku	べたつく	be sticky *(vb i)*
beto'byō	べと病	mildew *(n) [mycol]*
betsu'mei	別名	alias *[comput] [legal]*
bibun	微分；びぶん	differential; differentiation *[math]*
bibun-hōtei'shiki	微分方程式	differential equation *[math]*
bibun-kairo	微分回路	differentiating circuit *[elec]*
bibun-kaiseki-ki	微分解析器	differential analyzer *[comput]*
bibun-kanō na (no)	微分可能な(の)	differentiable *(adj) [math]*

bibun-ki	微分器	differentiator *[electr]*
bibun-kika'gaku	微分幾何学	differential geometry *[math]*
bibun-sekibun'gaku	微分積分学	calculus *[math]*
bibun suru	微分する	differentiate *{vb t} [math]*
bibun-tōji-ritsu	微分透磁率	differential permeability *[elecmg]*
bi'chōsei	微調整	fine-tuning *{n} [electr] [eng]*
bideo	ビデオ	video *[electr]*
bifū; soyokaze	微風	breeze *[meteor]*
bifun'ka	微粉化	atomization *[mech-eng]*
bifun-tan	微粉炭	pulverized coal *[geol]*
bi'hada-zai	美肌剤	skin beautifier *[pharm]*
bijō-kajo	尾状花序	catkin *[bot]*
bi'jutsu	美術；びじゅつ	art; fine arts
bijutsu'kan	美術館	art gallery; art museum *[arch]*
bijutsu-seidō	美術青銅	art bronze; statuary bronze *[met]*
bijutsu'teki na	美術的な	artistic; esthetic *{adj}*
bikan	尾管	tail pipe *[mech-eng]*
bikan-chiku	美観地区	esthetic area *[civ-eng]*
bi'kesshō	微結晶	microcrystal; microlite *[crys]*
"	微結晶	crystallite *[geol]*
bikō	尾鉱；尾鑛	tailings *[mining]*
bikō	備考	remark(s); note *{n} [pr]*
bikō	鼻孔	nostril *[anat]*
bi-kotsu	尾骨	coccyx (pl -coccyges, coccyxes) *[anat]*
bi'kōzō	微構造	fine structure *[atom-phys]*
"	微構造	microstructure *[sci-t]*
biman	瀰慢；びまん	pervasion *[med]*
bin	瓶；壜；びん	bottle; jar *{n} [eng]*
binkan	敏感	sensitivity *[physio]*
binnaga; binchō	鬢長；びんなが	albacore; Germo alalunga (fish) *[v-zoo]*
binrōji	檳榔子	betel nut *[bot]*
binrōju	檳榔樹	betel palm tree *[bot]*
binsen	便箋	letter paper; writing pad *[comm]*
bin'zume-hin	壜詰品	bottled goods *[food-eng]*
biru	ビル	building *{n} [arch]*
bīru	麦酒；ビール	beer *[cook]*
bīru-kōbo	麦酒酵母	brewing yeast; brewers' yeast *[cook]*
biryō	微量；びりょう	microdose *{n} [med]*
biryō-bunseki	微量分析	microanalysis *[an-ch]*
biryō-chūsha-hō	微量注射法	microinjection *[cyt]*
biryō-eiyō	微量栄養	micronutrient *[bioch]*
biryō-genso	微量元素	microelement; trace element *[geoch]*

biryō-seibun	微量成分	trace component *[ch]*
biryō-tenbin	微量天秤	microbalance *[eng]*
biryō-yōso	微量養素	micronutrient *[bioch]*
bi'ryūshi	微粒子	microparticle; fine particle *[poly-ch]*
bi'ryūshi-tantai	微粒子担体	microcarrier *[microbio]*
bi'ryūshi-un	微粒子雲	corpuscular cloud *[meteor]*
biryū'un	尾流雲	virga *{n}* *[meteor]*
bisai-ekusu'sen-shashin	微細エクス線写真	microradiograph *[an-ch]*
bisai-kōzō	微細構造	fine structure *[atom-phys]*
bisai-kōzō'gaku	微細構造学	leptology *[atom-phys]*
bisaku-rui	尾索類	tunicates; Tunicata *[i-zoo]*
bi'seibutsu	微生物	microbe; microorganism *[microbio]*
bi'seibutsu-bunseki	微生物分析	bioassay *[an-ch]*
bi'seibutsu'gaku	微生物学	microbiology *[bio]*
bi-seibutsu'gaku'teki-teiryō-hō	微生物学的定量法	microbioassay *[bio]*
bi'sen'i	微繊維	microfiber; fine fiber *[tex]*
bisha'chōseki	微斜長石	microcline *[miner]*
bishō	微小	minuteness *{n}* *[phys]*
"	微少	very small quantity *[math]*
bishō-	微小-	micro- *{prefix}*
bishō	微晶	microlite *[crys]*
bishō'butsu-sokkō-ki	微小物測光器	microphotometer *[eng]*
bishō-chūkū-kyū	微小中空球	microballoon *[mat]*
bishō-dan'menseki	微小断面積	infinitesimal cross section *[math]*
bishō-dōbutsu-rui	微小動物類	microfauna *[zoo]*
bishō-jishin	微小地震	microearthquake *[geophys]*
bishō na	微小な	minute *{adj}* *[math]*
bishō'seki	微晶石	microlite *[miner]*
bishō'shitsu-so'shiki	微小質組織	microcrystalline texture *[crys]*
bishō-shokubutsu-rui	微小植物類	microflora *[bot]*
bisumasu	ビスマス	bismuth (element)
bitamin	ビタミン	vitamin *[bioch]*
bitto	ビット	bit *[comput]*
biwa	枇杷；びわ	loquat (tree, fruit) *[bot]*
biwa	琵琶	Japanese lute *[music]*
bi'wakusei	微惑星	planitesimal; planetoid *{n}* *[astron]*
biyō-kenkō-ki	美容健康器	beauty and health instrument *[eng]*
Bizen-yaki	備前焼	Bizen stoneware *[cer]*
bō	望	full moon *[astron]*
bō	棒	rod; pole; stick *{n}* *[eng]*
bōbai-zai	防黴剤	fungicide; antiseptic *[bioch]*

boban	母盤	mother (of records) *{n} [acous]*
bōchō	膨張	expansion *[phys]*
bōchō-do	膨張度	dilatation *[phys]*
bōchō-kei	膨張計	dilatometer *[eng]*
bōchō'sei no	膨張性の	dilatant *{adj} [fl-mech]*
bōchō'tei	防潮提	seawall; tide embankment *[civ-eng]*
bōchō-uchū	膨張宇宙	expanding universe *[astrophys]*
bōchō-zai	膨張剤	blowing agent; foaming agent *[mat]*
bōchū	傍注	marginal note *[pr]*
bōchū-kakō	防虫加工	moth-proofing *{n} [tex]*
bōchū-seizai	防虫製剤	insect-proofing preparation *[pharm]*
bōei	防衛	defense *[mil]*
bōei-kikō	防衛機構	defense mechanism *[psy]*
bo'eki	母液	mother liquor *[ch-eng]*
bōeki	貿易	trade *{n} [econ]*
bōeki-fū	貿易風	monsoon; trade wind *[meteor]*
bōen-dai	望遠台	overlook; scenic overlook *[civ-eng]*
bōen'kyō	望遠鏡	telescope *{n} [opt] [eng]*
bōen-renzu	望遠レンズ	telephoto lens *[opt]*
bōen-sei	防炎性	flameproofness *[ind-eng]*
bōen-shashin'jutsu	望遠写真術	telephotography *[comm]*
bōfū	暴風	windstorm *[meteor]*
bōfū-garasu	防風ガラス	windshield *[eng]*
bōfū-hei	防風塀	windbreak *{n} [eng]*
bōfura; bōfuri	孑孑; ぼうふら	wiggler (mosquito larva) *[i-zoo]*
bōfū'u	暴風雨	rainstorm; storm *{n} [meteor]*
bōfu-zai	防腐剤	preservative; antiseptic *{n} [pharm]*
bōgai	防害	interference *{n} [comm] [phys]*
"	防害	jamming; disturbance *{n} [electr]*
"	防害; 妨害	hindrance; interference *{n}*
bōgai'butsu	妨害物	impediment; obstacle *{n}*
bōgai-ki	妨害機	jammer *[electr]*
bōgai-taikō	妨害対抗	antijamming *[electr]*
bōgen-sei	防眩性	glare resistance *[opt]*
bōgen-zai	防舷材	dock fender *[civ-eng]*
bo-gōkin	母合金	mother alloy *[met]*
bōgo-zai	防護材	fender; boom *[civ-eng] [eng]*
bōgo-zai	防護剤	protective agent *[ind-eng] [mat]*
bōgyo	防御; 防禦	defense; safeguard; protection *{n}*
bōgyū	犛牛; ぼうぎゅう	yak (animal) *[v-zoo]*
bōha'tei	防波堤	breakwater *[civ-eng]*
bōhyō'tei	防氷堤	ice barrier *[hyd]*

bo'in; bo'on	母音	vowel (sound) *[ling]*
bōjin-sei	防塵性	dustproofness *[eng]*
bōjun	膨潤；ぼうじゅん	swelling *{n} [phys]*
bō'kabi-zai	防黴剤	mildewproofing agent *[bioch]*
bōka-heki	防火壁	fire protection wall *[civ-eng]*
bokan	母艦	aircraft carrier *[nav-arch]*
bōkan no	防寒の	coldproof *{adj} [cl]*
bokashi-zome	暈し染め	ombre dyeing *{n} [tex]*
bokasu	暈す；ぼかす	shade off; gradate *{vb t} [graph]*
bokei-ban	母型盤	grid *[pr]*
boki	簿記	bookkeeping *[econ]*
bōkō	膀胱；ぼうこう	bladder *[anat]*
boku'chiku	牧畜；ぼくちく	livestock farming *[agr]*
boku'sō	牧草	meadow grass *[bot]*
"	牧草	pasturage; meadowland *[ecol]*
bōmatsu	防沫；ぼうまつ	splashproof *[ind-eng]*
bon	盆	tray *[cook] [furn]*
bonchi	盆地	basin *[geol]*
bonsai	盆栽；ぼんさい	dwarf tree culture *[bot]*
bon'seki	盆石	miniature stone landscape *[arch]*
bō'o	防汚；ぼうお	soiling prevention *[ind-eng]*
"	防汚	antifouling *{n} [met]*
bō'o-kakō	防汚加工	soil-resistant finish *[mat]*
bo'on	母音	vowel *[ling]*
bōon-ban	防音板	acoustical board *[acous]*
bōon-shitsu	防音室	soundproof chamber *[acous]*
bō'o-toryō	防汚塗料	antifouling coating *{n} [mat]*
bora	鯔；ぼら	gray mullet; Mugil cephalus (fish) *[v-zoo]*
bōru-gami	ボール紙	cardboard; paperboard *[mat] [paper]*
bōru-pen	ボールペン	ball-point pen *[graph]*
boruto	ボルト	volt; V *[elec]*
bōsei	防錆；ぼうせい	rust prevention *[met]*
bōseki	紡績；ぼうせき	spinning *{n} [tex]*
bōseki-shi	紡績糸	spun yarn *[tex]*
bosen	母線	bus; generating line *[comput] [math]*
bōsen	防染	resist printing *{n} [tex]*
bōsen-zai	防染剤	resist; resistance *{n} [tex]*
bōshi	防止	prevention *{n}*
bōshi	紡糸	spinning *{n} [tex]*
bōshi	帽子	hat; headgear *[cl]*
bōshi-eki	紡糸液；ぼうしえき	spinning solution *[tex]*
bōshi-kuchi'gane	紡糸口金	nozzle plate; spinneret *[eng]*

bōshin	傍心	excenter (of triangle) *[math]*
bō'shitsu-kessetsu	房室結節	atrioventricular node *[anat]*
bō'shitsu-sō	防湿層；防濕層	moisture barrier; vapor barrier *[civ-eng]*
bō'shiwa-sei	防皺性	crease resistance *[tex]*
bōshō	芒硝	mirabilite *[ch]*
bōshoku	防食	corrosion-proofing *[met]*
bōshoku-zai	防食剤	anticorrosive agent *[met]*
bo'shūdan	母集団	population *[math]*
bōshuku-sei	防縮性	shrinking resistance *[tex]*
bōshū-zai	防臭剤	antibromic *{n} [pharm]*
bōsō	暴走	runaway *{n} [ch]*
bosshokushi-san	没食子酸	gallic acid
bosshokushi'san-en	没食子酸塩	gallate *[ch]*
bosshokushi'san-tetsu	没食子酸鉄	iron gallate
bosshū	没収；没收	forfeiture *[legal]*
bosū	母数	parameter *[math]*
bōsui	紡錘	spindle *[text]*
bōsui'heki	防水壁	bulkhead *[nav-arch]*
bōsui-sei	防水性	waterproof(ness) *[ind-eng]*
bōsui-zai	防水剤	waterproofing agent *[mat]*
botai	母体	matrix (pl -trices, -trixes) *[bot]*
botan	釦；鈕；ボタン	button *{n} [cl]*
botan	牡丹；ぼたん	tree peony; Paeonia suffruticosa *[bot]*
botan-inko	牡丹鸚哥	lovebird (bird) *[v-zoo]*
botan-jime	釦閉め	button closure *[cl]*
boten	母点	generating point *[math]*
bōten	傍点	emphasis dot marks *[pr]*
botsu	没；ぼつ	setting; set *{n} [astron]*
bōun-zai	防曇剤	clouding preventive *[mat]*
bozai	母材	parent material; base material *[ind-eng]*
bōzai	防材	boom; fender *{n} [civ-eng]*
bō-zuhyō	棒図表；棒圖表	bar graph; bar chart *[stat]*
bu	分；ぶ	"bu" measure (tenth of one sun) = 0.119 in. *[math]*
"	部	division; section; group *{n}*
bu	部	part; class; department *{n}*
-bu	-部	-copy; -copies *{suffix} [pr]*
"	-部	-portion(s) *{suffix}*
bu'ai(-daka)	部合(高)	percentage; rate; ratio *[math]*
bubun	部分	segment *[comput]*
"	部分	part; portion *{n}*
bubun-bunsū	部分分数	partial fractions *[math]*

bubun'gun	部分群	subgroup *[math]*
bubun-jōken'bun	部分条件文	IF statement *[comput]*
bubun-keta'age	部分桁上げ	partial carry *{n}* *[comput]*
bubun-on	部分音	partial tone *[acous]*
bubun-retsu	部分列	substring *[comput]*
bubun-sekibun	部分積分	integration by parts *[math]*
bubun-shoku	部分食	partial eclipse *[astron]*
bubun-shūgō	部分集合	subset *[math]*
bubun-ten'i	部分転位	partial dislocation *[met]*
bubun-zu	部分図	partial view *[graph]*
budō	葡萄; ぶどう	grape (fruit) *[bot]*
budō'en	葡萄園	vineyard *[agr]*
budō('jō)-kyūkin	葡萄(状)球菌	staphylococcus (pl -i) *[microbio]*
bu-domari	歩留り	yield; yield rate *{n}* *[eng]*
budō'seki	葡萄石	prehnite *[miner]*
budō'shu	葡萄酒	wine *{n}* *[cook]*
budōshu'gaku	葡萄酒学	oenology *[cook]*
budōshu-jōzō'gyō'sha	葡萄酒醸造業者	vintner *[cook]*
budō'tō	葡萄糖	glucose; grape sugar; dextrose *[bioch]*
buhin	部品	component; part(s) *{n}* *[ind-eng]*
buhin-zu	部品図	part drawing *{n}* *[graph]*
buki	武器	weapon *[ord]*
bukka-shisū	物価指数	price index *[econ]*
bumon	部門	branch; department; section
bun	文; ぶん	statement; sentence *[comput]* *[gram]*
buna	山毛欅; ぶな	beech(nut), Japanese (tree) *[bot]*
bun'atsu	分圧	partial pressure *[phys]*
bun'atsu-kei	分圧計	potentiometer *[elec]*
bun'atsu-ki	分圧器; 分壓器	voltage divider *[elec]*
bun'atsu no hōsoku	分圧の法則	partial pressure, law of *[ch-eng]*
bunben	分娩; ぶんべん	parturition; childbirth *{n}* *[med]*
bun'betsu	分別	fractionation *[ch]*
bunbetsu-chinden	分別沈殿	fractional precipitation *[ch]*
bunbetsu-jōryū	分別蒸留	fractional distillation *[ch]*
bunbetsu-kesshō(-sayō)	分別結晶(作用)	fractional crystallization *[ch]*
bunbetsu-shōka	分別昇華	fractional sublimation *[ch]*
bunbo	分母; ぶんぼ	denominator *[math]*
bunbō'gu	文房具	stationery *{n}* *[mat]*
bunchin	文鎮; ぶんちん	paperweight *[furn]*
bundan	分断; 分斷	dividing *{n}* *[ind-eng]*
bundan-idenshi	分断遺伝子	split gene *{n}* *[gen]*
bundan suru	分断する	divide into pieces *{vb t}*

bundo-ki	分度器	protractor *[math]*
bun'eki-rōto	分液漏斗	separatory funnel *[ch]*
bunga	分芽	budding *{n} [bot]*
bungei	文芸；文藝	literary property *[pr]*
bungo	文語	literary (or written) language *[comm]*
bungo'bun	文語文	literary-style writing *[pr]*
bungyō	分業	division of labor *[bio]*
bunka	分化	differentiation *[bio] [cyt] [ling]*
"	分化	specialization *[bio]*
bunkai	分解	decomposition *[ch]*
"	分解	disassembly; takedown *{n} [eng]*
"	分解	degradation (of enzyme) *[microbio]*
"	分解	resolution *[opt] [spect]*
"	分解	cracking *{n} [org-ch] [petrol]*
bunkai	分塊	blooming *{n} [steel]*
bunkai-jōryū	分解蒸留	destructive distillation *[org-ch]*
bunkai-kōso	分解酵素	proteolytic enzyme *[bioch]*
bunkai-netsu	分解熱	heat of decomposition *[p-ch]*
bunkai-nō	分解能	resolving power; resolution *[opt]*
bunkai'sha	分解者	decomposer *[ecol]*
bunkai suru	分解する	decompose; disassemble; resolve *{vb}*
bunkai-ten	分解点	decomposition point *[ch]*
bunkai-yūkai	分解融解	incongruent melting *[p-ch]*
bunkai-zu	分解図；分解圖	exploded view *[graph]*
bunkaku	分画；分畫	fractionation *[bioch]*
bunkaku-enshin	分画遠心	differential centrifugation *[ch]*
bunka shita	分化した	specialized *{adj} [bio]*
bun'katsu	分割	cleavage *[bot]*
"	分割	resolution *[opt]*
"	分割	division *[pat]*
bunkatsu-hyō	分割表	contingency table *[math]*
bunkatsu-shutsu'gan	分割出願	divided application *[pat]*
bun'katsu suru	分割する	divide; split *{vb} [comput]*
bunka-zennō-sei	分化全能性	totipotency *[embryo]*
bunken	文献	bibliography; document; literature (the list) *[pr]*
bunki	分岐	divergence *[bio]*
"	分岐	branch; branching *{n} [ch] [comput]*
bunki-kaisen	分岐回線	multipoint line *[comm]*
bunki shita	分岐した	branched *{adj} [comput]*
"	分岐した	forked *{adj}*
bunki-ten	分岐点	branch point *[comput]*

bunkō	分光	spectral diffraction *[opt] [phys]*
bunkō-bunpu	分光分布	spectral distribution *[opt] [phys]*
bunkō-bunseki	分光分析	spectrochemical analysis *[ch]*
"	分光分析	spectrum analysis *[phys]*
bunkō'gaku	分光学	spectroscopy *[phys]*
bunkō-hansha-ritsu	分光反射率	spectral reflectance; spectro-reflectance *[an-ch]*
bunkō-kando	分光感度	spectral sensitivity *[electr]*
bunkō-kei	分光計	spectrometer *[spect]*
bunkō-ki	分光器	spectroscope *[spect]*
bunkō-kōdo-kei	分光光度計	spectrophotometer *[spect]*
bunkō-mitsudo	分光密度	spectral concentration *[spect]*
bunkō-netsu'ryō-kei	分光熱量計	spectrocalorimeter *[spect]*
bunkō-rensei	分光連星	spectroscopic binary star *[astron]*
bunkō-senkō-kei	分光旋光計	spectropolarimeter *[opt]*
bunkō-shashin-ki	分光写真器	spectrograph *[opt]*
bunkō-sokkō('gaku)	分光測光学	spectrophotometry *[spect]*
bunkō-taiyō-shashin-gi	分光太陽写真儀	spectroheliograph *[astron]*
bunkō-tōka-ritsu	分光透過率	spectral transmissivity *[opt]*
bunkō-tokusei	分光特性	spectral characteristic; spectro-scopic characteristic *[spect]*
bunkō-zōkan	分光増感	spectral sensitization *[photo]*
bunkyō-chiku	文教地区	educational district *[civ-eng]*
bun'kyoku	分局	branch office *[econ]*
bun'kyoku	分極	polarization *[elec] [phys]*
bunkyoku-denka	分極電荷	polarized charge *[elec]*
bunkyoku-ritsu	分極率	polarizability *[elec]*
bunkyū	分級	classifying; classification *{n}*
bunkyū suru	分級する	classify *{vb}*
bun'maki-dendō-ki	分巻電動機	shunt motor *[elec]*
bun'myaku	文脈	context *[comput] [pr]*
bun'on-fugō	分音符号	diacritical mark *[pr]*
bunpai	分配；ぶんぱい	division; distribution; sharing *{n}*
bunpai-ben	分配弁；分配瓣	distributing valve *[mech-eng]*
bunpai-hi	分配比	distribution ratio *[ch]*
bunpai-hōsoku	分配法則	distributive law *[math]*
bunpai-kansū	分配関数	partition function *[stat-mech]*
bunpai-keisū	分配係数	distribution coefficient; partition coefficient *[an-ch] [p-ch]*
bunpai no	分配の	distributive *{adj} [math]*
bunpai no hōsoku	分配の法則	partition, law of *[phys]*
bunpai-sei	分配性	distributive property *[math]*

bunpitsu	分泌；ぶんぴつ	secretion *[physio]*
bunpitsu	分筆	subdivision of lot *[civ-eng]*
bunpō	文法	grammar *[comm] [comput] [ling]*
bunpu	分布	distribution *[math] [stat]*
bunpu-kansū	分布関数	distribution function *[stat]*
bunpu-kyoku'sen	分布曲線	distribution curve *[math]*
bun'retsu	分裂；ぶんれつ	division *[cyt]*
"	分裂	splitting; fragmentation *{n}*
bunretsu'kin-rui	分裂菌類	Schizomycetes *[microbio]*
bunretsu-shisū	分裂指数	mitotic index *[cyt]*
bunretsu-so'shiki	分裂組織	meristematic tissue; meristem *[bot]*
bunri	分離；ぶんり	separation; division *{n}*
bunri-fukka; bunri-fukuka	分離複果	aggregate fruit *[bot]*
bunri-kaku	分離角	separation *[astron]*
bunri-keisū	分離係数	separation factor *[sci-t]*
bunri-kigō	分離記号	separator *[comput]*
bunri no hōsoku	分離の法則	segregation, law of *[gen]*
bunri suru	分離する	sort *{vb t} [comput]*
"	分離する	separate *{vb t}*
bunro	分路	shunt *{n} [elec]*
bunrui	分類；ぶんるい	classification; sort; sorting *[comput]*
bunrui'gaku	分類学；分類學	systematics *[bio]*
"	分類学	taxonomy *[syst]*
bunrui-ki	分類機	sorter *{n} [comput]*
bunrui-moku'roku	分類目録	classification index *[syst]*
bunrui-taikei	分類体系	classification system *[syst]*
bunrui-tan'i	分類単位	taxon (pl taxa, taxons) *{n} [syst]*
bun-ryoku	分力	component of force *[phys]*
bunryū	分流	tributary; distributary *{n}[hyd]*
bunryū-kan	分留管	fractionating column *[ch]*
bunryū-ki	分流器	shunt *{n} [elec]*
bunsan	分散；ぶんさん	variance *[stat]*
"	分散	dispersion *[phys]*
bunsan-bai	分散媒	dispersive medium *[elecmg]*
"	分散媒	dispersant; dispersion medium *[mat]*
bunsan-bunseki	分散分析	analysis of variance *[stat]*
bunsan-do	分散度	degree of dispersion *[ch]*
bunsan-jūgō	分散重合	dispersion polymerization *[poly-ch]*
bunsan-kei	分散系	disperse system *[ch]*
bunsan'shitsu	分散質	dispersoid *[ch]*
bunsan-sō	分散相	disperse phase *[ch]*
bunsan-zai	分散剤	dispersing agent; dispersant *[ch]*

bunsei'shi	分生子	conidium; conidiospore *[mycol]*
bunseishi-hei	分生子柄	conidiophore *[mycol]*
bunseki	分析	analysis *[ch-eng]*
bunseki-kagaku	分析化学	analytical chemistry *[ch]*
bun'setsu	分節	segment *[i-zoo]*
bun'setsu	文節	paragraph *[comput] [pr]*
bunshi	分子	molecule *[ch]*
"	分子	numerator *[math]*
bunshi-bun'kyoku	分子分極	molecular polarization *[p-ch]*
bunshi-dan	分子団；分子團	molecular cluster *[ch]*
bunshi-denba	分子電場	molecular (electric) field *[p-ch]*
bunshi-furui-karamu	分子篩いカラム	molecular sieve column *[an-ch]*
bunshi-haikō	分子配向	molecular orientation *[ch]*
bunshi-haikō-do	分子配向度	degree of molecular orientation *[ch]*
bunshi-iku'shu	分子育種	molecular breeding *{n} [ch]*
bunshi-jōryū	分子蒸留	molecular distillation *[ch]*
bunshi-jōryū-ki	分子蒸留機	molecular still *[ch]*
bunshi'kan-ryoku	分子間力	intermolecular force *[ch]*
bunshi'keisen	分至経線	colure *{n} [astron]*
bunshi-kyūkō-keisū	分子吸光係数	molecular extinction coefficient *[an-ch]*
bunshi'nai-ten'i	分子内転位	intramolecular rearrangement *[ch]*
bunshi'ryō	分子量	molecular weight *[ch]*
bunshi-seibutsu'gaku	分子生物学	molecular biology *[bio]*
bunshi-seiri'gaku	分子生理学	molecular physiology *[physio]*
bunshi-sen	分子線	molecular beam *[phys]*
bunshi-shiki	分子式	molecular formula *[ch]*
bunshi-sokudo-bunpu-soku	分子速度分布則	distribution of molecular velocities, law of *[stat-mech]*
bunshi-taiseki	分子体積	molecular volume *[ch]*
bunshi-tenmon'gaku	分子天文学	molecular astronomy *[astron]*
bunshi-yō	分子容	molecular volume *[ch]*
bunshi-zu	分子図；分子圖	molecular diagram *[ch]*
bunsho	文書	document; text *{n} [comput]*
bunshō	文章	composition; writing *{n} [comm]*
bunsho'ka	文書化	documentation *[comm]*
bunshū-kairo	分周回路	dividing circuit *[electr]*
bunshū-ki	分周器	frequency divider *[electr]*
bunshuku	分縮	partial condensation *[ch]*
bunsō-shidō-dendō-ki	分相始動電動機	split-phase-start motor *[elec]*
bunsō-shiji-yaku	分相指示薬	separate-phase indicator *[an-ch]*
bunsō-yūdō-dendō-ki	分相誘導電動機	split-phase induction motor *[elec]*
bunsū	分数	fraction *{n} [math]*

bunsui-kai	分水界	divide; watershed *{n} [geogr]*
bunten	分点	equinoctial point *[astron]*
bunten-getsu	分点月	tropical month *[astron]*
bun'ya	分野	field; area; subject matter *[pat]*
bun'yo	分与；分與	settlement *[legal]*
buppin	物品	article; goods; object *{n}*
buppōsō	仏法僧；佛法僧	broad-billed roller (bird) *[v-zoo]*
burashi	刷子；ブラシ	brush *{n} [eng]*
buri	鰤；ぶり	yellowtail; Seriola quinqueradiata (fish) *[v-zoo]*
buriki	ブリキ	tinplate *{n} [met]*
bussei	物性	physical properties *[ch] [phys]*
busshitsu	物質	material; matter; substance *{n} [ch] [phys]*
busshitsu-fu'metsu no hōsoku	物質不滅の法則	indestructibility of matter, law of *[phys]*
busshitsu-hozon no hōsoku	物質保存の法則	conservation of matter, law of *[phys]*
busshitsu-idō-sokudo	物質移動速度	mass-transfer rate *[phys]*
busshitsu-idō-yōryō-keisū	物質移動容量係数	coeffient of mass-transfer capacity *[phys]*
busshitsu-shūshi	物質収支	material balance *[ch-eng] [phys]*
buta	豚	pig; hog; swine (animal) *[v-zoo]*
butai	舞台	stage (in a theater) *{n} [arch]*
butō-gumo	舞踏蜘蛛	tarantula; Lycosa tarantula (spider) *[i-zoo]*
butō-shitsu	舞踏室	ballroom *[arch]*
butsuri'gaku	物理学	physics *[sci-t]*
butsuri-gakusha	物理学者	physicist *[phys]*
butsuri-henka	物理変化	physical change *[phys]*
butsuri'kagaku	物理化学	physical chemistry *[ch]*
butsuri-sei'kagaku	物理生化学	physical biochemistry *[bioch]*
butsuri-sokkō	物理測光	physical photometry *[opt]*
butsuri-tankō	物理探鉱	geophysical prospecting *{n} [eng]*
butsuri'teki-seishitsu	物理的性質	physical properties *[phys]*
butsu'ryū	物流	substance flow *[fl-mech]*
buttai	物体；物體	object; article; thing *{n} [phys]*
butten	物点	object point *[opt]*
buyo; buyu	蚋；ぶよ；ぶゆ	gnat; midge; sandfly; Simulium (insect) *[i-zoo]*
buzai	部材	member; member part *[ind-eng]*
byaku'dan	白檀	sandalwood (tree) *[bot]*
byaku'shin	柏槙；びゃくしん	juniper; Juniperus (tree) *[bot]*

byō	秒	second *{n} [math] [mech]*
byō	鋲	rivet; tack; thumbtack *{n} [eng]*
-byō	-秒	-second(s) *{suffix}*
-byō	-病	-disease *{suffix}*
byōbu	屏風	folding screen; screen *{n} [furn]*
byōdō	平等	equality *[math]*
byōdō'ka suru	平等化する	equalize; even; level *{vb t}*
byōdō na ba	平等な場	uniform field *[phys]*
byōga	描画; びょうが	drawing *{n} [graph]*
byōgai'chū	病害虫	noxious insect *[i-zoo]*
byōgai'chū-bōjo	病害虫防除	biological control (of insects) *[ecol]*
byōga suru	描画する	draw *{vb} [comput]*
byōgen-bi'seibutsu	病原微生物	pathogen *[med]*
byōgen'tai	病原体	pathogen *[med]*
byō'in	病院	hospital *[arch]*
byōjaku'sha	病弱者	invalid *{n} [med]*
byōjō	苗条; びょうじょう	shoot *{n} [bot]*
byōki	病気; 病氣	illness; disease; sickness *[med]*
byōri'gaku	病理学; 病理學	pathology *[med]*
byōsha	描写; 描寫	picture; depiction *{n} [graph]*
byō-yomi	秒読み; 秒讀み	countdown *{n} [aero-eng]*
byō'yomi-jikan	秒読み時間	countdown time *[aero-eng]*

C

cha	茶；ちゃ	tea *[bot] [cook]*
cha'bane-gokiburi	茶翅蜚蠊	German cockroach (insect) *[i-zoo]*
chadō; sadō	茶道	tea ceremony
cha-iro	茶色	brown (color) *[opt]*
cha'iro-ichi'monji	茶色一文字	viceroy (butterfly) *[i-zoo]*
chakku	チヤック	zipper *[cl]*
chaku'chi	着地	landing *{n} [navig]*
chaku'datsu-jizai	着脱自在	freely attachable and detachable *[mech-eng]*
chaku'hyō	着氷	icing; ice coating *{n} [hyd]*
chaku'hyō'sei-arashi	着氷性嵐	ice storm *[meteor]*
chaku'ji-kyoku	着磁極	magnetized pole *[elecmg]*
chak(u)kō-ryō	着香料	flavoring agent *[cook] [pharm]*
chaku'niku	着肉	inking *{n} [pr]*
chaku'riku	着陸	landing (of vehicle) *{n} [navig]*
chaku'riku suru	着陸する	land *{vb} [aero-eng]*
chaku'riku-tō	着陸灯；着陸燈	landing light *[aero-eng]*
chakusei-shokubutsu	着生植物	epiphyte *[bot]*
chaku'shoku	着色；ちゃくしょく	coloring; staining *{n} [opt]*
chaku'shoku-ryō	着色料	coloring material *{n} [pharm]*
chaku'shoku-ryoku	着色力	coloring strength (or power) *[opt]*
chaku'shoku-sei	着色性	coloring property *[opt]*
chaku'shoku-zai	着色剤	colorant; coloring agent *[mat]*
chakusō	着想	conception; idea *[psy]*
cha-no-yu	茶の湯	tea ceremony
cha o ireru	茶を入れる	brew tea *{vb t} [cook]*
cha'saji-ryō	茶匙量	teaspoonful *[cook]*
chasei	茶精	caffeine *[org-ch]*
cha-shitsu	茶室	tea ceremony room *[arch]*
chaso	茶素	caffeine *[org-ch]*
chawan	茶碗	rice bowl *[cer] [cook]*
chawan-mushi	茶碗蒸し	steamed egg custard *[cook]*
cha-zai	茶剤；茶劑	medicinal tea *[cook]*
cha'zuke	茶漬け	tea-doused cooked rice *[cook]*
chi	血	blood *[hist]*
chi	地；ち	earth *{n} [geol]*
chi	値；ち	value *{n} [math]*
chichi	乳；ちち	milk *{n} [physio]*
chichū-chō'onki	地中聴音機	geophone *[electr]*

chi-dai	血鯛；ちだい	crimson sea bream (fish) *[v-zoo]*
chi-denryū	地電流	terrestrial current; earth current *[geophys]*
chidori	千鳥；ちどり	plover (bird) *[v-zoo]*
chidori-nui	千鳥縫い	zigzag chain stitch *[cl]*
chidori-ue	千鳥植え	zigzag planting *{n} [agr]*
chi'eki-ken	地役権	easement *[legal]*
chi'en	遅延；遲延；ちえん	retardation; delay(ing) *{n}*
chi'en-jikan	遅延時間	delay time; retardation time *[electr]*
chi'en-keikō	遅延蛍光	delayed fluorescence *[atom-phys]*
chi'en-ōtō	遅延応答	delayed response *[comput]*
chi'en saseru	遅延させる	delay; retard *{vb t}*
chi'en-sayō	遅延作用	delayed action; delaying action *{n}*
chi'en-sen	遅延線	dealy line *[electr]*
chi'en-zai	遅延剤	retardation agent *[ch]*
chifusu-kin	チフス菌	typhoid bacillus *[microbio]*
chigai-dana	違い棚	alcove shelves, staggered *[arch]*
chi'gaku	地学	earth science; geoscience *[sci-t]*
chi'gire-gumo	千切れ雲	scattered clouds *[meteor]*
chihei'sen	地平線；ちへいせん	horizon *[astron]*
chihei-shisa	地平視差	horizontal parallax *[astron]*
chihō-chishi	地方地誌	chorography *[map]*
chihō'go	地方語	vernacular language *[ling]*
chihō-keikaku	地方計画	regional planning *{n} [civ-eng]*
chi'i	地衣	lichen *[bot]*
chi'iki	値域	range *{n} [math]*
chi'iki-keikaku	地域計画	regional planning *{n} [civ-eng]*
chi'iki-sei	地域制	zoning *{n} [civ-eng]*
chi'i-kai	地衣界	Lichenes; lichens *[bot]*
chi'i-shokubutsu	地衣植物	Lichenes; lichens *[bot]*
chi'i-tai	地衣帯；地衣帶	lichen zone *[bot]*
chi'jiki	地磁気	geomagnetism; terrestrial magnetism *[geophys]*
chi'jiki-arashi	地磁気嵐	geomagnetic storm *[geophys]*
chi-jiku	地軸	earth axis *[geod]*
chijima'nai ryūtai	縮まない流体	incompressible fluid *[fl-mech]*
chijimeru	縮める	shrink *{vb t} [tex]*
"	縮める	contract; shorten; shrink *{vb t}*
chijimi	縮み	shrinkage *[tex]*
chijimi-shiro	縮み代	shrinkage allowance *[plas] [tex]*
chijimu	縮む	shrink *{vb i} [tex]*
chijō-ha	地上波	ground wave *[comm]*

chijō-hōsō	地上放送	terrestrial broadcasting *[comm]*
chijō-kei	地上茎	terrestrial stem *[bot]*
chijō no	地上の	terrestrial *{adj} [sci-t]*
chijō-shisetsu-kan	地上敷設管	aboveground-laid pipe *[hyd]*
chijō-sui	地上水	surface water *[hyd]*
chijō-yusō	地上輸送	ground transportation *[trans]*
chijun	置閏; ちじゅん	intercalation *[astron]*
chikai	近い	close; nearby *{adj} {adv}*
chi'kai	地階	basement floor; cellar *[arch]*
chi'kaku	地殻; 地殼	crust; earth crust *[geol]*
chi'kaku	知覚	perception; sensation *[physio]*
chi'kaku-undō	地殻運動	crustal movement *[geol]*
chikan	置換; ちかん	displacement *[ch]*
"	置換	substitution *[ch] [math] [mol-bio]*
"	置換	permutation *[math]*
chikan-do	置換度	degree of substitution *[ch]*
chikan-ki	置換基	substituent *{n} [ch]*
chikan-ki; shikan-ki	弛緩期	diastole *{n} [physio]*
chika no	地下の	subterranean *{adj} [geogr]*
chikan-tai	置換体	substitution product *[ch]*
chikara	力; ちから	strength; power; force *[mech]*
chikara-keisū	力係数	force factor *[phys]*
chikara no bunkai	力の分解	resolution of forces *[math]*
chikara no kakudai-ritsu	力の拡大率	mechanical advantage *[mech-eng]*
chika-shitsu	地下室	basement *[arch]*
chika'sui	地下水	groundwater; underground water *[hyd]*
chika-suimen	地下水面	water table *[hyd]*
chika-tetsu(dō)	地下鉄(道)	subway *[trans]*
chikei	地形; ちけい	terrain *[eng]*
"	地形	topography *[geogr]*
chikei'gaku	地形学	geomorphology *[geol]*
chikei-zu	地形図	topographic map *[map]*
chikka	窒化; ちっか	nitriding *{n} [ch]*
chikka'butsu	窒化物	nitride *[ch]*
chikka-hōso	窒化硼素	boron nitride
chikka-keiso	窒化硅素	silicon nitride
chikka-rin	窒化燐	phosphorus nitride
chikka-tetsu	窒化鉄	iron nitride
chikō	地溝	rift valley; trough *[geol]*
chikō'sei-soku'shin-zai	遅効性促進剤	delayed-action accelerator *[rub]*
chi'kōsha	地向斜	geosyncline *[geol]*
chiku	地区; 地區; ちく	district *[geogr]*

chiku'denchi	蓄電池	storage battery; battery *[elec]*
chiku'ichi	逐一	in detail; one by one
chiku'ji	逐次；ちくじ	sequential(ness); serial; successive *{adj} {n} [math]*
chikuji-fuka-jūgō	逐次付加重合	consecutive (or successive) addition polymerization *[poly-ch]*
chikuji-hannō	逐次反応	consecutive reaction; successive reaction *[ch]*
chikuji-shori	逐次処理	sequential processing *[comput]*
chiku'san'butsu	畜産物	livestock *[agr]*
chiku'san'gaku	畜産学	animal husbandry *[agr]*
chiku'seki	蓄積；ちくせき	storage *[comput]*
"	蓄積	accumulation *{n}*
chiku'seki suru	蓄積する	store *{vb} [comput]*
"	蓄積する	accumulate *{vb}*
chiku'setsu-ninjin	竹節人参	Panax rhizome *[bot]*
chikyō	地峡；地峽	isthmus *[geog]*
chikyū	地球；ちきゅう	earth *{n} [astron]*
chikyū-butsuri'gaku	地球物理学	geophysics *[geol]*
chikyū'gai no	地球外の	extraterrestrial *{adj}* [astron]
chikyū-gi	地球儀	(terrestrial) globe *[map]*
chikyū-hōsha	地球放射	terrestrial radiation *[geophys]*
chikyū-jiba	地球磁場	geomagnetic field *[geophys]*
chikyū-kagaku	地球化学	geochemistry *[geol]*
chikyū-seishi-kidō	地球静止軌道	geostationary orbit *[aerosp]*
chikyū'shō	地球照	earthshine *{n} [astron]*
chikyū-taiki-sen	地球大気線	telluric line *[spect]*
chimei'teki-ayamari	致命的誤り	fatal error *[comput]*
chi'mitsu	緻密；ちみつ	close-packed; minute; fine *{n} [mat]*
chi'mitsu'sei no	緻密性の	close-packed *{adj} [mat] [mech]*
chin	狆；ちん	Japanese spaniel; Pekinese; pug dog (animal) *[v-zoo]*
chinami ni	因みに；ちなみに	in this connection; by the way *{adv}*
chinden	沈殿；ちんでん	precipitation *[ch]*
chinden'butsu	沈殿物；沈澱物	precipitate *{n} [ch]*
"	沈殿物	sediment *[geol]*
chinden'chi	沈殿池	settling pond (or basin) *[civ-eng]*
chinden-tekitei	沈殿滴定	precipitation titration *[ch]*
chinden-zai	沈殿剤	precipitant *[ch]*
chi-netsu	地熱	terrestrial heat *[geophys]*
chi'netsu-hatsuden'sho	地熱発電所	geothermal power plant *[mech-eng]*
chi'netsu no	地熱の	geothermal *{adj} [geophys]*

chingai-yaku (chingai-zai) 鎮咳薬（鎮咳剤） antitussive *{n} [pharm]*

chin'jō-yōgan 枕状溶岩 pillow lava *[geol]*

chin'jutsu'sho 陳述書 written statement *[legal]*

chinka 沈下 subsidence *[min-eng]*

chinka shita 沈下した submerged; settled; sunken *{adj}*

chinkei-yaku (chinkei-zai) 鎮痙薬（鎮痙剤） antispasmodic; spasmolytic *{n} [pharm]*

chinki-zai チンキ剤 tincture *{n} [pharm]*

chin'kō 沈降；ちんこう settling *{n} [eng] [met]*

" 沈降 sedimentation *[geol] [met] [sci-t]*

chinkō'butsu 沈降物 sediment *[geol]*

chinkō-den'i 沈降電位 sedimentation potential *[elec]*

chinkō-heikō 沈降平衡 sedimentation equilibrium *[an-ch]*

chinkō'so 沈降素 precipitin *{n} [immun]*

chinkō-sō 沈降槽 settling tank *[eng]*

chinkō-soku'shin-zai 沈降促進剤 sedimentation accelerator *[ch-eng]*

chin-kotsu; kinuta-bone 砧骨 anvil; incus (pl incudes) *[anat]*

chin'mi 珍味 delicacy *{n} [cook]*

chi'nō 知能 intelligence *[psy]*

chinō-gakushū 知能学習 intelligence learning *[comput]*

chinō-shisū 知能指数 intelligence quotient *[psy]*

chinpi 陳皮 dried orange peel *[pharm]*

chin'retsu-shitsu 陳列室 showroom *[arch]*

chinsei-kōkai 鎮静鋼塊 killed ingot *[met]*

chinsei-yaku (chinsei-zai) 鎮静薬（鎮静剤） sedative; tranquilizer *{n} [pharm]*

chinseki'do 沈積土 silt *{n} [geol]*

chinto-yaku (chinto-zai) 鎮吐薬（鎮吐剤） antiemetic *{n} [pharm]*

chintsū-yaku (chintsū-zai) 鎮痛薬（鎮痛剤） analgesic *{n} [pharm]*

chin'un-yaku (chin'un-zai) 鎮暈薬（鎮暈剤） antidinic *{n} [pharm]*

chin'yō-yaku (chin'yō-zai) 鎮痒薬（鎮痒剤） antipruritic *{n} [pharm]*

chirabari 散らばり dispersion *[math]*

chira-chira hikaru ちらちら光る shimmer; glitter *{vb i} [opt]*

chi'raku 地絡 ground fault *[elec]*

chirasu 散らす scatter; disperse; strew *{vb t}*

chira'tsuki ちらつき flicker *{n} [opt]*

chiri 塵 dust *[geol]*

chiri'gaku 地理学 geography *[sci-t]*

chirimen 縮緬；ちりめん crepe; crepe cloth *[tex]*

chiru 散る scatter *{vb i}*

chiryō 治療 treatment *[med]*

chi-ryoku 地力 fertility (of soil) *[agr]*

" 地力 terrestrial gravitation *[geophys]*

chiryō-shisū 治療指数 therapeutic index *[pharm]*

chiryō suru	治療する	treat; remedy *{vb} [med]*
chīsai	小さい	small *{adj} [math] [mech]*
chīsaku suru	小さくする	make small; reduce in size *{vb t}*
chisei	地勢	geographical features; terrain *[geogr]*
chisei-zu	地勢図	chorographic map *[graph]*
chisha	萵苣; ちしゃ	lettuce (vegetable) *[bot]*
chishi	地誌	topography *[geogr]*
chishi	智歯; 智齒	wisdom tooth *[dent]*
chishi-hen'i	致死変異	lethal mutation *[gen]*
chi'shiki	知識; ちしき	knowledge *{n} [psy]*
chishiki-kōgaku	知識工学	knowledge engineering *[comput]*
chishin-tenchō	地心天頂	geocentric zenith *[astron]*
chishin-zahyō	地心座標	geocentric coordinates *[astron]*
chishi-ryō	致死量	lethal dose *[med]*
chi'shitsu'gaku	地質学; ちしつがく	geology *[sci-t]*
chishitsu-jidai	地質時代	geological age *[geol]*
chishitsu-kōzō	地質構造	tectonics *[geol]*
chishitsu-nendai'gaku	地質年代学	geochronology *[geol]*
chishitsu-ondo-kei	地質温度計	geothermometer *[eng]*
chishitsu'teki-nendai-memori	地質的年代目盛り	geological time scale *[mech]*
chishitsu-zu	地質図	geological map *[graph]*
chisō	地層	earth stratum; layer *[geol]*
chisō-ruijū no hōsoku	地層累重の法則	superposition, law of *[geol]*
chisso	窒素; ちっそ	nitrogen (element)
chisso-hiryō	窒素肥料	nitrogenous fertilizer *[agr]*
chisso-junkan	窒素循環	nitrogen cycle *[bio]*
chi'sui-kōmori	血吸(い)蝙蝠	vampire bat (mammal) *[v-zoo]*
chitai	地帯; 地帶	zone *{n} [geog]*
chitai	遅滞; 遲滯; ちたい	delay; retardation *{n} [ind-eng]*
chi-tai-chi no	地対地の	surface-to-surface *{adj} [ord]*
chitai-ki	遅滞期	lag phase *[microbio]*
chi-tai-kū no	地対空の	surface-to-air *{adj} [ord]*
chi'taku	地卓	mesa *[geol]*
chitan	チタン	titanium (element)
chi'teki-shoyū'ken	知的所有権	intellectual property right *[legal]*
chitsu	膣	vagina *[anat]*
chitsu-senzai	膣洗剤	douche *{n} [pharm]*
chizu	地図; 地圖	map *{n} [graph]*
chizu'gaku	地図学	cartography *[graph]*
chizu-sakusei	地図作成	mapping *{n} [graph]*
chizu'sho	地図書	atlas (book) *[pr]*

chō	兆	trillion (U.S.) *[math]*
chō	腸	intestine *[anat]*
chō	蝶	butterfly (insect) *[i-zoo]*
chō	調	key *{n} [music]*
chō-	超-	ultra- *{prefix}*
chō'ai-shirabe	丁合い調べ	collating *{n} [pr]*
chō'biryō-bunseki	超微量分析	ultramicroanalysis *[an-ch]*
chō-bi'ryūshi	超微粒子	superfine grain *[mech] [phys]*
chō'bisai	超微細	ultrafineness *{n}*
chō'bisai-kakō	超微細加工	hyperfine processing *{n}*
chō'bisai'ka suru	超微細化する	render superfine *{vb t}*
chō'bisai-kōzō	超微細構造	ultrastructure *[mol-bio]*
"	超微細構造	hyperfine structure *[spect]*
cho'boku-chi	貯木池	timber pond *[ecol]*
chō'bunkai	超分解	hyperresolution *[comput]*
chōchin	提灯；ちょうちん	lantern (made of paper) *[opt]*
chō-chō	蝶蝶；蝶々	butterfly (insect) *[i-zoo]*
chō-chōha	超長波	very-low-frequency wave *[elecmg]*
chōchō-uo	蝶蝶魚	butterfly fish *[v-zoo]*
chō'dansei	超弾性	superelasticity *[mech]*
chōdatsu suru	調達する	procure *{vb}*; supply *{vb} [econ]*
chō'dendō	超伝導；超傳導	superconductivity *[sol-st]*
chō'dendō-ji'shaku	超伝導磁石	superconducting magnet *[elecmg]*
chō'dendō-ryōshi-kanshō-kei	超伝導量子干渉計	superconducting quantum interference meter *[eng]*
chō'dendō-zairyō	超伝導材料	superconducting material *[sol-st]*
chōdo	丁度	exactly *{adv}*
chōdo	調度	furnishings *[furn]*
chōdo	聴度；聽度	audibility *[acous]*
chō'dōden-jōtai	超導電状態	superconductive state *[sol-st]*
chō'dōki-dendō-ki	超同期電動機	supersynchronous motor *[eng]*
chō'en	長円；長圓	ellipse *{n} [math]*
chōen'men	長円面	ellipsoid *{n} [math]*
chō'enshin-bunseki	超遠心分析	ultracentrifugal analysis *[ch-eng]*
chō'enshin-ki	超遠心機	ultracentrifuge *[eng]*
chō'etsu-sū	超越数	transcendental number *[math]*
chō'fuku	重複；ちょうふく	duplication *[graph]*
chōfuku-jun'retsu	重複順列	repeated permutation *[math]*
chōfuku-jusei	重複受精	double fertilization *[bot]*
chōfuku suru	重複する	duplicate *{vb t} [graph]*
chōga	頂芽	terminal bud *[bot]*
chōga'gaku	蝶蛾学	lepidopterology *[bio]*

chōgata-ben	蝶形弁；蝶形瓣	butterfly valve *[eng]*
chōgata-ka	蝶形花	papilonaceous flower *[bot]*
chōga-yūsei	頂芽優勢	apical dominance *[bot]*
chō'gengo	超言語	metalanguage *[comput]*
chō'ginga-dan	超銀河団	supercluster of galaxies *[astron]*
chōgō	調号；ちょうごう	key signature *[music]*
chōgō	調合	mixing *{n}* *[pharm]*
chō'gōkin	超合金	superalloy *[met]*
chōgō-kōryō	調合香料	compounded spicery *[cook]*
chō'gun	超群；ちょうぐん	supergroup *[comm]*
chō-ha	長波	long wave *[comm]*
chō-ha	調波	harmonic wave *[phys]*
chōhatsu	徴発	requisition *{n}* *[mil]*
chō'himo-riron	超紘理論	superstring theory *[part-phys]*
chō-hōkaku	超方格	hyper square *[math]*
chōhō'kei	長方形	oblong; rectangle *{n}* *[math]*
chō'hōsha	超放射	superradiance *[opt]*
chōho-shiki	調歩式	start-stop system *[comput]*
chō-i	潮位	tide level *[geol]*
chō-iro'zōkan	超色増感	supersensitization *[photo]*
chōji	丁子；ちょうじ	clove (spice) *{n}* *[bot]* *[cook]*
chōji-kankei	調時関係	timed relationship *[ind-eng]*
chō-jiku	長軸	long axis; major axis *[math]*
chōjō	重畳；ちょうじょう	superposition *{n}* *[math]*
chōjō	頂上	zenith; crown; top *{n}*
chō'jōji-sei	超常磁性	superparamagnetism *[elecmg]*
chō'jū'genshi	超重原子	superheavy atom *[phys]*
chō'kachō-on	超可聴音	ultrasound *[acous]*
chōkai	潮解	deliquescence *[p-ch]*
chō-kaku	頂角；ちょうかく	vertical angle *[math]*
chō'kaku	聴覚；聽覺	audition; (sense of) hearing *[physio]*
chōkaku-ka	長角果	silizue *{n}* *[bot]*
chōkan-tai	潮間帯；潮間帶	intertidal zone *[geol]*
chōkan-zu	鳥瞰図	bird's eye view *[navig]*
chō'kasei-sei	超加成性	superadditivity *[photo]*
chō'kika-bunpu	超幾何分布	hypergeometric distribution *[math]*
chō'kika-kansū	超幾何関数	hypergeometric function *[math]*
chōki-yohō	長期予報	long-range forecast *{n}* *[meteor]*
chokkaku	直角；ちょっかく	right angle *[math]*
chokkaku'bun	直角分	quadrature component *[elec]*
chokkaku-chū	直角柱	right-angle cylinder; right prism *[math]*
chokkaku-jōgi	直角定規	try square *{n}* *[math]*

chokkaku-ri'kaku	直角離角	quadrature *[astron]*
chokkaku-sankak(u)kei	直角三角形	right triangle *[math]*
chokkaku-zahyō	直角座標	rectangular coordinates *[math]*
chokkan	直観；直觀	intuition *[psy]*
chokkei	直径；ちょっけい	diameter *[math]*
chokkei-sen no hōsoku	直径線の法則	rectilinear diameter, law of *[p-ch]*
chokketsu	直結	online *[comput]*
chokketsu-kei no	直結形の	direct-coupled *{adj} [elec]*
chokketsu suru	直結する	connect (or couple) directly *{vb t}*
chokkon	直根	taproot *[bot]*
chokkō-sei	直交性	orthogonality *[math]*
chokkō suru	直交する	cross orthogonally *{vb} [math]*
chokkō-zahyō	直交座標	rectangular coordinates *[math]*
chōkō	徴候；兆候	symptom; indications *[med]*
chō-kō'atsu	超高圧；超高壓	extra-high voltage *[elec]*
chō'kōbunshi	超高分子	superpolymer *[poly-ch]*
chō'kogata-denshi-kōgaku	超小形電子工学	microelectronics *[electr]*
chō-kogata'ka	超小形化	microminiaturization *[electr]*
chō'kogata-kōsei-bubun	超小形構成部分	microcomponent *[elec] [sci-t]*
chō'kogata-kōzō	超小形構造	microstructure *[sci-t]*
chōkō-gōkin	超硬合金	cemented carbide; sintered hard alloy *[met]*
chōkō-kōgu	超硬工具	(cemented) carbide tool *[eng]*
chōkoku	彫刻；ちょうこく	knurling *{n} [eng] [met]*
"	彫刻	engraving *{n} [graph] [pr]*
"	彫刻	sculpting *{n} [graph]*
chōkoku-tō	彫刻刀	burin *{n} [graph]*
chōkō-kyōdo	超高強度	ultra-high strength *{n} [phys]*
chō'kōon-shunkan-sakkin-hō	超高温瞬間殺菌法	ultra-high-temperature sterilizing method *[microbio]*
chō'kōshi-kōzō	超格子構造	superstructure *[sol-st]*
chōkō-shinkū	超高真空	ultra-high vacuum *{n} [phys]*
chōkō-shū'ha	超高周波	Extremely High Frequency; EHF *[comm]*
chō'kōsoku(do)-butsuri	超高速(度)物理	hypervelocity physics *[phys]*
chō'kōsoku-dōro	超高速道路	superhighway *[civ-eng]*
chō'kōsō-taiki-butsuri'gaku	超高層大気物理学	aeronomy *[geophys]*
chō'kōsō-taiki-kagaku	超高層大気化学	aeronomy *[geoch]*
choku'chō	直腸	rectum (pl -s, recta) *[anat]*
choku'dō no	直動の	direct-acting *{adj} [elec]*
choku-enchū	直円柱；直圓柱	right circular cylinder *[math]*
choku-ensui	直円錐	right circular cone *[math]*
chokuhō'tai	直方体	rectangular parallelepiped *[math]*

chō'kūki-riki'gaku	超空気力学	superaerodynamics *[fl-mech]*
choku'retsu	直列；ちょくれつ	serial(ness) *{n} [comput]*
"	直列	series *{n} [elec]*
chokuretsu-densō	直列伝送	serial transmission *[comput]*
chokuretsu-enzan	直列演算	serial arithmetic *[math]*
chokuretsu-hen'atsu-ki	直列変圧器	series transformer *[elecmg]*
chokuretsu'ka	直列化	serialization *[elec]*
chokuretsu'ka suru	直列化する	serialize *{vb} [comput]*
chokuretsu-kyōshin	直列共振	series resonance *[elec]*
choku'ryū	直流	direct current *[elec]*
chuku'ryū-denryoku	直流電力	direct-current power *[elec]*
choku-sa	直鎖；ちょくさ	normal chain; straight chain *[ch]*
choku'sa'jō	直鎖状	straight-chain form *[ch]*
choku'sa-tanka'suiso	直鎖炭化水素	straight-chain hydrocarbon *[org-ch]*
choku'sen	直線；ちょくせん	straight line; line *{n} [math]*
chokusen'gata-bunshi	直線形分子	linear molecule *[ch]*
chokusen-kasoku-sōchi	直線加速装置	linear accelerator (linac) *[nucleo]*
chokusen-zōfuku	直線増幅	linear amplification *[electr]*
choku'setsu	直接；ちょくせつ	directness *{n}*
chokusetsu-ben'eki	直接便益	direct benefit *[econ]*
chokusetsu-ki'oku	直接記憶	direct-access storage *[comput]*
choku'setsu ni	直接に	directly *{adv}*
chokusetsu-seigyo	直接制御	direct control *[comput]*
chokusetsu-senryō	直接染料	direct dye *[mat]*
chokusetsu-setsugō	直接接合	direct coupling *{n} [mech-eng]*
chokusetsu-shōmei	直接照明	direct illumination *[opt]*
choku'sha	直射	direct projection *[astron]*
choku'shi	直視	direct vision *[opt]*
chokushi-bunkō-ki	直視分光器	direct-vision spectroscope *[spect]*
chokushi-kan	直視管	direct-viewing tube *[electr]*
chokushi-rui	直翅類	Orthoptera *[i-zoo]*
chō'kussetsu	超屈折	superrefraction *[geophys]*
choku-teisū	直定数	literal *{n} [comput]*
choku'tsū	直通	direct line *[comm]*
chō-kyori-denwa	長距離電話	long-distance telephone *[comm]*
chō-kyori-kōhō	長距離航法	long-distance navigation *[navig]*
chō-kyori no	長距離の	long-distance; long-range *{adj}*
chō-kyori-senro	長距離線路	long-distance line *[comm]*
chō-kyori-sōden	長距離送電	long-distance transmission *[comm]*
chō-kyori-yobi'dashi	長距離呼出	long-distance call *[comm]*
chō'kyosei	超巨星	supergiant star *[astron]*
chō'kyōyaku	超共役	hyperconjugation *[p-ch]*

chō'kyūsoku-rei'kyaku	超急速冷却	ultra-rapid cooling *{n} [eng]*
chomei-chikei	著明地形	landmark *[eng] [navig]*
chomei na	著名な	famous; well-known *{adj}*
chōmen	帳面	notebook; account book *[pr]*
chōmen	彫面	facet *{n} [met]*
chōmi'ryō	調味料	seasoning *{n} [cook]*
chōna; te'ono	手斧	adze *[eng]*
chō-neji	蝶ねじ	wing nut *[mech-eng]*
chō-on	長音	long sound; long vowel *[ling]*
chō'on	超音；ちょうおん	ultrasound *[acous]*
chō'on-fugō	長音符号	prolonged-sound symbol *[comput]*
chō-onpa	超音波	ultrasonic wave *[acous]*
chō'onpa-chi'en-sen	超音波遅延線	acoustic delay line *[electr]*
chō'onpa-kenbi'kyō	超音波顕微鏡	scanning acoustic microscope *[opt]*
chō'onpa-kensa-hō	超音波検査法	ultrasonography *[eng]*
chō'onpa-kiroku	超音波記録	ultrasonogram *[eng]*
chō'onpa-senjō	超音波洗浄	ultrasonic cleaning *[eng]*
chō'onpa-shori	超音波処理	ultrasonication *[microbio]*
chō'onpa-tanshō-ki	超音波探傷器	ultrasonic flaw detector *[eng-acous]*
chō'onpa-yō'setsu	超音波溶接	ultrasonic welding *[plas]*
chō'on-pu	長音符	macron; prolong-sound mark *[pr]*
chō'onsoku no	超音速の	supersonic *{adj} [phys]*
chō'onsoku-ryokak(u)ki	超音速旅客機	supersonic transport; SST *[aero-eng]*
chōri	調理；ちょうり	cooking (of food) *{n} [cook]*
chōri-hō	調理法	recipe *{n} [cook]*
chō'ritsu	調律	tuning *{n} [acous]*
chō'rittai-eizō	超立体映像	hyperstereoscopy (map) *[graph]*
chō-rui	鳥類	Aves; birds *[v-zoo]*
chōrui'gaku	鳥類学；鳥類學	ornithology *[zoo]*
chōrui-sō	鳥類相	avifauna *[v-zoo]*
chō'ryoku	張力	tension *[mech]*
chō'ryoku	聴力；聽力	hearing *{n} [acous] [physio]*
chōryoku-kei	張力計	tension gage *[eng]*
chōryoku-kei	聴力計	audiometer *[eng]*
chōryū	潮流	tidal current *[ocean]*
chōsa	長鎖	heavy chain; long chain *[ch]*
chōsa	調査	survey *{n} [eng]*
"	調査	search *{n} [navig]*
chōsa	潮差	tide range *[ocean]*
chosak(u)ken	著作権	copyright; literary property *[legal]*
chōsan	超酸	superacid *[ch]*
chō(-sanji'gen)-kūkan	超(三次元)空間	hyperspace *[math]*

chōsei	調性	tonality *[music]*
chōsei	調整；ちょうせい	correction *[acous] [bio] [math]*
"	調整	tuning *{n} [comput] [electr]*
"	調整	justification *[graph]*
"	調整	adjustment; regulation *{n}*
chōsei-ben	調整弁；調整瓣	regulating valve *[mech-eng]*
chō-seidō	超青銅	super-bronze *[met]*
chōsei suru	調製する	manufacture; prepare *{vb} [ind-eng]*
chōsei-zai	調整剤	modifier; regulator *[ch]*
chōseki	長石；ちょうせき	feldspar *[miner]*
chōseki-fū	潮汐風	tidal wind *[meteor]*
chōseki-hyō	潮汐表	tidal table *[ocean]*
chōseki-ma'satsu	潮汐摩擦	tidal friction *[ocean]*
chōseki-sayō	潮汐作用	tidal action *[ocean]*
chōseki-shōsan	潮汐消散	tidal dissipation *[ocean]*
chōsetsu	調節；ちょうせつ	regulation *[bio]*
"	調節	adjustment; control *[eng] [sci-t]*
chōsetsu dekiru	調節出來る	adjustable *{adj}*
chōsetsu-ki	調節器	moderator (instrument) *[eng]*
chō-shaku	長尺	long-length *{n} [mech]*
chō'shikaku no	聴視覚の	audiovisual *{adj} [comm]*
chō'shin	長針	long hand (of clock); minute hand *[horol]*
chōshin-ki	聴診器；聽診器	stethoscope *[med]*
chō'shinsei	超新星	supernova (pl -novas,- novae) *[astron]*
chō'shitsu-atsu'en	調質圧延	temper rolling *{n} [met]*
chō'shitsu-tan	調湿炭	moisture-adjusted coal *[geol]*
chōsho	長所	advantage; merit; strong point *{n}*
chōshō	潮衝；ちょうしょう	riptide *[ocean]*
chōshoku	朝食；ちょうしょく	breakfast *{n} [cook]*
chōshoku	調色	toning *{n} [photo]*
chō-shūki	長周期	long period *[phys]*
chōsoku-ki	調速機	speed governor *[mech-eng]*
chō-sokushin-zai	超促進剤	ultra-accelerator *[poly-ch]*
chō'sosei	超塑性	superplasticity *[met]*
chō-sūgaku	超数学	metamathematics *[math]*
chosui'chi	貯水池；ちょすいち	reservoir *[civ-eng]*
chosui-tō	貯水塔	water tower *[civ-eng]*
chō'taishō-sei	超対称性	supersymmetry *[part-phys]*
chō'tajū'ka	超多重化	supermultiplication *[quant-mech]*
chō'tajū-kō	超多重項	supermultiplet *[quant-mech]*
chō-tanpa	超短波	very-high-frequency; VHF *[comm]*
chōtei	調停	arbitration; mediation; settlement

chō'teikū-hikō	超低空飛行	terrain following *{n} [navig]*
chō-tei'on	超低温	ultra-low temperature *{n} [phys]*
chōtei-shūha	超低周波	Extremely Low Frequency; ELF *[comm]*
chōten	長点; 長點	dash (Morse code) *{n} [comm]*
chōten	頂点	apex; vertex *[math]*
chō'tsugai	蝶番	hinge *{n} [mech-eng]*
chōwa-kansū	調和関数	harmonics *[math]*
chōwa-shindō'shi	調和振動子	harmonic oscillator *[electr]*
chōwa-sōchi	聴話装置	monitoring device *[comm]*
chōwa-sūretsu	調和数列	harmonic progression *[math]*
chōwa-yūkai	調和融解	congruent melting *{n} [p-ch]*
chōyō'sei no	腸溶性の	enteric *{adj} [pharm]*
chō-zame	蝶鮫; ちょうざめ	sturgeon; Acipenser mikadoi (fish) *[v-zoo]*
chozō	貯蔵; 貯藏	storage *[ind-eng]*
chōzō	彫像	statue *[furn]*
chozō-antei-sei	貯蔵安定性	shelf life; storage stability *[ind-eng]*
chozō-jumyō	貯蔵寿命	shelf life; storage life *[ind-eng]*
chō'zōkan	超増感	hypersensitization *[photo]*
chō'zume	腸詰め	sausage *[cook]*
chō'zume-chūdoku	腸詰め中毒	botulism *[med]*
chū	中	medium; middle *[math]*
chū	註; 注	note; remark *{n} [comput] [pr]*
chū	柱	cylinder *[math]*
-chū	-虫	-insect *{suffix}*
chūbō	厨房	galley (of ship) *[nav-arch]*
chū'buru no	中古の	used; secondhand *{adj}*
chūcho suru	躊躇する	hesitate; vacillate *{vb} [psy]*
chūdan	中断; 中斷	interruption; discontinuance *{n}*
chūdan suru	中断する	interrupt; suspend; discontinue *{vb}*
chūdo	稠度; ちゅうど	consistency *[mat]*
chūdo-kei	稠度計	consistometer *[eng]*
chūdoku	中毒	intoxication; poisoning *{n} [med]*
chū'ei	虫癭; ちゆうえい	gall; insect gall *[bot]*
chū'fusseki	中沸石	mesolite *[miner]*
chū-gurai no	中位いの	medium-degree *{adj}*
chūhaku-rō	虫白蠟	insect wax *[mat]*
chū'i	注意; ちゆうい	attention; caution; note; heeding *[psy]*
chūi-hō	注意報	alert signal *[comm]*
"	注意報	advisory *{n} [meteor]*
chū'i no-	中位の-	meso- *{prefix}*
chū'i-sū	中位数	median *{n} [math]*
chūjiku-kokkaku	中軸骨格	axial skeleton *[v-zoo]*

chū'jitsu-sei	忠実性；忠實性	fidelity *[comm]*
chūjō-shō	柱状晶	rod crystal *[crys]*
chūjō-zu(hyō)	柱状図(表)	histogram *[math]*
chū-kahi	中果皮	mesocarp *[bot]*
chūkai	仲介	intercession; (inter)mediation *{n}*
chūkai	厨芥；ちゅうかい	garbage *[civ-eng]*
chū-kakko	中括弧	brace(s) *[comput] [pr]*
chūkaku	中核	nucleus *[comput]*
chūkan	中間；ちゅうかん	intermediate; middle; midway *{n}*
chūkan'butsu	中間物	intermediate (product) *{n} [math]*
chūkan'chō	中間調	halftone *{n} [graph]*
chūkan-jōtai	中間状態	intermediate state *[quant-mech]*
chūkan-keitai	中間形態	mesomorphic state *[zoo]*
chūkan-ken	中間圏；中間圈	mesosphere *[meteor]*
chūkan (ni aru koto)	中間(にあること)	betweenness *[math]*
chūkan no	中間の	intermediate *{adj}*
chūkan no-	中間の-	mid- *{prefix}*
chūkan'shi	中間子	meson *[phys]*
chūkan'shitsu	中間質	intercalation *[geol]*
chūkan-shūha'sū	中間周波数	intermediate frequencies *[elec]*
chūkan'tai	中間体	intermediate (product) *{n} [ch]*
chūkei	中継；中繼	relaying *{n} [comm]*
chūkei	注型；ちゅうけい	casting *{n} [eng]*
chūkei-kairo	中継回路	junction circuit *[elec]*
chūkei-ki	中継器；中繼器	repeater *[electr]*
chūkei'yō-jushi	注型用樹脂	casting resin *[poly-ch]*
chūki	中期	metaphase *[bio]*
chūki	注記	annotation *[pr]*
chū'kibo-shūseki-kairo	中規模集積回路	medium-scale integrated circuit; MSI *[electr]*
chūki'jō	駐機場	apron (at an airport) *[trans]*
chūkō	昼光；晝光	daylight *[astron]*
chūkō-denkyū	昼光電球	daylight lamp *[elec]*
chūkō-dōbutsu	昼行動物	diurnal animal *[zoo]*
chūkō'hin	鋳鋼品	steel castings *[mat]*
chūkō-ritsu	昼光率	daylight factor *[opt]*
chūkū	中空；ちゅうくう	midair *{adv} {n} [navig]*
"	中空	hollowness *{n}*
chūkū-kō	中空鋼	cored steel *[steel]*
chūkū no	中空の	hollow *{adj} [tex]*
chūkū-shi	中空糸	hollow fiber; hollow yarn *[tex]*
chūkū-tanzō'hin	中空鍛造品	hollow forgings *[met]*

chū'mitsu-roppō-kōshi	稠密六方格子	dense hexagonal lattice *[crys]*
chūmon	注文	order; commission *{n}*
chūmon suru	注文する	commission; order; place an order *{vb t}*
chū'nikai	中二階	mezzanine *[arch]*
chūnyū	注入	injection *[mech-eng]*
chū'ō	中央；ちゅうおう	center *{n} [math]*
chūō-chi	中央値	median *{n} [math]*
chūō-kyoku	中央局	central office *[comm]*
chūō-mi'hiraki-pēji	中央見開き頁	centerfold; center spread *[pr]*
chū-on	中音	middle frequencies *[acous]*
chūō no	中央の	median *{adj} [math]*
"	中央の	central *{adj}*
chūō no-	中央の-	meso- *{prefix} [ch]*
chū'on-shokubutsu	中温植物	mesotherm *[bot]*
chūō-seigyo-sōchi	中央制御装置	central control unit *[comput]*
chūō-shori-sōchi	中央処理装置	central processing unit; CPU *[comput]*
chūrin'jō	駐輪場	bicycle parking area *[civ-eng]*
chūritsu	中立	neutrality
chūsai	仲裁	mediation
chūsei	中性；ちゅうせい	neutral; neutrality
chūsei-bishi	中性微子	neutrino *[phys]*
chūsei'chōseki	中性長石	andesine *[miner]*
chūsei-en	中性塩	neutral salt *[ch]*
chūsei-sen	中性線	neutral line *[meteor]*
chūsei'shi	中性子	neutron *[phys]*
chūseishi-bakudan	中性子爆弾	neutron bomb *[ord]*
chūseishi-boshi	中性子星	neutron star *[astron]*
chūseishi-dan'menseki	中性子断面積	neutron cross section *[nuc-phys]*
chūseishi-hokaku	中性子捕獲	neutron capture *[nuc-phys]*
chūseishi-kaisetsu-kei	中性子回折計	neutron diffractometer *[phys]*
chūseishi-keisū-ki	中性子計数器	neutron counter *[nucleo]*
chūseishi-ken'shutsu-ki	中性子検出器	neutron detector *[nucleo]*
chūseishi-sanran	中性子散乱	neutron scattering *[neucleo]*
chūseishi'sen-kōzō-kaiseki'gaku	中性子線構造=解析学	neutron radiography *[nucleo]*
chūseishi'sen-soku	中性子線束	neutron flux *[nucleo]*
chūseishi-sha'shutsu-ki	中性子射出器	neutron howitzer *[nucleo]*
chūsei-shokubutsu	中生植物	mesophyte *[bot]*
chūseki-do	沖積土	alluvial soil *[geol]*
chūseki-heichi	沖積平地	apron *[geol]*
chūseki-heiya	沖積平野	alluvial plain; floodplain *[geol]*
chū-sekki-jidai	中石器時代	Mesolithic Age *[archeo]*

chū-sen	中線	median line; midline *[math]*
chūsha	注射	injection *[med]*
chūsha'jō	駐車場	parking area *[civ-eng]*
chūsha-kinshi	駐車禁止	No Parking *[traffic]*
chūsha-ki	注射器	(hypodermic) syringe; injector *[med]*
chūshaku	注釈；注釋	annotation; comment *[comm] [pr]*
chūsha-tō	駐車灯	parking light *[elec]*
chūshi	中止	suspension; discontinuance
chū'shin	中心；ちゅうしん	center *{n} [math]*
chūshin'bu	中心部	kernel *{n}*
chūshin-ka	中心花	disk flower *[bot]*
chūshin-kaku	中心角	central angle *[math]*
chūshin-kaku	中心核	core *{n} [geol]*
chūshin-ken	中心圏；中心圈	centrosphere *[geol]*
chūshin-kiseki	中心軌跡	centroid *[math]*
chūshin-kyoku'gen-teiri	中心極限定理	central limit theorem *[math]*
chūshin no	中心の	central; middle; major; main *{adj}*
chūshin-sa	中心差	equation of the center *[astron]*
chūshin-shōtai	中心小体	centriole *[cyt]*
chūshi suru	中止する	discontinue; interrupt *{vb}*
chūshō-daisū'gaku	抽象代数学	abstract algebra *[math]*
chūshō-gei'jutsu	抽象芸術	abstract art
chūshō-kagaku	抽象科学	abstract science *[sci-t]*
chūshō-kigō	抽象記号	abstract symbol *[graph]*
chūshoku; chūjiki	昼食；晝食	midday meal; lunch *{n} [cook]*
chūshō-kūkan	抽象空間	abstract space *[math]*
chū'shoku'sei no	虫食性の	insectivorous *{adj} [v-zoo]*
chūshō-sei	抽象性	abstraction
chūshō'teki na	抽象的な	abstract *{adj}*
chū-shuku'shaku-zu	中縮尺図	medium-scale map *[map]*
chūshū no mangetsu	仲秋の満月	harvest moon *[astron]*
chūshū no meigetsu	仲秋の明月	harvest moon *[astron]*
chū'shutsu	抽出；ちゅうしゅつ	abstraction; extraction *{n} [ch]*
chūshutsu'butsu	抽出物	extract *{n} [ch]*
chūshutsu-jōryū	抽出蒸留	extractive distillation *[ch-eng]*
chūshutsu-meirei	抽出命令	extract instruction *[comput]*
chūshutsu-sei	抽出性	extractability *[ch]*
chūshutsu-sōchi	抽出装置	extractor *[eng]*
chū'shutsu suru	抽出する	extract *{vb t} [ch]*
chūshutsu'yō-yōzai	抽出用溶剤	extractant; extracting solvent *[ch]*
chūsui	虫垂	appendix (pl -dixes, -dices) *[anat]*
chūsū-shinkei-kei	中枢神経系	central nervous system *[anat]*

chū-tanpa	中短波	intermediate wave *[elecmg]*
chū-ten	中点	midpoint *[math]*
chū-tetsu	鋳鉄；鋳鐵	cast iron *[met]*
chūtō	柱頭	stigma *[bot]*
chūtō-do	中等度	moderate (-degree); medium-degree *{n}*
chūwa	中和；ちゅうわ	neutralization *[ch]*
chūwa-en	中和塩	neutralized salt *[ch]*
chūwa-ka	中和価	neutralization number *[an-ch]*
chūwa-netsu	中和熱	heat of neutralization *[p-ch]*
chūwa suru	中和する	neutralize *{vb}* *[ch]*
chūwa-zai	中和剤	neutralizing agent *[ch]*
chū-yu	中油	middle oil *[mat]*
chū-zahyō	柱座標	cylindrical coordinates *[math]*
chūzan'eki	抽残液	raffinate *[ch]*
chūzō-jushi	鋳造樹脂	casting resin *[poly-ch]*

D

dachō	駝鳥	ostrich (bird) *[v-zoo]*
da'doku; ja'doku	蛇毒	snake venom *[physio]*
da'eki	唾液	saliva *[physio]*
da'eki-sen	唾液腺	salivary gland *[physio]*
da'en	楕円；楕圓	ellipse *[math]*
daen-chū	楕円柱	elliptic cylinder *[math]*
daen-henkō	楕円偏光	elliptically polarized light *[opt]*
daen-men	楕円面	ellipsoid *{n}* *[math]*
daen-ritsu	楕円率	ellipticity *[electr]*
daen-sui	楕円錐	elliptic cone *[math]*
daen'tai	楕円体	ellipsoid *{n}* *[math]*
daen-zahyō	楕円座標	ellipsoidal coordinates *[math]*
da'fuku'jō no	蛇腹状の	snake-belly form *{adj}* *[graph]*
da'gakki-rui	打楽器類	percussion instruments *[music]*
dahei	舵柄；だへい	tiller *{n}* *[nav-arch]*
dai	代；だい	era *[geol]*
-dai	-台；-臺	-level(s); -vehicle(s); -age group *{suffix}*
dai-bakari	台秤	platform scale *[eng]*
dai-baku'hatsu-setsu	大爆発説	Big Bang theory *[astron]*
daichi	大地	earth *{n}* *[geol]*
daichi	台地	plateau; upland; tableland *[geol]*
daichi-den'i	大地電位	ground potential *[elec]*
daichi-teikō	大地抵抗	ground (or earth) resistance *[elec]*
daichō	大腸	large intestine *[anat]*
daichō-kin	大腸菌	coliforms; Escherichia coli *[microbio]*
daidai	橙；だいだい	bitter orange; bigarade (fruit) *[bot]*
dai'dokoro	台所；臺所	kitchen *[arch]* *[cook]*
dai'dokoro-yōhin	台所用品	kitchen utensil(s) *[cook]* *[eng]*
dai-dōmyaku	大動脈	aorta *[anat]*
dai-en	大円；大圓	great circle *[geod]*
daifūshi-san	大風子酸	hydnocarpic acid
daifūshi-yu	大風子油	chaulmoolgra oil *[mat]*
daihyō	代表	representative *{n}*
dai'ichi-en	第一塩	-ous salt; primary salt *[ch]*
dai'ichi-enka-suzu	第一塩化錫	stannous chloride
dai'ichi-kagō'butsu	第一化合物	-ous compound; primary compound *[ch]*
dai'ichi-rinsan-anmonyūmu	第一燐酸アンモニウム	ammonium primary phosphate

dai'ichi-rinsan-karushūmu	第一燐酸カルシウム	calcium primary phosphate
dai'ichi-rinsan-karyūmu	第一燐酸カリウム	potassium primary phosphate
dai'ichi-rinsan-natoryūmu	第一燐酸ナトリウム	sodium primary phosphate
dai'ichi-tanso-genshi	第一炭素原子	primary carbon atom *[ch]*
dai'ikkyū	第一級	primary *[ch]*
dai'isshu no ayamari	第一種の誤り	type one error *[math]*
dai'i suru	代位する	subrogate *{vb} [legal]*
daija	大蛇; だいじゃ	anaconda; big snake (reptile) *[v-zoo]*
dai-kajō	大過剰; 大過剩	great excess *{n} [ch]*
dai-kakko	大括弧	bracket(s) *[comput]*
dai-kaku	大核	macronucleus *[i-zoo]*
daikei	台形	frustum; trapezoid *[math]*
daikei-hizumi	台形歪	trapezoidal distortion *[electr]*
dai'kibo-shūseki'ka	大規模集積化	large-scale integration *[electr]*
dai'kibo-shūseki-kairo	大規模集積回路	large-scale integrated circuit; LSI *[electr]*
dai-kikō	大気候; 大氣候	macroclimate *[climat]*
daikon	大根	daikon radish (vegetable) *[bot]*
daikon-oroshi	大根下ろし	grated daikon (q.v.) *[cook]*
daiku	大工	carpenter *[eng]*
(dai-)kyū'shi	大臼歯	molar *{n} [anat]*
daimei	題名	title *[pr]*
daimei'shi	代名詞	pronoun *[gram]*
dai-moji	大文字	uppercase letter *[comput] [pr]*
dai'ni-en	第二塩	-ic salt; secondary salt *[ch]*
dai'ni-kagō'butsu	第二化合物	secondary compound *[ch]*
dai'ni-kyū	第二級	secondary *{n} [ch]*
dai'ni no	第二の	second (in order) *{adj} [math]*
dai'ni-rinsan-anmonyūmu	第二燐酸アンモ=ニウム	ammonium secondary phosphate
dai'ni-rinsan-karushūmu	第二燐酸カルシウム	calcium secondary phosphate
dai'ni-rinsan-karyūmu	第二燐酸カリウム	potassium secondary phosphate
dai'ni-rinsan-natoryūmu	第二燐酸ナトリウム	sodium secondary phosphate
dai'nishu no ayamari	第二種の誤り	type two error *[math]*
dai'ni-tanso-genshi	第二炭素原子	secondary carbon atom *[ch]*
dainō	大脳; 大腦	cerebrum (pl -s, -bra) *[anat]*
dainō-hi'shitsu	大脳皮質	cerebral cortex *[anat]*
dairi	代理; だいり	representation *{n}*
dairi'ken	代理権	representation right *[legal]*
dairi'nin	代理人	agent; deputy (a person)
dairi'seki	大理石	marble *{n} [petr]*
dairi'ten	代理店	agent (an establishment) *[econ]*

dai'san-kagōbutsu	第三化合物	tertiary compound *[ch]*
dai-sankaku-ho	大三角帆	lateen sail *[nav-arch]*
dai'san-kyū	第三級	tertiary *{n}* *[ch]*
dai'san no	第三の	third (in order) *{adj}* *[math]*
dai'san'sha	第三者	third person; third party *{n}*
dai'san-tanso-genshi	第三炭素原子	tertiary carbon atom *[ch]*
dai-seidō	大聖堂	cathedral *[arch]*
daisha	台車；臺車	flatcar; platform car *[rail]*
dai'shi(kyū)-kagō'butsu	第四(級)化合物	quaternary compound *[ch]*
dai'shi-tanso-genshi	第四炭素原子	quaternary carbon atom *[ch]*
daishoku-saibō	大食細胞	macrophage; histiocyte *[histol]*
dai-sōgen	大草原	savannah *[ecol]*
daisu	ダイス	die *{n}* *[eng]*
daisū'gaku	代数学	algebra *[math]*
daisū-shori-gengo	代数処理言語	algebraic language *[comput]*
daitai	大体	generally *{adj}*
daitai	大腿	thigh *[anat]*
daitai	代替；だいたい	alternate *{n}* *[comput]*
"	代替	substitution *{n}*
daitai'butsu	代替物	substitute *{n}* *[ind-eng]*
daitai-keiro	代替経路	alternate route *[comput]* *[navig]*
daitai-kī	代替キー	alternate key *[comput]*
daitai-kotsu	大腿骨	femur (bone) *[anat]*
dai-toshi-ken	大都市圏	metropolitan area *[civ-eng]*
daiyō'hin	代用品	substitute (an object) *{n}* *[ind-eng]*
dai'yon-hirei-kō	第四比例項	fourth proportional *[math]*
dai'yonkyū-amin	第四級アミン	quaternary amine *[org-ch]*
dai(-yōryō)-ki'oku	大(容量)記憶	mass storage *[comput]*
daiza	台座	pedestal *[arch]* *[civ-eng]* *[eng]*
dai-zentei	大前提	major premise *[logic]*
daizu	大豆；だいず	soybean *[bot]*
daizu-kasu	大豆粕	soybean meal *[agr]*
dajō-ki	打錠機	tableting machine *[eng]*
dakkai	脱灰	deashing; decalcification; deliming *[ch]*
"	脱灰	demineralization *[dent]*
dakki	脱気	deaeration *[mech-eng]*
dakki-sōchi	脱気装置	deaerator *[mech-eng]*
daku-do	濁度	turbidity *[an-ch]*
daku'fusseki	濁沸石	laumontite *[miner]*
dak(u)ka	濁化	turbidization *[ch]*
daku'on	濁音	sonant; voiced sound *[ling]*
daku'on-pu	濁音符	voiced sound mark *[pr]*

daku-ten	濁点	voiced sound mark *[pr]*
dame	駄目	no good
dan	段	step *{n} [arch] [ch-eng]*
"	段	stage; plate; tray *[ch-eng] [p-ch]*
"	段	column *[comput] [pr]*
"	段	level *{n}*
dan	壇	platform *[arch]*
danbō	暖房	heating *{n} [eng]*
danbōru	段ボール	corrugated fiberboard *[mat]*
dan'chigai no ie	段違いの家	split-level house *[arch]*
dandō	弾道；彈道	trajectory *[mech]*
dandō'gaku	弾道学	ballistics *[mech]*
dan'dori	段取り	preparatory plan *[civ-eng]*
dan'dori o kimeru	段取りを決める	setting priorities *[sci-t]*
dangai	断崖；斷崖	bluff; precipice *{n} [geol]*
dangan	弾丸	bullet *[ord]*
dango	団子	dumpling *[cook]*
dan'gumi	段組み	column setting *{n} [comput]*
dani	壁蝨；蜱；だに	mite; tick (insect) *[i-zoo]*
dani'gaku	壁蝨学；蜱学	acarology *[zoo]*
dankai	団塊；團塊	agglomerate; nodule *[geol] [sci-t]*
dankai	段階	stage; step; cycle; phase *[sci-t]*
dan-kō'ritsu	段効率	plate efficiency *[ch-eng]*
dankyū	段丘	terrace *[geol]*
dan'men	断面；斷面	profile *[math]*
dan'menseki	断面積	(cross-)sectional area *[math]*
danmen(-zu)	断面(図)	cross section *[geol]*
dannetsu	断熱；だんねつ	heat-insulating *{n} [civ-eng]*
"	断熱	adiabaticness *{n} [thermo]*
dannetsu-bōchō	断熱膨張	adiabatic expansion *[thermo]*
dannetsu-henka	断熱変化	adiabatic change *[thermo]*
dannetsu-katei	断熱過程	adiabatic process *[thermo]*
dannetsu-zai	断熱材	adiabatic material; heat-insulating material *[mat]*
danpen	断片	fragment *{n}*
danpen'ka	断片化	fragmentation *{n}*
dan'raku	段落	paragraph *[comput] [gram]*
dansa	段差	level difference; step difference *{n}*
dansei	男性	male (human); man *[bio]*
dansei	弾性；だんせい	elasticity *[mech]*
dansei-gendo	弾性限度	elastic limit *[mech]*
dansei-genkai	弾性限界	elastic limit *[mech]*

dansei-henkei	弾性変形	elastic deformation *[mech]*
dansei-hizumi	弾性歪	elastic strain *[mech]*
dansei-keisū	弾性係数	elastic coefficient; modular ratio *[mech]*
dansei-modori	弾性戻り	elastic recovery *[rub]*
dansei-ō'ryoku	弾性応力	elastic stress *[mech]*
dansei-ritsu	弾性率	modulus of elasticity *[mech]*
dansei-shi	弾性糸; だんせいし	elastic fiber *[tex]*
dansei-shindō	弾性振動	elastic oscillation *[phys]*
dansei-tai	弾性体	elastic material *[mat]*
dansei'teki na	男性的な	masculine (of humans) *{adj}* *[bio]*
dansei-yokō	弾性余効	elastic after-effect *[mech]*
dansen	断線; だんせん	disconnection *[elec]*
dansen-jumyō	断線寿命	burn-out life *[elec]*
dan-shoku	暖色; だんしょく	warm color *[opt]*
dan'shōmen-zu	断正面図	front sectional view *[graph]*
dansō	断層; 斷層	fault *[geol]*
dansō-hasai-tai	断層破砕帯	fractured zone *[geol]*
dansoku	弾速	bullet speed; projectile speed *[mech]*
dansō-satsu'ei-hō	断層撮影法	tomography; sectional radiology *[electr]*
dansō-sen	断層線	fault line *[geol]*
dansō-undō	断層運動	fault movement *[geol]*
dansū	段数; だんすう	number of plates; number of stages; number of steps *[p-ch]*
dantai	団体; 團體	association; group; organization
dantō	段塔	plate column *[ch-eng]*
dantō	弾頭	warhead *[ord]*
dantsū	段通; だんつう	cotton carpet; cotton rug *[furn]*
dan'zoku-shingō	断続信号	intermittent ringing; intermittent signals *[comm]*
dan'zumi	段積み; だんづみ	stacking *{n}*
dare	だれ	sag *{n}* *[rub]*
dare; tare	誰; だれ; たれ	who *{pron}*
dare no; tare no	誰の	whose *{pron}*
dareru	だれる	sag *{vb i}* *[rub]*
da'ryoku de susumu	惰力で進む	coast *{vb i}* *[navig]*
da'ryoku-unten	惰力運転	coasting *{n}* *[navig]*
dasei-hikō	惰性飛行	coasting flight *[navig]*
dashi	出し; だし	soup stock *[cook]*
dassan	脱酸; だっさん	deacidification *[ch]*
dassan(so)	脱酸(素)	deoxidation *[ch]*
dassanso-zai	脱酸素剤	deoxidizer; deoxidizing agent *[ch]*
dassei	脱錆; だっせい	derusting *{n}* *[ch-eng]*

dasshi	脱脂	degreasing *{n} [ch-eng]*
dasshi-daizu	脱脂大豆	defatted soybean *[cook]*
dasshi-men	脱脂綿	absorbent cotton; sanitary cotton *[tex]*
dasshin	脱心；脱芯	core removal *{n}*
dasshin-ki	脱進機	escapement *[horol]*
dasshi-nyū	脱脂乳	skim(med) milk *[cook]*
dasshitsu	脱湿；脱濕	dehumidification *[mech-eng]*
dasshutsu-sokudo	脱出速度	escape velocity *[aero-eng]*
dasshū-zai	脱臭剤	deodorant *{n} [mat] [pharm]*
dassui	脱水；だっすい	dehydration *[ch] [physio]*
"	脱水	desiccation *[hyd]*
dassui-do	脱水度	dehydration degree; deliquoring degree *[ch]*
dassui-ki	脱水機	dehydrator *[ch-eng]*
dassui-ritsu	脱水率	dehydration rate *[ch]*
dassuiso	脱水素	dehydrogenation *[ch]*
dassuiso-kōso	脱水素酵素	dehydrogenase *[bioch]*
dassui suru	脱水する	dehydrate *{vb t} [ch]*
dassui-zai	脱水剤	dehydrating agent *[ch]*
dasu	出す	remove; take out *{vb t}*
datchaku	脱着	desorption *[p-ch]*
datchisso	脱窒素	denitrification *[ch]*
datchitsu	脱窒	denitrification *[microbio]*
datchitsu-saikin	脱窒細菌	denitrifying bacteria *[microbio]*
datō-sei	妥当性	pertinence; validity *{n} [logic]*
datō'sei-kensa	妥当性検査	validation *[comput]*
datsu	啄長魚；だつ	needlefish; garfish (fish) *[v-zoo]*
datsu'aen	脱亜鉛	dezincification *[met]*
datsu-en	脱塩；脱鹽	demineralization; desalting *{n} [ch-eng]*
datsu'en-sui	脱塩水	desalted water *[ch-eng]*
datsu'hogo-zai	脱保護剤	deprotecting agent *[ch]*
datsu-ion-sui	脱イオン水	deionized water *[hyd]*
datsu'ji-kigō	脱字記号	caret *[comput] [pr]*
datsu'netsu-ki	脱熱器	heatsink (an instrument) *[elec]*
datsu'netsu-zai	脱熱剤	heatsink (a chemical) *[elec]*
datsu'raku	脱落	dropout; dropping out; omission *{n}*
datsu'reki	脱瀝	deasphalting *{n} [ch-eng]*
datsu'ri-hannō	脱離反応	elimination reaction *[org-ch]*
datsu'rin	脱燐；だつりん	dephosphorization *[ch] [met]*
datsu'rinsan'ka	脱燐酸化	dephosphorylation *[org-ch]*
datsu'ryū	脱硫	desulfurization; devulcanization *[ch-eng]*
datsuryū-zai	脱硫剤	desulfurizer *[ch-eng]*

dattan	脱炭	decarburization *[met]*
dattansan	脱炭酸	decarboxylation *[org-ch]*
dattansan-jūgō	脱炭酸重合	decarboxylation polymerization *[poly-ch]*
dattan-yaki'namashi	脱炭焼鈍	decarburization annealing *{n} [met]*
de	出；で	rising; rise *{n} [astron]*
de'guchi	出口	exit *{n} [arch] [comput]*
"	出口	outlet *{n}*
deiban'gan	泥板岩	shale *[geol]*
dei'do	泥土	mud; mire *[geol]*
deiryū	泥流	mudflow *[geol]*
deishō	泥漿；でいしょう	sludge *[geol]*
"	泥漿	slip; slurry *{n} [mat]*
deishō'ka	泥漿化	muddying; slurrying *{n} [ch-eng]*
deitan-sō	泥炭層	peat deposit *[geol]*
deka'guramu	瓧；デカグラム	decagram *[mech]*
deka'mētoru	籵；デカメートル	decameter *[mech]*
deko'boko	凸凹；でこぼこ	wavy surface *[cer]*
"	凸凹	bumps and dips; unevenness *[mech]*
de-mado	出窓	bay window; oriel window *[arch]*
denai; dewanai	でない；ではない	not *{adv}*
den'atsu	電圧；電壓	voltage *[elec]*
den'atsu-denryū-kei	電圧電流計	volt ammeter *[eng]*
den'atsu-kei	電圧計	voltmeter *[eng]*
den'atsu-kōka	電圧降下	voltage drop *[elec]*
den'atsu-ritoku	電圧利得	voltage gain *[elec]*
den'atsu-seiryū	電圧整流	voltage commutation *[elec]*
den'atsu-zōbai-ki	電圧増倍器	voltage multiplier *[electr]*
den'atsu-zōfuku-ki	電圧増幅器	voltage amplifier *[electr]*
denba	電場	electric field *[elec]*
denbun	伝聞	hearsay *{n} [legal]*
den'chaku	電着	electrocoating; electrodeposition; electroplating *[met]*
denchaku-ō'ryoku	電着応力	electrodeposition stress *[met]*
denchaku-tosō	電着塗装	electrodeposition coating *{n} [mat]*
denchi	電池	battery; cell *[elec]*
denchū	電鋳；電鑄	electroforming *{n} [met]*
denchū'hin	電鋳品	electrocast product *[mat]*
denchū-hō	電鋳法	galvanoplastics (method) *[met]*
dendō	伝導；傳導	transmission *[acous] [opt]*
"	伝導	conduction *[elec] [phys]*
dendō-do	伝導度	conductivity *[elec]*
dendō-ki	電動機	electric motor; motor *[elec]*

dendō-ritsu	伝導率	conductivity *[elec]*
dendō'shiki	電動式	electric-motor system *[elec]*
dengeki	電撃	electric shock *[physio]*
dengen	電源	power source *[elec]*
dengen-sōchi	電源装置	power supply *[elec]*
den'i	電位；でんい	electric potential *[elec]*
den'i-sa	電位差	potential difference *[elec]*
den'isa-kei	電位差計	potentiometer *[elec] [eng]*
den'isa-teki'tei	電位差滴定	potentiometric titration; potentiometry *[an-ch] [elec]*
denji-ba	電磁場	electromagnetic field *[elecmg]*
denji-ha	電磁波	electromagnetic wave *[elecmg]*
denji-hō	電磁砲；電磁砲	rail gun *[ord]*
denji-hōsha'sen	電磁放射線	electromagnetic radiation *[elecmg]*
denji-kai	電磁界	electromagnetic field *[elecmg]*
den'jiki	電磁気	electromagnetism *[phys]*
denjiki'gaku	電磁気学	electromagnetism (the discipline) *[phys]*
denji-ryoku	電磁力	electromagnetic force *[phys]*
denji-ryūtai-riki'gaku	電磁流体力学	magnetofluid dynamics; magneto-hydrodynamics; MHD *[phys]*
den'jishaku	電磁石	electromagnet *[elecmg]*
denji-yūdō	電磁誘導	electromagnetic induction *[elecmg]*
denjō-hakkō; denba-hakkō	電場発光	electroluminescence *[electr]*
denka	電化	electrification *[elec]*
denka	電荷	charge; electric charge *[elec]*
denkai	電界	electric field *[elec]*
denkai	電解；でんかい	electrolysis *[p-ch]*
denkai-atsu	電解圧	electrolytic potential *[p-ch]*
denkai-bunri	電解分離	electrolytic separation *[p-ch]*
denkai-chū'shutsu	電解抽出	electrolytic extraction *[p-ch]*
denka-idō	電荷移動	charge transfer *[p-ch]*
denkai-dō	電解銅	electrolytic copper *[met]*
denka-idō-jūgō	電荷移動重合	charge-transfer polymerization *[poly-ch]*
denka-idō-sakutai	電荷移動錯体	charge-transfer complex *[ch]*
denkai-eki	電解液	electrolyte *[p-ch]*
denkai-hakkō	電界発光	electroluminescence *[electr]*
denkai-hō'shutsu	電界放出	field emission *[electr]*
denkai-hō'shutsu-kenbi'kyō	電界放出顕微鏡	field-emission microscope *[electr]*
denkai-jōhatsu	電界蒸発	field evaporation *[elec]*
denkai-jūgō	電解重合	electrolytic polymerization *[poly-ch]*
denkai-kangen	電解還元	electrolytic reduction *[ch]*
denkai-kenma	電解研摩	electropolishing *{n} [met]*

denkai-kōka	電界効果	(electric) field effect *[electr]*
denkai-kondensa	電解コンデンサ	electrolytic capacitor *[elec]*
denkai-kyōdo-gensui'ryō	電界強度減衰量	field strength attenuation quantity *[elec]*
denkai-seiren	電解精錬	electrorefining *{n} [met]*
denkai-shitsu	電解質	electrolyte *[p-ch]*
denkai-sō	電解槽	electrolytic cell *[p-ch]*
denkai-tetsu	電解鉄	electrolytic iron *[met]*
denkai-yoku	電解浴	electrolytic bath *[met]*
denkei-hō	電型法	galvanoplastics method *[met]*
denken	電鍵；でんけん	key *{n} [elec]*
denki	電気；電氣；でんき	electricity *[phys]*
denki-aen'mekki	電気亜鉛めっき	electrogalvanizing *{n} [met]*
denki-bon	電気盆	electrophorus *[elec]*
denki-bunkai (denkai)	電気分解（電解）	electrolysis *[p-ch]*
denki-bunseki	電気分析	electroanalysis *[p-ch]*
denki-bunseki-kagaku	電気分析化学	electroanalytical chemistry *[p-ch]*
denki-chō	電気長	electrical length *[elecmg]*
denki-datsu'en	電気脱塩	electric desalting *{n} [ch-eng]*
denki-dendō-do	電気伝導度	electric conductivity *[elec]*
denki-eidō	電気泳動	electrophoresis *[p-ch]*
denki'gaku	電気学	electricity (the discipline) *[phys]*
denki-genshi'ka	電気原子価	electrovalence *[p-ch]*
denki-gishi	電気技師	electrical engineer *[elec]*
denki'han'jutsu	電気版術	electrotypy *[met]*
denki-hen'i	電気変位	electric displacement *[elec]*
denki-insei-do	電気陰性度	electronegativity *[p-ch]*
denki-ishi	電気石	tourmaline *[miner]*
denki-kagaku	電気化学	electrochemistry *[p-ch]*
denki'kagaku-hakkō	電気化学発光	electrochemiluminescence *[p-ch]*
denki'kagaku-retsu	電気化学列	electrochemical series; electromotive series *[p-ch]*
denki'kagaku-shinwa-ryoku	電気化学親和力	electrochemical affinity *[ch]*
denki'kagaku'teki-kōbai	電気化学的勾配	electrochemical gradient *[ch]*
denki'kagaku-tōryō	電気化学当量	electrochemical equivalent *[p-ch]*
denki-kairo	電気回路	electric circuit *[elec]*
denki-kami'sori	電気剃刀	electric shaver *[eng]*
denki-kanju-ritsu	電気感受率	electric susceptibility *[elec]*
denki-keisan-ki	電気計算機	electric computer *[comput]*
denki-kigu	電気器具	electric appliance *[eng]*
denki-kikai	電気機械	electric machine *[eng]*
denki-kikai-ketsu'gō-keisū	電気機械結合係数	electromechanical coupling factor *[mech-eng]*

denki-kikan'sha	電気機関車	electric locomotive *[mech-eng] [trans]*
denki'kō	電気工	electrician *[eng]*
denki-kō	電気鋼	electrical steel *[met]*
denki-kōgaku	電気工学	electrical engineering *[eng]*
denki-kōgaku	電気光学	electrooptics *[opt]*
denki'kōgaku-kōka	電気光学効果	electrooptic effect *[opt]*
denki-kōgyō	電気工業	electric industry *[eng]*
denki-kōtai	電気鋼帯	electrical steel band *[met]*
denki-mekki	電気めつき（鍍金）	electroplating *{n} [met]*
denki-mōkan-genshō	電気毛管現象	electrocapillarity *[phys]*
denki-nensei-kōka	電気粘性効果	electroviscous effect *[fl-mech]*
denki no	電気の	electric; electrical *{adj}*
denki no me	電気の目	electric eye *[electr]*
denki-onkyō'gaku	電気音響学	electroacoustics *[acous]*
denki-rikigaku	電気力学	electrodynamics *[elecmg]*
denki-rikisen	電気力線	electric line of force *[elec]*
denki-ro	電気炉；電氣爐	electric furnace *[eng]*
denki-ryoku	電気力	electric force *[elec]*
denki-seiri'gaku	電気生理学	electrophysiology *[med]*
denki'seki	電気石	tourmaline *[miner]*
denki'shi	電機子	armature *[elecmg]*
denki'shiki no	電気式の	electrically operated *{adj} [elec]*
denki'shiki-shitsudo-kei	電気式湿度計	electric hygrometer *[eng]*
denki-shintō	電気浸透	electroosmosis; electroendosmosis *[phys]*
denki-sōji-ki	電気掃除機	vacuum cleaner *[mech-eng]*
denki-sōkyokushi	電気双極子	electric dipole *[elec]*
denki-soryō	電気素量	elementary charge *[elec]*
denki-teikō	電気抵抗	electric resistance *[elec]*
denki-teikō-ritsu	電気抵抗率	electrical resistivity *[elec]*
denki'teki-seishitsu	電気的性質	electrical property *[phys]*
denki-tenbin	電気天秤	electrobalance *[an-ch]*
denki-tetsudō	電気鉄道	electric railway *[rail]*
denki-tōseki	電気透析	electrodialysis *[p-ch]*
denki-tsūshin('gaku)	電気通信（学）	telecommunications *[comm]*
denki-yōsetsu	電気溶接	electric welding *{n} [met]*
denki-zetsu'en	電気絶縁	electrical insulation *[elec]*
denki-zetsu'en-sei	電気絶縁性	electric insulating property *[mat]*
denki-zetsu'en-yu	電気絶縁油	electrical insulating oil *[mat]*
denkō	電光	lightning *{n} [geophys]*
den'kyoku	電極	electrode *[elec]*
denkyoku-den'i	電極電位	electrode potential *[p-ch]*
denkyū	電球	(incandescent) light bulb *[elec]*

dennetsu	伝熱	heat transmission *[thermo]*
dennetsu	電熱	electric heat *[thermo]*
dennetsu-yakin	電熱冶金	electrothermal metallurgy *[met]*
denpa	電波	electromagnetic wave; radio wave *[elecmg]*
denpa-bōenkyō	電波望遠鏡	radiotelescope *[eng]*
denpa-gen	電波源	radio source *[elecmg]*
denpa-ginga	電波銀河	radio galaxy *[astrophys]*
denpan	伝搬; でんぱん	propagation *[phys]*
denpan-chi'en	伝搬遅延	propagation delay *[electr]*
denpan suru	伝搬する	propagate *{vb} [acous] [phys]*
denpan-teisū	伝搬定数	propagation constant *[elecmg]*
denpa-shahei	電波遮蔽	radio screening *{n} [elecmg]*
denpa-tenmon'gaku	電波天文学	radio astronomy *[astron]*
denpa-tentai	電波天体	radio objects *[astron]*
denpō	電報	telegram *{n} [comm]*
denpun	澱粉	starch *{n} [bioch]*
denpun-ko	澱粉糊	starch paste *{n} [bioch]*
denrei	電鈴	electric bell *[comm] [elec]*
denri	電離; でんり	electrolytic dissociation; ionization *[ch]*
denri-bako	電離箱	ionization chamber *[nucleo]*
denri'sei-hōsha'sen	電離性放射線	ionizing radiation *[nucleo]*
denri-sō	電離層	ionosphere *[geophys]*
denryō-kei	電量計	coulometer; voltameter *[p-ch]*
den'ryoku	電力; でんりょく	electric power *[elec]*
denryoku-gentan'i	電力原単位	electric power consumption rate *[elec]*
denryoku-kēburu	電力ケーブル	electric cable; power cable *[elec]*
denryoku-kei	電力計	wattmeter *[eng]*
denryoku-sen	電力線	power line *[elec]*
denryoku-son	電力損	power loss *[electr]*
denryō-teki'tei	電量滴定	coulometric titration *[an-ch]*
denryū	電流; でんりゅう	electric current; current *[elec]*
denryū-jiki-kōka	電流磁気効果	galvanomagnetic effect *[elecmg]*
denryū-kei	電流計	ammeter *[eng]*
denryū-kō'ritsu	電流効率	current efficiency *[elec]*
denryū-mitsudo	電流密度	current density *[elec]*
denryū-sokutei-hō	電流測定法	galvanometry *[elec]*
denryū-teki'tei	電流滴定	amperometric titration *[p-ch]*
densan-ki	電算機	computer *[comput]*
densan-shoku'ji	電算植字	computerized typesetting *{n} [pr]*
denseki	電析	electrocrystallization *[crys]*
"	電析	electrodeposition *[met]*
densen	伝染	contagion *[med]*

densen	電線	electric wire; wire *[mat]*
densen'byō	伝染病	infectious (communicable) disease *[med]*
densen'sei no	伝染性の	contagious *{adj} [med]*
densha	電車	electric car; tram; trolley car *[trans]*
denshi	電子；でんし	electron *[phys]*
denshi-antei-ki	電子安定器	electronic ballast *[electr]*
denshi-boruto	電子ボルト	electronvolt (eV) *[phys]*
denshi-den'tatsu-kei	電子伝達系	electron transport system *[bioch]*
denshi-den'tatsu'tai	電子伝達体	electron carrier *[bioch]*
denshi-fukusha-ki	電子複写機	plain paper copier *[graph]*
denshi-haichi	電子配置	electronic configuration *[atom-phys]*
denshi-henshū	電子編集	electronic editing *{n} [comput]*
denshi-hō'shutsu	電子放出	electron emission *[phys]*
denshi-i'gaku	電子医学	medical electronics *[electr]*
denshi-jōji'sei-kyōmei	電子常磁性共鳴	electron paramagnetic resonance *[phys]*
denshi-jū	電子銃	electron gun *[electr]*
denshi-kaisetsu	電子回折	electron diffraction *[phys]*
denshi-kaku	電子殼	electron shell *[atom-phys]*
denshi-kan	電子管	electron tube *[electr]*
denshi-kanjō-hannō	電子環状反応	electrocycylic reaction *[p-ch]*
denshi-keisan-ki	電子計算機	electronic computer *[comput]*
denshi-kenbi'kyō	電子顕微鏡	electron microscope *[electr]*
denshi-kenbikyō-shashin	電子顕微鏡写真	electron micrograph *[graph]*
denshi-kidō	電子軌道	orbital *[atom-phys]*
denshi-kiki	電子機器	electronic instrument *[eng]*
denshi-kōgaku	電子工学	electronics *[phys]*
denshi-kōgaku	電子光学	electron optics *[electr]*
denshi-kōhō	電子航法	electronic navigation *[navig]*
denshin	伝信；でんしん	transmission *[comm]*
denshin	電信；でんしん	telegraph *{n} [comm]*
denshi-nadare	電子なだれ	electron avalanche *[electr]*
denshin-fugō	電信符号	telegraph code *[comm]*
denshin-hansō'ha	電信搬送波	telegraph carrier (wave) *[comm]*
denshi no	電子の	electronic *{adj}*
denshin-senro	電信線路	telegraph line *[comm]*
denshi-ōbun	電子オーブン	microwave oven *[cook] [eng]*
denshi-renji	電子レンジ	electronic range *[cook] [eng]*
denshi'renji-chōri-kanō	電子レンジ調理可能	microwavable *{adj} [cook] [electr]*
denshi'renji-chōri suru	電子レンジ調理する	microwave *{vb t} [cook]*
denshi-ryū	電子流	electron current *[elec]*
denshi-seikō-eki'teki	電子正孔液滴	electron hole droplets *[sol-st]*
denshi-sen	電子線	electron beam; electron ray *[electr]*

denshi-shashin	電子写真	electrophotography *[graph]*
denshi-shashin'shiki-insho-sōchi	電子写真式印書=装置	xerographic printer *[pr]*
denshi-shashin-sōchi	電子写真装置	electronograph *[electr]*
denshi'shiki-kinsen-tōroku-ki	電子式金銭登録機	electronic cash register *[electr]*
denshi'shiki-taku'jō-keisan-ki	電子式卓上計算機	electronic calculator *[electr]*
denshi-shindō	電子振動	electronic oscillation *[electr]*
denshi-shinwa-ryoku	電子親和力	electron affinity *[atom-phys]*
denshi-shōgeki-bunkō-hō	電子衝撃分光法	electron impact spectroscopy *[phys]*
denshi-shuppan	電子出版	electronic publishing *{n} [comput]*
denshi-sōsa	電子走査	electronic scanning *{n} [electr]*
denshi-techō	電子手帳；電子手帖	electronic notebook *[electr]*
denshi-tōryō	電子当量	electron equivalent *[phys]*
denshi-tōsha-yō'setsu	電子投射溶接	electron-beam welding *{n} [met]*
denshi'tsui-ketsugō	電子対結合	electron pair bond *[ch]*
denshi-yūbin	電子郵便	electronic mail *[comm]*
denshi-zōbai-kan	電子増倍管	electron multiplier *[electr]*
denshoku	電食	electrolytic corrosion *[met]*
densō	伝送；傳送；電送	transmission *[comm]*
densō'hin	電装品	electrical equipment (or parts) *[elec]*
densō-ki	伝送機	transmitter *[comm]*
densō-ki	電装器；電裝器	electrical equipment *[elec]*
densoku	電束	dielectric flux; electric flux *[elec]*
densō-sen(ro)	伝送線(路)	transmission line *[comm]*
densō-son'shitsu	伝送損失	transmission loss *[comm]*
densō suru	伝送する	transmit *{vb t} [comm]*
dentaku	電卓	desktop electronic calculator *[electr]*
den'tatsu	伝達	transfer *{n} [comm]*
den'tatsu-kansū	伝達関数	transfer function *[cont-sys]*
den'tatsu-ritsu	伝達率	transmissibility *[mech]*
dento	電鍍	electroplating *{n} [met]*
dentō	電灯；電燈	electric light *[elec]*
denwa	電話；でんわ	telephone; telephoning *{n} [comm]*
denwa-bangō	電話番号	telephone number *[comm]*
denwa'chō	電話帳	telephone book; telephone directory *[comm] [pr]*
den'wai	電歪	electrostriction *[mech]*
denwa-ki	電話機	telephone (instrument) *{n} [comm]*
denwa-kōkan-ki	電話交換機	telephone exchange *[comm]*
denwa o kakeru	電話を掛ける	telephone *{vb t} [comm]*

denwa-sen	電話線	telephone line *[elec]*
denwa suru	電話する	telephone *{vb t} [comm]*
deokishiribo-kakusan	デオキシリボ核酸	de(s)oxyribonucleic acid *[bioch]*
deru	出る	emerge (from) *{vb}*
de-sen	出線	outgoing line *[comm]*
deshi'guramu	瓰；デシグラム	decigram (unit of mass) *[mech]*
dētā-shori	データー処理	data processing *[comput]*
do	度	degree *[geogr] [math] [mech] [thermo]*
do	筌；ど	lobster trap *[eng]*
dō	道	district (e.g. Hokkaidō) *[geogr]*
dō	胴	torso; trunk (of body) *[anat]*
dō	銅	copper (element)
dō	どう；如何	how *{adv}*
do-atsu	土圧；土壓	earth pressure *[civ-eng]*
dō-atsu	動圧	dynamic pressure *[mech]*
do'atsu-keisū	土圧係数	coefficient of earth pressure *[civ-eng]*
dōbaku-sen	導爆線	detonating fuse *[eng]*
dōban-chōkoku	銅版彫刻	chalcography *[graph]*
do'bin	土瓶	earthenware teapot *[cer] [cook]*
doboku-kenchiku	土木建築	civil engineering and construction *[eng]*
doboku-kōgaku	土木工学	civil engineering *[eng]*
dobu	溝；どぶ	ditch; drain; sewer *[civ-eng]*
dobu-nezumi	溝鼠	sewer rat (animal) *[v-zoo]*
dō'butsu	動物；どうぶつ	animal *[zoo]*
dōbutsu-chiri'gaku	動物地理学	zoogeography *[bio]*
dōbutsu'en	動物園	zoo *[arch] [zoo]*
dōbutsu'gaku	動物学	zoology *[bio]*
dōbutsu-kagaku	動物化学	zoochemistry *[ch]*
dōbutsu-kai	動物界	Animalia *[syst]*
dōbutsu-ken	動物圏；動物圈	zoosphere *[ecol]*
dōbutsu-kisai-gakusha	動物記載学者	zoographer *[pr]*
dōbutsu-rikigaku	動物力学	zoodynamics *[bio]*
dōbutsu-rō	動物蠟	animal wax *[mat]*
dōbutsu-sen'i	動物纖維	animal fiber *[tex]*
dōbutsu'shi	動物誌	zoography *[pr]*
dōbutsu'shoku'sei no	動物食性の	zoophagous *{adj} [bio]*
dōbutsu-sō	動物相	fauna *[zoo]*
dōchi	同値	equivalent; equivalence *{n} [alg]*
dōchi no	同値の	equivalent *{adj} [logic]*
dōchō-kairo	同調回路	tuned circuit *[electr]*
dō'chūsei'shi-kaku	同中性子核	isotone *[nuc-phys]*
dō'chūsei'shi-tai	同中性子体	isotone *[nuc-phys]*

dodai 土台；土臺；どだい groundsill *[arch] [civ-eng]*
dō'dansei-keisū 動弾性係数 kinematic coefficient of elasticity *[mech]*
dōden-genshō 動電現象 electrokinetic phenomenon *[p-ch]*
dō-denki 動電気 dynamic electricity *[elec]*
dōden-ritsu 導電率 electric conductivity *[elec]*
dōden-sei 導電性 electroconductive property *[elec]*
dōden'sei-kōbunshi 導電性高分子 electroconductive polymer *[poly-ch]*
dōden-sō 導電層 electroconductive layer *[poly-ch]*
dōden-sokutei-hō 導電測定法 conductometry *[an-ch]*
dōden-zairyō 導電材料 electrically conductive material; electroconductive material *[mat]*

dōfū'butsu 同封物 enclosure; enclosed matter *{n}*
dōfun-dōbutsu 動吻動物 Kinorhyncha *[i-zoo]*
dōgen-hokaku'gan 同源捕獲岩 cognate xenolith *[petr]*
dōgo-hanpuku 同語反復 tautology *[logic]*
dogū 土偶；どぐう clay figure *[archeo]*
dōgu 道具；どうぐ implement; tool *[eng]*
dōgu-bako 道具箱 tool box *[eng]*
dōha 導波；どうは guided wave *[elecmg]*
dōha'kan 導波管 waveguide *[elecmg]*
dōhakan-kiri'kae-ki 導波管切換器 waveguide switch *[elecmg]*
dōhan-sū 同伴数 associate number *[math]*
dōha-ro 導波路 waveguide path *[elecmg]*
dō'i 同意；どうい approval; concurrence; consent
dōi-genso 同位元素 isotope *[nuc-phys]*
dōi'go 同意語 synonym *[gram]*
dōi-kaku 同位角；どういかく corresponding angles *[math]*
dōi'tai 同位体；同位體 isotope *[nuc-phys]*
dōitai-hyō'shiki 同位体標識 isotopic labeling *{n} [ch]*
dōitai-kōkan-hannō 同位体交換反応 isotopic exchange reaction *[ch]*
dōitai-tsui'seki'shi 同位体追跡子 isotopic tracer *[ch]*
dōitsu-kōzō'gata no 同一構造型の isotypic *{adj} [crys]*
dōitsu-ritsu 同一律 identity, law of *[logic]*
dō'itsu-sei 同一性 identification; sameness *{n}*
dōji 同時；どうじ at the same time; simultaneously *{adv}*
dōji-densō 同時伝送 simultaneous transmission *[comm]*
dōji-enshin 同時延伸 simultaneous stretching *[poly-ch]*
dōji-hannō 同時反応 simultaneous reaction *[ch]*
dōji-heikō no 同時平行の concurrent *{adj} [comput]*
dōji-henkan 同次変換 homogeneous transformation *[math]*
dōji-hōsō 同時放送 simultaneous broadcasting *[comm]*
dōji-keisū-kairo 同時計数回路 coincidence circuit *[electr]*

dō-jiku	動軸	driving axle *[mech-eng]*
dōjiku no	同軸の	coaxial *{adj} [elec] [electr]*
dōjiku-sen	同軸線	coaxial line *[elecmg]*
dōji no	同時の	simultaneous *{adj}*
dojō	土壌；土壤	soil *{n} [geol]*
dojō	泥鰌；土鰌	loach; Misgurnus fossilis (fish) *[v-zoo]*
dojō-danmen	土壌断面	soil profile *[geol]*
dojō'gaku	土壌学	soil science *[geol]*
dojō-hozen	土壌保全	soil conservation *[ecol]*
dōjū-kaku	同重核	isobar *[nuc-phys]*
dōjū'tai	同重体	isobar *[nuc-phys]*
dōka	同化	assimilation *[geol] [physio]*
dōka	同価；同價	equivalence *{n} [math]*
dō-kajū	動荷重	dynamic load *[civ-eng]*
do'kan	土管	earthenware pipe *[civ-eng]*
dōkan	導管	conduit; tube; pipe; duct *[eng]*
dōkan-so'shiki	導管組織	vascular tissue *[bot]*
dō'kansū	導関数	derivative *[math]*
dōka-sayō	同化作用	anabolism *[bioch]*
dōka-sen	導火線	fuse *{n} [eng]*
dōkei	同形	isomorphism *[math] [p-ch]*
dōkei	動径；動徑	radius vector *[astron]*
dōkei-haigū	同形配偶	isogamy *[bio]*
dōkei-ritsu	同形律	isomorphism, law of *[crys]*
do'ki	土器；土器	crockery; earthenware *[eng]*
"	土器	terra-cotta *[cer]*
dōki	同期	synchronism; synchronous *[phys]*
dō-ki	銅基	copper matrix *[met]*
dōki-dendō-ki	同期電動機	synchronous motor *[elec]*
dōki-dōsa	同期動作	synchronous operation *[comput]*
dōki-jōtai	同期状態	synchrony *[phys]*
dōki'ka	同期化	synchronization *[eng]*
doko	何処；何處；どこ	where *{adv}*
dōkō	瞳孔	pupil (of eye) *[anat]*
doku	毒；毒；どく	poison *{n} [mat]*
doku'butsu	毒物	poison *{n} [bioch]*
doku(-eki)	毒(液)	venom *{n} [physio]*
doku'jūseki	毒重石	witherite *[miner]*
dōkun-iji'go	同訓異字語	homonym: same kun (q.v.) but of different characters *[ling]*
doku-ninjin	毒人参	hemlock *[bot]*
doku'ritsu	独立；獨立	independentness *{n} [math]*

dokuritsu-eiyō-seibutsu	独立栄養生物	autotroph *[bio]*
dokuritsu-henkō	独立変項	argument *[logic]*
dokuritsu-hensū	独立変数	independent variable *[math]*
dokuritsu(-iden) no hōsoku	独立(遺伝)の法則	independent assortment, law of *[gen]*
dokuritsu-jishō	独立事象	independent events *[math]*
dokuritsu no hōsoku	独立の法則	independece, law of *[bio]*
dokuritsu-sei	独立性	independence *[math]*
dokuritsu-za'seki	独立座席	separate seat (of car) *[mech-eng]*
dokuro-mengata-suzume	髑髏面形天蛾	death's head moth (insect) *[i-zoo]*
doku-ryoku	毒力	virulence *[microbio]*
doku-sei	毒性	toxicity *[pharm]*
"	毒性	virulence *[microbio]*
dokusen'ken	独占権	monopoly; exclusive right *[econ] [legal]*
doku'so	毒素	toxin *[bioch]*
dokuso-sansei-sei	毒素産生性	toxigenicity *[microbio]*
dōkutsu	洞窟	cave; cavern *[geol]*
dōkutsu'gaku	洞窟学	speleology *[geol]*
doku-zeri	毒芹; どくぜり	water hemlock *[bot]*
dō-masatsu	動摩擦	kinematic friction *[mech]*
dōmen	導面	directrix (pl -trixes, -trices) *[math]*
dōmyaku	動脈	artery *[anat]*
do-nabe	土鍋	earthenware casserole *[cer] [cook]*
donburi	丼; どんぶり	bowl, often with cover *[cer] [cook]*
dō-nendo	動粘度	kinematic viscosity *[fl-mech]*
dō-nensei'ritsu; (-keisū)	動粘性率; (-係数)	kinematic viscosity (coefficient of) *[fl-mech]*
donguri	団栗; どんぐり	acorn *[bot]*
don'kaku	鈍角	obtuse angle *[math]*
donkaku-sankak(u)kei	鈍角三角形	obtuse triangle *[math]*
donsu	緞子; どんす	damask; satin damask *[tex]*
donten	曇点	cloud point *[ch-eng]*
donten	曇天	cloudy skies *[meteor]*
dōnyū	導入	introduction (into); guiding-in *{n}*
dōnyū-sen	導入線	leading-in wire *[elec]*
dōnyū suru	導入する	guide in *{vb}*
dō'on-hanpuku	同音反復	tautophony *[ling]*
dō'on-igi'go	同音異義語	homonym: same on (q.v.) but with different meaning *[ling]*
dō'on-iji	同音異字	homophone *[ling]*
dō'on-iji'go	同音異字語	homonym: same on (q.v.) but of different characters *[ling]*
dora	銅羅; どら	gong *[music]*

dō'rikigaku	動力学	dynamics; kinetics *[mech]*
dōrin	動輪	driving wheel *[mech-eng]*
doro	泥；どろ	mud *[geol]*
dōro	道路	road *[civ-eng]*
dōro-chizu	道路地図	road map *[map]*
dōro-kyō	道路橋	road bridge *[civ-eng]*
dōro-mō	道路網	highway net *[civ-eng]*
dōro-shōmei	道路照明	road lighting *[civ-eng]*
doru	弗；ドル	dollar *[econ]*
dōrui-kō	同類項	like terms *{n}* *[math]*
doryō'kō	度量衡	weights and measures *[sci-t]*
dō'ryoku	動力；どうりょく	motive power *[phys]*
dōryoku-dendō	動力伝導	power conduction *[elec]*
dōryoku-den'tatsu-sōchi	動力伝達装置	transmission *{n}* *[mech-eng]*
dōryoku-fuka	動力負荷	motive-power load *[mech-eng]*
dōryoku-kei	動力計	dynamometer *[eng]*
dōryoku-retsu	動力列	power train *[mech-eng]*
dōryoku-sū	動力数	power number *[fl-mech]*
dōsa	動作；どうさ	operation *[elec]* *[math]*
"	動作	action; motion; movement *[mech]*
dōsa-jikan	動作時間	operating time *[comput]*
dosei	土星	Saturn *[astron]*
dosei-zu	土性図	soil map *[civ-eng]*
doseki-ryū	土石流	boulder flow *[geol]*
dōshi	動詞	verb *[gram]*
dōshi'ku	動詞句	verb phrase *[gram]*
dōshin-en	同心円；同心圓	concentric circle *[math]*
dō'shite	如何して；どうして	why; how *{adv}*
dōshitsu-izō-kazō	同質異像仮像	paramorphism *[miner]*
do'shitsu-riki'gaku	土質力学	soil mechanics *[eng]*
dōshitsu-sei	同質性	homogeneity *[sci-t]*
dōshitsu-tazō	同質多像	polymorphism *[crys]*
dōshu-hakkō'sei no	同種発酵性の	homofermentative *{adj}* *[microbio]*
dōsō	同相	in phase *[phys]*
dōso'tai	同素体	allotrope *[ch]*
dosū	度数	frequency *[math]*
dosū-bunpu-zu	度数分布図	histogram *[math]*
dōsui-hankei	動水半径	hydraulic radius *[fl-mech]*
dōsui-kōbai	動水勾配	hydraulic gradient *[fl-mech]*
dosū-kansū	度数関数	probability mass function *[prob]*
dōtai	胴体	fuselage *[aerosp]*
dōtai	胴体	trunk; torso *{n}* *[anat]*

dōtai	導体	conductor *[elec]*
dōtai-kiroku-ki	動態記録器	kymograph *[med]*
dōtei	同定; どぅてい	identification of generic character *[bio]*
dōtei	同定	identification *[ch]*
dōtei suru	同定する	identify *{vb t}* *[bio]* *[ch]*
dōtei suru	同定する	establish identicalness (between substances) *[sci-t]*
dōteki-dansei-ritsu	動的弾性率	dynamic modulus *[mech]*
dōteki-heikō	動的平衡	dynamic equilibrium *[mech]* *[phys]*
dōteki-keikaku-hō	動的計画法	dynamic programming *[comput]*
dōteki-sei'shitsu	動的性質	dynamic properties *[electr]*
dōteki-shitsu'ryō-sayō no hōsoku	動的質量作用の=法則	kinetic mass action, law of *[ch]*
dō-tokusei	動特性	dynamic characteristic *[electr]*
dōtsū	導通	conducting; "on" *[elec]*
do'yōbi	土曜日	Saturday
dōyo'tai	同余体	isodiapheres *[nuc-phys]*
dōzen-pu	同前符: 々	Japanese ditto symbol *[pr]*
dōzoku-gengo	同族言語	cognate *[ling]*
dōzoku-kankei	同属関係	homology *[ch]*
dōzoku no	同族の	homologous *[bio]*
dōzoku'tai	同属体	homolog *[ch]*

E

e	絵; 繪; 画; 畫	picture; drawing; painting *{n} [graph]*
ebi	蝦; 海老; えび	shrimp (crustacean) *[i-zoo]*
ebi'cha-iro	蝦茶色	maroon (color) *[opt]*
e-chizu	絵地図; 繪地図	pictorial map *[graph]*
eda	枝; えだ	branch *{n} [bot]*
eda-mame	枝豆	green (young) soy bean *[bot]*
eda'nashi-sajō-bunshi	枝無し鎖状分子	unbranched chain molecule *[ch]*
eda-wakare	枝分れ	branching *{n} [bot]*
eda'zuno-reiyō	枝角羚羊	pronghorn antelope (animal) *[v-zoo]*
egaku	描く; 画く	draw; sketch; paint *{vb}*
e-gaso	絵画素	pixel *[electr]*
egonoki	斎墩果; えごのき	Perilla japonica *[bot]*
e'goma	荏胡麻; えごま	Perilla ocimoides *[bot]*
ego-yu	斎墩果油; えごゆ	styrax oil *[mat]*
eguru	抉る; 刳る	gouge *{vb t}*
e-ha'gaki	絵葉書	picture postcard *[comm]*
ei	鱏; えい	ray (fish) *[v-zoo]*
eibin'ka	鋭敏化	sensitization *[med]*
eidō	泳動	migration *[ch]*
eiga'kan	映画館	movie theater *[arch]*
eigyō	営業; えいぎょう	business; trade *{n} [econ]*
eigyō-himitsu	営業秘密	trade secret *[econ]*
eigyō-jikan	営業時間	business hours; office hours *[econ]*
eigyō'sho	営業所; 營業所	business office *[econ]*
ei'i	鋭意	zealously; assiduously; diligently *{adv}*
eiji	英字	alphabetical character; letter *[comput]*
eiji-retsu	英字列	alphabetic string *[comput]*
eikaku	鋭角	acute angle *[math]*
eikaku-sankak(u)kei	鋭角三角形	acute triangle *[math]*
eikan-zai	鋭感剤	sensitizer *[med]*
Eikoku	英国; えいこく	England; Great Britain *[geogr]*
Eikoku-netsu'ryō-tan'i	英国熱量単位	British thermal unit (Btu) *[thermo]*
eikyō	影響; 影響	influence *{n}*
eikyū	永久; えいきゅう	eternity; permanence *{n}*
eikyū-hizumi	永久歪み	permanent set *[mech]*
eikyū-hyōketsu'do	永久氷結土	permafrost *[geol]*
eikyū-jiba	永久磁場	permanent magnetic field *[elecmg]*
eikyū-ji'shaku	永久磁石	permanent magnet *[elecmg]*
eikyū ni	永久に	permanently *{adv}*

eikyū-nobi	永久伸び	tension set *{n} [mech] [poly-ch]*
eikyū-undō	永久運動	perpetual motion *[phys]*
ei'nen-henka	永年変化	secular variation *[astron]*
ei'nen-kasoku	永年加速	secular acceleration *[astron]*
ei'nen-setsu'dō	永年摂動	secular perturbation *[astrophys]*
ei'nen-shisa	永年視差	secular parallax *[astron]*
eiryū	癭瘤; えいりゅう	gall; cecidium (pl -dia) *[bot]*
eisei	衛生; 衞生	hygiene; sanitation *[med]*
eisei	衛星	satellite *[aerosp] [astron]*
eisei-chokka'ten	衛星直下点	subsatellite point *[astron]*
eisei-gaichū	衛生害虫	hygienically harmful insect *[i-zoo]*
eisei-keisan-ki	衛星計算機	satellite computer *[comput]*
eisei-tōki	衛生陶器	hygienic porcelain *[mat]*
eisei-tsūshin	衛星通信	satellite communication *[comm]*
eishin	映進	glide reflection *[crys]*
eishi-sei	曳糸性	spinnability; stringiness; thread-trailing property *[poly-ch] [tex]*
eisui'seki	鋭錐石	anatase *[miner]*
ei'sūji-shūgō	英数字集合	alphanumeric character set *[comp]*
eiyō	栄養; 営養; 營養	nutrition *[bio]*
eiyō-gakusha	栄養学者	nutritionist *[bio]*
eiyō-hanshoku	栄養繁殖	vegetative propagation *[bot]*
eiyō-ka	栄養価; 榮養價	nutritional value *[bio]*
eiyō-kyōsei	栄養共生	syntrophy; syntrophism *[bio]*
eiyō-saibō	栄養細胞	nutritive cell *[cyt]*
"	栄養細胞	trophocyte *[zoo]*
eiyō-seishoku	栄養生殖	vegetative reproduction; vegetative propagation *[bot]*
eiyō'shi	栄養士	dietician *[bio]*
eiyō-shitchō	栄養失調	malnutrition *[med]*
eiyō-shōgai	栄養障害	nutrition disorder *[med]*
eiyō'so	栄養素	nutrient *[bio]*
eiyō-yōkyū-kabu	栄養要求株	auxotroph *[gen]*
eiyō-yōkyū'sei-hen'i	栄養要求性変異	auxotrophic mutation *[gen]*
eiyō-yōkyū'sei-hen'i'tai	栄養要求性変異体	auxotroph *[gen]*
eizō	映像; えいぞう	projected image *[elec]*
eizō	影像	image *{n} [elec] [phys]*
eizō-ryoku	影像力	imaging power *[phys]*
eizō-shingō	映像信号	picture signals *[elec]*
eizō-shori	映像処理	image processing *{n} [comput]*
eizu	エイズ	AIDS *[med]*
eizu-kanren-shōkō'gun	エイズ関連症候群	AIDS-related complexes; ARC *[med]*

eki	液; えき	juice *{n} [bot]*
"	液	liquid *{n} [ch] [phys]*
"	液	liquor *[ch-eng] [pharm]*
eki	駅; 驛	station *[rail]*
eki'atsu-jidōsha-oshi'age-sōchi	液圧自動車押上装置	hydraulic car jack *[mech-eng]*
eki'atsu'shiki-chōsoku-ki	液圧式調速機	hydraulic governor *[mech-eng]*
eki-chō	益鳥; えきちょう	beneficial bird *[v-zoo]*
eki-chū	益虫; えきちゅう	beneficial insect *[i-zoo]*
eki'chū	液柱	jet *{n} [fl-mech]*
"	液柱	liquid column *[phys]*
eki-eki-chū'shutsu	液液抽出	liquid-liquid extraction *[ch-eng]*
eki'hō	液胞; 液胞	vacuole *[cyt]*
eki'jō-gomu	液状ゴム	liquid rubber *[poly-ch] [rub]*
eki'jō-sekken	液状石鹸	liquid soap *[mat]*
eki'ka	液化; えきか	liquefaction *[phys]*
eki'ka	液果	sap fruit; succulent fruit *[bot]*
eki'ka	腋窩	armpit *[anat]*
ekikai-den'i	液界電位	liquid junction potential *[p-ch]*
ekika-ryoku	液化力	liquefying power *[ch]*
ekika'sei no	液化性の	liquescent *{adj} [ch]*
eki-kōnai	駅構内	station premise(s) *[civ-eng]*
eki-maku	液膜	liquid film *[p-ch]*
eki'shin-tai'butsu-renzu	液浸対物レンズ	immersion objective *[opt]*
ekishō	液晶; えきしょう	liquid crystal *[p-ch]*
ekishō-hyōji	液晶表示	liquid crystal display *[electr]*
ekisō	液相	liquid phase *[ch]*
ekisu	エキス	extract *{n} [ch] [cook] [pharm]*
ekitai	液体; えきたい	liquid *{n} [phys]*
ekitai-chisso	液体窒素	liquid nitrogen *[mat]*
ekitai-eiyō-baichi	液体栄養培地	liquid nutrient medium *[microb]*
ekitai-junkatsu	液体潤滑	liquid lubrication *[mech-eng]*
ekitai-kūki	液体空気	liquid air *[phys]*
ekitai-nenryō	液体燃料	liquid fuel *[mat]*
ekitai-tansan	液体炭酸	liquid carbon dioxide *[mat]*
ekitai-zetsuen'butsu	液体絶縁物	liquid insulator *[mat]*
eki'teki	液滴	droplet *[ch] [meteor]*
ekkusu-jiku	エックス軸	x axis *[math]*
ekkusu'sen (ekusu'sen)	エックス線 (エクス線)	x-rays *[phys]*
ekkusu'sen-shōkaku-sanran	エックス線小角=散乱	x-ray small-angle scattering *[phys]*

ekkusu'sen-tōshi-hō	エックス線透視法	fluoroscopy *[eng]*
ekkusu-seppen	エックス切片	x intercept *[math]*
e-maki'mono	絵巻物；繪巻物	picture scroll *[graph]*
e-moji	絵文字	glyph; pictogram *[graph] [pr]*
en	円；圓；えん	yen (Japanese currency) *[econ]*
"	円；圓	circle *{n} [math]*
en	塩；鹽	salt *{n} [ch]*
-en	-円	-Yen (money) *{suffix}*
-en	-炎	-itis; -inflammation *{suffix} [med]*
-en	-園	-garden *{suffix}*
enbai	煙灰	flue cinder *[met]*
en'baku	燕麦；燕麥	oat *[bot]*
enban	円板	disc; disk *[math]*
enban-ben	円板弁；圓板瓣	disc valve *[mech-eng]*
enbun	塩分；えんぶん	salinity *[ocean]*
enbun-kei	塩分計；鹽分計	salinometer *[eng]*
enbun-yakusō	塩分躍層	halocline *{n} [ocean]*
enchi-ten	遠地点	apogee *[astron]*
enchō	延長	extension; prolongation *{n} [math]*
en'choku	鉛直；えんちょく	vertical *{n} [math]*
enchoku-hen'i	鉛直変位	vertical displacement *[mech]*
enchoku-kaku	鉛直角	vertical angle *[math]*
enchoku-ryoku	鉛直力	vertical force *[mech]*
enchoku-sen	鉛直線	plumb line *[eng]*
"	鉛直線	vertical line *[graph]*
enchō-sen	延長線	extension *[math]*
enchū	円柱	column *[arch]*
"	円柱	circular cylinder *[math]*
enchū'jō no	円柱状の	columnar; cylindrical *{adj} [math]*
enchū-rui	円虫類；圓蟲類	Nematoda *[i-zoo]*
endo	塩度	salinity *[ch]*
endō	煙道；煙道	flue *[eng]*
endo-kei	塩度計	salinimeter *[eng]*
en'eki-hō	演繹法	deduction *[logic]*
enerugī	エネルギー	energy *[phys]*
enerugī-hozon-soku	エネルギー保存則	conservation of energy, law of *[phys]*
enerugī-kinbun no hōsoku	エネルギー均分=の法則	equipartition of energy, law of *[stat-mech]*
engai	円蓋	canopy *[aero-eng]*
engai	塩害	salt damage *[ind-eng]*
engai	煙害；煙害	smoke damage *[ind-eng]*
engan	沿岸	coast; shore *{n} [geol]*

engan-chōryū	沿岸潮流	littoral current *[ocean]*
engan-dōro	沿岸道路	bund *{n}* *[civ-eng]*
engan-kai	沿岸海	marginal sea *[geogr]*
engan-su	沿岸洲	barrier beach; barrier bar *[geol]*
en'gawa	縁側	veranda *[arch]*
engei'gaku	園芸学; 園藝學	horticulture *[bot]*
en'ginga-chūshin-ten	遠銀河中心点	apogalacticon *[astron]*
en-i'sei	塩異性	salt isomerism *[ch]*
enji-iro	臙脂色	deep red (color) *[opt]*
en'jitsu-ten	遠日点	aphelion (pl aphelia) *[astron]*
enju	槐; えんじゅ	Japanese pagoda tree *[bot]*
en'juyō-ki	遠受容器	teleceptor *[physio]*
enka-aen	塩化亜鉛	zinc chloride
enka'butsu	塩化物	chloride *[ch]*
enka-chisso	塩化窒素	nitrogen chloride
enka-dai'ni-dō	塩化第二銅	cupric chloride
enka-dō	塩化銅	copper chloride
enka-gin	塩化銀	silver chloride
enka-hakkin	塩化白金	platinum chloride
enka'hakkin-san	塩化白金酸	chloroplatinic acid
enka'hakkin'san-en	塩化白金酸塩	chloroplatinate *[ch]*
enka-hiso	塩化砒素	arsenic chloride
enka-hōso	塩化硼素	boron chloride
enka-iō	塩化硫黄	sulfur chloride
enka-keiso	塩化硅素	silicon chloride
enka-kin	塩化金	gold chloride
enka-kinsan	塩化金酸	chloroauric acid
enkaku'butsu-kōdo-sokutei-ki	遠隔物光度測定器	telephotometer *[eng]*
enkaku-jiki'kishō-kei	遠隔自記気象計	telemeteograph *[eng]*
enkaku-jishin'gaku	遠隔地震学	teleseismology *[geophys]*
enkaku-kanchi	遠隔感知	remote sensing *{n}* *[eng]*
enkaku-seigyo	遠隔制御	remote control *[elecmg]*
enkaku'shiki-keiki	遠隔式計器	telegage *{n}* *[eng]*
enkaku-sōjū	遠隔操縦	remote control *[navig]*
enkaku-sōkan	遠隔相関	teleconnection *[meteor]*
enkaku-sokutei	遠隔測定	telemetering *{n}* *[eng]*
enkaku-tansa	遠隔探査	remote sensing *{n}* *[eng]*
enkan	円管	pipe *{n}* *[eng]* *[mat]*
enkan'men(-tai)	円環面(体)	toroid; torus *[math]*
enka-namari	塩化鉛	lead chloride
en-kansū	円関数	circular functions *[math]*

enka-rin	塩化燐	phosphorus chloride
enka-sanka-tetsu	塩化酸化鉄	iron chloride oxide
enka-shūso	塩化臭素	bromine chloride
enka-suigin	塩化水銀	mercury chloride
enka-suiso	塩化水素	hydrogen chloride
enka-suzu	塩化錫	tin chloride
enka-tetsu	塩化鉄；塩化鐵	iron chloride
en-kawase	円為替；圓爲替	yen exchange *[econ]*
enka-yōso	塩化沃素	iodine chloride
enki	延期	deferment; postponement *{n}*
enki	塩基；えんき	base *{n} [ch]*
enki-do	塩基度	basicity *[ch]*
enki-ka	塩基価	base number *[ch]*
en'kin'gahō	遠近画法	perspective *[graph]*
enkisei-a'ryūsan-namari	塩基性亜硫酸鉛	basic lead sulfite
enki'sei-en	塩基性塩	basic salt *[ch]*
enkisei-ryūsan-namari	塩基性硫酸鉛	basic lead sulfate
enkisei-sakusan-namari	塩基性酢酸鉛	basic lead acetate
enkisei-sanka'butsu	塩基性酸化物	basic oxide *[ch]*
enki-shokubai	塩基触媒	base catalyst *[ch]*
enki-tsui	塩基対	base pair *[mol-bio]*
enko	円弧	circle arc *[math]*
enko	塩湖	salt lake; saline lake *[hyd]*
enkō-kōdo-kei	炎光光度計	flame photometer *[spect]*
enko-sen	円弧線	circular arc *[math]*
enkyō	塩橋	salt bridge *[p-ch]*
en'kyori-	遠距離-	tele- *{prefix}*
enkyori-tsūshin-mō	遠距離通信網	telecommunications network *[comm]*
enmu	煙霧	fumes *{n} [ch]*
"	煙霧	haze *{n} [meteor]*
enmu'tai	煙霧体；煙霧體	aerosol *[ch]*
e-no-abura	荏の油	perilla oil *[mat]*
enoki	榎；えのき	hackberry; nettle tree (tree) *[bot]*
enoki-take	榎茸	enoki mushroom *[bot]*
en'paku	鉛箔	lead foil *[mat]*
enpei	掩蔽	occultation *[astron]*
enpi	鉛被	lead sheath *[mat]*
en'pitsu	鉛筆	pencil *{n} [eng]*
ensan	塩酸；鹽酸	hydrochloric acid
ensei	延性	ductility *[mat]*
ensei-hakai	延性破壊	ductile fracture *[mech]*
ensei no aru	延性のある	ductile *{adj} [mat]*

ensei-sen'i-ondo	延性遷移温度	ductility transition temperature *[met]*
ensei-shokubutsu	塩性植物	halophyte *[ecol]*
en'seki	塩析	salting out *[ch-eng]*
en'sekigai-sen	遠赤外線	far-infrared rays *[elecmg]*
enshi	遠視	farsightedness *[med]*
enshin	延伸	drawing; stretching *{n}* *[poly-ch]*
enshin-bai'ritsu	延伸倍率	draw ratio; stretch ratio *[poly-ch]* *[tex]*
enshin-bunri	遠心分離	centrifugation *[ch-eng]*
enshin-chūkei	遠心注型	centrifugal casting *{n}* *[poly-ch]*
enshin-dassui-ki	遠心脱水機	centrifugal dehydrator *[ch-eng]*
enshin-hi	延伸比	draw ratio; stretch ratio *[poly-ch]* *[tex]*
enshin-ki	遠心機	centrifuge *{n}* *[eng]*
enshin-ritsu	延伸率	draw ratio; stretch ratio *[poly-ch]* *[tex]*
enshin-roka-ki	遠心沪過機	centrifuge *{n}* *[eng]*
enshin-ryoku	遠心力	centrifugal force *[mech]*
enshin'sei no	遠心性の	centrifugal *{adj}* *[mech]*
enshō	延焼; 延燒	conflagration; spread of fire *[ch]*
en'shoku-hannō	炎色反応	flame reaction *[ch]*
enshū	円周; 圓周	circumference *[math]*
enshū	演習	exercise; practice *{n}* *[math]*
enshū-hizumi	円周歪	circumferential strain *[mech]*
enshū-ō'ryoku	円周応力	circumferential stress *[mech]*
enso	塩素; 鹽素	chlorine (element)
ensō	演奏	performance *[music]*
enso-ion	塩素イオン	chloride ion *[ch]*
enso'ka	塩素化	chlorination *[ch]*
enso-ryō	塩素量	chlorinity *[ocean]*
enso-san	塩素酸	chloric acid
ensō-shitsu	演奏室	concert hall *[arch]*
enso-shori	塩素処理	chlorination *[ch]*
enso-sui	塩素水	chlorine water *[ch]*
ensui	円錐; えんすい	circular cone; cone *[math]*
ensui'dai	円錐台	truncated cone *[math]*
ensui-funmu	塩水噴霧	salt-water spraying *[met]*
ensui-funmu-shiken	塩水噴霧試検	salt-spray test *[met]*
ensui-kajo	円錐花序	panicle *[bot]*
ensui-kei	円錐形	conical form *{n}* *[math]*
ensui-kyoku'sen	円錐曲線	conic sections *[math]*
ensui-kyoku'sen-kaiten'men'tai	円錐曲線回転面体	toroid *[math]*
entan	鉛丹	minium; red lead *[miner]*
entan	遠端	far end *{n}*

entarupī	エンタルピー	enthalpy *[thermo]*
en'tatsu-ryoku	遠達力	long-range force *[nuc-phys]*
en'tatsu-sayō	遠達作用	abscopal effect *[med]*
enten	延展	extending; spreading *{n}*
entō	煙筒	chimney; funnel *{n} [arch]*
entō	円筒；えんとう	cylinder *[math]*
entō'gata-enshin-bunri-ki	円筒形遠心分離機	tubular bowl centrifuge *[eng]*
entō'gata-tenjō	円筒形天井	vault *{n} [arch]*
entō-ha	円筒波	cylindrical wave *[elecmg]*
entō-kensaku	円筒研削	cylindrical grinding *{n} [mech]*
entoropī	エントロピー	entropy *[thermo]*
en'totsu	煙突；烟突	chimney *[arch]*
entō-zahyō	円筒座標	cylindrical coordinates *[math]*
en-undō	円運動	circular motion *[mech]*
en'yō	塩溶	salting in *[ch]*
en'yō	遠洋	deep sea; open sea *[geol]*
en'yō-kōro	遠洋海路	ocean lane *[navig]*
en'yoku-yaki'ire	塩浴焼入れ	salt-bath quenching *{n} [met]*
en'yō no	遠洋の	pelagic *{adj} [ocean]*
en'yō suru	援用する	cite; quote; invoke; claim *{vb}*
en'yō-teiki'sen	遠洋定期船	ocean liner *[nav-arch]*
enzan	演算；えんざん	calculation; operation *[comput] [math]*
enzan-jikan	演算時間	operation time *[comput]*
enzan-ki	演算器	arithmetic unit *[comput]*
enzan'shi	演算子	operator *[comput]*
enzan-sōchi	演算装置	arithmetic unit *[comput]*
enzan'sū	演算数	operand *[comput]*
enzan-zōfuku-ki	演算増幅器	operational amplifier *[electr]*
enzetsu	演説	speech; oral presentation *[comm]*
enzō	塩蔵；えんぞう	salting (preservation) *[cook] [food-eng]*
en'zui	延髄；延髓	medulla oblongata *[anat]*
era	鰓；えら	gill *[v-zoo]*
erabu	選ぶ；選ぶ；択ぶ	choose; select; prefer *{vb t}*
era-buta	鰓蓋	gill cover *[v-zoo]*
era'hiki-dōbutsu	鰓曳動物	Priapuloidea; Priapulida *[i-zoo]*
eri	襟；えり	collar *{n} [cl]*
eri'maki-raichō	襟巻き雷鳥	ruffed grouse (bird) *[v-zoo]*
eri'maki-tokage	襟巻き蜥蜴	frilled lizard *[v-zoo]*
erubyūmu	エルビウム	erbium (element)
erugu	エルグ	erg *[phys]*
esa	餌	bait *{n} [bio]*
eso	絵素；繪素；えそ	pixel *[electr]*

eso	壊疽；えそ	gangrene *[med]*
esuji'gata no	エス(S)-字形の	sigmoid *{adj}* *[bio]*
esuteru	エステル	ester *[org-ch]*
esuteru'ka	エステル化	esterification *[org-ch]*
ēteru	エーテル	ether *[org-ch]*
e-tsuke-seikei	絵付成形	foil decorating *{n}* *[poly-ch]*
etsu'nen'sei-shokubutsu	越年生植物	biennial plant *[bot]*
ettō	越冬	hibernation; wintering *[physio]*
ezo-matsu	蝦夷松	silver fir (tree) *[bot]*
ezo-negi	蝦夷葱	chive *[bot]*
ezo-raichō	蝦夷雷鳥	hazel grouse (bird) *[v-zoo]*
e-zu	絵図；繪圖	drawing; illustration; picture map *[graph]*

F

fanderuwārusu-ryoku	ファンデルワールス力	Van der Waals forces *[p-ch]*
ferumyūmu	フェルミウム	fermium (element)
fīto	呎；フイート	foot; feet (a measure) *[mech]*
fu	府	prefecture (e.g. Osaka-fu) *[geogr]*
fu	負；ふ	negative *[elec]*
"	負	minus *[math]*
fu	麩	wheat gluten bread *[cook]*
fu-	不-	non- *{prefix}*
fū	封	seal *{n} [mat]*
fū'ai	風合	feeling; handling *{n} [tex]*
fu'an	不安	uneasiness; anxiety; uncertainty *[psy]*
fu'antei na	不安定な	labile; unstable *{adj} [sci-t]*
fu'antei-sei	不安定性	instability *[phys]*
fu-atsu	負圧；負壓；ふあつ	negative pressure *[phys]*
fū-atsu	風圧；ふうあつ	wind pressure *[mech]*
fu'baku no	不爆の	unexploded *{adj} [ch] [ord]*
fubuki	吹雪；ふぶき	snowstorm *[meteor]*
fu'chaku	付着	accretion *[astron]*
fuchaku-kon	付着根	holdfast *{n} [bot]*
fuchaku'tai	付着体	hapten *[immun]*
fuchaku-tan	付着端	sticky end *[bioch]*
fuchi	縁；緣；ふち	margin; leaf edge *{n} [bot]*
"	縁	edge; rim; border *{n} [sci-t]*
fuchi	付値；賦値	valuation *{n} [math]*
fuchi-doru	縁取る	border; hem; fringe *{vb t} [cl]*
fuchi-ishi	縁石	curbstone *[civ-eng]*
fuchi-zori	縁反り	edge-curl *[mat]*
fū-chō	風鳥；ふうちょう	bird of paradise (bird) *[v-zoo]*
fuda	札；ふだ	tag *{n} [comm]*
fudan-gi	普段着；不断着	everyday wear; everyday clothes *[cl]*
fudan no	不断の；不斷の	incessant *{adj}*
fude	筆；ふで	brush *{n} [graph]*
fudō	浮動	floating *{n} [math]*
fūdō	風胴	wind tunnel *[eng]*
fūdo'byō	風土病	endemic disease *[med]*
fudō'ka	不動化	immobilization *{n}*
fudō-shōsū-ten	浮動小数点	floating decimal point *[math]*
fudō-shōsū'ten-sū	浮動小数点数	floating-point number *[comput]*

fudō-sō	不動層	immobile layer *[ch]*
fudō'tai	不導体	nonconductor *[elec]*
fudō-tai	不働態；不動態	passive state *[met]*
fudōtai'ka	不働態化	passivation *[met]*
fudō-ten	不動点	fixed point *[comput]*
fue	笛	flute; pipe; whistle *[music]*
fue-dai	笛鯛	snapper; Lutianus rivulatus (fish) *[v-zoo]*
fu'eiyō'ka	富栄養化	eutrophication *[ecol]*
fu'en	敷衍；ふえん	amplification; dilatation *[comm]*
fūfu-shin'shitsu	夫婦寝室	master bedroom *[arch]*
fu-genshi'ka	負原子価	negative valence *[ch]*
fu'gensū-bun'retsu	不減数分裂	ameiosis *[gen]*
fu'gensui-shindō	不減衰振動	undamped oscillation *[phys]*
fugō	符号；符號；ふごう	signal *{n}* *[comm]*
"	符号	sign; code *{n}* *[comput]* *[math]*
fugō'ka	符号化	coding *{n}* *[comput]*
fugō'ka-jusshin-hō	符号化十進法	coded decimal notation *[comput]*
fu'gōkaku	不合格	failing grade; rejected *[ind-eng]*
fugō'ka suru	符号化する	encode *{vb}* *[comm]*
fugō-ki	符号器；符號器	coder; encoder *[comm]* *[comput]*
fugō'tsuki-seisū	符号付き整数	signed integer *[comput]*
fugu	河豚；ふぐ	balloonfish; blowfish; globefish; puffer; swellfish *[v-zoo]*
fu'gu'ai	不具合	ill function *{n}* *[ind-eng]*
fuhai	腐敗	putrefaction; spoilage *[bioch]*
fu'haku	布帛；ふはく	woven fabric; cloth and silk *[tex]*
fu'hatsu	不発；不發	misfire *{n}* *[ch]*
fu'heikō	不平衡	unbalance *{n}*
fuhen	不変；不變	unchangeability *{n}*
fuhen	不偏；ふへん	impartiality; unbiased *[math]*
fuhen	普遍	universality *[math]*
fuhen-bunsan	不偏分散	unbiased variance *[math]*
fu-henchō	負変調	negative modulation *[electr]*
fuhen-men	不変面	invariable plane *[mech]*
fuhen-shūgō	普遍集合	universal set *[math]*
fuhen-suitei'ryō	不偏推定量	unbiased estimator *[math]*
fūhō'butsu	封包物；封包物	enclosure; encased matter *{n}*
fuhō-i'min	不法移民	illegal immigrant *[legal]*
fuhō-sei	不法性	unlawfulness *[legal]*
fu'hōwa	不飽和；不飽和	unsaturation *{n}* *[org-ch]*
fu'hōwa-shibō'san	不飽和脂肪酸	unsaturated fatty acid *[org-ch]*
fu'hōwa-tanka'suiso	不飽和炭化水素	unsaturated hydrocarbon *[org-ch]*

fuhyō	付表	appended table *[graph] [pr]*
fuhyō-kai	浮氷塊	ice floe *[ocean]*
fū'in	封印	seal; sealing *{n}*
fūin-kikai	封印機械	sealer *[eng]*
fuji	藤	Japanese wisteria (flower) *[bot]*
fuji-iro	藤色	lilac; lavender; mauve (color) *[opt]*
fūji'kome	封じ込め	containment *[bioch]*
-fujin	-夫人	Mrs. *{suffix}*
fujinka-i	婦人科医	gynecologist *[med]*
Fuji-san	富士山	Mount Fuji *[geogr]*
fuji'tsubo	富士壺；ふじつぼ	barnacle *[nav-arch]*
fu'jūbun na-	不充分な-	under- *{prefix}*
fujun'butsu	不純物	impurity *[sci-t] [sol-st]*
fuka	鱶；ふか	shark (large) (fish) *[v-zoo]*
fuka	不可	poor; unacceptable (grade) *[ind-eng]*
fuka	付加	addition *[ch]*
fuka	負荷	load *{n} [elec] [mech]*
fuka	富化	enrichment *[bio] [geol]*
fuka	賦課	levy *{n} [econ]*
fuka	孵化	incubation *[zoo]*
fūka	風化	weathering *{n} [geol]*
fu'kachō'on	不可聴音	infrasound *[acous]*
fu'kagyaku-hannō	不可逆反応	irreversible reaction *[ch]*
fu'kagyaku'sei no	不可逆性の	irreversible *{adj} [bio] [math]*
fukai	深い	deep *{adj}*
fūkai	風解	efflorescence *[ch]*
fukai-shisū	不快指数	discomfort index *[meteor]*
fuka-jūgō	付加重合	addition polymerization *[poly-ch]*
fu'kaketsu na	不可欠な	indispensable *{adj} {n}*
fu-kaku	俯角	dip *{n} [eng]*
"	俯角	angle of depression *[math]*
fukaku suru	深くする	deepen *{vb t}*
fu'kakutei na ba'ai	不確定な場合	ambiguous case *[math]*
fu'kakutei-sei	不確定性	indeterminancy *[math]*
fu'kakutei'sei-genri	不確定性原理	uncertainty principle *[phys]*
fukameru	深める	deepen; intensify *{vb t}*
fukami-sen	深海線	deep-sea cable *[elec]*
fu'kanbi	不完備	incomplete *{n} [math]*
fukan-chitai	不感地帯	blind zone *[comm]*
fukan-do	不感度	insensitiveness *{n}*
fukan-jikan	不感時間	dead time *[comput]*
fu'kanyū-sei	不可入性	impenetrability *{n} [mat]*

fu'kanzen-kesshō	不完全結晶	imperfect crystal *[crys]*
fu'kanzen'kin-rui	不完全菌類	Deuteromycetes *[mycol]*
fu'kanzen na-	不完全な-	under- *{prefix}*
fu'kanzen-nenshō	不完全燃焼	incomplete combustion *[ch]*
fuka-ritsu	負荷率	load factor *[civ-eng]*
fuka-sa	深さ	depth *[ocean]*
fuka-sakutai	付加錯体	addition complex *[ch]*
fu'kashi	不可視	invisibility *{n}* *[opt]*
fuka-shibori	深絞り	deep-drawing *{n}* *[eng]* *[steel]*
fuka'shibori-sei	深絞り性	deep-drawability *[steel]*
fukashi-garasu	不可視ガラス	invisible glass *[mat]*
fuka shita bakari no sakana (tori)	孵化したばかりの魚（鳥）	hatchling of fish (or bird) *[zoo]*
fuka-son	負荷損	load loss *[elec]*
fu'kassei	不活性	inert *[ch]*
fūka-tan	風化炭	effloresced coal; weathered coal *[geol]*
fukatsu	賦活	activation *[ch]*
fukatsu suru	賦活する	activate *{vb t}* *[bioch]*
fūkei	風景	scenery; landscape *[graph]*
fūkei-ga	風景画	landscape picture *[graph]*
fukei-zai	賦形剤	excipient; mass; vehicle *[pharm]*
fu'kenka'butsu	不鹼化物	unsaponifiable matter *[mat]*
fuki	蕗；ふき	butterbur; rhubarb (vegetable) *[bot]*
fuki	付記	supplementary note *[comm]* *[pr]*
fu'kihatsu'butsu	不揮発物	nonvolatile matter *[mat]*
fuki-ido	噴き井戸	artesian well *[geol]*
fu-kikan	負帰還	negative feedback *[electr]*
fu'kikkō-sogai	不拮抗阻害	uncompetetive inhibition *[bioch]*
fukin	布巾；ふきん	dishcloth; washcloth *[cook]* *[tex]*
fuki'nagashi	吹流し	streamer *{n}*
fu'kin'itsu na	不均一な	heterogeneous *{adj}*
fu'kin'itsu-sei	不均一性	heterogeneity; ununiformity *{n}*
fukin'ka	不均化	disproportionation *[ch]*
fuki no tō	蕗の薹	butterbur flower, stalk *[bot]*
fukin-sei	不均整	asymmetry *[ch]*
fu'kinshitsu	不均質	heterogeneous(ness) *{n}* *[ch]*
fu'kinshitsu-kei	不均質系	heterogeneous system *[sci-t]*
fu'kinshitsu-sei	不均質性	heterogeneity *[ch]*
fu'kisoku	不規則	irregularity *{n}*
fu'kisoku-henkō-sei	不規則変光星	irregular variable star *[astron]*
fu'kisoku'sei no	不規則性の	atactic *{adj}* *[poly-ch]*
fuki-zuna	吹き砂	blowsand *[geol]*

fukka-aen	弗化亜鉛	zinc fluoride
fukka'butsu	弗化物；ふっかぶつ	fluoride
fukka'butsu-tenka	弗化物添加	fluoridation *[eng]*
fukka-dō	弗化銅	copper fluoride
fukka-gin	弗化銀	silver fluoride
fukkaku	伏角；ふっかく	dip; inclination *[eng]*
"	伏角	angle of depression *[math]*
fukka-namari	弗化鉛	lead fluoride
fukka-rinsan-suzu	弗化燐酸錫	tin fluorophosphate
fukka-suiso	弗化水素	hydrogen fluoride
fukka-suiso-san	弗化水素酸	hydrofluoric acid
fukka-suzu	弗化錫	tin fluoride
fukka-tanso	弗化炭素	carbon fluoride
fukka-tetsu	弗化鉄	iron fluoride
fukka-yōso	弗化沃素	iodine fluoride
fukki	復帰；ふっき	(carriage) return; reversion *[comput]*
fukki-hen'i	復帰変異	back mutation; reversion mutation *[microbio]*
fukki suru	復帰する	be restored; revert to the original *{vb}* *[mech]*
fukki-totsu'zen-hen'i	復帰突然変異	reverse (or back) mutation *[microbio]*
fukkō	腹腔	abdominal cavity *[anat]*
fukō	浮鉱；浮鑛	float *{n}* *[geol]*
fūkō; kaza-muki	風向	wind direction *[meteor]*
fu'koku	布告	proclamation; ordinance; decree *{n}* *[legal]*
fuku	副	secondary *{adj}* *{n}*
fuku-	副-	co-; pro-; secondary-; sub- *{prefix}*
fuku-	複-	bi-; multi- *{prefix}*
fuku-ben	複瓣；ふくべん	compound petal *[bot]*
fuku'boku	副木	splint *{n}* *[med]*
fuku-bu	腹部	abdominal area *[anat]*
fuku-bunkai	複分解	double decomposition *[ch]*
fukuchō	復調；ふくちょう	demodulation *[comm]*
fukuchō	複調	bitonality *[music]*
fukuchō-ki	復調器	demodulator *[electr]*
fuku-dairi'nin	復代理人	subagent (a person) *[legal]*
fukudō-pisuton	複動ピストン	double-acting piston *[mech-eng]*
fuku-en	複塩；複鹽	double salt *[inorg-ch]*
fukugan	複眼	compound eye *[i-zoo]*
fuku'gen	復元	reversion *[steel]*
fukugen-ryoku	復原力；復元力	restoring force *[mech-eng]*
fukugen-sei	復元性	stability *[mech-eng]*

fukugen suru	復元する	restore *{vb} [comput]*
fuku'gō	複合；ふくごう	composite(ness)
fukugō'bun	複合文	compound statement *[comput]*
"	複合文	compound sentence *[math]*
fukugō'butsu	複合物	composite *{n} [mat]*
fukugō-kaku	複合核	compound nucleus *[nuc-phys]*
fukugō-kana'gata	複合金型	composite mold *[eng]*
fukugō-mekki	複合めっき	composite plating *[met]*
fukugō-on	複合音	compound tone *[acous]*
fukugō-sen'i	複合繊維	composite fiber *[tex]*
fukugō-shi'shitsu	複合脂質	compound lipid *[bioch]*
fukugō-soku	複合則	alligation *[math]*
"	複合則	law of mixture *[sci-t]*
fukugō-zai	複合材	composite material *[mat]*
fuku-hannō	副反応；副反應	side reaction *[ch]*
fuku-hon	副本	copy; additional copy; duplicate *{n} [graph]*
fuku'hyō	復氷	regelation *[hyd]*
fuku'i	復位	reinstatement *{n}*
fuku'jin	副腎；ふくじん	adrenal gland *[anat]*
fuk(u)ka	複果	compound fruit *[bot]*
fuku'kassha	複滑車	tackle; compound pulley *{n} [mech-eng]*
fuk(u)ki-totsu'zen-hen'i	復帰突然変異	back mutation; reverse mutation *[gen]*
fuku'kō	副虹	secondary rainbow *[opt]*
fuku'kōjō-sen	副甲状腺	parathyroid gland *[anat]*
fuk(u)kōkan-shinkei-kei	副交感神経系	parasympathetic nervous system *[anat]*
fuk(u)kon	匐痕	trail *{n} [paleon]*
fuku-kussetsu	複屈折	birefringence; double refraction *[opt]*
fuku-kyō	副鏡	secondary mirror *[opt]*
fuku'maku-kō	腹膜腔	peritoneal cavity *[anat]*
fukumō-dōbutsu	腹毛動物	Gastrotrichia *[i-zoo]*
fukura-hagi	脹ら脛	calf (pl calves, calfs) (of leg) *[anat]*
fukurami	脹らみ	swelling *{n} [physio]*
"	脹らみ	bulge; puff *{n}*
fukuramu	脹らむ	swell *{vb i}*
fukurashi-ko	脹らし粉	baking powder *[cook]*
fukure	ふくれ	blister *{n} [eng]*
fuku'ri	複利	compound interest *[econ]*
fukuro	袋	bag *{n} [eng]*
fukurō	梟；ふくろう	owl (bird) *[v-zoo]*
fukuro-kōji	袋小路	blind alley; cul de sac *[civ-eng]*
fukuro-neko	袋猫；ふくろねこ	dasyure (animal) *[v-zoo]*

fukurō-ōmu	梟鸚鵡	kakapo (bird) *[v-zoo]*
fukuryū-shiki	複流式	double-current system *[fl-mech]*
fukuryū-sui	伏流水	underflow water *[geol]*
fuku-sanbutsu	副産物	by-product *[eng]*
fuku-sayō	副作用	side effect *[comput] [pharm]*
fukusei	複製；ふくせい	replication *[bioch] [graph] [mol-bio]*
"	複製	duplicating *{n} [comput]*
fuku-seibun	副成分	accessory constituent *[ch]*
fuku-seibutsu	副生物	by-product *[eng]*
fukusei-kazan	複成火山	composite volcano *[geol]*
fuku-seisō	副精巣；副精巢	epididymis (pl -mides) *[anat]*
fuku'seisui-ben	副制水弁；副制水瓣	bypass valve *[eng]*
fukusei suru	複製する	reprint; duplicate reproduce *{vb t} [pr]*
fuku'sha	複写；複寫	ectype; copy; reproduction *[graph]*
fuku'sha	輻射；ふくしゃ	radiation *[phys]*
fukusha-ki	複写機	copying machine *[graph]*
fukusha-netsu	輻射熱	radiant heat *[thermo]*
fukusha-shōdo	輻射照度	irradiance *[elecmg]*
fukusha suru	複写する	copy; reproduce *{vb t} [comput]*
fukusha-(yō)shi	複写(用)紙	copying paper *[paper]*
fuku'shi	副詞	adverb *[gram]*
fuku'shoku	副食	side dish *[cook]*
fukuso-dansei-ritsu	複素弾性率	complex modulus of elasticity *[mech]*
fukuso-hensū	複素変数	complex variable *[math]*
fukuso-kan	複素環	heterocyclic ring *[org-ch]*
fukusokan'shiki-kagō'butsu	複素環式化合物	heterocyclic compound *[org-ch]*
fukuso-kussetsu-ritsu	複素屈折率	complex index of refraction *[opt]*
fukuso-kyōyaku	複素共役	complex conjugate *[math]*
fukusō-mekki	複層めっき	multiple-layer plating *[met]*
fuku-sosū	複素数	complex number *[math]*
fuku-sū	複数；ふくすう	plural *[gram]*
"	複数	plural number *[math]*
fuku'tan	復炭	recarburization *[steel]*
fuku'wa'jutsu	腹話術	ventriloquism *[comm]*
fuku-yō	複葉	compound leaf *[bot]*
fuku-yūten	複融点	double melting point *[thermo]*
fuku'zatsu na	複雑な；複雜な	complicated *{adj}*
fuku'zatsu-sa	複雑さ	complexity; complicatedness *{n}*
fuku'zatsu-seikei'hin	複雑成形品	complicated molded product *[ind-eng]*
fuku'zō	複像	multiple image *[phys]*
fu-kyoku	負極	negative electrode *[elec]*
fu'mekki	不めっき；不鍍金	unplated *{adj} [met]*

fūmi	風味	flavor *{n} [cook] [physio]*
fumi'kiri	踏切	grade crossing; railroad crossing *[civ-eng]*
fun	分；ふん	minute *{n} [geogr] [math] [mech]*
-fun; -pun	-分；-ふん；-ぷん	-minute(s) *{suffix}*
funa	鮒；ふな	crucian carp; Prussian carp; Carassius carassius (fish) *[v-zoo]*
funa'bin	船便；ふなびん	sea mail; surface mail *[comm]*
funa'ni-shōken	船荷証券	bill of lading *[econ]*
funa-yoi	船酔い	seasickness *[med]*
funa'zoko	船底	shipbottom *[nav-arch]*
funa'zumi	船積み	shipping; loading; lading *{n} [nav-arch]*
funben	糞便	feces *[physio]*
funben-o'sen	糞便汚染	fecal contamination *[ecol]*
fundō	分銅	weight; balance weight *[mech-eng]*
fune	舟	boat *{n} [nav-arch]*
fune	船	ship *{n} [nav-arch]*
fune'gata-haiza	舟形配座	boat conformation *[org-ch]*
fun'en	噴煙	volcanic smoke *[geol]*
fu'nenketsu-tan	不粘結炭	noncaking coal *[geol]*
fune no chūō'bu no	船の中央の	midship *{adj} [nav-arch]*
fu'nen-sei	不燃性；ふねんせい	incombustibility *[ch]*
fun'iki	雰囲気；雰圍気	ambience; aura; atmosphere *[psy]*
funiku'shoku'sei-dōbutsu	腐肉食性動物	carrion *{n} [zoo]*
funjō-hyō'setsu	粉状氷雪	neve *[geogr]*
funka	噴火	eruption (of volcano) *[geol]*
funkaseki	糞化石	guano *[mat]*
funki-kō	噴気孔	fumarole *[geol]*
fun'matsu	粉末；ふんまつ	powder *{n} [mat]*
funmatsu-do	粉末度	fineness of powder *[met]*
funmatsu-jushi	粉末樹脂	powdered resin *[resin]*
funmatsu-yakin	粉末冶金	powder metallurgy *[met]*
fun'mu	噴霧；ふんむ	spray *{n} [eng]*
funmu-hō	噴霧法	atomization method; spray method *[mech-eng]*
funmu-jūgō	噴霧重合	spray polymerization *[poly-ch]*
funmu'ka	噴霧化	nebulization *[mech-eng]*
funmu-kansō	噴霧乾燥	spray drying *{n} [paint]*
funmu-ki	噴霧器	atomizer; sprayer *[eng]*
funmu-ko	噴霧粉	atomized powder *[mat]*
funmu-yaki'ire	噴霧焼入れ	fog quenching *{n} [met]*
fu no	負の	negative *{adj} [math]*

fu(no)kaku	負(の)角	negative angle *[math]*
funō na hōtei'shiki	不能な方程式	inconsistent equations *[math]*
fu no nikō-bunpu-soku	負の二項分布則	negative binomial distribution, law of of *[bio]*
fu-nori	布海苔; ふのり	glue plant; Gliopeltis furcata *[bot]*
fu(no)sū	負(の)数	negative number *[math]*
funryū	噴流; ふんりゅう	jet *{n}* *[fl-mech]*
funryū-suishin	噴流推進	jet propulsion *[aero-eng]*
funsai	粉砕; ふんさい	comminution; grinding; pulverization *[mech-eng]*
funsai-hi	粉砕比; ふんさいひ	reduction ratio *[eng]*
funsai-hōsoku	粉砕法則	grinding, laws of *[mech-eng]*
funsai-ki	粉砕機; ふんさいき	crusher; shredder *[eng]*
funsai-sei	粉砕性	grindability *[mat]*
funsan	噴散; ふんさん	effusion *[p-ch]*
funseki	噴石; ふんせき	cinders *[geol]*
funseki-kyū	噴石丘	cinder cone *[geol]*
funsha-seikei	噴射成形	jet molding *{n}* *[eng]*
funsha-yaki'ire	噴射焼入れ	spray quenching *{n}* *[met]*
funsha-yaku	噴射薬	propellant *[mat]*
fun'shutsu	噴出; ふんしゅつ	eruption; gushing; spouting *{n}*
funshutsu'butsu	噴出物	ejecta *[geol]*
funsō	紛争; 紛爭	dispute; grievance; interference *[legal]*
funsui'sen	噴水泉	fountain *[arch]*
funtai-tosō	粉体塗装	powder coating *[mat]*
funtan	粉炭; ふんたん	powder coal *[geol]*
fūnyū-chūkei	封入注型	embedding (casting) *{n}* *[eng]*
furaisu-ban	フライス盤	milling machine *[eng]*
furan-ki	孵卵器; ふらんき	incubator *[agr]*
furanshūmu	フランシウム	francium (element)
fure	振れ; ふれ	deflection *[electr]* *[eng]*
fu'renzoku	不連続; 不連續	discontinuity *{n}* *[math]*
fu'renzoku no	不連続の	discontinuous *{adj}* *[math]*
fu'renzoku-ryū	不連続流	discontinuous flow *[fl-mech]*
fureru	振れる; ふれる	fluctuate; waver *{vb}*
fureru	触れる; 觸れる	touch; feel *{vb t}*
fureru koto	触れること	touching *{n}* *[physio]*
furiko	振り子; ふりこ	pendulum *[phys]*
fu'rikō	不履行; ふりこう	default *{n}* *[legal]*
furi-sode (no kimono)	振り袖(の着物)	long-sleeved (kimono) *[cl]*
fūrō	封蠟; ふうろう	sealing wax *[mat]*
fūrō	風浪; ふうろう	heavy seas *[ocean]*

furo'ba	風呂場	bathroom *[arch]*
furo'shiki	風呂敷；ふろしき	wrapping cloth square *[mat]*
fūro'sō	風露草	geranium (flower) *[bot]*
furu	振る	shake *{vb t}*
furu	降る	rain (upon); fall; descend *{vb}*
furue	震え	shivering *{n}* *[physio]*
furui	古い；旧い	old *{adj}*
furui	篩	sieve *{n}* *[eng]*
furui-zanbun	篩残分	sieve residue *[eng]*
furu-tori	旧鳥：隹；ふるとり	Chinese character radical denoting short-tailed bird *[pr]*
fu'ryoku	浮力	buoyancy; buoyant force *[fl-mech]*
fū-ryoku	風力	wind force *[meteor]*
fū'ryoku-hatsu'den'sho	風力発電所	wind power plant *[mech-eng]*
furyō-ritsu	不良率	fraction defective *{n}* *[ind-eng]*
fusa	房	tuft *{n}*
fūsa	封鎖	blockade; blocking *{n}* *[ch]*
fu'saku'i	不作爲	nonfeasance; omission *[legal]*
fusei	不整	dissymmetry; asymmetry *[ch]*
"	不整	asymmetry *[math]*
fusei	負性	negativeness *{n}* *[elec]*
fu'seigō	不整合	mismatching *{n}* *[elec]*
"	不整合	unconformity *[geol]*
fu'seigō-kōzō	不整合構造	incommensurate structure *{n}* *[sol-st]*
fusei-gōsei	不斉合成	asymmetric synthesis *[org-ch]*
fusei'ka	不斉化	disproportionation *[ch]*
fu'seikaku na	不正確な	inaccurate *{adj}*
fusei-kin	腐生菌	saprophyte; saprobe *[microbio]*
fusei-kō	不錆鋼	stainless steel *[steel]*
fusei('sei) no	腐生(性)の	saphrophytic *{adj}* *[bio]* *[bot]*
fusei-shokubutsu	腐生植物	saphrophytes *[bot]*
fūsei-sō	風成層	aeolian deposit *[geol]*
fusei suru	付勢する	energize *{vb}* *[elec]*
"	付勢する	activate *{vb}* *[eng]*
"	付勢する	trigger *{vb}*
fusei-tanso-genshi	不斉炭素原子	asymmetric carbon atom *[ch]*
fusen	浮選；浮選	flotation *[eng]*
fūsen	風船	balloon (toy)
fūsha; kaza-guruma	風車	windmill *[mech-eng]*
fushi	節；ふし	node *{n}* *[bot]*
"	節	knot *{n}* *[mat]*
fushi-ana	節穴	knot hole *[civ-eng]*

fushigi	不思議	strange(ness); wonder; marvel *{n}*
fu'shin	浮心	center of buoyancy *[fl-mech]*
fūshin	風疹	rubella; German measles *[med]*
fū'shinshi'kō	風信子鉱	zircon *[miner]*
fūshi-zai	封止剤	potting agent *[electr]*
"	封止剤	encapsulating agent *[mat]*
fu'shoku	腐食；腐蝕	corrosion *[met]*
fu'shoku	腐植	humus; ulmin *[geol]*
fū-shoku	風食	wind erosion *[geol]*
fu-shokubai-sayō	負触媒作用	negative catalysis *[ch]*
fushoku-fu	不織布	nonwoven fabric *[tex]*
fushoku-ganseki	腐食岩石	saprolite *[geol]*
fūshoku'reki	風触礫；風食礫	ventifact *[geol]*
fushoku-ritsu	腐食率	corrosion percent *[met]*
fushoku'sei no	腐食性の	saprophagous *{adj}* *[bio]*
"	腐食性の	caustic; corrosive *{adj}* *[ch]*
fushoku'shitsu	腐植質	humus *[geol]*
fu'soku	不足	inadequacy; insufficiency *{n}*
fusoku-rui	斧足類	bivalves *[i-zoo]*
fū-soku	風速	wind velocity *[meteor]*
fūsoku-kei	風速計	anemometer *[eng]*
fusseki	沸石	zeolite *[miner]*
fusso	弗素；ふっそ	fluorine (element)
fusso-jushi	弗素樹脂	fluoride resin *[poly-ch]*
fusso-shori (ha no)	弗素処理（歯の）	fluoridation (of teeth) *[dent]*
fusuma	麬；麸；ふすま	bran; wheat bran *[bot]* *[cook]*
fusuma	襖	sliding fusuma door *[arch]*
futa	蓋	cover; lid *{n}* *[eng]*
futa'go; sōsei'ji	双生児；ふたご	twins *[bio]*
futago-za	双子座；雙子座	Gemini; the Twins *[constel]*
fūtai	風袋	tare; packing *{n}* *[mech]*
futai-setsubi	付帯設備	supplemental facility *[ind-eng]*
futa'kobu-rakuda	二瘤駱駝	Bactrian camel (animal) *[v-zoo]*
futan	負担	charge *{n}* *[legal]*
futari	二人	two persons *{n}*
fu'tashika-sa	不確かさ	uncertainty *{n}* *[ind-eng]*
futatabi-	再-	re- *{prefix}*
futatsu no-	二つの-	bi- *{prefix}*
futatsu-ori(-ban)	二折（判）	folio *[pr]*
futei	不定；ふてい	undefined *{n}* *[logic]*
"	不定	uncertainty; indefiniteness *{n}*
futei-daimei'shi	不定代名詞	indefinite (or impersonal) pronoun *[gram]*

fu'teihi-kagō'butsu	不定比化合物	nonstoichiometric compound *[ch]*
futei-hō	不定法	infinitive mood *[gram]*
futei-kanshi	不定冠詞	indefinite article *[gram]*
fu'teiki no	不定期の	unscheduled *{adj} [transp]*
futei-sekibun	不定積分	antiderivative; indefinite integral *[math]*
futei'shi	不定詞	infinitive *[gram]*
fu'tekigō-sei	不適合性	incompatibility *[math]*
futen	付点	dot *{n} [music]*
fu'tenka	不点火	misfire *{n} [ch]*
futō	埠頭	wharf; quay; pier *[civ-eng]*
futō	不等	inequality *[math]*
fūto; futto	フート；呎	foot (pl feet) (length) *[mech]*
fūtō	封筒	envelope (stationery) *[comm]*
futo-aya	太綾；ふとあや	drill *{n} [tex]*
futō-eki	不凍液	antifreeze *[ch]*
futō'hen-sankak(u)kei	不等辺三角形	scalene triangle *[math]*
futoi	太い	thick *{adj} [mech]*
futo'ji	太字	bold face *[comput] [pr]*
fu'tōka-soshi	不等価素子	nonequivalence element *[comput]*
futo'maki-zushi	太巻き鮨	thick-rolled sushi (q.v.) *[cook]*
fu'tōmei-baitai	不透明媒体	opaque medium *[opt]*
fu'tōmei-do	不透明度	opacity *[opt]*
fu'tōmei na	不透明な	opaque *{adj}*
fu'tōmei-sa	不透明さ	opacity *{n}*
fu'tōmei-tai	不透明体	opaque body *[opt]*
futon	蒲団；蒲團	comforter; bedquilt *[furn]*
fūto-pondo	フートポンド；呎磅	foot-pound; ft-lb *[mech]*
fu'tōshi-iki	不透視域	zone of avoidance *[astron]*
futō-shiki	不等式	inequality *[math]*
fu'tōsui-sō	不透水層	impermeable layer *[geol]*
futsū	普通；ふつう	ordinaryness; normalness *{n}*
fu'tsui-denshi	不対電子	unpaired electron *[phys]*
futsū-kanten	普通寒天	nutrient agar *[mat]*
futsū no	普通の	ordinary *{adj}*
futten	沸点；沸點	boiling point *[p-ch]*
futten-jōshō-hō	沸点上昇法	ebullioscopy *[p-ch]*
futto	フット	foot (length) *[mech]*
futtō	沸騰；ふっとう	boiling *{n} [p-ch]*
futtō no teishi	沸騰の停止	defervescence *[p-ch]*
futtō-san(zai)	沸騰散(剤)	effervescent powder *[mat]*
futtō saseru	沸騰させる	boil *{vb t}*
futtō-sui	沸騰水	boiling water *[ch]*

futtō suru	沸騰する	boil *{vb i}*
fū'u	風雨	wind and rain *{n} [meteor]*
fuyakeru	ふやける	swell; become sodden *{vb i} [mat]*
fuyasu	殖やす；増やす	increase; add to *{vb t}*
fuyo	付与；付與	grant *{n} [pat]*
fuyō	芙蓉；ふよう	cotton rose; lotus (flower) *[bot]*
fuyu	冬	winter *{n} [astron]*
fuyū-baiyō	浮遊培養	suspension culture *[microbio]*
fuyū'butsu	浮遊物	suspended matter *[phys]*
fuyū-dōbutsu	浮遊動物	zooplankton *[ecol]*
fuyū-kokei'butsu	浮遊固形物	suspended solids; floating solids *[phys]*
fuyū no hōsoku	浮遊の法則	flotation, law of *[phys]*
fuyū-seibutsu	浮遊生物	plankton *[ecol]*
fuyū-senkō	浮遊選鉱；浮遊選鑛	flotation *[eng]*
fuyū-senkō-shiyaku	浮遊選鉱試薬	flotation reagent *[min-eng]*
fuyū-shokubutsu	浮遊植物	phytoplankton *[ecol]*
fuyū-yōryō	浮遊容量	stray capacitance *[electr]*
fuyū-zai	浮遊剤	flotation agent *[ch]*
fuzai	不在	absence *{n}*
fuzoku-sōchi	付属装置	attachment; accessory device *[eng]*
fu'zui'i-kin	不随意筋	involuntary muscle *[physio]*

G

ga	蛾；が	moth (insect); Heterocera *[i-zoo]*
gachō	鵝鳥；がちよう	goose (pl geese) (bird) *[v-zoo]*
ga-denryū	画電流；畫電流	picture current *[elec]*
gadorinyūmu	ガドリニウム	gadolinium (element)
gai-atsu	外圧；外壓	external pressure *[phys]*
gai-bibun	外微分；がいびぶん	exterior derivative *[math]*
gaibu-bunrui	外部分類	external sorting *[comput]*
gaibu-kansū	外部関数	external function *[comput]*
gaibu-kaso'ka	外部可塑化	external plasticization *[poly-ch]*
gaibu-kisei'chū	外部寄生虫	ectoparasite *[ecol]*
gaibu-kisei'sha	外部寄生者	ectoparasite *[ecol]*
gai'bunpitsu-sen	外分泌腺	exocrine gland *[anat]*
gaibu-shiji-yaku	外部指示薬	external indicator *[ch]*
gaibu-wari'komi	外部割(り)込み	external interrupt *[comput]*
gaibu-zatsu'on	外部雑音	external noise *[electr]*
gaichū	害虫	harmful (injurious) insect; pest *[i-zoo]*
gaigō	外合	superior conjunction *[astron]*
gaihai-sen	外擺線	epicycloid *[math]*
gai'haiyō	外胚葉	ectoderm; ectoblast *[embryo]*
gaihi	外皮；外被	integument *[anat] [zoo]*
"	外皮；外被	outer cover *[bot]*
gaihō-kōbutsu	外包鉱物；外包鑛物	perimorph *[miner]*
gai'in	外陰	vulva (pl vulvae) *{n} [anat]*
gai'in'sei no	外因性の	extrinsic *{adj}*
gai'in'teki-seishitsu	外因的性質	extrinsic property
gaiji	外字	foreign language; foreign character *[pr]*
gaika	外貨	foreign currency *[econ]*
gai'kahi	外果皮；がいかひ	epicarp; exocarp *[bot]*
gaikaku	外角	exterior angle; external angle *[math]*
gaikaku	外郭；がいかく	outer perimeter *[math]*
"	外郭	contour *[sci-t]*
gaikaku	外核	outer core *[geol]*
gaikei	外径；外徑	outside diameter *[math]*
gaikei	外景	exterior (view) *[arch]*
gaiken	外見	shaping *{n} [arch]*
"	外見	appearance *{n}*
gaiki-ken	外気圏；外氣圏	exosphere *[meteor]*
gaikō	外港；外港	outer harbor *[geogr]*
gaikō-dōbutsu	外肛動物	Ectoprocta *[i-zoo]*

gai'kokkaku	外骨格	exoskeleton *[i-zoo]*
gai'koku	外国; 外國	foreign country *[geogr]*
gaikoku'jin	外国人	foreigner; alien
gaikoku-kawase	外国為替	foreign exchange *[econ]*
gai'kōso	外酵素	exoenzyme *[bioch]*
gai'kotsu	骸骨	skeleton *[anat]*
gaimen no	外面の	superficial *{adj}*
ga'in	画因; 畫因	motif *[graph]*
gai'nen	概念	concept; idea *[math] [psy]*
gai'nichi-rizumu	概日リズム	circadian rhythm *[bio]*
gairai'go	外来語; 外來語	forign-origin word *[ling]*
gairai-han'dōtai	外来半導体	extrinsic semiconductor *[electr]*
gairai-kesshō	外来結晶	xenocryst *[crys]*
gairin'sen	外輪船	paddleboat *[nav-arch]*
gairo	街路; がいろ	street *[civ-eng] [traffic]*
gairo-ju	街路樹	shade tree; street tree *[bot] [ecol]*
gairo-keitō	街路系統	street system *[civ-eng]*
gai'ryaku	概略	schema *{n} [comput]*
gairyaku no	概略の	approximate *{adj} [math]*
gai-ryoku	外力	external force *[mech]*
gaisan	概算	approximate calculation *[math]*
gai'setsu	外接	circumscription *[math]*
gai'setsu	概説	general account; general statement; outline; review *[pr]*
gaisetsu-en	外接円; 外接圓	circumcircle *[math]*
gaisetsu shita	外接した	circumscribed (-ing) *{adj} [math]*
gaisetsu'sho	概説書	general information manual *[pr]*
gaisetsu-takak(u)kei	外接多角形	circumscribed polygon *[math]*
gaishi	外資	foreign capital *[econ]*
gaishi	碍子	insulator *[elec]*
gai'shin	外心	circumcenter (of triangle) *[math]*
gaishō	外傷	trauma *[med]*
gai'shoku-hen	外植片	explant *[cyt]*
gaishū	外周; がいしゅう	periphery *[math]*
gaishū-men	外周面	peripheral surface *[math]*
gaisō	外挿	extrapolation *[math]*
gaisō	外装; 外裝	outer packaging *[ind-eng]*
gaisō	鎧装	armoring *{n} [rub]*
gaisō'hin	外装品	trim (of automobile) *{n} [mat]*
gaisō'ji	外装地	covering fabric *[tex]*
gaisō suru	外装する	sheathe; armor; wrap *{vb t} [eng]*
gaisui	崖錐	talus *{n} [geol]*

gaiten suru	外添する	externally add *{vb t}* *[ch]*
gaitō	外筒	outer tube; outer cylinder *[mech-eng]*
gaitō	外套	coat; mantle *{n}* *[cl]*
gaitō	街灯；街燈	street light *[civ-eng]*
gaitō-jikō	該当事項	pertinent data; relevant data *[math]*
gai-wakusei	外惑星	outer planet; superior planet *[astron]*
gaiyō'sei no	外洋性の	pelagic; of the ocean *{adj}* *[geol]*
gaiyō-yaku	外用薬；外用藥	medicine for external application *[pharm]*
gaiyō-zu	概要図；概要圖	schematic diagram *[graph]*
gaka	画架；畫架；がか	easel *[graph]*
ga-kaku	画角	field angle *[opt]*
gake	崖；がけ	cliff; precipice; bluff *[geogr]*
gake-kuzure	崖崩れ	earth fall; landslide; landslip *[geol]*
gakkai	学会；學會	scientific society *[sci-t]*
gakki	楽器；樂器	musical instrument *[music]*
gakkō	学校；學校	school *{n}* *[arch]*
gaku	額	frame; plaque *{n}* *[furn]*
-gaku	-学；-學	study of; science of *{suffix}*
gaku'dō	学童	pupil; school-age child
gaku'fu	楽譜	music (sheet); score *{n}* *[music]*
gaku'hen	萼片；がくへん	sepal *[bot]*
gaku'jutsu	学術；がくじゅつ	learning; scholarship; science *[sci-t]*
gakujutsu'teki na	学術的な	scientific *{adj}*
gaku'mei	学名	scientific name *[comm]*
(gakumei-)meimei-hō	(学名-)命名法	nomenclature *[sci-t]*
gakumon-ryō'iki	学問領域	discipline (of study) *[sci-t]*
gaku'on	楽音；樂音	musical tone *[acous]*
gaku'shū	学習	learning; study *{n}*
gakushū-katei	学習過程	learning process *[psy]*
gama	がま	vug *[petr]*
gama	蒲；がま	cattail; Typha latifolia *[bot]*
gama; hiki-gaeru	蝦蟇；がま	toad; Buff vulgaris *[v-zoo]*
gama-doku	蝦蟇毒	toad poison *[pharm]* *[physio]*
gamen	画面	picture *[comput]*
gamen-idō	画面移動	scroll; scrolling *{n}* *[comput]*
gan	雁；がん	wild goose (bird) *[v-zoo]*
gan	癌；がん	cancer *[med]*
ganban	岩板	plate; rock plate *[geol]*
ganban	岩盤	bedrock *[geol]*
gan-dare	雁垂：厂；がんだれ	kanji (q.v.) radical for wild goose *[pr]*
gan'en	岩塩	halite; rock salt *[miner]*
gan'en-haisha	岩塩背斜	salt anticline *[geol]*

gan'en-kō	岩塩坑	salt mine *[min-eng]*
gangi'ei	雁木鱏；がんぎえい	skate; thornback (fish) *[v-zoo]*
gan-idenshi	癌遺伝子	oncogene *[gen]*
ganhyō'shō	含氷晶	cryohydrate *[ch]*
gan'i	含意；がんい	implication *[logic]*
gan'i suru	含意する	imply; mean *{vb i}* *[psy]*
ganjō na	頑丈な	sturdy *{adj}*
ganka'kiseki	頑火輝石	enstatite *[miner]*
ganken; ma'buta	眼瞼	eyelid *[anat]*
ganken-sei	頑健性	robustness *[ind-eng]*
gankin	眼筋；がんきん	eye muscle *[anat]*
ganki-ritsu	含気率	porosity *[phys]*
gankyū	眼球	eyeball; bulbus oculi *[anat]*
ganma-sen	ガンマ線	gamma rays *[nuc-phys]*
ganmen	岩綿；がんめん	mineral wool; rock wool *[mat]*
gan-modoki	雁擬き；がんもどき	tofu-vegetable deep-fried cake *[cook]*
ganpeki	崖壁；がんぺき	quay; quaywall; palisade *[civ-eng]*
ganpi-shi	雁皮紙	unsized silk paper *[mat]*
ganryō	含量	content (quantity) *[ch]*
ganryō	顔料；がんりょう	pigment *[biochem]* *[mat]*
ganryū-ken	岩流圏；岩流圏	asthenosphere *[geol]*
gan'sai; ganshi	岩滓	scoria (pl -riae) *[geol]*
gan'sanso'kō-seisei-genso	含酸素鉱生成元素	lithophile; oxyphile *{n}* *[geoch]*
gansei-shokubutsu	岩生植物	lithophyte *[ecol]*
ganseki-chikyū-kagaku	岩石地球化学	lithogeochemistry *[geoch]*
ganseki'gaku	岩石学	petrology; lithology *[geol]*
ganseki-ken	岩石圏	lithosphere *[geol]*
ganseki-kisai'gaku	岩石記載学	petrography *[geol]*
ganseki-sei'in'gaku	岩石成因学	petrogeny *[geol]*
gansetsu	岩屑；がんせつ	detritus *[geol]*
gan'shin	含浸	impregnation *[ch-eng]* *[rub]*
ganshin'shō	眼振症	nystagmus *[med]*
ganshin-zai	含浸剤	impregnant; impregnating agent *[poly-ch]*
gansho	願書；がんしょ	application *[pat]*
ganshō	岩漿；がんしょう	magma *[geol]*
ganshō	岩礁	reef *[geol]*
gansui-baku'yaku	含水爆薬	slurried explosive; water gel *[ord]*
gansui-ritsu	含水率	water content *[ch-eng]*
gansui-ryō	含水量	moisture content; water content *[ch-eng]*
gansui-sei	含水性	hydrousness *[ch]*
gantai	岩帯；岩帯	zone *[geol]*
ganten	眼点	eyespot; stigma *[i-zoo]*

gan-yaku	丸薬	pill; medicinal pill *{n} [pharm]*
gan'yu-jushi	含油樹脂	oleoresin *[mat]*
gan'yu-ritsu	含油率	oil content *[ch-eng]*
gan'yū-ryō	含有量	content (quantity) *[ch]*
gappei	合併	union *[math]*
gara	柄；がら	pattern *{n} [graph]*
garagara-hebi	がらがら蛇	rattlesnake (reptile) *[v-zoo]*
garakuta	がらくた	debris *[ecol] [geol]*
garasu	硝子；ガラス	glass *[mat]*
garasu-chōkoku-gihō	硝子彫刻技法	hyalography *[graph]*
garasu'ga-gihō	硝子画技法	hyalography *[graph]*
garasu'jō no	硝子状の	hyaline; glassy *{adj} [geol] [i-zoo]*
garasu'ka-ishi	硝子化石	vitrified rock *[geol]*
garasu'ka-seishitsu	硝子化性質	vitrescence *[mat]*
gare'ba	がれ場	scree *[geol]*
ga'rin	芽鱗	bud scale *[bot]*
garyūmu	ガリウム	gallium (element)
gaso	画素	picture element (pel, pixel) *[electr]*
gasshō'dan	合唱団	chorus *[music]*
gasu	瓦斯；ガス	gas *[phys]*
gatchi	合致	agreement; matching; coincidence *{n}*
gatsu; getsu	月；がつ；げつ	month *[astron]*
-gatsu	-月	-month (of year) *{suffix} [astron]*
gattai suru	合体する	coalesce *{vb}*
gazami	蝤蛑；がざみ	blue crab *[i-zoo]*
gāze	ガーゼ	gauze *[mat]*
gazō	画像；がぞう	picture; picture image *[comm] [electr]*
gazō-kōgaku	画像工学	picture image engineering *[electr]*
gazō-shori	画像処理	image processing *{n} [comput]*
ge'doku	解毒；げどく	detoxification *[bioch]*
gedoku-yaku	解毒薬	antidote *{n} [pharm]*
gei'jutsu	芸術；藝術	art; arts *{n}*
gei-kaku	迎角	angle of attack *{n} [aero-eng]*
gei'rō	鯨蠟；げいろう	spermaceti *[mat]*
geishu; kujira-hige	鯨鬚	baleen *[v-zoo]*
gei-yu	鯨油	whale oil *[mat]*
geji-geji	蚰蜒；げじげじ	millipede (insect) *[i-zoo]*
geka'i	外科医；外科醫	surgeon *[med]*
geki'hen	激変	cataclysm *[geol]*
gekihen'gata-henkō-sei	激変型変光星	cataclysmic variable star *[astron]*
geki'jō	劇場	theater *[arch]*
geki-ryoku	撃力	impulsive force; impact *[mech]*

geki'ryū	激流	torrent *[hyd]*
geki-yaku	劇薬	powerful medicine *{n} [pharm]*
gekkaku-sa	月角差	parallactic inequality *[astron]*
gekkan'shi	月刊誌	monthly periodical *[pr]*
gekkei	月経; 月經	menstruation *[physio]*
gekkei'ju	月桂樹	bay tree; laurel (tree) *[bot]*
gen	元; げん	element *[math]*
gen	源; 原; げん	origin *[comput]*
gen	弦	chord *[math]*
gen	絃	string; chord *[music]*
gen'atsu	減圧; 減壓	reduced pressure *[ch-eng]*
gen'atsu-jōryū	減圧蒸留	vacuum distillation *[ch-eng]*
gen'atsu-shitsu	減圧室	decompression chamber *[eng]*
genba	現場	field; in the field; jobsite; on site; worksite *[ind-eng]*
genbu'gan	玄武岩	basalt *[miner]*
genbu'gan'shitsu-yōgan	玄武岩質溶岩	basaltic lava *[miner]*
genbun	原文	original text *[graph] [pr]*
genbun	減分	decrement *[comput] [math]*
genchi-kōnyū	現地購入	local purchase *[econ]*
gendai	現代	modern times; the present *{n}*
gendai-sūgaku	現代数学	modern mathematics *[math]*
gendo	限度	limit *{n} [math]*
gendō-jiku	原動軸	driving shaft *[mech-eng]*
gendō-ki	原動機	prime mover *[mech-eng]*
gendō'sho	原動所	power plant *[mech-eng]*
gen-eiyō'kabu	原栄養株; 原榮養株	prototroph *[microbio]*
gen'eki	原液	dope; stock solution; undiluted solution *{n} [ind-eng]*
ge'netsu-zai	解熱剤	antifebrile *{n} [pharm]*
gengai no	言外の	implied; unexpressed *{adj} [comm]*
gengai-kenbi'kyō	限外顕微鏡	ultramicroscope *[opt]*
gengai-roka	限外沪過	ultrafiltration *[ch-eng]*
gengai-roka-ki	限外沪過器	ultrafilter *{n} [eng]*
gengai-roka-maku	限外沪過膜	ultrafiltration membrane *[ch-eng]*
gen'gakki-rui	絃楽器類	stringed instruments *[music]*
gengan-keiseki	原岩形跡	palimpsest *[geol]*
gengi	原義	primary meaning *[comm]*
gengo	言語	language; speech *[comm] [ling]*
gengo'gaku	言語学	linguistics; philology *[comm]*
gengo'on	言語音	speech sound *[physio]*
gen'in	原因; げんいん	cause; grounds; reason *{n} [sci-t]*

genji	減磁	demagnetization *[elecmg]*
genji-ryoku	減磁力	demagnetizing force *[elecmg]*
genka	原価；原價	cost; prime cost *[econ]*
"	現価	present value *[econ] [math]*
gen-kabu	原株；げんかぶ	original strain *[microbio]*
genkai	限界；げんかい	margin *[graph] [pr] [sci-t]*
"	限界	boundary; limit *{n} [math]*
"	限界	limitation *{n}*
genkai-chi	限界値	threshold value *[math]*
genkai-genka	限界原価	marginal cost *[econ]*
genkai-kakaku	限界価格	price ceiling *[econ]*
genkai-kensa	限界検査	marginal checking *[electr]*
genkai-shiken	限界試験	critical-limit test *[mech-eng]*
genkai-tokusei-chi	限界特性値	threshold characteristic value *[sci-t]*
genkaku-saibō	原核細胞	prokaryolytic cell *[cyt]*
genkaku-seibutsu	原核生物	prokaryote *[cyt]*
genkaku-zai	幻覚剤	hallucinogen; psychedelic drug *{n} [pharm]*
genkan	玄関	entrance *[arch]*
genkan-zai	減感剤	desensitizer *[photo]*
genka-shō'kyaku	原価消却	depreciation of cost *[econ]*
genkei	原型	prototype *[eng]*
genkei'shitsu	原形質	protoplasm *[cyt]*
genkei'shitsu-bunri	原形質分離	plasmolysis *[physio]*
genkei'shitsu-maku	原形質膜	plasma membrane; plasmalemma *[cyt]*
genkei'shitsu-tai	原形質体	protoplast *[cyt]*
genki	元気	vitality; vigor; stamina *{n}*
genki	元期	epoch *[astron]*
genki	減輝	decalescence *[met]*
genkin	現金	cash; money *[econ]*
genkō	原鉱；原鑛	mineral *[geol]*
genkō	減光	dimming (of light) *{n} [elec]*
"	減光	extinction *[astron] [opt]*
genkō-ki	減光器	dimmer *[elec]*
genkyoku	減極	depolarization *[elec]*
genmai	玄米	unpolished rice *[bot] [cook]*
genmaku-hōshi	原膜胞子	chlamydospore *[mycol]*
genmen	原綿	raw cotton *[bot] [tex]*
"	原綿	raw stock *[tex]*
genmō	原毛	raw wool; raw hair *[tex]*
gen'no'shōko	現の証拠	Geranium thunbergii *[bot]*
genomu	ゲノム	genome *[gen]*
genpa-fugō	現波符号	cable code *[comm]*

genpō	減法	subtraction (method) *[math]*
genpō-kon'shoku	減法混色	subtractive color mixing *[opt]*
genri	原理	principle *[sci-t]*
genryō	原料	raw material *[ind-eng]*
genryō	減量	loss (of weight) *[mech]*
genryō-kakō	減量加工	weight-reduction processing *[ind-eng]*
genryō-ritsu	減量率	weight-loss rate *[mech]*
gensei-dōbutsu	原生動物	Protozoa *[i-zoo]*
gensei-seibutsu	原生生物	Protista; protists *[bio]*
gen-sen'i	原繊維；原纖維	fibril *[bio]*
gen'shaku	原尺	full scale *[graph]*
genshaku-zu	原尺図；原尺圖	full-scale (-size) drawing *[graph]*
genshi	原子；げんし	atom *[ch]*
genshi	原始	original *[comput]* *[pr]*
genshi	原紙	base paper; paper stock; untrimmed paper *[mat]* *[paper]*
genshi-bakudan	原子爆弾	atom bomb *[ord]*
genshi-baku'hatsu	原子爆発	atomic explosion *[phys]*
genshi-bangō	原子番号	atomic number *[nuc-phys]*
genshi-bunritsu	原子分率	atomic fraction *[phys]*
genshi-butsuri'gaku	原子物理学	atomic physics *[phys]*
genshi-dokei	原子時計	atomic clock *[horol]*
genshi-gengo	原始言語	source language *[comput]*
genshi-hankei	原子半径	atomic radius *[p-ch]*
genshi'ika no	原子以下の	subatomic *{adj}* *[phys]*
genshi-jika-ritsu	原子磁化率	atomic susceptibility *[elecmg]*
genshi'ka	原子価	valence *[ch]*
genshi'ka-denshi	原子価電子	valence electron *[atom-phys]*
genshi-kaku	原子核	atomic nucleus *[nuc-phys]*
genshi(-kaku)-dantō	原子(核)弾頭	nuclear warhead *[ord]*
genshi'kaku-kōgaku	原子核工学	nucleonics *[eng]*
genshi'kaku-nyūzai	原子核乳剤	nuclear emulsion *[nucleo]*
genshi'ka-ryoku no ba	原子価力の場	valence force field *[ch]*
genshi'ka-yōdō	原子価揺動	valence fluctuation *[ch]*
genshi-kigō	原子記号	atomic symbol *[ch]*
genshi no	原始の	primitive *{adj}* *[bio]*
genshi no-	原始の-	proto- *{prefix}*
genshi-ro	原子炉	nuclear reactor; reactor *[nucleo]*
genshi'ro-yōki	原子炉容器	reactor vessel *[nucleo]*
genshi'ryō	原子量	atomic weight *[ch]*
genshi-ryoku	原子力	atomic energy; atomic power *[nucleo]*
genshi'ryoku'sen	原子力船	atomic powered ship *[nav-arch]*

genshi'ryoku-suishin	原子力推進	atomic propulsion *[mech]*
genshi-sei	原始星	protostar *[astron]*
genshi-sen	原子線	atomic beam *[phys]*
genshi-shitsu'ryō	原子質量	atomic mass *[phys]*
genshi-shitsuryō-sū	原子質量数	atomic mass number *[phys]*
genshi-shitsuryō-tan'i	原子質量単位	atomic mass unit *[phys]*
genshi'teki na	原始的な	primordial *{adj}* *[bio]*
genshi-wakusei	原始惑星	protoplanet *[astron]*
genshi-yō	原子容	atomic volume *[p-ch]*
genshō	現象	phenomenon *[math]*
genshō	減少	reduction *[math]*
"	減少	abatement; diminution *{n}*
genshō-kansū	減少関数	decreasing function *[math]*
gen'shoku	原色	primary color(s) *[opt]*
genshoku-hō no genshoku	減色法の原色	subtractive primary color *[opt]*
genshō-ryō	減少量	decrement *[math]*
genso	元素	element *[ch]*
gensō	舷窓	porthole *[nav-arch]*
genso-bunseki	元素分析	ultimate analysis; elementary analysis *[an-ch]*
genso-kōbutsu	元素鉱物	native element(s) *[geol]*
gensoku	舷側	broadside *{n}* *[nav-arch]*
gensoku	原則	principle; rule *[sci-t]*
gensoku	減速；げんそく	deceleration; speed reduction *[mech]*
gensoku-ha'guruma	減速歯車	reduction gear *[mech-eng]*
gensoku-ki	減速機	decelerator; reduction gear *[eng]*
gensoku-ritsu	減速率	deceleration rate *[mech]*
gensoku-zai	減速材	moderator *[nucleo]*
genso-shūki-hyō	元素周期表	periodic table of the elements *[ch]*
gensū	減数	subtrahend *[math]*
gensū-bun'retsu	減数分裂	meiosis *[cyt]*
gensui	原水	raw water *[civ-eng]* *[hyd]*
gensui	減衰	attenuation; damping *{n}* *[phys]*
gensui-hi	減衰比	damping ratio *[phys]*
gensui-ki	減衰器	attenuator *[electr]*
gensui-shindō	減衰振動	damped oscillation *[phys]*
gentai	減退	reduction *[bio]*
"	減退	abatement *[med]*
gentan	原反	material cloth *[tex]*
gen-tan'i	原単位	basic unit *[econ]*
gentei	限定	limitation *[legal]*
gentei'shi	限定詞	qualifier *[gram]*

genten	原点	origin *[math]*
gen'waku	眩惑	dazzlement; bewilderment *{n} [psy]*
gen'yō'go	現用語	living language *[ling]*
gen'yō no	現用の	currently in use *{adj} [pat]*
gen-yu	原油	crude oil *[geol]*
genzai no	現在の	current; present *{adj}*
genzan-ki	減算器	subtracter *[electr]*
genzō	現像	development; developing *{n} [photo]*
genzō-eki	現像液	developer *[photo]*
genzō-shori	現像処理	processing *{n} [photo]*
genzō-zai	現像剤	developer; developing agent *[photo]*
geppō	月報	monthly report *[pr]*
geri	下痢	diarrhea *[med]*
geru	ゲル	gel *[ch]*
gerumanyūmu	ゲルマニウム	germanium (element)
geshi	夏至; げし	summer solstice *[astron]*
gesshoku	月食; 月蝕	lunar eclipse *[astron]*
ge'sui	下水	sewage *[civ-eng]*
gesui'dō	下水道	sewer *[civ-eng]*
geta	下駄	clogs (footwear) *[cl]*
getchō'seki	月長石	moonstone *[miner]*
getsu'men-jishin	月面地震	moonquake *[astron]*
getsu'men-kakō'jō-kubomi	月面火口状窪み	lunar crater *[astron]*
getsu'men-riku'chi	月面陸地	terra (of moon) *{n} [astron]*
getsu'men'shi	月面誌	selenography *[astron]*
getsu'men-zu	月面図	lunar map; selenograph *[astron]*
getsu'yōbi	月曜日	Monday
gi; waza	技	craft; skill; art *{n}*
gi-	偽-; 僞-	false- *{prefix}*
gi-	擬-	pseudo- *{prefix}*
-gi	-儀	instrument for astonomical measurements *{suffix}*
gi'dansei	擬弾性	anelasticity *[mech]*
gi'fusei	擬不斉	pseudoasymmetry *[p-ch]*
gigi	疑義	doubt *{n} [psy]*
gi'heikō	偽平衡	false equilibrium *[phys]*
gi'henkei'tai	偽変形体	pseudoplasmodium *[i-zoo]*
gi-idenshi	偽遺伝子	pseudogene *[gen]*
giji	擬似	artificiality; falseness *{n}*
giji-fugō	擬似符号	false signal *[comm]*
giji-fuka	擬似負加	artificial load *[mech-eng]*
giji-henkan	擬似変換	affine transformation *[math]*

giji-hōkō'zoku-sei	擬似芳香族性	pseudoaromaticity *[org-ch]*
giji no-	擬似の-	pseudo- *{prefix}*
gi'jinsei	擬靭性	pseudotoughness; pseudotenacity *[mech]*
giji-ransū-retsu	擬似乱数列	pseudorandom number sequence *[comput]*
giji-shūki	擬似周期	pseudoperiod *{n}* *[geophys]*
giji-taishō	擬似対称	pseudosymmetry *[crys]*
giji-tōhō	擬似等方	quasi-isotropic(ness) *[phys]*
giji-yūsei'teki-kōzatsu	擬似有性的交雑	parasexual hybridization *[gen]*
gi'jutsu	技術	technology *[sci-t]*
gijutsu-kōgaku	技術工学	technology *[sci-t]*
gijutsu-suijun	技術水準	state of the art; technical level *[ind-eng]*
gijutsu-tantō'sha	技術担当者	technical personnel; technical person in charge *[ind-eng]*
gijutsu'teki-hyōjun	技術的標準	technical standard *[ind-eng]*
gijutsu-yōgo	技術用語	technical term *[comm]*
gimon-fu	疑問符	question mark *[comput]* *[gram]* *[pr]*
gin	銀；ぎん	silver (element)
gin'biki	銀引き	silvering *{n}* *[met]*
gin-dara	銀鱈；ぎんだら	sablefish; Anoplopoma fimbria *[v-zoo]*
gindei	銀泥	silver paint *[mat]*
gin'en	銀塩	silver salt *[ch]*
ginga	銀河	galaxies; Milky Way *[astron]*
gingakei'gai-sei'un	銀河系外星雲	extragalactic nebula *[astron]*
ginga-kyoku	銀河極	galactic poles *[astron]*
ginga-zahyō	銀河座標	galactic coordinates *[astron]*
gin-masu	銀鱒	silver salmon (fish) *[v-zoo]*
ginmen	銀面	grain side (of skin) *[v-zoo]*
ginnan	銀杏	ginkgo nut *[bot]* *[cook]*
gin-nezu	銀ネズ	silver gray (color) *[opt]*
ginpo	銀宝；ぎんぽ	gunnel; Enedrias nebulosus (fish) *[v-zoo]*
gin-rō'zuke	銀蠟付	silver soldering *{n}* *[met]*
ginsei'seki	銀星石	wavellite *[miner]*
ginshi	銀糸	silver thread *[tex]*
ginshu	銀朱	vermillion; mercuric sulfide *[met]*
gin-teki'tei	銀滴定	argentometry *[an-ch]*
gin-zame	銀鮫	ratfish; chimera; Chimaera phantasma (fish) *[v-zoo]*
gi'on	擬音	imitation sound; sound effect *[acous]*
giratsuki	ギラツキ	dazzle; glare *{n}* *[opt]* *[physio]*
gisan	蟻酸	formic acid
gi-san	擬酸	pseudo acid *[ch]*

gisan-en	蟻酸塩	formate *[ch]*
gishi	義肢	artificial limb *[eng]*
gisō	擬装；ぎそう	camouflage *{n} [bio] [mil]*
gisō'butsu	擬装物	dummy *{n} [math]*
gi'sosei-ryūdō	擬塑性流動	pseudoplastic flow *[fl-mech]*
gi'sosei-ryūdō'tai	擬塑性流動体	pseudoplastic fluid *[fl-mech]*
gi-suijun	擬水準	dummy level *[math]*
gizō	偽造	counterfeiting; forging *{n} [legal]*
gizō	偽像	false image *[opt]*
go	五	five *{n} [math]*
go	語	computer word *[comput]*
"	語	word *[gram]*
gō	合	conjunction *[astron]*
gō	号；號	issue; number *{n} [pr]*
gō	濠	trench *[geogr]*
gōban	合板	plywood *[mat]*
goban'me-shiken	碁板目試験	crosscut adhesion test *[eng]*
gobō	牛蒡；ごぼう	burdock; cocklebur(r) (vegetable) *[bot]*
gobu	五分	fifty percent; one-half *[math]*
gobu-gobu	五分五分	evenness; even parts; a draw
go'bun no ichi	五分の一	one-fifth *[math]*
gō'chaku suru	合着する	coalesce *{vb} [phys]*
gōdō	合同	congruent *[math]*
gōdō-sankak(u)kei	合同三角形	congruent triangles *[math]*
goei'tai	護衛隊	convoy *{n} [ord]*
go'enka-rin	五塩化燐	phosphorus pentachloride
gofun	胡粉	chalk *{n} [mat]*
gōgai	号外；號外	extra edition *[pr]*
go'gatsu	五月	May (month)
gogo	午後	afternoon; postmeridian *[astron]*
gohan	御飯	boiled rice; meal *[cook]*
gōhan	合板	plywood *[mat]*
gohen	互変；互變	enantiotropy *[ch]*
gohen dekiru	互変出来る	interchangeable *{adj}*
gohen-i'sei	互変異性	tautomerism *[ch]*
gohen-isei'tai	互変異性体	tautomer *[ch]*
gohō	語法	expression; usage *[ling]*
gōhō no	合法の	within the law *{adj} [legal]*
go'i	語彙	glossary; vocabulary *[pr]*
gōi	合意	agreement *[legal]*
go'in	五音	pentatonic scale *[music]*
gō'itsu-ha	合一波	composite wave *[phys]*

go-jikan	語時間	word time *[comput]*
gojo-hō	互除法	algorithm *[comput]*
"	互除法	mutual division *[math]*
gojū'kara	五十雀	nuthatch (bird) *[v-zoo]*
gojū-kō	五重項	quintet *{n}* *[quant-mech]*
gojun	語順	order *{n}* *[gram]*
gojū-sen	五重線	quintet *[spect]*
goka-genso	五価元素	pentad *[ch]*
gōkaku	合格	acceptance *[ind-eng]*
gokaku-chū	五角柱	pentagonal prism *[math]*
go'kaku-jūni'men'tai	五角十二面体	pyritohedron *[crys]*
gokak(u)kei	五角形	pentagon *[math]*
gokaku-sui	五角錐	pentagonal pyramid *[math]*
gokan	互換	transposition *[math]*
gokan	語間	spacing (between words) *{n}* *[comp]* *[pr]*
gokan	語感	linguistic sense; word feel *[ling]*
goka no	五価の	pentavalent *{adj}* *[ch]*
gokan-sei	互換性	interchangeability *[comput]*
gōka'sen	豪華船	luxury ship *[nav-arch]*
gōkei	合計；ごうけい	sum; total *{n}* *[math]*
goki	誤記	mistake (in notation) *{n}* *[comm]*
gokiburi	蜚蠊；ごきぶり	cockroach (insect) *[i-zoo]*
gōkin	合金；ごうきん	alloy *[met]*
gōkin'ka-mekki-kōban	合金化めっき鋼板	alloy-plated steel sheet *[steel]*
gōkin-kō	合金鋼	alloy steel *[steel]*
gōkin-kōgu-kō	合金工具鋼	alloy tool steel *[steel]*
goko-hito'kumi	五個一組	pentad *[climat]*
goku'raku-chō	極楽鳥	bird-of-paradise (bird) *[v-zoo]*
gokuraku'chō-ka	極楽鳥花	bird-of-paradise flower *[bot]*
goma	胡麻；ごま	sesame; sesame seed *[bot]*
gomame	鱓；田作；ごまめ	dried small sardines *[cook]*
gōmei-gaisha	合名会社	partnership *[econ]*
gomi	塵；芥；ごみ	dust *{n}* *[geol]*
gōmō	剛毛	bristle; seta (pl setae) *{n}* *[bio]*
gomoku-gohan	五目御飯	rice with various ingredients *[cook]*
gomu	ゴム；護謨	rubber *[org-ch]*
gon-ben	言扁；言偏：訁	kanji (q.v.) radical denoting speech *[pr]*
gondō-kujira	巨頭鯨；ごんどう鯨	Globicephala whale *[v-zoo]*
gōnen-shi	合撚糸	double-twisted yarn *[tex]*
gonzui	権瑞；ごんずい	plotosid; Plotosus anguillaris (fish) *[v-zoo]*
gōri'ka	合理化	rationalization *[legal]*

gōri-sei	合理性	rationality *{n}*
gōri'teki na	合理的な	rational; reasonable *{adj}*
goro; koro	頃	at about (time) *{adv}*
gō-ryoku	合力	resultant of forces *[mech]*
gōryū	合流	combined flow; confluence *[hyd]*
gōryū'gata-chōkika-kansū	合流型超何幾関数	confluent hypergeometric function *[math]*
gosa	誤差	error *[math]*
gosa-bunsan	誤差分散	error variance *[math]*
go'sadō	誤作動	malfunction; wrong operation *[ind-eng]*
go'sanka-ni'chisso	五酸化二窒素	dinitrogen pentoxide
go'sanka-rin	五酸化燐	phosphorus pentoxide
gosa-ritsu	誤差率	relative error *[math]*
gōsei	合成；ごうせい	synthesis *[ch] [poly-ch]*
gō-sei	剛性	rigidity *[mech]*
gōsei-fū	合成風	resultant wind *[meteor]*
gōsei-hi'kaku	合成皮革	synthetic leather *[poly-ch]*
gōsei-jushi	合成樹脂	synthetic resin *[poly-ch]*
gōsei-jushi-kagaku	合成樹脂化学	plastics (the field) *[poly-ch]*
gōsei-kagaku'sha	合成化学者	synthetic chemist *[ch]*
gōsei-kansū	合成関数	composite function *[math]*
gōsei-kōso	合成酵素	ligase *[bioch]*
gōsei-meidai	合成命題	compound statement *[logic]*
gosei no	互生の	alternate *{adj} [bot]*
gōsei-on	合成音	composite tone *[acous]*
gōsei-ritsu	剛性率	modulus of rigidity; shear modulus *[mech]*
gōsei-shi	合成紙	synthetic paper *[poly-ch]*
gōsei-sū	合成数	composite number *[math]*
go-sen	語線	word line *[comput]*
gōsen	合繊；合纖	synthetic fiber *[tex]*
gōsen-ori'mono	合繊織物	synthetic-fiber fabric *[tex]*
go'shiki-hiwa	五色鶸；ごしきひわ	goldfinch (bird) *[v-zoo]*
go-shin	五心	five centroids (of triangle) *[math]*
goshin-hō	五進法	quinary *{n} [comput]*
gō'tai	剛体	rigid body *[mech]*
go'yaku suru	語訳する	translate for the words *{vb} [comm]*
go'yaku suru	誤訳する	mistranslate *{vb} [comm]*
go'yō-ron	語用論	pragmatics *[ling]*
goza	茣蓙；御座；ござ	rush mat, bound at edges *[mat]*
gozen	午前	ante meridian (a.m.); forenoon *[astron]*
go'zoku	語族	phylum (pl phyla) *[ling]*
gu	具；ぐ	ingredient(s) *[cook]*
gubijin'sō	虞美人草	corn poppy; field poppy *[bot]*

guchi	ぐち	croaker (fish) *[v-zoo]*
gūhatsu'teki na	偶発的な	accidental *{adj} [math]*
gū-kansū	偶関数	even function *[math]*
gūki-sei	偶奇性	parity *[phys]*
gumi	茱; 胡頽子; ぐみ	oleaster *[bot]*
gun	群	group *{n} [bio] [syst]*
gunbi	軍備	armament *[ord]*
gundan	群団; 群團	alliance *[bot]*
gunjō	群青; ぐんじょう	ultramarine blue (color) *[inorg-ch]*
gunju'hin	軍需品	ordnance *[eng]*
"	軍需品	munitions *[ord]*
gunkan	軍艦	battleship; warship *[nav-arch]*
gunkan-dori	軍艦鳥	frigate bird *[v-zoo]*
gun'san-kyōdō'tai	軍産共同体	military-industrial complex *[econ]*
gunshū	群集	community *[bio] [ecol]*
gun-sokudo	群速度	group velocity *[phys]*
guntai-seibutsu	群体生物	colonial organism *[bio]*
guntō	群島	archipelago *[geogr]*
gun'wake	群分け	grouping *{n} [comm]*
gurabiya-in'satsu	グラビヤ印刷	rotogravure printing *[pr]*
gurafu	グラフ	graph *[math]*
gurai; kurai	位	about *[math]*
guramu	瓦; グラム	gram *[mech]*
guramu-shiki'ryō	グラム式量	gram-formula weight *[ch]*
gurikōgen	グリコーゲン	glycogen *[bioch]*
gū-ryoku	偶力	couple of forces *{n} [phys]*
gūsū	偶数	even number *[math]*
gūsū-pēji	偶数ページ	verso (pl versos) *[pr]*
gūzen	偶然	accident; chance; fortuity *[math]*
gūzen-gosa	偶然誤差	accidental error *[math]*
gūzen-sei	偶然性	contingency; eventuality *[math]*
gyaku	逆; ぎゃく	converse *{n} [logic]*
"	逆	inverse; reverse *{n} [math]*
gyaku-atsu	逆圧; 逆壓	back pressure *[mech]*
gyaku-denryū	逆電流	reverse current *[elec]*
gyaku-fū	逆風	unfavorable winds *{n} [navig]*
gyaku-gyō'retsu	逆行列	inverse matrix *[math]*
gyaku-hannō	逆反応	reverse reaction *[ch]*
gyaku-heikō no (na)	逆平行の(な)	antiparallel *{adj} [bioch] [gen] [phys]*
gyaku-hei'retsu-setsu'zoku-kairo	逆並列接続回路	antiparallelly connected circuit *[elec]*
gyaku-hi	逆比	inverse ratio *[math]*

gyaku-hōkō	逆方向	reverse (opposite) direction *[math]*
gyaku'hōkō-in'satsu	逆方向印刷	backward printing *{n}* *[comput]*
gyaku-jikan	逆時間	inverse hour; inhour *[nucleo]*
gyaku'jū no	逆従の; 逆從の	obsequent *{adj}* *[geol]*
gyaku-kaiten	逆回転; 逆回轉	counterrotation *[mech]*
gyaku-kansū	逆関数	inverse function *[math]*
gyaku-kiden'ryoku	逆起電力	counterelectromotive force *[elecmg]*
gyaku-kō; gyakkō	逆行	retrograde motion *[astron]*
gyaku-kongō	逆混合	back mixing *{n}* *[ch-eng]*
gyak(u)kōsen-shashin	逆光線写真	shadowgraph *[graph]*
gyak(u)kō-undō	逆行運動	retrograde motion *[astron]*
gyaku-nijō no hōsoku	逆二乗の法則	inverse square law *[phys]*
gyaku no	逆の	reciprocal *{adj}* *[alg]*
gyaku no-	逆の-	counter- *{prefix}*
gyaku'ryū	逆流	reflux *[ch]*
"	逆流	reverse current *[electr]*
gyaku-sankaku-kansū	逆三角関数	inversely trigonometric functions *[math]*
gyaku-senpū	逆旋風	anticyclone; high pressure; high *[meteor]*
gyaku-shintō	逆浸透	reverse osmosis *[ch-eng]* *[p-ch]*
gyaku-shintō'maku	逆浸透膜	reverse-osmosis membrane *[p-ch]*
gyaku-sōhan-sei	逆相反性	antireciprocity *[math]*
gyaku'sū	逆数	reciprocal *{n}* *[alg]*
gyaku-taiden'atsu	逆耐電圧	peak inverse voltage *[electr]*
gyaku-taishō	逆対称; 逆對稱	antisymmetry *[math]*
gyaku-teki'tei	逆滴定	back titration *[ch]*
gyaku'ten	逆転	reverse (driving) *{n}* *[mech-eng]*
"	逆転	inversion *[meteor]*
gyakuten saseru	逆転させる	reverse; invert *{vb t}*
gyaku-tensha-kōso	逆転写酵素	reverse transcriptase *[microbio]*
gyakuten-sō	逆転層; 逆轉層	inverted layer *[meteor]*
gyakuten suru	逆転する	reverse; go into reverse *{vb}*
gyaku'tokei-mawari	逆時計回り	counterclockwise rotation *[mech]*
-gyo	-魚	-fish *{suffix}*
gyō	行; ぎょう	line *{n}* *[comput]* *[pr]*
"	行	row *{n}* *[comput]* *[math]*
gyō	業	vocation; calling *{n}*
gyō-bangō	行番号	line number *[comput]*
gyo'dō	魚道	fishway *[civ-eng]*
gyofun	魚粉	fish meal *[mat]*
gyōga'i	仰臥位	horizontal supine position *[med]*
gyogan	魚眼	fisheye *[mat]*
gyogan'seki	魚眼石	apophyllite *[miner]*

gyogun-tanchi	魚群探知	fish detection *[ocean]*
gyo'gyō	漁業；ぎょぎょう	fishing industry *[ocean]*
gyogyō-tori'shimari'sen	漁業取締船	fisheries patrol boat *[nav-arch]*
gyō-inji-ki	行印字機	line printer *[comput]*
gyo'jō	漁場	fishing grounds *[navig]*
gyōkai'gan	凝灰岩	tuff *[geol]*
gyokai-rui	魚介類	seafood; fish and shellfish *[cook]*
gyōkaku	仰角	angle of elevation *{n} [eng] [math]*
gyōkan	行間	spacing (between lines) *{n} [comp] [pr]*
gyō'ketsu	凝血；ぎょうけつ	coagulation (clotting) of blood *[physio]*
"	凝血	coagulated blood *[physio]*
gyō'ketsu	凝結	curdling *{n} [bioch]*
"	凝結	coagulation *[ch]*
"	凝結	setting *{n} [civ-eng]*
gyōketsu-ten	凝結点	congealing point *[ch]*
gyōketsu-zai	凝結剤	coagulant *[ch]*
gyokō	漁港；漁港	fishing port *[geogr]*
"	漁港	fishery harbor *[ocean]*
gyōko	凝固；ぎょうこ	coagulation; curdling *{n} [ch]*
"	凝固	solidification *[phys]*
gyōko'so	凝固素	rennin *[bioch]*
gyōkō'tai	凝膠体	jelly *[ch] [geol]*
gyōko-ten	凝固点	solidifying point *[phys]*
gyōko-yoku	凝固浴	coagulating bath *[ch]*
gyoku'mō	玉毛	pill *{n} [tex]*
gyoku'ro	玉露	tea, green superior grade *[cook]*
gyoku'zui	玉髄；玉髓	chalcedony *[miner]*
gyomō	漁網	fishing net *[eng] [tex]*
gyomō-shi	漁網糸	fishing twine *[tex]*
gyōmu	業務	business *[econ]*
gyō'nin-ben	行人扁；行人偏：彳	kanji (Chinese character) radical denoting a small walk *[pr]*
gyōnyū	凝乳	curd *[cook]*
gyōnyū-kōso	凝乳酵素	rennen *[bioch]*
gyorai	魚雷	torpedo *{n} [ord]*
gyoran('jō)-sekkai'gan	魚卵(状)石灰岩	oolitic limestone *[petr]*
gyoran'seki	魚卵石	oolite; oolith *[petr]*
gyō'retsu	行列；ぎょうれつ	queue; queueing *{n} [comput] [eng]*
"	行列	matrix (pl matrices) *[math]*
gyō'retsu-shiki	行列式	determinant *[math]*
gyōri	凝離	segregation *[ch]*
gyo'rui	魚類	fish; Pisces *[v-zoo]*

gyorui'gaku	魚類学；魚類學	ichthyology *[zoo]*
gyōsei	行政；ぎょうせい	administration
gyōsei-chūshin'chi	行政中心地	administrative center *[civ-eng]*
gyōsho-tai	行書体	cursive-style writing *[comm]*
gyōshū	凝集；ぎょうしゅう	agglutination *[ch] [immun]*
"	凝集	aggregation (of powder) *[phys]*
"	凝集	cohesion *[phys] [sci-t]*
gyōshū-hakai	凝集破壊	cohesive failure *[adhes]*
gyō'shuku	凝縮；ぎょうしゅく	condensation *[ch] [mech]*
gyōshuku'butsu	凝縮物	condensate *[mat]*
gyōshuku-eki	凝縮液	concentrated solution *[ch]*
gyōshuku-ki	凝縮器	condenser *[ch-eng]*
gyōshuku-men	凝縮面	condensation plane *[phys]*
gyōshuku-ryoku	凝縮力	cohesion; cohesive force *[sci-t]*
gyōshū-ryoku	凝集力	flocculating force *[ch]*
"	凝集力	cohesive strength *[mech]*
"	凝集力	cohesion *[phys]*
gyōshū saseru	凝集させる	agglutinate; condense *{vb t}*
gyōshū-zai	凝集剤	flocculant; coagulant *[ch]*
gyo'son	漁村；ぎょそん	fishing village *[geogr]*
gyo-yu	魚油；ぎょゆ	fish oil *[mat]*
gyū'kyaku-yu	牛脚油	neatsfoot oil *[mat]*
gyū'nyū	牛乳；ぎゅうにゅう	milk (of cow) *{n} [cook] [physio]*
gyūshi	牛脂；ぎゅうし	beef tallow *[mat]*
gyūshin'ri	牛心梨	custard apple (fruit) *[bot]*

H

ha	歯；齒；は	tooth (pl teeth) *[anat]*
"	歯	gear tooth *[mech-eng]*
ha	葉；は	leaf (pl leaves) *{n} [bot]*
ha'aku'ryoku no aru	把握力のある	prehensile *{adj} [v-zoo]*
haba	幅；巾；はば	amplitude *[math]*
"	幅；巾	width; breadth; range *{n}*
haba'dashi	幅出し；はばだし	tentering *{n} [tex]*
haba'ki	幅木；はばき	baseboard; plinth *[arch]*
habu	波布；偽匙倩；はぶ	Trimeresurus flavoviridis; habu (reptile) *[v-zoo]*
ha-burashi	歯刷子；歯ブラシ	toothbrush *[dent]*
ha'butae	羽二重；はぶたえ	habutae silk *[tex]*
hachi	八；はち	eight *{n} [math]*
hachi	鉢；はち	bowl; basin; pot *[eng]*
hachi	蜂；はち	bee (insect) *[i-zoo]*
-hachi	-鉢	-pot(s) *{suffix} [hort]*
hachi'bun no ichi	八分の一	one-eighth *{n} [math]*
hachi'chū-rui	鉢虫類	Scyphozoa *[i-zoo]*
hachi-doku	蜂毒；はちどく	bee toxin *[physio]*
hachi-dori	蜂鳥；はちどり	hummingbird (bird) *[v-zoo]*
hachi'gatsu	八月；はちがつ	August (month)
hachigū'shi	八隅子；はちぐうし	octet *[atom-phys]*
hachi'hen-kei	八辺形	octagon *[math]*
hachi'ji-gata	八字形；はちじがた	figure eight *{n} [math]*
hachijū'shi	八重子	octet *[atom-phys]*
ha'chiku	淡竹；はちく	black bamboo; Phyllostachys nigra *[bot]*
hachi'maki	鉢巻；鉢巻	headband *[cl]*
hachi'men'tai	八面体	octahedron (pl -s, -hedra) *[math]*
hachi-mitsu	蜂蜜；はちみつ	honey *[cook] [i-zoo]*
hachi no su	蜂の巣；蜂の巢	honeycomb *[i-zoo]*
hachi'no'su-yō no	蜂の巣様の	beehive-like *{adj}*
hachi-rō	蜂蠟；はちろう	bee glue; propolis *{n} [pharm]*
hachi'ue	鉢植；はちうえ	houseplant; potted plant *[bot]*
hachō	波長；はちょう	wavelength *[phys]*
hachū'gaku	爬虫学	herpetology *[bio]*
hachū-rui	爬虫類	reptiles; Reptilia *[v-zoo]*
hachū'rui-gakusha	爬虫類学者	herpetologist *[bio]*
hada	肌；膚；はだ	skin; the body *{n} [anat]*
hada-are	肌荒れ；はだあれ	surface roughness *{n} [met]*

hada-are	肌荒れ；はだあれ	roughening *{n} [plas]*
hada-iro	肌色	flesh color (color) *[opt]*
hadaka	裸；はだか	naked; nude; bare *{n}*
hadaka-sen	裸線	bare wire *[elec]*
hadan	破談	cancellation; breaking off *{n}*
hadan-hizumi	破断歪；破斷歪	set at break; strain at fracture *[mech]*
hadan-kyōdo	破断強度	breaking tensile strength *[mech]*
hadan-men	破断面；はだんめん	broken-out section *[graph]*
"	破断面	fractured surface *[mech-eng]*
hadan-nobi	破断伸び	breaking elongation *[mech]*
hadan-sen	破断線	break line *[graph]*
hadan-shindo	破断伸度	breaking elongation *[mech]*
hadan-ten	破断点	fracture point *[mech]*
hada-yaki(-shori)	肌焼き（処理）	case hardening *[met]*
hada-zawari	肌ざわり	feel to touch *{n} [physio]*
hadō	波動；はどう	wave motion *[phys]*
hadō-hōtei'shiki	波動方程式	wave equation *[phys]*
hadō-riki'gaku	波動力学	wave mechanics *[quant-mech]*
hadō suru	波動する	undulate; fluctuate *{vb}*
hae; haya	鮠；はえ；はや	dace (pl dace) (fish) *[v-zoo]*
hae	蠅；はえ	fly (insect) *[i-zoo]*
hae'tori-gumo	蠅取蜘蛛	jumping spider (insect) *[i-zoo]*
ha'fuku	波腹	antinode *[phys]*
hafunyūmu	ハフニウム	hafnium (element)
ha'gaki	葉書	postcard *[comm]*
hagane; kō	鋼；はがね；こう	steel *[met]*
hagare	剝れ	peeling; scaling *{n} [mech] [paint]*
ha-gata	歯形；齒形	tooth profile *[mech]*
ha'gatsuo	歯鰹；はがつお	bonito; Sarda orientalis (fish) *[v-zoo]*
hage-taka	禿鷹；はげたか	condor; vulture (bird) *[v-zoo]*
hage-washi	禿鷲；はげわし	cinerous vulture (bird) *[v-zoo]*
hagi	脛；はぎ	leg *[anat]*
hagi	萩	bush clover (herb) *[bot]*
ha-giri	歯切り	gear cutting *{n} [mech]*
ha'giri-ban	歯切り盤	gear generator *[mech]*
hagi'tori	剝ぎ取り	grazing *{n} [v-zoo]*
Hagi-yaki	萩焼	Hagi ware *[cer]*
hagu	剝ぐ；はぐ	peel; strip *{vb}*
ha'guchi	羽口	tuyere *[steel]*
ha'guki	歯茎；齒莖	gingiva (pl -vae); gum *[anat]*
ha-guruma	歯車；はぐるま	gear; toothed gear *[mech]*
ha'guruma-hi	歯車比	gear ratio *[mech]*

ha'guruma-retsu	歯車列	gear train *[mech]*
ha-haba	歯幅	face width *[mech]*
hahen (doki no)	破片(土器の)	shard *{n} [archeo]*
-hai; -pai; -bai	-杯	-cup(s) *{suffix}*
hai	灰; はい	ash(es) *[ch]*
hai	胚	embryo *[bot]*
hai	肺	lung *[anat]*
hai-atsu	背圧; 背壓	back pressure *[mech]*
hai-baiyō	胚培養	embryo culture *[microbio]*
haibun	灰分	ash content *[mat]*
haibun	配分	allocation *{n}*
hai'butsu	廃物; 廢物	refuse; waste *{n} [ecol]*
hai'butsu-saisei-riyō	廃物再生利用	recycling *{n} [ecol] [ind-eng]*
haichi	配置; はいち	configuration *[ch] [math] [phys]*
"	配置	layout *{n} [graph]*
"	配置	arrangement; placement; layout *[math]*
"	配置	disposition *[mech-eng]*
"	配置	deployment *[mil]*
haichi suru	配置する	arrange; dispose; place *{vb} [mech-eng]*
haichi-zu	配置図	layout *{n} [graph]*
haichū-ritsu	排中律	excluded middle, law of *[logic]*
haiden	配電; はいでん	distribution of power *[elec]*
haiden-ban	配電盤	switchboard *[elec]*
haiden-den'atsu	配電電圧	distribution voltage *[elec]*
haiden-ki	配電器	power-distribution unit *[elec]*
hai-dōmyaku	肺動脈	pulmonary artery *[anat]*
hai'eki-sō	排液槽	liquid-evacuation vessel *[ch]*
haiga	胚芽	embryo bud; germ *[bot]*
haigō	配合; はいごう	mixing *{n} [mech-eng]*
haigō suru	配合する	compound *{vb t} [ind-eng]*
haigō-tan	配合炭	compounded coal *[geol]*
haigō-zai	配合剤	compounding ingredient *[mat]*
haigū'sha	配偶者	spouse *[bio]*
haigū'shi	配偶子	gamete *[bio]*
haigūshi-nō	配偶子嚢	gametangium (pl -gia) *[bio]*
haigū'tai	配偶体	gametophyte *[bot]*
haigyo	肺魚	lungfish (fish) *[v-zoo]*
haigyō suru	廃業する	go out of business *{vb} [econ]*
haihan no	排反の	disjoint *{adj} [math]*
haihō	肺胞; 肺胞	alveolus (pl -li) *[anat]*
hai'i-ion	配位イオン	coordinating ion *[ch]*
hai'i-i'sei	配位異性	coordination isomerism *[ch]*

hai'i-kagō'butsu	配位化合物	coordination compound *[ch]*
hai'i-ketsugō	配位結合	coordinate bond *[ch]*
hai'i-kōbunshi	配位高分子	coordination polymer *[poly-ch]*
hai-iro	灰色	gray (color) *[opt]*
hai'i'shi	配位子	ligand *[ch]*
hai'i'shi-ba	配位子場	ligand field *[ch]*
hai'i-sū	配位数	coordination number *[phys]*
hai'i-sui	配位水	coordinated water *[ch]*
haijo	排除；はいじょ	abatement *[legal]*
hai'jiku (no)	背軸(の)	abaxial *{adj}* *[bio]*
hai-jōmyaku	肺静脈	pulmonary vein *[anat]*
hai'ka	灰化	incineration *[ch]*
haikan	配管；はいかん	piping; plumbing *{n}* *[civ-eng]*
haikan-kei	配管系	pipeline *[eng]*
haikan'kō	配管工	plumber (a person) *[eng]*
haikei	背景	background *[comput]* *[graph]*
hai-keisei	胚形成	embryogenesis *[embryo]*
haikei-shori	背景処理	background processing *{n}* *[comput]*
haiki	排気	exhaust air; exhaust gas *[mech-eng]*
haiki	廃棄；はいき	discarding *{n}* *[ecol]*
"	廃棄	abandonment *[legal]*
haiki-ben	排気弁；排氣瓣	exhaust valve *[mech-eng]*
haiki'butsu	廃棄物	wastes; discards *{n}* *[ecol]*
haiki'butsu-shori	廃棄物処理	waste disposal *[ecol]*
haiki-en	排気煙；排氣煙	exhaust smoke *[eng]*
haiki-kan	排気管	exhaust pipe *[mech-eng]*
haiki-ryō	排気量	displacement *{n}* *[fl-mech]*
haikō	配向；はいこう	orientation *[crys]* *[p-ch]*
haikō-do	配向度	degree of orientation; degree of preferred orientation *[crys]*
haikō-kyoku'sen	配光曲線	light distribution curve *[opt]*
haikō-seichō	配向成長	epitaxy *[crys]*
haiku	俳句；はいく	Japanese poem of 17 syllables *[pr]*
haimen-kō	背面光	backlight *{n}* *[graph]* *[opt]*
haimen-zu	背面図	rear view *[graph]*
hainyū'tai	胚乳体	xeniophyte *[bot]*
hai'retsu	配列；はいれつ	arrangement *[ch]*
"	配列	array; rank *{n}* *[comput]* *[math]*
hairetsu-do	配列度	degree of orientation *[org-ch]*
hairetsu-kettei	配列決定	sequencing *{n}* *[bioch]* *[gen]*
hairetsu-mei	配列名	array name *[comput]*
hairu	入る；はいる；いる	enter *{vb}*

haisen	配線	wiring *{n} [elec]*
haisen	擺線；はいせん	cycloid *[math]*
haisen-ban	配線盤	plugboard *[comput]*
haisen-zu	配線図	wiring diagram *[elec]*
haisetsu'butsu	排泄物	excrement; excreta *[physio]*
hai'setsu-kei	排泄系	excretory system *[anat]*
haisha	背斜	anticline *{n} [geol] [min-eng]*
hai'shoku	配色	color scheme *[opt]*
haishu	胚珠；はいしゅ	ovule *[bot]*
hai'shutsu	排出	excretion *[physio]*
haisui	背水	backwater *[hyd]*
haisui	排水	drainage water *[hyd]*
haisui	廃水；廢水	waste water *[ind-eng]*
haisui	配水	water supply *[hyd]*
haisui-kan	排水管	drain pipe *[civ-eng]*
haisui-kō	排水孔	weep hole *[civ-eng]*
haisui-men	排水面	drainage surface *[hyd]*
haisui-ryō	排水量	displacement (of water) *[fl-mech]*
ha'ita; shitsū	歯痛	toothache *[med]*
hai'ta	排他；はいた	exclusion *[comput]*
haita-ritsu	排他律	exclusion principle *[quant-mech]*
haita'teki-sanshō	排他的参照	exclusive reference *[comput]*
haitō'tai	配糖体	glucoside; glycoside *[bioch]*
haiyō	胚葉	germ layers *[bio]*
haiza	配座	conformation *[ch]*
haizō	肺臓；肺臟	lung *[anat]*
hajiki	弾き；彈き；はじき	beading (of a paint) *{n} [poly-ch]*
hajiki-tsubo	弾き壺	dashpot *[mech-eng]*
hajiku	弾く	repel; flip away *{vb t}*
haji suru	把持する	hold *{vb t}*
hakai	破壊；破壞	failure *{n} [adhes]*
"	破壊；はかい	destruction; demolition *[civ-eng]*
"	破壊	breaking; rupture *{n} [mech]*
hakai-den'atsu	破壊電圧	breakdown voltage *[elec]*
hakai-jinsei	破壊靭性	breaking toughness *[mech]*
hakai-kajū	破壊荷重	breaking load *[mech]*
hakai-keisū	破壊係数	modulus of rupture *[mech]*
hakai-ō'ryoku	破壊応力	breaking stress *[mech]*
hakai-rikigaku	破壊力学	fracture mechanics *[phys]*
hakai-shiken	破壊試験	destruction test *[mech]*
hakai-tsuyo-sa	破壊強さ	breaking strength *[mech]*
haka-jikan	破過時間	breakthrough time *[ch-eng]*

hakama	袴；はかま	Japanese trousers *[cl]*
hakari	秤	scale; balance (weighing device) *[eng]*
hakari-bin	量り瓶	weighing bottle *[an-ch]*
hakaru	計る；測る；はかる	measure; gage; survey *{vb t}*
hakaru	量る；秤る	weigh *{vb}*
ha-kazu	歯数	number of teeth (gear) *[mech]*
hake	刷毛；はけ	brush *{n}* *[eng]*
ha-kei; nami-gata	波形	waveform *{n}* *[phys]*
hake'nuri	刷毛塗り	brushing *{n}* *[paint]*
haki'dashi-kan	吐出し管	discharge pipe *[civ-eng]*
haki'mono	履き物	footwear *[cl]* *[eng]*
ha'kiri-ban; ha'giri-ban	歯切り盤	gear cutter (machine) *[mech]*
hakka	白化	etiolation *[bot]*
"	白化	clouding; whitening *{n}* *[plas]*
hakka	薄荷；はっか	mint; peppermint *[bot]* *[cook]*
hakkaku-kei	八角形	octagon *[math]*
hakka no	八価の；八價の	octavalent *[ch]*
hakka'nō	薄荷脳	menthol *[org-ch]*
hakka-ondo	発火温度	ignition temperature; kindling temperature *[thermo]*
hakka-ten	発火点；發火點	ignition point *[ch]*
hakka-yu	薄荷油	peppermint oil *[cook]* *[mat]*
hakkekkyū	白血球	leukocyte; white blood cell *[hist]*
hakken	発見；はっけん	discovery *[min-eng]*
hakken-hō	発見法	heuristics *[psy]*
hakken'sha	発見者	discoverer *[pat]*
hakken suru	発見する	discover; detect *{vb}* *[pat]*
hakkin	白金；はっきん	platinum (element)
hakkin-ji	白金耳	platinum-wire loop *[ch]*
hakkin-san	白金酸	platinic acid
hakkin'san-en	白金酸塩	platinate *[ch]*
hakkō	発光	luminescence *[phys]*
hakkō	発行	issue; issued; publication *[pat]* *[pr]*
hakkō	発効	become effective; coming into force
hakkō	発酵；はっこう	fermentation; zymosis *[microbio]*
hakkō-bi	発行日	issuance date; publication date *[pat]*
hakkō-bunkō-bunseki	発光分光分析	emission spectral analysis *[spect]*
hakkō-daiōdo	発光ダイオード	light-emitting diode *[electr]*
hakkō-dan	発光団；發光團	luminophore group *[phys]*
hakkō-dan	発香団	osmophore group *[phys]*
hakkō-ho'kōso	発酵補酵素	cozymase *[bioch]*
hakkō-kagaku	発酵化学	zymochemistry *[ch]*

hakkō-kei	発酵計	zymometer *[microbio]*
hakkō-kōgaku	発酵工学	fermentation technology *[microbio]*
hakkō-kō'ritsu	発光効率	luminous efficiency *[opt]*
hakkō'sei no	発酵性の	fermentative; fermentable *{adj} [microbio]*
hakkō-sei'un	発光星雲	emission nebula *[astron]*
hakkō suru	発行する	publish *{vb t} [pr]*
hakkō suru	発酵する	ferment *{vb i} [microbio]*
hakkō-toryō	発光塗料	luminous paint *[mat]*
hakkyoku-kan	八極管	octode *[electr]*
hako	箱	box *{n} [eng]*
-hako	-箱	-box(es) *{suffix}*
hakō	波高	pulse height *[electr]*
"	波高	wave height *[ocean]*
hakobe	繁縷; はこべ	chickweed; Stellaria media *[bot]*
hakobu	運ぶ	carry; transport *{vb t}*
hakō-dōbutsu	爬行動物	reptile *[v-zoo]*
hako-fugu	箱河豚; はこふぐ	boxfish (fish) *[v-zoo]*
hako-game	箱亀; はこがめ	box turtle (reptile) *[v-zoo]*
hakō-ritsu	波高率	crest factor; peak factor *[phys]*
hakō-ten	波高点	wave crest *[phys]*
hako-yaki'namashi	箱焼鈍し	box annealing *{n} [met]*
haku	箔; はく	foil *{n} [mat]*
haku	吐く	expel; spew; vomit *{vb}*
haku	穿く; 履く	wear on feet *{vb}*
haku	掃く	sweep *{vb}*
haku'a	白亜; 白堊; はくあ	chalk *{n} [mat] [petr]*
haku'a'ka	白亜化	chalking *{n} [rub]*
haku'a'shitsu	白亜質	cementum (pl -ta) *[dent]*
haku-ban	薄板	thin sheet *[mat]*
haku'butsu'gaku	博物学	natural history *[sci-t]*
haku'butsu'kan	博物館	museum *[arch]*
hakuchō	白鳥	swan (bird) *[v-zoo]*
hakuchō-za	白鳥座	Cygnus; the Swan *[constel]*
haku'datsu-sayō	剥脱作用	exfoliation *[geol]*
haku'do	白土	clay; china clay *[geol]*
"	白土	terra alba *[mat]*
haku-do	白度	whiteness *[opt]*
haku'dō	白道	moon's path *[astron]*
haku'dō	白銅	cupro-nickel *[met]*
haku-eki	白液	white liquor *[paper]*
haku'enkō	白鉛鉱; 白鉛鑛	cerusite *[miner]*
haku'han	白斑	facula (pl -ae) *[astron]*

haku'hen'jō no	薄片状の	thin-leaf- shaped *{adj} [sci-t]*
haku'ka	白華；はくか	efflorescence *[miner]*
haku-maku	薄膜	thin film *{n} [electr]*
hakumei	薄明	twilight *[astron]*
haku-netsu	白熱	incandescence *[opt]*
"	白熱	white heat *{n} [thermo]*
hakuran'kai	博覧会	exposition; fair *{n}*
haku'ri	剝離；はくり	shivering *{n} [cer]*
hakuri-sei	剝離性	releasability *{n} [adhes]*
"	剝離性	fissility *{n} [geol]*
hakuri-sei (no)	剝離性(の)	peelable; detachable; strippable; releasable *{adj} [adhes]*
hakuri-shi	剝離紙	release paper *[adhes]*
haku-rō	白蠟；はくろう	white Japan wax *[mat]*
hakuryū'seki	白榴石	leucite *[miner]*
hakusai	白菜	celery cabbage; Chinese cabbage (vegetable) *[bot]*
haku-sekirei	白鶺鴒	pied wagtail (bird) *[v-zoo]*
haku-sen	白銑	white pig iron; white iron *[mat]*
haku'sha	拍車	spur *{n} [transp]*
haku-shoku	白色；はくしょく	white (color) *[opt]*
haku'shoku-do	白色度	whiteness *{n} [opt]*
haku'shoku-fukyū	白色腐朽	white rot *[bot]*
haku'shoku-kō	白色光	white light *[opt]*
haku'shoku-waisei	白色矮星	white dwarf *[astron]*
haku'shoku-zatsu'on	白色雑音	white noise *[phys]*
haku'sui	白水	white water *[ocean]*
haku'tekkō	白鉄鉱；白鐵鑛	marcasite *[miner]*
haku'yō	白楊	white poplar (tree) *[bot]*
ha'kyoku'teki na	破局的な	catastrophic *{adj}*
hama	浜；濱；はま	beach *{n} [geol]*
hamachi	魬；はまち	yellowtail (young fish) *[v-zoo]*
hamaguri	蛤；はまぐり	clam: cherry-stone (shellfish) *[i-zoo]*
hama-nasu	浜茄子；はまなす	sweet brier *[bot]*
hama-suge	浜菅；はますげ	coco grass; nut grass *[bot]*
hame'ai	嵌め合	fit *{n} [eng]*
hame'ba	嵌め歯	cog *[eng]*
ha'men	波面	wavefront; wave surface *[phys]*
ha-men	破面	fracture(d) surface *[mech-eng]*
ha-men	歯面	tooth flank (gear) *[mech]*
hameru	嵌める	insert; set in; snap in *{vb t}*
ha-mizo	歯溝；はみぞ	blade channel; gullet (gear) *[mech]*

hamo	鱧；はも	pike conger; sea eel; Muraenesox cinereus (fish) *[v-zoo]*
ha'mono	刃物；刄物	cutlery *[cook] [food-eng]*
han	版	block; edition; plate *{n} [pr]*
han	斑；はん	plaque (bacteria) *[microbio]*
han-	反-	anti- *{prefix}*
han-	半-	semi- *{prefix}*
han-	汎-	pan- *{prefix}*
hana	花	blossom; flower *{n} [bot]*
hana	鼻	nose *{n} [anat]*
hanabi	花火	fireworks *{n} [eng]*
hana'bira	花弁；花瓣；花びら	petal *[bot]*
hana-guma	鼻熊	coati; coatimundi (animal) *[v-zoo]*
hanahada	甚だ	very *{adv}*
hanareru	離れる	separate; detach from *{vb i}*
hanasu	話す	speak; talk *{vb}*
hanasu	離す；はなす	separate; detach; disconnect *{vb t}*
hanasu koto	話すこと	speaking; talking *{n} [physio]*
hana-yasai	花野菜	cauliflower (vegetable) *[bot]*
hanbai	販売；販賣	sale; sales; selling *{n} [econ]*
hanbai'ten	販売店	shop; store *{n} [arch]*
han'betsu-shiki	判別式	discriminant *[math]*
hanbun	半分	half *{n} [math]*
han-bunsū	繁分数	complex fraction *[math]*
han'busshitsu	反物質	antimatter *[phys]*
hanchi-haba	半値巾；半値幅	half-width *[math]*
hanchō	反跳；はんちょう	recoil *{n} [mech]*
hanchō	半潮	half tide *[ocean]*
hanchō-hannō	反跳反応	rebound reaction *[ch]*
han-chōkei	半長径	semimajor axis *[math]*
hanchū	範疇；はんちゅう	category *[ling]*
han'chūsei'shi	反中性子	antineutron *[phys]*
han-daku'on	半濁音	p sound *[ling]*
han'dakuon-pu	半濁音符	p-sound mark *[pr]*
handa'zuke	半田付け	soldering (soft) *{n} [eng]*
handa'zuke-sei	半田付け性	solderability *{n} [eng]*
han-denchi	半電池	half-cell *[p-ch]*
han'denchi-den'i	半電池電位	half-cell potential *[p-ch]*
handō	反動	counteraction; reaction *[phys]*
han'dōkō	斑銅鉱	bornite *[miner]*
handoku-ritsu	判読率	readability *[comm]*
handō'sei-kōbunshi	半導性高分子	semiconductive polymer *[poly-ch]*

han'dōtai	半導体；半導體	semiconductor *[sol-st]*
handōtai'ban	半導体板	wafer *[electr]*
handōtai'hen	半導体片	chip *[electr]*
handōtai-ken'shutsu-ki	半導体検出器	semiconductor detector *[nucleo]*
handōtai-rēzā	半導体レーザー	semiconductor laser *[opt]*
handō'teki na-	反動的な-	counter- *{prefix}*
hane	羽；はね	wing(s) *[aero-eng] [v-zoo]*
"	羽	blade; fan; paddle *{n} [eng]*
"	羽	plumage *[v-zoo]*
hane'age-do	跳ね上げ戸	trapdoor *[arch]*
hane-bashi	跳ね橋	drawbridge *[civ-eng]*
hane-guruma	羽根車	impeller; runner *[mech-eng]*
han'ei	半影	penumbra (pl -brae, -bras) *[astron]*
hane-kaeru	跳ね返る	rebound *{vb}*
han'en	半円；半圓	semicircle *{n} [math]*
han'en-kei	半円形	semicircular shape *[math]*
hanga	版画；版畫	print (woodblock) *{n} [graph]*
hangan	斑岩；はんがん	porphyry *[petr]*
hangen-ki	半減期	half-life; half-value period *[ch] [nucleo]*
han'genzan-ki	半減算器	half-subtracter *[electr]*
hangen-sō	半減層	half-value thickness *[phys]*
han'getsu	半月；はんげつ	half moon *[astron]*
han'getsu'gata-koya	半月型小屋	quonset hut *[arch]*
hango	半語	halfword *[comput]*
han'gōsei-kōbunshi	半合成高分子	semisynthetic polymer *[poly-ch]*
han'gōsei-sen'i	半合成繊維	semisynthetic fiber *[poly-ch] [tex]*
hangun	半群	semigroup *{n} [math]*
han-hachō	半波長	half-wavelength *[elecmg]*
han-heimen	半平面	half plane *[math]*
han'hōkō'zoku-sei	反芳香族性	antiaromaticity *[ch]*
han'i	範囲；はんい	extent; range; scope *{n} [math]*
han'i'go	反意語	antonym *[gram]*
hanji-ba	反磁場；はんじば	demagnetizing field *[elecmg]*
han'jika-ritsu	反磁化率	diamagnetic susceptibility *[elecmg]*
hanji-sei	反磁性	diamagnetism *[elecmg]*
hanji'sei no	反磁性の	diamagnetic *{adj} [elecmg]*
hanjisei-tai	反磁性体	diamagnetic substance *[elecmg]*
hanka-haba	半価巾；半價幅	half-value width; half-width *[math]*
hankaku	半角	en *[comput] [pr]*
han'kakushi	反核子	antinucleon *[part-phys]*
han'kanmen'tai no	半完面体の	hemihedral *{adj} [crys]*
han'kansei-yu	半乾性油	semidrying oil *[mat]*

han'kansetsu-shōmei	半間接照明	semiindirect lighting *{n} [eng]*
han'kansū-bibun	汎関数微分	functional derivative *[math]*
han'kasan-ki	半加算器	half-adder *[electr]*
han'kaseki-jushi	半化石樹脂	semifossil resin *[resin]*
hanka-sō	半価層	half-value layer *[phys]*
hankei	半径；半徑	radius (pl radii) *[math]*
hankei-hōkō no	半径方向の	radial (direction) *{adj} [math]*
hankei-hōkō-sokudo	半径方向速度	radial velocity *[mech]*
han'ketsu	判決	trial decision *[legal] [pat]*
hanki-kan	半規管；はんきかん	semicircular canal *[anat]*
han'kinzoku	半金属	metalloid; semimetal *{n} [ch]*
han'kōka-yu	半硬化油	partially hydrogenated oil *[mat]*
hankon	瘢痕；はんこん	scar *{n} [bio]*
han'kōshitsu-sen'i'ban	半硬質繊維板	semihardboard *[mat]*
han'kotai	半固体	semisolid *{n} [phys]*
hankyō	反響；反響	echo (pl echoes); resonance *{n} [acous]*
han'kyōji-sei	反強磁性	antiferromagnetism *[sol-st]*
han'kyōji'sei-tai	反強磁性体	antiferromagnetic material *[sol-st]*
hankyō-kensa	反響検査	echo check *{n} [comput]*
han'kyōshin	反共振	antiresonance *[elec]*
han'kyōshin-shūha'sū	反共振周波数	antiresonance frequency *[elec]*
hankyō-shitsu	反響室	echo chamber *[acous]*
hankyō-yokusei-ki	反響抑制器	echo suppressor *[electr]*
han'kyōyūden-tai	反強誘電体	antiferroelectric (pl -s) *{n} [sol-st]*
hankyū	半球；はんきゅう	hemisphere *[geogr] [math]*
hanmen-zō	半面像	hemihedral form *[crys]*
hanmō-ki	反毛機	garnett *[text]*
han'nichi-shūchō	半日周潮	semidiurnal tide *[ocean]*
han-nijū-tsūshin	半二重通信	half-duplex communication *[comm]*
hannō	反応；反應	reaction *[ch]*
hannō-bōsō	反応暴走	runaway reaction *[ch]*
hannō-chūkan'tai	反応中間体	reaction intermediate *[ch]*
hannō-do	反応度；はんのうど	reactivity *[ch]*
hannō-eki	反応液	reaction (-ed, -ing) solution *[ch]*
hannō-genri	反応原理	reaction principle *[miner]*
hannō-jisū	反応次数	order of reaction; reaction order *[ch]*
hannō-kei	反応系	reaction system *[ch]*
hannō-keiro	反応経路（-径路）	reaction path *[ch]*
hannō-ki	反応期	reacting period *[ch]*
hannō-ki	反応器；反應器	reactor *[ch]*
hannō-netsu	反応熱	heat of reaction *[p-ch]*
hannō-ritsu	反応率	conversion *[ch-eng]*

hannō-sei	反応性	reactivity *[ch]*
hannō-seigyo-zai	反応制御剤	reaction-controlling agent *[ch]*
hannō'sei-kaso-zai	反応性可塑剤	reactive plasticizer *[mat]*
hannō-seisei'butsu	反応生成物	reaction product; resultant *[ch]*
hannō'sei-senryō	反応性染料	reactive dye *[mat]*
hannō-sha'shutsu-seikei	反応射出成形	reaction injection molding; RIM *[eng]*
hannō-shitsu	反応室	reaction chamber *[ch-eng]*
hannō-sokudo-shiki	反応速度式	rate equation *[ch]*
hannō suru	反応する	react *{vb}* *[ch]*
hannō'tai	反応体	reactant(s) *[ch]*
ha no aru	歯のある	toothed *{adj}* *[bio]*
hanpa-kangen-den'i	半波還元電位	halfwave reduction potential *[p-ch]*
hanpa-seiryū	半波整流	half-wave rectification *[electr]*
hanpatsu-dansei	反発弾性；反撥彈性	rebound resiliency; rebound elasticity *[mech]*
hanpatsu-ryoku	反発力	repulsive force; repulsion *[mech]*
hanpatsu-sei	反発性	repellency *{n}* *[mech]*
han'patsu suru	反発する；反撥する	repel *{vb}*
han-pirei no	反比例の	inversely proportional *{adj}* *[math]*
hanpuku	反復；はんぷく	recursion; repetition *[comput]*
"	反復	(re)iteration; replication *[math]*
hanpuku-dōsa	反復動作	repetitive movement *[ind-eng]*
hanpuku'mei	反復名	tautonym *[bio]*
hanpuku no	反復の	iterative *{adj}* *[math]*
hanpuku'on-pu	反復音符	repeat-sound mark *[pr]*
hanran'gen	氾濫原	floodplain *[geol]*
hanrei; bonrei	凡例	legend *[map]*
hanrei'gan	斑糲岩	gabbro (pl gabbros) *[miner]*
han-ryoku	反力	reactive force *[mech]*
han'ryūshi	反粒子	antiparticle *[part-physics]*
han'sabaku	半砂漠；はんさばく	semidesert *[ecol]*
hansai-sō	反彩層	reversing layer *[astrophys]*
hansaku-dōbutsu	半索動物	Hemichordata *[syst]*
han'sayō	反作用	counteraction; reaction *[phys]*
hansei	伴星；はんせい	companion star *[astron]*
hansei-iden	伴性遺伝	sex-linked inheritance *[gen]*
han'sen; ho-bune	帆船	sailboat *[nav-arch]*
hansha	反射；はんしゃ	reflection *[phys]*
"	反射	reflex *[physio]*
hansha-bōenkyō	反射望遠鏡	reflecting telescope *[opt]*
hansha-jittai'kyō	反射実体鏡	reflection stereoscope *[opt]*
hansha-kaku	反射角	angle of reflection *[phys]*

hansha-kenbi'kyō 反射顕微鏡 reflecting microscope *[opt]*
hansha-kōgaku-kei 反射光学系 catoptric system *[opt]*
hansha'kyō 反射鏡 speculum (pl -la, -lums) *[opt]*
hansha-men 反射面 plane of reflection *[phys]*
hansha no hōsoku 反射の法則 reflection, law of *[phys]*
hansha-ritsu 反射律 reflexive property *[math]*
hansha-ritsu 反射率; はんしゃ=りつ reflectance; reflection factor; reflectivity *[opt] [phys]*
hansha-ro 反射炉; 反射爐 reverberatory furnace *[eng]*
hansha-sei'un 反射星雲 reflection nebula *[astron]*
hansha'teki na 反射的な reflexive *{adj}*
hansha'teki-seishitsu 反射的性質 reflexive property *[math]*
hanshō 反証; はんしょう refutation; counterevidence *[legal]*
hanshoku 繁殖; はんしょく propagation *[bio] [microbio]*
hanshoku'jo 繁殖所 rookery *{n} [zoo]*
hanshoku-zōkan 反色増感 antisensitization *[photo]*
hanshu 播種; はんしゅ broadcast seeding *{n} [agr]*
hansō-bu 搬送部 carrier part *[comm]*
hansō-denryū 搬送電流 carrier current *[comm]*
hansō-ha 搬送波; はんそうは carrier wave *[comm]*
hanson-shu 汎存種 ubiquitous species *[bio]*
hansū 半数; はんすう haploid number *[gen]*
hansui-sekkō 半水石膏 hemihydrate gypsum *[inorg-ch]*
hansū'tai 半数体; 半數體 haploid *[gen]*
hantai 反対; はんたい contradiction; opposition *[legal]*
hantai'go 反対語; 反對語 antonym *[gram]*
hantai no 反対の opposite *{adj} [math]*
hantai no- 反対の- counter- *{prefix}*
han'taisū-hōgan-shi 半対数方眼紙 semilog graph paper *[math]*
hantai suru 反対する object; contradict *{vb}*
han-tankei 半短径 semiminor axis *[math]*
hantei 判定 judgment *[legal]*
hanten 反転; 反轉 reverse; reversal *{n} [ch]*
" 反転; はんてん inversion *[ch] [math]*
" 反転 reflection *[math]*
hanten 斑点; はんてん speck; speckle; spot *{n} [bio] [opt]*
hanten-bunpu 反転分布 population inversion *[atom-phys]*
hanten-genzō-shori 反転現像処理 reversal processing *{n} [photo]*
hanten-in'satsu 反転印刷 reverse printing *{n} [pr]*
hanten saseru 反転させる reverse; turn around *{vb t}*
hanten suru 反転する invert *{vb} [comput]*
" 反転する reverse *{vb i}*

hantō	半島	peninsula *[geogr]*
han'tokei-mawari	反時計回り	counterclockwise rotation *[mech]*
hantō-maku	半透膜	semipermeable membrane *[phys]*
han'tōmei-tai	半透明体	translucent body *[opt]*
han'yō no	汎用の; はんようの	universally-used *{adj}*
han'yō-sei	汎用性	universal-use property *[sci-t]*
han'yōshi	反陽子	antiproton *[part-phys]*
ha'ori	羽織; はおり	Japanese half-coat *[cl]*
happi	法被; 半被	Japanese livery half-coat *[cl]*
happō-seikei	発泡成形; 發泡成形	expansion molding *{n} [eng] [plas]*
happō-setchaku-zai	発泡接着剤	foam glue *[mat]*
happō'shu	発泡酒	sparkling wine *[cook]*
happyō	発表	publication *[pat] [pr]*
hara	原	field *{n} [ecol]*
hara	腹	abdomen *[anat]*
hara-bire	腹鰭; はらびれ	ventral fin *[v-zoo]*
haran	葉蘭; はらん	aspidistra *[bot]*
harau	払う	pay *{vb t} [econ]*
"	払う	clear away; sweep (away) *{vb t}*
hare	晴	fine (weather) *{n} [meteor]*
hare	腫	swelling *{n} [med]*
ha'retsu	破裂; はれつ	bursting *{n} [paper]*
"	破裂	rupture; explosion *{n}*
haretsu-tsuyo-sa	破裂強さ	bursting strength *[mech]*
hari	針	spine *{n} [bot]*
"	針	needle *{n} [eng]*
hari	梁	beam; joist *[arch]*
hari	鍼; はり	acupuncture *[med]*
hari-ari	針蟻; はりあり	fireant (insect) *[i-zoo]*
hari-bako	針箱	sewing box *[furn]*
hari'chōseki	玻璃長石	sanidine *[miner]*
hari'dashi	張出し	bulging *{n}*
hari'dashi-mado	張出し窓	bay window *[arch]*
hari'dashi-shichū	張出し支柱	boom *{n} [aero-eng]*
hari-dasu	張り出す	strut *{vb i}*
hari'gane	針金; はりがね	wire *{n} [elec] [met]*
hari'gane-jiki-roku'on	針金磁気録音	wire recording *{n} [elecmg]*
hari'gane-gōshi	針金格子	wiregrating *{n} [elecmg]*
hari-i	鍼医; はりい	acupuncture doctor *[med]*
hari'jō no	針状の	acicular *{adj}*
hariken	ハリケン	hurricane *[meteor]*
hari-nezumi	蝟; はりねずみ	hedgehog (animal) *[v-zoo]*

hari'tekkō	針鉄鉱; 針鐵鑛	goethite *[miner]*
harō	波浪	billow (of waves); surge *{n} [ocean]*
harogen	ハロゲン	halogen(s) *[ch]*
haru	春	Spring (season) *[astron]*
haru'same	春雨; はるさめ	cellophane noodles *[cook]*
"	春雨; はるさめ	spring rain *[meteor]*
hasai	破砕; 破碎	breaking *{n} [mech]*
"	破砕	crushing *{n} [min-eng]*
hasai-ki	破砕機	crusher *[eng]*
hasami	鋏; はさみ	scissors *[eng]*
hasami-gane	挟み金	shim (of metal) *{n} [eng] [met]*
hasami-ita	挟み板	shim (of wood) *{n} [eng]*
hasamu	挟む; 挾む	sandwich *{vb t}*
hasan suru	破算する	clear *{vb t} [comput]*
hasen	破線	broken line *[graph]*
ha'setsu	波節	wave node; node *[phys]*
hashi	端	edge *{n} [math]*
hashi	箸; はし	chopsticks *[cook]*
hashi	橋	bridge *{n} [civ-eng]*
hashi-bako	箸箱	chopstick box *[furn]*
hashibami	榛; はしばみ	hazel (tree) *[bot]*
hashibami no mi	榛の実	hazelnut (nut) *[bot]*
hashi'biro-gamo	嘴広鴨	shoveler (bird) *[v-zoo]*
hashi'buto-gara	嘴太雀	marsh tit (bird) *[v-zoo]*
hashi-dai	橋台; 橋臺	abutment (of a bridge) *[civ-eng]*
hashi'go	梯子; はしご	ladder *[eng]*
hashigo-kōzō	梯子構造	ladder structure *[poly-ch]*
hashika	疹; 麻疹; はしか	measles *[med]*
hashi'kake	橋架け; はしかけ	bridging *{n} [ch]*
"	橋架け	cross-linking *{n} [poly-ch]*
hashi'kake-ketsu'gō	橋架け結合	crosslinkage *[ch]*
hashi'kake-kōzō	橋架け構造	bridged structure *[ch]*
hashike	艀; はしけ	barge *{n} [nav-arch]*
hashira	柱	column; post *{n} [arch]*
hashiri	走り	first of the season *{n} [cook]*
hashiri'dokoro	走野老; はしり=どころ	Scopolia japonica (herb) *[pharm]*
hashiru	走る	run *{vb i}*
hashiru koto	走ること	running *{n} [physio]*
hashi-watashi	橋渡し	bridging; mediation *{vb t}*
hashō'fū	破傷風	tetanus *[med]*
ha-soku	波束	wave packet *[phys]*

hassan	発散；發散	transpiration *[bio]*
"	発散；はっさん	divergence *[math] [opt]*
"	発散	dissipation *[phys]*
hassan'butsu	発散物	emanation *{n} [nuc-phys]*
hassan-kyūsū	発散級数	divergent series *[math]*
hassan-renzu	発散レンズ	diverging lens *[opt]*
hassan-ritsu	発散率	emissivity *[thermo]*
hassan-sen	発散線	divergence line *[math]*
hassei	発生；はっせい	development *[bio]*
"	発生	evolution; nascency *[ch]*
hassei'gaku	発生学	embryology *[bio]*
hassei-hō	発声法；發聲法	diction *[ling]*
hassei-ki	発生期	nascent state *[ch]*
hassei-ki	発生器；發生器	generator *[electr]*
hassei-ro	発生炉	producer (a furnace) *[ch]*
hassei-ryō	発生量	yield *{n}*
hassei suru	発生する	generate *{vb} [electr]*
hassha	発射；はっしゃ	launch(ing); firing *{n} [aero-eng] [eng]*
"	発射	emission *[elecmg]*
hassha-dai	発射台	launch pad *[aerosp]*
hassha-seibi'tō	発射整備塔	gantry *[aerosp]*
hassha'tai	発射体	projectile *[ord]*
hasshin	八進	octal *{n} [comput]*
hasshin	発信；はっしん	sending *{n} [comm]*
hasshin	発振	oscillation *[phys]*
hasshin-denpa	発振電波	oscillating radio wave *[elecmg]*
hasshin-hyōki	八進表記	octal notation *[math]*
hasshin-ki	発振器	oscillator *[comm] [electr]*
hasshin-kikō	発振機構	earthquake mechanism *[geophys]*
hasshin-sū	八進数	octal number *[math]*
hasshin suru	発信する	send *{vb} [comm]*
hasshoku-dan	発色団；發色團	chromophore *[ch]*
hasshoku-genzō	発色現像	chromogenic development *[photo]*
hasshoku-sokudo	発色速度	color development speed *[photo]*
hasshū-zai	発臭剤	odorant *{n} [mat]*
hassō-bi	発送日	date of mailing *[pat]*
hassui-kakō	撥水加工	water-repellent finishing *{n} [mat]*
hassui-sei	撥水性	water repellency *[mat]*
hassui-zai	撥水剤	water repellent *{n} [mat]*
hassuru	発する；發する	discharge; emit; emanate *{vb}*
hasu	斜；はす	diagonal; oblique *{n} [math]*
hasu	蓮	lotus *[bot]*

hasū	波数	wave number *[phys]*
hasu-ba	斜歯；はすば	helical tooth *[mech]*
hasuba-ha'guruma	斜歯歯車	helical gear *[mech]*
hasu no mi	蓮の実	lotus seed *[bot] [cook]*
hata	旗	flag *{n} [eng]*
hata	機；はた	loom *{n} [tex]*
hatahata	鰰；はたはた	sandfish; Arctoscopus japonicus *[v-zoo]*
hatake	畑；はたけ	field (plowed, cultivated); farm *[agr]*
hataki	叩き；はたき	duster *[eng]*
hata-nezumi	畑鼠；はたねずみ	vole (animal) *[v-zoo]*
hata'ori-dori	機織鳥；はたおり鳥	weaverbird (bird) *[v-zoo]*
hataraki	働き；はたらき	working; work; labor *{n}*
hataraki-ari	働き蟻	worker ant (insect) *[i-zoo]*
hataraki-bachi	働き蜂	worker bee (insect) *[i-zoo]*
hataraku	働く；仂く	work; toil; labor *{vb}*
hatei	波堤	dike (or levee) break *[civ-eng]*
hato	鳩；はと	pigeon; dove (bird) *[v-zoo]*
hatō	波頭	wavefront; whitecap *[ocean]*
hatō	波濤；はとう	waves; billows; surges *{n} [ocean]*
hato'ba	波止場	quay *[civ-eng]*
hato'me	鳩目	eyelet; grommet *[eng]*
hato-mugi	鳩麦；鳩麥	adlay; adlai *[bot]*
hatsu'den-ki	発電機	dynamo; generator *[elec]*
"	発電機	engine *[mech-eng]*
hatsu'den'sho	発電所	generating station; power plant *[mech-eng]*
hatsu'en-ryūsan	発煙硫酸	fuming sulfuric acid
hatsu'en-shōsan	発煙硝酸	fuming nitric acid
hatsu'ga	発芽；はつが	germination; sprouting *{n} [bot]*
hatsu'gan-busshitsu	発癌物質	carcinogen *[med]*
hatsu'gan'sei no	発癌性の	carcinogenic *{adj} [med]*
hatsu'iku suru	発育する	develop; grow *{vb i} [physio]*
hatsu'mei	発明；發明	invention *[pat]*
hatsumei no meishō	発明の名称	name (title) of invention *[pat]*
hatsumei no shōsai na setsumei	発明の詳細な説明	detailed explanation of invention *[pat]*
hatsumei'sha	発明者	inventor *[pat]*
hatsu-netsu	発熱；はつねつ	heat generation *[thermo]*
hatsu'netsu-hannō	発熱反応	exothermic reaction *[ch]*
hatsu'netsu'sei no	発熱性の	exothermic *{adj} [p-ch] [phys]*
hatsu'netsu-tai	発熱体	heat-emitting body *[thermo]*
hatsu'on	発音；はつおん	pronunciation *[ling]*
hatsu-take	初茸；はつたけ	Lactarius hatsudake (mushroom) *[mycol]*

hatsu'yō	発葉	foliation *[bot]*
hatsu'yu-sei	撥油性	oil repellency *{n} [eng]*
hatten	発展	progress; development; growth *{n}*
hatten-hōtei'shiki	発展方程式	evolution equation *[quant-mech]*
hau	這う	crawl; creep *{vb}*
hayabusa	隼；はやぶさ	falcon (bird) *[v-zoo]*
hayai	早い；速い；はやい	fast; quick; rapid *{adj} {n} [phys]*
hayaku	早く；速く	rapidly *{adv}*
hayaku(-kara)	早く(から)	early *{adv}*
hayaku suru	早くする	accelerate; expedite *{vb t}*
hayameru	早める	accelerate; hasten; speed *{vb t}*
haya'modori-kikō	早戻り機構	quick-return mechanism *[mech-eng]*
haya'okuri-zenshin	早送り前進	fast-feed advance *{n} [mech-eng]*
haya-sa	速さ	rate; speed; rapidity *{n} [mech]*
haya'se	早瀬；早瀨	rapids *[hyd]*
hayashi	林	small forest; grove; thicket *[ecol]*
haya'te	疾風；早手；はやて	squall *{n} [meteor]*
hazama	狭間；はざま	ravine; gorge; defile; narrows *[geogr]*
ha'zao	歯桿；はざお	rack *{n} [mech-eng]*
haze	沙魚；鯊；はぜ	goby (pl gobies, goby) (fish) *[v-zoo]*
hazu	巴豆；はず	croton (plant) *[bot]*
hazumi	弾み；彈み；はずみ	spring; rebound; momentum; chance *[mech]*
hazumi-guruma	勢車；弾車	flywheel *[mech-eng]*
hazureru	外れる	depart from; miss *{vb}*
hazusu	外す	detach; remove; unfasten *{vb t}*
hazu-yu	巴豆油；はずゆ	croton oil *[mat]*
hebi	蛇	snake (reptile) *{n} [v-zoo]*
hechima	糸瓜；へちま	dishcloth gourd; luffa; loofah *[bot]*
hei	閉；へい	on *{n} [elec]*
hei	塀	fence; wall *{n} [arch]*
heiban	餅盤；へいばん	laccolith; laccolite *[geol]*
heiban-baiyō	平板培養	plate culture *[microbio]*
heiban-hanpuku-hō	平板反復法	replica plating method *[bioch]*
heiban-in'satsu-ki	平板印刷機	lithographic press *[pr]*
heiban'jō no	平板状の	tabular *{adj} [crys]*
heichi	平地	flat ground *[geol]*
heichi	並置	juxtaposition *{n} [math]*
heichō-kaizan	平頂海山	guyot *[geol]*
heigen	平原	plain *{n} [geol]*
heigō	併合	merge *{n} [comput]*
heigō suru	併合する	merge *{vb} [comput]*
hei'hatsu-hannō	並発反応	simultaneous reaction *[ch]*

heihō-kon	平方根	square root *[math]*
heihō-sei	閉包性；閉包性	closure property *[math]*
heijō-gantai	餅状岩体	laccolith *[geol]*
heika	閉果；へいか	indehiscent fruit *[bot]*
hei'kaku	平角	straight angle *[math]*
heikatsu-do	平滑度；へいかつど	smoothness *{n} [sci-t]*
heikatsu'ka	平滑化	leveling; smoothing; flattening *{n}*
heiki	兵器	ordnance *[ord]*
heikin(-chi)	平均(値)	average; mean value *{n} [math] [stat]*
heikin-chi no teiri	平均値の定理	mean value theorem *[math]*
heikin-fuka-moru-sū	平均付加モル数	mean number of moles added *[ch]*
heikin-gosa	平均誤差	mean error *[math]*
heikin-ji	平均時；へいきんじ	mean time *[astron]*
heikin-jiyū-jikan	平均自由時間	mean free time *[acous] [phys]*
heikin-jiyū-kōtei	平均自由行程	mean free path *[acous] [phys]*
heikin-jumyō	平均寿命	mean life *[phys]*
heikin-kaimen	平均海面	mean sea level *[ocean]*
heikin-mitsudo	平均密度	mean density *[mech]*
heikin-nijō-gosa	平均二乗誤差	mean square error *[cont-sys]*
heikin no hōsoku	平均の法則	averages, law of *[math]*
heikin-tairyū-jikan	平均滞留時間	mean residence time *[nucleo]*
heikin-yobi'dashi-jikan	平均呼出時間	average access time *[comput]*
heikō	平行；へいこう	overlap *{n} [comput]*
"	平行	parallel *{n} [comput] [math]*
heikō	平衡；へいこう	balance *{n} [ch] [physio]*
"	平衡	equilibrium (pl -riums, -ria) *[ch]*
heikō-idō	平行移動	translation *[math]*
heikō-jōtai	平衡状態	trim *{n} [aero-eng]*
"	平衡状態	equilibrium condition *[ind-eng]*
heikō'kō-nōdo	平行光濃度	specular density *[photo]*
heikō-men	平行面	parallel planes *[math]*
heikō na	平行な	parallel *{adj} [math]*
heikō-rokumen'tai	平行六面体	parallelepiped *[math]*
heikō-sen	平行線	parallel lines *[math]*
heikō-shihen'kei	平行四辺形	parallelogram *[math]*
heikō-shihen'kei no hōsoku	平行四辺形の法則	parallelogram, law of *[math]*
heikō-shori	平行処理	parallel processing *{n} [comput]*
heikō'sui	平衡錘	counterweight *{n} [mech-eng]*
heikō-teisū	平衡定数	equilibrium constant *[ch]*
hei-kukan	閉区間；へいくかん	closed interval *[math]*
hei-kyoku'sen	閉曲線	closed curve *[math]*
heimen	平面；へいめん	plane *{n} [math]*

heimen-ha	平面波；へいめんは	plane wave *[phys]*
heimen-henkō	平面偏光	plane-polarized light *[opt]*
heimen-hizumi	平面歪	plane strain *[mech]*
heimen-kaku	平面角	plane angle *[math]*
heimen-kika'gaku	平面幾何学	plane geometry *[math]*
heimen-kōsa	平面交差	grade crossing *{n}* *[traffic]*
heimen-kyō	平面鏡	plane mirror *[opt]*
heimen'teki na	平面的な	two-dimensional; planar *{adj}* *[math]*
heimen-zu	平面図	ichnograph; plan view *{n}* *[graph]*
hei'nen	平年	common year *[astron]*
heinyū-gan	迸入岩	intrusion rocks *[geol]*
hei-ō no	平凹の	planoconcave *{adj}* *[opt]*
hei'retsu	並列；へいれつ	parallel; in parallel *{n}* *[elec]*
heiretsu-densō	並列伝送	parallel transmission *[comput]*
heiretsu-kairo	並列回路	parallel circuit *[elec]*
heiretsu-kyōshin	並列共振	parallel resonance *[elec]*
heiro	平炉；平爐；へいろ	open-hearth furnace *[steel]*
heiro	閉路	closed circuit *[elec]*
heiro-hō	平炉法	open-hearth process *[steel]*
heiryū	並流	parallel flow *[elec]*
heisa-junkan-kei	閉鎖循環系	closed circulatory system *[anat]*
heisa-kei	閉鎖系	closed system *[eng]*
Heisei	平成；へいせい	Heisei (era name) (1989-)
heishi-ben	閉止弁；閉止瓣	shut-off valve *[mech-eng]*
hei'shin	並進	translation *[math]*
heishin-undō	並進運動	translational motion *[mech]*
heisho-kyōfu'shō	閉所恐怖症	claustrophobia *[psy]*
hei'soku	閉塞；へいそく	choking; clogging *{n}* *[fl-mech]*
"	閉塞	obstruction *[med]*
"	閉塞	occlusion *[meteor]*
"	閉塞	blockage; blockade *{n}* *[rail]*
heisoku-kūkan	閉塞空間	block section *[rail]*
heitan	平坦	smooth *{n}*
hei-totsu no	平凸の	planoconvex *{adj}* *[opt]*
heiwa	平和	peace *[mil]*
heiya	平野	plain *{n}* *[geol]*
hei'yō suru	併用する	use in combination *{vb}*
heki-ga	壁画	mural; wall painting *[graph]*
heki'gan	壁龕；へきがん	niche; nexus *[geol]*
heki'gyoku	碧玉	jasper *[petr]*
heki'kai	劈開	cleavage *[crys]*
hekomi	凹み；へこみ	dent; depression; indentation *[math]*

hekuto'guramu	瓸；ヘクトグラム	hectogram *[mech]*
hen	辺；邊；へん	edge; side *{n}* *[math]*
hen	扁；偏	left-hand radical of a kanji (Chinese character) *{n}* *[pr]*
-hen; -ben	-扁；-偏	-left-hand radical of a kanji *{suffix}*
hen'atsu-ki	変圧器；變壓器	transformer *[elecmg]*
hen-bibun	偏微分	partial differentiation *[math]*
hen'bō-kan'kyaku	偏旁冠脚	Chinese character radicals *{n}* *[pr]*
hen'chiku	扁蓄；扁竹	knotweed; knotgrass (herb) *[bot]*
henchō	変調；へんちょう	modulation *[comm]*
henden'sho	変電所	substation *[elec]*
hendō-hi	変動費	variable expenses (or costs) *[ind-eng]*
hen-dō'kansū	偏導関数	partial derivative *[math]*
hendō-keisū	変動係数	coefficient of variation *[math]*
hen'en-fūsa-sei	辺縁封鎖性	peripheral sealing property *[mat]*
hen'fukuchō-sōchi	変復調装置（モデム）	modem (modulator-demodulator) *{n}* *[electr]*
hengan	片岩；へんがん	schist *[geol]*
hen'i	変位	displacement *[comput]* *[elec]* *[phys]*
hen'i	変異	variation *[bio]*
"	変異	mutation *[gen]*
"	変異	untoward event *{n}*
hen'i	変移	conversion; transmutation *[ch]*
hen'i	偏位	deflection *[eng]*
hen'i	偏移	deviation *[eng]*
hen'i	偏倚；へんい	bias *{n}* *[electr]*
hen'i-kabu	変異株	mutant strain *[microbio]*
hen'i-ken'shutsu-ki	変位検出器	displacement detector *[eng]*
hen'i-shu	変異種	mutant *[gen]*
hen'i-yūdō'gen	変異誘導源	mutagen *[microbio]*
henji	返事	reply; answer; response *{n}* *[comm]*
henjō-koku'en	片状黒鉛	flake graphite *[miner]*
henka	変化；變化；へんか	change *{n}* *[math]*
henkaku	偏角	declination *[astron]*
"	偏角	deflection angle *[geod]*
"	偏角	amplitude; argument *[math]*
"	偏角	angle of deviation *[opt]*
henkaku-kōdo	変角光度	deformation luminosity *[opt]*
henkan	変換；へんかん	change; conversion *[comput]* *[math]*
"	変換	translation *[comput]* *[math]*
henkan-ki	変換器；變換器	converter *[comput]*
"	変換器	transducer *[eng]*

henkan suru	変換する	convert; translate *{vb} [comput]*
henka shita-	変化した-	meta- *{prefix}*
henkei	変形	metamorphosis (pl -phoses) *[bio]*
"	変形	transformation *[comput] [math]*
"	変形	deformation *[mech]*
henkei-dōbutsu	扁形動物	flatworm; Plat(y)helminthes *[i-zoo]*
henkei'kin-rui	変形菌類	slime molds; Myxomycota *[mycol]*
henkei shi'uru	変形しうる	deformable *{adj} [math]*
henkei suru	変形する	transform *{vb} [comput]*
henkei'tai	変形体; 變形體	plasmodium *[microbio]*
henkō	変更	alteration; change *{n} [math]*
henkō	偏向	deflection *[electr] [eng]*
henkō	偏光; へんこう	polarization; polarized light *[opt]*
henkō-ban	偏向板	deflection plate *[electr]*
henkō-ban	偏光板	polarizing plate *[opt]*
henkō-bunkō	偏光分光	polarization spectroscopy *[spect]*
henkō-do	偏光度	degree of polarization *[opt]*
henkō-idenshi	変更遺伝子	modifier gene *[gen]*
henkō-kaiseki	偏光解析	ellipsometry *[opt]*
henkō-kenbi'kyō	偏光顕微鏡	polarizing microscope *[opt]*
henkō-ki	偏光器	polariscope *[opt]*
henkō-sei	変光星	variable star *[astron]*
henkō'shi	偏光子	polarizer *[opt]*
henkyoku	編曲	adaptation; arrangement *[music]*
henkyoku-ten	変曲点	inflection point *[math]*
henma'gan	片麻岩; へんまがん	gneiss *[petr]*
henniku	偏肉	thickness deviation *[eng]*
hen'on-dōbutsu	変温動物	poikilotherm *[zoo]*
henpa	偏波	polarized wave *[electr]*
henpei-ritsu	偏平率	oblateness *{n} [astron]*
"	扁平率	ellipticity *[electr]*
henreki-denshi	遍歴電子	itinerant electron *[phys]*
henri-kyōsei	片利共生	commensalism *[ecol]*
hen-ritsu	偏率; へんりつ	flattening *{n} [astron]*
hensa	偏差	variation *[astron]*
"	偏差	deviation *[bio] [eng] [math]*
hen'sankaku'men-tai	偏三角面体	scalenohedron *[crys] [geom]*
hensei	変成; 變成	denaturation *[ch]*
"	変成	modification; transformation *[mech-eng]*
hensei	変性	modification; transmutation *[plas] [rub]*
hensei-arukōru	変性アルコール	denatured alcohol *[ch]*
hensei-fū	偏西風	westerly; westerly wind *{n} [meteor]*

hensei-gan	変成岩	metamorphic rock *[geol]*
hensei-kenki'sei-kin	偏性嫌気性菌	obligate anaerobe *[bio]*
hensei-ki	変成器；へんせいき	transformer *[elec]*
hensei-kōki'sei-kin	偏性好気性菌	obligate aerobe *[bio]*
hensei-sayō	変成作用	metamorphism *[geol]*
hensei suru	変性する	modify *{vb t}*
hensei-zai	変性剤；變性劑	denaturant; modifying agent *[mat]*
henseki	偏析；へんせき	segregation *[met]*
henseki'un	片積雲	fractocumulus; cumulus fractus *[meteor]*
hen'senkō	変旋光	mutarotation *[ch]*
hen'shikaku'men-tai	偏四角面体	trapezohedron *[crys] [math]*
henshin	偏心；へんしん	eccentricity *[math] [mech]*
henshin-kyori	辺心距離	apothem *[math]*
henshin no	偏心の	eccentric *{adj} [sci-t]*
henshin-rin	偏心輪	eccentric wheel *{n} [eng]*
henshin-ritsu	偏心率	eccentricity *[mech]*
hen'shitsu	変質；へんしつ	degeneration; quality-change *[mat]*
hen'shitsu-kazō	変質仮像；變質假像	alternation pseudomorphism *[miner]*
hen'shoku	変色	color-change; discoloration *[opt]*
henshu	変種	variety *[bio] [syst]*
henshū	編集；へんしゅう	editing *{n} [comput] [pr]*
henshū'sha	編集者	editor (a person) *[pr]*
henshū (suru)	編集(する)	edit *{vb t} [comput] [pr]*
henso	編組	braid *{n} [tex]*
hensō	変奏	playing of a variation *[music]*
henso-hannō	変素反応	transmutation *[ch]*
hensoku-ki	変速機；へんそくき	transmission; gearbox *[mech-eng]*
hensoku-sei	変則性	anomaly *[sci-t]*
hensoku-sōchi	変速装置	variable speed gear *[mech-eng]*
hensō-shōgō	返送照合	loop checking *{n} [comput]*
hensū	変数；變數	variable *{n} [math] [comput]*
hentai	変態；へんたい	modification; transformation *{n} [bio]*
hentai-ten	変態点	transformation point *[met]*
hentō	返答	answer *{n} [comm]*
hentō	扁桃；へんとう	almond (tree, flower) *[bot]*
hentō-sen	扁桃腺	tonsils *[anat]*
hen'yō	偏揺；偏搖	yaw *{n} [mech]*
henzai suru	偏存する	exist displaced *{vb i} [math]*
hera	箆；へら	spatula; scoop *{n} [cook] [eng]*
hera-jika	箆鹿；へらじか	elk (animal) *[v-zoo]*
herasu	減らす	decrease; reduce *{vb t}*
heri	縁；へり	border *{n} [math]*

heri'tsugi-setchaku	縁継ぎ接着	edge-jointing *{n} [adhes]*
heryūmu	ヘリウム	helium (element)
heso	臍; へそ	navel; umbilicus *[anat]*
heso no o	臍の緒	umbilical cord *[aero-eng] [embryo]*
hetari	へたり	sagging *{n} [mech]*
heya	部屋; 室	room *{n} [arch]*
hi	日; 陽; ひ	day; sun *{n} [astron]*
hi	火	fire *{n} [ch]*
hi	比	ratio (pl ratios) *[math]*
hi	卑; ひ	base; less noble *{n} [ch]*
hi	杼	shuttle *{n} [tex]*
hi	緋	scarlet (color) *[opt]*
hi-	否-; 非-	a-; non-; un- *{prefix}*
hi-asshuku-kyōdo	比圧縮強度	specific compressive strength *[mech]*
hi'asshuku-ritsu	非圧縮率	incompressibility *[mech]*
hibachi	火鉢	brazier for charcoal *[furn]*
hibaku	被曝	exposure *{n} [nucleo]*
hi-bana	火花	spark *{n} [elec]*
hi'bana-hōden	火花放電	spark discharge *[elec]*
hibari	雲雀; ひばり	lark; skylark (bird) *[v-zoo]*
hibi	罅; ひび	crazing *{n} [eng]*
hibiki	響き; 響き; ひびき	echo; resounding *{n} [acous]*
hibiku	響く	resound; echo *{vb} [acous]*
hi-bi'nenketsu'sei-resshitsu-tan	非微粘結性劣質炭	nonfinely-coking low-grade coal *[miner]*
hibi'ware	罅割れ; ひび割れ	crazing *{n} [cer] [met]*
hi-bunsan	比分散	specific dispersion *[ch]*
hi'bunsan-gata	非分散型	nondispersive (-type) *[math]*
hi-bunseki'tai	被分析体	analyte *[an-ch]*
hi'byōkin('sei) no	非病菌(性)の	saprophytic *{adj} [bio]*
hichaku-ryō	被着量	adhering quantity *[adhes]*
hichaku'tai	被着体	adherend *[adhes]*
hichi'riki	篳篥; ひちりき	flageolet-like musical instrument *[music]*
hi'chokkaku-sankak(u)kei	非直角三角形	oblique triangle *[math]*
hi'chokketsu	非直結	offline *[comput]*
hi'chokusen-sei	非直線性	nonlinearity *[math]*
hi'chōwa-shindō'shi	非調和振動子	anharmonic oscillator *[electr]*
hida	襞; ひだ	fold *{n} [cl] [med]*
hidaku-hō	比濁法	turbidimetry *[an-ch]*
"	比濁法	nephelometry *[opt]*
hidaku-kei	比濁計	turbidimeter *[an-ch]*
"	比濁計	nephelometer *[opt]*

hi'dansei	非弾性	anelasticity *[mech]*
hi-dansei-ritsu	比弾性率	specific modulus *[mech]*
hi'dansei-sanran	非弾性散乱	inelastic scattering *[phys]*
hi'dansei-shōtotsu	非弾性衝突	inelastic collision *[mech]*
hidari	左; ひだり	left; the left *{n} [math]*
hidari'gawa no	左側の	sinistral (of the left side) *{adj} [med]*
hidari-giki; hidari-kiki	左利き	left-handed *{n} [math] [physio]*
hidari'maki no	左巻の	sinistrorse *{adj} [bot]*
hidari'muki no	左向きの	sinistral (leftward facing) *{adj} [med]*
hidari-sokumen-zu	左側面図	left side view *{n} [graph]*
hidari'te no hōsoku	左手の法則	left-hand rule *[elecmg]*
hidari-tsume	左詰め	left-justify *[comput] [graph]*
hidari-yose	左寄せ	left-justify *[comput] [graph]*
hi-dendō'do	比電導度	specific conductivity *[elec]*
hi'denkai'shitsu	非電解質	nonelectrolyte *[mat]*
hi-denki-teikō	比電気抵抗	resistivity *[elec]*
hideri	日照り; 旱	drought *[agr]*
hi'dōji'sei no	非同時性の	asynchronous *{adj} [phys]*
hidō'ka	非働化; ひどうか	inactivation *{n}*
hi-dokei	日時計	sundial *[horol]*
hi'dōki-densō	非同期伝送	asynchronous transmission *[comput]*
hi'dōki no	非同期の	asynchronous *{adj} [comput] [phys]*
hidoro'chū-rui	ヒドロ虫類	Hydrozoa *[i-zoo]*
hidorokishiru	ヒドロキシル	hydroxyl *{n} [ch]*
hidoronyūmu-ion	ヒドロニゥムイオン	hydronium ion *[inorg-ch]*
hidō'uran'kō	砒銅ウラン鉱	zeunerite *[miner]*
hi-enzan'sū	被演算数	operand *[math]*
hifu	皮膚; ひふ	skin; integument *{n} [anat]*
hi'fuku	被覆; ひふく	coating; covering *{n}*
hifuku-āku-yōsetsu	被覆アーク溶接	covered-arc welding *{n} [met]*
hifuku-kakō	被覆加工	cladding *{n} [eng]*
hifuku-ryoku	被覆力	covering power (by paint) *[eng]*
hi-fukusha-genkō	被複写原稿	matter being copied *{n} [opt]*
hifu-nanka-yaku	皮膚軟化薬	emollient *{n} [pharm]*
hifu-ō'hen	皮膚黄変	xanthism *[bio]*
hifu-shijō-kin	皮膚糸状菌	dermatophyte *[mycol]*
higan	彼岸	equinoctial week *[astron]*
higan-jio	彼岸潮	equinoctial tide *[ocean]*
higashi	東; ひがし	east (direction) *{n} [geod]*
higashi-kaze; kochi	東風	easterly wind; easterlies *[meteor]*
hi'gata	干潟	tideland *[geogr]*
"	干潟	mudflat *[geol]*

hige	鬚; 髭; 髯; ひげ	mustache; beard; whiskers *[anat]*
hige-kesshō	鬚結晶; ひげ結晶	crystal whisker *[crys]*
hige-ne	鬚根	fibrous root *[bot]*
hi-gensū	被減数	minuend *[math]*
hige-zenmai	髭薇; ひげ発条	hairspring *[horol]*
hi-gōsei	比剛性	ratio of rigidity *[mech]*
hi-gōshi	火格子	fire grate *[eng]*
higurashi	蜩; ひぐらし	cicada; evening cicada (insect) *[i-zoo]*
hi'hakai-kensa	非破壊検査	nondestructive testing *{n}* *[eng]*
hi'hakai no	非破壊の	nondestructive *{adj}*
hi'hakai-yomi'dashi	非破壊読出し	nondestructive readout *[comput]*
hi-haretsu-kyōdo	比破裂強度	specific bursting strength *[mech]*
hi-hen	日扁; 日偏: 日	kanji (q.v.) radical denoting sun *[pr]*
hi-hen	火扁; 火偏: 火	kanji (q.v.) radical denoting fire *[pr]*
hihi	狒狒; 狒々; ひゝ	baboon (animal) *[v-zoo]*
hi-hōsha'nō	比放射能	specific radioactivity *[nuc-phys]*
hihyō	批評; ひひょう	commentary; critique *[pr]*
hi-hyō'menseki	比表面積	specific surface area *[eng]*
hi'iki	顧客; ひいき	favoring customer *[econ]*
hi-in'satsu-tai	被印刷体	printing substrate *[pr]*
hi'ion-kaimen-kassei-zai	非イオン界面=活性剤	nonionic surfactant; nonionic surface-active agent *[mat]*
hiji	肘; ひじ	elbow *{n}* *[anat]*
hiji'kake	肘掛	armrest *{n}* *[furn]*
hi-jika-ritsu	比磁化率	specific susceptibility *[p-ch]*
hijiki	鹿尾菜; ひじき	hijiki algae; Hizikia fusiforme *[bot]*
hi'jikkō'bun	非実行文	nonexecutable statement *[comput]*
hi'jikō-sei	非時効性	nonaging property *[eng]*
hi'jikyū-sei	非持久性	volatility *[comput]*
hi'jisei-kō	非磁性鋼	nonmagnetic steel *[met]*
hi'jisei no	非磁性の	nonmagnetic *{adj}* *[elecmg]*
hijō	非常; ひじょう	contingency; emergency
hijō-guchi	非常口	emergency exit *[arch]*
hijō(-ji)	非常(時)	emergency; crisis time
hijō ni	非常に	very *{adv}*
hijo'sū	被除数	dividend *[math]*
hijō'sū	被乗数; 被乗數	multiplicand *[math]*
hijū	比重; ひじゅう	specific gravity *[mech]*
hijū-bin	比重瓶	pycnometer *[eng]*
hijū-kei	比重計	densimeter; hydrometer *[eng]*
hi'junkan'teki-kōrinsan'ka	非循環的光燐酸化	noncyclic photophosphorylation *[bioch]*
hika-aen	砒化亜鉛	zinc arsenide

hika'butsu	砒化物；ひかぶつ	arsenide *[ch]*
hikae	控え	brace; stay; shore; strut *{n} [civ-eng]*
"	控え	memo; note; entry *{n} [comm]*
hikae-kabe	控え壁	buttress *{n} [arch]*
hikage-no-kazura	日陰の葛；石松；ひかげのかずら	buck grass; club moss; coral evergreen; ground pine *[bot]*
hikage'no'kazura-rui	ひかげの蔓類	Lycopsida *[bot]*
hi'kagyaku-sei	非可逆性	irreversibility *{n} [bio] [math]*
hi-kaihei'sū	被開平数	radicand *[math]*
hi'kaiten-ryū	非回転流	irrotational flow *[fl-mech]*
hi-kakō'butsu	被加工物	workpiece material *[mech-eng]*
hikaku	比較；ひかく	relation *[comput]*
"	比較	comparison *[math]*
hi'kaku	皮殻	incrustation *[geol]*
hikaku-kairo	比較回路	comparator circuit *[electr]*
hikaku-ki	比較器	comparator *[comput] [eng]*
hikaku suru	比較する	compare *{vb}*
hikan	悲観；悲觀	pessimism *[psy]*
hi'kanjō no	非環状の	acyclic *{adj} [org-ch]*
hi'kanshō'sei-sanran	非干渉性散乱	incoherent scattering *{n} [phys]*
hikan suru	悲観する	be pessimistic *{vb} [psy]*
hikari	光；ひかり	light *{n} [opt]*
hikari- also see kō-		
hikari-bunkai	光分解	photolysis *[p-ch]*
hikari-denki'kagaku-denchi	光電気化学電池	photovoltatic cell *[electr]*
hikari-doku'ritsu-eiyō-seibutsu	光独立栄養生物	photoautotroph *[bio]*
hikari-gōsei	光合成	photosynthesis *[bioch]*
hikari-gōsei-saikin	光合成細菌	photosynthetic bacteria *[microbio]*
hikari-denki	光電気；光電氣	photoelectricity *[electr]*
hikari-denri	光電離	photoionization *[p-ch]*
hikari-isei'ka	光異性化	photoisomerization *[p-ch]*
hikari-jūgō	光重合	photopolymerization *[poly-ch]*
hikari-kagaku-kei	光化学系	photosystem *[bioch]*
hikari-kaihen	光壊変；光壞變	photodisintegration *[nuc-phys]*
hikari-kairi	光解離	photodissociation *[p-ch]*
hikari-ki'denryoku-kōka	光起電力効果	photovoltatic effect *[electr]*
hikari-ki'denryoku-seru	光起電力セル	photovoltatic cell *[electr]*
hikari-kōon-kei	光高温計	optical pyrometer *[eng]*
hikari no kyōdo	光の強度	luminous intensity *[opt]*
hikari-onkyō-bunkō	光音響分光	photoacoustic spectroscopy *[spect]*
hikari-ryōshi; kō'ryōshi	光量子	light quantum; photon *[quant-mech]*

hikari-sanka	光酸化	photooxidation *[ch]*
hikari-shū(ki)-sei	光周(期)性	photoperiodism *[physio]*
hikari-shūsei; kō-shusei	光周性	photoperiodism *[physio]*
hikari-teko	光挺子; ひかりてこ	optical lever *[opt]*
hikari-tsūshin	光通信	optical communication *[comm]*
hi-kassei	比活生	specific activity *[microbio]*
hi-kasū	被加数	augend *[math]*
hika-suzu	砒化錫	tin arsenide
hika-tetsu	砒化鉄	iron arsenide
hike	引け	shrink mark; sink mark *[met]*
hi-kenpa-ki	比検波器	ratio detector *[electr]*
hi'kesshō-ryō'iki	非結晶領域	amorphous region *[ch]*
hiki	引き	pull; tug *{n} [eng]*
hiki-ami	引網; ひきあみ	seine net *[eng]*
-hiki (-biki)	-匹	-unit(s) of animal(s) *{suffix}*
hiki-bune	引舟	tugboat; towboat *[nav-arch]*
hiki'dashi	引出し	drawer (of chest) *[furn]*
hiki'dashi-sen	引出(し)線	outgoing line *[elec]*
hiki-do	引戸	sliding door *[arch]*
hikido'shiki-yane	引戸式屋根	sliding roof (of car) *[mech-eng]*
hiki'gaeru	蟇蛙; 蟆	toad (amphibian); bufo *[v-zoo]*
hiki'gane	引金	trigger *{n} [ord]*
hi-ki'hatsu-do	比揮発度	relative volatility *[thermo]*
hi'kikkō'teki-sogai	否拮抗的阻害	noncompetetive inhibition *[bioch]*
hiki'komi-guchi	引込(み)口	service entrance *[elec]*
hiki-nami	引き波	undertow *[ocean]*
hiki-nobashi	引伸し	enlarging; enlargement *{n} [photo]*
hiki'nuki-hannō	引抜き反応	abstraction reaction *[ch]*
hi'kinzoku	非金属	nonmetal *[mat]*
hi-kinzoku	卑金属; ひきんぞく	base metal *[met]*
hi'kinzoku-kaizai'butsu	非金属介在物	nonmetallic inclusion *[mech-eng]*
hiki-okoshi	引起; ひきおこし	Amethysanthus japonicus *[bot] [pharm]*
hiki'saki	引裂き	tearing *{n} [tex]*
hiki'saki-tsuyo-sa	引裂強さ	tear(ing) strength *[mech]*
hiki-saku	引き裂く	tear; rip open *{vb t}*
hiki-sū	引き数	argument *[comput] [math]*
hiki'tsuri	引き攣り; ひきつり	drawing up; cramping *{n}*
hiki-watasu	引渡す	surrender (give to) *{vb t} [legal]*
hiki-zan	引き算	subtraction *[math]*
hiki-zuna	引き綱	tow rope *[eng]*
hiki'zuri	引き摺り	drag *{n}*
hikizuru	引き摺る	drag *{vb t}*

hikkakaru	引っ掛かる	catch on; be caught on *{vb}*
hikkake-tsuyo-sa	引掛強さ	loop strength *[tex]*
hikkaki-kata-sa	引掻き硬さ	scratch hardness *[mech]*
hikkaku	引っ掻く	scratch; claw; maul *{vb t}*
hikki-sei	筆記性；ひっきせい	writability *[graph]*
hikkonda	引っ込んだ	recessed *{adj} [arch]*
hikkuri-kaeru	引っくり返る	overturn *{vb i}*
hikkuri-kaesu	引っくり返す	overturn; upset; capsize *{vb t}*
hikō	飛行；ひこう	flight; flying *{n} [aero-eng]*
hikō-kanban	飛行甲板	flight deck *[aero-eng]*
hikō-keiro	飛行径路	flight path *[navig]*
hikōki	飛行機	airplane *[aerosp]*
hikō-ki'roku-ki	飛行記録器	flight recorder *[eng]*
hiko-kyoku	被呼局	called station *[comm]*
hikō'sen	飛行船	airship *[aerosp]*
hi'kōsoku-oshi'dashi	非拘束押出し	unrestricted extrusion *[met]*
hi-kotsu	腓骨；ひこつ	fibula (bone) *[anat]*
hi'kōyu	非鉱油；非鑛油	nonpetroleum oil *[mat]*
hiku	引く	subtract *{vb} [math]*
"	引く	pull; tug *{vb} [mech]*
hikui	低い	low *{adj} [math]*
hi'kui-dori	火喰鳥；ひくいどり	cassowary (bird) *[v-zoo]*
hikumeru	低める	lower; lessen the height *{vb t}*
hi-kussetsu	比屈折	specific refraction *[mech]*
hi-kyōdo	比強度	specific strength *[mech]*
hi'kyokusei-kagō'butsu	非極性化合物	nonpolar compound *[ch]*
hi'kyōyaku no	非共役の	nonconjugated *{adj} [org-ch]*
hi'kyūmen-kyō	非球面鏡	aspherical mirror *[opt]*
hima	蓖麻；箆麻；ひま	castor-oil plant; Ricinus communis *[bot]*
himaku	飛膜	patagium *[v-zoo]*
himaku	被膜	film; coating film; covering film *[mat]*
himashi-yu	蓖麻子油	castor oil *[mat]*
hi'matsu-kansen	飛沫感染	droplet infection *[med]*
hi'mawari	向日葵；ひまわり	sunflower (flower) *[bot]*
hime-aka-tate'ha	姫赤立羽	painted lady (butterfly) *[i-zoo]*
himei-kenkyū	碑銘研究	epigraphy *[comm]*
hi'mitsu	秘密；ひみつ	secret *{n} [ind-eng]*
himitsu-hogo	秘密保護	security *[comput]*
himitsu-sei	秘密性	confidentiality *{n} [ind-eng]*
himo	紐；ひも	cord; string *{n} [mat]*
himō	被毛	covering fur (animal) *[mat]*
himo'gata-dōbutsu	紐形動物	Nemertinea *[i-zoo]*

hina	雛	chick; fledgling (bird) *[agr] [v-zoo]*
hina'geshi	雛芥子；ひなげし	corn poppy; red poppy *[bot]*
hinan'jo	避難所	refuge; sanctuary; shelter *[arch]*
hindo	頻度	frequency *[math]*
hi-nendo	比粘度	specific viscosity *[fl-mech]*
hi'nensei-ryūtai	非粘性流体	inviscid fluid *[fl-mech]*
hineri-shindō	捻り振動	twisting vibration *[spect]*
hineru	捻る；撚る；拈る	twist *{vb t}*
hi-netsu	比熱	specific heat *[thermo]*
hingan	玢岩；ひんがん	porphyrite *[miner]*
hin-gasu	貧ガス	lean gas *[ch]*
hin'i	品位	grade; quality *{n} [ind-eng]*
hi'nin-gu	避妊具	contraceptive (device) *[pharm]*
hi'nin-yaku; (-zai)	避妊薬(剤)	contraceptive (medicine) *[pharm]*
hinkei	品型	form *{n} [bio]*
hinmei	品名	article name; name of item *{n}*
hinmoku	品目	item; a list of articles *{n} [pr]*
hinmō-rui	貧毛類	Oligochaeta *[i-zoo]*
hi no-	火の-	pyro- *{prefix} [ch]*
hi-no-de	日の出	sunrise *[astron]*
hinoki	檜；ひのき	cypress; white cedar (tree) *[bot]*
hinpan na	頻繁な	frequent *{adv} [math]*
hin'shi	品詞	part(s) of speech *{n} [gram]*
hin'shitsu	品質；ひんしつ	quality *[ind-eng]*
hinshitsu-hoshō	品質保証	quality assurance *[ind-eng]*
hinshitsu-kanri	品質管理	quality control *[math]*
hinshitsu-kijun	品質基準	quality standard *[ind-eng]*
hinshu	品種	breed; race; variety *{n} [agr] [bio]*
"	品種	form *{n} [bio] [syst]*
hin-yōbai	貧溶媒	bad solvent; poor solvent *[ch]*
hi'nyō'ki-kei	泌尿器系	urinary system *[anat]*
hi'ōgi	檜扇；ひおうぎ	blackberry lily; leopard flower; Belamcanda *[bot]*
hi'onkyō-han'i	非音響範囲	zone of silence *[acous]*
hippari-bane	引張りばね	tension spring *{n} [mech-eng]*
hippari-dansei-ritsu	引張弾性率	tensile modulus *[mech]*
hippari-hadan-nobi	引張破断伸	tensile breaking elongation; tensile fracture elongation *[mech]*
hippari-kyōdo	引張強度	tensile strength *[mech]*
hippari-ō'ryoku	引張応力	tensile stress *[mech]*
hippari-ryoku	引張力	tension *[mech]*
hippari-sendan-ō'ryoku	引張剪断応力	tensile shear stress *[mech]*

hippari-shiken	引張試験	tensile test; tension test *[mech]*
hippari-tsuyo-sa	引張強さ	tensile strength *[mech]*
hippari-zanryū-ō'ryoku	引張殘留応力	tensile residual stress *[mech]*
hipparu	引っ張る	pull; tug; draw; stretch *{vb}*
hi'puroton'sei-yōbai	非プロトン性溶媒	aprotic solvent *[ch]*
hira-gana	平假名; ひらがな	kana (q.v.), cursive style *{n} [pr]*
hīragi	柊; ひいらぎ	holly *[bot]*
hira-ha'guruma	平歯車	spur gear *[mech]*
hirai-ki	避雷器	lightning arrester *[elec]*
hirai-shin	避雷針	lightning conductor; lightning rod *[elec]*
hiraita	開いた	open *{adj} [elec]*
hiraita uchū	開いた宇宙	open universe *[astron]*
hira'kezuri-ban	平削り盤	planing machine *[eng]*
hiraki-mado	開き窓	casement window *[arch]*
hira-maki-e	平蒔絵	gilt flat lacquerware *[eng] [furn]*
hira-mame	平豆	lentil (legume) *[bot]*
hirame	比目魚; ひらめ	flatfish; flounder; halibut *[v-zoo]*
hirameki	閃き	flash *{n} [opt]*
hira-ori	平織	plain weave *{n} [tex]*
hiratai	平たい	flat *{adj} [math]*
hira-take	平茸; ひらたけ	agaric; Pleurotus ostreatus *[bot]*
hira-yasuri	平鑢; ひらやすり	flat file *[eng]*
hire	鰭; ひれ	fin *[v-zoo]*
hire-ashi	鰭足	flipper *[v-zoo]*
hire'hari'sō	鰭玻璃草	comfrey; Symphytum officinale *[bot]*
hirei	比例; ひれい	proportion *[math]*
hirei-chūkō	比例中項	mean proportional *{n} [math]*
hirei-gendo; (-genkai)	比例限度; (-限界)	limit of proportionality *{n} [mech]*
hirei shite	比例して	proportionally *{adv} [math]*
hi-ritsu	比率	percentage; rate; ratio *{n} [math]*
hiro	尋; ひろ	fathom *{n} [ocean]*
hirō	疲労; 疲勞; ひろう	fatigue; tiring *{n} [physio]*
hiro'ba	広場	place; square; plaza; circle *{n} [civ-eng]*
hirogeru	広げる	broaden; spread; widen; expand *{vb t}*
hiroi	広い; 廣い; ひろい	wide; broad *{adj}*
hiro'ji	広路	avenue; boulevard *[civ-eng]*
hirō-kiretsu	疲労亀裂	fatigue cracking *{n} [met]*
hiro'kuchi-bin	広口瓶	wide-mouth bottle *[an-ch]*
hiroku suru	広くする	make wide *{vb t}*
hiromaru	広まる	widen; broaden; spread *{vb i}*
hiromeru	広める	widen; broaden; spread *{vb t}*
hiro-sa	広さ	breadth; width *[mech]*

hiru	昼；晝；ひる	daytime; midday; noon *[astron]*
hiru	蛭	leech *[i-zoo]*
hiru-ishi	蛭石；ひるいし	vermiculite *[miner]*
hiru'mushiro	蛭蓆；ひるむしろ	pickerelweed *[bot]*
hiryō	肥料	fertilizer *[agr] [mat]*
hi'saku-sei	被削性	machinability *{n} [mech]*
hisan	砒酸；ひさん	arsenic acid
hisan	飛散	scatter; spatter *{n}*
hisan-dō	砒酸銅	copper arsenate
hisan-en	砒酸塩	arsenate *[ch]*
hisan-hōso	砒酸硼素	boron arsenate
hisashi	庇；ひさし	eaves *[arch]*
hi'seibutsu'teki-kankyō	非生物的環境	abiotic environment *[ecol]*
hi'seishoku-nuno	非製織布	unwoven cloth *[tex]*
hiseki	飛跡	track (of flying object) *{n} [aero-eng]*
hi-sekibun-kansū	被積分関数	integrand *[math]*
hi'senkei-hadō	非線形波動	nonlinear wave *[phys]*
hi'senkei-keikaku-hō	非線形計画法	nonlinear programming *{n} [comput]*
hi'senkei-sei	非線形性	nonlinearity *[phys]*
hi-senkō-do	比旋光度	specific rotation *[opt]*
hi'senkō-sei	非旋光性	nonrotatory(ness) *{n} [opt]*
hi-sesshoku'tai	被接触体	catalysant *[ch]*
hi-setchaku	比接着	specific adhesion *[adhes]*
hi'setchaku'sei no	非接着性の	abhesive; nonadhesive *{adj} [adhes]*
hishi	菱；ひし	water chestnut (fruit) *[bot]*
hishi	榧子；ひし	Japanese nutmeg seed *[bot]*
hishi-gata	菱形	rhombus *{n} [math]*
hishi'gata no	菱形の	rhomboid *{adj} [math]*
hi-shigeki-sei	被刺激性	irritability *[bio] [physio]*
hi'shimen'dōkō	砒四面銅鉱	tennantite *[miner]*
hi'shin-do	非心度	noncentrality degree *[math]*
hishin'kei no	皮針形の	lanceolate *{adj} [bot]*
hishi no mi	菱の実	water caltrop; water chestnut *[bot]*
hishi-okō	皮脂汚垢	sebaceous grime *[physio]*
hishi-sen	皮脂腺	sebaceous gland *[physio]*
hishi-shokubutsu	被子植物	angiosperm *[bot]*
hi-shitsu	比湿；比濕	specific humidity *[meteor]*
hi'shitsu	皮質	cortex (pl cortices, cortexes) *[anat]*
hi'shōgeki'shiki-inji-sōchi	非衝撃式印字装置	nonimpact printer *[pr]*
hishoku-kan	比色管	color-comparison tube *[an-ch]*
hishō-shitsu	非晶質	amorphous matter *[phys]*
hishō'shitsu-gōkin	非晶質合金	amorphous alloy *[met]*

hi'shūki'teki na	非周期的な	aperiodic *{adj}* *[phys]*
hiso	砒素；ひそ	arsenic (element)
hisō	皮層	cortex (pl cortex) *[bot]*
hi'sōtai'teki-kinji	非相対的近似	nonrelative approximation *[phys]*
hisō'teki na	皮相的な	superficial *{adj}*
hissu-jōken	必須条件	mandatory clause *[legal]*
hissu-shibō-san	必須脂肪酸	essential fatty acid *[org-ch]*
hissu-yōken	必須要件	necessary condition *[math]*
hisui	翡翠；ひすい	imperial jade; jade; nephrite *[miner]*
hisui'teki na	非水的な	nonaqueous *{adj}* *[ch]*
hisui-yōbai	非水溶媒	nonaqueous solvent *[ch]*
hitai	額；ひたい	forehead *[anat]*
hi'taiō-kōgen	非対応抗原	heterologous antigen *[immun]*
hi-taiseki	比体積；ひたいせき	specific volume *[mech]*
hi'taishō	非対称；非對稱	asymmetry *{n}* *[sci-t]*
hi'taishō-do	非対称度	skewness *[sci-t]*
hi'taishō-ha	非対称波	asymmetrical wave *[phys]*
hitashi(-mono)	浸し(物)	boiled greens in soy sauce dip *[cook]*
hitasu	浸す	immerse; soak; steep *{vb t}*
hitei	否定；ひてい	negation *[comput]* *[logic]*
"	否定	NOT; negative *{n}* *[comput]*
"	否定	denial; disavowal; negation *{n}*
hitei	飛程；ひてい	range (of flight) *{n}* *[navig]*
hi'teijō-ryū	非定常流	unsteady flow *{n}* *[fl-mech]*
hi-teikō	比抵抗	resistivity; specific resistance *[elec]*
hitei-ōtō-moji	否定応答文字	negative acknowledge character *[comput]*
hitei-ronri-seki	否定論理積	NOT-AND
hitei-ronri'seki-enzan	否定論理積演算	NOT-AND operation *[comput]*
hitei-ronri'seki-soshi	否定論理積素子	NOT-AND element *[comput]*
hitei-ronri-wa	否定論理和	NOT-OR
hitei suru	否定する	negate *{vb}* *[comput]*
hi'tekkō	砒鉄鉱；砒鐵鑛	loellingite *[miner]*
hiten-shūsa	非点収差	anastigmatic aberration *[opt]*
hito	人；ひと	man; mankind; person *[bio]*
hito'de	人手	manpower *[ind-eng]*
hitode	海星；ひとで	starfish *[i-zoo]*
hitode-rui	海星類	Asteroidea; starfishes *[i-zoo]*
hito'e	一重；ひとえ	one layer *{n}* *[tex]*
hito-hiro	一尋；ひとひろ	one fathom (= 6 feet) *{n}* *[math]*
hi'tōjiku no	非等軸の	anisometric *{adj}* *[crys]*
hi-tōji-ritsu	比透磁率	specific permeability *[mech]*
hito-ka	人科	Hominidae (people) *[bio]*

hito'ka-dōbutsu	人科動物	hominids *[v-zoo]*
hito'kobu-rakuda	一瘤駱駝	dromedary (animal) *[v-zoo]*
hito-ma	一間	one room *{n} [arch]*
hito-maki	ひと巻き	whorl (of a shell) *{n} [i-zoo]*
hito-men'eki-fuzen-uirusu	人(ヒト)免疫不全=ウイルス	human immunodeficiency virus; HIV *[bioch]*
hitomi	瞳; ひとみ	pupil (of eye) *[anat]*
hito-mori	一盛り	serving; one serving *{n} [cook]*
hitori	一人; ひとり	one person *{n}*
hito'sashi-yubi	人差指; 食指	forefinger; index finger *[anat]*
hito-seichō-horumon	人成長ホルモン	human growth hormone *[bioch]*
hitoshī	等しい	equal *{adj} [math]*
hitoshikunai	等しくない	unequal *{adj} [math]*
hito'soroi no hōseki	一揃いの宝石	parure *{n} [lap]*
hito'suji-shima-ka	一筋縞蚊	Asian tiger mosquito; Aedes albopictus (insect) *[i-zoo]*
hito-zoku	人族; ヒト族	Homo *[bio]*
hitsu	櫃; ひつ	coffer; chest (wooden) *[furn]*
hitsuji	羊; ひつじ	sheep (pl sheep) (animal) *[v-zoo]*
hitsu'jun	筆順	order of strokes (of kana, kanji) *[pr]*
hitsu'jū no	必従の; 必従の	consequent *{adj} [geol]*
hitsu'yō	必要; ひつよう	necessary *{n} [math]*
hitsu'yō-jōken	必要条件	necessary condition *[math]*
hitsuzen'teki na	必然的な	necessary *{adj} [logic]*
hi'uchi-ishi	燧石; ひうちいし	flint *[miner]*
hiyasu	冷す	cool; chill *{vb t}*
hiyō	比容	specific volume *[mech]*
hiyō	費用	cost(s); expense(s); expenditure *[econ]*
hiyodori	鵯; ひよどり	bulbul (bird) *[v-zoo]*
hi'yoke(-ita)	日除け(板)	sun visor *[mech-eng]*
hiyoku'ka	肥沃化	fertilization *[agr]*
hiyosu	菲沃斯; ひよす	henbane; stinking nightshade *[bot]*
hi-yūden-ritsu	比誘電率	specific inductive capacity *[elec]*
hiyū'dōkō	砒黝銅鉱	tennantite *[miner]*
hiza	膝; ひざ	lap *{n} [anat]*
hiza'gashira	膝頭	knee *{n} [anat]*
hizō	脾臓; 脾臟; ひぞう	spleen *{n} [anat]*
hi-zō'shoku-sokudo	比増殖速度	specific multiplication rate *[bio]*
hi'zuke	日付	date; dating *{n} [comput] [pat]*
hi'zuke-henkō-sen	日付変更線	international date line *[astron]*
hizume	蹄; ひづめ	hoof (pl hooves) *[v-zoo]*
hizumi	歪; ひずみ	distortion; strain *{n} [phys]*

hizumi-chizu	歪地図; 歪地圖	crustal strain map *[geol]*
hizumi-jikō	歪時効	strain aging *{n} [met]*
hizumi-kei	歪み計	strain gage *[eng]*
hizumi-kōka	歪硬化	strain hardening *{n} [met]*
hizumi-ō'ryoku	歪応力	strain stress *[mech]*
hizumi-ritsu	歪み率	distortion factor *[comm]*
hizumi-taishō no	歪対称の	skew-symmetric *{adj} [math]*
hizumi'tori-yaki'namashi	歪取焼鈍	stress-relieving annealing *{n} [met]*
ho	帆	sail *{n} [nav-arch]*
ho	穂; 穗; ほ	spike; ear *{n} [bot]*
hō	法; ほう	law *[legal]*
"	法	method *[math]*
hō	砲; 砲; ほう	cannon *[ord]*
hō	頬; ほお	cheek *{n} [anat]*
hōan'kō	方安鉱; 方安鑛	senarmontite *[miner]*
hobo	略; ほぼ; ほゞ	about *{adv}*
hōbō	魴鮄; ほうぼう	gurnard; sea robin; Chelidonichthys kumu (fish) *[v-zoo]*
hobu-ban	ホブ盤	hobbing machine *[mech]*
hobu-kakō	ホブ加工	hobbing *{n} [mech]*
hobu-kiri	ホブ切	hobbing *{n} [mech]*
ho-bune	帆船	sailboat *[nav-arch]*
hō'butsu'men	放物面	paraboloid *{n} [math]*
hōbutsu'men-hansha-ki	放物面反射器	parabolic reflector *[elecmg]*
hō'butsu-sen	放物線	parabola *[math]*
hōbutsu'sen'jō no	放物線状の	parabolic *{adj} [math]*
hōbutsu'tai	放物体	projectile *[ord]*
hō-byōri'gaku	法病理学	forensic pathology *[med]*
hōchi suru	放置する	let stand; leave as is *{vb t}*
hōchō	庖丁	cook's knife (pl knives) *[cook]*
hochō-ki	補聴器; 補聽器	hearing aid *[acous]*
hodai	補題	lemma (pl lemmas, lemmata) *[logic]*
hōden	放電; ほうでん	discharge *{n} [elec]*
hōden-bako	放電箱	spark chamber *[nucleo]*
hōden-hakkō	放電発光	electroluminescence *[electr]*
hōden-kan	放電管	discharge tube *[elec]*
hōden-ritsu	放電率	discharge rate *[elec]*
hōden-tō	放電灯	discharge lamp *[electr]*
hodo	程; ほど	about; extent *{n}*
hodo	歩度	rate (of pace) *{n} [physio]*
hodō	歩道	sidewalk *[civ-eng]*
hodō	舗道	pavement *[civ-eng]*

hōdō	報道；ほうどう	report; news *{n} [comm]*
hōdō-kikan	報道機関	news media *[comm]*
hodoku	解く；ほどく	undo; unfasten; loosen *{vb t}*
hō'enkō	方鉛鉱；方鉛鑛	galena *[miner]*
hō'fukka-aen	硼弗化亜鉛	zinc borofluoride
hō'fukka-dō	硼弗化銅	copper borofluoride
hō'fukka-gin	硼弗化銀	silver borofluoride
hō'fukka-namari	硼弗化鉛	lead borofluoride
hō'fukka-suiso-san	硼弗化水素酸	borofluoric acid
hō'fukka-suzu	硼弗化錫	tin borofluoride
hō'fukka-tetsu	硼弗化鉄	iron borofluoride
ho'fuku	匍匐；ほふく	creep *{n} [plas]*
hofuku-shi	匍匐枝	runner *[bot]*
hōfutsu'seki; hōfusseki	方沸石	analcite; analcime *[miner]*
ho'gai	補外	extrapolation *[math]*
hōgaku	方角	direction *[eng]*
"	方角	bearing *{n} [navig]*
hōgaku	邦楽	Japanese music *[music]*
hō'gaku	法学；法學	jurisprudence; law *[legal]*
hōgan	方眼；ほうがん	plotting *{n} [graph]*
hōgan-kukaku-zumen	方眼区画図面	graticule *[map] [opt]*
hōgan-shi	方眼紙	graph paper *[graph]*
hōgen	方言	vernacular language *[ling]*
hogo	保護；ほご	protection *{n} [comput]*
hōgō	抱合；抱合	conjugation *{n} [pharm]*
hōgō(-sen)	縫合(線)	suture *{n} [med]*
hogo-fuku	保護服	protective clothing *[cl]*
hōgō-ryoku	抱合力	cohesive force *[mech]*
hogo-shoku	保護色	protective coloration *[zoo]*
hogo-sō	保護層	protective coating *[photo]*
hogo suru	保護する	protect *{vb t}*
hōgō'tai	抱合体；抱合體	conjugate *{n} [bioch]*
hogo-te'bukuro	保護手袋	protective gloves *[cl]*
hogo-tosō	保護塗装	protective coating *[ind-eng]*
hogo'yō	保護葉	bract *[bot]*
hogo-zai	保護剤	protective agent *[mat]*
hogusu	解す；ほぐす	disentangle; untangle; loosen *{vb}*
hōgyoku	宝玉；寶玉	gem *[miner]*
hōhō	方法；ほうほう	method *[math]*
hō'hōseki	方硼石	boracite *[miner]*
ho'i	補遺	addendum; appendix; supplement *[pr]*
hōi	方位；ほうい	azimuth *[astron]*

hō'i	方位	direction *[eng]*
"	方位	orientation *[math]*
"	方位	bearing *{n} [navig]*
hō'i-hai'retsu	方位配列	preferred orientation *[petr]*
hō'i-kaku	方位角	azimuth *[astron] [geod]*
"	方位角	bearing angle *[astron]*
ho'inshi	補因子	cofactor *[bioch]*
hoji	保持；ほじ	retention *[ch-eng]*
hoji-buzai	保持部材	retaining member *[eng]*
hoji-jikan	保持時間	retention time *[an-ch]*
"	保持時間	retention period *[comput]*
"	保持時間	maintenance time *{n}*
hōjin	法人	juridical person; juristic person *[legal]*
hoji-ryoku	保持力	holding power *{n} [adhes]*
hoji-ryoku	保磁力	coercive force *[elecmg]*
hōji-shingō	報時信号	time signal *[comm]*
hoji suru	保持する	hold; retain *{vb t} [adhes]*
hoji-yōryō	保持容量	retention volume *[an-ch]*
hojo	補助；ほじょ	backup *{n} [comput]*
"	補助	auxiliary; secondary *{n}*
hojo-denkyoku	補助電極	auxiliary electrode *[elec]*
hojo-teiri	補助定理	lemma (pl lemmas, lemmata) *[logic]*
hojo'teki-shudan	補助的手段	supplementary procedure *[ind-eng]*
hojo-yoku	補助翼	aileron *[aero-eng]*
hojo-zu	補助図	auxiliary view *[graph]*
hojū	補充	replenishment *{n}*
hoka	外；他；ほか	elsewhere; outside; other *{adv} {n}*
hōka	法貨	legal tender; lawful money *[econ]*
hōka'butsu	硼化物	boride *[ch]*
hōka-dō	硼化銅	copper boride
hō-kagaku	法化学	forensic chemistry *[ch]*
hō-kagaku	法科学	forensic science *[sci-t]*
hōkai	崩壊；ほうかい	(radioactive) decay *{n} [nuc-phys]*
hō'kaiseki	方解石	calcite; iceland spar *[miner]*
hōkai suru	崩壊する	disintegrate *{vb i} [nuc-phys]*
"	崩壊する	collapse *{vb i}*
hōkai-teisū	崩壊定数	decay constant *[nuc-phys]*
hōkai-zai	崩壊剤；崩壞劑	disintegrating agent *[pharm]*
ho-kaku	補角	supplementary angle *[math]*
hokaku'gan	捕獲岩；ほかくがん	xenolith *[petr]*
hokaku-kesshō	捕獲結晶	xenocryst *[crys]*
hokaku no	補角の	supplementary (angular) *{adj} [math]*

hokaku sareta hikari	捕獲された光	trapped light *[opt]*
hokan	補間	interpolation *[math]*
hokan	保管；ほかん	save *{n} [comput]*
"	保管	custody; safekeeping *{n} [legal]*
hōkan	砲艦；砲艦	gunboat *[nav-arch]*
hokan-jumyō	保管寿命	storage life *[eng]*
hokan'sha	保管者	custodian; trustee *[legal]*
hōka-tetsu	硼化鉄	iron boride
hō'keiseki	方珪石	cristabolite *[miner]*
hoken	保険；ほけん	insurance *[econ]*
hoken	保健	health (preservation of) *[med]*
hoken'jo	保健所	health center *[arch] [med]*
hoken-kin	保険金	insurance money *[econ]*
hōki	箒；ほうき	broom *[eng]*
hōki	法規	rules and regulations *{n} [legal]*
hōki	放棄	waiver; abandonment *[legal]*
hōki-boshi	箒星；ほうきぼし	comet *[astron]*
hōki'mushi-dōbutsu	箒虫動物	Phoronidea *[i-zoo]*
hōkin	砲金	gunmetal *[met]*
hōki suru	放棄する	abandon; waive *{vb t} [legal]*
hokki-gai	北寄貝；ほっきがい	surf clam (shellfish) *[i-zoo]*
hokkyoku	北極；ほっきょく	North Pole *[geogr]*
hokkyoku-guma	北極熊	polar bear (animal) *[v-zoo]*
hokkyoku-kai	北極海	Arctic Ocean *[geogr]*
hokkyoku-ken	北極圏；北極圏	Arctic Circle *[geod]*
hokkyoku-kō	北極光	aurora borealis *[geophys]*
hokkyoku-sei	北極星	Polaris *[astron]*
hokō	歩行	locomotion; walking *{n} [physio]*
hōkō	方向；ほうこう	direction *[eng]*
hōkō'da	方向舵	rudder; vertical rudder *[aero-eng]*
hōkō-kaku	芳香核	aromatic nucleus *[org-ch]*
hōkoku	報告；ほうこく	report *{n} [comm] [pr]*
hōkoku'sho-sakusei	報告書作成	report generation *[comput]*
hoko no	補弧の	supplementary *{adj} [math]*
hōkō-sei	方向性	directionality *{n}*
hōkō-shiji-ki	方向指示器	turn signal (automobile) *[mech-eng]*
ho'kōso	補酵素；ほこうそ	coenzyme *[bioch]*
hōkō-tenkan	方向転換	veering *{n} [meteor]*
hōkō-yogen	方向余弦	direction cosines *[math]*
hōkō-zai	芳香剤	aromatic agent; perfume *[mat] [org-ch]*
hōkō'zoku-kagō'butsu	芳香族化合物	aromatic compound *[org-ch]*
hōkō'zoku-sei	芳香族性	aromaticity *{n} [org-ch]*

hoku'to-sei	北斗星；ほくとせい	Big Dipper; Great Dipper *[astron]*
hokuto-shichi'sei	北斗七星	Big Dipper; Great Dipper *[astron]*
hokyō suru	補強する	reinforce; strengthen *{vb} [eng]*
hokyō-zai	補強剤	reinforcing agent *[mat]*
hōkyū	俸給	salary *[econ]*
hokyū'sen	補給船	supply ship; tender *[nav-arch]*
hōmatsu	泡沫；泡沫	bubbles; foam; froth *{n} [ch]*
homeru	褒める	praise; commend *{vb t}*
hōmon-gi	訪問着；ほうもんぎ	visiting kimono; gala kimono *[cl]*
hōmu	ホーム	platform *[civ-eng] [rail]*
hon	本；ほん	book *[pr]*
-hon; -bon; -pon	-本；-ほん；-ぼん；-ぽん	-unit(s) of a thin-cylinder-like object *{suffix}*
hō'naga-suzume'bachi	頬長雀蜂	yellow jacket (insect) *[i-zoo]*
hon-bako	本箱；ほんばこ	bookcase *[furn]*
honbun; hon'mon	本文	body *[pat]*
hon-dana	本棚	bookshelf (pl -shelves) *[arch]*
hondawara	馬尾藻；ほんだわら	sargasso (pl -s); gulfweed *[bot]*
hone; kotsu	骨	bone *{n} [anat] [hist]*
hone'gumi	骨組み；ほねぐみ	skeleton *[anat]*
hon'ei	本影	umbra (pl umbras, umbrae) *[astron]*
hō'netsu-ki	放熱器	radiator *[eng]*
honnō	本能；ほんのう	instinct *[psy]*
honrai no basho ni	本来の場所に	in site *{adv} [sci-t]*
honsen	本線	main line *[rail]*
honshin	本震	main shock (earthquake) *[geophys]*
honshitsu'teki-henkō-sei	本質的変光星	intrinsic variable star *[astron]*
honshitsu'teki na	本質的な	intrinsic *{adj}*
hontai	本体	mainframe *[comput]*
honten	本店	main store *[arch] [econ]*
ho-nuno	帆布；ほぬの	canvas *{n} [tex]*
hon-yaki	本焼；本燒	glost firing *{n} [cer]*
hon'yaku	翻訳；飜譯	translation *{n} [comm] [mol-bio] [pr]*
"	翻訳；ほんやく	compilation *[comput]*
hon'yaku'sha	翻訳者	translator; interpreter *[comm]*
ho'nyū-dōbutsu	哺乳動物	mammal *[v-zoo]*
ho'nyū-rui	哺乳類	mammals; Mammalia *[v-zoo]*
hō'ō'boku	鳳凰木	poinciana (tree) *[bot]*
ho'on-zai	保温剤；保溫劑	heat-insulating material *[mat]*
hoppō-shinyō'ju-rin	北方針葉樹林	taiga; boreal coniferous forest *[ecol]*
hoppu	忽布；ほっぷ	hop *{n} [bot]*
hora-gai	法螺貝；ほらがい	conch; triton; trumpet shell *[i-zoo]*

hōraku-sen	包絡線; 包絡線	envelope *{n} [comm] [math]*
horei (suru)	保冷(する)	maintaining cooled *{vb}*
hōrei	放冷	still-air cooling; leaving to cool *{n} [ch] [ch-eng]*
hōrei	法令; ほうれい	act; ordinance *{n} [legal]*
horei-zai	保冷剤	cold insulator *[mat]*
hōren'sō	菠薐草; 鳳蓮草; = ほうれんそう	spinach (vegetable) *[bot]*
hori	堀; 濠; 壕	ditch; canal; moat *{n} [civ-eng]*
hori-ido	掘井戸	dug well *{n} [eng]*
hori-nezumi	掘り鼠	pocket gopher; gopher *[v-zoo]*
hō'ritsu	法律; ほうりつ	law *[legal]*
horō	歩廊	corridor; gallery *[arch]*
horohoro-chō	ほろほろ鳥	guinea fowl (bird) *[v-zoo]*
hōrō'shitsu	琺瑯質	enamel *{n} [mat]*
horumon	ホルモン	hormone *[bioch]*
horumyūmu	ホルミウム	holmium (element)
horyo'gan	捕虜岩; ほりょがん	xenolith *[petr]*
ho'ryoku	補力	intensification *[photo]*
hōryū	放流	discharge *{n} [hyd]*
horyū-jikan	保留時間	holding time *[comput]*
horyū suru	保留する	retain *{vb t}*
hōsan	硼酸; ほうさん	boric acid
hōsan-aen	硼酸亜鉛	zinc borate
hōsan-en	硼酸塩	borate *[ch]*
hōsan-namari	硼酸鉛	lead borate
hosei	補正; ほせい	amendment; correction; revision *[pat]*
hosei-ban	補正板	correcting plate *[opt]*
hosei-kyakka	補正却下	dismissal of revision *[pat]*
hosei-ritsu	補正率	correction factor *{n}*
hosei('sho)	補正(書)	amendment; correction *[pat]*
hosei suru	補正する	correct *{vb}*
hōseki	宝石; 寶石	gem; jewel; precious stone *[miner]*
hōseki'gaku	宝石学	gemology *[miner]*
hōseki-gakusha	宝石学者	lapidarist *[lap]*
hōseki-genseki	宝石原石	boule *[crys]*
hōseki-jiku'uke	宝石軸受	jewel bearing *{n} [horol]*
hōseki-saiku'jutsu	宝石細工術	lapidary *[sci-t]*
hōseki-saiku'nin	宝石細工人	lapidist *[lap]*
hōsen	法線	normal *{n} [math]*
hōsen-kin	放線菌	ray fungus; Actinomyces *[microbio]*
hōsen-ō'ryoku	法線応力	normal stress *[mech]*

hōsha	硼砂	borax *[miner]*
hōsha	放射；ほうしや	emission; radiation *[phys]*
hōsha	報謝	recompense; payment; remuneration *{n}*
hōsha'chū-rui	放射虫類	Radiolaria *[i-zoo]*
hōsha-do	放射度；ほうしゃど	emissive power *[thermo]*
hōsha-hōkai	放射崩解	decay (radioactive) *{n} [nuc-phys]*
hōsha-kagaku	放射化学	radiochemistry *[ch]*
hōsha-kei	放射計	radiometer *[electr]*
hōsha-kō	放射光	irradiated light *[opt]*
hōsha-kō'ritsu	放射効率	radiant efficiency *[opt]*
hōsha-kyōdo	放射強度	emissive power; emittance *[thermo]*
hōsha'nō	放射能	radioactivity *[nuc-phys]*
hōsha'nō-bunseki	放射能分析	radioassay *[an-ch]*
hōsha-ritsu	放射率	emissivity *[thermo]*
hōshasei-busshitsu	放射性物質	radioactive substance *[nuc-phys]*
hōshasei-chiri	放射性塵	radioactive dust *[nuc-phys]*
hōshasei-dō'i-genso	放射性同位元素	radioisotope *[nuc-phys]*
hōshasei-genso	放射性元素	radioactive element *[ch] [phys]*
hōshasei-hakkō	放射性発光	radioluminescence *[phys]*
hōshasei-hōkai	放射性崩壊	radioactive decay *[nuc-phys]*
hōshasei-kōka'butsu	放射性降下物	radioactive fallout *[nucleo]*
hōshasei-nendai-sokutei	放射性年代測定	radioactive dating *{n} [nucleo]*
hōsha'sei no	放射性の	radioactive *{adj} [nuc-phys]*
hōshasei-tanso-nendai-sokutei	放射性炭素年代=測定	radiocarbon dating *{n} [nucleo]*
hōshasei-tsuiseki'shi	放射性追跡子	radioactive tracer *[nucleo]*
hōsha'sen	放射線	ray *{n} [math]*
"	放射線	radioactive rays *[nuc-phys]*
hōshasen-dōi'tai	放射線同位体	radioisotope *[nuc-phys]*
hōshasen-i'gaku	放射線医学	radiology *[med]*
hōshasen-jūgō	放射線重合	radiation polymerization *[poly-ch]*
hōshasen-kagaku	放射線化学	radiation chemistry *[nucleo]*
hōshasen-kei	放射線計	dosage meter; dosimeter *[nucleo]*
hōshasen-kiken-sei	放射線危険性	radiation hazard *[med]*
hōshasen-men'eki-kentei-hō	放射線免疫検定法	radioimmunoassay *[immun]*
hōshasen-ruiji'sayō-busshitsu	放射線類似作用物質	radiomimetic substance *[ch]*
hōshasen-seibutsu'gaku	放射線生物学	radiobiology *[bio]*
hōshasen-sokutei-bunseki	放射線測定分析	radiometric analysis *[an-ch]*
hōshasen-sonshō	放射線損傷	radiation damage *[nucleo]*
hōsha-soku	放射束	radiant flux *[opt]*
hōsha-sōshō	放射相称	radial symmetry *[bot] [sci-t]*

hōsha'tai	放射体	emitter *[electr]*
hōsha-teki'tei	放射滴定	radiometric titration *[an-ch]*
hōsha-tōkyū	放射等級	bolometric magnitude *[astron]*
hoshi; sei	星；ほし；せい	star *{n} [astron]*
hōshi	胞子；胞子；ほうし	spore *[bio]*
hoshi-budō	干葡萄；乾し葡萄	raisin *[cook]*
hōshi'chū-rui	胞子虫類	Sporozoa *[i-zoo]*
hoshi'gata-ha'guruma	星形歯車	star gear *[mech]*
hoshi'gata-setsu'zoku	星形接続	star connection *[elec]*
hoshi-jirushi	星印	asterisk *[comput] [pr]*
hō'shiki	方式	mode; method; system *[comput] [math]*
hoshi-muku-dori	星椋鳥；ほしむく=どり	starling; Sturnus vulgaris (bird) *[v-zoo]*
hōshi'nō	胞子嚢；胞子嚢	sporangium (pl -gia) *[bot]*
hōshi'nō'hei	胞子嚢柄	sporangiophore *[bot]*
hoshin suru	歩進する	step and advance *{vb}*
hōshi'tai	胞子体	sporophyte *[bot]*
ho'shitsu-zai	保湿剤；保濕劑	humectant *[ch]*
hoshi-zame	星鮫；ほしざめ	spotted shark (fish) *[v-zoo]*
hoshō	保証	guarantee *{n}*
hoshō	補償；ほしょう	compensation; restitution *[legal]*
hoshō	堡礁；保礁	barrier reef *[geol]*
hōshō	包晶；包晶	peritectic *{n} [p-ch]*
hōshō	泡鐘；泡鐘	bubble cap *[ch-eng]*
hoshō-ban	補償板	compensator *[opt]*
hoshō-hō	補償法	compensation method; indemnification method *[econ]*
hōshō-kin	報奨金	payment *[econ]*
ho-shoku	補色	complementary color *[opt]*
hoshoku'sei no	捕食性の	predatory; predaceous *{adj} [zoo]*
hoshoku'sha	捕食者	predator *[zoo]*
hoshō'nin	保証人	guarantor *[legal]*
hoshu	保守	maintenance *[comput] [ind-eng]*
hoshū	補修	repair *{n} [ind-eng]*
hōshū	報酬；ほうしゅう	fee; recompense *[econ]*
hoshu-jikan	保守時間	maintenance time *[ind-eng]*
hoshu-sei	保守性	maintainability *[eng]*
hō'shutsu	放出	discharge *{n} [fl-mech]*
hōshū-zengen no hōsoku	報酬漸減の法則	diminishing return, law of *[bio]*
hoshū-zai	捕収剤；捕收劑	collector *[bioch]*
"	捕収剤	scavenger *[nucleo]*
hosō	舗装；ほそう	pavement; paving *{n} [civ-eng]*

hōso	硼素; ほうそ	boron (element)
hōsō	包装; 包裝	packaging; package; packing *{n} [ind-eng]*
hōsō	放送	broadcast *{n} [comm]*
hōsō	疱瘡	smallpox *[med]*
hō'sōda'seki	方曹達石	sodalite *[miner]*
hosoi	細い; ほそい	narrow; thin *{adj} {n} [math]*
ho'soku	補足	supplement *{n} [legal]*
hōsoku	法則; ほうそく	law; principle; rule *[legal] [sci-t]*
hosoku suru	細くする	attenuate *{vb t}*
hosomeru	細める	thin (narrow) down *{vb t} [mech-eng]*
hoso-michi (sakanakaba no)	細道 (坂半ばの)	berm *[civ-eng]*
hoso-nagai	細長い	long and narrow *{adj} [math]*
hoso'o-raichō	細尾雷鳥	sharp-tailed grouse (bird) *[v-zoo]*
hosō-ro	舗装路	paved road *[civ-eng]*
hosu	乾す; 干す; ほす	dry; desiccate; air *{vb t}*
hosū	補数	complement *{n} [math]*
hosū-enzan'shi	補数演算子	complementary operator *[comput]*
ho'sui-ryoku	保水力	water-holding capacity *[hyd]*
ho'sui-sei	保水性	water retentivity *[hyd]*
hosū no tei	補数の底	complement base *[comput] [math]*
hōtai	繃帯; 包帯; 包帶	bandage *{n} [med] [pharm]*
hotaru	蛍; 螢; ほたる	firefly (insect) *[i-zoo]*
hotaru-ishi	蛍石	fluorite; fluorspar *[miner]*
ho'tate-gai	帆立貝	scallop (shellfish) *{n} [i-zoo]*
hōtei	法廷	courtroom *[legal]*
hōtei	方庭	quadrangle *[civ-eng]*
hōtei-keiryō-tan'i	法定計量単位	statutory units *[phys]*
hōtei'shiki	方程式	equation *[math]*
hōtei'shiki-kei	方程式系	system of equations *[math]*
hōtei'shiki-ron	方程式論	theory of equations *[math]*
hoten-zai	補填剤	filler *[mat]*
hotō	堡島; 保島; ほとう	barrier island *[geol]*
hōtō	砲塔; 砲塔	turret *[ord]*
hotoke-no-za	仏の座; 佛の座	bee nettle *[bot]*
hotondo	殆ど; ほとんど	nearly; nearly all *{adv}*
hototogisu	杜鵑; 時鳥; 不如帰	cuckoo (bird) *[v-zoo]*
hototogisu; hototogisu'sō	杜鵑草	toad lily (flower) *[bot]*
hōwa	飽和; 飽和; ほうわ	saturation *[phys]*
hōwa-ji'soku-mitsudo	飽和磁束密度	saturation magnetic flux density *[elecmg]*
hōwa-jōki	飽和蒸気	saturated steam *[phys]*
hōwa-jōki-atsu	飽和蒸気圧	saturation vapor pressure *[thermo]*
hōwa-yōeki	飽和溶液	saturated solution *[ch]*

hoya	海鞘；老海鼠；ほや	sea squirt *[i-zoo]*
hōyō	包葉；包葉	bract *[bot]*
hōyū'butsu	包有物	inclusion *[crys] [met] [petr]*
hozen	保全	preservation; conservation *[ecol]*
hozen(-sei)	保全(性)	integrity *[comput]*
hozo'ana-setsu'gō	枘穴接合	mortise joint *[eng]*
hozon	保存；ほぞん	storage *[comput]*
"	保存	preservation; conservation *[ecol]*
"	保存	maintenance *[ind-eng]*
hozon-ketsu'eki	保存血液	whole (preserved) human blood *[hist]*
hozon-ki'oku	保存記憶	archival memory *[comput]*
hozon-sei	保存性	shelf life *[ind-eng]*
hozon-shoku	保存食	keepable foods *[cook] [food-eng]*
hozon suru	保存する	save *{vb t} [comput]*
hozon-zai	保存剤	preservative *[mat]*
hozo to hozo-ana	枘と枘穴	tenon and mortise *[eng]*
hozo-tsugi	枘継ぎ；ほぞつぎ	tenon joint *[adhes]*
hōzuki	酸漿；鬼灯	Chinese lantern plant *[bot]*
hyakka-jiten	百科事典	encyclopedia *[pr]*
hyakka'ten	百科店	department store *[arch]*
hyaku	百；ひゃく	hundred *{n} [math]*
hyaku'bun-hi	百分比	percent; percentage *[math]*
hyakubun'i-sū	百分位数	percentile *[math]*
hyaku'bun-ritsu	百分率	percent; percentage *[math]*
hyaku'man	百万	million *[math]*
hyaku'man'chō	百万兆	trillion (England) *[math]*
hyaku'nen-kan	百年間	century (one hundred years)
hyaku'nichi-zeki	百日咳	whooping cough; pertussis *[med]*
hyō	表	table; list *{n} [comput] [math]*
hyō	俵	bale *{n} [ind-eng]*
hyō	豹；ひょう	leopard; panther (animal) *[v-zoo]*
hyō	雹	hail; hailstone *{n} [meteor]*
hyōdai	表題	caption; heading; title *{n} [pr]*
hyōdo	表土	regolith; topsoil *[geol]*
hyōdō	秤動	libration *[astron] [phys]*
hyōfudo	漂布土	fuller's earth *[geol]*
hyōga	氷河；ひょうが	glacier *[hyd]*
hyōga-chi'shitsu'gaku	氷河地質学	glaciogeology *[geol]*
hyōga'gaku	氷河学	glaciology *[geol]*
hyōga-gakusha	氷河学者	glaciologist *[geol]*
hyōgai	雹害；ひょうがい	hail damage *{n} [ecol]*
hyōga-ki	氷河期	glacial epoch *[geol]*

hyōga-sayō	氷河作用	glaciation *[geol]*
hyōgen	表現；ひょうげん	expression *{n}* *[comput]*
hyōgen-do	表現度	expressivity *[gen]*
hyōgen'kei	表現型	phenotype *[gen]*
hyōhaku	標白	bleaching *{n}* *[cook]* *[tex]*
hyōhi	表皮	skin (of foam) *{n}* *[ch]*
"	表皮	cuticle; epidermis *[hist]*
hyōhi-akkon	表皮圧痕	surface-skin impression *[physio]*
hyōhin	標品	specimen *[bioch]*
hyōhon	標本；ひょうほん	sample; specimen *[math]* *[sci-t]*
hyōhon-chū'shutsu	標本抽出	sampling *{n}* *[math]*
hyōhon'ka	標本化	sampling *{n}* *[sci-t]*
hyō'i-moji	表意文字	ideogram; ideograph *[pr]*
hyōji	表示；ひょうじ	display *{n}* *[comput]* *[electr]*
"	表示	indication; representation *[comput]*
"	表示	expression *{n}*
hyōji-chi	表示値	displayed (indicated) value *[electr]* *[math]*
hyōji-(ga)men	表示(画)面	screen *{n}* *[comput]*
hyōji-kan	表示管	display tube *[electr]*
hyōji-sōchi	表示装置	visual display *[comput]*
"	表示装置	display (apparatus) *[electr]*
hyōji-sōsa'taku	表示操作卓	display console *[comput]*
hyōjun	標準；ひょうじゅん	standard *{n}* *[comput]* *[phys]*
hyōjun-denkyoku	標準電極	standard electrode *[elec]*
hyōjun-hensa	標準偏差	standard deviation *[math]*
hyōjun-ji	標準時	standard time *[astron]*
hyōjun-jōtai	標準状態	standard state; standard conditions *[phys]*
hyōjun'ka	標準化	standardization *[eng]*
hyōjun'ka suru	標準化する	standardize *{vb}* *[sci-t]*
hyōjun-ki'atsu	標準気圧	standard atmospheric pressure *[meteor]*
hyōjun-kyōdo	標準強度	proof (alcohol) *[cook]*
hyōjun no ichi	標準の位置	standard position *[math]*
hyōjun-sanka'kangen-den'i	標準酸化還元電位	standard oxidation-reduction potential *[p-ch]*
hyōjun-so'shiki	標準組織	normal structure *[sci-t]*
hyōjun-(yō)eki	標準(溶)液	standard solution *[ch]*
hyōka	評価	evaluation *[math]*
hyōkai	氷塊；ひょうかい	calf; calved ice *[geol]*
hyōkan	氷冠	ice cap *[hyd]*
hyōketsu-bōshi	氷結防止	deicing *{n}* *[eng]*
hyō'ketsu suru	氷結する	freeze *{vb t}*
hyōki	氷期	Ice Age *[geol]*

hyōki-hō	表記法	notation; notation method *[math]*
hyōkō	氷縞；ひょうこう	varve *{n} [geol]*
hyōkō	標高	elevation *[eng]*
hyō'men	表面；ひょうめん	surface *{n} [math]*
hyōmen-chō'ryoku	表面張力	surface tension *[fl-mech]*
hyōmen-den'i	表面電位	surface potential *[elec]*
hyōmen-hifuku-zai	表面被覆剤	surface-covering agent *[mat]*
hyōmen-kassei-zai	表面活性剤	surface-active agent; surfactant *[mat]*
hyōmen-kōka-sō	表面硬化層	surface-hardened layer *[met]*
hyō'menseki	表面積	surface area *[math]*
hyōmen-teikō	表面抵抗	surface electrical resistance *[elec]*
hyōmen-teikō-ritsu	表面抵抗率	surface resistivity *[elec]*
hyōmu	氷霧	ice fog *[meteor]*
hyōnō	氷嚢；ひょうのう	ice bag; ice pack *[med]*
hyō'on-moji	表音文字	phonogram *[pr]*
hyōryō	秤量	weighing *{n} [eng]*
hyōryū	漂流	drift *{n} [navig]*
hyōryū'sen	漂流船	derelict ship *[navig]*
hyōsaku-san	氷酢酸	glacial acetic acid
hyōsen	標線	bench mark; marked line *[eng]*
hyō'setsu-kajū	氷雪荷重	ice load *[eng]*
hyō'setsu-shokubutsu	氷雪植物	cryophyte *[ecol]*
hyōshi	拍子	beat; measure *{n} [music]*
hyō'shiki	標識；ひょうしき	tag *{n} [comput]*
"	標識	sign; mark; indicator *{n} [navig]*
"	標識	labeling *{n} [nucleo]*
hyōshiki-genshi	標識原子	tagged atom *[phys]*
hyōshiki-zu'an	標識図案	logotype *[graph]*
hyō-shitsu	氷室	freezer; ice compartment *[mech-eng]*
hyōshō	氷晶	cryohydrate; ice crystal *[ch] [p-ch]*
hyōshō	氷床	ice sheet *[hyd]*
hyō'shōseki	氷晶石	cryolite *[miner]*
hyōshō'u	氷晶雨	ice crystal rain *{n} [meteor]*
hyō'taiseki	氷堆石	moraine *[geol]*
hyōtan	瓢箪；ひょうたん	gourd *[bot]*
hyōtei	標定	standardization *[ch]*
hyō'ten	氷点	freezing point *[p-ch]*
hyō'ten	標点	mark *{n} [navig]*
hyōten-hō	氷点法	cryoscopy *[an-ch]*
"	氷点法	freezing-point method *[p-ch]*
hyōten-kōka-zai	氷点降下剤	freezing-point depressant *[p-ch]*
hyōten-kyori	標点距離	gage length *[mech]*

hyōtō	氷島; ひょうとう	ice island *[ocean]*
hyōtō	氷塔	serac *{n}* *[hyd]*
hyōyū-denji'kai	漂遊電磁界	stray electromagnetic field *[elecmg]*
hyōyū-ji'soku	漂遊磁束	stray flux *[elecmg]*
hyōyū-yōryō	漂遊容量	stray capacitance *[electr]*
hyōzai-dōbutsu	表在動物	epifauna *[bio]*
hyōzai-seibutsu	表在生物	epibion *[bio]*
hyōzai-teisei-seibutsu	表在底生物	epibenthos *[bio]*
hyōzan	氷山	iceberg *[ocean]*
hyō-zetsu	氷舌; ひょうぜつ	glacial tongue *[hyd]*

I

i	位	place *{n} [math]*
i	胃	stomach *{n} [anat]*
ī; ii	良い	good *{adj}*
i'bari	铸ばり；鑄	fin; flash *{n} [eng] [met]*
ibitsu	歪；いびつ	distortion; warpage *[mech]*
ibo	疣；いぼ	wart *[med]*
ibo-dai	疣鯛；いぼだい	harvest fish; Psenopsis anomala *[v-zoo]*
ibo-inoshishi	疣猪；いぼいのしゝ	warthog (animal) *[v-zoo]*
ibuki-bōfū	伊吹防風	seseli *[bot]*
ibuki-jakō'sō	伊吹麝香草	Thymus quinquecostatus (herb) *[bot]*
ibushi	燻し	smoking; barbequeing *{n} [cook]*
ibusu	燻す	smoke; fumigate *{vb t}*
i'butsu	異物	foreign matter *[sci-t]*
ichi	一；壱	one *{n} [math]*
ichi	位置	location *{n} [comput]*
"	位置	position *{n} [comput] [math] [navig]*
ichiba	市場	market; marketplace *{n} [econ]*
ichi-bosū	位置母数	location parameter *[math]*
ichi-bu	一分；いちぶ	one "bu" (= 0.119 in.) *{n} [math]*
ichi-bu	一部	one part; one section *[math]*
ichi-chōsei	位置調整	justification *[comput] [pr]*
ichi-chōsetsu	位置調節	positioning *{n} [navig]*
ichi'dan-karyū	一段加硫	one-step cure *{n} [rub]*
ichi'ekisei-setchaku-zai	一液性接着剤	one-component adhesive *[poly-ch]*
ichi-enerugī	位置エネルギー	potential energy *[mech]*
ichi'enka-iō	一塩化硫黄	sulfur monochloride
ichi'enka-keiso	一塩化珪素	silicon monochloride
ichi'enka-tanso	一塩化炭素	carbon monochloride
ichi'enka-yōso	一塩化沃素	iodine monochloride
ichi'enki-san	一塩基酸	monobasic acid *[ch]*
ichi'gatsu	一月	January (month)
ichi'gen-haichi	一元配置	one-way layout *{n} [math]*
ichi'gime-jikan	位置決め時間	positioning time *{n} [comput]*
ichigi'teki	一義的	primarily *{adv}*
ichigo	苺；莓；いちご	strawberry (fruit) *[bot]*
ichi-gō	一合	one "go" (= 0.381 pint) *{n} [math]*
ichi-hyōji-kikō	位置表示機構	cursor *[comput]*
ichi'i	一位；水松；いちい	Japanese yew; Taxus cuspidata (tree) *[bot]*
ichi'ji	一次	primary *{n} [ch] [elec] [math]*

ichi'ji	一次	first-step *{n} [eng]*
"	一次	linear *{n} [math]*
"	一次	first-order *{n} [p-ch]*
ichi'ji	一時	one o'clock (time) *{n}*
"	一時	temporary; a time; transient *{n}*
ichiji-baiyō	一次培養	primary culture *[microbio]*
ichiji-denshi	一次電子	primary electron *[phys]*
ichiji-haiyō	一次胚葉	primary germ layer *[bot]*
ichiji-hannō	一次反応	first-order reaction *[p-ch]*
ichiji-hōtei'shiki	一次方程式	linear equation *[math]*
ichiji-kai	一時解	transient solution *[math]*
ichiji-kaiki-shiki	一次回帰式	linear regression equation *[math]*
ichi-jikan	一時間	one hour *{n}*
ichiji-kansū	一次関数	linear function *[math]*
ichiji-ki'oku	一次記憶	primary storage *[comput]*
ichiji-ki'oku	一時記憶	temporary memory *[comput]*
ichiji-kōzō	一次構造	primary structure *[aero-eng]*
ichijiku	無花果; いちじく	fig (fruit) *[bot]*
ichi'jiku-enshin	一軸延伸	monoaxial stretching; uniaxial stretching *{n} [poly-ch]*
ichi'jiku-haikō	一軸配向	uniaxial orientation *[crys]*
ichi'jiku'sei-kesshō	一軸性結晶	uniaxial crystal *[crys]*
ichiji-makisen	一次巻線	primary winding *{n} [elecmg]*
ichi'ji no	一次の	primary *{adj}*
ichi'ji no	一時の	transient *{adj}*
ichiji-sukēru	一次スケール	primary scale *[met]*
ichiji-taisha-sanbutsu	一次代謝産物	primary metabolite *[bioch]*
ichiji-teishi	一時停止	pause *{n} [comput]*
ichiji-teishi-hyō'shiki	一時停止標識	stopsign *[traffic]*
ichiji-teishi-meirei	一時停止命令	pause instruction *[comput]*
ichiji'teki na	一時的な	temporary *{adj} [math]*
ichiji('teki)-ki'oku	一時(的)記憶	temporary storage *[comput]*
ichiji'teki ni	一時的に	temporarily; for a while *{adv}*
ichi-jō	一丈	one "jō" (= 3.324 yds.) *{n} [math]*
ichi'jū-kō	一重項	singlet *[quant-mech]*
ichi-kankaku	位置感覚	position sense *[physio]*
ichi'matsu-moyō	市松模様	checkerboard pattern *[graph]*
ichi'monji-chō	一文字蝶	white admiral (butterfly) *[i-zoo]*
ichi'nen no	一年の	annual; yearly *{adj}*
ichi'nen-sei	一年生	annual *{n} [bot]*
ichi'nen'sei-shokubutsu	一年生植物	annual plant; therophyte *{n} [bot]*
ichi no en	位置の円	circle of position *{n} [navig]*

ichi no hosū	一の補数	one's complement *[comput]*
ichi'ran-hyō	一覧表	table *{n} [comput]*
"	一覧表	chart *{n} [sci-t]*
ichi'ren	一連	continuous; sequence *{n} [comput]*
ichi'ren-bangō	一連番号	serial number *[math]*
ichi-riron'dan ni tōka na	一理論段に等価な	height equivalent to a theoretical plate *[ch-eng]*
ichi'ritsu-kijun	一律基準	uniform standard *[phys]*
ichi'san-enki	一酸塩基	monoacidic base *[ch]*
ichi-suisan-hyō	位置推算表	ephemeris *[astron] [pr]*
ichi-tenkan	位置転換	transposition *{n} [comm]*
ichi-tenmon'gaku	位置天文学	position astronomy *[astron]*
ichi-toku'i-sei	位置特異性	position specificity *[bio]*
ichi-yoku-hō	一浴法	single-bath process *[ch]*
ichi'zuke suru	位置付けする	position; rank *{vb t}*
i'chō	胃腸	stomach and intestines *{n} [anat]*
ichō	銀杏；公孫樹	ginkgo (tree, nut) *[bot]*
i'chō-senshoku-sei	異調染色性	metachromasia *[ch]*
i'den	遺伝；遺傳；いでん	heredity; inheritance *[bio]*
iden-angō	遺伝暗号	genetic code; codon *[mol-bio]*
iden'gaku	遺伝学	genetics (the discipline) *[bio]*
iden'shi	遺伝子	gene *{n} [gen]*
idenshi-chizu	遺伝子地図	genetic map *[gen]*
idenshi-dō'nyū-dōbutsu	遺伝子導入動物	transgenic animal *[gen] [v-zoo]*
idenshi-gata	遺伝子型	genotype *{n} [gen]*
idenshi'gata-hindo	遺伝子型頻度	genotypic frequency *[gen]*
idenshi-ginkō	遺伝子銀行	gene bank *[gen]*
idenshi-gun	遺伝子群	genome *[gen]*
idenshi-kakusan	遺伝子拡散	gene flow *[gen]*
idenshi-kōgaku	遺伝子工学	gene(tic) engineering *{n} [gen]*
idenshi-kyūgen	遺伝子給源	gene pool *[gen]*
idenshi-sōsa	遺伝子操作	gene manipulation *{n} [gen]*
idenshi-zairyō	遺伝子材料	gene stock *[gen]*
iden'teki-fudō	遺伝的浮動	genetic drift *[gen]*
iden'teki-kumi'kae	遺伝的組換え	genetic recombination *[gen]*
ido	井戸	well; water-well *{n} [eng]*
ido	緯度	latitude *[geod]*
i'dō	異同	dissimilarity *{n}*
i'dō	移動；いどう	migration *[ch] [petrol]*
"	移動	shift *{n} [comput]*
"	移動	movement *{n}*
idō-do	移動度；易動度	mobility *[phys]*

idō-heikin	移動平均	moving average *[comput]*
idō-shō	移動床	moving bed *[ch-eng]*
idōshō'shiki-sesshoku-bunkai-hō	移動床式接触=分解法	moving-bed catalytic cracking process *[ch-eng]*
idō-sō	移動相	mobile phase *[ch-eng]*
idō-sō	移動層	moving bed *[ch-eng]*
idō suru	移動する	move *{vb t}* *[comput]*
idō-undō	移動運動	locomotion *[sci-t]*
ie	家	house *{n}* *[arch]*
ie-dani	家蜱；いえだに	house tick (insect) *[i-zoo]*
i'en	以遠	farther than; more distant *{n}* *[navig]*
i'en-ken	以遠権；以遠權	beyond right *{n}* *[legal]*
ie-suzume	家雀；いえすずめ	house sparrow (bird) *[v-zoo]*
iga	毬；いが	burr (as of chestnut) *{n}* *[bot]*
i-ga	衣蛾；いが	case-making clothes moth *[i-zoo]*
i'gai no-	以外の-	extra- *{prefix}*
i'gaku	医学；醫学	medicine (the discipline) *[med]*
i'gata	鋳型；いがた	mold; ingot mold *{n}* *[cer]* *[eng]*
"	鋳型	template *[mol-bio]*
i'gata-hannō	鋳型反応	template reaction *[met]*
i'gata-yu	鋳型油	casting oil *[mat]*
i-genshi	異原子	hetero atom *[ch]*
i'geta	井桁	parallel crosses *[graph]*
igi	異議	dissent *{n}* *[pat]*
igi	意義	meaning; significance *{n}* *[comm]*
igi-mōshi'tate	異議申立	statement of opposition *[pat]*
Igirisu	イギリス	England *[geogr]*
igo no-	以後の-	post- *{prefix}*
i(-gusa)	藺(草)	rush *{n}* *[bot]*
i-hada	鋳肌	cast(ing) surface *[eng]*
i'han	違反	breach; infraction *{n}* *[legal]*
i'heki	囲壁；圍壁	enclosure wall *[ind-eng]*
ihō	違法	illegal *{n}* *[legal]*
ihō-moji	違法文字	illegal character *[comput]*
ihō no	違法の	illegal *{adj}* *[legal]*
i'hō-sei	異方性	anistropy *[phys]*
i'in-kai	委員会；委員會	committee *{n}*
iji	維持	maintenance *[comput]* *[ind-eng]*
ijō	以上	at least; more than *{n}* *[math]*
ijō	異常；いじょう	abnormality; fault *{n}* *[comput]* *[math]*
"	異常	anomaly *[fl-mech]* *[opt]*
ijō-denpan	異常伝搬	anomalous propagation *[acous]*

ijō-kei	異常形	aberrant *{n} [bio]*
ijō-kōsen	異常光線	extraordinary (light) ray *[opt]*
ijō na	異常な	abnormal *{adj}*
ijū'tai	異重体	heterobar *[ch]*
ika	以下	at most; less than; fewer than *{n} [math]*
ika	烏賊; いか	cuttlefish; inkfish; squid *[i-zoo]*
i'kaku-setsugō'tai	異核接合体	heterokaryon *[mycol]*
ikanago	玉筋魚; いかなご	sand eel; lance (fish) *[v-zoo]*
ika ni	如何に	how *{adv}*
ika no-	以下の-	infra- *{prefix}*
i'kansoku	維管束	vascular bundle *[bot]*
i'kansoku-shokubutsu-rui	維管束植物類	vascular plants; Tracheophyta *[bot]*
ikari	錨; 碇	anchor *{n} [nav-arch]*
ikari'sō	碇草; いかりそう	barrenwort; Epimedium Herba *[bot]*
ika(-sayō)	異化(作用)	catabolism *[bioch]*
ike	池	pond *[geogr]*
ike-bana	生け花; 活け花	flower arrangement *{n}*
ikei-haigū	異形配偶	anisogamy *[bot]*
ikei-haigū'shi	異形配偶子	anisogamete *[bot]*
iken	意見	advice; opinion *{n}*
ike'su	生け洲; 生け簀	fish preserve; fishery *[ecol]*
iki	息	breath *{n} [physio]*
iki(-chi)	閾(値)	threshold *[math]*
"	閾(値)	limen *[psy]*
ikka	一価; 一價	monovalence; univalence *[ch]*
ikka-arukōru	一価アルコール	monohydric alcohol *[org-ch]*
ikka-genso	一価元素	monad *[ch]*
ikkaku	一角	narwhal(e); Monodon *[v-zoo]*
ikkaku'jū-za	一角獣座	Unicorn, the *[constel]*
ikka no	一価の	monovalent *{adj} [ch]*
ikka'sei no	一過性の	temporary *{adj} {n} [math]*
ikkatsu-shori	一括処理	batch processing *{n} [ind-eng]*
ikken	一間; いっけん	one "ken" (= 5.965 ft.) *{n} [math]*
ikken	一軒	one unit of house *{n} [arch]*
ikō	衣桁	clothing stand *[cl] [furn]*
ikō	移行	conversion *[comput]*
"	移行	transference *[math]*
"	移行	migration *[poly-ch]*
"	移行	transferral *{n}*
ikōkan'shiki-kagō'butsu	異項環式化合物	heterocyclic compound *[org-ch]*
i'komi	铸込み	pouring; casting *{n} [eng] [met]*
i'komi-seikei	铸込成形	casting; slip casting *{n} [cer] [eng]*

ikō suru	移項する	transpose *{vb t} [math] [stat]*
iku; yuku	行く	go *{vb}*
iku'byō	育苗；いくびょう	seedling culture *[agr]*
ikura	イクラ	salmon roe (individually) *[cook]*
i'kyoku-kagō'butsu	異極化合物	heteropolar compound *[ch]*
i'kyoku-ketsu'gō	異極結合	heteropolar bond *[ch]*
i'kyoku'kō	異極鉱；異極鑛	hemimorphite *[miner]*
i'kyoku-sei	異極性	heteropolarness *[elecmg]*
i'kyoku-zō	異極像	hemimorphic form *[crys]*
ima	今；いま	now *{adv}*
i'ma	居間；いま	living room; sitting room *[arch]*
Imari-yaki	伊万里焼	Imari porcelain *[cer]*
imi	意味；いみ	meaning; sense *{n} [comm]*
i'min	移民	emigrant; immigrant *[legal]*
imi no aru sa	意味のある鎖	sense strand *[mol-bio]*
imi-ron	意味論	semantics *[comm] [comput] [ling]*
imi suru	意味する	imply; mean *{vb} [psy]*
imo	芋	potato (pl -toes) (vegetable) *[bot]*
imo-gai	芋貝	cone shell (shellfish) *[i-zoo]*
imo-gayu	芋粥；いもがゆ	rice porridge with sweet potato *[cook]*
imo-mushi	芋虫	caterpillar (green); hornworm *[i-zoo]*
imono'ba	鋳物場	foundry *[eng]*
imono'yō-jushi	鋳物用樹脂	foundry resin *[poly-ch]*
imori	井守；いもり	newt; eft; water lizard *[v-zoo]*
ina'bikari	稲光	lightning *{n} [geophys]*
inada	鰍；いなだ	yellowtail (young fish of) *[i-zoo]*
inago	蝗；いなご	locust *[i-zoo]*
i'nai-shōka	胃内消化	gastric digestion *[physio]*
inari-zushi	稲荷鮨；いなりずし	flavored rice in fried tofu (q.v.) pocket *[cook]*
ina'zuma	稲妻；いなずま	lightning *{n} [geophys]*
ina'zuma-yoko'bai	稲妻横這；いな=ずまよこばい	zigzag-striped leafhopper (insect) *[i-zoo]*
inchi	吋；インチ	inch (unit of length) *{n} [mech]*
in-denka	陰電荷	negative charge *[elec]*
in-denshi	陰電子	negatron *[phys]*
Indo-sokei	印度素馨	frangipani; Plumeria rubra *[bot]*
Indo-yō	印度洋；インド洋	Indian Ocean *[geogr]*
ine	稲；稲；いね	rice plant *[bot]*
in'ei	陰影	shadowing; shading *{n} [graph] [opt]*
inga	印画	print *{n} [photo]*
inga-ritsu	因果律	causality (causation), law of *[phys]*

inga(-yō)shi	印画(用)紙	photographic paper; printing paper *[paper] [pr]*
inga-zō	陰画像	negative image; negative picture *[photo]*
i'nin	委任; いにん	authorization *[legal]*
i'nin'jō	委任状	letter of attorney *[pat]*
in-ion	陰イオン	anion; negative ion *[ch] [phys]*
inji	印字	print; printing *{n} [comput]*
inji-shutsu'ryoku	印字出力	printout *{n} [comput]*
inji-sōchi	印字装置	printer *[comput]*
injūmu	インジウム	indium (element)
inka-den'atsu	印加電圧	applied voltage; impressed voltage *[elec]*
in'kansū-bibun-hō	陰関数微分法	implicit differentiation *[calc]*
inka-ten	引火点	flash point *[ch]*
inkei	陰茎	penis (pl penes, penises) *{n} [anat]*
inko	鸚哥; いんこ	cockatoo; macaw; parakeet (bird) *[v-zoo]*
inkō	咽喉; いんこう	throat *{n} [anat]*
inku-chaku'niku-sei	インク着肉性	ink-attachment property *[pr]*
in-kyoku	陰極; いんきょく	cathode; negative electrode *[elec]*
inkyoku-denryū	陰極電流	cathode current *[elec]*
inkyoku-eki	陰極液	catholyte *[ch]*
inkyoku-sen	陰極線	cathode rays *[electr]*
inkyoku'sen-kan	陰極線管	cathode-ray tube; CRT *[electr]*
inmen	陰面; 隠面	hidden surface *[comput]*
in-nyō	辶繞: 辶; いん=にょう	kanji (Chinese character) radical denoting to go *[pr]*
inochi	命	life (pl lives) *[bio]*
inokozuchi	牛膝; いのこずち	Achyranthes japonica *[bot]*
inondo	莳蘿; いのんど	dill *[bot] [cook]*
inoshishi	猪; いのしし	wild boar (animal) *[v-zoo]*
inpei-ryoku	隠蔽力	covering power; hiding power *[eng]*
inpei suru	隠蔽する	conceal; hide *{vb t}*
inpīdansu	インピーダンス	impedance; electrical impedance *[elec]*
in'retsu-zuyo-sa	引裂強さ	tearing strength *[mech]*
inrō	印籠; いんろう	medicine case; pillbox *{n}* (antique)
inryō	飲料	beverage(s); drink(s) *[cook]*
in-ryoku	引力	attractive force; gravity *[mech] [phys]*
in'ryoku no hōsoku	引力の法則	gravitation, law of *[mech]*
inryō-sui	飲料水	drinking water *[cook] [hyd] [sci-t]*
in'satsu	印刷; いんさつ	print; printing *{n} [comput] [pr]*
insatsu-denshin-ki	印刷電信機	teleprinter *[comm]*
insatsu-gyō	印刷行	print line *{n} [comput] [pr]*
insatsu'jutsu	印刷術	printing (the technology) *[comm] [graph]*

insatsu-kairo	印刷回路	printed circuit *[electr]*
insatsu-ki	印刷機	printing press *[pr]*
insatsu-kikō	印刷機構	print station *[comput]*
insatsu-senmei-do	印刷鮮明度	print contrast ratio *[graph]*
insatsu-shutsu'ryoku	印刷出力	printout *{n} [comput]*
in'satsu suru	印刷する	print *{vb t} [comput] [pr]*
insatsu-tekisei	印刷適性	printability *[pr]*
in-sayō	飲作用	pinocytosis (pl -toses) *[cyt]*
insei-shokubutsu	陰生植物	shade plant *[bot]*
inseki	隕石；いんせき	meteorite *[astron]*
inseki-kō	隕石孔	meteorite crater *[astron]*
insen	陰線	hidden line *[comput]*
insha	印写	imaging *{n} [graph]*
inshi	因子	factor *{n} [gen] [math]*
insho	印書	print *{n} [comput] [graph] [pr]*
inshō	印章	seal; stamp *{n} [graph]*
inshō	印象	impression *[dent] [graph]*
insho-gyō	印書行	print line *{n} [comput] [pr]*
in'shoku'butsu	飲食物	eatables and drinkables; food and drink *[cook]*
in'shoku no	飲食の	dietary *{adj} [cook] [med]*
inshō-zai	印象剤	impression agent *[graph]*
inshurin	インシュリン	insulin *[bioch]*
insū	因数；因數	factor *{n} [math]*
insū-bunkai	因数分解	factorization; factoring *{n} [math]*
insū-teiri	因数定理	factor theorem *[math]*
intāchenji	インターチエンジ	interchange *{n} [traffic]*
in'taizō'butsu	隠退蔵物；隱退藏物	cache *{n} [comput]*
in'tetsu	隕鉄	meteoric iron *[astron]*
"	隕鉄	siderite *[miner]*
intō	咽頭	pharynx (pl pharynges, pharynxes) *[anat]*
intō no	咽頭の	pharyngeal *{adj} [anat]*
inu	犬	dog (animal) *{n} [v-zoo]*
inu-hakka	犬薄荷	catnip; Nepata cataria *[bot]*
inu-udo	犬独活；いぬうど	angelica (herb) *[bot]*
in'yō	引用	citation; quotation; reference *[pr]*
in'yō-bunken	引用文献	cited reference *[pr]*
in'yō-fu	引用符：「」，" "	quotation marks *[comput] [gram] [pr]*
in'yō suru	引用する	cite *{vb} [pr]*
in'zei	印税	royalty *[legal]*
iō	硫黄；いおう	sulfur (element)
iō'ka	硫黄華	flowers of sulfur *[pharm]*

iō-kakyō	硫黄架橋	sulfur bridge; sulfur cross-linking *[rub]*
ion	イオン	ion *[ch]*
ion-dokuritsu-idō no hōsoku	イオン独立移動の=法則	independent ionic mobilities, law of *[p-ch] [phys]*
ion-genshi'ka	イオン原子価	electrovalence *[p-ch]*
ion-hankei	イオン半径	ionic radius *[p-ch]*
ion-ka	イオン価	valence *[ch]*
ion-ketsu'gō	イオン結合	ionic bond(ing) *{n} [p-ch]*
ion-kōkan-jushi	イオン交換樹脂	ion-exchange resin *[poly-ch]*
ion-kyōdo	イオン強度	ionic strength *[p-ch]*
i'on-sei	異温性	heterothermy *[physio]*
ion-tsui	イオン対	ion pair *[nucleo]*
iō-saikin	硫黄細菌	Thiobacillus *[microbio]*
ippan-jōtai	一般状態	general condition; general state *{n}*
ippan-kai	一般解	general solution *[math]*
ippan-mei	一般名	generic name (for medicine) *[pharm]*
ippan-yō	一般用	general purpose *[ind-eng]*
ippan-zu	一般図	general drawing; general view *[graph]*
ippon-seki	一本石	monolith *[mat]*
ippō-tsūkō-ro	一方通行路	one-way road *[traffic]*
ippun	一分; いっぷん	one minute *{n} [mech]*
i'rei	異例	unusual example *{n}*
ire'ko	入れ子	nest; nesting *{n} [comput]*
ire'mono	入れ物	container *[eng]*
ireru	入れる	insert; place into *{vb t}*
iri	入; いり	setting; set *{n} [astron]*
iri'e	入江	inlet; sound *{n} [geogr]*
iri'e-game	入江亀	terrapin (tortoise) *[v-zoo]*
iri'guchi; iri'kuchi	入口	entry *{n} [arch] [comput]*
iri'guchi	入口	inlet *{n}*
iri'guchi-jōken	入口条件	entry condition *[comput]*
iri'guchi no ma	入口の間	foyer; vestibule *[arch]*
iri'guchi-ten	入口点	entry point; entrance *[comput]*
iri-sen	入り線	incoming wire *[elec]*
iro	色	color *{n} [opt]*
iro-awase	色合せ	color matching *{n} [opt] [poly-ch]*
iro-buchi	色縁	color fringe *[opt]*
iro-chi'kaku	色知覚	color perception *[physio]*
irodoru	彩る; 色取る	color; paint; dye *{vb t}*
iro-kankaku	色感覚	color sensation *[physio]*
iro-kenrō-do	色堅牢度	colorfastness *{n} [eng]*
iro-kūkan	色空間	color space *[opt]*

iro-mei	色名	color name *[opt]*
iro-mura	色斑；いろむら	mottling *{n} [opt]*
iro-nagare	色流れ	dye bleeding *{n} [tex]*
iro-ondo	色温度	color temperature *[stat-mech]*
iro-rittai	色立体	color solid *[opt] [poly-ch]*
iro-shisū	色指数	color index *[astron]*
iro-shūsa	色収差；色收差	chromatic aberration *[opt]*
iru	入る	enter *{vb}*
iru	居る	be present *{vb i}*
iru	煎る；炒る；いる	roast *{vb t} [cook]*
i'rui	衣類	clothes; clothing *{n} [cl]*
iruka	海豚；いるか	dolphin (mammal) *[v-zoo]*
i'ryō	医療；醫療	medical treatment *[med]*
iryō-hojo'in	医療補助員	paramedic (a person) *[med]*
i'ryū	移流	advection *[meteor] [ocean]*
i'ryūdō-sei	易流動性	easy-flowing property *[fl-mech]*
i'san	遺産	estate; inheritance *[legal]*
i'san-hōsoku	遺産法則	legacy, law of *[math]*
ise-ebi	伊勢蝦；いせえび	lobster (spiny) *[i-zoo]*
i'sei	異性；いせい	opposite sex *[bio]*
"	異性	isomerism *[ch]*
isei'ka	異性化	isomerization *[ch]*
isei'ka-jūgō	異性化重合	isomerization polymerization *[poly-ch]*
isei'ka'tō	異性化糖	isomerized sugar *[bioch]*
isei'tai	異性体	isomer *[ch]*
i'sen	移染	migration *[tex]*
i'sen-sei	移染性	migrating property *[tex]*
i'sen-sei	異染性	metachromasia *[ch]*
i'sha	医者；醫者	doctor; physician (a person) *{n} [med]*
ishi	石	stone *{n} [geol]*
"	石	rock *{n} [petr]*
i'shi	医師	doctor (a person) *{n} [med]*
ishi	意思	intention *[legal] [psy]*
ishi-bashi	石橋	stone bridge *[civ-eng]*
ishi-dōrō	石燈籠	stone lantern *[arch]*
ishi-kettei	意思決定	decision-making *{n} [psy]*
i'shiki	意識	consciousness *[psy]*
ishi'kiri'ba	石切場	quarry; stone quarry *{n} [eng]*
ishi'ku	石工	stonemason (a person) *[civ-eng]*
ishimochi	石首魚；石持；いしもち	croaker; Argyrosomus argentatus (fish) *[v-zoo]*
ishin-kō	囲心腔；圍心腔	pericardial cavity *[anat]*

i'shitsu	異質；いしつ	heterogeneity *{n} [bio] [gen]*
i'shitsu-dōkei	異質同形	homomorphy *[bio]*
"	異質同形	homeomorphism *[crys]*
i'shitsu-ken	異質圏；異質圈	heterosphere *[meteor]*
i'shitsu-setsu'gō	異質接合	heterojunction *[electr]*
ishi'wata	石綿	asbestos *[miner]*
i'shō	意匠	design *{n} [graph] [pat]*
i'shoku	移植；いしょく	grafting; transplantation *[bio] [bot]*
ishoku-hen	移植片	implant *{n} [med]*
ishoku-sei	移植性	portability *[comput]*
ishoku'tai	移植体	transplant *{n} [med]*
i'shōsei-sei	易焼成性	burnability *[cer]*
ishō-tōroku-negai	意匠登録願	application for design registration *[pat]*
i'shu	異種；いしゅ	different variety *[bot]*
i'shū	異臭	off odor; off flavor *{n} [cook]*
ishu-hakkō(-sei)	異種発酵(性)	heterofermentative(ness) *[microbio]*
ishu-i'shoku	異種移植	xenotransplantation *[immun]*
ishu'ishoku-hen	異種移植片	xenograft *[immun]*
ishu-kōgen	異種抗原	xenoantigen *[immun]*
ishu-kōhai	異種交配	hybridization; interbreeding *{n} [gen]*
i'shuku	萎縮	atrophy *{n} [med]*
ishu-nijū-rasen	異種二重螺旋	heteroduplex *[gen]*
ishu-setsugō'tai	異種接合体	heterokaryon *[mycol]*
iso	磯	beach; shore *{n} [geol]*
isō	位相；いそう	phase *{n} [astron] [elec] [math] [phys]*
isō	移送	sending; transfer; transport(ation) *{n} [trans]*
"	移送	migration *{n}*
isō	移相	phase shift *[electr]*
isō-ben'betsu	位相弁別；位相辧別	phase discrimination *[electr]*
iso-chidori	磯千鳥；いそちどり	beach plover (bird) *[v-zoo]*
iso-gani	磯蟹；いそがに	sand crab (crustacean) *[i-zoo]*
isogaseru	急がせる	hurry; expedite *{vb t}*
iso'ginchaku	磯巾着	sea anemone (marine coelenterate) *[i-zoo]*
isogu	急ぐ	hurry *{vb i}*
isō-henchō	位相変調	phase (or pulse) modulation *[electr]*
isō-hen'i	位相偏移	phase deviation *[comm]*
isō-kaku	位相角	phase angle *[phys]*
isō-ki	移相器；いそうき	phase shifter *[elec]*
isō-kika'gaku	位相幾何学	topology *[math]*
isō-kiri'kae-kanshō-kei	位相切換干渉計	phase-switching interferometer *[opt]*
i'son-kankei	依存関係	dependence *[math]*

isō'sa-kenbi'kyō 位相差顕微鏡 phase-contrast microscope *[opt]*
isō-sokudo 位相速度 phase velocity *[phys]*
isō-sūgaku 位相数学 topology *[math]*
isō suru 移送する move; send *{vb} [trans]*
isō-yugami 位相歪み phase distortion *[comput]*
issaku-jitsu; ototoi 一昨日 day before yesterday *{n}*
issan 逸散；いっさん dissipation *[phys]*
issanka'butsu 一酸化物 monoxide *[ch]*
issanka-chisso 一酸化窒素 nitrogen monoxide; nitric oxide
issanka-iō 一酸化硫黄 sulfur monoxide
issanka-namari 一酸化鉛 lead monoxide; litharge *[ch]*
issanka-ni'chisso 一酸化二窒素 dinitrogen monoxide
issanka-ni'iō 一酸化二硫黄 disulfur monoxide
issan-sei 逸散性 fugacity *[ch]*
isseiki 一世紀；いっせいき one century *{n}*
isshaku 一尺 one "shaku" (= 0.994 ft.) *{n} [math]*
isshi-tayū 一雌多雄 polyandry *[bio]*
isshō 一升；いっしょう one "shō" (= 0.447 US gallon) *{n} [math]*
issun 一寸 one "sun" (= 1.193 in.) *{n} [math]*
isu 椅子；いす chair *{n} [furn]*
isū 位数 digit *[comput] [math]*
i'suikan-kei 胃水管系 gastrovascular system *[i-zoo]*
isuka 鶍；いすか crossbeak; crossbill (bird) *[v-zoo]*
isū-sei 異数性 heteroploidy *[gen]*
ita 板；いた plate *{n} [eng]*
ita-bane 板発条；板ばね leaf spring *{n} [mech-eng]*
itachi 鼬；いたち weasel (animal) *[v-zoo]*
ita-gami 板紙 cardboard; paperboard *[mat] [paper]*
ita-ishi 板石 flagstone *[geol]*
i'taku 委託；いたく consignment; charge; commission; trust *[econ]*
itaku'sha 委託者 client; requestor (a person) *{n} [econ]*
itameru 炒める；煠める frizzle; fry *{vb t} [cook]*
itameru 痛める；傷める damage; hurt *{vb t}*
itami 痛み；いたみ pain *{n} [physio]*
itamu 痛む hurt *{vb} [physio]*
itamu 傷む spoil; be damaged *{vb} [ind-eng]*
ita-sa 痛さ painfulness *[physio]*
ita-ware 板割れ；いたわれ sheet cracking *{n} [eng]*
itchi 一致；いっち coincidence; matching *{n} [math]*
itchi-suitei'ryō 一致推定量 consistent estimator *[stat]*
itchi suru 一致する agree; match *{vb}*

itchō	一兆	billion (British system) *{n} [math]*
"	一兆	trillion (American system) *{n} [math]*
iten	移転；移轉	transfer *{n}*
ite-za	射手座	Sagittarium; the Archer *[constel]*
ito	糸；絲	string; thread; yarn *{n} [tex]*
ito	意図；意圖	intention *[psy]*
ito-gire	糸切れ	yarn breakage *[tex]*
ito-guruma	糸車	spinning wheel *[tex]*
ito'hiki	糸引き	cobwebbing (of paint) *{n} [org-ch]*
"	糸引き	stringiness *{n} [adhes] [poly-ch]*
ito'hiki-aji	糸引鯵	cobblerfish; Alectis ciliaris *[v-zoo]*
ito-kuzu	糸屑；いとくず	lint *{n} [tex]*
ito'mari	糸まり	coiling *{n} [ch] [tex]*
itomari'jō-bunshi	糸毬状分子	coiled molecule *[ch]*
ito-michi	糸道	thread guide *[tex]*
ito-mura	糸斑；いとむら	yarn unevenness *[poly-ch] [tex]*
ito-mushi	糸蒸し	yarn steaming *[tex]*
ito-neri	糸練り	yarn scouring *[tex]*
ito-sabi	糸錆；糸さび	filiform corrosion *[met]*
ito-takak(u)kei	糸多角形	funicular polygon *[mech]*
itoyo	糸魚；棘魚；いとよ	three-spined stickleback; Gaseterosteus aculeatus (fish) *[v-zoo]*
ito'zome-ori'mono	糸染め織物	yarn-dyed fabric *[tex]*
itsu	何時；いつ	when *{adv} {n}*
itsu'datsu-ken	逸脱圏；逸脱圏	exosphere *[meteor]*
itsudemo	何時でも	anytime *{adv}*
itsumo	何時も；いつも	always *{adv}*
itsu'ryū	溢流	overflow *{n} [civ-eng]*
itsutsu'go	五つ子；いつつご	quintuplets *[bio]*
itsuwari	偽り	lie; falsehood; deceit *{n}*
itsuwari no	偽りの	false; fake *{adj}*
itsuwari no-	偽りの-	pseudo- *{prefix}*
ittai'teki ni	一体的に	integrally; as a single body *{adv}*
itterubyūmu	イッテルビウム	ytterbium (element)
ittoryūmu	イットリウム	yttrium (element)
iwa	岩	boulder; large rock *[geol]*
iwa-kan	違和感	sense of incompatibility *[psy]*
iwana	岩魚；いわな	char(r) (fish) *[v-zoo]*
iwashi	鰯；いわし	sardine (fish) *[v-zoo]*
iwashi-yu	鰯油	sardine oil *[mat]*
iwa-tsubame	岩燕	house marten (bird) *[v-zoo]*
i'yaku	医薬；醫藥；いやく	medicament *[med]*

i'yaku-bugai'hin	医薬部外品	quasi drug *{n} [med]*
iyaku'hin	医薬品	pharmaceuticals *{n} [pharm]*
iyaku'hin-kagaku	医薬品化学	medicinal chemistry *[ch]*
i'yaku suru	意訳する	translate for meaning; translate freely *{vb} [comm]*
iya'mi	嫌味；いやみ	bad flavor; disagreeableness *[cook]*
iyū-gōkin	易融合金	fusible alloy *[met]*
izari-uo	躄魚；いざりうお	frogfish (fish) *[v-zoo]*
i'zen no-	以前の-	ante-; pre- *{prefix}*
izumi	泉；いずみ	fountain; spring *{n} [hyd]*
i-zuna	鋳砂	casting sand *[eng]*
i'zutsu-kiso	井筒基礎	well foundation *[civ-eng]*

J

ja'guchi	蛇口；じゃぐち	faucet *[eng]*
jakō	麝香；じゃこう	musk *[physio]*
jakō-jika	麝香鹿	musk deer (animal) *[v-zoo]*
jakō-neko	麝香猫	cacomistle; civet (animal) *[v-zoo]*
jakō-ushi	麝香牛	musk ox (animal) *[v-zoo]*
jaku-denkai'shitsu	弱電解質	weak electrolyte *[p-ch]*
jaku'den-kiki	弱電機器	light electrical appliance *[elec]*
jaku-enki	弱塩基	weak base *[ch]*
jaku'ji-sei	弱磁性	feeble magnetism *[phys]*
jaku-san	弱酸；じゃくさん	weak acid *[ch]*
jama-ban	邪魔板	baffle board *[eng]*
jamon'gan	蛇紋岩	serpentinite *[petr]*
jamon'seki	蛇紋石	serpentine *[miner]*
ja'no'me-chō	蛇の目蝶	satyr (butterfly) *[i-zoo]*
jari	砂利；じゃり	gravel *[geol]*
jetto-kiryū	ジエット気流	jet stream *{n}* *[meteor]*
ji	字	character *[gram]*
ji	地；じ	ground *{n}* *[geol]* *[tex]*
"	地；じ	matrix (pl matrices, matrixes) *[met]*
ji	痔；じ	hemorrhoid *[med]*
ji-	自-	self- *{prefix}*
ji-	次-	next-; following-; sub- *{prefix}*
-ji	-時	-o'clock *{suffix}* *[mech]*
ji'a-	次亜-；次亞-	hypo--ous; hypo--ite *{suffix}* *[ch]*
jia'enso-san	次亜塩素酸	hypochlorous acid
jia'enso'san-en	次亜塩素酸塩	hypochlorite *[ch]*
jia'harogen'san-en	次亜ハロゲン酸塩	hypohalogenous acid salt *[ch]*
ji'ai	地合(い)	formation *[paper]*
jia'rinsan	次亜燐酸	hypophosphorous acid
jia'rinsan-en	次亜燐酸塩	hypophosphite *[ch]*
jia'rinsan-natoryūmu	次亜燐酸ナトリウム	sodium hypophosphite
jia'ryūsan	次亜硫酸	hyposulfurous acid
jia'ryūsan-en	次亜硫酸塩	hyposulfite *[ch]*
jia'shōsan	次亜硝酸	hyponitrous acid
jia'shōsan-en	次亜硝酸塩	hyponitrite *[ch]*
jia'shūso-san	次亜臭素酸	hypobromous acid
jia'shūso'san-en	次亜臭素酸塩	hypobromite *[ch]*
jia'yōso-san	次亜沃素酸	hypoiodous acid
jiba	磁場；じば	magnetic field *[elecmg]*

jiban	地盤	ground base *{n} [civ-eng]*
ji'biki	字引	dictionary *[pr]*
ji'biki-ami	地引網；地曳網	seine net *{n} [eng]*
jibun	自分	one's self *{n}*
ji-bun'katsu	時分割	time sharing *{n} [comput]*
ji'chaku	自着	autohesion *[org-ch]*
jidō	自動；じどう	automatic(ness) *{n} [eng]*
jidō-bakari	自動秤	automatic scale *[eng]*
jidō-dēta-shori	自動データ処理	automatic data processing *{n} [comp]*
jidō'gaku	児童学；兒童學	pedology *[med]*
jidō-jisho	自動辞書	automatic dictionary *[comput]*
jidō'ka	自動化	automation *[comput]*
jidō-kensa	自動検査	automatic check *[comput]*
jidō-kikai	自動機械	automaton *[comput]*
jidō-kōen	児童公園	children's park *[civ-eng]*
jidō-ningyō	自動人形	automaton *[eng]*
jidō no	自動の	auto- *{adj}*
jidō-ondo'chōsetsu-sōchi	自動温度調節装置	thermostat *[eng]*
jidō-ri'toku-seigyo	自動利得制御	automatic gain control *[electr]*
jidō-ritsu	自同律	identity, law of *[logic]*
jidō-ryōmen-fukusha-ki	自動両面複写機	automatic two-side copier *[opt]*
jidō-sai'shidō	自動再始動	automatic restart *[comput]*
jidō-seigyo	自動制御	automatic control *[cont-sys]*
jidō-seizu-ki	自動製図機	automatic drawing instrument *[graph]*
jidō'sha	自動車；じどうしゃ	automobile; car *[mech-eng]*
jidōsha-denwa	自動車電話	car telephone *[comm]*
jidōsha-gasorin	自動車ガソリン	motor gasoline *[mat]*
jidōsha-kankei no	自動車関係の	automotive-related *{adj} [mech-eng]*
jidōsha-sangyō	自動車産業	automotive industry *[mech-eng]*
ji-dōshi	自動詞	intransitive verb *[gram]*
jidō-sōda-sōchi	自動操舵装置	autopilot (of a ship) *{n} [navig]*
jidō-sōjū-sōchi	自動操従装置	autopilot (of a plane) *{n} [navig]*
jidō-sōsa'ka	自動操作化	automation *[eng]*
jidō-suishin no	自動推進の	automotive *{adj} [mech-eng]*
jidō-teishi	自動停止	automatic stop *[comput]*
ji-gane	地金；じがね	ground metal *{n} [met]*
jigen	字源	word origin *[comm]*
jigen	次元	dimension(s) *[math] [phys]*
jigen-bun'retsu-zukei	次元分裂図形	fractal *{n} [math]*
jigen-shinkan	時限信管	time fuse *{n} [eng]*
jigi	字義	word meaning *{n} [comm]*
ji-giri	地霧	ground fog *[meteor]*

jigo-bunseki	事後分析	postmortem *[comput]*
jigu	治具；じぐ	jig *{n} [mech-eng]*
jigyō	事業	business; enterprise; undertaking *[econ]*
jihada-men	地肌面	ground surface *[geol]*
ji'hatsu-bun'kyoku	自発分極	spontaneous polarization *[elec]*
ji'hatsu-jika	自発磁化	spontaneous magnetization *[elecmg]*
ji-heki	磁壁	magnetic wall *[phys]*
ji-hoku	磁北	magnetic north *[geophys]*
ji'i	磁位	magnetic potential *[elecmg]*
ji'ichi-sokutei-sōchi	自位置測定装置	loran *[navig]*
ji'jitsu	事実；事實；じじつ	actuality; fact *{n}*
ji'jitsu no	事実の	factual *{adj} [math]*
ji'jō	二乗；二乘；自乗	square *{n} [math]*
jijō	事情	circumstances; situation *{n}*
jika	時価；時價	current price; market value *[econ]*
jika	磁化	magnetization *[elecmg]*
ji-kai	磁界	magnetic field *[elecmg]*
ji-kaku	時角	hour angle *[astron]*
ji'kaku-shōjō	自覚症状	subjective symptom *[med]*
jikan	時間；じかん	hour; duration *{n} [mech]*
"	時間	time *{n} [phys]*
-jikan	-時間	-hour(s) *{suffix} [mech]*
jikan-bōchō	時間膨張	time dilation *[phys]*
jikan ga aru	時間がある	have time
jikan ga nai	時間がない	have no time
jikan-hō'shutsu	時間放出	time-release; timed release *{n}*
jikan-me'mori	時間目盛	time scale *[comput]*
jikan no tan'i	時間の単位	unit of time *[math] [phys]*
jikan-seibutsu'gaku	時間生物学	chronobiology *[bio]*
jikan-sen'yū-ritsu	時間占有率	time occupancy *{n}*
jikan-tai	時間帯；時間帶	time zone *[astron]*
jika-ritsu	磁化率	magnetic susceptibility *[elecmg]*
jika-tabi	地下足袋	workman's rubber tabi (q.v.) *[cl]*
jika-yōi'jiku	磁化容易軸	axis of easy magnetization *[elecmg]*
jikei	字形	typeface; face *{n} [comput] [graph] [pr]*
jikei	自形	automorphic; euhedral; idiomorphic *[petr]*
jikei-men	自形面	idiomorphic surface *[math]*
jiken	時圏；時圈	hour circle *[astron]*
jiki	時期	season; time *{n} [astron]*
jiki	磁気；磁氣；じき	magnetism *[phys]*
jiki	磁器；磁器	china; porcelain *[cer] [mat]*
jiki-arashi	磁気嵐	magnetic storm *[geophys]*

jiki-chi'en-sen	磁気遅延線	magnetic delay line *[electr]*
jiki-denki	磁気電気	magnetoelectricity *[elecmg]*
jiki'denki-kōka	磁気電気効果	magnetoelectric effect *[elecmg]*
jiki-fujō	磁気浮上	magnetic levitation *[mech-eng]*
jiki'gaku	磁気学	magnetics (the discipline) *[elecmg]*
jiki-haku'maku	磁気薄膜	magnetic thin film *[sol-st]*
jiki-henkaku	磁気偏角	magnetic declination *[geophys]*
jiki-hentai	磁気変態	magnetic transformation *[elecmg]*
jiki-hizumi	磁気歪；じきひずみ	magnetic strain; magnetostriction *[elecmg]*
jiki-hokkyoku	磁気北極	magnetic north *[geophys]*
jiki-in'satsu	磁気印刷	magnetic-ink printing *{n} [pr]*
jiki-kaiten-kōka	磁気回転効果	gyromagnetic effect *[elecmg]*
jiki-kaku'undōryō-hi	磁気角運動量比	magnetomechanical ratio *[phys]*
jiki-ken	磁気圏；磁氣圈	magnetosphere *[geophys]*
jiki'ken-bi	磁気圏尾	magnetotail *[geophys]*
jiki'ken-kaimen; (-sen)	磁気圏界面；(-線)	magnetopause *[geophys]*
jiki-kōgaku	磁気光学	magnetooptics *[opt]*
jiki'kōgaku no	磁気光学の	magnetooptical *{adj} [opt]*
jiki-kondensa	磁器コンデンサ	ceramic capacitor *[elec]*
jiki-kyōmei	磁気共鳴	magnetic resonance *[phys]*
jiki-kyōmei-eizō	磁気共鳴映像	magnetic-resonance imaging; MRI *[phys]*
jiki'more-hen'atsu-ki	磁気漏れ変圧器	leakage transformer *[elecmg]*
jiki no	磁気の	magnetic *{adj} [phys]*
jiki-ryō	磁気量	quantity of magnetism *[phys]*
jiki-sekidō	磁気赤道	magnetic equator *[geophys]*
jiki-shigo'sen	磁気子午線	magnetic meridian *[geophys]*
jiki-tan'kyoku	磁気単極	magnetic monopole *[elecmg]*
jiki-teikō	磁気抵抗	reluctance *{n} [elecmg]*
jiki-yūdō	磁気誘導	magnetic induction *[elecmg]*
jikkai	実開	Utility Model Early Disclosure (abbrev in the Japanese) *[pat]*
jikkaku-kei	十角形	decagon *[math]*
jikka no; jukka no	十価の	decavalent *{adj} [ch]*
jikken	実験；じっけん	experiment *{n} [sci-t]*
jikken-kekka	実験結果	experiment result *[ch]*
jikken (o) suru	実験(を)する	experiment *{vb} [ch]*
jikken-shiki	実験式；實驗式	empirical formula *[ch]*
jikken-shitsu	実験室	laboratory *[ch]*
jikkō	実公	Utility Model Gazette (abbrev) *[pat]*
jikkō	実行	execution; run *{n} [comput]*
jikkō'bun	実行文	executable statement *[comput]*
jikkō-chi	実効値	effective current *[elec]*

jikkō-chi	実効値; じっこうち	root-mean-square (rms) value *[elec]*
jikkō-den'atsu	実効電圧; 實効電壓	effective voltage *[elec]*
jikkō-denka	実効電荷	effective charge; net charge *[elec]*
jikkō-jikan	実行時間	execution time; running time *[comput]*
jikkō-kanō-sei	実行可能性	viability; ability to execute *{n}*
jikkō-menseki	実効面積	effective area *[mech]*
jikkō saseru	実行させる	run *{vb t} [comput]*
jikkō suru	実行する	execute *{vb t} [comput]*
jikkō'teki na	実効的な	effective *{adj}*
jiko	事故; じこ	breakdown; fault *{n} [elec]*
"	事故	failure *[elec] [mech]*
"	事故	trouble *{n} [ind-eng]*
"	事故	accident *{n} [ind-eng] [traffic]*
jikō; mimi-aka	耳垢	earwax; cerumen *[physio]*
jikō	事項	article(s); matter(s); fact(s) *{n}*
jikō	時効; じこう	aging *{n} [eng]*
"	時効	period of limitation *[pat]*
jiko-bunkai	自己分解	autodegradation *[microbio]*
jiko-chakka	自己着火	autoignition *[phys]*
jikō-fū; chikō-fū	地衡風	geostrophic wind *[meteor]*
jikō-kōka	時効効果	aging effect *[poly-ch]*
jikō-kōka	時効硬化	age hardening *{n} [met]*
jiko'kōka-setchaku-zai	自己硬化接着剤	self-curing adhesive *[poly-ch]*
ji'koku	時刻; じこく	time instant; time; hour *{n} [phys]*
ji'koku-hyō	時刻表	time table *[trans]*
jiko-men'eki	自己免疫	autoimmunity *[immun]*
jiko-men'eki'ka	自己免疫化	autoimmunization *[immun]*
jiko-nenchaku	自己粘着	autohesion *[rub]*
jikō-rekka	時効劣化	aging degradation *[eng]*
jiko-ritsu	事故率	accident rate *[ind-eng]*
jikō-sei	自硬性	self-hardening property *[steel]*
jiko-shōka	自己消化	autolysis *[bioch] [microbio]*
jiko-shōka	自己消火	self-fire-extinguishing *{n} [mat]*
jiko-suihei'sei no	自己水平性の	self-leveling *{adj} [eng]*
jiko-suishin no	自己推進の	self-propelling *{adj} [mech-eng]*
jiko-yūdō	自己誘導	self-induction *{n} [elecmg]*
jiko-yūkai	自己融解	autolysis *[microbio]*
jiku	軸; じく	axis (pl axes) *{n} [math] [mech]*
"	軸	axle; shaft *[mech-eng]*
jiku	磁区; 磁區; じく	magnetic domain *[sol-st]*
jikū	時空	space-time *{n} [rela]*
jiku-bako	軸箱	axle box; journal box *[eng]*

jiku-gumi	軸組；じくぐみ	wall framing *{n} [arch]*
jiku-hi	軸比	axial ratio *[crys] [electr]*
jiku'hyō	軸標	parameter *[crys]*
jik(u)ka sareru	軸架される	be axially hung *{vb} [mech-eng]*
jik(u)ka suru	軸架する	hang axially *{vb t} [mech-eng]*
ji'kū-renzoku'tai	時空連続体	space-time continuum *[rela]*
jiku-ritsu	軸率	axial ratio *[crys]*
jiku-ryū	軸流	axial flow *[fl-mech]*
jiku-shin	軸芯	shaft core *[mech-eng]*
jiku'soku-tōzō-hō	軸測投像法	axonometry *[graph]*
jiku'tsui(-kotsu)	軸椎(骨)	axis (bone) *[v-zoo]*
jiku-uke	軸受	bearing *{n} [mech-eng]*
jiku'uke-dai	軸受台	bearing stand; pedestal *[mech-eng]*
jiku'uke-kō	軸受鋼	bearing steel *[met]*
ji-kyoku	磁極	magnetic pole *[elecmg] [geophys]*
jimu	事務	office work *[econ]*
jimu-kiki	事務機器	office instrument *[eng]*
jimu-shori no kikai'ka	事務処理の機械化	office automation *[comput]*
jin	仁；じん	kernel *[bot]*
-jin; -nin	-人	-person *{suffix}*
jin'ai	塵埃；じんあい	dust *{n} [geol]*
jin'ai-sokutei-ki	塵埃測定器	konimeter *[eng]*
ji'nari	地鳴り	brontides *[geophys]*
jinbē-zame	甚兵衛鮫	whale shark *[v-zoo]*
jinchōge	沈丁花	daphne; Daphne odora *[bot]*
jindō'kyō	人道橋	footbridge; pedestrian bridge *[civ-eng]*
ji-nezumi	地鼠	shrew; shrewmouse (animal) *[v-zoo]*
jin'i no	人為の	artificial *{adj}*
jinkan	腎管	nephridium (pl -ia) *[i-zoo]*
jinken-hi	人件費	personnel expenses *[econ]*
jinkō	人口	population *[bio]*
jinkō	沈香；じんこう	agalloch; lignaloe (incense) *[bot]*
jinkō-chinō	人工知能	artificial intelligence; AI *[comput]*
jinkō'gaku	人口学	demography *[ecol]*
jinkō-gengo	人工言語	artificial language *[comput]*
jinkō-hōsoku	人口法則	law of progression *[bio]*
jinkō-i'butsu	人工遺物	artifact *[archeo]*
jinkō-jikō	人工時硬	artificial aging *{n} [met]*
jinkō-kajō	人口過剰	overpopulation *[bio]*
jinkō no	人工の	artificial; manmade *{adj} [ind-eng]*
jinkō-shikon	人工歯根	artificial tooth root *[mat]*
jinkō-shushi	人工種子	synthetic seed *[bot]*

jinkō-sōgu	人工装具	prosthesis (pl -theses) *{n} [med]*
jinkō-taiyō-kōsen	人工太陽光線	artificial sunlight *[elec]*
jinkō-tsūki	人工通気	artificial ventilation *[min-eng]*
jinkō-zunō'gaku	人工頭脳学	cybernetics *[sci-t]*
jinmei'bo	人名簿	directory *[pr]*
jinmei'roku	人名録	directory *[pr]*
jinpi-sen'i	靭皮繊維	bast fiber *[tex]*
jinrui	人類；じんるい	Homo sapiens; mankind *[bio]*
jinrui-chikei'gaku	人類地形学	anthropogeomorphology *[bio]*
jinrui'gaku	人類学；人類學	anthropology *[bio]*
jinsei	靭性；じんせい	toughness *[mech]*
"	靭性	tenacity *[tex]*
jinshu	人種	race *{n} [bio]*
jintai	靭帯；靭帶	ligament *[hist]*
jintai-kō'shitsu-so'shiki	人体硬質組織	hard human tissue *[hist]*
jinzō	人造	artificial; manmade *{n} [ind-eng]*
jinzō	腎臓；じんぞう	kidney *[anat]*
jinzō-gomu	人造ゴム	artificial rubber; manmade rubber *[poly-ch] [rub]*
jippen-kei	十辺形；十邊形	decagon *[math]*
jippun; juppun	十分	ten minutes *{n}*
jirai	地雷	land mine *[ord]*
jirei	事例	case; example *{n}*
jirei-shindō	自励振動	self-induced vibration *[mech]*
ji'rinsan	次燐酸	hypophosphoric acid
ji'rinsan-en	次燐酸塩	hypophosphate *[ch]*
ji'ritsu no	自立の	self-sustaining *{adj} [mech]*
ji'ritsu-shinkei-kei	自律神経系	autonomic nervous system *[anat]*
jirukonyūmu	ジルコニウム	zirconium (element)
ji-ryoku	磁力；じりょく	magnetic force *[elecmg]*
jiryoku-kei	磁力計	magnetometer *[eng]*
jiryoku-kiroku	磁力記録	magnetogram *[elecmg]*
jiryoku-kiroku-ki	磁力記録機	magnetograph *[elecmg]*
jiryoku-sen	磁力線	magnetic lines of force *[elecmg]*
jiryū'tekkō	磁硫鉄鉱	pyrrhotite *[miner]*
jisa	時差	time difference *[astron]*
ji'sage	字下げ	indentation *[comput]*
ji'satsu-idenshi	自殺遺伝子	suicide gene *[gen]*
ji'satsu-ki'shitsu	自殺基質	suicide substrate *[bioch]*
ji-sei	磁性	magnetic property; magnetism *[phys]*
jisei-handōtai	磁性半導体	magnetic semiconductor *[sol-st]*
jisei-ryūtai	磁性流体	magnetic fluid *[mat]*

ji'shaku	磁石; じしゃく	magnet *[elecmg]*
jishi	磁子	magneton *[phys]*
ji'shin	磁針	magnetic needle *[elecmg]*
ji'shin	地震; じしん	earthquake *[geophys]*
ji'shin	磁心; 磁芯	magnetic core *[electr]*
jishin-ha	地震波	seismic waves *[geophys]*
jishin-hōkō	磁針方向	compass direction *[navig]*
jishin-katsud-do	地震活動度	seismicity *[geophys]*
jishin-kei	地震計	seismograph *[eng]*
ji'shin-ki'oku	磁心記憶	core storage *[comput]*
jishin no-	自身の-	auto- *{prefix}*
jishin('sei) no	地震(性)の	seismic *{adj} [geophys]*
jisho	辞書; 辭書; じしょ	dictionary *[pr]*
jishō	事象	event *{n} [comput] [math]*
jishō-bun	自消分	self-consumed part *{n}*
ji'shoku'teki na	自触的な	autocatalytic *{adj} [ch]*
jishō-sei	自消性	self-extinguishing property *[mat]*
jishu	耳珠	tragus *[anat]*
ji-soku	磁束; じそく	magnetic flux *[elecmg]*
ji'soku-mitsudo	磁束密度	magnetic flux density *[elecmg]*
jissai no	実際の; 實際の	actual *{adj} [math]*
jissai no shōsū-ten	実際の小数点	actual decimal point *[math]*
jisseki	実積	showings; business showings; actual results *[ind-eng]*
jisshaku	実尺	true measure *{n} [sci-t]*
jisshi	実施; じっし	practice; working *{n} [pat]*
jisshi-bumon	実施部門	area of use *{n} [pat]*
jisshi'ken	実施権	license; right of exploitation *[pat]*
jisshin-hō	十進法	decimal numeration *[math]*
jisshin no	十進の	decimal *{adj} [math]*
jisshin-shiki	十進式	decimal system *[math]*
jisshin-shōsū	十進小数	decimal fraction *[math]*
jisshin-sū	十進数	decimal number *[math]*
jisshi-rei	実施例; 實施例	example; example of application *[pat]*
jisshi-rensei	実視連星	visual binaries *[astron]*
jisshi-taiyō	実施態様	mode of working *{n} [pat]*
jisshi-tōkyū	実視等級	visual magnitude *[astron]*
jisshitsu	実質	matter; substance; essence *{n}*
jisshō	実証; 實證	actual proof *[legal]*
jissō	実装	mounting *{n} [elec]*
jissoku	実測; じっそく	actual measurement *[sci-t]*
jissū	実数	real numbers *[math]*

jisū	次数	degree; order *{n} [math]*
ji-suberi	地滑べり	landslide *[geol]*
jisupuroshūmu	ジスプロシウム	dysprosium (element)
jitai	字体	font *[comput] [graph]*
ji-teisū	時定数	time constant *[phys]*
ji'tekkō	磁鉄鉱; 磁鐵鑛	magnetite *[miner]*
jiten	自転; 自轉; じてん	rotation (around own axis) *{n} [astron] [mech]*
jiten	事典	cyclopedia *[pr]*
ji-ten	時点	time-point; point in time *[mech]*
jiten	辞典	dictionary; lexicon *[pr]*
jiten'sha	自転車; 自轉車	bicycle *[mech-eng]*
ji'ten suru	自転する	turn on own axis *{vb i} [astron] [mech]*
ji-tetsu	地鉄; 地鐵	iron matrix; matrix *[met]*
jitsu-dan'yaku	実弾薬; 實彈藥	live ammunition *[ord]*
jitsu'eki	実益	actual profit *[econ]*
jitsu'en	実演	performance; live performance *{n}*
jitsu-jikan	実時間	real time *[comput]*
jitsu-jikan-keisan	実時間計算	real-time calculation *[comput]*
jitsu'rei	実例; じつれい	example *{n} [pat]*
"	実例	illustration; actual example *{n}*
jitsu-teisū	実定数	real constant *[comput]*
jitsu'yō-sei	実用性	practicability; practical utility *[pat]*
jitsuyō-shin'an	実用新案	utility model *[pat]*
jitsuyō-shin'an-kōhō	実用新案公報	Utility Model Gazette *[pat]*
jitsuyō-shin'an-tōroku	実用新案登録	utility model registration *[pat]*
jitsuyō-shin'an-tōroku-negai	実用新案登録願	application for utility model registration *[pat]*
jitsuyō'teki-gigei	実用的技芸	practical art *{n}*
jitsuyō-yōryō	実用容量	practical capacity *[sci-t]*
jitsu-zō	実像	real image *[opt]*
jittai'kyō	実体鏡	stereoscope *[opt]*
jittai no	実体の; 實體の	substantive *{adj}*
ji'wai	磁歪	magnetostriction *[elecmg]*
ji'yaku	字訳	transliteration *[comput]*
jiyū	自由; じゆう	freedom; liberty *{n}*
jiyū-denshi	自由電子	free electron *[phys]*
jiyū-denshi-rēzā	自由電子レーザー	free-electron laser *[opt]*
jiyū-do	自由度	degree of freedom *[mech] [p-ch] [stat]*
jiyū-kōro	自由行路	free path *[phys]*
jiyū-kōtei	自由行程	free path *[phys]*
jiyū-kūkan-hachō	自由空間波長	free-space wavelength *[phys]*

jiyū na	自由な	free *(adj)*
jiyū-rakka	自由落下	free fall *(n) [mech]*
jiyū-sui	自由水	free water *[ch]*
jizai-kagi	自在鉤	pothook *[eng]*
jizai-tsugi'te	自在継ぎ手	universal joint *[eng]*
jizoku-sei	持続性；持續性	durability *[eng]*
jō	条；條；じょう	article *[pat] [pr]*
jō	帖	quire (paper) *[mat]*
jō	錠	lock *(n) [eng]*
"	錠	tablet *(n) [pharm]*
jō'bibun-hōtei'shiki	常微分方程式	ordinary differential equation *[math]*
jō'bitaki	尉鶲；上鶲	redstart (bird) *[v-zoo]*
jōbu	上部	top part *(n) [math]*
jōbu-kōzō	上部構造	superstructure *[civ-eng] [nav-arch]*
jōbun	条文；條文	provision *[legal]*
jōbu na	丈夫な	sturdy; healthy; robust *(adj)*
jō-bunsan	常分散	normal dispersion *[phys]*
jō'chaku	蒸着；じょうちゃく	vacuum deposition; vacuum evaporation; vaporization *[eng]*
jōchō-kensa	冗長検査	residue check *[comput]*
jōchō-sei	冗長性	redundancy *(n) [comput] [math]*
jōchū	常駐	permanent resident *[comput]*
jōchū-rui	条虫類；條蟲類	tapeworms; Cestoda *[i-zoo]*
jo'enso-zai	除塩素剤	antichlor *[ch-eng]*
jogai	除外	exception; exclusion *[legal]*
jōge	上下	above and below; up and down *[math]*
jō-gen	上限	upper limit *[math]*
jō-gen	上弦	first quarter (of moon) *[astron]*
jōgi	定規；じようぎ	ruler *[eng]*
jō-haku	上膊	upper arm; brachium *(n) [anat]*
jō-hassan	蒸発散	evapotranspiration *[hyd]*
jō'hatsu	蒸発；じょうはつ	evaporation; vaporization *[thermo]*
jōhatsu'gan	蒸発岩；蒸發岩	evaporite *[geol]*
jōhatsu-kanko	蒸発乾固	evaporating to dryness *(n) [phys]*
jōhatsu-kei	蒸発計	atmometer; evaporimeter *[eng]*
jōhatsu-ki	蒸発器	evaporator *[ch-eng] [mech-eng]*
jōhatsu-netsu	蒸発熱	heat of vaporization *[thermo]*
jōhatsu-sennetsu	蒸発潜熱	latent heat of vaporization *[thermo]*
jōhatsu-zanryū'butsu	蒸発殘留物	evaporated residue *[ch-eng]*
jōhatsu-zara	蒸発皿	evaporating dish *[ch]*
jo'hensū	助変数	parameter *[math]*
jōhi-soshiki	上皮組織	epithelial tissue *[hist]*

johō	除法	division *{n} [math]*
jōhō	上方	upper side *[math]*
jōhō	乗法；乘法	multiplication *{n} [math]*
jōhō	情報；じょうほう	intelligence *{n} [comm]*
"	情報	information *[comput]*
jōhō-gen	情報源	information source *[comput]*
jōhō-gijutsu'gaku	情報技術学	information technology *[comm]*
jōhō-kensaku	情報検索	information retrieval *[comput]*
jōhō-shori	情報処理	information processing *{n} [comput]*
jōhō-yōryō	情報容量	information capacity *[photo]*
jōji-sei	常磁性	paramagnetism *[elecmg]*
jōji'sei no	常磁性の	paramagnetic *{adj} [elecmg]*
jōjisei-kesshō	常磁性結晶	paramagnetic crystal *[elecmg]*
jōjisei-kyōmei	常磁性共鳴	paramagnetic resonance *[phys]*
jōjisei-tai	常磁性体	paramagnetic substance *[mat]*
jōjō-kaisha	上場会社	listed company *[econ]*
jōka	上科	family group *[bot] [syst]*
"	上科	superfamily *[syst] [zoo]*
jōka	浄化；淨化	decontamination *[eng]*
jōkai	蒸解	digestion *[ch-eng] [paper]*
jōkai-jozai	蒸解助剤	digestion assistant *[ch-eng]*
jōka-ki	浄化器；淨化器	clarifier *[eng]*
jōka-sō	浄化槽	septic tank; water-purifying tank *[civ-eng]*
jōken	条件；じょうけん	condition *{n} [math]*
jōken-hansha	条件反射	conditioned reflex *[psy]*
jōken-meidai	条件命題	conditional statement *[logic]*
jōken-shiki	条件式；條件式	conditional expression *[comput]*
jōken'teki-kenki'sei-seibutsu	条件的嫌気性生物	facultative anaerobe *[microbio]*
jōken'teki na	条件的な	facultative *{adj} [microbio]*
jōken'tsuki; jōken'zuki	条件付	conditional *{n} [math]*
jōken'tsuki-futō-shiki	条件付不等式	conditional inequality *[math]*
jōken'tsuki-kaku'ritsu	条件付確率	conditional probability *[math]*
jōken'tsuki no	条件付きの	conditional *{adj}*
jōken-zai	条件剤	modifier *[mat]*
jōki	蒸気；じょうき	steam; (water) vapor *{n} [phys] [thermo]*
jōki-atsu	蒸気圧；蒸氣壓	steam pressure *[mech]*
"	蒸気圧	vapor pressure *[thermo]*
jōki'atsu-kei	蒸気圧計	vaporimeter *[eng]*
jōki-kikan	蒸気機関	steam engine *[mech-eng]*
jōki-kikan'sha	蒸気機関車	steam locomotive *[rail]*

jōki-mitsudo	蒸気密度	vapor density *[ch]*
jōki-mu	蒸気霧	steam fog *[meteor]*
jōkō	上綱	superclass *[bio] [syst]*
jōkō	条項；條項	clause *[legal]*
jōkō'jō	乗降場；乘降場	platform *[rail]*
jōkoku	上告	appeal *{n} [legal]*
jōkon	条痕	streak *{n} [miner]*
jōkon	乗根	root *{n} [math]*
jō-kōsen	常光線	ordinary ray *[opt]*
jo'kōso	助酵素；じょこうそ	coenzyme *[bioch]*
jōkū-ha	上空波	sky wave *[elecmg]*
jō'kyaku	乗客；乘客	passenger (traveler) *[trans]*
jōkyō	状況；狀況	status; state; condition *{n}*
jōkyō-hōkoku	状況報告	advisory *{n} [meteor]*
jō'mae'ya	錠前屋	locksmith *[arch]*
jōmen-zu	上面図；上面圖	top view *[graph]*
jōmoku	上目	superorder *[bio] [syst]*
jō'myaku	静脈；靜脈	vein *[anat]*
jo'netsu	除熱	heat removal *[thermo]*
jō'on	上音	overtone *[acous]*
jō'on	常温；常溫	normal temperature *[thermo]*
jō'on-kaku'yūgō	常温核融合	cold fusion *[nuc-phys]*
jō'on-kōka	常温硬化	cold setting; room-temperature cure *[poly-ch]*
jōran	擾乱；じょうらん	disturbance *{n}*
jorei	徐冷	slow cooling *{n} [thermo]*
jōrei	常例	standard example; usual example *{n}*
jō'riku	上陸	landing *{n} [navig]*
jō-ritsu	乗率；乘率	multiplying factor *[math]*
joron	序論	introduction; preface *[pr]*
jōryō-bunseki	常量分析	macroanalysis *[an-ch]*
jōryoku-ju	常緑樹	evergreen tree *[bot]*
jōryū	蒸留；じょうりゅう	distillation *[ch]*
jōryū-gawa	上流側	upstream side *[hyd]*
jōryū-ki	蒸留器；蒸溜器	still (an apparatus) *{n} [ch-eng]*
jōryū-kan	蒸留管	distilling tube *[ch]*
jōryū'shu	蒸留酒	distilled liquor; spirit *[cook]*
jōryū-sōchi	蒸留装置	distillation apparatus *[ch]*
jōryū-sui	蒸留水	distilled water *[ch]*
jōryū-tō	蒸留塔	distillation column (or tower) *[ch]*
josai	除滓	skimming *{n} [steel]*
jō-saibō	娘細胞；孃細胞	daughter cell *[cyt]*

jōsan	蒸散；じょうさん	transpiration *[bio]*
josei	女性	female (human); woman; women *{n} [bio]*
josei'teki na	女性的な	feminine *{adj}*
josen	除染	decontamination *[eng]*
josen-shisū	除染指数	decontamination index *[nucleo]*
jo'setsu-ki	除雪機	snowplow *{n} [eng]*
jōsha-ken	乗車券；乘車劵	passenger ticket *[trans]*
jōshin'sho	上申書	petition *{n} [pat]*
jo'shitsu	除湿；除濕；じょしつ	dehumidification; demoisturization *[mech-eng]*
jō-shitsu no	上質の	high-quality; fine-quality *{adj}*
jō-shitsu-shi	上質紙	woodfree paper *[mat]*
jōshō	上昇	ascension; ascent *[aerosp]*
jōshō-kiryū	上昇気流	updraft *[fl-mech]*
jo'shokubai	助触媒；助觸媒	promoter *[ch]*
joshoku'dan	助色団；助色團	auxochrome *[ch]*
josō'ro	助走路	runway *[civ-eng]*
josō'zai	除草剤；除草劑	herbicide; weedkiller *[mat]*
josū	序数；じょすう	ordinal number *[math]*
josū	除数	divisor *[math]*
jōsū	条数；條數	number of threads *[tex]*
jōsū	乗数；乘數	multiplier *[math]*
jō-suidō	上水道	waterworks; water supply *[hyd]*
jōsui'jō	浄水場	water purifying plant *[civ-eng]*
jōtai	状態；狀態	condition; state *{n} [phys] [sci-t]*
jōtai	常態；じょうたい	normality; ordinary state *{n}*
jōtai-chōsetsu	状態調節	conditioning *{n} [poly-ch]*
jōtai(-hōtei)-shiki	状態(方程)式	equation of state *[p-ch]*
jōtai-zu	状態図	phase diagram *[p-ch]*
jo'tanpaku	除蛋白	deproteinization *[org-ch]*
jōto	譲渡；讓渡	assign *{n} [pat]*
jōto-kanō na (no)	譲渡可能な(の)	transferable *{adj} [pat]*
jōtō na (no)	上等な(の)	high-grade; first-class *{adj}*
jōto suru	譲渡する	cede *{vb t} [pat]*
jō-wan	上腕	upper arm *{n} [anat]*
jō'wan-kotsu	上腕骨	humerus (bone) *[anat]*
jōyaku	条約；條約	agreement; convention; treaty *[legal]*
jōyo	剰余；剩餘	remainder *[math]*
jōyō-hakumei	常用薄明	civil twilight *[astron]*
jōyō-ji	常用時	civil time *[astron]*
jōyo-kō	剰余項；剩餘項	remainder *[math]*
jōyō-mei	常用名	trivial name *[bio]*

jōyō-nen	常用年	civil year *[astron]*
jōyo no teiri	剰余の定理	remainder theorem *[math]*
jōyō'sha	乗用車; 乘用車	passenger car; automobile *[mech-eng]*
jōyō-taisū	常用対数; 常用對數	common logarithm *[math]*
jō-yūden'tai	常誘電体	normal dielectric *[mat]*
jozai	助剤	assistant; auxiliary; dyeing auxiliary *[ch-eng]*
jōzai	錠剤	pellet *{n}* *[pharm]*
jōzō	醸造; じょうぞう	brewing *{n}* *[cook]*
jōzō'gaku	醸造学; 釀造學	zymurgy *[microbio]*
jōzō'shu	醸造酒	brewage *{n}* *[cook]*
jōzō suru	醸造する	brew *{vb t}* *[cook]*
jū; tō	十	ten *{n}* *[math]*
jū'aryūsan-en	重亜硫酸塩	bisulfite *[ch]*
jū'aryūsan-sōda	重亜硫酸ソーダ	sodium bisulfite
jū-bako	重箱; じゅうばこ	food boxes, stacked, for picnic *[furn]*
ju'baku-yaku	受爆薬	explosion acceptor *[ord]*
jūbun	十分; 充分	sufficient; adequate *{n}*
jūbun na	十分な	adequate; sufficient *{adj}*
jū'bun no ichi	十分の一	one-tenth *{n}* *[math]*
jū-bunretsu	縦分裂; 縱分裂	longitudinal division *[bio]*
jūdan suru	縦断する	cut vertically *{vb t}*
jūden	充電	charging; charge *{n}* *[elec]*
jūdō	従動	following *{n}*
jūdō	縦動; じゅうどう	vertical motion *[mech]*
judō no	受動の	passive *{adj}* *[math]*
judō-tai	受動態	passive voice *{n}* *[gram]*
judō-yusō	受動輸送	passive transport *[cyt]*
ju'eki	樹液	sap *{n}* *[bot]*
jū'fuka	重付加	polyaddition *[poly-ch]*
jū'fuku-taisū no hōsoku	重複対数の法則	iterated logarithms, law of *[math]*
jufun	受粉	pollination (pollen-receiving) *[bot]*
jufun	授粉	pollination (pollen-giving) *[bot]*
jū'gatsu	十月	October (month)
jū-genshi	重原子	heavy atom *[ch]*
jūgō	重合; じゅうごう	polymerization *[poly-ch]*
jūgō-bōshi-zai	重合防止剤	(polymerization) inhibitor *[poly-ch]*
jūgō-do	重合度	degree of polymerization; polymerization degree *[poly-ch]*
jūgō-kaishi-zai	重合開始剤	(polymerization) initiator *[poly-ch]*
jūgō-kinshi-zai	重合禁止剤	(polymerization) inhibitor *[poly-ch]*
jūgō-ritsu	重合率	conversion *[poly-ch]*

jūgō'tai	重合体	polymer *[poly-ch]*
jūgō-teishi-zai	重合停止剤	shortstopping agent; shortstop *[poly-ch]*
jūgō-yokusei-zai	重合抑制剤	(polymerization) retarder *[poly-ch]*
jūgō-yu	重合油	polymerized oil *[mat]*
jūgyō'sha	従業者	employee *[ind-eng]*
juhi	樹皮；じゅひ	bark *{n} [bot]*
"	樹皮	cascara *{n} [pharm]*
juhyō	樹氷	soft rime *[meteor]*
jūichi'gatsu	十一月	November (month)
jūi'gaku	獣医学；獸醫學	veterinary medicine *[med]*
jū-i'shi	獣医師；じゅういし	veterinary doctor *[med]*
jūji'kei no	十字形の	cruciform *{adj} [sci-t]*
jūji'seki	十字石	staurolite *[miner]*
jūji-sen	十字線	cross hair; cross wire *{n} [eng]*
jujō'sei no	樹上性の	arboreal *{adj} [zoo]*
jujō-tokki	樹状突起	dendrite *[anat]*
jū'jūji'fusseki	重十字沸石	harmotome *[miner]*
jū-kaiki	重回帰；重回歸	multiple regression *[math]*
jū'kaiki-bunseki	重回帰分析	multiple regression analysis *[math]*
jukan	樹冠；じゅかん	crown (of a tree) *{n} [bot]*
jū'kansen	重感染	superinfection *[microbio]*
jūken	重圏；重圈	barysphere *[geol]*
jū-kinzoku	重金属	heavy metal *[met]*
jukkaku-kei; jikkaku-kei	十角形	decagon *[math]*
jukō-bu	受光部	receptor unit *[opt]*
jū-kōgyō	重工業	heavy industry *[ind-eng]*
jukō-kaku	受光角	angle of reception of light *{n} [opt]*
jukō-shibori	受光絞り	receptor field stop *[opt]*
juku'go	熟語；じゅくご	kanji (q.v.) compound word *[comm]*
juku'ren	熟練；熟練	skill *{n}*
jukuren'kō	熟練工	skilled workman *[ind-eng]*
jukusei	熟成	maturing; maturation; ripening *{n} [bio]*
"	熟成	aging *{n} [ch] [eng]*
jukushita	熟した	ripe *{adj} [bot]*
jūkyo-chiku	住居地区	residential sector *[civ-eng]*
jūman-tai	充満帯；充滿帶	full band *[electr]*
ju'moku	樹木；じゅもく	tree *[bot]*
jumoku'en	樹木園	arboretum *[bot]*
jumoku'gaku	樹木学	dendrology *[for]*
jumoku-genkai-sen	樹木限界線	timberline *[ecol]*
jumyō	寿命；壽命	lifetime; predestined lifespan *{n}*
jun	順	sequence; sequential *[comput]*

jun-	準-	quasi- *{prefix}*
jū'nan-sei	柔軟性	softness *[mat]*
jū'nan-taiō'ryoku	柔軟対応力	flexible response *[mil]*
jun-antei na (no)	準安定な(の)	metastable *{adj}* *[sci-t]*
jun'antei-sō	準安定相	metastable phase *[p-ch]*
jūnan-zai	柔軟剤	softening agent *[mat]*
jun'atsu	順圧; 順壓	barotropy *[phys]*
jun-baiyō	純培養	pure culture *[microbio]*
junbaku	殉爆	sympathetic detonation *[ch]*
junbi	準備; じゅんび	preparation *{n}*
junchō ni	順調に	satisfactorily; favorably *{adv}*
jun'danmen	純断面	net section *[graph]*
jun'dansei-kō'sanran	準弾性光散乱	quasi-elastic light scattering *{n}* *[opt]*
jun'dansei-sanran	準弾性散乱	quasi-elastic scattering *{n}* *[mech]*
jun-dōkei	準同形	homomorphs *[ch]*
jūnen'kan	十年間	decade (ten years) *{n}*
jun'gengo	準言語	sublanguage *[ling]*
jun-gūzen	純偶然	pure chance *[math]*
jun-heigen	準平原	base-leveled plain; peneplain *[geol]*
jun-hōkō	順方向	forward direction *{n}* *[electr]*
jun'hōkō-den'atsu	順方向電圧	forward voltage *[electr]*
jun-hōseki	準宝石; 準寶石	semiprecious stone *[miner]*
jun'i	順位	sequence; precedence *{n}* *[comput]* *[eng]*
"	順位	order; rank(ing); succession *{n}* *[math]*
jūni'gatsu	十二月	December (month)
jūni'men'tai	十二面体	dodecahedron *[math]*
jūnishi'chō	十二指腸	duodenum (pl -dena, -denums) *[anat]*
jūni'shin-hō	十二進法	duodecimal numeration *[math]*
jun'i-soku	順位則	sequence rule *[math]*
jun(ji)-yobi'dashi	順(次)呼出し	sequential access *[comput]*
junjo	順序; じゅんじょ	sequence *{n}* *[comput]*
junjo-dōsa	順序動作	sequential operation *[comput]*
junjo-kairo	順序回路	sequential circuit *[elec]*
junjo-sū	順序数	ordinal number *[math]*
junjo-tsui	順序対	ordered pair *[math]*
junjo'zuke; junjo'tsuke	順序付け	ordering; sequencing *{n}* *[math]*
junka	純化	purification *[eng]*
junka	馴化	acclimatization *[bio]*
junka	順化; 馴化	habituation *[bio]*
jun-kagaku'teki-hōhō	純化学的方法	purely chemical method *[ch]*
jun-kagaku'teki-hōhō	準化学的方法	quasi-chemical method *[ch]*
junkan	循環; じゅんかん	wraparound *{n}* *[comput]*

junkan-kei	循環系	circulatory system *[anat]*
junkan-ki'oku	循環記憶	cyclic storage *[comput]*
junkan-ryūdō	循環流動	circulation *[physio]*
junkan-shōsū	循環小数	repeating decimal *[math]*
junkan'teki-kō'rinsan'ka	循環的光燐酸化	cyclic photophosphorylation *[bioch]*
jun'katsu-yu	潤滑油	lubricating oil *[mat]*
jun'katsu-zai	潤滑剤	lubricant *[mat]*
jun-kesshō	準結晶	quasi-crystal *[crys]*
jun'kesshō-sō	準結晶相	quasi-crystalline phase *[crys]*
junkō	巡航；じゅんこう	cruise (by ship) *{n} [trans]*
junkō-undō	順行運動	direct motion *[astron]*
jun(kō)sei('jō)-denpa'gen	準(恒)星(状)電波源	quasi-stellar radio source (quasar) *[astron]*
jun-kyosei	準巨星	subgiant *{n} [astron]*
jun(n)ō; jun'ō	順応	adaptation *[physio] [psy]*
jūnō	十能；じゅうのう	fire shovel *[eng]*
jū no hosū	十の補数	ten's complement *[comput]*
jun'on	純音	pure tone; simple tone *[acous]*
junpū	順風	favorable wind; fair wind *[navig]*
jun'retsu	順列	permutations *[math]*
junryū	順流	favorable current; fair current *[navig]*
jun-ryūshi	準粒子	quasi-particle *[phys]*
junsai	蓴菜；じゅんさい	water shield; Brasenia schreberi *[bot]*
junsen	準線	directrix (pl -trixes, -trices) *[math]*
junshi'sen	巡視船	patrol boat *[nav-arch]*
jun-shūki'teki na	準周期的な	quasi-periodic *{adj} [sci-t]*
junsui-baiyō	純粋培養	pure culture *[microbio]*
junsui-busshitsu	純粋物質	pure substance *[phys]*
junsui no	純粋の	pure *{adj} [math]*
jun'teitai-zensen	準停滞前線	quasi-stationary front *[meteor]*
jun-waisei	順矮星	subdwarf star *[astron]*
jun'yō	順養；じゅんよう	acclimatization; domestication *{n}*
jun'yō-baiyō	順養培養	acclimating cultivation *[microbio]*
jun'yō'kan	巡洋艦	cruiser *[nav-arch]*
jun'yō-kōkai	巡洋航海	cruise (by ship) *{n} [trans]*
jun'yō'sen	巡洋船	cruise ship *[nav-arch]*
junzoku'teki-fukusei	順続的複製	sequential replication *[mol-bio]*
junzuru	準ずる	conform to; correspond to; is patterned after *{vb}*
jūō-hi	縦横比；縱橫比	aspect ratio *[eng]*
jū-oku	十億	billion (American system) *[math]*
jūrai no gi'jutsu	従来の技術	prior technology *[pat]*

jū'retsu	縦列	column *[comput]*
juri	受理	acceptance *[pat]*
jūroku'shin-hō	十六進法	hexadecimal notation *[comput]*
jū-rui	獣類；獸類	Theria; animals *[v-zoo]*
juryō	受領	receipt; acceptance *{n}*
jūryō	重量；じゅうりょう	weight *[mech]*
jūryō-bunseki	重量分析	gravimetric analysis *[an-ch]*
jūryō'heikin-bunshi'ryō	重量平均分子量	weight-average molecular weight *[ch]*
jūryō-hyaku'bun-ritsu	重量百分率	weight percentage *[mech]*
jū-ryoku	重力；じゅうりょく	gravity; gravitation *[mech] [phys]*
jūryoku'ba	重力場	gravitational field *[mech]*
jūryoku-ha	重力波	gravitational wave *[rela]*
jūryoku-hen'i	重力偏移	gravitational shift *[astron]*
jūryoku-hō	重力法	gravitational method *[phys]*
jūryoku-ka'sokudo	重力加速度	acceleration of gravity; acceleration of free fall *[mech]*
jūryoku-kei	重力計	gravimeter *[eng]*
jūryoku-kussei	重力屈性	geotropism *[bot]*
jūryoku-ryōshi	重力量子	graviton *[phys]*
jūryoku-shitsu'ryō	重力質量	gravitational mass *[phys]*
jūryoku-teisū	重力定数	gravitational constant *[mech]*
jūryō-moru-nōdo	重量モル濃度	molality *[ch]*
jūryō-sokutei	重量測定	gravimetry *[eng]*
jūryō-teki'tei	重量滴定	gravimetric titration *[an-ch]*
jū'ryūsan-en	重硫酸塩	bisulfate *[ch]*
jū'ryūshi	重粒子	baryon *[part-phys]*
jusei	受精；授精	fertilization *[physio]*
jushi	樹脂；じゅし	resin *[org-ch] [poly-ch]*
jūshi	獣脂；じゆぅし	tallow *{n} [mat]*
jushi-ganshin-asshuku-mokuzai	樹脂含浸圧縮木材	compregnated wood *[mat]*
jushi-ganshin-shi	樹脂含浸紙	impregnated paper *[poly-ch]*
jushi'jō no	樹枝状の	arborescent *{adj} [bio]*
jushi'jō-seichō	樹枝状成長	dendritic growth *[sci-t]*
jushi'jō-shō	樹枝状晶	dendrite *[crys]*
jushi'ka	樹脂化	resinification *[org-ch] [poly-ch]*
jushi-kakō-zai	樹脂加工剤	resin-finishing agent; resin-treatment agent *[mat]*
jushi-karyū	樹脂加硫	resin cure *{n} [rub]*
jushin	受信	reception; receiving *{n} [comm] [electr]*
jūshin	重心；じゅうしん	barycenter (of triangle) *[math]*
"	重心	center of gravity *[mech]*

jushin-ki	受信機	receiver *[electr]*
jushin-ki	受振機	geophone *[electr]*
jushi-san	樹脂酸	disproportionated rosin; resin acid *[rub]*
jū'shitsu	鞣質; じゅうしつ	tannin *[org-ch]*
jūsho	住所	address *{n} [comm]*
"	住所	domicile *[legal]*
jūshō'seki	重晶石	barite *[miner]*
jū'shukugō	重縮合	condensation polymerization; polycondensation *[poly-ch]*
jū'shuseki'san-en	重酒石酸塩	bitartrate *[ch]*
jūsō (jūtansan-sōda)	重曹（重炭酸曹達）	baking soda (an abbreviation) *[cook]*
jū-sōkan	重相関; 重相關	multiple correlation *[math]*
jūsō-kōka	重層効果	interlayer effect *[photo]*
jusshin-hō; jisshin-hō	十進法	decimal system *[math]*
jusshin-shiki	十進式	decimal system *[math]*
jūsui	重水; じゅうすい	heavy water *[inorg-ch]*
jū'suiso	重水素	deuterium; heavy hydrogen *[ch]*
jutai	受体	acceptor *[sol-st]*
jūtai	獣帯; 獸帶	girdle; zodiac *{n} [astron]*
jūtaku	住宅	dwelling; residence *{n} [arch]*
jūtaku-chi(ku)	住宅地(区)	residential sector *[civ-eng]*
jūtaku-denwa	住宅電話	residence telephone *[comm]*
jū'tansan-en	重炭酸塩	bicarbonate *[ch]*
jū'tansan-sōda	重炭酸曹達	sodium bicarbonate; baking soda *[cook]*
jūten	重点	emphasis; important point *{n}*
jūten(-butsu)	充填(物)	filling *{n} [dent]*
jūten'butsu	充填物	packing *{n} [mat]*
jūten-tō	充填塔	packed tower; packed column *[ch-eng]*
jūten-zai	充填剤	filler *[ch-eng]*
jūtō	充当	appropriation *{n}*
jutsu'bu	述部	predicate *{n} [gram]*
jutsu'go	術語	term; terminology *{n} [comm]*
juwa-ki	受話機	receiver (of telephone) *[comm]*
juyō	受容	acceptance *[psy]*
juyō	需用; じゅよう	demand *{n} [econ]*
jūyō	重要; じゅうよう	importance *{n}*
juyō'ka	需用家	consumer *[econ]*
juyō-kyōkyū no hōsoku	需要供給の法則	supply and demand, law of *[econ]*
juyō-nō'ryoku	受容能力	competence *[immun]*
jūyō-sangyō	重要産業	major industry *[econ] [ind-eng]*
jūyō-sei	重要性	importance; weight; significance *{n}*
jū'yōshi	重陽子	deuteron *[ch]*

juyō'tai	受容体	acceptor *[ch]*
jūyu	重油; じゅうゆ	fuel oil; heavy oil *[mat] [petrol]*
jūyu-antei-zai	重油安定剤	fuel oil stabilizer *[petrol]*
jūyu-kikan	重油機関	heavy-oil engine *[mech-eng]*
juzō-kan	受像管	picture tube *[electr]*
jūzoku	従属; 從屬	subordination *[bio]*
"	従属; じゅうぞく	dependency *[comput]*
jūzoku-eiyō-seibutsu	従属栄養生物	heterotroph *[bio]*
jū(zoku)-hensū	従(属)変数	dependent variable *[math]*
jūzoku-tokkyo	従属特許	dependent patent *[pat]*

K

ka	可	fair (grading) *{n} [ind-eng]*
ka	価；價；か	value *{n} [math]*
ka	科	family (pl -lies) *[bio] [syst]*
"	科	group (viruses) *{n} [bio] [syst]*
ka	蚊	mosquito (insect) (-es, -s) *[i-zoo]*
ka'ansokukō-san	過安息香酸	perbenzoic acid
kaba	樺	birch (tree) *[bot]*
kaba	河馬；かば	hippopotamus (animal) *[v-zoo]*
kabe	壁；かべ	wall *[arch]*
kabe-dokei	壁時計	wall clock *[horol]*
kabe-gami	壁紙	wallpaper *{n} [mat]*
kabe-ita	壁板	wallboard *[mat]*
kabe-konsento	壁コンセント	wall outlet *[elec]*
kabe-tsuchi-iro	壁土色	wall-mud color (color) *[opt]*
kabi	黴；かび	mildew; mold *{n} [mycol]*
kabi-doku	黴毒	fungal toxin; mycotoxin *[mycol]*
kabin	花瓶	vase; flower vase *[furn]*
kabin-sei	過敏性	hypersensitivity *[immun]*
kabocha	南瓜；かぼちゃ	pumpkin (vegetable) *[bot]*
kabu	株	stock *{n} [econ]*
"	株	strain *{n} [microbio]*
kabu no-	下部の-	infra- *{prefix}*
ka'bunsū	仮分数；假分數	improper fraction *[math]*
kabura; kabu	蕪青；蕪	turnip (vegetable) *[bot]*
kabure	気解；かぶれ	rash (skin) *[med]*
kaburi	カブリ	fog *{n} [photo]*
kaburi'mono	被り物	headgear *[eng]*
kabuto-gani	兜蟹；鱟	horseshoe crab; helmet crab *[i-zoo]*
kabuto-mushi	兜虫	beetle (insect) *[i-zoo]*
kachi	価値；價値；かち	value *{n} [math] [sci-t]*
"	価値	merit; worth *{n}*
ka'chiku	家畜	livestock *[agr]*
kachō	河潮	fluvial tide *[hyd]*
kachō-han'i	可聴範囲	audible range *[acous]*
kachō'ka no	可聴下の	subaudio; subaudible *{adj} [acous]*
kachō-keihō	可聴警報	audible alarm *[comput]*
kachō-shūha-hasshin-ki	可聴周波発振器	audio frequency oscillator *[electr]*
kachō-shūha'sū	可聴周波数	audio frequencies *[comm] [phys]*
kachū	花柱	style *{n} [bot]*

kadai-tōkaku-kensa	課題統覚検査	thematic apperception test; TAT *[psy]*
kadan	花壇	flower bed *[bot]*
ka-den'atsu	過電圧；過電壓	overvoltage *[elec]*
kaden-ryūshi	荷電粒子	charged particle *[part-phys]*
ka-denshi	価電子	valence electrons *[atom-phys]*
ka'denshi-tai	価電子帯	valence band *[sol-st]*
kado	角；かど	corner *{n}* *[arch]* *[civ-eng]*
kado	門；かど	gate; door *[arch]*
kadō	火道	vent (volcano) *{n}* *[geol]*
kadō	華道	floral art *{n}*
kadō	稼動	operation; work *{n}*
kado-gamae	門構：門	kanji (q.v.) character radical for gate *[pr]*
kadō-jikan	稼動時間	operating time *[comput]*
kadō-koa	可動コア	movable core *[elecmg]*
ka'doku-do	可読度；可讀度	legibility *[pr]*
kadomyūmu	カドミウム	cadmium (element)
kadō-ritsu	稼動率	rate of operation *{n}* *[ind-eng]*
kaede	楓；かえで	maple (tree) *[bot]*
ka'en	火炎	flame *{n}* *[ch]*
ka'enso-san	過塩素酸	perchloric acid
ka'enso'san-en	過塩素酸塩	perchlorate *[ch]*
kaeri	逆鉤；かえり	burr *{n}* *[met]*
kaeru	蛙；かえる	frog (amphibian) *[v-zoo]*
kaeru	反る；かえる	warp *{vb i}* *[mech]*
kaeru	変える；變える	change; alter *{vb t}*
kaeru	帰る；歸る	return; come (or go) back *{vb i}*
kaeru	替える；換える；代える	change; substitute; convert *{vb t}*
kaeru	孵る	hatch; be hatched *{vb i}* *[agr]*
kaesu	返す	return; give back *{vb t}*
kaesu	帰す；歸す	send back; dismiss *{vb t}*
kafu	下付	grant; issue *{n}* *[legal]*
ka'fuka	過負荷	overload *{n}* *[electr]*
ka'fukka-tanka'suiso	過弗化炭化水素	perfluorohydrocarbon *[org-ch]*
kafun	花粉	pollen *[bot]*
kafun-baiyō	花粉培養	pollen culture *[bot]*
kafun'gaku	花粉学	palynology *[paleon]*
kafun-ryū	花粉粒	pollen grain *[bot]*
kaga	花芽	flower bud *[bot]*
kagaku	化学；化學	chemistry *[sci-t]*
kagaku	科学；科學	science *[sci-t]*
kagaku-bunseki	化学分析	chemical analysis *[ch]*

kagaku-chōmi'ryō	化学調味料	chemical seasonings (flavorings) *[cook]*
kagaku-doku'ritsu-eiyō-seibutsu	化学独立栄養生物	chemoautotroph *[microbio]*
kagaku-gi'jutsu	科学技術	technology *[sci-t]*
kagaku-gōsei-bakuteria	化学合成バクテリア	chemosynthetic bacteria *[microbio]*
kagaku-hannō	化学反応	chemical reaction *[ch]*
kagaku-henka	化学変化	chemical change *[ch]*
kagaku-hōtei'shiki	化学方程式	chemical equation *[ch]*
kagaku-jō'chaku	化学蒸着	chemical vapor deposition *[met]*
kagaku-juyō-ki	化学受容器	chemoreceptor *[physio]*
kagaku-kanwa	化学緩和	chemical relaxation *[ch]*
kagaku-kenma	化学研摩	chemical polishing *{n} [met]*
kagaku-ketsu'gō	化学結合	chemical bond *[ch]*
kagaku-kigō	科学記号	scientific symbol *[sci-t]*
kagaku-kisō-seichō-hō	化学気相成長法	chemical vapor deposition *[met]*
kagaku-kisō-seki'shutsu-hō	化学気相析出法	chemical vapor deposition *[met]*
kagaku-kōgaku	化学工学	chemical engineering *[eng]*
kagaku-kyūchaku	化学吸着	chemisorption *[p-ch]*
kagaku-muki'eiyō-seibutsu	化学無機栄養生物	chemolithotroph *[microbio]*
kagaku'ryō-ron	化学量論	stoichiometry *[p-ch]*
kagaku'ryōron-shiki	化学量論式	stoichiometric equation *[p-ch]*
kagaku'ryōron'teki na	化学量論的な	stoichiometric *{adj} [p-ch]*
kagaku'ryō-sū	化学量数	stoichiometric number *[p-ch]*
kagakusen'gaku	化学線学	actinology *[phys]*
kagaku-sen'i	化学繊維	chemical fiber; manmade fiber *[tex]*
kagaku'sha	化学者	chemist *[ch]*
kagaku'sha	科学者	scientist *[sci-t]*
kagaku-shiki	化学式	chemical formula *[ch]*
kagaku-shinwa'ryoku	化学親和力	chemical affinity *[ch]*
kagaku-shōten	化学焦点	actinic focus *[opt]*
kagaku-shūshoku	化学修飾	chemical modification *[ch]*
kagaku-teisū	化学定数	chemical constant *[ch]*
kagaku'teki-hōhō	科学的方法	scientific method *[sci-t]*
kagaku'teki-keikaku-hō	科学的計画法	scientific programming *{n} [comput]*
kagaku'teki-kisū-hō	科学的記数法	scientific notation *[math] [sci-t]*
kagaku'teki-sanso-yōkyū-ryō	化学的酸素要求量	chemical oxygen demand (COD) value *[org-ch]*
kagaku'teki-seishitsu	化学的性質	chemical property *[ch]*
kagaku-tōryō	化学当量	chemical equivalent *[ch]*
kagami	鏡	mirror *{n} [opt]*
kagami-seidō	鏡青銅	mirror bronze *[met]*
kagayaki	輝き	shine; sparkle; glint *{n} [opt]*

kage	影	shadow *{n} [opt]*
kage	陰；蔭	shade *{n} [opt]*
kage-e	影絵；影繪	shadow picture *[graph]*
kagen	下限	lower limit *[math]*
kagen	下弦	last quarter (of moon) *[astron]*
kagen-teikō-ki	加減抵抗器	rheostat *[elec]*
kagerō	蜉蝣；かげろう	dayfly; dragonfly; mayfly (insect) *[i-zoo]*
-kagetsu	-箇月	-month(s) *{suffix}*
kagi	鉤	hook *{n} [eng]*
kagi	鍵；かぎ	key *{n} [comput] [eng]*
kagi-ana	鍵穴	keyhole *[eng]*
kagi-bashira	鉤柱	davit *{n} [nav-arch]*
kagi-genshi	鍵原子	key atom *[ch]*
kagi'jō no	鍵状の	hamate *{adj} [bio]*
kagi-kakko	鉤括弧：［　］；⌞⌝	hooked punctuation mark (in Japanese) *[pr]*
kagi-nawa	鉤縄；鉤繩	hook-tipped rope *[eng]*
kagi-no-te no	鍵の手の	right-angled *{adj} [arch]*
ka'gisan	過蟻酸	performic acid
ka'gisan-en	過蟻酸塩	performate *[ch]*
kagi-zume	鉤爪	claw *{n} [anat]*
kago	籠	basket *[eng]*
kagō	化合	combination *[ch]*
kagō'butsu	化合物	compound *[ch]*
kago'gata-kaiten'shi	籠形回転子	cage rotor *[elec]*
kagu	家具	furniture *[arch]*
kagu'shi	家具師	cabinetmaker *[furn]*
ka'gyaku-hannō	可逆反応	reversible reaction *[ch]*
ka'gyaku-jūgō	可逆重合	reversible polymerization *[ch]*
ka'gyaku no	可逆の	reversible *{adj} [ch]*
ka'gyaku-sei	可逆性	reversibility *[math]*
kagyū	蝸牛	cochlea (pl cochleae, -s) *[anat]*
kahan-kyō	可搬橋	portable bridge *[civ-eng]*
kahei	貨幣	money *[econ]*
kahei'gaku	貨幣学	numismatics *{n}*
kahen-dengen	可変電源	variable power source *[elec]*
kahen-kigō	可変記号	variable symbol *[comput]*
kahen-kondensa	可変コンデンサ	variable capacitor *[elec]*
kahen'sei no	可変性の	variable *{adj} [ch] [math]*
ka'hensei-sayō	過変成作用	ultrametamorphism *[petr]*
kahen-yōryō	可変容量	variable capacitance *[elec]*
kahi	果皮	pericarp; rind (of fruit) *[bot]*

kahi'hen	花被片	tepal *[bot]*
kahō	下方	lower side *{n} [math]*
kahō	加法	addition (method) *[math]*
kahō-kōshiki	加法公式	additive formula *[math]*
kahō no-	下方の-	under- *{prefix}*
ka'hōsan	過硼酸	perboric acid
ka'hōsan-en	過硼酸塩	perborate *[ch]*
ka'hōwa	過飽和; 過飽和	supersaturation *[p-ch]*
ka'hōwa-koyō'tai	過飽和固溶体	supersaturated solid solution *[p-ch]*
ka'hōwa-yōeki	過飽和溶液	supersaturated solution *[ch]*
kai	貝	shell; shellfish *[i-zoo]*
kai	界	kingdom *[bio] [syst]*
"	界	field *{n} [elec]*
kai	開; かい	off; open *{n} [elec]*
kai	解	solution; integral *[math]*
kai	櫂	oar; paddle *{n} [eng] [nav-arch]*
-kai	-回	-time(s) *{suffix}*
-kai	-階	-floor(s) *{suffix}*
kai-bai	貝灰	shell lime *[agr]*
kaiban'zakuro-ishi	灰礬柘榴石	grossularite *[miner]*
kai-bashira	貝柱; かいばしら	scallop; scallop eyes *[cook]*
kai'batsu-kōdo	海抜高度	height above sea level *{n} [eng]*
kaibō'gaku	解剖学	anatomy *[bio]*
kai'bōjun	解膨潤	deswelling *{n} [ch]*
kaibun	灰分	ash; ash content *[ch]*
kaibun-baiyō	回分培養	batch culture *[microbio]*
kaibun-sōsa	回分操作	batch process *[eng]*
kaichō	海鳥	seabird *[v-zoo]*
kaichō	階調	gradation *[graph]*
kaichō-do	階調度	gradient *[photo]*
kai'chōseki	灰長石	anorthite *[miner]*
kaichū-dentō	懐中電灯; 懷中電燈	flashlight *[eng]*
kaichū-saibai	海中栽培	mariculture *[agr]*
kaichū-saikō	海中採鉱	undersea mining *{n} [min-eng]*
kai'chūseki	灰柱石	meionite *[miner]*
kaidai-moku'roku	解題目録	annotated catalog *[pr]*
kaidan	階段	stairway *[arch]*
"	階段	step *{n} [arch] [comput]*
kai-dashi	買出し	shopping wholesale; laying-in *[econ]*
kaido	海土	sea mud *[geol]*
kaido	開度	aperture *[opt]*
kaidoku-ki	解読器; 解讀器	decoder *[comm]*

kai'en	海塩	sea salt *[geol]*
kaifū	海風	sea breeze *[meteor]*
kaifuku	回復	recovery *[comput] [med] [met] [poly-ch]*
kaifuku-ritsu	回復率	recovery rate *[nuc-phys]*
kaifuku-shisū	回復指数	recovery exponent *[nuc-phys]*
kai'fusseki	灰沸石	scolecite *[miner]*
kaiga	絵画；繪畫	pictorial art *[graph]*
kaigan	海岸；かいがん	beach; shore; seashore *{n} [geol]*
kaigan-dankyū	海岸段丘	marine terrace *[geol]*
kaigan-sen	海岸線	shoreline; coastline *[geol]*
kaigan-shokusei	海岸植生	coast vegetation *[bot]*
kaigan-shōtaku'chi	海岸沼沢地	marine swamp *[ecol]*
kaigan-taiki	海岸大気	seaside atmosphere *[meteor]*
kai'gara	貝殻；かいがら	shell *{n} [zoo]*
kaigara-mushi	貝殻虫；貝殼蟲	scale (insect) *[i-zoo]*
kaigi-shitsu	会議室；會議室	conference room *[arch]*
kaigō	会合	association *[ch]*
"	会合	meeting; assembly *{n}*
kaigō'sei-hannō	会合性反応	associative reaction *[ch]*
kaigō-shūki	会合周期	synodic period *[astron]*
kaigun	海軍	navy *{n} [mil]*
kaigyō	改行；かいぎょう	line feed *[comput]*
kaigyō-fukki	改行復帰	carriage return *[comput]*
kaigyō-fukki-moji	改行復帰文字	new-line character *[comput]*
kaigyū	海牛	manatee; sea cow (mammal) *[v-zoo]*
kai'hatsu	開発	development *[sci-t]*
kaihei-ben	開閉弁；開閉瓣	closing valve *[mech-eng]*
kaihei-ki	開閉器	switch *{n} [elec]*
kaihi	回避	refrainment *[legal]*
"	回避	avoidance; aversion *[psy]*
kaihō	解法	solution (method) *[math]*
kaihō	解放	release *[comput]*
kaihō'bun	開放文	open sentence *[math]*
kaihō-junkan-kei	開放循環系	open circulatory system *[bio]*
kaihō-kei	開放系	open system *[eng]*
kai'hōkō	灰硼鉱；灰硼鑛	colemanite *[miner]*
kai'i	回位	disclination *[crys]*
kaiji	界磁	field magnet *[elecmg]*
kaiji	開示	disclosure *[pat]*
kaiji-eisei	海事衛星	maritime satellite *[aero-eng]*
kaiji-hō	海事法	maritime law *[legal]*
kaiji no	海事の	nautical *{adj} [ocean]*

kaiji-senrin	界磁線輪	field coil *[elecmg]*
kaiji suru	開示する	disclose *{vb} [pat]*
kaijō	階乗; かいじょう	factorial *[math]*
kaijō-jūgō	塊状重合	block polymerization; bulk polymerization; mass polymerization *[poly-ch]*
kaijō-shūseki	塊状集積	agglomeration *[met]*
kaijō-tan	塊状炭	lump coal; massive coal *[geol]*
kaijō-tei'haku-shi'setsu	海上停泊施設	sea berth *[nav-arch]*
kai'jūgō	解重合	depolymerization *[poly-ch]*
kai'jūseki	灰重石	scheelite *[miner]*
kaika	開花; かいか	blossoming; inflorescence *{n} [bot]*
kaikan-hannō	開環反応	ring-opening reaction *[poly-ch]*
kaikan-jūgō	開環重合	ring-opening polymerization; scission polymerization *[poly-ch]*
kaikei('gaku)	会計(学)	accounting *{n} [econ]*
kaikei	塊茎; 塊莖	tuber *[bot]*
kaikei-ki	会計機; かいけいき	accounting machine *[comput] [eng]*
kaikei-nendo	会計年度	fiscal year *[econ]*
kaikei'shi	会計士; 會計士	accountant (a person) *[econ]*
ka'i-keta'afure	下位桁溢れ	underflow *{n} [comput]*
kaiki	回帰; 回歸; かいき	regression *[stat]*
kaiki-chitai	皆既地帯	zone of totality; zone of total eclipse *[astron]*
kaiki(-hō)	回帰(法)	recursion *[logic]*
kaiki-nen	回帰年	tropical year *[astron]*
kaiki-sen	回帰線	tropic *{n} [astron] [geod]*
kaiki-shoku	皆既食	total eclipse *[astron]*
kaiko	蚕; かいこ	silkworm *[i-zoo]*
kaikō	海溝	trench; trough (ocean) *{n} [geol]*
kaikō	開口	aperture *[opt]*
kaikō	解膠; かいこう	peptization *[ch]*
"	解膠	deflocculation *[ch-eng]*
kaiko-ga	蚕蛾; かいこが	silkworm moth; Bombyx (insect) *[i-zoo]*
kaikō-sū	開口数	numerical aperture *[opt]*
kaikō-tō	灰光灯	limelight *[opt]*
kaikō-zai	解膠剤	deflocculant *[ch-eng]*
kai-kukan	開区間; 開區間	open interval *[math]*
kaikyō	海峡; 海峽	channel (ocean) *{n} [civ-eng] [navig]*
"	海峡; かいきょう	narrows; strait (ocean) *{n} [geogr]*
kaikyū	階級	class; rank *{n} [stat]*
kaimen	界面	boundary *{n} [ch] [phys]*
"	界面	interface *{n} [ch] [p-ch] [sci-t]*

kaimen	海面；かいめん	sea surface *[ocean]*
kaimen	海綿	sponge *{n}* *[i-zoo]*
kaimen-chō'ryoku	界面張力	interfacial tension *[phys]*
kaimen-denki'kagaku	界面電気化学	interfacial electrochemistry *[p-ch]*
kaimen-dōbutsu	海綿動物	Porifera *[i-zoo]*
kaimen-dōden-genshō	界面動電現象	interfacial electrokinetic phenomenon *[p-ch]*
kaimen-dōden'i	界面動電位	electrokinetic potential *[phys]*
kaimen-hakai	界面破壊	interfacial failure *[adhes]*
kaimen-jūgō	界面重合	interfacial polymerization *[ch]*
kaimen-jū'shukugō	界面重縮合	interfacial polycondensation *[poly-ch]*
kaimen-kagaku	界面化学	surface chemistry *[p-ch]*
kaimen-kanshō-kei	海面干渉計	sea interferometer *[opt]*
kaimen-kassei-zai	界面活性剤	surface-active agent; surfactant *[poly-ch]*
kaimen-ki'atsu	海面気圧	sea-level pressure *[meteor]*
kaimen-setchaku	界面接着	interfacial bond *[adhes]*
kaimen-tetsu	海綿鉄	sponge iron *[met]*
kai no shūgō	解の集合	solution set *[math]*
kai'nyū'ka	解乳化	demulsification *[ch-eng]*
kainyū'ka-zai	解乳化剤	demulsifier *[ch-eng]*
kai'ō-sei	海王星	Neptune *[astron]*
kai'retsu	開裂	cleavage *[bio]* *[geol]*
kai'retsu-hannō	開裂反応	cleavage reaction *[crys]*
kairi	海里	nautical mile *[navig]*
kairi	解離；かいり	dissociation *[p-ch]*
kairi-do	解離度	degree of dissociation *[p-ch]*
kairi-teisū	解離定数	dissociation constant *[p-ch]*
kairo	回路；かいろ	network *{n}* *[comm]*
"	回路	circuit *[elec]*
kairo	海路	seaway *[navig]*
kairo	開路	open circuit *[elec]*
kairo	懐炉；懷爐	body-warmer stick; pocket heater *[eng]*
kairoban'jō no jikken	回路板上の実験	breadboarding *{n}* *[electr]*
kairo-mō	回路網	network *[comm]* *[elec]*
(kairo-)shadan-ki	(回路)遮断器	circuit breaker *[elec]*
kairo-zu	回路図；回路圖	circuit diagram *[elec]*
kai-rui	貝類	shellfish *[i-zoo]*
kairui'gaku	貝類学；貝類學	conchology *[i-zoo]*
kairyō	改良	improvement *{n}*
kairyū	海流	ocean current *[ocean]*
kaisai-parupu	解砕パルプ	broken pulp *[paper]*
kaisaku-kō	快削鋼	free-cutting steel *[met]*

kaisan'butsu	海産物	marine products *[ocean]*
kaisei	快晴	clear (weather) *{n} [meteor]*
kaisei-rui	海星類	Asteroidea *[i-zoo]*
kaisei-sō	海成層	sea layer *[geol]*
kaiseki	解析；かいせき	analysis (pl -yses) *[math]*
kaiseki-kika'gaku	解析幾何学	analytic geometry *[math]*
kaiseki-riki'gaku	解析力学	analytical dynamics *[mech]*
kaiseki-ryōri	会席（懐石）料理	individual-tray (simplistic) dinner *[cook]*
kaisen	回線；かいせん	circuit; line *{n} [elec]*
kaisen	開繊；開纖	opening (of fibers) *{n} [tex]*
kaisen-fuka	回線負荷	circuit load; line load *[elec]*
kaisen'jō no	回旋状の	convolute *{adj} [bot]*
kaisen-kōkan	回線交換	circuit (or line) switching *{n} [comm]*
kaisen-sokudo	回線速度	line speed *[comm]*
kaisen-zu	回線図	circuit diagram *[elec]*
kaisetsu	解説	explanation; interpretation *[comm] [pr]*
kaisetsu	回折；かいせつ	diffraction *[phys]*
kaisetsu-gōshi	回折格子	diffraction grating *[spect]*
kaisetsu-kaku	回折角	diffraction angle *[phys]*
kaisetsu-zukei	回折図形	diffraction pattern *[crys]*
kaisha	会社；會社	company (-nies) *[econ]*
kaishaku	解釈；解釋	interpretation *[comm]*
kaishaku suru	解釈する	interpret *{vb} [comm]*
kaishi	開始	start(ing)-up *{n} [ind-eng]*
kaishi-hannō	開始反応	initiation reaction *[poly-ch]*
kaishin	海震	sea shock; sea quake *[geophys]*
kaishi-shingō	開始信号	start signal *[comput]*
kai'shitsu-zai	改質剤	modifier *[poly-ch]*
kaishi-zai	開始剤	initiator *[poly-ch]*
kai'shoku-shokubutsu	灰色植物	Glaucophyta *[bot]*
kaishoku'tai	灰色体	graybody *[thermo]*
kaisho-tai	楷書体	script (print) style kanji writing *[pr]*
kaishū	回収；回收	recovery; salvage *{n} [ecol]*
"	回収	harvesting *{n} [microbio]*
"	回収	collection *{n}*
kaisō	海草	marine algae; seaweed *[bot]*
kaisō	海葱；かいそう	sea onion; squill *[bot]*
kaisō	階層	hierarchy *[math]*
kaisō'chōseki	灰曹長石	oligoclase *[miner]*
kaisō-kōzō	階層構造	hiearchical structure *[comput]*
kaisū	回数；回數	frequency; number of times *[phys]*
kaisū	階数	rank *{n} [math]*

kaisui	海水	seawater; salt water *[ocean]*
kai'te	買手	buyer *[econ]*
kaitei	改訂	revision *[pr]*
kaitei	海底	ocean floor *[ocean]*
kaitei	開廷	court session *[legal]*
kaitei-ban	改訂版	revision; revised edition *[pr]*
kaitei-funka	海底噴火	submarine eruption *[geophys]*
kaitei-kazan	海底火山	submarine volcano *[geol]*
kaitei'tai	海底体；海底體	magma (pl -s, -ta) *[geol]*
kaiten	回転；回轉	rotation; revolution *[math] [mech]*
kaiten-antei-ki	回転安定器	gyrostabilizer *[eng]*
kaiten-ban	回転盤	turntable *[eng-acous]*
kaiten-bane	回転羽根	moving blade; moving vane *[mech-eng]*
kaiten-bin	回転瓶	roller bottle *[cyt] [eng]*
kaiten-danmen-zu	回転断面図	revolved section *{n} [graph]*
kaiten-hankei	回転半径	turning radius *[mech-eng]*
kaiten-hanten (kaihan)	回転反転（回反）	rotatory inversion *[math]*
kaiten-isei'tai	回転異性体	rotamer; rotational isomer *[ch]*
kaiten-jiku	回転軸	rotating axis *[mech]*
kaiten-ki(ki)	回転機(器)	rotary machine *[mech-eng]*
kaiten-kyō'ei	回転鏡映	rotatory reflection *[opt]*
kaiten-ro	回転炉；回轉爐	rotary kiln *[eng]*
kaiten-seikei	回転成型	rotational molding *{n} [poly-ch]*
kaiten'shi	回転子；かいてんし	rotor *[elec]*
kaiten-shindō	回転振動	rotational vibration *[spect]*
kaiten-shintō-baiyō	回転振盪培養	rotary shaking culture *[microbio]*
kaiten-sokudo-kei	回転速度計	tachometer *[eng]*
kaiten suru	回転する	rotary; rotating *{adj} [math]*
"	回転する	revolve *{vb i} [math]*
kaiten-tobira	回転扉	revolving door *[arch]*
kaiten-ude	回転腕	crank *{n} [mech-eng]*
kaiten-undō	回転運動	rotational motion *[mech]*
kaitetsu'kiseki	灰鉄輝石	hedenbergite *[miner]*
kaitetsu'zakuro-ishi	灰鉄柘榴石	andradite *[miner]*
kaitō	解凍	thawing *{n} [eng]*
kaitō	解糖	glycolysis *[bioch]*
kaitō'jō no	海島状の	sea-island form *{adj} [geol]*
kaitō suru	解答する	solve *{vb t} [math]*
kaitsuburi	鸊鷉；かいつぶり	grebe (bird) *[v-zoo]*
kai-uke	櫂受け	oarlock *{n} [nav-arch]*
kai'un	海運	sea shipping *{n} [trans]*
kaiwa	会話；會話	conversation *[comm]*

kaiwa-gata	会話形; 会話型	conversational (interactive) mode *[comput]*
kaiwa'gata-shori	会話型処理	interactive processing *{n} [comput]*
kaiwa-hō'shiki	会話方式	conversational mode *[comput]*
kaiyō	海洋	ocean *[geogr]*
kaiyō	潰瘍; かいよう	ulcer *[med]*
kaiyō'gaku	海洋学	oceanography *[geophys]*
kaiyō-kikō	海洋気候	marine climate *[climat]*
kaiyō-seibutsu'gaku	海洋生物学	marine biology *[bio]*
kaiyō'tei-kaku'dai-setsu	海洋底拡大説	sea floor spreading theory *[geol]*
kaizai'butsu	介在物	inclusion *{n} [crys] [met] [petr]*
kaizan	海山	seamount *[geol]*
kaizō	改造	rebuilding; remodeling *{n} [arch]*
kaizō(-do)	解像(度)	resolution (degree of) *[opt]*
kaizō-ryoku	解像力	resolving power *[opt]*
kaizu	海図; 海圖	hydrographic chart *[map]*
kaizu-yōshi	海図用紙	chart paper *[graph] [paper]*
kaji	舵; かじ	rudder *[eng]*
kaji	火事	conflagration; fire *{n} [ch]*
kajika	鰍; かじか	bullhead; Cottus pollus (fish) *[v-zoo]*
kajiki	旗魚; かじき	swordfish (fish) *[v-zoo]*
ka'jikō	過時効	overaging treatment *[met]*
kajiru	齧る; かじる	gall *{vb i} [met]*
"	齧る	gnaw; nibble; bite *{vb}*
kaji'tori-ha'guruma	舵取り歯車	steering gear *[mech-eng]*
kaji'tori-sōchi	舵取り装置	steering system *[mech-eng]*
ka'jitsu	果実; 果實	fruit *{n} [bot]*
kajitsu'gaku	果実学	carpology *[bot]*
kajitsu'jo	果実序	infructescence *[bot]*
kajitsu-senbetsu-ki	果実選別機	fruit grader; fruit sorter *[agr]*
kajitsu'shu	果実酒	fruit wine *[cook]*
kaji'ya	鍛冶屋	blacksmith (a person) *[met]*
kajo	花序	inflorescence *[bot]*
kajō	箇条; 箇條	item *{n}*
kajō	過剰; 過剩	excess *{n}*
kajō-jūten	過剰充填	overpacking *{n} [poly-ch]*
kajō-seisan	過剰生産	overproduction *[ind-eng]*
kajō-shintan	過剰浸炭	excess carburizing *{n} [met]*
kaju	果樹	fruit tree *[bot]*
kajū	果汁	fruit juice *[bot] [cook]*
kajū	荷重	load *{n} [mech]*
kaju'en	果樹園	orchard; fruit orchard *[agr]*
kajū-heikin	加重平均	weighted average *[math]*

kajū-shibori'gu	果汁絞り具	reamer; juicer *[eng]*
kakan	花冠	corolla *[bot]*
kakan-sei	可換性	commutative property *[math]*
ka'kanshō-sei	可干渉性	coherence *[phys]*
ka'kansoku-sei	可観測性	observability *[cont-sys]*
ka'karyū	過加硫	over-vulcanization *{n} [rub]*
kakato	踵; かかと	heel (of foot) *{n} [anat]*
kakato'zuke	踵付け	heeling (a shoe) *{n} [eng]*
kake	欠け; 缺け; かけ	chip; chipping *{n} [cer]*
kake-e	掛け絵	hanging picture (ornament) *{n} [graph]*
kakei'shiki no	可傾式の	tiltable *{adj} [eng]*
kake-jiku	掛け軸	hanging scroll (ornament) *[graph]*
kake-mono	掛け物	hanging scroll (ornament) *[graph]*
kakera	欠片; かけら	shard *{n} [archeo]*
kakeru	掛ける	multiply *{vb} [math]*
"	掛ける	hang; suspend *{vb}*
kakeru	駆ける; 驅ける	run; race *{vb}*
kakete yuku tsuki	欠けてゆく月	waning moon *[astron]*
kake-zan	掛け算	multiplication *[math]*
kaki	垣	fence; fencing *{n} [arch]*
kaki	柿	persimmon (fruit, tree) *[bot]*
kaki	牡蠣; かき	oyster (shellfish) *[i-zoo]*
kaki	夏期	summer; summertime *[astron]*
kaki'dashi	書き出し	writing; writing-out *[comput]*
kaki-dasu	書き出す	write *{vb} [comput]*
Kaki'emon-yaki	柿右衛門焼	Kakiemon porcelain *[cer]*
kaki-engei	花卉園芸	floriculture *[bot]*
kaki'komi	書き込み	writing; writing-in *{n} [pr]*
kaki'komi-hogo	書き込み保護	write protection *{n} [comput]*
kaki-komu	書き込む	write *{vb} [comput]*
kaki'maze-bō	掻き混ぜ棒	stirring rod *[an-ch]*
kaki'naoshi	書き直し	rewriting; transcription *{n} [comput]*
ka'kisei	過寄生	hyperparasitism *[ecol]*
kakitsubata	燕子花; 杜若	iris (flower) *[bot]*
kakkazan	活火山	active volcano *[geol]*
kakki'teki na	画期的な; 畫期的な	epochal; revolutionary *{adj}*
kakko	括弧	parenthesis (pl -ses) *[math] [pr]*
kakkō	恰好; 格好	appearance; shape; form *{n}*
kakkō	郭公; かっこう	cuckoo, Japanese (bird) *[v-zoo]*
kakō	下降	descension *[math]*
kakō	下綱	infraclass *[bio] [syst]*
kakō	火口	crater *[geol]*

kakō	加工；かこう	finishing; processing *{n} [eng]*
"	加工	machining *{n} [met]*
kakō	河口	estuary *[geogr]*
kakō-denpun	化工澱粉	modified starch *[bioch]*
kakō'gan	花崗岩；かこうがん	granite *[petr]*
kakō-gen	火口原	crater basin (of a volcano) *[geol]*
kakō-hentai	加工変態	work transformation *[met]*
kakō-jozai	加工助剤	processing aid *[poly-ch] [rub]*
kakō'jutsu	火工術	pyrotechnics *[eng] [mat]*
ka'kōka	過硬化	overcure *{n} [poly-ch] [rub]*
kakō-kei	加工径；加工徑	working diameter *[math]*
kakō-kiryū	下降気流	downdraft *{n} [phys]*
kakō-ko	火口湖	crater lake *[hyd]*
kakō-kōka	加工硬化	work hardening *{n} [met]*
kakō-konseki	加工痕跡	worked traces *[met]*
kakon	仮根；假根	rhizoid *{n} [bot]*
kakō-ō'ryoku	加工応力	working stress *[met]*
kakō-ryū	下向流	downward flow *[fl-mech]*
kakō-sei	加工性	forming property; processibility; workability *[eng] [mat]*
kakō-shi	加工糸	finished yarn *[text]*
kakō-shi	加工紙	converted paper *[mat] [paper]*
kakō suru	加香する	add for fragrance *{vb} [food-eng]*
kakō-toppū	下降突風	downgust *[phys]*
ka'kotsu	仮骨；假骨	callus *{n} [med]*
kakō-yūki no	加工誘起の	strain-induced *{adj} [mech]*
kaku	角	angle *{n} [math]*
kaku	画	stroke (of a Chinese character) *[pr]*
kaku	格	case *{n} [gram]*
kaku	核	nucleus (pl nuclei, -es) *[astron] [cyt] [nucleo] [nuc-phys] [sci-t]*
"	核	kernel *[atom-phys] [bot]*
"	核	core *{n} [geol]*
kaku; kara	殼；殻	shell *{n} [ch] [eng]*
kaku	書く	write *{vb}*
kakuban'gan	角蛮岩；角蠻岩	breccia *[miner]*
kaku'bun	画分	fraction *[sci-t]*
kaku-bun'retsu	核分裂	mitosis (pl -toses) *[cyt]*
(kaku-)bunretsu	(核)分裂	nuclear fission *[nuc-phys]*
kaku-bunretsu-rensa-hannō	核分裂連鎖反応	fission chain reaction *[nuc-phys]*
kaku-bunretsu-ro	核分裂炉	fission reactor *[nucleo]*
kaku-bunretsusei-busshitsu	核分裂性物質	fissionable material *[nucleo]*

kaku-butsuri'gaku	核物理学	nuclear physics *[phys]*
kaku'chi	画地；畫地	lot; plot *{n} [civ-eng]*
kaku'chō	角頂；かくちょう	vertex (pl -es, vertices) *[math]*
kaku'chō	拡張；擴張	escape *{n} [comput]*
kakuchō-kanō	拡張可能	open-ended *[comput]*
kakuchō-kanō-gengo	拡張可能言語	extensible language *[comput]*
kakuchō-ki'oku	拡張記憶	extended storage *[comput]*
kaku'chū	角柱	prism *[crys] [math] [opt]*
kakudai-kyō	拡大鏡	magnifying glass *[opt]*
kakudai-ritsu	拡大率	magnifying power *[opt]*
kaku'dai suru	拡大する	expand *{vb} [comput]*
"	拡大する	enlarge *{vb} [eng]*
kakū-densen	架空電線	overhead wire *[elec]*
kakudo	角度	angle *{n} [math]*
kaku'eki-teisha no	各駅停車の	local (not express) *[rail] [trans]*
kaku-enerugī	核エネルギー	nuclear energy *[nucleo]*
kaku'enkō	角鉛鉱；角鉛鑛	phosgenite *[miner]*
ka'kuen-san	過枸櫞酸	percitric acid
kaku'gata	角形	square *{n} [math]*
kaku'gata	核型	karyotype *[cyt]*
kaku'ginkō	角銀鉱	cerargyrite *[miner]*
kaku'han	攪拌；かくはん	agitation; stirring *{n} [mech-eng]*
kakuhan-bō	攪拌棒	stirring rod *[mat]*
kakuhan-ki	攪拌機	stirrer; agitator *[ch-eng]*
kaku-hannō	核反応	nuclear reaction *[nuc-phys]*
kakuhan-sōchi	攪拌装置	agitator (apparatus) *[ch-eng]*
kakuhan-yoku	攪拌翼	impeller *[mech-eng]*
kaku'heki	隔壁	bulkhead *[aero-eng] [nav-arch]*
kaku'hi	核皮	cascara *[pharm]*
kakū-hiki'komi	架空引込	aerial lead-in *[elec]*
kaku'itsu'teki na	画一的な	uniform; standardized *{adj}*
kaku-jiki-kyōmei	核磁気共鳴	nuclear magnetic resonance *[phys]*
kaku'ka	核果	drupe *[bot]*
kaku-kagaku	核化学	nuclear chemistry *[atom-phys]*
kaku'kan-kyori	核間距離	internuclear distance *[p-ch]*
kaku-keisei	核形成	nucleation *[ch]*
kaku-kō	角鋼	square (steel) bar *[civ-eng]*
kaku-kōgaku	核工学	nucleonics *[eng]*
kaku koto	書く事	writing; to write *{n} [comm]*
kaku-kōzō	核構造	shell structure *[nuc-phys]*
kaku'kyaku-rui	核脚類	Tylopoda (camels, etc.) *[v-zoo]*
kaku'maku	角膜	cornea *[anat]*

kaku'maku	核膜	nuclear membrane *[cyt]*
kaku'maku	隔膜；かくまく	diaphragm *[anat] [phys]*
kaku-nenryō	核燃料	nuclear fuel *[nucleo]*
kaku'nin	確認	confirmation; identification *{n}*
kakū no	架空の	imaginary; fanciful *{adj} [psy]*
kakunō'ko	格納庫	hangar (for airplanes) *[civ-eng]*
kaku'nō-shitsu	格納室	bay *{n} [civ-eng]*
kaku'nō suru	格納する	house; contain *{vb t}*
kaku no tan'i	角の単位	unit of angle *[math]*
kakure-gani	隠れ蟹	oyster crab; pea crab (crustacean) *[i-zoo]*
kaku'reki	角礫；かくれき	rubble *{n} [geol]*
kaku'reki'gan	角礫岩	breccia *[miner]*
kakureta	隠れた	hidden; invisible *{adj} [graph] [opt]*
kaku'ri	隔離	isolation *[microbio]*
kaku'ritsu	確率；かくりつ	probability *{n} [math]*
kaku'ritsu-hensū	確率変数	random variable *[math]*
kaku'ritsu-katei	確率過程	stochastic process *[math]*
kaku'ritsu-mitsudo-kansū	確率密度関数	probability density function *[math]*
kaku-ryō	角稜	angle edge *{n} [math]*
kaku-ryoku	核力	nuclear force *[nuc-phys]*
kaku'san	拡散；擴散	diffusion *[opt] [phys]*
kaku-san	核酸；かくさん	nucleic acid *[bioch]*
kakusan-ban	拡散板	deflector *[eng]*
kakusan-bunkai-kōso	核酸分解酵素	nuclease; nucleolytic enzyme *[bioch]*
kakusan-en	核酸塩	nucleate *{n} [bioch]*
kakusan-hansha	拡散反射	diffuse reflection *[opt]*
kakusan-keisū	拡散係数	diffusivity; diffusion coefficient *[phys]*
kakusan-rissoku-denryū	拡散律速電流	diffusion-controlled current *[elec]*
kakusan-ritsu	拡散率	diffusivity *[phys]*
kakusan-sō	拡散層	diffusion layer *[phys]*
kakusan-tōseki	拡散透析	diffusion dialysis *[p-ch]*
kakusei-ki	拡声器；擴聲器	loudspeaker *[acous]*
kaku-seisei	核生成	nucleation *[ch]*
kakusei-sōchi	拡声装置	public address system *[acous]*
kakusei-yaku	覚醒薬；覺醒藥	antihypnotic *{n} [pharm]*
kaku'sen	画線	streak *{n} [microbio]*
kaku'senseki	角閃石	hornblende *[miner]*
kaku'shi	核子	nucleon *[phys]*
kakushi'ba	隠し場	cache *{n} [comput]*
kakushin	革新	innovation; reform; renovation *{n}*
kakushin	確信	confidence *{n} [psy]*
kaku-shindō'sū	角振動数	angular frequency *[phys]*

kaku'shitsu	角質；かくしつ	keratin *[bioch]*
kakushu	核種	nuclide *[nuc-phys]*
kaku'shū no	隔週の	every other week; fortnightly *{adv}*
kaku-shutsu'ryoku	核出力	nuclear yield *[nucleo]*
kaku-sokudo	角速度	angular velocity *[mech]*
kakusu	隠す	conceal; hide *{vb} [comput]*
kaku'sui	角錐；かくすい	pyramid *[math]*
kakusui-dai	角錐台	truncated pyramid *[math]*
kakutoku-keishitsu	獲得形質	acquired character *[bio]*
kakutoku-men'eki	獲得免疫	acquired immunity *[immun]*
kakutoku-men'eki-fuzen'shō	獲得免疫不全症	acquired immune deficiency syndrome; AIDS *[med]*
kaku'tsuke; kaku'zuke	格付	grading *{n} [ind-eng]*
kaku-undō	角運動	angular motion *[mech]*
kaku-undō'ryō	角運動量	angular momentum *[mech]*
kaku'undō'ryō-hozon-soku	角運動量保存則	conservation of angular momentum, law of *[mech]*
kaku-yūgō	核融合	nuclear fusion *[nuc-phys]*
kaku-zai	角材	square timber *{n} [civ-eng]*
kaku-zatō	角砂糖；かくざとう	sugar cube *[cook]*
kaku-zōshoku-ro	核増殖炉	nuclear breeder reactor *[nucleo]*
kakyō	架橋；かきょう	cross-linking *{n} [org-ch]*
kakyō-mitsudo	架橋密度	crosslinking density *[org-ch]*
kakyō-saku'tai	架橋錯体	bridged complex *[ch]*
ka'kyōseki-shō	過共析晶	hypereutectoid *[met]*
ka'kyōshō	過共晶	hypereutectic *[met]*
kakyō-ten	架橋点	cross-link site *[poly-ch]*
kakyō-zai	架橋剤	cross-linking agent *[poly-ch]*
kakyū	火球	bolide *[astron]*
"	火球	fireball *[astron] [nucleo]*
kakyū	過給	supercharging *{n} [mech-eng]*
kakyū-ki	過給機	supercharger *[mech-eng]*
kama	釜；かま	cauldron; kettle; pot *{n} [cook]*
kama	窯	furnace; kiln *[cer] [eng]*
"	窯	oven *[cook]*
kama	鎌	sickle *[agr] [eng]*
kamaboko	蒲鉾；かまぼこ	boiled fish pudding dish *[cook]*
kamae	構え	kanji (q.v.) encasement radical *[pr]*
kamakiri	蟷螂；かまきり	praying mantis (insect) *[i-zoo]*
ka'mangan-san	過マンガン酸	permanganic acid
ka'mangan'san-en	過マンガン酸塩	permanganate *[ch]*
kamasu	梭魚；魳；かます	saury-pike; barracuda (fish) *[v-zoo]*

kama'zan	釜残	still residue *[ch-eng]*
kama'zan-eki	釜残液	residue solution *[ch-eng]*
kame	瓶；甕；かめ	crock; earthenware jug; pot; urn *[cer]*
kame	亀；龜	turtle (reptile) *[v-zoo]*
kame-mushi	亀虫；椿象	stinkbug (insect) *[i-zoo]*
kamen-zu	下面図	bottom view *[graph]*
kame-rui	亀類；龜類	Chelonia (tortoises, turtles) *[v-zoo]*
kami	紙	paper *{n} [mat] [paper]*
kami	髪；髮	hair (of head) *[anat]*
kami'ai	嚙(み)合(い)	intermeshing *{n} [mech-eng]*
kami'ai-seido	嚙合精度	intermeshing precision *[mech-eng]*
kami'awase-yō'setsu	嚙合せ溶接	seam welding *{n} [met]*
kami-kazari	髪飾り	hair ornament *[lap]*
kami-kōgaku	紙工学	paper technology *[paper]*
ka'min	夏眠	aestivation *[physio]*
kaminari	雷；かみなり	thunder *{n} [geophys]*
kaminari-hōden	雷放電	lightning discharge *[geophys]*
kami no	上の	higher; upper *{adj} [math]*
kami no ke	髪の毛	hair (of head) *[anat]*
kami'okuri-kikō	紙送り機構	carriage *[comput]*
"	紙送り機構	paper feed mechanism *[comput]*
ka'mitsu	果蜜	nectar *[bot]*
kami-yasuri	紙鑢；紙やすり	sandpaper *{n} [mat]*
kami-zaiku	紙細工	papercrafted ware *[furn]*
kamo	鴨；かも	duck (bird, wild) *[v-zoo]*
kamome	鷗；かもめ	gull; seagull (bird) *[v-zoo]*
kamo-nanban	鴨南蛮；鴨南蠻	duck soup noodles *[cook]*
kamonohashi	鴨嘴；かものはし	platypus *[v-zoo]*
ka'motsu	貨物	railway freight *[trans]*
kamotsu-jidōsha	貨物自動車	truck *{n} [mech-eng]*
kamotsu'sen	貨物船	cargo vessel; freighter *[nav-arch]*
kamotsu-shitsu	貨物室	hold *{n} [nav-arch]*
kamu	嚙む；咬む	chew; bite *{vb}*
kan	勘	hunch *[psy]*
kan	桿	rod *{n} [eng]*
-kan	-巻；-卷	-scroll(s) *{suffix}*
kana	仮名；假名；かな	Japanese alphabet *[comm] [pr]*
kana-ami	金網	wire netting; wire gauze *{n} [ch]*
Kanada	カナダ；加奈陀	Canada *[geogr]*
kanae	鼎；かなえ	tripod; tripod kettle *[met]*
kana-gata	金型	die; mold *{n} [eng] [poly-ch]*
kana'gu	金具	hardware *[eng]*

kaname	要；かなめ	pivot *{n} [mech]*
kana'mono	金物	hardware *[comput] [eng]*
kana'shiki	鉄敷き	anvil *[met]*
kana'shiki-dai	鉄敷台	anvil block *[met]*
kana'toko-gumo	鉄床雲	incus; anvil cloud; thunderhead *[meteor]*
kan'atsu-shi	感圧紙	pressure-sensitive paper *[paper]*
kana'zuchi	金槌；かなづち	hammer *{n} [eng]*
kana-zukai	仮名遣い	kana (q.v.) orthography *[comm]*
kanba; kaba	樺	birch (tree) *[bot]*
kanban; kōhan	甲板	deck (of a ship) *{n} [nav-arch]*
kanban	看板	signboard; sign *{n} [econ]*
kanbatsu	旱魃	drought *[climat]*
kanbatsu	間伐	thinning (of trees) *{n} [ecol] [for]*
kan'botsu saseru	陥没させる	sink; let collapse *{vb t}*
kanben-sa	簡便さ	convenience *{n}*
kanbu	患部	afflicted part; affected part *[med]*
kanchaku	嵌着；かんちゃく	attaching, snapped (fitted) in *[eng]*
kanchi	換地	replotting *{n} [civ-eng]*
kanchi-ki	感知器	sensor *[eng]*
kanchō	干潮	ebb tide; low tide *[ocean]*
kanchō-zai	灌腸剤；浣腸剤	enema *{n} [pharm]*
kanchū-eki	灌注液；浣注液	douche; irrigation *{n} [pharm]*
kandan-kei	寒暖計	thermometer *[eng]*
kanden	感電	electric shock; electrification *[elec]*
kan'denchi	乾電池	dry cell *[elec]*
kando	感度	sensitivity *[electr] [sci-t]*
"	感度	sensitiveness *[sci-t]*
kane	金；かね	money (colloquial) *[econ]*
"	金	metal *[mat]*
kane	鉦	chime; gong *{n} [music]*
kane	鐘	bell *[music]*
kane-hen	金扁；金偏：金	kanji (Chinese character) radical denoting metal *[pr]*
ka'nen	加撚	twisting *{n} [tex]*
kan'en	肝炎	hepatitis *[med]*
ka'nen'butsu	可燃物	combustible matter *[mat]*
ka'nen no	可燃の	flammable *{adj} [ch]*
kane no shita	鐘の舌	clapper (of bell) *[music]*
ka'nen-sei	可燃性	combustibility *[ch]*
kane-seidō	鐘青銅	bell bronze *[met]*
ka'netsu	加熱；かねつ	heating *{n} [thermo]*

ka'netsu-do	過熱度；かねつど	degree of superheating *[thermo]*
ka'netsu-jōki	過熱蒸気	superheated vapor *[thermo]*
ka'netsu suru	加熱する	heat *{vb t} [thermo]*
ka'netsu suru	過熱する	superheat *{vb t} [thermo]*
kangae	考え；かんがえ	thought; opinion; idea *{n} [psy]*
kangaeru	考える	think *{vb i} [psy]*
kangaeru koto	考えること	thinking *{n} [psy]*
kangai	灌漑；かんがい	irrigation; watering *{n} [agr]*
kangai no	環外の	exocyclic *{adj} [org-ch]*
kan-gakki	管楽器；管樂器	wind instruments *[music]*
kangeki	間隙；かんげき	gap *{n} [bot] [elec]*
"	間隙	crevice; interstice *[sci-t]*
kangeki-busshitsu	間隙物質	interstitial matter *[sci-t]*
kangeki-ritsu	間隙率	porosity *[phys]*
kangen	還元；かんげん	reduction *[ch] [comput]*
kangen-den'i	還元電位	reduction potential *[p-ch]*
kangen'en	還元炎	reducing flame *[ch]*
kangen-hyō	乾舷標	Plimsoll mark *[nav-arch]*
kangen-nendo	還元粘度	reduced viscosity *[poly-ch]*
kangen-nyūtō	還元乳糖	lactitol *[bioch]*
kangen suru	還元する	reduce; deoxidize *{vb t} [ch]*
kangen'tai	還元体	reductant; reduction product *[ch]*
kangen-tō	還元糖	reducing sugar *[org-ch]*
kangen-zai	還元剤	reducing agent; reductant *[ch]*
kan-gezai	緩下剤	laxative *{n} [pharm]*
kangō	嵌合；かんごう	engagement; fit; fitting-together *[eng]*
kango'fu	看護婦；かんごふ	nurse (woman) *{n} [med]*
kango'shi	看護士	nurse (man) *{n} [med]*
kangō suru	嵌合する	fit together; snap together *{vb} [eng]*
kan'hentō-yu	甘扁桃油	sweet almond oil *[mat]*
kani	蟹；かに	crab (crustacean) *{n} [i-zoo]*
kan'i	簡易	spreadsheet *{n} [comput]*
kani'kui-zaru	蟹喰い猿	crab-eating macaque (animal) *[v-zoo]*
kan'i-meidai	換位命題	converse *{n} [logic]*
kan-i'sei	環異性；かんいせい	ring isomerism *[ch]*
kani-sei'un	蟹星雲	Crab Nebula *[astron]*
kani-za	蟹座；かにざ	Cancer; the Crab *[constel]*
kanji	感じ	feeling(s); sensibility; sensation *{n} [physio]*
kanji	漢字；かんじ	Chinese character *[pr]*
kanji-fundō	感じ分銅	sensing weight *{n} [mech]*
kanjiki-usagi	樏兎；檋兎；かん=じきうさぎ	snowshoe hare (animal) *[v-zoo]*

kanjiru	感じる	feel; experience; suffer *{vb i} [physio]*
kanjō	桿状；かんじょう	rod form *{n} [microbio]*
kanjō	勘定	count; computation; tally *{n} [math]*
kanjō	感情；かんじょう	emotion; feeling *{n} [psy]*
kanjō	環状；環狀	cyclic form *{n} [electr] [org-ch]*
kanjō-kōsa-ro	環状交差路	rotary intersection *[traffic]*
kanjō-kōzō	環状構造	annulation *{n} [i-zoo]*
kanjō'men(-tai)	環状面(体)	toroid; torus *[math]*
kanjō no	環状の	annular *{adj} [anat] [astron] [trans]*
kanjō o suru	勘定をする	pay the bill; settle accounts *{vb} [econ]*
kanjō-sei'un	環状星雲	Ring Nebula *[astron]*
kanjō-sukima	環状隙間	annular clearance *[eng] [mech-eng]*
kanjō suru	勘定する	count; compute *{vb t} [math]*
kanjō'tai	桿状体；杆状体	rod *{n} [anat] [microbio]*
kanjō-tanka'suiso	環状炭化水素	cyclic hydrocarbon *[org-ch]*
kanju-sei	感受性	susceptibility *[physio]*
kanka	環化；かんか	cyclization *[org-ch]*
kanka-jūgō	環化重合	cyclopolymerization *[poly-ch]*
kan'kaku	間隔；かんかく	space; spacing; gap; pitch *{n} [comput] [pr]*
kan'kaku	感覚；感覺	sensation *[physio]*
kankaku-keiji-kikō	間隔計時機構	interval timer *[eng]*
kankaku-ki(kan)	感覚器(官)	sense organ *[physio]*
kankaku o hedateta	間隔をへだてた	leaving a space; with space left *{adj}*
kankaku-saibō	感覚細胞	sensory cell *[physio]*
kankaku-settei	間隔設定	designation of spacing *{n} [comput]*
kankaku-shinkei	感覚神経	sensory nerve *[physio]*
kan'karyū-soku'shin-zai	緩加硫促進剤	slow accelerator *[rub]*
kankei	関係；かんけい	relation; relationship *[math] [sci-t]*
kankei-dōbutsu	環形動物	Annelida; earthworms *[i-zoo]*
kan-keisei	環形成	ring formation *[org-ch]*
kankei-shiki	関係式；關係式	relational expression *[math]*
kan'ketsu	間欠；かんけつ	intermittency *{n} [eng] [mech-eng]*
kanketsu'bun	完結文	sentence *{n} [gram]*
kanketsu-(fun)sen	間欠(噴)泉	geyser *[hyd]*
kanketsu-rokō-kōka	間欠露光効果	intermittency effect *[photo]*
kanki	刊記；かんき	imprint *{n} [pr]*
kanki	換気；換氣	ventilation *[eng]*
kanki-ki	換気機	ventilator *[eng]*
kankin	桿菌；かんきん	bacillus (pl -li) *[microbio]*
kan-kiri	缶切り；罐切り	can opener *[eng]*
kankitsu	柑橘；かんきつ	citrus (fruit) *[bot]*

kankō 甘汞；かんこう calomel *[miner]*
kankō-do 感光度 sensitivity *[photo]*
kan-kōgen 乾荒原 desert *{n}* *[geogr]*
kankō-genshi 感光原紙 sensitizing paper *[paper]*
kankō-jidōsha 観光自動車 sightseeing car *[mech-eng]*
kankō-kaku 感光核 sensitivity speck *[photo]*
kankō-kei 感光計 sensitometer *[photo]*
kankō-kyaku 観光客；觀光客 tourist *[trans]*
kankō-sei 感光性 photosensitiveness *[photo]*
kankō'sei-jushi 感光性樹脂 photopolymer *[poly-ch]*
kankō-shi 感光紙 sensitized paper *[photo]*
kankō suru 感光する respond optically *{vb}* *[opt]*
kankō'tai 乾膠体 xerogel *[ch]*
kankō-tai 感光体 (light-)sensitive material *[photo]*
kankyō 環境；かんきょう environment *[comput]* *[ecol]*
kankyō 艦橋 bridge *{n}* *[nav-arch]*
kankyō-sōsa-denshi'kenbikyō 環境走査電子顕微鏡 environmental scanning electron microscope; ESEM *[electr]*
kankyū-hō 環球法 ring and ball method *[ch-eng]*
kankyū-ondo-kei 乾球温度計 dry-bulb thermometer *[eng]*
kankyū-sokudo 緩急速度 tempo (pl -s, tempi) *[music]*
kanman-nenshō 緩慢燃焼 slow combustion *[ch]*
kanmen-zō 完面像 holohedral form; holohedry *[crys]*
kan'mi; ama'mi 甘味；かんみ sweetness *{n}* *[cook]* *[physio]*
kanmi-ka'jitsu'shu 甘味果実酒 fortified wine *[cook]*
kanmi-ryō; (-zai) 甘味料；（-剤） sweetening agent *[cook]* *[pharm]*
kan-mon'myaku-kei 肝門脈系 hepatic portal system *[anat]*
kanmuri 冠；かんむり crest *{n}* *[arch]* *[cl]* *[nav-arch]*
" 冠 crown *{n}* *[mech]*
" 冠 crown part of a Chinese character *[pr]*
kanmuri-ha'guruma 冠歯車 crown gear *[mech]*
kanmuri-nokogiri 冠鋸 crown saw (a tool) *[eng]*
kanmuri-zuru 冠鶴 crowned crane (bird) *[v-zoo]*
kanna 鉋；かんな plane (a tool) *{n}* *[eng]*
kannai no 環内の intraannular *{adj}* *[ch]*
kanna-kuzu 鉋屑 wood shavings *[mat]*
kannetsu-setchaku-zai 感熱接着剤 heat-sensitive adhesive agent *[adhes]*
kannetsu-shi 感熱紙 thermosensitive paper *[mat]*
kannō-kensa 官能検査 sensory test *[physio]*
kannō-ki 官能基；かんのうき functional group *[org-ch]*
kannō-kōryō 感応光量 responding-light quantity *[opt]*
kannō-sei 官能性 functionality (the property) *[poly-ch]*

kannō'sū	官能数	functionality (the numeral) *[poly-ch]*
kannuki	閂；かんぬき	bolt (for a door); latch *{n} [eng]*
kannuki-dome	閂止め	crow's foot *[cl]*
kannuki-nui	閂縫い	bartack *{n} [cl] [eng]*
kannyū-do	貫入度	quantity penetrated *[civ-eng]*
kannyū-shiken	貫入試験	penetration test *[civ-eng]*
kanoko'sō	鹿の子草；かのこ草	valerian; Valeriana fauriei *[bot]*
kanoko-zome	鹿の子染	dappled cloth *[tex]*
kan'on-soshi	感温素子	temperature-sensitive element *[electr]*
kanō-sei	可能性	possibility *[math]*
kanpa	寒波	cold wave *[meteor]*
kanpan	乾板	dry plate; photographic dry plate *[photo]*
kanpō	官報	official gazette *[pat]*
kanpō	漢方	Chinese medicine (the field) *[med]*
kanpo-dōbutsu	緩歩動物	Tardigrada *[i-zoo]*
kanpō-yaku	漢方薬	Chinese medicinal drug *[med]*
kanpyō	干瓢；かんぴょう	dried gourd strips (or shavings) *[cook]*
kanpyō-ki	間氷期	interglacial period *[geol]*
kanrak(u)kō	陥落孔	sinkhole *[geol]*
kanran'seki	橄欖石	olivine *[miner]*
kanrei-zensen	寒冷前線	cold front *[meteor]*
kanren	関連；關連	relationship *[math]*
kanren-sei	関連性	relevancy *[math]*
kanren-seigyo-kei	関連制御系	related control system *[eng]*
kanri	管理；かんり	administration; management *{n} [ind-eng]*
kanri-kōgaku	管理工学	management engineering *[eng]*
kanri'sha	管理者	manager (a person) *[ind-eng]*
kanro	管路	conduit; duct *[mech-eng]*
kanro-ni	甘露煮	sweet-stewed dish *[cook]*
kanryaku-ki'oku-kigō	簡略記憶記号	mnemonic symbol *[comput]*
kanryō	完了	completion; finish *{n}*
kanryō	乾量	dry weight *[mech]*
kanryū	乾留	carbonization; dry distillation *[ch]*
kanryū	環流；かんりゅう	gyre *[ocean]*
kanryū	還流	reflux; refluxing *{n} [ch-eng]*
kanryū-hi	還流比	reflux ratio *[ch-eng]*
kanryū-rei'kyaku-ki	還流冷却機	reflux condenser *[ch-eng]*
kansa	監査	auditing *{n} [comput] [econ]*
kan-saibō	幹細胞；幹細胞	stem cell *[cyt]*
kansan-hyō	換算表	conversion table *[math]*
kansan suru	換算する	convert (mathematically) *{vb} [math]*
kan'satsu	観察	observation *{n}*

kan'satsu 鑑札 license *{n} [legal]*
kansatsu-chi 観察値 observed value *[math]*
kansei 慣性；かんせい inertia *[mech]*
kansei-keitai 乾性形態 xeromorphism *[bot]*
kansei no hōsoku 慣性の法則 inertia, law of *[mech]*
kansei-ryoku 慣性力 inertial force *[mech]*
kansei(-sen'i)-kei'retsu 乾性(遷位)系列 xerosere *{n} [ecol]*
kansei-shitsu'ryō 慣性質量 inertial mass *[mech]*
kansei-shokubutsu 乾性植物 xerophyte *[ecol]*
kansei suru 完成する complete; finish *{vb}*
kansei-tō 管制塔 control tower *[navig]*
kansei-yu 乾性油 drying oil *[mat]*
kansei-yūdō 慣性誘導 inertial guidance *[navig]*
kanseki 缶石；罐石 boiler scale; scale *{n} [ch]*
kansen 汗腺 sweat gland *[physio]*
kansen 感染；かんせん infection *[med]*
kansen-genzō 感染現像 infectious printing *[photo]*
kansen'shō 感染症 infectious disease *[med]*
kan'setsu 間接；かんせつ indirect(ness) *{n}*
kan'setsu 関節；關節 joint *{n} [anat]*
kan'setsu 環節 segment *{n} [i-zoo]*
kansetsu-bunseki 間接分析 indirect analysis *[an-ch]*
kansetsu-meirei 間接命令 indirect instruction *[comput]*
kansetsu ni 間接に indirectly *{adv}*
kansetsu-satsu'ei 間接撮影 fluorography *[graph]*
kansetsu-shi 関節肢 jointed appendage *[anat]*
kansetsu-shōmei 間接証明 indirect proof *[logic]*
kansetsu-sokutei 間接測定 indirect measurement *{n} [math]*
kan'shamon-ori 緩斜文織 reclined twill *[tex]*
kanshi 冠詞 article *[gram]*
kanshi 監視 monitoring *{n} [ind-eng]*
kanshi-eisei 監視衛星 surveillance satellite *[aero-eng]*
kanshiki-bōshi 乾式紡糸 dry spinning *{n} [tex]*
kanshiki-ekusu'sen-satsu'ei 乾式エクス線撮影 xeroradiography *[graph]*
kanshiki-hō 乾式法 dry process *{n}*
kanshiki-hōshasen-shashin-jutsu 乾式放射線写真術 xeroradiography *[graph]*
kan'shiki-kagō'butsu 環式化合物 cyclic compound *[org-ch]*
kan'shikkyū-shitsudo-kei 乾湿球湿度計 wet and dry bulb hygrometer *[eng]*
kanshin 感震 vibration-sensitive *[geol]*
kanshi-tan'matsu 監視端末 monitor station *[comput]*
kan'shitsu-kei 乾湿計；乾濕計 psychrometer *[eng]*

kan'shitsukyū-shitsudo-kei	乾湿球湿度計	wet-and-dry-bulb hygrometer *[eng]*
kan'shitsu-zō	乾漆像	dry-lacquered image (ornament)
kansho	甘藷; かんしょ	sweet potato (vegetable) *[bot]*
kanshō	干渉; かんしょう	interference *[phys]*
kanshō	緩衝	buffer *{n} [comput]*
kanshō	環礁	atoll *[geogr]*
kanshō-bunkō-hōhō	干渉分光方法	interference spectroscopy *[spect]*
kansho-denpun	甘藷澱粉	sweet potato starch *[mat]*
kanshō-eki	緩衝液	buffer solution *[ch]*
kanshō-jima	干渉縞	interference fringes *[opt]*
kanshō-kei	干渉計	interferometer *[opt]*
kanshō-ki	緩衝器	buffer; shock absorber *{n} [mech-eng]*
kan-shoku	寒色	cool color *[opt]*
kanshoku	間食	snack *{n} [cook]*
kanshoku	感触; 感觸	texture *[physio]*
kanshō-sei	干渉性	coherence *[phys]*
kanshō'sei-sanran	干渉性散乱	coherent scattering *{n} [elecmg]*
kansho'tō	甘蔗糖	cane sugar *[org-ch]*
kanshō-zai	緩衝剤	buffer; buffering agent *{n} [ch]*
kanshū	監修	editorialship *[pr]*
"	監修	supervision *{n}*
kan'shutsu-eki	缶出液; 罐出液	bottoms *{n} [petrol]*
kansō	乾燥; かんそう	dry; drying *{n} [sci-t]*
kansō-genryō	乾燥減量	loss by drying *{n} [mech]*
kansō-hōshasen-shashin	乾燥放射線写真	xeroradiography *[graph]*
kansō-jōtai	乾燥状態	aridity *[climat]*
kansō-jūryō	乾燥重量	dry weight *[mech]*
kansō-ki	乾燥器	desiccator *[ch-eng]*
kansō-ki	乾燥機	dryer; drier *{n} [eng]*
kansoku	観測; かんそく	observation *[math]*
kansoku'jo	観測所	observatory *[arch]*
kansō(-sayō)	乾燥(作用)	desiccation *[ch-eng]*
kansō suru	嵌挿する	insert, fitting *{vb} [eng]*
kansō suru	嵌装する	mount, fitted *{vb} [eng]*
kansō-tai	乾燥帯; 乾燥帶	arid belt *[climat]*
kansō-zai	乾燥剤; 乾燥劑	desiccant; drying agent *[ch]*
kansū	関数; 關數; 函数	function *{n} [math]*
kansū-hassei-ki	関数発生器	function generator *[electr]*
kansū-hyō	関数表	function table *[comput]*
kansui	灌水; かんすい	irrigation *[civ-eng]*
kantai	寒帯; 寒帶	frigid zone; polar zone *[climat]*
kantai	環帯	clitellum *[i-zoo]*

kantai	艦隊	fleet (of ships) *{n} [ord]*
kan'taiheiyō-kazan-tai	環太平洋火山帯	circumpacific volcanic zone *[geol]*
kan-tai-kan no	艦対艦の	surface-to-surface *{adj} [ord]*
kan-tai-kū no	艦対空の	surface-to-air *{adj} [ord]*
kantan'fu	感嘆符	exclamation point *{n} [comput] [pr]*
kantan na	簡単な	simple *{adj} [math]*
kantei	鑑定; かんてい	appraisal; expert opinion *{n}*
kanten	寒天	agar; agar-agar *[mat]*
kanten	乾点	dry point *[an-ch]*
kantō'shi	間投詞	interjection *{n} [gram]*
kantsū suru	貫通する	penetrate; perforate *{vb t} [sci-t]*
kanwa	緩和; かんわ	relaxation *[mech]*
kanwa-hensū	緩和変数	slack variable *[math]*
kanwa-jikan	緩和時間	relaxation time *[phys]*
kanwa-ritsu	緩和率	relaxation ratio *[mech]*
kan'yaku-hōsoku	簡約法則	cancellation rule *[math]*
kan'yō'go	慣用語	idiom *[ling]*
kan'yō-mei	慣用名	common-usage name *[sci-t]*
kan-yu	肝油	liver oil *[mat]*
ka'nyū-denshin	加入電信	telegraph-exchange (Telex) *[comm]*
ka'nyūsan	過乳酸	perlactic acid
kanzai	寒剤; 寒劑	freezing mixture *[p-ch]*
"	寒剤	cryogen *[phys]*
kanzai'nin	管財人	trustee *[econ]*
kanzei	関税	tariff *[econ]*
kanzen-kesshō	完全結晶	perfect crystal *[crys]*
kanzen-kitai	完全気体	ideal gas *[thermo]*
kanzen na	完全な	perfect; complete; ideal *{adj}*
kanzen-nenshō	完全燃焼	complete combustion *[ch]*
kanzen-ryūtai	完全流体	inviscid fluid; perfect fluid *[phys]*
kanzen-sei	完全性	integrity *[comput]*
"	完全性	perfectness *{n}*
kanzō	甘草; かんぞう	licorice (herb) *[bot]*
kanzō	肝臓	liver *[anat]*
kan'zume'hin	缶詰品; 罐詰品	canned goods *[cook] [food-eng]*
kan'zume-seizō'sho	缶詰製造所	cannery *[food-eng]*
kao	顔	face *{n} [anat]*
kaori	香り	flavor; fragrance *{n} [physio]*
kappan-in'satsu	活版印刷	letterpress printing; typographic printing *{n} [pr]*
kappan(-in'satsu)'jutsu	活版(印刷)術	typography *[pr]*
kappan'saku	括帆索	gasket *[nav-arch]*

kara	空；から	null *{n} [math]*
"	空	empty(ness) *{n}*
kara	殼；殼	shell; husk *{n} [eng] [mat]*
karada	体；體；からだ	body (pl bodies) *{n} [anat]*
karada no	体の	somatic *{adj} [bio]*
-kara (hanareta)	-から(離れた)	-a; -ab *{prefix}*
Karafuto-inu	樺太犬；からふと犬	Saghalien (Sakhalin) dog *[v-zoo]*
kara-kami	唐紙	sliding door; bamboo paper *[paper]*
kara-kusa	唐草；からくさ	arabesque; foliage; scrollwork *[graph]*
kara-matsu	唐松；落葉松	Japanese larch *[bot]*
karami	鍰；からみ	slag *[met]*
kara'mi	辛味；からみ	pungency; saltiness *[cook] [physio]*
karami'ai	絡み合	interlocking; intertwining *{n} [tex]*
karami-ori	絡み織り	gauze fabric *[tex]*
kara'oke	空オケ	music tape for sing-along *[music]*
karashi	枯し	seasoning *{n} [steel]*
karashi	芥子；からし	mustard (condiment) *[bot]*
karashi-jimo	枯らし霜	killing frost *[meteor]*
karashi'na	芥菜	mustard plant (vegetable) *[bot]*
karashi-yu	芥子油	mustard (-seed) oil *[org-ch]*
karasu	烏	blackbird; crow (bird) *[v-zoo]*
karasu	枯らす	let wither *{vb t}*
karasu-garei	烏鰈；からすがれい	halibut (fish) *[v-zoo]*
karasu-uri	烏瓜	snake cucumber; snake gourd; Trichosanthe cucumeroides *[bot]*
karatachi	枳殻；枳；からたち	trifoliate orange *[bot]*
Karatsu-yaki	唐津焼	Karatsu ware *[cer]*
karē	カレー	curry (condiment) *{n} [cook]*
karei	鰈；かれい	flatfish; turbot (fish) *[v-zoo]*
karei	過冷	supercooling *{n} [thermo]*
kareru	枯れる	wither; dry up *{vb}*
kareta	枯れた	dry; dried up *{adj} [bot]*
kari	雁；かり	wild goose (bird) *[v-zoo]*
kari	借り	borrow(ing) *{n} [math]*
kari-bunsū	仮分数；假分數	improper fraction *[math]*
kari'gan'en	加里岩塩	sylvite *[miner]*
karihorunyūmu	カリホルニウム	californium (element)
kari-hyōjun	仮標準	tentative standard *[ind-eng]*
kari'ire	刈り入れ	harvesting; reaping *{n} [agr]*
kari-kikaku	仮規格	tentative standard *[sci-t]*
kari no-	仮の-；假の-	pseudo- *{prefix}*
ka'rinsan-bunkai	加燐酸分解	phosphorolysis *[bioch]*

ka'rinsan-sekkai	過燐酸石灰	calcium superphosphate
kari-nui	仮縫い	basting *{n} [cl]*
kariru	借りる	borrow *{vb} [math]*
kari-sū	借り数	borrow digit *[math]*
kari-yori	仮撚り	false twist *[tex]*
karokon	括楼根; 瓜呂根; かろ(う)こん	Trichosanthes Radix (root) *[bot]*
karu-gamo	軽鴨; 輕鴨	spot-billed duck (bird) *[v-zoo]*
karui	軽い; 輕い	light; lightweight *{adj}*
karu-ishi	軽石	pumice *[geol]*
karushūmu	カルシウム	calcium (element)
ka-ryoku	火力	fire power *[phys]*
ka'ryoku-hatsu'den	火力発電	thermal (steam) power generation *[elec]*
ka'ryoku-hatsuden'sho	火力発電所	thermal power plant *[mech-eng]*
karyū	加硫; かりゅう	cure; vulcanization *{n} [ch-eng] [rub]*
karyū	顆粒; 果粒	granule *[bio]*
karyū-fusoku	加硫不足	undervulcanization *[rub]*
karyū-gawa	下流側	downstream-side *[hyd]*
karyū-kan	加硫缶; 加硫罐	vulcanization pan *[rub]*
karyūmu	カリウム	potassium (element)
karyū-soku'shin-zai	加硫促進剤	vulcanization accelerator *[ch-eng]*
karyū-yu	加硫油	vulcanized oil *[mat]*
karyū-zai	加硫剤	vulcanizing agent *[ch-eng]*
karyū-zai	顆粒剤	granule *[pharm]*
kasa	笠	hat (of bamboo or sedge) *[cl]*
kasa	傘	umbrella *[eng]*
kasa	嵩	bulk *[pr]*
kasa	暈; かさ	halo (pl -s, -es) *[astron]*
kasa'gi	笠木; かさぎ	coping *{n} [arch]*
kasago	笠子; かさご	scorpion fish *[v-zoo]*
kasa-ha'guruma	傘歯車	bevel gear *[mech]*
kasa-hijū	嵩比重	bulk density *[eng] [poly-ch]*
"	嵩比重	bulk specific gravity *[mech]*
kasai-hōchi-ki	火災報知機	fire alarm *[eng]*
kasai-hoken	火災保険	fire insurance *[econ]*
kasa-keisū	嵩係数	bulk factor *[eng] [poly-ch]*
ka'sakusan	過酢酸	peracetic acid
kasa-mitsudo	嵩密度	bulk density *[eng] [poly-ch]*
ka-san	過酸	peracid; peroxy acid *[ch]*
kasanari	重なり	overlap *{n} [math] [met]*
kasanari-kaku	重なり角	overlap angle *[mech-eng]*
kasane-awase	重ね合せ	superimposition; superposition *[math]*

kasane-ita'bane	重ね板ばね	laminated spring *{n} [eng]*
kasaneru	重ねる	stack; heap; superimpose *{vb t}*
kasane-tsugi	重ね継ぎ	lap joint *[arch]*
ka'sanka-aen	過酸化亜鉛	zinc peroxide
ka'sanka'butsu	過酸化物	peroxide *[ch]*
ka'sanka-gin	過酸化銀	silver peroxide
ka'sanka-kohaku-san	過酸化琥珀酸	succinic acid peroxide
ka'sanka-kōso	過酸化酵素	peroxidase *[bioch]*
ka'sanka-suiso	過酸化水素	hydrogen peroxide
kasan-ki	加算器	adder *[comput]*
kasasagi	鵲; かささぎ	magpie, Korean (bird) *[v-zoo]*
kase	桛; かせ	reel *{n} [tex]*
kase	綛; かせ	hank; skein *{n} [tex]*
kase'age	桛揚げ	reeling *{n} [tex]*
kasei	化成	chemical synthesis; transformation *[ch]*
"	化成	formation *[p-ch]*
kasei	火星	Mars *[astron]*
kasei-den'atsu	化成電圧	formation voltage *[elec]*
kasei-do	苛性度; かせいど	causticity *[ch]*
kasei-gan	火成岩	igneous rocks *[petr]*
kasei'hin	化成品	chemically-synthesized product *[ch]*
kasei-hyōmen'gaku	火星表面学	areography *[astron]*
kasei no	苛性の	caustic *{adj} [ch]*
kasei-ritsu	加成率; かせいりつ	additivity *[ch]*
kasei-sei	加成性	additivity *[ch]*
kasei-shori-sei	化成処理性	forming treatability *[eng]*
kasei-sōda	苛性ソーダ	caustic soda; sodium hydroxide *[ch]*
kasei-soku	加成則	additivity rule *[ch]*
kase'jime-ki	綛締め機	yarn bundling machine *[eng]*
ka'seki	化石; かせき	fossil *[paleon]*
kaseki-jushi	化石樹脂	fossil resin *[poly-ch]*
kaseki-nenryō	化石燃料	fossil fuel *[geol]*
kaseki-sayō	化石作用	petrifaction *[geol]*
ka'sen	下線	underline *{n} [pr] [comput]*
kasen	化繊; 化纖	chemical fiber *[tex]*
kasen-hyō	河川氷	river ice *[hyd]*
kasen no ryūryō	河川の流量	river discharge *[hyd]*
kasetsu	仮説; 假說	hypothesis *[logic] [sci-t]*
kasetsu-kentei	仮説検定	hypothesis testing *{n} [math]*
kasetsu no	仮説の	hypothetical *{adj} [sci-t]*
kasetsu suru	架設する	construct; install *{vb t}*
kasha	貨車; かしゃ	freight car *[rail]*

kashi	樫；かし	oak (tree) *[bot]*
kashi	河岸；かし	riverbank; waterfront *[geol]*
kashi	菓子	confectionery; sweets; desserts *[cook]*
kashi-do	華氏度	degree Farenheit *[thermo]*
kashi-dori	樫鳥；かしどり	jay (bird) *[v-zoo]*
kashi-en	可視炎	visible flame *[ch]*
kashi'ka	可視化	visualization: rendering visible *[opt]*
kashi-kō	可視光	visible light *[opt]*
kashime	加締め；かしめ	caulk *{n}* *[eng]*
kashi-pan	菓子パン	sand dollar *[i-zoo]*
kashira-moji	頭文字	initial(s) *[pr]*
kashira(-mo)jigo	頭(文)字語	acronym *[comm]*
ka'shitsu-ki	加湿器；加濕器	humidifier *[mech-eng]*
kashiwa	柏；槲；かしわ	oak (tree) *[bot]*
kashō	煆焼	calcination *[eng]*
kashoku-hō no genshoku	加色法の原色	additive primary color *[opt]*
kashoku'sei no	果食性の	frugivorous *{adj}* *[zoo]*
kashō-yaku; yakedo-gusuri	火傷薬	antipyrotic; burn medicine *{n}* *[pharm]*
kasō	下層	substratum (pl -strata) *[geol]*
kaso'butsu	可塑物；かそぶつ	plastic substance *[mat]* *[poly-ch]*
kaso-do	可塑度	degree of plasticity; plasticity number *[mech]*
kaso'ka	可塑化	plasticization *[eng]* *[poly-ch]*
kasō-kairo	仮想回路	virtual circuit *[elec]*
ka'soku	加速；かそく	acceleration *[mech]*
ka'soku	仮足；假足	pseudopod(ium) (pl -podia) *[cyt]*
kasoku-den'atsu	加速電圧	acceleration voltage *[elec]*
kasoku-do	加速度	acceleration *[math]* *[mech]*
kasoku(do)-kei	加速(度)計	accelerometer *[eng]*
kasoku-sōchi	加速装置	accelerator *[mech-eng]*
kasoku-undō	加速運動	accelerated motion *[phys]*
kaso-sei	可塑性	plasticity *[poly-ch]*
kasō-sen	仮想線	imaginary line *[math]*
kasō-shitsu'ryō	仮想質量	virtual mass *[mech]*
kasō'un	下層雲	low clouds *[meteor]*
kaso-zai	可塑剤	plasticizer; flexibilizer *[poly-ch]*
kassei	活性；かっせい	activity *[ch]*
kassei-bu'i	活性部位	active site *[bioch]*
kassei-chūshin	活性中心	active center *[bioch]*
kassei-haku'do	活性白土	activated clay *[mat]*
kassei'ka	活性化	activation *[ch]*
kassei'ka-enerugī	活性化エネルギー	activation energy *[p-ch]*

kassei'ka-netsu	活性化熱	heat of activation *[p-ch]*
kassei-ki	活性基	active group *[ch]*
kassei-o'dei	活性汚泥	activated sludge *[civ-eng]*
kassei-seibun	活性成分	active component *[elec]*
kassei-sen	活性線	actinic ray *[opt]*
kassei-shu	活性種	active species *[ch]*
kassei-sui	活性水	activated water *[hyd]*
kassei-tan	活性炭	activated charcoal *[mat]*
kassei-ten	活性点	active center; active site *[astron]*
kassei-zai	活性剤	activator *[ch]*
kasseki	滑石; かっせき	talc *[miner]*
kassen	活栓	stopcock *[eng]*
kassen	活線	live wire *[elec]*
kassen	括線	vinculum (pl -a) *[math]*
kassen	割線	secant *[math]*
kassetsu	滑節	hinged joint *[mech-eng]*
kassha	滑車; かっしゃ	block; pulley(s) *{n}* *[eng]*
kasshi	滑子	cursor *[comput]*
kasshoku	褐色; 褐色	brown (color) *[opt]*
kasshoku-shokubutsu	褐色植物	Chromophyta *[bot]*
kassō	褐草	brown algae (sing alga) *[bot]*
kassō-ro	滑走路	landing strip; airstrip; runway *[aero-eng]*
kassō-shokubutsu-kō	褐藻植物綱	Phaeophyta *[bot]*
kassō'so	褐藻素	fucoxanthin *[bioch]*
kassui-ki	渇水期	dry season *[climat]*
kasu	滓; かす	dregs; refuse *[met]*
kasu	借す	lend *{vb t}*
kasū	加数	addend *[math]*
kasū	仮数; 假數	mantissa *[math]*
kasugai	鎹; かすがい	clamp; U-nail *{n}* *[eng]*
ka'sui	花穂; かすい	spike *{n}* *[bot]*
kasui-bunkai	加水分解	hydrolysis *[ch]*
kasui'bunkai-kōso	加水分解酵素	hydrolase *[bioch]*
kasumi	霞; かすみ	haze; mist *{n}* *[meteor]*
kasumi-gakari	霞がかり	hazing *{n}* *[meteor]*
kasumi-ishi	霞石	nepheline *[miner]*
kasunda	霞んだ	hazy *{adj}* *[meteor]*
kasureru	掠れる	become hoarse *{vb i}* *[physio]*
kasuri	絣; かすり	splashed pattern *[tex]*
kasu-zame	粕鮫; かすざめ	angelfish; monk fish; shark ray; Squatina japonica *[v-zoo]*
kata	形	forma *{n}* *[bio]*

kata	肩	shoulder *{n} [anat]*
kata	型	form *{n} [bio] [syst]*
"	型; 形; かた	type *{n} [comput] [pr]*
"	型	die; mold; model *{n} [eng]*
kata	潟	lagoon *[geogr]*
katabami	酢漿草; かたばみ	oxalis; wood sorrel *[bot]*
kata'ban; kata-ita	型板	template *[eng]*
kata'buri-kajū	片振(り)荷重	pulsating load [mech-eng]
katachi	形; かたち	form; shape *{n} [math] [phys] [sci-t]*
kata-chijimi	型縮み	mold shrinkage *[rub]*
kata-gami	型紙	pattern paper *[paper]*
kata'gami-nasen	型紙捺染	stencil printing *{n} [tex]*
kata'gawa-kentei	片側検定	one-tailed test *[math]*
kata'gawa no-	片側の-	semi- *{prefix}*
kata'gi	堅木	hardwood *[mat]*
kata'ha-baito	片刃バイト	knife tool *[eng]*
katai	固い; 硬い; 堅い	hard *{adj} [phys]*
katai enki	かたい塩基	hard base *[ch]*
katai san	かたい酸	hard acid *[ch]*
kata-ita	型板	template *[resin]*
kata'ji	肩字	superscript *[comput] [pr] [sci-t]*
kata'jime	型締め	clamping; closing of mold; locking of mold *{n} [mech-eng]*
kata'jime-atsu'ryoku	型締圧力	clamping (or locking) pressure *[mech-eng]*
kata-kana	片假名; カタカナ	kana (q.v.), script style *[pr]*
kata-karyū	型加硫	mold curing *{n} [rub]*
kata'kuchi-iwashi	片口鰯	anchovy; Engraulis japonica (fish) *[v-zoo]*
katamari	固まり; 塊	lump; mass; clod *{n}*
katamen-gōkin'ka-kōban	片面合金化鋼板	one-side-alloyed steel sheet *[mat] [steel]*
kata'mimi-chō	片耳聴	monoaural hearing *{n} [acous]*
kata'mochi-bari	片持梁	cantilever *{n} [eng]*
kata-moji	肩文字	superscript *[comput]*
kata'mono	硬物	hardware *[comput]*
katamuki	傾き; かたむき	gradient; inclination; slope *[geol] [math]*
katamuki no chūshin	傾きの中心	metacenter *[fl-mech]*
katana	刀	Japanese sword *[ord]*
katan-chū-tetsu	可鍛鋳鉄	malleable cast iron *[met]*
ka'tansan	過炭酸	percarbonic acid
ka'tansan-en	過炭酸塩	percarbonate *[ch]*
katan-sei	可鍛性	malleability *[met]*
katan suru	加炭する	carburize *{vb t} [met]*
kata-sa	硬さ	hardness *[eng]*

katasa-sū	硬さ数	hardness number *[eng]*
kata'taisū-memori	片対数目盛	semilogarithmic paper *[math]*
kata'te-nokogiri	片手鋸	handsaw *{n} [eng]*
kata'tsumuri	蝸牛；かたつむり	snail (gastropod mollusk) *[i-zoo]*
katayori	偏り；片寄り	bias; deviation *{n} [eng] [math]*
katayori-yure	偏揺れ；偏搖れ	yaw; yawing *{n} [mech]*
kata'yure-kei	片揺れ計	yawmeter *[eng]*
kata-zukeru	片付ける	tidy; clear away *{vb t}*
katei	仮定；假定；かてい	assumption; hypothesis *[math] [sci-t]*
katei	過程	course; process; stage *{n}*
katei'yō'hin	家庭用品	household utensil(s) *[eng]*
katen	火点；火點	fire point *[ch]*
katen	果点	dot (of fruit) *{n} [bot]*
kato	過渡	transition *{n}*
katō	果糖	fructose; fruit sugar *[bioch]*
katō-gōban	可撓合板	flexible plywood *[mat]*
kato-jōtai	過渡状態	transient state *[elec]*
kato-mō	家兎毛	rabbit hair *[mat] [tex]*
kato-ō'tō	過渡応答	transient response *[phys]*
katō-rennyū	加糖練乳	sweetened condensed milk *[cook]*
ka-tori-senkō	蚊取り線香	mosquito-repellent incense *[mat]*
katō-sei	可撓性	flexibleness; flexibility *[mech]*
katsu	かつ	and *{conj} [logic]*
katsu'dan-sō	活断層；活斷層	active fault *[geol]*
katsudō-do	活動度	activity *[p-ch]*
katsudō-keisū	活動係数	activity coefficient *[p-ch]*
katsu'enkō	褐鉛鉱；褐鉛鑛	vanadinite *[miner]*
katsu'ji	活字；かつじ	type *{n} [pr]*
katsu'ji-chūzō-ki	活字鋳造機	typecasting machine *[pr]*
katsuji-gata; katsuji-kei	活字形	type font *[graph]*
katsuji-gōkin	活字合金	type metal *[met]*
katsuo	鰹；かつお	bonito; Katsuwonus pelamis (fish) *[v-zoo]*
katsuo'bushi	鰹節	bonito, dried *[cook]*
katsuo'bushi-mushi	鰹節虫	museum beetle; bacon beetle *[i-zoo]*
katsuo no eboshi	鰹の烏帽子	Portugese man o'war *[i-zoo]*
katsura	桂	katsura (tree) *[bot]*
katsura; kazura	鬘；かつら；かずら	wig *[cl]*
katsu'renseki	褐簾石	allanite *[miner]*
katsurō	括楼；括樓；栝樓	gualou (fruit, seed); Trichosanthes kirilowii Maxim *[bot]*
katsurō-kon	括楼根	Trichosanthes Radix (root) *[bot]*
katsu'ryō	活量	activity *[thermo]*

katsuyō'ka	活用化	activation *[comput]*
katsu-zai	滑剤; 滑劑	lubricant *[mat]*
kattan	褐炭	brown coal *[geol]*
kattekkō	褐鉄鉱; 褐鐵鑛	limonite *[miner]*
kau	買う	buy; purchase *{vb t}*
kau	飼う	keep; raise (animals, birds, fish) *{vb t}*
kawa	川; 河	river *[hyd]*
kawa	皮	hide; skin *{n} [cl]*
kawa	革	leather *[cl]*
kawa	鈹; かわ	matte *{n} [met]*
kawa'bari	皮張り	skinning (of ink) *{n} [pr]*
kawa-bune	川船	river boat *[nav-arch]*
kawa'doko	川床	riverbed *[geol]*
kawahagi	皮剝ぎ	filefish; leatherfish; Stephanolepis cirrhifer *[v-zoo]*
kawaī; kawaii	可愛い	loveable; sweet *{adj}*
kawairashī	可愛らしい	cute; charming *{adj}*
kawaita	乾いた	dry *{adj}*
kawa-kamasu	河梭魚; かわかます	pike (fish) *[v-zoo]*
kawa'kami	川上	upstream *{n} [hyd]*
kawakasu	乾かす	dry *{vb t}*
kawaki (nodo no)	渇き(喉の)	thirst *{n} [physio]*
kawaku	乾く	dry *{vb i}*
kawara	瓦; かわら	tile *{n} [mat]*
kawasemi	翡翠; 川蟬; 鴗	kingfisher (bird) *[v-zoo]*
kawase-rēto	為替レート	exchange rate *[econ]*
kawa-shimo	川下	downstream *[hyd]*
kawa'uso	川獺; かわうそ	otter (animal) *[v-zoo]*
kaya	萱; 茅	cogon-grasses; eularias *[bot]*
kaya	蚊帳; かや	mosquito net *[mat]*
ka'yaku-rui	火薬類; 火藥類	explosives *[ord]*
kaya-nezumi	萱鼠	harvest mouse (animal) *[v-zoo]*
ka-yōbai-bunkai	加溶媒分解	solvolysis *[ch]*
ka'yōbi	火曜日	Tuesday
kayō'ka	可溶化	solubilization *[ch]*
ka'yōso-san	過沃素酸	periodic acid
ka'yōso'san-en	過沃素酸塩	periodate *[ch]*
kayu	粥; かゆ	rice porridge; rice gruel *[cook]*
kayu'mi	痒み	itchiness *[physio]*
kayu'mi-dome	痒み止め	antipruritic *{n} [pharm]*
kayu-sa	痒さ	itchiness *[physio]*
kayū-sei	可融性	fusibility *[thermo]*

kaza-guruma	風車；かざぐるま	windmill *[mech-eng]*
kaza'kami	風上	windward *{n} [meteor]*
kaza'kami no	風上の	windward *{adj} [ocean]*
kaza-muki	風向き	wind direction *[meteor]*
kaza'muki-kei	風向計	yawmeter *[meteor]*
kazan	火山；かざん	volcano (pl -es, -s) *[geol]*
kaza-nami	風波	wind wave *[meteor]*
kazan-bai	火山灰	volcanic ash *[geol]*
kazan'gaku	火山学	volcanology *[geol]*
kazan-gan'shi	火山岩滓	scoria (pl -riae) *[geol]*
kazan-tō	火山島	volcanic island *[geogr]*
kazari-hige	飾り鬚；かざりひげ	kern *[pr]*
kazari-ita	飾り板	plaque *[furn]*
kazari(-mono)	飾り(物)	ornament *[furn]*
kazaru	飾る	decorate; adorn *{vb}*
kaza'shimo	風下	downwind *[navig]*
"	風下	lee *[sci-t]*
kaza'shimo no	風下の	leeward *{adj} [ocean]*
kazasu	翳す；かざす	hold over head, eyes; shade *{vb}*
kaze	風	wind *{n} [meteor]*
kaze	風邪；かぜ	common cold *[med]*
kaze'bie-shisū	風冷え指数	wind-chill index *[meteor]*
kaze-gusuri	風邪薬；かぜぐすり	cold medicine *[pharm]*
kazoeru	数える	count *{vb t} [math]*
ka'zoku	家族	family (pl -lies) *[bio]*
kazu	数；數；かず	number *{n} [math]*
kazu-heikin-bunshi'ryō	数平均分子量	number-average molecular weight *[ch]*
kazu-no-ko	数の子	herring roe (fish egg) *[cook]*
ke	毛	hair *[anat]*
"	毛	fur *[mat] [v-zoo]*
keba	毛羽；毳；けば	fluff; fuzz *[mat]*
"	毛羽；毳	nap; pile *{n} [tex]*
"	毛羽；毳	feather *{n} [v-zoo]*
keba'dachi	毛羽立ち	fluff formation; napping *[tex]*
ke-bari	毛鉤	fly (for fishing) *{n} [eng]*
kedamono; kemono; jū	獣；獸	animal *[zoo]*
ke'gaki	罫書き	marking-off; scoring *{n} [eng]*
ke'gaki-setsu'dan	罫書き切断	scribe-cutting *{n} [eng]*
kegawa	毛皮	fur *{n} [v-zoo]*
kei	系	series *[bio] [ch] [syst]*
"	系	system *[geol] [sci-t]*
"	系	corollary (pl -laries) *[logic]*

kei	計	total *{n} [math]*
kei	径; 徑; けい	diameter *[math]*
kei'atsu-sei	傾圧性	baroclinity *[phys]*
keibi na	軽微な; 輕微な	slight; mild *{adj}*
keidai-baiyō	継代培養	subculture *[microbio]*
keiden-ki	継電器; 繼電器	relay *{n} [elec]*
keido	経度; 經度	longitude *[geod]*
kei-dōmyaku	頸動脈	carotid artery *[anat]*
keido'fū	傾度風	gradient wind *[meteor]*
kei'ei	経営; 經營	operation; management *[ind-eng]*
kei'ei-soshiki	経営組織	management organization *[econ]*
kei'fukka-aen	珪弗化亜鉛	zinc silicofluoride
kei'fukka'butsu	珪弗化物	silicofluoride *[ch]*
kei'fukka-dō	珪弗化銅	copper silicofluoride
kei'fukka-namari	珪弗化鉛	lead silicofluoride
kei'fukka-suiso-san	珪弗化水素酸	hydrosilicofluoric acid
kei'fukka-suzu	珪弗化錫	tin silicofluoride
kei'fukka-tetsu	珪弗化鉄	iron silicofluoride
keigan	珪岩	quartzite *[petr]*
kei-gashira	⺕頭: 彐, ⺕, ⺔	kanji radical denoting a pig's head *[pr]*
keigō	係合	meshing *{n} [eng]*
keigō suru	係合する	mesh(ing) together *{vb} [eng]*
keihi	経費; 經費	expenditure *[econ]*
keihi-san	桂皮酸; けいひさん	cinnamic acid
keihi'san-en	桂皮酸塩	cinnamate *[ch]*
keihi-yu	桂皮油	cinnamon oil *[mat]*
keihō	警報; けいほう	alarm; siren; warning *{n} [electr]*
kei'i-gi	経緯儀; 經緯儀	altazimuth; theodolite *[eng]*
keiji-ban	掲示板; 揭示板	bulletin board *[comm]*
keiji-henka	経時変化	change with passage of time *{n} [ind-eng]*
keiji-kikō	計時機構	timer *[electr]*
keijō-keisū	形状係数	shape factor *[fl-mech]*
keijō-ki'oku-gōkin	形状記憶合金	shape-memory alloy *[met]*
keijō-seigyo	形状制御	shape control *{n} [eng]*
keijō-tōketsu-sei	形状凍結性	shape-freezability *[eng]*
keika	珪化	silicification *[geol]*
keika'butsu	珪化物; けいかぶつ	silicide *[ch]*
keikai'seki	珪灰石	wollasynonite *[miner]*
keika-kiroku	経過記録	log *{n} [comm] [comput] [navig]*
keikaku	計画; けいかく	schedule *{n} [comput]*
"	計画; 計畫	plan; scheme; project *{n} [ind-eng]*
keikaku	傾角	inclination *[geol] [math] [sci-t]*

keikaku'do-ban	傾角度板	inclinometer *[eng]*
keikaku-hoshu	計画保守	scheduled maintenance *[ind-eng]*
keikaku-zu	計画図；計畫圖	scheme drawing *(n) [graph]*
keikan'seki	鶏冠石	realgar *[miner]*
keiken	経験；けいけん	experience *(n) [psy]*
keiken-ron	経験論；經驗論	empiricism *[sci-t]*
keiken'teki na	経験的な	empirical *(adj) [sci-t]*
keiki	計器；けいき	gage; gauge *(n) [eng]*
keiki-ban	計器板	gage board *[electr]*
keiki-chaku'riku	計器着陸	instrument approach *[navig]*
keiki-chaku'riku-hōshiki	計器着陸方式	instrument landing system *[navig]*
keiki-hikō	計器飛行	instrument flight *[navig]*
kei-kinzoku	軽金属；輕金屬	light metal *[met]*
keikō	傾向；けいこう	tendency; trend *(n) [math]*
keikō	蛍光；けいこう	fluorescence *[atom-phys]*
keikō-ganryō	蛍光顔料	fluorescent pigment *[ch]*
kei-kōgyō	軽工業	light industry *[ind-eng]*
keikō-kenbikyō-hō	蛍光顕微鏡法	fluorescence microscopy *[opt]*
keikoku-shoku	警告色	warning coloration *[bio]*
keikoku-tō	警告灯；警告燈	warning light *[navig] [opt]*
keikō-shiji-yaku	蛍光指示薬	fluorescent indicator *[ch]*
keikō-supekutoru	蛍光スペクトル	fluorescence emission spectrum *[spect]*
keikō'tai	蛍光体；螢光体	luminophor; phosphor; fluorescent substance *[phys]*
keikō-teiryō-hō	蛍光定量法	fluorimetry *[an-ch]*
keikō-tō	蛍光灯	fluorescent lamp *[electr]*
keikō-toryō	蛍光塗料	fluorescent paint *[ch] [mat]*
kei-kotsu	脛骨；けいこつ	tibia (bone) *[anat]*
kei-kōzō	軽構造	light structure *[poly-ch]*
keikō-zō'haku	蛍光増白	fluorescent brightening *(n) [ch]*
keikō-zōhaku-zai	蛍光増白剤	optical brightening agent *[ch]*
keimu'sho	刑務所	prison *[arch]*
kei'nen-henka	経年変化	secular variation *[astron]*
keiran	鶏卵；けいらん	hen's egg *[agr]*
keiri'shi	計理士	public accountant *[econ]*
ke-iro	毛色；けいろ	color of hair; color of fur *[bio]*
keiro	経路；けいろ	path; pathway *[navig]*
"	経路	course; channel; route; process *(n)*
keiro-sentaku	経路選択；經路選擇	routing *(n) [comput]*
keiryō	計量；けいりょう	measuring; weighing *(n) [sci-t]*
keiryō('gaku)	計量(学)	metrics *[math]*
keiryō-garasu	計量ガラス	measuring glass *[eng]*

keiryō-kappu	計量カップ	measuring cup *[eng]*
keiryō-saji	計量匙	measuring spoon *[eng]*
keiryū	渓流；溪流	torrent *[hyd]*
kei'ryūshi	軽粒子；輕粒子	lepton *[part-phys]*
keisan	計算；けいさん	calculation; tally *{n}* *[math]*
keisan	珪酸；けいさん	silicic acid
keisan-aen	珪酸亜鉛	zinc silicate
keisan'aenkō	珪酸亜鉛鉱	willemite *[miner]*
keisan-dansō'zō-hō	計算断層像法	comput(eriz)ed tomography; CT *[med]*
keisan-en	珪酸塩	silicate *[ch]*
keisan-jaku	計算尺	slide rule *[math]*
keisan-kagaku	計算化学	computational chemistry *[ch]*
keisan-ki	計算機；けいさんき	computer; calculator *[comput]*
keisanki-enyō-seizō	計算機援用製造	computer-aided manufacturing; CAM *[comput]*
keisanki-enyō-sekkei	計算機援用設計	computer-aided design; CAD *[comput]*
keisanki-gengo	計算機言語	computer language *[comput]*
keisanki-kagaku	計算機科学	computer science *[comput]*
keisanki-mō	計算機網	computer network *[comput]*
keisan-namari	珪酸鉛	lead silicate
keisan suru	計算する	calculate; compute; count *{vb}* *[math]*
keisan-tetsu	珪酸鉄	iron silicate
keisan-zuhyō	計算図表	nomograph *[math]*
keisatsu'sho	警察所	police station *[arch]*
keisei	形成；けいせい	formation *{n}*
keisei-sō	形成層	cambium (pl -s, -bia) *[bot]*
keiseki	珪石；けいせき	silica *[miner]*
keisen-bui	繋船ブイ	mooring buoy *[eng]*
keisen'chi	繋船池；係船池	marina *[civ-eng]*
keisen'chū	繋船柱	bollard *[nav-arch]*
keisen-ichi	繋船位置	berthing (for ship) *{n}* *[nav-arch]*
keisen'seki	珪線石	sillimanite *[miner]*
keisha	珪砂；けいしゃ	quartz sand; silica sand *[geol]*
keisha	傾瀉	decantation *[eng]*
keisha-do	傾斜度	pitch *{n}* *[math]* *[sci-t]*
keisha-gi	傾斜儀	clinometer *[geol]*
keisha-hi	傾斜比	rake ratio *[nav-arch]*
keisha-kaku	傾斜角	tilt angle *[elecmg]*
"	傾斜角	inclination *[math]* *[sci-t]*
keisha-kei	傾斜計	clinograph *[geol]*
keisha shite iru	傾斜している	sloping *{adj}* *[math]*
keishi	係止；けいし	stop(ping), linked *{n}* *[eng]*

keishi	罫紙；けいし	line-ruled paper *[mat]*
keishiki	形式	expression *[comput]*
kei-shin	傾心；けいしん	center of inclination *[math]*
kei'shitsu-dō'nyū	形質導入	transduction *[microbio]*
kei'shitsu-saibō	形質細胞	plasma cell *[hist]*
kei'shitsu-tenkan	形質転換	transformation *[cyt]*
kei'shitsu-tenkan-saibō	形質転換細胞	transformant *[cyt]*
kei('shitsu)-yu	軽(質)油	light oil *[mat]*
keiso	珪素；硅素；ケイ素	silicon (element)
keisō	計装；けいそう	instrumentation *[eng]*
keisō-bunseki	計装分析	instrumental analysis *[eng]*
keisō'do	珪藻土；けい=そうど	diatomaceous earth; kieselguhr; loam *[geol]*
keiso-jushi	珪素樹脂	silicone resin *[poly-ch]*
keiso-kō	珪素鋼	silicon steel *[steel]*
kei'soku	計測；けいそく	instrumentation *[eng]*
keisoku'gaku	計測学	metrology *[phys]*
keisoku-kiki	計測機器	measuring instrument *[eng]*
keisō-shokubutsu	珪藻植物	diatom *[i-zoo]*
keisū	係数；係數	coefficient; modulus; factor *[math]*
keisū'gata no	計数型の	digital *{adj}* *[comput]*
keisū-hō	計数法	counting method *[math]*
keisui	軽水；けいすい	light water *[nucleo]*
keisui-ro	軽水炉；輕水爐	light-water reactor *[nucleo]*
keisū'ka	計数化；けいすうか	digitization *[comput]*
keisū-kan	計数管	counter *[nucleo]*
keisū'ka suru	計数化する	digitize *{vb}* *[comput]*
keisū-ki	計数器	counter *[comput]*
keitai'gaku	形態学	morphology *[bio]*
keitai-ji	経帯時；經帶時	zone time *[astron]*
keitai-keisei	形態形成	morphogenesis *[embryo]*
keitai-sei	携帯性	portability *[comput]*
keitai'yō no	携帯用の	portable *{adj}* *[eng]*
kei'tetsu	継鉄；繼鉄	yoke *{n}* *[elecmg]*
ke-ito	毛糸；けいと	wool yarn *[tex]*
keitō	系統；けいとう	line; pedigree *[gen]*
"	系統	system *[math]*
"	系統	strain *{n}* *[microbio]*
keitō'gaku	系統学	systematics *[bio]*
keitō-hassei	系統発生	phylogeny (pl -nies) *[bio]*
keiyaku	契約；けいやく	contract; agreement *{n}* *[econ]* *[legal]*
keiyō'shi	形容詞；けいようし	adjective *[gram]*

kei-yu	軽油; 輕油; けいゆ	gas oil; light oil *[mat]*
keizai'gaku	経済学; 經濟學	economics *[ind-eng]*
ke'jusu	毛繻子; けじゅす	satinet *[tex]*
ke-kabi	毛黴; 毛かび	mucor *[mycol]*
kekka	結果; けっか	result *{n}* *[math]*
kekkan	欠陥; 缺陥	defect; flaw; fault *{n}* *[sci-t]*
kekkan	血管; けっかん	blood vessel *[anat]*
kekkan-kaku'chō-yaku	血管拡張薬	vasodilator *[physio]*
kekka no	結果の	consequent *{adj}*
kekkan-shū'shuku-yaku	血管収縮薬	vasoconstrictor *[pharm]*
kekkan-zōei-hō	血管造影法	angiography *[med]*
kemono-hen	獣偏: 犭	kanji (q.v.) radical for dog *[pr]*
kemuri	煙; けむり	smoke *{n}* *[eng]*
kemuri-seki'ei	煙石英	smoky quartz *[miner]*
kemuri-ten	煙り点	smoke point *[eng]*
ke-mushi	毛虫	caterpillar (black) *[i-zoo]*
ken	県; 縣; けん	prefecture *[geogr]*
ken	圏; 圈	zone *{n}* *[meteor]*
ken	腱	tendon *[anat]*
-ken	-件	-case(s) *{suffix}* *[legal]*
-ken	-県; -縣	-Prefecture *{suffix}*
-ken	-軒	-unit(s) of house(s) *{suffix}*
-ken	-圏	sphere of influence *{suffix}*
ken'atsu-hō	検圧法; 檢壓法	manometry *[eng]*
kenban	鍵盤	keyboard *[comput]* *[music]*
kenbi-bunkō-bunseki	顕微分光分析	microspectrophotometry *[spect]*
kenbi'kyō	顕微鏡; 顯微鏡	microscope *[opt]*
kenbikyō-kensa-hō	顕微鏡検査法	microscopy *[opt]*
kenbikyō-shashin	顕微鏡写真	photomicrograph *[graph]*
kenboku	堅木	hardwood *{n}* *[mat]*
kenchiku	建築; けんちく	building *{n}* *[arch]*
kenchiku'gaku	建築学	architecture (the discipline) *[eng]*
kenchiku'ka	建築家	architect (a person) *[arch]*
kenchiku-zairyō	建築材料	building materials *[arch]* *[civ-eng]*
ken'daku	懸濁; けんだく	suspension *[ch]*
kendaku-bunri-hō	懸濁分離法	elutriation *[ch-eng]* *[min-eng]*
kendaku(-eki)	懸濁(液)	suspension *{n}* *[ch]*
kendaku-jūgō	懸濁重合	suspension polymerization *[poly-ch]*
kendaku-sei	懸濁性	suspendability *[ch-eng]*
kenden-ki	検電器; 檢電器	electroscope; voltage detector *[eng]*
ken'eki	検疫	quarantine *{n}* *[med]*
ken'eki-byōchi	検疫錨地	quarantine anchorage *[civ-eng]*

ken'eki'sen	検疫船	quarantine ship *[nav-arch]*
ken'etsu	検閲	censorship; inspection
kengen	権限；權限	authority
ken'in'ryoku	牽引力	traction *[mech]*
ken'i-yaku	健胃薬	stomachic *{n}* *[pharm]*
kenjō	健常	healthy and normal *[bio]*
kenjō no	剣状の；劍狀の	xiphoid *{adj}* *[anat]*
kenjū	拳銃	handgun *[ord]*
kenka	堅果	nut *[bot]*
kenka	鹼化；けんか	saponification *[ch]*
kenkai-men	圏界面；圈界面	tropopause *[meteor]*
kenka-ka	鹼化価	saponification number *[an-ch]*
kenka-shokubutsu	顕花植物	flowering plant *[bot]*
kenki-bunkai	嫌気分解	anaerobic decomposition *[bioch]*
kenki-kin	嫌気菌	anaerobe *[bio]*
kenki'sei-kin	嫌気性菌	anaerobe *[bio]*
kenkō	健康；けんこう	health; healthy *{n}* *[med]*
kenkō'do-sō	堅硬土層	hardpan; caliche *[geol]*
kenkō-kotsu	肩甲骨	shoulder blade; scapula *[anat]*
kenkoku	圏谷	cirque *[geol]*
kenkō na	健康な	healthy *{adj}* *[bio]*
kenkō'shi	検光子	analyzer *[opt]*
kenkyū	研究；けんきゅう	research; study *{n}* *[sci-t]*
kenkyū'jo	研究所	laboratory; research institute *[sci-t]*
kenkyū'sha	研究者	researcher (a person)
kenkyū-shitsu	研究室	laboratory (pl -ries) *[sci-t]*
kenkyū suru	研究する	research; study *{vb}*
kenma	研摩	polishing *{n}* *[ch-eng]*
kenma-men	研摩面	polished surface *[eng]*
kenma-sei	研摩性	abrasiveness *[mat]*
kenma-zai	研摩剤	abrasive *{n}* *[mat]*
kennetsu	顕熱；顯熱	sensible heat *[thermo]*
kenpa	検波；檢波	detection *[elec]*
kenpō	憲法	constitution *[legal]*
kenri	権利；權利	right; privilege *{n}* *[legal]*
kenrō-do	堅牢度	fastness *[ind-eng]*
kenryō-sen	検量線	calibration curve *[eng]*
kenryū-kei	検流計	galvanometer *[eng]*
kensa	検査；檢查；けんさ	checking *{n}* *[comput]*
"	検査	inspection; testing *{n}* *[ind-eng]*
"	検査	verification *{n}*
kensaku	検索	retrieval *[comput]*

kensaku-ban	研削盤	grinding machine *[eng]*
kensaku-kakō	研削加工	grinding *{n} [mech-eng]*
kensaku-kinō	検索機能	retrieval function *[comput]*
kensaku-men	研削面	ground surface *[mech-eng]*
kensaku-sei	研削性	grindability *[mat]*
kensaku suru	研削する	grind *{vb t} [mech-eng]*
kensaku suru	検索する	retrieve *{vb} [comput]*
kensaku-ware	研削割れ	grinding cracks *[mat]*
kensaku-yake	研削焼	grinding burn *[mech-eng]*
kenseki'un	巻積雲；絹積雲	cirrocumulus *[meteor]*
kenshi	犬歯；犬齒	canine(s) *[dent]*
kenshi-sen	絹糸腺	silk gland *[i-zoo]*
kenshō	検証；けんしょう	verification *[comput]*
ken'shoku-zai	顕色剤；顯色劑	color developer; developer *[opt]*
kenshō-suru	検証する	identify; verify *{vb} [legal]*
kenshū'in	研修員	research study member *{n}*
kenshuku	捲縮	crimp; curl *{n} [tex]*
ken'shutsu	検出；檢出	detection *[eng]*
ken'shutsu-bu	検出部	sensing element; sensor *[eng]*
ken'shutsu-ki	検出器	sensor (an instrument) *[eng]*
kensō'un	巻層雲；絹層雲	cirrostratus cloud *[meteor]*
kensui-sei	懸垂性	suspending property *[ch-eng]*
kensui-sen	懸垂線	catenary (pl -naries) *[math]*
kentei	検定；檢定	assay; verification *{n} [an-ch]*
"	検定	testing *{n} [ind-eng]*
kentei-tōkei'ryō	検定統計量	test statistic *[math]*
kentō-kei	検糖計	saccharimeter *[eng]*
ke-nuki	毛抜き	tweezers *[eng]*
ken'un	巻雲；絹雲	cirrus *[meteor]*
ken'yaku-ritsu	倹約律；儉約律	parsimony, law of *[sci-t]*
kenzai	建材	building material *[arch]*
kenzan	検算	verification *[math]*
ke-ori'mono	毛織物	woolen cloth *[tex]*
kerubin-ondo-memori	ケルビン温度目盛	Kelvin temperature scale *[thermo]*
keshi	罌粟；芥子；けし	poppy (pl -poppies) (herb) *[bot]*
keshi-gomu	消しゴム	eraser *[eng]*
keshi-in	消印	cancellation mark; postmark *[comm]*
keshi no mi	罌粟の実	poppy seed *[bot] [cook]*
keshi-tsubu	罌粟粒；芥子粒	poppy seed *[bot] [cook]*
keshi-yu	罌粟油；けしゆ	poppy seed oil *[mat]*
keshō-ban	化粧板	decorative laminate *[poly-ch]*
keshō'hin-rui	化粧品類	cosmetics; toiletry *[mat]*

keshō-sekken	化粧石鹸	toilet soap *[mat]*
keshō-shitsu	化粧室	powder room *[arch]*
kessei	血清；けっせい	serum (pl -s, sera) *[physio]*
kessei'gaku	血清学	serology *[bio]*
kesseki	欠席；缺席	default *{n} [legal]*
kessen	結線	connection *{n} [elec]*
kessen suru	結線する	connect *{vb t} [elec]*
kesshi-dōbutsu	齧歯動物	rodent *[v-zoo]*
kesshikiso	血色素	blood pigment; hemoglobin *[bioch]*
kesshō	血漿	blood plasma; plasma *[hist]*
kesshō	結晶；けっしょう	crystal *[crys]*
kesshō-ban	血小板	platelet *[hist]*
kesshō-denseki	結晶電析	electrocrystallization *[p-ch]*
kesshō-do	結晶度	crystallinity *[crys]*
kesshō'gaku	結晶学	crystallography *[phys]*
kesshō-gata	結晶型	crystal form *[crys]*
kesshō-hengan	結晶片岩	crystalline rock; schist *[geol]*
kesshō-hiki'age-hō	結晶引上げ法	crystal-pulling method *[crys]*
kesshō-hō'i	結晶方位	crystal orientation *[crys]*
kesshō-jiku	結晶軸	crystallographic axis *[crys]*
kesshō'ka	結晶化	crystallization; crystallizing *{n} [crys]*
kesshō'ka-do	結晶化度	degree of crystallinity *[crys]*
kesshō'ka suru	結晶化する	crystallize *{vb i} [ch]*
kesshō-kei(tai)	結晶形態	crystal form *[crys]*
kesshō-kōshi	結晶格子	crystal lattice *[crys]*
kesshō-kōzō	結晶構造	crystal structure *[crys]*
kesshō-ryūdo	結晶粒度	grain-size number *[steel]*
kesshō'sei-jūgō'tai	結晶性重合体	crystalline polymer *[poly-ch]*
kesshō'sei no	結晶性の	crystalline *{adj} [crys]*
kesshō'shi	結晶子	crystallite *[geol]*
kesshō-shūgō-soshiki	結晶集合組織	grain texture *[crys]*
kesshō-sui	結晶水	water of crystallization *[ch]*
kessoku-chi	欠測値	missing value *[math]*
kesu	消す	erase; extinguish *{vb}*
kesu dōgu	消す道具	eraser *[eng]*
keta	桁；けた	spar *{n} [aero-eng]*
"	桁	column; digit; order of magnitude *[math]*
keta'age	桁上げ	carry *{n} [math]*
ke'tai	懈怠；けたい	default *{n} [legal]*
keta-idō	桁移動	shift; shifting *{n} [comput]*
keta no atai	桁の値	place value *[math]*
keta-okuri	桁送り	shift *{n} [comput]*

ketchaku	結着	binding *{n} [ch-eng]*
ketchaku-zai	結着剤	binding agent *[mat]*
ketchō	結腸	colon *[anat]*
keton	ケトン	ketone *[org-ch]*
ketsu'atsu	血圧；血壓	blood pressure *[physio]*
ketsu'atsu-kei	血圧計	sphygmomanometer *[med]*
ketsu'eki	血液；けつえき	blood *[hist]*
ketsu'eki-daiyō-zai	血液代用剤	blood substitute *[pharm]*
ketsu'eki-gata	血液型	blood type *[immun]*
ketsu'eki-gyōko-soku'shin-yaku	血液凝固促進薬	styptic *{n} [pharm]*
ketsu'eki-shikiso	血液色素	blood pigment *[bioch]*
ketsu'gan	頁岩；けつがん	shale *[petr]*
ketsugan-yu	頁岩油	shale oil *[mat]*
ketsu'gō	結合；けつごう	binding *{n} [bio]*
"	結合	association; bond; bonding *{n} [ch]*
"	結合	coupling *{n} [elec]*
"	結合	connection *[mech-eng]*
ketsugō-enerugī	結合エネルギー	bond energy *[p-ch]*
ketsugō-hōsoku	結合法則	associative law *[math]*
ketsugō-i'sei	結合異性	linkage isomerism *[ch]*
ketsugō-kassei	結合活性	avidity *[immun]*
ketsugō-keisū	結合係数	coefficient of coupling; coupling constant *[phys]*
ketsugō-ryoku	結合力	bonding strength *[mech]*
ketsugō'shi	結合子	connector *[electr]*
ketsugō-sui	結合水	bonding water; bound water *[ch]*
ketsugō-zai	結合剤	binder; bonding agent *[adhes] [poly-ch]*
ketsu'gyokuzui	血玉髄；血玉髓	bloodstone; heliotrope *[miner]*
ketsu'jo	欠除；欠如	lack; deficiency *{n}*
ketsu'maku	結膜	conjunctiva (pl -s, -tivae) *[anat]*
ketsu'men-zō	欠面像	merohedrism *[crys]*
ketsu'ro	結露；けつろ	dew condensation *[meteor]*
ketsu'ron	結論	conclusion *[logic]*
ketsu'zō	結像	image formation; imaging *{n} [graph]*
kettei	決定；けってい	decision *{n}*
kettei-keisū	決定係数	coefficient of determination *[math]*
kettei'ken	決定権；決定權	right of decision; sayso *{n} [legal]*
ketten	欠点	defect; shortcoming *{n} [sci-t]*
kettō	血統	pedigree *[gen]*
ke'u-genso	希有元素；けう元素	rare element *[ch]*
ke'u-kitai	希有気体；けう気体	noble gas; rare gas *[ch]*

ke-ware	毛割れ	hair(line) cracks *[met]*
keyaki	欅；けやき	kiaki; Zelkova serrata (tree) *[bot]*
kezuru	削る；けずる	shave *{vb t}* *[eng]*
ki	木	tree *[bot]*
"	木	wood *[mat]*
ki	生；き	raw(ness); pure(ness) *{n}*
ki	季	season *{n}* *[climat]*
ki	紀	period *[geol]*
ki	基	base; group; radical *{n}* *[ch]*
ki	期	age; era; period; time *[geol]*
ki	貴	high (in rank); nobleness *{n}* *[ch]*
ki	黄	yellow (color) *[opt]*
ki	樹	tree *[bot]*
ki	機	opportunity; chance; occasion *{n}*
-ki	-器；-器	-instrument *{suffix}*
-ki	-機	-unit(s) of machinery *{suffix}*
"	-機	-unit(s) of aircraft *{suffix}*
"	-機	machine(s) *{suffix}*
ki'ankō	輝安鉱；きあんこう	antimonite; stibnite *[miner]*
ki'atsu	気圧；氣壓	atmospheric pressure; barometric pressure *[phys]*
ki'atsu-kei	気圧計	barometer *[eng]*
ki'atsu no tani	気圧の谷	trough *[meteor]*
ki'atsu-zōdai-hō	気圧増大法	pressurization method *[eng]*
kiba	牙	tusk; fang *[zoo]*
ki'baku	気曝	aeration *[eng]*
kibaku-sōchi	起爆装置	fuse *{n}* *[eng]*
kibaku-yaku	起爆薬	initiating explosive; initiator *[ord]*
kiban	基板	substrate *[electr]*
kiban	基盤	bedrock; base *[geol]*
kiban-kōzō	基盤構造	infrastructure *[civ-eng]*
ki'bashiri	木走り	tree creeper (bird) *[v-zoo]*
ki'batan	黄巴旦；きばたん	sulfur-crested cockatoo (bird) *[v-zoo]*
ki'beri-tate'ha	黄縁立羽	mourning cloak (butterfly) *[i-zoo]*
kibi	黍；きび	millet *[bot]*
kibo	規模	scale; scope; plan *{n}*
kichin	キチン	chitin *[bioch]*
kichinto	きちんと	accurately; exactly; neatly *{adv}*
kichi-sū	既知数	known quantity *[math]*
kidai	基台；基臺	base; base stand *{n}* *[eng]*
kidan	気団；氣團	air mass *[meteor]*
kiden	起電	electric generation *[elec]*

kiden-ban	起電盤	electrophorus *[elec]*
kiden'gaku	機電学	mechatronics *[electr]*
kiden-ryoku	起電力	electromotive force; e.m.f. *[p-ch]*
kido	輝度；きど	luminance *[opt]*
kidō	気道；氣道	respiratory tract *[anat]*
kidō	軌道；きどう	orbit; track *{n} [astron] [phys] [rail]*
"	軌道	locus (pl loci) *[math]*
kidō	起動	starting; start *{n} [comput] [elec]*
kidō	機胴	fuselage *[aero-eng]*
kidō-denshi	軌道電子	orbital electron *[atom-phys]*
kidō(-kansū)	軌道(関数)	orbital *[atom-phys]*
kidō-kettei	軌道決定	orbit determination *[astron]*
kidō-ki	起動機	starter *[elec]*
ki'dōkō	輝銅鉱	chalcocite *[miner]*
kidō-kyoku'ten	軌道極点	apsis *[astron]*
kido'rui-genso	希土類元素	rare earth element *[ch]*
kidō saseru	起動させる	activate *{vb t} [comput]*
ki'en	輝炎	luminous flame *[ch]*
kifu	基布	base fabric *[poly-ch]*
ki'fusseki	輝沸石	heulandite *[miner]*
ki-gasu	希ガス	rare gas *[ch]*
kigasu-kagō'butsu	希ガス化合物	rare (inert) gas compound *[ch]*
kigasu-kōzō	希ガス構造	rare (inert) gas structure *[ch]*
kigasu'rui-genso	希ガス類元素	rare (inert) gas element *[ch]*
kigen	期限	term; time limit; deadline *{n}*
ki'ginkō	輝銀鉱	argentite *[miner]*
kigō	記号；記號；きごう	symbol *{n} [ch] [sci-t]*
kigō-ron	記号論	semiotics *[comm]*
kigu	器具；きぐ	instrument *[eng]*
kigū-kensa	奇偶検査	even parity check *[comput]*
"	奇偶検査	odd-even check *[comput]*
kigū-sei	奇偶性	parity *[comput]*
kigu-seizu	器具製図	instrument drawing *[graph]*
kigyō-kimitsu	企業機密	industrial secret *[ind-eng]*
kigyō-sei	企業性	commercial potential *[econ]*
ki'hada(-maguro); kiwada	黄肌(鮪)	yellowfin tuna (fish) *[v-zoo]*
ki'hada	黄檗；きはだ	Chinese cork tree *[bot]*
kihak(u)'ka	希薄化	dilution; rarefaction *[acous] [ch]*
ki'haku na	希薄な	rare; rarefied *{adj} [fl-mech]*
kihaku-yōeki	希薄溶液	dilute solution *[ch]*
ki'hatsu	揮発；揮發	volatilization *[thermo]*
kihatsu-bun	揮発分	volatile matter *[mat]*

kihatsu-do	揮発度；きはつど	volatility *[thermo]*
kihatsu-horyū-zai	揮発保留剤	fixative *[mat]*
kihatsu'sei no	揮発性の	volatile *{adj}* *[ch]*
kihatsu'sei-yōzai	揮発性溶剤	volatile solvent *[ch]*
ki-hen	木扁；木偏：木	kanji (q.v.) radical for tree or wood *[pr]*
kihi-zai	忌避剤	repellent *[bioch]* *[mat]*
kihō	気泡；氣泡	bubble (air or gas) *{n}* *[phys]*
kihō	記法	notation *[math]*
kihō	起泡	foaming; frothing *{n}* *[eng]*
kihon	基本；きほん	master *{n}* *[comput]*
kihon-kai	基本解	fundamental solution *[math]*
kihon-on	基本音	fundamental tone *[acous]*
kihon-sei	基本星	fundamental star *[astron]*
kihon-shindō	基本振動	fundamental vibration *[mech]*
kihon-shoku	基本色	base color *[opt]*
kihon-shūha'sū	基本周波数	fundamental frequency *[phys]*
kihon-tan'i	基本単位	base unit; fundamental unit *[phys]*
kihon-tan'i-kumi'awase-gata no	基本単位組合せ型の	modular *{adj}* *[arch]*
kihō-sei	起泡性	foaming property *[eng]*
kihō-tō	気泡塔	bubble tower *[ch-eng]*
kihō-zai	起泡剤	foaming agent; frother *[poly-ch]*
ki-ichigo	木苺；きいちご	raspberry; Rubus palmatus (fruit) *[bot]*
ki-iro	黄色；きいろ	yellow (color) *[opt]*
ki'iro-mushi'kui	黄色虫食い	yellow warbler (bird) *[v-zoo]*
ki-ito	生糸；きいと	raw silk *[tex]*
kiji	木地	wood grain *[mat]*
kiji	生地	ground *{n}* *[mat]*
"	生地	cloth *[tex]*
kiji	素地	body *[cer]*
kiji	雉子；きじ	pheasant (bird) *[v-zoo]*
kiji-hō	記示法	notation *[comput]*
kiji-ryoku	起磁力	magnetomotive force *[elecmg]*
kijō-shuppan	机上出版	desktop publishing *{n}* *[comput]*
kijū-ki	起重機	crane; jack *{n}* *[mech-eng]*
kijun	基準；きじゅん	base; reference *{n}* *[comput]* *[sci-t]*
kijun	規準	criterion (pl -ria); norm; standard
kijun-denkyoku	基準電極	reference electrode *[p-ch]*
kijun-fundō	基準分銅	standard weight *[mech]*
kijun'ka	基準化	scaling *{n}* *[comput]* *[eng]*
kijun'ka-insū	基準化因数	scale factor *[eng]*
kijun-sen	基準線	reference line; zero line *[mech]* *[navig]*

kijun-shindō	基準振動	normal vibration *[phys]*
kijun-ten	基準点	reference point *[navig]*
ki'jutsu	記述; きじゅつ	description *[comput]*
kijutsu'kō	記述項	entry *{n}* *[comput]*
kijutsu'shi	記述子	descriptor *[comput]*
kijutsu-tōkei'gaku	記述統計学	descriptive statistics *[math]*
kika	気化; 氣化	gasification *[ch-eng]*
kika'gaku	幾何学; きかがく	geometry *[math]*
kika-heikin	幾何平均	geometric mean *[math]*
kika-heikin-soku	幾何平均則	geometric mean rule *[math]*
kikai	機械; きかい	machine; machinery; equipment *[mech-eng]*
kikai-(gen)go	機械(言)語	machine language *[comput]*
kikai'ka	機械化	mechanization *[eng]*
kikai-kakō	機械加工	machining *{n}* *[met]*
kikai-kanki	機械換気	mechanical ventilation *[eng]*
kikai'kō	機械工	mechanic (a person) *[mech-eng]*
kikai-kōgaku	機械工学	mechanical engineering *[eng]*
kikai-kotoba	機械言葉	machine language *[comput]*
kika-i'sei	幾何異性	geometrical isomerism *[p-ch]*
kikai-shūri'kō	機械修理工	mechanic (a person) *[mech-eng]*
kikai-son	機械損	mechanical loss *[eng]*
kikai-sōsa	機械操作	machine operation *[comput]*
kikai'teki-hin'shitsu-keisū	機械的品質係数	mechanical quality factor *[ind-eng]*
kika-ki	気化器; 氣化器	carburetor *[ch-eng]*
kika-kōgaku	幾何光学	geometrical optics *[opt]*
ki'kaku	規格; きかく	specification; technical standard *[ind-eng]*
kikaku-chi	規格値	technical standard value *[ind-eng]*
kikaku'ka	規格化	standardization *[eng]*
kikaku'ka-keisū	規格化係数	normalization factor *[photo]*
kikaku-nuno	規格布	standard cloth *[tex]*
kikaku suru	規格する	standardize {vb t} *[eng]*
kika-kyūsū-soku	幾何級数則	geometric progression, law of *[bio]*
kikan	気乾; 氣乾	air drying *{n}* *[eng]*
kikan	気管	trachea (pl -cheae) *[anat]*
kikan	汽缶; 汽罐	boiler *[mech-eng]*
kikan	帰還; 歸還; きかん	feedback *{n}* *[comput]* *[electr]* *[sci-t]*
kikan	基幹	stem *{n}* *[ling]*
kikan	期間	period of time *[phys]*
kikan	器官; 噐官	organ *[bio]*
kikan	機関; 機關	engine; machine *[mech-eng]*
"	機関	authority; agency; organ *{n}*

kikan-baiyō	器官培養	organ culture *[bot]*
kikan-dai	機関台	engine bed *[mech-eng]*
kika-netsu	気化熱	evaporation heat; heat of vaporization *[thermo]*
kikan-hassei	器官発生	organogenesis *[embryo]*
kikan-jū	機関銃	machine gun *[ord]*
kikan'sei no	起寒性の	frigorific *{adj}* *[thermo]*
kikan'sha	機関車	engine; locomotive *[rail]*
kikanshi	気管支; きかんし	bronchus (pl -i) *[anat]*
kikan-shitsu	機関室	engine room *[nav-arch]*
ki-kansū	奇関数	odd function *[math]*
kikan-zai	起寒剤	freezing mixture *[p-ch]*
kikei'gaku	奇形学	teratology *[med]*
kiken	危険; 危險; きけん	danger; hazard *[ind-eng]*
kiken'butsu	危険物	dangerous object; hazardous material *[mat]*
kiken-do	危険度	risk *{n}* *[math]*
kiken-ritsu	危険率	level of significance *[math]*
"	危険率	risk factor *{n}*
kiki	危機	crisis (pl crises)
kiki	機器; きき	instrument; machinery and tools *[eng]*
kiki-bunseki	機器分析	instrumental analysis *[ind-eng]*
ki-kinzoku	貴金属	noble metal; precious metal *[met]*
kikkō-sayō	拮抗作用	antagonism *[bio]*
kikkō-sei	拮抗性	antagonism; competitiveness *[bio]*
kikkō'seki	亀甲石; 龜甲石	septarium *{n}* *[geol]*
kikkō-sogai-zai	拮抗阻害剤	competitive inhibitor *[bioch]*
kikō	気孔	pore *{n}* *[bio]*
kikō	気候; きこう	climate *[climat]*
"	気候	weather *{n}* *[meteor]*
kikō	機構	mechanism; machinery *[mech-eng]*
"	機構	structure; function; organization
kikoeru	聞こえる	audible *{adj}*
kikō'gaku	気候学; 氣候學	climatology *[meteor]*
kikō-gakusha	気候学者	climatologist *[climat]*
kikori	樵夫; きこり	woodcutter; logger (a person)
kikō-ritsu	気孔率	porosity *[bio]*
kiku	菊	chrysanthemum (flower) *[bot]*
kiku	利く	be effective; work *{vb}*
kiku	聞く; 聴く	hear; listen *{vb}* *[physio]*
ki'kui-mushi	木食虫	wood borer (insect) *[i-zoo]*
kiku-ishi	菊石	ammonite *[paleon]*
kiku-itadaki	菊戴; きくいただき	goldcrest (bird) *[v-zoo]*

kiku'me-ishi	菊目石	star coral *[i-zoo]*
kiku'me-so'shiki	菊目組織	chrysanthemum structure *[mech-eng]*
kiku'na	菊菜	edible chrysanthemum leaves *[cook]*
kiku-niga'na	菊苦菜；きくにがな	chicory (herb) *[cook]*
ki-kurage	木耳	Auricularia auricula (mushroom) *[bot]*
kiku-san	菊酸	chrysanthemic acid
ki'kyaku	棄却	discard; discarding *{n}*
kikyaku-iki	棄却域	critical region *[math]*
kikyō	桔梗；ききょう	balloonflower; Chinese bellflower *[bot]*
kikyū	気球	aerostat; balloon *[aero-eng]*
kime	肌理；きめ	texture (of skin) *[physio]*
kimi	黄身	yolk (of egg) *[bioch]*
kimitsu	機密	privacy *[comput]*
kimitsu-do	気密度	airtightness *[eng]*
kimo	肝	liver *[anat]*
ki'mono	着物；きもの	clothing; Japanese kimono *[cl]*
kimu-kasetsu	帰無仮説；歸無假說	null hypothesis *[math]*
kin	金；きん	gold (element)
kin	菌	fungus (pl -i) *[mycol]*
-kin	-筋	-muscle *{suffix}* *[anat]*
kina	規那；きな	quinine; cinchona *[org-ch]*
kinbaku suru	緊縛する	bind tightly *{vb t}*
kinban	勤番	worker on duty *[ind-eng]*
kinchi-ten	近地点	perigee *[astron]*
kinchō-bōshi	緊張紡糸	stretch spinning *{n}* *[tex]*
kindei	金泥	gold paint *[mat]*
kin'en	禁煙；禁煙	Smoking Prohibited *[ecol]*
kin'enka'suiso-san	金塩化水素酸	hydrogen chloroaurate
kingaku	金額	amount of money *[econ]*
kin'gaku	菌学；きんがく	mycology *[bot]*
kin'ginga-chūshin-ten	近銀河中心点	perigalacticon *[astron]*
kingyo	金魚	goldfish (fish) *[v-zoo]*
kingyo'sō	金魚藻	hornwort *[bot]*
kin-hon'i-seido	金本位制度	gold standard *[econ]*
kin'i no	近位の	proximal *{adj}* *[med]*
kin-iro; kon'jiki	金色	gold (color) *[opt]*
kin'itsu-hannō	均一反応	homogeneous reaction *[ch]*
kin'itsu-jūgō'butsu	均一重合物	homogeneous polymer *[poly-ch]*
kin'itsu na	均一な	homogenous *{adj}* *[ch]* *[sci-t]*
kinji-chi	近似値	approximate value *[math]*
kinjin	菌蕈；きんじん	mushrooms *{n}* *[bot]*
kin'jisa	均時差	equation of time *[astron]*

kin'jitsu-ten	近日点	perihelion (pl -helia) *[astron]*
kin-kabu	菌株; きんかぶ	bacterial strain; fungal strain *[mycol]*
kinkai	近海	adjoining seas; neighboring seas *[geogr]*
kinkai	金塊	bullion *[met]*
kinkai	菌界	Mycota; Eumycetes *[mycol]*
kinkan	金柑; きんかん	kumquat (fruit) *[bot]*
kinkan-shoku	金環食	annular eclipse *[astron]*
kinki	菌器	mycetome *[i-zoo]*
kinkon	菌根	mycorrhiza (pl -zae, -s) *[bot]*
kin'kōseki	金紅石	rutile *[miner]*
kinko-shitsu	金庫室	vault (as in a bank) *{n}* *[arch]*
kinkō-ten	均衡点	equilibrium point *[ch]*
kinkyū-hoshu	緊急保守	emergency maintenance *[comput]* *[ind-eng]*
kinkyū-shingō	緊急信号	urgency signal *[comm]*
kinme-dai	金眼鯛; きんめだい	Beryx splendens (fish) *[v-zoo]*
kin'mitsuda	金密陀	massicot *[miner]*
kin-mokusei	金木犀	fragrant olive *[bot]*
kinnetsu	均熱	soaking *{n}* *[steel]*
kinniku	筋肉; きんにく	muscle *{n}* *[anat]*
kin(niku)-sen'i	筋(肉)繊維	muscle fiber *[hist]*
kin(niku)-so'shiki	筋(肉)組織	muscular system *[anat]*
kinō	昨日; きのう	yesterday *{adv}*
kinō	帰納	induction *[logic]*
kinō	機能	function; capability; facility *[comput]*
ki no atai	生の値	pure value *[math]*
kinoko	茸; きのこ	mushroom *{n}* *[mycol]*
ki-no-me	木の目	prickly ash sprouts *[cook]*
ki no sa	生の差	pure difference *{n}* *[math]*
kinō-sei	機能性	functionality *[comput]*
kinō'sei-kōbunshi	機能性高分子	functional polymer *[org-ch]*
kinō-teishi	機能停止	breakdown; failure *{n}* *[mech]*
kinō-zairyō	機能材料	functional materials *[mat]*
kinpaku	金箔; きんぱく	gold leaf; gold foil *[mat]*
kinri	金利	interest *{n}* *[econ]*
kinrin-kōen	近隣公園	neighborhood park *[civ-eng]*
kin-rui	菌類	fungus (pl fungi) *[mycol]*
kinrui'gaku	菌類学; 菌類學	mycology *[bot]*
kin'ryokuseki	金緑石	chrysoberyl *[miner]*
kinsan	金酸	auric acid
kinsan-en	金酸塩	aurate *[ch]*
kinsei	金星; きんせい	Venus (planet) *[astron]*
kinsei	均整; 均斉	balance; uniformness *[ch]*

kinsei'seki	菫青石	cordierite *[miner]*
kinsei-sen	禁制線	forbidden line *[atom-phys]*
kin-sekigai-ryō'iki	近赤外領域	near-infrared region *[elecmg]*
kin-sekigai-sen	近赤外線	near-infrared radiation *[elecmg]*
kinsen	金銭; 金銭	money *[econ]*
kinsen-sei	均染性	leveling property *[ch]*
kin'setsu-ki	近接基; きんせつき	vicinal group *[org-ch]*
kin'setsu-kōka	近接効果	proximity effect *[elec]*
kinshi	近視	myopia; nearsightedness *[med]*
kinshi	金糸	gold thread *[mat]*
kinshi	菌糸; きんし	hypha (pl hyphae) *[mycol]*
kinshi	禁止	inhibit *[comput]*
kin-shigai'sen	近紫外線	near-ultraviolet radiation; near-ultraviolet rays *[elecmg]*
kinshi-hyō'shiki	禁止標識	prohibition sign *[civ-eng]*
kinshi-kei	菌糸形	mycelial form *[mycol]*
kinshi-kumi'awase	禁止組合せ	forbidden combination *[comput]*
kinshi-meirei	禁止命令	restraining order *[legal]*
kinshi suru	禁止する	inhibit; prohibit *{vb}*
kinshi'tai	菌糸体	mycelium (pl -a) *[mycol]*
kinshi-tai	禁止帯; 禁止帯	forbidden band *[ch]*
kin'shitsu	均質; きんしつ	homogeneous; homogeneity *[sci-t]*
kin'shitsu-gyūnyū	均質牛乳	homogenized milk *[cook]*
kin'shitsu-ken	均質圏; 均質圏	homosphere *[meteor]*
kin-shiya-zō	近視野像	near-field pattern *[opt]*
kinshu	菌種	bacterial type *[mycol]*
kin'shuku-sei	緊縮性	tautness *[mat]*
kinsō'gaku	金相学	metallography *[met]*
kin-sū	金数	gold number *[an-ch]*
kintai	菌体	bacterial cell; fungal cell *[mycol]*
kin'tatsu-ryoku	近達力	short-range force *[phys]*
kin-ten	金点	gold point *[thermo]*
kinten-nen	近点年	anomalistic year *[astron]*
kinton	金団; きんとん	chestnut-sweet potato confection *[cook]*
kinu	絹; きぬ	silk *[tex]*
kinu'goshi-dōfu	絹漉し豆腐	fine-grained tofu (q.v.) *[cook]*
kinu-ito	絹糸	silk yarn *[tex]*
kin-un'mo	金雲母	phlogopite; bronze (brown) mica *[miner]*
kinu-ori'mono	絹織物	silk fabric *[tex]*
kinuta-bone; chin-kotsu	砧骨	anvil (bone) *{n} [anat]*
kin'yōbi	金曜日	Friday
ki'nyū	記入	entry *{n} [comput]*

kin'zoku	金属; きんぞく	metal *[mat]*
kinzoku-ion-hōsa-zai	金属イオン封鎖剤	sequestering agent *[ch]*
kin(zoku-)kai	金(属)塊	bullion *[met]*
kinzoku'kan-kagō'butsu	金属間化合物	intermetallic compound *[ch]*
kinzoku-kenbi'kyō	金属顕微鏡	metallurgical microscope *[opt]*
kinzoku-ketsu'gō	金属結合	metallic bond *[p-ch]*
kinzoku-kōtaku	金属光沢; 金屬光澤	metallic luster *[opt]*
kinzoku no hirō	金属の疲労	metal fatigue *[met]*
kinzoku'ryū ni suru	金属粒にする	prill *{vb t}* *[ch-eng]*
kinzoku-sekken	金属石鹼	metallic soap *[org-ch]*
kinzoku-soshiki'gaku	金属組織学	metallography *[met]*
ki'oku	記憶; きおく	memory; storage *[comput]*
"	記憶	memory *[psy]*
ki'oku-hō	記憶法	mnemonic system *[psy]*
ki'oku-hogo	記憶保護	memory protection *[comput]*
ki'oku-ichi	記憶位置	store location *[comput]*
ki'oku-sōchi	記憶装置	memory; storage *[comput]*
ki'oku-soshi	記憶素子	memory (device); storage element *[comput]*
ki'oku suru	記憶する	store *{vb t}* *[comput]*
"	記憶する	remember *{vb}* *[psy]*
ki'oku-yōryō	記憶容量	storage capacity *[comput]*
kira-kira	キラキラ	glitter *{n}* *[opt]*
kirameki	煌; きらめき	sheen *{n}* *[opt]*
kirazu	雪花菜; きらず	bean curd refuse *[cook]*
-kire	-切	-piece(s); -slice(s) *{suffix}*
kire-aji	切れ味	cutting quality; sharpness *[eng]*
kire-ha	切れ刃	cutting edge *[eng]*
kirei na	綺麗な; 奇麗な	beautiful; clean *{adj}*
kirēto-shiyaku	キレート試薬	chelating agent *[org-ch]*
ki'retsu	亀裂	crack *{n}* *[sci-t]*
kiri	桐	empress tree; paulownia (tree) *[bot]*
kiri	錐; きり	drill; awl; gimlet *{n}* *[eng]*
kiri	霧	fog *{n}* *[meteor]*
kiri-bako	霧箱	cloud chamber *[nucleo]*
kiri-ban	切り板	cut-to-order sheet *[mat]*
kiri'fuki	霧吹き	spraying *{n}* *[eng]*
kirigirisu	螽蟖; きりぎりす	grasshopper (long-horned); katydid (insect) *[i-zoo]*
kiri-ha	切り刃	cutting blade *[eng]*
kiri-hanasu	切り離す	disconnect *{vb t}* *[elec]* *[eng]*
kiri'ho	切穂; きりほ	cutting *{n}* *[bot]*
kiri'kae	切り替え	switch; switching *{n}* *[comput]*

kiri'kake-bu	切欠(け)部	notched part *[eng]*
kiri'kaki	切欠き；切缺き	notch *{n} [eng]*
kiri'kiseki	錐輝石	acmite; aegerine *[miner]*
kiri'komi(-fuka-sa)	切込み(深さ)	depth of cut; cut *{n} [eng]*
kiri'kuzu	切り屑	chips *{n} [mat]*
kiri'me	切り目	kerf *[eng]*
kiri-mizo	切り溝	kerf *[eng]*
kirin	麒麟；きりん	giraffe (animal) *[v-zoo]*
"	麒麟；きりん	kylin (mythical animal)
kiri'same	霧雨；きりさめ	drizzle *{n} [meteor]*
kiri'sute	切り捨て	truncation; rounding down *[math]*
kiri'sute-gosa	切り捨て誤差	truncation error *[comput]*
kiri-suteru	切り捨てる	round down *{vb} [math]*
kiro'guramu	瓩；キログラム	kilogram *[mech]*
kiro'karorī	キロカロリー	kilocalorie *[thermo]*
ki'roku	記録；記錄	archiving *{n} [comput]*
"	記録；きろく	record; recording *{n} [comput] [pr]*
kiroku-baitai	記録媒体	recording medium *[electr]*
kiroku'chō	記録帳	docket *[legal]*
kiroku-den'ryoku-kei	記録電力計	recording wattmeter *[eng]*
kiroku(-hokan)	記録(保管)	archiving *{n} [comput]*
kiroku-hokan'jo	記錄保管所	archive *[comput]*
kiroku-kei	記録計	recorder *[eng]*
kiroku-kikō	記録機構	recording mechanism *[eng]*
kiroku-men	記録面	recording surface *[eng]*
kiroku-mitsudo	記録密度	recording density *[comput]*
kiroku-ondo-kei	記録温度計	thermograph *[eng]*
kiroku suru	記録する	record *{vb t}*
kiro'mētoru	粁；キロメートル	kilometer *[mech]*
kiro'rittoru	竏；キロリットル	kiloliter *[mech]*
kiro'watto-ji	キロワット時	kilowatt-hour; kWh *[elec]*
kiru	切る	cut *{vb t}*
kiru	着る	wear *{vb}*
kiru koto	切ること	truncation *[comput]*
kisa	器差	instrumental error *[eng]*
ki'saku suru	機削する	machine-shave *{vb t} [met]*
kisan	希酸；稀酸；きさん	dilute acid *[ch]*
kisan-rui	希酸類	rare acids *[ch]*
kisasage	木豇豆；楸；きささげ	Japanese catalpa *[bot]*
kisei	寄生	parasitism *[ecol]*
kisei-fuku	既製服	ready-to-wear clothing *[cl]*

kisei no	既製の	ready-made *{adj} [ind-eng]*
"	既製の	established; accepted; completed *{adj}*
kisei-sei	寄生性；きせいせい	parasitic property *[bio]*
kisei-seibutsu	寄生生物	parasite *[bio]*
kisei suru	規正する	regulate; readjust; control *{vb t}*
kisei suru	規制する	control; regulate *{vb t}*
kisei'tai	寄生体	parasite *[bio]*
kisei'un	輝星雲	bright nebula (pl -s, -lae) *[astron]*
kiseki	軌跡；きせき	locus (pl -ci) *[math]*
kiseki	輝石	augite *[miner]*
kisen	汽船	steamship; steamer *[nav-arch]*
kisen	軌線	trajectory *[mech]*
kisen	基線	base line *[navig] [sci-t]*
kisen	輝線	emission line *[spect]*
kisenon	キセノン	xenon (element)
kisen-shōkyo	帰線消去	blackout (of TV) *{n} [electr]*
ki'setsu	季節；きせつ	season *[climat]*
kisetsu'fū	季節風	monsoon *[meteor]*
kisetsu-hendō	季節変動	seasonal variation *[geoph]*
ki'shaku	希釈；希釋	dilution *[ch]*
kishaku-do	希釈度；きしゃくど	dilution, degree of *[ch]*
kishaku-netsu	希釈熱	heat of dilution *[p-ch]*
kishaku-ritsu	希釈率	dilution ratio *[ch]*
kishaku suru	希釈する	dilute; thin *{vb t}*
kishaku-zai	希釈剤	diluent *[ch]*
kishin	晷針；きしん	gnomon *[eng]*
ki'shitsu	基質	matrix (pl matrices, -es) *[bio] [geol]*
"	基質	substrate *[bioch]*
ki-shitsu	機室	cabin (aircraft, seacraft) *[arch]*
ki'shitsu-toku'i-sei	基質特異性	substrate specificity *[bioch]*
kishō	気象；きしょう	weather conditions *[meteor]*
kishō'gaku	気象学	meteorology *[sci-t]*
kishō-tsū'hō	気象通報	weather report *[meteor]*
kishu	寄主	host *{n} [bio]*
kishu	機種	machine type *[mech-eng]*
kishu-choku'hō'i	機首直方位	true heading *{n} [navig]*
kishu-ji'hō'i	機首磁方位	magnetic heading *{n} [navig]*
kishuku'sha	寄宿舎	dormitory (pl -ries) *[arch]*
kiso	起訴	indictment; prosecution *[legal]*
kiso	基礎；きそ	fundamentals *[sci-t]*
kiso	機素	machine element *[eng]*
kisō	気相	gas(eous) phase; vapor phase *[ch]*

kiso-dojō-gaku	基礎土壌学	pedology *[geol]*
kisō-jūgō	気相重合	gas-phase polymerization *[poly-ch]*
kisoku	規則；きそく	rule *{n}* *[math]*
"	規則	regulations *{n}*
kisoku'sei-kōbunshi	規則性高分子	regular polymer *[poly-ch]*
kiso-seisan	基礎生産	primary production *[bio]*
kiso-zu	基礎図	foundation drawing *[graph]*
kissō'kon	吉草根	valerian *[bot]*
kissō-san	吉草酸	valeric acid
kissō'san-en	吉草酸塩	valerate *[ch]*
kissui	喫水；きっすい	draft *{n}* *[nav-arch]*
kissui-sen	喫水線	waterline *[nav-arch]*
kisu	鱚；きす	Sillago sihama (fish) *[v-zoo]*
kisū	奇数；奇數	odd number *[math]*
kisū	基数	base; cardinal number; fundamental number; radix *[math]*
kisū-hō	基数法	number system *[math]*
kisui'enkō	輝水鉛鉱	molybdenite *[miner]*
kisū no hosū	基数の補数	radix complement *[math]*
kisū-pēji	奇数頁	recto (pl -s) *[pr]*
kita	北；北；きた	north *[geod]*
kitai	気体；氣體；きたい	gas (pl -es) *{n}* *[phys]*
kitai	基体	substrate *[eng]*
kitai	機体	airframe *[aero-eng]*
kitai-chi	期待値	expectation *[math]*
kitai-funsha	気体噴射	gas injection *[mech-eng]*
kitai-hannō no hōsoku	気体反応の法則	gaseous reaction, law of *[ch]*
kitai-kakusan no hōsoku	気体拡散の法則	diffusion of gas, law of *[phys]*
kitai no hō'soku	気体の法則	gas law; law of gases *[thermo]*
kitai-ryū'shutsu no hōsoku	気体流出の法則	effusion of gas, law of *[phys]*
kitai-shibori	気体絞り	gas wiper *[eng]*
kitai-teisū	気体定数	gas constant *[thermo]*
kita-jūji-za	北十字座	Northern Cross *[constel]*
kita-kaiki-sen	北回帰線	tropic of Cancer *[geod]*
kita-kaze	北風	northerlies; northerly wind *[meteor]*
kitaku	寄託	deposition *[microbio]*
kitaku suru	寄託する	deposit at *{vb t}* *[microbio]*
kita'makura	北枕	four-saddle puffer (fish) *[v-zoo]*
kitanai	汚ない	dirty; soiled; messy *{adj}*
kitchiri to	きっちりと	punctually; snugly *{adv}*
kitei	基底	base *{n}* *[comput]*
kitei	既定	predefined *{n}*

kitei	規定；きてい	provision(s); definition *{n}*
kitei-do	規定度	normality *[ch]*
kitei-eki	規定液	normal solution *[ch]*
kitei-jōtai	基底状態	ground state *[quant-mech]*
kitei suru	規定する	stipulate; ordain *{vb} [legal]*
kiteki	汽笛；きてき	steam whistle *[eng]*
kiten	起点；起點	origin *[math]*
"	起点	fiducial point *[opt]*
kiten	基点	origin; radix point *[comput]*
"	基点	cardinal points *[geod]*
kiten-gengo	起点言語	source language *[ling]*
kitsu'en suru	喫煙する	smoke *(e.g., a cigarette) {vb t}*
kitsui	きつい	tight; strong; powerful *{adj}*
kitsuku suru	きつくする	make tight; tighten *{vb t}*
kitsune	狐；きつね	fox (animal) *[v-zoo]*
kitsune-zaru	狐猿	lemur (animal) *[v-zoo]*
ki'tsutsuki	啄木鳥；きつつき	woodpecker (bird) *[v-zoo]*
kitte	切手；きって	stamp *{n} [comm]*
kitte-shūshū	切手収集	philately *[comm]*
ki'yaku	規約	protocol *[comput]*
kiyomeru	清める	clean; purify *{vb}*
Kiyomizu-yaki	清水焼	Kiyomizu ware *[cer]*
kiyu	基油；きゆ	base oil *[mat]*
kizai	器材	material for tools *[mech-eng]*
kizai	基材；きざい	base; base composition; base material *{n} [ch] [eng] [pharm] [poly-ch]*
"	基材	substrate *[eng]*
kizai	基剤	base composition *[ch]*
kizai	機材	machine material *[mech-eng]*
kizami	刻み；きざみ	serration *[bot]*
"	刻み	knurl *{n} [eng]*
"	刻み	pitch *{n} [mech]*
kizamu	刻む	chip; dice; mince *{vb t}*
kizu	疵；瑕；きず	defect; flaw; fault *{n} [ind-eng]*
kizu	傷；創	wound; injury *{n} [med]*
kizu-ato	傷跡	scar *{n} [med]*
ki'zuisen	黄水仙	jonquil (flower) *[bot]*
ki'zuta	木蔦；きずた	ivy *[bot]*
ko	子	child *[bio] [comput]*
"	子	member *{n} [comput]*
ko	弧	arc *{n} [astron] [math]*
ko	個；こ	piece; unit *{n}*

ko	箇	piece; unit *{n}*
kō	香；こう	incense *{n} [mat]*
kō	項	term *{n} [comput] [math] [spect]*
"	項	paragraph *[pat]*
"	項	item *[pr]*
kō	綱	class *{n} [bio] [syst]*
-ko	-箇	-unit(s) of object(s) *{suffix}*
kō-	抗-	anti- *{prefix}*
kō- see also hikari	光-	light-; photo- *{prefix}*
kō'aen'kō	紅亜鉛鉱	zincite *[miner]*
ko-ago; shō'gaku	小顎	maxilla (pl -illae) *[anat]*
kōan	考案；こうあん	contrivance; devisement; device *{n} [pat]*
kōan suru	考案する	devise *{vb}*
kō'atsu	高圧；高壓	high voltage *[elec]*
"	高圧；こうあつ	high pressure *[phys]*
kōatsu-butsuri'gaku	高圧物理学	high-pressure physics *[phys]*
kōatsu-densen	高圧電線	high-voltage cable *[elec]*
kōatsu-gaishi	高圧碍子	high-voltage insulator *[elec]*
kōatsu-hō	高圧法	high-pressure process *[ch-eng]*
kōatsu-kagaku	高圧化学	high-pressure chemistry *[p-ch]*
kōatsu-kagaku	高圧科学	high-pressure science *[phys]*
kōatsu-mekkin-ki	高圧滅菌器	autoclave *[eng]*
kōatsu no	高圧の	high pressure *{adj} [phys]*
kōatsu-seikei	高圧成形	high-pressure molding *[resin]*
kōatsu-sekisō	高圧積層	high-pressure laminating *[poly-ch]*
kōatsu-zai	降圧剤	hypotensor *{n} [pharm]*
kōbai	勾配；こうばい	pitch *{n} [arch] [sci-t]*
"	勾配	gradient *[geol] [math]*
"	勾配	incline *{n} [sci-t]*
kōbai	購買	purchasing *{n} [econ]*
kōban	交番	police box; police stand *[arch]*
kōban	鋼板	steel sheet; steel plate *[mat]*
kōban-ji'kyoku	交番磁極	alternating magnetic pole *[elecmg]*
kōban-kyoku'sei	交番極性	alternating polarity *[phys]*
koban-zame	小判鮫	shark sucker (fish) *[v-zoo]*
kobaruto	コバルト	cobalt (element)
kōbashī; kanbashī	芳しい	fragrant; delicious-smelling *{adj} [cook]*
Kōbe	神戸；こうべ	Kobe *[geogr]*
ko'betsu	個別	separate; individually *{n}*
kōbo	酵母；こうぼ	yeast *[mycol]*
kōbō	光芒	beam of light; shaft of light *[opt]*
kōbo-ekisu	酵母エキス	yeast extract *[mycol]*

kōbo-kin	酵母菌；こうぼきん	yeast *[mycol]*
kōboku-genkai-sen	高木限界線	tree line *[ecol]*
kobore	零；こぼれ	spill *{n} [eng]*
"	零；こぼれ	overflow *{n} [sci-t]*
kobosu	零す；溢す	spill *{vb t}*
kōbu	後部	hind part; rear (or back) part *{adv}*
kōbun(-hō)	構文(法)	syntax *[comput] [ling]*
kō'bunkai	光分解	photolysis *[p-ch]*
kō'bunretsu	光分裂	photofission *[nuc-phys]*
kōbun'ron	構文論	syntactics *[ling]*
kō'bunshi	高分子；こうぶんし	high-molecular compound; high polymer; macromolecule; polymer *[poly-ch]*
kōbunshi-denkai'shitsu	高分子電解質	polyelectrolyte *[poly-ch]*
kōbunshi-gōkin	高分子合金	polymer alloy *[met] [poly-ch]*
kōbunshi-gyōshū-zai	高分子凝集剤	polymeric flocculant *[poly-ch]*
kōbunshi'ka	高分子化	conversion to high polymer; high polymerization *[poly-ch]*
kōbunshi-kagaku	高分子化学	macromolecular chemistry; polymer chemistry *[ch]*
kōbunshi-kagō'butsu	高分子化合物	high-molecular compound; polymer *[poly-ch]*
kōbunshi-kaimen-kassei-zai	高分子界面活性剤	high-molecular surfactant *[poly-ch]*
kōbunshi-kaso-zai	高分子可塑剤	polymeric plasticizer *[poly-ch]*
kōbunshi no	高分子の	macromolecular *{adj} [poly-ch]*
ko-buta	小豚	piglet (animal) *[v-zoo]*
kō'butsu	好物；こうぶつ	favorite dish *[cook]*
kō'butsu	鉱物；こうぶつ	mineral *[geol]*
kōbutsu'gaku	鉱物学；鑛物學	mineralogy *[inorg-ch]*
kōbutsu-rō	鉱物蠟	mineral wax; ozocerite *[geol]*
kōbutsu-sen'i	鉱物繊維	mineral fiber *[tex]*
kō-byō	光秒	light second *{n} [mech]*
kōcha	紅茶	black tea *[cook]*
ko'chaku	固着	adherence; fastening; sticking *{n} [adhes]*
kōchaku-sōchi	降着装置	landing gear; undercarriage *[aero-eng]*
kōchaku suru	膠着する	agglutinate (onto) *{vb} [ch]*
kochaku-zai	固着剤	binder; sticking agent *[mat]*
kochi	鯒；牛尾魚；こち	flathead; Platycephalus indicus (fish) *[v-zoo]*
kochi	東風；こち	easterly winds *[meteor]*
kōchi	公知	publication *{n} [pat]*
kōchi	高地	upland *[geogr]*
ko-chi'jiki'gaku	古地磁気学	paleomagnetics *[geophys]*
kōchō	紅潮	flushing *{n} [physio]*

kōchō	高潮	high water *[ocean]*
kōchō-dōbutsu	腔腸動物	Coelenterata *[i-zoo]*
kōchō-eki	高張液	hypertonic solution *[ch]*
kōchō-ha	高調波	harmonics; higher harmonics *[acous]*
kōchō-ryoku	抗張力	tensile strength *[mech]*
kō-chō'ryoku-kō	高張力鋼	high-tensile-strength steel *[met]*
kōchō-seki	抗張積	tensile product *[mech]*
kōchū	甲虫; こうちゅう	beetle (insect) *[i-zoo]*
kōchū	鉤虫	hookworm *[i-zoo]*
kōchū'seki	紅柱石	andalusite; zincite *[miner]*
kodai-murasaki	古代紫	purple of the ancients; Tyrian purple (color) *[opt]*
kodama	木霊; こだま	echo (pl -es) *{n}* *[acous]*
kō'dansei; hikari-dansei	光弾性	photoelasticity *[opt]*
kōden-bunkō-kōdo-kei	光電分光光度計	photoelectric spectrophotometer *[spect]*
kō'denchi	光電池	photocell *[electr]*
kō'dendō-seru	光伝導セル	photoelectric cell; photocell *[electr]*
kōden-hansha-ritsu	光電反射率	photoelectric reflectivity *[electr]*
kōden-kan	光電管	photo(electric) tube *[electr]*
kō'denki; hikari-denki	光電気	photoelectricity *[electr]*
kōden-kōdo-kei	光電光度計	photoelectric photometer *[eng]*
kōden-kōka	光電効果	photoelectricity *[electr]*
kō'denryū; hikari-denryū	光電流	photocurrent *[p-ch]*
kōden-sei	向電性	galvanotropism *[bio]*
kō'denshi-hō'shutsu	光電子放出	photoemission *[electr]*
kō'denshi-kōgaku	光電子工学	optoelectronics *[electr]*
kō'denshi-shūseki-kairo	光電子集積回路	optoelectronic integrated circuit *[electr]*
kō'denshi-zōbai-kan	光電子増倍管	multiplier phototube *[electr]*
kōdo	光度; こうど	luminosity; luminous intensity *[opt]*
kōdo	耕土	arable soil *[agr]*
kōdo	高度	altitude *[eng]*
kōdō	公道	public road; highway *[civ-eng]*
kōdō	行動; こうどう	action; behavior; conduct *{n}* *[psy]*
kōdō	坑道	gallery (pl -leries) *[min-eng]*
kōdō	黄道	ecliptic; girdle *{n}* *[astron]*
kōdō	講堂	auditorium (pl -s, -ria) *[arch]*
ko-dōbutsu'gaku	古動物学	paleozoology *[paleon]*
kodo-hō	弧度法	radian measure *[math]*
kōdō-jūni'kyū	黄道十二宮	zodiac *[astron]*
kōdō-kagaku	行動科学	behavioral science *[psy]*
kōdo-kei	光度計	photometer *[eng]*
kōdo-kei	高度計	altimeter *[eng]*

kōdo-kei 硬度計 hardness meter *[eng]*

kōdō-kō 黄道光 zodiacal light *[astron] [geophys]*

kō'doku'so 抗毒素；こうどくそ antitoxin *[immun] [pharm]*

kōdo-kyoku'sen 光度曲線 light curve *[astrophys]*

kodomo-beya 子供部屋 child's (children's) room *[arch]*

kōdo no 高度の high degree of *{adj}*

kōdo-soku'ryō-ki 高度測量器 altimeter *[eng]*

kōdo-sū 硬度数 hardness number *[eng]*

koe 声；聲；こえ voice *{n} [comm] [physio]*

ko'eki-chū'shutsu 固液抽出 solid-liquid extraction *[p-ch]*

kōen 公園；こうえん park *{n} [civ-eng]*

kōen 紅炎 prominence *[astron]*

kōen 後縁 trailing edge *[aero-eng]*

koendoro 胡荽；こえんどろ coentro *[bot]*

kōen-kin 好塩菌 halophile *[bio]*

kōen'kō 紅鉛鉱 crocoite *[miner]*

kōen'kō 黄鉛鉱 wulfenite *[miner]*

kōen-sei 好塩性 halophilism *[bio]*

kō-faibā 光ファイバー optical fiber *[opt]*

kōfu 甲布 upper (of shoe) *{n} [cl]*

kō'fuku 降伏；こうふく yielding; yield *{n} [mech]*

kōfuku-chi 降伏値 yield value *[mech]*

kōfuku-ka 降伏価 yield value *[mech]*

kōfuku-kyōdo 降伏強度 yield strength *[mech]*

kōfuku-ō'ryoku 降伏応力 yield stress *[mech]*

kōfuku-ten 降伏点 yield point *[mech]*

kōfun 興奮；こうふん excitement; excitation *[psy]*

kōfun-yaku 興奮薬 stimulant *{n} [pharm]*

ko'gai 戸外；こがい outdoors *{n} [geogr]*

kōgai 公害；こうがい environmental pollution; public hazard *[ecol]*

kōgai 郊外 suburbs *[geogr]*

kōgai-kōgyō'chi 郊外工業地 suburban industrial land *[civ-eng]*

ko-gaisha 子会社；子會社 subsidiary company *[econ]*

kōgaku 工学；こうがく engineering; technology *[sci-t]*

kōgaku 光学；こうがく optics *[phys]*

kōgaku-i'sei 光学異性 optical isomerism *[p-ch]*

kōgaku-isei'tai 光学異性体 enantiomorph; optical isomer *[ch]*

kōgaku-jundo 光学純度 optical purity *[ch]*

kōgaku-kaiyō'gaku 光学海洋学 optical oceanography *[ocean]*

kōgaku-kassei 光学活性 optically active; optical activity *[opt]*

kōgaku-kassei-jūgō'tai 光学活性重合体 optically active polymer *[poly-ch]*

kōgaku-kassei-tai	光学活性体	optically active substance *[mat]*
kōgaku-kei	光学系	optical system *[opt]*
kōgaku-keisan	光学計算	optical computing *{n}* *[comput]* *[opt]*
kōgaku-keisan-ki	光学計算機	optical computer *[comput]* *[opt]*
kōgaku-kenbi'kyō	光学顕微鏡	optical microscope *[opt]*
kōgaku-kenbikyō-shashin	光学顕微鏡写真	optical (photo)micrograph *[opt]*
kōgaku-kikai	光学機械	optical instrument *[opt]*
kōgaku-kusabi	光学楔	optical wedge *[opt]*
kōgaku-kyūshū	光学吸収	optical absorption *[opt]*
kōgaku-nōdo	光学濃度	absorbance *[p-ch]*
kōgaku-onkyō'gaku	工学音響学	engineering acoustics *[acous]*
kōgaku-sen'i	光学繊維	optical fiber *[opt]*
kōgaku(shiki)moji'ninshiki	光学(式)文字認識	optical character recognition *[comput]*
kōgaku(shiki)moji-yomitori-sōchi	光学(式)文字読取=装置	optical character reader *[comput]*
kōgaku'teki-fuka-sa	光学的深さ	optical depth *[opt]*
kōgaku'teki-sō'antei-sei	光学的双安定性	optical bistability *[opt]*
kōgaku'teki-taishō'tai	光学的対称体	enantiomer *[ch]*
kōgaku-tokui'sei	光学特異性	optical characteristic *[opt]*
kōgan	睾丸	testicles *[anat]*
ko'gane-gokiburi	黄金蜚蠊	Oriental cockroach (insect) *[i-zoo]*
ko-gara	小雀; こがら	willow tit (bird) *[v-zoo]*
ko'garashi	凩; 木枯らし	wind: wintry blast *[meteor]*
ko-garasu	小鴉	jackdaw (bird) *[v-zoo]*
ko-gata	小形	small size
kogata-erebēta	小型エレベーター	dumbwaiter *[eng]*
kogata'ka	小型化	miniaturization *[electr]*
"	小型化	scaling down (of size, shape) *{vb}*
ko-gatana	小刀; こがたな	small knife (pl knives) *{n}* *[eng]*
kogata-senpaku	小型船舶	small ship *[nav-arch]*
koge	焦げ; こげ	scorch; scorching; singe *{n}* *[cook]*
koge'cha-iro	焦茶色	umber (color) *[opt]*
kō'gedoku-zai	抗解毒剤	antiantibody *[immun]*
kōgei	工芸; 工藝	technology *[sci-t]*
kōgen	光源; こうげん	light source *[opt]*
kōgen	抗原	antigen *[immun]*
kōgen	荒原	desert *{n}* *[geogr]*
kō'genchū-sei	抗原虫性	antiprotozoic property *[microbio]*
kōgen-hokyō-zai	抗原補強剤	adjuvant *[immun]*
kōgen-sei	抗原性	antigenicity *[immun]*
kogi-fune	漕ぎ舟	rowboat *[nav-arch]*
kōgin	光銀; こうぎん	photosilver *[photo]*

kōgo	口語；こうご	colloquial speech; spoken language *[comm]*
kōgo	交互	alternate *{n} [math]*
kōgō	咬合	occlusion *{n} [anat]*
kōgo'bun	口語文	colloquial-style writing *[comm]*
kōgo-inka	交互印加	alternating application *[elec]*
kōgo-kyōjūgō'tai	交互共重合体	alternating copolymer *[poly-ch]*
kōgo no	交互の	alternative *{adj}*
kōgo-sayō	交互作用	interaction *[phys] [math]*
kō'gōsei	光合成	photosynthesis *[bioch]*
kōgo'tai	口語体	colloquial style *[gram]*
kōgo'teki-hyōgen	口語的表現	colloquialism *[gram]*
kogu	漕ぐ	row (a boat) *{vb}*
kōgu	工具	implement; tool *{n} [eng]*
ko-guchi	小口	edge *{n} [pr]*
kōgu-kō	工具鋼；こうぐこう	tool steel *[met]*
ko-guma-za	小熊座	Ursa Minor *[astron]*
kōgyō	工業；こうぎょう	industry (pl -tries)
kōgyō	鉱業；鑛業	mining industry *[eng]*
kōgyō'ka	工業化	industrialization *[ind-eng]*
kōgyō-kagaku	工業化学	industrial chemistry *[ch]*
kōgyō'ken	鉱業権	mining right *[legal]*
kōgyoku	硬玉	jadite *[miner]*
kōgyoku	鋼玉	corundum *[miner]*
kō'gyokuzui	紅玉髄；紅玉髓	sard *[miner]*
kōgyō-senyō-chiku	工業専用地区	exclusive industrial district *[civ-eng]*
kōgyō-shoyū-ken	工業所有権	industrial property right *[legal]*
kōgyō-yōsui	工業用水	industrial water *[ch]*
kōha	光波	light wave *[opt]*
ko'hada	小鰭；こはだ	gizzard shad (medium) (fish) *[v-zoo]*
ko-ha'guruma	小歯車	pinion *{n} [mech-eng]*
ko'haku	琥珀；こはく	topaz *[miner]*
kohaku-iro	琥珀色	amber (color) *[opt]*
ko'haku-san	琥珀酸	succinic acid
kohaku'san-en	琥珀酸塩	succinate *[ch]*
kohaku'san-tetsu	琥珀酸鉄	iron succinate
kōhan; kanban	甲板	deck (of ship) *{n} [nav-arch]*
kōhei	公平	impartial(ity) *{n}*
kōhei-sa	公平さ	impartiality *{n}*
kōhen; ō'hen	黄変；黄變	yellowing (fiber, paint) *{n} [opt]*
kōhen	鋼片	billet; bloom; slab (of steel) *{n} [steel]*
kōhen-do	黄変度	degree of yellowing *{n} [opt]*
kōhi	公比；こうひ	common ratio *[math]*

kōhi	甲皮	carapace *[i-zoo]*
kōhī	珈琲; コーヒー	coffee *[bot] [cook]*
ko-hitsuji no ke	子羊の毛	lamb's wool *[tex]*
kōhō	公報	gazette *[pat]*
kōhō	後方	rear; back *{n}*
kōhō	航法	navigation *[eng]*
kōhō e magatta	後方へ曲った	recurved *{adj} [sci-t]*
ko'hone	河骨; 川骨; こほね	candock; spatterdock *[bot]*
kōhō-sanran	後方散乱	backscattering *{n} [phys]*
kōhyō	公表	description *[pat]*
kō'hyōhaku-shiyaku	光漂白試薬	photobleaching reagent *[opt]*
koi	鯉; こい	carp; Cyprinus carpio (fish) *[v-zoo]*
kō'i	後位	rear position *{n}*
kō'i	高位	high order; high level *{n} [math]*
ko-i'ga	小衣蛾; こいが	webbing clothes moth (insect) *[i-zoo]*
kō'iki-hyūzu	広域ヒューズ	general purpose fuse *[elec]*
koi'kuchi-shōyu	濃口醤油	dark-colored soy sauce *[cook]*
ko'i ni	故意に	deliberately; intentionally *{adv} [psy]*
kō'i no	高位の	high-grade; high-level *{adj}*
koi-nobori	鯉幟; こいのぼり	carp streamer *[furn]*
ko-inu	小犬	puppy (animal) *[v-zoo]*
kō'i-shitsu	更衣室	dressing room *[arch]*
kōji	麹; 糀; こうじ	malted rice; malt; koji *[cook] [mycol]*
koji-akeru	抉じ開ける	pry open *{vb t}*
kōji-hō	麹法	koji (q.v.) process *[microbio]*
kōji-kabi	麹黴	koji mold; Aspergillus *[mycol]*
ko'jika-iro	仔鹿色; 子鹿色	fawn (color) *[opt]*
kō-jiki	硬磁器	true porcelain *[cer]*
kōji-kin	麹菌	koji mold *[mycol]*
kōji-kōzō	高次構造	higher-order structure *[mol-bio]*
kō-jiku	光軸	optical axis *[opt]*
kō'jiku (no)	向軸(の)	adaxial *{adj} [bio]*
ko-jima	小島	small island; key *{n} [geol]*
kojin	個人	individual; person *{n}*
kojin-gosa	個人誤差	individual error; personal error *[math]*
ko-jinrui'gaku	古人類学	paleoanthropology *[bio]*
kojin-sa	個人差	individual (or personal) difference *[math]*
kojireru	拗れる; こじれる	twisted; go wrong *{vb i}*
kojiru	抉る; こじる	gouge *{vb}*
kōji-san	麹酸	kojic acid *[ch]*
kōjō	工場	factory; plant; workshop *[ind-eng]*
kōjō-haisui	工場排水	industrial waste water *[ecol]*

kojō-rettō	弧状列島	arcuate islands *[geol]*
kōjō-sei	恒常性	homeostasis *[bio]*
kōjō-sen	甲状腺	thyroid gland *[anat]*
kōjōsen'shu	甲状腺腫	goiter *[med]*
kōjō-zai	向上剤	improver (a chemical) *[mat]*
kō'jūgō	光重合	photopolymerization *[poly-ch]*
kōjūgō'tai	高重合体	high polymer *{n}* *[poly-ch]*
kōjun	公準	postulate *{n}* *[logic]*
kōjun	降順；こうじゅん	descending order *[math]*
kōjun-chū'shutsu'butsu	高純抽出物	quintessence (extraction) *{n}* *[ch]*
kōjū-rui	後獣類；後獸類	Metatheria *[v-zoo]*
kō-jūten-haigō	高充填配合	heavily-loaded mix *[ch-eng]*
koka	固化	caking; solidification *{n}* *[eng]*
koka	糊化	gelatinization *[eng]*
kōka	効果；こうか	effect *{n}* *[math]* *[sci-t]*
kōka	降下	descent *{n}* *[aero-eng]*
kōka	硬化	curing *{n}* *[org-ch]*
"	硬化	hardening; solidification *{n}* *[phys]*
kōka	硬貨	coin *{n}* *[econ]*
kōka	鉱化；鑛化	mineralization *[geol]*
kō'kabi-kōsei-busshitsu	抗黴抗生物質	antifungal antibiotics *[microbio]*
kō'kabi'sei no	抗黴性の	antimycotic *{adj}* *[pharm]*
kōka'butsu	降下物	fallout (atomic) *{n}* *[nucleo]*
kōka-dōro'kyō	高架道路橋	overpass *{n}* *[traffic]*
kō'kagaku	光化学	photochemistry *[p-ch]*
kō'kagaku-dai'ichi-hōsoku	光化学第一法則	photochemistry, first law of *[p-ch]*
kō'kagaku-dai'ni-hōsoku	光化学第二法則	photochemistry, second law of *[p-ch]*
kō'kagaku-denchi	光化学電池	photochemical cell *[electr]*
kō'kagaku-kyūshu no hōsoku	光化学吸収の法則	photochemical absorption, law of *[p-ch]*
kō'kagaku no	光化学の	photochemical *{adj}* *[p-ch]*
kō'kagaku-tōryō	光化学当量	photochemical equivalent *[p-ch]*
kō'kagaku-tōryō no hōsoku	光化学当量の法則	photochemical equivalent, law of *[p-ch]*
kōkai	公海	high seas; international waters; open sea *{n}* *[ocean]*
kōkai	公開	early disclosure; unexamined *[pat]*
kōkai	叩解；こうかい	beating (of pulp) *{n}* *[eng]*
kōkai	航海	voyage *{n}* *[navig]*
kōkai	鋼塊	steel ingot *[met]*
kōkai-do	叩解度	degree of beating *[eng]*
kōkai-hyō	航海表	nautical table(s) *[navig]*
kōkai-nisshi	航海日誌	log book *[navig]* *[pr]*
kōkai-reki	航海暦；航海曆	nautical almanac *[navig]* *[pr]*

kōkai sareta	公開された	laid open to public inspection *[pat]*
kōkai suru	公開する	disclose *{vb}* *[pat]*
kōkai-tō	航海灯	navigation light; running light *[nav-arch]*
kō'kaiyō-zai	抗潰瘍剤	ulcer preventive *{n}* *[pharm]*
kōka-kazan-saisetsu'butsu	降下火山砕屑物	tephra *{n}* *[geol]*
kō-kaku	交角	intersection angle *[civ-eng]*
kōkaku-kō no	甲殻綱の	crustacean *{adj}* *[i-zoo]*
kōkaku-rui	甲殻類；甲殼類	Crustacea *[i-zoo]*
kōkan	光冠；光環	corona *[astron]*
kōkan	交換；こうかん	interchange *{n}* *[comput]*
"	交換	switching *{n}* *[elec]*
"	交換	exchange; exchanging *{n}* *[math]*
kōkan-atsukai'sha	交換扱者	exchange operator *[comm]*
kōkan-denryū	交換電流	exchange current *[elec]*
kōkan-hōsoku	交換法則	commutative law *[math]*
kōkan-ki	交換器	switchboard *[comm]*
kōkan-ki	交換機	switching equipment *[eng]*
kōka no hōsoku	効果の法則	effect, law of *[psy]*
kōkan'sei no	好乾性の	xerophilic *{adj}* *[bio]*
ko-kan'setsu	股関節；股關節	hip joint *[anat]*
ko'kansetsu-bu	股関節部	hip *{n}* *[anat]*
kōkan'shi	交換子	commutator *[math]*
kōkan-shinkei-kei	交感神経系	sympathetic nervous system *[anat]*
kōkan suru	交換する	exchange *{vb t}*
kōka-shokubai	硬化触媒	curing catalyst *[ch]*
kōka-yu	硬化油；こうかゆ	hardened oil; hydrogenated oil *[mat]*
koka-zai	固化剤	hardening agent; solidifying agent *[mat]*
kōka-zai	降下剤	depressant *[ch]*
kōka-zai	硬化剤	curing agent; hardener *[poly-ch]*
kōka-zai	鉱化剤；鑛化劑	mineralizer *[geol]*
koke	苔	moss *[bot]*
koke'gaku	苔学	muscology *[bot]*
kōkei	口径	aperture *[opt]*
kokei-bun	固形分	solids content; solid components *{n}*
kokei'butsu	固形物	solid matter *[phys]*
kōkei-hi	口径比；こうけいひ	relative aperture *[opt]*
kō'keiso-kō	高珪素鋼	high-silicon steel *[met]* *[steel]*
kō'kekkaku-zai	抗結核剤	antituberclous drug; tuberculostatic drug *{n}* *[pharm]*
koke-momo	苔桃；こけもも	cowberry; mountain cranberry *[bot]*
kōken	後件	consequent *{n}* *[logic]*
kō-kenchi	光検知	light detection *[opt]*

kōken-so'shiki	光顕組織	optical microstructure *[opt]*
kokera-ita	柿板；コケラ板	shingle *{n} [mat]*
kokera-yane	柿屋根；コケラ屋根	shingle roof *[arch]*
koke-rui	苔類	Bryophyta *[bot]*
kō'kessei	抗血清	antiserum *{n} [immun]*
koketsu	固結	concretion *[geol]*
kō-ketsu'atsu	高血圧；高血壓	hypertension *[med]*
kōki	光輝	brilliance *[opt]*
kōki	香気	aroma; fragrance *[physio]*
kōki	後期	anaphase *[bio]*
kō-ki'atsu	高気圧	high; high pressure *{n} [meteor]*
kō'kiden-kōka	光起電効果	photovoltatic effect *[electr]*
kō'kiden'ryoku-seru	光起電力セル	photovoltatic cell *[electr]*
kōki-kin	好気菌；こうききん	aerobe *[bio]*
ko-kikō'gaku	古気候学	paleoclimatology *[geol]*
kōki-kokyū	好気呼吸	aerobic respiration *[bio]*
kōkin-ryoku	抗菌力	antimicrobial activity *[microbio]*
kōkin-sei	抗菌性	antibacterial property *[microbio]*
kōkin-zai	抗菌剤	antibacterial agent *[microbio]*
kōki-sei	好気性	aerobicness *{n} [bio]*
kōki'sei-bi'seibutsu	好気性微生物	aerobic bacteria *[microbio]*
kōki'sei-shokubutsu	好気性植物	aerobe *[bot]*
kōki-yaki'namashi	光輝焼鈍	bright annealing *{n} [met]*
kokka	骨化	ossification *[physio]*
kokka	黒化；黑化	blackening *{n} [opt]*
kokkai-giji'dō	国会議事堂	capitol (Japanese); diet building *[arch]*
kokkaku	骨格；こつかく	skeleton; physique *[anat]*
kokka suru	骨化する	ossify *{vb} [physio]*
kokkei na	滑稽な	comical; funny; humorous *{adj}*
koko	此処；此所；こゝ	here *{adv} {n} {pron}*
kokō	糊膏	liniment *{n} [pharm]*
kōkō	航行	navigation *[eng]*
kōkō	鉱坑；鑛坑	mine *{n} [min-eng]*
kōkō-eisei	航行衛星	navigation satellite *[aero-eng]*
kōko'gaku	考古学	archeology *[sci-t]*
kō'kōka-sei	光硬化性	photohardening property *[mat] [opt]*
kōkoku	公告；こうこく	open for public inspection; publication *[pat]*
kōkoku	広告；廣告	advertisement *[comm]*
kōkoku-tō	広告塔	advertisement tower *[arch]*
kō'kokyū	光呼吸	photorespiration *[bioch]*
kokoro	心；こころ	mind; mentality *{n} [psy]*

kokoromi	試み; こころみ	trial; attempt *{n}*
kō'kōsa	光行差	aberration *[astron]*
kōkō-sei	向光性	heliotropism *[bio]*
kō'kōso	抗酵素	antienzyme *{n} [bioch]*
kō'kōtai	抗抗体	antiantibody *[immun] [pharm]*
kōkō-ten	降交点	descending node *[astron]*
kōkotsu'gyo-rui	硬骨魚類	Osteichthyes; bony fishes *[v-zoo]*
kōkotsu-rui	硬骨類	Teleostei *[v-zoo]*
kōkō-zai	硬膏剤	plaster *{n} [pharm]*
-koku	-国; -國	-country *{suffix}*
kōkū	航空; こうくう	aeronautical; aviation *[aero-eng]*
kokuban	黒板; 黒板	blackboard *[comm]*
kōkū'bin	航空便	airmail *{n} [comm]*
(kōkū-)bokan	(航空-)母艦	aircraft carrier *[nav-arch]*
koku'chō	黒鳥; こくちょう	black swan (bird) *[v-zoo]*
koku-deido	黒泥土	muck soil *{n} [geol]*
kōkū-denshi-kōgaku	航空電子工学	aviation electronics; avionics *[eng]*
koku-do	黒度	blackness *{n} [opt]*
koku-dō	国道	national highway *[civ-eng]*
koku'dōkō	黒銅鉱; 黒銅鑛	tenorite *[miner]*
koku-eki	黒液	black liquor *[mat] [paper]*
koku'en	黒鉛; こくえん	graphite *[miner]*
koku'en'ka	黒鉛化	graphitization *[met]*
kōkū'gaku	航空学	aeronautics *[fl-mech]*
koku'haku-eiga	黒白映画	black-and-white movie *[opt]*
kōkū-hyō'shiki	航空標識	radio beacon *[navig]*
koku'ji	告示	announcement; notification *[comm]*
koku'ji	国字; 國字; こくじ	Japanese kanji characters *[pr]*
kōkū-junkatsu-yu	航空潤滑油	aviation oil *[mat]*
kōkū-kaisha	航空会社	airline company *[econ]*
kōkū-kamotsu	航空貨物	air cargo; air freight *[trans]*
kōkū-kamotsu-yusō	航空貨物輸送	air freighting *{n} [trans]*
kōkū-kansei	航空管制	flight control *{n} [aero-eng]*
kōkū'ki	航空機; こうくうき	aircraft; airplane *[aero-eng]*
kōkūki-seigyo	航空機制御	aircraft control *[navig]*
kōkū-kōtsu	航空交通	air traffic *[navig]*
kōkū-kōtsū-kansei	航空交通管制	air traffic control *[navig]*
koku-kujira	克鯨; こくくじら	gray whale *[v-zoo]*
koku'motsu	穀物; こくもつ	cereal; grain *[bot]*
kōkū-nenryō	航空燃料	aviation fuel *[mat]*
kōkū-riki'gaku	航空力学	aerodynamics *[fl-mech]*
koku'ritsu-kōen	国立公園	national park *[civ-eng]*

kōkū'ro-tōdai	航空路灯台	airway beacon *[navig]*
koku'ryū	穀粒	grain particle *[bot]*
koku'sai	国際；國際	international *{adj} {n}*
kokusai-chōsa	国際調査	international search *[pat]*
kokusai-chōsa-kikan	国際調査機関	international searching authority *[pat]*
kokusai-hi'zuke-henkō-sen	国際日付変更線	international date line *[astron]*
kokusai-kikaku	国際規格	international standard *[phys]*
kokusai-shutsu'gan-bangō	国際出願番号	international filing number *[pat]*
kokusai-shutsu'gan-bi	国際出願日	international filing date *[pat]*
kokusai-tan'i	国際単位	international unit *[bio]*
kokusai-tokkyo-bunrui	国際特許分類	international patent classification *[pat]*
koku'setsu suru	刻設する	provide engraved *{vb} [graph]*
koku-shiki'so	黒色素	melanin *[bioch]*
koku'shoku-ka'yaku	黒色火薬	gunpowder *[ord]*
kō'kussetsu'sei-hi'senkei-kōgaku	光屈折性非線形光学	photorefractive nonlinear optics *[opt]*
kō'kussetsu'sei-kesshō	光屈折性結晶	photorefractive crystal *[crys]*
kō'kussetsu'sei-kōbunshi	光屈折性高分子	photorefractive polymer *[poly-ch]*
koku'tai	黒体；黑體	blackbody *[thermo]*
kokutai-hōsha-netsu	黒体放射熱	blackbody-radiation heat *[thermo]*
koku'tan	黒檀；こくたん	ebony (tree) *[bot]*
kokuteki	黒滴	black drop *[astron]*
koku'ten-shūki	黒点周期	sunspot cycle *[astron]*
kōkū-tō	航空灯；航空燈	running light *[aero-eng] [navig]*
kōkū'uchū-denshi-kōgaku	航空宇宙電子工学	aerospace electronics *[electr]*
kōkū'uchū-kōgaku	航空宇宙工学	aerospace engineering *[eng]*
kōkū-uchū no	航空宇宙の	aerospace; airspace *{adj} [meteor]*
kokuyō'seki	黒曜石	obsidian *[geol]*
kokuzō-mushi	穀象虫；こくぞう虫	rice weevil (insect) *[i-zoo]*
kōkyō-dantai	公共団体	public body *{n}*
kōkyō'kyoku	交響曲	symphony *[music]*
kōkyō-sui'iki	公共水域	public water domain *[ind-eng]*
kokyū	呼吸；こきゅう	breathing; respiration *{n} [physio]*
kōkyū	光球	photosphere *[astron]*
kōkyū	高級	higher (in grade)
kokyū'ki-kei	呼吸器系	respiratory system *[anat]*
kokyū-kōso	呼吸酵素	respiratory enzyme *[enz]*
kokyū-shō	呼吸商	respiratory quotient *[physio]*
kō-kyūsui'sei-kōbunshi	高吸水性高分子	super-(water-)absorbent polymer *[poly-ch]*
koma	駒	piece (chess or game) *{n} [eng]*
"	駒	frame (of movie) *{n} [opt]*
koma	独楽；こま	top (toy) *{n} [eng]*

kōma	黄麻; こうま	jute *[bot]*
koma-dori	駒鳥; こまどり	Japanese robin (bird) *[v-zoo]*
komakai	細かい	finely divided *{adj}* *[mech-eng]*
komaka-sa	細かさ	fineness *[mech-eng]*
ko'maku	鼓膜	eardrum; tympanic membrane *[anat]*
koma-tsugumi	駒鶫; こまつぐみ	robin (bird) *[v-zoo]*
komatsu'na	小松菜	Brassica Rapa var. pervidis (vegetable) *[bot]*
kome	米; こめ	rice (raw) *[bot]* *[cook]*
kome-nuka	米糠	rice bran *[bot]* *[cook]*
kome'nuka-rō	米糠蠟	rice bran wax *[mat]*
kome'tsuki-mushi	米搗虫	click beetle (insect) *[i-zoo]*
kōmi	香味	flavor; flavor and taste *{n}* *[physio]*
kōmi'sō	香味草	herbs (for seasoning) *[bot]* *[cook]*
komo	菰; 薦; こも	straw mat; rush mat
kōmoku	項目	item; entry *[comput]* *[math]*
kōmon	肛門	anus *[anat]*
kōmori	蝙蝠; こうもり	bat (mammal) *[v-zoo]*
kōmori-ga	蝙蝠蛾	Japanese swift moth (insect) *[i-zoo]*
komori-guma	子守り熊	koala *[v-zoo]*
ko'mugi	小麦; こむぎ	wheat *[bot]*
ko'mugi-denpun	小麦澱粉	wheat starch *[cook]*
ko'mugi no baku'ga	小麦の麦芽	wheat germ *[bot]*
kō-myaku	鉱脈; 鑛脈	(mineral) vein; streak *{n}* *[min-eng]*
kōmyō'tan	光明丹	minium *[miner]*
kon	根	root *{n}* *[dent]* *[math]*
kon	紺; こん	navy blue *[opt]*
kona	粉; こな	flour; powder *[cook]* *[mat]*
kona-furui	粉篩; 粉ふるい	flour sifter; sifter *[cook]* *[eng]*
kōnai-tanmatsu	構内端末	local terminal *[comput]*
kona'jirami	粉虱; こなじらみ	whitefly; white-fly (insect) *[i-zoo]*
kona-kaigara-mushi	粉貝殻虫	mealy-bug (insect) *[i-zoo]*
konbō-shi	混紡糸	blended yarn *[tex]*
konbu	昆布; こんぶ	kelp; tangle (seaweed) *{n}* *[bot]*
konchū	昆虫; こんちゅう	insect *[i-zoo]*
konchū'gaku	昆虫学	entomology *[i-zoo]*
konchū-rō	昆虫蠟	insect wax *[mat]*
konchū-rui	昆虫類	Insecta *[i-zoo]*
kon'date(-hyō)	献立(表)	menu *[cook]*
kondensā	コンデンサー	capacitor *[elec]*
kō-nen	光年	light-year *[astrophys]*
koneru	捏る; こねる	knead; mix; work *{vb-t}* *[sci-t]*

kongō	混合；こんごう	mixing *{n} [sci-t]*
kongō'butsu	混合物	mixture *[sci-t]*
kongō-genshi'ka	混合原子価	mixed valence *[ch]*
kongō-kan	混合管	mixer tube *[electr]*
kongō'seki	金剛石	diamond *[miner]*
kongō'sha	金剛砂	emery *[miner]*
kongō-shiji-yaku	混合指示薬	mixed indicator *[ch-eng]*
kongō shinai	混合しない	immiscible *{adj} [ch]*
kō'nichi-sei	向日性	heliotropism *[bio]*
kōnichi'sei-shokubutsu	好日性植物	heliophyte *[ecol]*
ko-ni'motsu	小荷物	small package; parcel *[trans]*
kōnin-kaikei'shi	公認会計士	certified public accountant *[econ]*
kon'jiki; kin-iro	金色	golden (color) *[opt]*
konjō	紺青；こんじょう	Prussian blue (color) *[opt]*
konkan	根管	root canal *[dent]*
konkei	根茎；根莖	rhizome; rootstock *{n} [bot]*
konki	根基	radical *{n} [math]*
konnyaku	蒟蒻；こんにゃく	devil's tongue *[cook]*
kono	此の；斯の；この	this *{adj}*
kōnō	効能	efficacy; effect *[pharm]*
konoha-zuku	木の葉木菟	Japanese scops owl (bird) *[v-zoo]*
kō-no-mono	香の物	pickle(s) *{n} [cook]*
konoshiro	鮗；鰶；このしろ	gizzard shad (fish) *[v-zoo]*
kōnotori	鸛；こうのとり	Japanese stork (bird) *[v-zoo]*
konpi	根被	velamen *[bot]*
konpūtā'ka-taijiku-dansō-zōhō	コンピュータ化=体軸断層像法	computerized axial tomography (CAT) *[med]*
konran-jōtai	混乱状態	chaos; chaotic state *{n}*
konren	混練；こんれん	kneading; mixing *{n} [eng] [poly-ch]*
"	混練；混練	mulling *{n} [eng]*
konren suru	混練する	knead *{vb t}*
kōryō	光量	quantity of light *[opt]*
konryū	根粒	root nodule *[bot]*
konsei	混成	mixing *{n} [eng]*
"	混成	hybridization *[gen]*
konsei'gata-konpūtā	混成型コンピュータ	hybrid computer *[comput]*
konsei'shu	混成酒	liqueur *[cook]*
kon'seki	痕跡	trace *{n} [sci-t]*
konseki-kikan	痕跡器官	vestigial organ *[bio]*
konseki no	痕跡の	vestigial *{adj} [bio]*
konsento	コンセント	electric outlet *[elec]*
konshin	混信	radio interference *[comm] [phys]*

konshō	混晶	mixed crystal *[crys]*
kon-shōmō	根小毛	rootlet *[bot]*
konshu'butsu	混種物	hybrid *{n}* *[gen]*
konsū	混数	mixed number *[math]*
konsui-fushoku	混水腐食	brackish water corrosion *[met]*
kontena	コンテナ	container *[rail]*
konwa-sei	混和性	miscibility *[ch]* *[sci-t]*
konwa-zai	混和剤	blending agent *[mat]*
kon'yū	混融; こんゆう	mixed melting *{n}* *[thermo]*
kon'yū-shiken	混融試験	mixed examination; mixed melting point test *{n}* *[an-ch]*
kon'yū-ten	混融点	mixed melting point *[thermo]*
konzai'butsu	混在物	adulterant; mixed-present matter *[ch]*
kon'zetsu saseru	根絶させる	eradicate *{vb t}* *[eng]*
kō'on-kei	高温計	pyrometer *[eng]*
kō'onkyō-bunkō-hō	光音響分光法	photoacoustic spectroscopy *[spect]*
kō'on-sei	恒温性	homoiothermy *[physio]*
kō'on-sō	恒温槽	thermostat *[eng]*
koppai-jiki	骨灰磁器	bone china *[cer]* *[mat]*
kōra	甲羅; こうら	carapace *[v-zoo]*
Korai-yaki	高麗焼	Koryo ware *[cer]*
kōraku	交絡	confounding *{n}* *[math]*
kōran	高欄; 高欄	handrail; railing *{n}* *[eng]*
kore	此れ; 是; 之; これ	this *{n}*
korera-kin	コレラ菌	cholera bacillus *[microbio]*
kori	梱; こり	bale *{n}* *[ind-eng]*
kōri	氷; こおり	ice *{n}* *[p-ch]*
kōri	公理	axiom *[logic]*
kōri	行李	wicker trunk; luggage *[transp]*
kōri-makura	氷枕	ice bag *[med]*
kōri-mizu	氷水	ice water *[ch]* *[cook]*
kōrin	紅燐	scarlet phosphorus *[ch]*
kōrin-kudō	後輪駆動	rear wheel drive *[mech-eng]*
kō'ri'nyō-zai	抗利尿剤	antidiuretic *{n}* *[pharm]*
ko'ritsu	孤立	loneness; solitaryness *{n}*
kō'ritsu	工率	rate of production *[ind-eng]*
kō'ritsu	効率	efficiency *[ind-eng]*
ko'ritsu-denshi-tsui	孤立電子対	lone electron pair *[p-ch]*
kōri-zatō	氷砂糖	crystal sugar *[cook]*
koro	転; 轉; ころ	roller *[eng]*
kōro	光路	optical path *[opt]*
kōro	香炉	incense burner *[eng]*

kōro	高炉; 高爐; こうろ	blast furnace *[met]*
kōro	航路	fairway; sea route *[navig]*
kōrō	硬蠟	silver solder *[met]*
kōro-chō	光路長	optical path length *[opt]*
korogari	転がり; 轉がり	rolling *{n}* *[mech]*
korogari-ma'satsu	転がり摩擦	rolling friction *[mech]*
korogari-yasu-sa	転がり易さ	rollability *[mech]*
korogaru	転がる; ころがる	roll *{vb i}*
korogasu	転がす	roll *{vb t}*
kōrogi	蟋; こおろぎ	cricket (insect) *[i-zoo]*
koroha	胡盧巴; ころは	fenugreek; Trigonella foenumgraecum *[bot]*
kōro-hyō'shiki	航路標識	navigation mark *[navig]*
koroido	コロイド	colloid *[ch]*
koroido-tōryō	コロイド当量	colloid equivalence *[p-ch]*
koromo	衣; ころも	clothes; garment *[cl]*
"	衣	dough coating for deep frying *[cook]*
koromo-hen	衣扁; 衣偏: ネ; ころもへん	kanji (Chinese character) radical denoting clothing *[pr]*
kōro-sa	光路差	optical path difference *[opt]*
kōro'seki	光鹵石; こうろせき	carnallite *[miner]*
kōrui	項類	category *[math]*
kōryan	高粱; コーリヤン; カオリヤン	kaoliang; sorghum; koaliang *[bot]*
kōryan-yu	高粱油	kaoliang oil *[mat]*
koryō	糊料; こりょう	thickener *[mat]*
kōryō	光量	light quantity *[opt]*
kōryō	香料	fragrance; perfume *{n}* *[mat]*
kōryō	恒量	constant weight *[mech]*
kōryoku	抗力	drag *{n}* *[aero-eng]* *[fl-mech]*
kōryoku	効力	potency; efficacy *[sci-t]*
kōryoku-kō	高力鋼	high-tensile steel *[met]* *[steel]*
kō'ryoshi	光量子	light quantum; photon *[quant-mech]*
kōryū	向流	countercurrent *{n}* *[fl-mech]* *[sci-t]*
kōryū	交流; こうりゅう	alternating current *[elec]*
kōryū	光流	flux of light *[phys]*
kōryū	後流	slipstream; wash *{n}* *[aero-eng]*
"	後流	wake *{n}* *[fl-mech]*
kōryū-bunpai-hō	向流分配法	countercurrent distribution method *[ch]* *[ch-eng]*
kōryū-denryoku	交流電力	alternating-current power *[elec]*
kō-ryūdō-sei	高流動性	high fluidity; high liquidity *[fl-mech]*
kōryū'seki	紅榴石	pyrope *[miner]*

kōsa	公差；こうさ	tolerance *[eng]*
kōsa	交叉	crossing-over; cross *{n} [math]*
kōsa	交差	intersection *[comput]*
kōsa	光差	equation of light; light time *[astron]*
kōsa	黄砂	yellow sand *[geol]*
kōsa-fūsa	交叉封鎖	cross-sealing *{n} [eng]*
kōsai	公債	public loan *[econ]*
kōsai	虹彩；こうさい	iris (pl -es, irides) *[anat] [opt]*
kōsai; kōshi	鉱滓；鑛滓	slag *[met]*
kōsai-shibori	虹彩絞り	iris diaphragm *[opt]*
kōsa-kanwa	交差緩和	crossrelaxation *[phys]*
kō'saku	鋼索	cable (of steel) *{n} [eng]*
kōsaku'ba	工作場	workshop *[ind-eng]*
kōsaku'butsu	工作物	workpiece *[eng]*
kōsaku-kikai	工作機械	machine tool *[eng]*
kōsaku-tetsudō	鋼索鉄道	funicular railway *[rail]*
kōsan	鉱酸；こうさん	inorganic acid; mineral acid *[ch]*
kō-sanran	光散乱	light scattering *{n} [opt]*
kōsan-sei	抗酸性	acid-fastness *{n} [mat]*
kōsan'sei-busshitsu	好酸性物質	acidophil; oxyphil *{n} [bio]*
kōsa-seki	交差積	cross product *[math]*
kōsa-taishō-sei	交差対称性	crossing symmetry *[part-phys]*
kōsa-ten	交差点	intersection *[traffic]*
kōsatsu	考察；こうさつ	consideration; discussion *[pr]*
kōsei	向性	tropism *[bio]*
kōsei	校正	proofreading *{n} [pr]*
kōsei	恒星	fixed star *[astron]*
kōsei	較正；校正	calibration *[sci-t]*
kōsei	構成；こうせい	configuration *[ch] [comput] [mech]*
"	構成	constitution; make-up *{n} [mech]*
"	構成	construction *[pat]*
kōsei-bu'hin	構成部品	component *[mech-eng]*
kōsei-busshitsu	抗生物質	antibiotics *[microbio] [pharm]*
ko-seibutsu'gaku	古生物学	paleontology *[bio]*
kōsei-dōbutsu	後生動物	Metazoa *[zoo]*
kō-seido no	高精度の	highly precise *{adj}*
kō'seido-tokei	高精度時計	chronometer *[horol]*
kōsei'fū	恒星風	stellar wind *[astron]*
kosei-haku'a	湖生白亜	travertine *{n} [geol]*
kōsei-ichi'shi	恒星位置誌	uranometry *[astron]*
kōsei-ji	恒星時	sidereal time *[astron]*
kōsei-jitsu	恒星日	sidereal day *[astron]*

kōsei'jō-ginga	恒星状銀河	quasi-stellar galaxy *[astron]*
kōsei'jō-tentai	恒星状天体	quasar *[astron]*
kōsei-kyoku'sen	較正曲線	calibration curve *[eng]*
kosei-matsuba'ran-rui	古生松葉蘭類	Psilopsida *[bot]*
kō'seisaido-terebi(jon)	高精細度テレ=ビジョン	high-definition television *[electr]*
kō'seisan-sei	高生産性	high productivity *[ind-eng]*
kō-seishin'byō-yaku	抗精神病薬	antipsychotic *{n}* *[pharm]*
kōsei-shinka	恒星進化	stellar evolution *[astrophys]*
kōsei-shūdan	恒星集団	star cloud *[astron]*
kōsei-shūki	恒星周期	sidereal period *[astron]*
kōsei'teki na	後成的な	epigenetic *{adj}* *[embryo]*
kōsei-tōkei'gaku	恒星統計学	stellar statistics *[math]*
kōsei-yōso	構成要素	component; constituent *[sci-t]*
kōsei-zu	構成図；構成圖	block diagram *[eng]*
kōseki	鉱石；鑛石	mineral; ore *[geol]*
kōseki'un	高積雲	altocumulus *[meteor]*
kōseki'un	航跡雲	condensation trail; contrail *[meteor]*
kō'sekkō	硬石膏	anhydrite *[miner]*
kōsen	光線；こうせん	light rays *[opt]*
kōsen	鉱泉	mineral spring *[hyd]*
kōsen	鋼線	steel wire *[mat]*
kosen'gaku	古銭学；古錢學	numismatics *{n}*
kōsetsu	降雪	snowfall *[meteor]*
kōsetsu-ryoku	抗折力	flexural strength *[mech]*
kōsha	向斜	syncline *{n}* *[geol]* *[min-eng]*
koshi	腰；こし	hip *{n}* *[anat]*
"	腰	stiffness *[mech]*
ko-shi	故紙	waste paper *[mat]*
kōshi	光子	photon *[opt]*
"	光子	light quantum *[quant-mech]*
kōshi	格子；こうし	lattice *[crys]*
"	格子	grating; grid *[spect]*
kōshi	鉱滓	slag *[met]*
kōshi-bunkō-kei	格子分光計	grating spectrometer *[spect]*
koshi'kake	腰掛	bench; seat *{n}* *[furn]*
kōshi-kekkan	格子欠陥	lattice defect *[crys]*
kōshi-kenpa	格子検波	grid detection *[electr]*
kō'shiki	公式；こうしき	formula (pl -s, -lae) *[math]*
kōshiki'ka	公式化	formulation *[math]*
kōshin	光心	optical center *[opt]*
kōshin	更新	renewal *{n}*

kōshin-ha	後進波；こうしんは	retrograding wave *[elec]*
kōshin-ka'sokudo	向心加速度	centripetal acceleration *[mech]*
koshi no tsuyo-sa	腰の強さ	nerve *{n} [poly-ch] [rub]*
kōshin'ryō	香辛料	condiment; seasoning; spice *{n} [cook]*
kōshin-ryoku	向心力	centripetal force *[mech]*
kōshin-shi'enai shigen	更新しえない資源	nonrenewable resources *[sci-t]*
kōshin-shiki	恒真式	tautology *[logic]*
kōshin-shi'uru shigen	更新しうる資源	renewable resources *[sci-t]*
kōshin suru	更新する	update *{vb} [comput]*
ko-shio	小潮；こしお	neap tide *[ocean]*
koshi o kakeru	腰を掛ける	sit down *{vb}*
koshiraeru	拵える；こしらえる	make; manufacture; fashion *{vb t}*
kōshi-sanran	格子散乱	lattice scattering *{n} [sol-st]*
kōshi-sui	格子水	lattice water *[ch]*
kōshi-teisū	格子定数	lattice constant *[crys]*
kōshi-ten	格子点	lattice point *[math]*
kōshitsu-gomu	硬質ゴム	hard rubber *[mat]*
kōshitsu-jiki	硬質磁器	hard porcelain *[mat]*
kōshitsu-sen'iban	硬質繊維板	hardboard *[mat]*
kōshitsu-tan	硬質炭	anthracite; hard coal *[miner]*
koshō	故障	failure; fault *{n} [elec]*
koshō	胡椒；こしょう	pepper (condiment) *{n} [cook]*
koshō	湖床	lakebed *[geol]*
koshō	湖沼	lakes and marshes *[hyd]*
kōshō	公称；公稱	nominal (value) *{n} [ind-eng]*
kōshō	鉱床；鑛床	mineral deposit; ore deposit *[geol]*
koshō-kensa'nin	故障検査人	troubleshooter (a person) *{n}*
kōshoku	孔食；孔蝕	pitting corrosion *[met]*
kō'shokubai-hannō	光触媒反応	photocatalytic reaction *[ch]*
ko-shokubutsu'gaku	古植物学	paleobotany *[paleon]*
kōshoku-do	黄色土	ocher *[miner]*
kōshoku-kaburi	黄色カブリ	yellow fog *[photo]*
kōshoku'kei-chakushoku-zai	黄色系着色剤	yellow colorants *[mat]*
kōshoku-ori'mono	交織織物	union cloth *[text]*
kōshokusei-busshitsu	光色性物質	photochromics *{n} [ch]*
kōshoku-shokubutsu	紅色植物	red algae; Rhodophyta *[bot]*
kōshō-ō'ryoku	公称応力	nominal stress *[met]*
kōshū-eisei	公衆衛生	public health *[med]*
kōshu'gaku	耕種学	agronomy *[agr]*
kō-shūha no	高周波の	high-frequency *{adj} [comm]*
kō'shūha-yūdō-ka'netsu	高周波誘導加熱	high-frequency induction heating *{n} [eng]*
kosō	固相	solid phase *[phys]*

kōso	酵素；こうそ	enzyme *[bioch]*
kōsō	紅藻	red algae *[bot]*
kōsō	構想	design *{n}* *[sci-t]*
kōsō-biru	高層ビル	high-rise building *[arch]*
kōso-bunkai	酵素分解	enzyme degradation; zymolysis *[bioch]*
kōsō-denryū	高層電流	electrojet *{n}* *[geophys]*
kōso'gaku	酵素学	enzymology *[bioch]*
kosō-hannō	固相反応	solid-phase reaction *[ch]*
kosō-jūgō	固相重合	solid state polymerization *[poly-ch]*
kōso-ketsugō-men'eki-kyūchaku'zai-kentei(-hō)	酵素結合免疫=吸着剤検定(法)	enzyme-linked immunosorbent assay *[immun]*
kosō-kin	枯草菌	Bacillus subtilis *[microbio]*
kōso-ki'shitsu-fukugō'tai	酵素基質複合体	enzyme-substrate complex *[bioch]*
kō-soku	光束；こうそく	beam of light; luminous flux *[opt]*
kō-soku	光速	velocity of light *[opt]*
kōsoku	拘束	constraint; restraint *[mech]*
kō-soku	高速	high speed *[mech]* *[phys]*
kōsoku	梗塞	stoppage; blocking *{n}* *[med]*
kōsoku-dōro	高速道路	expressway *[civ-eng]*
kō'sokudo-sei	高速度星	high-velocity stars *[astron]*
kōsoku-hassan-do	光束発散度	luminous emittance; luminous radiance *[opt]*
kōsoku-keta'age	高速桁上げ	high-speed carry *{n}* *[comput]*
kōsoku-shibori	光束絞り	diaphragm; iris (pl -es, irides) *[opt]*
kōso-men'eki-sokutei-hō	酵素免疫測定法	enzyme immunoassay; enzyme-linked immunosorbent assay *[immun]*
kosō-netsu	枯草熱	hay fever *[med]*
kōso no kokusai-tan'i	酵素の国際単位	international unit of enzyme activity *[bioch]*
kōsō'un	高層雲	altostratus *[meteor]*
kōso-zenku'tai	酵素前駆体	zymogen *[bioch]*
kossetsu	骨折；こっせつ	fracture (of bone) *{n}* *[med]*
kosu	越す	cross; pass (over, across) *{vb}*
kosu	漉す	filter; strain *{vb t}*
kosu-	越す-	over- *{prefix}*
kōsū	恒数	constant *{n}* *[math]*
kōsui	香水	perfume *{n}* *[mat]*
kōsui	硬水	hard water *[ch]*
kōsui	鉱水；鑛水	mineral water *[hyd]*
kō-suijun-gengo	高水準言語	high-level language *[comput]*
kōsui-nanka-zai	硬水軟化剤	water softener *[mat]*
kōsui'ryō	降水量	precipitation (amount of) *[meteor]*

kōsui-sei	向水性	hydrotropism *[bio]*
kosuru	擦る；こする	rub; scour; chafe *{vb t}*
kotae	答え	reply; answer *{n} [comm]*
kotai	固体；こたい	solid *{n} [math] [phys]*
kōtai	抗体	antibody *[immun]*
kōtai	後退	retrograde movement *[astron]*
"	後退	backspace *[comput]*
kōtai	鋼帯；鋼帶	steel band *[mat]*
kotai-butsuri'gaku	固体物理学	solid-state physics *[phys]*
kotai-hassei	個体発生	ontogeny *[embryo]*
kō-tai'iki	広帯域；廣帶域	broadband; wideband *[comm]*
kōtai'iki-kūchū-sen	広帯域空中線	wideband antenna *[elecmg]*
kotai'kairo-keisan-ki	固体回路計算機	solid state computer *[comput]*
kōtai-kyūsū	交代級数	alternating series *[math]*
kōtai-moji	後退文字	backspace character *[comput]*
kotai-ronri-gi'jutsu	固体論理技術	solid-logic technology *[comput]*
kōtai saseru	後退させる	retract *{vb t}*
kotai-soshi	固体素子	solid-state component *[comput]*
kotai-suishin-yaku	固体推進薬	solid propellant *[mat]*
kōtai suru	交代する；交替する	alternative; substituting *{adj}*
kōtai suru	後退する	recede; retreat *{vb i}*
kotai-tansan (dorai-aisu)	固体炭酸（ドライ＝アイス）	solid carbon dioxide (dry ice) *[inorg-ch]*
kō'taku	光沢；光澤	luster *{n} [opt]*
kōtaku-do	光沢度；こうたくど	glossiness *[opt]*
kōtaku'do-kei	光沢度計	glossmeter *[opt]*
kōtaku-hoji-ritsu	光沢保持率	gloss retention *[ind-eng]*
kōtaku-kei	光沢計	glossmeter; glossimeter *[eng]*
kōtaku-sei	光沢性	glossiness *[opt]*
kōtaku-zai	光沢剤	brightener *[met]*
kote	鏝；こて	trowel *{n} [eng]*
kotei	固定	fixing; securing *{n} [eng]*
kōtei	工程；こうてい	(production) process *[ind-eng]*
kōtei	行程	stroke *{n} [mech-eng]*
"	行程	distance; journey *[navig]*
kōtei	肯定	acknowledgment; assertion *{n}*
kōtei	後庭	backyard *[civ-eng]*
kōtei	高低	high and low; undulations; unevenness *{n}*
kotei-hi	固定費	fixed expenses *[econ]*
kotei'ka-kōso	固定化酵素	immobilized enzyme *[bioch]*
kōtei-kanri	工程管理	process control *{n} [eng]*
kotei-ki'oku	固定記憶	fixed storage; permanent storage *[comput]*

kotei-kondensa	固定コンデンサ	fixed capacitor *[elec]*
kōtei no	公定の；こうていの	official *{adj}*
kōtei-sen	航程線	rhumb line *[map]*
kotei'shi	固定子；こていし	stator *[elec]*
kotei-shō	固定床	fixed bed *[ch-eng]*
kotei-shōsū'ten-sū	固定小数点数	fixed-point number *[comput]*
kotei-sō	固定相	fixed phase; stationary phase *[ch-eng]*
kotei-sō	固定層	fixed bed *[ch-eng]*
kotei suru	固定する	fix; secure *{vb} [eng]*
kotei-tanso	固定炭素	fixed carbon *[ch]*
kōtei'teki na	肯定的な	positive; assertive *{adj}*
kōtei-zu	工程図；工程圖	process chart *[graph] [ind-eng]*
kōten	公転；公轉	revolution (common axis) *{n} [astron]*
kōten	光点；光點	light spot; radiant *{n} [phys]*
kōten'getsu	交点月	nodical month *[astron]*
kōten suru	公転する	revolve about common axis *{vb} [astron]*
kote-nuri	鏝塗り	trowel coating *{n} [civ-eng]*
koto	琴；こと	Japanese harp *[music]*
koto	事	matter; thing; particulars *{n}*
kōtō	喉頭；こうとう	voice box; larynx (pl larynges, -es) *[anat]*
kotoba	言葉；ことば	word *[gram] [ling]*
kōtō'chū-rui	鉤頭虫類	Acanthocephala *[i-zoo]*
kōtō-gen	恒等元	identity element *[math]*
kōtō-gyō'retsu	恒等行列	identity matrix *[math]*
kotonaru-	異る-	allo- *{prefix}*
koto ni yoru	事による	depends on matters
kōtō ni yoru	口頭による	by word of mouth *{adv} [legal]*
kōtō-shiki	恒等式	identity *{n} [math]*
-kotsu	-骨	-bone *{suffix} [anat]*
kōtsū	交通；こうつう	traffic *{n} [eng]*
kotsu'ban	骨盤；こつばん	pelvis (pl -es, pelves) *[anat]*
kōtsū-kōgaku	交通工学	traffic engineering *{n} [civ-eng]*
kotsu'maku	骨膜	periosteum (pl -tea) *[anat]*
kōtsū-seigyo	交通制御	traffic control *[traffic]*
kōtsū-shingō	交通信号	traffic signal *[traffic]*
kōtsū-shingō-tō	交通信号灯	traffic light *[traffic]*
kotsu-yu	骨油；こつゆ	bone oil *[mat]*
kotsu-zai	骨材	filler *[cer]*
"	骨材	aggregate *[mat]*
kotsu-zui	骨髄；骨髓	bone marrow *[hist]*
kotsuzui'shu-saibō	骨髄腫細胞	myeloma cell *[cyt] [med]*

kottō'hin	骨董品	antiques; curios *[furn]*
kō'u'ryō	降雨量	precipitation; rainfall *[meteor]*
ko-ushi	子牛；犢	calf (pl calves; calfs) (animal) *[v-zoo]*
kō'utsu-yaku	抗鬱薬；こぅうつ薬	antidepressant *[pharm]*
kowa-meshi	強飯；こわめし	glutinous rice with red beans *[cook]*
kōwan-chiku	港湾地区；港灣地區	harbor district *[civ-eng]*
koware'me	毀れ目；こわれ目	break; crack location *{n}*
kowa-sa	剛さ；こわさ	rigidity; stiffness *[mech]*
kowasu	毀す；壊す；壞す	break; destroy; smash *{vb t}*
kōya-dōfu	高野豆腐	freeze-dried tofu (q.v.) *[cook]*
ko-yama	小山	knoll *[geol]*
kōyō	効用	utility *{n} [comput]*
kōyō'ju	広葉樹；廣葉樹	hardwood; broadleaf tree *[bot]*
ko-yōkai-do	固溶解度	solid solubility *[phys]*
koyō'ka-netsu-shori	固溶化熱処理	solid-solution treatment *[phys]*
kō'yokusei-yaku	抗抑制薬	counterdepressant *{n} [pharm]*
koyomi	暦；曆；こよみ	almanac; calendar *[astron] [pr]*
kōyō-shisetsu	公用施設	public utility *[ind-eng]*
koyō-tai	固溶体；固溶體	solid solution *[phys]*
kō'yu	鉱油；鑛油；こうゆ	mineral oil *[mat]*
kōyū	公有	public domain *[legal]*
ko-yubi	小指	little (fifth) finger *[anat]*
koyū-chi	固有値	characteristic value; eigenvalue *[math]*
koyū-jikan	固有時間	proper time *[astron]*
koyū-nendo	固有粘度	intrinsic viscosity *[p-ch]*
koyū'on	固有音；こゆうおん	eigentone *[acous]*
koyū-sei	固有性	characteristic property *[p-ch]*
koyū-teikō	固有抵抗	electrical resistivity; specific resistance *[elec]*
koyū-undō	固有運動	proper motion *[astron]*
kozai	糊剤	paste *{n} [mat]*
kōzan	鉱山；鑛山	mine *{n} [min-eng]*
kōzan-kōgaku	鉱山工学	mining engineering *{n} [eng]*
kōzan-shokubutsu	高山植物	alpine plant *[bot]*
kozato-hen	阜扁；阜偏：阝；こざとへん	kanji (Chinese character) radical denoting a small community *[pr]*
kō'zatsu	交雑；交雜	hybridization *[gen]*
kōzatsu-iku'shu	交雑育種	hybridization breeding *{n} [gen]*
ko-zeni	小銭；小錢	small change *[econ]*
kōzo	楮；こうぞ	paper mulberry *[bot]*
kōzō	構造；こうぞう	construction; structure *[ch] [sci-t]*
kōzō'butsu	構造物	structure *[civ-eng]*

kōzō-buzai	構造部材	structural member *[mat]*
kōzō-chi'shitsu'gaku	構造地質学	structural geology; tectonics *[geol]*
kōzō'gaku	構造学	tectonics *[civ-eng]*
kō'zōkan-zai	光増感剤	photosensitizer *[opt]*
kōzō-kanzen-sei	構造完全性	structural integrity *[eng]*
kōzō-kōgaku	構造工学	structural engineering *[civ-eng]*
kōzō-nensei	構造粘性	structural viscosity *[fl-mech]*
kōzō-seizu	構造製図	structural drawing *{n} [graph]*
kōzō-shiki	構造式	structural formula *[ch]*
kōzō-tsuyo-sa	構造強さ	structural strength *[mech]*
kōzō'yō-kō(zai)	構造用鋼(材)	structural steel *[steel]*
kōzō'yō-setchaku-zai	構造用接着剤	structural adhesive agent *[adhes]*
kōzō-zu	構造図；こうぞうず	construction plan *{n} [graph]*
kōzui	洪水；こうずい	flood *{n} [hyd]*
ku; kyū	九；く	nine *{n} [math]*
ku	区；區	ward (of a city) *{n} [geogr]*
ku	句	phrase *{n} [ling]*
kū-baitai	空媒体	empty medium *[comput]*
kubai-yaku (kubai-zai)	駆梅薬（駆梅剤）	antiluetic; antisyphilitic *{n} [pharm]*
kubi	首；頸；くび	neck *{n} [anat]*
kubi-kazari	首飾り	necklace *[lap]*
kubomi	窪み；凹み；くぼみ	indentation *{n} [math] [mech-eng]*
kūbō-sō	空乏層	depletion layer *[electr]*
kubun	区分；區分	segment; partition *{n} [comput]*
"	区分；くぶん	section *{n} [math]*
ku'bun no ichi	九分の一	one-ninth *[math]*
kuchi	口；くち	mouth *{n} [anat]*
"	口	opening *{n}*
kuchibashi	嘴；くちばし	beak; bill (of a bird) *{n} [v-zoo]*
kuchibiru	唇；くちびる	lip *{n} [anat]*
kuchi'dashi-sen	口出線	lead wire *{n} [eng]*
kuchiku'kan	駆逐艦	destroyer *[nav-arch]*
kuchi'nashi	山梔子；くちなし	cape jasemine; gardenia (flower) *[bot]*
kuchi'saki	口先；くちさき	proboscis *[zoo]*
kūchū-chisso-kotei	空中窒素固定	atmospheric nitrogen fixation *[ch-eng]*
kūchū-fuyō	空中浮揚	levitation *[phys]*
kūchū-kairo	空中回路	air corridor *[navig]*
kūchū-seibutsu'gaku	空中生物学	aerobiology *[bio]*
kūchū-sen	空中線	antenna (pl -nae, -s) *[elec]*
kuchū-zai	駆虫剤；驅蟲劑	anthelminthic; vermicide *{n} [pharm]*
kuda	管；くだ	pipe; tube; quill *{n} [eng]*
kudake-nami	砕け波；碎け波	breaker(s) *[ocean]*

kudakeru	砕ける；くだける	break; be broken *{vb i}*
kudake-yasu-sa	砕け易さ	friability *[mat]*
kudaku	砕く	crush; shatter; smash *{vb t}*
kuda'mono	果物；くだもの	fruit *{n} [bot]*
kudari	下り；くだり	descent; descension *{n} [navig]*
kudari-kōbai	下り勾配	declivity *[geol]*
kudaru; oriru	下る	descend *{vb} [navig]*
kūden	空電；くうでん	static *{n} [comm]*
"	空電	atmospherics *[geophys]*
kūdō	空洞；くうどう	cavity (pl -ties) *[bio]*
"	空洞	cavitation *[fl-mech]*
kudō-gen	駆動源	driving source *[mech-eng]*
kūdō-genshō	空洞現象	cavitation *[fl-mech]*
kudō-guruma	駆動車	driving wheel *[mech-eng]*
kudō-jiku	駆動軸；くどうじく	drive shaft *[eng]*
kudo'kanran'seki	苦土橄欖石	forsterite *[miner]*
kudō-kikō	駆動機構	drive *{n} [comput]*
kūdō-kyōshin-ki	空胴共振器	cavity resonator *[elecmg]*
kūdō-renga	空胴煉瓦	hollow brick *[mat]*
kudō-retsu	駆動列	drive train *[mech]*
kūdō-ritsu	空洞率	void content *[geol] [poly-ch]*
ku'en-san	枸櫞酸；くえんさん	citric acid
kuen'san-en	枸櫞酸塩	citrate *[ch]*
kuen'san-san'anmonyūmu	枸櫞酸三アンモ=ニウム	tribasic ammonium citrate
ku'ensan-tetsu	枸櫞酸鉄	iron citrate
kugai'sō	九蓋草；くがいそう	Culver's root; Veronica sibirica *[bot]*
ku'gatsu	九月	September (month)
kūgeki	空隙；くうげき	void *{n} [mech-eng]*
"	空隙	crevice *[sci-t]*
kūgeki'bu	空隙部	pore *{n} [met]*
kūgeki-bun	空隙分	void component *{n} [mat]*
kūgeki-ritsu	空隙率	void ratio *[geol]*
"	空隙率	void volume *[poly-ch]*
kugi	釘；くぎ	nail *{n} [mat]*
kugi-kakushi	釘隠し；くぎかくし	nailhead cover *[eng]*
kugiri-kigō	区切り記号	delimiter *[comput]*
kugiri-ten	区切り点	breakpoint *[comput]*
kūgun	空軍	air force *[mil]*
kūhaku	空白；くうはく	null *{n} [math]*
"	空白	blank *{n}*
ku'hentō-yu	苦扁桃油	bitter almond oil *[mat]*

kūhō	空胞；空胞	vacuole *[cyt]*
kui	杭；くい	pile(s) *{n} [eng]*
kū'i	空位；くうい	vacancy *[crys]*
kui'chigai	食違い	stagger *{n} [mech-eng]*
kui'chigai-ten	食違い点	jog *{n} [photo]*
ku'iki-danbō	区域暖房	zone heating *{n} [eng]*
kuina	水鶏；くいな	rail; waterrail (bird) *[v-zoo]*
kui'uchi-ki	杭打ち機	pile driver *[mech-eng]*
kujaku	孔雀；くじゃく	peacock (bird) *[v-zoo]*
kujaku'seki	孔雀石	malachite *[miner]*
kujira	鯨；くじら	whale (mammal) *[v-zoo]*
kujira-rui	鯨類	Cetacea (whales, dolphins, etc.) *[v-zoo]*
kujiru	抉る；くじる	gouge; prize; wrench; pry *{vb} [mech]*
kū-jishō	空事象	empty event *[prob] [stat]*
kūji-shū'ritsu	空時収率；空時收率	space-time yield *[ch-eng]*
ku'kaiseki	苦灰石	dolomite *[miner]*
ku'kaku	区画；區畫	partition *{n} [comput]*
ku'kak(u)kei	九角形	nonagon *[math]*
kukaku-shitsu	区画室；くかくしつ	compartment *{n} [nav-arch]*
kukaku suru	区画する	demarcate; divide; partition *{vb}*
kūkan	空間；くうかん	space *{n} [astron] [comput]*
kūkan-denka-kōka	空間電荷効果	space-charge effect *[electr]*
kūkan-denka-seigen-denryū	空間電荷制限電流	space-charge-limited current *[electr]*
kūkan-gun	空間群	space group *[crys]*
kūkan-haichi	空間配置	spatial arrangement *[elec]*
kūkan-kōkō'gaku	空間航行学	astronautics *[aero-eng]*
kūkan-kōshi; kūkan-gōshi	空間格子	space lattice *[crys]*
kūkan no	空間の	spatial *{adj} [phys]*
kūkan-ritsu	空間率	voidage *[ch-eng]*
kūkan-sokudo	空間速度	space velocity *[astron] [ch-eng]*
kukei	矩形；くけい	rectangle *[math]*
kukei-dōha'kan	矩形導波管	rectangular waveguide *[elecmg]*
kuki	茎；莖；くき	stalk; stem *{n} [bot]*
kūki	空気；空氣；くうき	air *{n} [ch]*
kūki-bukuro	空気袋	airbag *[mech-eng]*
kūki-chōwa	空気調和	air conditioning *[mech-eng]*
kūki-hi	空気比	air ratio *[eng]*
kūki-kai	空気塊；くうきかい	cell *[meteor]*
kūki-kansō	空気乾燥	air drying *{n} [eng]*
kūki-o'sen	空気汚染	air pollution *[ecol]*
kūki-riki'gaku	空気力学	aerodynamics *[fl-mech]*
kūki-sanka	空気酸化	air oxidation *[ch]*

kūki-seijō-ki	空気清浄器	air cleaner *[eng]*
kūki-tori'ire-guchi	空気取入口	air intake *[aero-eng]*
kūki-yoku	空気浴	air bath *[thermo]*
kukkyoku-kiretsu	屈曲亀裂	flex cracking *{n} [poly-ch]*
kuko	枸杞；くこ	Chinese wolfberry; Lycium chinense *[bot]*
kūkō	空孔	vacancy *[ch]*
"	空孔	hole *[sol-st]*
kūkō	空港；空港	airport *[civ-eng]*
kū'kōshi-ten	空格子点	vacancy *[crys]*
kukuri-zome	括り染め	tie-dyeing *{n} [tex]*
kuma	熊；くま	bear (animal) *[v-zoo]*
kuma'de	熊手	rake *{n} [eng]*
kuma'dori	隈取り	shadowing *{n} [graph]*
kumi	組	set *{n} [math]*
kumi'awase	組合せ	merge; merging; combination *{n} [comput]*
"	組合せ	association; combinations *[math]*
kumi'awase-kairo	組合せ回路	combinational circuit *[electr]*
kumi'kae	組換え；くみかえ	recombination *[gen]*
kumikae-DNA	組換えDNA	recombinant DNA *[bioch]*
kumikae-iden'shi	組換え遺伝子	recombination gene *[gen]*
kumikae'tai	組換え体	recombinant *{n} [gen]*
kumi'komi-kinō	組込み機能	built-in function *[comput]*
kumi-komu	組込む	incorporate *{vb}*
kumi'tate	組立	assembly *[mech-eng]*
kumi'tate-johō	組立除法	synthetic division *[math]*
kumi-tateru	組(み)立る	assemble *{vb}*
kumi'tate-zu	組立図	assembly drawing *{n} [graph]*
kumi-yaku	苦味薬	amara; bitters *{n} [pharm]*
kumo	雲	cloud *{n} [meteor]*
kumo	蜘蛛；くも	spider (insect) *[i-zoo]*
kumo('gata)-rui	蛛(形)類	Arachnida *[i-zoo]*
kumo-hitode	蜘蛛人手	brittle star *[i-zoo]*
kumo-no-su	蜘蛛の巣	spider web *[i-zoo]*
kumo'no'su-kabi	蜘蛛の巣かび	Rhizopus *[microbio]*
kumori	曇り；くもり	tarnish (silver) *{n} [met] [miner]*
"	曇り	cloudy *[meteor]*
"	曇り	bloom; chill (of paint) *{n} [poly-ch]*
"	曇り	haze; hazing *{n} [poly-ch]*
"	曇り	dulling *{n}*
kumori-ka	曇り価	haze value *{n}*
kumori-ten	曇り点	cloud point *[ch-eng]*
kumori-zora	曇り空	overcast sky *[meteor]*

kumo-zaru	蜘蛛猿; くもざる	spider monkey (animal) *[v-zoo]*
kun	訓; くん	Japanese phonetics of meaning for Chinese characters *[ling]*
kun'en	燻煙	smoking *{n} [ch] [cook]*
kuni	国; 國	country *[geogr]*
kuni-gamae	国構: 囗; くに=がまえ	Chinese character radical of encasement *[pr]*
kunjō	燻蒸	fumigation *[eng]*
kunjō-shōdoku	燻蒸消毒	fumigation sterilization *[eng]*
kunren	訓諫; 訓練	training *{n}*
kunsei	燻製; くんせい	smoking *{n} [cook]*
kunsei-niku	燻製肉	smoked meat *[cook]*
kuōku	クォーク	quark *[part-phys]*
kura	鞍; くら	saddle *{n} [eng] [transp]*
kuraberu	比べる; 較べる	compare *{vb}*
kurage	水母; 海月; くらげ	jellyfish; medusa (pl -dusae) *[i-zoo]*
kurai; gurai	位; くらい; ぐらい	about; location *[math]*
kurai	暗い	dark *{adj}*
kurai'dori	位取り	scaling *{n} [comput]*
kura'jō-kōshō	鞍状鉱床	saddle reef *[geol]*
kuraku naru	暗くなる	darken; become dark *{vb i}*
kuraku suru	暗くする	darken; make dark *{vb t}*
kura'yami	暗闇	darkness *{n} [astron]*
kūrei	空冷	air-cooling *{n} [mech-eng]*
kūretsu	空列	null string *[comput]*
kuri	栗; くり	chestnut (tree, nut) *[bot]*
kuri'age	繰上げ	carry(ing) up *[math]*
kuri'kaeshi	繰(り)返し	repetition *[math]*
kuri'kaeshi-enzan	繰返し演算	repetitive operation *[comput]*
kuri'kaeshi-kigō	繰返し記号: 々	repeat symbol for kanji *[pr]*
kuriputon	クリプトン	krypton (element)
kuri'sage	繰下げ	carry(ing) down *[math]*
Kurita-yaki	栗田焼	Kurita ware *[cer]*
kuro	黒; 黑; くろ	black (color) *{n} [opt]*
kūro	空路	air route; airway *[navig]*
kuro-ari	黒蟻	black ant (insect) *[i-zoo]*
kuro-dai	黒鯛	black porgy; Mylio macrocephalus *[v-zoo]*
kuro-denki'seki	黒電気石	schorlite; schorl *[miner]*
kuro'goke-gumo	黒後家蜘蛛	black widow (spider) *[i-zoo]*
kuro-guwai	黒慈姑; くろぐわい	water chestnut (fruit) *[bot]*
kuro-hyō	黒豹; くろひょう	black leopard (animal) *[v-zoo]*
kuroi	黒い; 黑い	black *{adj} [opt]*

kuro-ichigo	黒苺; 黒莓	blackberry (fruit) *[bot]*
kuro'ji	黒字	black-ink balance *[econ]*
kuro-kabi	黒黴; くろかび	bread mold; Aspergillus niger *[bot]* *[mycol]*
kuro-kajiki	黒梶木; 黒旗魚	black marlin (fish) *[v-zoo]*
kuro-kōji-kin	黒麹菌	black koji mold *[mycol]*
kuro'kō-kōshō	黒鉱鉱床	black ore deposit *[geol]*
kuro-kurumi	黒胡桃	black walnut (tree) *[bot]*
kuroku suru	黒くする	blacken *{vb t}*
kuro-maguro	黒鮪; くろまぐろ	bluefin tuna; Thunnus thynnus *[v-zoo]*
kuro-maku	黒膜	black film *[p-ch]*
kuro-mame	黒豆	black soybean *[bot]* *[cook]*
kuro-matsu	黒松	Japanese black pine *[bot]*
kuro-moji	黒文字; くろもじ	lindera; spicebush *[bot]*
kuromu	クロム	chromium (element)
kurorofuruorokābon	クロロフルオロ=カーボン	chlorofluorocarbon *[mat]*
kuro-shio	黒潮	black stream *[ocean]*
kuro'tama-tan	黒玉炭	jet *{n}* *[miner]*
kuro-ten	黒貂; くろてん	marten; sable (animal) *[v-zoo]*
kuro-ume'modoki	黒梅擬	buckthorne *[bot]*
kuro-unmo	黒雲母	biotite; black mica *[miner]*
kurubushi	踝; くるぶし	ankle (of foot) *{n}* *[anat]*
kuru'byō	佝僂病	rickets *[med]*
kuruma	車; くるま	wheel *{n}* *[eng]*
"	車	automobile; car; vehicle *[mech-eng]* *[trans]*
kuruma-ebi	車海老	prawn (crustacean) *[i-zoo]*
kuruma-isu	車椅子	wheelchair *[eng]*
kurumi	胡桃; くるみ	walnut (tree, nut) *[bot]*
kurumi-wari(-ki)	胡桃割り(器)	nutcracker *[eng]*
kurushimu	苦しむ	suffer; agonize *{vb i}*
kusa	草	grass *[bot]*
kusabi	楔; くさび	wedge *{n}* *[eng]* *[meteor]*
kusabi-bunkō-shashin-ki	楔分光写真器	wedge spectrograph *[spect]*
kusabi'gata-moji	楔形文字	cuneiform character *[comm]*
kusabi'jō no	楔状の	sphenoid *{adj}* *[geom]*
"	楔状の	cuneiform *{adj}* *[ling]*
kusa'chi	草地	meadow *[ecol]*
kusa-hibari	草雲雀; くさひばり	grass cricket (insect) *[i-zoo]*
kusai	臭い	malodorous; bad-smelling *{adj}*
kusa-iro	草色	green (dark) (color) *[opt]*

kusa-kanmuri	草冠: 艹, 艹 ; くさかんむり	kanji (Chinese character) radical denoting grass *[pr]*
kusa'kari-ki	草刈機	mowing machine *[eng]*
ku'sanka-shi'yōso	九酸化四沃素	tetraiodine enneaoxide
kusa o kū	草を食う	grazing *{adj} [v-zoo]*
kusare	腐れ	decay *{n} [bio] [mat]*
kusari	腐り	spoilage *[bio] [mat]*
kusari	鎖; くさり	chain *{n} [mat]*
kusari-ami	鎖編み	chain stitch *[tex]*
kusari-ha'guruma	鎖歯車	sprocket *[eng]*
kusari-hebi	鎖蛇	viper (reptile) *[v-zoo]*
kusaru	腐る; くさる	decay; rot *{vb i}*
kuse	癖; くせ	habit *[psy]*
kūseki-ritsu	空席率	vacancy rate *{n} [ch] [nucleo]*
kushami; kusami	嚔; くしゃみ; = くさみ	sneeze *{n} [physio]*
kushami suru	嚔する	sneeze *{vb} [physio]*
kushi	串	skewer; spit *{n} [cook]*
kushi	櫛; くし	comb *{n} [eng]*
kū-shiken; kara-shiken	空試験; 空試驗	blank test *[ind-eng]*
kushi'kezuru	梳る; くしけずる	comb (hair) *{vb}*
kū-shūgō	空集合	empty set; null set *[math]*
kūsō	空槽	empty tank *[ch-eng]*
kussaku-deisui	掘鑿泥水	excavation mud *[geol]*
kussaku-ki	掘鑿機; くっさくき	excavator *[eng]*
kussei	屈性	tropism *[bio] [bot]*
kussetsu	屈折; くっせつ	refraction *[phys]*
kussetsu-bōenkyō	屈折望遠鏡	refracting telescope *[opt]*
kussetsu-do	屈折度	degree of refraction *[phys]*
kussetsu-ha	屈折波	refracted wave *[phys]*
kussetsu-kei	屈折計	refractometer *[eng]*
kussetsu-kōgaku-kei	屈折光学系	dioptric system *[opt]*
kussetsu no hōsoku	屈折の法則	refraction, law of *[phys]*
kussetsu-ritsu	屈折率	index of refraction *[phys]*
kussetsu-ryoku	屈折力	refractive power *[phys]*
kussetsu-ten	屈折点	inflection point *[math]*
kusshin	屈伸	expansion and contraction *[mech]*
kusuguru	擽る; くすぐる	tickle *{vb t}*
kusu(-noki)	楠; 樟; くす(のき)	camphor tree; Cinnamomum camphora *[bot]*
kusuri	薬; 藥; くすり	medicine *{n} [pharm]*
kusuri-dansu	薬簞笥	medicine chest *[furn]*
kusuri-yubi	薬指	ring (fourth) finger *{n} [anat]*

kūtai	空帯; 空帶	empty band *[chem]*
Kutani-yaki	九谷焼	Kutani porcelain *[cer]*
kuten	句点; くてん	full stop *[comput] [pr]*
"	句点	period (punctuation mark) *[gram] [pr]*
kūten suru	空転する	idle; run idle *{vb} [mech-eng]*
kutō-hō	句読法; 句讀法	punctuation *[gram]*
kutō-moji	句読文字	punctuation character *[comput]*
kutō-ten	句読点	punctuation mark *[gram] [pr]*
kutsu	靴	shoe *{n} [cl]*
kutsū	苦痛	pain *{n} [physio]*
kutsu-bera	靴篦; くつべら	shoehorn *[cl] [eng]*
kutsu-gata	靴型	last (of shoe) *{n} [cl]*
kutsu'shita	靴下	socks *{n} [cl]*
kuttsuku	くっつく	stick together *{vb} [adhes]*
kuwa	桑	mulberry *[bot]*
kuwa	鍬	hoe *{n} [agr]*
kuwaeru	加える	add *{vb} [math]*
kuwae-zan	加え算	addition *[math]*
kuwai	慈姑; くわい	arrowhead *[bot]*
kuwanomi'mo	桑実藻; くわのみ藻	Pandorina *[bot]*
kuwashī	詳しい	detailed; be familiar *{adj}*
ku-yakusho	区役所; 區役所	ward office *[arch]*
kūyu no	空輸の	airborne *{adj} [aero-eng]*
kuzu	屑; くず	rubbish; scraps; trash *{n} [ecol]*
kuzu	葛; くず	arrowroot; kudzu vine *[bot]*
kuzu'kago	屑籠	wastebasket *[eng]*
kuzu-ke	屑毛	scrap wool *[tex]*
kuzureru	崩れる	crumble; collapse *{vb i}*
kuzuri	屈狸; くずり	wolverine (animal) *[v-zoo]*
kuzusu	崩す	break (money) *{vb t}*
"	崩す	demolish; level *{vb t}*
kyabetsu	キャベツ	cabbage (vegetable) *[bot]*
kyakka	却下	dismissal *[pat]*
kyakkan'teki na	客観的な	objective *{adj}*
kyaku	客	customer *[econ]*
kyaku	脚	leg (of an article) *{n}*
kyaku'chū	脚注; 脚註	footnote *[comput] [pr]*
kyaku'sha	客車	passenger car *[rail] [transp]*
kyaku-shitsu	脚室	landing-gear well *[aero-eng]*
kyaku-tobira	脚扉	landing-gear door *[aerosp]*
kyapashitansu	キャパシタンス	capacitance; C *[elec]*
kyara	伽羅; きゃら	aloeswood (incense) *[bot]*

kyōboku	喬木; きょうぼく	big tree *[bot]*
kyō'bunsan	共分散	covariance *[stat]*
kyō'bunsan-bunseki	共分散分析	analysis of covariance *[stat]*
kyōchiku'tō	夾竹桃; きょう=ちくとう	oleander (flower) *[bot]*
kyō'chin	共沈; きょうちん	coprecipitation *[ch]*
kyōchin-zai	共沈剤	coprecipitator *[bioch]*
kyōchō	強調; きょうちょう	enhancement *[comput]*
kyōdai	鏡台; 鏡臺	mirror stand; dressing table *[furn]*
kyodai-bunshi	巨大分子	macromolecule *[org-ch] [poly-ch]*
kyodai-bunshi no	巨大分子の	macromolecular *{adj} [poly-ch]*
kyodai-jūgō	巨大重合	macropolymerization *[poly-ch]*
kyodai-kesshō	巨大結晶	macrocrystal *[crys] [petr]*
kyodaku	許諾; きょだく	permission
kyodan	鋸断; 鋸斷	sawing (apart or through) *{n} [eng]*
kyō'dansei-tai	強弾性体	ferroelastic (pl -s) *{n} [mat]*
kyō-denkai'shitsu	強電解質	strong electrolyte *[p-ch]*
kyōdo	強度; きょうど	strength *[mech]*
"	強度	intensity *[phys]*
kyōdō	協働	cooperation *[elec]*
kyōdō-hatsumei'sha	共同発明者	coinventor; inventorship *[pat]*
kyōdō no-	共同の-	co- *{prefix}*
kyōdō'tai	共同体	consortium (pl -tia) *[bot]*
kyō'ei-men	鏡映面	reflection plane *[opt]*
kyō-enki	強塩基	strong base *[ch]*
kyōfū	強風; きょうふう	gale *{n} [meteor]*
kyō'futsu	共沸; きょうふつ	azeotropy *[ch]*
kyōfutsu-jōryū	共沸蒸留	azeotropic distillation *[ch]*
kyōfutsu-kongō'butsu	共沸混合物	azeotrope; azeotropic mixture *[ch]*
kyō'futten	共沸点	azeotropic point *[ch]*
kyogi	虚偽	falsity *[math]*
kyōgi	協議; きょうぎ	consultation; conference *[comm]*
kyōgō	競合; きょうごう	competition *[ecol] [gen]*
kyōgō suru	競合する	compete *{vb}*
kyōgō'teki-fuka	競合的付加	competitive addition *[ch]*
kyojaku	虚弱; きょじゃく	fragility; weakness *{n}*
kyōjin-chū-tetsu	強靭鋳鉄	high-tensile cast iron *[met]*
kyōjin-kō	強靭鋼	tough and hard steel *[steel]*
kyōjin-sei	強靭性	toughness *[mech]*
"	強靭性	tenacity *[tex]*
kyōji-sei	強磁性	ferromagnetism *[sol-st]*
kyōji'sei-kesshō	強磁性結晶	ferromagnetic crystal *[sol-st]*

kyō'jisei-tai	強磁性体	ferromagnetic material *[sol-st]*
kyōji suru	挟持する	hold, sandwiched *{vb t} [eng]*
kyōji suru	教示する	instruct; teach *{vb t}*
kyō'jūgō	共重合	copolymerization *[poly-ch]*
kyōjūgō'tai	共重合体	copolymer *[poly-ch]*
kyoka	許可	granted *[pat]*
"	許可	permission *{n}*
kyōka	強化	reinforcing *{n} [civ-eng]*
"	強化	strengthening *{n} [mech]*
kyōka	莢果；きょうか	legume *[bot]*
kyōka-baiyō	強化培養	enrichment culture *[microbio]*
kyōka-boku	強化木	compregnated wood *[poly-ch]*
kyōka-garasu	強化ガラス	tempered glass *[mat]*
kyōkai	境界；きょうかい	boundary *[comput] [geol] [sci-t]*
kyōkai-jōken	境界条件	boundary condition *[math]*
kyōkai-sen	境界線	border line *[math]*
kyōkai-sō	境界層	boundary layer *[fl-mech] [phys]*
kyōkaku	共角	coplanar *[math]*
kyōkaku	夾角；きょうかく	included angle *{adj} [math]*
kyōka-nenryō	強化燃料	enriched fuel *[mat]*
kyōka-rojin	強化ロジン	fortified rosin *[mat]*
kyō-karyū	共加硫	covulcanization *[poly-ch] [rub]*
kyōka-sen'i-shīto	強化繊維シート	reinforced fiber sheet *[tex]*
kyōka'sho	教科書	textbook *[pr]*
kyōka-zai	強化剤	enrichment agent *[cook]*
"	強化剤	reinforcing agent *[ind-eng]*
kyōken'byō	恐犬病	rabies; hydrophobia *[med]*
kyokkō	極光	aurora polaris *[geophys]*
kyōkō	胸腔	thoracic cavity *[anat]*
kyōkoku	峡谷；峽谷	canyon; gorge; gulch *{n} [geogr]*
kyō-kotsu	胸骨	breast bone; sternum *[anat]*
kyoku	極；きょく	pole *{n} [elec] [geogr] [math]*
kyoku'atsu-tenka-zai	極圧添加剤	extreme-pressure additive *[mat]*
kyoku'biryō-kagaku	極微量化学	ultramicrochemistry *[ch]*
kyokubu-denryū	局部電流	local current *[elec]*
kyokuchō-onsoku	極超音速	hypersonic speed *[fl-mech]*
kyokuchō-onsoku-kōkūki	極超音速航空機	hypersonic transport; HST *[trans]*
kyokuchō-onsoku no	極超音速の	hypersonic *{adj} [acous]*
kyokuchō-onsoku-ryū	極超音速流	hypersonic flow *[fl-mech]*
kyokuchō-tai'kaku	極頂対角	parallactic angle *[astron]*
kyokuchō-tanpa	極超短波	ultra-high-frequency; UHF *[comm]*
"	極超短波	ultrashort waves *[comm]*

kyokuchō-tanpa	極超短波	microwave *{n} [elecmg]*
kyoku'dai	極大；きょくだい	maximum *[math]*
kyokudai-chi	極大値	maximum value *[math]*
kyoku'do ni-	極度に-	super- *{prefix}*
kyoku'gen	極限；きょくげん	limit *{n} [math]*
kyoku'gen-kyōdo	極限強度	ultimate strength *[mech]*
kyoku'gen-nendo	極限粘度	intrinsic viscosity (number) *[p-ch]*
kyoku'hi-dōbutsu	棘皮動物	Echinodermata *[i-zoo]*
kyoku-jiku	極軸	polar axis *[eng]*
kyoku-kan; kyokkan	極冠	polar cap *[astron] [hyd]*
kyoku-kōfun no hōsoku	極興奮の法則	polar excitation, law of *[bio]*
kyoku'men	曲面	curved surface; surface *{n} [math]*
kyoku'ritsu	曲率；きょくりつ	curvature *[math]*
kyoku'ritsu-hankei	曲率半径	radius of curvature *[math]*
kyoku'ritsu-men	曲率面	surface of curvature *[math]*
kyoku-sei	極性	polarity *[math] [phys]*
kyokusei-bunshi	極性分子	polar molecule *[p-ch]*
kyokusei-kagō'butsu	極性化合物	polar compound *[ch]*
kyokusei-ki	極性基	polar group *[org-ch]*
kyoku'sen	曲線；きょくせん	curve; curved line *[math]*
kyokusen-zahyō	曲線座標	curvilinear coordinates *[math]*
kyoku'setsu-kakō	曲折加工	bending processing *{n} [ind-eng]*
kyokusha-hō	曲射砲；曲射砲	howitzer *[ord]*
kyoku-shi	局紙	Japanese vellum *[paper]*
kyoku-shigai	極紫外	extreme ultraviolet *[phys]*
kyoku'shō	極小；きょくしょう	minimum (pl -ma) *{n} [math]*
kyoku'shō-chi	極小値	minimum value *[math]*
kyoku'sho-masui-yaku	局所麻酔薬	local anesthetic *{n} [pharm]*
kyokushō no-	極小の-	mini- *{prefix}*
kyokusho'teki na	局所的な	local *{adj} [comput]*
kyoku-ten	極点	highest point; extreme point *{n} [math]*
kyoku'ten-zu	極点図	pole figure *[graph]*
kyoku-zahyō	極座表	polar coordinates *[math]*
kyō'kyaku	橋脚	bridge pier; pier *[civ-eng]*
kyōkyū-den'atsu	供給電圧	service voltage *[elec]*
kyōkyū-genryō	供給原料	feedstock *[agr]*
kyōkyū-kō	供給口	supply orifice *[mech-eng]*
kyōmaku	莢膜；きょうまく	capsule *[microbio]*
kyōmaku	強膜；鞏膜	sclera (pl -s, sclerae) *[anat]*
kyōmei	共鳴；きょうめい	resonance *[phys] [quant-mech]*
kyōmei-keikō	共鳴蛍光	resonance fluorescence *[atom-phys]*
kyōmei-sanran	共鳴散乱	resonance scattering *{n} [nuc-phys]*

kyōmen	鏡面；きょうめん	mirror-finished surface *[eng]*
"	鏡面	mirror surface *{n} [opt]*
kyōmen-kōtaku	鏡面光沢	specular gloss *[opt]*
kyōmen no	共面の	coplanar *{adj} [geom]*
kyōmen-shūsei	鏡面修正	figuring *{n} [opt]*
kyōmi	興味	interest; zest *[psy]*
kyon	羌；きょん	muntjak (animal) *[v-zoo]*
kyō'nenketsu-tan	強粘結炭	coking coal *[geol]*
kyō'netsu-genryō	強熱減量	ignition loss *[ch]*
kyō'netsu-zanbun	強熱残分	ignition residue *[ch]*
kyō'nin	杏仁；きようにん	apricot stone *[pharm]*
kyo'reki	巨礫	boulder *[geol]*
kyori	距離；きょり	distance *{n} [math] [mech]*
kyō'riki-ko	強力粉	strong flour *[cook]*
kyori-shisū	距離指数	distance modulus *[astron]*
kyori-sokutei	距離測定	range-finding *{n} [eng]*
kyōryō	橋梁；きょうりょう	bridges *{n} [civ-eng]*
kyōryoku-jinken	強力人絹	high-tenacity rayon *[tex]*
kyōryoku-sayō	協力作用	synergism *[ecol]*
kyōryū	恐竜；恐龍	dinosaur *[v-zoo]*
kyō-san	強酸	strong acid *[ch]*
kyosei	巨星	giant star *[astron]*
kyōsei	共生	symbiosis (pl -oses) *[ecol]*
kyōsei-junkan	強制循環	forced circulation *[mech-eng]*
kyōsei-junkan'gata-jō'hatsu-kan	強制循環形蒸発缶	forced-circulation evaporator *[eng]*
kyōsei-kankei	共生関係	symbiosis *[ecol]*
kyōsei-kansō	強制乾燥	forced drying *{n} [mech-eng]*
kyōsei-shindō	強制振動	forced oscillation *[mech]*
kyōseki'shō	共析晶	eutectoid *[p-ch]*
kyōsen	共線	collinear *[math]*
kyō-sen	胸腺	thymus (pl -es, thymy) (gland) *[anat]*
kyoshi	鋸歯；きょし	serration *[bot]*
kyoshi'jō no	鋸歯状の	serrate *{adj} [bot]*
kyōshin	共振	resonance *[elec] [phys]*
kyōshin-kairo	共振回路	resonance circuit *[elec]*
kyōshin'shi	共振子	resonator *[phys]*
kyōshin-shūha'sū	共振周波数	resonance frequency *[phys]*
kyōshin-yaku	強心薬	cardiac; cordial *{n} [pharm]*
kyoshi'teki na	巨視的な	macroscopic *{adj} [sci-t]*
kyō-shitsu	教室	classroom *[arch]*
kyoshō	裾礁；きょしょう	fringing reef *[geol]*

kyōshō	共晶；きょうしょう	eutectic *{n} [met]*
kyōshō	橋床	bridge floor; bridge deck *[civ-eng]*
kyō'shokubai	共触媒；共觸媒	cocatalyst *[ch]*
kyō'shuku'gō	共縮合	copolycondensation *[poly-ch]*
kyō'shukugō-jushi	共縮合樹脂	cocondensation resin *[poly-ch]*
kyō'shukugō'tai	共縮合体	copolycondensate *[poly-ch]*
kyōshū-zai	矯臭剤	odor-correcting agent *[pharm]*
kyōsō	競争；競爭	competition *[ecol]*
kyōsō-hannō	協奏反応	concerted reaction *[org-ch]*
kyōsō-hannō	競争反応	competitive reaction *[ch]*
kyōsō'kyoku	協奏曲	concerto *[music]*
kyōson-tentai	共存天体	symbiotic objects *[astron]*
kyōsō-zai	強壮剤	tonic *{n} [pharm]*
kyosū	虚数；きょすう	imaginary number *[math]*
kyōtai	供体	donor *[ch]*
kyō-tai'iki	狭帯域；狹帶域	narrowband *[comm]*
kyotan-yaku (kyotan-zai)	去痰薬（去痰剤）	expectorant *{n} [pharm]*
kyōtei	協定	agreement
kyōten	共点	concurrent *{n} [math]*
Kyoto	京都；きょうと	Kyoto *[geogr]*
kyōtō'i	橋頭位	bridgehead position *[ch]*
kyōtsū-gengo	共通言語	common language *[comput]*
kyōtsū-iki	共通域	common area *[comput]*
kyōtsū no-	共通の-	co- *{prefix}*
Kyō-yaki	京焼	Kyoto ware *[cer]*
kyō'yaku	共役；きょうやく	conjugate; conjugation *[ch] [math]*
kyōyaku-fukuso'sū	共役複素数	complex conjugate *[math]*
kyōyaku-kaku	共役角	conjugate angle *[math]*
kyōyaku-kei	共役系	conjugated system *[org-ch]*
kyōyaku na	共役な	conjugate *{adj}*
kyōyaku-nijū-ketsu'gō	共役二重結合	conjugated double bond *[p-ch]*
kyoyō-denryū-mitsudo	許容電流密度	allowable current density *[elec]*
kyoyō-gan'yū-ryō	許容含有量	permissible content *[ind-eng]*
kyoyō-gendo	許容限度	allowed limit *[ind-eng]*
kyoyō-genkai	許容限界	tolerance limits *[eng] [math]*
kyoyō-gosa	許容誤差	allowable error *[math]*
kyōyō-ronri	共用論理	shared logic *[comput]*
kyoyō-sa	許容差	tolerance *[math]*
kyoyō-senryō	許容線量	permissible dose *[nucleo]*
kyōyo'tai	供与体；供與体	donor *[ch]*
kyoyō-tai'netsu-ondo	許容耐熱温度	allowable heat-resisting temperature *[thermo]*

kyō'yūden-sei	強誘電性	ferroelectricity *[sol-st]*
kyō'yūden'sei-kesshō	強誘電性結晶	ferroelectric crystal *[sol-st]*
kyō'yūden-tai	強誘電体	ferroelectric (pl -s) *{n}* *[sol-st]*
kyōyū-genshi'ka	共有原子価	covalence *[ch]*
kyōyū-ketsugō	共有結合	covalent bond *[ch]*
kyōyū-kongō'butsu	共融混合物	eutectic mixture *[p-ch]*
kyōyū-ten	共融点	eutectic point *[p-ch]*
kyōzatsu'butsu	夾雑物	foreign matter *[sci-t]*
kyo'zetsu-hannō	拒絶反応	rejection reaction *[ch]*
kyo'zetsu-satei	拒絶査定	decision to reject *[pat]*
kyozō	虚像; きょぞう	virtual image *[opt]*
kyōzō	鏡像; きょうぞう	mirror image *[opt]*
kyōzō-i'sei	鏡像異性	enantiomorphism; mirror-image isomerism *[ch]*
kyōzō(-isei)'tai	鏡像(異性)体	enantiomorph *[ch]*
kyū	灸; きゅう	moxibustion *[med]*
kyū	級	class; grade *{n}* *[syst]*
kyū	球	sphere *{n}* *[math]*
kyū'chaku	吸着; きゅうちゃく	adsorption *[ch]*
kyūchaku-bai	吸着媒	adsorbent *[ch]*
kyūchaku-shitsu	吸着質	adsorbate; adsorptive *{n}* *[ch]*
kyūchaku-tōon-shiki	吸着等温式	adsorption isotherm *[p-ch]*
kyūchaku-zai	吸着剤	adsorbent *[ch]*
kyūden(-keitō)	給電(系統)	power supply (system) *[elec]*
kyūden-sen	給電線	feeder line *[elec]*
kyū'denshi-sei	求電子性	electrophilic property *[p-ch]*
kyū'denshi-shi'yaku	求電子試薬	electrophilic reagent *[p-ch]*
kyūden-sōchi	給電装置	power supply *[electr]*
kyūgen	給源	supply source *[ind-eng]*
kyū'in-bin	吸引瓶	suction bottle *[an-ch]*
kyū'in-roka	吸引沪過	suction filtration *[sci-t]*
kyūji	給餌	feeding (of animal) *{n}* *[agr]*
kyūjō-koku'en	球状黒鉛	spheroidal graphite *[geol]*
kyūjō-seidan	球状星団	globular star cluster *[astron]*
kyūka	球果; 毬果; 球花	cone *[bot]*
kyūka	球顆; きゅうか	variole *[geol]*
kyū-kaimen	球界面	spherical interface *[math]*
kyū-kaiten	急回転	spinning *{n}* *[mech]*
kyū('ka)-kansetsu	球(窩)関節	ball-and-socket joint *[anat]*
kyū'kaku	嗅覚; 嗅覺	olfaction; sense of smell *[physio]*
kyū'kaku-kei	嗅覚計	odorimeter; olfactometer *[eng]*
kyūkaku-shiyaku	求核試薬	nucleophilic reagent *[an-ch]*

kyū-kandan-kei	球寒暖計	bulb thermometer *[eng]*
kyūka no	九価の	nonavalent *{adj} [ch]*
kyū-kazan	休火山	dormant volcano *[geol]*
kyūkei	休憩；きゅうけい	rest; pause *{n}*
kyūkei	球形	spherical form *[math]*
kyūkei	球茎；球莖	corm *[bot]*
kyūkei (no)	球形(の)	globular; spherical *[math]*
kyūketsu	急結	quick-setting (cement) *[eng]*
kyūki-ben	吸気弁；吸氣瓣	intake valve *[mech-eng]*
kyūkin	球菌	coccus (pl -i) *[microbio]*
kyūkō-do	吸光度	absorbance; extinction *[p-ch]*
kyūkō-kōdo-bunseki	吸光光度分析	absorption spectrophotometry; molecular absorptiometric analysis *[spect]*
kyūkō-kōdo-kei	吸光光度計	absorptiometer *[an-ch] [eng]*
kyūkon	球根	bulb *[bot]*
kyūkō-ressha	急行列車	express train *[trans]*
kyūkō-ritsu	吸光率	absorptivity *[an-ch]*
kyūko-zai	急硬剤	quick-hardening admixture *[mat]*
kyū'kyoku-hatsu'gan-busshitsu	究極発癌物質	ultimate carcinogen *{n} [pharm]*
kyūkyoku-totsuzen-hen'i'gen	究極突然変異原	ultimate mutagen *{n} [pharm]*
kyūkyū'sha	救急車	ambulance *[med]*
kyūmei-dō'i	救命胴衣	life jacket *[nav-arch]*
kyūmei'gu	救命具	life preserver *[eng]*
kyūmei'tei	救命艇	lifeboat *[nav-arch]*
kyūmen	球面；きゅうめん	spherical surface *[eng] [math]*
kyūmen-kyō	球面鏡	spherical mirror *[opt]*
kyūmen-sankak(u)kei	球面三角形	spherical triangle *[math]*
kyūmen-shūsa	球面収差	spherical aberration *[opt]*
kyūmen-tenmon'gaku	球面天文学	spherical astronomy *[astron]*
kyūmin	休眠	dormancy *[bot]*
"	休眠	diapause *[physio]*
kyūnan'sen	救難船	rescue ship *[nav-arch]*
kyū'netsu-hannō	吸熱反応	endothermic reaction *[p-ch]*
kyū'netsu'sei no	吸熱性の	endothermic *{adj} [p-ch]*
kyū no hosū	九の補数	nine's complement *[comput]*
kyūnyū-zai	吸入剤	inhalant *{n} [pharm]*
kyū'on-ritsu	吸音率	acoustic absorptivity; sound absorption coefficient *[acous]*
kyūrei	急冷；きゅうれい	quenching *{n} [met]*
kyūrei-haku'tai	急冷薄帯	quenched thin strip *[met]*
kyūrei-koyō-shori	急冷固溶処理	quenching solution treatment *[met]*

kyūri	胡瓜；きゅうり	cucumber (vegetable) *[bot]*
kyūryō	給料	pay; salary; wages *{n} [econ]*
kyūryō-chitai	丘陵地帯	foothills *[geogr]*
kyūryū	急流	rapids *[hyd]*
kyūryū	穹窿；きゆぅりゆぅ	vault *{n} [arch]*
kyuryūmu	キユリウム	curium (element)
kyū-saibō	嗅細胞	olfactory cell *[physio]*
kyūsei no	急性の	acute; of sudden onset *{adj} [med]*
kyūseki	求積；きゅうせき	quadrature *[math]*
kyūseki-hō	求積法	planimetry; stereometry *[math]*
"	求積法	mensuration *[sci-t]*
kyū-sekki-jidai	旧石器時代	Paleolithic Age *[archeo]*
kyū-shamen	急斜面	escarpment *[geol]*
kyūshi	休止	rest; pause *{n} [music]*
"	休止	quiescence *[zoo]*
kyūshi	臼歯	molar *[dent]*
kyūshi	給糸	yarn-feeding *{n} [text]*
kyūshi-jōtai	休止状態	dormant state *[comput]*
kyūshin-ryoku	求心力	centripetal force *[mech]*
kyūshin'sei no	求心性の	centripetal *{adj} [mech]*
kyū'shitsu	吸湿；吸濕	moisture absorption *[ch-eng]*
kyūshitsu-sei	吸湿性	hygroscopicity *[bot] [ch]*
kyūshō	球晶	spherulite *[geol]*
kyūshū	吸収；きゅうしゅう	absorption *[ch]*
kyūshū-inshi	吸収因子	absorption factor *[acous] [phys]*
kyūshū-keisū	吸収係数	absorption coefficient *[acous] [phys]*
kyūshū-netsu	吸収熱	heat of absorption *[thermo]*
kyūshū-ritsu	吸収率	absorption factor *[acous] [phys]*
"	吸収率	absorptance *[phys]*
kyūshū-ryoku	吸収力	absorptivity *[an-ch]*
kyūshū'sei no	吸収性の	absorptive *{adj} [ch]*
kyūshū-zai	吸収剤；吸收劑	absorbent *[mat]*
kyūsoku-kidō	急速起動	quick start *{n} [mech-eng]*
kyū'soku no	急速の	express *{adj} [trans]*
kyūsoku-tōketsu	急速凍結	quick freezing *{n} [eng]*
kyūsō-sokudo	給送速度	feeding speed *[tex]*
kyūsu	急須；きゅうす	teapot (small) *[cer]*
kyūsū	級数	series (pl series) *[math]*
kyūsui-do	吸水度	water-absorbing capacity *[ch] [tex]*
kyūsui-honkan	給水本管	water main *[civ-eng]*
kyūsui-ritsu	給水率	water absorption *[ch]*
kyūsui-sen	給水栓	faucet; water tap *[eng]*

kyūsui-yōryō	吸水容量	water absorption capacity *[mat]*
kyūtai	球体	sphere *[math]*
kyūtai	給体	donor *[sol-st]*
kyūyo	給与; 給與	allowance *[econ]*
kyūyō	休養; きゅうよう	rest *{n}* *[med]*
kyūyu	給油	oiling *{n}* *[mech-eng]*
kyūyu-do	吸油度	oil absorptiveness *[mech-eng]*
kyū-zahyō	球座標	spherical coordinates *[math]*
kyūzō	吸蔵; 吸藏	occlusion *[phys]*
kyūzō	急増; きゅうぞう	sudden increase *{n}*
kyūzō-kagō'butsu	吸蔵化合物	occlusion compound [ch]
kyūzō no	急造の	hurriedly constructed *{adj}* *[arch]*

M

ma	間；ま	room *{n} [arch]*
ma; aida	間；ま；あいだ	interval (of time) *{n}*
ma'aji	真鯵；まあじ	horse mackerel; Trachurus trachurus (fish) *[v-zoo]*
mabara na	疎らな	sparse; scattered; sporadic *{adj}*
mabataki	瞬き；まばたき	blink; blinking *{n} [phyiso]*
ma'biki	間引き	thinning *{n} [agr]*
ma-biku	間引く	thin (as seedlings); cull *{vb}*
mabushi-sa	眩しさ	glare *{n} [opt]*
ma'buta	瞼；眼瞼；目蓋	eyelid *{n} [anat]*
machi	町；街	city; town; street *[geogr]*
machi	襠；まち	gusset; gore; inset *[cl]*
ma'chigai	間違い	mistake; error *{n}*
machigau	間違う	err; make a mistake *{vb}*
machi-gyō'retsu	待ち行列	queue *{n} [comput] [math]*
machi-jikan	待ち時間	latency; waiting time *[comput]*
machin	馬銭；まちん	nux vomica; poison-nut tree *[bot]*
ma'dai	真鯛；まだい	red sea bream; porgy; Chrysophrys major (fish) *[v-zoo]*
ma'dake	真竹；まだけ	bamboo; Phyllostachys bambusoides *[bot]*
madara	斑；まだら	mottle; mottled; mottling *{n} [bio]*
"	斑	mottles *{n} [plas]*
ma-dare	麻垂：广；まだれ	kanji (q.v.) radical denoting roof *[pr]*
mado	窓；まど	window *[arch]*
mado-garasu	窓硝子	window pane *[arch] [mat]*
ma-dori	間取り	room-arrangement plan *[graph]*
mado-ue'waku	窓上枠	yoke *{n} [arch]*
mado-waku	窓枠	window frame *[arch]*
mae	前；まえ	ahead; in front of *{adv} {n}*
"	前	before *{adv} {prep}*
mae'ba	前歯；前齒	front teeth; incisors *[anat] [dent]*
mae-baiyō	前培養	preculture *{n} [microbio]*
mae-barai	前払い；前拂い	prepayment *[econ]*
mae-kake	前掛け	apron *[cl]*
mae-kōka	前硬化	precure; precuring *{n} [plas]*
mae-muki no	前向きの	forward-facing *{adj}*
mae no-	前の-	ante-; pre- *{prefix}*
mae no-; mae e-	前の-；前へ	pro- *{prefix}*
mae-rokō	前露光	pre-exposure *[photo]*

mae-shori	前処理；前處理	conditioning; pretreatment *{n} [eng]*
mae-suberi	前滑り	forward slip (in rolling) *{n} [mech]*
magai no	紛いの；擬いの	artificial; counterfeit; spurious *{adj}*
ma'gamo	真鴨；まがも	mallard (bird) *[v-zoo]*
magari	曲(が)り	warp *{n} [cer]*
"	曲(が)り	tortuosity *[mech]*
magari'ba-kasa-ha'guruma	曲り歯傘歯車	spiral bevel gear *[mech]*
magatta	曲った	curved *{adj} [math]*
mage	曲げ；まげ	bend; bending *{n} [eng]*
mage-dansei-ritsu	曲げ弾性率	flexural modulus *[mech]*
mage-gōsei	曲げ剛性	flexural rigidity *[mech]*
mage-hirō-kyōdo	曲げ疲労強度	bending fatigue strength *[mech]*
mage-ki'retsu	曲げ亀裂	flex cracking *{n} [mech]*
mage-kowa-sa	曲げ強さ	flexural rigidity *[mech]*
mage-ō'ryoku	曲げ応力	flexural stress; bending stress *[mech]*
mageru	曲げる	bend; curve *{vb t} [mech-eng]*
mage-sei	曲げ性	bendability *[mech]*
mage-tsuyo-sa	曲げ強さ	bending strength; flexural strength *[mech]*
magirawashī	紛らわしい	confusing; misleading *{adj}*
magire-komu	紛れ込む	stray in *{vb}*
maguneshūmu	マグネシウム	magnesium (element)
maguro	鮪；まぐろ	tuna (fish) *[v-zoo]*
maguro-kanyu	鮪肝油	tuna liver oil *[mat]*
mahi-yaku	麻痺薬；まひやく	paralyzant *{n} [pharm]*
mahō-bin	魔法瓶；まほうびん	vacuum bottle; thermos bottle *[eng]*
mahō-jin	魔法陣	magic square *[math]*
mai-	毎-；毎-；まい-	every-; each-: apiece- *{prefix}*
-mai	-枚	-sheet(s) (of paper) *{suffix} [paper]*
mai'botsu'butsu	埋没物	implant *{n} [med]*
mai'botsu suru	埋没する	bury; embed *{vb t}*
mai-byō	毎秒；まいびょう	per second *[mech]*
mai-fun	毎分	per minute *[mech]*
mai-ji	毎時	per hour *[mech]*
maikuro-ha	マイクロ波	microwave *{n} [elecmg]*
maikuro-konpūta	マイクロコン=ピュータ	microcomputer *[comput]*
maimai-ga	舞舞蛾；舞々蛾	gypsy moth (insect) *[i-zoo]*
mai'nen; mai-toshi	毎年	every year; annually *{adv}*
mai-nichi	毎日	every day *{adv}*
mairu	哩；マイル	mile (unit of length) *[mech]*
ma'iruka	真海豚；まいるか	dolphin (mammal) *[v-zoo]*
mai-shū	毎週	weekly *{adv} {n}*

mai-toshi	毎年；まいとし	every year; yearly *{n} {adv}*
maiyō-shi	枚葉紙	sheet *{n} [pr]*
maizō	埋蔵；埋藏	embedding *{n} [sci-t]*
maizō'ryō	埋蔵量	reserve(s) *{n} [min-eng]*
majiwari	交わり	intersection *[math]*
majiwari-sū	交わり数	intersection number *[math]*
ma'jutsu-sū	魔術数	magic number *[math]*
ma'kajiki	真旗魚；まかじき	marlin; spearfish; Makaira mitsukurii (fish) *[v-zoo]*
maki	槙；まき	Chinese black pine; podocard (tree) *[bot]*
maki; takigi	薪；まき；たきぎ	firewood *[mat]*
maki'age-ki	巻揚げ機	winch *{n} [mech-eng]*
maki-ageru	巻き上げる	roll up; wind up *{vb t}*
maki-e	蒔き絵；蒔き繪	lacquer; lacquering *{n} [furn]*
maki-hige	巻鬚；まきひげ	tendril *[bot]*
maki-jaku	巻尺；卷尺	tape measure *[eng]*
maki-keitai	巻形態	wound profile *[tex]*
maki'modoshi	巻き戻し	rewinding; unwinding *{n} [mech-eng]*
maki-modosu	巻き戻す	rewind *{vb t}*
maki'mono	巻物；まきもの	scroll; rolled book *{n} [graph] [pr]*
maki'sen	巻線	winding *{n} [elec]*
maki'sen-teikō-ki	巻線抵抗器	wire-wound resistor *[elec]*
maki'sū-hi	巻数比	turns ratio *{n} [elec]*
ma-kita; shin'poku	真北；真北	true north *[navig]*
maki-tesshin	巻鉄心；卷鐵芯	rolled core; wound iron core *[elecmg]*
maki'tori	巻取り	take up; taking up; coiling *{n} [tex]*
maki'tori-chōryoku	巻取張力	take-up tension *[tex]*
maki'tori-gami	巻取紙	web *{n} [paper]*
makkō	末項	last term *{n} [math]*
makkō	抹香；まっこう	incense *{n} [mat]*
makkō'kujira-yu	抹香鯨油	sperm oil *[mat]*
maku	幕	curtain *{n} [furn]*
maku	膜；まく	film; membrane *[ch-eng] [mat]*
maku	巻；巻く；捲く	wind; coil; roll *{vb t}*
maku	蒔く	sow; plant seeds *{vb t} [agr]*
maku-den'i	膜電位	membrane potential *[physio]*
maku-denkō	幕電光	sheet lightning *{n} [geophys]*
maku-denryū	膜電流	membrane current *[physio]*
maku-heikō	膜平衡	membrane equilibrium *[ch-eng]*
makura	枕；まくら	pillow *{n}*
makura-gai	枕貝	olive shell (shellfish) *[i-zoo]*
makura'gi	枕木	tie; sleeper *{n} [rail]*

makuri	海人草；まくり	corsican weed *[bot]*
maku-roka-ki	膜沪過器；膜濾過器	membrane filter *[sci-t]*
maku'shi-moku	膜翅目	Hymenoptera *[i-zoo]*
maku-teikō	膜抵抗	membrane resistance *[ch-eng]*
makuwa-uri	真桑瓜	makuwa melon; Cucumis melon (fruit) *[bot]*
maku-yōryō	膜容量	membrane capacitance *[elec]*
mamako	継粉；繼粉；ままこ	lump *{n}* *[cook]*
mame	豆	pea (sometimes bean) *[bot]*
mame-ishi	豆石	pisolite *[petr]*
mame-ka	豆科	Leguminosae *[bot]*
mame-kogane-mushi	豆黄金虫	Japanese beetle; Popilla japonica (insect) *[i-zoo]*
ma-mizu	真水；淡水；まみず	fresh water *[hyd]*
mamō	摩耗	abrasion; wear; wearing *{n}* *[eng]* *[geol]*
mamō-genryō	摩耗減量	abrasion loss *[plas]*
mamoru	守る；護る	guard *{vb}*
mamushi	蝮；まむし	pit viper (reptile) *[v-zoo]*
man	万；萬	ten thousand *{n}* *[math]*
mana-gatsuo	真名鰹；まながつお	butterfish; harvest fish (fish) *[v-zoo]*
mana-ita	俎；まないた	chopping board *[cook]*
manbō	翻車魚；まんぼう	ocean sunfish; Mola mola (fish) *[v-zoo]*
manchō	満潮；滿潮	high tide *{n}* *[ocean]*
mandai	万鯛；まんだい	kingfish; moonfish; Lampris regius *[v-zoo]*
mane	真似	mimicry *[bio]*
maneru	真似る	imitate; copy *{vb}*
maneshi-tsugumi	マネシ鶫	mockingbird (bird) *[v-zoo]*
mangan	マンガン	manganese (element)
mange'kyō	万華鏡	kaleidoscope *[opt]*
mangetsu	満月	full moon *{n}* *[astron]*
ma ni au	間に合う	be on time
ma ni awanai	間に合わない	miss; be not on time
ma-ni-awase	間に合わせ	makeshift *{n}*
ma ni awaseru	間に合わせる	make do *{vb}*
manjū	饅頭	bun with bean jam filling *[cook]*
mannen	万年；萬年	ten thousand years *{n}*
mannen-gusa	万年草；まんねん草	sedum *[bot]*
mannen'hitsu	万年筆	fountain pen *[graph]*
mannenrō	迷送香；まんねん=ろう	rosemary (pl -maries) *[bot]*
mannen'setsu	万年雪	firn *[geol]* *[hyd]*
mannen-setsu'gen	万年雪原	neve *[geogr]*
man'riki	万力；まんりき	vise; vice *{n}* *[eng]*

manryō-bi	満了日；滿了日	expiration date *[econ]*
mansei	慢性	chronicity *{n} [med]*
mansei no	慢性の	chronic *{adj} [med]*
manzoku saseru	満足させる	satisfy *{vt t}*
ma'ō	麻黄	ephedra; mahuang *[bot]*
mare na	稀な；希な	rare; infrequent *{adj}*
mari	毬	ball (a toy) *[eng]*
-maru	-丸	name suffix for e.g. ships, swords
maru-atama	丸頭	buttonhead (screw) *[eng]*
marui	丸い；円い；圓い	round *{adj} [math]*
maru-kō	丸鋼	round bar of steel, rod *[civ-eng]*
marume	丸め	round(ing) off *{n} [comput] [math]*
marume no gosa	丸めの誤差	rounding error *[math]*
marumeta sū	丸めた数	round number *[math]*
maru'mi	丸味	mellowness (flavor) *{n} [cook]*
"	丸味	roundness *{n} [math]*
maru-tenjō	丸天井	dome (ceiling) *{n} [arch]*
maru-yane	丸屋根	dome (roof) *{n} [arch]*
masai	摩砕	grinding; milling *{n} [mech-eng]*
ma'satsu	摩擦；まさつ	friction *[mech]*
masatsu-butsuri'gaku	摩擦物理学	tribophysics *[phys]*
masatsu-denki	摩擦電気	frictional electricity; triboelectricity *[elec]*
masatsu'gaku	摩擦学	tribology *[phys]*
masatsu-guruma	摩擦車	friction wheel *[mech]*
masatsu-keisū	摩擦係数	friction coefficient *[mech]*
masatsu-kenrō-do	摩擦堅牢度	abrasion fastness *[mat] [tex]*
masatsu no hōsoku	摩擦の法則	friction, law of *[mech]*
masatsu-yō'setsu	摩擦溶接	friction welding *[plast]*
mashi-bun	増分；增分	increment *{n} [math]*
massatsu	抹殺	cancellation; erasure; obliteration *[math]*
masshō	抹消	erasure; obliteration *[math]*
masshō-shinkei-kei	末梢神経系	peripheral nervous system *[anat]*
masshō suru	抹消する	delete; erase *{vb} [comput]*
massugu	真っ直	straight *{n} [math]*
massugu na	真っ直な	straight *{adj} [math]*
massugu ni suru	真っ直ぐにする	make straight; straighten *{vb t}*
masu	鱒；ます	trout; Oncorhynchus masou (fish) *[v-zoo]*
masui-yaku	麻酔薬	anesthetic *{n} [pharm]*
masu-no-suke	鱒の介；ますのすけ	chinook salmon; king salmon (fish) *[v-zoo]*
mata	叉；また	tine *[acous]*
matatabi	木天蓼；またたび	silvervine *[bot]*

matcha	抹茶	powdered tea *[cook]*
mato	的	target *{n} [phys]*
mato-dai	的鯛; まとだい	dory; Zeus japonicus (fish) *[v-zoo]*
matsu	松	pine (tree) *{n} [bot]*
matsu	待つ	wait *{vb}*
matsu-ba	松葉	pine needle *[bot]*
matsu'ba-yu	松葉油	pine-needle oil *[mat]*
matsuba'zue	松葉杖	crutches *[med]*
matsu'ge	睫毛	eyelash(es) *{n} [anat]*
matsu-kasa	松毬; 松笠	pinecone *[bot]*
matsu'kasa-uo	松毬魚; まつかさ=うお	pinecone fish; Monocentris japonica *[v-zoo]*
matsu'mo	松藻	hornwort; horned liverwort *[bot]*
matsu no mi	松の実; 松の實	pine nut; pignolia *[bot]*
matsuri'ka	茉莉花; まつりか	jasmine *[bot]*
matsuri'ka-cha	茉莉花茶	jasmine tea *[cook]*
matsu-take	松茸	matsudake; pine mushroom *[bot]*
matsu-yani	松脂; まつやに	pine resin; colophony *[mat]*
mattan-ki	末端基	end group; terminal group *[ch]*
mattan-ten'i-kōso	末端転移酵素	terminal transferase *[bioch]*
mawari	回り; 廻り	turning; revolution *{n}*
mawari	周り	surroundings *{n}*
mawari-michi	回り道	detour *{n} [civ-eng] [traffic]*
mawari-sunpō	回り(周り)寸法	girth *{n} [mech]*
mawaru	回る; 廻る	revolve; rotate; spin; turn *{vb i}*
mawasu	回す	revolve; rotate; spin; turn *{vb t}*
mawata	真綿	floss silk; floss *{n} [tex]*
ma'yaku	麻薬; 痲藥	narcotic(s) *{n} [pharm]*
mayaku-chūdoku	麻薬中毒	narcotic intoxication; narcotism *[med]*
mayu	眉	eyebrow *{n} [anat]*
mayu	繭; まゆ	cocoon *[i-zoo]*
maze-ori	交ぜ織	blended weave; mixed weave *[tex]*
mazui	不味い; まずい	unpalatable *{adj} [cook] [physio]*
me	目; 眼	eye *{n} [anat]*
"	目	grain (of wood, cloth) *[mat]*
"	目	stitch *{n} [tex]*
me	芽	bud *{n} [bot]*
me'bachi	眼撥; めばち	bigeye tuna (fish) *[v-zoo]*
me'baru	眼張; めばる	rock cod; rockfish (fish) *[v-zoo]*
me'bōki	目箒; めぼうき	sweet basil; Ocimum basilicum *[bot]*
me-bunryō	目分量	eye measure *{n} [sci-t]*
medaka	目高	cyprinodont (fish) *[v-zoo]*

me'dama-moyō	目玉模様	eyespot *{n} [i-zoo]*
me'dama-yaki	目玉焼	fried eggs *[cook]*
medo	目処；めど	goal; prospect (colloquial) *{n}*
me'gane	眼鏡；めがね	eyeglasses; glasses *[opt]*
megane-guma	眼鏡熊	spectacled bear (animal) *[v-zoo]*
megane-zaru	眼鏡猿	tarsier (animal) *[v-zoo]*
meguru	回る；繞る；めぐる	surround; come round *{vb}*
me-gusuri; gan'yaku	眼薬	eyewash; eye lotion *{n} [pharm]*
mei'an-hō	明暗法	shading *{n} [graph]*
mei'an-kaisen	明暗界線	terminator *[astron] [geophys]*
meiban	銘盤	nameplate *[comm]*
mei-cha	銘茶；めいちゃ	tea of well-known brand *[cook]*
meidai	命題	proposition *[logic]*
meido	明度	lightness; color value; value *{n} [opt]*
meido-nōdo	明度濃度	luminous density *[photo]*
meigara	銘柄；めいがら	brand; brand name; make *{n} [econ]*
meigetsu	明月	bright moon *[astron]*
meigi de	名義で	in the name of *[legal]*
meigi-henkō	名義変更	name change *[pat]*
meihaku na	明白な	evident *{adj}*
Meiji	明治；めいじ	Meiji (era name) (1868-1912)
meiji no	明示の	explicit *{adj}*
meiji-sengen	明示宣言	explicit declaration *[comput]*
mei-junnō	明順応	light adaptation *[physio]*
meikaku'ka	明確化	clarification; explanation *[comm]*
meikaku na	明確な	clear-cut; explicit; definite *{adj}*
meikō	迷光	stray light *[opt]*
meikoku	銘刻	inscription *[graph]*
meikyū	迷宮；めいきゅう	labyrinth *{n}*
meimei-hō	命名法	nomenclature *[sci-t]*
mei'metsu	明滅	blinking *{n} [comm]*
"	明滅	flicker *{n} [opt]*
meimoku-sokudo	名目速度	nominal speed; rated speed *[comput]*
mei'on	鳴音	singing *{n} [elec]*
mei'ō-sei	冥王星	Pluto (planet) *[astron]*
meirei	命令；めいれい	command; instruction; order *{n} [comput]*
meirei'bun	命令文	statement *[comput]*
meirei-jikkō	命令実行	execution; instruction execution *[comput]*
meirei-tori'dashi	命令取出し	instruction fetch *[comput]*
mei'ro	迷路	maze *{n} [psy]*
meiryō-do	明瞭度	articulation *[comm]*
meisai-ki'nyū-seikyū'sho	明細記入請求書	invoice *{n} [econ]*

meisai'sho	明細書	specification(s) *[pat]*
meishi	名刺	business card; calling card *[pr]*
meishi	名詞	noun *[gram]*
meishi-kyori	明視距離	least distance of distinct vision *[navig]*
mei-shiya	明視野	bright field *[opt]*
meishō	名称; 名稱	nomenclature; name *[sci-t]*
meisō-denryū	迷走電流	stray current *[elec]*
me'ita-garei	目板鰈; めいた=がれい	frog flounder; Pleuronichthys cornutus (fish) *[v-zoo]*
meji	目地	joint *{n}* *[eng]*
mejina	眼仁奈; めじな	opaleye; Girella punctata (fish) *[v-zoo]*
me'jiro	目白; めじろ	white-eye; silvereye (bird) *[v-zoo]*
mejiro-zame	目白鮫	requiem shark; Carcharhinus gangeticus (fish) *[v-zoo]*
meji-zai	目地剤	joint compound; joint mixture *[civ-eng]*
me'kajiki	目旗魚; めかじき	broadbill; Xiphias glacius; sword-fish *[v-zoo]*
mekki; tokin	鍍金	plating; galvanizing *{n}* *[met]*
mekkin	滅菌	sterilization *[biol]*
mekkin-ki	滅菌機	sterilizer *[eng]*
mekkin-sōchi	滅菌装置	sterilizer *[eng]*
mekki-yoku	鍍金浴; めっき浴	plating bath *[met]*
mekura	盲	blind *{n}* *[med]*
mekura-ana	盲穴; めくら穴	blind hole *[eng]* *[met]*
mekuru	捲る; めくる	turn over (a page) *{vb t}* *[pr]*
me'mai	目眩い	vertigo; dizziness *[physio]*
me'mori	目盛; めもり	graduation; scale *{n}* *[phys]*
memori-sen	目盛線	scale mark; graduation line *[graph]*
memori'tsuki no	目盛付きの	graduated *{adj}* *[ch]*
men	面	face *{n}* *[anat]* *[crys]* *[math]*
"	面	disk *[astron]*
"	面	mask; face mask *{n}* *[cl]*
"	面	facet *[mat]* *[met]*
"	面	plane; surface *{n}* *[math]*
men; wata	綿	cotton *[tex]*
menada	赤目魚; めなだ	Japanese mullet; Liza haematocheila (fish) *[v-zoo]*
menbō	麺棒	rolling pin *[cook]* *[eng]*
menderebyūmu	メンデレビウム	mendelevium (element)
mendori	雌鳥	hen; female bird *[v-zoo]*
me-neji	雌螺子; めねじ	internal thread *[mech-eng]*
men'eki	免疫; めんえき	immunity (pl -ties) *[immun]*

men'eki'gaku	免疫学	immunology *[bio]*
men'eki-gen	免疫原	immunogen *[immun]*
men'eki-guroburin	免疫グロブリン	immune globulin *[immun]*
men'eki-iden'gaku	免疫遺伝学	immunogenetics *[med]*
men'eki'ka	免疫化	immunization *[immun]*
men'eki-kagaku	免疫化学	immunochemistry *[ch]*
men'eki-kentei-hō	免疫検定法	immunoassay *[immun]*
men'eki-kyūchaku-zai	免疫吸着剤	immunoadsorbent *[immun]*
men'eki-sei	免疫性	immunity *[immun]*
mengai-hen'kaku-shindō	面外変角振動	out-of-plane deformation vibration *[p-ch]*
mengai-shindō	面外振動	out-of-plane vibration *[p-ch]*
men'jitsu-yu	綿実油	cottonseed oil *[mat]*
men-kaku	面角	interfacial angle *[crys]*
men'kaku-fuhen no hōsoku	面角不変の法則	constant interfacial angle, law of
men'kaku-ittei no hōsoku	面角一定の法則	constant interfacial angle, law of; constancy of facial angle, law of *[crys]*
men-kankaku	面間隔	interplanar spacing; lattice spacing *{n} [crys]*
menkyo	免許	license *{n} [legal]*
men'mō	綿毛	wool hair *[bot]*
mennai-ihō'sei	面内異方性	in-plane anisotropy *[crys]*
mennai-shindō	面内振動	in-plane vibration *[crys]*
me'nō	瑪瑙; めのう	agate *[miner]*
men-rō	綿蠟	cotton wax *[mat]*
men-rui	麵類; 麺類	noodles; noodle-type food; pasta *[cook]*
menseki	免責	escape *[comput]*
menseki	面積; めんせき	area *[math]*
menseki no tan'i	面積の単位	unit of area *[math]*
menseki-ritsu	面積率	area percentage *[math]*
menshi	綿糸	cotton yarn *[tex]*
menshin-kōshi	面心格子	face-centered lattice *[crys]*
menshin-rippō-kōshi	面心立方格子	face-centered cubic lattice *[crys]*
menshin-sanpō-kōshi	面心三方格子	face-centered rhombohedral lattice *[crys]*
menshin-seihō-kōshi	面心正方格子	face-centered tetragonal lattice *[crys]*
me'nuke	目抜; めぬけ	rockfish; ocean perch (fish) *[v-zoo]*
me-nuki	目貫き	swort hilt ornament *[eng]*
menzei	免税	duty exemption; tax exemption *[econ]*
menzei'hin	免税品	duty-free goods; tax-exempt goods *[econ]*
mesu	雌	female *[zoo]*
mesu-buta	雌豚	sow (animal) *{n} [v-zoo]*
mesu-hitsuji	雌羊	ewe (animal) *{n} [v-zoo]*
mesu-ko'uma	雌子馬	filly (pl fillies) (animal) *[v-zoo]*

mētoru	米；メートル	meter (unit of length) *[mech]*
me'tsuke	目付；めつけ	grammage; weight per area *[mech]*
me-ushi	雌牛	cow (animal) *{n} [v-zoo]*
me'yasu	目安；めやす	criterion (pl -ria) *{n} [math]*
"	目安	aim *{n}*
me'zamashi-dokei	目覚まし時計	alarm clock *[horol]*
me'zameru	目覚める	wake; awake *{vb i}*
me'zumari	目詰まり	loading (machining) *{n} [met]*
"	目詰まり	mesh-clogging *{n} [tex]*
mi	実；實；み	fruit *{n} [bot]*
mi	身	body (pl bodies) *{n} [anat]*
mi'au	見合う	counterbalance; offset *{vb}*
mi-bae	実蠅；みばえ	fruit fly (insect) *[i-zoo]*
michi	道	road; way; street; path *[civ-eng]*
mi'chi	未知	unknown (value) *[math]*
michibiku	導く	guide *{vb}*
michi-ita	道板	gangplank *[nav-arch]*
michi-sū	未知数	unknown (number) *{n} [math]*
michite yuku tsuki	満ちてゆく月	waxing moon *[astron]*
michi-yanagi	路柳；道柳	knotweed; knotgrass (herb) *[bot]*
midare	乱れ；亂れ；みだれ	disruption *[crys]*
"	乱れ	turbulence *[fl-mech]*
midare-bako	乱れ箱	clothes tray *[cl] [furn]*
mi'dashi	見出し	heading; header *{n} [comput] [pr]*
mi-dashinami	身嗜み	personal care *{n}*
midori	緑；綠	green (color) *[opt]*
midori-mushi	緑虫；みどりむし	euglena *[i-zoo]*
midori'mushi-shokubutsu	緑虫植物	Euglenophyta *[bot]*
midori'zakuro-ishi	緑柘榴石	grossular *[miner]*
mie'nai	見えない	invisible *{adj} [opt]*
migaki	磨き；摩き；みがき	polish; polishing *{n} [mech-eng]*
migaki-garasu	磨き硝子	frosted glass *[mat]*
migaki-kin	磨き金	burnished gold *[met]*
migaki-ko	磨(き)粉	polishing powder *[mat]*
migaku	磨く	polish; buff; burnish *{vb} [eng]*
migi	右；みぎ	right (direction) *{n} [math]*
migi'gawa no	右側の	dextral *{adj} [med]*
migi-kiki	右利き	right-handed *{n} [math]*
migi'maki no	右巻きの	dextrorse *{adj} [bot]*
migi'muki no	右向きの	dextral *{adj} [med]*
migi-neji	右ねじ	right-handed screw *[mech-eng]*
migi-sokumen-zu	右側面図	right side view *[graph]*

migi'te no hōsoku	右手の法則	right hand rule *[elecmg]*
migi-tsume	右詰め	right-justify(ing) *[comput] [pr]*
mi'hiraki	見開き	spread *{n} [pr]*
mi'hiraki-tobira	見開き扉	double-spread title page *[pr]*
mi'hiraki-zuban	見開き図版	double plate *[pr]*
mihon	見本	sample *{n} [ind-eng]*
mijika'de-hōkō	短手方向	shortish direction *[math]*
mijikai	短い	short *{adj} {n} [math]*
mijin'ko	微塵子；みじんこ	water flea; Daphnia pulex *[i-zoo]*
mi'juku na	未熟な	immature; unripe *{adj} [bio]*
mi'kake	見掛け	apparent(ness) *{n} [math]*
mikake-hijū	見掛比重	apparent specific gravity *[mech]*
mikake-nendo	見掛粘度	apparent viscosity *[fl-mech]*
mi'kaku	味覚；味覺	gustation; sense of taste *{n} [physio]*
mikan	蜜柑	mandarin orange; tangerine (fruit) *[bot]*
(mikan-)kona'jirami	(蜜柑)粉虱	white-fly (insect) *[i-zoo]*
mi'kansei no	未完成の	unfinished *{adj}*
mika-zuki	三日月	crescent moon *[astron]*
mike-neko	三毛猫	tortoiseshell cat (animal) *[v-zoo]*
miki	幹	trunk (of tree) *{n} [bot]*
mi'kōkai	未公開	unpublished *{n} [pat]*
mi'kōkai no	未公開の	unpublished *{adj} [pat]*
mi-kosu	見越す	anticipate *{vb}*
mikuron	ミクロン	micron (pl -s, micra) *[mech]*
mimei	未明	early dawn *{n} [astron]*
mimi	耳；みみ	ear *{n} [anat]*
"	耳	lug *{n} [eng]*
"	耳	selvage; trimmings *{n} [tex]*
mimi-aka	耳垢	earwax; cerumen *[physio]*
mimi-nobi	耳延び	selvage stretch *{n} [tex]*
mimi-sen	耳栓	earplug *[eng]*
mimi-tabu	耳朶；耳たぶ	earlobe *{n} [anat]*
mimi-ware	耳割れ	edge cracking *{n} [plas]*
mimizu	蚯蚓；蚓；ミミズ	earthworm *[i-zoo]*
mimizuku	木菟；みみずく	horned owl; eared owl (bird) *[v-zoo]*
mina; minna	皆；みな；みんな	all *{adv} {n}*
minami	南；みなみ	south *{n} [geod]*
minami-jūji-za	南十字座	Southern Cross *[constel]*
minami-kaiki-sen	南回帰線	tropic of Capricorn *[geod]*
minami-kaze	南風	southerly wind; southerlies *[meteor]*
minamoto	源；みなもと	source *{n} [sci-t]*
minato	港；港	harbor *{n} [geogr]*

mine-sen	峰線	crest line *[geol]*
mi'nikui	見難い；見悪い	difficult to see *{adj}*
min'ji no	民事の	civil *{adj} [legal]*
minna; mina	皆	all *{adv} {n}*
mi'nogashi-ayamari-ritsu	見逃し誤り率	residual error ratio *[comput]*
mino-kasago	簑笠子；みのかさご	lionfish; Pterois lunulata (fish) *[v-zoo]*
mino-mushi	簑虫	bagworm; fagot worm *[i-zoo]*
minpō	民法	civil law *[legal]*
min'zoku	民族	race *{n} [bio]*
minzoku-seibutsu'gaku	民族生物学	ethnobiology *[bio]*
mippei-gata	密閉型	closed mold *[eng]*
mippei-ki	密閉器；密閉器	stoppered container *[ind-eng]*
"	密閉器	sealed (sealable) container *{n}*
mippū	密封	sealing *{n} [eng]*
mippū suru	密封する	seal; close air-tight *{vb t}*
mirai	味蕾	taste bud *{n} [anat]*
miri'bāru	ミリバール	millibar *[mech]*
miri'guramu	瓱；ミリグラム	milligram *[mech]*
miri'mētoru	粍；ミリメートル	millimeter *[mech]*
miri'mikuron	ミリミクロン	millimicron *[mech]*
mirin	味醂；みりん	sweet sake (q.v.) *[cook]*
miru	見る；観る	see; look at *{vb}*
miru	水松；海松；みる	thick-haired codium *[bot]*
miru-gai	海松貝；水松貝	trough shell; Tresus keenae *[i-zoo]*
miru koto	見ること	seeing *{n} [physio]*
miru'kui	海松食；水松食	trough shell; Tresus keenae *[i-zoo]*
misago	鶚；みさご	fish hawk; osprey (bird) *[v-zoo]*
misairu	ミサイル	missile *{n} [ord]*
misaki	岬	cape; promontory *[geol]*
mise	店	store; shop *{n} [econ]*
mi'seijuku	未成熟	immaturity *{n} [bio]*
miseru	見せる	show; exhibit *{vb t}*
mishima-okoze	三島虎魚；みしま=おこぜ	stargazer (fish) *[v-zoo]*
mi'shiyō-baitai	未使用媒体	virgin medium *[comput]*
miso	味噌；みそ	soybean paste *[cook]*
misosazai	鷦鷯；みそさざい	wren (bird) *[v-zoo]*
mi'soshiki-tanso	未組織炭素	unorganized carbon *[ch]*
miso-shiru (omiotsuke)	味噌汁；（御御御=付；おみおつけ）	soybean paste soup *[cook]*
mitchaku	密着	adherence *[eng] [phys]*
mitchaku-atsu'en	密着圧延	tight rolling *{n} [met]*

mitchaku-ori'mage	密着折曲げ	tight-adhesion bending *{n} [eng]*
mitchaku-sei	密着性	tight-adhesiveness *{n} [phys]*
mitchaku-yaki'tsuke	密着焼付け	contact printing *{n} [photo]*
mitei-keisū-hō	未定係数法	undetermined coefficient method *[math]*
mi'tori-zu	見取図	sketch *{n} [graph]*
mitsu	蜜; みつ	honey *[cook]*
mitsu'ba	三葉	honewort; trefoil; wild chervil *[bot]*
mitsu'bachi	蜜蜂	honeybee (insect) *[i-zoo]*
mitsu'dasō	密陀僧	litharge *[miner]*
mitsu'do	密度; みつど	density (pl -ties) *[mech]*
"	密度	count *{n} [tex]*
mitsudo'ha-riron	密度波理論	density wave theory *[astrophys]*
mitsudo-kansū	密度関数	density function *[math]*
mitsudo-kei	密度計	densimeter *[eng]*
mitsudo-kōbai-kan	密度勾配管	density gradient tube *[an-ch]*
mitsudo-ryū	密度流	density current *[ocean]*
mitsudo-yakusō	密度躍層	pycnocline *[geophys]*
mitsu'go	三つ子; 三生児	triplets *{n} [bio]*
mitsu'gumi	三つ組	triad *[math]*
mitsu'gumi no	三つ組の	ternary *{adj} [sci-t]*
mitsukaru	見付かる	be found; be found out *{vb i}*
mitsukeru	見付ける	discover; find; sight *{vb t}*
mitsu'mame	蜜豆	dessert of fruit, agar, syrup *[cook]*
mi'tsumori	見積り	estimation *[math]*
mitsu'rin	密林	jungle *[ecol]*
mitsu'rō	蜜蠟	(yellow) beeswax *{n} [mat]*
mitsu'yubi-namake'mono	三つ指樹懶	three-toed sloth (animal) *[v-zoo]*
mi-yasui	見易い	easy to see *{adj}*
mizo	溝	channel *{n} [civ-eng]*
"	溝	groove; slot *{n} [eng]*
mizore	霙; みぞれ	sleet *{n} [meteor]*
mizu	水; みず	water *{n} [ch]*
mizu no	水の	aqueous; of water *{adj} [ch]*
mizu no hozen	水の保全	water conservation *[ecol]*
mizu no junkan	水の循環	hydrologic cycle *[hyd]*
mizu-ame	水飴	millet jelly (pl jellies) *[cook]*
mizu'ame'jō	水飴状	starch-syrup state *[mech]*
mizu-ba'riki	水馬力	water horsepower *[hyd]*
mizu-bōsō	水疱瘡	chicken pox; varicella *[med]*
mizu-bunsan-kei	水分散系	aqueous dispersed system *[ch]*
mizu-denryō-kei	水電量計	eudiometer *[eng]*
mizu-dokei	水時計	water clock *[horol]*

mizu-dōryoku-kei	水動力計	hydraulic dynamometer *[eng]*
mizu'game-za	水瓶座	Aquarius; the Water Bearer *[constel]*
mizu-gire	水切れ	water break *[paper]*
mizu'goke	水苔	bog moss; peat moss; sphagnum *[bot]*
mizu'kaki	水掻き	web (of bird's foot) *[v-zoo]*
mizu'kiri-ban	水切り板	drainboard (kitchen) *[arch] [cook]*
mizu'kiri-dai	水切り台	drainboard (kitchen) *[arch] [cook]*
mizu'nagi-dori	水凪鳥	shearwater (bird) *[v-zoo]*
mizu-nira	水韮; みずにら	quillwort; Isosetes *[bot]*
mizu-nomi'ba	水呑場	drinking fountain *[arch]*
mizu'saki(-annai)'nin	水先(案内)人	pilot (a person) *{n} [navig]*
mizu'saki'sen	水先船	pilot boat *[navig]*
mizu-shūshi	水収支; 水收支	water balance *[ind-eng]*
mizu'suki-gami	水漉紙	waterleaf paper *[paper]*
mizu-sumashi	水澄まし	whirligig beetle (insect) *[i-zoo]*
mizu-tama; mizu-dama	水玉	water droplet; dewdrop *[hyd]*
mizu-togi	水研ぎ	wet sanding; wet rubbing *{n} [eng]*
mizu-tori	水鳥; 水禽	waterfowl (bird) *[v-zoo]*
mizu-tōryō	水当量	water equivalent *[meteor]*
mizu'umi	湖; みずうみ	lake *[hyd]*
mizu-yaki'ire	水焼入れ	water quenching *{n} [eng]*
mizu-yuki	水雪	slush; melting snow *[meteor]*
mo	藻; も	alga (pl -ae) *[bot]*
mō	網	network *{n} [comput]*
moba	藻場	seaweed bed *[bot] [ecol]*
mochi	持ち	wear; durability *{n} [ind-eng]*
mochi	餅; もち	glutinous-rice cake; rice cake *[cook]*
mochi	黐; もち	ilex (tree) *[bot]*
mochi-ageru	持上げる	lift; raise *{vb t}*
mochi ga i'i	持ちが良い	long-lasting; wears well
mochi-gome	糯米; もちごめ	glutinous rice (cereal) *[cook]*
mochi-yasui	持ち易い	easy to hold *[eng]*
mōchō	盲腸	appendix (pl -es, -pendices) *[anat]*
mōdo	猛度	brisance *[ord]*
-modoki	-擬; -もどき	-like; in the style of *{suffix}*
modori	戻り; 戾り	return *{n} [comput]*
modoru	戻る	return; go back; revert *{vb i}*
modosu	戻す	return; give back; restore *{vb t}*
moeru	燃える	burn *{vb i}*
mōfu	毛布	blanket *{n} [furn] [tex]*
mo-fuku	喪服	mourning clothes *[cl]*
mō'gaku-dōbutsu	毛顎動物	Chaetognatha *[i-zoo]*

mogura	土竜；もぐら	mole (animal) *[v-zoo]*
moguru	潜る；もぐる	burrow *{vb} [zoo]*
mogusa	艾；もぐさ	moxa *{n} [bot]*
mō'hatsu	毛髪；毛髮	hair *{n} [anat]*
mō'hitsu	毛筆	brush (for writing, painting) *{n} [graph]*
mohō	模倣	imitation *{n}*
moji	文字；もじ	character; letter *[comput] [graph] [pr]*
moji-fugō	文字符号	character code *[comput]*
moji-hassei-ki	文字発生器	character generator *[comput]*
moji-hōtei'shiki	文字方程式	literal equation *[math]*
moji-kankaku	文字間隔	character space *[comput]*
moji-kensa	文字検査	character check *[comput]*
moji-nin'shiki	文字認識	character recognition *[comput]*
moji no kumi	文字の組	character set *[comput]*
moji-retsu	文字列	character string *[comput]*
moji-shiki	文字式	character expression *[comput]*
mōkan	毛管	capillary tube *[eng]*
mōkan-genshō	毛管現象	capillarity (pl -ties) *[fl-mech]*
mokei	模型	pattern; model *{n} [graph] [sci-t]*
mōkin-rui	猛禽類	Raptores (birds of prey) *[v-zoo]*
mokkōgu	木工具	woodworking tool *[eng]*
mokkō-seihin	木工製品	wood product *[mat]*
mōkō	毛鉱；毛鑛	jamesonite *[miner]*
moku	目；もく	order *{n} [bio] [syst]*
moku'bu	木部	xylem *[bot]*
mokubu-hōsha-so'shiki	木部放射組織	xylem ray *[bot]*
mokubu-jū'so'shiki	木部柔組織	xylem parenchyma *[bot]*
mokubu-sen'i	木部纖維	xylem fiber *[bot]*
moku-fun	木粉	wood flour; wood meal *[mat]*
mokuhan	木版	wood-block printing *{n} [graph]*
moku'hon-shokubutsu	木本植物	woody plant *[bot]*
moku'hyō	目標	target; goal *{n} [sci-t]*
mokuhyō-gengo	目標言語	target language *[ling]*
mokuhyō-tō	目標塔	pylon *[civ-eng]*
moku'ji	目次	table of contents *[pr]*
moku'ji	黙字；默字	mute letter *[ling]*
moku'ji no	黙示の	implicit *{adj}*
mok(u)ka	木化	lignification *[bot]*
moku-kyō	木橋	wooden bridge *[civ-eng]*
moku'me	木目	moire (wood) *[bot]*
moku'me-zuke	木目付け	wood graining *{n} [eng]*
moku-neji	木螺子；木ねじ	wood screw *[mech-eng]*

moku'ren	木蓮；もくれん	magnolia (tree, flower) *[bot]*
moku-rō	木蠟；もくろう	Japan wax; vegetable wax *[mat]*
moku'roku	目録；もくろく	catalog *{n} [comput]*
moku'roku-henkō	目録変更	recataloging *{n} [pr]*
moku'roku-sagyō	目録作業	cataloging *{n} [comput]*
moku'saku-eki	木酢液	pyroligneous acid; wood vinegar *[org-ch]*
moku'saku-san	木酢酸	pyroligneous acid
mokusei	木星	Jupiter (planet) *[astron]*
moku'sei	木精；もくせい	methanol; methyl alcohol; wood spirit *[org-ch]*
mokusei-kōgei	木製工芸	woodcraft *[furn]*
moku-sen	木栓	wood plug *[eng]*
moku'sen	木船	wooden ship *[nav-arch]*
moku'shaku-seki	木錫石	wood tin *[miner]*
mokushi-kensa	目視検査	visual inspection *[ind-eng]*
mokushi-kyori	目視距離	visibility (pl -ties) *[meteor]*
moku'shitsu'so	木質素；もくしつそ	lignin *[bioch]*
moku-soku	目測	eye estimation *{n} [physio]*
mokutan	木炭；もくたん	charcoal *[mat]*
mokuteki	目的；もくてき	aim; end; object; objective; purpose; target *{n}*
mokuteki'chi	目的地	destination *[navig]*
mokuteki-gengo	目的言語	object language *[comput]*
mokuteki'go	目的語	object *{n} [gram]*
mokuteki-kaku	目的格	objective case *[gram]*
mokuteki-kansū	目的関数	objective function *[math]*
mokuteki'ron	目的論	teleology *[sci-t]*
moku'tō	木糖	wood sugar; xylose *[bioch]*
moku'yōbi	木曜日；もくようび	Thursday
mokuzai-bōfu-yu	木材防腐油	wood-preservative oil *[mat]*
mokuzai-fukyū	木材腐朽	wood decay *[bot]*
mokuzai-fukyū-kin	木材腐朽菌	wood-rotting fungus *[mycol]*
mokuzai-kanryū	木材乾留	carbonization of wood *[ch]*
mokuzai-kōgyō	木材工業	lumber industry *[eng]*
mokuzai-sen'i	木材繊維	wood fiber *[bot]*
mōmaku	網膜	retina (pl -s, -inae) *{n} [anat]*
momen	木綿；もめん	cotton *[bot] [tex]*
momen-dōfu	木綿豆腐	cotton-strained tofu (q.v.) *[cook]*
momi	籾；もみ	unhulled rice (cereal) *[bot]*
momi	樅；もみ	fir (tree) *[bot]*
momi	紅絹；もみ	red silk cloth *[tex]*
momi-gara	籾殻	rice chaff; chaff *[bot]*

momiji	紅葉；もみじ	Japanese maple (leaves, tree) *[bot]*
momo	股	thigh *{n} [anat]*
momo	桃	peach (fruit, tree) *[bot]*
momo-iro	桃色	pink; rose (color) *[opt]*
momo'iro-sango	桃色珊瑚	pink coral *[i-zoo]*
momoke	ももけ	pilling *{n} [tex]*
momu	揉む	rub; crumble *{vb t}*
mon	門；もん	gate *[arch] [electr]*
"	門	phylum (pl phyla) *[bio] [syst]*
"	門	division *{n} [bot] [syst]*
monazu-ishi	モナズ石	monazite *[miner]*
mondai	問題	problem; question *{n} [math]*
mondai-ten	問題点	moot point; problem point *{n}*
mon'gara-kawa'hagi	紋柄皮剝	spotted triggerfish *[v-zoo]*
mon-myaku	門脈	portal vein *[anat]*
mono	物	thing; object; article *{n}*
mono	者	person *{n}*
mono'oki	物置	closet *[arch]*
mon-ori'mono	紋織物	figured cloth *[tex]*
monshi	門歯	incisor *[dent]*
moraru-yōeki	モラル溶液	molal solution *[ch]*
mōra suru	網羅する	collect exhaustively *{vb t}*
more	漏れ	leak; leakage *{n} [eng]*
more-denryū	漏れ電流	leakage current *[elec]*
more'dome	漏れ止め	seal *{n} [mech-eng]*
mori	森	forest *[ecol]*
mori	銛；もり	harpoon *{n} [eng]*
mori-azami	森薊；もりあざみ	Circium dipsacolepsis *[bot]*
moribuden	モリブデン	molybdenum (element)
mori'wake-ryō	盛り分け量	serving size *[cook]*
moroko	諸子；もろこ	chub; roach (fish) *[v-zoo]*
moro-sa	脆さ	brittleness; fragility *{n} [mech]*
moru	モル	mole (abbreviated mol) *[ch]*
moru	盛る	heap *{vb}*
moru	漏る；洩る	leak *{vb}*
moru-eki	モル液	molar solution *[ch]*
moru-shitsu'ryō	モル質量	molar mass *[p-ch]*
mōsai-kekkan	毛細血管	capillary (pl -laries) *{n} [anat]*
mōsen-ga	毛氈蛾；もうせんが	carpet moth (insect) *[i-zoo]*
mosha	模写；模寫	copy; replica; reproduction *{n}*
mosha-densō	模写電送	facsimile; FAX *{n} [comm]*
moshiki'shu	模式種	genotype *{n} [syst]*

moshiki'teki-gai'ryaku-zu	模式的概略図	model(-wise) schematic *[graph]*
moshiki-zu	模式図；模式圖	type diagram *[graph]*
mōsō'chiku	孟宗竹	bamboo: Phyllostachys pubescens *[bot]*
mō-ten	盲点	blind spot *[physio]*
moto	元；もと	base; origin; original *{n}*
moto-gata	元型	master mold *[eng]*
motomeru	求める	seek *{vb}*
moto no	元の	original *{adj}*
moto-nori	元糊；もとのり	stock paste *[mat]*
moto-zu	元図	original drawing *{n} [graph]*
motsu	持つ	hold; carry *{vb t}*
motsu	もつ	organ(s) of birds, animals, etc. *[cook]*
motsureru	縺れる	tangle; snarl *{vb i}*
motto	もっと	more *{adv} {n}*
moya	靄；もや	mist *{n} [meteor]*
moyai-zuna	舫い綱	mooring rope *[nav-arch]*
moyashi	萌し；もやし	bean sprouts *[bot] [cook]*
moyasu	燃やす	burn *{vb t}*
moyau	舫う	moor (a boat) *{vb t}*
moyō	模様；模樣	pattern *{n} [graph]*
moyō-shi	模様紙	pattern paper *[mat]*
mozō	模造	imitation *{n} [eng]*
mozō-gin	模造銀	mock silver *[met]*
mozō'hin	模造品	imitation; counterfeit *{n} [legal]*
mozō-kin	模造金	mock gold *[met]*
mozu	百舌鳥；鵙；鴃	shrike (bird) *[v-zoo]*
mozuku	水雲；もずく	Nemycystus decipiens (seaweed) *[bot]*
mozu'modoki	百舌擬；鵙擬	vireo (bird) *[v-zoo]*
mu-	無-	a-; non- *{prefix}*
mu'atsu-seikei	無圧成形	zero-pressure molding *{n} [eng]*
mu'chitsu'jo	無秩序	disorder; randomness *{n} [math]*
muchō-ongaku	無調音楽	atonal music *[acous]*
muchō-ten	無潮点	amphidromic point *[map]*
muda	無駄；むだ	waste(fulness); useless(ness) *{n} [ind-eng]*
muda-jikan	無駄時間	dead time *[eng]*
mu'denkai-mekki	無電解めっき	electroless plating *{n} [met]*
mu'denwa'sen-denwa-ki	無電話線電話機	cordless telephone *[comm]*
mu'doku'sei no	無毒性の	nontoxic *{adj} [med]*
mu'en-kon	無縁根	extraneous root *[math]*
mu'en-tan	無煙炭	anthracite coal *[geol]*
mu'fukusha no	無輻射の	nonradiant *{adj} [phys]*
mu'gaku-rui	無顎類	jawless fishes *[v-zoo]*

mugen	無限；むげん	infinity; infinitude *{n} [math]*
"	無限	unlimited(ness) *{n} [meteor]*
mugen-dai	無限大	infinity; infinite quality *{n} [math]*
mugen'dai no	無限大の	infinite *{adj} [math]*
mugen'en-shōten no	無限遠焦点の	afocal *{adj} [opt]*
mugen-kyūsū	無限級数	infinite series *[math]*
mugen no	無限の	infinite; limitless *{adj}*
mugen-shō	無限小	infinitesimal *{n} [math]*
mugen-shūgō	無限集合	infinite set *[math]*
mugi	麦；麥；むぎ	wheat (cereal) *[bot]*
muhei no	無柄の	sessile *{adj} [bot]*
mu-hōi'kaku-sen	無方位角線	agonic line *[geophys]*
mu'hōkō'sei no	無方向性	nondirectional; nonoriented; unoriented *{adj} [steel]*
mu'hyō	霧氷	rime *{n} [meteor]*
muji	無地	plain; solid-color; unfigured *{n} [tex]*
mu'jigen-kansū	無次元関数	dimensionless function *[math]*
mu'jigen-kō	無次元項	dimensionless term *[math]*
mu'jigen-ryō	無次元量	dimensionless quantity *[math]*
mujin-hyō'teki-ki	無人標的機	target drone *[aero-eng]*
mujin-ki	無人機	unmanned aircraft *[aerosp]*
muji-zome	無地染め	solid dyeing *{n} [tex]*
mu'jōken-bun	無条件文	unconditional statement *[comput]*
mu'jōken-meirei	無条件命令	imperative statement *[comput]*
mujun	矛盾；むじゅん	contradiction; conflict; incompatibility; inconsistency *{n} [logic]*
mujun-ritsu	矛盾律	contradiction, law of *[logic]*
mu'jūryō-jōtai	無重量状態	weightlessness; zero gravity *[mech]*
muka	霧化	atomization *[mech-eng]*
mukade	百足；むかで	centipede (insect) *[i-zoo]*
mukae-kaku	迎え角	attack angle *[aero-eng]*
mukago	零余子；むかご	bulfil; propagule *[bot]*
mukai-awase-rareru-oya-yubi	向かい合わせ=られる親指	opposable thumb *[anat]*
mukai-tan	無灰炭	ashless coal *[geol]*
mu'kessei-baichi	無血清培地	serum-free medium *{n} [microbio]*
mu'ketten-undō	無欠点運動	zero defects; ZD *{n} [ind-eng]*
muki	無機；むき	inorganic *{n} [ch] [mat]*
muki-eiyō	無機栄養	mineral nutrient *[bot]*
muki-eiyō-seibutsu	無機栄養生物	autotrophs; lithotrophs *[bio]*
muki-ganryō	無機顔料	inorganic pigment *[mat]*
muki-kagaku	無機化学	inorganic chemistry *[ch]*

muki-kagō'butsu	無機化合物	inorganic compound *[ch]*
muki-ka'sanka'butsu	無機過酸化物	inorganic peroxide *[ch]*
muki-kōbunshi	無機高分子	inorganic polymer *[poly-ch]*
muki-kokyū	無気呼吸	anaerobic respiration *[bio]*
mukin-shitsu	無菌室	sterile chamber *[microbio]*
muki'shitsu	無機質	inorganic matter *[ch] [geol]*
"	無機質	mineral *[geol]*
mukō	無効	invalid; invalidity *{n} [legal]*
mukō	霧虹；むこう	fogbow; bow *[meteor]*
mukō no	無効の	null; void *{adj} [legal]*
mukō-shinpan	無効審判	trial of invalidation *[legal]*
mukō-zuchi	向う槌	sledge hammer *[eng]*
mukō'zune	向こう脛	shin (of leg) *{n} [anat]*
muku	向く	turn toward; face *{vb}*
muku	剥く；むく	peel *{vb}*
muku-dori	椋鳥；むくどり	gray starling (bird) *[v-zoo]*
mukuge	木槿；槿；むくげ	rose-of-sharon *[bot]*
mukuiru	報いる	reward; recompense *{vb t}*
mu'kyoku'sei-yōbai	無極性溶媒	nonpolar solvent *[mat]*
mukyō-shitsu	無響室	anechoic room *[acous]*
muna-bire	胸鰭	pectoral fin *[v-zoo]*
mune	胸	thorax (pl thoraxes, thoraces); breast; chest *{n} [anat]*
mune'aka-hiwa	胸赤鶸	linnet (bird) *[v-zoo]*
mu'netsu-yōeki	無熱溶液	athermal solution *[ch]*
mu'on-ha'guruma	無音歯車	silent gear *[mech-eng]*
mura	村	village *[geogr]*
mura	斑；むら	unevenness; ununiformity *{n}*
murasaki	紫；むらさき	gromwell *[bot]*
"	紫	purple (color) *{n} [opt]*
murasaki-gai	ムラサキ貝	purplish Washington clam *[i-zoo]*
murasaki-i'gai	紫貽貝	mussel (shellfish) *[i-zoo]*
murasaki-seki'ei	紫石英	amethyst; violet quartz *[miner]*
murasaki-uma'goyashi	紫馬肥	alfalfa *[bot]*
muri-sū	無理数	irrational number *[math]*
muro-aji	室鰺；むろあじ	mackerel scad (fish) *[v-zoo]*
mu'saibō	無細胞；無細胞	cell-less *[bio]*
musai-shoku	無彩色	achromatic color *[opt]*
mu'saku'i	無作為；むさくい	random *{n} [math]*
mu-saku'i'ka	無作為化	randomization *[math]*
mu'sansei no	無酸性の	anoxic *{adj} [ch]*
mu'sanso-dō	無酸素銅	oxygen-free copper *[met]*

musasabi	鼯鼠; 鼯; むささび	flying squirrel (animal) *[v-zoo]*
mu'seibutsu no	無生物の	inanimate *{adj} [sci-t]*
musei-hōden	無声放電	silent discharge *[elec]*
musei-seishoku	無性生殖	asexual reproduction *[bio]*
mu'sekitsui-dōbutsu	無脊椎動物	invertebrates *[i-zoo]*
mu'sekitsui-dōbutsu'gaku	無脊椎動物学	invertebrate zoology *[bio]*
musen-denshin	無線電信	radiotelegraphy *[comm]*
musen-denshin'kyoku	無線電信局	radiotelegraph station *[comm]*
musen-denwa	無線電話	radiotelephony *[comm]*
musen-kōkō	無線航行	radio navigation *[navig]*
musen-kyoku	無線局	radio station *[comm]*
musen-shashin-densō	無線写真電送	radiophototelegraphy; radiophoto; radiophotography *[comm] [graph]*
musen-shūha	無線周波	radio frequency *[elecmg]*
mu'setten-suitchi	無接点スイッチ	contactless switch *[elec]*
mushi	虫; 蟲; むし	bug; insect; worm *[i-zoo]*
mushi'ba	虫歯; 齲歯	carious (decayed) tooth *[dent]*
mushi-bun'retsu	無糸分裂	amitosis *[cyt]*
mushi-ki	蒸し器	steamer *[cook] [eng]*
mushi-kobu	虫瘤	gall *{n} [bot]*
mushi'kui	虫食い	willow wren; chiffchaff (bird) *[v-zoo]*
mushi-megane	虫眼鏡; 虫めがね	magnifying glass *[opt]*
mushi'mono	蒸し物	steamed dishes *[cook]*
mushi-ni	蒸し煮	steam cooking *{n} [cook] [food-eng]*
mushiri-toru	毟り取る	pluck off *{vb}*
mushiro	蓆; 筵; むしろ	straw mat(ting) *{n} [mat]*
mushi'tori	虫取り	debugging (eliminating insects) *{vb}*
mushi-yaki	蒸し焼き	casserole baking *{n} [cook]*
mu'shūsa-ten	無収差点; 無收差點	stigmatic point *[opt]*
musu	蒸す	steam *{vb t}*
musubi'me	結び目	knot *{n} [tex]*
mu'sui	無水; むすい	anhydrous *{n} [ch]*
musui-ansokkō-san	無水安息香酸	benzoic anhydride
musui'butsu	無水物	anhydride *[ch]*
musui-futaru-san	無水フタル酸	phthalic anhydride
musui-keisan	無水珪酸	silicic anhydride
musui-kohaku-san	無水琥珀酸	succinic anhydride
musui-marein-san	無水マレイン酸	maleic anhydride
musui-merito-san	無水メリト酸	mellitic anhydride
musui-rakusan	無水酪酸	butyric anhydride
musui-ryūsan	無水硫酸	sulfuric anhydride
musui-sakusan	無水酢酸	acetic anhydride

musui-san	無水酸	acid anhydride *[ch]*
musui-sekkō	無水石膏	anhydrous gypsum *[miner]*
musui-shūsan	無水蓚酸	oxalic anhydride
mutan'jō no	無端状の	endless *{adj} [mech-eng]*
mu'teigi-gai'nen	無定義概念	undefined concept *[math]*
mu'teigi-yōgo	無定義用語	undefined term *[logic]*
mu'tei'i	無定位	astatic *{n} [phys]*
mu'teikei	無定形	amorphous *{n} [phys]*
mu'teikei-kōbunshi	無定形高分子	amorphous polymer *[poly-ch]*
mu'teki	霧笛	foghorn *[navig]*
mutō	霧灯；霧燈	foglight *[navig]*
mutsu	鯥；むつ	Japanese bluefish; Scombrops boops (fish) *[v-zoo]*
mu'yōzai'gata-setchaku-zai	無溶剤型接着剤	solventless adhesive *[adhes]*
myaku	脈；みゃく	vein *[geol] [min-eng]*
myaku'dō	脈動	pulsation *[elec]*
myaku'dō-sei	脈動星	pulsating star *[astron]*
myaku'haku	脈拍	pulse *{n} [physio]*
myaku'raku-maku	脈絡膜	choroid *[anat]*
myaku-seki	脈石	gangue *[geol]*
myaku-sō	脈相	venation *[bot]*
myōban	明礬	alum *[inorg-ch] [miner]*
myōban-nameshi	明礬鞣し	tawing *{n} [eng]*
myōga	茗荷	Japanese ginger; Zingiber mioga *[bot]*
myōgo-nichi	明後日	day after tomorrow *{n}*
myōji	苗字	last name *{n}*
myō-myōgo-nichi	明々後日	three days from now *{n}*
myō'nichi	明日	tomorrow *{n}*

N

na	名	name *{n} [comput]*
na	菜	greens (vegetables) *{n} [bot]*
nabe	鍋	pan; pot; saucepan *{n} [cook] [eng]*
nabe-rui (nabe-mono)	鍋類（鍋物）	pot-cooked dishes *[cook]*
Nabeshima-yaki	鍋島焼	Nabeshima porcelain *[cer]*
nada	灘	ocean with high waves *{n} [ocean]*
nadare	雪崩；なだれ	avalanche *{n} [hyd]*
nadeshiko	撫子；なでしこ	pink; wild pink *{n} [bot]*
nae	苗	seedling; sapling *[bot]*
nae'doko	苗床	seedbed *[agr]*
na'fuda	名札	label *{n} [comput]*
nagai	長い	long *{adj} [math]*
naga-imo	長芋	Chinese yam *[bot]*
naga'kame-mushi	長椿象；長亀虫	milkweed bug (insect) *[i-zoo]*
nagare	流れ；ながれ	flow *{n} [comput] [fl-mech]*
"	流れ	current; stream *{n} [hyd]*
nagare'gi	流れ木	driftwood *{n}*
nagare-kaiseki	流れ解析	flow analysis *[comput]*
nagare-mizu	流れ水	runoff *{n} [hyd]*
nagare no muki	流れの向き	flow direction *[eng]*
nagare-sen	流れ線	flowline *[eng]*
nagare-tokusei	流れ特性	flow characteristics *[mech]*
nagare-zu	流れ図	flow chart; flow diagram *[comput] [eng]*
naga-sa	長さ	length *{n} [math]*
nagasa-hōkō-jika	長さ方向磁化	longitudinal magnetization *[eng-acous]*
nagasa no tan'i	長さの単位	unit of length *[math]*
nagashi	流し	sink *{n} [arch] [cook]*
nagashi-ami	流し網	drift net *[eng]*
nagashi-ita	流し板	drainboard *[cook]*
nagashi-nuri	流し塗り	flow coating *{n} [eng]*
nagasu	流す	drain; flush; let flow *{vb t}*
naga'te-hōkō	長手方向	lengthwise direction; longitudinal direction; longish direction *[eng]*
nage'komi-dennetsu-ki	投込電熱器	immersion heater *[eng]*
nagi	凪；なぎ	calms *{n} [meteor]*
nagisa	渚	beach; waterside *{n} [geol]*
Nagoya	名古屋	Nagoya *[geogr]*
nagu	凪ぐ	becalm; calm down *{vb} [meteor]*
nagu	薙ぐ	mow; cut down *{vb t}*

nai-atsu	内圧；内壓	internal pressure *[phys]*
nai'bu	内部；ないぶ	interior part; the inside *{n}*
nai(bu)-hyōjun-hō	内(部)標準法	internal standard *[an-ch] [spect]*
naibu-kakusan	内部拡散	internal diffusion *[ch-eng]*
naibu-kio'ku	内部記憶	internal memory *[comput]*
naibu-kisei'chū	内部寄生虫	endoparasite *[ecol]*
naibu-kisei'sha	内部寄生者	endoparasite *[ecol]*
nai'bun'pitsu-kei	内分泌系	endocrine system *[physio]*
nai'bun'pitsu-sen	内分泌腺	endocrine gland *[physio]*
naibu-shiji'yaku	内部指示薬	internal indicator *[ch]*
naibu-teikō	内部抵抗	internal resistance *[elec]*
naibu-te'tsuzuki	内部手続	internal procedure *[comput]*
nai'fuku-yaku	内服薬	medicine for internal use *[med]*
naigō	内合	inferior conjunction *[astron]*
naiha	内破	implosion *[phys]*
nai'haiyō	内胚葉	endoderm; endoblast *[embryo]*
naihō-kōbutsu	内包鉱物；内包鑛物	endomorph *[miner]*
nai-hyōjun	内標準	internal standard *[spect]*
nai'kahi	内果皮	endocarp *[bot]*
naikai	内海	inland sea *[ocean]*
naika-i	内科医	internal medicine specialist *[med]*
nai-kaku	内角	interior angle; internal angle *[math]*
nai-kaku	内核	inner core *[geol]*
naikei	内径；内徑	inside diameter *[math]*
naikō	内腔	lumen *[sci-t]*
naikō-dōbutsu	内肛動物	Entoprocta; Endoprocta *[i-zoo]*
nai'kōso	内酵素；ないこうそ	endoenzyme *[bioch]*
nai'nen-kikan	内燃機関	internal combustion engine *[mech-eng]*
nai'riku	内陸	inland *{n} [geogr]*
nai'riku-koku	内陸国；内陸國	landlocked country *[geogr]*
nai-ryoku	内力	internal force *[mech]*
naisei-dōbutsu	内生動物	infauna *[zoo]*
naisei-hōshi	内生胞子	endospore *[bio]*
naisen	内線	extension (telephone) *{n} [comm]*
nai'setsu	内接	inscription (circle) *[math]*
naisetsu-en	内接円	incircle *{n} [math]*
naisetsu suru	内接する	inscribed; inscribing *{adj} [math]*
naisetsu-takak(u)kei	内接多角形	inscribed polygon *[math]*
nai'shin	内心	incenter *{n} [math]*
nai'shintō	内浸透	endosmosis (pl -moses) *[physio]*
naisō	内相	inside appearance *{n}*
naisō	内挿	interpolation *[math]*

naisō	内装	trim; interior furnishing *{n} [furn]*
naisō	内層	inner layer *{n} [geol]*
naiten suru	内添する	internally add *{vb}*
naitō	内筒	inner cylinder; inner tube *[eng]*
nai-wakusei	内惑星	inferior planet; inner planet *[astron]*
naiyō	内容	content(s); substance *{n}*
najimu	馴染む	become attuned; attune *{vb i}*
naka	中	inside *{n} [math]*
"	中	interior *{n}*
naka'dane	中種；なかだね	sponge (bread) *{n} [cook]*
naka'go	中子	tang *{n} [eng]*
naka-goro	中頃	about center; middle, more or less; midway, more or less *{n} [math]*
naka-guro	中黒；中黑	black-filled period *[pr]*
naka-niwa	中庭	atrium; courtyard *[arch]*
naka-shin	中芯	corrugating medium *[paper]*
naka-toji	中綴じ	saddle stitching (bookmaking) *{n} [pr]*
naka'yasumi-dan	中休み段	landing (stairway) *{n} [arch]*
naka-yubi	中指	middle finger *[anat]*
naku koto	泣くこと	crying *{n} [physio]*
nama	生；なま	raw; uncooked *{n} [cook]*
nama-dēta	生データ	raw data *[comput]*
na'mae	名前	name *{n} [comput]*
nama-ge'sui	生下水	raw sewage *[civ-eng]*
nama-gomu	生ゴム	crude rubber; raw rubber *[org-ch]*
nama-ito	なま糸	gray yarn *[tex]*
nama-jōki	生蒸気	live steam *[phys]*
nama-kawa	生皮；なまかわ	rawhide *{n} [mat]*
namake'mono	樹懶；なまけもの	sloth (animal) *[v-zoo]*
namako	海鼠；生子；なまこ	sea slug; sea cucumber; trepang *[i-zoo]*
nama-kōbo	生酵母	live yeast *[mycol]*
namako-ita	海鼠板；ナマコ板	corrugated iron sheet *[met]*
nama-konkurīto	生コンクリート	unhardened concrete *[mat]*
nama no	生の	raw; rare; uncooked *{adj} [cook]*
namari	鉛；なまり	lead (element)
namari-san	鉛酸	plumbic acid
namari'san-en	鉛酸塩	plumbate *[ch]*
namari'tetsu'myōban	鉛鉄明礬	plumbojarosite *[miner]*
nama-shīto	生シート	raw sheet *[mat]*
namasu	膾；なます	Japanese fish salad *[cook]*
nama-zaimoku	生材木	green lumber *{n} [bot]*
namazu	鯰；なまず	catfish; Parasilurus asotus *[v-zoo]*

name'ko	滑子；なめこ	Pholiota nameko (mushroom) *[bot]*
namekuji	蛞蝓；なめくじ	slug *{n} [i-zoo]*
nameraka	滑らか	smooth *{n} [mech]*
nameraka na	滑らかな	smooth *{adj}*
nameraka-sa	滑らかさ	lubricity; smoothness; slickness *[mat]*
nameshi-gawa	鞣し皮	leather *[mat]*
nameshi(-hō)	鞣し(法)	tanning *{n} [eng]*
nami	波	wave *{n} [fl-mech] [ocean] [phys]*
namida	涙；泪	tear fluid *{n} [physio]*
nami-gumo	波雲	billow cloud *[meteor]*
nami'hin	並品	average-quality article *[ind-eng]*
nami-kubo	波窪；なみくぼ	wave trough *[phys]*
nami-kyū	並級	ordinary grade *[ind-eng]*
nami no mine	波の峰	wave crest *[phys]*
nami no soko'biki	波の底引き	undertow *[ocean]*
nami'sū	並数	mode *{n} [math]*
naname	斜；ななめ	diagonal; oblique *{n} [math]*
naname ni suru	斜めにする	slant *{vb t}*
naname no	斜の	diagonal; oblique *{adj} [math]*
nan'boku	南北	south and north *{n} [geod]*
nanchū	南中	culmination *[astron]*
nando	納戸；なんど	closet *{n} [arch]*
nani	何；なに	what *{pron}*
nan'ji	何時；なんじ	what time *{n}*
nanji-sei	軟磁性	soft magnetic property *[phys]*
nanka	軟化	softening *{n} [phys]*
nanka-ten	軟化点	softening point *[phys]*
nanka-zai	軟化剤	softener; softening agent *[mat]*
nan-kinzoku	軟金属	soft metal *[met]*
nankō	軟膏	ointment; unguent *{n} [pharm]*
nankō	軟鋼	mild steel; soft steel *[met]*
nan'kotsu	軟骨	cartilage *[hist]*
nankotsu'gyo	軟骨魚	cartilaginous fish; selachian *[v-zoo]*
nankotsu'gyo-rui	軟骨魚類	Chondrichthyes *[v-zoo]*
nan'kyoku	南極；なんきょく	South Pole *[geod]*
nankyoku-kai	南極海	antarctic ocean *[geogr]*
nankyoku-ken	南極圏	antarctic circle *[geod]*
nankyoku-kō	南極光	aurora australis *[geophys]*
nan'kyoku no	南極の	antarctic *{adj}*
nannen-kakō	難燃加工	flame retardant finish *[mat]*
nannen-kakō-zai	難燃加工剤	flame retardancy agent *[ch]*
nannen'ka suru	難燃化する	render fire-resistant *{vb} [civ-eng]*

nannen-sei	難燃性	flame resistance *[ch]*
nannen-zai	難燃剤	fire-retarding material *[mat]*
na-no-hana	菜の花	rape blossom *[bot]*
na'nori	名乗り; 名乘り	name; full name (of a person) *{n}*
nanpu'byō	軟腐病	soft rot (plant disease) *[bot]*
nan'shiki-kogata-hikō'sen	軟式小型飛行船	blimp *[aero-eng]*
nan'shitsu-gomu	軟質ゴム	soft rubber *[rub]*
nan'shitsu-tan	軟質炭	soft coal *[geol]*
nansui	軟水	soft water *[ch]*
nantai-dōbutsu	軟体動物	mollusks; Molluska *[i-zoo]*
nanten	南天	nandin *[bot]*
Nan'yō	南洋	South Seas *[geogr]*
Nan'yō-sugi	南洋杉	araucaria (tree) *[bot]*
naoru	治る; なおる	heal *{vb i}* *[med]*
naosu	直す; なおす	fix; correct; repair; mend *{vb t}* *[eng]*
Nara	奈良; なら	Nara *[geogr]*
nara	楢; なら	Japanese oak (tree) *[bot]*
naraberareta	並べられた	ordered *{adj}* *[math]*
naraberu	並べる	juxtapose; put side by side *{vb}*
narabu	並ぶ	line up; be in a row; rank with *{vb}*
narai	倣い; ならい	profiling *{n}* *[eng]*
narasu	鳴らす	ring; sound *{vb t}* *[acous]*
narau	習う	learn *{vb}* *[psy]*
nare	慣れ	habituation *[bio]*
naru	鳴る	ring *{vb i}* *[acous]*
naruko-yuri	鳴子百合	sealwort; Polygonum falcatum (flower) *[bot]*
nasen; nassen	捺染	textile printing *{n}* *[graph]*
nasen-nori	捺染糊	printing paste *[tex]*
nashi	梨; なし	pear (fruit, tree) *[bot]*
nashi'ji	梨地	aventurine; matte *{n}* *[mat]*
nashi'ji-shi'age	梨地仕上	satin finish *[mat]*
nashi'jō no	梨状の	pear-shaped *{adj}* *[math]*
nasu	茄子; なす	eggplant (vegetable) *[bot]*
na'tane-abura	菜種油	rapeseed oil; canola oil *[mat]*
natoryūmu	ナトリウム	sodium (element)
natoryūmu-hōsuika'butsu	ナトリウム硼水化物	sodium borohydride
natsu	夏; なつ	summer *{n}* *[astron]*
natsu'in	捺印	sealing; imprinting *{n}* *[graph]*
natsu-jikan	夏時間	daylight saving time *[astron]*
natsume	棗; なつめ	Chinese date tree; jujube tree *[bot]*
"	棗	tea powder container *[furn]*
natsume-yashi	棗椰子	date palm (tree) *[bot]*

natsu-shiro'giku	夏白菊	feverfew; Chrysanthemum parthenium *[bot]*
nattō	納豆；なっとう	fermented soybeans *[cook]*
nawa	縄；繩；なわ	rope *{n} [mat]*
nawa'bari-kōdō	縄張(り)行動	territoriality; territorial behavior *[zoo]*
nawa-bashigo	縄梯子	rope ladder *[eng]*
naze	何故；なぜ	why *{adv}*
nazoru	擦る；なぞる	trace; follow *{vb}*
nazuna	薺；なずな	shepherd's purse *[bot]*
ne	根	root *{n} [bot]*
nebari-zuyo-sa	粘り強さ	tack; tackiness; tenacity *{n} [mech]*
"	粘り強さ	persistence *[psy]*
neda	根太；ねだ	joist *[arch]*
nedan	値段	price; rate *{n} [econ]*
ne-dana	寝棚；寢棚	bunk *{n} [arch] [furn]*
negi	葱；ねぎ	Welsh (Spanish) onion *[bot]*
Negoro-nuri	根来塗	red and black lacquered ware *[furn]*
ne'iro	音色；ねいろ	timbre; tone color *[acous]*
neji	ねじ；螺子；捻子；捩子；螺止	screw *{n} [eng]*
neji-ha'guruma	螺子歯車	screw gear; crossed helical gear *[mech-eng]*
neji-mawashi	ねじ回し	screw driver *[eng]*
nejire	捩れ；ねじれ	torsion; tortuosity; twist *{n} [mech]*
nejire-kaku	捩れ角	angle of twist *[mech]*
nejire-kiri	捩れ錐	twist drill *[eng]*
nejire-purizumu	捩れプリズム	antiprism *[math]*
nejireru	捩れる	be twisted; twist *{vb i} [eng]*
nejiri	捩り；ねじり	torsion *[mech]*
nejiri-bō	捩り棒	torsion bar *[mech-eng]*
nejiri-gōsei	捩り剛性	torsional rigidity *[mech]*
nejiri-ito	捩り糸	twisted thread *[tex]*
nejiri-kowa-sa	捩り剛さ	torsional rigidity *[mech]*
nejiri-shindō	捩り振動	torsional oscillation; torsional vibration *[mech] [phys]*
nejiri-tsuyo-sa	捩り強さ	torsional strength *[mech]*
nejiru	捩じる	twist; screw; wrench *{vb}*
neji'tate	ねじ立て	tapping *{n} [mech-eng]*
neji-yama	ねじ山	screw thread *{n} [eng]*
neji'zuno-reiyō	捩じ角羚羊	koodoo; kudu (animal) *[v-zoo]*
nekaseru	寝かせる	lay on side; send to sleep *{vb t}*
nekashi	寝かし；ねかし	maturation *[resin] [rub]*
nekiri'mushi	根切り虫	cutworm (insect) *[i-zoo]*

nekkan-atsu'en	熱間圧延	hot rolling *{n} [met]*
nekkan-kakō	熱間加工	hot working *{n} [met]*
nekkan-kōka-setchaku-zai	熱間硬化接着剤	hot-setting adhesive *[adhes]*
nekkan-mage-kakō	熱間曲げ加工	hot bending *{n} [met]*
nekkan-zeisei	熱間脆性	hot shortness *[met]*
nekken; netsu-ken	熱圏; 熱圏	thermosphere *[meteor]*
neko	猫	cat (animal) *[v-zoo]*
neko(no)me-ishi	猫眼石	cat's eye *[miner]*
neko-zame	猫鮫; ねこざめ	horn(ed) shark; Port Jackson shark; Heterodontus japonicus *[v-zoo]*
ne'maki	寝巻き; 寝間着	nightclothes *[cl]*
nemu'ke	眠気; ねむけ	sleepiness *{n} [physio]*
nemuri	眠り	sleep *{n} [physio]*
nemuru	眠る; 睡る	sleep; slumber *{vb i} [physio]*
nen; toshi	年	year *[astron]*
-nen	-年	-year(s) *{suffix}*
nenban'gan	粘板岩	slate *{n} [petr]*
nenchak'ka-sei	粘着化性	tackifying property *[mech]*
nenchaku-bōshi-zai	粘着防止剤	abherent *[mat]*
nenchaku-fuyo-zai	粘着付与剤	tackifier *[mat]*
nenchaku-ryoku	粘着力	adhesive strength *[eng]*
nenchaku-sei	粘着性	stickiness; tackiness; pressure-sensitive adhesiveness *{n} [mat]*
nenchaku'sei-busshitsu	粘着性物質	adhesive(s) *[mat]*
nenchaku'sei-fuyo-zai	粘着性付与剤	tackifier *[mat]*
nenchaku-tēpu	粘着テープ	adhesive tape; sticky tape *[mat]*
nenchaku-zai	粘着剤	pressure-sensitive adhesive *[adhes]*
nendai'gaku	年代学	chronology (pl -gies) *[sci-t]*
nendai-sokutei	年代測定	dating; estimation of date *[archeo]*
nendan-sei	粘弾性	viscoelasticity (pl -ties) *[mat]*
nendo	粘土	clay *[geol]*
nendo	粘度; ねんど	viscosity (pl -ties) *[fl-mech]*
nendo-fuyo-zai	粘度付与剤	thickener *[mat]*
nendo-heikin-bunshi'ryō	粘度平均分子量	viscosity-average molecular weight *[ch]*
nendo-kai	粘土塊	clay lump *[geol]*
nendo-kei	粘度計	viscometer *[eng]*
nendo-shisū	粘度指数	viscosity index *[ch-eng]*
nendo-sū	粘度数	viscosity number *[ch-eng] [poly-ch]*
nen'eki	粘液	mucus *[physio]*
nen'eki-kin	粘液菌	slime bacteria *[microbio]*
nen'eki-san	粘液酸	mucic acid
nen'eki-saikin-moku	粘液細菌目	Myxobacterales *[microbio]*

nen'eki'sei no	粘液性の	mucous *{adj} [physio]*
nen'eki-sō	粘液層	slime layer *[microbio]*
nen-fuka-kyoku'sen	年負荷曲線	yearly load curve *[ind-eng]*
nen-fuka-ritsu	年負荷率	yearly load factor *[ind-eng]*
nengappi	年月日	date: year, month, day *{n}*
nengō	年号	year name; name of era *{n}*
nenkai	捩回	twisting *{n} [eng]*
nenkan	年鑑	yearbook; almanac *[pr]*
nen'ketsu-sei	粘結性	coking property *[ch-eng]*
nenketsu'sei-hoten-zai	粘結性補填剤	binding filler *[mat]*
nenketsu-tan	粘結炭	caking coal *[geol]*
nenketsu-zai	粘結剤	binder *[mat]*
nenmaku	粘膜	mucous membrane *[hist]*
nenpō	年報	annual report *[econ] [pr]*
nenrei	年齢	age (in years) *{n} [bio]*
nenrin-nendai'gaku	年輪年代学	dendrochronology *[geol]*
nenryō	燃料；ねんりょう	fuel *{n} [mat]*
nenryō-denchi	燃料電池	fuel cell *[elec]*
nenryō-kei	燃料計	fuel gage *[eng]*
nenryō-yu	燃料油	fuel oil *[mat]*
nensei	粘性	viscosity (pl -ties) *[fl-mech]*
nensei-gensui-keisū	粘性減衰係数	viscous damping coefficient *[mech]*
nensei-hakai	粘性破壊	viscous fracture *[fl-mech]*
nensei-keisū	粘性係数	coefficient of viscosity *[fl-mech]*
nensei-ritsu	粘性率	coefficient of viscosity *[fl-mech]*
nensei-ryūdō	粘性流動	viscous flow *[fl-mech]*
nensei-ryūtai	粘性流体	viscous fluid *[fl-mech]*
nensei-teikō	粘性抵抗	viscous drag *[fl-mech]*
nen'shikai-kankō'butsu	年四回刊行物	quarterly publication *[pr]*
nenshi-ki	撚糸機	yarn twisting machine *[eng]*
nen'shitsu	粘質	mucilage *{n} [mat]*
nen'shitsu no	粘質の	viscid *{adj} [bot]*
nenshō	燃焼；燃燒	combustion *[ch]*
"	燃焼；ねんしょう	burning *{n} [eng]*
nenshō-kanryō	燃焼完了	burnout *[aero-eng]*
nenshō-netsu	燃焼熱	heat of combustion *[p-ch]*
nenshō-sei	燃焼性	flammability *[mat]*
nenshō-shitsu	燃焼室	combustion chamber *[mech-eng]*
nenshō-shūryō	燃焼終了	burnout *[aero-eng]*
nenso	粘素	mucin *[bioch]*
nensō	年層	varve *[geol]*
nenso-sei	粘塑性	viscoplasticity (pl -ties) *[mech]*

neojimu	ネオジム	neodymium (element)
neon	ネオン	neon (element)
neppa	熱波	heat wave *[meteor]*
neputsunyūmu	ネプツニウム	neptunium (element)
neri	煉り; 練り	milling *{n} [mech-eng] [pr] [rub]*
"	錬り; 鍊り	tempering *{n} [met]*
neri-ginu	練り絹	glossed silk *[tex]*
neri-ha'migaki	練歯磨	toothpaste; dentifrice *{n} [pharm]*
neri-ito	練り糸	scoured yarn *[tex]*
neru	寝る; 寢る	sleep; slumber *{vb i} [physio]*
neru	練る	knead *{vb}*
nessen	熱線	heat rays *[thermo]*
nesshin-mekki	熱浸鍍金	hot-dip plating *{n} [met]*
nessui	熱水	hot water *[ch]*
nessui-kayō-sei	熱水可溶性	hot-water solubility *[p-ch]*
nessui-kōshō	熱水鉱床	hydrothermal deposit *[p-ch]*
netsu	熱; ねつ	fever *{n} [med]*
"	熱	heat *{n} [thermo]*
netsu-antei-do	熱安定度	thermal stability *[phys]*
netsu-antei-sei	熱安定性	thermal stability *[phys]*
netsu'bai	熱媒	heat medium *[phys]*
netsu-bōchō	熱膨張	thermal expansion *[phys]*
netsu-bōchō-keisū	熱膨張係数	coefficient of thermal expansion *[phys]*
netsu-bōchō-ritsu	熱膨張率	coefficient of thermal expansion *[phys]*
netsu-bunkai	熱分解	thermal decomposition; thermal cracking; pyrolysis *[ch]*
netsu-bunseki	熱分析	thermal analysis *[an-ch]*
netsu-chūsei'shi	熱中性子	thermal neutron *[nucleo]*
netsu-dendō-do	熱伝導度	thermal conductivity *[thermo]*
netsu-dendō-ritsu	熱伝導率	coefficient of thermal conductivity; heat conduction coefficient *[thermo]*
netsu'den-keiden-ki	熱電継電器	thermorelay *[eng]*
netsu'den-kenryū-kei	熱電検流計	thermogalvanometer *[eng]*
netsu-denki	熱電気; 熱電氣	thermoelectricity *[phys]*
netsu'den no	熱電の	thermoelectric *{adj} [phys]*
netsu'den-retsu	熱電列	thermoelectric series *[met]*
netsu-denshi	熱電子	thermion; thermoelectron *[electr]*
netsu-den'tatsu-keisū	熱伝達係数	heat-transfer coefficient *[thermo]*
netsu-den'tsui	熱電対	thermocouple; thermoelement *[electr]*
netsu-den'tsui-retsu	熱電対列	thermopile *[eng]*
netsu'en-junkan	熱塩循環	thermohaline circulation *[ocean]*
netsu-ganryō	熱含量	heat content *[thermo]*

netsu-hassei	熱発生	thermogenesis *[thermo]*
netsu-hatsu'den-soshi	熱発電素子	thermoelectric generating element; thermal converter *[electr]*
netsu-heikō	熱平衡	thermal equilibrium *[thermo]*
netsu-henkei	熱変形; 熱變形	heat distortion *[phys]*
netsu-hensei	熱変性	heat denaturation *[ch]*
netsu'hensei-eki'shō	熱変性液晶	thermotropic liquid crystal *[p-ch]*
netsu-hirō	熱疲労	heat fatigue *[mech]*
netsu-hizumi	熱歪; ねつひずみ	thermal strain *[mech]*
netsu-hōsha	熱放射	thermal emission *[therm]*
netsu-jūgō	熱重合	thermal polymerization *[poly-ch]*
netsu'jūgō-kinshi-zai	熱重合禁止剤	thermopolymerization inhibitor *[mat]*
netsu-ka	熱価; 熱價	heat value; heat rate *[mech-eng]*
netsu-kagaku	熱化学	thermochemistry *[p-ch]*
netsu'kagaku-hōtei'shiki	熱化学方程式	thermochemical equation *[p-ch]*
netsu'kaku-hannō	熱核反応	thermonuclear reaction *[nuc-phys]*
netsu-kakusan-ritsu	熱拡散率	thermal diffusivity *[thermo]*
netsu-karyū	熱加硫	hot-vulcanization *[rub]*
netsu-kaso'ka	熱可塑化	heat plasticization *[eng]*
netsu-kaso-sei	熱可塑性	thermal plasticity *[poly-ch]*
netsu'kaso'sei-jushi	熱可塑性樹脂	thermoplastic resin *[mat]*
netsu'kaso'sei no	熱可塑性の	thermoplastic *{adj}* *[poly-ch]*
ne'tsuke	根付	ornamental carving (antique) *{n}*
netsu'ke	熱気	feverishness *{n}* *[med]*
netsu-kikan	熱機関	heat engine *[mech-eng]* *[poly-ch]*
netsu-kōkan-ki	熱交換器	heat exchanger *[eng]*
netsu'kōka-sei	熱硬化性	thermosetting property *[mat]*
netsu'kōka'sei-jushi	熱硬化性樹脂	thermosetting resin *[mat]*
netsu-kō'ritsu	熱効率	thermal efficiency *[ch-eng]*
netsu no-	熱の-	pyro- *{prefix}* *[thermo]*
netsu-ō'hen	熱黄変	heat yellowing *{n}* *[opt]*
netsu-ō'ryoku	熱応力; 熱應力	thermal stress *[mech]*
netsu-rekka	熱劣化	thermal degradation *[ch]*
netsu-riki'gaku	熱力学	thermodynamics *[phys]*
netsu-rikigaku-dai'ichi-hōsoku	熱力学第一法則	thermodynamics, first law of *[thermo]*
netsu-rikigaku-dai'ni-hōsoku	熱力学第二法則	thermodynamics, second law of *[thermo]*
netsu-rikigaku-dai'san-hōsoku	熱力学第三法則	thermodynamics, third law of *[thermo]*
netsu'rikigaku'teki-kōritsu	熱力学的効率	thermodynamic efficiency *[thermo]*
netsu-ri'reki	熱履歴; 熱履歴	heat history *[phys]*

netsu-rōka	熱老化	thermal aging *{n} [ch]*
netsu-ryō	熱量；ねつりょう	quantity of heat *[thermo]*
netsu'ryō-kei	熱量計	calorimeter *[eng]*
netsu-ryū	熱流	heat flow *{n} [thermo]*
netsu-seikei	熱成形	thermoforming *{n} [poly-ch]*
netsu-seki'un	熱積雲	thermal cumulus *[meteor]*
netsu-setto-sei	熱セット性	heat-setting property *[tex]*
netsu-shūshi	熱収支；熱收支	heat balance *[thermo]*
netsu-son'shitsu	熱損失	heat loss *[phys]*
netsu-tairyū	熱対流；熱對流	thermal convection *[meteor]*
netsu-tan'i	熱単位	thermal unit *[phys]*
netsu-teichaku	熱定着	heat fixing *{n} [eng]*
netsu-tenbin	熱天秤	thermobalance *[eng]*
netsu-tōryō	熱当量；熱當量	thermal equivalent *[thermo]*
netsu-yōryō	熱容量	heat capacity; thermal capacity *[thermo]*
netsu-yōyū'gata-setchakuzai	熱溶融型接着剤	hot-melt adhesive *[adhes] [poly-ch]*
netsu-zatsu'on	熱雑音；熱雜音	thermal noise *[electr]*
nettai	熱帯；熱帶	torrid zone; tropical zone *[climat]*
nettai-bōfū	熱帯暴風	tropical storm *[meteor]*
nettai-chihō	熱帯地方	tropics *[climat]*
nettai no	熱帯の	tropical *{adj} [climat]*
nettai-shima-ka	熱帯縞蚊；ねったいしまか	yellow fever mosquito; Aedes aegypti (insect) *[i-zoo]*
nettai-ta'u-rin	熱帯多雨林	tropical rainforest *[ecol]*
nettai-tei'ki'atsu	熱帯低気圧	tropical depression *[meteor]*
nettai-yu	熱帯油	tropical oil *[mat]*
nettō	熱湯；ねっとう	boiling (or hot) water *[ch]*
ne'uchi	値打ち	value *{n} [sci-t]*
nezumi	鼠；ねずみ	mouse (pl mice) (animal) *[v-zoo]*
nezumi-ka no	鼠科の	murine *{adj} [v-zoo]*
nezumi-iruka	鼠海豚	porpoise (mammal) *[v-zoo]*
nezumi-tori	鼠取り	mousetrap *[eng]*
nezumi-zame	鼠鮫；ねずみざめ	porbeagle; mackerel shark (fish) *[v-zoo]*
ni	二；弐	two *{n} [math]*
ni'bai	二倍；にばい	double *{n} [math]*
nibai-kaku no kō'shiki	二倍角の公式	double-angle formula *[math]*
nibai no	二倍の	double *{adj} [math]*
nibai-sei	二倍性	diploidy *[gen]*
nibai-seido	二倍精度	double precision *[comput]*
nibai-sū	二倍数；二倍數	diploid number *[gen]*
nibai'tai	二倍体；二倍體	diploid *{n} [gen]*
nibe	鰾膠；にべ	isinglass; fish glue *[mat]*

ni-bitashi	煮浸し	stewed fish or vegetables *[cook]*
ni-boshi	煮干し	dried small sardines *[cook]*
nibui	鈍い；にぶい	dull *{adj}*
nibun-hō	二分法	dichotomy (pl -mies) *[bio] [comput]*
"	二分法	bisecting method *[math]*
nibun-keisen	二分経線	equinoctial colure *[astron]*
nibun no ichi	二分の一	one-half; half *{n} [math]*
ni-bun'retsu	二分裂	binary fission *[bio]*
ni'bunshi-hannō	二分子反応	bimolecular reaction *[ch]*
nibun-tansaku	二分探索	dichotomizing search *[comput]*
nichi'botsu	日没；にちぼつ	sunset *[astron]*
nichi'getsu-saisa	日月歳差	lunisolar precessions *[astrophys]*
nichi-heikin-ki'on	日平均気温	daily mean temperature *[meteor]*
nichi-henka	日変化	diurnal variation *[astron]*
nichi'men-zahyō	日面座標	heliographic coordinates *[astron]*
nichi'nichi'sō	日日草；日々草	periwinkle (groundcover) *[bot]*
nichi'yōbi	日曜日	Sunday
nidan-gumi	二段組み	double-column setting *{n} [pr]*
ni'eki'sei-setchaku-zai	二液性接着剤	two-component adhesive *[adhes]*
ni'enka-butsu	二塩化物	dichloride *[ch]*
ni'enka-iō	二塩化硫黄	sulfur dichloride
ni'enki-san	二塩基酸	dibasic acid *[ch]*
ni-fuda	荷札	tag; luggage tag *{n} [trans]*
niga'mi	苦味	bitter flavor; bitterness *[physio]*
nigami-yaku; nigami-gusuri	苦味薬	bitters *{n} [pharm]*
nigami-zake	苦味酒	bitters *[cook]*
nigari	苦汁；にがり	bittern *[ch-eng]*
niga-sa	苦さ	bitterness *[cook] [physio]*
nigashi-ben	逃し弁；逃し瓣	relief valve *[mech-eng]*
ni'gatsu	二月	February (month)
niga'yomogi	苦艾；にがよもぎ	wormwood; Artemesia absinthium *[bot]*
nige'men-mamō	逃げ面摩耗	flank wear *{n} [mech-eng]*
nigen-gōkin	二元合金	binary alloy *[met]*
nigen-jōtai	二元状態	binary condition *[sci-t]*
nigen-keiden-ki	二元継電器	two-element relay *[elec]*
nigiri	握り	knob *[eng]*
nigiri-zushi	握り鮨；にぎりずし	hand-shaped sushi (q.v.) *[cook]*
nigiru	握る	grasp *{vb t}*
nigi'teki ni	二義的に	secondarily *{adv}*
ni'goi	似鯉；にごい	Hemibarbus barbus (a carp type) (fish) *[v-zoo]*
nigori-do	濁り度	turbidity *[an-ch]*

ni'goshin(-hō)-fugō	二-五進(法)符号	biquinary code *[math]*
Nihon (see also Nippon)	日本; にほん	Japan *[geogr]*
Nihon'go	日本語	Japanese (the language) *[ling]*
Nihon-jika	日本鹿	Japanese deer; sika (animal) *[v-zoo]*
Nihon-kai	日本海	Sea of Japan *[geogr]*
Nihon-ma	日本間	Japanese-style room *[arch]*
Nihon-nōen	日本脳炎	Japanese encephalitis *[med]*
nihon'sa-mattan	二本鎖末端	blunt end; flush end *[bioch]*
Nihon-san	日本酸	Japanic acid *[ch]*
Nihon'shu	日本酒	sake; Japanese wine *[cook]*
Nihon-tei'en	日本庭園	Japanese garden *[bot]*
Nihon-zaru	日本猿	Japanese monkey; Macaca fuscata *[v-zoo]*
niji	虹; にじ	rainbow *[opt]*
ni'ji	二次	quadratic; secondary; second-order *{adj} {n} [math]*
niji-denchi	二次電池	secondary cell *[elec]*
niji-gen	二次元	two dimensions *[sci-t]*
nijigen-hen'kaku-hansha	二次元変角反射	two-dimensional deformation reflection *[phys]*
niji-gen no	二次元の	two-dimensional *{adj} [sci-t]*
niji-hannō	二次反応	second-order reaction *[ch]*
niji-hōtei'shiki	二次方程式	quadratic equation *[math]*
niji-ion-shitsu'ryō-bunseki	二次イオン質量分析	secondary ion mass spectroscopy *[phys]*
niji-iro no	虹色の	iridescent *{adj} [opt]*
niji-kairo	二次回路	secondary circuit *[elec]*
niji-kansū	二次関数	quadratic function *[math]*
niji-keikaku-hō	二次計画法	quadratic programming *[comput]*
niji-kei'shiki	二次形式	quadratic form *[math]*
ni'jiku-enshin	二軸延伸	biaxial stretching *{n} [mech] [poly-ch]*
ni'jiku-haikō	二軸配向	biaxial orientation *[poly-ch]*
ni'jiku-ō'ryoku	二軸応力	biaxial stress *[mech]*
nijiku'sei-kesshō	二軸性結晶	biaxial crystal *[crys]*
niji-kyoku'men	二次曲面	quadratic surface *[math]*
niji-maki'sen	二次巻線	secondary winding *{n} [elecmg]*
niji-masu	虹鱒; にじます	rainbow trout (fish) *[v-zoo]*
niji-medaka	虹目高	guppy (pl guppies) (fish) *[v-zoo]*
nijimi	滲み; にじみ	bleeding *{n} [tex]*
nijimi-dashi	滲み出し	sweating-out *{n} [plas]*
ni'ji no	二次の	two-dimensional *{adj} [math]*
nijisseiki; nijusseiki	二十世紀	nijisseiki pear (fruit) *[bot]*
"	二十世紀	twentieth century *{n}*
niji-sukēru	二次スケール	secondary scale *[met]*

niji-taisha-sanbutsu	二次代謝産物	secondary metabolite *[bioch]*
niji'teki-	二次的-	meta- *{prefix}*
niji-ten'i-ten	二次転移点	second-order transition point *[p-ch]* *[thermo]*
ni'jō	二乗; 二乘	square *{n}* *[math]*
nijō-kon	二乗根	square root *[math]*
nijū	二十	twenty *{n}* *[math]*
nijū-densō	二重伝送	duplex transmission *[comm]*
nijū-isseiki	二十一世紀	twenty-first century *{n}*
nijū-ketsu'gō	二重結合	double bond *[p-ch]*
nijū-kō	二重項	doublet *[quant-mech]*
nijū-kōdo	二重光度	dual brightness *[opt]*
nijū'men'tai	二十面体	icosahedron *[math]*
nijū-mōken-hō	二重盲検法	double-blind test method *[ind-eng]*
nijū no	二重の	dual; double *{adj}*
nijū-rokō	二重露光	double exposure *[photo]*
nijū-sei	二重星	binary star *[astron]*
nijū-seki'bun	二重積分	double integral *[math]*
nijūshi'men'tai	二十四面体	icositetrahedron; trisoctahedron *[crys]* *[math]*
nijū'sō	二重奏	duet *[music]*
nijū-sō	二重層	double layer *[phys]*
nijū-yoji	二十四時	twenty-four hours (time: midnight) *{n}*
nijūyo'jikan-sei	二十四時間制	twenty-four hour system *[mech]*
nijū-yuka	二重床	false floor *[arch]*
nika	二価	bivalence; divalence *[ch]*
nika-arukōru	二価アルコール	dihydric alcohol *[org-ch]*
nikai-u'jō-fuku'yō	二回羽状複葉	bipinnately compound leaf *[bot]*
ni'kaku-kyōson'tai	二核共存体	dikaryon; dicaryon *[mycol]*
nikaku-tai	二核体	binuclear compound *[ch]*
nika no	二価の	bivalent; divalent *{adj}* *[ch]*
ni'kannō-sei	二官能性	bifunctionality *{n}* *[ch]*
ni'kannō'sei no	二官能性の	bifunctional *{adj}* *[ch]*
ni'kan'shiki-kagō'butsu	二環式化合物	bicyclic compound *[org-ch]*
nikawa	膠; にかわ	glue *{n}* *[mat]*
nikin'sa	二均差	variation (of moon) *[astron]*
ni-kitō	二気筒	two cylinder *{n}* *[mech-eng]*
nikkei	肉桂; にっけい	cinnamon (spice) *[bot]*
nikkeru	ニッケル	nickel (element)
nikki	日記	diary (pl -ries) *[comm]* *[pr]*
nikkō	日光	daylight; sunlight *[astron]*
nikkō-hansha-kei	日光反射計	heliotrope *[eng]*

nikkō-setsu'yaku-jikan	日光節約時間	daylight saving time *{n} [astron]*
nikō-bunpu	二項分布	binomial distribution *[math]*
nikō-enzan	二項演算	dyadic operation *[comput]*
nikō-kyūsū	二項級数	binomial series *[math]*
nikō-shiki	二項式	binomial *[math]*
nikō-teiri	二項定理	binomial theorem *[math]*
nikō-tenkai	二項展開	binomial expansion *[math]*
niku	肉	flesh *{n} [anat]*
"	肉	meat *[cook]*
niku'gan-so'shiki	肉眼組織	macrostructure *[met]*
niku'jō'tai	肉状体; 肉狀體	callus *[bot]*
niku'jū	肉汁	bouillon; meat broth *[microbio]*
niku'mori-yō'setsu	肉盛溶接	buildup welding; padded welding *{n} [met]*
niku'shitsu(-chū)-rui	肉質(虫)類	Sarcodina *[i-zoo]*
niku'shoku-dōbutsu	肉食動物	carnivorous animal *[zoo]*
niku-zuki	肉月: ⺼; にくづき	kanji (q.v.) radical denoting flesh *[pr]*
niku'zuku	肉豆; にくずく	nutmeg (spice) *[bot]*
ni'kyoku-kan	二極管	diode *[electr]*
ni(mei)mei-hō	二(命)名法	binomial nomenclature *[syst]*
nimen-kaku	二面角	dihedral angle *[math]*
ni'mono	煮物	simmered dish *[cook]*
ni'motsu	荷物	baggage; luggage *[trans]*
ni'motsu-azukari'jo	荷物預り所	baggage room *[transp]*
ni'motsu-guruma	荷物車	baggage car *[rail]*
-nin; -jin	-人	-person(s) *{suffix}*
nin-ben	人扁; 人偏: 亻; にんべん	kanji (q.v.) radical denoting person or man *[pr]*
nindō (suikazura)	忍冬; にんどう	Japanese honeysuckle *[bot]*
ni'nen'goto no	二年毎の	biennial *{adj} [bot]*
nin'gen	人間; にんげん	human being; Homo sapiens *[bio]*
ningen'gaku	人間学	anthropology *[bio]*
ningen-kōgaku	人間工学	ergonomics *[ind-eng]*
ningen-seitai'gaku	人間生態学	human ecology *[ecol]*
nin-getsu	人月	man-month *{n} [econ]*
ningyō	人形	doll (toy) *[eng]*
nin'i	任意	arbitrary *[comput] [math]*
nin'i-hōki	任意放棄	waiver *{n} [legal]*
nin'i no	任意の	discretionary; arbitrary *{adj} [math]*
"	任意の	optional *{adj} [comput] [math]*
nin'i-sentaku	任意選択; 任意選擇	option *[comput] [math]*
nin'i-teishi-meirei	任意停止命令	optional stop instruction *[comput]*
nin'i-teisū	任意定数	arbitrary constant *[math]*

ninjin	人参；にんじん	carrot (vegetable) *[bot]*
"	人参	ginseng (root) (herb) *[bot]*
ninka	認可	approval *[pat]*
ninka-bangō'satsu	認可番号札	license plate *[legal]*
ninmu-butai	任務部隊	task force *[ord]*
ninmu-kantai	任務艦隊	task fleet *[ord]*
nin-nen	人年	man-year *{n} [econ]*
ninniku	大蒜；にんにく	garlic (bulb) *[bot]*
ni no hosū	二の補数	two's complement *[comput]*
nin'shiki	認識；にんしき	recognition *[comput] [psy]*
"	認識	knowledge; awareness *[psy]*
ninshō	認証	authentication *[comm]*
"	認証	certification; validation *{n}*
ninshō suru	認証する	certify *{vb} [legal]*
nin-shū	人週	man-week *{n} [econ]*
nintei	認定	acknowledgment *{n}*
niobu	ニオブ	niobium (element)
nioi	匂い；香い；におい	odor; smell (pleasant) *{n} [physio]*
nioi	臭い	odor; smell (unpleasant) *{n} [physio]*
nioi o hanatsu	香い(臭い)を放つ	odoriferous *{adj} [physio]*
nioi'zuke	臭い付け	odorizing *{n} [ch-eng]*
niou	匂う；香う	smell; give off a pleasant fragrance *{vt} [physio]*
niou	臭う	smell (unpleasant) *{vb i} [physio]*
Nippon (see also Nihon)	日本；にっぽん	Japan *[geogr]*
Nippon'go; Nihon'go	日本語	Japanese (the language) *{n} [ling]*
Nippon-hyōjun-ji	日本標準時	Japanese standard time *[astron]*
Nippon Kōgyō Kikaku	日本工業規格	Japan Industrial Standards *[ind-eng]*
Nippon no; Nihon no	日本の	Japanese *{adj}*
Nippon Yakkyoku-hō	日本薬局方	Japanese Pharmacopoeia *[pharm]*
nira	韮；韭；にら	leek; scallion; shallot (vegetable) *[bot]*
nire	楡；にれ	elm (tree) *[bot]*
nire-ki'kui-mushi	楡木食虫	elm bark beetle (insect) *[i-zoo]*
ni'rinsan-suiso-natoryūmu	二燐酸水素=ナトリウム	disodium hydrogen phosphate
nirin'sha	二輪車	two-wheeler *{n} [mech-eng]*
ni'ritsu-haihan	二律背反	antinomy
niru	似る	resemble *{vb}*
niru	煮る	boil; cook *{vb t}*
niryō'ka	二量化	dimerization *[poly-ch]*
niryō'tai	二量体	dimer *[poly-ch]*
ni'ryūka-keiso	二硫化硅素	silicon disulfide

ni'ryūka-tanso	二硫化炭素	carbon disulfide
nisan-enki	二酸塩基	diacid base *[ch]*
ni'sanka-chisso	二酸化窒素	nitrogen doxide
ni'sanka-enso	二酸化塩素	chlorine dioxide
ni'sanka-iō	二酸化硫黄	sulfur dioxide; sulfurous acid gas
ni'sanka-keiso	二酸化硅素	silicon dioxide; silica
ni'sanka-namari	二酸化鉛	lead dioxide
ni'sanka-tanso	二酸化炭素	carbon dioxide
nise-gusuri	偽薬；僞藥	placebo *{n} [pharm]*
ni-seibun kara naru	二成分から成る	binary *{adj} [sci-t]*
ni-seibun-kei	二成分系	two-component system *[ch]*
nise'mono	偽物；にせもの	fake; sham *{n}*
nisen-setchi	二線接地	two-line ground *{n} [elec]*
nisen'shiki-kaisen	二線式回線	two-wire circuit *[elec]*
nishi	西	west *{n} [geod]*
nishi-kaze	西風	westerly wind; westerlies *[meteor]*
nishi-keisen	二至経線	solstitial colure *[astron]*
nishiki-hebi	錦蛇；にしきへび	python (reptile) *[v-zoo]*
nishiki-ori	錦織り	brocade *[tex]*
nishin	鯡；鰊；にしん	herring; Clupea pallasi (fish) *[v-zoo]*
nishin-enzan	二進演算	binary arithmetic operation *[comput]*
nishin-hō	二進法	binary notation; binary system *[math]*
nishin-sū	二進数	binary number *[comput]*
nishin-sūji	二進数字	binary digit; bit *[comput]*
nishin-yōso-retsu	二進要素列	binary element string *[comput]*
ni'shoku-kaburi	二色カブリ	dichroic fog *[photo]*
ni'shoku-kō'on-kei	二色高温計	two-color pyrometer *[eng]*
ni'shoku-sei	二色性	dichroism *[opt]*
ni'shoku'sei-hi	二色性比	dichroic ratio *[opt]*
nishū ni ikkai no	二週に一回の	fortnightly; once every two weeks *{adv}*
nisō-kyō	二層橋	double-deck bridge *[civ-eng]*
nissha	日射；にっしゃ	insolation *[astron]*
nissha-kei	日射計	actinometer *[eng]*
nisshi	日誌	journal; diary *[comm]*
nisshin-zahyō	日心座標	heliocentric coordinates *[astron]*
nisshō-chōsei	日照調整	solar control *[meteor]*
nisshō-kei	日照計	heliograph *[eng]*
nisshoku	日食	solar eclipse *[meteor]*
nisshū-chō	日周潮	diurnal tide *[ocean]*
nisshū-hyōdō	日周秤動	diurnal libration *[astron]*
nisshū-shisa	日周視差	diurnal parallax *[astron]*
nisshū-undō	日周運動	diurnal motion *[astron]*

nisu	ニス	varnish *[mat]*
ni-sugata	荷姿；にすがた	type of packing *{n} [ind-eng]*
ni-sui	二水：冫；にすい	kanji (q.v.) radical denoting ice *[pr]*
nitchū	日中	daytime *[astron]*
nitō	二糖；にとう	disaccharide *[bioch]*
ni-tōbun	二等分	bisection *[math]*
ni'tōbun-sen	二等分線	bisectrix *[crys]*
ni'tōbun-sen; (-men)	二等分線；(-面)	bisector *[math]*
ni'tōbun suru	二等分する	bisect *{vb} [math]*
ni'tōhen-sankak(u)kei	二等辺三角形	isosceles triangle *[math]*
niwa	庭	garden; yard *[arch] [bot]*
niwa-tori	鶏；雞；にわとり	chicken; barn fowl (bird) *[v-zoo]*
niwa-yanagi	庭柳	knotweed; knotgrass (herb) *[bot]*
ni'yaku	荷役；にやく	cargo work(er) *[aerosp] [nav-arch]*
niza-dai	仁座鯛；にざだい	surgeon fish (fish) *[v-zoo]*
ni'zukuri-ki	荷造り機	packing (baling) machine *[eng]*
no	の	of *{Japanese particle}*
no	野	field *{n} [geogr]*
nō	脳；腦；のう	brain *{n} [anat]*
no aida no-	の間の-	inter- *{prefix}*
no ato-	の後-	meta- *{prefix}*
nobasu	延ばす；伸ばす	lengthen; extend *{vb t}*
nōberyūmu	ノーベリウム	nobelium (element)
nobi	伸び；のび	elongation; extension *[mech]*
"	伸び	stretch *{n} [mech] [physio]*
nobi	延び	mileage *[pr]*
nobi-kei	伸び計	extensimeter *[eng]*
nobi-ritsu	伸び率	elongation rate (or percentage) *[mech]*
no-biru	野蒜；のびる	wild rocambole; Allium grayi *[bot]*
nobiru	延びる；伸びる	lengthen; stretch *{vb i}*
nobi-sei	伸び性	extensibility *[mech]*
nobori	幟；のぼり	flag; banner; streamer *{n}*
nobori	登り；昇り；上り	ascension *[navig]*
nobori-keisha	上り傾斜	acclivity *[geol]*
noboru	登る	ascend; climb *{vb} [navig]*
nodo	喉；のど	throat *{n} [anat]*
nōdo	濃度；のうど	concentration *[ch]*
"	濃度	power *[math]*
"	濃度	density (pl -ties) *[photo]*
nōdō	能動	active *{n} [gram]*
nodo-aki	喉開き；のどあき	gutter *[pr]*
nodo'bue	喉笛	windpipe; trachea *{n} [anat]*

nōdo-kōbai	濃度勾配	concentration gradient *[ch]*
nodo'kusari	喉腐り	dragonet (fish) *[v-zoo]*
nodo'shiro-mushi'kui	喉白虫食い	whitethroat (bird) *[v-zoo]*
nōdō-tai	能動態	active voice *{n} [gram]*
nōdō'teki na	能動的な	active; aggressive *{adj}*
nōdō-yusō	能動輸送	active transport *[cyt] [physio]*
nō'gaku	農学	agriculture *[agr] [bio]*
nōgei-kagaku	農芸化学	agricultural chemistry *[agr] [ch]*
nogi-hen	ノ木扁; ノ木偏: 禾; のぎへん	kanji (Chinese character) radical for grain *[pr]*
nogisu	ノギス	vernier caliper *[eng]*
nō'gyō	農業; のうぎょう	agriculture; farming *{n} [agr]*
nōgyō-dojō'gaku	農業土壌学	edaphology *[ecol]*
nōgyō'ka	農業家	farmer *[agr]*
nōgyō-shiken'jō	農業試験場	agricultural experiment station *[agr]*
nō'ha	脳波; 腦波; のうは	brain wave *[physio]*
nō'ha-kiroku-hō	脳波記録法	electroencephalography; EEG *[med]*
ho'hara-tsugumi	野原鶫	fieldfare (bird) *[v-zoo]*
nōhō	嚢胞; 囊胞	cyst *[hist]*
nō-jiki'zu-kiroku-hō	脳磁気図記録法	magnetoencephalography *[med]*
nōjō	農場	farm *{n} [agr]*
nō-kan	脳幹	brain stem *{n} [anat]*
no'kanzō	野萱草; のかんぞう	daylily; Homerocallis longituba *[bot]*
nō'kasui'tai	脳下垂体	hypophysis; pituitary (gland) *{n} [anat]*
noki	軒	eaves *[arch]*
nōki	農機	agricultural machinery *[agr] [eng]*
nōkō'ginkō	濃紅銀鉱	pyrargyrite *[miner]*
nokogiri	鋸; のこぎり	saw (a tool) *{n} [eng]*
nokogiri-ha	鋸波	sawtooth wave *[electr]*
nokogiri'sō	鋸草	Siberian yarrow; Achillea sibirica *[bot]*
nokogiri-zame	鋸鮫	saw shark; Pristiophorus japonicus *[v-zoo]*
nōkō'ka	濃厚化	inspissation *[ch] [geoch]*
"	濃厚化	thickening *{n} [ch-eng]*
nōkō-kongō-ki	濃厚混合気	rich mixture *[ch]*
nomi	蚤; のみ	flea (insect) *[i-zoo]*
nomi	鑿; のみ	chisel *{n} [eng]*
nomi'mono	飲物	drinkables; drinks *{n} [cook]*
nomu	飲む; 呑む	drink; swallow *{vb} [physio]*
nōmu	濃霧	dense fog *[meteor]*
nomu-noni-tekishita	飲むのに適した	drinkable; potable *{adj} [cook]*
no no hana	野の花	wildflower (flower) *[bot]*
noren	暖簾; のれん	shop door curtain *{n} [furn]*

nori	海苔; のり	laver (sea weed) *{n} [bot]*
nori	糊; のり	starch; paste *{n} [bioch]*
nori'kumi'in	乗組員; 乗組員	crew *{sing, pl} [trans]*
nori'nuki	糊抜き	desizing *{n} [tex]*
nori'ochi-sei	糊落性	easy washing (property) *{n} [tex]*
nori-tobi	糊飛び	paste skipping *{n} [tex]*
nō'ritsu	能率; のうりつ	efficiency *[eng] [phys]*
"	能率	performance *[ind-eng]*
nori'zuke; nori'tsuke	糊付け	sizing; starching *{n} [tex]*
noro-noro-unten	のろのろ運転	stop-and-go driving *{n} [traffic]*
noru	乗る; 乗る	ride; mount *{vb}*
nō'ryoku	能力	capacity; capability *{n}*
nō-sango	脳珊瑚; のうさんご	brain coral *[i-zoo]*
noseru	乗せる	mount (upon); let mount *{vb t}*
noshi	熨; のし	iron (for ironing clothes) *{n} [eng]*
nōshi	囊子; のうし	cyst *[med]*
nōshoku-kōka	濃色効果	hyperchromic effect *[p-ch]*
nō'shuku	濃縮	concentration; concentrating *{n} [ch]*
nō'shuku-bai'ritsu	濃縮倍率	concentration scale factor; enrichment scale factor *{n} [ch]*
nō'shuku-nenryō	濃縮燃料	enriched fuel *[mat]*
nōtan-bun'kyoku	濃淡分極	concentration polarization *[p-ch]*
no'usagi	野兎	jackrabbit (animal) *[v-zoo]*
nō-yaku	農薬	agricultural chemical; pesticide *[mat]*
nōzen'kazura	凌霄花; のうぜん=かずら	great trumpet flower; Campsis grandiflora (flower) *[bot]*
nozoki-ana	覗き穴; 覘き穴	peephole *[mech-eng]*
nozoku	覗く	peek in; peep; look; snoop *{vb}*
nozoku	除く	remove; eliminate *{vb}*
nozomu	臨む	face; confront *{vb}*
no'zura-ishi	野面石; のづら石	quarry stone *[mat]*
nū; nu'u	縫う	sew; stitch *{vb t}*
nuka	糠; ぬか	rice bran *[bot] [cook]*
nuka-ka	糠蚊	biting midge (insect) *[i-zoo]*
nuka-miso	糠味噌	salted rice-bran paste *[cook]*
nukarumi	泥濘; ぬかるみ	mud; muddy place *[ecol]*
nuki-dasu	抜き出す	extract *{vb} [comput]*
nuki'kōbai	抜き勾配	draft *{n} [eng]*
nuki'tori-kensa	抜取検査	sampling inspection *[ind-eng]*
nuku	抜く; 抜く	extract; pull out *{vb t}*
numa'chi	沼地	marsh; swamp *{n} [ecol]*
numa-mamushi	沼蝮; ぬままむし	water moccasin (reptile) *[v-zoo]*

numeri	ヌメリ	sliminess; slipperiness *{n} [mat]*
nuno	布	cloth *[tex]*
nuno'kizai-sekisō'hin	布基材積層品	fabric-base laminate *[mat]*
nuno'me	布目	texture (of cloth) *[tex]*
nuno-yasuri	布鑢	emery cloth *[mat]*
nure	濡れ	wetting *{n} [ch]*
nure-sei	濡れ性	wetting property *[ch]*
nureta	濡れた	wet *{adj}*
nurude	白膠木; ぬるで	sumac (tree) *[bot]*
nuta	饅; ぬた	vinegar-soy paste salad of fish and vegetables *[cook]*
nyō	尿	urine *[physio]*
nyō	繞; にょう	kanji (Chinese character) radical which surrounds the base *[pr]*
nyōdō	尿道	urethra *{n} [anat]*
nyōsan	尿酸	uric acid
nyōsan-en	尿酸塩	urate *[ch]*
nyōso	尿素; にょうそ	urea *[org-ch]*
nyōso-jushi	尿素樹脂	urea resin *[poly-ch]*
nyōso-jushi-toryō	尿素樹脂塗料	urea resin coating *[met]*
nyōso-sen'i	尿素繊維	polyurea fiber *[tex]*
nyū	乳; にゅう	milk *{n} [physio]*
nyū'bachi	乳鉢	mortar *[sci-t]*
nyūbō	乳棒	pestle *[sci-t]*
nyū'daku-eki	乳濁液	emulsion *[ch]*
nyūdaku'shitsu	乳濁質	emulsoid *{n} [ch]*
nyudō-gumo	入道雲	cumulonimbus (cloud) *[meteor]*
nyūhaku-do	乳白度	opacity *[opt]*
nyūhaku-zai	乳白剤	opaquefier; opalizer *[plas]*
nyūjū	乳汁	milk *{n} [bioch]*
nyūka	入荷	arrival of goods *{n} [ind-eng]*
nyūka	乳化	emulsification *[ch]*
nyūka-jūgō	乳化重合	emulsion polymerization *[poly-ch]*
nyūka-zai	乳化剤	emulsifying agent; disperser *[mat]*
nyūkō	乳光	opalescence *[opt]*
nyūkō	乳香	mastic; olibanum; frankincense *[mat]*
nyūkyū'shi	乳臼歯	deciduous teeth and molars *[anat] [dent]*
nyūro	入路	adit *[civ-eng]*
nyū'ryoku	入力; にゅうりょく	enter; input *[comput] [electr]*
"	入力	work input *[mech]*
nyū'ryoku-jun	入力順	entry sequence *[comput]*
nyū'ryoku-saki	入力先	destination *[comput]*

nyū'ryoku-sōchi	入力装置	input device; input unit *[comput] [eng]*
nyū'ryoku suru	入力する	enter; input *{vb} [comput]*
nyūsan	乳酸；にゅうさん	lactic acid
nyūsan-en	乳酸塩	lactate *[ch]*
nyūsan-kin	乳酸菌	lactic acid bacteria *[microbio]*
nyūsan'kin-gyūnyū	乳酸菌牛乳	acidophilous milk *[cook]*
nyū'satsu	入札	bid *{n} [econ]*
nyū'satsu suru	入札する	bid *{vb} [civ-eng] [econ]*
nyū-seihin	乳製品	dairy product *[agr]*
nyū-seki'ei	乳石英	milky quartz *[miner]*
nyū-sen	乳腺	mammary gland; milk gland *[physio]*
nyūsha-ha	入射波	incident wave *[electr]*
nyūsha-kaku	入射角	angle of incidence; incident angle *[opt]*
nyūsha-men	入射面	plane of incidence *[opt]*
nyūshi	乳脂；にゅうし	milk fat *[bioch] [cook]*
nyū-shibō	乳脂肪	butterfat *[cook]*
nyūshi-kei	乳脂計	lactobutyrometer; lactometer *[eng]*
nyūshō	乳漿	whey *[cook]*
nyū'shutsu-ryoku	入出力	input/output *[comput]*
nyū'tō	乳糖	lactose; milk sugar *[bioch]*
nyūyō'ji	乳幼児；乳幼兒	infants and toddlers *[bio]*

O

o 尾 train *{n} [astron]*
" 尾 tail *{n} [v-zoo]*
ō 凹；おう concave *{n} [math]*
ō-ago; daigaku 大顎 mandible *{n} [anat]*
ō-ao'sagi 大青鷺 great blue heron (bird) *[v-zoo]*
Ō-aza 大字；おゝあざ Major Section (of village) *[geogr]*
ōbā オーバー overcoat *{n} [cl]*
ōba-gibōshi 大葉擬宝珠 large-leaved plantain lily *[bot]*
ō-bako 大葉子；車前草 plantain; Plantago asiatica *[bot]*
ō-ban 凹板 intaglio *[lap]*
obi 帯；帶；おび sash (for kimono) *{n} [cl]*
" 帯 belt; band; strap; strip *{n} [eng]*
obi'jō-hanmon 帯状斑紋 zonation *[geol]*
obi-kō 帯鋼；おびこう band steel *[met]*
o-bire 尾鰭 caudal fin *[v-zoo]*
oboeru 覚える；覺える learn; commit to memory *{vb} [psy]*
ō-bu 凹部 concave part *[math]*
ō-bun'retsu 横分裂 transverse fission *[bio]*
ō'buntai-keisei 横分体形成 strobilation *[i-zoo]*
oburāto オブラート medicinal wafer *{n} [pharm]*
ochiru 落ちる drop; fall *{vb i}*
ō-chō'zame 大蝶鮫 beluga (fish) *[v-zoo]*
ō-dai 王台；王臺 queen cell *[i-zoo]*
o'daku 汚濁 pollution *[ecol]*
o'daku-shisū 汚濁指数 pollution index *[ecol]*
odamaki 苧環；おだまき columbine (flower) *[bot]*
ō'dan 黄疸；おうだん jaundice *[med]*
ō'dan 横断；横斷 crossing; intersection *{n} [traffic]*
ōdan-hodō 横断歩道 crosswalk *[civ-eng]*
ōdan-men 横断面 cross section *[math]*
ōdan-ryū 横断流 transverse flow *[poly-ch]*
ōdan-sen 横断線 transversal line *{n} [math]*
ōdan-sen no 横断線の transversal *{adj} [math]*
ōdan suru 横断する cross; cut horizontally *{vb t}*
o'dei 汚泥 sludge *[civ-eng]*
oden 御田；おでん stewed miscellaneous vegetables *[cook]*
ō'do 黄土 loess *[geol]*
" 黄土 Chinese yellow; ocher *[miner]*
ō'dō 黄銅；おうどう brass *[met]*

ō'dō'kō	黄銅鉱	chalcopyrite *[miner]*
ō-dōkutsu	大洞窟	cavern *[geol]*
ō-dōri	大通り	main street; highway *[traffic]*
odori'ba	踊り場	landing (stairway) *{n} [arch]*
odoroki	驚き	surprise *{n}*
odoroku	驚く	be surprised; be astonished *{i vb} [psy]*
ō-fubuki	大吹雪	blizzard *[meteor]*
ō'fuku'dō suru	往復動する	reciprocate *{vb} [mech-eng]*
ō'fuku-kippu	往復切符	round-trip ticket *[trans]*
ō'fuku-undō	往復運動	reciprocating motion *[mech-eng]*
ōga'i	横臥位	on one's side *{n} [med]*
ō-gama	大釜	cauldron *[cook]*
ō-gama	大鎌	scythe *{n} [agr]*
ō-gata-senpaku	大型船舶	large ship *[nav-arch]*
ōgi	扇	fan; folding fan *{n} [eng]*
ōgi-gata	扇形	sector *{n} [math]*
ō'gon	黄金；おうごん	gold *[met]*
ōgon-bunkatsu	黄金分割	golden cut *[math]*
ōgon-hi	黄金比	golden ratio *[math]*
ōgon'shoku-shokubutsu	黄金色植物	Chrysophyta *[bot]*
ō'gyoku	黄玉	topaz *[miner]*
ohagi	お萩	rice dumpling in azuki (q.v.) jam *[cook]*
o hanarete-	を離れて-	ab- *{prefix}*
ō'han-in'satsu	凹版印刷	intaglio printing *{n} [pr]*
ō-hashi	大嘴	toucan (bird) *[v-zoo]*
ō'hen	黄変；黄變	yellowing *{n} [poly-ch] [tex]*
ō-hige'mawari	大鬚回	volvox *[bot] [i-zoo]*
o'hitsuji-za	雄羊座	Aries; the Ram *[constel]*
ō-hō	王蜂	queen bee (insect) *[i-zoo]*
o'hyō	大鮃；おひょう	Pacific halibut (fish) *[v-zoo]*
ōi	多い；おおい	abundant; much *{adj} [math]*
ōi-	多い-	multi- *{prefix}*
ōi	覆い	cover; covering *{n} [eng]*
oi-kanmuri	老冠：耂	kanji (q.v.) radical for "old age" *[pr]*
oi'kawa	追河；おいかわ	Zacco platypus (a carp-type fish) *[v-zoo]*
oi-kaze	追風	tailwind *[meteor]*
ōi'mono	覆い物	covering *{n} [mat]*
oishī	おいしい	delicious; tasty *{adj} [cook] [physio]*
oishī mono	おいしい物	delicacy (pl -cies) *[cook]*
ō-itadori	大虎杖	giant knotweed *[bot]*
ōji	欧字；歐字	letter; alphabet letter *{n} [pr]*
ojiya	おじや	rice gruel (thick) *[cook]*

oka	丘；岡	hill *[geogr]*
ō'ka	黄化	etiolation *[bot]*
ō'kaku'maku	横隔膜	diaphragm *[anat]*
ōkami	狼；おおかみ	wolf (pl wolves) (animal) *[v-zoo]*
o'kan	悪寒；惡寒	chill *{n} [med]*
ō'kan	王冠	crown *{n} [lap]*
ō'kan no	王冠の	coronal *{adj} [astron]*
okara	雪花菜；おから	bean curd lees (dregs) *{n} [cook]*
okashī	可笑しい；おかしい	funny; amusing; odd; illogical *{adj}*
okazu	御数；おかず	side dish(es) *[cook]*
oke	桶	bucket *[eng]*
okera	朮；おけら	Atractylodes japonica *[bot]*
(o)kera	螻蛄；(お)けら	mole cricket (insect) *[i-zoo]*
ō-ketsu; kame-ana	甌穴	pothole *[geol]*
ō'ketsu-kari	黄血カリ	yellow prussiate of potash *[ch]*
ō'ketsu-sōda	黄血ソーダ	yellow prussiate of soda *[ch]*
ōki	嘔気	nausea *[med]*
ōkī	大きい；おゝきい	large; big *{adj} [math]*
ōkī-	大きい-	macro- *{prefix}*
oki'ai	沖合	offshore *{n} [geol]*
oki'ai-gyogyō	沖合漁業	offshore fisheries *[ecol]*
oki'ai-tō	沖合島	offshore island *[geogr]*
oki'kae	置き換え	substitute (n) *[comput]*
oki-kaeru	置き換える	replace *{vb} [comput]*
ōkiku suru	大きくする	enlarge *{vb t}*
okiru	起きる	wake up; awaken *{vb i}*
ōki-sa	大きさ	dimension; magnitude; size *{n} [math]*
oki-ware	置割れ	season crack *{n} [met]*
ō'kō	横坑	adit *[civ-eng]*
o koeta-	を越えた-	meta- *{prefix}*
o koete-	を越えて-	trans-; ultra- *{prefix}*
okonomi-yaki	お好み焼	Japanese pancake *[cook]*
okoze	虎魚；おこぜ	stonefish (fish) *[v-zoo]*
oku	奥；奧	depth; back; inner part *{n}*
oku'ba	奥歯；奧齒	back tooth; molar *[anat] [dent]*
ō-kubo'chi	大窪地	caldera *[geol]*
oku'gai-bakuro	屋外暴露	outdoor exposure *[meteor]*
okujō	屋上	rooftop *[arch]*
oku'nai-haisen	屋内配線	house wiring; inside wiring *{n} [elec]*
oku'nai-shōten'gai	屋内商店街	indoor shopping mall *[arch]*
okure	遅れ；遲れ	lag; time lag *{n} [comput] [phys]*
okuri	送り	shift *{n} [comput]*

okuri-gana	送りがな	kana (q.v.) phonetics beside Chinese characters *[ling] [pr]*
okuri'jō	送り状	invoice *{n} [econ]*
okuri-kan	送り杆	feed rod *[mech-eng]*
okuri-kan	送り管	feed pipe *[mech-eng]*
okuri'komi	送り込み	feed in; feeding in *{n} [mech-eng]*
okuri-natto	送りナット	feed nut *[mech-eng]*
okuri-sōchi	送り装置	feeding apparatus *[mech-eng]*
okuri-undō	送り運動	feed motion *[mech-eng]*
okuru	送る	send *{vb t}*
okutan-ka	オクタン価	octane number *[eng]*
ōkute	多くて	at most *[math]*
oku'yuki	奥行き	depth *[arch]*
oku'zuke	奥付け	colophon *[pr]*
ōkyū-te'ate	応急手当	first aid *[med]*
ō'men-kyō	凹面鏡	concave mirror *[opt]*
omiotsuke	御御御付；おみお=つけ	soybean paste soup *[cook]*
omocha; gangu	玩具	toy *{n} [eng]*
omoi	思い	thought *{n} [psy]*
omoi	重い	heavy *{adj} [mech]*
omoi-dasu	思い出す	remember; recall *{vb} [psy]*
omo'mi	重み	significance *[math]*
omo'mi'zuke-ki	重み付け器	weighting instrument *[eng]*
omo-sa	重さ	heaviness; weight *{n} [mech]*
omo'sa no tan'i	重さの単位	unit of weight *[math]*
omoshi	重し	weight *{n} [mech]*
omoshiroi	面白い	interesting; amusing *{adj}*
omoshiro'mi	面白み	interest; amusingness; funness *{n}*
omoshiro-sa	面白さ	interest; amusingness; funness *{n}*
omote	表；おもて	front *{n} [arch]*
"	表	surface; right side *[eng]*
"	表	outside; out front; outdoors *{n}*
omote-ito	表糸	needle thread; upper thread *[cl] [tex]*
omote'me	表目	knit (knitting stitch) *{n} [tex]*
omote-saku	表作	main season crop; summer crop *[agr]*
omoto	万年青；おもと	Rhodea japonica *[bot]*
omou	思う	think *{vb} [psy]*
omo'waku	思惑；思わく	speculation; opinion *{n} [psy]*
omo'yu	重湯	rice gruel (thin) *[cook]*
ōmu	鸚鵡；おうむ	parrot (bird) *{n} [v-zoo]*
ōmu-gai	鸚鵡貝；おうむがい	nautilus (shellfish) *[i-zoo]*

ō-mugi	大麦; 大麥	barley *[bot]*
ō'mugi-nuka	大麦糠	barley bran *[agr]*
on	音; おん	Japanese phonetics of sound for kanji (q.v.) *[ling] [pr]*
o'naga-dori	尾長鶏	long-tailed cock (bird) *[v-zoo]*
o'naga-gara	尾長雀; おながガら	long-tailed tit (bird) *[v-zoo]*
o'naga-zame	尾長鮫; おながざめ	thresher shark; thrasher; fox shark; Alopias pelagicus (fish) *[v-zoo]*
o'naga-zaru	尾長猿	ceropith; Ceropithecus (monkey) *[v-zoo]*
onaji	同じ	same *{adj} {n}*
onamomi	苓; おなもみ	cockebur(r) *[bot]*
on-atsu	音圧; 音壓	acoustic pressure; sound pressure *[acous]*
onba; onjō	音場	sound field *[acous]*
onbin	音便; おんびん	euphonic change; euphony *[ling]*
onchi	温置; おんち	incubation *[microbio]*
onchō	温調	thermoregulation *[physio]*
onchō no takasa	音調の高さ	pitch of sound *[acous]*
ondan-zensen	温暖前線	warm front *[meteor]*
ondo	温度; 溫度; おんど	temperature *[thermo]*
ondo-gyaku'ten	温度逆転	temperature inversion *[meteor]*
ondo-hyōjun	温度標準	temperature scale *[thermo]*
ondo-i'zon-sei	温度依存性	temperature dependency *[sci-t]*
ondo-jōshō-keisū	温度上昇係数	temperature-rise coefficient *[mech-eng]*
ondo-kanju'sei-hen'i	温度感受性変異	temperature-sensitive mutation *[gen]*
ondo-kei	温度計	thermometer *[eng]*
ondo-keisū	温度係数	temperature coefficient *[phys]*
ondo-kiroku-zu	温度記録図	thermogram *[eng]*
ondo-kōbai	温度勾配	temperature gradient *[thermo]*
ondo-reikō	温度冷光	thermoluminescence *[atom-phys]*
ondori	雄鳥; おんどり	rooster; male bird (bird) *[v-zoo]*
ondo-teki'tei	温度滴定	thermometric titration *[an-ch]*
o'ne	尾根; おね	ridge (of pressure) *{n} [meteor]*
o-neji	雄螺子; おねじ	external thread *[mech-eng]*
o'ne-sen	尾根線	ridge line *[meteor]*
ō-netsu	黄熱; おうねつ	yellow fever *[med]*
on'gaku	音楽; おんがく	music *[music]*
ongak(u)'ka	音楽家	musician (a person) *[music]*
ongak(u)kai'jō	音楽会場	music hall *[arch]*
ongaku-onkyō'gaku	音楽音響学	musical acoustics *[acous]*
oni-geshi	鬼罌粟; おにげし	Oriental poppy (flower) *[bot]*
o'nigiri	お握り	cooked-rice balls, hand-shaped *[cook]*
oni-nezumi	鬼鼠; おにねずみ	bandicoot (animal) *[v-zoo]*

oni-ukogi	鬼五加；おにうこぎ	Acanthopanax spinosus *[bot]*
oni-yuri	鬼百合	tiger lily (flower) *[bot]*
on-kagaku	音化学	sonochemistry *[ch]*
onkai	音階	scale *{n} [music]*
onkan-atsu'en	温間圧延	warm rolling *{n} [met]*
on'ketsu-dōbutsu	温血動物	warm-blooded animal *[zoo]*
onkō	音高	pitch *{n} [acous]*
on'kyō	音響；おんきょう	sound *{n} [acous]*
onkyō-denshi-kōgaku	音響電子工学	acoustoelectronics *[acous]*
onkyō'gaku	音響学；音響學	acoustics *[phys]*
onkyō-ketsugō-sōchi	音響結合装置	acoustic coupler *[eng-acous]*
onkyō-kōgaku	音響光学	acoustooptics *[opt]*
onkyō-kō'ritsu	音響効率	acoustic efficiency *[acous]*
onkyō-soku'shin	音響測深	echo sounding *{n} [eng]*
onkyō-sokushin-ki	音響測深機	echo sounder; sonic depth finder *[eng]*
onkyō-sokushin-kiroku-zu	音響測深記録図	echogram *[eng]*
onkyō'teki-chūjitsu-sei	音響的忠実性	acoustic fidelity *[acous]*
onna	女	woman (pl women) *[bio]*
onna-hen	女扁；女偏：女；おんなへん	kanji (Chinese character) radical denoting woman *[pr]*
ono	斧	axe *{n} [eng]*
ono-ishi	斧石；おのいし	axinite *[miner]*
onpa	音波	acoustic wave; sound wave *[acous]*
onpa-kensa-hō	音波検査法	sonography *[eng]*
onpa-shori	音波処理	sonication *[acous] [cyt] [microbio]*
onpu	音符	note; musical notation *{n} [music]*
"	音符	diacritical mark *[pr]*
onpū-danbō	温風暖房	hot-air heating *{n} [mech-eng]*
on-ritsu	音律	temperament *[music]*
on-ryō	音量	loudness; sound volume; volume *[acous]*
onryō-kei	音量計	sound level meter *[eng]*
on'sa	音叉	tuning fork *[acous]*
onsei	音声；おんせい	speech; voice *{n} [physio]*
onsei'gaku	音声学；音聲學	phonics *[ling]*
onsei-gōsei	音声合成	speech synthesis *[comput]*
onsei-jibo	音声字母	phonetic alphabet *[ling]*
onsei-kanshi-sōchi	音声監視装置	bug *{n} [electr]*
onsei-kei	音声計	phonometer *[acous]*
onsei-kiroku-sōchi	音声記録装置	voice recorder *[eng]*
onsei-nin'shiki	音声認識	speech recognition *[acous]*
onsei-ōtō-sōchi	音声応答装置	audio response unit *[comput]*
onsei-tai'iki	音声帯域	voice-band *[comm]*

onsei-tan'i	音声単位	voice unit *[acous]*
on'sekimen	温石綿	chrysotile (asbestos) *[miner]*
onsen	温泉	hot springs *[hyd]*
on'setsu	音節	syllable *[ling]*
onshi	音子	phonon *[sol-st]*
on-shitsu	音質	tone quality *[acous]*
on-shitsu	温室	greenhouse; hothouse *[arch] [bot]*
on'shitsu-kōka	温室効果	greenhouse effect *[ecol] [meteor]*
on-shitsu-shisū	温湿指数	temperature-humidity index *[meteor]*
onshō; on'doko	温床	hotbed; frame *[bot]*
on-shoku	音色; おんしょく	timbre; tone quality *[acous]*
onso	音素	phoneme *[ling]*
on-soku	音速	acoustic velocity; sonic speed *[acous]*
on'soku-shōheki	音速障壁	sonic barrier *[aero-eng]*
onsu	オンス	ounce; oz (unit of mass) *[mech]*
ontai	温帯; 溫帶	Temperate Zone *[climat]*
ontai-kikō	温帯気候	temperate climate *[climat]*
ontei	音程	interval (music) *[acous]*
ontei-ritsu	音程律	octaves, law of *[ch]*
ontō	温湯	warm water *[hyd]*
on'yaku	音訳; 音譯	transliteration *[comput] [ling]*
ō'ren	黄連; 黄蓮	goldthread; Coptis japonica *[bot]*
ori	檻; おり	cage *{n} [eng]*
Oribe-yaki	織部焼	Oribe ware *[cer]*
ori'chō	折り丁	section; signature *{n} [pr]*
ori-gami	折紙; おりがみ	paper folding *{n} [paper]*
ori-mageru	折曲げる	bend; fold *{vb t}*
ori'mage-tsuyo-sa	折曲げ強さ	bending strength *[mech]*
ori'me	織り目	between threads of a fabric *{n} [tex]*
ori'mono	織物; おりもの	fabric; textile; woven fabric *[tex]*
ori'mono-kōgaku	織物工学	textiles (the engineering field) *[tex]*
ori'mono-shi'age-ki	織物仕上機	textile finishing machine *[eng]*
ō'rin	黄燐	white phosphorus; yellow phosphorus *[ch]*
ori-nuno	織布	woven cloth *[tex]*
oriru	下りる; 降りる	descend; alight *{vb}*
ori-taosu	折り倒す	fold and fell *{vb}*
ori'tatami-isu	折り畳み椅子	folding chair *[furn]*
ori'tatami-jaku	折り畳み尺	zigzag rule; folding ruler *{n} [eng]*
ori-tatamu	折り畳む	fold *{vb}*
oroshi	下し; おろし	grated daikon *[cook]*
oroshi	颪; おろし	fall wind; katabatic wind *[meteor]*
oroshi'uri	卸売; 卸賣	wholesale *{n} [econ]*

orosu	下ろす; おろす	grate (as vegetable) *{vb} [cook]*
orosu	下ろす	slice meat *{vb t} [cook]*
orosu	降ろす; 下ろす	borrow *{vb} [math]*
"	降ろす; 下ろす	lower; bring (let) down; discharge (a passenger) *{vb t}*
oru	折る	bend; fold; break *{vb t} [mech-eng]*
ō'ryoku	応力; おうりょく	stress *{n} [mech]*
ō'ryoku	横力	lateral force *[mech]*
ōryoku-fuku'kussetsu	応力複屈折	stress birefringence *[poly-ch]*
ōryoku-fu'shoku-ware	応力腐食割れ	stress corrosion cracking *{n} [met]*
ōryoku-kanwa	応力緩和	stress relaxation *[met]*
ōryoku-ki'retsu	応力亀裂	stress cracking *{n} [poly-ch]*
ōryoku-shūchū	応力集中	stress concentration *[mech]*
ōryoku-yūki	応力誘起	stress induction *[mech]*
ōryoku-zu	応力図; 應力圖	stress diagram *[graph] [met]*
ō'ryū	横流; おうりゅう	crosscurrent *{n} [fl-mech]*
osa	筬; おさ	reed (for weaving) *{n} [tex]*
osa-game	長亀; おさがめ	leatherback (turtle) *[v-zoo]*
osae	押え	weight *{n} [mech]*
osae-ita	押え板	pressure plate *[mech-eng]*
osae-neji	押えねじ	cap screw *[eng]*
ō-saji	大匙	tablespoon *[cook] [eng]*
Ōsaka	大阪; おおさか	Osaka *[geogr]*
osameru	収める; 納める	store (within); house (in) *{vb t}*
osa-mushi	歩行虫; おさむし	ground beetle (insect) *[v-zoo]*
o'sen	汚染	contamination; pollution *[ecol]*
o'sen'butsu; (-busshitsu)	汚染物; (-物質)	contaminant; pollutant *[ecol]*
ō'setsu-ma	応接間	drawing room; reception room *[arch]*
ō'shakkō	黄錫鉱	stannite *[miner]*
oshi	唖; 啞; おし	mute (a person) *[physio]*
oshi'age-manriki	押上げ万力	jack *[mech-eng]*
oshi-bō	押し棒; おしぼう	push rod *[mech]*
oshi-botan	押釦; 押ボタン	push button *[electr]*
o'shida	雄羊歯; おしだ	male fern *[bot]*
oshi'dashi	押出し	extrusion *[poly-ch]*
oshi'dashi-atsu'ryoku	押出し圧力	extrusion pressure *[poly-ch]*
oshi'dashi-ki	押出機; おしだしき	extruder *[eng] [poly-ch]*
oshi'dashi-nagare	押出し流れ	piston flow; plug flow *[fl-mech]*
oshi'dashi-seikei	押出し成形	extrusion molding *{n} [met] [poly-ch]*
oshi'dashi-sokudo	押出し速度	extrusion rate *[poly-ch]*
oshi'dashi-zai	押出し材	extruded material *[mat]*
oshi-dori	鴛鴦; おしどり	mandarin duck (bird) *[v-zoo]*

oshi-hirogeru	押し広げる	widen; spread *{vb t} [mech-eng]*
oshi-ire	押し入れ	bedding closet; wall closet *[arch]*
oshi-ita	押し板	pressure plate *[mech-eng]*
oshi'komi-kōdo	押込み硬度	indentation hardness *[met]*
oshi'komi-tsūfū	押込通風	forced draft *[mech-eng]*
oshi-mugi	押麦; 押麥	pressed barley *[agr]*
oshi'noke-ryō	押除け量	displacement *[fl-mech]*
ō-shio	大潮; おおしお	spring tide *[ocean]*
oshi-sageru	押下げる	depress; press down *{vb t}*
oshi'sugi	押過ぎ	overcuring (of resin) *{n} [ch-eng]*
oshi-tsubusu	押し潰す	collapse *{vb t} [mech-eng]*
oshi'tsuke-ryoku	押付け力	clamping force *[mech]*
oshi-yasu-sa	押し易さ	ease of pushing; pushability; rollability *[mech]*
ō'shoku-do	黄色土	yellow earth *[agr] [geol]*
ō'shoku-sohō	黄色素胞	xanthophore *[cyt]*
Ōshū	欧州; おうしゅう	Europe *[geogr]*
osoi	遅い; 遲い	late; slow *{adj} [mech] [phys]*
osoku suru	遅くする	slow down *{vb t}*
osomeru	遅める	make slow(er); slow down *{vb}*
o'son	汚損	fouling; soiling *{n} [ecol]*
osoraku	恐らく	probably *{adv}*
osu	雄; 牡	male *[bio]*
osu-bachi	雄蜂; おすばち	drone (insect) *[i-zoo]*
osu-hitsuji	雄羊	ram (animal) *[v-zoo]*
ō-sui	王水	aqua regia *[ch]*
osui'dame	汚水溜	cesspool *[civ-eng]*
osumyūmu	オスミウム	osmium (element)
osu-uma	雄馬	stallion (animal) *[v-zoo]*
osu-ushi	雄牛	bull (animal) *[v-zoo]*
o'tafuku-kaze	お多福風	mumps; parotitis *[med]*
o'tafuku-mame	お多福豆	broad bean (large); fava bean *[bot]*
ō-taki	大滝; 大瀧	cataract; great waterfall *[hyd]*
o'tama'jakushi	お玉杓子	ladle; soup ladle *{n} [cook]*
o'tama'jakushi	お玉杓子	tadpole *[v-zoo]*
ō'tekkō	黄鉄鉱; 黃鐵鑛	iron pyrite; pyrite *[miner]*
o'ten	汚点	stain *{n} [ecol]*
ō'ten	横転	roll *{n} [aerosp]*
oto	音	sound; tone; noise *{n} [acous]*
oto no ōki-sa	音の大きさ	loudness *[acous]*
oto no taka-sa	音の高さ	sound pitch *[acous]*
oto no tsuyo-sa	音の強さ	sound intensity *[acous]*

ōtō	応答；應答	answer *{n} [comput]*
ō'tō	応答；おうとう	response *[bio] [comm] [comput]*
otogai	頤；おとがい	chin *{n} [anat]*
otogiri'sō	弟切草；おとぎり草	Saint John's wort *[bot]*
ō'tō-jikan	応答時間	response time *[comput]*
ō'tō-ki	応答機	transponder *[comm]*
otoko	男	man (pl men) *[bio]*
otome-za	乙女座	Virgo; the Virgin *[constel]*
otori	囮；おとり	decoy *{n} [mil]*
otoshi-do	落とし戸	trapdoor *[arch]*
otoshi-hi	落とし樋	chute *[eng]*
otosu	落とす	drop *{vb t}*
ototoi	一昨日；おととい	day before yesterday *{n}*
ō-totsu	凹凸；おうとつ	unevenness *[mat]*
"	凹凸	hills and valleys *{n}*
ō-totsu no	凹凸の	concavo-convex *{adj} [opt]*
ottosei	膃肭獣；おっとせい	eared seal; seal; fur seal *[v-zoo]*
o'ushi-za	牡牛座	Taurus; the Bull *[constel]*
oya	親；おや	parent *{n} [bio] [comput]*
"	親	host; owner *{n} [comput]*
oya-bashira	親柱	newel post *[civ-eng]*
oyabitcha	おやびっちゃ	sergeant major; Abudefduf vaigiensis (fish) *[v-zoo]*
oya-gengo; oya-kotoba	親言語	host language *[comput]*
oya-kabu	親株	parent strain *[microbio]*
ōyake no	公の	official; public *{adj}*
oyako-donburi	親子丼	rice in donburi (q.v.) with chicken and eggs *[cook]*
oyako-shūgō	親子集合	set *{n} [comput]*
oya-moji	親文字	parent character; root character *[pr]*
ō-yamori	大守宮；おゝやもり	tokay (reptile) *[v-zoo]*
oya-neji	親ねじ	lead screw *{n} [eng]*
oya-neri	親練り；親練り	master batch *[poly-ch] [rub]*
oya-saibō	親細胞；親細胞	parent cell *[cyt]*
oya-yubi	親指	big toe; thumb *{n} [anat]*
oya-zuna	親綱	main rope *[civ-eng]*
ōyō	応用；應用	application(s) *[sci-t]*
oyobi	及び	and *{conj} [logic]*
ōyō-butsuri'gaku	応用物理学	applied physics *[phys]*
oyogaseru	泳がせる	cause to swim *{vb t}*
oyogu	泳ぐ	swim *{vb i}*
ōyō-jōhō-gi'jutsu	応用情報技術	applied information technology *[comm]*

ōyō-kagaku	応用化学	applied chemistry *[ch]*
ōyō-kagaku	応用科学	applied science *[sci-t]*
ōyō-kenkyū	応用研究	application study *[sci-t]*
ōyō-rikigaku	応用力学	applied mechanics *[mech]*
oyoso	凡そ	about *{adv}*
ō-yumi	弩；大弓	crossbow *[ord]*
ō-zato	邑：阝；おおざと	kanji (q.v.) radical denoting a large community *[pr]*
ozon	オゾン	ozone *[ch]*
ōzuru	応ずる；應ずる	respond *{vb}*

P

-pa (-ba, -wa)	-羽	-unit(s) of bird(s) *{suffix}*
pafe	パフェ	parfait *[cook]*
painto	パイント	pint (unit of volume) *[mech]*
pakkin	パッキン	packing *{n} [mat]*
pāmu'kaku-yu	パーム核油	palm kernel oil *[mat]*
pāmu-yu	パーム油	palm oil; palm butter *[mat]*
pan	パン； 麺麭； 麺包	bread *[cook]*
pan-dane	パン種	leaven *{n} [cook]*
panfuretto	パンフレット	pamphlet *[pr]*
pan-kiji	パン生地	bread dough *[cook]*
pan-ko	パン粉	breadcrumb *[cook]*
pan-kōbo	パン酵母	baker's yeast *[cook]*
panku	パンク	puncture; rupture (of tire) *{n}*
pan'ya	パン屋	bakery (pl -eries) *[arch] [cook]*
pan-yaki	パン焼き	baking *{n} [food-eng]*
papain	パパイン	papain *[bioch]*
papirusu	パピルス	papyrus *[bot]*
pappu-zai	巴布剤	cataplasm; poultice *{n} [pharm]*
parafin-retsu	パラフィン列	paraffin series *[org-ch]*
parafin-rō	パラフィン蠟	paraffin wax *[mat]*
parafin-yu	パラフィン油	paraffin oil *[mat]*
paragomu no ki	パラゴムの木	para rubber tree *[bot]*
para'i	パラ位	para position *[org-ch]*
parajūmu	パラジウム	palladium (element)
para'mitsu	波羅蜜； ぱらみつ	jackfruit *[bot]*
para-suiso	パラ水素	parahydrogen *[atom-phys]*
para'tsuki	ぱらつき	unevenness; irregularity *{n}*
Pari	パリ	Paris *[geogr]*
pāru	パール	pearl *{n} [mat]*
parusu	パルス	pulse *{n} [physio]*
pāseku	パーセク	parsec *[astron]*
pāsento	パーセント	percent *[math]*
paseri	パセリ	parsley *[bot]*
pasokon	パソコン	personal computer *[comput]*
pasu	パス	calliper *[eng]*
"	パス	run *{n} [ind-eng]*
pasukaru	パスカル	pascal (unit of pressure) *[mech]*
patān-nin'shiki	パターン認識	pattern recognition *[comput]*
pate	パテ	putty *[mat]*

pate'tsuke パテ付け puttying (in painting) *[eng]*
paundaru パウンダル poundal (unit of strength) *[mech]*
peinto ペイント paint *{n} [mat]*
pēji 頁；ページ page *{n} [pr]*
penchi ペンチ cutting pliers; pincers *[eng]*
penishirin ペニシリン penicillin *[microbio]*
penki ペンキ paint *{n} [mat]*
pepuchido-ketsu'gō ペプチド結合 peptide bond *[org-ch]*
piezo-denki ピエゾ電気 piezoelectricity *[sol-st]*
pika-pika ピカピカ sparkle; glitter *{n} [opt]*
piko- ピコ- pico-; micromicro- *{prefix} [math]*
pin-ana ピン穴 pinhole *[opt]*
pin-ha'guruma ピン歯車 pinwheel *[mech-eng]*
pinsetto ピンセット tweezers *[eng]*
pinto ピント focus (pl -foci, focuses) *{n} [opt]*
pin-yasuri ピン鑢；ピンやすり pin file *[eng]*
pirafu ピラフ pilaf (a rice dish) *[cook]*
piro-denki ピロ電気 pyroelectricity *[sol-st]*
pirubin-san ピルビン酸 pyruvic acid *[bioch]*
pisuton ピストン piston *[mech-eng]*
pītan 皮蛋；ピタン green (preserved) egg *[cook]*
pitō-kan ピトー管 pitot tube *[eng]*
pittari ぴったり exactly *{adv}*
poazu ポアズ poise; P (unit of viscosity) *[fl-mech]*
poji ポジ positive *{n} [photo]*
ponchi ポンチ punch *{n} [eng]*
pondo 封度；ポンド pound (unit of mass) *{n} [mech]*
pondo 磅；ポンド pound (currency) *{n} [econ]*
ponpu ポンプ pump *{n} [mech-eng]*
ponpu-suisha ポンプ水車 reversible pump turbine *[mech-eng]*
poribiniru-arukōru (pobāru) ポリビニルアルコール（ポバール） polyvinyl alcohol; PVA; PVOH *[poly-ch]*
pori'esuteru ポリエステル polyester *[poly-ch]*
pori'ēteru ポリエーテル polyether *[poly-ch]*
porimā ポリマー polymer *[poly-ch]*
porimerāze ポリメラーゼ polymerase *[bioch]*
pori-rinsan-en ポリ燐酸塩 polyphosphate *[inorg-ch]*
pori-sakusan-biniru ポリ酢酸ビニル polyvinyl acetate (PVAc) *[poly-ch]*
pori-san ポリ酸 polyacid *[ch]*
poronyūmu ポロニウム polonium (element)
posuto ポスト mailbox *[comm]*
potensharu ポテンシャル potential *[elec] [phys]*

potsu	ぽつ	dot *{n} [pr]*
purachinaito	プラチナイト	platynite *[miner]*
purachinoido	プラチノイド	platinoid *{adj} {n} [met]*
puragu	プラグ	plug *{n} [elec] [sci-t]*
puraseojimu	プラセオジム	praseodym (element)
puraseojimyūmu	プラセオジミウム	praseodymium (element)
purasuchikku	プラスチック	plastic; plastics *[mat]*
puratō	プラトー	plateau *[electr] [geogr] [geol]*
purau'kō	プラウ耕	plowing *{n} [agr]*
purazuma	プラズマ	plasma *[phys]*
purazuma-butsuri'gaku	プラズマ物理学	plasma physics *[phys]*
purazuma-kakō	プラズマ加工	plasma etching *{n} [graph]*
purazuma-uchū-ron	プラズマ宇宙論	plasma cosmology *[astron]*
purin	プリン	pudding *[cook]*
purinto-haisen-ban	プリント配線板	printed circuit board *[electr]*
purinto-kairo	プリント回路	printed circuit *[electr]*
purinto-kiban	プリント基板	printed (wiring) board *[electr]*
purizumu	プリズム	prism *[crys] [math] [opt]*
puroguramu	プログラム	program *{n} [comput]*
puro-kōso	プロ酵素	proenzyme *[bioch]*
puromechūmu	プロメチウム	promethium (element)
puropera	プロペラ	propeller *[mech-eng]*
purotoakuchinyūmu	プロトアクチニウム	protoactinium; protactinium (element)
puroton-ji'ryoku-kei	プロトン磁力計	proton magnetometer *[elecmg]*
puroton'sei-yōbai	プロトン性溶媒	protic solvent *[ch]*
purutonyūmu	プルトニウム	plutonium (element)
putomain-chū'doku	プトマイン中毒	ptomaine poisoning *[med]*
pyūma	ピューマ	puma; cougar (animal) *[v-zoo]*

R

raba	騾馬；らば	mule (animal) *[v-zoo]*
raden	螺鈿；らでん	lamina of mother of pearl *[i-zoo]*
ragō suru	螺合する	screw-mesh *{vb} [mech]*
raiden	雷電	thunderbolt *[geophys]*
raigeki	雷撃	thunderstroke *[meteor]*
rai-getsu	来月；來月	next month *{n}*
rai-hōden	雷放電	lightning discharge *[geophys]*
raikan	雷管	cap (explosive); detonator *[ord]*
raimei	雷鳴	thunder *{n} [meteor]*
raimu'byō	ライム病	Lyme disease *[med]*
rai-mugi	ライ麦	rye (cereal) *[bot]*
rai-nen	来年；來年	next year *{n}*
raisan	雷酸	fulminic acid
raisan-en	雷酸塩	fulminate *[org-ch]*
raisan-suigin	雷酸水銀	mercury fulminate
rai-shū	来週	next week *{n}*
rai'u	雷雨	thunderstorm *[meteor]*
rai'un	雷雲	thunder cloud *[meteor]*
rajūmu	ラジウム	radium (element)
rakkan	楽観；樂觀	optimism *[psy]*
rakka no tei'ritsu	落下の定律	falling, law of *[mech]*
rakkan suru	楽観する	be optimistic *{vb} [psy]*
rakka-san	落下傘	parachute *{n} [aero-eng]*
rakka'sei-yu	落花生油	peanut oil *[mat]*
rakka-shiken	落下試検	drop test *[mech]*
rakkyō	辣韮；らっきょう	scallion; Allium chinense *[bot]*
rakkyō-zuke	辣韮漬	pickled scallions *[cook]*
rakkyū-kyōdo	落球強度	falling-ball strength *[mech]*
rakkyū-nendo-kei	落球粘度計	falling-sphere viscometer *[eng]*
rakkyū-shōgeki-shiken	落球衝撃試験	falling-ball impact test *[mech]*
rakuda	駱駝；らくだ	camel (animal) *[v-zoo]*
rakudai	落第	failure *[ind-eng]*
rakuda-mō	駱駝毛	camel hair *[tex]*
raku-men	落綿	cotton flock; waste cotton *[tex]*
raku-nō	酪農	dairy farming *{n} [agr]*
rakurai	落雷	thunderbolt *[meteor]*
raku'sa	落差	head *{n} [fl-mech]*
raku-san	酪酸；らくさん	butyric acid
rakusan-en	酪酸塩	butyrate *[ch]*

raku'sui-shōgeki-shiken	落錘衝撃試験	falling-weight impact test *[mech]*
raku-tai	落体	falling body *[mech]*
Raku-yaki	楽焼；樂焼	Raku ware *[cer]*
raku'yō'sei no	落葉性の	deciduous *{adj}* *[bot]*
raku'yō-zai	落葉剤	defoliant *[mat]*
rāmen	老麺；拉麺	Chinese noodles *[cook]*
ran	卵	egg *[cyt]*
ran	蘭；蘭；らん	orchid *[bot]*
ran	欄；欄	column; field *{n}* *[comput]*
ranchō	乱調；亂調	hunting; racing *{n}* *[cont-sys]* *[elec]*
"	乱調	rough idling *{n}* *[mech-eng]*
randō'kō	藍銅鉱	azurite *[miner]*
rangyoku	藍玉	aquamarine *[miner]*
ran-hansha	乱反射	diffuse reflection *[phys]*
ran'katsu	卵割	cleavage *[bio]*
rankei no	卵形の	ovate *{adj}* *[math]*
rankin-rui	卵菌類	Oomycetes *[mycol]*
ran'ma	欄間	fanlight; transom *[arch]*
ran'ō	卵黄；らんおう	yolk *[bioch]*
ran'ō-nō	卵黄嚢	yolk sac *[embryo]*
ranpaku	卵白	albumen; eggwhite *[bioch]*
ranpō	卵胞；卵胞	follicle *[bio]*
ranryū	乱流；らんりゅう	turbulent flow *[fl-mech]*
ranryū-kakusan	乱流拡散	eddy diffusion *[fl-mech]*
ranryū-ken	乱流圏；亂流圈	turbosphere *[meteor]*
ranryū'ken-kaimen	乱流圏界面	turbopause *[meteor]*
ranryū-nensei	乱流粘性	turbulent viscosity *[fl-mech]*
ran-saibō	卵細胞；卵細胞	ovule *[cyt]*
ranshi	乱視	astigmatism *[opt]*
ran'shoku-shokubutsu	藍色植物	Cyanophyta *[bot]*
ran'shōseki	藍晶石	cyanite; kyanite *[miner]*
ransō	卵巣；卵巢	ovary (pl -ries) *{n}* *[anat]*
ransō	藍草	blue-green algae *[bot]*
ransō'un	乱層雲	nimbostratus *[meteor]*
ransū	乱数；亂數	random numbers *[math]*
ransū-hai'retsu	乱数配列	random number sequence *[math]*
ransui'enka	藍水鉛華	ilsemannite *[miner]*
ran-taisei	卵胎生	ovoviviparism *[v-zoo]*
rantan	ランタン	lanthanum (element)
rantanido-retsu	ランタニド列	lanthanide series *[ch]*
ran'tekkō	藍鉄鉱；藍鐵鑛	vivianite *[miner]*
ran'un	乱雲	nimbus (pl nimbi, -es) *[meteor]*

rappa	喇叭；らっぱ	trumpet; bugle *{n} [music]*
rappa-kan	喇叭管	Fallopian tube *[anat]*
rappa'zuisen	喇叭水仙	daffodil (flower) *[bot]*
ra'retsu suru	羅列する	enumerate *{vb} [math]*
rasai-rui	裸鰓類	Nudibranchia *[i-zoo]*
rasemi-tai	ラセミ体	racemic modification *[org-ch]*
rasen	螺旋；らせん	helix (pl helices, -es); spiral *{n} [math]*
rasen-jiku	螺旋軸	screw axis *[crys]*
rasen'jō-ten'i	螺旋状転位	screw dislocation *[photo]*
rasen-kaidan	螺旋階段	corkscrew staircase *[arch]*
rasen-undō	螺旋運動	spiral motion *[mech]*
rasha	羅紗；らしゃ	rhaxa; raxa *[tex]*
rashin-ban	羅針盤	compass *[eng]*
rashin-gi	羅針儀	compass *[eng]*
rashin'gi-bako	羅針儀箱	binnacle *[nav-arch]*
rashin-hō'i	羅針方位	compass bearing *[navig]*
rashi-shokubutsu	裸子植物	Gymnospermae *[bot]*
rei	令；齢；齡	instar *[i-zoo]*
rei	例	example *[sci-t]*
rei	零	zero (pl zeros, zeroes) *[math]*
reibai	冷媒	coolant; refrigerant *[mat]*
reibaku-yaku	励爆薬；勵爆藥	explosion donor *[ord]*
reibō	冷房	air conditioning; cooling *{n} [mech-eng]*
reichō-moku	霊長目；靈長目	Primates *[bio] [v-zoo]*
rei-den'i	零電位	zero potential *[phys]*
rei'en	冷炎	cool flame *[ch]*
rei-enshin	冷延伸	cold drawing; cold stretching *{n} [plas]*
rei-fuku	礼服；禮服	formal clothes *[cl]*
rei'gai	例外	exception (to the rule) *{n}*
rei'hatsu-den'atsu	励発電圧	excitation voltage *[elec]*
rei'i-chōsei	零位調整	zero adjustment *[eng]*
rei'i-hosei	零位補正	zero correction *[math]*
rei-in'kyoku-hō'shutsu	冷陰極放出	cold-cathode emission *[electr]*
rei'ji	励磁	magnetic excitation *[elecmg]*
reiji-hannō	零次反応	zero-order reaction *[p-ch]*
reiji suru	励磁する	magnetize *{vb} [elecmg]*
reika	零下	below zero *[thermo]*
reikan-atsuzō-sei	冷間圧造性	cold-headability *[met]*
reikan-kakō	冷間加工	cold working *{n} [met]*
reikan-tanzō'hin	冷間鍛造品	cold-forged product *[met]*
reikei	零系	null system *[math]*
rei'ketsu-dōbutsu	冷血動物	cold-blooded animal *[zoo]*

reiki	冷気	cold; chill; cold weather *{n} [meteor]*
reiki	励起	excitation *[elecmg]*
reiki-ichi'jū-kō	励起一重項	excited singlet *[atom-phys]*
reiki-jōtai	励起状態	excited state *[atom-phys]*
reiki-kikan	励起期間	induction period *[p-ch]*
reiki-kō	励起光	excited light *[opt]*
reiki'shi	励起子	exciton *[sol-st]*
reikō	冷光	cold light *[opt]*
rei'kyaku	冷却; れいきゃく	cooling *{n} [eng]*
reikyaku-eki	冷却液	coolant liquid; cooled liquid *[mat]*
reikyaku-ki	冷却器	condenser; cooler *[ch] [mech-eng]*
reikyaku-nō	冷却能	cooling power *[thermo]*
reikyaku-sui	冷却水	cooling water; cooled water *[ch]*
rei-ritsu	捩率	torsion *[mech]*
rei-seikei	冷成形	cold molding *{n} [plast]*
rei-sen	零線	zero line *[geophys]*
reishi	茘枝; れいし	lichi; litchi (tree, fruit) *[bot]*
reishin	励振	excitation *[phys]*
rei-shūgō	零集合	null set *[math]*
rei'shuku	冷縮	condensation (of gas) *[ch]*
reisui-kai	冷水塊	cold eddy *{n} [geol]*
reisui-kayō-sei	冷水可溶性	cold-water solubility *[ch]*
rei-ten	零点	zero point *[math]*
reitō	冷凍; れいとう	refrigeration *[mech-eng]*
reitō-ki	冷凍機	refrigerating machine *[eng]*
reitō-shoku'hin	冷凍食品	frozen food *[cook]*
reitō-zai	冷凍剤	cryogen *[phys]*
reiyō	羚羊; れいよう	antelope (animal) *[v-zoo]*
reizō'ko	冷蔵庫; 冷藏庫	refrigerator *[mech-eng]*
reizō'sha	冷蔵車	refrigerator car *[rail]*
reki'gan	礫岩; れきがん	rudite *[geol]*
"	礫岩	conglomerate *[petr]*
reki'sei	瀝青; れきせい	asphalt; bitumen *[mat]*
rekisei(-busshitsu)	瀝青(物質)	pitch *{n} [mat]*
rekisei-sagan	瀝青砂岩	tar sand *[petr]*
rekisei-tan	瀝青炭	bituminous coal *[geol]*
rekishi	歴史; 歷史	history (pl -ries)
rekishi'ka	歴史家	historian (a person)
rekka	劣化	degradation; deterioration *[ch] [eng]*
rekkai-ka	裂開果	dehiscent fruit *[bot]*
rekkai-sei	裂開性	fissility *[geol]*
rekkaku	劣角	inferior angle; minor angle *[math]*

rekko	劣弧	inferior arc; minor arc *[math]*
rekkyo	列挙	enumeration *[math]*
rekkyo suru	列挙する	enumerate; list *{vb} [pr]*
ren	連; れん	tribe *[bot] [syst]*
"	連	string *{n} [eng]*
"	連	ream (paper) *{n} [mat]*
renbō-ki	練紡機	roving frame *[tex]*
renchū	連铸; 連鑄	continuous casting *{n} [met]*
rendō	連動	interlocking *{n} [mech-eng]*
rendō-sōchi	連動装置	interlocking device *[mech-eng]*
rendō suru	連動する	move interlocked (interlinked) *{vb}*
renga	連火: 灬; れんが	kanji (q.v.) radical denoting fire *[pr]*
renga	煉瓦	brick *[arch] [mat]*
rengen	連言	conjunction *[logic]*
renge'sō	蓮華草; れんげそう	Chinese milk vetch *[bot]*
ren-gōkin	錬合金	wrought alloy *[met]*
renjaku	連雀; れんじゃく	waxwing (bird) *[v-zoo]*
renji'fu	連字符	hyphen *[comput] [pr]*
ren'jitsu; hasu no mi	蓮実; 蓮實	lotus seed *[bot] [cook]*
ren'ketsu	連結; れんけつ	concatenation; link; linkage *{n} [comput]*
renketsu-ban	連結板	connecting plate *[mech-eng]*
renketsu-bu	連結部	connecting-link portion *[mech-eng]*
renketsu-hannō	連結反応	ligation *[mol-bio]*
renketsu-i'sei	連結異性	linkage isomerism *[ch]*
renketsu-kan	連結杆	connecting rod *[mech-eng]*
renketsu-ki	連結器	coupler *[eng]*
renketsu-sei	連結性	connectivity *[comput]*
renketsu suru	連結する	concatenate *{vb} [comput]*
renkō	錬鋼; 鍊鋼	wrought steel *[met]*
renkon	蓮根	lotus root *[bot] [cook]*
renniku	練肉; 練肉	grinding; milling *{n} [mech-eng]*
rennyū	練乳	condensed milk *[cook]*
renraku	連絡	communication; liaison *[comm]*
ren'ritsu-futō'shiki	連立不等式	system of inequalities *[math]*
ren'ritsu-hōtei'shiki	連立方程式	simultaneous equations *[math]*
rensa	連鎖; れんさ	chain; chaining *{n} [comput]*
rensa-baku'hatsu	連鎖爆発	chain explosion *[ch]*
rensa-hannō	連鎖反応	chain reaction *[nucleo] [poly-ch]*
rensa-idō-zai	連鎖移動剤	chain transfer agent *[poly-ch]*
rensa('jō)-kyūkin	連鎖(状)球菌	streptococcus (pl -i) *[microbio]*
rensa-jūgō	連鎖重合	chain polymerization *[poly-ch]*
rensa-ritsu	連鎖律	chain rule *[math]*

rensa-seichō-hannō	連鎖成長(生長)反応	chain propagation reaction *[poly-ch]*
rensa-tantai	連鎖担体	chain carrier *[poly-ch]*
rensa-teishi-hannō	連鎖停止反応	chain termination reaction *[poly-ch]*
rensa-teishi-zai	連鎖停止剤	chain stopper *[poly-ch]*
rensei	連星	binary star; couple *{n} [astron]*
ren'setsu-bō	連接棒	connecting rod *[mech-eng]*
renshi	連糸	string *{n} [comput]*
renshin'ro-kōhō	連針路航法	traverse sailing *[navig]*
rensō	連想	association *[psy]*
rentai-kei	連体形	participial adjective *[gram]*
rentan	錬炭	briquet coal *[mat]*
ren-tetsu	錬鉄；錬鐵	wrought iron *[met]*
rentsū-ro	連通路	connecting passageway *[arch]*
renyūmu	レニウム	rhenium (element)
renzoku	連続；連續	consecutive *[comput]*
"	連続；れんぞく	sequence *[eng]*
"	連続	continuous *[math]*
"	連続	continuum *[math] [spect]*
"	連続	train *{n} [phys]*
renzoku-baiyō	連続培養	continuous culture *[microbio]*
renzoku-bangō	連続番号	consecutive numbers *[math]*
renzoku-chōhyō	連続帳票	continuous form *[comput]*
renzoku-chūzō	連続鋳造	continuous casting *{n} [met]*
renzoku-jūgō	連続重合	continuous polymerization *[poly-ch]*
renzoku-kaku'ritsu-hensū	連続確率変数	continuous random variable *[math]*
renzoku no hōsoku	連続の法則	succession, law of *[bio]*
"	連続の法則	continuity, law of *[math]*
renzoku-oshi'dashi-seikei	連続押出成形	continuous extrusion molding *{n} [eng]*
renzoku-ro	連続炉	continous furnace *[met]*
renzoku-sei	連続性	continuity *[civ-eng]*
renzoku'tai	連続体	continuum *[math] [phys]*
renzu	レンズ	lens *[opt]*
reppen	裂片	lobe *[bot]*
ressei	劣性	recessive *{n} [gen]*
ressei no	劣性の	recessive *{adj} [gen]*
ressei-so'shitsu	劣性素質	recessive trait *[gen]*
ressha	列車	train *{n} [rail] [transp]*
retsu	列；れつ	column; string *{n} [math]*
"	列	train *{n} [mech-eng]*
retsu'dan-chō	裂断長	breaking length *[paper]*
retsu'ri	列理	grain effect *[rub]*
retteru	レッテル	label *{n} [paper]*

rēzā	レーザー	laser *[opt]*
ribo-kaku'san	リボ核酸	ribonucleic acid; RNA *[bioch]*
richūmu	リチウム	lithium (element)
ri'eki	利益	gain; profit *{n}* *[econ]*
ri'eki	離液	syneresis *[ch]*
ri'eki-jun'retsu	離液順列	lyotropic series *[ch]*
rihan suru	離反する	deflect *{vb}* *[psy]*
rikai	理解	understanding *{n}* *[psy]*
rikai	離解	maceration *[paper]* *[pharm]* *[zoo]*
rikan saseru	離間させる	separate; space apart *{vb t}*
rikan suru	離間する	space apart *{vb}* *[pr]*
rikei-sei	離型性	mold releasability *[mech-eng]*
rikei-shi	離型紙	release paper *[plas]*
rikei-zai	離型剤	mold lubricant; mold releasing agent; releasing agent *[mat]* *[met]* *[rub]*
"	離型剤	parting compound; parting agent *[met]*
riki'gaku	力学; りきがく	dynamics *[mech]*
"	力学	mechanics *[phys]*
rikigaku-ji	力学時	dynamical time *[astron]*
rikigaku-shisa	力学視差	dynamical parallax *[astron]*
riki-jō	力場	force field *[mech]*
riki-ka	力価; 力價	titer *[ch]*
"	力価	potency (pl -cies) *[pharm]*
riki-ritsu	力率	power factor *[elec]*
riki'ritsu-kei	力率計	power factor meter *[eng]*
riki-seki	力積	impulse *[mech]*
riki-sen	力線	line of force *[phys]*
rikkyaku-ten	立脚点	footing; ground; position *{n}*
riku	陸; りく	land *{n}* *[geogr]*
rikuchi	陸地	land *{n}* *[geogr]*
rikuchi-shinnyū-hyo'shiki	陸地進入標識	landfall mark *[navig]*
rikuchi-sho'nin	陸地初認	landfall *[navig]*
riku-dana	陸棚	continental shelf *[geol]*
riku'gen-taiseki'butsu	陸源堆積物	terrigenous sediment *[geol]*
riku-gun	陸軍	army (pl armies) *[mil]*
riku-hyō	陸標	landmark *[eng]* *[navig]*
riku-kaze; riku'fū	陸風	land breeze *[meteor]*
rik(u)kyō	陸橋	viaduct *[civ-eng]*
"	陸橋	land bridge *[geogr]*
riku-nanpū	陸軟風	land breeze *[meteor]*
rikusei-dōbutsu	陸生動物; 陸棲動物	land animal *[zoo]*
rikusei-shokubutsu	陸生植物; 陸棲植物	land plant *[bot]*

riku-sui	陸水	inland water *[geogr]*
rikusui'gaku	陸水学	limnology *[ecol]*
"	陸水学	hydrology *[geophys]*
rin	燐；りん	phosphorus (element)
rinban'seki	燐礬石	amblygonite *[miner]*
rindō	竜胆；龍膽	autumn bellflower; gentian (flower) *[bot]*
rin'dōkō	燐銅鉱	libethenite *[miner]*
rin'gaku	林学	forestry *[ecol]*
ringo	林檎；りんご	apple (fruit, tree) *[bot]*
ringo no shibori-kasu	林檎の絞り滓	pomace *{n} [agr]*
ringo-san	林檎酸	malic acid
ringo'san-en	林檎酸塩	malate *[ch]*
rinji no	臨時の	contingent *{adj}*
rinjō-koku'en	鱗状黒鉛	flaky graphite *[geol]*
rinka-aen	燐化亜鉛	zinc phosphide
rinkai	臨界；りんかい	criticality *[nucleo]*
"	臨界	threshold *[phys]*
rinkai	臨海	coast *{n} [geogr] [ocean]*
rinkai-atsu	臨界圧	critical pressure *[fl-mech]*
rinkai-kaku	臨界角	critical angle *[math]*
rinkai-miseru-nōdo	臨界ミセル濃度	critical micelle concentration *[p-ch]*
rinkai-mitsudo	臨界密度	critical density *[astron]*
rinkai no	臨海の	coastal *{adj} [geogr]*
rinkai-ondo	臨界温度	critical temperature *[p-ch]*
rinkai'seki	燐灰石	apatite *[miner]*
rinkai-shindō-sū	臨界振動数	critical frequency *[electr]*
rinkai-shisū	臨界指数	critical index *[phys]*
rinkai-shitsu'do	臨界湿度	critical humidity *[ch-eng]*
rinkai-shitsu'ryō	臨界質量	critical mass *[nucleo]*
rinkai-ten	臨界点	critical point *[math] [p-ch]*
rin'kaku	輪郭；りんかく	outline; profile *{n} [graph]*
"	輪郭	contour; delineation *{n} [math] [sci-t]*
rinkaku-sen	輪郭線	outline; border line *{n} [graph]*
rinkan	輪環	ring *{n} [eng]*
rinka-tetsu	燐化鉄	iron phosphide
rinkei-dōbutsu	輪形動物	Rotifera *[i-zoo]*
rin'keiseki	鱗珪石	tridymite *[miner]*
rinkō	燐光	phosphorescence *[atom-phys]*
rin'kōseki	燐鉱石	rock phosphate *[geol]*
rinkō-shi'setsu	臨港施設	terminal facilities *[civ-eng]*
rinkō'tai	燐光体	luminophor; phosphor; phosphorescent substance *[phys]*

rinpa	リンパ	lymph *[hist]*
rinpa-kei	リンパ系	lymphatic system *[anat]*
rinpen'jō no	鱗片状の	fish-scale form *{adj}* *[graph]*
rinsan	燐酸；りんさん	phosphoric acid
rinsan-aen	燐酸亜鉛	zinc phosphate
rin-sanbutsu	林産物	forest product(s) *[ind-eng]*
rinsan-dō	燐酸銅	copper phosphate
rinsan-en	燐酸塩	phosphate *[ch]*
rinsan-ichi-karyūmu	燐酸一カリウム	monobasic potassium phosphate
rinsan-ichi'suiso-namari	燐酸一水素鉛	lead monohydrogen phosphate
rinsan-ion	燐酸イオン	phosphate ion *[ch]*
rinsan'ka	燐酸化	phosphorylation *[org-ch]*
rinsan'ka-kōso	燐酸化酵素	kinase *[bioch]*
rinsan-namari	燐酸鉛	lead phosphate
rinsan-ni-karyūmu	燐酸二カリウム	dibasic potassium phosphate
rinsan-ni'suiso-anmonyūmu	燐酸二水素アンモ=ニウム	ammonium dihydrogen phosphate
rinsan-ni'suiso-karushūmu	燐酸二水素カル=シウム	calcium dihydrogen phosphate
rinsan-ni'suiso-karyūmu	燐酸二水素カリウム	potassium dihydrogen phosphate
rinsan-ni'suiso-namari	燐酸二水素鉛	lead dihydrogen phosphate
rinsan-ni'suiso-natoryūmu	燐酸二水素ナト=リウム	sodium dihydrogen phosphate
rinsan-san'natoryūmu	燐酸三ナトリウム	tribasic sodium phosphate
rinsan-suiso-ni'anmonyūmu	燐酸水素二アンモ=ニウム	diammonium hydrogen phosphate
rinsan-suiso-ni'karyūmu	燐酸水素二カリウム	dipotassium hydrogen phosphate
rinsan-suiso-ni'natoryūmu	燐酸水素二ナト=リウム	disodium hydrogen phosphate
rinsan-tetsu	燐酸鉄	iron phosphate
rinsei	輪生	verticillation *[bot]*
rin'seidō	燐青銅	phosphor bronze *[met]*
rinsetsu-kaku	隣接角	adjacent angles *[math]*
rinshi'gaku	鱗翅学	lepidopterology *[zoo]*
rin-shi'shitsu	燐脂質	phospholipid *[bioch]*
rin-tanpaku'shitsu	燐蛋白質	phosphoprotein *[bioch]*
ri'nyō-yaku (ri'nyō-zai)	利尿薬（利尿剤）	diuretic *{n}* *[pharm]*
rinzu	綸子；りんず	figured satin *[tex]*
ripo-tanpaku'shitsu	リポ蛋白質	lipoprotein *[bioch]*
rippō	立方；りっぽう	cube *{n}* *[math]*
rippō-hachi'men'tai	立方八面体	cuboctahedron *[math]*
rippō'kei no	立方形の	cuboidal *{adj}* *[math]*

rippō-kon	立方根	cube root *[math]*
rippō-saimitsu-kōzō	立方最密構造	cubic closest-packed structure *[crys]*
rippō-shō	立方晶	cubic crystal *[crys]*
rippō-shōkei	立方晶系	cubic system; isometric crystal system *[crys]*
rippō'tai	立方体	cube *{n}* *[math]*
ripuru-den'atsu	リプル電圧	ripple voltage *[elec]*
ri'reki(-genshō)	履歴(現象)	hysteresis (pl -eses) *[phys]*
ri'reki-son	履歴損	hysteresis loss *[phys]*
riron	理論; りろん	theory (pl -ries) *[math]* *[sci-t]*
riron-butsuri'gaku	理論物理学	theoretical physics *[phys]*
riron-dansū	理論段数	theoretical number of plates *[ch-eng]*
riron-sūgaku	理論数学	theoretical mathematics *[math]*
risan-hensū	離散変数	discrete variable *[math]*
risan-kaku'ritsu-hensū	離散確率変数	discrete random variable *[math]*
risan no	離散の	discrete *{adj}* *[math]*
risan'teki-hyōgen	離散的表現	discrete representation *[comput]*
ri'setsu	離接	disjunction *[logic]*
rishi	利子	interest *[econ]*
rishin-kinten-kaku	離心近点角	eccentric anomaly *[astron]*
rishin-ritsu	離心率	eccentricity (pl -ties) *[math]*
rishō	離昇	lift-off *{n}* *[aero-eng]*
rishō	離漿	syneresis (pl -eses) *[ch]*
risō	理想; りそう	ideal *{n}* *[math]* *[psy]*
risō-kitai	理想気体	ideal gas *[thermo]*
risō-kitai no hōsoku	理想気体の法則	ideal gas law *[thermo]*
risō-kitai no shiki	理想気体の式	ideal gas equation *[thermo]*
ri'soku	利息	interest *[econ]*
risō-ryūtai	理想流体	ideal fluid *[fl-mech]*
risō'teki na	理想的な	ideal; perfect *{adj}* *[math]*
risō-yōeki	理想溶液	ideal solution *[ch]*
risshin-ben	立心扁; 立心偏: 忄; りっしんべん	kanji (Chinese character) radical denoting the mind *[pr]*
risshū	立秋	first day of autumn *[astron]*
risshun	立春	first day of spring *[astron]*
rissoku	律速	rate-controlling *{n}* *[ch]*
rissoku-dankai	律速段階	rate-determining step *[ch]*
rissoku-katei	律速過程	rate-determining process *[ch]*
rissuru	律する	regulate *{vb t}*
risu	栗鼠; りす	squirrel (animal) *[v-zoo]*
risu-zaru	栗鼠猿	squirrel monkey (animal) *[v-zoo]*
ritan-yaku; ritan-zai	利胆薬; 利胆剤	cholagogue *{n}* *[pharm]*

ritoku	利得	gain *{n} [econ] [elec]*
ritomasu	リトマス	litmus *[mat]*
ritsu	率；りつ	ratio (pl ratios) *[math]*
"	率	rate *{n} [sci-t]*
-ritsu	-律；りつ	-law *{suffix} [sci-t]*
-ritsu	-率	-rate; -percent(age) *{suffix} [math]*
ritsu'men-zu	立面図；立面圖	elevation view *[graph]*
rittai	立体；りったい	cube; solid *{n} [math]*
rittai-haichi	立体配置	configuration *[ch] [math]*
rittai-haiza	立体配座	conformation; constellation *[org-ch]*
rittai-hizumi	立体歪；立體歪	steric strain *[ch]*
rittai-hōsō	立体放送	stereophonic broadcast *[comm]*
rittai-inshi	立体因子	steric factor *[p-ch]*
rittai-i'sei	立体異性	stereoisomerism *[org-ch]*
rittai-isei'tai	立体異性体	stereoisomer *[org-ch]*
rittai-jūgō'tai	立体重合体	space polymer *[poly-ch]*
rittai-kagaku	立体化学	stereochemistry *[p-ch]*
rittai-kaku	立体角	solid angle *[math]*
rittai-kika'gaku	立体幾何学	solid geometry *[math]*
rittai-kisoku-do	立体規則度	stereoregularity; stereospecificity *[org-ch]*
rittai-kisoku-sei	立体規則性	stereoregularity; stereospecificity *[org-ch]*
rittai-kisoku'sei-jūgō	立体規則性重合	stereoregular (stereospecific) polymerization *[poly-ch]*
rittai'kōsa-jūji'ro	立体交差十字路	cloverleaf (pl -leafs, -leaves) *[traffic]*
rittai'kyō	立体鏡	stereoscope *[opt]*
rittai-onkyō-kei	立体音響系	stereophonic sound system *[acous]*
rittai-onkyō-kōka	立体音響効果	stereophony *[acous]*
rittai'on'teki na	立体音的な	stereophonic *{adj} [acous]*
rittai-saimitsu-kōzō	立体最密構造	cubic closest-packed structure *[crys]*
rittai-shōgai	立体障害	steric hindrance *[ch]*
rittai'teki na	立体的な	three-dimensional *{adj} [math]*
rittai-tokui'sei-jūgō	立体特異性重合	stereospecific polymerization *[poly-ch]*
rittai-zō	立体像	stereoscopic image *[opt]*
rittō	立刀：刂；りっとう	kanji (q.v.) radical denoting a sword *[pr]*
rittoru	立；リットル	liter *[mech]*
riyō	利用	use; utilization; application *{n}*
riyō'sha	利用者	user *[comput]*
riyū	理由；りゆう	reason; cause; grounds *{n}*
riyū'sho	理由書	document giving reason *[legal]*
ro	炉；爐；ろ	furnace; kiln; oven *[eng]*

ro	艪；艫；櫓；ろ	ore; paddle *{n} [eng]*
rō	蠟；ろう	wax *{n} [mat]*
rō'a	聾唖；聾啞；ろうあ	deafness and dumbness; deaf mute *[med]*
roba	驢馬；ろば	donkey; ass (animal) *[v-zoo]*
roban-zai	路盤材	roadbed material *[mat]*
rō-biki	蠟引き	waxing *{n} [ind-eng]*
robotto	ロボット	robot *[cont-sys]*
rō-chōkoku('jutsu)	蠟彫刻(術)	cerography *[graph]*
rōden	漏電	leak *{n} [elec]*
rōden-shadan-ki	漏電遮断器	ground fault interrupter *[elec]*
rōdō	労働；労仂；勞働	labor; work *{n}*
rōdō-hi	労働費	labor cost *[econ]*
rō'ei-ji'soku	漏洩磁束	leakage flux *[elecmg]*
ro'eki	沪液；濾液；ろえき	filtrate *[sci-t]*
ro'en	路縁	berm *[civ-eng]*
rofu	沪布；濾布	filter cloth *[mat]*
rofu-sokudo	沪布速度	filter cloth speed *[mech]*
rogai-chūsha	路外駐車	offstreet parking *{n} [civ-eng]*
rō-gami	蠟紙	wax paper *[mat]*
rō-gata	蠟型	wax impression; wax pattern *[eng]*
rōhai'butsu	老廃物；老廢物	waste product (of body) *{n} [physio]*
roha-kairo	沪波回路	filtration circuit *[elec]*
roha-ki	沪波器；濾波器	filter *{n} [eng]*
rōhi	浪費	wastefulness *[ecol]*
rōhi suru	浪費する	waste *{vb t}*
rohō	沪胞；濾胞	follicle *[bio]*
rojūmu	ロジウム	rhodium (element)
roka	沪過；濾過；ろか	filtration *[sci-t]*
rōka	廊下	corridor; passageway; hallway *[arch]*
roka-jozai	沪過助剤；濾過助劑	filter aid *[mat]*
roka-ki	沪過器	filter *{n} [eng]*
roka-sei	沪過性	filterability *[eng]*
rōka-sei	老化性	aging property *[mat]*
roka-sesshoku'sha	沪過摂食者	filter feeder *[zoo]*
roka-sokudo	沪過速度	filtering speed *[mech]*
roka suru	沪過する	filter *{vb t}*
ro-kata	路肩	road shoulder *[civ-eng]*
rō'ketsu-zome	臈纈染め；臘纈染め	batik *[tex]*
rokkaku'ana'tsuki-neji	六角穴付きねじ	hexagon socket screw *[eng]*
rokkaku-kei	六角形	hexagon *[math]*
rokka no	六価の	hexavalent *{adj} [ch]*
rokkotsu	肋骨	rib bone *[anat]*

rokkyoku-kan	六極管	hexode *[electr]*
rokō	露光	exposure *[photo]*
rokō-kei	露光計	exposure meter *[eng]*
roku	六	six *{n} [math]*
roku'bun-gi	六分儀	sextant *[eng]*
roku'bun no ichi	六分の一	one-sixth *{n} [math]*
roku'fukka-iō	六弗化硫黄	sulfur hexafluoride
roku'gatsu	六月	June (month)
roku'jū-shinpō	六十進法	sexagesimal scale *[math]*
rokumen-kōzō	六面構造	hexagonal structure *[crys]*
rokumen'tai	六面体	hexahedron (pl -s, -hedra) *[math]*
roku'on	録音; 錄音	sound recording *{n} [acous]*
"	録音; ろくおん	recording *{n} [sci-t]*
roku'on-ki	録音機	sound recorder *[eng]*
roku'on suru	録音する	record *{vb} [acous]*
roku'on'tai	録音体	recording *{n} [acous]*
roku'rin'sha	六輪車	six wheeler; six-wheel car *[mech-eng]*
roku'ro	轆轤; ろくろ	potter's wheel; pulley *[eng]*
roku'shō	緑青; ろくしょう	patina (pl -s, -inae) *[met]*
"	緑青; 綠青	verdigris *[org-ch]*
rokutan'tō	六炭糖	hexose *[bioch]*
rokuzai	肋材	rib *{n} [arch]*
romen-haisui	路面排水	surface drainage *[civ-eng]*
ronbun	論文	article; thesis *[comm] [pr]*
ronri	論理; ろんり	logic *[electr]*
ronri-enzan	論理演算	logic(al) operation *[comput]*
ronri-enzan'shi	論理演算子	logical operator *[comput]*
ronri'gaku	論理学	logic *[math]*
ronri-seki	論理積	AND *[comput]*
ronri'seki-kairo	論理積回路	AND circuit *[electr]*
ronri-sekkei	論理設計	logical design *[comput]*
ronri-shiki	論理式	logical expression *[comput]*
ronri-soshi	論理素子	logic element *[comput]*
ronri-tan'i	論理単位	logical unit *[comput]*
ronri-wa	論理和	logical sum *[comput]*
"	論理和	OR *[comput]*
ronri'wa-kairo	論理和回路	OR circuit *[electr]*
ronshō	論証	argument *[logic]*
ronsō	論争; 論爭	argument; controversy *{n}*
roppō-kōshi	六方格子	hexagonal lattice *[crys]*
roppō-sai'mitsu-kōzō	六方最密構造	hexagonal closest-packed structure *[crys]*
roppō-shōkei	六方晶形	hexagonal system *[crys]*

rōrenshūmu	ローレンシウム	lawrencium (element)
rōseki	蠟石	agalmatolite *[miner]*
roshi	沪紙；濾紙	filter paper *[mat]*
rōshi	老視	presbyopia *[med]*
ro-shin	炉心；爐芯	reactor core *[nucleo]*
roshin-yōyū	炉心溶融	meltdown *{n}* *[nucleo]*
ro-shitsu	沪室	filter chamber *[ch]*
ro-shō	沪床；ろしょう	filter bed *[ch]*
ro'shutsu	露出	exposure *[civ-eng]* *[min-eng]*
ro'soku	路側	roadside *{n}* *[civ-eng]*
rōsoku	蠟燭	candle *[mat]*
rosui-do	沪水度	freeness *{n}* *[paper]*
rosui-sei	沪水性	degree of freeness *[paper]*
roten	露点	dew point *[ch]* *[meteor]*
roten-kōhan	露天甲板	weather deck *[nav-arch]*
roten-saikō	露天採鉱	open pit mining; strip mining *[min-eng]*
rōto	漏斗；ろうと	funnel *{n}* *[eng]*
rōto-gumo	漏斗雲	funnel cloud *[meteor]*
rōto-kan	漏斗管	funnel tube *[ch-eng]*
rōto-kō	漏斗孔	crater *[geol]*
rōwa	漏話	crosstalk *[comm]*
royō	沪葉	filter leaf *[ch]*
ro'zoku	蘆票；ろぞく	sweet sorghum *[bot]*
rō-zuke	蠟付け	soldering (hard) *{n}* *[met]*
rubijūmu	ルビジウム	rubidium (element)
rui	類；類	class; genus (pl genera) *{n}* *[syst]*
"	類；るい	type; sort; species; variety *{n}*
rui-betsu	類別	classification *[syst]*
rui'en-kankei	類縁関係	relationship *[bio]*
ruigi'go-jiten	類義語辞典	thesaurus (pl -sauri, -es) *[pr]*
ruigo	類語	synonym *[gram]*
ruigo'shū	類語集	thesaurus *[pr]*
rui'ji no-	類似の-	quasi- *{prefix}*
ruiji-sei	類似性	resemblance; similarity *{n}*
ruiji'tai	類似体；類似體	analogue *[bio]*
ruijō	累乗；累乘	power; radical *[math]*
rui'kinzoku-genso	類金属元素	metalloid element *[ch]*
ruisan-ki	累算器	accumulator *[comput]*
rui'seki-bunpu-kansū	累積分布関数	cumulative distribution function *[math]*
rui'seki-kyokusen	累積曲線	cumulative curve *[math]*
rui-sen	涙腺；涙腺	lacrimal gland *[anat]*
rui'senkei-dōbutsu	類線形動物	Nematomorpha *[i-zoo]*

ruiteki	涙滴	teardrop *{n} [lap]*
ruiwa	累和	cumulative sum *[math]*
ruri	瑠璃	lapis lazuli *[petr]*
ruri-jisa	瑠璃萵苣；るりじさ	borage (herb) *[bot]*
ruri-zora	瑠璃空	azure sky *[astron]*
rusu'ban-denwa	留守番電話	telephone answering machine *[comm]*
rutenyūmu	ルテニウム	ruthenium (element)
rutsubo	坩堝	crucible *[sci-t]*
ryakki-hō	略記法	abbreviation *[comput]*
ryaku'fu	略符；りゃくふ	logogram *[graph]*
ryaku'go	略語	code *{n} [comput]*
"	略語	abbreviation *[pr]*
ryakumei	略名	abbreviation; abbreviated name *{n}*
ryakushō	略称；略稱	abbreviation *[comm]*
ryaku-zu	略図；略圖	schematic view *{n} [graph]*
ryō	良	good (grade) *[ind-eng]*
ryō	量；りょう	quantity; amount; volume *[math]*
"	量	mass *[mech]*
ryō	稜	ridge; edge *{n} [geol] [math]*
ryō'aen'kō	菱亜鉛鉱	smithsonite *[miner]*
ryō-chokkaku-sankak(u)kei	両直角三角形	birectangular triangle *[math]*
ryō'dōtai	良導体	good conductor *[elec]*
ryō'fusseki	菱弗石	chabazite *[miner]*
ryō'gawa-kentei	両側検定	two-tailed test *[math]*
ryō'gawa ni-	両側に-；兩側に-	amphi- *{prefix}*
ryō'gawa-sōshō	両側相称	bilateral symmetry *[bio]*
ryō-heikin-bunshi'ryō	量平均分子量	weight-average molecular weight *[ch]*
ryōhen-kyōyū no kaku	両辺共有の角	coterminal angles *[math]*
ryōhō	両方	both *{pron}*
ryō'iki	領域；りょういき	region *[comput]*
"	領域	area; domain; territory *[math]*
ryōji'kan	領事館	consulate *[arch]*
ryō'ka	量化	quantification *[sci-t]*
ryōkai	領海	territorial waters *[geogr]*
ryōkai-do	了解度	intelligibility *[psy]*
ryo'kaku; ryo'kyaku	旅客	passenger; tourist *[trans]*
ryō-kaku	稜角	dihedral angle *[math]*
ryokak(u)ki	旅客機	passenger airplane *[aero-eng]*
ryokaku'sen	旅客船	passenger ship *[nav-arch]*
ryokaku-setsubi	旅客設備	passenger facilities *[trans]*
ryokan	旅館；りょかん	Japanese inn *[arch]*
ryoken	旅券	passport *[trans]*

ryōkin	料金	charge; fee; fare; rate *{n} [econ]*
ryokō	旅行；りょこう	travel; trip *{n} [trans]*
ryokō-annai'jo	旅行案内所	travel bureau *[trans]*
ryokō-ni'motsu	旅行荷物	baggage; luggage *[trans]*
ryokō-nittei	旅行日程	itinerary (pl -aries) *[trans]*
ryoku-cha	緑茶；りょくちゃ	green tea *[cook]*
ryoku'chi-tai	緑地帯；緑地帶	green belt *[civ-eng]*
ryokuchū'seki	緑柱石	beryl *[miner]*
ryokudei'seki	緑泥石	chlorite *[miner]*
ryō'kudo'kō	菱苦土鉱	magnesite *[miner]*
ryō'kudo'seki	菱苦土石	magnesite *[miner]*
ryoku-eki	緑液	green liquor *[paper]*
ryoku'en'dōkō	緑塩銅鉱	atacamite *[miner]*
ryoku'enkō	緑鉛鉱	pyromorphite *[miner]*
ryokuhi'dōkō	緑砒銅鉱	olivenite *[miner]*
ryoku'nai'shō	緑内症	glaucoma *[med]*
ryoku'senseki	緑閃石	actinolite *[miner]*
ryoku'shoku-shokubutsu	緑色植物	Chlorophyta *[bot]*
ryoku-sō	緑藻	green algae *[bot]*
ryokyaku	旅客	passenger *[trans]*
ryōkyaku-ki	両脚規	divider; compass *[eng] [math]*
ryōkyoku-sei	両極性	binary transmission *[comm]*
ryōmen'tai	菱面体	rhombohedron *[math]*
ryōmen'tai-shōkei	菱面体晶系	rhombohedral system *[crys]*
ryō'mimi-chō	両耳聴	biaural hearing *{n} [acous]*
ryō'mochi-bari	両持梁	simple beam *[eng]*
ryō-ō no	両凹の	biconcave; concavo-concave *{adj} [opt]*
ryōri	料理；りょうり	cookery; cooking *{n} [food-eng]*
ryōri'ten	料理店	restaurant *[arch]*
ryōsan	量産	mass production *[ind-eng]*
ryōsei	両性；兩性	amphoteric *[ch]*
ryōsei-denkai'shitsu	両性電解質	ampholyte; amphoteric substance *[ch]*
ryōsei-dōbutsu	両性動物	hermaphrodite *[bio]*
ryōsei-ion	両性イオン	ampho-ion *[ch]*
ryōsei-rui	両棲類；兩生類	Amphibia *[v-zoo]*
ryōshi	量子；りょうし	quantum (pl quanta) *[quant-mech]*
ryōshi-hi'yaku	量子飛躍	quantum leap; quantum jump *[phys]*
ryōshi'ka	量子化	quantization *[quant-mech] [sci-t]*
ryōshi-kagaku	量子化学	quantum chemistry *[ch]*
ryōshi-kō'ritsu	量子効率	quantum efficiency *[electr]*
ryō'shinbai-sei	両親媒性	amphiphilic property; amphipatic property *[org-ch]*

ryōshi-rikigaku	量子力学	quantum mechanics *[phys]*
ryōshi-sū	量子数	quantum number *[quant-mech]*
ryō-shitsu no	良質の	good-quality *{adj}*
ryō'soku-sōshō	両側相称	bilateral symmetry *[bio]*
ryōsui(-kei)	両錐(形)	bipyramid *[math]*
ryō'taisū-memori	両対数目盛	logarithmic paper *[math]*
ryotei	旅程	itinerary (pl -aries) *[trans]*
ryotei-nikki	旅程日記	itinerary *[trans]*
ryō'te-kiki	両手利き	ambidextrality *[physio]*
ryō'tekkō	菱鉄鉱; 菱鐵鑛	siderite *[miner]*
ryō-totsu no	両凸の	biconvex *{adj} [opt]*
"	両凸の	convexo-convex *{adj} [opt]*
ryū	留; りゆう	stationary point *[astron]*
ryū	竜; 龍; りゅう	dragon (mythical animal)
ryū'an	硫安	ammonium sulfate (abbrev in Japanese)
ryūan'dōgin'kō	硫安銅銀鉱	polybasite *[miner]*
ryūdo	粒度; りゅうど	grain size; particle size *[mech]*
ryūdō	流動; りゅうどう	drift; flow; flowing *{n} [fl-mech]*
ryūdo-bunpu	粒度分布	particle size distribution *[mech]*
ryūdo-bunpu-hōsoku	粒度分布法則	size distribution (of particles), law of *[eng]*
ryūdō-den'i	流動電位	flow potential *[elec]*
ryūdō-do	流動度	fluidity *[fl-mech]*
ryūdō'gaku	流動学	rheology *[mech]*
ryūdō-hensei-sayō	流動変成作用	rheomorphism *[petr]*
ryūdō'ka	流動化	fluidization *[ch-eng] [fl-mech]*
ryūdō'ka-zai	流動化剤	fluidizing agent *[pharm]*
ryūdō-kōgaku	流動光学	rheooptics *[opt]*
ryūdō-ō'ryoku	流動応力	flow stress *[mech]*
ryūdō-parafin	流動パラフィン	liquid paraffin *[mat]*
ryūdō-ritsu	流動率	fluidity *[fl-mech]*
ryūdō-saibō-keisoku-hō	流動細胞計測法	flow cytometry *[cyt]*
ryūdō-sei	流動性	fluidity *[fl-mech]*
ryūdō-shō	流動床	fluidized bed *[eng]*
ryūdō-sō	流動層	fluidized bed *[eng]*
ryūdō-sokudo	流動速度	drift velocity *[sol-st]*
ryūdō-ten	流動点	pour point *[fl-mech]*
ryūgan	竜眼; 龍眼	longan (tree, fruit) *[bot]*
ryūgū no tsukai	竜宮の使い	oarfish (fish) *[v-zoo]*
ryūhi'dōkō	硫砒銅鉱	enargite *[miner]*
ryūhi'tekkō	硫砒鉄鉱	mispickel; arsenopyrite *[miner]*
ryūhyō	流氷; りゅうひょう	ice flow *{n} [hyd]*

ryū'iki	流域；りゅういき	basin; valley *[geol]*
"	流域	watershed *{n} [hyd]*
ryūjō-do	粒状度	granularity *[photo]*
ryūjō-fun	粒状粉	granular powder *[mat]*
ryūjō-hyō'setsu	粒状氷雪	neve *[hyd]*
ryūjō-jūgō'tai	粒状重合体	bead polymer *[poly-ch]*
ryūjō-sei	粒状性	graininess *[photo]*
ryūjō-so'shiki	粒状組織	granulation *[astron]*
"	粒状組織	granular texture *[petr]*
ryūka-aen	硫化亜鉛	zinc sulfide
ryūka'butsu	硫化物	sulfide *[ch]*
ryūka-chisso	硫化窒素	nitrogen sulfide
ryūka-dai'ichi-suigin	硫化第一水銀	mercurous sulfide
ryūka-dai'ni-suigin	硫化第二水銀	mercuric sulfide
ryūka-dō	硫化銅	copper sulfide
ryūka-gin	硫化銀	silver sulfide
ryūka-hiso	硫化砒素	arsenic sulfide
ryūkai	粒界；りゅうかい	grain boundary; grain interface *[crys]*
ryūka-keiso	硫化硅素	silicon sulfide
ryūka-ki	硫化機	xanthating machine; xanthator *[eng]*
ryūka-kin	硫化金	gold sulfide
ryūka-koku'hen	硫化黒変	sulfide staining *{n} [met]*
ryūka-namari	硫化鉛	lead sulfide
ryūka-rin	硫化燐	phosphorus sulfide
ryūka-suigin	硫化水銀	mercury sulfide
ryūka-suiso	硫化水素	hydrogen sulfide
ryūka-suzu	硫化錫	tin sulfide
ryūka-tanso	硫化炭素	carbon sulfide
ryūka-tetsu	硫化鉄	iron sulfide
ryūkei	粒径；粒徑	grain (particle) diameter *[crys] [geol]*
ryūki	隆起；隆起	protuberance; uplift; upheaval *[geol]*
ryūkō	流光	streamer *[geophys]*
ryūkō'byō	流行病	epidemic *{n} [med]*
ryūkō'sei-kanbō	流行性感冒	influenza; flu *[med]*
ryūkō'sei-nō'en	流行性脳炎	epidemic encephalitis *[med]*
ryū'kotsu	竜骨；龍骨	keel *[nav-arch]*
ryūmon'gan	流紋岩	rhyolite *[miner]*
ryūnō	竜脳；龍腦	borneol *[org-ch]*
ryūnyū-eki	流入液	influent *{n} [sci-t]*
ryūnyū no	流入の	incurrent *{adj} [i-zoo]*
ryū'ryō	流量；りゅうりょう	flow rate *[fl-mech]*
ryūryō-kei	流量計	flow meter *[eng]*

ryūryō-keisū	流量係数	discharge coefficient *[fl-mech]*
ryū'sa	流砂	quicksand *[geol]*
ryūsan	硫酸; りゅうさん	sulfuric acid
ryūsan-aen	硫酸亜鉛	zinc sulfate
ryūsan-bando	硫酸礬土	aluminum sulfate
ryūsan-dai'chi-tetsu	硫酸第一鉄	ferrous sulfate
ryūsan-dai'ni-dō	硫酸第二銅	cupric sulfate
ryūsan-dai'ni-tetsu	硫酸第二鉄	ferric sulfate
ryūsan-dō	硫酸銅	copper sulfate
ryūsan-en	硫酸塩	sulfate *[ch]*
ryūsan'en-kō	硫酸塩鉱	anglesite *[miner]*
ryūsan-gin	硫酸銀	silver sulfate
ryūsan-haibun	硫酸灰分	sulfated ash *[ch]*
ryūsan-ion	硫酸イオン	sulfate ion *[ch]*
ryūsan'ka	硫酸化	sulfation *[ch]*
ryūsan-ki	硫酸基	sulfate group *[ch]*
ryūsan-kon	硫酸根	sulfate radical *[ch]*
ryūsan-namari	硫酸鉛	lead sulfate
ryūsan-shi	硫酸紙	vegetable parchment *[mat]*
ryūsan-suigin	硫酸水銀	mercury sulfate
ryūsan-suiso-natoryūmu	硫酸水素ナトリウム	sodium hydrogen sulfate
ryūsan-tetsu	硫酸鉄	iron sulfate
ryūsei	流星; りゅうせい	meteor; shooting star *[astron]*
ryūsei'ate-toppatsu-tsūshin	流星宛突発通信	meteor burst communications *[comm]*
ryūsei'jin	流星塵	micrometeorite *[astron]*
ryūsei'u	流星雨	meteor shower *[astron]*
ryūseki-sen	流跡線	trajectory *[meteor]*
ryūsen	流線	line of flow; streamline *[fl-mech]*
ryūsen'kei no	流線形の	streamlined *{adj}* *[aero-eng]*
ryūshi	粒子; りゅうし	grain; particle *{n}* *[mat]* *[phys]*
ryūshi'jō	粒子状	particulate *[phys]*
ryūshi-ka'soku-ki	粒子加速器	particle accelerator *[nucleo]*
ryūshi-setsu	粒子説	corpuscular theory *[opt]*
ryū-shi'shitsu	硫脂質	sulfolipid *[bioch]*
ryū'shutsu	流出	runoff *{n}* *[hyd]*
ryū'shutsu'butsu	流出物	distillate *[ch]*
ryūshutsu'butsu-ryū	流出物流	effluent *[fl-mech]*
ryū'shutsu-eki	流出液	effluent *[ch]*
ryū'shutsu no	流出の	excurrent *{adj}* *[bio]*
ryū-soku	流速; りゅうそく	flow velocity *[ch]* *[geol]*
"	流速	flux *{n}* *[phys]*

ryūsoku-mitsudo	流速密度	flux density *[phys]*
ryūsui	流水	running water *[hyd]*
ryūtai	流体；りゅうたい	fluid *{n}* *[phys]*
ryūtai-funryū'giri	流体噴流切	liquid-jet cutting *{n}* *[mech-eng]*
ryūtai-jiki'gaku	流体磁気学	hydromagnetics *[phys]*
ryūtai-hensoku-sōchi	流体変速装置	hydraulic gear *[mech-eng]*
ryūtai-ma'satsu	流体摩擦	fluid friction *[fl-mech]*
ryūtai-rikigaku	流体力学；流體力學	hydrodynamics; hydromechanics; fluid mechanics *[fl-mech]*
ryūtai-sei'rikigaku	流体静力学	hydrostatics *[fl-mech]*
ryūtai-sei'rikigaku-heikō	流体静力学平衡	hydrostatic equilibrium *[fl-mech]*
ryūtai-tsugi'te	流体継手	fluid coupling; hydraulic coupling *{n}* *[mech-eng]*
ryūtsū-kikō	流通機構	distribution structure (of goods) *[econ]*
ryūzan	流産	abortion *[med]*
ryūzen'kō	竜涎香	ambergris (perfume) *{n}* *[physio]*
ryūzetsu'ran	竜舌蘭；龍舌蘭	agave *[bot]*

S

sa	差；さ	inequality *[astron]*
"	差	difference *[math]*
sa	鎖	chain *{n}* *[ch]*
saba	鯖；さば	mackerel (fish) *[v-zoo]*
sabaku	裁く	judge; pass judgment *{vb t}*
sabaku	捌く	handle; deal with; sell *{vb t}*
sabaku	砂漠；さばく	desert *{n}* *[geogr]*
sabaku-shokubutsu	砂漠植物	desert plant *[bot]*
sabi (sei)	錆；さび（せい）	rust *{n}* *[met]*
sabi (shū)	銹；鏽（しゆう）	rust *{n}* *[met]*
sabi'byō	銹病；さびびょう	rust (wheat disease) *{n}* *[bot]*
sabi'dome	錆止め	rustproofing *{n}* *[met]*
sabi-kobu	錆瘤；さびこぶ	tubercle *[met]*
sabō	砂防	erosion control *[civ-eng]*
saboten	仙人掌；サボテン	cactus (pl cacti, cactus, -es) *[bot]*
sa'bun	差分	finite difference *[math]*
sa-chō	鎖長	chain length *[poly-ch]*
sadō	作動；さどう	functioning; working *{n}* *[ind-eng]*
sadō	茶道	tea ceremony
sadō-den'i-kei	差動電位計	differential electrometer *[eng]*
sadō-denryū-kei	差動電流計	differential ammeter *[eng]*
sadō-eki	作動液；さどうえき	hydraulic fluid *[mat]*
sadō saseru	作動させる	actuate; operate *{vb t}* *[ind-eng]*
sadō-shudan; sadō-te'date	差動手段	differential means *[mech-eng]*
sadō-sōchi	差動装置	differential gear *[mech-eng]*
sadō'teki ni	作動的に	functionally *{adv}* *[ind-eng]*
sadō-yu	作動油	hydraulic oil *[mat]*
sadō-zōfuku-ki	差動増幅器	differential amplifier *[electr]*
saegiru	遮る；さえぎる	obstruct; intercept; block *{vb t}*
sagan	砂岩；さがん	sandstone *[petr]*
sagasu	探す	search; search for *{vb}*
sagen	左玄；さげん	port (direction from ship) *{n}* *[navig]*
sageru	下げる	hang; dangle; lower *{vb}*
sage-yoku	下げ翼	flap (aircraft) *{n}* *[aero-eng]*
sagi	鷺；さぎ	egret; heron (bird) *[v-zoo]*
saguru	探る	explore; look for *{vb}*
sagyō	作業；さぎょう	work; operation *{n}*
sagyō-iki	作業域	work area *[comput]*
sagyō-machi'gyōretsu-kōmoku	作業待行列項目	work queue entry *[comput]*

sagyō-sei	作業性	operability; working property *[ind-eng]*
sagyō-shitsu	作業室	work room *[arch]*
sagyō-shitsu	作業質	work quality *[ind-eng]*
sagyō suru	作業する	work *{vb}*
sagyō-tan'matsu	作業端末	workstation *[comput]*
sa'hen	左辺	left-hand side *{n}*
sai	犀；さい	rhinoceros (animal) (pl -es, -eri) *[v-zoo]*
-sai	-才；-歳	-year(s) old *{suffix}*
saibai	栽培	cultivation; growing; raising *{n} [agr]*
saibai-henshu-shokubutsu	栽培変種植物	cultivar; cultigen *[bot]*
saiban	裁判	adjudication; trial *[legal]*
saiban'sho-chōsha	裁判所庁舎	courthouse *[arch] [legal]*
saibō; saihō	細胞；さいぼう	cell *[bio]*
saibō-bunka	細胞分化	cell specialization (differentiation) *[cyt]*
saibō-bun'retsu	細胞分裂	cell division *[cyt]*
saibō'gai-kōso	細胞外酵素	exoenzyme *[bioch]*
saibō'gaku	細胞学；細胞學	cytology *[bio]*
saibō-ginkō	細胞銀行	cell bank *[cyt]*
saibō-heki	細胞壁	cell wall *[cyt]*
saibō-kagaku	細胞化学	cytochemistry *[ch]*
saibō-kikan	細胞器官	organelle *[cyt]*
saibō-kokyū	細胞呼吸	cell respiration *[bio]*
sai'boku	砕木；碎木	groundwood *{n} [paper]*
saibō-maku	細胞膜	cell membrane *[cyt]*
saibō'nai-kōso	細胞内酵素	endoenzyme *[bioch]*
saibō-senbetsu-ki	細胞選別器	cell sorter *[eng]*
saibō-shitsu	細胞質	cytoplasm *[cyt]*
saibō'shitsu-bun'retsu	細胞質分裂	cytokinesis *[cyt]*
saibō-shūhen-kō	細胞周辺腔	periplasmic space *[microbio]*
saibun'ka	細分化	fine-dividing *{n}*
sai-bunrui	細分類	detailed classification *[syst]*
sai'chin	再沈	reprecipitation *{n} [ch]*
saichi suru	載置する	place upon *{vb}*
saichō	犀鳥	hornbill (bird) *[v-zoo]*
saidai	最大	maximum (pl -ma); greatest; largest *[math]*
saidai-chi	最大値	greatest value *[math]*
saidai-kyoyō-ryō	最大許容量	maximum permissible amount *[ind-eng]*
saidai-shigoto	最大仕事	maximum work *[mech] [thermo]*
saido	彩度	chroma; saturation *[opt]*
sai'en	菜園	vegetable garden *[bot]*
sai'enshin	再延伸	restretching *{n} [plas]*
saifun	細粉；さいふん	fine powder *[mat]*

sai'futsu-ki	再沸器	reboiler *[eng]*
saigai	災害	accident; disaster *{n}*
saigai; era-buta	鰓蓋	operculum; gill cover *[v-zoo]*
saigai-bōshi	災害防止	disaster prevention *[ind-eng]*
saigan'sen	砕岩船	rock-cutter (ship) *[nav-arch]*
saigeki	細隙；さいげき	slit *{n} [eng]*
saigen-sei	再現性	repeatability; reproducibility *[math]*
saigo no	最後の	last; final *{adj}*
sai'haibun	再配分	reallocation *{n}*
sai'haichi	再配置	relocation *{n}*
sai'haichi suru	再配置する	relocate *{vb t} [comput]*
sai'hensei	再編成	regrouping *{n} [math]*
sai'hensei suru	再編成する	reorganize *{vb} [comput]*
saihin-chi	最頻値	mode *{n} [math]*
saihō; saibō	細胞	cell *[bio]*
saihō	裁縫；さいほう	sewing *{n} [cl] [tex]*
sai'hosō	再舗装	resurfacing *{n} [civ-eng]*
saihyō-ki	砕氷機	ice breaker *[eng]*
sai'in-yaku	催淫薬	aphrodisiac *{n} [pharm]*
sai'jikkō	再実行	rerun *{n} [comput]*
saijin-kei	細塵計	dust counter *[eng]*
saijō'i no	最上位の	uppermost *{adv} [math]*
saikachi	皂莢；皁莢	honey locust (tree) *[bot]*
saika'i no	最下位の	lowest *{adv} [math]*
sai'kaishi	再開始	restart *{n} [comput]*
sai'kakō	再加工	rework *{n} [ind-eng]*
saikaku-chi	最確値	most probable value *[math]*
sai'kesshō	再結晶	recrystallization *[crys]*
sai'ketsugō	再結合	rebonding *{n} [ch]*
sai'ketsugō-katei	再結合過程	recombination process *[astron]*
saiki	再輝	recalescence *[met]*
sai'kikei'sei-busshitsu	催奇形性物質	teratogen *[med]*
saikin	採金	gold mining *{n} [min-eng]*
saikin	細菌；さいきん	bacterium (pl -ria); germs; microbes *[microbio]*
saikin	最近	recently; lately *{adv} {n}*
saikin'gaku	細菌学	bacteriology *[microbio]*
saikin-hatsu'iku-soshi	細菌発育阻止	bacteriostasis *[microbio]*
saikin-hatsu'iku-soshi-zai	細菌発育阻止剤	bacteriostat *[mat]*
saikin no	細菌の	bacterial *{adj} [microbio]*
saikin no	最近の	the most recent; the latest *{adj}*
saiki no	再帰の；再歸の	reflexive *{adj} [gram]*

saikin-shiken	細菌試検	bacteriological test *[bact]*
saiki-shinsei	再帰新星	recurrent nova *[astron]*
saiki'teki na	再帰的な	recursive *{adj} [comput]*
saiki'teki-kansū	再帰的関数	recursive function *[math]*
saikō	細孔；さいこう	pore *{n} [bio] [met]*
"	細孔	micropore *[geol]*
saikō	採光	natural illumination *[opt]*
saikō	採鉱；採鑛	mining *{n} [min-eng]*
sai'kōbi no	最後尾の	rearmost *{adv}*
saikō-kachō'gen	最高可聴限	upper audible limit *{n} [acous]*
saikō-ki	砕鉱機；碎鑛機	jaw crusher *[eng]*
saikon	細根	feeder root *[bot]*
saikō'saitei-ondo-kei	最高最低温度計	maximum-minimum thermometer *[thermo]*
sai'kōsei	再構成	reconfiguration; restructuring *{n} [comput]*
saikō-sokudo	最高速度	top speed *{n} [nav-arch] [phys]*
saikō-ten	最高点	vertex (pl -es, vertices) *[math]*
sai-kotsu'zai	細骨材	fine aggregate *[mat]*
saikurotoron	サイクロトロン	cyclotron *[nucleo]*
saikuru	サイクル	cycle *[sci-t]*
sai'kutsu	採掘	digging; mining *{n} [min-eng]*
saimatsu'ka	細末化	comminuition *[ch-eng]*
saimei-shoku	最明色	optimal color *[opt]*
saimin-zai	催眠剤	hypnotic *{n} [pharm]*
sai'mitsu-hai'retsu	最密配列	densest array; hexagonal array *[crys]*
saimitsu-jūten	最密充填	closest packing *{n} [crys]*
saimitsu-kōshi	最密格子	closest-packed lattice *[crys]*
saimitsu-kōzō	最密構造	densest structure; closest-packed structure *[crys]*
saimoku	細目	item *{n}*
sai'myaku	細脈	veinlet *[bot]*
sai'nan	災難；さいなん	disaster; calamity *{n}*
sai'neri	再煉り；再練り	remilling *{n} [rub]*
saireki'gan	細礫岩	granule conglomerate *[geol]*
sai'retsu	鰓裂	branchial cleft; gill slit *[v-zoo]*
sai'riyō-gōsei	再利用合成	salvage synthesis *[bioch]*
sai-rui	菜類；菜類	leafy vegetables *[cook]*
sairui-sei	催涙性；催淚性	tearing property *[physio]*
saisan ga toreru	採算が取れる	profitable *[econ]*
saisa-undō	歳差運動	precession *[mech]*
saisei	再生；さいせい	regeneration *[bio] [ch] [comput] [elec]*
"	再生	playback *{n} [eng-acous]*
saisei-gomu	再生ゴム	reclaimed rubber *[rub]*

saisei-ki	再生機	regenerator; repeater *[eng]*
saisei-riyō suru	再生利用する	recycle *{vb} [ecol] [eng]*
saisei-sen'i	再生繊維；再生纖維	regenerated fiber *[tex]*
saisei-sen'i'so	再生繊維素	regenerated cellulose *[tex]*
saisei'shiki-chūkei-ki	再生式中継器	regenerating repeater *[comm]*
saisei-zai	再生剤	reclaiming agent *[rub]*
sai'seki	採石；さいせき	quarrying *{n} [eng]*
saisetsu'butsu	砕屑物	sediment (clastics; detritus) *[geol]*
sai'shidō	再始動	restart *{n} [comput]*
sai'shiki-kin	彩色金	mozaic gold *[mat]*
sai'shikō	再試行	retry *{n} [comput]*
saishin	再審	retrial *{n} [legal]*
saishin'shiki no	最新式の	newest; state of the art *{adj} [ind-eng]*
sai'shiyō-funō	再使用不能	nonreusable *[comput]*
sai'shiyō-kanō	再使用可能	reusable *[comput]*
saisho	最初；さいしょ	first; earliest *{n}*
sai'shō	最小；さいしょう	mininum (pl -ma) *{n} [math]*
saishō-baichi	最小培地	minimal medium *[microbio]*
saishō-chi	最小値	least value *[math]*
saishō-chishi-ryō	最小致死量	minimum lethal dose *[med]*
saishō'gen-shiryō	最小限資料	minimum documentation *[pr]*
saishō-jikan no genri	最小時間の原理	least time, principle of *[opt]*
saishō-kachō-chi	最小可聴值	threshold of audibility *{n} [acous]*
saishō-kōbai'sū	最小公倍数	least common multiple *[math]*
saishō-kō'bunbo	最小公分母	least common denominator *[math]*
sai'shokurin	再植林	reforestation *[ecol] [for]*
saishō-nijō-hō	最小二乗法	method of least squares *[math]*
saishō-nijō-suitei'ryō	最小二乗推定量	least squares estimator *[math]*
saisho no-	最初の-	proto- *{prefix}*
saishō no	最小の	minimal: smallest *{adj}*
saishō no	最少の	minimal: least *{adj}*
saishō-riron-dansū	最小理論段数	minimum theoretical number of plates *[ch-eng]*
saishō-ritsu	最少律	minimum, law of *[bio]*
saishō-sayō no hōsoku	最小作用の法則	least action, law of *[mech]*
saishō-shigoto	最小仕事	minimum work *[mech] [thermo]*
saishū	採収；採收	harvesting *{n} [agr]*
saishū	採集；さいしゅう	collecting; picking; gathering *{n}*
saishu-ritsu	採取率	extraction rate *[ch]*
"	採取率	recovery *{n} [met]*
saishū-riyō'sha	最終利用者	end user *[comput]*
saisō	彩層	chromosphere *[astron]*

saita	最多	most abundant; most numerous *{n} [math]*
saitai; seitai	臍帯; 臍帶	umbilical cord *[embryo]*
saitei	裁定	arbitral decision *[legal]*
saitei-tori'hiki	裁定取引	arbitrage *[legal]*
saiteki'ka	最適化	optimization *[math] [eng]*
saiteki na	最適な	optimum *{adj} [math]*
saiten	採点	grading *{n} [ind-eng]*
saito-yaku (saito-zai)	催吐薬（催吐剤）	nauseant *{n} [pharm]*
sai-yūsen	最優先	highest priority *{n}*
saiyū-suitei'ryō	最尤推定量	maximum likelihood estimator *[math]*
saji	匙; さじ	spoon *{n} [cook] [eng]*
sa'jiki	桟敷; 棧敷; さじき	seating area *[arch]*
sajin-arashi	砂塵嵐	dust storm *[meteor]*
sajō-bunshi	鎖状分子	chain molecule *[p-ch]*
sajō-jūgō'tai	鎖状重合体	chain polymer; linear polymer *[poly-ch]*
sakaki	榊; 賢木; さかき	sakaki (tree) *[bot]*
sakamata	逆叉; さかまた	killer whale; orc; grampus *[v-zoo]*
saka-mushi	酒蒸し	sake (q.v.)-steamed dishes *[cook]*
sakana	魚; さかな	fish (pl fish, -es) *[v-zoo]*
sakana'tsuri	魚釣	fishing *{n}*
saka-noboru	溯る	go upstream; trace back to; retroact *{vb}*
sakan'yō-shō'sekkai	左官用消石灰	plasterer's lime *[mat]*
sakari'ba	盛り場	amusement district *[civ-eng]*
sakasa(ma)	逆さ(ま)	inverted; overturned; upside down *[math]*
sakasa(ma) ni suru	逆さ(ま)にする	invert; turn upside down *{vb t}*
saka'te-hobu-kakō	逆手ホブ加工	reverse hobbing *{n} [mech]*
sakazuki	杯; 坏; 盃	cup for Japanese wine *[cer] [cook]*
sakazuki'jō no	坏状の	winecup-shaped *{adj} [math]*
sake	酒; さけ	Japanese rice wine *[cook]*
sake	鮭; さけ	salmon; Oncorhynchus keta (fish) *[v-zoo]*
sake-kasu	酒粕; 酒糟	sake lees *[cook]*
sake'me	裂け目	crevasse *[geol]*
"	裂け目	crack *{n}*
sake-rui	酒類; 酒類	alcoholic beverages *[cook]*
saki	先	ahead *{adv} {n}*
"	先	tip (location) *{n}*
saki'boso ni suru	先細にする	taper (down) *{vb t}*
saki'boso-yoku	先細翼	tapered wing *[aero-eng]*
sa-kin	砂金	gold dust *[met]*
saki no-	先の-	pre- *{prefix}*
saki-ototoi	一昨々日; さき=おととい	three days ago *{n}*

saki-ototoshi	一昨々年; さき=おととし	three years ago *{n}*
saki'togari-penchi	先尖りペンチ	needle-nose pliers *[eng]*
sakkaku	錯角	alternating interior angles *[math]*
sakkaku	錯覚	illusion *[psy]*
sakkarin	サッカリン	saccharin *[org-ch]*
sakkin	殺菌; さっきん	sterilization *[microbio]*
sakkin suru	殺菌する	sterilize *{vb}* *[microbio]*
sakkin-tō	殺菌灯; 殺菌燈	sterilizing lamp *[opt]*
sakkin-zai	殺菌剤	bactericide; bacteriocide *{n}* *[mat]*
sak'kiseichū-yaku	殺寄生虫薬	parasiticide *{n}* *[pharm]*
sakkyoku'ka	作曲家	composer (a person) *[music]*
sa-kotsu	鎖骨	collar bone; clavicle *{n}* *[anat]*
saku	柵	fence *{n}* *[arch]*
saku	朔	new moon *[astron]*
saku	蒴; 朔	capsomere *[bot]*
saku	咲く	bloom; blossom; flower *{vb i}* *[bot]*
saku	裂く	tear; rip; crack *{vb t}*
saku'bō	朔望	syzygy *{n}* *[astron]*
saku'dō	索道	cableway *[mech-eng]*
"	索道	ropeway *[min-eng]*
saku'en	錯塩; 錯鹽	complex salt *[inorg-ch]*
sakugan-ki	鑿岩機	rock drill *[eng]*
sakugo	錯誤; さくご	error; mistake *{n}* *[sci-t]*
saku'haku	削剥	denudation *[geol]*
saku'hyō	作表	tabulating; tabulation *{n}* *[math]*
saku'hyō	索表	table lookup *{n}* *[comput]*
saku'hyō suru	作表する	tabulate *{vb}* *[math]*
saku'in	索引; さくいん	index (pl -es, indices) *{n}* *[pr]* *[comput]*
saku'in-chizu	索引地図	index map *[comput]*
saku'in-zuke	索引付け	indexing *{n}* *[comput]*
saku-ion	錯イオン	complex ion *[ch]*
saku'jitsu	昨日	yesterday *{n}*
sakujo	削除	deletion *[comput]*
"	削除	cancellation *[pr]*
saku'jō-sui	索状水	funicular water *[ch]*
sakujo suru	削除する	delete *{vb}* *[comput]*
saku-ka	蒴果	capsomere; capsule *[bot]*
saku-kagō'butsu	錯化合物	complex compound *[ch]*
saku-ki	錯基	complex radical *[ch]*
sakuma	削摩	ablation *[geol]*
saku'motsu	作物; さくもつ	crops; farm products *[agr]*

saku'motsu'gaku	作物学	crop science *[agr]*
sakura	桜；櫻；さくら	cherry (tree, wood, blossom) *[bot]*
sakura-bana; ōka	桜花	cherry blossom (flower) *[bot]*
sakura-ebi	桜蝦	spotted shrimp (crustacean) *[i-zoo]*
sakura-mochi	桜餅	rice cake wrapped with cherry leaf *[cook]*
sakuran	錯乱；錯亂	confusion *[psy]*
sakuranbō	桜ん坊	cherry (pl cherries) (fruit) *[bot]*
sakura no hana	桜の花	cherry blossom (flower) *[bot]*
sakusan	酢酸；さくさん	acetic acid
sakusan-dō	酢酸銅	copper acetate
sakusan-en	酢酸塩	acetate *[ch]*
sakusan-kinu	柞蚕絹	tussah silk *[tex]*
sakusan-namari	酢酸鉛	lead acetate
sakusei	作製	fabrication *[ind-eng]*
sakushi	錯視	optical illusion *[physio]*
sakutai	錯体；錯體	complex *{n}* *[ch]*
saku'ya	昨夜；さくや	last night; yesterday evening *{n}*
saku'yu-ki	搾油機	oil press *[agr]* *[eng]*
sakuzu-hō	作図法；作圖法	graphic determination method *[graph]*
sakuzu-sōchi	作図装置	plotter *[graph]*
saku'zu suru	作図する	plot *{vb}* *[graph]*
sakyū	砂丘	sand dune *[geol]*
samaryūmu	サマリウム	samarium (element)
same	鮫；さめ	shark (fish) *[v-zoo]*
same-kan'yu	鮫肝油	shark liver oil *[mat]*
sāmetto	サーメット	cermet *[cer]*
samui	寒い	chilly; cold *{adj}* *[physio]*
samu-sa	寒さ	coldness (of weather) *{n}* *[meteor]*
san	三；参；參	three *{n}* *[math]*
san	酸；さん	acid *[ch]*
-san; -zan; -yama	-山	-mountain *{suffix}*
sanagi	蛹；さなぎ	chrysalis; pupa (pl pupae) *[i-zoo]*
san-arai	酸洗い	pickling *{n}* *[met]*
"	酸洗い	souring *{n}* *[tex]*
sanbai	三倍	triple *{n}*
sanbai ni naru	三倍になる	triple *{vb i}*
sanbai ni suru	三倍にする	triple *{vb t}*
sanbai no	三倍の	triple; treble *{adj}* *[math]*
sanbai-zu	三倍酢	sake, soy sauce, and vinegar sauce *[cook]*
san'bashi	桟橋；棧橋	pier; quay *[civ-eng]*
sanbō-ben	三方弁；三方瓣	three-way valve; cross valve *[mech-eng]*
sanbō-hen'sankaku'men-tai	三方偏三角面体	trigonal scalenohedron *[math]*

sanbō-hen'shikaku'men-tai	三方偏四角面体	trigonal trapezohedron *[math]*
sanbō-ryōsui	三方両錐	trigonal bipyramid *[math]*
sanbō-shōkei; sanpō-shōkei	三方晶形	trigonal system *[crys]*
sanbō-sui	三方錐	trigonal pyramid *[math]*
sanbun	酸分	acid content *[ch]*
sanbun-hō	三分法	trisecting method; trichotomy *[math]*
sanbun-hōsoku	三分法則	trichotomy, law of *[math]*
san'bun no ichi	三分の一	one-third; third *{n}* *[math]*
san'butsu	産物；さんぶつ	product; fruition; outcome *[ind-eng]*
sanchi	山地	mountainous land *[geogr]*
sandan	散弾；散彈	shot *{n}* *[ord]*
sandan-ronpō	三段論法	syllogism *[logic]*
sando	酸度	acidity *[ch]*
sandō-yaku	散瞳薬	mydriatic *{n}* *[pharm]*
san'eki-kei	三液系	three-component system *[poly-ch]*
san-enka'butsu	酸塩化物	acid chloride *[ch]*
san'enka-hiso	三塩化砒素	arsenic trichloride
san'enka-hōso	三塩化硼素	boron trichloride
san'enka-rin	三塩化燐	phosphorus trichloride
san'enka-yōso	三塩化沃素	iodine trichloride
san'enki-rensa	三塩基連鎖	triplet *[bioch]*
san'enki-san	三塩基酸	tribasic acid *[ch]*
san'fukka-enso	三弗化塩素	chlorine trifluoride
san'fukka-hiso	三弗化砒素	arsenic trifluoride
san'fukka-hōso	三弗化硼素	boron trifluoride
san'gaku	山岳	mountain; hill *[geogr]*
san'gatsu	三月	March (month)
sangen-gōkin	三元合金	ternary alloy *[met]*
sangen-haichi	三元配置	ternary arrangement *[math]*
sangen-kyōjūgō'tai	三元共重合体	terpolymer *[poly-ch]*
sangen-kyōshō	三元共晶	ternary eutectic *[met]*
sango	珊瑚；さんご	coral *[i-zoo]*
sango-hebi	珊瑚蛇；さんごへび	coral snake (reptile) *[v-zoo]*
sango-ju	珊瑚珠	coral beads *[lap]*
sango-shō	珊瑚礁	coral reef *[geol]*
sangyō	産業；さんぎょう	industry (pl -tries) *[econ]*
sangyō-chūdoku	産業中毒	industrial poisoning *{n}* *[med]*
sangyō-kikai	産業機械	industrial machinery *[eng]*
sangyō-kōgai	産業公害	industrial pollution *[ecol]*
san'itsu	散逸	dissipation *[phys]*
san'itsu-kansū	散逸関数	dissipation function *[p-ch]*
sanji-gen	三次元	three dimensions *[sci-t]*

sanji'gen-hyōji-sōchi	三次元表示装置	three-dimensional display *[comput]*
sanji'gen-kōbunshi	三次元高分子	three-dimensional polymer *[poly-ch]*
sanji-gen no	三次元の	three-dimensional *{adj} [sci-t]*
sanji-hannō	三次反応	third-order reaction *[p-ch]*
san'jiku-ō'ryoku	三軸応力	triaxial stress *[mech]*
sanji-kyoku'sen	三次曲線	cubic curve *[math]*
sanji no	三次の	three-dimensional *{adj} [math] [sci-t]*
sanjō	三乗；三乘	cube *{n} [math]*
sanjō-kon	三乗根	cube root *[math]*
san'jū	三十	thirty *{n} [math]*
san'jū	三重	triple *{n}*
sanjū-fukugō'tai	三重複合体	ternary complex *[bioch]*
sanjū-ketsugō	三重結合	triple bond *[org-ch] [p-ch]*
sanjū no	三重の	triple *{adj}*
sanjū'sō	三重奏	trio *{n} [music]*
sanjū'suiso	三重水素	tritium *[nuc-phys]*
sanjū-ten	三重点	triple point *[p-ch]*
san'jutsu	算術；さんじゅつ	arithmetic *{n} [math]*
sanjutsu-heikin	算術平均	arithmetic mean *[math]*
sanjutsu-keta-afure	算術桁溢れ	arithmetic overflow *[comput]*
sanjutsu-kyūsū	算術級数	arithmetic series *[math]*
sanjutsu-meirei	算術命令	arithmetic instruction *[comput]*
sanjutsu-shiki	算術式	arithmetic expression *[comput]*
sanjū'yōshi	三重陽子	triton *[nuc-phys]*
sanka	酸化；さんか	oxidation *[ch]*
sanka	酸価；酸價	acid value *[ch]*
sanka-aen	酸化亜鉛	zinc oxide
sanka-bōshi-zai	酸化防止剤	antioxidant *[ch]*
sanka'butsu	酸化物	oxide *[ch]*
sanka'butsu-en	酸化物塩	oxy salt; oxide salt *[ch]*
sanka-chisso	酸化窒素	nitrogen oxide
sanka-dai'ni-suigin	酸化第二水銀	mercuric oxide
sanka-den'i	酸化電位	oxidation potential *[p-ch]*
sanka-dō	酸化銅	copper oxide
sanka-en	酸化炎	oxidizing flame *[ch]*
sanka-enka-rin	酸化塩化燐	phosphorus oxide chloride
sanka-enso	酸化塩素	chlorine oxide
sanka-fusso	酸化弗素	fluorine oxide
sanka-gin	酸化銀	silver oxide
sanka-hakkin	酸化白金	platinum oxide
sanka-himaku	酸化皮膜	oxide film *[ch]*
sanka-hiso	酸化砒素	arsenic oxide

sanka-hōso	酸化硼素	boron oxide
sanka-iō	酸化硫黄	sulfur oxide
sankai-seidan	散開星団	open cluster *[astron]*
sanka-jūgō	酸化重合	oxidative polymerization *[poly-ch]*
sanka'kangen-den'i	酸化還元電位	oxidation-reduction potential; redox potential *[p-ch]*
sanka'kangen-hannō	酸化還元反応	oxidation-reduction reaction; redox reaction *[ch]*
sanka'kangen-kōso	酸化還元酵素	oxidation-reduction enzyme; oxidoreductase *[bioch]*
sanka'kangen-shiji-yaku	酸化還元指示薬	oxidation-reduction indicator *[an-ch]*
sanka-keiso	酸化硅素	silicon oxide
sanka-kōso	酸化酵素	oxidase *{n} [bioch]*
san'kaku	三角；さんかく	triangle *[math]*
sankaku-chū	三角柱	trigonal (triangular) prism *[math]*
sankaku-hi	三角比	trigonometric ratio *[math]*
sankaku-hō	三角法	trigonometry *[math]*
sankaku-jaku	三角尺	triangle scale *[graph]*
sankaku-jōgi	三角定規	triangle (a tool) *{n} [eng]*
sankaku'jō no	三角状の	deltoid *{adj} [geom]*
sankaku-kansū	三角関数	trigonometric function *[trig]*
sankaku-kei	三角形	triangle *{n} [math]*
sankak(u)kei no	三角形の	triangular; trigonal *{adj} [math]*
sankaku-kōtō-shiki	三角恒等式	trigonometric identity *[math]*
sankaku-ryōsui	三角両錐	trigonal bipyramid *[crys] [math]*
sankaku-shisa	三角視差	trigonometric parallax *[astron]*
sankaku-soku'ryō	三角測量	triangulation *{n} [eng] [navig]*
sankaku-su	三角洲	delta *{n} [geog]*
sankaku-sui	三角錐	trigonal (triangular) pyramid *[math]*
sankaku-zahyō	三角座標	triangular coordinates *[math]*
sankaku-zuhyō	三角図表	triangular diagram *[ch-eng] [geol]*
sanka-maku	酸化膜	oxide film *[ch]*
sanka-namari	酸化鉛	lead oxide
sanka no	三価の	trivalent *{adj} [ch]*
sankan'shiki-kagō'butsu	三環式化合物	tricyclic compound *[org-ch]*
sanka-rin	酸化燐	phosphorus oxide
sanka-shūso	酸化臭素	bromine oxide
sanka-sū	酸化数	oxidation number *[ch]*
sanka-suigin	酸化水銀	mercury oxide
sanka suru	酸化する	oxidize *{vb} [ch]*
sanka-suzu	酸化錫	tin oxide
sanka'teki-rinsan'ka	酸化的燐酸化	oxidative phosphorylation *[bioch]*

sanka-tetsu	酸化鉄；酸化鐵	iron oxide
sanka-yōso	酸化沃素	iodine oxide
sanka-zai	酸化剤	oxidizer; oxidizing agent *[ch]*
sankei	三形	trimorphism *[bio]*
sankei-kajo	散形花序	umbel *[bot]*
sanki'shiki-bakki-hōhō	散気式暴気方法	diffuser method *[eng]*
sankō	参考；さんこう	reference *{n}*
sankō	散光	diffused light *[opt]*
sankō-sen	参考線	reference line *[graph]*
sankō-shiki	三項式	trinomial *{n}* *[math]*
sankō suru	鑽孔する	perforate *{vb}* *[eng]*
san'kyaku	三脚	tripod *[eng]*
sankyoku(-shinkū)-kan	三極(真空)管	triode *[electr]*
sanma	秋刀魚；さんま	mackerel pike (fish) *[v-zoo]*
sanmi	酸味	sourness *{n}* *[cook]* *[physio]*
sanmi-fuyo'zai	酸味付与剤	acidulant *[cook]* *[mat]*
san-musui'butsu	酸無水物	acid anhydride *[ch]*
sanmyaku	山脈	mountain range *[geogr]*
san-ni'musui'butsu	酸二無水物	acid dianhydride *[ch]*
san'nin	三人	three persons *{n}*
san-ni'san'ka-tetsu	三二酸化鉄	iron sesquioxide
sannyū	酸乳	sour milk *[cook]*
sa'nō	砂嚢；さのう	gizzard *[v-zoo]*
sanpai	酸敗	rancidification *[cook]*
sanpai-do	酸敗度	rancidity *[cook]*
sanpan	舢板；三板	sampan *{n}* *[nav-arch]*
sanpen-sokuryō	三辺測量	trilateration *[eng]*
sanpō	算法	algorithm *[comput]*
sanpō-gengo	算法言語	algorithmic language *[comput]*
sanpō-shōkei; sanbō-shōkei	三方晶形	trigonal system *[crys]*
sanpō-shōkei no	三方晶形の	trigonal *{adj}* *[crys]*
sanpu	散布	spraying *{n}* *[agr]*
san'puku	山腹	hillside; mountainside *{n}* *[geogr]*
sanpu-zai	散布剤	dusting powder *[mat]* *[pharm]*
sanpu-zu	散布図	scatter diagram *[math]*
sanran	散乱；散亂	scattering *{n}* *[phys]*
sanran-ha	散乱波	scattered wave *[phys]*
sanran-kaku	散乱角	scattering angle *[phys]*
sanran-kan	産卵管	ovipositor *[i-zoo]*
sanran-kō	散乱光	scattered light *[opt]*
sanran-kyōdo-bunpu	散乱強度分布	scattering intensity distribution *[opt]*
sanrin'seki	三燐石	triphylite *[miner]*

sanrin'sha	三輪車	tricycle *[mech-eng]*
san'roku-chitai	山麓地帯	piedmont *[geol]*
san'roku-kansha'men	山麓緩斜面	pediment *[geol]*
sanryō	山稜	arete *[geol]*
sanryō'tai	三量体; 三量體	trimer *[poly-ch]*
san'ryūka-rin	三硫化燐	phosphorus trisulfide
sansa-kyoku'sen	三叉曲線	trident *{n} [math]*
san'san-enki	三酸塩基	triacidic base *[ch]*
san'sanka-hiso	三酸化砒素	arsenic trioxide
san'sanka-hōso	三酸化硼素	boron trioxide
san'sanka-iō	三酸化硫黄	sulfur trioxide
san'sanka-ni'chisso	三酸化二窒素	dinitrogen trioxide
san'sanka-ni'iō	三酸化二硫黄	disulfur trioxide
san'sanka-rin	三酸化燐	phosphorus trioxide
sansei	酸性; さんせい	acidity; acidic *{n} [ch]*
sansei-do	酸性度	acidity *[ch]*
sansei-en	酸性塩	acid salt *{n} [ch]*
sansei-hakudo	酸性白土	acid clay; fuller's earth *[geol]*
sansei-hisan-namari	酸性砒酸鉛	acidic lead arsenate
sansei'ji (mitsu'go)	三生児 (三つ子)	triplets *{n} [bio]*
sansei-kō	酸性鋼	acid steel *[steel]*
sansei-sanka'butsu	酸性酸化物	acidic oxide *[ch]*
sansei-senryō	酸性染料	acid dye *[org-ch]*
sansei'u	酸性雨; さんせいう	acid rain *[meteor]*
sansen-shiki	三線式	three-wire system *[elec]*
sansha-shōkei	三斜晶形	triclinic system *[crys]*
san'shigeki-chi	三刺激值; 三刺戟値	tristimulus values *[opt]*
sanshin	三進	ternary *{n} [logic] [math]*
sanshin-hō	三進法	ternary notation *[math]*
sanshō	山椒	Japanese pepper; prickly ash (leaves) Xanthoxylum piperitum (herb) *[bot]*
sanshō	参照	reference *{n} [comput]*
sanshoku-hyōji	三色表示	trichromatic expression *[opt]*
sanshoku-kei	三色系	trichromatic system *[opt]*
sanshoku-keisū	三色係数	trichromatic coefficient *[opt]*
sanshoku-sumire	三色菫	pansy (flower) *[bot]*
sanshō suru	参照する	access; look up *{vb} [comput]*
sanshō'uo	山椒魚	salamander (amphibian) *[v-zoo]*
sanshō-yōeki	参照溶液	reference solution *[an-ch]*
san'shūka-hōso	三臭化硼素	boron tribromide
san'shutsu no	三出の	ternate *{adj} [bot]*
san'shutsu suru	算出する	calculate; compute *{vb} [math]*

sanso	酸素	oxygen (element)
sansō-dendō-ki	三相電動機	three-phase motor *[elec]*
sansō-kairo	三相回路	three-phase circuit *[elec]*
sanso-san	酸素酸	oxoacid; oxygen acid *[ch]*
sanso-ten	酸素点	oxygen point *[thermo]*
sansū	算数	arithmetic *{n}* *[math]*
sansui-sōchi	散水装置	sprinkler; water sprinkler *[eng]*
san'suiso-en	酸水素炎	oxyhydrogen flame *[ch]*
santan'tō	三炭糖	triose *[bioch]*
san'tōbun	三等分	trisection *[math]*
san'tōbun suru	三等分する	trisect *{vb}* *[math]*
santō-rui	三糖類	trisaccharide *[bioch]*
san'tōshi	酸通し	souring *{n}* *[tex]*
san-tsukuri	彡旁; さんつくり	kanji (q.v.) radical denoting color *[pr]*
san-yaku	散薬; 散藥	powdered medicine *{n}* *[pharm]*
sanzui	三水: 氵; さんずい	kanji (q.v.) radical denoting water *[pr]*
sao-bakari	棹秤; さおばかり	beam scale *[eng]*
sa'on	差音	difference tone *[acous]*
sappa	鯯; 拶双魚; さっぱ	pilchard; Harengula zunasi (fish) *[v-zoo]*
sara	皿	dish; plate *{n}* *[cer]* *[cook]*
sara'arai-ki	皿洗機	dishwasher *[eng]*
sarasa	更紗; さらさ	calico (pl -coes, -cos) *[tex]*
sarashi-ko	晒し粉	bleaching powder *[mat]*
saru	猿; さる	monkey (pl monkeys) (animal) *[v-zoo]*
saru-men'eki-fuzen-uirusu	猿免疫不全ウィルス	simian immunodeficiency virus; SIV *[immun]*
saru no	猿の	simian *{adj}* *[v-zoo]*
saru no koshi'kake	猿の腰掛け	polypore; shelf fungus *[mycol]*
saru-ogase	猿麻桛; 猿おがせ	usnea *[bot]*
saru'ogase'modoki	猿麻桛擬	Spanish moss *[bot]*
saru-suberi	百日紅; 猿滑; さるすべり	crape myrtle; crepe myrtle; Lagerstroemia indica (tree, flower) *[bot]*
sasa	笹; 篠	bamboo grass *[bot]*
sasae	支え	prop; support *{n}* *[eng]*
sasae-sen	支え線	guy; guy wire *{n}* *[eng]*
sasage	豇豆; さゝげ	black-eyed pea *[bot]*
sasa no ha	笹の葉	bamboo leaf *[bot]*
sasayaki	囁き; ささやき	whisper *{n}* *[acous]* *[physio]*
sasayaki no kairō	囁きの回廊	whispering gallery *[acous]* *[arch]*
sasayaku	囁く	whisper *{vb}*
sa-sen	鎖線	chain line *[graph]*
sasen-daen-henpa	左旋楕円偏波	left-hand polarized wave *[elecmg]*

sa-sen(kō)'sei no	左旋(光)性の	levorotatory *{adj} [opt]*
sasen'tō	左旋糖	levulose *{n} [bioch]*
sa-setsu	左折；させつ	left turn *{n} [navig]*
sa'setsu-shasen	左折車線	left-turn lane *[traffic]*
sa'shi	砂嘴	sandbank *{n} [geol]*
sashi-e	挿絵；挿繪	illustration; cut-in; cut *{n} [graph]*
sashi-gane	差し金	carpenter's square *[eng]*
sashi'ki-doko	挿木床	cutting bed (nursery) *[bot]*
sa'shiki-kagō'butsu	鎖式化合物	chain compound *[ch]*
sashi'ko-ori	刺し子織り	quilt weave *[tex]*
sashi'mi	刺身；さしみ	sliced raw fish *[cook]*
sashi-mono	差物	addition metal *[met]*
sashō	査証	visa *[trans]*
sa'soku-tsūkō	左側通行	Keep to the Left (traffic)
sasori	蝎；蠍；蠆；さそり	scorpion (insect) *[i-zoo]*
sasori-za	蝎座；さそり座	Scorpio; the Scorpion *[constel]*
sassenchū-zai	殺線虫剤	nematocide *{n} [pharm]*
sasso-zai	殺鼠剤	rodenticide *{n} [mat]*
sassuru	察する	surmise; estimate *{vb}*
sasu	刺す	pierce *{vb}*
sasu	指す	point at; indicate *{vb}*
sasu	螫す	sting (by an insect) *{vb t} [i-zoo]*
sa'su	砂洲	reef; sand bar *[geol]*
satchū-zai	殺虫剤	insecticide *{n} [mat]*
satei	査定	assessment *[sci-t]*
sa-tetsu	砂鉄；砂鐵	iron sand; magnetite sand; magnetic sand *[geol] [miner]*
satō	砂糖；さとう	sugar *{n} [bioch] [cook]*
satō-daikon	砂糖大根	sugarbeet *[bot]*
sato-imo	里芋	taro (pl taros) (vegetable) *[bot]*
satō-kibi	砂糖黍	sugarcane *[bot]*
satō'kibi-shibori'gara	砂糖黍搾り殻	bagasse *[cook]*
satō'morokoshi	砂糖蜀黍	sweet sorghum *[bot]*
satsu	冊；册	volume; a copy *{n} [pr]*
satsu	刷	impression; issue *{n} [pr]*
-satsu	-冊	-copy (-ies) of bound text *{suffix}*
satsu'ei-ki	撮影機	motion picture camera *[photo]*
Satsuma-age	薩摩揚げ	fried fish balls *[cook]*
Satsuma-imo	薩摩芋；さつま=いも	sweet potato; Ipomoea batatas (vegetable) *[bot]*
Satsuma-yaki	薩摩焼	Satsuma ware *[cer]*
satsu-zai	擦剤；さつざい	liniment *{n} [pharm]*

satsu'zō-kan	撮像管	television camera tube; camera tube; pickup tube *[electr]*
sawa-gani	沢蟹；澤蟹	river crab (crustacean) *[i-zoo]*
sawara	鰆；さわら	Spanish mackerel; Scomberomorus niphonius (fish) *[v-zoo]*
sawaru	触る；觸る	touch; feel *{vb t}*
sawaru	障る	hinder; interfere with *{vb t}*
saya	莢；さや	pod; husk *[bot]*
saya	鞘	sheath; scabbard *{n} [ord]*
saya	匣鉢；サヤ	sagger *{n} [cer]*
saya-endō	莢豌豆；さや豌豆	garden pea (legume) *[bot]*
saya-ingen	莢隠元	kidney bean (legume) *[bot]*
sayō	作用；さよう	action; effect *{n} [ch] [phys]*
"	作用	function *{n}*
sayō-han'sayō no hōsoku	作用反作用の法則	action and reaction, law of *[phys]*
sa-yoku	砂浴	sand bath *[med]*
sayori	針魚；鱵；さより	halfbeak; hemiramph (fish) *[v-zoo]*
sayō'ryoku	作用力	effort *{n} [phys]*
sayō-sen	作用線	line of action *{n} [mech-eng]*
sayō suru	作用する	function; work; act upon *{vb}*
sayū	左右；さゆう	left and right *{n} [math]*
sayū-sōshō	左右相称	bilateral symmetry *[bio]*
"	左右相称	zygomorphy *[bot]*
sazae	栄螺；さざえ	turban shell; turbo (pl -s, turbines) *[i-zoo]*
saza'nami	漣；小波；さざなみ	ripple *{n} [ocean]*
sazan'ka	山茶花；さざんか	sasanqua (shrub, flower) *[bot]*
se	背	back *{n} [anat]*
se	瀬；瀨	rapids; a current *[hyd]*
sebamaru	狭まる；狹まる	narrow: become narrower *{vb i}*
sebameru	狭める	narrow: make narrower *{vb t}*
se-bire	背鰭	dorsal fin *[v-zoo]*
se'biro	背広；背廣	suit (man's) *[cl]*
se-bone	背骨	backbone; spine *{n} [anat]*
se'chō	背丁	signature *[pr]*
sedai	世代	generation *{n} [bio] [comput]*
se'gawa	背革	quarter binding *{n} [pr]*
sei	正	positive *{n} [elec] [math] [sci-t]*
sei	世	epoch *[geol]*
sei	性	sex; gender *{n} [bio]*
sei	姓；せい	surname; last name; family name *{n}*
sei	星；せい	star *{n} [astron]*

sei-atsu	静圧；静壓	static force *[mech]*
sei'betsu	性別	classification by sex *{n} [bio]*
seibi	整備；せいび	maintenance; servicing *{n} [ind-eng]*
"	整備	preparation *{n}*
seibi-in	整備員	ground crew *[aero-eng]*
seibō	精紡	fine spinning; spinning *{n} [tex]*
seibun	成分；せいぶん	component *[ch] [elec] [math] [sci-t]*
"	成分	ingredient *{n} [pharm]*
sei'bunkai	生分解	biodegradation *[bioch]*
sei'bunkai-sei	生分解性	biodegradability *[mat]*
sei'butsu	生物；せいぶつ	organism; living organism *[bio]*
seibutsu-bunkai-kanō no	生物分解可能の	biodegradable *{adj} [mat]*
seibutsu-bunkai-sei	生物分解性	biodegradability *[mat]*
seibutsu-butsuri'gaku	生物物理学	biophysics *[sci-t]*
seibutsu-chikyū-kagaku	生物地球化学	biogeochemistry *[geoch]*
seibutsu-chikyū'kagaku-teki-junkan	生物地球化学的=循環	biogeochemical cycles *[geochem]*
seibutsu-chiri'gaku	生物地理学	biogeography *[ecol]*
seibutsu-dokei	生物時計	biological clock *[physio]*
seibutsu'gaku	生物学	biology *[sci-t]*
seibutsu-gakusha	生物学者	biologist *[bio]*
seibutsugaku'teki-teiryōhō	生物学的定量法	bioassay *[an-ch]*
seibutsu(-genson)'ryō	生物（現存）量	biomass *[ecol]*
seibutsu-gi'jutsu	生物技術	biotechnology *[eng]*
seibutsu-gi'jutsu-sangyō	生物技術産業	bioindustry *[eng]*
seibutsu-gunkei	生物群系	biome *[bio]*
seibutsu-hakkō	生物発光	bioluminescence *[bio]*
seibutsu-i'gaku	生物医学	biomedicine *[med]*
seibutsu-kagaku-kenchi-ki	生物化学検知器	biosensor *[eng]*
seibutsu-kagaku-soshi	生物化学素子	biochip *[bio]*
seibutsu-ken	生物圏；生物圈	biosphere *[ecol]*
seibutsu-kentei	生物検定	bioassay *[an-ch]*
seibutsu-kōgaku	生物工学	bioengineering; biotechnology *[eng]*
"	生物工学	ergonomics *[ind-eng]*
seibutsu ni yoru kondakuka	生物による混濁化	bioturbation *[ocean]*
sei'butsu no-	生物の-	bio- *{prefix}*
seibutsu-onkyō'gaku	生物音響学	bioacoustics *[bio]*
seibutsu-sen(sō)	生物戦（争）	biological warfare *[mil]*
seibutsu-sō	生物相	biota *[bio]*
seibutsu-sokutei'gaku	生物測定学	biometrics *[stat]*
seibutsu-tai	生物帯；生物帶	biome *[ecol]*
seibutsu'tai-ryō	生物体量	biomass *[ecol]*

seibutsu'teki-bōjo	生物的防除	biological control *[ecol]*
seibutsu'teki-kankyō	生物的環境	biotic environment *[ecol]*
seibutsu-tōkei'gaku	生物統計学	biostatistics *[math]*
seibutsu-zai	生物剤；生物劑	biological agent *[ord]*
seibyō	性病	venereal disease *[med]*
seichi-baiyō	静置培養	static (or stationary) culture *[microbio]*
seichō	成長；生長	growth *[physio]*
seichō-hannō	生長反応	growth reaction; propagation reaction *[org-ch]*
seichō-inshi	生長因子；成長因子	growth factor *[physio]*
seichō'ka	清澄化	clarification (of a liquid) *[mech-eng]*
seichō-roka	清澄沪過	clarifying filtration *[eng]*
sei'chōseki	正長石	orthoclase *[miner]*
seichō-setsu'gō	成長接合	grown junction *[electr]*
seichō-yokusei-zai	生長抑制剤	growth retardant *[bot]*
seichō-zai	清澄剤；清澄劑	finings *[cook]*
seichō-zai	清澄剤；せいちょうざい	clarifying agent; clarifier; clarificant *[mat]*
seichō-zai	整腸剤	intestine medicine *{n}* *[pharm]*
seichū-bire	正中鰭	median fin *[v-zoo]*
seidan	星団；星團	cluster; star cluster *{n}* *[astron]*
seidei	青泥；せいでい	blue mud *[geol]*
seiden-han'patsu	静電反発	electrostatic repulsion *[elec]*
sei-den'i	静電位	electrostatic potential *[elec]*
sei-denkai	静電界	electrostatic field *[elec]*
sei-denki	静電気；靜電氣	static electricity *[elec]*
seiden-ryoku	静電力	electrostatic force *[elec]*
seiden-shashin-hō	静電写真法	xerography *[graph]*
seiden'shiki-inshō-sōchi	静電式印書装置	electrostatic printer *[electr]* *[pr]*
seiden'shiki-ki'oku	静電式記憶	electrostatic storage *[comput]*
seiden-yōryō	静電容量	electrostatic capacity *[elec]*
seiden-yūdō	静電誘導	electrostatic induction *[elec]*
seido	精度；せいど	accuracy; precision *[math]* *[sci-t]*
seidō	青銅；せいどう	bronze *{n}* *[met]*
seidō	制動	braking *{n}* *[mech-eng]*
"	制動	damping *{n}* *[phys]*
seidō-ki	制動機	trigger *{n}* *[eng]*
seidō'ki-jidai	青銅器時代	Bronze Age *[archeo]*
seidō-ryoku	制動力	braking force *[mech]*
seidō'shi	制動子	damper *[mech-eng]*
sei-eki	精液	seminal fluid; semen (pl -semina) *[physio]*
sei-en	正塩；正鹽	normal salt *[ch]*

sei'enkō	青鉛鉱	linarite *[miner]*
seifuku	制服	uniform *{n} [cl]*
seifun-ki	製粉機	flour mill; mill *[agr]*
seigan	西岸	west coast *[geogr]*
seigen	正弦；せいげん	sine *[math]*
seigen'ha	正弦波	sine wave; sinusoidal wave *[phys]*
seigen-hōsoku	正弦法則	sines, law of *[math]*
seigen-inshi	制限因子	limiting factor *[bio]*
seigen-ki	制限器	limiter *[electr]*
seigen-kōso	制限酵素	restriction enzyme; restriction endonuclease *[bioch]*
sei'genso	生元素	bioelement *[ch] [bio]*
seigō	正号；正號	positive sign *{n} [math]*
seigō	整合；せいごう	matching *{n} [comput]*
"	整合	adjustment *[eng]*
seigō-ki	整合器	impedance-matching box *[elec]*
seigo ni	正後に	dead astern (ship) *{adv} [eng]*
sei'gōsei	生合成	biosynthesis (pl -theses) *[bioch]*
seigō suru	整合する	adjust; coordinate *{vb}*
seigyo	制御；せいぎょ	control *{n} [comput] [math] [sci-t]*
seigyo-ban	制御盤	control panel *[comput]*
seigyo-bō	制御棒	control rod *[nucleo]*
seigyo'bun	制御文	control statement *[comput]*
seigyo'go	制御語	control word *[comput]*
seigyo-kei	制御系	control system *[eng]*
seigyo-keiden-ki	制御継電器	control relay *[comput] [elec]*
seigyo'kei'gaku	制御系学	control systems *[eng]*
seigyo-moji	制御文字	control character *[comput]*
seigyo-shirei	制御指令	control command *[comput]*
seigyo-sōchi	制御装置	control unit; controller *[comput]*
seigyo-sōsa	制御操作	control operation *[comput]*
seigyo-taku	制御卓	operator console *[comput]*
sei-hachi'men'tai	正八面体	regular octahedron *[math]*
seihaku-mai	精白米	polished rice *[agr] [cook]*
sei-hannō	正反応；正反應	forward reaction *[ch]*
sei-hansha	正反射	mirror reflection; regular reflection; specular reflection *[opt]*
seihin	製品；せいひん	commercial product; manufactured product; finished product; product *[ind-eng]*
seihō	製法	production method *[ind-eng]*
seihō-gyō'retsu	正方行列	square matrix *[math]*
seihō-hai'retsu	正方配列	square array *[math]*

seihō-hen'i	青方偏移	blue shift *[astrophys]*
seihō'kei	正方形	square *{n} [math]*
seihon	製本；せいほん	bookbinding *{n} [pr]*
seihō-shō	正方晶	tetragonal crystal *[crys]*
seihō'shō-kei	正方晶形	tetragonal system *[crys]*
seihyō	星表	star catalog *[astron]*
seihyō	製氷；せいひょう	ice-making *{n} [eng]*
seihyō-ki	製氷機	ice machine *[eng]*
seihyō-ki	製表機	tabulator *{n} [comput]*
seihyō suru	製表する	tabulate *{vb} [comput]*
sei'iku	生育	growth; development *[bio] [physio]*
sei'iku'chi	生育地；せいいくち	habitat *[ecol]*
sei-inshi	性因子	sex factor *[microbio]*
seiji	青磁；せいじ	celadon *[cer]*
sei'jiki no	正磁気の	paramagnetic *{adj} [elecmg]*
seijin	成人	adult (person) *[bio]*
seijō	正常；せいじょう	normalcy; normalness *{n}*
seijō-bunsan	正常分散	normal dispersion *[phys]*
seijō-do	清浄度	cleanliness *[met]*
seijō no	正常の	normal *{adj} [sci-t]*
seijō-ryū	正常粒	normal grain *[phys]*
seijō-saibō	星状細胞	astrocyte *[cyt]*
seijun-neji	整準ねじ	leveling screw *[eng]*
seijū-rui	正獣類；正獸類	Eutheria *[v-zoo]*
sei'kagaku	生化学；せいかがく	biochemistry *[ch]*
seikagaku'teki-sanso-yōkyū'ryō	生化学的酸素要求量	biochemical oxygen demand; BOD *[microbio]*
seika-hō	青化法	cyaniding process *[met]*
seikaku(-sa)	正確(さ)	accuracy; accurateness *[math] [sci-t]*
sei-kakuchū	正角柱	regular prism *[math]*
seikaku-do	正確度；せいかくど	accuracy *{n} [sci-t]*
seikaku na	正確な	accurate; correct *{adj} [math]*
seikan-kūkan	星間空間	interstellar space *[astron]*
seikan-yaku (seikan-zai)	制汗剤（制汗劑）	antiperspirant *{n} [pharm]*
seikatsu	正割；せいかつ	secant *[an-geo]*
seikatsu-kan	生活環	life cycle *[bio]*
seikatsu-ryoku	生活力	viability *[bio]*
seikatsu-shi	生活史	life history *[bio]*
seikei	成形；せいけい	molding; shaping; fashioning *{n} [eng]*
"	成形	forming *{n} [rub]*
seikei	整形；せいけい	fairing *{n} [aero-eng]*
"	整形	trueing; truing *{n} [mech-eng]*

seikei-kana'gata	成形金型	shaping mold *[eng]*
seikei-ronri-shiki	正形論理式	well-formed formula *{n} [logic]*
seikei-sei	成形性	formability; moldability *[mat]*
seikei-tan	成型炭	briquet coal *[mat]*
sei-kettei	性決定	sex determination *[gen]*
seiki	世紀; せいき	century *{n}*
seiki	正規	normality *{n} [ch]*
seiki	生気	animation; life; vitality *[bio]*
seiki	精気	energy; spirit; essence *[bio]*
seiki-bunpu	正規分布	normal distribution *[math]*
seiki'ka	正規化	normalization *[sci-t]*
sei-kikan	正帰還; 正歸還	positive feedback *[electr]*
seiki'ka suru	正規化する	normalize *{vb} [comput] [sci-t]*
sei'kikō'gaku	生気候学	bioclimatology *[climat]*
seiki-kyoku'sen	正規曲線	normal curve *[stat]*
seikin-sayō	静菌作用	bacteriostasis *[microbio]*
seikin'sei no	静菌性の	bacteriostatic *{adj} [microbio]*
seikin'seki	青金石	lapis lazuli; lazurite *[miner]*
seikin-zai	静菌剤	bacteriostat; bacteriostatic agent *[pharm]*
sei'kishō'gaku	生気象学	biometeorology *[meteor]*
seikō	性交	coitus; sexual intercourse *[bio]*
seikō	製鋼	steelmaking *{n} [met]*
seikō-dōbutsu	星口動物	Sipunculoidea; Sipunculida *[i-zoo]*
sei-kyoku	正極	positive electrode *[elec]*
seikyū	請求; せいきゅう	demand; request *{n} [pat]*
seikyū'sho	請求書	bill; statement of account *[econ]*
sei-masatsu-keisū	静摩擦係数	coefficient of static friction *[mech]*
seimei	生命; せいめい	life (pl lives) *[bio]*
seimei-gi'jutsu	生命技術	biotechnology *[eng]*
seimei-iji-keitō	生命維持系統	life-support system *[eng]*
seimei-kagaku	生命科学	life sciences *[bio]*
seimei-kōgaku	生命工学	biotechnology *[eng]*
seimei no-	生命の-	bio- *{prefix}*
sei'mitsu-sa	精密さ	precision *[math]*
sei-mon	声紋; 聲紋	voice print *[eng-acous]*
sei-nen'gappi	生年月日	date of birth *{n} [bio]*
sei'nen-ki	青年期	adolescence; youth period *[psy]*
sei-netsu	青熱	blue heat *[thermo]*
sei'netsu-moro-sa	青熱脆さ	blue brittleness *[met]*
sei'netsu-sosei'kakō-hō	青熱塑性加工法	blue-heat working *{n} [met]*
seinō	性能; せいのう	efficiency; performance; capacity; power; properties *[ind-eng] [sci-t]*

sei no sū 正の数；せいのすう positive number *[math]*
seinō-teika-insū 性能低下因数 degradation factor *[comput]*
sei-on 清音 voiceless sound *[ling]*
sei'on-ki 整温器；整溫器 thermostat *[eng]*
sei-raku'sa 静落差 static head *[fl-mech]*
seiren 精練；精練 boiling off; degumming; scouring *{n} [tex]*
seiren 精錬；精鍊 refining *{n} [eng]*
seiren 製錬；せいれん smelting *{n} [met]*
seiren'sho; seiren'jo 精錬所 refinery *[met]*
sei'retsu 整列；せいれつ lineup *{n} [ch]*
" 整列 sort *[comput]*
" 整列 alignment *[eng]*
seiri-bangō 整理番号 reference number *[sci-t]*
seiri'gaku 生理学；せいりがく physiology *[bio]*
seiri-kagaku 生理化学 physiological chemistry *[ch]*
sei-riki'gaku 静力学 statics (the discipline) *[mech]*
seirin'kō 青燐鉱 lazulite *[miner]*
seiri-shoku'en-sui 生理食塩水 normal saline; physiological saline *[physio]*
seiri suru 整理する rearrange *{vb} [math]*
sei'ritsu-zō 正立像 erect image *[opt]*
sei-roku'men'tai 正六面体 cube *{n} [math]*
seiryō-inryō'sui 清涼飲料水 soft drink *[cook]*
seiryō-kan 清涼感 refreshing feel *[physio]*
sei'ryoku 勢力；せいりょく influence; power; strength *{n}*
sei'ryoku 精力 energy; vigor; vitality *[bio]*
seiryoku-ken 勢力圏；勢力圈 sphere of activity *[astron]*
sei'ryoku-shoku 青緑色 cyan (color) *[opt]*
seiryū 精留；精溜 rectification *[ch]*
seiryū 整流；せいりゅう rectification *[elec]*
" 整流 commutation *[elecmg]*
seiryū 整粒 screening *{n} [eng]*
" 整粒 granulation *[sci-t]*
seiryū-kairo 整流回路 rectifier circuit *[electr]*
seiryū-ki 整流器 rectifier *[elec]*
seiryū-ki 整粒機 granulator *[eng]*
seiryū-kōka 整粒効果 particle-unifying effect *[phys]*
seiryū'shi 整流子 commutator *[elecmg]*
seiryū suru 精流する rectify *{vb} [elec]*
seiryū-tō 精留塔 rectifying column; rectifying tower *[ch-eng]*
sei-saibō 性細胞；性細胞 sex cell(s) *[bio]*

seisan	生産；せいさん	production *[ind-eng]*
seisan	青酸	hydrocyanic (or prussic) acid (obsolete)
seisan-gishi	生産技師	product engineer *[ind-eng]*
seisan-jikan	生産時間	production time *[comput]*
sei-sankak(u)kei	正三角形	equilateral triangle *[math]*
seisan-kanri	生産管理	production control *[ind-eng]*
seisan'koku-mei	生産国名	country of origin *[ind-eng]*
seisan-ryoku	生産力	productivity *[ind-eng]*
seisan-sei	生産性	productivity *[ind-eng]*
seisan-setsu'bi	生産設備	production facilities *[ind-eng]*
seisan'sha	生産者	producer *[ecol]* *[econ]*
seisan suru	生産する	manufacture; produce *{vb}*
seisan-taisei	生産体制	production structure *[ind-eng]*
seisan-zai	制酸剤	antacid *{n}* *[pharm]*
seisei	生成	generation *[comput]*
seisei	精製；せいせい	purification; refinement *[eng]*
seisei'butsu	生成物	product *[ind-eng]*
seisei'jo; seisei'sho	精製所	refinery (pl -eries) *[ch-eng]*
seisei-netsu	生成熱	heat of formation *[p-ch]*
seisei-sui	精製水	purified water *[hyd]*
seisei suru	生成する	generate *{vb}* *[comput]*
seisei'tō	精製糖	refined sugar *[bioch]*
seisei-yu	精整油	raffinate *[ch-eng]*
seiseki	成積	result; showing; record *{n}* *[econ]*
seiseki-keisū	成積係数	performance coefficient *[thermo]*
sei'sekkai	生石灰	quicklime *[inorg-ch]*
seisen	精選；精選	cleaning (of ore) *{n}* *[miner]*
sei-senmō	性繊毛；性線毛	sex pilus (pl -pili) *[gen]*
seisen-shoku'ryō('hin)	生鮮食料(品)	perishable foods *[cook]*
sei-senshoku'tai	性染色体	sex chromosome *[gen]*
sei'setsu	正接；せいせつ	tangent *[math]*
sei'setsu-hōsoku	正接法則	tangents, law of *[math]*
seishi	生死	life and death *{n}* *[bio]*
seishi	制止	control; check; restraint *{n}*
seishi	精子	sperm cell *[bio]*
seishi	製糸	silk reeling *{n}* *[tex]*
seishi	製紙	papermaking *{n}* *[paper]*
seishi	静止	stillness; repose *{n}*
seishi-keisei	精子形成	spermatogenesis *[physio]*
seishiki	正式	formal; official; proper *{n}*
sei-shiki	整式	polynomial *[math]*
seishi-ki	静止期	interphase; resting stage *{n}* *[cyt]*

seishin; seishi	静振	seiche *[fl-mech]*
seishin	精神；せいしん	mind *{n}* *[psy]*
seishin-antei-yaku	精神安定薬	tranquilizer *{n}* *[pharm]*
seishin'byō	精神病	mental disease *[med]*
seishin-ijō-hatsu'gen-busshitsu	精神異常発現物質	psychedelic substance *{n}* *[pharm]*
seishin-sokutei'gaku	精神測定学	psychometrics *[psy]*
seishi shita	静止した	quiescent *{adj}*
seishi-shitsu'ryō	静止質量	rest mass *[rela]*
sei'shitsu	性質；せいしつ	character; nature; properties *[sci-t]*
seisho	成書	published volume *[pr]*
seisho	清書	clean copy; fair copy *[pat]* *[pr]*
seishoku	生殖	procreation; reproduction *[bio]*
seishoku	星食	lunar occultation; occultation *[astron]*
seishoku	製織	weaving *{n}* *[tex]*
seishok(u)ki	生殖器	reproductive organ *[anat]*
seishoku-nō'ryoku	生殖能力	reproductive potential *[bio]*
seishoku-nuno	製織布	woven cloth *[tex]*
seishoku-sen	生殖腺	gonad *[bio]*
seishō'shi	生松脂	turpentine *[mat]*
seishu	清酒	Japanese rice wine; sake *[cook]*
seisō	清掃	cleaning; scavengery *{n}* *[civ-eng]*
seisō	整相	phasing *{n}* *[electr]*
seisō	精巣；精巣	testis (pl testes); testicle(s) *[anat]*
seisō-dōbutsu	清掃動物	scavenger (animal) *[zoo]*
seisō-jōtai	精巣上体	epididymis (pl epididymides) *[bio]*
seisō-kagō'butsu	成層化合物	lamellar compound *[crys]*
seisō-ken	成層圏；成層圏	stratosphere *[meteor]*
seisoku'chi	生息地；棲息地	habitat *[ecol]*
seisō-tesshin	成層鉄心	laminated (iron) core *[elecmg]*
seisū	整数；せいすう	integer; integral number; whole number *[math]*
sei-sui'atsu	静水圧；静水壓	hydrostatic pressure *[fl-mech]*
"	静水圧	static water pressure *[hyd]*
seisū-keikaku-hō	整数計画法	integer programming *{n}* *[comput]*
seisū-seigyo	整数制御	integer control *[comput]*
seitai	生体；生體	living body *[bio]*
seitai	声帯；聲帶	vocal cord *[anat]*
seitai-dansei'tai	生体弾性体	bioelastic (pl -s) *{n}* *[mech]*
seitai-denki	生体電気	bioelectricity *[physio]*
seitai-gai de	生体外で	in vitro *{adj}* *{adv}* *[bio]*
seitai'gaku	生態学	ecology *[bio]*

seitai-kaibō	生体解剖	vivisection *[bio]*
seitai-kei	生態系	ecosystem *[ecol]*
seitai-kikō	生態気候	ecoclimate *[bio]*
seitai-kōbunshi	生体高分子	biopolymer *[bioch]*
seitai-kōgaku	生体工学	biological engineering *{n} [sci-t]*
seitai-maku	生体膜	biomembrane *[cyt]*
seitai'nai-bunkai-sei	生体内分解性	biodegradability *[mat]*
seitai-nai de	生体内で	in vivo *{adj} {adv} [bio]*
seitai-sen'i	生態遷移	ecological succession *[ecol]*
sei(tai)-shokubai	生（体）触媒	biocatalyst *[bioch]*
seitai'teki-chi'i	生態的地位	ecological niche *[ecol]*
seitai-zairyō	生体材料	biomaterial *[mat]*
sei-takak(u)kei	正多角形	regular polygon *[math]*
sei'taka-shigi	背高鴫	stilt (bird) *[v-zoo]*
sei-ta'men'tai	正多面体	regular polyhedron *[math]*
seitan	成端	terminating *{n} [elec]*
seitan	精炭	clean coal *{n} [geol]*
seitei	制定	enactment; institution *[legal]*
sei-teisū	整定数	integer constant *[math]*
sei'teki na	静的な；靜的な	static *{adj} [comput]*
sei'teki-nikei	性的二形	sexual dimorphism *[bio]*
seiten	晴天；せいてん	fair weather *[meteor]*
sei-tōki	精陶器	fine earthenware *[cer]*
sei-tokusei	静特性	static characteristic *[electr]*
seito-yaku (seito-zai)	制吐薬（制吐剤）	antiemetic *{n} [pharm]*
sei'uchi	海馬；海象；セイ=ウチ	walrus (marine mammal) *[v-zoo]*
sei'un	星雲	nebula (pl -las, -lae) *[astron]*
seiyō	西洋；せいよう	Occident; West *[geogr]*
seiyō-kyū	西洋球	Western hemisphere *[geogr]*
seiyō-ninjin	西洋人参	American ginseng; Panax quinquefolium *[bot] [pharm]*
seiyō-ryōri	西洋料理	Western food; Western dishes; Occidental cuisine *[cook]*
seiyō-yama'hakka	西洋山薄荷	melissa (herb) *[bot]*
seiyu	精油	essential oil *[mat]*
seiyu'jo	製油所	oil refinery *[ch-eng]*
seiza	星座	constellation *[astron]*
seizai'jō	製材場	sawmill *[ind-eng]*
seiza-tōei-ki	星座投影機	planetarium (pl -s, -ia) *[astron]*
seizen ni	正前に	dead ahead (ship) *{adv} [navig]*
seizō	製造；せいぞう	manufacture; production *{n} [ind-eng]*

seizō-gaisha	製造会社	manufacturing company *[ind-eng]*
seizō-gyōsha	製造業者	manufacturer *[ind-eng]*
seizō-koku	製造国；製造國	country of manufacture *[ind-eng]*
seizō-moto	製造元	manufacturer; maker; producer *[ind-eng]*
seizon-ken	生存圏	biosphere *[ecol]*
seizon-nō'ryoku	生存能力	viability *[bio]*
seizō-sōchi	製造装置	producing device *[eng]*
seizu	星図；星圖	star atlas; celestial map *[astron]*
seizu	製図	cartography; drafting; drawing *{n}* *[graph]*
seizu-ban	製図板	drawing board *[graph]*
sekai	世界；せかい	world *[geogr]*
sekai-chi'teki-shoyū'ken	世界知的所有権	world intellectual property *[pat]*
sekai-ji	世界時	universal time *[astron]*
seki	咳	cough *{n}* *[physio]*
seki; ato	跡	trace *{n}* *[math]*
seki	堰	dam *{n}* *[civ-eng]*
seki	積	product (of multiplication) *[arith]*
sekiban-in'satsu	石版印刷	lithography *[pr]*
seki'boku	石墨；石墨	graphite *[miner]*
seki'bun	積分；せきぶん	integral calculus *[math]*
sekibun-hō	積分法	integration method *[math]*
sekibun-ki	積分器	integrator *[electr]*
sekibun-kairo	積分回路	integration (-ing) circuit *[elec]*
sekibun-shaku'do	積分尺度	integral scale *[math]*
sekibun suru	積分する	integrate *{vb t}* *[math]*
sekibun-tōkyū	積分等級	integrated magnitude *[astron]*
seki'chū	脊柱	backbone; spine; spinal column; vertebral column *[anat]*
seki'dei	赤泥	red mud *[geol]*
seki'dō	赤道；せきどう	equator *[geod]*
sekidō-gi	赤道儀	equatorial telescope *[eng]*
seki'dōkō	赤銅鉱	red copper ore *[miner]*
sekidō no	赤道の	equatorial *{adj}* *[geod]*
sekidō-zahyō	赤道座標	equatorial coordinates *[astron]* *[geod]*
seki'ei	石英；せきえい	quartz *[miner]*
seki'ei'gan	石英岩	quartzite *[petr]*
seki'ei-hangan	石英斑岩	quartz porphry *[miner]*
seki'ei-hengan	石英片岩	quartz schist *[miner]*
seki'ei'shitsu no	石英質の	siliceous *{adj}* *[miner]*
sekigai-bunkō'gaku	赤外分光学	infrared spectroscopy *[spect]*
sekigai-bunkō-ki	赤外分光器	infrared spectrometer *[spect]*
sekigai-bunkō-kōdo-kei	赤外分光光度計	infrared spectrophotometer *[spect]*

sekigai-sen	赤外線	infrared radiation (rays) *[elecmg]*
sekigai'sen-bunkō-kōdo-hō	赤外線分光光度法	infrared spectrophotometry *[spect]*
sekigai'sen-jidō-tsui'bi	赤外線自動追尾	infrared homing *{n} [eng]*
sekigai'sen-tenmon'gaku	赤外線天文学	infrared astronomy *[astron]*
sekihan	赤飯	glutinous rice with red beans *[cook]*
sekihō-hen'i	赤方偏移	red shift *[astrophys]*
seki'i	赤緯；せきい	declination *[astron]*
seki'i-jiku	赤緯軸	declination axis *[eng]*
seki'i-kan	赤緯環	declination circle *[eng]*
seki'jun	石筍	stalagmite *[geol]*
sekika	石果	drupe *[bot]*
seki'ko	潟湖	lagoon *[geogr]*
sekimen; ishi-wata	石綿	asbestos *[miner]*
seki-netsu	赤熱	red heat *[thermo]*
seki'nin	責任	liability; responsibility *[legal]*
sekiran'un	積乱雲；積亂雲	cumulonimbus *[meteor]*
sekiri	石理	fabric; texture *{n} [geol]*
sekiri-kin	赤痢菌	dysentery bacillus; Shigella *[microbio]*
seki-rin	赤燐	red phosphorus *[ch]*
seki'ryō	積量	tonnage *[nav-arch]*
seki-ryoku	斥力	repulsion; repulsive force *[mech]*
sekisai	積載	carrying; loading *{n} [trans]*
seki'saku	脊索；せきさく	notochord *[v-zoo]*
seki'saku-dōbutsu	脊索動物	chordates *[zoo]*
seki'saku-dōbutsu-mon	脊索動物門	Chordata *[zoo]*
sekisan	積算	addition (cumulative) *[math]*
"	積算	integration *[eng] [math]*
sekisan-nōdo	積算濃度	integral density *[photo]*
sekisei-inko	背黄青鸚哥	budgerigar (bird) *[v-zoo]*
seki'shitsu-inseki	石質隕石	aerolite *[geol]*
seki'shoku-kyosei	赤色巨星	red giant *[astron]*
sekishō'mo	石菖藻	tape grass; eel grass *[bot]*
seki'shutsu	析出	deposition *[ch]*
"	析出	precipitation *[met]*
sekisō	積層	laminating; lamination *{n} [sci-t]*
sekisō-ban	積層板	laminate; laminated sheet *[mat]*
sekisō'butsu	積層物	laminate; laminated article *[mat]*
seki'tan	石炭；せきたん	coal *[geol]*
sekitan-bukuro	石炭袋	Coalsack *[astron]*
sekitan'ka-do	石炭化度	rank of coalification *[geol]*
sekitan-ro	石炭炉；石炭爐	coal furnace *[eng]*
sekitan-san	石炭酸	carbolic acid; phenol

seki'tekkō	赤鉄鉱; 赤鐵鑛	hematite; red iron ore *[miner]*
seki'tetsu-inseki	石鉄隕石	siderolite *[geol]*
seki'tsui	脊椎; せきつい	vertebra (pl -brae) *[anat]*
sekitsui-dōbutsu	脊椎動物	vertebrates; Vertebrata *[v-zoo]*
sekitsui-dōbutsu'gaku	脊椎動物学	vertebrate zoology *[bio]*
seki'un	積雲	cumulus *[meteor]*
seki-wa	積和	sum of products *[math]*
seki'yu	石油; せきゆ	petroleum *[mat]*
sekiyu-bi'seibutsu	石油微生物	petroleum microorganism *[microbio]*
sekiyu-hakkō	石油発酵	petroleum fermentation *[microbio]*
sekiyu-jushi	石油樹脂	petroleum resin *[org-ch]*
sekiyu'kagaku-kōgyō	石油化学工業	petrochemical industry *[org-ch]*
sekiyu'kagaku-seihin	石油化学製品	petrochemical product *[mat]*
sekiyu'kei-hōkō'zoku-tanka'suiso	石油系芳香族=炭化水素	petroleum aromatics *[org-ch]*
sekiyu-kikan	石油機関	petroleum engine *[mech-eng]*
sekiyu-kōkusu	石油コークス	petroleum coke *[mat]*
sekiyu-sai'kutsu	石油採掘	oil-well digging (drilling) *{n} [petrol]*
sekiyu-seisei	石油精製	petroleum refining *{n} [ch-eng]*
seki-zui	脊髄; 脊髓	spinal cord *[anat]*
sekka	石化	petrifaction *[geol]*
sekkai	石灰; せっかい	lime *[inorg-ch]*
sekkai-chisso	石灰窒素	lime nitrogen; nitrolime *[inorg-ch]*
sekkai-dō	石灰洞	lime grotto *[geol]*
sekkai'gan	石灰岩	limestone *[petr]*
sekkai'ka	石灰化	calcification *[geoch] [physio]*
sekkai'ka	石灰華	tufa *[geol]*
sekkai-kō	石灰光	limelight *[eng]*
sekkai-nyū'eki	石灰乳液	lime emulsion *[mat]*
sekkai'seki	石灰石	limestone *[petr]*
sekkai'shitsu no	石灰質の	calcareous *{adj} [sci-t]*
sekkaku	接角	adjacent angles *[math]*
sekka-sayō	石化作用	petrifaction *[geol]*
sekka'sekkō	雪花石膏	alabaster *[miner]*
sekkei	赤経; 赤經	right ascension *[astron]*
sekkei	設計; せっけい	plan *{n} [graph]*
"	設計	design *{n} [sci-t]*
sekkei	雪渓; 雪溪	snow valley; snow gorge *[geol]*
sekkei-gishi	設計技師	design engineer *[eng]*
sekkei-kōgaku	設計工学	design engineering *{n} [eng]*
sekkei-zu	設計図; 設計圖	design drawing *{n} [graph]*
sekkekkyū	赤血球	erythrocyte; red blood cell *[hist]*

sekken	石鹼；せっけん	soap *{n} [mat]*
sekken-sui	石鹼水	soapwater *[hyd]*
sekki	炻器；石器	stoneware *[cer]*
sekki	石基	groundmass; matrix *[petr]*
sekki-jidai	石器時代	Stone Age *[archeo]*
sekkin suru	接近する	approach; near *{vb}*
sekkō	石膏；せっこう	gypsum *[miner]*
sekkyoku'shi	接極子	armature *[elecmg]*
sekkyoku'teki na	積極的な	aggressive; positive; vigorous *{adj}*
sekō	背甲	carapace *[v-zoo]*
sekō	施工	application; execution (of work) *[civ-eng]*
sekō	施行	carrying out *{n}*
seku	咳く	cough *{vb} [physio]*
se'madara-kogane-mushi	背斑黄金虫；せま=だらこがねむし	Asiatic beetle; Oriental beetle (insect) *[i-zoo]*
semai	狭い；狹い	narrow *{adj}*
semaku suru	狭くする	narrow down *{vb t}*
semaru	迫る	approach *{vb}*
semi	蟬；せみ	cicada (insect) *[i-zoo]*
sen	千	thousand *{n} [math]*
sen	栓	cock; plug; stopper; tap *{n} [ch]*
sen	腺	gland *[anat]*
sen	銭；錢	sen (Japanese money) *[econ]*
sen	線	line *{n} [electr] [math]*
sen	線	ray *[math] [opt]*
-sen	-船	-ship *{suffix} [nav-arch]*
-sen	-腺	-gland *{suffix} [anat]*
-sen	-銭	-sen (money) *{suffix} [econ]*
sen'aen'kō	閃亜鉛鉱	sphalerite; zincblende *[miner]*
se'naka	背中	back *{n} [anat]*
senbai	専売；專賣	monopoly *[econ]*
senbai'hin	専売品	proprietary article *[ind-eng]*
senbai'ken	専売権	monopoly *[econ]*
senbei	煎餅；せんべい	Japanese rice cracker *[cook]*
sen'betsu	選別；選別	selection; selecting *{n} [comput]*
"	選別	sorting *{n} [met]*
"	選別	separation *[min-eng]*
senbi	船尾	stern (of ship) *{n} [nav-arch]*
senbi-kōhan	船尾甲板	quarter deck *[nav-arch]*
senbin	洗瓶	washing bottle *[an-ch]*
senbi no hō ni	船尾の方に	abaft *{prep} [nav-arch]*
senbi-suisen	船尾垂線	after perpendicular *{n} [nav-arch]*

senbi'yoku	先尾翼	canard *[aero-eng]*
sen-bōchō-keisū; (-ritsu)	線膨張係数；（-率）	coefficient of linear expansion *[thermo]*
senbō'kyō	潜望鏡	periscope *[nav-arch] [opt]*
senbun	線分	segment; line segment *[math]*
senburi	千振；せんぶり	Japanese green gentian *[bot]*
sencha	煎茶	green tea (medium grade) *[cook]*
sen'chaku	染着；せんちゃく	dyeing *{n} [ch-eng]*
senchaku-do	染着度	degree of dyeing power *[tex]*
senchaku-ritsu	染着率	degree of exhaustion *[tex]*
senchaku-sei	染着性	dye affinity *[tex]*
senchi'guramu	甅；センチグラム	centigram (unit of mass) *[mech]*
senchi'mētoru	糎；センチメートル	centimeter (unit of length) *[mech]*
senchō	千兆	quadrillion (U.S.) *[math]*
senchō	船長	ship's captain; commander; master
senchū	線虫	nematode *[i-zoo]*
senchū-kujo-zai	線虫駆除剤	nematicide; nematocide *{n} [mat]*
sendan	栴檀；せんだん	Japanese bead tree *[bot]*
sendan	剪断；剪斷	shear *{n} [mech]*
sendan-kōfuku-ten	剪断降伏点	yield point in shear *[mech]*
sendan-ō'ryoku	剪断応力	shearing stress *[mech]*
sendan-sayō	剪断作用	shearing *{n} [mech]*
sendan-sokudo	剪断速度	shear rate *[fl-mech]*
sendarai-ko	銑ダライ粉	iron castings; lathe scraps *[met]*
sendo	鮮度	freshness *[cook]*
sendo	繊度；纖度	fineness (thread) *[tex]*
sen'ei-do	鮮鋭度	sharpness *[photo]*
sengai-ki	船外機	outboard motor *[eng]*
sengan-hatsu'mei	先願発明	prior (-applied) invention *[pat]*
sengan-yaku	洗眼薬	eyewash; collyrium *{n} [pharm]*
sengen	選言；選言	disjunction *[logic]*
sengen'bun	宣言文	declarative statement *[comput]*
sengen suru	宣言する	declare; proclaim *{vb}*
sen-getsu	先月	last month *{n}*
sengyō	専業；專業	specialty (of business) *[econ]*
sen'i	遷移	transition *[comput]*
sen'i	繊維；せんい	fiber *[tex]*
sen'i-aen'kō	繊維亜鉛鉱	wurtzite *[miner]*
sen'i-ga'saibō	繊維芽細胞	fibroblast *[hist]*
sen'i-kinzoku	遷移金属	transition metal *[ch]*
sen'i-kyōka-kinzoku	繊維強化金属	fiber-reinforced metal *[met]*
sen'i-nori-zai	繊維糊剤	fiber thickening agent *[mat]*
sen'i'shitsu	繊維質	fibrous matter *[mat]*

sen'i'shitsu-shoku'hin	繊維質食品	roughage *{n} [cook]*
sen'i'so	繊維素; せんいそ	cellulose; fibrin *[bioch]*
sen'i'so-gen	繊維素原	fibrinogen *[bioch]*
sen'i-soku	繊維束	fiber bundle *[tex]*
senjiru	煎じる	decoct *{vb t} [cook] [food-eng]*
senjō	洗浄; 洗滌	washing *{n}*
senjō	線状; せんじょう	linearness *{n} [math]*
senjō	線条; 線條	filament *[astron] [elec]*
senjō-bunshi	線状分子	linear molecule *[p-ch]*
senjō-chi	扇状地	alluvial fan *[geol]*
senjō-jūgō'tai	線状重合体	linear polymer *[poly-ch]*
senjō-ki	洗浄機; 洗滌機	washer; washing machine *[eng]*
senjō-kōbunshi	線状高分子	linear polymer *[poly-ch]*
senjō-ryoku	洗浄力; 洗滌力	detergency *[mat]*
senjō-zai	洗浄剤	detergent *[mat]*
sen'jutsu	戦術; 戰術	tactics *[mil]*
sen'jutsu'yō-kōkū-ki	戦術用航空機	tactical aircraft *[aerosp]*
senkai	旋回; せんかい	turning; whirl *{n} [mech-eng]*
senkai-ken	旋回圏	turning circle *[navig]*
senkei	線形	linear form *[math]*
senkei-dōbutsu	線形動物	roundworm; Nematoda *[i-zoo]*
senkei-keikaku(-hō)	線形計画(法)	linear programming *{n} [math]*
senketsu	潜穴	burrow *{n} [paleon]*
senkō	先行	precedence *[comput] [math]*
"	先行	antecedent *{n}*
senkō	穿孔; せんこう	hole *{n} [comput]*
"	穿孔	punching *{n} [eng]*
"	穿孔	bore *{n} [paleon]*
senkō	閃光	flash *{n} [opt] [meteor]*
senkō	選鉱; 選鑛	beneficiation; mineral dressing *[min-eng]*
senkō	線香	incense stick *[mat]*
senkō-bunken	先行文献	earlier document *[pr]*
senkō-do	旋光度	optical rotation; angle of rotation *[opt]*
senkō-kei	旋光計	polarimeter *[opt]*
senkō-ki	穿孔機	boring machine *[eng]*
senkō-kō'bunkai	閃光光分解	flash photolysis *[p-ch]*
senkō-moji	先行文字	leading graphics *[comput]*
senkō-sei	閃光星	flare star *[astron]*
senkō-sei	旋光性	optical rotatory power; optical activity *[opt]*
senkō'sei-tai	旋光性体	optically active substance *[mat]*
sen-kotsu	仙骨; せんこつ	sacrum (bone) *{n} [anat]*

senku-busshitsu	先駆物質	precursor *[ch]*
senku-hōden	先駆放電	leader stroke *[meteor]*
senkyō	船橋	bridge (of ship) *{n} [nav-arch]*
senmei-do	鮮明度	visibility *[electr]*
"	鮮明度	definition; resolution *[photo]*
senmen'jo	洗面所	lavatory; rest room *[arch]*
senmen-ki	洗面器	lavatory basin *[cer]*
senmō	選毛	wool sorting *{n} [tex]*
senmō	繊毛；纖毛	cilia (sing cilium) *[i-zoo]*
"	繊毛；せんもう	fimbria (pl fimbriae); pilus (pl pili) *[microbio]*
senmō'chū-kō	繊毛虫綱	Ciliata *[i-zoo]*
senmon	専門；專門	specialty *{n}*
senmon'ka	専門家	expert; professional person; specialist
senmon'ten	専門店	specialty store *[arch] [econ]*
sen-nen	先年	former year(s) *{n}*
sennen'kan	千年間	millenium (a thousand years) *{n}*
sennetsu	潜熱	latent heat *[thermo]*
sen-netsu'bōcho-ritsu	線熱膨張率	coefficient of linear expansion *[thermo]*
sen-nuki	栓抜き	bottle opener; corkscrew *[eng]*
sennyū	潜入	immersion *[astron]*
sen'on-soku	遷音速	transonic speed *[fl-mech] [phys]*
sen'paku	船舶	ship *{n} [nav-arch]*
sen'paku-denwa	船舶電話	maritime mobile radiotelephone *[comm]*
sen'paku('yō) no	船舶(用)の	marine *{adj} [nav-arch]*
senpū	旋風	cyclone; windspout *[meteor]*
senpū-ki	扇風機	fan (electric) *{n} [eng]*
sen'puku-ki	潜伏期	incubation period; latent period *[med]*
sen'puku shita	潜伏した	cryptic *{adj} [zoo]*
sen'ritsu	旋律	melody *[acous] [music]*
senro	線路	railroad track *[rail]*
senro-den'atsu-kōka	線路電圧効果	line drop *[elec]*
senro-son	線路損	line loss *[elec]*
sen'ryaku	戦略	strategy (pl -gies) *[mil]*
senryō	染料；せんりょう	dye; dyestuff *{n} [ch]*
senryō	線量	dose (of radiation) *{n} [med]*
senryō-chō'shoku	染料調色	dye toning *{n} [photo]*
senryō-kei	線量計	dosimeter *[eng]*
sen-ryoku	潜力	latent force *[mech]*
senryoku'gan	閃緑岩	diorite *[miner]*
senryō-ritsu	線量率	dosage *[med]*
"	線量率	dose rate *[nucleo]*

senryū	栓流	plug flow *[fl-mech]*
senryū	潜流	undercurrent *[ocean]*
sensaku	旋削	turning *{n}* *[mech-eng]*
sensei-kyō'jutsu'sho	宣誓供述書	affidavit *[legal]*
senseki-ritsu	占積率	space factor *[elecmg]*
"	占積率	occupancy rate *{n}*
sensha	戦車	military tank *[ord]*
senshi-baiyō	穿刺培養	stab culture *[microbio]*
sen-shitsu	船室	cabin (in a ship) *[nav-arch]*
sen'shoku	染色；せんしょく	dyeing; staining *{n}* *[bioch]* *[ch]*
sen'shoku	染織	dyeing and weaving *{n}* *[tex]*
senshoku-buntai	染色分体	chromatid *[cyt]*
senshoku-idō	浅色移動	hypsochromic shift *[p-ch]*
senshoku-kenrō-do	染色堅牢度	colorfastness; dyefastness *[ch]* *[tex]*
senshoku'shitsu	染色質	chromatin *[bioch]*
senshoku'tai	染色体	chromosome *[cyt]*
senshoku'tai-sū	染色体数	chromosome number *[cyt]*
senshō'seki	尖晶石	spinel *[miner]*
senshu	船首	bow; prow; stem (of ship) *[nav-arch]*
senshu-ha	船首波	bow wave *[nav-arch]*
senshu-kissui	船首喫水	forward draft *[nav-arch]*
senshu-suisen	船首垂線	forward perpendicular *[nav-arch]*
senshu-zō	船首像	figurehead *[nav-arch]*
sensō	船倉	hold (of ship) *[nav-arch]*
sensō	戦争；戰爭	war *{n}* *[mil]*
sen-sokudo	線速度	linear velocity *[mech]*
sensu	扇子；せんす	fan; folding fan *{n}* *[eng]*
sensui	潜水	diving *{n}* *[ocean]*
sensui'fu	潜水夫	diver (a person) *[ocean]*
sensui-fuku	潜水服	diving suit *[nav-arch]*
sensui'kan	潜水艦	submarine *{n}* *[nav-arch]*
sensui suru	潜水する	submerge *{vb}* *[nav-arch]*
sentai	船体；船體	hull (of a ship) *[nav-arch]*
sentai	蘚苔	moss(es); bryophyte *[bot]*
sentai-chūō(-bu)	船体中央(部)	amidship *{n}* *[nav-arch]*
sentai-chūō'bu ni	船体中央部に	amidships *{adv}* *[nav-arch]*
sentai'gaku	蘚苔学	bryology *[bot]*
sen'taku	洗濯；せんたく	laundry; laundering *{n}*
sen'taku	選択；選擇	select(ing); selection *{n}* *[sci-t]*
sentaku-ki	洗濯機	laundry machine; washing machine; washer *[eng]*
sentaku-kyūshū	選択吸収；選擇吸收	selective absorption *[elecmg]*

sentaku-ritsu	選択率	selectivity *[electr]*
sentaku-sei	選択性	selectivity *[an-ch]*
sentan	先端	tip (location) *{n}*
sentei	剪定；せんてい	pruning *{n} [bot]*
sentei	選定	selection; selecting *{n} [comput]*
senten	先点	cusp *[astron]*
sen-tetsu	銑鉄；銑鐵	pig iron *[mat]*
sentō-bu	尖頭部；先頭部	tip part *{n}*
sen'uran'kō	閃ウラン鉱	uraninite *[miner]*
sen-yaku	煎薬；せんやく	decoction; infusion *{n} [pharm]*
sen'yō-kaisen	専用回線	dedicated line *[comput]*
sen'yō-keisan-ki	専用計算機	special-purpose computer *[comput]*
sen'yō-shitsu (fune no)	専用室（船の）	stateroom (of a ship) *[nav-arch]*
sen'yū-menseki-ritsu	占有面積率	coverage *[graph]*
senzai	線材	wire rod *[mat]*
senzai-sei	潜在性	latency *[med] [physio]*
senzai shita	潜在した	cryptic *{adj} [zoo]*
senzan'kō	穿山甲	pangolin; scaly anteater (animal) *[v-zoo]*
senzo	先祖	ancestor(s) *[bio]*
senzō	潜像；せんぞう	latent image *[graph]*
senzō-ho'ryoku	潜像補力	latensification *[photo]*
senzō-kaku	潜像核	latent-image nuclei *[photo]*
sen-zu	線図；線圖	diagram *[graph]*
se'ou	背負う	carry on back; shoulder; bear *{vb}*
seppan-shusshi	折半出資	equally-shared investment *[econ]*
seppen	切片；せっぺん	intercept *{n} [math]*
seren	セレン	selenium (element)
seri	芹；せり	Japanese parsley; water dropwort; Oenanthe javanica (herb) *[bot]*
seri'uke-dai	迫受台；迫受臺	abutment *[arch]*
seron-chōsa	世論調査	public opinion survey *[comm] [psy]*
serurōsu	セルロース	cellulose *[bioch]*
seryūmu	セリウム	cerium (element)
seshūmu	セシウム	cesium (element)
sessaku	切削	cutting *{n} [met]*
sessaku-kakō	切削加工	cutting working *{n} [met]*
sessaku-sei	切削性	machinability *[met]*
sessaku-yu	切削油；せっさくゆ	cutting oil *[mat]*
sesseki	榍石；せっせき	sphene *[miner]*
sessen	接栓	contact plug *[eng]*
sessen	接線；せっせん	tangent *[math]*
sessen-jū'danmen	接線縦断面	tangential section *[bot]*

sessen-ō'ryoku	接線応力	tangential stress *[mech]*
sessen-shoku	接線食	grazing occultation *[astron]*
sessen-sokudo	接線速度	tantengial velocity *[mech]*
sesshi-do	摂氏度；攝氏度	degree Celsius; degree centigrade *[thermo]*
sesshi-ondo-me'mori	摂氏温度目盛	Celsius temperature scale *[thermo]*
sesshi suru	楔止する	stop, wedged *{vb} [eng]*
sesshoku	雪食	nivation *[geol]*
sesshoku'gata-setchaku-zai	接触型接着剤	contact adhesive *[poly-ch]*
sesshoku-hannō	接触反応	catalytic reaction; contact catalysis *[ch] [ch-eng]*
sesshoku'hannō-seisei'butsu	接触反応生成物	catalysate *[ch]*
sesshoku-jūgō	接触重合	catalytic polymerization *[poly-ch]*
sesshoku-kaku	接触角	contact angle *[fl-mech]*
sesshoku-kidō	接触軌道	osculating orbit *[astron]*
sesshoku-soshi	接触阻止	contact inhibition *[cyt]*
sesshu	接取	ingestion; intake *[bio]*
sesshu	接種；せっしゆ	inoculation *[bio]*
sesshu'butsu	接種物	inoculum (pl -la) *[microbio]*
sesshu suru	接種する	inoculate *{vb} [bio]*
sesshu-zairyō	接種材料	inoculum (pl -la) *[microbio]*
sessoku-dōbutsu	節足動物	Arthropoda *[i-zoo]*
sētā	セーター	sweater *[cl]*
setchaku	接着；せっちゃく	adhesion; bonding; cementing; glu(e)ing *{n} [adhes] [mech] [poly-ch]*
"	接着	splice (rope) *{n} [eng]*
setchaku-men	接着面	adherend *[adhes]*
setchaku-ryoku	接着力	bonding force *[mech] [poly-ch]*
setchaku-sei	接着性	adhesiveness property *[adhes]*
setchaku-zai	接着剤	adhesive (agent); bonding agent; cement; glue *[adhes] [mat]*
setchi	接地	grounding *{n} [elec]*
setchi	設置	installation *[comput] [ind-eng]*
setchi-ban	接地板	earth plate *[elec]*
setchi suru	設置する	install *{vb} [ind-eng]*
seto'biki	瀬戸引き	porcelain enamel; vitreous enamel *[mat]*
seto'mono	瀬戸物；瀨戸物	earthenware; porcelain *[eng] [mat]*
seto'mono-rui	瀬戸物類	chinaware *[mat]*
Seto-Naikai	瀬戸内海	Inland Sea (of Japan) *[geogr]*
Seto-yaki	瀬戸焼	Seto ware *[cer]*
setsu	説	theory (pl -ries) *[sci-t]*
setsu	節；せつ	section *{n} [bio] [comput] [syst]*
"	節	clause *[gram]*

setsu	節	link {n} [mech-eng]
setsubi	設備	facility (pl -ties) {n} [eng]
setsubi'bu	接尾部	suffix part [gram]
setsubi'go	接尾語	suffix [gram]
setsubi'ji	接尾辞	suffix [gram]
setsu'dan	切断； 切斷	abscission [bot]
"	切断； せつだん	scission [ch]
"	切断	cutting {n} [mech-eng]
setsudan-kāfu	切断カーフ	kerf [eng]
setsudan-men	切断面	cutting plane [graph]
setsudan-shingō	切断信号	disconnect signal [comm] [elec]
setsu'dan suru	切断する	disconnect {vb} [elec]
setsu'gan-renzu	接眼レンズ	eyepiece; ocular [opt]
setsu'gō	接合； せつごう	conjugation [bot] [i-zoo]
"	接合	joining; jointing {n} [civ-eng]
"	接合	junction [electr]
setsugō-ban	接合盤	connection box [comput]
setsugō'bu	接合部	junction [elec]
"	接合部	joint [eng]
setsugō-hōshi	接合胞子	zygosperm; zygospore [embryo]
setsugō'kin-rui	接合菌類	Zygomycetes [mycol]
setsugō-men	接合面	joining surface [adhes]
setsu'go-sen	摂護腺	prostate (gland) [anat]
setsugō-shi	接合子	zygote [embryo]
setsugō-ten	接合点	junction {n} [elec]
setsugō-yū'hatsu	接合誘発	zygotic induction [microbio]
setsu'mei	説明； せつめい	clarification; explanation [comm]
setsumei-zu	説明図； 說明圖	explanatory diagram [graph]
setsu'zoku	接続	connection [comput]
setsu'zoku-bako	接続箱	junction box [eng]
setsu'zoku-ki	接続器； 接續器	connector [elec]
setsu'zoku'shi	接続詞	conjunction; connective {n} [gram]
settei-hizumi	設定歪	set strain {n} [mech]
settei suru	設定する	set {vb} [comput]
"	設定する	establish; institute {vb}
setten	接点	contact; contact point {n} [math]
setten	節点	node [astron] [electr] [eng]
setten-idō	節点移動	joint displacement [civ-eng]
setten-teikō	接点抵抗	contact resistance [elec]
settō-bu	接頭部	prefix part [gram]
settō'go	接頭語	prefix {n} [gram]
settō'ji	接頭辞	prefix {n} [gram]

settō'tai	切頭体; 切頭體	frustum (pl -s, frustra) *[math]*
se'yore	背凭れ; せよれ	backrest *[furn]*
seyū	施釉; せゆう	glazing; enameling *{n} [cer]*
-sha	-車	-vehicle *{suffix}*
shabu-shabu	しゃぶしゃぶ	soup dish with meat and vegetables, cooked at the table *[cook]*
shachi	鯱; しゃち	killer whale; Orcinus orca *[v-zoo]*
shachi	車地	capstan (ship) *[nav-arch]*
shachi'hoko	鯱; しゃちほこ	dolphinlike fish (mythical)
shadai	車台; 車臺	chassis (pl chassis) *[mech-eng]*
shadan	遮断; しゃだん	shutdown *{n} [comput]*
shadan-ki	遮断器; 遮斷器	circuit breaker *[elec]*
shadan-shūha'sū	遮断周波数	cutoff frequency *[electr]*
sha'ei	射影	projection *[graph] [math]*
sha'futsu	煮沸; しゃふつ	boiling *{n} [p-ch]*
sha'futsu-ki	煮沸器	boiler *[mech-eng]*
shahei	遮蔽	shielding *{n} [elecmg]*
"	遮蔽	masking *{n} [electr]*
"	遮蔽	screening; covering *{n} [sci-t]*
shahen	斜辺; 斜邊; しゃへん	hypotenuse; hypothenuse (of a right triangle) *[math]*
shahō-jūni'men'tai	斜方十二面体	rhombic dodecahedron *[crys]*
shahō-kei	斜方形	rhombus (pl -es, rhombi) *[math]*
shahō-shōkei	斜方晶形	orthorhombic system *[crys]*
sha'jiku	車軸	axle *[mech-eng]*
sha'jiku-yu	車軸油	axle oil *[mat]*
sha-kaku	斜角	bevel; oblique angle *{n} [math]*
sha-kakuchū	斜角柱	oblique prism *[math]*
shakkotsu	尺骨; しゃっこつ	ulna (bone) *[anat]*
shakkuri; shakuri	噦; 吃逆	hiccups *{n} [physio]*
shakkuri suru	しゃっくりする	hiccup *{vb} [physio]*
shako	車庫	garage (automobile) *[arch]*
shako	蝦蛄; しゃこ	mantis crab; squilla (pl -s, squillae) *[i-zoo]*
shakō	射光	streamer *[geophys]*
shakō	斜坑	incline; inclined shaft *{n} [min-eng]*
sha-kōsen	斜光線	oblique ray (of light) *[opt]*
shakotsu'kō	車骨鉱	bournonite *[miner]*
shakō-zahyō	斜交座標	cartesian coordinates; oblique coordinates *[math]*
shaku	尺; しゃく	Japanese foot (unit of length) = 0.994 ft.) *[mech]*

shaku	勺；しゃく	shaku (unit of volume) measure = 0.038 U.S. pint *[mech]*
shaku	杓	ladle *{n} [met]*
shaku-do	尺度	scale *{n} [eng]*
shaku'dō	赤銅	copper-gold Japanese alloy *[met]*
shaku'do-tan'i	尺度単位	unit of scale *[math]*
shaku'netsu	灼熱	ignition; strong heating *{n} [ch]*
shaku'tori-mushi	尺取り虫	inchworm *[i-zoo]*
shaku'yaku	芍薬	peony (pl -nies) (herbaceous plant); Paeonia lactiflora *[bot]*
sha'men	斜面；しゃめん	ramp; slope *{n} [geol]*
"	斜面	inclined plane *[math]*
shamen-baiyō	斜面培養	slant culture *{n} [microbio]*
shamon-ori	斜紋織り	twill weave *[tex]*
shanpen-cha	香片茶	jasmine tea *[cook]*
sharin	車輪	wheel *{n} [eng]*
sharyō	車両；車兩	vehicle *[trans]*
shasen	射線	ray *[astron] [math]*
shasen	斜線	slash *{n} [comput]*
sha'shin	写真；しゃしん	photograph *{n} [graph]*
shashin-ban	写真版；寫真版	photo plate *[graph]*
shashin-chōkoku	写真彫刻	photoengraving *{n} [graph]*
shashin-densō	写真電送	phototelegraphy; telephotography *[comm]*
shashin-hantei-kesshō	写真判定決勝	photo finish (in competition) *{n}*
shashin-heihan	写真平版	photolithography *[pr]*
shashin'jutsu	写真術	photography (the technique) *[graph]*
shashin-ki	写真機	camera *[opt]*
shashin no	写真の	photographic *{adj} [photo]*
shashin-nōdo	写真濃度	photographic density; photographic intensity *[opt]*
shashin-nyūzai	写真乳剤	photographic emulsion *[photo]*
shashin-satsu'ei	写真撮影	photography; photographing *{n} [graph]*
shashin-shoku'ji-ki	写真植字機	photocomposing machine *[graph]*
shashin-sokkō-tōryō	写真測光当量	photometric equivalent *[photo]*
shashin-soku'ryō-hō	写真測量法	photogrammetry *[eng]*
shashin-tenchō-tō	写真天頂筒	photographic zenith tube *[opt]*
shashō'sha	車掌車	caboose *[rail]*
shashu	車種	automobile type; car type *[trans]*
sha'shutsu	射出；しゃしゅつ	ejection *[eng]*
shashutsu-atsu'ryoku	射出圧力	injection pressure *[eng]*
shashutsu-fuki'komi-seikei	射出吹込成形	injection blow molding *{n} [eng]*
shashutsu-hitomi	射出瞳	exit pupil *[opt]*

shashutsu-ritsu	射出率	injection rate *[eng]*
shashutsu-seikei	射出成形	injection molding *{n} [eng] [poly-ch]*
shasui-ro	射水路; しゃすいろ	chute *[hyd]*
shatai	車体	vehicle body; car body *[trans]*
shatai	斜体; しゃたい	inclined letter *[comput]*
shatsu	シャツ	shirt *[cl]*
shayū'renseki	斜黝簾石	clinozoisite *[miner]*
shazō	写像; しゃぞう	mapping *{n} [comput] [math]*
shazō suru	写像する	map *{vb t} [comput] [math]*
shi	氏	Mister; Mr.
shi; yon	四	four *{n} [math]*
shi	市	city (pl cities) *[geogr]*
shi	史	history (pl -ries) *{n}*
shi	死	death *[bio]*
shi	糸; し	ten-thousandth of one *{n} [math]*
shi	肢	limb *[bio]*
shi	詩	poetry; verse *{n} [pr]*
shī	椎; しい	chinquapin (tree) *[bot]*
-shi	-氏	Mr.; Mister *{suffix}*
shi'agari'hin	仕上り品	end product *[ind-eng]*
shi'age	仕上; 仕上げ	finish; finishing *{n} [ind-eng]*
shi'age'yō-kōgu-kō	仕上用工具鋼	finishing steel *[steel]*
shi'age-zai	仕上剤	finishing agent *[mat]*
shian('ka'butsu)-ion	シアン(化物)イオン	cyanide ion *[ch]*
shiba	柴; しば	brushwood; firewood *[mat]*
shiba'fu	芝生; しばふ	lawn *[arch] [bot]*
shiba'kari-ki	芝刈機	lawn mower *[eng]*
shiban	師板; 篩板	sieve plate *[ch-eng]*
shiba-tsuchi	芝土	turf *[geol]*
shibi	鮪; しび	tuna (fish) *[v-zoo]*
shibire-ei	痺れ鱏; しびれえい	electric ray (fish) *[v-zoo]*
shibireru	痺れる	become numb *{vb i} [physio]*
shibire-take	痺れ茸	Psilocybe venetana (poison mushroom) *[bot]*
shibo	皺; しぼ	crimp *{n} [eng] [tex]*
"	皺; しぼ	grain (leather) *[mat]*
"	皺; しぼ	craping *{n} [tex]*
shibō	子房	ovary (pl -ries) *[bot]*
shibō	脂肪; しぼう	fat *{n} [bioch] [physio]*
shibomu	萎む	droop; shrivel; wilt *{vb i}*
shibori	絞り; しぼり	iris (diaphragm); stop *{n} [opt] [photo]*
shibori-ben	絞り弁; 絞り瓣	throttle valve *[mech-eng]*
shibori-ritsu	絞り率	wringing rate *[tex]*

shibori-seikei 絞り成形 draw forming; stretch forming *{n} [eng]*
shibō-ritsu 死亡率 mortality (rate) *[med]*
shibori-zome 絞り染め tie-dyeing *{n} [tex]*
shiboru 絞る；搾る；しぼる press; squeeze; wring; extract *{vb}*
shibō-san 脂肪酸；しぼうさん fatty acid *[org-ch]*
shibō-so'shiki 脂肪組織 adipose tissue; fat tissue *[hist]*
shibo'tsuke 皺付け；しぼ付け embossing *{n} [eng]*
shibō'zoku-kagō'butsu 脂肪族化合物 aliphatic compound *[org-ch]*
shibu 渋；澁；しぶ persimmon tannin; tan *{n} [org-ch]*
shibu 師部；篩部 phloem *[bot]*
shibu-gami 渋紙 tanned paper *[paper]*
shibui 渋い astringent *{adj} [physio]*
shibu'mi 渋味 astringency *[physio]*
" 渋味 subdued refinement; severity *{n}*
shi'bun-chi 四分値 quartile *[math]*
shi'bun-en 四分円；四分圓 quadrant *[math]*
shibun'i-hensa 四分位偏差 quartile deviation *[math]*
shibun'i-sū 四分位数 quartile *[math]*
shi'bun no ichi 四分の一 one-quarter; one-fourth; quarter *[math]*
shi'bun-no-ichi hachō'ban 四分の一波長板 quarter-wave plate *[opt]*
shi'bun no san 四分の三 three-quarters; three-fourths *[math]*
shi-bunshi 四分子 tetrad *[bioch]*
shichi 七；しち seven *{n} [math]*
shichi'bun no ichi 七分の一 one-seventh *{n} [math]*
shichi'gatsu 七月 July (month)
shichi'hen-kei 七辺形；七邊形 heptagon *[math]*
shichi'jū-kō 七重項 septet *[quant-mech]*
shichi'jū-sen 七重線 septet *[spect]*
shichi'kak(u)kei 七角形 heptagon *[math]*
shichi-ka no 七価の；七價の heptavalent *{adj} [ch]*
shichi'men-chō 七面鳥 turkey (bird) *[v-zoo]*
shichi'mi-tō'garashi 七味唐辛子 seven-spice pepper (condiment) *[cook]*
shichō-hasshin-ki 弛張発振器 relaxation oscillator *[electr]*
shichū 支柱 stay (aircraft) *{n} [aero-eng]*
" 支柱 fulcrum; strut; support *{n} [eng]*
shida 羊歯；歯朶；しだ fern *[bot]*
shida-rui 羊歯類；歯朶類 Pteropsida *[bot]*
shidō 市道 city road *[civ-eng]*
shidō 始動 start; start-up *{n} [comput] [ind-eng]*
shi'enka-keiso 四塩化硅素 silicon tetrachloride
shi'enka-tanso 四塩化炭素 carbon tetrachloride
shi'fukka-keiso 四弗化硅素 silicon tetrafluoride

shi'fukka-tanso	四弗化炭素	carbon tetrafluoride
shiga	歯牙	tooth (pl teeth) *{n} [anat] [dent]*
shiga-hassei	歯牙発生	dentition *[dent] [physio]*
shigai	市街	city streets *[civ-eng]*
shigai-chi	市街地	urban area *[civ-eng]*
shigai-densha	市街電車	trolley (pl -s, trollies) *[mech-eng]*
shigai-sen	市外線	toll line (telephone) *[comm]*
shigai'sen	紫外線；しがいせん	ultraviolet radiation; ultraviolet rays *[elecmg]*
shigai'sen-kenbi'kyō	紫外線顕微鏡	ultraviolet microscope *[opt]*
shigai'sen-kōka	紫外線硬化	ultraviolet curing *{n} [ch]*
shigai'sen-kyūshū-zai	紫外線吸収剤	ultraviolet (ray) absorber *[mat]*
shi'gatsu	四月	April (month)
shigeki	刺激；刺戟；しげき	stimulus (pl -uli) *[physio]*
shigeki-zai	刺激剤	irritant; stimulant *[mat]*
shigen	資源	resource(s) *[sci-t]*
shigen-kanri	資源管理	resource management *[comput]*
shigen-kyōjūgō'tai	四元共重合体	quadripolymer *[poly-ch]*
shigi	鴫；鷸；しぎ	longbill; snipe (bird) *[v-zoo]*
shigi-yaki	鴫焼き	fried eggplant with miso sauce *[cook]*
shigo-kan	子午環	transit circle *[eng]*
shigo-kōchoku	死後硬直	rigor mortis *[path]*
shigoku	扱く；しごく	squeeze (through hands); strip *{vb}*
shigo-sen	子午線	meridian *[astron] [geod]*
shigō suru	嚙合する	gear-mesh; tooth-mesh (with) *{vb} [mech-eng]*
shi'goto	仕事；しごと	work *{n} [mech]*
shigoto'ba	仕事場	workshop *[mech-eng]*
shigoto-kansū	仕事関数	work function *[sol-st]*
shigoto-ritsu	仕事率	rate of production *[ind-eng]*
"	仕事率	power *{n} [phys]*
shigoto suru	仕事する	work *{vb}*
shigure	時雨；しぐれ	rain shower (late autumn to early winter) *[meteor]*
shigyō	始業	starting of work *{n} [ind-eng]*
shigyō	紙業	paper industry *[ind-eng] [paper]*
shi'hanmen-sei	四半面性	tetartohedrism *[crys]*
shi'hanmen-zō	四半面像	tetartohedry *{n} [crys]*
shihan suru	市販する	market; sell *{vb} [econ]*
shi'harai	支払い；支拂い	payment *[econ]*
shihei	紙幣	paper money *[econ]*
shihen	試片	workpiece; testpiece *[ind-eng]*

shihen-kei	四辺形	quadrilateral *{n} [math]*
shihen'kei no	四辺形の	quadrate *{adj} [math]*
shihō-ben	四方弁; 四方瓣	four-way valve *[mech-eng]*
shihō-dōbutsu	刺胞動物	Cnidaria *[i-zoo]*
shihon	資本	capital; funds *[econ]*
shihon-gōdō	資本合同	consortium *[econ]*
shihō'sho	示方書	specifications *[ind-eng]*
shihyō	指標; しひょう	index (pl -es, indices) *[comput] [math]*
"	指標	characteristic *{n} [math]*
"	指標	indicator *{n}*
shi'iku	飼育	breeding; raising *{n} [agr]*
shi'in; shi'on	子音	consonant *[ling]*
shiji	支持	support *{n} [arch] [math]*
shiji	指示	command; instruction *{n} [comput]*
shiji-dai	支持台	support stand *[eng]*
shiji-denkai'shitsu	支持電解質	supporting electrolyte *[p-ch]*
shi-jiku	支軸	supporting axle (shaft, axis) *[mech-eng]*
shijimi	蜆; しじみ	corbicula (shellfish) *[i-zoo]*
shiji-saibō-sō	支持細胞層	feeder layer *[microbio]*
shiji'tai	支持体; 支持體	carrier; support *{n} [eng]*
shi'jitsu-tai	子実体; 子實體	fruiting body *[bot]*
shiji-waku	支持枠	support frame *[eng]*
shiji-yaku	指示薬	indicator *[an-ch]*
shiji'yaku-shiken-shi	指示薬試験紙	indicator paper *[an-ch]*
shijō	市場	market; marketplace *{n} [econ]*
shijō	糸条; 糸條	thread strand *[tex]*
shijō-kin	糸状菌	filamentous fungus; mold fungus *[microbio]*
shijō'muke-yasai-nōjō	市場向け野菜農場	truck farm; vegetable farm *[agr]*
shijō no	歯状の; 齒狀の	dentate *{adj}*
shijō-sei	市場性	marketability *{n} [econ]*
shijū	死重	dead weight *[mech]*
shijū'kara	四十雀	great tit (Japanese bird) *[v-zoo]*
shijū-kō	四重項	quartet *[quant-mech]*
shijū'kyoku	四重極	quadrupole *[elecmg] [math]*
shijū-ten	四重点	quadruple point *[p-ch]*
shijun	視準	collimation *[phys]*
shijun-ki	視準器	collimator *[opt]*
shijun-sen	視準線	collimation line *[opt]*
shijū-sei	四重性	quadruplicity *{n}*
shijū'sō	四重奏	quartet *[music]*
shika	鹿; しか	deer (stag, doe) (pl deer) *[v-zoo]*
shika-genso	四価元素	tetrad *[ch]*

shika-hotetsu	歯科補綴	dental prosthesis *[eng]*
shika-i'gaku	歯科医学	dentistry *[med]*
shi'kake'hin	仕掛け品	partially-finished goods *[ind-eng]*
shi-kaku	死角	dead angle *[eng]*
shi-kaku	視角	visual angle *[opt]*
shi'kaku	視覚	sense of sight; vision *{n} [physio]*
shikaku-chū	四角柱	square column *[math]*
shikaku-ha	四角波	square wave *[elec]*
shikakui	四角い; しかくい	square *{adj} [math]*
shikaku-kei	四角形	square (shape) *{n} [math]*
shikaku-kensa	視覚検査	sight check *[comput]*
shikaku-kōka	視覚効果	optoaccoustic effect *[phys]*
shikaku-nōdo	視覚濃度	visual density *[photo]*
shikan; chikan	弛緩	slackening; loosening *{n}*
shikan	師管	sieve tube *[bot]*
shikan	歯冠; 齒冠	tooth crown; corona dentis *[dent]*
shikan'shiki-kagō'butsu	脂環式化合物	alicyclic compound *[org-ch]*
shikan-shitsu	士官室	wardroom *[nav-arch]*
shikan-tetsu'dō	市間鉄道	interurban railway *[rail]*
shika-sei	資化性	assimilating property *[physio]*
"	資化性	resource-converting property *{n}*
shika suru	資化する	assimilate *{vb} [microbio]*
shi-kazan	死火山; しかざん	extinct volcano *[geol]*
shike	時化; しけ	storm *{n} [meteor]*
shiken	試験; 試驗; しけん	test; testing; trial *{n} [ind-eng]*
shiken'hen	試験片	testpiece; specimen *[ind-eng] [sci-t]*
shiken-kan	試験管	test tube *[an-ch]*
shiken'kan-tate	試験管立	test tube stand *[an-ch]*
shiken-kigu	試験器具	testing apparatus *[eng]*
shiken-shi	試験紙	test paper *[ch]*
shi'ketsu-zai	止血剤	hemostatic; styptic *{n} [pharm]*
shiki	式	formula (pl -s, -mulae) *[ch] [math]*
"	式	equation; expression *[math]*
"	式	system *[sci-t]*
shiki	紙器; しき	container made of paper *[mat]*
shiki'betsu-bangō	識別番号	identification number *[comput] [math]*
shiki'betsu-nō	識別能	discerning (discriminating, distinguishing) power *[psy]*
shiki'betsu suru	識別する	identify *{vb} [psy]*
shiki-do	色度; しきど	chromaticity *[opt]*
shiki'do-zu	色度図; 色度圖	chromaticity diagram *[opt]*
shiki-hyō	色票	color chart *[opt]*

shiki'hyō-kei	色票系	color order system *[opt]*
shiki'i	閾；敷居	threshold *[math] [phys] [physio]*
shiki'i	敷居	doorsill *[arch]*
shiki'i-chi	閾値；敷居値	threshold (value) *[comput] [phys]*
shiki'iki'ka no	識閾下の	subliminal *{adj} [psy]*
shiki-ita	敷板	planking *{n} [civ-eng]*
shiki'kaku	色覚；色覺	color vision *[physio]*
shiki'kōdo	色光度	chrominance *[opt]*
shikimi	樒；しきみ	Japanese star anise (herb) *[bot]*
shikimi-san	樒酸	shikimic acid *[ch]*
shiki'mō	色盲	color blindness *[med]*
shiki'mono	敷物	floor covering *{n} [arch] [mat]*
shikin	試金	assay *{n} [an-ch]*
shikin	資金	capital; funds *[econ]*
shiki-nōdo	式濃度	formality *[ch]*
shiki-ryō	式量	formula weight *[ch]*
shiki-sa	色差	color difference *[opt]*
shiki'sai	色彩；しきさい	color *{n} [opt]*
shikisai-kei	色彩計	colorimeter *[opt]*
shikisai-shashin	色彩写真	color photograph *[photo]*
shikisai-shōhyō	色彩商標	color trademark *[econ]*
shikisa-kei	色差計	color-difference meter *[opt]*
shiki-shi	色紙；しきし	paper square, colored *[comm]*
shiki'so	色素；しきそ	coloring matter; pigment *[mat]*
shiki'sō	色相	hue *[opt]*
shikiso-haijo-shiken	色素排除試験	dye exclusion test *[cyt]*
shikiso'hō	色素胞；色素胞	chromatophore *[hist]*
shiki-sū	色数	color number *[opt]*
shikka	失火	misfire *{n} [ch] [mech-eng]*
shikkan	疾患	disease *[med]*
shikkatsu	失活	deactivation *[ch]*
shikki; shikke	湿気；濕氣	moisture; humidity *[hyd] [meteor]*
shikki	漆器	lacquerware *[furn]*
shikki no aru	湿気のある	damp; humid; moist *{adj} [hyd]*
shikkui	漆喰；しっくい	mortar; plaster; stucco *{n} [mat]*
shikkyū-ondo-kei	湿球温度計	wet-bulb thermometer *[eng]*
shikō	志向	inclination (of mind) *[psy]*
shikō	思考	thought *{n} [psy]*
shikō	指向	direction *[eng]*
shikō	歯孔	tooth cavity *[dent]*
shikō	歯垢	dental plaque *[dent]*
shikō	試行	trial *[ind-eng]*

shikō	篩孔	sieve pore *[eng]*
shikō-dōbutsu	趾行動物	digitigrade animal *[v-zoo]*
shikō'hin	嗜好品	favorite foods *[cook]*
shikō-hōsoku	思考法則	thought, laws of *[logic]*
shikon	紫根	Lithospermi Radix *[bot] [pharm]*
shikon	紫紺	bluish purple (color) *[opt]*
shikon	歯根	root (of tooth) *{n} [dent]*
shikō-sakugo	試行錯誤	trial and error (behavior) *{n} [psy]*
shikō-sei	指向性	directionality; directivity *[comm]*
shiku	敷く; 布く	spread; lay down (as paper) *{vb}*
shi'kumi	仕組み	arrangement; setup *{n}*
shi'kyokushi-kyōmei	四極子共鳴	quadrupole resonance *[elecmg]*
shikyū	子宮	uterus *[anat]*
shikyū	至急	urgent *{n}*
shikyū'ka	四級化	quaternization *[ch]*
shima	島	island *[geogr]*
shima	縞	striation *[geol]*
"	縞	streak *{n} [miner] [opt]*
"	縞	stripe *{n} [tex]*
shima-aji	島味	garganey (bird) *[v-zoo]*
shima-aji	縞鯵; しまあじ	hardtail; yellowtail; Caranx leptolepis (fish) *[v-zoo]*
shima-gatsuo	縞鰹; しまがつお	pomfret; Lepidotus brama (fish) *[v-zoo]*
shima-guni	島国; 島國	island country *[geogr]*
shima'jō-kōzō	縞状構造	banded structure *[met] [petr]*
shima'jō-sekitan	縞状石炭	banded coal *[geol]*
shima-ka	縞蚊	striped mosquito; Stegomyia fasciata (insect) *[i-zoo]*
shima'menō	縞瑪瑙; しまめのう	onyx *[miner]*
shima-ō'dan-hodō	縞横断歩道	zebara crossing *{n} [traffic]*
shimau	仕舞う	put away; finish *{vb}*
shima'uma	縞馬	zebra (animal) *[v-zoo]*
shime'gu	締め具	clamp *{n} [eng]*
shimei'roku	氏名録	directory (pl -ries) *[comm] [pr]*
shimeji	湿地; 占地; しめじ	champignon (mushroom) *[bot]*
shi'men	紙面	paper space *[pr]*
shime-neji	締めねじ	bolt *{n} [mech-eng]*
shime(neji)-kugi	締め(ねじ)釘	bolt *{n} [mech-eng]*
shimen-kōzō	四面構造	tetragonal structure *[crys]*
shimen'tai	四面体	tetrahedron (pl -s, -hedra) *[math]*
shimerasu	湿らす; 濕らす	moisten *{vb t}*
shimeru	締める	shut; close; tighten *{vb}*

shimesu-hen	示扁; 示偏: ネ; しめすへん	kanji (Chinese character) radical denoting religion *[pr]*
shime-tsukeru	締め付ける	tighten (onto) *{vb t}*
shi'metsu-ki	死滅期	death phase *[microbio]*
shime-yaki	締焼	biscuit firing *{n} [cer]*
shimi	染み; しみ	stain spot *{n} [tex]*
shimi	紙魚; しみ	bookworm; silverfish (insect) *[i-zoo]*
shimo	霜	frost *{n} [hyd]*
shimo-bashira	霜柱	frost column *[hyd]*
shimo'goe	下肥	manure *[agr]*
shi'mon	指紋	fingerprint *{n} [anat] [graph]*
shimon-saishu	指紋採取	fingerprinting *{n} [legal]*
shimo-otoshi	霜落し	defrosting *{n} [eng]*
Shimo'tsuke-sō	下野草	meadowsweet; Filipendura multijuga *[bot]*
shimo'yake	霜焼け; 霜焼け	frostbite *[med]*
shi'myaku	翅脈	vein; nervure *[i-zoo]*
shin	芯; 心	core *{n} [sci-t]*
shinabita	萎びた; しなびた	wilted *{adj} [bot]*
shina'gire	品切れ	shortage; out of stock; sold out *{n}*
shi'nai	市内	in-city; municipal; urban; local
shi-nai	為ない; 爲ない	do (or does) not do *{vb}*
shinanoki	科木; しなのき	Japanese linden (tree) *[bot]*
shinayaka	撓やか; 靭	supple; pliant *{n} [mat]*
shinbi-sei	審美性	aesthetic quality
shinbō	心房	atrium (pl atria) *{n} [anat]*
shinbu-baiyō	深部培養	submerged culture *[microbio]*
shinbun	新聞; しんぶん	newspaper *[comm] [pr]*
shin-bunsū	真分数	proper fraction *[math]*
shinbun-yōshi	新聞用紙	newsprint *[mat] [paper]*
shin'chin'taisha	新陳代謝	metabolism *[physio]*
shinchō	伸長	elongation; extension; stretch *{n} [mech]*
shinchō-ki	伸長器	expander *[electr]*
shin'choku-do	真直度	straightness *[math]*
shinchō-ritsu	伸長率	elongation ratio *[mech]*
shinchū	真鍮; しんちゅう	brass *[met]*
shindai	寝台; 寢台	bed *{n} [furn]*
shindai-setsubi	寝台設備	berthing (for sleeping) *{n} [nav-arch]*
shindai'sha	寝台車	sleeping car *[rail]*
shindan	診断; 診斷	diagnostics *[comput]*
shindan suru	診断する	diagnose *{vb} [med]*
shindo	伸度	ductility *[mat]*
shindo	震度	seismic intensity *[geophys]*

shindō	振動；しんどう	quake; tremor *{n} [geophys]*
"	振動	shock; vibration *{n} [mech]*
"	振動	oscillation *[phys]*
shindō-ban	振動板	diaphragm *[eng-acous] [phys]*
shindō-genso	親銅元素	chalcophile element *[geol]*
shindō-jun'i	振動準位	vibrational level *[p-ch]*
shindō-kei	振動計	vibrometer; vibrograph *[eng]*
shindo-ki'roku	深度記録	fathogram (by echo sounder) *[eng]*
shindō-ryōshi-sū	振動量子数	vibrational quantum number *[p-ch]*
shindō'shi	振動子	vibrator *[elec] [mech-eng]*
shindō suru	振動する	oscillate; shake; vibrate *{vb} [phys]*
shin'eki-sei	親液性	lyophilicness *[ch]*
shin'en	深淵	abyss *[geol]*
shin'en-do	真円度；真圓度	roundness *[math]*
shin'en-rui	真猿類	Anthropoidea (monkeys, etc.) *[v-zoo]*
shingai	侵害	infringement *[legal]*
shingai suru	侵害する	prejudice *{vb} [legal]*
shin-gane	芯金	core; core bar; core metal *{n} [met]*
shingen	震源	hypocenter (of earthquake); seismic focus *[geophys]*
shin-getsu	新月	new moon *[astron]*
shingi	真偽；真僞	authenticity; truth or falsehood *{n}*
shingō	信号；信號	signal *{n} [comm]*
shingō-kō'en	信号紅炎	red flare *{n} [comm]*
shingō-saisei	信号再生	signal regeneration *[comm]*
shingō-shori	信号処理	signal processing *[comput]*
shin-hijū	真比重	true specific gravity *[mech]*
shi'niku	歯肉	gum *{n} [anat]*
shi-nikui	為難い；しにくい	difficult to do
shi'nise	老舗；しにせ	store of long standing *[econ]*
shin'jitsu	真実；真實	truth; reality; fact *{n}*
shin'jō-kesshō	針状結晶	needle crystal *[crys]*
shinju	真珠；しんじゅ	pearl *[mat]*
shinju-kōtaku	真珠光沢	pearly luster *[miner]*
shinju-sō	真珠層	mother of pearl; nacre *[i-zoo]*
shinju'un	真珠雲	nacreous cloud *[meteor]*
shinka(-ron)	進化(論)	evolution (theory) *[bio]*
shinkai	深海	abyss; deep sea *[ocean]*
shin-kaihyō	新海氷	young ice *[hyd]*
shinkai no	深海の	benthic; deep-sea *{adj} [bio] [ocean]*
shinkai-shō	深海床	deep-sea floor *[geol]*
shinkaku-saibō	真核細胞	eukaryotic cell *[bio]*

shinkaku-seibutsu	真核生物	eukaryote; eucaryote *[bio]*
shinkan-sen	新幹線	Bullet Train Line *[rail]*
shinkei	神経；しんけい	nerve *{n} [anat]*
shinkei-dentatsu-busshitsu	神経伝達物質	neurotransmitter *[bioch]*
shinkei-kei	神経系；神經系	nervous system *[anat]*
shinkei-mō	神経網	nerve net *[i-zoo]*
shinkei-saibō	神経細胞	nerve cell; neuron *[hist]*
shinkei'sei-juyō'tai	神経性受容体	neuroreceptor *[bioch]*
shinkei-sen'i	神経纖維	nerve fiber *[cyt]*
shinkei'setsu	神経節	ganglion *[anat]*
shinkei-shūmatsu	神経終末	nerve ending *{n} [anat]*
shinkei-seibutsu'gaku	神経生物学	neurobiology *[bio]*
shinkei-seibutsu-gakusha	神経生物学者	neurobiologist *[bio]*
shinkei-tan'i	神経単位	neuron *[hist]*
shinkei-yakuri'gaku	神経薬理学	neuropharmacology *[pharm]*
shinki-genso	親気元素	atmophile element *[meteor]*
shin-kin	心筋	cardiac muscle *[hist]*
shinkin-doku'so	真菌毒素	mycotoxin *[microbio]*
shinki no	新規の	new; fresh *{adj}*
shinki no gōsei	新規の合成	de novo synthesis *[bioch]*
shinki'rō	蜃気楼；蜃氣樓	mirage *[opt]*
shinki-sei	新規性	novelness *{n} [pat]*
shinkō-shoku	深紅色	lake; dark red (color) *[opt]*
shinku	真紅；深紅	crimson; magenta (color) *[opt]*
shinkū	真空；しんくう	vacuum (pl -s, vacua) *{n} [phys]*
shinkū-do	真空度	degree of vacuum *[phys]*
shinkū-jō'chaku	真空蒸着	metallizing; vacuum evaporation *{n} [eng]*
"	真空蒸着	vacuum plating *{n} [met]*
shinkū-kan	真空管	vacuum tube *[electr]*
shinkū-nai de	真空内で	in vacuo *{adv} [phys]*
shinkū-seikei	真空成形	vacuum forming *{n} [eng]*
shinkū-yōkai-ro	真空溶解炉	vacuum melting furnace *[met]*
shinkyū-do	真球度	sphericity *[math]*
shin'kyū-taihi	新旧対比	new and old contrast *{n}*
shinmai	新米	new-crop rice *[bot]*
"	新米	beginner; novice *{n}*
shinmi-busshitsu	辛味物質	acrid substance *[cook]*
shin-mitsudo	真密度	true density *[mech]*
shin no hosū	真の補数	true complement *[math]*
shinnyō	之繞：辶；辵；しんにょう	kanji (q.v.) radical denoting to advance *[pr]*
shinnyū	浸入；しんにゅう	infiltration; intrusion *[geol]*

shinnyū	浸入	invasion *[med] [mil]*
shinnyū-do	浸入度	penetration; degree of penetration *[eng]*
shinnyū'do-hi	浸入度比	penetration ratio *[mech-eng]*
shinnyū'do-kei	針入度計	penetrometer *[eng]*
shi'nō	子嚢；しのう	ascus (pl asci) *[mycol]*
shin'ō	震央；しんおう	epicenter; seismic center *[geophys]*
shinobi	忍び	stealth *{n}*
shino-bue	篠笛；しのぶえ	bamboo flute *[music]*
shinō'ka	子嚢果	ascocarp *[mycol]*
shinō'kin-rui	子嚢菌類	Ascomycetes *[mycol]*
shī no mi	椎の実；椎の實	sweet acorn; chinquapin nut *[bot]*
shin'on	震音	warble tone *[acous]*
shinonome	東雲；しののめ	daybreak; dawn *{n} [astron]*
shin-ō'ryoku	真応力；真應力	actual stress *[mech]*
shin'oshi-dai	心押台；心押臺	tailstock *[mech-eng]*
Shino-yaki	志野焼	Shino ware *[cer]*
shinpai	心配	worry; concern *{n} [psy]*
shinpan	審判	trial *[legal]*
shinpi	真皮；しんぴ	dermis *[anat]*
shinpo-sei	進歩性	inventiveness *[pat]*
shinpuku	振幅；しんぷく	amplitude *[phys]*
shinpuku-henchō	振幅変調	amplitude modulation; AM *[electr]*
shinpuku-hizumi	振幅歪	amplitude distortion *[electr]*
shinpuku-tōka-ritsu	振幅透過率	amplitude transmittance *[elecmg]*
shinrai-do	信頼度；しんらいど	reliability *[comput] [eng] [math]*
shinrai-genkai	信頼限界	confidence limits *[math]*
shinrai-kukan	信頼区間	confidence interval *[math]*
shinrai-sei	信頼性	reliability *[comput] [eng] [math]*
shinrai-suijun	信頼水準	confidence level *[ind-eng]*
shinrai suru	信頼する	trust; believe; rely on *{vb}*
shinrei-bunri	深冷分離	cryogenic separation *[eng]*
shinri	心理	psychology *[bio]*
shinri-chi	真理値	truth value *[logic]*
shinri'chi-hyō	真理値表	truth table *[logic]*
shinri'gaku	心理学；しんりがく	psychology (the discipline) *[bio]*
shin'rin	森林	forest; woodland; timberland *[ecol]*
shinrin-bassai	森林伐採	deforestation *[for]*
shinrin-ōkami	森林狼	timber wolf (animal) *[v-zoo]*
shinrō-kei	浸漏計	lysimeter *[eng]*
shinrui	親類；しんるい	relative; kinsman *{n} [bio]*
shinryū	浸硫	sulfurizing *{n} [ch]*
shinsa	審査；しんさ	review *{n} [legal] [pat]*

shinsai	震災	earthquake disaster *[geophys]*
shinsa'kan	審査官	examiner *[pat]*
shinsa-seikyū	審査請求	demand for examination *[pat]*
shinsei	申請	application; petition *{n} [pat]*
shinsei	新星	nova (pl novas, novae) *[astron]*
shinsei-saikin-moku	真正細菌目	Eubacteriales *[microbio]*
shinsei-genso	親生元素	biophile element *[bioch]*
shinsei no	真性の	intrinsic; inborn *{adj}*
shinsei'sho	申請書	petition; written application *{n} [legal]*
shinsei'teki-seishitsu	真性的性質	intrinsic property *[mat]*
shinseki	浸漬；しんせき	dipping *{n} [met]*
"	浸漬	immersion *[sci-t]*
shinseki	親戚	relative; kinsman *{n} [bio]*
shinseki-dasshi	浸漬脱脂	dipping degreasing *{n} [ch-eng] [met]*
shinseki-fu'shoku	浸漬腐食	dip corrosion *[met]*
shinseki-genso	親石元素	lithophile element *[geoch]*
shinseki-tosō	浸漬塗装	dip coating *{n} [eng]*
shin-sekki-jidai	新石器時代	Neolithic Age *[archeo]*
shinsen	心線	core; core wire *{n} [elec]*
shinsen	伸線	wiredrawing *{n} [met]*
shinsen	浸染	dip dyeing *{n} [tex]*
shinsen-do	新鮮度；しんせんど	freshness *{n} [cook]*
shinsen na	新鮮な	fresh *{adj} [cook]*
shinsen-sei	伸線性	drawability; wire drawing property *[met]*
shinsha	辰砂；しんしゃ	cinnabar *[miner]*
shin'shi'sei no	親脂性の	lipophilic *{adj} [ch]*
shin'shitsu	心室	chamber (of heart); ventricle *{n} [anat]*
shin-shitsu	寝室；寢室	bedroom *[arch]*
shin-shōgo	真正午	apparent noon *[astron]*
shin'shoku	浸食；浸触	erosion *[geol]*
shin'shoku	侵食；浸触	corrosion *[met]*
shin'shoku-idō	深色移動	bathochromic shift *[p-ch]*
shin'shoku-kanō-sei	侵食可能性	erodibility *[geol]*
shin'shuku'shiki-antena	伸縮式アンテナ	telescopic antenna *[elecmg]*
shin'shutsu	浸出；しんしゅつ	leaching *{n} [ch-eng] [geoch]*
"	浸出	seeping out *{n} [fl-mech]*
"	浸出	exosmosis *[physio]*
shin'shutsu'butsu	浸出物	leachate *{n} [ch]*
"	浸出物	exudate *{n} [med]*
shin'shutsu-eki	浸出液	leached solution *[ch]*
"	浸出液	exudate *{n} [geol]*
"	浸出液	percolate liquid

shinsū	真数；真數	antilogarithm *[math]*
shinsui	進水	launching (of a ship) *{n} [nav-arch]*
shinsui	浸水	retting *{n} [ch-eng] [tex]*
shinsui-do	親水度	hydrophilicity *[ch]*
shinsui-ki	親水基	hydrophilic group *[ch]*
shinsui-sei	親水性	hydrophilic property *[ch]*
shintai	身体；身體	body (pl bodies) *{n} [anat]*
shintai no	身体の	somatic *{adj} [bio]*
shintai-shōgai	身体傷害	bodily injury; disability *[med]*
shintai-shōgai'sha	身体傷害者	disabled person *[med]*
shintai suru	進退する	advance and retreat; move *{vb}*
shintan	浸炭	carburization; cementation *[met]*
shintan-chikka-ro	浸炭窒化炉	carbonitriding furnace *[met]*
shintan-shin'chitsu	浸炭浸窒	carbonitriding *{n} [met]*
shinten-sei	伸展性	extensibility *[mech]*
shin'tetsu-genso	親鉄元素	siderophile element *[ch]*
shintō	浸透；しんとう	osmosis (pl -ses); permeation *[ch] [p-ch]*
shintō	親等	degree of consanguinity *{n} [gen]*
shintō-atsu	浸透圧	osmotic pressure *[p-ch]*
shintō'atsu-kei	浸透圧計	osmometer *[an-ch]*
shintō-baiyō	振盪培養	shake culture *{n} [microbio]*
shintō-keisū	浸透係数	osmotic coefficient *[p-ch]*
shintō-sei	浸透性	permeability *[fl-mech]*
shintō-zai	浸透剤	penetrating agent *[mat]*
shinwa-ryoku	親和力	affinity (pl -ties) (the strength) *[ch]*
shinwa-sei	親和性	affinity (the nature) *[ch]*
shin'yō'ju	針葉樹	conifer; softwood *[bot]*
shi'nyō-o'dei	屎尿汚泥	raw sludge; sewage sludge *[civ-eng]*
shin-yu	浸油	oil soaking *{n} [met]*
shin'yu-do	親油度	oleophilicity *[ch]*
shin'yu'sei no	親油性の	lipophilic *{adj} [ch]*
shinzai	心材	heartwood *[bot]*
shinzai	浸剤	infusion *{n} [ch]*
shinzō	心臓；しんぞう	heart *{n} [anat]*
shinzō'kei no	心臓形の	cardioid *{adj} [math]*
shin(-zō)-kekkan no	心(臓)血管の	cardiovascular *{adj} [anat]*
shio	塩；鹽；しお	salt *{n} [ch]*
shio	潮；汐	tide *{n} [ocean]*
shio-damari	潮溜り	tide pool *[ocean]*
shio-kara	塩辛	salted fish guts *[cook]*
shio-kara-sa	塩辛さ	saltiness *{n} [cook] [physio]*
shio-maneki	潮招き	fiddler crab (crustacean) *[i-zoo]*

shi'on	子音	consonant *[ling]*
shi'on-ki	止音器	damper *[music]*
shiore	萎れ；しおれ	wilting; wilt *{n} [bot]*
shio'sai	潮騒；しおさい	ocean sound; ocean roar *{n} [acous]*
shio-tsunami	潮津波	tidal bore *{n} [ocean]*
shio'zuke	塩漬	salting; salt pickling *{n} [food-eng]*
shippei-bunrui'gaku	疾病分類学	nosology *[med]*
shippō-yaki	七宝焼；七寶焼	cloisonne ware *[cer]*
shiraberu	調べる	examine; investigate *{vb}*
shira'ko	白子	albinism; albino *[bio]*
shirami	蝨；虱；しらみ	louse (pl lice) *[i-zoo]*
shira-sagi	白鷺	egret (bird) *[v-zoo]*
shirasu-boshi	白簀乾；白子乾	dried young sardines *[cook]*
shira-taki	白滝；白瀧	konnyaku noodles *[cook]*
shira'uo	白魚；しらうお	whitebait; icefish; Salanx microdon *[v-zoo]*
shirei	指令	command; instruction; order *{n} [comput]*
shirei-gengo	指令言語	command language *[comput]*
shi'retsu-kyōsei	歯列矯正	orthodontics *[med]*
shiri	尻	buttocks *{n} [anat]*
shiritai	知りたい	wish to know
shirizoku	退く	retreat; recede *{vb i}*
shiro	白	white (color) *{n} [opt]*
shiro	代；しろ	allowance; margin; material *[tex]*
shiro-ari	白蟻	termite (insect) *[i-zoo]*
shiro-bō	白棒	white rouge *[mat]*
shiro-budō'shu	白葡萄酒	white wine *[cook]*
shiro-garashi	白芥子；白がらし	white mustard (condiment) *[bot]*
shiro-goshō	白胡椒	white pepper (condiment) *[bot]*
shiro-hata	白旗	white flag *[comm]*
shiroi	白い	white *{adj}*
shiro-iruka	白海豚；しろいるか	beluga (marine mammal) *[v-zoo]*
shiro-kabi	白黴；白かび	mildew *{n} [mycol]*
shiro-katsuo-dori	白鰹鳥	gannet (bird) *[v-zoo]*
shiro-kin	白金；しろきん	white gold *[met]*
shiro-kuma	白熊	polar bear (animal) *[v-zoo]*
shiro-miso	白味噌	white miso (q.v.) *[cook]*
shiro-nagasu-kujira	白長須鯨	blue whale *[v-zoo]*
shiroppu	シロップ	syrup *[cook] [pharm]*
shiro-sango	白珊瑚	white coral *[i-zoo]*
shiro-unmo	白雲母；しろうんも	muscovite; white mica *[miner]*
shiro-uri	白瓜	white muskmelon (fruit) *[bot]*

shiro-zake	白酒	white sake (q.v.) *[cook]*
shiru	知る；識る	know *{vb}* *[psy]*
shiru	汁	juice *{n}* *[bot]*
shiruko	汁粉；しるこ	azuki (q.v.) confection soup *[cook]*
shirushi	印；しるし	mark *{n}* *[comm]*
shirusu	印す	mark *{vb}* *[graph]* *[pr]*
shiryō	紙料	paper stock; stock *[mat]* *[paper]*
shiryō	資料	data (sing datum) *[math]*
shiryō	試料；しりょう	sample; specimen *{n}* *[ch]*
shiryō	飼料	feed; feedstuff *{n}* *[agr]*
shiryō-dai	試料台	specimen table; specimen stand *{n}*
shiryō'hen	試料片	testpiece *{n}* *[sci-t]*
shiryō'kan	資料館	library (pl -braries) *[arch]*
shi-ryoku	視力	visual acuity *[physio]*
shi'ryoku no-	視力の-	opto- *{prefix}*
shi'ryoku-zōkyō-zai	紙力増強剤	paper-strength enhancer *[mat]* *[paper]*
shi'ryoku-zu	示力図	force diagram *[graph]* *[mech-eng]*
shiryū	支流	tributary (pl -taries) *[hyd]*
shi'ryūka-shi'hiso	四硫化四砒素	tetrarsenic tetrasulfide
shiryū'tai	糸粒体	mitochondrion (pl -dria) *[cyt]*
shisa	視差	parallax *[math]* *[opt]*
shisai	仔細；しさい	meaning; details *{n}* *[comm]*
shi'saku	試作	trial production *[ind-eng]*
shisa-kussetsu-kei	示差屈折計	differential refractometer *[eng]*
shisan	資産	asset; property *[econ]*
shisan	試算	trial calculation *[math]*
shi'sanka-ni'chisso	四酸化二窒素	dinitrogen tetroxide
shi'sanka-san'namari	四酸化三鉛	trilead tetroxide
shi'san-sanka-tetsu	四三酸化鉄	triiron tetroxide
shisei	姿勢	posture *{n}*
shi-seibun-kei	四成分系	four-component system *{n}* *[eng]*
shisei-shiki	示性式	rational formula *[ch]*
shisei-sū	示性数	figure of merit *{n}* *[math]*
shiseki	歯石	dental calculi; tartar *[dent]*
shisen	視線	line of sight *{n}* *[sci-t]*
shisen-sokudo	視線速度	radial velocity *[mech]*
shi'setsu	施設	equipment; facility *{n}* *[eng]*
shisetsu-kairo	私設回路	private line *[comm]*
shi'sha-go'nyū	四捨五入	rounding; rounding off *{n}* *[math]*
shisha-zai	止瀉剤	antidiarrheal *{n}* *[pharm]*
shishi	四肢	limbs; the four limbs *[anat]*
shishi	獅子；しし	lion (animal) *[v-zoo]*

shishi-dōbutsu	四肢動物	Tetrapoda (four-limbed animals) *[v-zoo]*
shishin	指針	indicator; pointer *[eng]*
shi'shitsu	脂質	lipid *[bioch]*
shi'shitsu	歯質; 齒質	dentin *[dent]*
shishi-udo	猪独活; ししうど	angelica (herb) *[bot]*
shishi-za	獅子座	Leo; the Lion *[constel]*
shishō	支承	support; bearing *{n} [civ-eng]*
shishō	視床	thalamus (pl -ami) *{n} [anat]*
shishōka'bu	視床下部	hypothalamus *{n} [anat]*
shishō suru	支承する	receive and support *[mech-eng]*
shishū	刺繡; ししゅう	embroidery (pl -deries) *{n} [tex]*
shi'shūka-keiso	四臭化硅素	silicon tetrabromide
shi'shūka-tanso	四臭化炭素	carbon tetrabromide
shishū-so'shiki	歯周組織	periodontium (pl -dontia) *[med]*
shiso	紫蘇; しそ	beefsteak plant; perilla *[bot] [cook]*
shisō	思想	thought; idea *{n} [psy]*
shiso'kiseki	紫蘇輝石	hypersthene *[miner]*
shi'soku-rui	四足類	quadrupeds *[v-zoo]*
shi'son	子孫	progeny *{n} [bio]*
shissei-shokubutsu	湿性植物	hygrophyte *[bot]*
shisshiki-funsai	湿式粉砕	wet grinding *{n} [mech-eng]*
shisshiki-kensaku	湿式研削	wet grinding *{n} [mech-eng]*
shisshoku	湿食; 濕蝕	wet corrosion *[met]*
shissoku	失速	stall; stalling *{n} [aero-eng] [mech-eng]*
shisū	指数; 指數	index *{n} [crys] [math]*
"	指数; しすう	exponent *{n} [math]*
shisū-hōsoku	指数法則	exponents, law of *[math]*
shisū-hōtei'shiki	指数方程式	exponential equation *[math]*
shisui-kō	試錐孔	borehole *[min-eng]*
shisū'ka	指数化	exponentiation *[math]*
shisū-kansū	指数関数	exponential function *[math]*
shisutemu-kōgaku	システム工学	systems engineering *{n} [eng]*
shita	下	below; beneath; down(ward) *{adv} [math]*
shita	舌	tongue *{n} [anat]*
shita'bae	下生え	undergrowth *[ecol]*
shita'biki-sō	下引き層	substratum (pl -strata) *[photo]*
shita-birame	舌平目; したびらめ	sole (fish) *{n} [v-zoo]*
shita-e	下絵; 下繪	draft *{n} [graph]*
shita-enogu	下絵具	underglaze color *[cer]*
shita-gata	下型	drag *{n} [met]*
shita-goshirae	下拵え	preparation; prearrangement *{n}*
shita-gusuri	下薬; したぐすり	underglaze *{n} [cer]*

shita'ji	下地	foundation; groundwork *{n}*
shī-take	椎茸；しいたけ	shiitake mushroom; Cortinellus shiitake *[bot]*
shi'taku	支度；仕度	preparations; arrangements *{n}*
shita'muki-keisha	下向き傾斜	declivity (pl -ties) *[geol]*
shita'muki-tsūfū	下向き通風	downdraft *{n}* *[phys]*
shita ni ō	下に凹	downwardly concave *[opt]*
shita ni totsu	下に凸	downwardly convex *[opt]*
shita no-	下の-	infra-; sub-; under- *{prefix}*
shitan'tō	四炭糖	tetrose *[bioch]*
shita-nuno	舌布	tongue fabric (shoe) *[cl]*
shita'nuri	下塗り	base coat; undercoat(er) *[mat]*
shita-sen	下線	underline *{n}* *[comput]* *[pr]*
shitatari	滴り	drip *{n}* *[hyd]*
shita'tsugi	舌継ぎ；舌繼ぎ	whipgraft *[bot]*
shita'tsuki-kigō	下付き記号	subscript symbol *[pr]*
shita'tsuki-moji	下付き文字	subscript character *[pr]*
shita'tsuki-sūji	下付き数字	subscript numeral *[pr]*
shita-yori	下撚り	primary twist *[tex]*
shita-zawari	舌触り；したざわり	feel to the palate *{n}* *[cook]*
shitchi	湿地；濕地；しっち	wetlands; damp ground *[ecol]* *[geol]*
shitei	指定	specification; designation *{n}*
shitei	視程	visibility (pl -ties) *[meteor]*
"	視程	visibility range *[navig]*
shiteki	至適	optimum (pl -tima) *{n}* *[math]*
shiteki-ondo	至適温度	optimum temperature *[physio]*
shiteki suru	指摘する	indicate; point out *{vb}*
shiten	支店	branch store; branch office
shiten	支点	fulcrum; supporting point *[mech]*
shiten	死点	dead center; dead point *{n}* *[mech-eng]*
shiten	至点	solstices *[astron]*
shitsu	質；しつ	quality *[ind-eng]*
shi'tsū; ha-ita	歯痛；齒痛	toothache *[med]*
shitsu-ban	湿板	photographic wet plate *[photo]*
shitsu-denchi	湿電池	wet cell *[elec]*
shitsu-do	湿度；濕度；しつど	humidity (pl -ties) *[meteor]*
shitsu'do-kei	湿度計	hygrometer *[eng]*
shitsū-dome	歯痛止め	antidontalgic *{n}* *[pharm]*
shitsu'do-shisū	湿度指数	humidity index *[climat]*
shitsu'gai-kotsu	膝蓋骨	kneecap; patella (pl patellae) *[anat]*
shitsu'jun-kyōdo	湿潤強度	wet strength *[met]* *[paper]*
shitsu'jun-sei	湿潤性	wettability *[ch]*

shitsu'jun-zai	湿潤剤	moistening agent; humectant; wetting agent *[mat]*
shitsu'mon	質問	query; question *{n} [comm]*
shitsu'mu-ritsu	悉無律; しつむりつ	naught or one, law of *[math]*
shitsu'nai	室内	interior; in the room *[arch]*
shitsu-on	室温; 室溫	room temperature *[thermo]*
shitsu'ryō	質量; しつりょう	mass *{n} [mech]*
shitsuryō-bunseki'gaku	質量分析学	mass spectrometry *[an-ch]*
shitsuryō-bunseki-kei	質量分析計	mass spectrometer *[eng]*
shitsuryō-bunseki-ki	質量分析器	mass spectrograph *[eng]*
shitsuryō-hi	質量比	mass ratio *[astrophys]*
shitsuryō-hozon no hōsoku	質量保存の法則	conservation of mass, law of *[phys]*
shitsuryō-kyūshū-keisū	質量吸収係数	mass absorption coefficient *[phys]*
shitsuryō-sayō no hōsoku	質量作用の法則	mass action, law of *[p-ch]*
shitsuryō-sū	質量数	mass number *[nuc-phys]*
shittō	失透	devitrification *[ch]*
shiwa	皺; しわ	crease; pucker; wrinkle *{n} [tex]*
shiwa-do	皺度	rugosity *[bio]*
shi'ya	視野	visual field *[physio]*
shi'yaku	試薬	reagent *[an-ch]*
shi'yaku-bin	試薬瓶	reagent bottle *[an-ch]*
shi-yakusho	市役所	city hall *[arch]*
shi-yasui	為易い; しやすい	easy to do
shiyō	子葉	cotyledon *[bot]*
shi'yō(-gaki)	仕様(書)	(technical) specification *[ind-eng]*
shiyō-kajū	使用荷重	working load *[eng]*
shi'yōka-keiso	四沃化硅素	silicon tetraiodide
shiyō-kan	使用感	feel in use
shiyō-kanō	使用可能	enabled *[comput]*
shi'yōka-tanso	四沃化炭素	carbon tetraiodide
shiyō-ryō	使用量	quantity used *[sci-t]*
shiyō-sei	使用性	usability *[comput]*
shiyō-sei	脂溶性; しようせい	fat-soluble(ness); lipophilicity *[ch]*
shiyō'sha	使用者	user (person) *[comm] [comput]*
shiyō'sho	仕様書; 仕樣書	specifications *[ind-eng]*
shiyō suru	使用する	use *{vb}*
shiyō'zumi-nenryō	使用済み燃料	spent fuel *[nucleo]*
shiyū-dōtai-genshō	雌雄同体現象	hermaphroditism *[bio]*
shiza-hai'i'shi	四座配位子	quadridentate ligand *[ch]*
shizai	資材	material(s) *[sci-t]*
shizen	自然; しぜん	nature; mother nature *{n}*
shizen-gengo	自然言語	natural language *[comput]*

shizen-gyōko	自然凝固	spontaneous coagulation *[rub]*
shizen-hakka	自然発火	spontaneous ignition *[ch]*
shizen-hō'shutsu	自然放出	spontaneous emission *[electr]*
shizen-hōsoku	自然法則	nature, law of *[bio]*
shizen-kagaku	自然科学	natural science *[sci-t]*
shizen-kin	自然金	native gold *[geol]*
shizen-nenshō	自然燃焼	spontaneous combustion *[ch]*
shizen-ni'shin	自然二進	natural binary *[math]*
shizen-rōka	自然老化	natural aging *{n}* *[poly-ch]*
shizen-sentaku	自然選択	natural selection *[bio]*
shizen-shi	自然史	natural history *[sci-t]*
shizen-sū	自然数	natural numbers *[math]*
shizen-taisū	自然対数；自然對數	natural logarithm *[math]*
shizen-tōta	自然淘汰	natural selection *[bio]*
shizen-totsu'zen-hen'i	自然突然変異	spontaneous mutation *[gen]*
shi-zui	歯髄；齒髓	dental pulp *[dent]* *[hist]*
shi'zui; me-shibe	雌蕊	pistil *[bot]*
shizuka	静か；靜か；しずか	quietness; silence *{n}* *[acous]*
shizuka na	静かな	quiet; silent *{adj}* *[acous]*
shizuka na taiyō	静かな太陽	quiet sun *[astrophys]*
shizuke-sa	静けさ	silence *{n}* *[acous]*
shizume-bori(-zaiku)	沈め彫(細工)	intaglio (pl -glios) *{n}* *[lap]*
shizumeru	沈める	sink *{vb t}*
shizumeru	静める	quiet; calm down *{vb t}*
shizumu	沈む；しずむ	sink; be submerged *{vb i}*
shō	升；しょう	sho (unit of volume) = 0.477 U.S. gallon *[mech]*
shō	商	quotient *[math]*
shō	章	chapter *[pr]*
shō	笙	Japanese mouth organ *[music]*
shō	衝	opposition *[astron]*
shō	礁	reef *[geol]*
shō-	小-	micro- *{prefix}*
-shō	-省	-Ministry *{suffix}*
shō'an	硝安	abbreviation for ammonium nitrate
shōbai	商売；商賣	business; trade; commerce *{n}* *[econ]*
shōbō(-jidō)sha	消防(自動)車	fire engine *[mech-eng]*
shōbō'sho	消防署	fire station *[arch]*
shōbu	菖蒲；しょうぶ	iris; sweet flag; Acorus calamus *[bot]*
shobun	処分；處分	disposal; dealing; disposition *{n}*
shōchō	小腸	small intestine *{n}* *[anat]*
shōchō	象徴；しょうちょう	symbol *{n}*

shōchū	焼酎	low-grade distilled spirit *[cook]*
shodai-baiyō	初代培養	primary culture *[microbio]*
shōdaku	承諾	consent *{n} [legal]*
shō'denki	焦電気；焦電氣	pyroelectricity *[sol-st]*
shō-denryoku'ka	省電力化	conservation of electric power *[elec]*
shodō	書道	calligraphy *[comm]*
shōdo	照度	illuminance; illumination *{n} [opt]*
shōdō	章動	nutation *[astron]*
shōdō	晶洞	druse *[geol]*
shōdō	衝動；しょうどう	impulse; drive *{n} [phys] [psy]*
"	衝動	impact *{n} [phys]*
shōdo-kei	照度計	illuminance meter *[eng]*
shōdoku	消毒	disinfection *{n} [med]*
shōdoku suru	消毒する	disinfect; sanitize *{vb} [med]*
shōdoku-yaku	消毒薬；消毒藥	disinfectant *{n} [med]*
shōdon-ro	焼鈍炉；燒鈍爐	annealing furnace *[met]*
shōdō'seki	晶洞石	geode *[geol]*
shō-en	小円	small circle *[geod]*
shō'en-ki	消煙器	smoke consumer *[eng]*
shō'en-zai	消炎剤	antiphlogistic *{n} [pharm]*
shōga	生薑；しょうが	ginger; Zingiber officinale (herb) *[bot]*
shōgai	障害；しょうがい	fault *{n} [eng]*
shōgai'butsu	障害物	obstacle; obstruction; impedance *{n}*
shōgai-tankyū	障害探究	troubleshooting *{n} [comput]*
shōgai-tankyū suru	障害探究する	troubleshoot *{vb} [comput]*
shōgai-tsuikyū	障害追究	troubleshooting *{n} [comput]*
shōgan	床岩	bedrock *[geol]*
shō-gan'yaku	小丸薬	pellet *{n} [pharm]*
Shōgatsu	正月	January; the New Year *{n}*
shō'geki	衝撃；しょうげき	impact; shock *{n} [mech]*
"	衝撃	percussion *[mech-eng]*
"	衝撃	impulse *[phys]*
shōgeki'ha	衝撃波	impulse wave; shock wave *[phys]*
shōgeki-hakai	衝撃破壊	impact fracture *[mech]*
shōgeki-ō'ryoku	衝撃応力	impact stress *[mech]*
shōgeki'shiki-insho-sōchi	衝撃式印書装置	impact printer *[graph]*
shōgeki'shiki-seifun-ki	衝撃式製粉器	hammer mill *[eng]*
shōgeki-tsuyo-sa	衝撃強さ	impact strength *[mech]*
shogen	諸元	various dimensions; various factors *[math]*
shogen	緒言	introduction; foreword *[pr]*
shōgen	象限；しょうげん	quadrant *{n} [math]*
shōgen no	象限の	quadrantal *{adj} [navig]*

shōgen-soku	消滅則	extinction rule *[crys]*
shōgo	正午; しょうご	noon *[astron]*
shōgō	商号; 商號	commercial name *[ind-eng]*
"	商号	trade name *[pat]*
shōgō	照合	collating *{n}* *[comput]*
shōgō-ki	照合機	collator *[comput]*
shōgō suru	照合する	collate *{vb}* *[comput]*
shōgun	小群	pod *{n}* *[v-zoo]*
shōgyō	商業; しょうぎょう	business *[econ]*
shōgyō-chi'iki	商業地域	commercial district *[civ-eng]*
shōgyō'ka	商業化	commercialization *[econ]*
shō-gyō'retsu-shiki	小行列式	minor; minor determinant *[math]*
shōgyō-sei	商業性	commerciality *[econ]*
shō-ha'guruma	小歯車; 小齒車	pinion *{n}* *[mech-eng]*
shō'heki	晶癖	crystal habit *[crys]*
shō-henkō	消偏光	quenching of polarized light *{n}* *[opt]*
shōhi	消費	expenditure *[econ]*
shōhin	商品	merchandise *{n}* *[ind-eng]*
shōhi'sha	消費者	consumer (a person) *[econ]*
shōhi-zai	消費財	consumer goods *[econ]*
shohō	処方; 處方	recipe; formula *{n}* *[ch]* *[pharm]*
"	処方; しょほう	prescription *[med]* *[pharm]*
shōhon	抄本	extract; abstract *{n}* *[pr]*
shohō-sen	処方箋	prescription *[med]* *[pharm]*
shohō'sen ni yoru yaku'hin	処方箋による薬品	ethical drug *[pharm]*
shō'hōshi	焦胞子; 焦胞子	smut spore *[mycol]*
shohō'shū	処方集	formulary (pl -laries) *{n}* *[pharm]*
shō-hōsoku	商法則	quotient, law of *[math]*
shohō suru	処方する	prescribe *{vb}* *[med]*
shōhō'tai	小胞体	endoplasmic reticulum; ER *[cyt]*
shohō-yaku	処方薬	prescription (medicine) *[pharm]*
shōhō-zai	消泡剤; 消泡劑	antifoamer; antifoam(ing) agent; defoaming agent *[mat]*
shōhyō	商標	trademark *{n}* *[ind-eng]* *[pat]*
shōhyō-mei	商標名	propietary name (medicine) *[pharm]*
shōhyō-tōroku	商標登録	trademark registration *[pat]*
shōji	消磁	demagnetization *[elecmg]*
shōji	障子	sliding shoji door *[arch]*
shō-jiten	小辞典; 小字典	glossary (pl -ries) *[pr]*
shōjō	猩猩; しょうじょう	orangutang (animal) *[v-zoo]*
shōjō-fuku'yō	掌状複葉	palmately compound leaf *[bot]*
shōjō-kōkan-chō	猩々紅冠鳥	cardinal (bird) *[v-zoo]*

shōka	昇華；しょうか	sublimation *[thermo]*
shōka	消化	slaking *{n} [cer]*
"	消化	digestion *[physio]*
shōka	硝化	nitrification *[microbio]*
"	硝化	nitration *[org-ch]*
shōka	漿果；しょうか	berry (pl berries) *[bot]*
shōka'butsu	昇華物	sublimate *{n} [ch]*
shōka-eki	消化液	digestive fluid *[physio]*
shō-kahei	小花柄	pedicel *[bot]*
shōkai	照会；照會	inquiry (pl -ries) *[comput] [legal]*
shōka-iō	昇華硫黄	sublimed sulfur; flowers of sulfur *[pharm]*
shōka-kan	消化管	digestive tract *[anat]*
shōka-kei	消化系	digestive system *[anat]*
shō-kaki	小火器	small arm(s) *[ord]*
shōka-ki	消火器；しょうかき	fire extinguisher *[eng]*
shōka-kikan	消化器官	digestive organ *[anat]*
shō-kakko	小括弧	parenthesis (pl -ses) *[math] [pr]*
shōka-kōso	消化酵素	digestive enzyme *[bioch]*
shōkaku	小核	micronucleus *[i-zoo]*
shōka-men	硝化綿；しょうかめん	cellulose nitrate; nitrocellulose; nitrocotton *[org-ch]*
shōka-saikin	硝化細菌	nitrifier (bacteria) *[microbio]*
shōka-sayō	硝化作用	nitrification *[microbio]*
shōka-sen	消火栓	hydrant *[civ-eng]*
shōka suru	昇華する	sublime *{vb t} [thermo]*
shōka-zai	消火剤	fire-extinguishing agent *[mat]*
shōkei	小計	subtotal *{n} [math]*
shōkei	晶系；しょうけい	crystal system *[crys]*
shōkei-moji	象形文字	hieroglyph *[comm] [pr]*
shoken	所見；しょけん	finding(s) *{n} [med]*
shō-kessetsu	小結節	nodule *{n} [anat]*
shō'ketsu	焼結；しょうけつ	sintering *{n} [met]*
shōketsu-kō	焼結鉱；焼結鑛	sintered ore *[geol]*
shōketsu-sei	焼結性	sinterability *[met]*
shoki	初期	initial period *[sci-t]*
shō'kibo no-	小規模の-	mini- *{prefix}*
shoki-chi	初期値；しょきち	initial value *[math]*
shoki-jōken	初期条件	initial-period condition *[ch]*
shō-kikō	小気候	microclimate *[climat]*
shōkin	渉禽；しょうきん	wading bird *[v-zoo]*
shoki no	初期の	incipient *{adj} [med]*
shoki-setsubi-shitsu	諸機設備室	utility room (in a home) *[arch]*

shokkaku	触覚; 觸覺	sense of touch; tactile sense *[physio]*
shokka-rui	食果類	Frugivora *[v-zoo]*
shokki	食器; しょっき	tableware; dinnerware *[cer] [eng]*
shokki-dana	食器棚	cupboard; sideboard *[furn]*
shokkō	燭光	candlepower *[opt]*
shōko	証拠; 證據	evidence; proof *{n} [legal]*
shōko	礁湖	lagoon *[geogr]*
shōkō	小孔	pore *{n} [bio]*
shōkō	昇汞; しょうこう	corrosive sublimate *[inorg-ch]*
shōkō	消光	quenching *{n} [sol-st]*
shōkō-busshitsu	消光物質	quencher (of light) *[opt]*
shōkō'da	昇降舵	elevator (aircraft) *[aero-eng]*
shokō-dōbutsu	蹠行動物	plantigrade animal *[v-zoo]*
shōkō-guchi	昇降口	companionway *[nav-arch]*
shōkō'gun	症候群	syndrome *[med]*
shōkō-ki	昇降機	elevator *[arch]*
shōko-shitsu	消弧室	arc extinguishing chamber *[electr]*
shōkō'yō-hashigo	昇降用梯子	trap *{n} [aero-eng] [nav-arch]*
shōkō-zai	消光剤	quencher *[mat]*
shoku	食; 蝕; しょく	eclipse *{n} [astron]*
shoku	燭; しょく	candlepower *[opt]*
shoku'bai	触媒; しょくばい	catalyst *[ch]*
shokubai-doku	触媒毒; 觸媒毒	anticatalyzer; anticatalyst; catalytic poison *[ch]*
shokubai-hannō	触媒反応	catalytic reaction *[ch]*
shokubai-kassei'ka-zai	触媒活性化剤	catalyst activator *[ch]*
shoku'bun	食分	eclipse *{n} [astron]*
shoku'butsu	植物; しょくぶつ	plant; vegetation; flora *{n} [bot]*
shokubutsu'en	植物園	botanical garden *[bot]*
shokubutsu'gaku	植物学	botany *[bio]*
shokubutsu-gunraku	植物群落	plant community *[bot]*
shokubutsu-kagaku	植物化学	plant chemistry; phytochemistry *[bot]*
shokubutsu-kai	植物界	Plantae; plant kingdom *[bot]*
shokubutsu-rō	植物蠟	vegetable wax *[mat]*
shokubutsu-seichō-chōsei-busshitsu	植物生長調整物質	plant-growth substance *{n} [bot]*
shokubutsu'sei-gensei-seibutsu	植物性原生生物	plantlike protista *[bot]*
shokubutsu-sen'i	植物繊維	vegetable fiber *[tex]*
shokubutsu-sō	植物相	flora *[bot]*
shokubutsu-tokkyo	植物特許	plant patent *[pat]*
shokubutsu-yu	植物油	vegetable oil *[mat]*

shoku-chūdoku	食中毒	food poisoning *{n} [med]*
shoku'chū-rui	食虫類；食蟲類	Insectivora *[v-zoo]*
shoku'chū-shokubutsu	食虫植物	carnivorous (or insectivorous) plant *[bot]*
shoku'dai	燭台；燭臺	candlestand *{n} [furn]*
shoku'dō	食堂	dining room *[arch]*
shoku'dō	食道	alimentary canal; esophagus; gullet *[anat]*
shokudō'sha	食堂車	dining car *[rail]*
shoku'en	食塩；食鹽	table salt; common salt; halite; salt *{n} [ch]*
shoku'en-sui	食塩水	salt water *[ch] [cook]*
"	食塩水	brine *[mat] [ocean]*
shoku'fu; ori-nuno	織布	fabric cloth *[tex]*
shoku'gyō'byō	職業病	occupational disease *[med]*
shoku'hin	食品；しょくひん	food *[cook] [food-eng]*
shokuhin-eisei'gaku	食品衛生学	food hygiene *[med]*
shokuhin-gyōkai	食品業界	food industry world *[food-eng]*
shokuhin-kōgyō	食品工業	food industry *[food-eng]*
shokuhin-senryō	食品染料	food color *[food-eng] [mat]*
shokuhin-tenka'butsu	食品添加物	food additive *[food-eng] [mat]*
shoku'ji	食事	a meal *[cook]*
shoku'ji	植字	typesetting *{n} [pr]*
shokuji-ryōhō'gaku	食餌療法学	dietetics *[med]*
shokuji-shitsu	食事室	dining room *[arch]*
shokuji suru	植字する	typeset *{vb} [pr]*
shoku'mei	職名	occupation; title of office *{n}*
shokumō	植毛	flocking *{n} [tex]*
shoku'motsu	食物；しょくもつ	food *[bio]*
shokumotsu-kōgaku	食物工学	food engineering *{n} [eng]*
shokumotsu-mō	食物網	food web *[ecol]*
shokumotsu-rensa	食物連鎖	food chain *[ecol]*
shokumotsu-sen'i	食物繊維	dietary fiber *[cook]*
shoku-nen	食年	eclipse year *{n} [astron]*
shoku'niku no	食肉の	predatory *{adj} [zoo]*
shoku-rensei	食連星	eclipsing binary *[astron]*
shokuryō-asari	食糧漁り	foraging *{n} [v-zoo]*
shokuryō-kōgaku	食糧工学	food engineering *{n} [eng]*
shokuryō-kōgyō	食糧工業	food industry *[econ]*
shoku-saibō	食細胞；食細胞	phagocyte *[cyt]*
shoku-sayō	食作用	phagocytosis (pl -toses) *[cyt]*
shoku'shin-narai-kikō	触針倣い機構	tracking profile mechanism *[eng]*
shoku'shu	触手；觸手	antenna (pl -nae, -s); feeler; tentacle *[i-zoo]*

shoku-zu	食酢	vinegar; table vinegar *{n} [cook]*
shoku'wan	触腕	tentacle (of squid) *[i-zoo]*
shokuzen'shu	食前酒	aperitif *{n} [cook]*
shokuyō-abura; shokuyō'yu	食用油	edible oil; food oil *[cook]*
shoku'yō-kōryō	食用香料	food flavor *[cook]*
shoku'yoku-kōshin-yaku	食欲亢進薬	aperitive *{n} [pharm]*
shoku'yō'shi; shoku'yō-abura	食用脂	edible fat *[cook]*
shoku'yō-shikiso	食用色素	food color *[cook]*
shō'kyaku-ro	焼却炉；焼却爐	incinerator *[eng]*
shōkyo	消去；しょうきょ	elimination *[math]*
"	消去	clear; erase; erasure *[comput]*
shōkyo-kanō-ki'oku	消去可能記憶	erasable storage *[comput]*
shōkyoku-sayō	消極作用	depolarization *[elec]*
shōkyoku'teki na	消極的な	passive; negative *{adj}*
shōkyo suru	消去する	eliminate; erase *{vb}*
shōkyū	小球	pellet *{n} [sci-t]*
shokyū no-	初級の-	proto- *{prefix} [ch]*
shomei	書名	book title *[pr]*
shomei	署名	signature *[legal]*
shōmei	証明；證明	certification; proof *[logic]*
shōmei	照明	illumination; lighting *{n} [opt]*
shōmei-dan	照明弾	flare bomb *[ord]*
shōmei-kigu	照明器具	lighting fixture *[opt]*
shōmei'sho	証明書	certificate *[legal]*
shomei suru	署名する	sign *{vb} [legal]*
shōmei suru	証明する	certify *{vb} [legal]*
shōmei suru	照明する	illuminate *{vb} [opt]*
shōmen	正面；しょうめん	facade *[arch]*
"	正面	front *{adv} {n}*
shōmen-danmen-zu	正面断面図	front sectional view; front view cross section *{n} [graph]*
shōmen-genkan	正面玄関	main entrance *[arch]*
shōmen no	正面の	frontal *{adj}*
shōmen-zu	正面図；正面圖	front elevation; front view *[graph]*
shōmetsu	消滅；しょうめつ	annihilation *[part-phys]*
"	消滅	quenching *{n} [quant-mech]*
"	消滅	dissipation; expiation *{n}*
shōmi	正味	net *{n} [econ]*
shōmi-jūryō	正味重量	net weight *[mech]*
shōmō'hin	消耗品	expendable article *[ecol] [ind-eng]*
shō-moji	小文字	lowercase letter *[comput] [pr]*
shōmoku'sei no	焦木性の	pyroligneous *{adj} [ch-eng]*

shōmō suru	消耗する	consume; dissipate; exhaust *{vb}*
sho'motsu	書物；しょもつ	book; volume *{n} [pr]*
shōni	小児；小兒	child (pl children) *[bio]*
shō-nijū'men'tai	正二十面体	icosahedron *[math]*
shōnin	承認	acknowledgment; approval *{n} [pat]*
shō'nin	証人；證人	witness *{n} [legal]*
shōnō	小脳；小腦	cerebellum (pl -s, -bella) *{n} [anat]*
shōnō	樟脳；しょうのう	camphor *[org-ch]*
shōnō-san	樟脳酸	camphoric acid
shōnō-yu	樟脳油	camphor oil *[mat]*
shōnyū'seki	鐘乳石	stalactite *[geol]*
shō'on-ki	消音器	muffler; silencer *[eng]*
shōrei suru	奨励する	encourage *{vb}*
shori	処理；處理；しょり	treatment; processing *[ind-eng] [comput]*
shori-nō'ryoku	処理能力	throughput *[comput]*
shori'ryō	処理量	throughput *[comput]*
shori suru	処理する	treat; process *{vb t} [ind-eng]*
shōro	松露	truffle *[bot] [cook]*
shōrō	鐘楼；鐘樓	bell tower *[arch]*
shōroku	抄録；抄錄	abstract *{n} [pr]*
shorui	書類	documents *[comm] [legal] [pr]*
shō'ryaku	省略	omission; abridgment *{n}*
shōryaku'ji-kaishaku	省略時解釈	default assumption *{n} [comput]*
shōryaku-kigō	省略記号	ellipsis (pl ellipses) *[pr]*
shōryō-bunseki	小量分析	semimicroanalysis *[an-ch]*
shōryoku no	省力の	labor-saving *{adj} [ind-eng]*
shōryoku-tōshi	省力投資	labor-saving investment *[econ]*
shosai	書斎	study; library (a room) *{n} [arch]*
shōsai	詳細；しょうさい	details; particulars *{n}*
shōsai na setsu'mei	詳細な説明	detailed explanation *[pat]*
shōsai-zu	詳細図	detail drawing *{n} [graph]*
shōsan	硝酸；しょうさん	nitric acid
shōsan-dai'ichi-tetsu	硝酸第一鉄	ferrous nitrate
shōsan-dai'ni-tetsu	硝酸第二鉄	ferric nitrate
shōsan-dō	硝酸銅	copper nitrate
shōsan-en	硝酸塩	nitrate *[ch]*
shōsan-gin	硝酸銀	silver nitrate
shōsan-kin	硝酸菌	Nitrobacter *[microbio]*
shōsan-namari	硝酸鉛	lead nitrate
shōsan-sanka-gin	硝酸酸化銀	silver oxide nitrate
shōsan-suigin	硝酸水銀	mercury nitrate
shōsan-tetsu	硝酸鉄	iron nitrate

shōsei	焼成；燒成	baking *{n} [eng] [met]*
"	焼成；しょうせい	firing; burning; heat treating *{n} [eng]*
"	焼成	sintering *{n} [met]*
"	焼成	calcination; roasting *{n} [min-eng]*
shoseki	書籍	books; publications *[pr]*
shōseki	晶析	crystallization *[crys]*
shōseki	硝石	niter *[inorg-ch]*
"	硝石	potassium nitrate; saltpeter *[inorg-ch]*
shō'sekkai	消石灰	calcium hydroxide; slaked lime *[inorg-ch]*
shōsen	商船	commercial ship; merchant ship *[nav-arch]*
shōsen	焦線	caustic line *[mat] [opt]*
shōsha	商社	business firm *[econ]*
shōsha	照射；しょうしゃ	irradiation *[eng]*
shōsha'ryō	照射量	dose *{n} [med]*
shōsha(-sen)'ryō	照射(線)量	exposure *[nucleo]*
shōsha'senryō-ritsu	照射線量率	exposure rate *[nucleo]*
shōshi'jō no	小歯状の	denticulate *{adj} [zoo]*
sho'shiki	書式	form; format *{n} [comput] [pr]*
shoshiki-okuri	書式送り	form feed *[comput]*
shoshiki-sakusei	書式作成	formatting *{n} [comput]*
shō-shitsu	晶質	crystalloid *{n} [bot]*
shoshō	初晶	primary crystal *[met]*
shō-shokudō	小食堂	dinette *[arch]*
shōshoku-sokudo	消色速度	decolorization speed *[opt]*
shō-shuku'shaku-zu	小縮尺図	small-scale map *[map]*
shō'shutsu	晶出	crystallizing out *[ch-eng]*
"	晶出	crystallization *[crys]*
shō'shutsu-kaku	晶出核	crystallization nucleus *[ch-eng]*
shōshū-zai	消臭剤	deodorant *{n} [pharm]*
sho'soku(do)	初速(度)	initial velocity *[mech]*
shōsū	小数；小數	decimals *[math]*
shōsū no-	少数の-	oligo-; olig- *{prefix} [sci-t]*
shōsū no hōsoku	少数の法則	small numbers, law of *[math]*
shōsū-ten	小数点；小數點	decimal point; radix point *[math]*
shōtai	晶帯	zone *[crys]*
shōtai-jiku	晶帯軸；晶帶軸	zone axis *[crys]*
shōtaku'chi	沼沢地；沼澤地	swamp *{n} [ecol]*
shoten	書店	bookstore *[arch] [pr]*
shōten	消点；消點	vanishing point *[navig] [opt]*
shōten	焦点；しょうてん	focus (pl foci) *{n} [math] [opt]*
shōten'gai	商店街	shopping center; shopping district *[civ-eng]*

shōten-hi	焦点比	focal ratio *[opt]*
shōten-kyori	焦点距離	focal length *[opt]*
shōten-men	焦点面	focal plane *[opt]*
shotō	蔗糖；しょとう	sucrose; cane sugar *[bioch]*
shōtō	小塔	turret *[arch]*
shōtō	少糖	oligosaccharide *[bioch]*
sho'toku	所得	earnings; income *[econ]*
shotoku-zei	所得税	income tax *[econ]*
shō'totsu	衝突；衝突	collision; impact *{n} [mech] [phys]*
shōtotsu'gata-kasoku-ki	衝突型加速器	colliding-beam accelerator *[electr]*
shōtotsu-ki	衝突器	collider *[phys]*
Shōwa	昭和；しょうわ	Shōwa (era name) (1926-1989)
shō-wakusei	小惑星	asteroid *[astron]*
shō-yaku	生薬；生薬	crude drug *{n} [pharm]*
shōyaku'gaku	生薬学	pharmacognosy *[pharm]*
shōyō	小葉	leaflet *[bot]*
shōyō	商用	commercial *{n} [econ]*
shoyō no	所要の	necessary *{adj}*
shōyō-shikin-hō	焼溶試金法	scorification *[met]*
shōyō-shūha'sū	商用周波数	commercial frequency *[elecmg]*
shōyō suru	賞用する	favor in use *{vb}*
shōyu	松油	pine oil *[mat]*
shōyu	醤油；しょうゆ	soy sauce *[cook]*
shoyū-kaku	所有格	possessive case *[gram]*
shoyū'sha	所有者	owner *[comput]*
shozai	所在	location *{n}*
shō-zentei	小前提	minor premise *[logic]*
shōzō-ga	肖像画	portrait *[graph]*
shozoku	所属	affiliation; position; post *{n}*
shozoku suru	所属する	belong to; affiliated with *{vb}*
shōzō-sui	抄造水	papermaking water *[eng]*
shōzuku	小豆蔻；しょうずく	cardamon; cardamum (herb) *[bot]*
shu	主；しゅ	master *{n} [pr]*
shu	朱	cinnabar *[miner]*
"	朱	vermillion (color) *[opt]*
"	朱	red mercuric sulfide
shu	種；しゅ	species (pl species) *[bio] [syst]*
shū	州	state *{n} [geogr]*
shū	洲	continent *[geogr]*
shū	週	week *{n} [astron]*
-shu	-種	-type(s) *{suffix}*
-shū	-州	-State (U.S.) *{suffix}*

shubo	酒母；しゅぼ	yeast culture *[mycol]*
shu-bōdo	主ボード	mother board *[comput]*
shūbun	秋分	autumnal equinox *[astron]*
shūchaku	収着；收着	sorption *[p-ch]*
shūchaku-eki	終着駅；終着驛	terminal station *[rail]*
shuchi	主値	principal value *[math]*
shūchū-gō'u	集中豪雨	locally severe rainstorm *[meteor]*
shūchū-seigyo	集中制御	centralized control *[comput]*
shūchū-teisū-tōka-kairo	集中定数等価回路	lumped-constant equivalent circuit *[elec]*
shudai-zu	主題図	thematic map *[graph]*
shūdan	集団；集團	group *{n} [bio] [gen]*
shūdan-iden'gaku	集団遺伝学	population genetics *[gen]*
shūden-ki	集電器	collector *[electr]*
shūdō-ha'guruma	摺動歯車	sliding gear *[mech-eng]*
shūdō-jizai no (na)	摺動自在の(な)	freely slidable *{adj} [eng]*
shudō-kei	手動系	manual system *[comput]*
shūdo-kei	臭度計	odorometer *[eng]*
shūdomonasu	シユードモナス	Pseudomonas *[microbio]*
shu-dōmyaku	主動脈	main artery *{n} [anat]*
shudō'shiki	手動式	manually operated *[ind-eng]*
shudō(-sō)sa	手動(操)作	manual operation *[comput]*
shū'en	周縁	rim *{n} [eng]*
"	周縁	periphery *{n} [math]*
shūen-kōsen	周縁光線	marginal ray *[opt]*
shufu	首府	capital (city) *[geogr]*
shūfuku	修復	repair; restoration *{n} [eng] [sci-t]*
shūfuku suru	修復する	restore *{vb} [eng]*
shugei	手芸；手藝	handicraft
shu-genshi'ka	主原子価	principal valence *[ch]*
shū'ginkō	臭銀鉱	bromyrite *[miner]*
shugo	主語	subject *{n} [gram]*
shūgō	集合；しゅうごう	set *{n} [comput] [math]*
"	集合	class *{n} [gram] [math]*
"	集合	aggregation *[phys]*
shūgō-so'shiki	集合組織	texture *[crys]*
shūgō'tai	集合体	consortium (pl consortia) *[bot]*
"	集合体	aggregate *{n} [geol] [mat]*
shū'goto no	週毎の	weekly *{adv}*
shūgyō	終業	end of work *{n} [ind-eng]*
shūha	周波；しゅうは	wave cycle *[elec]*
shu-hachō	主波長	dominant wavelength *[phys]*
shūha'sū	周波数	frequency (pl -cies) *[phys]*

shūhasū-benbetsu-ki	周波数弁別器	frequency discriminator *[electr]*
shūhasū-henchō	周波数変調	frequency modulation; FM *[comm]*
shūhasū-henkan	周波数変換	frequency conversion *[electr]*
shūhasū-ō'tō	周波数応答	frequency response *[eng]*
shūhasū-tai'iki	周波数帯域	frequency band *[phys]*
shūhasū-teibai-ki	周波数逓倍器	frequency multiplier *[electr]*
shūhasū-teikō-ki	周波数逓降器	frequency demultiplier *[electr]*
shu-heimen	主平面	principal plane *[opt]*
shūhen	周辺; 周邊	limb *[astron]*
"	周辺; しゅうへん	perimeter *[math]*
shūhen	終辺	terminal side *[math]*
shūhen-chō	周辺長	peripheral length *[math]*
shūhen-genkō	周辺減光	limb darkening *{n} [astrophys]*
shūhen-ka	周辺花	ray flower *[bot]*
shūhen-kaku'ritsu	周辺確率	marginal probability *[math]*
shūhen-kiki	周辺機器	peripherals; peripheral devices *[comput]*
shūhen-sōchi	周辺装置	peripherals *[comput]*
shūhen-zōkō	周辺増光	limb brightening *{n} [astrophys]*
shuhi	珠皮	integument *[bot]*
shuhi	種皮	testa *[bot]*
shū-hō	週報	weekly report *[pr]*
shū'i	周囲; 周圍	surroundings *{n} [ecol] [eng] [phys]*
"	周囲	perimeter *[math]*
shū'i no-	周囲の-	amphi- *{prefix}*
shu-jiku	主軸	principal axis *[math] [mech] [opt]*
shujiku-dai	主軸台	headstock *[mech-eng]*
shūjin	集塵;しゅうじん	dust collection *[eng]*
shūka-aen	臭化亜鉛	zinc bromide
shūka'butsu	臭化物	bromide *[ch]*
shūka-chisso	臭化窒素	nitrogen bromide
shūka-dō	臭化銅	copper bromide
shūka-gin	臭化銀	silver bromide
shūkai	集塊	agglomerate *[bot] [geol]*
shūkaidō	秋海棠	begonia (flower) *[bot]*
shūkai-giji'dō	州会議事堂	capitol (state building) *[arch]*
shūka-iō	臭化硫黄	sulfur bromide
shūkai-sōgō-shisetsu	集会総合施設	convention center *[arch]*
shūkai-uchū'sen	周回宇宙船	orbiter; space orbiter *[aero-eng]*
shūka-keiso	臭化硅素	silicon bromide
shukaku	主格	nominative case *[gram]*
shūkaku	収獲; 收獲	harvest; harvesting *{n} [agr]*
shūkan	習慣; しゅうかん	habit *[psy]*

-shūkan	-週間	-week(s) *{suffix}*
shukan-kaku'ritsu	主観確率	subjective probability *[math]*
shukan-seigyo-ki	主幹制御器	master controller *[comput]*
shūkan'shi	週刊誌	weekly periodical *[pr]*
shukan'teki na	主観的な; 主觀的な	subjective *{adj}* *[psy]*
shūka-rin	臭化燐	phosphorus bromide
shūka-suiso	臭化水素	hydrogen bromide
shūka-suzu	臭化錫	tin bromide
shūka-yōso	臭化沃素	iodine bromide
shūkei	集形	combination *[crys]*
shu-kei'retsu	主系列	main sequence *[astron]*
shu'keiretsu-sei	主系列星	main sequence stars *[astron]*
shū'ketsu	終結	termination *[comput]*
shūketsu-shiki	終結式	resultant *{n}* *[mech]*
shūki	周期; しゅうき	period *{n}* *[ch]* *[math]* *[phys]*
"	周期	cycle *{n}* *[sci-t]*
shūki	秋期	autumn *[astron]*
shūki	臭気	odor *[physio]*
shūki	秋季	autumn season *[astron]*
shūki	終期	telophase *[cyt]*
shūki-kansū	周期関数	periodic function *[math]*
shuki-kenchi-ki	酒気検知器	Breathalyzer *[eng]*
shu-ki'oku(-kikō)	主記憶(機構)	main storage; main memory *[comput]*
shūki-ritsu	周期律	periodic law; law of periodicity *[ch]*
shūki'ritsu-hyō	周期律表	periodic table of elements *[ch]*
shūki'ritsu(hyō)-bunrui	周期律(表)分類	periodic classification *[ch]*
shūki'teki na	周期的な	periodic *{adj}* *[math]*
shūki-undō	周期運動	periodic motion *[mech]*
shukka	出荷	shipping *{n}* *[ind-eng]*
shukkō	出鋼	tapping *{n}* *[steel]*
shukō	主港; 主港	major port; major harbor *[geogr]*
shūkō	集光	condensing *{n}* *[opt]*
shu-kōka	主効果	main effect *[math]*
shukon	主根	taproot *[bot]*
shūkō-renzu	集光レンズ	converging lens *[opt]*
shūkō-ryoku	集光力	light-gathering power *[opt]*
shu-kōsen	主光線	principal ray *[opt]*
shuku'gō	縮合; しゅくごう	condensation *[ch]*
shukugō'butsu; (-tai)	縮合物; (-体)	condensate *{n}* *[mat]*
shuku(gō)-jūgō	縮(合)重合	condensation polymerization; polycondensation *[poly-ch]*
shukugō-zai	縮合剤	coupling agent *[ch]*

shuku'sha-ki	縮写機；縮寫機	reduction printer *[graph]*
shuku-shaku	縮尺	reduced scale *[graph]*
shuku'shō suru	縮小する	reduce; curtail; contract *{vb}*
shuku'shu; yado-nushi	宿主	host *{n} [bio]*
shuku'shu-saibō	宿主細胞	host cell *[gen]*
shuku'tai	縮退	degeneracy *[phys]*
shukutai-sei	縮退星	degenerate star *[astron]*
shuku'zu-ki	縮図器；縮圖器	pantograph *[graph]*
shu-kyō	主鏡	primary mirror; main mirror *[opt]*
shū'kyoku	褶曲；しゅうきょく	bend; fold; warp *{n} [geol]*
shūkyoku-sei	周極星	circumpolar star *[astron]*
shumi	趣味	hobby (pl hobbies)
shu'moku-zame	撞木鮫	hammerhead shark; Sphyrna zygaena *[v-zoo]*
shūmō'sei no	周毛性の	peritrichous *{adj} [microbio]*
shunbun	春分	vernal (spring) equinox *[astron]*
shun'giku	春菊；しゆんぎく	Chrysanthemum coronarium (vegetable) *[bot]*
shū-ni'kai no	週二回の	biweekly *{adv}*
shun'ji-chi	瞬時値	instantaneous value *[phys]*
shunji-den'atsu	瞬時電圧	instantaneous voltage *[elec]*
shunkan	瞬間；しゅんかん	instant *{n} [phys]*
shunkan-sokudo	瞬間速度	instantaneous velocity *[mech]*
shun'ka-shū'tō	春夏秋冬	seasons (of the year) *{n} [astron]*
shunki	春期	spring *{n} [astron]*
shunki-hatsu'dō-ki	春機発動期	puberty *[physio]*
shun-maku	瞬膜	nictitating membrane *[v-zoo]*
shūnō-bako	収納箱；收納箱	receiving box *[traffic]*
shun'setsu-ki	浚渫機	dredging machine *[eng]*
shūnyū-inshi	収入印紙	revenue stamp *[econ]*
shuppan	出版；しゅっぱん	publication *{n} [pr]*
shuppan suru	出版する	publish *{vb} [pr]*
shuppatsu	出発；出發	departure *[trans]*
shuppatsu-yotei-jikan	出発予定時間	expected time of departure *{n} [trans]*
shūraku	集落	colony (pl -nies) *[bio] [microbio]*
shūrei	週齢	age in weeks *{n} [bio]*
shūren	収斂；しゅうれん	convergence *[math] [phys]*
shūren-sei	収斂性；收斂性	astringency *[physio]*
shūren-zai	収斂剤	astringent *{n} [pharm]*
shūri	修理	repair; servicing *{n} [eng]*
shūri-hoshu	修理保守	corrective maintenance *[eng]*
shū'ritsu	収率；しゅうりつ	yield; yield in percentage *{n} [eng]*
shuro	棕櫚	hemp palm; Trachycarpus fortunei *[bot]*
shūroku suru	収録する	record *{vb}*

shurui	種類；種類	type; sort; variety *{n} [sci-t]*
shuryō	狩猟	hunting *{n}*
shūryō	収量；しゅうりょう	yield; yield in quantity *{n} [eng]*
shūryō	終了	termination *[comput]*
shūryō-jikoku	終了時刻	finish time *[ind-eng]*
shu-ryōshi-sū	主量子数	principal quantum number *[atom-phys]*
shuryō-zuki	狩猟月	hunter's moon *[astron]*
shuryū-dan	手榴弾	hand grenade *[ord]*
shūsa	収差；收差	aberration *[opt]*
shūsan	蓚酸；しゅうさん	oxalic acid
shūsan-en	蓚酸塩	oxalate *[org-ch]*
shūsan-tetsu	蓚酸鉄	iron oxalate
shusei	酒精	alcohol; spirits of wine *[cook]*
shūsei	修正	retouching *{n} [photo]*
shu-seibun	主成分	main component *[ch] [math]*
shusei-inryō	酒精飲料	alcoholic beverage *[cook]*
shusei-zai	酒精剤	spirit *{n} [pharm]*
shūsei-zai	集成材	laminated wood *[mat]*
shu'seki	酒石	tartar *[org-ch]*
shūseki	集積；しゅうせき	accumulation *[elec] [math]*
shūseki-baiyō	集積培養	accumulative cultivation *[microbio]*
shūseki-do	集積度	degree of accumulation *{n}*
shuseki'ei	酒石英	cream of tartar *[cook]*
shūseki'ka	集積化	integration; integrating *{n} [electr]*
shūseki-kairo	集積回路	integrated circuit; IC *[electr]*
shuseki-san	酒石酸	tartaric acid
shuseki'san-en	酒石酸塩	tartrate *[org-ch]*
shūseki suru	集積する	archive *{vb} [comput]*
shū'setsu suru	摺接する	slide contacting against *{vb} [mech-eng]*
shushi; tane	種子	seed; pit; stone *{n} [bot]*
shushi	趣旨	meaning; opinion; aim *{n} [psy]*
shūshi'fu	終止符	period; full stop *[gram] [pr]*
shushi-shokubutsu	種子植物	seed plant(s) *[bot]*
shushi-yu	種子油	seed oil *[mat]*
shu'shoku	主食	staple food *[cook]*
shūshoku	修飾	modification *[sci-t]*
shūshoku'go	修飾語	modifier; qualifier *[gram]*
shūshoku'shi	修飾子	qualifier *[comput] [gram]*
shu-shōten	主焦点；主焦點	principal focus (of lens) *[opt]*
shū'shuku	収縮；しゅうしゅく	contraction; shrinkage *{n} [mech-eng]*
shūshuku-hō	収縮胞；收縮胞	contractile vacuole *[cyt]*
shūshuku-keisū	収縮係数	contraction coefficient *[fl-mech]*

shūshuku-ki	収縮期	systole *{n} [physio]*
shūshuku-kō	収縮孔	shrinkage cavity *[met]*
shūshuku suru	収縮する	shrink *{vb} [mech]*
shūso	臭素; しゅうそ	bromine (element)
shūso-ion	臭素イオン	bromide ion *[ch]*
shūso'ka	臭素化	bromination *[ch]*
shūso-ka	臭素価	bromine number *[an-ch]*
shū'soku	収束	convergence *[math]*
shū'soku	集束	focusing *{n} [opt]*
shū-sokudo	周速度	peripheral velocity *[mech]*
shū-soku(do)	終速(度)	final velocity *[mech]*
shūsoku-kyūsū	収束級数	convergent series *[math]*
shūsoku-renzu	収束レンズ	converging lens *[opt]*
shūsoku-zai	集束剤	fiber size *{n} [mat] [tex]*
shūso-san	臭素酸	bromic acid
shūso'san-en	臭素酸塩	bromate *[ch]*
shūso-sui	臭素水	bromine water *[ch]*
shussa	出差	evection *[astrophys]*
shussei	出生	birth *{n} [bio]*
shussei-ritsu	出生率	birthrate *[bio]*
shussei-shibō-katei	出生死亡過程	birth and death process *[math]*
shussen'kō	出銑口	taphole *[met]*
shussen suru	出銑する	tap *{vb} [steel]*
shusshō-ritsu	出生率	birthrate *[bio]*
shusu-ori	朱子織り; 繻子織り	satin weave *{n} [tex]*
shutai	主体	subject *{n} [comput]*
shūtan	終端; しゅうたん	terminal *[elec]*
"	終端	end *{n}*
shūtan-bako	終端箱	terminal box *[elec]*
shūtan(-sōchi)	終端(装置)	terminal *{n} [comput]*
shūtan-sokudo	終端速度	terminal velocity *[phys]*
shutchō'jo	出張所	branch office *[econ]*
shuten	主点	principal point *[opt]*
shūten	終点	end point; terminal point *[math]*
"	終点	last stop; terminal *[trans]*
shutō	種痘	vaccination *[immun]*
shutsu'ga	出芽	budding *{n} [bot]*
shutsu'gan	出願; しゅつがん	application *[pat]*
shutsugan-bi	出願日	filing date *[pat]*
shutsugan-kōkoku	出願公告	publication of application *[pat]*
shutsugan'nin	出願人	applicant *[pat]*
shutsugan no bun'katsu	出願の分割	division of application *[pat]*

shutsu'gen	出現	emersion *[astron]*
shutsu'nyūkō-kinshi	出入港禁止	embargo (pl -goes) *{n} [econ]*
shutsu'ryoku	出力；しゅつりょく	output *[comput] [electr] [sci-t]*
"	出力	work output *[mech]*
shutsu'ryoku-sōchi	出力装置	output unit *[comput]*
shū'u	驟雨；しゅうう	shower; sudden rain *{n} [meteor]*
shūyaku	集約	collecting together *{n}*
shūyaku'teki na	集約的な	intensive *{adj} [agr]*
shuyō	腫瘍；しゅよう	tumor *[med]*
shūyō	収容；收容	accomodation; containment; capacity
shuyō-hassei	腫瘍発生	oncogenesis *[med]*
shuyō-idenshi	腫瘍遺伝子	oncogene *[gen]*
shu'yoku	主翼	wing (aircraft) *{n} [aero-eng]*
shuzai	主剤	base resin (adhesive) *[org-ch]*
shūzen	修繕	repair *{n} [eng]*
shu'zoku	種族	stellar population *[astron]*
"	種族	race; tribe; family *{n} [bio]*
so	素	principle *{n} [ch]*
so	疎	sparse *{n} [math]*
sō	双；雙；そう	congruence *[math]*
"	双	pair; both *{n}*
sō	相	phase *{n} [ch] [math] [phys]*
"	相	facies (pl facies) *[geol]*
sō	層；層	layer *{n} [geol] [geophys] [meteor]*
"	層	ply (pl plies) *{n} [mat]*
sō	槽	tank; vat *[ch] [eng]*
sō-	双-	bi- *{prefix}*
sō-antei (no)	双安定(の)	bistable *{adj} [sci-t]*
sō'antei-soshi	双安定素子	bistable element *[ch]*
soba	側；傍；そば	side *{n}*
soba	蕎麦；そば	buckwheat noodles *[cook]*
sō'batsu-jizai	挿抜自在	freely insertable and extractable *[mech-eng]*
sōbi	装備；裝備	equipment *[ind-eng]*
sobō	粗紡	roving *{n} [tex]*
so'byō	素描	sketch *{n} [graph]*
sōchaku	装着；裝着	attach; attaching; attachment *{n} [eng]*
sōchi	装置；そうち	apparatus; device; equipment; unit *[eng]*
sōchi-doku'ritsu	装置独立	device independence *[comput]*
sōchi-i'zon	装置依存	device dependence *[comput]*
sōchi-seigyo-moji	装置制御文字	device control character *[comput]*
sōchi suru	装置する	equip with; fit with *{vb} [ind-eng]*

sōchō	双潮	double tide *[ocean]*
sō'chōseki	曹長石	albite *[miner]*
sōda	曹達；ソーダ	soda (sodium carbonate)
sōda'fusseki	曹達沸石	natrolite *[miner]*
sōda-ki	操舵機	steering gear *[mech-eng]*
sōdan	相談	consultation; conference *[comm]*
sō-danmen	総断面；總斷面	gross section *[graph]*
sōda-sui	ソーダ水	carbonated water *[cook]*
sode	袖	sleeve *[cl]*
sōden	送電	power transmission *[elec]*
sōden-mō	送電網	transmission network *[comm]*
sōden-sen(ro)	送電線(路)	transmission line *[elec]*
sodō	素銅	elemental copper *[ch]*
sōdō	相同	homology *[bio] [ch] [org-ch]*
sōdō-kikan	相同器官	homologous organ *[bio]*
sōdō no	相同の	homologous *{adj} [sci-t]*
sōdō no hōsoku	相同の法則	homology, law of *[bio]*
soe'ji	添字；そえじ	subscript *[comput] [pr] [sci-t]*
so'eki-sei	疎液性	lyophobicness *[ch]*
sō'en	蒼鉛；そうえん	bismuth (element)
sōen-zai	掃鉛剤	scavenger *[ch]*
sōfū	送風	air blowing; fanning *{n} [ch-eng]*
sogai	阻害	inhibition *[sci-t]*
sogai-busshitsu	阻害物質	inhibitor *[bioch] [ch]*
sogai suru	疎外する	alienate *{vb}*
sōgan-jittai-kenbi'kyō	双眼実体顕微鏡	stereoscopic microscope *[opt]*
sōgan'kyō	双眼鏡	binoculars *[opt]*
sōgen	草原	grassy plain; prairie *[geogr]*
sōgen-raichō	草原雷鳥	prairie chicken (bird) *[v-zoo]*
sogi-hagi	殺ぎ接ぎ	sypher joint *[eng]*
sōgo-henchō-rōwa	相互変調漏話	intermodulation crosstalk *[electr]*
sōgo-hirei no hōsoku	相互比例の法則	reciprocal proportions, law of *[ch]*
sōgo-hōsoku	相互法則	law of reciprocity *[math]*
sōgo'jōken'teki na	相互条件的な	biconditional *{adj} [logic]*
sōgo-kakusan	相互拡散	interdiffusion *[p-ch]*
sōgo-kankei	相互関係	interaction *[sci-t]*
sōgo-kantsū-ami'me	相互貫通網目	interpenetrating network *{n}*
sōgō-kō'ritsu	総合効率	overall efficiency *[ind-eng]*
sōgo-sayō	相互作用	interaction *[phys]*
sōgo'teki na	相互的な	reciprocal *{adj}*
sōgo-tsū'shin-hō'shiki	相互通信方式	intercommunicating system; intercom *[comm]*
sōgo-yūdō	相互誘導	mutual induction *[elecmg]*

sōgo-yūsei	相互優勢	codominance *[bio]*
sōgun	層群; 層群	terrane *[geol]*
so-gyōshū'tai	疎凝集体	rough aggregate *[geol] [mat]*
sōgyō suru	操業する	operate *{vb} [ind-eng]*
sō-hai'shutsu-kō	総排出腔	cloaca *[v-zoo]*
sōhaku-hi	桑白皮	mulberry bark *{n} [bot] [pharm]*
sōhan no	相反の	reciprocal *{adj} [math]*
so-hannō	素反応; 素反應	elementary reaction *[p-ch]*
sōhan-sei	相反性	reciprocity (pl -ties) *[math]*
sōhan-soku	相反則	reciprocity law *[graph]*
sōhō	総包; 總包	involucre *[bot]*
sōhon	草本; そうほん	herb; herbage *[bot]*
sōho-sei	相補性	complementation *[gen]*
sō'i'gaku	層位学	stratigraphy *[geol]*
sō'in	掃引; そういん	sweep *{n} [electr]*
sō'in-kairo	掃引回路	sweep circuit *[electr]*
so-inshi	素因子	prime factors *[math]*
sō'in suru	掃引する	sweep *{vb} [comput]*
soji	素地	groundwork; foundation *{n}*
sōji	走時	travel time *[geophys]*
sōji	相似; そうじ	analogy; similarity {n} *[bio] [math]*
sōji	掃除	cleaning *{n}*
sōji-hōsoku	相似法則	similitude, law of *[phys]*
sōji no	相似の	analogous *{adj} [sci-t]*
sōji-tai	草字体	cursive-style writing *[comm] [pr]*
sōjō-heikin	相乗平均	geometric mean *[math]*
sōjō-kajo	総状花序	raceme *[bot]*
sō'jōken-bun	双条件文	biconditional sentence *[math]*
sōjō-kōka	相剰効果	synergistic effect *[bioch]*
sōjō-sayō	相剰作用; 相乗作用	synergism *[ecol]*
sōjō-zai	相乗剤; 相乗劑	synergist *[ch]*
sōjū	操縦; そうじゅう	steering *{n} [navig]*
sōjū-kan	操縦桿	control lever *[aero-eng]*
sōjū-sei	操縦性	maneuverability *[navig]*
sōjū'sha (sōjū'shi)	操縦者 (操縱士)	pilot (a person) *{n} [aero-eng]*
sō'jushin	送受信	transmission *[electr]*
sōjū-shitsu	操縦室; 操縱室	cockpit *[aero-eng]*
soka	蔬果; そか	vegetables and fruits *[cook] [food-eng]*
sōka	痩果; そうか	achene *[bot]*
sōka-heikin	相加平均	arithmetic mean *[math]*
sōkai'chōseki	曹灰長石	labradorite *[miner]*
sōkai'hōkō	曹灰硼鉱	ulexite *[miner]*

sōkai'shinseki	曹灰針石	pectolite *[miner]*
sōkai'tei	掃海艇	minesweeper *[nav-arch]*
sōka-kaisen	装荷回線	loaded circuit *[elec]*
sōkaku-rui	双殻類；雙殼類	bivalve *[i-zoo]*
sōkan	相関；相關	correlation *[math]*
sōkan	層間	between layers; inter-layer
sōkan-keisū	相関係数	correlation coefficient *[math]*
sōkan-kikō'gaku	総観気候学	synoptic climatology *[climat]*
sōkan-kishō'gaku	総観気象学	synoptic meteorology *[meteor]*
sōkan-sei	相関性	correlation; correlative property *[math]*
sōkan-shiki	相関式	correlation equation *[math]*
sōkan-zu	相関図	correlation diagram *[math]*
sōka suru	装架する	hang, mounted; mount, hanging *{vb} [eng]*
sokei	素馨；そけい	jasmine (shrub, flower) *[bot]*
sōkei	総計；總計	final total *[math]*
sokei'bu	鼠径部	groin *{n} [anat]*
sōkin-kawase	送金為替	remittance *[econ]*
sōkin-ko'gitte	送金小切手	remittance check *[econ]*
sokki	速記	shorthand; stenography *[comm]*
sokkō	測光；そっこう	photometry *[opt]*
sokkō-ki	測光器	photometer *[eng]*
sokkon	足痕	track *{n} [paleon]*
sokkō-saishoku	測高彩色	hypsometric tinting *{n} [graph]*
sokkō'sei-hiryō	速効性肥料	quick-acting fertilizer *[agr]*
soko	其処；其所；そこ	there *{n} {adv}*
soko	底	bottom *{n}*
sōko	倉庫	storehouse; warehouse *[arch] [ind-eng]*
"	倉庫	magazine *[ord]*
sōkō	草稿	draft *{n} [pr]*
sōkō	走向	running *{n}*
sōkō-jikan	走行時間	transit time *[trans]*
sōkō-kyori-kei	走行距離計	odometer *[eng]*
soko-ni	底荷	ballast *{n} [aero-eng] [nav-arch]*
"	底荷	bottom cargo *[nav-arch]*
soko o miyo	そこを見よ	quod vidae; q.v. *[pr]*
sōkō-sei	走光性	phototaxis (pl -taxes) *[bio]*
so-kotsu'zai	粗骨材	coarse aggregate *[geol]*
-soku (-zoku)	-足	-pair(s) (of footwear) *{suffix}*
-soku	-則	-rule *{suffix}*
sokubaku-denshi	束縛電子	bound electron *[atom-phys] [phys]*
sokubaku-jōken	束縛条件	restraining condition *[math]*
soku'baku suru	束縛する	bind *{vb t}*

sokubaku-undō	束縛運動	constrained motion *[mech]*
sokuchi'gaku	測地学; 測地學	geodesy *[geophys]*
sokuchi-gakusha	測地学者	geodesist (a person) *[geophys]*
sokuchi-sankak(u)kei	測地三角形	geodetic triangle *[geod]*
sokuchi-sen	測地線	geodesic line *[geod]*
soku'do	速度; そくど	rate; speed; velocity (pl -ties) *{n}* *[math] [mech] [phys]*
soku'do	測度	measure *{n}* *[math]*
sokudo-bunpu-soku	速度分布則	velocity distribution, law of *[stat-mech]*
sokudo-henchō	速度変調	velocity modulation *[electr]*
sokudo-kei	速度計	speedometer *[eng]*
sokudo-kōbai	速度勾配	velocity gradient *[fl-mech]*
sokudo-ron	速度論	theory of rate process *[ch] [phys]*
"	速度論	kinetics *[mech]*
sokudo-seigen	速度制限	speed limit *[traffic]*
sokudo-seigyo-sōchi	速度制御装置	speed control apparatus *[eng]*
sokudo-shiki	速度式	rate equation *[math]*
sokudo-suitō	速度水頭	velocity head *[fl-mech]*
sokudo-teisū	速度定数	rate constant *[p-ch]*
soku'en	側縁	lateral edge *[math]*
soku'fusseki	束沸石	stilbite *[miner]*
soku-genshi'ka	側原子価	auxiliary valence *[ch]*
soku'hō	速報; そくほう	bulletin *[pr]*
soku'hō	側方	side *{n}*
sokuhō-junnō	側方順応	lateral adaptation *[photo]*
soku'itsu-sei	束一性	colligative property *[p-ch]*
soku'ji (ni)	即時(に)	immediately *{adv}*
sokuji'gaku	測時学	horology *[sci-t]*
sokuji-shirei	即時指令	immediate command *[comput]*
sokuji-yobi'dashi	即時呼び出し	random access *[comput]*
sokuji-yobi'dashi-ki'oku	即時呼び出し記憶	random access memory; RAM *[comput]*
soku'kaku-ki; sokkakki	測角器	goniometer *[eng]*
soku'men no	側面の	lateral *{adj}* *[math]*
sokumen-zu	側面図	lateral view; side view *{n}* *[graph]*
soku'on	促音	double (long) consonant *[ling]*
soku'on	側音	sidetone *[comm]*
soku'on-pu	促音符; そくおんぷ	double consonant mark; tsu-sound mark *[pr]*
soku'ō suru	即応する; 即應する	conform *{vb}*
soku-ro	側路	bypass *{n}* *[civ-eng]*
soku'ryō	測量	survey (pl surveys) *{n}* *[eng]*
sokuryō suru	測量する	survey; measure *{vb t}*
soku'sa	側鎖; そくさ	side chain *[org-ch]*

soku'seki no	即席の	impromptu; instant; improvized *{adj}*
soku-sen	側線	lateral line *[v-zoo]*
sokushin-gi	測深儀	depth finder *[eng]*
sokushin'jō no	束針状の	bundled-needle state *{adj}*
soku'shin suru	促進する	accelerate; promote *{vb}*
sokushin-zai	促進剤	accelerator; promoter *[ch]*
soku'shoku	測色；そくしょく	colorimetry *[opt]*
sokutai-ha	側帯波；側帶波	sideband wave *[elecmg]*
sokutei	側庭	side yard *[arch] [civ-eng]*
sokutei	測定；そくてい	measurement *[sci-t]*
sokutei'shi	測定子	gage element *[eng]*
soku'tō-kotsu	側頭骨	temporal bone *{n} [anat]*
sōkyoku-men	双曲面；雙曲面	hyperboloid *{n} [math]*
sōkyoku-sen	双曲線	hyperbola (pl -las, -lae) *[math]*
sōkyoku'sen-kansū	双曲線関数	hyperbolic function *[math]*
sōkyoku'shi	双極子	dipole *[elecmg]*
sōkyū-kin	双球菌	Diplococcus *[microbio]*
sōmen	素麺；索麺	noodles; vermicelli *[cook]*
sōmen	創面	injury surface *[bio]*
so'men'gan	粗面岩	trachyte *{n} [petr]*
someru	染める	dye; tint *{vb t}*
sōmoku	草木	plant; herb *{n} [bot]*
sō-mokuroku	総目録	general catalog *[pr]*
somō-shi	梳毛糸	worsted yarn *[tex]*
son; mura	村	village *[geogr]*
son'eki	損益	profit and loss *[econ]*
sōnetsuryō-hozon no hōsoku	総熱量保存の法則	constant heat summation, law of *[p-ch]*
songai	損害	damage *{n} [ind-eng]*
songai-baishō	損害賠償	damage compensation; indemnity *[legal]*
songai-kyō'i	損害脅威	damage threat *[mil]*
sono	其の；その	that; those *{adj}*
so'nō	嗉嚢；そのう	crop *{n} [v-zoo]*
sonota	其の他	other; others *{n}*
son-ritsu	損率	loss factor *[elec]*
sonshi	損紙	broke *{n} [paper]*
son'shitsu	損失	loss *{n}*
son'shitsu-dansei-keisū	損失弾性係数	loss modulus *[mech]*
son'shitsu-dansei-ritsu	損失弾性率	loss modulus *[mech]*
sonshō	損傷	damage *{n} [sci-t]*
sō-nyō	走繞：⻍；そう=にょう	kanji (Chinese character) radical denoting to go *[pr]*
sōnyū	挿入	insertion *[comput] [mol-bio]*

sōnyū	装入；そうにゅう	charging *{n} [eng] [met]*
sōnyū'butsu	装入物	charge; charging material *[met]*
sōnyū'hō-bunrui	挿入法分類	insertion sort *[comput]*
sōnyū-kagō'butsu	挿入化合物	intercalation compound *[ch]*
sōnyū-shikkatsu	挿入失活	insertional inactivation *[microbio]*
sōnyū suru	挿入する	insert *{vb}*
sonzai	存在	existence *{n}*
sonzai-kigō	存在記号	existential quantifier *[logic]*
sonzoku	存続；存續	continuance; persistence *{n}*
sō'on	騒音；騷音	noise *[acous]*
sō'on-kōgai	騒音公害	noise pollution *[ecol]*
sora	空	sky (pl skies) *[astron]*
sora-iro	空色	sky blue (color) *[opt]*
sora-mame	空豆	broad bean; fava; horsebean (legume) *[bot]*
sōran	総覧；總覧	conspectus; comprehensive survey *[pr]*
sore	其れ；それ	that; those *{pron}*
sorenaraba	それならば	then *{conj} [math]*
sori	反り	camber *{n} [eng]*
sori	橇；そり	sled *{n} [eng]*
sōri-kyōsei	相利共生	symbiosis (pl -bioses) *[ecol]*
sō-ritsu	相律	phase rule *[ch]*
sōro	走路	runway *[civ-eng]*
soroban	十露盤；そろばん	abacus (pl -es, abaci) *[math]*
soroe	揃え	justification *[graph]*
soroeru	揃える	arrange in order; unify *{vb t}*
soru	反る；そる	warp; bend backward *{vb i} [mech]*
soru	剃る	shave (as a beard) *{vb}*
sō-rui	藻類；そうるい	algae *[bot]*
sōryō	総量；總量	gross weight *[mech]*
sōryōji'kan	総領事館	consulate general *[arch]*
soryū	粗粒	grit *{n} [mat]*
sōryū	層流；層流	laminar flow *[fl-mech]*
so'ryūshi	素粒子	elementary particle *[part-phys]*
so'ryūshi-butsuri'gaku	素粒子物理学	particle physics *[phys]*
sōsa	走査；そうさ	scan; scanning *{n} [comput] [electr]*
sōsa	操作	operation *[comput]*
sōsa-denshi-kenbi'kyō	走査電子顕微鏡	scanning electron microscope; SEM *[electr]*
sōsa-denshi'kenbikyō-shashin	走査電子顕微鏡写真	scanning electron micrograph *[electr]*
sōsa'gata-tonneru-kenbikyō	走査型トンネル=顕微鏡	scanning tunneling microscope *[electr]*
sōsa-hensū	操作変数	operating variable *[math]*

sōsa'in	操作員	operator (a person) *[comput]*
sōsa-ki	走査器	scanner *[comm] [comput]*
sō'saku suru	創作する	create *{vb}*
sōsa-seigyo-ban	操作制御盤	operator control panel *[comput]*
sōsa-sōchi	走査装置	scanner *[electr]*
sōsa-sokudo	操作速度	scanning rate *[electr]*
sōsa'taku	操作卓	console *[comput]*
sō'satsu	相殺	mutual cancellation *[math]*
sosei	組成；そせい	composition *[ch]*
sosei	塑性；そせい	plasticity *[mech]*
sōsei	双生	twins *[bot]*
sōsei	走性	taxis *{n} [physio]*
sōsei	創成	generating; generation *{n}*
sosei-henkei	塑性変形	plastic deformation *[mech]*
sosei-hizumi	塑性歪	plastic strain *[mech]*
sōsei-ion	双性イオン	ampho-ion *[ch]*
sōsei'ji; futa'go	双生児	twins *[bio]*
sosei-kakō	塑性加工	plastic forming *{n} [eng]*
sosei-nagare	塑性流れ	plastic flow *[phys]*
sosei-nendo	塑性粘度	plastic viscosity *[fl-mech]*
sōseki'un	層積雲	stratocumulus *[meteor]*
sosen	素線	strand; wire strand *{n} [eng]*
sosen	祖先	ancestor(s) *[bio]*
sōsen'kei no	双線形の	bilinear *{adj} [math]*
sōsetsu	総説；綜説	general account; outline *{n} [pr]*
so-sha	粗砂	coarse sand *[petr]*
so'shaku	咀嚼；そしゃく	chewing *{n} [physio]*
soshi	素子	element *[elec]*
sōshi	双糸；雙絲	two-ply yarn *[tex]*
so'shiki	組織；そしき	tissue *[bio]*
"	組織	fabric *[geol]*
"	組織	organization *[ind-eng]*
"	組織	texture *[met]*
"	組織	structure; system *[sci-t]*
soshiki-baiyō	組織培養	tissue culture *[cyt]*
soshiki'gaku	組織学	histology *[anat]*
soshiki-kagaku	組織化学	histochemistry *[ch]*
soshiki'teki-meimei-hō	組織的命名法	nomenclature *[sci-t]*
soshiki'teki na	組織的な	systematic *{adj}*
sōshin	送信	sending; transmission *{n} [electr]*
sōshin-ki	送信機	transmitter *[comm] [electr]*
sōshin suru	送信する	send *{vb t} [comm]*

soshō	訴訟	appeal; litigation *(n) [legal]*
sōshō	双晶；そうしょう	twin *(n) [crys]*
sōshō	相称；相稱	symmetry (pl -tries) *[bio]*
sōshō	創傷	wound *(n) [med]*
sōshō	総称；總稱	generic name *[sci-t]*
sō'shoku	装飾	ornamentation; decor *[furn]*
sōshoku-dōbutsu	草食動物	herbivore *[v-zoo]*
sōshoku'hin	装飾品	accessories (sing accessory) *[cl] [furn]*
sōshoku-katsu'ji	装飾活字	dingbat; ornament *[comput] [pr]*
sō'shoku'sei no	草食性の	herbivorous *(adj) [v-zoo]*
sōshō-mei	総称名	generic name *[sci-t]*
sosogu	注ぐ；灌ぐ	pour (into, onto); sprinkle *(vb)*
sosū	素数	prime number *[math]*
so'sui	疏水	canal *[civ-eng]*
sosui-do	疎水度	hydrophobicity *[ch]*
sosui-ki	疎水基	hydrophobic group *[ch]*
sosui-sei	疎水性	hydrophobic property *[ch]*
sōsui-sei	走水性	hydrotaxis *[bio]*
sōtai-enzan	双対演算	dual operation *[comput]*
sōtai-gosa	相対誤差	relative error *[math]*
sōtai-nendo	相対粘度	relative viscosity *[fl-mech]*
sōtai-ryōshi-kō'ritsu	相対量子効率	relative quantum efficiency *[electr]*
sōtai-sei(-genri)	相対性(原理)	relativity (pl -ties) *[phys]*
sōtai-shitsu-do	相対湿度	relative humidity *[meteor]*
sōtai'teki-bunkō-bunpu	相対的分光分布	relative spectral distribution *[opt]*
so-tanpaku'shitsu	粗蛋白質	crude protein *[bioch]*
sōtei	装丁；装訂；装幀	bookbinding design(ing) *(n) [pr]*
sōtei	装蹄	horseshoeing *(n) [trans]*
sō-ten'i	相転移	phase transition *[phys]*
so'tetsu	蘇鉄；蘇鐵；そてつ	cycad; Japanese sago palm *[bot]*
sotō	粗糖	raw sugar *[bioch]*
sōtō	相等	equality (pl -ties) *[math]*
soto'bari-kōkan	外張鋼管	outer-lined steel pipe *[mat]*
soto-bun'maki	外分巻；外分巻	long shunt *[elec]*
soto-gawa; gai'soku	外側	outer side *(n)*
soto-ita	外板	planking *(n) [nav-arch]*
soto'maki no	外巻きの	revolute *(adj) [bot]*
soto'muki-maku-denryū	外向き膜電流	outward membrane current *[physio]*
soto'muki-nagare	外向き流れ	outward flow *[fl-mech]*
soto no-	外の-	extra- *(prefix)*
soto'nori(-sunpō)	外法(寸法)	outside measurement *[math]*
sō'un	層雲；そううん	stratus (pl strati) *[meteor]*

sōwa-kigō	総和記号	summation notation *[math]*
soyo-kaze	そよ風	breeze *[meteor]*
sōyō-sei	相溶性；相容性	compatibility *[ch]*
sōyō-sei aru	相溶性ある	miscible *{adj} [ch]*
sozai	素材	raw material; copy *{n} [pr]*
"	素材	elemental material *[ind-eng]*
sōzai	総菜；惣菜	everyday dish *[cook]*
sōzoku'nin	相続人；相續人	heir *[legal]*
sōzō(-ryoku)	想像(力)	imagination *[psy]*
sōzōshī	騒騒しい	noisy *{adv} [acous]*
su	州；洲	shoal; bar *{n} [geol]*
su	巣；巢	cavity (casting) (pl -ties) *[met]*
"	巣	nest *{n} [met] [v-zoo]*
su	酢	vinegar *[cook]*
sū	数；數；すう	number; numeral *[math]*
su-basu	酢蓮；すばす	vinegared lotus root *[cook]*
suberasu	滑らす	let slip; let slide *{vb} [mech-eng]*
suberi	滑り	glide; slide; slip *{n} [crys]*
suberi-bō	滑り棒	slide bar *[eng]*
suberi'dome-kakō	滑り止め加工	nonslip finish; slipproof finish *[rub]*
suberi-hiyu	滑り莧；すべりひゆ	purslane; protulaca *[bot]*
suberi-ma'satsu	滑り摩擦	sliding friction *[mech]*
suberi-men	滑り面	slip plane *[crys]*
suberi-ritsu	滑り率	slip factor; slippage *{n} [eng]*
suberi-tsugi'te	滑り継手	slip joint *[eng]*
suberi-yasui	滑り易い	slippery *{adj} [mat]*
suberu	滑る；辷る	slide; glide; slip *{vb i} [mech]*
subete	総て；凡て；全て	all *{n}*
subete no-	総ての-	omni- *{prefix}*
subomeru	窄める	make narrower; pucker *{vb}*
su-boshi	素乾し	drying in the shade *{n}*
sū'chaku suru	枢着する	attach pivotally *{vb} [eng]*
sūchi-hensū	数値変数	variable *{n} [math]*
sūchi-seigyo	数値制御	numerical control *[comput]*
sūchi-seigyo-kei	数値制御系	numerical control system *[cont-syst]*
sūchi-sekibun	数値積分	numerical integration *[math]*
sū-choku'sen	数直線	number line *[math]*
su'dachi	酸橘；すだち	Japanese citron *[bot]*
sudare	簾；すだれ	bamboo (reed, rattan) blind *[furn]*
sudare-ori	簾織	cord fabric; tire fabric *[tex]*
sue	末	descendant *[bio]*
"	末	end; closing; tip *{n}*

sueru	据える；すえる	put in place; install *{vb t}*
sueru	饐える	turn sour; spoil; go bad *{vb i}*
sue-tsuke	据え付け	installation *[ind-eng]*
su-gaki	酢牡蠣；すがき	vinegared oysters *[cook]*
sū'gaku	数学；數學	mathematics *[sci-t]*
sūgaku'teki-kensa	数学的検査	mathematical check *[comput]*
sūgaku'teki-kinō-hō	数学的帰納法	mathematical induction *[logic]*
sugi	杉	cedar; cryptomeria (tree) *[bot]*
sugi-yani	杉脂；すぎやに	cryptomeria resin *[mat]*
sugu'ba-kasa-ha'guruma	直刃傘歯車	straight bevel gear *[mech-eng]*
sū-heikin-bunshi'ryō	数平均分子量	number-average molecular weight *[ch]*
sū-heikin-jūgō'do	数平均重合度	number-average degree of polymerization *[poly-ch]*
sui	錐；すい	cone; pyramid *[math]*
sui	錘	spindle *[tex]*
sui'aen'dōkō	水亜鉛銅鉱	aurichalcite *[miner]*
sui'aen'kō	水亜鉛鉱	wulfenite *[miner]*
sui-atsu	水圧；水壓	water pressure; hydraulic pressure *[mech]*
sui'atsu-ekitai	水圧液体	hydraulic fluid *[mat]*
sui'atsu-kōgaku	水圧工学	hydraulic engineering *{n} [civ-eng]*
sui'atsu no	水圧の	hydraulic *{adj} [eng]*
suibai-hō	水媒法	water medium method *[ch]*
suiban'dokō	水礬土鉱	bauxite *[miner]*
suibun	水分；すいぶん	water content; moisture *[ch] [hyd]*
suibun-gan'ryō	水分含量	moisture content *[agr] [ch]*
suibun-junkan	水分循環	hydrologic cycle *[hyd]*
suibun-kassei	水分活性	water activity *[hyd]*
suibun-ritsu	水分率	regain *{n} [tex]*
suibun-taisha	水分代謝	water metabolism *[bio]*
suichoku-antei'ban	垂直安定板	vertical stabilizer; fin *[aero-eng]*
suichoku-bi'yoku	垂直尾翼	vertical tail *[aero-eng]*
suichoku-jiku	垂直軸	vertical axis *[math]*
suichoku-kan	垂直環	vertical circle *[astron]*
suichoku-men	垂直面	vertical plane *[math]*
suichoku na	垂直な	perpendicular; vertical *{adj} [math]*
suichoku-sei	垂直性	verticality *[math]*
suichoku-sōsa	垂直走査	vertical scanning *{n} [eng]*
suichū	水柱；すいちゅう	water column *[mech-eng]*
suichū-baku'rai	水中爆雷	depth charge *[ord]*
suichū-chō'on-ki	水中聴音器	hydrophone *[eng-acous]*
suichū-jū	水中銃	underwater gun *[ord]*
suichū no	水中の	submerged; in-water *{adj}*

suichū'onpa-tanchi-ki	水中音波探知機	sonar *{n} [eng]*
suichū-seibutsu-baiyō	水中生物培養	aquiculture *[bio]*
suichū'yoku	水中翼	hydrofoil *[nav-arch]*
sui-chū-yu no	水中油の	oil-in-water *{adj} [ch]*
sui'dōkō	翠銅鉱	dioptase *[miner]*
suidō-sui	水道水	tap water *[hyd]*
suigai	水害	flood damage *[ecol]*
suigan	燧岩；すいがん	chert *[geol]*
suigin	水銀；すいぎん	mercury (element)
suigin-chū	水銀柱	mercury column *[phys]*
suigin-ki'atsu-kei	水銀気圧計	mercury barometer *[eng]*
suigin-ki'oku	水銀記憶	mercury storage *[comput]*
suigin-setten-keiden-ki	水銀接点継電器	mercury-wetted contact relay *[electr]*
suigin-tō	水銀灯	mercury vapor lamp *[electr]*
suigyū	水牛	water buffalo (animal) *[v-zoo]*
suihan-ki	炊飯器	rice cooker *[cook] [eng]*
suihei	水平；すいへい	horizontality *{n} [sci-t]*
suihei-bi'yoku	水平尾翼	horizontal tail plane *[aero-eng]*
suihei-da	水平舵	diving rudder *[aero-eng]*
suihei-henkō	水平偏向	horizontal deflection *[electr]*
suihei na	水平な	flat; horizontal *{adj} [math]*
suihei-sen	水平線	horizon *[astron]*
suihei-sōsa	水平走査	horizontal scanning *{n} [eng]*
suihen-chiku	水辺地区	waterfront area *[civ-eng]*
suihi	水簸	elutriation; levigation *[ch-eng]*
sui'i	水位	water level; water table *[hyd]*
sui'i-kei	水位計	water gage *[eng]*
sui'i'teki na	推移的な	transitional; transitive *{adj}*
sui'i'teki na seishitsu	推移的な性質	transitive property *[math]*
suijō-hikōki	水上飛行機	hydroplane *{n} [nav-arch]*
suijō-kajo	穂状花序	spike *{n} [bot]*
sui'jōki	水蒸気	steam; water vapor *{n} [phys]*
sui'jōki-hōwa-shitsu	水蒸気飽和室	humidor *[ch-eng]*
sui'jōki-jōryū	水蒸気蒸留	steam distillation *[ch-eng]*
suijō'tai	錐状体；錐狀體	cone *[hist] [math]*
suijun	水準	level *{n} [comput] [math]*
suijun-ki	水準器	level (tool) *[eng]*
suika	水化	hydration *[ch]*
suika	西瓜；すいか	watermelon (fruit) *[bot]*
suika'butsu	水化物	hydrate *{n} [ch]*
suikai (kasui-bunkai)	水解（加水分解）	hydrolysis *[ch]*
suikai'butsu	水解物	hydrolysate; hydrolyzate *[ch] [geol]*

sui'kasseki	水滑石	brucite *[miner]*
suikei-densen'byō	水系伝染病	water-borne infectious disease *[med]*
suiken	水圏；水圏	hydrosphere *[hyd]*
suikō	遂行	performance; executing *{n}*
suikō-hō	水攻法	waterflooding *{n} [petr]*
suikō-hō	水耕法	hydroponics *[bot]*
suikō-nōjō	水耕農場	hydroponic farm *[agr] [bot]*
suimin	睡眠；すいみん	sleep; slumber *{n} [physio]*
suimin'byō-byōgen'chū	睡眠病病原虫	Trypanosoma *[i-zoo]*
sui'mitsu no	水密の	watertight *{adj} [nav-arch]*
sui'mitsu'tō	水蜜桃	white peach (fruit) *[bot]*
sui-mon	水門	floodgate; sluice; water gate *[civ-eng]*
suimon'gaku	水文学	hydrology *[geophys]*
suimon-kishō'gaku	水文気象学	hydrometeorology *[meteor]*
sui'mono	吸物	clear soup *[cook]*
sui'on-yakusō	水温躍層	thermocline *[geophys]*
suirai	水雷	torpedo *{n} [ord]*
suirei-rōru	水冷ロール	water-cooled roll(er) *[ind-eng]*
suiren	水蓮；すいれん	water lily (flower) *[bot]*
suiren	吹錬；吹錬	blowing *{n} [steel]*
suiri'gaku	水理学	hydraulics *[fl-mech]*
sui'riku-ryōsei no	水陸両性の	amphibious *{adj} [bio]*
suiri-ryoku	推理力	reasoning power *[psy]*
suiro	水路	water channel; watercourse *[civ-eng]*
suiron	推論	deduction; reasoning *{n} [math] [logic]*
suiro-sokuryō'jutsu	水路測量術	hydrography *[geogr] [navig]*
suiryō-kei	水量計	water meter *[eng]*
sui'ryoku	水力	hydraulic power; waterpower *[mech]*
sui'ryoku	推力	thrust; propulsion *{n} [mech]*
suiryoku-hatsu'den	水力発電	hydroelectric power generation *[mech-eng]*
suiryoku-kō'ritsu	水力効率	hydraulic efficiency *[mech]*
sui'ryoku no	水力の	hydraulic *{adj} [eng]*
suiryoku-saikō	水力採鉱	hydraulic mining *{n} [min-eng]*
suisai-enogu	水彩絵具	watercolor *{n} [mat]*
suisan'butsu	水産物	aquatic product; marine product *[ocean]*
suisan'gyō	水産業	fisheries *[ecol]*
suisan'ka	水酸化；すいさんか	hydroxylation *[org-ch]*
suisan'ka-aen	水酸化亜鉛	zinc hydroxide
suisan'ka'butsu	水酸化物	hydroxide *[ch]*
suisan'ka'butsu-en	水酸化物塩	hydroxide salt *[ch]*
suisan'ka-dō	水酸化銅	copper hydroxide
suisan'ka-hakkin	水酸化白金	platinum hydroxide

suisan'ka-kin	水酸化金	gold hydroxide
suisan'ka-namari	水酸化鉛	lead hydroxide
suisan'ka-tetsu	水酸化鉄	iron hydroxide
suisan-ki	水酸基	hydroxyl group *[ch]*
suisan-shiken'jo	水産試験所	fisheries experiment station *[ecol]*
suisei	水星	Mercury (planet) *[astron]*
suisei	彗星；すいせい	comet *[astron]*
suisei-dōbutsu	水生動物；水棲動物	aquatic animal *[bio]*
suisei'en	水生園	aquatic garden *[bio]*
suisei-gan	水成岩	sedimentary rock *[petr]*
suisei-gasu	水性ガス	water gas *[mat]*
suisei no	水生の	aquatic *{adj}* *[bio]*
suisei-seibutsu'gaku	水生生物学	hydrobiology *[bio]*
suisei-shokubutsu	水生植物；水棲植物	aquatic plant *[bot]*
suisei-toryō	水性塗料	water(-base) paint *[mat]*
suisen	水仙	narcissus (pl -s, -cissi, narcissus) (flower) *[bot]*
suisen	水洗	rinsing; washing (with water) *{n}*
suisen	水線	waterline *[geol]*
suisen	垂線	perpendicular *{n}* *[math]*
suisen-benjo	水洗便所	flush toilet *[eng]*
suisha	水車	water turbine; waterwheel *[mech-eng]*
sui'shin	垂心	orthocenter (triangle) *[math]*
suishin-jiku	推進軸	propeller shaft *[mech-eng]*
suishin'ryoku	推進力	propulsion; thrust *{n}* *[mech]*
suishin-yaku	推進薬	propellant *[mat]*
sui-shitsu	水質	water quality *[hyd]*
sui'shitsu-o'daku	水質汚濁	water pollution *[ecol]*
suishō	水床	water bed *[geol]*
suishō	水晶；すいしょう	rock crystal *[miner]*
suishō-bunkō-shashin-ki	水晶分光写真器	quartz spectrograph *[spect]*
suishō-dokei	水晶時計	crystal clock; quartz clock *[horol]*
sui'shoku	水食；水蝕	water erosion *[ecol]*
suishō-shindō'shi	水晶振動子	crystal resonator (or vibrator) *[electr]*
suiso	水素；すいそ	hydrogen (element)
suisō	水槽	water tank *[eng]*
suiso-bakudan	水素爆弾	hydrogen bomb *[ord]*
suiso-ion-nōdo	水素イオン濃度	hydrogen ion concentration *[ch]*
suiso'ka	水素化	hydrogenation *[org-ch]*
suiso'ka-bunkai	水素化分解	hydrocracking *{n}* *[petrol]*
suiso'ka'butsu	水素化物	hydride *[inorg-ch]*
suiso'ka-hiso	水素化砒素	arsenic hydride

suiso'ka-hōso	水素化硼素	boron hydride
suiso'ka-keiso	水素化硅素	silicon hydride
suiso'ka-namari	水素化鉛	lead hydride
suiso'ka-rin	水素化燐	phosphorus hydride
suiso'ka-suzu	水素化錫	tin hydride
suiso-ketsu'gō	水素結合	hydrogen bond *[p-ch]*
suiso-san	水素酸	hydroacid *[ch]*
suiso-shisū	水素指数	hydrogen exponent *[ch]*
suiso-ten'i-jūgō	水素転移重合	hydrogen-transfer polymerization *[poly-ch]*
suiso-tenka (suiten)	水素添加（水添）	hydrogenation *[org-ch]*
suiso-tenka-bunkai	水素添加分解	hydrocracking *{n}* *[petrol]*
suiso-tōryō	水素当量	hydrogen equivalent *[ch]*
sui'tanban	水胆礬	brochantite *[miner]*
suitei	水底	bottom of the sea *[ocean]*
suitei	推定; すいてい	assumption; estimation *[sci-t]*
suitei ni sumu	水底に住む	benthonic *{adj}* *[bio]* *[ocean]*
suitei'ryō	推定量	estimator *[math]*
suiten	水添	hydrogenation *[org-ch]*
suitō	水痘	chicken pox *[med]*
sui'un	水運	water transport *[trans]*
suiwa	水和; すいわ	hydration *[ch]*
suiwa'butsu	水和物	hydrate *[ch]*
suiwa-netsu	水和熱	heat of hydration *[p-ch]*
suiwa-sū	水和数	hydration number *[ch]*
suiwa-sui	水和水	hydrated water; water of hydration *[ch]*
sui-yaku	水薬; 水藥	liquid medicine *[pharm]*
sui'yōbi	水曜日	Wednesday
sui-yōeki	水溶液	aqueous solution; water solution *[ch]*
sui-yoku	水浴	water bath *[ch]*
suiyō'sei-jushi	水溶性樹脂	water-soluble resin *[poly-ch]*
suizai	水剤; 水劑	solution *{n}* *[pharm]*
suizō	膵臓; すいぞう	pancreas (pl pancreata) *{n}* *[anat]*
suizok(u)kan	水族館	aquarium (pl -iums, -ia) *[arch]*
suji	筋; すじ	muscle; tendon *{n}* *[anat]*
sūji	数字; 數字; すうじ	numeric character; numeric *[comput]*
"	数字	digit; number; numeral *[math]*
-suji	-筋	-strand(s) *{suffix}*
sūji'go	数字語	numeric word *[comput]*
sūji-hōtei'shiki	数字方程式	numerical equation *[math]*
sūji'ka	数字化	digitization *[math]*
suji'ko	筋子	salmon roe (en masse) *[cook]*
sukanjūmu	スカンジウム	scandium (element)

suki	隙	aperture; opening *{n}*
suki	鋤	plough *{n} [agr]*
suki'ire	漉入；すき入れ	watermark *[paper]*
suki'ma	隙間；すきま	hiatus (pl -es) *[crys]*
"	隙間	clearance; crevice; gap; aperture *[eng]*
sukima-gēji	隙間ゲージ	thickness gage; clearance gage *[eng]*
suki'ma-kaze	隙間風	draft *{n} [fl-mech]*
suki'yaki	鋤焼き；鋤焼き	sukiyaki (Japanese dish) *[cook]*
sukkiri	すっきり	clean-cut *{adv}*
sukoppu	スコップ	shovel; scoop *{n} [eng]*
sukoshi	少し	small quantity; a bit *{n}*
suku	漉く	manufacture paper *{vb} [paper]*
sukui-kaku	掬い角；抄い角	rake angle *{n} [eng]*
sukunai	少い	few; scanty *{adj} [math]*
"	少い	little; limited in amount *{adj} [math]*
sukunaku'tomo	少くとも	at least *{adv} [math]*
sukū	掬う	scoop (up, out); ladle *{vb t}*
sumasu	済ます；濟ます	finish *{vb t}*
sumasu	澄ます	clarify (as soup) *{vb}*
sumi	炭	charcoal *[mat]*
sumi	墨；墨；すみ	Japanese ink; india ink *[mat]*
sumi-e	墨絵；墨繪；すみえ	black and white painting; sumi (q.v.) painting; ink-brush painting *[graph]*
sumi-ire	墨入れ	inking *{n} [civ-eng] [graph]*
sumi-ishi	隅石	quoin *{n} [arch]*
sumi'komi-kyōsei	住み込み共生	inquilinism *[zoo]*
sumi-niku	隅肉	fillet *{n} [eng]*
sumire	菫；すみれ	violet (flower) *{n} [bot]*
sumire-iro	菫色	violet (color) *{n} [opt]*
sumomo	李；すもも	Japanese plum (fruit) *[bot]*
suna	砂	sand *{n} [geol]*
suna-arashi	砂嵐	sandstorm *[meteor]*
suna-dokei	砂時計	hourglass; sandglass *[horol]*
suna-fuki	砂吹き	sandblasting *{n} [eng]*
sundō-sōchi	寸動装置	inching device *[eng]*
sune	脛	shin (of leg) *{n} [anat]*
su'neri	素練り	mastication *[ch-eng] [rub]*
sunome-shi	簀の目紙	laid paper *[mat]*
su-no-mono	酢の物	vinegared dishes; salads *[cook]*
sunpō	寸法	dimension(s); size *{n} [math] [phys]*
sunpō-antei-sei	寸法安定性	dimensional stability *[mat]*
supana	スパナ	wrench; spanner *{n} [eng]*

supekutoru	スペクトル	spectrum *[math] [phys]*
supin'ha-ryōshi	スピン波量子	magnon *[sol-st]*
supirohētā-moku	スピロヘータ目	Spirochaetales *[microbio]*
suppon	鼈; すっぽん	soft-shelled turtle; snapping turtle (reptile) *[v-zoo]*
sū-retsu	数列	sequence *{n} [math]*
suri-bachi	擂り鉢	mortar of earthenware *[cer] [cook]*
suriko'gi	擂り粉木; すりこぎ	pestle of wood *[cook] [eng]*
sūri-keikaku-hō	数理計画法	mathematical programming *{n} [comput]*
suri'komi	刷り込み	imprinting *{n} [bio]*
suru	為る; する	do; perform *{vb}*
suru	刷る	print *{vb}*
suru	摩る; 擦る	file; rub; chafe *{vb}*
suru	擂る	grind; mash *{vb} [mech-eng]*
surudoi	鋭い; するどい	sharp *{adj} [eng]*
surudo-sa	鋭さ	acuity (pl -ities) *{n} [bio]*
"	鋭さ	sharpness *[opt]*
surume	鯣; するめ	dried cuttlefish; dried squid *[cook]*
surume-ika	鯣烏賊: するめいか	sagittated squid (marine mollusk) *[i-zoo]*
sūryō	数量	quantity (pl -ties) *[math]*
sūryō'ka	数量化	quantification *[math]*
sūryō'shi	数量詞	quantifier *[gram]*
sushi	鮨; 寿司; 壽司; 鮓; すし	vinegared rice with fish *[cook]*
sūshi	数詞	numeral *[gram]*
sū-shiki	数式	numerical expression (or formula) *[math]*
suso-moyō	裾模様	kimono with design at skirt bottom *[cl]*
susu	煤; すす	soot *[mat]*
susugu	濯ぐ; 嗽ぐ; 雪ぐ	rinse; wash *{vb t}*
susuki	薄; すすき	eulalia *[bot]*
susumeru	進める	advance; move forward *{vb t}*
susumi-kaku	進み角	lead angle *[phys]*
susumu	進む	advance; progress *{vb i}*
sutareta	廃れた; 癈れた	obsolete *{adj}*
suteru	捨てる; 棄てる	discard; throw away *{vb t}*
sutoronchūmu	ストロンチウム	strontium (element)
suwaru	座る; 坐る	sit *{vb i}*
su'yaki	素焼; 素焼	biscuit firing *{n} [cer]*
suyaki'ban; suyaki-ita	素焼板	clay plate; porous sheet *[ch]*
suyaki-gama	素焼窯	biscuit kiln *[eng]*
suzu	錫; すず	tin (element)
suzu'biki	錫引き	tinning *{n} [met]*

suzu-gamo	鈴鴨; すずがも	scaup duck (bird) *[v-zoo]*
suzu-gōkin	錫合金	tin alloy *[met]*
suzu-ishi	錫石	cassiterite *[miner]*
suzuki	鱸; すずき	sea bass; Lateolabrax japonicus (fish) *[v-zoo]*
suzume	雀; すずめ	sparrow (bird) *[v-zoo]*
suzume-bachi	雀蜂; すずめばち	wasp (insect) *[i-zoo]*
suzume-ga	雀蛾; すずが	sphinx moth; hawk moth (insect) *[i-zoo]*
suzu-mekki	錫めっき	tin plating *[met]*
suzu'na	菘; 鈴菜; すずな	turnip (vegetable) *[bot]*
suzu'ran	鈴蘭; 鈴蘭	lily of the valley (flower) *[bot]*
suzuri	硯; すずり	inkstone *[comm]*
suzu-sakutai	錫錯体; 錫錯體	tin complex *[ch]*
suzu-san	錫酸	stannic acid
suzushi-sa	涼しさ; すずしさ	coolness (of weather) *{n}* *[meteor]*

T

ta	田；た	field: rice or water field *[agr]*
ta-	多-	multi- *{prefix} [sci-t]*
-taba	-束	-bundle(s) *{suffix}*
ta'bai-seido	多倍精度	multiple-precision *[comput]*
tabako	煙草；たばこ	tobacco (pl -cos) *[bot]*
tabaneru	束ねる	bundle; bunch; sheave *{vb t}*
tabe'mono	食べ物	food; edibles *{n} [bio]*
tabe'rareru	食べられる	is eaten *{vb}*
"	食べられる	is edible *{adj} [cook]*
taberu	食べる	eat *{vb} [physio]*
taberu koto	食べること	eating *{n} [physio]*
tabi	足袋；たび	Japanese socks *[cl]*
tabun	多分	probably *{adv}*
ta'bunsan	多分散	polydispersity *[poly-ch]*
ta'bunsan-kei	多分散系	polydisperse system *[poly-ch]*
ta'bunshi-sei	多分子性	polymolecularness *[org-ch]*
tachi'agari	立ち上り	rise *{n} [rub]*
tachi'agari-bubun	立ち上り部分	rise part *[comm]*
tachi'agari-jikan	立ち上り時間	buildup time; rise time *[comm] [cont-sys]*
tachibana	橘；たちばな	mandarin orange (fruit) *[bot]*
tachi'gare'byō	立枯れ病	wilt (plant disease) *{n} [bot]*
tachi'iri-kinshi-ku'iki	立入り禁止区域	Restricted Area *[civ-eng] [ord]*
tachi'jakō'sō	立ち麝香草	thyme; Thymus vulgaris (herb) *[bot]*
tachi'ki	立木；たちき	timber; living tree *[bot]*
tachi'sagari	立下り	fall *{n} [comm]*
tachi-uo	太刀魚；たちうお	cutlass fish; hairtail; scabbard fish; Trichiurus lepturus (fish) *[v-zoo]*
ta'chō	多調	polytonality *[music]*
tada	只；ただ	free; gratuitous *{n}*
tada-ima	只今；ただいま	now; right now *{adv} {n}*
tadan-ryūdō-sō	多段流動層	multiple fluidized bed *[eng]*
tadan'shiki no	多段式の	multistage *{adj} [eng]*
ta'dantō-kakko-mokuhyō-sai'totsunyū-dan	多弾頭各個目標再=突入弾	multiple independently targetable reentry vehicle; MIRV *[ord]*
tadan-zōfuku-ki	多段増幅器	cascade (multistage) amplifier *[electr]*
tadashī	正しい	right; proper; correct; true *{adj}*
tadashi'gaki	但し書	proviso (pl -sos, -soes) *{n} [legal]* "
"	但し書	provision *{n} [legal]*
tadashi-sa	正しさ	correctness; properness; legality *{n}*

tade	蓼；たで	jointweed; Polygonum hydropiper *[bot]*
tade-ka	蓼科	Polygonaceae *[bot]*
ta-dōshi	他動詞	transitive verb *[gram]*
ta'enki-san	多塩基酸	polybasic acid *[ch]*
taeru	耐える	endure; tolerate *{vb}*
taezu	絶えず	constantly; incessantly *{adv}*
ta'fuhōwa-shibō-san	多不飽和脂肪酸	polyunsaturated fatty acid *[org-ch]*
tagane	鏨；たがね	chisel *{n} [eng]*
tagayasu	耕す；たがやす	till *{vb} [agr]*
tagen-haichi	多元配置	multiway layout *[math]*
tagen-hōtei'shiki	多元方程式	plural equation *[math]*
ta'genshi'ka-genso	多原子価元素	polygen; polyvalent element *[ch]*
tagen'sū	多元数	hypercomplex number; quaternion *[math]*
tageri	田鳧；たげり	lapwing (bird) *[v-zoo]*
tagui	類；類；たぐい	kind; sort; variety; class *{n}*
"	類	match; equal *{n}*
taguru	手繰る	pull hand over hand *{vb t}*
tahen'kei	多辺形；多邊形	polygon *[math]*
tahō-sei	多泡性；多泡性	frothiness *{n} [mat]*
tahō-taiki	多方大気	polytropic atmosphere *[meteor]*
tai	対；對；たい	versus *{prep} [math]*
"	対	compared to; compared with *[math]*
"	対	opposite *{n}*
tai	体；體；たい	field *{n} [math]*
tai	帯；帶；たい	band *{n} [crys]*
"	帯	zone *[geogr]*
tai	鯛	porgy; sea bream (fish) *[v-zoo]*
tai-	対-	anti- *{prefix}*
taiban	胎盤；たいばん	placenta (pl -centas, -centae) *[embryo]*
tai-bōchō-keisū	体膨張係数	coefficient of volume expansion; coefficient of cubic expansion *[thermo]*
tai-bōchō-ritsu	体膨張率	coefficient of volume expansion *[thermo]*
tai'butsu'kyō	対物鏡	objective; object lens *[opt]*
tai'butsu-renzu	対物レンズ	objective lens *[opt]*
taichi-den'atsu	対地電圧；對地電壓	voltage to ground *{n} [elec]*
taichi-sokudo	対地速度	ground speed *[aero-eng]*
taichō-kaku	対頂角	opposite angle *[math]*
taichū-sei	耐虫性	insect resistance *[bot]*
tai-dai'ka no hōsoku	体大化の法則	increase in size, law of *[bio]*
taiden	帯電；帶電	electrification *[elec]*
tai'den'atsu	耐電圧	withstand voltage *[elec]*
taiden-bōshi	帯電防止	static charge prevention *[poly-ch]*

taiden'bōshi-kakō	帯電防止加工	antistatic finish *[eng]*
taiden'bōshi-zai	帯電防止剤	antistatic agent *[mat]*
tai-denkyoku	対電極	counter-electrode *[elec]*
taiden-tai	帯電体	charged body *[ch] [elec]*
tai-eki	体液；體液	bodily fluid *{n} [physio]*
tai'en'ka	耐炎化	flameproofing *{n} [eng]*
tai'enka-shori	耐炎化処理	flame-resisting treatment *[eng]*
tai'en-sei	耐塩性；耐鹽性	salt tolerance *[agr]*
tai'ensui-sei	耐塩水性	salt-water resistance *[mat]*
taifū	台風；颱風；大風	typhoon *[meteor]*
tai'fūka-sei	耐風化性	weathering resistance *[mat]*
taifū no me	台風の目	eye of a hurricane *[meteor]*
taigan-renzu	対眼レンズ	eyepiece; ocular *{n} [opt]*
tai-gen'yu'sei-kō	耐原油性鋼	crude-oil-resistant steel *[steel]*
taigū-meidai	対偶命題	contrapositive *{n} [logic]*
Taihei-yō	太平洋	Pacific Ocean *[geogr]*
taihen	対辺；對邊	opposite side *[math]*
tai'henshoku-sei	耐変色性	colorfastness *{n} [tex]*
taihi	対比	contrast *{n} [opt]*
taihi	堆肥；たいひ	compost *{n} [agr]*
tai'hikkaki-sei	耐引っ掻き性	scratch resistance *[mat]*
taihō	大砲；大砲	cannon; artillery (en masse) *[ord]*
tai'i-hō	対位法	counterpoint *[music]*
tai'iki	帯域；たいいき	zone *[elec]*
tai'iki-haba	帯域巾；帯域幅	bandwidth *[comm]*
tai'in-chō	太陰潮	lunar tide *[ocean]*
tai'in-getsu	太陰月	lunar month; lunation *[astron]*
tai'in-jitsu	太陰日	lunar day *[astron]*
tai'inkyoku	対陰極	anticathode *[electr]*
tai'in-nen	太陰年	lunar year *[astron]*
tai'in-taiyō-reki	太陰太陽暦	lunisolar calendar *[astron]*
tai-ion	対イオン	counterion *[ch]*
taiji	胎児；胎兒；たいじ	fetus *[embryo]*
taiji no	耐磁の	antimagnetic *{adj} [mat]*
taijin-sei	耐塵性	dust resistance *[mat]*
tai'jitsu'shō	対日照	counterglow *{n} [astron]*
taijō-fū	帯状風；帯狀風	zonal wind *[meteor]*
tai'jō no	苔状の	mossy *{adj}*
taijū	体重	body weight *[physio]*
taijū-bakari	体重秤	bathroom scale *[eng]*
taika	退化	degeneration; degeneracy *[bio]*
taika	袋果；たいか	follicle *[bot]*

tai(-ka)'atsu-ki	耐(加)圧器	autoclave *[eng]*
taika-busshitsu	退化物質	degenerate matter *[astron]*
taika'butsu	耐火物	refractory (pl refractories) *{n} [mat]*
tai'kachi-sen'ryaku	対価値戦略	countervalue strategy *[mil]*
taika-do	耐火度	refractoriness *{n} [cer]*
tai'kajū-nō	耐荷重能	load-carrying capacity *[trans]*
taikaku	体格；たいかく	build (of body); physique *[med]*
taikaku	対格	accusative case *[gram]*
taikaku-sen	対角線	diagonal line *[math]*
taikaku'sen-gyō'retsu	対角線行列	diagonal matrix *[math]*
taika-nendo	耐火粘土	fireclay; refractory clay *[geol]*
taikan-kikō	体感気候	sensible climate *[climat]*
taikan-sei	耐乾性	drought resistance *[agr]*
taikan-sei	耐寒性	cold resistance *[agr] [mat]*
taika-sei	耐火性	fireproofness; fire resistance *[mat]*
taikei	大系	outline (of history) *{n} [pr]*
taikei	大計	long-range plan *{n}*
taikei	体系	system *[sci-t]*
taikei	体形	body shape *[bio]*
taiken	大圏；大圏	great circle *[geod]*
taiken	体験；體驗	experience *{n} [psy]*
taiken-kōro	大圏航路	great circle route *[navig]*
taiki	大気；大氣；たいき	atmosphere *[meteor]*
taiki	待機	standby *{n} [comput]*
(tai-)ki'atsu	(大)気圧	atmospheric pressure; barometric pressure *[phys]*
taiki-chō	大気潮	atmospheric tide *[meteor]*
taiki'gai-taiyō-hōsha	大気外太陽放射	extraterrestrial radiation *[astron]*
taiki-jikan	待機時間	standby time *[comput]*
taiki-jōken	大気条件	ambient conditions *[phys]*
taiki-ken	大気圏	atmosphere, the *[meteor]*
taiki'kō	大気光	airglow *{n} [geophys]*
taiki no	大気の	atmospheric *{adj} [meteor]*
taiki-o'sen	大気汚染	air pollution *[ecol]*
taiki-shitsu	大気質	air quality *[meteor]*
taiki-sokudo	対気速度	airspeed *[aero-eng]*
taiko	太古	ancient times; prehistory *[archeo]*
taiko	太鼓；たいこ	drum *{n} [music]*
taikō	大綱	general rules; basic principles; outline
taikō	対向；對向	opposition *{n}*
taikō	対抗	antagonism *[bio]*
taikō	退行	regression *[psy]*

taikō	退行	retrograding *{n} [med]*
taikō-antei-sei	耐光安定性	light-resisting stability *[opt]*
tai(kō)-denkyoku	対(向)電極	counter-electrode *[elec]*
taikō-hannō	対向反応	opposing reaction *[ch]*
taikō-kenrō-do	耐光堅牢度	colorfastness to light *[tex]*
tai'kōon-sanka-sei	耐高温酸化性	high-temperature-oxidation resistance *[ch]*
taikō-sei	耐光性	light resistance *[electr] [mat]*
taikō-sei	耐航性	seaworthiness *[nav-arch]*
taikō-sei	耐候性	weatherability; weather resistance *[mat]*
taikō-shasen	対向車線	opposing lane *[traffic]*
taikō suru	対抗する	countervail *{vb}*
taikyū-do	耐久度	permanence *[mat]*
taikyū-gendo	耐久限度	endurance limit; fatigue limit *[mech]*
taikyū-sei	耐久性	durability *[mat]*
taikyū-shiken	耐久試験	endurance test *[ind-eng]*
taikyū(-shōhi)-zai	耐久(消費)財	durable goods *[econ]*
taima	大麻	hemp; Indian hemp *[bot]*
taimai	玳瑁; 瑇瑁; たいまい	hawksbill turtle (reptile) *[v-zoo]*
tai'mamō-chū-tetsu	耐摩耗鋳鉄	abrasion-resistant cast iron *[met]*
tai'mamō-sei	耐摩耗性	abrasion resistance *[mat]*
taimatsu	松明; 炬火	torch *{n} [opt]*
tai'nai-jusei	体内受精	internal fertilization *[physio]*
tai'nen-sei	耐燃性	burning resistance *[mat]*
tai-netsu	体熱	body heat *[bio]*
tai'netsu	耐熱; たいねつ	heatproof *[mat]*
tai'netsu-do	耐熱度	heat resistance (degree) *[mat]*
tai-netsu'kiretsu-sei	耐熱亀裂性	thermal cracking resistance *[mat]*
tai'netsu-sei	耐熱性	heat resisting property *[mat]*
tai'netsu-shokki	耐熱食器	heat-resistant food utensil *[cer] [cook]*
tai'nin no hōsoku	耐忍の法則	toleration, law of *[bio]*
ta'inkan	多員環	many-membered ring *[org-ch]*
tai'ō	対応; 對應	correspondence *{n} [math] [phys]*
taiō-hen	対応辺	corresponding sides *[math]*
taiō-jōtai no genri	対応状態の原理	corresponding states, law of *[ch]*
tai'on-kei	体温計	clinical thermometer *[med]*
taiō suru	対応する	corresponding *{adj} [math]*
taiō suru	対応する	correspond to *{vb}*
tairagi	玉珧; たいらぎ	fan shell; pen shell; razor shell *[i-zoo]*
taira na	平な	flat *{adj} [math]*
tai'riku	大陸; たいりく	continent *[geogr]*
tairiku-bunsui'kai	大陸分水界	continental divide *[geol]*

tairiku-dana	大陸棚	continental shelf *[geol]*
tairiku-hyō'i	大陸漂移	continental drift *[geol]*
tairiku'kan-dandō-dan	大陸間弾道弾	intercontinental ballistic missile; ICBM
tairiku'kan no	大陸間の	intercontinental *{adj} [geogr]*
tairiku-kidan	大陸気団	continental air mass *[meteor]*
tairiku-kikō	大陸気候	continental climate *[climat]*
tairiku-shamen	大陸斜面	continental slope *[geol]*
tai'ritsu	対立；對立	opposition; contrast *{n}*
tairitsu-iden'shi	対立遺伝子	allele *[gen]*
tairitsu-inshi	対立因子	allele *[gen]*
tairitsu-ka'setsu	対立仮説	alternative hypothesis *[math]*
tai-rui	苔類；たいるい	liverwort; Hepaticae *[bot]*
tai-ryoku	体力；體力	physical fitness *[med]*
"	体力	body strength *[physio]*
tai-ryoku	耐力	yield strength *[mech] [met]*
"	耐力	proof stress *[met]*
tai'ryoku-heki	耐力壁	load-bearing wall *[arch]*
tai'ryoku-kyōdo	耐力強度	yield strength *[mech]*
tairyō-seisan	大量生産	mass production *[ind-eng]*
tairyū	対流；たいりゅう	convection *[ocean] [phys]*
tairyū-jikan	滞留時間	residence time *[ch-eng] [nucleo]*
"	滞留時間	retention time *[ch-eng] [poly-ch]*
tai-ryūkai-fushoku-sei	耐粒界腐食性	intergranular corrosion resistance *[met]*
tairyū-ken	対流圏	troposphere *[meteor]*
tairyū-saibō	対流細胞	convection cell *[meteor]*
tai-saibō	体細胞；體細胞	somatic cell *[bio] [cyt]*
tai'saibō-kōzatsu	体細胞交雑	somatic hybridization *[gen]*
tai'sanka-sei	耐酸化性	oxidation resistance *[mat]*
tai'san-sei	耐酸性	acidproofness; acid resistance *[mat]*
tai'san-sei no	耐酸性の	acid-resistant *{adj} [mat]*
taisei	大勢	general trend *{n}*
taisei	体制	organization *[bot]*
taisei	耐性	tolerance; resistance *[agr] [med]*
taisei	胎生	viviparity *[physio]*
taisei	態勢	attitude *[psy]*
taisei'gaku	胎生学	embryology *[bio]*
taisei-kankaku	体性感覚	somatic sensation *[physio]*
taisei no	対生の	opposite *{adj} [bot]*
taisei-shinkei-kei	体性神経系	somatic nervous system *[physio]*
taisei-shushi	胎生種子	viviparous seed *[bot]*
Taisei-yō	大西洋	Atlantic Ocean *[geogr]*
tai'seki	体積；たいせき	volume *[math]*

tai'seki	堆石；たいせき	moraine *[geol]*
tai'seki	滞積；滯積	deposit; silting *{n} [civ-eng]*
taiseki-asshuku-keisū	体積圧縮係数	coefficient of volume compressibility *[mech-eng]*
taiseki-bōchō-keisū	体積膨張係数	coefficient of cubical expansion *[thermo]*
taiseki'butsu	堆積物	sediment; deposit *{n} [geol]*
taiseki'chi	対蹠地	antipodes *[geod]*
taiseki-dansei-keisū	体積弾性係数	bulk modulus *[mech]*
taiseki-dansei-ritsu	体積弾性率	bulk modulus *[mech]*
taiseki'gaku	堆積学	sedimentology *[geol]*
taiseki-gan	堆積岩	sedimentary rock *[petr]*
taiseki-hyaku'bun-ritsu	体積百分率	volume percent(age) *[math]*
taiseki-kei	体積計	volumeter *[eng]*
taiseki-koyū-teikō	体積固有抵抗	volume resistivity; volume specific resistance *[elec]*
taiseki-nagare	体積流れ	volume flow *[fl-mech]*
taiseki-nensei	体積粘性	volume viscosity *[fl-mech]*
taiseki-ritsu	体積率	volume fraction *[math]*
taiseki-ryoku	体積力	volume force *[phys]*
taiseki suru	堆積する	accumulate; heap *{vb}*
taiseki-ten	対蹠点	antipodal point *[bot]*
tai'sen-saku'sen	対潜作戦	antisubmarine warfare *[mil]*
taisetsu	体節	segment *[zoo]*
taisetsu na	大切な	important; weighty; significant *{adj}*
tai'setsu-tsuyo-sa	耐折強さ	folding endurance; folding strength *[paper]*
taisha	代謝；たいしゃ	metabolism *[physio]*
taisha	堆砂	sediment (silting) *{n} [geol]*
taisha-busshitsu	代謝物質	metabolite *[bioch]*
taisha'kaiten-sū	代謝回転数	turnover number (enzyme) *[bioch]*
taisha-kikkō-busshitsu	代謝拮抗物質	antimetabolite *{n} [pharm]*
taisha-san'butsu	代謝産物	metabolic product *[bioch]*
taishi'kan	大使館	embassy *[arch]*
tai-shin	耐振	shake resistance *{n} [mat]*
tai-shin	耐震；たいしん	shock resistance *{n} [eng]*
tai'shin-kō	耐震鋼	shock-resistant steel *[steel]*
taishin-kōshi	体心格子	body-centered lattice *[crys]*
taishin-rippō-kōshi	体心立方格子	body-centered cubic lattice *[crys]*
taishin-rippō-shō	体心立方晶	body-centered cubic crystal *[crys]*
taishin-sekkei	耐震設計	earthquake-resistant design *[arch]*
tai-shitsu	体質	body type *[physio]*
tai'shitsu-ganryō	体質顔料	body; extender pigment *{n} [ch] [mat]*

tai'shiwa-sei	耐皺性	creaseproofness *{n} [mat]*
tai'shiwa'sei no	耐皺性の	creaseproof *{adj} [mat]*
Taishō	大正；たいしょう	Taishō (era name) (1912-1926)
taishō	対称；對稱	symmetry (pl -tries) *[math]*
taishō	対象	object *{n} [math] [psy]*
"	対象	subject; target *{n} [sci-t]*
taishō	対照	control *{n} [bio] [cont-sys]*
"	対照	contrast *{n} [math]*
"	対照	reference *{n} [opt] [sci-t]*
taishō-bunpu	対称分布	symmetric distribution *[math]*
taishō-eki	対照液	contrast solution; control solution; reference solution *[an-ch]*
tai'shōgeki-sei	耐衝撃性	impact resistance; shock resistance *[mat]*
taishō-gengo	対象言語	target language *[ling]*
taishō-hyō	対照表	cross-reference table *[pr]*
taishō-jiku	対称軸	symmetry axis *[math] [mech]*
tai'shōka-sei	耐消化性	slaking resistance *[cer]*
tai-shoku	体色；體色	body color *[zoo]*
tai'shoku	退色；たいしょく	fading *{n} [tex]*
tai'shoku-sei	耐食性	corrosion resistance *[mat]*
taishō-men	対称面	symmetry plane *[opt]*
taishō-sei	対称性	symmetry; symmetric property *[ch] [math]*
taishō-sei	対掌性	chirality *[ch]*
taishō-shu	対称種	symmetry species *[spect]*
taishō'tai	対掌体	antipode *[ch]*
taishō'teki na	対称的な	symmetric *{adj} [math]*
taishū-sei	耐銹性	rust resistance *[mat] [met]*
taisō	体操；たいそう	exercise (physical fitness) *{n}*
taisū	対数；たいすう	logarithm *[math]*
taisū-gensui-ritsu	対数減衰率	logarithmic decrement *[math] [phys]*
taisū-heikin	対数平均；對數平均	logarithmic mean *[math]*
taisui-sei	耐水性	water resistance; waterproofness *[mat]*
taisui'sō	帯水層；帶水層	aquifer *[geol]*
taisū-kyūsū-soku	対数級数則	logarithmic series, law of *[bio]*
taisū-me'mori	対数目盛	logarithmic scale *[math]*
taisū-nendo	対数粘度	logarithmic viscosity *[fl-mech]*
taisū no hōsoku	大数の法則	large numbers, law of *[math]*
taisū-seiki-soku	対数正規則	log normal distribution, law of *[bio]*
taisū-zō'shoku	対数増殖	logarithmic growth *[microbio]*
taitō na	対等な	equivalent *{adj} [logic]*
taitō-sei	耐冬性	winter hardiness *[agr]*
taitō-sei	耐凍性	freezing hardiness *[agr]*

taiwa	対話；對話；たいわ	conversation; dialog *[comm]*
"	対話	interaction *[comput]*
taiwa'gata-zukei	対話型図形	interactive graphics *[comput]*
taiya	タイヤ	tire (of a vehicle) *{n} [eng]*
tai'yakuhin-sei	耐薬品性	chemicals resistance *[mat]*
tai'yō	太陽；たいよう	sun *{n} [astron]*
taiyō-chō	太陽鳥	sunbird (bird) *[v-zoo]*
taiyō-chō	太陽潮	solar tide *[ocean]*
taiyō-chū	太陽柱	sun pillar *[meteor]*
taiyō-denchi	太陽電池	solar cell *[electr]*
taiyō-fū	太陽風	solar wind *[geophys]*
taiyō-ha'guruma	太陽歯車	sun wheel *[eng]*
taiyō-ji	太陽時	solar time *[astron]*
taiyō-jishin'gaku	太陽地震学	helioseismology *[astron]*
taiyō-katsudō	太陽活動	solar activity *[astron]*
taiyō-kei	太陽系	solar system *[astron]*
taiyō-ken	太陽圏	heliosphere *[geophys]*
taiyō-koku'ten	太陽黒点	sunspot *[astron] [meteor]*
taiyō-kōsen	太陽光線	sunlight *[astron]*
taiyō-kōten	太陽向点	solar apex *[astron]*
taiyō'kyō	太陽鏡	helioscope; siderostat *[opt]*
taiyō'men-baku'hatsu	太陽面爆発	solar flare *[astron] [meteor]*
taiyō-netsu-ryoku	太陽熱力	solar thermal power *[phys]*
taiyō-ro	太陽炉；太陽爐	solar furnace *[eng]*
taiyō-shisa	太陽視差	solar parallax *[astron]*
taiyō-teisū	太陽定数	solar constant *[meteor]*
taiyō-tō	太陽灯；太陽燈	heat lamp; sunlamp *[elec]*
tai'yu no	耐油の	oilproof; oil resistant *{adj} [mat]*
tai'yu-sei	耐油性	oilproofness *{n} [mat]*
taiza	胎座	placenta (pl -centas, -centae) *[embryo]*
ta'jigen	多次元；たじげん	multidimensional(ness) *{n} [math]*
ta'jigen-gengo	多次元言語	multidimensional language *[comput]*
ta'jiku-oshi'dashi-ki	多軸押出機	multiscrew extruder *[eng]*
tajū-	多重-	multi- *{prefix} [sci-t]*
tajū-do	多重度；たじゅうど	multiplicity *[math]*
tajū-ketsu'gō	多重結合	multiple bond *[ch]*
tajū-kō	多重項	multiplet (term) *[quant-mech]*
ta'jūkyoku-hōsha	多重極放射	multipole radiation *[phys]*
tajū-sanran	多重散乱	multiple scattering *{n} [phys]*
tajū-sei	多重星	multiple star *[astron]*
tajū-sen	多重線	multiplet (line) *[spect]*
tajū-sōsa	多重操作	multiplex operation *[comput]*

tajū-zō	多重像	ghost image *[electr]*
taka	鷹; たか	hawk (bird) *[v-zoo]*
taka	多価; 多價	polyvalence *{n} [ch]*
taka'ashi-gani	高脚蟹; たかあし蟹	giant Japanese spider crab *[i-zoo]*
taka-denkai'shitsu	多価電解質	polyelectrolyte *[org-ch]*
taka-fenōru	多価フェノール	polyhydric phenol; polyphenol *[org-ch]*
takai	高い	expensive; costly *{adj} [econ]*
"	高い	high; tall; elevated; superior *{adj}*
taka'ka	多花果; たかか	syncarp; multiple fruit *[bot]*
taka-kōtai	多価抗体; 多價抗體	multivalent antibody; polyvalent antibody *[bio]*
ta'kaku'chū	多角柱	prism *[math]*
ta'kaku-dantō	多核弾頭	multiple warhead *[ord]*
ta'kak(u)kei	多角形	polygon *[math]*
ta'kaku-saibō	多核細胞	multinucleate cell *[cyt]*
ta'kaku-sakutai	多核錯体	polynuclear complex *[org-ch]*
ta'kaku-sui	多角錐; たかくすい	polygonal cone *[math]*
ta'kaku-zu	多角図	polygonal graph *[graph]*
taka-maki-e	高蒔絵; たかまきえ	gilt lacquer(ware), embossed *[furn]*
takameru	高める	raise higher; heighten; promote *{vb}*
ta'kannō no	多官能の	multifunctional *{adj} [org-ch]*
ta'kannō-sei	多官能性	polyfunctionality *[org-ch]*
taka-no-me-ishi	鷹眼石	hawk's eye *[miner]*
takan'shiki	多環式	polycyclic *{n} [org-ch]*
takan'shiki no	多環式の	polycyclic *{adj} [org-ch]*
takan'shiki-tanka'suiso	多環式炭化水素	polycyclic hydrocarbon; polynuclear hydrocarbon *[org-ch]*
takara-gai	宝貝; 寶貝	cowrie (shellfish) *[i-zoo]*
taka-sa	高さ	altitude; height *[eng] [geogr] [math]*
taka-shio	高潮	floodtide; high tide *[ocean]*
take	丈; たけ	height (of body); tallness *{n} [anat]*
take	竹	bamboo *[bot]*
take	茸; たけ	mushroom *{n} [bot]*
takei-henka	多形変化	polymorphic change *[bio] [crys]*
takei-sei	多形性	polymorphism *[bio] [crys]*
takei-shu	多型種	polytypic species *[ch]*
ta'keitai-sei	多形態性	pleomorphism *[crys]*
take-kanmuri	竹冠: ⺮	kanji (q.v.) radical for bamboo *[pr]*
takenoko	筍; 竹の子	bamboo shoot *[bot]*
takenoko-bane	竹の子ばね	volute spring *[mech-eng]*
takenoko-gai	筍貝	anger shell; terebra (shellfish) *[i-zoo]*
ta'kesshō'sei-koku'en	多結晶性黒鉛	polycrystalline graphite *[miner]*

ta'kesshō-shitsu	多結晶質	polycrystalline substance *{n} [crys]*
take-zaiku	竹細工	bamboo-crafted ware *[furn]*
taki	滝；瀧；たき	waterfall; cataract *[hyd]*
taki'komi-gohan	炊込御飯	rice steamed with ingredients *[cook]*
ta'kitō-kikan	多気筒機関	multiple-cylinder engine *[mech-eng]*
takkō	卓効	excellent effect *{n}*
tako; benchi	胼胝	callus *[med]*
tako	蛸；章魚；たこ	octopus (pl -puses, -pi) *[i-zoo]*
tako	凧；たこ	kite (toy) *[eng]*
ta'kōdo-kei	多孔度計	porosimeter *[eng]*
ta'kokuseki-gaisha	多国籍会社	multinational company *[econ]*
takō-sei	多孔性	porosity *[phys]*
takō-shiki	多項式	multinomial; polynomial *{n} [math]*
takō-shitsu	多孔質	porous matter *[mat]*
taku'an(-zuke)	沢庵(漬け)	pickle of daikon (q.v.) *[cook]*
taku'etsu-fū	卓越風	prevailing wind *[meteor]*
takujō'gata-kensa	卓上型検査	desk checking *{n} [comput]*
takujō-kaizan	卓状海山	guyot *[geol]*
taku'men	卓面；たくめん	pinacoid *{n} [crys]*
ta'kyoku-hōsha	多極放射	multipole radiation *[phys]*
ta'kyoku-kan	多極管	multielectrode tube *[electr]*
tama	玉；球；珠；たま	bead; ball; gem; sphere *{n} [mat]*
tama	弾（弾丸）；彈	bullet; shot; shell *{n} [ord]*
tama'gata-ben	玉形弁；玉形瓣	globe valve *[mech-eng]*
tamago	卵；たまご	egg *[cook] [cyt]*
tamago-dake	卵茸	egg mushroom *[bot]*
tamago-dōfu	卵豆腐	steamed egg custard *[cook]*
tamago-senbetsu-ki	卵選別機	oometer *[agr] [eng]*
tamago-toji	卵綴じ；たまごとじ	egg-drop soup *[cook]*
tamago-yaki	卵焼き	eggroll *[cook]*
tama-jiku'uke	玉軸受	ball-and-socket joint *[mech-eng]*
tama-mushi	玉虫	buprestid (beetle) *[i-zoo]*
tama'mushi-iro	玉虫色	iridescent color *[opt]*
tama'mushi(-zome)	玉虫(染め)	fluorescent dyeing *{n} [tex]*
tama-negi	玉葱；たまねぎ	onion (vegetable) *[bot]*
tamari-jōyu	溜まり醬油	refined, concentrated soy sauce *[cook]*
tama-tsugi'te	玉継手；玉繼手	ball-and-socket joint *[mech-eng]*
tamen-hatsu'gen-sei	多面発現性	pleiotropy *[gen]*
tamen'tai	多面体	polyhedron *[math]*
tamen'teki-kōka	多面的効果	pleiotropic effect *[gen]*
tameru	溜める；貯める	accumulate; save; amass *{vb t}*
tameru	撓める	bend; train *{vb t}*

tameshi-yaki	試し焼き	proof print *{n} [photo]*
tamesu	試す；ためす	test; try; sample *{vb} [ind-eng]*
ta'mokuteki-eisei	多目的衛星	multipurpose satellite *[aero-eng]*
tan	炭	char *{n} [mat]*
tan	痰	phlegm; sputum *[physio]*
tan	端；たん	edge *{n} [math]*
tana	棚	shelf *{n} [furn]*
tana-dan	棚段	tray *{n} [an-ch]*
tana'dan-tō	棚段塔	plate column *[ch-eng]*
tanago	鰱；たなご	Japanese bitterling (fish) *[v-zoo]*
tan'an	炭安	abbreviation for ammonium carbonate *[ch]*
tan'antei-kairo	単安定回路	monostable circuit *[electr]*
tanban	単板；單板	veneer *{n} [mat]*
tanban	胆礬；膽礬	blue vitriol; cupric sulfate *[mat]*
"	胆礬；たんばん	chalcanthite *[miner]*
tanbo	田圃；たんぼ	field: rice or water field *[agr]*
tanbun	単文；單文	simple sentence *[gram]*
tan'bunsan-sei	単分散性	monodispersing property *[poly-ch]*
tan'bunsan-kei	単分散系	monodisperse system *[poly-ch]*
tan-bunshi	単分子；たんぶんし	single molecule *[ch]*
tan'bunshi-hannō	単分子反応	unimolecular reaction *[ch]*
tan'bunshi-maku	単分子膜	monolayer; monomolecular film *[p-ch]*
tan'bunshi-sa	単分子鎖	monomolecular chain *[ch]*
tan'bunshi-sō	単分子層	monolayer *[p-ch]*
tanchi	探知	ascertainment; detection *{n}*
tanchō-sei	単調性	monotonic property *[sci-t]*
tanchō-zuru	丹頂鶴	Japanese crane (bird) *[v-zoo]*
tan'deki	耽溺；たんでき	addiction *[med]*
tan'deki-yaku	耽溺薬	addictive drug *{n} [pharm]*
tanden	炭田	coalfield *[min-eng]*
tandoku-hatsu'mei	単独発明；單獨發明	sole invention *[pat]*
tandoku-jūgō	単独重合	homopolymerization *[poly-ch]*
tandoku-jūgō'tai	単独重合体	homopolymer *[poly-ch]*
tane	種；たね	pit, stone (of fruit); seed *{n} [bot]*
tane-ha'guruma	種歯車	master gear *[mech-eng]*
tane-kesshō	種結晶	seed crystal *[crys]*
tane-kōji	種麹；たねこうじ	seed malt *[cook]*
tane'nashi-budō	種無し葡萄	seedless grape (fruit) *[bot]*
ta'nen'sei no	多年生の	perennial *{adj} [bot]*
ta'nensei-shokubutsu	多年生植物	perennial plant; herb *[bot]*
ta'nensei-sōhon	多年生草本	perennial herb *[bot]*
tane-shō	種晶	seed crystal *[crys]*

tan'fu'hōwa-shibō-san	単不飽和脂肪酸	monounsaturated fatty acid *[bioch]*
tan-furiko	単振子；たんふりこ	simple pendulum *[mech]*
tan'genshi-bunshi	単原子分子	monoatomic molecule *[ch]*
tan'genshi no	単原子の	monoatomic *{adj}* *[ch]*
tango	単語；たんご	word *[comput]* *[gram]*
tangusuten	タングステン	tungsten (element)
tanhen	単変；單變	monotropy *[fl-mech]*
tan'hōkō'sei-kairo-mō	単方向性回路網	unilateral network *[comm]*
tani	谷；たに	rill *{n}* *[astron]*
"	谷	valley *[geogr]* *[geol]*
"	谷	trough (of pressure) *[meteor]*
tan'i	単位；單位；たんい	unit *[phys]*
tan'i-gen	単位元	identity element *[math]*
tani-giri	谷霧；たにぎり	valley fog *[meteor]*
tan'i-hizumi	単位歪	unit strain *[mech]*
tan'i-ka'kaku-hyōji	単位価格表示	unit pricing *{n}* *[econ]*
tani-kaze	谷風；たにかぜ	valley breeze; valley wind *[meteor]*
tan'i-kōshi	単位格子	unit cell *[crys]*
tan'i-ō'ryoku	単位応力	unit stress *[mech]*
tan'i-seishoku	単為生殖	parthenogenesis *[i-zoo]*
tani-sen	谷線	trough line *[meteor]*
ta'nishi	田螺；たにし	mud snail; pond snail *[i-zoo]*
tan'itsu no	単一の	simple; single *{adj}* *[math]*
tan'itsu-hannō	単一反応	single reaction *[ch]*
tan'itsu-sei	単一性	unity (pl -ties) *{n}*
tan'itsu-shoku	単一色	pure color *[opt]*
tan-jiku	短軸；たんじく	minor axis; short axis *[math]*
tan'jiku-kesshō	単軸結晶	uniaxial crystal *[crys]*
tan'jitsu-shokubutsu	短日植物	short-day plant *[bot]*
tanjō	誕生	birth *{n}* *[bio]*
tan-jōryū	単蒸留；單蒸溜	simple distillation *[ch]*
tanjū	膽汁；たんじゅう	bile; gall *{n}* *[physio]*
tanjū-maki	単重巻；單重卷	simplex winding *{n}* *[elec]*
tan'jun	単純；たんじゅん	simplicity *{n}*
tanjun-kikai	単純機械	simple machine *[mech-eng]*
tanjun-kōshi	単純格子	primitive lattice; simple lattice *[crys]*
tanjun na	単純な	simple; plain *{adj}*
tanjun-shi'shitsu	単純脂質	simple lipid *[bioch]*
tanjū-san	胆汁酸	bile acid *[bioch]*
tanka	単価；単價	unit cost *[econ]* *[ind-eng]*
tanka	担架；擔架	stretcher *[med]*
tanka	炭化；たんか	carbonization *[ch]* *[geoch]*

tanka'butsu	炭化物	carbide *[inorg-ch]*
tanka-hōso	炭化硼素	boron carbide
tanka-keiso	炭化硅素	silicon carbide
tan'kaku-rui	単殻類；單殼類	univalve *[i-zoo]*
tan'kaku-saku'tai	単核錯体	mononuclear complex *[ch]*
tan'kannō no	単官能の	monofunctional *{adj}* *[ch]*
tankan'shiki-kagō'butsu	単環式化合物	monocyclic compound; mononuclear compound *[org-ch]*
tanka'suiso	炭化水素	hydrocarbon *[org-ch]*
tanka'suiso-ki	炭化水素基	hydrocarbon radical *[org-ch]*
tanka'suiso-shika-kin	炭化水素資化菌	hydrocarbon-assimilating bacteria *[microbio]*
tanka-tetsu	炭化鉄；炭化鐵	iron carbide
tan-keisha(-sō)	単傾斜(層)	monocline *{n}* *[geol]*
tanken	探検；探檢	exploration; expedition *{n}*
tanken	短剣；短劍	short sword; dagger; dirk *[ord]*
tan-kesshō	単結晶	single crystal; simple crystal *[crys]*
tan-ketsu'gō	単結合	single bond *[ch]*
tanki-yohō	短期予報	short-range forecast *{n}* *[meteor]*
tanko	淡湖	freshwater lake *[hyd]*
tankō	単項	monadic(ness) *{n}* *[comput]*
tankō	炭坑；たんこう	coal mine; coalpit *[min-eng]*
tankō	炭鉱；炭鑛	colliery (pl -lieries) *[min-eng]*
tankō	探鉱	prospecting *{n}* *[min-eng]*
tankō	鍛工	metalworker; smith (a person) *[met]*
tankō	鍛鋼	forged steel *[steel]*
tankō-enzan	単項演算	monadic (or unary) operation *[comput]*
tankō-enzan'shi	単項演算子	monadic operator; unary operator *[comput]*
tankō'ginkō	淡紅銀鉱	proustite *[miner]*
tankō-moku	単孔目	Monotremata *[v-zoo]*
tankō-shiki	単項式	monomial *{n}* *[math]*
tankō-tsū'shin	単向通信	one-way communication; simplex communication *[comm]*
tan'kyoku-hatsu'den-ki	単極発電機	homopolar generator *[electr]*
tan'kyoku'sei no	単極性の	unipolar *{adj}* *[elec]*
tan'kyori-ryoku	短距離力	short-range force *[phys]*
tankyū	探究；たんきゅう	research; investigation; inquiry *{n}*
tankyū-kin	単球菌	monococcus (pl -cocci) *[microbio]*
tan'matsu-riyō'sha	端末利用者	terminal user *[comput]*
tan'matsu(-sōchi)	端末(装置)	terminal *{n}* *[elec]*
tanmen-zu	端面図；端面圖	end view *{n}* *[graph]*
tan-nankō	単軟膏	simple ointment *{n}* *[pharm]*

tan naru	単なる	mere; simple; sheer *{adj}*
tan ni	単に	merely; simply; solely; only *{adv}*
tannō	胆嚢；膽嚢	gall bladder *[anat]*
tanpa	短波	short wave *[comm]*
tanpaku'kō	蛋白光	opalescence *[opt]*
tanpaku-kōgaku	蛋白工学	protein engineering *{n} [bioch]*
tanpaku'shitsu	蛋白質	protein *[bioch]*
tanpaku('shitsu)-bunkai-kōso	蛋白(質)分解酵素	protease *[bioch]*
tanpaku'shitsu-bunkai-ryoku	蛋白質分解力	proteolytic power *[bioch]*
tanpaku'shitsu-hensei	蛋白質変性	protein denaturation *[bioch]*
tanpaku'shitsu-jushi	蛋白質樹脂	protein resin *[poly-ch]*
tanpaku'shitsu-kasui-bunkai'butsu	蛋白質加水分解物	protein hydrolysate *[bioch]*
tanpaku'shitsu no sei'gōsei	蛋白質の生合成	protein biosynthesis *[bioch]*
tanpopo	蒲公英；たんぽぽ	dandelion *[bot]*
tan'raku	短絡	short circuit *{n} [civ-eng] [elec]*
tanren	鍛錬；鍛鍊	forging; temper *{n} [met]*
"	鍛錬；たんれん	training *{n}*
tanri	単離	isolation *[ch]*
tanrin'shi-yōsei	担輪子幼生	trochophore larva *[i-zoo]*
tanri suru	単離する	isolate *{vb} [ch]*
tanryō'tai	単量体	monomer *[poly-ch]*
tanryō'tai-hannō'sei-hi	単量体反応性比	monomer reactivity ratio *[poly-ch]*
tanryū'shiki-densō	単流式伝送	neutral transmission *[comm]*
tansa	短鎖；たんさ	short chain *[ch]*
"	短鎖	light chain *[immun]*
tan'saibō-seibutsu	単細胞生物	unicellular organism *[bio]*
tan'saibō-tanpaku'shitsu	単細胞蛋白質	single-cell protein *[bioch]*
tansa-ki	探査機	probe *{n} [aero-eng]*
tansa-ki	探査器	sensor *{n} [eng]*
tansaku	探索	search *{n} [comput]*
tansan	炭酸；たんさん	carbonic acid
tansan-aen	炭酸亜鉛	zinc carbonate
tansan-dō	炭酸銅	copper carbonate
tansan-dōka	炭酸同化	carbon dioxide assimilation *[bio]*
tansan-en	炭酸塩	carbonate *[ch]*
tansan-gin	炭酸銀	silver carbonate
tansan-namari	炭酸鉛	lead carbonate
tansan-suiso-en	炭酸水素塩	hydrogencarbonate *[ch]*
tansan-suiso-natoryūmu	炭酸水素ナトリウム	sodium hydrogencarbonate
tansan-tetsu	炭酸鉄	iron carbonate

tan-seido	単精度；たんせいど	single-precision *[comput]*
tan'seki	胆石；膽石	gallstone *[med]*
tan-sen'i	単繊維	single fiber *[tex]*
tan'setsu	鍛接；たんせつ	forge welding *{n} [met]*
tansha	単斜；たんしゃ	monocline *[min-eng]*
tansha-shō	単斜晶	monoclinic crystal *[crys]*
tansha-shōkei	単斜晶系	monoclinic system *[crys]*
tanshi	単糸	monofilament; single yarn *[tex]*
tanshi	短枝	short shoot *{n} [bot]*
tanshi	端子；たんし	terminal *{n} [comput] [elec]*
tanshi-ban	端子板	terminal board; terminal strip *[elec]*
tanshi-den'atsu	端子電圧	terminal voltage *[elec]*
tanshi-gire	単糸切れ	monofilament breakage; single yarn breakage *[tex]*
tanshi-hōshi	担子胞子	basidiospore *[mycol]*
tanshi-ki	担子器；たんしき	basidium (pl -ia) *[mycol]*
tanshi'ki'ka	担子器果	basidiocarp *[mycol]*
tanshi'kin-rui	担子菌類	Basidiomycetes *[mycol]*
tan'shin	短針；たんしん	hour hand; short hand (of clock) *[horol]*
tan-shindō	単振動	simple harmonic motion *[mech]*
tanshin-kaisen	単信回線	simplex circuit *[comm]*
tanshin-tsū'shin	単信通信	simplex communication *[comm]*
tan-shitsu	炭質	coal property *[mat]*
tansho	短所	defect; shortcoming; demerit *{n}*
tanshō	単晶；たんしょう	single crystal *[crys]*
tanshō-ki	探傷機	flaw detector *[eng]*
tanshokkō	単色光	monochromatic light *[opt]*
tan-shoku	単色	monochrome *{n} [opt]*
tan-shoku	淡色	light color; pale color; pallidness *[opt]*
tanshoku-kei	単色計	monochromator *[spect]*
tanshoku-kōka	淡色効果	hypochromism; hypochromic effect *[p-ch]*
tanshoku-sei	単色性	monochromaticity *[opt]*
tanshō'tō	探照灯	searchlight *[opt]*
tanshuku suru	短縮する	abbreviate; contract; shorten *{vb}*
tanso	炭素；たんそ	carbon (element)
tansō	単層	monolayer *[p-ch]*
tansō-baiyō	単層培養	monolayer culture *[microbio]*
tansō-den'ryoku	単相電力	single-phase power *[elec]*
tan-sōkan	単相関	simple correlation *[math]*
tansokan'shiki-kagō'butsu	単素環式化合物	homocyclic compound *[org-ch]*
tanso-kō	炭素鋼	carbon steel *[steel]*
tansoku	単速	single speed *[phys]*

tan-sokuha'tai-densō	単側波帯伝送	single-sideband transmission *[comm]*
tanso-nendai-sokutei	炭素年代測定	carbon dating *{n} [nucleo]*
tanso-sasshi	炭素刷子	carbon brush *[elec]*
tanso-sei	炭素星	carbon star *[astron]*
tanso-sen'i	炭素繊維	carbon fiber *[mat] [tex]*
tanso'shitsu no	炭素質の	carbonaceous *{adj} [sci-t]*
tanso-shūshi	炭素収支	carbon balance *[geoch]*
tanso-takan-kagō'butsu	炭素多環化合物	carbopolycyclic compound *[ch]*
tanso-tankan-kagō'butsu	炭素単環化合物	carbomonocyclic compound *[ch]*
tanso-tanso-fukugō'butsu	炭素炭素複合物	carbon-carbon composite *[mat]*
tanso'zoku-genso	炭素族元素	carbon family element *[ch]*
tansu	箪笥; たんす	chest (of drawers) *[furn]*
tansū	単数; 單數	singular *{n} [gram]*
"	単数	singular number *[math]*
tansui	淡水	fresh water *[hyd]*
tansui'gyo	淡水魚	freshwater fish *[v-zoo]*
tansui'ka'butsu	炭水化物	carbohydrate *[bioch]*
tantai	担体; 擔體	carrier *[ch] [phys] [sol-st]*
tantai	単体	simple substance *[ch]*
tantai-hō	単体法	simplex method *[math]*
tantaru	タンタル	tantalum (element)
tanten	短点	dot (Morse code) *{n} [comm]*
tantō	単糖	monosaccharide *[bioch]*
tantō-rui	単糖類	monosaccharides *[bioch]*
tanuki	狸; たぬき	raccoon dog (animal) *[v-zoo]*
tan'yō	単葉	simple leaf *[bot]*
tan-yu	短油	short oil *[mat]*
tanza-hai'i'shi	単座配位子	monodentate (unidentate) ligand *[ch]*
tan-zaku	短冊; たんざく	paper rectangle (for poetry) *[paper] [pr]*
tanzō	鍛造	forging *{n} [met]*
tan-zome	反染め	piece dyeing *{n} [tex]*
taosu	倒す	fell; lay; bring down *{vb t}*
tara	鱈; たら	cod; Gadus macrocephalus (fish) *[v-zoo]*
taraba-kani	鱈場蟹	King crab *[i-zoo]*
tara-kan'yu	鱈肝油	cod-liver oil *[mat]*
tara-no-ki	楤の木; たらのき	Japanese angelica tree; Aralia elata *[bot]*
tare	垂れ; たれ	marinade; sauce *{n} [cook]*
tare-kazari	垂れ飾り	pendant *{n} [lap]*
tareru	垂れる	droop; hang; dangle; drip *{vb i}*
tare'sagari	垂れ下がり	sagging (of paint) *{n} [poly-ch]*
ta'retsu-yō	多裂葉	multifid leaf *[bot]*
tarinai	足りない	be insufficient; is lacking

tariru	足りる	suffice; be adequate *{vb}*
taru	樽；たる	barrel; cask; keg *[eng]*
tāru	タール	tar *{n} [mat]*
tarumi	弛み；たるみ	sag; slack *{n} [eng]*
taru'shi	樽師	cooper (a person) *[eng]*
taryō'tai	多量体	multimer *[poly-ch]*
ta'ryūka'butsu	多硫化物	polysulfide *[org-ch]*
ta'ryūka-gomu	多硫化ゴム	polysulfide rubber *[rub]*
taryūmu	タリウム	thallium (element)
taryū-sei	多粒性	graininess *[mat]*
tasai	多彩	multicolor *{n} [opt]*
ta'saibō no	多細胞の	multicellular *{adj} [bio]*
ta'san-enki	多酸塩基	polyacidic base *[ch]*
ta'seibun-shokubai	多成分触媒	multicomponent catalyst *[ch]*
ta'shasen-dōro	多車線道路	multilane road *[civ-eng] [trans]*
tashika	確か；たしか	certainty; definiteness *{n}*
"	確か	perhaps; probably *{adv}*
tashikameru	確かめる	ascertain; confirm; verify *{vb}*
tashika ni	確かに	certainly; surely; definitely *{adv}*
tashi-kōsa	多枝交差	multiple intersection *[traffic]*
tashi-sei	多雌性	polygyny *[bio]*
tashi-zan	足し算	addition *[math]*
tashi-zui	多雌蕊；たしずい	polygyny *[bot]*
ta'shoku-sei	多色性	pleochroism *[opt]*
ta'shoku'sei no	多色性の	pleochroic *{adj} [opt]*
tashu'tayō	多種多様	diverse; diversity *{n} [comm]*
tasō	多相	polyphase; multiphase *{n} [elec]*
tasō-ban	多層板	multilayer sheet *[mat]*
tasogare	黄昏；曛；たそがれ	twilight *[astron]*
tasō-kairo-kiban	多層回路基板	multilayered circuit board *[elec]*
tasō-kōryū	多相交流	polyphase current *[elec]*
tasō-oshi'dashi	多層押出し	multilayer extrusion *[eng]*
tasu	足す	add *{vb t} [math]*
tasū-hadan	多数破断	multiple fracture *[mech] [sci-t]*
tasukeru	助ける	help; aid; save (others) *{vb t}*
tasū'ketsu	多数決；たすうけつ	majority (decision) *{n} [comput]*
tasū'ketsu-enzan	多数決演算	majority operation *[comput]*
tasū no-	多数の-	multi- *{prefix}*
tatai-sei	多態性	pleomorphism *[bio]*
tataki	叩き；敲き；たたき	pounded (minced) fish, poultry, or meat dish *[cook]*
tataku	叩く	strike; beat; spank *{vb t}*

tatami	畳；疊；たたみ	flooring of straw mat *{n} [arch]*
tatami-komu	畳み込む	fold in; turn into *{vb t}*
tatamu	畳む	fold; furl *{vb t}*
tate	盾；楯；たて	shield *{n} [ord]*
tate	縦；縱	height; length; vertical *{n} [mech]*
tate'biki	縦引き	ripsawing *{n} [mech-eng]*
tate-dansei-keisū; (-ritsu)	縦弾性係数；(-率)	modulus of longitudinal elasticity; Young's modulus *[mech]*
tategami	鬣；鬐；髦；騣	mane *[zoo]*
tategami-hitsuji	鬐羊	aoudad; Barbary sheep (animal) *[v-zoo]*
tate'gata-funsai-ki	縦型粉砕機	vertical mill *[eng]*
tate'gata-sha'shutsu-seikei-ki	縦型射出成形機	vertical injection molding machine *[eng]*
tate'gu	建具	fittings; fixtures; furnishings *[arch]*
tate-hizumi	縦歪み；たてひずみ	longitudinal strain *[mech]*
tate-hōkō	縦方向	longitudinal direction; machine direction *[eng]*
tate-ito	经糸；經糸；縦糸	warp (thread) *{n} [tex]*
tate-jiku	縦軸	axis of ordinates *[math]*
tate'jō-kazan	楯状火山；盾状火山	shield volcano *[geol]*
tate'mono	建物	building *{n} [arch]*
tate-nami	縦波	longitudinal wave *[phys]*
tate ni suru	縦にする	stand upright *{vb t}*
tateru	立てる；建てる	build; construct; erect; raise *{vb} [arch]*
tate-sen	縦線	vertical line *[comput] [math]*
tate-sendan	縦剪断	longitudinal shear *[mech]*
tate-shindō	縦振動	longitudinal vibration *[mech]*
"	縦振動	longitudinal oscillation *[phys]*
tate-waku	竪枠	jamb *{n} [arch]*
tate'yure	縦揺れ；縱搖れ	pitch *{n} [mech]*
tate-zahyō	縦座標	axis of ordinates; ordinate *[math]*
tate-zome	建染め；縦染め	vat dyeing *{n} [tex]*
tate'zome-senryō	建染め染料	vat dye *[tex]*
tatō	多糖	polysaccharide *[bioch]*
tatoeba	例えば；たとえば	for example; e.g. (exemplii gratia) *{adv}*
tatō-rui	多糖類	polysaccharides *[bioch]*
tatsu'maki	竜巻；龍巻；たつまき	waterspout (upon water); landspout (upon land) *[meteor]*
tatsu no otoshi'go	竜の落(と)し子	seahorse (fish) *[i-zoo]*
tatta	たった	only; merely *{adv}*
tawami	撓み；たわみ	deflection; flexure; slack *{n} [mech]*
tawami-kei	撓み計	deflectometer *[eng]*

tawami-ren'ketsu-ki	撓み連結器	flexible coupler *[mech-eng]*
tawami-sei	撓み性	flexibility *[mech]*
tawara	俵；たわら	straw sack; straw bag *[agr]*
tawashi	束子；たわし	scrub brush *[cook]*
tayō-sei	多様性	diversity *{n} [bio]*
ta'yōso-mokei	多要素模型	multielement model *[eng]*
tayō'tai	多様体；多樣體	manifold *{n} [math]*
ta'yūzui	多雄蕊；たゆうずい	polyandry *[bot]*
taza-hai'i'shi	多座配位子	polydentate ligand *[ch]*
ta'zuna	手綱	bridle; reins *{n} [trans]*
te	手	hand *{n} [anat]*
te-ami	手編み	hand-knitting *{n} [tex]*
te-arai	手洗い；てあらい	rest room; toilet; lavatory *[arch]*
"	手洗い	hand-washing *{n} [tex]*
te-bako	手箱	lacquered case; box; etui *{n} [furn]*
te-bata	手機；てばた	hand loom *{n} [tex]*
te'biki'sho	手引書	manual (a document) *{n} [pr]*
te-bori	手彫り	hand-engraving *{n} [graph] [tex]*
tēburu	卓子；テーブル	table *{n} [furn]*
te'chō	手帳；手帖	notebook; memo pad *[pr]*
te-fuigo	手吹子；手ふいご	hand bellows *[met]*
te-fuki	手吹き	hand-blowing (of glass) *{n} [cer]*
te-gaki	手書き	hand-writing *{n} [comm]*
te-gaki	手描(き)	hand-painting *{n} [graph] [tex]*
tegaki-moji	手書き文字	hand-written character *[comm]*
tegami	手紙	letter *{n} [comm]*
tegusu	天蚕糸；てぐす	silk gut *[i-zoo] [mat]*
te-hen	手扁；手偏：扌	kanji (q.v.) radical denoting hand *[pr]*
tei	底；てい	base *{n} [math]*
tei'an	提案	proposal; suggestion *{n}*
tei-atsu	低圧；低壓	low tension; low voltage *[elec]*
"	低圧；ていあつ	low pressure *{n} [meteor] [phys]*
tei'atsu-densen	低圧電線	low-voltage cable *[elec]*
tei-atsu no	低圧の	low voltage *{adj} [elec]*
"	低圧の	low pressure *{adj} [phys]*
tei'atsu-seikei	低圧成形	low-pressure molding *{n} [eng]*
tei'bairitsu-kenbikyō-shashin	低倍率顕微鏡写真	photomacrograph *[opt]*
teiban	底盤	batholith *[geol]*
teibō	堤防	embankment *[civ-eng]*
tei'boku	低木	shrub *[bot]*
tei-bunshi'ryō-kagō'butsu	低分子量化合物	low-molecular-weight compound *[ch]*

tei'chaku	定着；ていちゃく	fixation; fixing *{n} [photo]*
teichaku-eki	定着液	fixing solution; fixer *[photo]*
teichaku-sei	定着性	setting property *[eng] [mat]*
teichaku suru	定着する	take root *{vb i} [bot]*
"	定着する	fix *{vb t} [civ-eng] [photo]*
teichō-eki	低張液	hypotonic solution *[physio]*
teiden	停電	power failure; service interruption *[elec]*
tei'en	庭園	garden *{n} [bot]*
teigi	定義；ていぎ	definition *[ling] [math]*
teigi-zumi	定義済み	predefined *[comput]*
tei'haku-tō	停泊灯；碇泊燈	anchor light *[opt]*
teihaku-yochi	停泊余地	berth *{n} [nav-arch]*
teihen	底辺；底邊	base *{n} [math]*
tei-hirei no hōsoku	定比例の法則	definite composition, law of *[ch]*
"	定比例の法則	definite proportions, law of *[ch]*
tei'iki-roha-kairo	低域沪波回路	low-pass filtration circuit *[elec]*
tei'ji	丁字；ていじ	T (the letter)
teiji	低次	low-order *[comput] [math]*
teijō-denryū	定常電流	stationary current *[elec]*
teijō-fū	定常風	steady wind *[meteor]*
teijō-ha	定常波	standing wave; stationary wave *[phys]*
teijō-hannō	定常反応	steady reaction *[ch]*
teijō-jōhō'gen	定常情報源	stationary information source *[comput]*
teijō-jōtai	定常状態	steady state *[phys]*
"	定常状態	stationary state *[quant-mech]*
teijō-ki	定常期	stationary phase *[microbio]*
teijō-ryū	定常流	stationary flow *[fl-mech]*
teijō-sei	定常性	regularity; constancy; steadiness *{n}*
teijō-shindō	定常振動	steady-state vibration *[mech]*
teijō'tai	蹄状体	ungulla *[math]*
tei-jūgō'tai	低重合体	low grade polymer; low polymer *[poly-ch]*
teikaku	定格；ていかく	rating *{n} [elec]*
tei-kaku	底角	base angle *[math]*
teikaku-den'atsu	定格電圧	rated voltage *[elec]*
teikaku-shutsu'ryoku	定格出力	rated output *[mech-eng]*
teikaku-sokudo	定格速度	rated speed *[comput]*
tei-kanshi	定冠詞	definite article *[gram]*
teikei	提携；ていけい	cooperation; affiliation *{n}*
teikei-shorui	定型書類	formal documents *[pr]*
tei'ketsu	締結	fastening; tightening *{n}*
tei-ki'atsu	低気圧；低氣壓	low; low pressure *{n} [meteor]*
teiki-hozen	定期保全	periodic maintenance *[ind-eng]*

teiki-kankō'butsu	定期刊行物	periodical *(n) [pr]*
teiki'sen	定期船	liner (a ship) *[nav-arch]*
teiki-shiken	定期試験	routine test *[ind-eng]*
teikō	抵抗；ていこう	resistance *[elec] [mech] [phys]*
"	抵抗	drag *(n) [fl-mech]*
teikō-keisū	抵抗係数	drag coefficient *[fl-mech]*
teikō-ki	抵抗器	resistor *[elec]*
teikō-ritsu	抵抗率	resistivity; specific resistance *[elec]*
teikō-sei	抵抗性	resistance *[bot]*
teikō-soshi	抵抗素子	resistance element *[elec]*
teikō'tai	抵抗体	resistor *[elec]*
teimen-zu	底面図	bottom view *(n) [graph]*
teimi-ryoku	呈味力	flavor-generating power *[cook]*
teimi-sei	呈味性；ていみせい	tasteableness; flavor-manifesting property *[cook] [physio]*
teimi-seibun	呈味成分	tasteable component *[cook]*
tei-on	低音	low frequencies *[acous]*
tei-on	低温；ていおん	low temperature *[thermo]*
teion(-butsuri)'gaku	低温(物理)学	cryogenics *[phys]*
teion-dōbutsu	定温動物	homeotherm *[physio]*
teion'gaku	低温学	cryogenics *[phys]*
teion-kakusei-ki	低音拡声器	woofer *[eng-acous]*
teion-kei	低温計	cryometer *[thermo]*
teion-kōka-sei	低温硬化性	low-temperature-hardening property *[mat]*
teion-sakkin(-hō)	低温殺菌(法)	pasteurization *[food-eng]*
teion-zeisei	低温脆性	cold brittleness *[mech]*
teiri	定理	theorem *[logic] [math]*
teiryō	定量；ていりょう	determination *[an-ch]*
teiryō-bunseki	定量分析	quantitative analysis *[an-ch]*
teiryō-hō	定量法	assay *(n) [an-ch]*
teiryō'teki na	定量的な	quantitative *(adj) [ch]*
tei'satsu	偵察；ていさつ	reconnaissance *[mil]*
teisei	訂正	amendment; correction *[pat]*
teisei-bunseki	定性分析	qualitative analysis *[an-ch]*
teisei-seibutsu	底生生物	benthos *(n) [ecol]*
teisei'teki na	定性的な	qualitative *(adj) [ch]*
tei-sekibun	定積分	definite integral *[math]*
teisetsu	定説	established theory *[sci-t]*
teisha	停車	parking; stopping (of vehicle) *[traffic]*
teishi	停止	halt *(n) [comput]*
tei-shibō'bun no	低脂肪分の	low-fat *(adj) [cook]*
tei'shibō-nyū	低脂肪乳	lowfat milk *[cook]*

teishi-hannō	停止反応	termination reaction *[org-ch]*
teishi-jikan	停止時間	downtime *{n}* *[ind-eng]*
teishin-kōshi	底心格子	base-centered lattice *[crys]*
teishi-shingō	停止信号	stop signal *[comput]*
teishi'tō	停止灯；停止燈	stoplight *[traffic]*
teishi-zai	停止剤	terminator; terminating agent *[mat]*
teishoku	抵触；ていしょく	conflict; infringement *{n}* *[legal]*
teishoku-hannō	呈色反応	color reaction *[ch]*
teishoku-i'sei	呈色異性	chromoisomerism; chromotropy *[opt]*
teisoku	定速；ていそく	constant speed *{n}* *[mech]* *[phys]*
tei'sokudo-satsu'ei-shashin	低速度撮影写真	time-lapse photography *[photo]*
teisū	定数；ていすう	constant *{n}* *[comput]* *[math]*
teisū-kansū	定数関数；定數關數	constant function *[math]*
teisun-sōchi	定寸装置	tru(e)ing apparatus *[eng]*
teitai	停滞；停滯	stagnation; delay *{n}*
teitai-sui	停滞水	stagnant water *[hyd]*
teiten	定点；定點	fixed point *[ch]* *[math]* *[meteor]*
tei'yōshiki-hyōji	定様式表示	formatted display *[comput]*
teizai-ha	定在波；ていざいは	standing wave *[phys]*
teizai'ha-hi	定在波比	standing wave ratio *[phys]*
tejō	手錠	handcuffs *{n}* *[eng]*
te'jun	手順	procedure; routine *{n}* *[comput]*
teki	滴；てき	drop; droplet (of liquid) *{n}* *[phys]*
teki	適	approved (an abbreviation) *[pat]*
-teki	-的	-tic; -tical; -like *{suffix}*
tekigō saseru	適合させる	cause to conform (or agree) with *{vb}*
tekigō-sei	適合性	compatibility; conformity *{n}*
tekigō suru	適合する	conform with *{vb}*
teki-ka	摘果；てきか	fruit thinning *{n}* *[agr]*
tekikaku; tekkaku	適格	eligibility; competence *{n}*
tekika-rōto	滴下漏斗	dropping funnel; drop funnel *[ch]*
tekika suru	滴下する	add dropwise; drip; trickle *{vb}* *[ch]*
teki'ō	適応；適應	adaptation *[gen]* *[physio]*
"	適応；てきおう	adjustment *[psy]*
tekiō-keisū	適応係数	accomodation coefficient *[stat-mech]*
tekiō-kōso	適応酵素	adaptive enzyme *[bioch]*
tekiō-sei	適応性	adaptability *[physio]*
teki'ryō	適量	appropriate quantity; sufficient amount *{n}* *[math]*
tekisei	適正	proper *{n}*
tekisei	適性	aptitude *[psy]*
"	適性	appropriateness; properness *{n}*

tekisei-kensa	適性検査	aptitude test *[psy]*
tekisei-sei	適正性	appropriateness; validity *{n}*
teki'setsu na	適切な	appropriate; adequate *{adj}*
teki'sha-sei'zon	適者生存	survival of the fittest *[bio]*
tekisū-kei	滴数計	stalagmometer *[eng]*
teki'tei	滴定；てきてい	titration *{n} [an-ch]*
tekitei-bunseki	滴定分析	titrimetric analysis *[an-ch]*
tekitei-eki	滴定液	titrant (solution) *[an-ch]*
tekitei-hi	滴定比	titration ratio *[an-ch]*
tekitei-kyoku'sen	滴定曲線	titration curve *[an-ch]*
tekitei-shi'yaku	滴定試薬	titrant (reagent) *[an-ch]*
teki'yō	摘要	abstract; synposis *{n} [comm] [pr]*
teki'yō	適用	application *[sci-t]*
tekiyō-sei	適用性	applicability *{n}*
tekkai	撤回	cancellation; withdrawal *{n}*
tekken	鉄圏；鐵圏	siderosphere *[geophys]*
tekki-jidai	鉄器時代	Iron Age *[archeo]*
tekkin-konkurīto	鉄筋コンクリート	reinforced concrete *[mat]*
tekkō'jō	鉄工場	iron shop *[met]*
tekkō'sho	鉄工所	iron works *[met]*
tekkyo	撤去	evacuation; withdrawal *{n}*
teko	梃子；てこ	lever *[eng]*
te-kōgu	手工具	hand tool *[eng]*
teko-hi	梃子比	leverage *[mech]*
te-kubi	手首	wrist *{n} [anat]*
tekunechūmu	テクネチウム	technetium (element)
tema	手間；てま	time and labor *{n} [ind-eng]*
te'mochi-sakugan-ki	手持ち鑿岩機	jackhammer *{n} [mech-eng]*
ten	天	sky; heavens *[astron]*
ten	点；點；てん	dot; speck; spot *{n} [graph]*
"	点	point *{n} [math]*
ten	貂；てん	ermine; marten (animal) *[v-zoo]*
te'naga-ebi	手長蝦；てながえび	prawn; Palaemon nipponensis *[i-zoo]*
te'naoshi	手直し	debugging *{n} [comput]*
"	手直し	correction; later adjustment; readjustment; rework *[ind-eng]*
te-nassen; te-nasen	手捺染	hand-printing *{n} [tex]*
tenbi	天火	oven *[cook] [eng]*
tenbi; tenpi	天日	sun light; sun heat *[cook]*
ten'bin	天秤；てんびん	balance; pair of scales *[eng]*
tenbin-bō	天秤棒	shouldering pole *[eng]*
tenbin-za	天秤座	Libra; the Balance *[constel]*

tenbi-zai	点鼻剤；てんびざい	collunarium; nosedrop *{n} [pharm]*
tenbi-zarashi	天日晒し	sun bleaching *{n} [tex]*
tenbō'sha	展望車	observation car *[rail]*
tenbō-tō	展望塔	observation tower *[arch]*
tenchaku	展着；てんちゃく	spreading *{n} [adhes]*
tenchaku	貼着	gluing on *{n} [adhes]*
tenchaku-zai	展着剤	spreading agent *[mat]*
tenchi suru	転置する	transpose *{vb} [gram] [math]*
tenchō	天頂；てんちょう	zenith *[astron]*
tenchō-gi	天頂儀	zenith telescope *[opt]*
tenchō-in'ryoku	天頂引力	zenith attraction *[astron]*
tenchō-kyori	天頂距離	zenith distance *[astron]*
ten-denka	点電荷；てんでんか	point charge *[elec]*
ten-denshi-shiki	点電子式	dot electron formula *[atom-phys]*
tendō	転動；轉動	rolling *{n} [mech]*
tendō-zōryū-sōchi	転動造粒装置	tumbling granulator *[mech-eng]*
ten'en-sei	展延性	extensibility *[mech]*
ten'en-tosō	展延塗装	spread coating *{n} [paint]*
tengai	天蓋	canopy (pl -pies) *[tex]*
tengan-eki	点眼液	collyrium; eyewash *{n} [pharm]*
tengi	転義；てんぎ	figurative meaning; transferred meaning *[comm]*
tengun	点群；點群	point group *[crys]*
ten-gusa	天草；石花菜	agar agar *[mat]*
tengu-take	天狗茸	death cup; Amanita pantherina (poison mushroom, fly killer) *[bot]*
tengu-zaru	天狗猿	proboscis monkey (animal) *[v-zoo]*
ten-hen'i	点変異	point mutation *[gen]*
ten'i	転位；轉位；てんい	translocation *[bioch]*
"	転位	rearrangement *[ch]*
"	転位	dislocation *[crys] [photo]*
"	転位	transposition *{n}*
ten'i	転移	transfer *{n} [bio]*
"	転移	transition *[met] [thermo]*
ten'i-kōso	転位酵素	transferase *[bioch]*
ten'i-RNA	転移RNA	transfer ribonucleic acid; tRNA *[mol-bio]*
ten'i-ten	転移点	transition point *[thermo]*
tenji	点字；點字；てんじ	Braille *[comm]*
tenji	展示	display; exhibition *{n}*
tenji; soe'ji	添字	subscript *[math]*
tenjiku-nezumi	天竺鼠;てんじく鼠	guinea pig (animal) *[v-zoo]*
tenji-shitsu	展示室	showroom; display room *[arch]*

tenji suru	展示する	display *{vb}*
tenji-yaku	点耳薬	eardrop *{n} [pharm]*
tenjō	天井；てんじょう	ceiling *[arch]*
tenjō-daka	天井高	ceiling height *[arch]*
ten-jō'seki	点乗積；點乗積	dot product; inner product *[math]*
tenjō-senpū-ki	天井扇風機	ceiling fan *[eng]*
tenjū-sei	塡充性	extension *[phys]*
tenka	点火	ignition *[ch]*
tenka	添加	addition *[ch]*
tenka	転化	conversion *[ch-eng]*
"	転化	transformation *{n}*
tenka'butsu	添加物	additive *[mat]*
tenka-fun	天花粉；天瓜粉	baby powder; nursery powder *[pharm]*
tenkai	展開；てんかい	development; expansion *[math]*
"	展開	evolution *{n}*
tenkai-shiki	展開式	expanded expression *[math]*
tenkai-zai	展開剤	developer; developing agent *{n}*
tenkai-zu	展開図	development (drafting) *{n} [graph]*
"	展開図	expansion plan *{n} [graph]*
tenkan	転換；てんかん	conversion *[ch] [math] [nuc-phys]*
"	転換	transmutation *[ch]*
"	転換	transformation *[cyt]*
tenkan-kō'ritsu	転換効率	transformation efficiency *[cyt]*
tenkan-ten	転換点	turning point; switching point *{n}*
tenka-ondo	点火温度	ignition temperature; kindling temperature *[thermo]*
tenka-ritsu	転化率	conversion *[org-ch]*
tenka-ro	転化炉	converter reactor *[nucleo]*
tenka-sei	点火性	ignitability *[petr]*
tenka suru	点火する	ignite *{vb} [ch]*
tenka suru	添加する	add *{vb t} [ch] [sci-t]*
tenka'tō	転化糖	invert sugar; invertose *[cook]*
tenka-zai	添加剤；てんかざい	additive; admixture *{n} [mat]*
"	添加剤	adjuvant *{n} [mat] [pharm]*
tenkei'teki na	典型的な	typical *{adj}*
tenki	天気	weather *{n} [meteor]*
tenki suru	転記する	transcribe; copy *{vb} [comput]*
tenki-yohō	天気予報	weather forecast *[meteor]*
tenki-zu	天気図	weather map *[meteor]*
tenkō-shadō	転向車道	turning roadway (or lane) *[traffic]*
tenkyō-gi	転鏡儀	transit (for surveying) *{n} [eng]*
tenkyū	天球	celestial sphere *[astron]*

tenkyū-seki'dō	天球赤道	celestial equator *[astron]*
ten-mado	天窓	skylight *[arch]*
tenmaku	天幕	awning; tent; cloth pavilion *[arch]*
ten'metsu	点滅；てんめつ	flashing on and off *{n}* *[navig]*
tenmon-dai	天文台；天文臺	observatory *[arch]* *[astron]*
tenmon'gaku	天文学；天文學	astronomy *[sci-t]*
tenmon-gakusha	天文学者	astronomer *[astron]*
tenmon-hakumei	天文薄明	astronomical twilight *[astron]*
tenmon-kōhō	天文航法	astronavigation *[navig]*
tenmon-sankak(u)kei	天文三角形	astronomical triangle *[astron]*
tenmon-suihei	天文水平	celestial horizon *[astron]*
tenmon-tan'i	天文単位	astronomical unit *[astron]*
ten'nen	天然；てんねん	nature *{n}*
tennen'butsu	天然物	natural product *[mat]*
tennen-chaku'shoku-ryō	天然着色料	natural coloring material *[cook]*
tennen-gasu	天然ガス	natural gas *[mat]*
tennen-jushi	天然樹脂	natural resin *[mat]*
tennen-kagō'butsu	天然化合物	native compound *[mat]*
tennen-kinen'butsu	天然記念物	natural monument *{n}*
tennen no	天然の	natural *{adj}*
tennen-sen'i	天然繊維	natural fiber *[tex]*
tennen-senryō	天然染料	natural dye *[mat]*
tennen-shigen	天然資源	natural resource *[mat]*
tennen-shigen-hogo	天然資源保護	conservation of natural resources *[ecol]*
tennen-shiki'so	天然色素	natural pigment *[mat]*
tennen-tanpaku'shitsu	天然蛋白質	native protein *[bioch]*
tennen'tō	天然痘	smallpox *[med]*
tennō-sei	天王星	Uranus (planet) *[astron]*
ten no seki'dō	天の赤道	celestial equator *[astron]*
ten no shigo-sen	天の子午線	celestial meridian *[astron]*
te-no-hira	手の平；掌	palm (of hand) *{n}* *[anat]*
te-no-kō	手の甲	back of hand *{n}* *[anat]*
ten-on'gen	点音源	point source *[acous]*
tenpo-chiku	店舗地区	shopping district *[civ-eng]*
tenpu; chōfu	貼付	gluing on *{n}*
tenpura	天麩羅；てんぷら	tempura: Japanese deep-fried dish *[cook]*
tenpu-shorui	添付書類	attached (accompanying) document *[pat]* *[pr]*
tenran'seki	天藍石	lazulite *[miner]*
tenrin-rashin-gi	転輪羅針儀	gyrocompass *[navig]*
tenro	転炉；轉爐；てんろ	converter *[met]*
tenro-kō	転炉鋼	converter steel *[steel]*
tenrō-sei	天狼星	Sirius *[astron]*

tenryō	填料；てんりょう	filler; loading material *[mat]*
tenryū	転流	commutation *[electr]*
tensai	天災	natural disaster; act of God *{n}*
tensai	甜菜；てんさい	sugarbeet *[bot]*
tensai'tō	甜菜糖	beet sugar *[cook]*
tensei	展性	malleability *[met]*
tensei'seki	天青石	celestite *[miner]*
tensen	点線	dotted line *[graph]*
ten'setsu suru	転接する	revolvingly contact *{vb} [mech-eng]*
tensha	転写；轉寫	transcription; transfer *{n} [graph]*
"	転写	transfer printing *{n} [graph] [pr]*
"	転写	transcription *[mol-bio]*
tensha-shi	転写紙	transfer paper *[graph]*
tensha suru	転写する	transcribe *{vb} [comput]*
ten'shoku	点食；點蝕	pitting (by corrosion) *{n} [met]*
tenshoku-zai	展色剤	vehicle *[graph] [pr]*
tensō	転相	phase inversion *[ch]*
tensō-ritsu	転送率	transfer rate *[comput]*
tensō-sokudo	転送速度	transfer rate; transfer speed *[comput]*
tensō suru	転送する	move; transfer *{vb t} [comput]*
tensū	点数	marks; merit marks *{n} [ind-eng]*
tensū	添数	suffix *[math]*
ten-tai	天体；天體	celestial body; heavenly body *[astron]*
tentai-bōenkyō	天体望遠鏡	astronomical telescope *[opt]*
tentai-butsuri'gaku	天体物理学	astrophysics *[astron]*
tentai-ichi-hyō	天体位置表	ephemeris (pl ephemerides) *[astron] [pr]*
tentai-kaku-butsuri'gaku	天体核物理学	nuclear astrophysics *[astron]*
tentai-reki	天体暦	celestial almanac; ephemeris *[astron] [pr]*
tentai-sekkin-hikō	天体接近飛行	flyby *{n} [navig]*
tentai'zu'gaku	天体図学	uranography *[astron]*
tentei	天底	nadir *[astron]*
tentei	点綴；てんてい	plotting *{n} [math]*
ten-teki	天敵	natural enemy *[bio]*
tenteki-roshō	点滴沪床	trickling filter bed *[civ-eng]*
ten'teki suru	点滴する	administer dropwise *{vb t}*
tenteki-yaku	点滴薬	drops; drop medicine *{n} [pharm]*
tentō	転倒	overturning; upsetting *{n}*
tentō-mushi	天道虫；瓢虫	ladybug (insect) *[i-zoo]*
tentō suru	点灯する	turn the light on *{vb} [elec]*
te'nugui	手拭い	towel; hand towel *{n} [tex]*
ten-yaku	填薬	loading (of weapon) *{n} [ord]*
ten-yō'setsu	点溶接	spot welding *{n} [met]*

tenza	転座；轉座；てんざ	translocation *[gen]*
te'ono; chōna	手斧	adze; hatchet *{n}* *[eng]*
te'ori no	手織りの	homespun; handwoven *{adj}* *[tex]*
te'oshi-guruma	手押し車	hand truck *[eng]*
te o utsu	手を打つ	adopt a measure; take action *{vb}*
teppan-yaki	鉄板焼	griddle-roasted food *[cook]*
teppō	鉄砲；てっぽう	gun; firearm *[ord]*
teppō-uo	鉄砲魚；鉄砲魚	archerfish *[v-zoo]*
terepin-yu	テレピン油	turpentine oil *[mat]*
teri'yaki	照焼；照り焼き	soy-broiled fish, chicken, etc. *[cook]*
terubyūmu	テルビウム	terbium (element)
terupen-rui	テリペン類	terpenes *[ch]*
teruru	テルル	tellurium (element)
te'sage-tō	手提げ灯	portable lamp *[eng]*
te-saguri	手探り	trial and error *[math]*
te-sagyō	手作業	manual operation *[ind-eng]*
te-sentaku	手洗濯	hand-washing *{n}* *[tex]*
tessa	轍叉；てっさ	frog *[rail]*
tessai; tesshi	鉄滓	slag *[met]*
tesshin	鉄心；鐵芯	iron core *[elecmg]*
tesson	鉄損；てっそん	iron loss *[elecmg]*
te'suki-gami	手抄き紙；手漉き紙	handmade paper; handsheet; vat paper *[paper]*
te'suki-ki	手抄機	sheet machine *[eng]* *[paper]*
te'suri	手摺り	handrail; banister; railing *[arch]*
tesū'ryō	手数料	handling fee; commission *[econ]*
tetsu	鉄；鐵；てつ	iron (element)
tetsuban'zakuro-ishi	鉄礬柘榴石	almandine *[miner]*
tetsu'dō	鉄道	railroad; railway *[rail]*
tetsudō-renraku'sen	鉄道連絡船	railway ferry *[nav-arch]*
tetsudō-shingō	鉄道信号	railway signal *[rail]*
tetsu'gaku	哲学	philosophy (pl -phies)
tetsu-guro	鉄黒；鐵黒	iron black *[ch]*
tetsu'jūseki	鉄重石	ferberite *[miner]*
tetsu'kanran'seki	鉄橄榴石	fayalite *[miner]*
tetsu-kōseki	鉄鉱石	iron ore *[miner]*
tetsu no wata	鉄の綿	steel wool *[mat]*
tetsu-ritsu	鉄率	iron modulus *[ch]*
tetsu-saku'tai	鉄錯体	iron complex *[ch]*
tetsu-unmo	鉄雲母；てつうんも	lepidomelane *[miner]*
te'tsuzuki	手続き；てつづき	procedure *[comput]* *[pat]*
tettō	鉄塔	pylon *[civ-eng]*

tettsui	鉄槌	hammer (made of iron) *{n} [eng]*
te-uchi	手打ち	hand-made *{n} [cook] [food-eng]*
te'uchi-soba	手打ち蕎麦	hand-made buckwheat noodles *[cook]*
te-yō'setsu	手溶接	manual welding *{n} [met]*
te-yumi	手弓	handbow *{n} [ord]*
te-zukuri no	手造りの	hand-formed; hand-shaped; handmade *{adj}*
to	戸	door *[arch]*
to	都	metropolis *[geogr]*
tō; jū	十	ten *{n} [math]*
tō	塔	tower *{n} [arch]*
tō	筒	cylinder *[math]*
tō	糖	sugar *[bioch]*
tō	薹；とう	pedicel *[bot]*
-tō	-島	-island *{suffix}*
tō-asa	遠浅；遠淺	shoaling beach *[ocean]*
tō'atsu-men	等圧面	isopiestic surface *[phys]*
tō'atsu-sen	等圧線	isobar *[meteor]*
tobaseru	飛ばせる	fly; let fly *{vb t}*
tobasu	飛ばす	skip (over); jump (over) *{vb}*
tōben	答弁；答辯	reply *{n} [comm]*
tōben'sho	答弁書	response; written response *[comm]*
tobi	鳶；とび	kite; black-eared kite (bird) *[v-zoo]*
tobi'dashi-sokudo-teisū	飛出し速度定数	elutriation velocity constant *[eng]*
tobi-dasu	飛出す	pop out; leap out *{vb}*
tobi-ei	鳶鱝；とびえい	eagle ray; Myliobatus tobijei (fish) *[v-zoo]*
tobi-gaeru	飛び蛙	flying frog (amphibian) *[v-zoo]*
tobi-guchi	鳶口	fire hook *[eng]*
tobi-hi	飛び火	jump fire *{n}*
tobi-iro	鳶色；とびいろ	auburn; brown (color) *[opt]*
tobi-ishi	飛石	stepping stone *[arch]*
tobi'koshi	飛越し	jump; skip(ping) *{n} [comput]*
"	飛越し	overreach *{n}*
tōbi-kōzō	頭尾構造	head-to-tail structure *[poly-ch]*
tobi-nezumi	跳び鼠	jerboa (animal) *[v-zoo]*
tobira	扉；とびら	door *[arch]*
"	扉	title page *[pr]*
tobi-tokage	飛び蜥蜴	flying lizard *[v-zoo]*
tobi-uo	飛び魚	flying fish; Prognichthys agoo *[v-zoo]*
tōbō-kessetsu	洞房結節	sinoatrial node *{n} [anat]*
tobu	飛ぶ	fly *{vb i}*
tō'bunpai	等分配	equipartition *[ch]*

tō'bunpai-soku	等分配則	equipartition law *[stat-mech]*
tō'bunsan-sei	等分散性	homoscedasticity *[math]*
tōbyō-kōka	投錨効果	anchor effect *[adhes]*
tochaka	トチャカ	Irish moss *[bot]*
tōchaku	到着	arrival *[trans]*
tochi'ba-ninjin	栃葉人参	Panax japonicus; Japanese ginseng *[bot]*
tochi-kaihatsu	土地開発	land development *[civ-eng]*
tochi no	土地の；とちの	local; of the locality *{adj} [geogr]*
tochinoki; tochi	栃；とち(のき)	buckeye; horse chestnut tree (tree) *[bot]*
tochi-shiyō-seigen-hō	土地使用制限法	zoning law *[civ-eng] [legal]*
tōchō	塔頂	tower summit *[ch-eng]*
tōchō	盗聴；とうちょう	tapping; wiretap *{n} [comm]*
tōchō-eki	等張液	isotonic solution *[p-ch] [physio]*
tōchō-gyō'retsu	等長行列	isometric matrix *[math]*
tōchō-shazō	等長写像	isometry *[math]*
tōchō-sōchi	盗聴装置	wiretap *{n} [comm]*
tōchū	灯柱	lamp post *[civ-eng]*
tō'chūshin	等中心	isocenter *[graph]*
todaeru	途絶える	cease; halt; come to a stop *{vb}*
tōdai	灯台；燈臺	lighthouse *[arch] [navig]*
to'dana	戸棚	cupboard; pantry; wardrobe *[arch]*
tō'den'i	等電位	equipotential *[elec]*
tō'den'i-men	等電位面	equipotential surface *[elec]*
tōden no	等電の	isoelectric *{adj} [elec]*
tōden-ten	等電点	isoelectric point *[p-ch]*
tōdo	陶土	china clay; kaolin; kaolinite *[miner]*
tōdo	糖度	sugariness *{n} [cook]*
to'dō-fu'ken	都道府県	metropolis and districts; urban and rural prefectures (in Japan) *[geogr]*
todoke'sho	届け書；届け書	written notification *[legal]*
todomari	止まり	rest; stopping *{n}*
tōdo-tai	凍土帯	tundra *[ecol]*
tō'ei	投影；とうえい	projection *[map] [math]*
tō'ei-ki	投影機	projector *[eng] [opt]*
tō'ei-zu	投影図	projection drawing *{n} [graph]*
tōfu	豆腐；とうふ	bean curd *[cook]*
tō-fu'hyō	灯浮標	light buoy *[eng]*
tō'fukkaku-sen	等伏角線	isoclinic line *[geophys]*
tōgai	凍害	frost damage *[cer]*
"	凍害	freezing damage *[meteor]*
tōgai; zugai	頭蓋；とうがい；ずがい	cranium; skull *[anat]*

tōgai-kotsu; zugai-kotsu	頭蓋骨	cranium; skull *[anat]*
tōgan	冬瓜；とぅがん	wax gourd (vegetable) *[bot]*
tōgan	東岸	east coast *[geogr]*
tō'garashi	唐辛；とうがらし	cayenne (red) pepper; capsicum *[bot]*
togaru	尖る	be pointed; taper to a point *{vb i}*
togatta	尖った	acuate; pointed *{adj} [bio]*
toge	刺；棘；莿；とげ	thorn *[bot]*
tōge	峠；とうげ	mountain ridge *[geol]*
tōgen	糖原	glycogen *[bioch]*
toge no aru	刺のある	barbed *{adj} [bot] [mat]*
togi'age	研ぎ上げ	honing *{n} [mech-eng]*
tōgō	等号；等號	equality; equal sign *{n} [math]*
tōgō	統合	consolidation; integration *{n}*
tō-goma	唐胡麻	castor oil plant; Ricinus communis *[bot]*
tō'goma no mi	唐胡麻の実	castor bean *[bot]*
tō'gorō-iwashi	藤五郎鰯	silversides (fish) *[v-zoo]*
tōgo-ron(pō)	統語論(法)	syntax *[ling]*
togu	研ぐ；磨ぐ；とぐ	sharpen; whet; polish; burnish; wash rice *{vb t}*
tōgyō'sha	当業者；當業物	one skilled in the art *{n}*
tō'hatsu	頭髪	hair (of the head) *[anat]*
tō'hen no	等辺の	equilateral *{adj} [math]*
tōhi	逃避；とうひ	escape; evasion *{n}*
tōhi	頭皮	scalp *{n} [anat]*
tōhi-chūkō	等比中項	geometric mean *[math]*
tōhi-kyūsū	等比級数	geometric series *[math]*
tōhi'kyūsū-soku	等比級数則	geometric series, law of *[bio]*
tō-hishin'kei no	倒皮針形の	oblanceolate *{adj} [bot]*
tōhi-sū'retsu	等比数列	geometric progression *[math]*
tōhi-yu	橙皮油	bitter orange oil *[cook] [mat]*
tō-hō'i'kaku-sen	等方位角線	isogonic line *[geophys]*
tōhon	謄本；とうほん	copy (pl copies) *{n} [legal]*
tōhō-sei	等方性	isotropy; isotropism *[phys]*
tōhō'sei no	等方性の	isotropic *{adj} [phys]*
toi	樋	drain; gutter *{n} [arch]*
"	樋	spout *{n} [met]*
toi	問い；とい	question *{n} [comm]*
tōi	遠い；とおい	distant; far; remote *{adj} {adv} {n}*
tō'i	糖衣；とうい	sugar coating *{n} [pharm]*
toi-awase	問合せ	enquiry; inquiry; query *{n} [comput]*
tō'i-jō	糖衣錠	sugar coated tablet *{n} [pharm]*
to'ishi	砥石；といし	grindstone; whetstone *[eng]*

to'ishi-dai	砥石台	wheel spindle stock *[eng]*
tō'itsu	統一	unity *[bio]*
toji	綴じ；とじ	sewing and stitching (bookbinding) *[pr]*
tōji	冬至	winter solstice *[astron]*
tōji'go	頭辞語	acronym *[comm]*
tō'jikan'sen-zu	等時間線図	time contour map *[map]*
tōji'ki	陶磁器	pottery (pl -teries) *[cer] [mat]*
tō'jiku-shō	等軸晶	isometric crystal *[crys]*
tō'jiku-shōkei	等軸晶系	isometric system *[crys]*
tōji-ritsu	透磁率；とうじりつ	(magnetic) permeability *[elecmg]*
tōji'ritsu-kei	透磁率計	permeameter *[eng]*
tojiru	閉じる	close; shut *{vb}*
tōji-ryoku	透磁力	magnetic permeating force *[elecmg]*
tōji-sei	等時性	isochronism *[mech]*
tōji-sei	透磁性	magnetic permeability *[elecmg]*
tōji-sen	等時線	isochrone *[phys]*
tōji'sha	当事者	concerned party *{n}*
tojita-uchū	閉じた宇宙	closed universe *[astron]*
tōjō	凍上	frost heaving *{n} [geol]*
tōjō	搭乗；塔乗	boarding (a plane) *{n} [trans]*
tōjō-ben	頭上弁；頭上瓣	overhead valve *[mech-eng]*
tō'jū'ryoku-senzu	等重力線図	isogonal map *[min-eng]*
tōka	投荷	jettison *{n} [eng]*
tōka	透化	vitrification *[crys]*
tōka	透過	permeation *[ch]*
"	透過	transmission *[opt]*
tōka	等化	equalization *[electr]*
tōka	等価；等價；とうか	equivalence *[ch] [math]*
"	等価	parity *[math]*
tōka	頭化	cephalization *[zoo]*
tōka	頭花	head *{n} [bot]*
tōka	糖化	saccharification *[bioch]*
to'kachi-ishi	十勝石	obsidian *[geol]*
tōka-chūsei-yaki'tsuke-nōdo	等価中性焼付け濃度	equivalent neutral printing density *[photo]*
tōka'gata-denshi-kenbi'kyō	透過型電子顕微鏡	transmission electron microscope *[electr]*
tokage	蜥蜴；とかげ	lizard (reptile) *[v-zoo]*
tō'kaika'bi-sen	等開花日線	isoanthesic line *[meteor]*
tōka-keisū	透過係数	permeability coefficient *[fl-mech]*
"	透過係数	transmission coefficient *[opt]*
tōka-kō	透過光	transmitted light *[opt]*
tōka'kō-kenbikyō-hō	透過光顕微鏡法	transmitted light microscopy *[opt]*

tōka-kōso	透過酵素	permease *[bioch]*
tōka-kōso	糖化酵素	diastatic enzyme *[bioch]*
tō'kaku	頭殻；とうかく	capsid *[microbio]*
tōkaku'senseki	透角閃石	tremolite *[miner]*
tōkaku-takak(u)kei	等角多角形	equiangular polygon *[geom]*
tōkaku-tō'ei-zu	等角投影図	isometric drawing *{n}* *[graph]*
tōkan	陶管	pottery tube *[mat]*
tō-kankaku-chū'shutsu-hō	等間隔抽出法	equal-interval sampling method *[math]*
tōka no	等価の	equivalent *{adj}* *[logic]*
tōka-ritsu	透過率；とうかりつ	percent transmission; transmittance *[an-ch] [opt] [phys]*
"	透過率	permeability *[ch]*
"	透過率	transmissivity *[elecmg]*
"	透過率	transmission factor *[phys]*
tōka-sei	透過性	permeability; permeating property *[ch]*
tō-ka'sokudo	等加速度	uniform acceleration *[mech]*
tokasu	溶かす	melt; dissolve *{vb t}* *[ch]*
tōka-teikō	等価抵抗	equivalent resistance *[elec]*
tōka-yu	橙花油	orange flower oil; neroli oil *[mat]*
tokei	時計；とけい	clock; timepiece *{n}* *[comput] [horol]*
tōkei	統計；とうけい	statistics *[math]*
tokei'gaku	時計学；時計學	horology *[sci-t]*
tōkei'gaku	統計学	statistics (the discipline) *[math]*
tōkei'gaku'teki-jūritsu	統計学的重率	statistical weight *[math]*
tokei-ji'kake	時計仕掛	clockwork; clock mechanism *[horol]*
tokei-mawari	時計回り	clockwise rotation *[mech]*
tōkei-riki'gaku	統計力学	statistical mechanics *[phys]*
tōkei'ryō	統計量	statistic *{n}* *[math]*
tōkei-sen	等傾線	isoclinic line *[geophys]*
tō'keisha-sen	等傾斜線	isoclinal *{n}* *[geophys]*
tokei'sō	時計草	passionflower; Passiflora coerulea (flower) *[bot]*
tōkei'teki-suisoku	統計的推測	statistical inference *[math]*
tokei-zara	時計皿	watch glass *[ch]*
tōken	刀剣；刀劍	sword(s) *[ord]*
tokeru	溶ける	soluble *{adj}* *[ch]*
"	溶ける	dissolve; melt *{vb i}* *[ch]*
tōketsu-hogo-zai	凍結保護剤	cryoprotectants *[microbio]*
tōketsu-hozon	凍結保存	cryopreservation *[eng]*
tōketsu-kansō	凍結乾燥	lyophilization *[ch-eng]*
"	凍結乾燥	freeze drying *{n}* *[food-eng]*
tōketsu-ki	凍結機；凍結器	freezer *[eng]*

toki	斎; 齋; とき	meal served a priest *{n} [cook]*
toki	時	time *{n} [phys]*
toki	鴇; 朱鷺; とき	Japanese crested ibis (bird) *[v-zoo]*
tōki	冬期	winter; winter season *{n} [astron]*
tōki	当換; とうき	angelica (herb) *[bot]*
tōki	陶器	ceramics; earthenware *[cer]*
tōki	登記	registration *[pat]*
toki-bun'katsu-hō'shiki	時分割方式	time division system *[comm]*
tōki-do	透気度	air permeability *[fl-mech]*
tōki-in'satsu	陶器印刷	ceramic printing *{n} [cer] [pr]*
toki-iro	鴇色	pink (color) *{n} [opt]*
toki-kei'retsu	時系列	time series *[math]*
tō'kiseki	透輝石	diopside *[miner]*
toki-teisū	時定数	time constant *[phys]*
tokiwa-iro	常磐色; ときわいろ	malachite green (color) *[opt]*
tokkai	特開; とっかい	patent early disclosure (abbreviation) *[pat]*
tokken	特権; 特權	privilege; special rights *[legal]*
tokken-meirei	特権命令	privileged instruction *[comput]*
tokkyo	特許; とっきょ	patent *{n} [legal]*
tokkyo-bangō	特許番号	patent number *[pat]*
Tokkyo-chō	特許庁; 特許廳	Patent Office of Japan *[pat]*
Tokkyo'chō-Chōkan	特許庁長官	Director-General of the Patent Office *[pat]*
tokkyo-funsō	特許紛争	patent interference *[pat]*
tokkyo-genbo	特許原簿	patent register *[pat]*
tokkyo'hin	特許品	patented product; patented article *[pat]*
tokkyo-kanō'sei	特許可能性	patentability *[pat]*
tokkyo'ken	特許権; 特許權	patent right *[pat]*
tokkyo'ken'sha	特許権者	patentee *[pat]*
tokkyo-kōhō	特許公報	patent gazette *[pat]*
tokkyo-kyō'ryoku-jōyaku	特許協力条約	patent cooperation treaty *[pat]*
tokkyo-sei	特許性	patentability *[pat]*
tokkyo-seikyū no han'i	特許請求の範囲	Claim(s) *{n} [pat]*
tokkyo-shinsa'kan	特許審査官	examiner of patents *[pat]*
tokkyo-shutsu'gan-kōkoku	特許出願公告	patent application publication *[pat]*
tokkyo-yaku	特許薬; 特許藥	patent medicine *{n} [pharm]*
tokkyū	特急	super-express (train) *{n} [trans]*
toko	床; とこ	bed *{n} [furn]*
tō'kōdo-sen	等光度線	isophotal contour *[opt]*
toko-jirami	床蝨	bedbug (insect) *[i-zoo]*
tokon	吐根; とこん	ipecac; Cephaelis ipecacuanha *[bot] [pharm]*

tō'kon	刀根	tang *{n} [eng] [ord]*
tō'kon	痘痕	blister *{n} [eng]*
Toko'name-yaki	常滑焼	Tokoname ware *[cer]*
toko-no-ma	床の間	alcove (in room for ornament) *[arch]*
tokoroten	心天；ところてん	gelidium jelly *[cook]*
tōkō-sen	等高線	contour line *[map] [meteor]*
tōkō'sen-ue	等高線植え	contour planting *{n} [agr]*
tokō-shi	塗工紙	coated paper *[mat]*
tō-kotsu	橈骨；とうこつ	radius (pl radii) (bone) *[anat]*
tō-kotsu; abumi-bone	鐙骨；とうこつ	stirrup; stapes (bone) *{n} [anat]*
toku	解く；とく	solve *{vb} [math]*
toku'betsu	特別；とくべつ	special *{n} [sci-t]*
toku'betsu-kigō	特別記号；特別記號	special symbol *[comput] [pr]*
toku'betsu-kigō-katsu'ji	特別記号活字	dingbat *[comput] [pr]*
toku'betsu-kyūkō-ressha	特別急行列車	super-express train *[trans]*
toku'betsu-moku'teki	特別目的	specific purpose *{n}*
toku'betsu na	特別な	special *{adj}*
tokuchō	特徴	characteristic; feature *{n} [sci-t]*
toku'i-hannō	特異反応	specific reaction *[ch]*
toku'i-kai	特異解；とくいかい	singular solution *[math]*
toku'i'kei-kesshō	特異形結晶	idiomorphic crystal *[crys]*
toku'i na	特異な	peculiar; specific *{adj}*
toku'i-sei	特異性	specificity *[bio]*
toku'i-setsu'dō	特異摂動；特異攝動	singular perturbation *[math]*
toku'i'teki na	特異的な	singular; specific *{adj}*
toku'i-ten	特異点；特異點	singularity *{n} [math]*
toku'mei	匿名；とくめい	anonymity *{n}*
tokumei no	匿名の	anonymous *{adj}*
tōku ni-	遠くに-	ab- *{prefix}*
tokuri; tokkuri	徳利；徳利	sake (q.v.) server; wine pourer *[cer]*
tokusa-rui	木賊類；砥草類；とくさるい	Sphenosida *[bot]*
toku'sei	特性；とくせい	characteristic; peculiarity *{n} [sci-t]*
tokusei-chi	特性値	characteristic value *[math]*
tokusei-kansū	特性関数	characteristic function *[math]*
tokusei-kyoku'sen	特性曲線	characteristic curves *[math]*
tokusei-ondo	特性温度	characteristic temperature *[sol-st]*
toku'shitsu	特質	characteristic(s) *{n}*
toku'shu	特殊；とくしゅ	special *{n}*
tokushu-kai	特殊解	particular solution *[math]*
tokushu-kigō	特殊記号	special character *[comput]*
tokushu-kikō	特殊機構	special feature *[comput]*

tokushu-kinō	特殊機能	special function *[comput]*
tokushu-kō	特殊鋼	special alloy steel *[steel]*
tokushu-moji	特殊文字	special character *[comput] [pr]*
tokushu no	特殊の	special *{adj}*
toku'tei	特定；とくてい	specially designated; specific *{n}*
tokutei-baichi	特定培地	defined medium *[microbio]*
tokutei-hatsu'mei	特定発明	specified invention *[pat]*
tokutei-kagaku-busshitsu	特定化学物質	specified chemical substance *[mat]*
tokutei no	特定の	specific; specified *{adj}*
Tōkyō	東京；とうきょう	Tokyo *[geogr]*
tō'kyoku-kesshō	等極結晶	homopolar crystal *[crys]*
tō'kyoku-ketsugō	等極結合	homopolar bond *[p-ch]*
tōkyū	等級；とうきゅう	magnitude *[astron] [math]*
"	等級	grade *{n} [math]*
tō'kyūshū-ten	等吸収点	isosbestic point *[ch]*
to'maku	塗膜	coating film; paint film *[mat]*
tomato	蕃茄；トマト	tomato (pl -toes) (vegetable) *[bot]*
tome-gane	留め金	clasp; fastener; latch; hook *{n} [eng]*
tōmei	透明；とうめい	transparent *{n} [phys]*
tōmei-do	透明度	transparency (degree of) *{n} [phys]*
tōmei-ga	透明画	transparency *{n} [graph]*
tōmei na	透明な	transparent; pellucid *{adj}*
tōmei-tai	透明体	transparent body *[opt]*
tomeru; todomeru	止める	stop; halt; arrest *{vb t}*
tōmin	冬眠	hibernation *[physio]*
tō-mitsu	糖蜜；とうみつ	molasses *[cook]*
tō'mitsudo-sen	等密度線	isopycnic line *[meteor]*
to-mō	兎毛	rabbit fiber; rabbit hair *[tex]*
tō'morokoshi	玉蜀黍	corn (grain) *[bot]*
tō'morokoshi-denpun	玉蜀黍澱粉	corn starch *[cook] [mat]*
tō'morokoshi-yu	玉蜀黍油	corn oil *[cook] [mat]*
tomo'sen-bin	共栓瓶	bottle with ground stopper *[ch]*
tomosu	点す；點す；ともす	turn the light on *{vb}*
ton	瓲；噸；トン	ton (unit of volume, weight, mass) *[mech]*
tonakai	馴鹿；となかい	reindeer (animal) *[v-zoo]*
tōnan-keihō-ki	盗難警報器	burglar alarm *[eng]*
tonbo	蜻蛉；とんぼ	dragonfly (insect) *[i-zoo]*
tō'nensei-jōtai	等粘性状態	isoviscous state *[fl-mech]*
toneriko	梣；秦皮；とねりこ	ash (tree) *[bot]*
tō'nisshō-sen	等日照線	isohel *[meteor]*
ton-katsu	豚カツ	pork cutlet, Japanese style *[cook]*
tonneru	隧道；トンネル	tunnel *{n} [eng]*

tonosama-batta	殿様蝗	Asiatic locust (insect) *[i-zoo]*
tonshi	豚脂	lard; hog fat *{n} [mat]*
ton'ya	問屋; とんや	wholesaler *[econ]*
tōnyō'byō	糖尿病	diabetes mellitus *[med]*
tōnyū	豆乳	soybean milk *[cook]*
tōnyū-kasu	豆乳粕	soybean milk lees (dregs) *[cook]*
tōnyū-kō	投入口	input port (orifice) *[eng]*
tō'on	等温; とうおん	isothermal *{n} [thermo]*
tō'on-jika-ritsu	等温磁化率	isothermal magnetic susceptibility *[elecmg]*
tō'on-jōryū	等温蒸留	isothermal distillation *[ch]*
tō'on-sen	等温線	isotherm *{n} [geophys]*
toppan	凸版; とっぱん	relief plate *[pr]*
toppan-in'satsu	凸版印刷	letterpress printing *{n} [pr]*
toppan-nasen	凸版捺染	relief printing *{n} [tex]*
toppan-rinten-ki	凸版輪転機	rotary letterpress machine *[pr]*
toppatsu-koshō	突発故障	sudden failure *[ind-eng]*
toppū	突風; 突風	gust (of wind); squall *{n} [meteor]*
topputsu	突沸	bumping *{n} [ch]*
tora	虎; とら	tiger (animal) *[v-zoo]*
toraeru	捕える; 捉える	catch; capture; snare *{vb t}*
tora-fugu	虎河豚; とらふぐ	Fugu rubripes rubripes (fish) *[v-zoo]*
tora-gisu	虎鱚; とらぎす	Parapercis pulchella (fish) *[v-zoo]*
tōrai-ha	到来波	incoming wave *[phys]*
tōrai-kaku	到来角	arrival angle *[navig]*
tō'rai'u-sen	等雷雨線	isoceraunic line *[meteor]*
tora-kanmuri	虎かんむり; 虍	kanji (q.v.) radical for tiger crown *[pr]*
tōran'kei no	倒卵形の	obovate *{adj} [bio] [math]*
tora-no-me-ishi	虎眼石	tiger's eye; tigereye *[miner]*
tori	鳥; とり	bird *[v-zoo]*
tori	雞	chicken; hen; rooster *[v-zoo]*
tori'atsukai-setsumei'sho	取扱説明書	instruction manual; user's manual *[pr]*
tori'be	取瓶; とりべ	ladle *{n} [met]*
tori'dashi	取出し	fetch *{n} [comput]*
tori-dasu	取出す	fetch *{vb} [comput]*
tori-gai	鳥貝	cockle (shellfish) *[i-zoo]*
tori'hada	鳥肌	gooseflesh *[physio]*
tori-hen	鳥扁; 鳥偏: 鳥; とりへん	kanji (Chinese character) radical at left for long-tailed bird *[pr]*
tori'hiki	取り引き	transaction *[comput] [econ]*
tori'i	鳥居	torii (Shinto archway) *{n} [arch]*
tori-kabuto	鳥兜; とりかぶと	aconite; wolfsbane *[bot]*
tori'kae	取替え	exchange; replacement *[comput]*

tori-kawasu	取り交わす	interchange *{vb}*
tori'keshi	取(り)消し	cancel; cancellation *[comput]*
"	取消し	revocation *{n}*
tori'keshi-moji	取消し文字	cancel character *[comput]*
tori-kesu	取消す	cancel *{vb}*
tori-kyaku	鳥脚: 灬; とり=きゃく	kanji (Chinese character) radical at bottom denoting bird *[pr]*
tori-mochi	鳥黐; とりもち	birdlime *[mat]*
tori-nabe	取り鍋	ladle *{n} [met]*
tori-niku	鶏肉	chicken meat *[cook]*
tori'sage	取下げ	withdrawal *[pat]*
tori'te	取手	acceptor; recipient; taker *{n}*
tori'tsuke	取付け	mounting; attachment *{n} [eng]*
tori'tsuke-buhin	取付部品	fittings *{n} [eng]*
tori'tsuke-kana'gu	取付け金具	mounting bracket *[eng]*
tori'tsuke-mono	取付け物	fittings *{n} [furn]*
tori-tsukuri	鳥旁: 鳥; とり=つくり	kanji (q.v.) radical at right denoting long-tailed bird *[pr]*
tō'ritsu-zō	倒立像	inverted image *[opt]*
toro	とろ	tuna abdomen flesh *[cook]*
tōrō	灯籠; 燈籠	garden lantern; lantern *[furn]*
tō'roku	登録; とうろく	registration *{n}*
tōroku'bo	登録簿	directory (pl -ries) *[comput] [pr]*
tōroku-shōhyō	登録商標	registered trademark *[pat]*
tōroku suru	登録する	register *{vb} [comput] [pat]*
tōron	討論	discussion; debate *[comm]*
tororo	薯蕷; とろゝ	grated Japanese yam *[cook]*
tororo'aoi	黄蜀葵; とろろ=あおい	hibiscus *[bot]*
toru	取る	take *{vb t}*
tō-rui	豆類	legume *[bot]*
tō-rui	糖類	saccharides *[bioch]*
tō-rui	籐類	cane; rattan *{n} [furn] [mat]*
toruko-ishi	トルコ石	turquois; turquoise *{n} [miner]*
toryō	塗料	coating material; paint *{n} [mat]*
tōryō	当量; とうりょう	equivalent; equivalency *[ch] [mat]*
"	当量; 當量	equivalent weight *[ch]*
toryō-bake	塗料刷毛	paintbrush *[eng]*
tōryō-dendō-do; (-ritsu)	当量伝導度; (-率)	equivalent conductivity *[ch]*
tōryō-hannō	当量反応	equivalent reaction *[ch]*
tōryō-hi	当量比	equivalence ratio *[p-ch]*
tō'ryoku-yu	冬緑油	wintergreen oil *[org-ch]*

toryū	砥粒	abrasive grain *[mech-eng]*
toryūmu	トリウム	thorium (element)
tōsa	踏査	reconnaisance *{n} [eng]*
tosaka	鶏冠; 鳥冠; とさか	cockscomb; comb *{n} [v-zoo]*
tōsaku	盗作	piracy *[comput]*
tōsan	糖酸	saccharic acid
tōsan'do	逃散度	fugacity *[thermo]*
tōsa-sū'retsu	等差数列	arithmetic progression; arithmetic sequence *[math]*
to'satsu-zai	塗擦剤	embrocation; liniment *{n} [pharm]*
tōseki	凍石	soapstone; steatite *[miner]*
tōseki	透析; とうせき	dialysis (pl -yses) *[p-ch]*
tōseki	等積	equivalent (in area) *{n} [math]*
tōseki	陶石	porcelain stone; pottery stone *[cer]*
tōseki'butsu	透析物	dialyzate *[p-ch]*
tō'sekkō	透石膏	selenite *[miner]*
tō'senryō-zu	等線量図	isodose chart *[nucleo]*
tō'senseki	透閃石	tremolite *[miner]*
tō'setsu suru	当接する; 當接する	abut *{vb}*
tō'shaku'sei-shū'shuku	等尺性収縮	isometric contraction *[physio]*
tōsha suru	投射する	eject; project *{vb}*
toshi; nen	年; とし; ねん	year *[astron]*
tōshi	投資	investment *[econ]*
tōshi	陶歯	dental procelain *[mat]*
tōshi-bangō	通し番号; とおしばんごう	consecutive numbers; serial number; sequence number *[math]*
tōshi-do	透視度	translucency; transparency *[opt]*
toshi'gaku	都市学	urbanology *[civ-eng]*
tōshi-gaku	投資額; とうしがく	amount of money invested *{n} [econ]*
toshi-gasu	都市ガス	city gas *[mat]*
toshi-kōtsu	都市交通	urban traffic *[traffic]*
toshi-mondai	都市問題	urban problem *[civ-eng]*
to'shin	都心	civic center *[civ-eng]*
"	都心	urban center *[geogr]*
tō'shin	灯心	wick *{n} [tex]*
tōshin-sen	等深線	depth contour; isobath *[ocean]*
tōshin-sen	等震線	isoseismal line *[geophys]*
tō'shi'shitsu	糖脂質	glycolipid *[bioch]*
tō('shitsu)	糖(質)	carbohydrate; glucide; sugar *{n} [bioch]*
tō'shitsu-do	透湿度; 透濕度	water-vapor permeability *[fl-mech]*
tō'shitsu-sei	透湿性	moisture permeability *[fl-mech]*
tō'shitsu-taisha	糖質代謝	carbohydrate metabolism *[bioch]*

tōshi-zu	透視図	perspective view *[graph]*
tōsho	頭書	headnote *[pat]*
tōshō; shimo-yake	凍傷	frostbite *{n} [med]*
tōshō	糖漿；とうしょう	sugar syrup *[cook]*
tōshō-do	等照度	isolux *{n} [opt]*
tōshō'do-kyoku'sen	等照度曲線	equiluminous curve *[opt]*
toshō'jitsu	杜松実	juniper berry *[bot]*
tosho'kan	図書舘；圖書館	library (pl -braries) *[arch]*
tō'shoku	橙色	orange (color) *{n} [opt]*
tōshoku-sen	等色線	isochromatic lines *[opt]*
tōshō-sen	等照線	isophot *[opt]*
toshu'seki	吐酒石	tartar emetic; potassium antimonyl tartrate *[org-ch] [tex]*
to'shutsu	吐出；としゅつ	extrusion *[eng] [tex]*
to'shutsu-atsu	吐出圧	extrusion pressure *[mech]*
to'shutsu-ryū	吐出流	discharge flow *{n} [fl-mech]*
tosō	塗装	coating (of paint) *{n} [mat] [poly-ch]*
tōsō	痘瘡；とうそう	smallpox *[med]*
tōsoku-kō no	頭足綱の	cephalopod *{adj} [i-zoo]*
tōsoku-ondo	等速温度	isokinetic temperature *[meteor]*
tōsoku-sen	等速線	isotach *[meteor]*
tō'soku-undō	等速運動	uniform motion *[mech]*
tosō-shi	塗装紙	coated paper *[paper] [resin] [rub]*
toso('shu)	屠蘇(酒)	spiced sake (q.v.) *[cook]*
tosshi	凸子	tappet *[mech-eng]*
tōsui-sei	透水性	water permeability *[fl-mech]*
tōsui-sō	透水層	permeable layer *[geol]*
tōta	淘汰；とうた	selection *[gen]*
totan-ita	トタン板	galvanized sheet iron; sheet zinc *[met]*
tō-tanpaku'shitsu	糖蛋白質	glycoprotein *[bioch]*
to'tate-gumo	戸閉蜘蛛	trapdoor spider *[i-zoo]*
tō'ten	読点；讀點	comma *[gram] [pr]*
tōtō-kōzō	頭頭構造	head-to-head structure *[poly-ch]*
totsu	凸	convex; convexity *{n} [math] [sci-t]*
totsu-bu	凸部	convex part *[mech]*
totsu-jō	突条；突條	protruding strip *[mech]*
totsu'men-kyō	凸面鏡	convex mirror *[opt]*
totsu'nen	突燃	deflagration *[ch]*
totsu'nyū suru	突入する	project into; thrust into *{vb t}*
totsu-ō no	凸凹の；とつおうの	convexo-concave *{adj} [opt]*
totsu-takak(u)kei	凸多角形	convex polygon *[math]*
totsu'zen-hen'i	突然変異	mutation *[gen]*

totsu'zen-hen'i-yūhatsu-busshitsu	突然変異誘発物質	mutagen *[gen]*
totsu'zen no	突然の	abrupt; sudden *{adj}*
totte	把手；とって	doorknob; grip; pull *{n} [arch] [eng]*
"	把手	handle; handhold *{n} [eng]*
tottei	突堤	groin; jetty (pl jetties) *[civ-eng]*
tō'u-sen	等雨線	isohyet *[meteor]*
tōwa-shoku	等和色	secondary color *[opt]*
tōyō	東洋；とうよう	Asia; The East; Orient *[geogr]*
tōyō-dantsū	東洋段通	Oriental rug *[furn]*
tōyō-jūtan	東洋絨毯	Oriental carpet *[furn]*
tōyō-kanji	当用漢字	Chinese characters (kanji) for daily use in Japan *[comm] [pr]*
tōyō-kyū	東洋球	Eastern Hemisphere *[geogr]*
tō-yu	灯油；燈油	kerosene; lamp oil *[mat]*
tō-yu	桐油	China wood oil; tung oil; wood oil *[mat]*
tō-yu	桃油	peach oil *[mat]*
tōzai	東西；とうざい	east and west *[geogr]*
tōzai	陶材	porcelain *[cer] [mat]*
tōzai-ken	東西圏	prime vertical *[astron]*
tō'zai-nan'boku	東西南北	directions (of compass) *[geod]*
tōzai-shisū	東西指数	zonal index *[meteor]*
tozan	登山	mountaineering; mountain climbing *{n}*
tōzō	倒像	inverted image *[opt]*
tō'zoku-kamome	盗賊鴎	skua (bird) *[v-zoo]*
tsu	津	harbor; port *{n} [geogr]*
-tsū	-通	-item(s) of correspondence *{suffix}*
tsuba	唾	saliva *{n} [physio]*
tsuba	鍔；つば	flange *{n} [eng]*
"	鍔	guard; hilt (of sword) *{n} [ord]*
tsubaki	椿；つばき	camellia (flower) *[bot]*
tsubaki-yu	椿油	camellia oil *[mat]*
tsubakuro-ei	燕鱝；つばくろえい	butterfly ray; Gymnura japonica (fish) *[v-zoo]*
tsubame	燕；つばめ	swallow (bird) *[v-zoo]*
tsubame-uo	燕魚	batfish; Platax pinnatus (fish) *[v-zoo]*
tsubasa	翼；つばさ	wing *{n} [aerosp] [zoo]*
tsubo	坪	tsubo (Japanese area) (= 3.954 sq. yds.) *[mech]*
tsubo	壺	jar; pot; crock; urn *{n} [cer] [furn]*
tsubomi	蕾；つぼみ	bud *{n} [bot]*
tsubo'ryō	坪量；つぼりょう	basis weight; grammage *[mech]*

tsubo-sū	坪数	number of tsubo; acreage *[math]*
tsubu	粒	particle; grain *[mat] [phys]*
tsubushi-henkei	潰し変形	collapsing deformation *[mech]*
tsubusu	潰す; つぶす	crush; smash; mash *{vb}*
tsuchi	土	earth; soil; loam *{n} [agr] [geol]*
tsuchi	槌; つち	hammer; mallet *{n} [eng]*
tsūchi	通知	notification; notice *{n} [comm]*
tsuchi-bone; tsui-kotsu	槌骨	hammer (ear bone) *{n} [anat]*
tsuchi-buta	土豚; つちぶた	aardvark (animal) *[v-zoo]*
tsuchi-fumazu	土踏まず	arch (of foot) *{n} [anat]*
tsuchi-hen	土扁; 土偏; 扌	kanji (q.v.) radical denoting earth *[pr]*
tsūden-yōryō	通電容量	current-carrying capacity *[elec]*
tsūdō	通洞	tunneling *{n} [eng]*
"	通洞	adit *[min-eng]*
tsūfū	痛風	gout *[med]*
tsūfū-kansō-ki	通風乾燥機	draft dryer *[eng]*
tsūfū-kei	通風計	draft gage *[eng]*
tsūfū-kō	通風孔	vent; vent hole *{n} [eng]*
tsūfū suru	通風する	air; ventilate *{vb t}*
tsuge	黄楊; つげ	box tree (tree) *[bot]*
tsugi	次	next *{n}*
tsugi'awase-nui	継ぎ合せ縫い	zigzag seaming *{n} [cl] [tex]*
tsugi-ki	接木; つぎき	grafting (of branch) *{n} [bot]*
tsugi'me	継ぎ目	joint *{n} [eng]*
tsugi no	次の	next; following *{adj}*
tsugi'te	継手; 繼手	coupling; joint *{n} [eng]*
tsugu	注ぐ	pour (e. g. tea) *{vb t}*
tsugu	接ぐ	join; graft *{vb t}*
tsugumi	鶫; つぐみ	thrush (bird) *[v-zoo]*
tsūhō	通報	information *[comm]*
tsui	対; 對; つい	pair *{n} [math]*
tsui'ka	追加; ついか	addition; supplement *[pat]*
tsuika no	追加の	additional; added *{adj}*
tsuika-roku'on	追加録音	dub; dubbing *{n} [electr]*
tsuika-tokkyo	追加特許	patent of addition; additional patent *[pat]*
tsui-kotsu	椎骨	vertebra (pl -brae) *{n} [anat]*
tsui-kotsu; tsuchi-bone	槌骨	hammer (ear bone) *{n} [anat]*
tsui-seisei	対生成	pair creation *[sci-t]*
tsui'seki	追跡; ついせき	pursuit *[comput]*
tsui'seki'shi	追跡子	tracer *[ch]*
tsui'seki suru	追跡する	trace *{vb} [comput]*

tsui-shōgen	対消滅	pair annihilation *[sci-t]*
tsui'shu	堆朱	red lacquerware with relief pattern *[furn]*
tsui'tate	衝立；ついたて	screen *{n}* *[furn]*
tsuji'tsuma	辻褄；つじつま	consistency (of facts) *{n}* *[logic]*
tsuka	柄；つか	hilt; sword guard; haft *[ord]*
tsūka	通貨	currency; current money *[econ]*
tsukai-yasui	使い易い	easy to use
tsūka-kō	通過口	passage orifice *[eng]*
tsukami	摑み；つかみ	chuck; clamp; grip; gripper *{n}* *[eng]*
"	摑み	gripping *{n}* *[physio]*
tsukami-bō	摑み棒	grab bar *[eng]*
tsukami'gu	摑み具	gripper *[eng]*
tsukamu	摑む	seize; grasp; grab *{vb t}*
tsukaneru	束ねる	bundle; bunch; sheave *{vb t}*
tsūkan-kō	通関港；通關港	port of entry *[trans]*
tsukare	疲れ；つかれ	fatigue *{n}* *[mech-eng]* *[physio]*
tsukare-hason	疲れ破損	fatigue fracture *[mech]*
tsukare-shiken	疲れ試験	endurance test *[eng]*
tsukare-tsuyo-sa	疲れ強さ	fatigue strength *[mat]*
tsūka-ryo'kyaku	通過旅客	transit passenger *[trans]*
tsukau	使う	use; employ *{vb t}*
tsuke-bashira	付柱	pilaster *[arch]*
tsūkei	通径；通徑	latus rectum *[math]*
tsuke'komi	漬け込み	soaking *{n}* *[tex]*
tsuke'mono	漬け物	pickle(s) *{n}* *[cook]*
tsuki	月；つき	moon *[astron]*
tsuki no de	月の出	moonrise *[astron]*
tsuki no hyōdō	月の秤動	libration of moon *[astron]*
tsuki no iri	月の入り	moonset *[astron]*
tsuki no kasa	月の暈；つきのかさ	lunar halo *[meteor]*
tsuki no sentan	月の尖端	lunar cusp *[astron]*
tsuki no umi	月の海	mare (pl maria) *[astron]*
tsuki'age-dansō	突上げ断層	thrust fault *[geol]*
tsuki-awaseru	突合わせる	match *{vb t}* *[comput]*
"	突合わせる	butt against (one another) *{vb}*
tsuki'awase-tsugi'te	突合わせ継(ぎ)手	butt joint *{n}* *[eng]*
tsuki'awase-yō'setsu	突合わせ溶接	butt weld *{n}* *[met]*
tsuki'gaku	月学	selenology *[astron]*
tsuki-heikin-ki'on	月平均気温	monthly mean temperature *[meteor]*
tsuki-hen	月扁；月偏：月	kanji (q.v.) radical denoting moon *[pr]*
tsuki-keitai'gaku	月形態学	selenomorphology *[astron]*
tsuki'mawari	付き廻り	following; pursuing *{n}*

tsuki'mawari-sei	付き廻り性	throwing power *[met] [paint]*
tsūkin-ken	通勤圏	commutable area *[civ-eng]*
tsuki-no-wa-guma	月の輪熊	Japanese bear (animal) *[v-zoo]*
tsuki'nuki'sō	突貫草	feverroot; Triosteum perfoliatum *[bot]*
tsūki-sei	通気性	gas permeability *[fl-mech]*
tsuki-shūkai-uchū'sen	月周回宇宙船	lunar orbiter *[aero-eng]*
tsuki-tansa-ki	月探査機	lunar probe *[aero-eng]*
tsuki-yama	築山；つきやま	hill (artificial) *[arch]*
tsūkō'ken	通行権	right-of-way *{n} [traffic]*
tsūkoku	通告	notification; notice *{n} [comm]*
tsūkō'ryō-chōshū'jo	通行料徴収所	tollgate *[civ-eng]*
tsūkō'ryō-ryōkin'jo	通行料料金所	tollgate *[civ-eng]*
tsuku	突く；突く	strike; thrust; pierce *{vb}*
tsuku	着く；付く；つく	adhere; stick to *{vb} [mech]*
"	着く	arrive at *{vb} [trans]*
tsukuda'ni	佃煮；つくだに	soysauce-reduced food *[cook]*
tsukue	机	desk *[furn]*
tsukuri	旁；つくり	right-hand radical of a Chinese character
tsukuroi	繕い；つくろい	repair; patch *{n}*
tsukuroi-nuri	繕い塗	touchup (painting) *{n} [eng]*
tsukurou	繕う	repair; mend; darn *{vb t}*
tsukuru	作る；造る	make; create; construct *{vb t}*
tsukushi	土筆；つくし	field horsetail *[bot]*
tsumami'mono	摘まみ物	relish; snack *{n} [cook]*
tsumami'tori	撮まみ取り	picking *{n} [v-zoo]*
tsumamu	抓む；摘む；撮む	pinch; pick *{vb t}*
tsumari	詰(ま)り	in other words; in short *{adv}*
tsuma'yōji	爪楊子	toothpick *[eng]*
tsume	爪；つめ	nail *{n} [anat]*
"	爪	pawl *{n} [mech-eng]*
"	爪	claw; talon *{n} [zoo]*
tsume	詰め	justification *[graph]*
tsume-guruma	爪車	ratchet *[eng]*
tsume-kiri	爪切り	nail clippers *[eng]*
tsume-kusa	爪草	pearlwort; Sagina japonica *[bot]*
tsumeru	詰める	pack; cram; stuff; crowd *{vb t}*
tsumetai	冷い；つめたい	cold; chill(y); cool *{adj}*
tsumeta-sa	冷たさ	coldness (sensation) *{n} [physio]*
tsumi'kae	積替；つみかえ	transshipment *[trans]*
tsumi'kasane	積み重ね	stack(ing) *{n}*
tsumi-ki	積み木	building blocks (a toy) *[eng]*
"	積み木	piled lumber *{n} [mat]*

tsumi'ki'shiki no	積み木式の	modular *{adj} [arch]*
tsumi'ni	積荷；つみに	cargo (pl cargoes, cargos) *[trans]*
tsumu	剪む；つむ	clip; cull; pluck; pick *{vb t}*
tsumu	摘む	pick; pluck; cull *{vb}*
tsumu	積む	stack; heap; pile *{vb}*
tsumugi	紬；つむぎ	pongee *[tex]*
tsumugi-guruma	紡ぎ車	spinning wheel *[tex]*
tsumugu	紡ぐ	spin; make yarn *{vb} [tex]*
tsumuji-kaze	旋風；飄風	cyclone; whirlwind *[meteor]*
tsuna	綱；つな	rope *{n} [mat]*
tsunagi	繋ぎ；つなぎ	nexus (pl nexuses, nexus) *[comm]*
tsu'nami	津波；つなみ	seismic sea wave; tsunami *[ocean]*
tsunbo	聾；つんぼ	deafness *[med]*
tsuno	角；つの	antler(s); horn(s) *[zoo]*
tsuno'jō no	角状の	corneous *{adj} [graph]*
tsuno-tokage	角蜥蜴；つのとかげ	horned toad (reptile) *[v-zoo]*
tsuno-zame	角鮫	spiny dogfish; Squalus mitsukurii (fish) *[v-zoo]*
tsupparu	突っ張る	thrust out; stretch; insist on *{vb}*
tsuranuku	貫く；つらぬく	penetrate; pierce through *{vb}*
tsurara	氷柱；つらら	icicle *[hyd]*
tsurara-ishi	氷柱石；つららいし	stalactite *[geol]*
tsuri	釣り	fishing (for fish) *{n}*
tsuri'age-sōchi	吊上げ装置	lifting gear *[mech-eng]*
tsuri'ai	釣り合い	equilibrium; equalizing *{n} [mech]*
"	釣り合い	balance *{n} [mech-eng]*
tsuri'ai-omori	釣合い錘	counterweight *{n} [mech-eng]*
tsuri'ai-sōchi	釣合い装置	equalizer *[eng]*
tsuri-awase	釣り合わせ	balancing *{n}*
tsuri-bakari	吊り秤；つりばかり	hanging scale *[eng]*
tsuri-bari	釣(り)針；釣鉤	fishhook *[eng]*
tsuri-bashi	吊り橋	suspension bridge *[civ-eng]*
tsuri-sen	吊り線	suspension wire *[comm]*
tsuri-sen (o-tsuri)	釣り銭（おつり）	change (money) *{n} [econ]*
tsuri'te	吊り手；釣手	hanger; hanging implement *[eng]*
tsuri-zao	釣り竿；つりざお	fishing rod *[eng]*
tsūro	通路	aisle; passageway *[arch] [traffic]*
tsuru	蔓；つる	vine *[bot]*
tsuru	鶴；つる	crane (bird) *[v-zoo]*
tsurugi	剣；劍；つるぎ	sword *[ord]*
tsurugi-medaka	剣目高	swordtail (fish) *[v-zoo]*
tsuru'hashi	鶴嘴；つるはし	pick; pickaxe; mandrel *{n} [eng]*

tsuru-hebi	蔓蛇；つるへび	vine snake *[v-zoo]*
tsuru-koke'momo	蔓苔桃	cranberry (fruit) *[bot]*
tsuru'maki-bane	蔓巻ばね	helical spring *[eng]*
tsuru'maki-kaku	蔓巻角	helix angle *[math]*
tsuru'maki-sen	蔓巻線	helix (pl helixes, helices) *[math]*
tsuru-reishi	蔓茘枝；つるれいし	bitter gourd; Momordica charantia *[bot]*
tsuryūmu	ツリウム	thulium (element)
tsūsei-kenki'sei-kin	通性嫌気性菌	facultative anaerobe *[microbio]*
tsūsen-ro	通船路	navigation pass (for ships) *[navig]*
tsū'shin	通信；つうしん	communication *[comm]*
tsūshin'bun	通信文	message *[comm]*
tsūshin-eisei	通信衛星	communications satellite *[aero-eng] [comm]*
tsūshin-jōhō-den'tatsu	通信情報伝達	communications *[comm]*
tsūshin-kaisen	通信回線	(tele)communication line *[comm]*
tsūshin-ki	通信機；つうしんき	transmitter *[comm]*
tsūshin-kinō	通信機能	telecommunication facility *[comm]*
tsūshin-mō	通信網	communication network *[comm]*
tsūshin-riron	通信理論	communication theory *[comm]*
tsūshi-sei	通紙性；つうしせい	paper-passage property *[eng]*
tsūsō-shi	通草紙；つうそうし	rice paper *[mat] [paper]*
tsuta	蔦；つた	ivy (pl ivies) *[bot]*
tsuta-urushi	蔦漆	poison ivy *[bot]*
tsutsu	筒；つつ	sleeve; barrel *{n} [eng]*
"	筒	cylinder; tube *{n} [math]*
tsutsuga-mushi	恙虫；つつがむし	deerfly; gadfly (insect) *[i-zoo]*
tsutsuji	躑躅；つつじ	azalea; rhododendron (flower, bush) *[bot]*
tsutsumu	包む；包む	encase; envelope; wrap *{vb t}*
tsūwa	通話	(telephone) call; message *[comm]*
tsuwabuki	石蕗；つわぶき	silveredge; leopard plant *[bot]*
tsūwa-jikan	通話時間	duration of a phone call *[comm]*
tsūwa-kairo	通話回路	speaking (or talking) circuit *[comm] [elec]*
tsūwa-ro	通話路	channel *{n} [comm] [elec]*
tsuya	艶；つや	luster; shine; sheen *{n} [opt]*
tsuya'dashi	艶出し；つやだし	polishing *{n} [ch-eng]*
"	艶出し	glazing; lustering *{n} [tex]*
tsuya'dashi-zai	艶出し剤	lustering agent *[mat] [tex]*
tsuya'keshi	艶消し；つやけし	delustering *{n} [opt]*
tsuya'keshi-garasu	艶消し硝子	frosted glass *[mat]*
tsuya'keshi-hifuku-maku	艶消し被覆膜	matte coating film *[mat]*
tsuya'keshi-hōrō	艶消し琺瑯	matte enamel *{n} [mat]*
tsuya'keshi no	艶消しの	matted *{adj} [mat]*
tsuya'keshi-zai	艶消剤	delustering agent; matting agent *[mat]*

tsū'yaku	通訳；通譯	interpreter (a person); interpreting *[comm]*
tsū'yaku suru	通訳する	interpret *{vb} [comm]*
tsuyoi	強い	strong *{adj}*
tsuyo-sa	強さ	strength *[mech]*
tsuyo-sa (hikari no)	強さ(光の)	intensity (pl -ties) (of light) *[opt]*
tsuyu	汁；つゆ	broth; soup *[cook]*
tsuyu	露	dew *{n} [meteor]*
tsuyu; bai'u	梅雨；つゆ；ばいう	rainy season (Japanese) *[meteor]*
tsuyu-kusa	露草	day-flower; Commelina communis *[bot]*
tsuzumi	鼓；つづみ	hand drum *[music]*
tsuzura	葛籠；つづら	wicker chest, trunk, basket *[furn]*

U

u	鵜；う	cormorant (bird) *[v-zoo]*
ubu'gi	産衣；産着；うぶぎ	swaddling clothes; baby clothes *[cl]*
uchi	内；うち	inside; interior; one's home *{n}*
uchi'age	打上げ	launch *{n} [aero-eng]*
uchi'age'kanō-jikan-tai	打上げ可能時間帯	launch window *[aerosp]*
uchiba-ha'guruma	内歯歯車	internal gear *[mech-eng]*
uchi-bari	内張	inner lining *{n} [mat]*
uchi'bari-kōkan	内張鋼管	inner-lined steel pipe *[met]*
uchi'bari-renga	内張煉瓦	lining brick *[mat]*
uchi-bike	内引け	internal shrinkage *[met]*
uchi-bun'maki	内分巻；内分卷	short shunt *{n} [elec]*
uchi-gawa	内側；うちがわ	inner side *{n} [math]*
uchi'gawa ni-	内側に-	intra- *{prefix}*
uchi'gawa ni magatta	内側に曲った	recurved *{adj} [sci-t]*
uchi-ha'guruma	内歯車	internal gear *[mech-eng]*
uchi-ha'guruma-kudō	内歯車駆動	internal gear drive *[mech-eng]*
uchi-ha'mono-kō	打刃物鋼	wrought tool steel *[mat]*
uchi'kake	打(ち)掛け；裲襠	overgarment kimono *[cl]*
uchi-kami'ai	内噛み合	inner gearing *[mech-eng]*
uchi'kiri	打切り	truncation *[comput]*
uchi'kiri-gosa	打切り誤差	truncation error *[comput]*
uchi-kiru	打切る	abort *{vb} [comput]*
uchi'kizu	打ち疵；うちきず	bruise *{n} [met]*
uchi-ko	打粉	dusting powder *[mat]*
uchi'komi-(hon)sū	打込(本)数	count *{n} [tex]*
uchi'maki no	内巻きの	involute *{adj} [bot]*
uchi-mono	打物	wrought tool *[eng]*
uchi-muki	内向き	directed inwardly
uchi'muki-maku-denryū	内向き膜電流	inward membrane current *[physio]*
uchi ni-	内に-	intra- *{prefix}*
uchi-nobasu	打ち延ばす	beat thin; hammer thin *{vb t}*
uchi-nori	内法；うちのり	inside measure *{n} [math]*
uchi'nuki	打抜き	punching; blanking *{n} [eng]*
uchi'nuki-kakō	打抜き加工	blanking *{n} [eng]*
uchi'nuki-kakō-sei	打抜き加工性	punching quality *[eng]*
uchi'nuki-ryoku	打抜き力	blanking force *[eng]*
uchi-nuku	打ち抜く	blank; punch out *{vb} [eng]*
uchiwa	団扇；團扇；うちわ	fan (round, hand-held) *{n} [eng]*
uchi'wake'sho	内訳書；内譯書	detailed bill *[econ]*

u'chū	宇宙；うちゅう	universe *[astron]*
uchū-butsuri'gaku	宇宙物理学	astrophysics *[astron]*
uchū-fuku	宇宙服	space suit *[eng]*
uchū-genri	宇宙原理	cosmological principle *[astron]*
uchū-hikō'shi	宇宙飛行士	astronaut *[aerosp]*
uchū-jin	宇宙塵	cosmic dust *[astron]*
uchū-ki	宇宙機	spacecraft *[aero-eng]*
uchū-kūkan	宇宙空間	space *{n} [astron]*
uchū-kyoku	宇宙局	space station *[aero-eng]*
uchū-renraku'sen	宇宙連絡船	space shuttle *[aero-eng]*
uchū-ron	宇宙論	cosmology *[astron]*
uchū-seibutsu'gaku	宇宙生物学	exobiology; space biology *[bio]*
uchū-sei'in-ron	宇宙成因論	cosmogony *[astrophys]*
uchū'sen	宇宙船	space vehicle *[aero-eng]*
uchū-sen	宇宙線	cosmic rays *[nuc-phys]*
uchū'sen-bōen'kyō	宇宙線望遠鏡	cosmic-ray telescope *[eng]*
uchū-shinka-ron	宇宙進化論	cosmogony *[astrophys]*
uchū-tsui'bi	宇宙追尾	space tracking *{n} [aero-eng]*
uchū-yoi	宇宙酔い	spacesickness *[aerosp] [med]*
uchū-yū'ei	宇宙遊泳	space walk *{n} [aero-eng]*
ude	腕	arm *{n} [anat]*
ude-dokei	腕時計；うでどけい	watch; wristwatch *{n} [horol]*
ude'gi	腕木	crossarm; crosspiece; semaphore *[rail]*
ude'gi-shingō	腕木信号	semaphore *[comput]*
ude'gi-shingō-ki	腕木信号機	semaphore signal *[comm]*
ude'wa	腕輪	bracelet; bangle *[lap]*
udo	独活；獨活；うど	Aralia cordata (vegetable) *[bot]*
udon	饂飩；うどん	Japanese noodles *[cook]*
ue	上	above; up; upward *{n} [math]*
ue	筌；うえ	lobster trap *[eng]*
ue	飢え	hunger; starvation *[physio]*
ue'ba-kōgu	植刃工具	inserted tool *[eng]*
ue'ki	植木	potted plant *[bot]*
ue'komi	植込み	shrubbery *[agr] [bot]*
ue ni-	上に-	supra- *{prefix}*
ue ni ō	上に凹	upwardly concave *[math] [opt]*
ue ni totsu	上に凸	upwardly convex *[math] [opt]*
ue no-	上の-	over-; super-; supra- *{prefix}*
ueru	飢える；餓える	starve *{vb} [physio]*
ueru	植える；うえる	plant; set (a plant) *{vb t}*
ue'sen	上線	overline *{n} [pr]*
ue'tsuke	植付け	planting *{n} [agr]*

ugai	含嗽；うがい	gargle *{n} [physio]*
u'gai	雨害	rain damage *[ecol]*
ugai-yaku	含嗽薬	collutorium; gargle; mouthwash *{n} [pharm]*
u-gan	右岸	right bank *[geogr]*
ugatsu	穿つ	dig; drill; excavate *{vb t}*
u'gen	右玄	starboard *[navig]*
ugokanai	動かない	immobile; unmoving *{adj}*
"	動かない	does not move
ugokasu	動かす	move *{vb t}*
ugoki-uru	動き得る	motile; movable *{adj} [bio]*
ugoku	動く；うごく	move; budge *{vb i}*
"	動く	moving; mobile; working *{adj}*
ugoku-hodō	動く歩道	moving sidewalk; moving walkway *[civ-eng]*
ugui	鯎；石班魚；うぐい	dace; chub (fish) *[v-zoo]*
uguisu	鶯；うぐいす	bush warbler (bird) *[v-zoo]*
uguisu-iro	鶯色	brownish green (color) *[opt]*
u-hen	右辺；右邊	right-hand side *[math]*
u'hyō	雨氷	clear ice; glaze *{n} [hyd]*
uikyō	茴香；ういきょう	fennel; Foeniculum vulgare (herb) *[bot]*
uiroido	ウイロイド	viroid *{n} [microbio]*
uirusu	ウイルス	virus *[microbio]*
uirusu'gaku	ウイルス学	virology *[microbio]*
uirusu-ryūshi	ウイルス粒子	virus particles *[microbio]*
uji	蛆；うじ	maggot *[i-zoo]*
ujiku'kon	羽軸根	quill *[v-zoo]*
u'jō-fuku'yō	羽状複葉	pinnately compound leaf *[bot]*
u'jō no	羽状の	pinnate *{adj} [bot]*
u'ka	羽化；うか	emergence *[i-zoo]*
ukabu	浮かぶ	float *{vb}*
u'kai	迂回	detour; bypass *{n} [trans]*
u-kai; u-gai	鵜飼い	cormorant fishing *{n}*
ukai-ro	迂回路	bypass; detour road *{n} [civ-eng]*
u-kanmuri	ウ冠：宀；うかん=むり	kanji (Chinese character) radical denoting roof *[pr]*
uke'guchi	受口	socket *[elec]*
uke'ire	受入れ	acceptance; receiving *[ind-eng]*
uke'ire-kensa	受入検査；受入檢查	acceptance inspection *[ind-eng]*
uke'ire-shiken	受入試験	acceptance test *[ind-eng]*
uke-kana'mono	受金物	brace *{n} [eng]*
uke'ki	受け器	receiver; receptacle *[ch]*
uke'mi	受身	passive *{n} [microbio] [psy]*
uke'mi-men'eki	受身免疫	passive immunity *[immun]*

uke'oi-kōji 請負工事 contract work *[civ-eng]*
uke'oi'nin 請負人 contractor (a person) *[civ-eng]*
ukeru 受ける receive *{vb}*
uke'te 受け手 recipient *{n}*
uke'tori 受取り receiving *{n} [comput]*
uke(-to)ru 受(取)る receive; accept *{vb}*
uke-yōseki 受け容積 content volume *[ind-eng]*
uke-zara 受け皿 receiving dish; saucer *[cer] [cook]*
uki 浮き lifting *{n} [adhes]*
" 浮き float *{n} [eng]*
u-ki 雨季; うき rainy season *[meteor]*
uki-ami 浮き編 float stitch *[tex]*
uki-bakari 浮秤; うきばかり hydrometer *[eng]*
uki-bashi; uki-hashi 浮橋 pontoon bridge *[civ-eng]*
uki-bori 浮彫 relief; relief engraving *{n} [lap]*
uki'bori-zaiku 浮彫細工 cameo (pl -eos) *{n} [lap]*
uki-bukuro 浮き袋 float *{n} [eng]*
" 鰾; 浮き袋 air bladder; swim bladder (fish) *[v-zoo]*
uki'dashi-in'satsu-ki 浮出し印刷機 embossing press *[pr]*
uki-kasu 浮滓 dross; scum *[met]*
uki-kusa 浮草 duckweed; floating weed *[bot]*
uki-sanbashi 浮桟橋 floating pier *[civ-eng]*
uki'yo-e 浮世絵; 浮世繪 Japanese genre painting *[graph]*
u'kon 鬱金; 欝金; うこん curcuma; tumeric *[bot]*
u'kon-iro 鬱金色 saffron (yellow color) *{n} [opt]*
uku 浮く float *{vb i}*.
uma 馬; うま horse (animal) *[v-zoo]*
uma'mi 旨味 tastiness; savoriness *{n} [cook] [physio]*
uma-ni 甘煮 fish and vegetables stewed in soy sauce and sugar *[cook]*
uma'nori-fundō 馬乗り分銅 rider *[ch-eng] [eng]*
uma-rui 馬類 Equoidea (horses, etc.) *[v-zoo]*
uma'ya 厩; 廐; 馬屋 stable (for horses) *{n} [arch]*
umaya-goe; kyūhi 厩肥 manure *[agr]*
ume 梅 Japanese apricot *[bot]*
ume-bi; uzume-bi 埋め火 banked fire *[ch]*
ume-boshi 梅干; うめぼし pickled ume (q.v.) *[cook]*
ume'komi-moji 埋め込み文字 pad character *[comput]*
umeru 埋める bury *{vb t}*
ume'tate 埋立て reclamation *[civ-eng]*
ume'tate-chi 埋立地 reclaimed land *[civ-eng]*
ume-zake; ume'shu 梅酒 ume (q.v.) brandy *[cook]*

ume-zu	梅酢	ume (q.v.) vinegar; ume juice *[cook]*
umi	海; 海; うみ	ocean; sea *[geogr]*
umi	膿	pus *[med]*
umi-azami	海薊	xenia *[bot]*
umi-chiki	海地気	sea ground *[elec]*
umi-era	海鰓; うみえら	sea pen *[i-zoo]*
umi-game	海亀; 海龜	sea turtle (reptile) *[v-zoo]*
umi-giri	海霧	sea fog *[meteor]*
umi-jari	海砂利; うみじゃり	sea gravel *[geol]*
umi-kaze	海風	sea breeze *[meteor]*
umi'kaze-zensen	海風前線	sea breeze front *[meteor]*
umi-nari	海鳴り	oceanic noise *[acous]*
umi-neko	海猫	black-tailed gull (bird) *[v-zoo]*
umi no	海の	marine; oceanic *{adj}* *[ocean]*
umi-shio; kai'en	海塩; 海鹽	sea salt *[cook]* *[ocean]*
umi-tsubame	海燕	(stormy) petrel (bird) *[v-zoo]*
umi-uchiwa	海団扇; うみうちわ	sea fan *[i-zoo]*
umō	羽毛	feather *{n}* *[v-zoo]*
umō'jō no	羽毛状の	plumous *{adj}* *[bot]*
unagi	鰻; うなぎ	eel (fish) *[v-zoo]*
unagi no kaba'yaki	鰻の蒲焼	eel, broiled glazed with soy sauce *[cook]*
unaji	項; うなじ	nape (of neck) *[anat]*
unari	呻り	buzzing; hum *{n}* *[i-zoo]*
unari	唸り; うなり	beat; beats *{n}* *[phys]*
unari-shūha'sū	唸り周波数	beat frequency *[electr]*
un'bo; un'mo	雲母	mica *[miner]*
un'bo-hen	雲母片	mica flake *[miner]*
un'chin	運賃	shipping charge; cartage *[econ]* *[ind-eng]*
un'chin-hyō	運賃表	fare table; tariff table *[trans]*
un-chō	雲頂	cloud top *[meteor]*
undō	運動; うんどう	motion *{n}* *[mech]*
"	運動	exercise; locomotion; athletics *{n}*
undō-denji-yūdō	運動電磁誘導	motional induction *[elecmg]*
undō-enerugī	運動エネルギー	kinetic energy *[mech]*
undō-fuku	運動服	sports clothes *[cl]*
undō'gaku	運動学	kinematics *[mech]*
undō-gutsu	運動靴	sports shoes *[cl]*
undō-hōtei'shiki	運動方程式	equation of motion *[fl-mech]*
undō'jō	運動場	playground *[civ-eng]*
undō-kenbi'kyō	運動顕微鏡	traveling microscope *[opt]*
undō-nendo	運動粘度	kinematic viscosity *[fl-mech]*
undō no hōsoku	運動の法則	motion, laws of *[mech]*

undō-ondo	運動温度	kinetic temperature *[phys]*
undōron'teki-hōtei'shiki	運動論的方程式	kinetic equation *[mech]*
undō'ryō	運動量	momentum (pl -s, momenta) *[mech]*
undō'ryō-hozon-soku	運動量保存則	conservation of momentum, law of *[mech]*
undō-saibō	運動細胞	motor cell *[physio]*
undō-sei	運動性	motility *[bio]*
undō-seidan	運動星団	moving star cluster *[astron]*
undō-shinkei	運動神経	motor nerve *[physio]*
undō-shitsu	運動室	gymnasium (pl -s, -nasia) *[arch]*
undō-yokusei	運動抑制	immobilization *[bioch]*
une	畝；うね	ridge; furrow; groove *{n} [agr] [geol]*
un'ei	運営；運營	administration; operation *[econ]*
une-ori	畝織り	rib weave *[tex]*
uneri	うねり	wave (in glass); waviness *{n} [cer] [plas]*
"	うねり	undulation *[geol]*
"	うねり	swell *{n} [ocean]*
uneru	うねる	undulate *{vb}*
un'ga	運河；うんが	canal; waterway *[civ-eng]*
unga-kyō	運河橋	canal bridge *[civ-eng]*
unga-shōkō-ki	運河昇降機	canal elevator *[civ-eng]*
uni	雲丹；うに	sea urchin *[i-zoo]*
un-kei	雲形	cloud form *{n} [meteor]*
unkō-ro	運航路	shipping lane *[navig]*
un-kyō	雲鏡	cloud mirror *[meteor]*
un'mo	雲母	mica *[miner]*
un'nun	云々；うんぬん	et cetera (abbreviated etc.)
u-no-hana	卯の花	bean curd refuse *[cook]*
unpan	運搬	carrying; haulage; transport *{n} [trans]*
unpan'sha	運搬車	truck *{n} [eng]*
un-ryō	雲量	cloud amount *[meteor]*
un'ryō-kei	雲量計	nephelometer *[meteor]*
un'ryō-ritsu	雲量率	cloud cover *[meteor]*
un'shin-sū	運針数	stitch number per inch *[tex]*
un-shoku	暈色；うんしょく	iridescence *{n} [miner] [opt]*
Unshū-mikan	温州蜜柑	tangerine; Citrus unsiu (fruit) *[bot]*
untan-ki	運炭機	coal conveyor *[min-eng]*
un'ten	運転；運轉	operation; driving; running *{n} [ind-eng]*
unten-dokei	運転時計	drive clock *[mech-eng]*
unten-menkyo('jō)	運転免許(状)	driving license *[trans]*
unten'sha	運転者	driver (a person) *[trans]*
unten'shu	運転手	cabdriver; chauffeur *[trans]*
unten-teishi	運転停止	shutdown *{n} [comput]*

un'yō	運用	use; working; employment *{n}*
un'yō-shinrai-do	運用信頼度	operational reliability *[ind-eng]*
un'yō-yūkō-sei	運用有効性	operational effectiveness *[ind-eng]*
unzai	運材	log hauling *{n} [agr]*
unzan	運算	operation *[math]*
uo; sakana	魚；うお；さかな	fish (pl fish, fishes) *{n} [v-zoo]*
uo-gashi	魚河岸	fish market *[econ]*
uo-hen	魚扁；魚偏：⻥；うおへん	kanji (Chinese character) radical denoting fish *[pr]*
uo-jōyu (gyo'shō)	魚醤油（魚醤）	fish sauce *[cook]*
uo-za	魚座	Pisces; the Fishes *[constel]*
ura	浦	creek; inlet; seashore; beach *[geogr]*
ura	裏；うら	obverse; reverse *{n} [math]*
"	裏	back side; other side *[paper] [tex]*
"	裏	undersurface *{n}*
ura'ate-zai	裏当て材	backing material *[tex]*
ura'bari	裏張	lining *{n} [tex]*
ura-dōri	裏通り	backstreet *[civ-eng]*
ura'dōshi	裏通し	penetration *[tex]*
ura'gaki	裏書き	endorsement *{n} [econ]*
ura-gane	裏金	backplate *{n} [arch]*
ura'goshi	裏漉し；うらごし	strainer *[cook] [eng]*
ura-ito	裏糸	bobbin thread; under thread *[tex]*
ura'ji	裏地	lining *{n} [cl]*
ura'me	裏目	purl *{n} [tex]*
uran	ウラン	uranium (element)
ura-nuke	裏抜け；裏脱け	strike-through *{n} [pr]*
ura-saku	裏作；うらさく	winter crop; off season crop *[agr]*
ura'uchi	裏打ち	lining *{n} [mat]*
ura-utsuri	裏移り	offset *{n} [pr]*
ure'yuki	売れ行；賣れ行	demand (for merchandise); sales *{n} [econ]*
uri	瓜；うり	melon (fruit) *[bot]*
uri'age-genka	売上原価；賣上原價	cost of goods sold *[econ]*
u'rin	雨林	rainforest *[ecol]*
uri-te	売手	seller *{n} [econ]*
uroko	鱗；うろこ	scale (of fish) *{n} [v-zoo]*
uroko-gata	鱗形	imbrication *[graph]*
uroko-unmo	鱗雲母	lepidolite *[miner]*
ūron-cha	烏龍茶；ウーロン茶	oolong tea *[bot] [cook]*
uru	売る；賣る	sell *{vb}*
uru	得る	gain; obtain *{vb}*
urū-bi	閏日；うるうび	intercalary day *[astron]*

urū-byō	閏秒	leap second; intercalary second *[astron]*
uruchi-mai	粳米；うるちまい	nonglutinous rice *[bot] [cook]*
urū-doshi	閏年	leap year *[astron]*
urushi	漆；うるし	Japanese lacquer; japan *{n} [mat]*
urushi-ito	漆糸	urushi yarn *[tex]*
urushi-nuri	漆塗り	japanning *{n} [met]*
urū-zuki	閏月	intercalary month *[astron]*
u'ryō	雨量	rainfall; precipitation amount *[meteor]*
u'ryō-kei	雨量計	rain gage; udometer *[eng]*
u'ryoku-jurin	雨緑樹林	rain green forest *[ecol]*
usagi	兎；兎；うさぎ	rabbit (animal) *[v-zoo]*
usagi-rui	兎類	Lagomorpha (rabbits, etc.) *[v-zoo]*
u'sen-da'en-henpa	右旋楕円偏波	right-handed elliptically polarized wave *[elecmg]*
u'sen(kō)-sei	右旋(光)性	dextrorotatory *[opt]*
u'sen'tō	右旋糖	dextrose *[bioch]*
u-setsu	右折；うせつ	right turn *{n} [traffic]*
u'setsu-kinshi	右折禁止	No Right Turn *[traffic]*
u'setsu-shasen	右折車線	right-turn lane *[traffic]*
ushi	牛；うし	bull; cow; cattle (animal) *[v-zoo]*
ushi-gaeru	牛蛙	bull frog (amphibian) *[v-zoo]*
ushi-hen	牛扁；牛偏：牜	kanji (q.v.) radical for cow or bull *[pr]*
ushinau	失う	lose *{vb t}*
ushinawareta	失われた	lost *{adj}*
ushi no	牛の	bovine *{adj} [v-zoo]*
ushi-no-shita	牛の舌	tongue sole (fish) *[v-zoo]*
ushio	汐；うしお	tide: evening tide *{n} [ocean]*
ushio	潮；うしお	tide: morning tide *{n} [ocean]*
ushiro	後；うしろ	behind; back; rear *{n}*
ushiro-ita	後板	backplate *{n} [arch]*
ushiro-muki no	後向きの	backward-facing *{adj}*
ushiro no	後の；うしろの	posterior *{adj} [zoo]*
ushi-taiji-kessei	牛胎児血清	fetal calf serum *[microbio]*
ushi-zoku no	牛属の	bovine *{adj} [v-zoo]*
u'shoku	雨食；雨蝕	rainwash *{n} [geol]*
u'shoku	齲食；齲蝕	dental caries; tooth decay *[dent]*
u'soku-tsūkō	右側通行	Keep to the Right *[traffic]*
usu	臼	mortar *[sci-t]*
usu-akari	薄明り	pale light *[astron]*
usu-ban	薄板	sheet *{n} [mat]*
usu-dōshi	薄通し	tight milling *{n} [rub]*
usu-gumori	薄曇り	slightly overcast skies *{n} [meteor]*

usui 薄い dilute {adj} [ch]

" 薄い pale; thin; weak {adj}

u'sui 雨水 rainwater [hyd]

u'sui'kō 雨水溝 storm sewer [civ-eng]

usu-ita'gane 薄板金 sheet metal [met]

usu-kinzoku'ban 薄金属板 sheet metal [met]

usu-kōhan 薄鋼板 sheet iron [elec] [met]

usu'kuchi-shōyu 薄口醬油 light-colored soy sauce [cook]

usume-eki 薄め液 thinner; reducer (for paint) {n} [mat]

usumeru 薄める dilute; thin {vb t} [ch]

usu-mono 薄物 lightweight fabric [tex]

usu-murasaki 薄紫 lavender (color) {n} [opt]

usu-sa 薄さ sheerness; thinness {n} [tex]

usu-sha 薄紗；うすしゃ tarlatan [tex]

usu'yō-shi 薄葉紙 thin paper; tissue paper [mat] [paper]

uta 歌；唄；うた song; poem {n} [pr]

utagai no nai 疑いのない unquestionable {adv}

utagawashī 疑わしい doubtful; questionable {adj} {adv}

utagawashī atai 疑わしい値 suspicious value [math]

uta-tsugumi 歌鶇；うたつぐみ song thrush (bird) [v-zoo]

utau 歌う；唄う sing; recite; chant {vb}

u'teki 雨滴 raindrop [hyd]

u-ten 雨天 rainy weather [meteor]

utsu 打つ hit; strike {vb t}

utsubo 鱓；うつぼ moray; Gymnothorax kidako (fish) [v-zoo]

utsubo-gusa 靫草；うつぼぐさ selfheal; Prunella vulgaris [bot]

utsubuse no; utsubushi no 俯伏せの；俯伏しの prone; on one's face {adj}

utsukushī 美しい beautiful {adj}

utsuro na 空ろな；うつろな hollow {adj}

utsushi 写し；寫し copy (pl copies) {n} [graph]

utsushi-zome 写し染め direct printing {n} [tex]

utsusu 写す copy; map {vb}

utsusu 移す shift; transfer {vb}

utsuwa 器；噐；うつわ vessel; container [eng]

uwabami'sō (mizu'na) 蟒草（水菜） potherb mustard; Brassica japonica [bot]

uwa'bari 上張り facing {n} [eng]

uwa-biki 上引き upper coating {n} [rub]

uwa-enogu 上絵具；上繪具 overglaze color {n} [cer]

uwa'gasane-in'satsu 上重ね印刷 overprint {n} [pr]

uwa-gusuri 釉薬；釉藥 glaze {n} [cer]

uwa-iro 上色 overtone (color); top color [opt]

uwa-ito 上糸；うわいと needle thread [tex]

uwa-jiku	上軸	arm shaft *[mech-eng]*
uwa-muki no	上向きの	upwardly directed *{adj}*
uwa'muki-ryū	上向流	upward flow *[fl-mech]*
uwa'muki-tsū'fū	上向き通風	updraft *{n}* *[fl-mech]*
uwa'muki-toppū	上向き突風	upgust *{n}* *[meteor]*
uwa-nuri	上塗り；うわぬり	final coat; finishing coat; overcoat *[mat]*
uwa'nuri-toryō	上塗り塗料	top coat *[paint]*
uwa-yori	上撚り	final twist *[tex]*
uwa-zori no	上反りの	dished *{adj}* *[plas]*
uwa'zumi-eki	上澄み液	supernatant; supernatant liquid *[ch]*
uzu; uzumaki	渦；うず；うずまき	vortex (pl vortices, -es) *[astron]* *[fl-mech]*
"	渦	eddy; swirl *{n}* *[fl-mech]*
uzu	烏頭；うず	aconite root *[bot]* *[pharm]*
uzu-denryū	渦電流	eddy current *[elec]*
uzu-do	渦度	vorticity *[fl-mech]*
uzu'gata	渦形	spiral; whorl *{n}* *[math]*
uzu-ito	渦糸	vortex filament *[fl-mech]*
uzu'maki	渦巻；渦巻	swirl; whirl *{n}* *[mech-eng]*
"	渦巻	whirlpool *[ocean]*
uzu'maki-benmō'chū-rui	渦鞭毛虫類	Dinoflagellata *[i-zoo]*
uzu'maki-benmō-shokubutsu	渦鞭毛植物	Dinophyta *[bot]*
uzu'maki-denryū	渦巻き電流	eddy current *[elec]*
uzu'maki-nensei	渦巻(き)粘性	eddy viscosity *[fl-mech]*
uzu'maki-sei'un	渦巻星雲	spiral nebula *[astron]*
uzu'maki-sen	渦巻線	spiral line *[math]*
uzumeru	埋める	bury *{vb t}*
uzu-nensei	渦粘性	eddy viscosity *[fl-mech]*
uzura	鶉；うずら	quail (bird) *{n}* *[v-zoo]*
uzura-mame	鶉豆	mottled kidney beans (legume) *[bot]*
uzu-retsu	渦列	vortex street *[fl-mech]*
uzu-sen	渦線	vortex line *[fl-mech]*
uzu-shitsu	渦室	whirl chamber *[mech-eng]*
uzu-sō	渦層	vortex sheet *[fl-mech]*
uzu-tsui	渦対；渦對	vortex pair *[fl-mech]*
uzu-wa	渦環	vortex ring *[fl-mech]*

W

wa	和；わ	sum; total *[math]*
wa	輪	loop; ring; circle; wheel *{n} [eng]*
wa	環	ring *{n} [astron]*
"	環	loop *{n} [math]*
-wa; -pa	-把	-bundle(s) *{suffix}*
wa-bane	輪ばね	ring spring *[mech-eng]*
wa'bun	和分	summation *[math]*
wa'bun	和文	Japanese text; Japanese writing *[pr]*
wa'chū	話中	busy (telephone) *[comm]*
wa'chū-on	話中音	busy signal *[comm]*
wa'chū-sen	話中線	engaged line *[comm]*
wa'ei no	和英の	Japanese (and) English *{adj}*
wa-enogu	和絵具；和繪具	Japanese colors *[opt]*
wa-fuku	和服	Japanese clothing *[cl]*
wa-gama	輪窯；わがま	ring kiln *[cer] [eng]*
wa'gata-ben	輪形弁；輪形瓣	ring valve *[mech-eng]*
wa'gata-ji'shaku	輪形磁石	ring magnet *[elecmg]*
wa'gata-setsu'zoku	輪形接続	ring connection *[elec]*
wa'gen	和弦	chord *[acous]*
wagō-sei	和合性	compatibility *[bot]*
wa-gusari	輪鎖；わぐさり	endless chain *[mech-eng]*
wa-hagane	和鋼；わはがね	Japanese steel *[steel]*
wa-ha'guruma	輪歯車；輪齒車	ring gear *[mech-eng]*
wa-hō	和方	Japanese medicine (the field) *[med]*
wa-hon	和本	Japanese binding *{n} [pr]*
wai-do	歪度	skewness *[math]*
wai-jiku	ワイ軸	y axis *[math]*
waika-zai	矮化剤	growth retardant *[agr]*
wai'kyoku	歪曲；わいきょく	distortion *[photo]*
wai-ritsu	歪率	distortion factor *[comm]*
wai-ryoku	歪力	stress; tension *{n} [mech]*
wai-sei	矮性	dwarfness *[med]*
wai-sei	矮星	dwarf star *[astron]*
wai-seppen	ワイ切片	y intercept *[math]*
wa'ji	和字	Japanese (native coined) character *[comm]*
waka	和歌	Japanese poem of 31 syllables *[pr]*
waka'ba o kū (koto)	若葉を食う(こと)	browsing *{adj} [bio]*
waka-eda	若枝	sprig *[bot]*
waka'gaeri	若返り	rejuvenation; rejuvenescence *{n}*

waka-gi	若木；わかぎ	sapling; young tree *{n} [bot]*
wakai	若い	young *{adj}*
wakame	若布；和布；わかめ	wakame seaweed *[bot]*
waka-mushi	若虫	nymph (of insect) *[i-zoo]*
wa-kanmuri	ワ冠：冖；わかん=むり	kanji (Chinese character) radical denoting lid *[pr]*
wa'kan no	和漢の	Japanese (and) Chinese *{adj}*
waka'na	若菜	young greens (vegetable) *[bot]*
waka-nae	若苗	young seedling (rice) *[bot]*
waka-neri	若湅り；若練り	premature scouring *{n} [tex]*
wa'kan-yaku	和漢薬；わかんやく	Japanese and Chinese medicines; Oriental medicines *[pharm]*
wakaranai	分からない；解ら=ない	do not understand *{vb} [comm]*
wakare-zuna	別れ砂	parting sand *[met]*
wakari-yasui	分かり易い	easy to understand *[comm]*
wakaru	分かる；解る；判る	understand; comprehend *{vb} [psy]*
waka'sagi	若鷺；鰙；公魚	pond smelt (fish) *[v-zoo]*
wakasu	沸かす	boil (as water) *{vb} [cook]*
wakatsu	分かつ	divide; separate *{vb}*
wake	訳；譯；わけ	meaning; sense *{n} [comm]*
wake'gi	分葱；わけぎ	welsh onion (vegetable) *[bot]*
wakeru	分ける	part; subdivide *[math]*
waki	脇；わき	side *{n} [anat]*
waki'aka-tsugumi	脇赤鶫	redwing (bird) *[v-zoo]*
waki-bara	脇腹	flank; side (of body) *[anat]*
waki'dashi	湧き出し；涌き出し	source *[hyd]*
waki-michi	脇道；わきみち	side road *[civ-eng]*
waki-mizu	湧水	spring water; subsoil water *[hyd]*
waki-no-shita	脇の下；腋の下	armpit; underarm *{n} [anat]*
waki'zashi	脇差	short sword; auxiliary sword *[ord]*
waku	枠；わく	collar; yoke *[eng]*
"	枠	frame *{n} [math]*
wakuchin	ワクチン	vaccine *[med]*
waku'gome(-hō)	枠込(法)	flask molding *{n} [met]*
waku'sei	惑星；わくせい	planet *[astron]*
wakusei-butsuri'gaku	惑星物理学	planetophysics; planetary physics *[astrophys]*
wakusei'gaku	惑星学	planetology *[astron]*
wakusei'jō-sei'un	惑星状星雲	planetary nebula *[astron]*
wakusei'kan-busshitsu	惑星間物質	interplanetary material *[astron]*
wakusei'kan-kūkan	惑星間空間	interplanetary space *[astron]*

wakusei-undō no hō'soku	惑星運動の法則	laws of planetary motion *[astron]*
waku-sen	枠線	frame wire *[met]*
waku-tsugi'te	枠継手	yoke joint *[mech-eng]*
wa-mei	和名	Japanese name *{n}*
wa'mushi-rui	輪虫類	Rotifera *[i-zoo]*
wan	椀	bowl (of wood or lacquer) *[cook]*
wan	碗	bowl (of chinaware) *[cer] [cook]*
wan	湾; 灣; わん	bay; gulf *{n} [geogr]*
-wan	-椀	-bowls *{suffix}*
wana	罠; わな	trap; snare *{n} [eng]*
wana	輪奈	loop pile *[tex]*
wana-kairo	罠回路	trap circuit *[elec]*
wan'gake	椀掛け	panning *{n} [min-eng]*
wan'gata-antena	椀形アンテナ	dish antenna *[elecmg]*
wani	鰐; わに	crocodile; alligator (reptile) *[v-zoo]*
wani'gawa-ware	鰐皮割れ	alligatoring; crocodiling *{n} [mat] [met]*
wani-hada	鰐肌	alligator skin *[met]*
wanira	ワニラ	vanilla; Vanilla planifolia *[bot]*
wani-rui	鰐類; わに類	Crocodilia *[v-zoo]*
wanisu	ワニス	varnish *{n} [mat]*
wan'kyoku	湾曲; 灣曲	bend; curve; curvature *[geol] [math]*
wan'kyoku'bu	湾曲部	bight *{n} [geol]*
wan'kyoku-kesshō	湾曲結晶	curved crystal *[crys]*
wan'kyoku saseru	湾曲させる	incurvate *{vb t}*
wan'nyū	湾入	embayment *{n} [geol]*
wa no hō'soku	和の法則	sum rule *[math]*
wan-ryū	湾流; わんりゅう	Gulf Stream *[ocean]*
wanshō	腕章	armband *[cl]*
wansoku-dōbutsu	腕足動物	Brachiopoda *[i-zoo]*
wa'o-kitsune-zaru	輪尾狐猿	ringtail lemur (animal) *[v-zoo]*
wa'on	和音	accord *{n} [acous]*
"	和音	chord *[music]*
wa'on-kai	和音階	Japanese scale *{n} [music]*
wāpuro	ワープロ	word processor *[comput]*
wara	藁; わら	straw *[agr] [bot]*
wara-bai	藁灰	straw ashes *[agr] [ch]*
warabi	蕨; わらび	bracken; Pteridium aquilinum *[bot]*
wara-iro	藁色	straw color *[opt]*
warai	笑い	laughter *[physio]*
warai-take	笑い茸	Panaeolus papilionaceus *[bot]*
waraji-mushi	草鞋虫; わらじ虫	pill bug; sow bug *[i-zoo]*
wara'tekkō	藁鉄鉱; 藁鐵鑛	utahite *[miner]*

ware	割れ；われ	crack *{n} [sci-t]*
ware'kuchi; ware'guchi	割れ口	fracture *{n} [miner]*
ware'me	割れ目	crack; hiatus *{n} [geol]*
ware'me-fun'shutsu	割れ目噴出	fissure eruption (volcano) *[geol]*
waremokō	我亦紅；われもこう	burnet; Sanguisorba officinalis *[bot]*
wari	割	one-tenth *{n} [math]*
wari'ai	割合；わりあい	proportion; rate; ratio *{n} [math]*
wari'ate	割当；わりあて	allocation; assignment *{n}*
wari'ate-hō	割当法；割當法	quota method *[math]*
wari'ate-ryō	割当量	quota *[math]*
wari'ate-shūha'sū-tai	割当周波数帯	assigned frequency band *[comm]*
wari'biki	割引き	discount *{n} [econ] [ind-eng]*
wari'biki-ritsu	割引率	discount rate *[econ]*
wari'dashi	割出し	dividing; indexing *{n} [mech-eng]*
wari'dashi-dai	割出し台	index head; dividing head *[mech-eng]*
wari'dashi-ha'guruma-sōchi	割出し歯車装置	index gear mechanism *[mech-eng]*
wari-dasu	割出す	index *{vb}*
wari'furi	割振り	allocation *[comput]*
wari'furi-kaijo	割振解除	deallocation *[comput]*
wari-gata	割り型	split mold *[eng] [met]*
wari-ha'guruma	割(り)歯車	split gear *[mech-eng]*
wari-jiku'uke	割り軸受	split bearing *{n} [eng]*
wari-kireru	割り切れる	divisible *{adj} [math]*
wari-kireru koto	割り切れること	divisibility *[math]*
wari'komi	割(り)込み	interrupt; interruption; trap *[comput]*
wari'komi-ichi	割り込み位置	interstitial position *[sci-t]*
wari'komi-seigyo-kairo	割り込制御回路	interruption control circuit *[elec]*
wari'komi-yōkyū	割り込み要求	interrupt request *[comput]*
wari-komu	割り込む	interrupt *{vb t} [comput]*
wari'modoshi	割(り)戻し	rebate *{n} [econ]*
wari-pin	割ピン	locking pin; cotter pin *[eng]*
wari'tsuke	割(り)付	allocation; assignment *{n}*
wari-zan	割り算	division *[math]*
wari'zan-kigō	割り算記号	division symbol *[math]*
waru	割る	divide; drop below (a value) *{vb} [math]*
"	割る	split; crack apart *{vb t}*
warui	悪い；惡い；わるい	bad; wrong; injurious *{adj}*
waru-sa	悪さ	badness *[ind-eng]*
wasabi	山葵；わさび	Japanese horseradish *[cook]*
wasabi-oroshi	山葵下ろし	grated horshradish *[cook]*
wa'sai	和裁	Japanese sewing *{n} [cl] [tex]*
wasan	和算	Japanese mathematics *[math]*

wase	早生；わせ	early-maturing rice plant *[bot]*
wasei	和声；和聲	harmony *[music]*
wasei'hin	和製品	Japanese product *[ind-eng]*
waserin	ワセリン	vaseline *[mat] [pharm]*
wa'sha	話者	speaker; talker *[comm] [comput]*
washi	鷲；わし	eagle (bird) *[v-zoo]*
wa'shi	和紙	Japanese paper *[mat] [paper]*
washi-mimizuku	鷲木菟	eagle owl (bird) *[v-zoo]*
wa-shitsu	和室；わしつ	Japanese room *[arch]*
wa-shoku	和食	Japanese-style food *[cook]*
wa-shūgō	和集合	union (sets) *[math]*
wa-shūha'sū	和周波数	sum frequency *[phys]*
wasure'na-gusa	勿忘草；わすれな草	forget-me-not (flower) *[bot]*
wasureru	忘れる	forget *{vb} [psy]*
wata	綿；わた	cotton *{n} [bot] [tex]*
wata-abura-mushi	綿油虫	cotton aphid (insect) *[i-zoo]*
wata-gashi	綿菓子	cotton candy *[cook]*
wata'ge	綿毛	down (of bird) *{n} [v-zoo]*
wata-hagi-ki	綿剥ぎ機	cotton stripper *[eng]*
wata-kuri-ki	綿繰り機	cotton gin *[eng]*
wata-kuzu	綿屑	lint *{n} [tex]*
wata-maki-ki	綿蒔き機	cotton planter *[eng]*
wata'mi'hana-zō'mushi	綿実花象虫	boll weevil (insect) *[i-zoo]*
watari	渡り；わたり	transition *[comm]*
"	渡り	migration *[v-zoo]*
watari-dori	渡り鳥	migrating bird *[v-zoo]*
watashi-bune	渡し船	ferry *{n} [nav-arch]*
wata-tsumi-ki	綿摘み機	cotton picker *[eng]*
wa-tetsu	和鉄；和鐵；わてつ	Japanese iron *[met]*
watto	ワット	watt (unit of power); W *[phys]*
watto-ji	ワット時	watt-hour (unit of energy) *[elec]*
wa-yaku	和薬；和藥	Japanese medicine *[pharm]*
wa'yō no	和様の；和様の	Japanese-style *{adj}*
waza	技；わざ	craft; skill; art *{n}*
waza-mono	業物	fine Japanese sword *[ord]*
wazan	和算	Japanese mathematics *[math]*
wazato	わざと	deliberately *{adv}*
wazuka na	僅かな；わずかな	scant; a few; slight *{adj}*
wazurawashī	煩わしい	troublesome *{adj}*

Y

ya	矢	arrow *[ord]*
ya	野；や	field; plain *{n}* *[geogr]*
yabu	藪；やぶ	thicket; bush; brush; bamboo grove *[ecol]*
yabu-ka	藪蚊	striped mosquito (insect); Aedes *[i-zoo]*
yabu-kanzō	藪萱草	tawny day-lily; Hemerocallis fulva *[bot]*
yabure'me	破れ目	tear; rupture; breach *{n}*
yādo; yāru	碼；ヤード；ヤール	yard (unit of length) *[mech]*
yado-kari	宿借；やどかり	hermit crab *[i-zoo]*
yado-nushi	宿主	host *{n}* *[bio]*
yadori'gi-tsugumi	寄生木鶫	mistle thrush (bird) *[v-zoo]*
yae'zaki no	八重咲きの	double-flowered *{adj}* *[bot]*
yae-zakura	八重桜；八重櫻	double-flowered cherry tree *[bot]*
ya'gai	野外	open air; outdoor *{n}* *[meteor]*
yagen	薬研；藥研；やげん	crusher; mortar *{n}* *[pharm]*
yagi	山羊；やぎ	goat (animal) *[v-zoo]*
yagi-mō	山羊毛	goat hair *[tex]*
yagi-nyū	山羊乳	goat's milk *[cook]*
yagi-za	山羊座	Capricorn; the Goat *[constel]*
yagu	夜具	bedding *{n}* *[furn]*
yagura	櫓；矢倉	tower; turret; scaffold *{n}* *[arch]*
ya'han	夜半	midnight; dead of night *[astron]*
ya'hazu-moyō	矢筈模様	herringbone *[cl]* *[graph]*
yaiba	刃；刄；やいば	blade (of knife) *[ord]*
yaiba'jō-ten'i	刃状転位	edge dislocation *[photo]*
yaiba'uke	刃受け	knife bearing *[mech-eng]*
ya'iro-chō	八色鳥	fairy pitta (bird) *[v-zoo]*
ya-ita	矢板	sheet piling *{n}* *[civ-eng]*
ya-jirushi	矢印	arrow; arrow mark *[comput]* *[graph]*
ya'jirushi-tō	矢印灯	arrow signal; arrow light *[traffic]*
yajū	野獣；野獸	wild animal *[v-zoo]*
yakamashī	喧しい	noisy *{adj}* *{n}* *[acous]*
yakan	夜間	nighttime *[astron]*
yakan	薬缶；藥罐；やかん	kettle (of copper or brass) *[cook]*
yakan-hōsha	夜間放射	nocturnal radiation *[geophys]*
yakan-tsū'shin	夜間通信	night transmission *[comm]*
yakan('yō)-sōgan'kyō	夜間(用)双眼鏡	night glasses *[opt]*
yake	焼け；燒け；やけ	burning; yellowing *{n}* *[met]*
"	焼け	burn mark *[plas]*
"	焼け	scorch; scorching *{n}* *[rub]*

yakedo	火傷；やけど	burn (of flesh); scald *{n} [med]*
yakedo-gusuri	火傷薬	antipyrotic; burn medicine *{n} [pharm]*
yaki-bame	焼嵌め；焼き嵌め	shrinkage fit *[mech-eng]*
yaki'ire	焼入；焼き入れ	quenching; hardening; tempering *{n} [met]*
yaki'ire-jikō	焼入時効	quench aging *{n} [met]*
yaki'ire-jōtai-zu	焼入状態図	quenching diagram *[met]*
yaki'ire-ō'ryoku	焼入応力	quenching stress *[met]*
yaki'ire-sei	焼入れ性	hardenability *[met]*
yaki'ko	焼粉；やきこ	chamotte *[cer]*
yaki'modoshi	焼き戻し	tempering *{n} [met]*
yaki'modoshi-iro	焼き戻し色	temper color *[met]*
yaki'modoshi-moro-sa	焼き戻し脆さ	temper brittleness *[met]*
yaki'modoshi-zeisei	焼き戻し脆性	temper brittleness *[met]*
yaki'mono	焼き物	ceramics; pottery *[cer]*
"	焼き物	broiled dishes; roasted dishes *[cook]*
yaki-myōban	焼(き)明礬	burnt alum *[inorg-ch] [pharm]*
yaki'namashi	焼鈍(し)	annealing *{n} [eng]*
yaki-namasu	焼鈍す；やきなます	anneal *{vb} [eng]*
yaki'narashi	焼準(し)	normalizing *{n} [met]*
yakin'gaku	冶金学；やきんがく	metallurgy *[sci-t]*
yaki'sugi	焼き過ぎ	overheating *{n} [met]*
yaki-tori	焼(き)鳥	grilled (or broiled) chicken *[cook]*
yaki'tsuke	焼付け	baking *{n} [cer] [paint]*
"	焼付け	annealing (metal, alloy, glass) *[eng]*
"	焼付け	printing *{n} [photo]*
"	焼付け	burning-on *{n}*
yaki'tsuke-kōka	焼付硬化	baking hardening *{n} [paint]*
yaki'tsuke-toryō	焼付塗料	baking finish *[paint]*
yaki'tsuki	焼付き	seizure *{n} [met]*
yaki-tsuku	焼付く	gall *{vb} [met]*
yaki-ware	焼割れ	quench(ing) crack *[met]*
yaki-zakana	焼き魚	grilled or broiled fish *[cook]*
yaki-zatō	焼砂糖；やきざとう	caramel *{n} [cook]*
yakko-dōfu	奴豆腐	tofu (bean curd), in cubes *[cook]*
yakkō'gaku	薬効学；藥効学	pharmacodynamics *[pharm]*
yakkyoku	薬局	pharmacy *[arch] [pharm]*
yakkyoku-hō	薬局方	pharmacopoeia *[pharm] [pr]*
yakō	夜光	nightglow *[geophys]*
yakō'chū	夜光虫	Noctiluca *[i-zoo]*
yakō-dōbutsu	夜行動物	nocturnal animal *[v-zoo]*
yakō-toryō	夜光塗料	luminous paint *[mat]*
yakō'un	夜光雲	luminous cloud; noctilucent cloud *[meteor]*

yaku	約	approximately; about *{adv} [math]*
yaku	訳	translation *[comm] [pr]*
yaku	葯；やく	anther *[bot]*
yaku	焼く	burn; bake; fire; roast; grill *{vb t}*
yaku-baiyō	葯培養	anther culture *[bot]*
yaku'bun	約分	reduction of a fraction *{n} [math]*
yaku'butsu	薬物；藥物	drugs; medicines *[pharm]*
yaku'butsu-dōtai'gaku	薬物動態学	pharmacokinetics *[pharm]*
yaku'butsu-taisha	薬物代謝	drug metabolism *[physio]*
yakuchū-pompu	薬注ポンプ	chemical injection pump *[eng]*
yaku-eki	薬液	chemical solution *[ch]*
"	薬液	liquid medicine *[pharm]*
yaku'gai	薬害	chemical injury; spray injury *[agr]*
yaku'gaku	薬学	pharmaceutical science *[pharm]*
yaku'hin	薬品；やくひん	chemical(s) *{n} [ch]*
"	薬品	medicine(s) *[pharm]*
yakuhin-kōgai	薬品公害	chemical pollution *[ecol]*
yakuhin-seizō-kōgaku	薬品製造工学	pharmaceutical engineering *[pharm]*
yakuhin-zakka'ten	薬品雑貨店	drugstore *[pharm]*
yakuji-hō	薬事法	pharmaceutical affairs law *[legal] [pharm]*
yaku'hō	薬包；藥包	cartridge *[ord]*
yak(u)kyō-ō'dō	薬莢黄銅	cartridge brass *[met]*
yaku'mi	薬味	condiment(s); spice(s) *{n} [cook]*
yakumi'shu	薬味酒	medicated wine *[cook]*
yakuri'gaku	薬理学	pharmacology *[ch]*
yakuri-iden'gaku	薬理遺伝学	pharmacogenetics *[gen]*
yakuri-kagaku	薬理化学	pharmaceutical chemistry *[ch-eng]*
yaku'rikigaku'teki-kōka	薬力学的効果	pharmacodynamic effect *[pharm]*
yakuryō'gaku	薬量学	posology *[med]*
"	薬量学	dosimetry *[nucleo] [pharm]*
yakuryō-kei	薬量計	dosimeter *[nucleo] [pharm]*
yaku'sō	薬草	medicinal herb; medicinal plant *[pharm]*
"	薬草	simple *{n} [pharm]*
yakusō	躍層；やくそう	thermocline *[geophys]*
yaku'ten	薬店	drugstore *[arch] [pharm]*
yaku'yō-saku'motsu	薬用作物	medicinal crop *[agr]*
yaku'yō-sekken	薬用石鹼	medicated soap *[mat] [pharm]*
yaku'yō-shokubutsu	薬用植物	medicinal plant *[bot] [pharm]*
yaku'zai	薬剤；やくざい	drug; medicine *{n} [pharm]*
yakuzai'gaku	薬剤学；藥劑學	pharmaceutics *[pharm]*
yakuzai-josō	薬剤除草	chemical weeding *{n} [agr]*
yakuzai'shi	薬剤師	pharmacist *[pharm]*

yakuzai-taisei	薬剤耐性	drug resistance *[med]*
yama	山；やま	crest (screw thread) *{n} [eng]*
"	山	mountain; peak; hill *[geogr]*
"	山	crest (of a wave) *{n} [ocean]*
yama-arashi	山荒し	porcupine (animal) *[v-zoo]*
yama'ba-ha'guruma	山歯歯車	double helical gear *[mech-eng]*
yamabiko	山彦	echo (pl -es) *{n} [acous]*
yamabuki	山吹；やまぶき	Japanese rose; globeflower *[bot]*
yama'dori-zenmai	山鳥薇	cinnamon fern; Osmunda cinnamoneum *[bot]*
yama'gara	山雀；やまがら	titmouse (bird) *[v-zoo]*
yama'gata-kō	山形鋼	angle steel *[arch]*
yama'gata-zai	山形材	angle bar *[arch]*
yama-giri	山霧	mountain fog *[meteor]*
yama-gobō	山牛蒡；やまごぼう	dyers grape; pokeweed *[bot]*
yamai	病	illness; disease *[med]*
yamai-dare	病垂：疒；やまい=だれ	kanji (Chinese character) radical denoting illness *[pr]*
yama-jiso	山紫蘇；やまじそ	Mosla japonica *[bot]*
yama-kaji	山火事	forest fire *[for]*
yama-kashū	山何頸烏	cat-brier; Smilax oldhamii *[bot]*
yama-kaze	山風	mountain breeze *[meteor]*
yama-kuzure	山崩れ	earth avalanche; earthflow *[geol]*
yama'me	山女；やまめ	Oncorhynchus masou (trout-type fish) *[v-zoo]*
yama-narashi	山鳴らし	aspen; poplar (tree) *[bot]*
yama-nari	山鳴り	mountain rumble *[geol]*
yama'ne	山鼠；やまね	dormouse; marmot (animal) *[v-zoo]*
yama-neko	山猫	lynx; wildcat (animal) *[v-zoo]*
yama-no-imo	山の芋；薯蕷	Japanese yam; Dioscorea japonica *[bot]*
yama no taka-sa	山の高さ	thread height (screw) *[eng]*
yama'se	山背	yamase *[meteor]*
yama-tsu'nami	山津波	earthflow *[geol]*
yama-yuri	山百合；やまゆり	goldband lily; Lilium auratum *[bot]*
yameru	止める	stop; cease; discontinue; end *{vb}*
yami	闇；暗；やみ	darkness; dark *{n} [astron]*
yami-yo	闇夜	pitch-dark night *[astron]*
ya'mori	守宮；家守；やもり	gecko; wall lizard (reptile) *[v-zoo]*
yamō'shō	夜盲症	night-blindness; nyctalopia *[med]*
yanagi	柳；やなぎ	willow (tree) *[bot]*
yanagi-hakka	柳薄荷	hyssop (herb) *[bot]*
yanagi-kurage	柳水母	chrysaora *[i-zoo]*
yanagi-tade (ma'tade)	柳蓼（真蓼）	smartweed; Polygonum hydropiper *[bot]*
yane	屋根；やね	roof *{n} [arch]*

yane-beya	屋根部屋	attic; attic room *[arch]*
yane'buki-zai(ryō)	屋根葺き材(料)	roofing material *[arch] [mat]*
yane-gawara	屋根瓦	roof tile; roofing tile *[arch] [mat]*
yani	脂；やに	resinous exudate *[bot]*
"	脂；やに	tree resin *[bot] [resin]*
ya'on	夜温	night temperature *[thermo]*
yao'ya	八百屋；やおや	grocery store *[arch]*
yarappa-kon	药刺巴根；やらっぱ=こん	jalapa root; jalap; Ipomoea purga *[bot]*
yari	槍	spear; javelin; lance *{n} [ord]*
yari'naoshi	遣り直し	rerun; redoing *{n} [comput]*
yāru	碼；ヤール	yard (unit of length) *[mech]*
yasai	野菜；やさい	vegetable *[agr] [bot]*
yasai-saibai	野菜栽培	vegetable growing *{n} [agr]*
yasan'ken	野蚕繭	wild cocoon *[i-zoo]*
yasan-shi	野蚕糸	wild silk *[tex]*
yasashī	易しい	easy (not difficult); simple *{adj}*
yase-chi	瘦せ地	infertile land; barren soil *[agr]*
yasei-kabu	野生株	wild strain *[gen]*
yasei-kōbo	野生酵母	wild yeast *[mycol]*
yasei no	野生の	naturally growing; wild *{adj} [bot]*
yasei no	野性の	savage; wild; feral; untamed *{adj} [bio]*
yasei-seibutsu	野生生物	wildlife *[zoo]*
ya-sen	矢線	arrow; arrow line *[graph]*
yasen-zu	矢線図；矢線圖	arrow diagram *[graph]*
yashi	椰子；やし	palm; palmetto (pl -tos, -toes) *[bot]*
yashi-guma	椰子熊	sun bear (animal) *[v-zoo]*
yashi no mi	椰子の実	coconut (fruit) *[bot]*
ya'shio-ji	八潮路	long sea route *[navig]*
yashiro	社；やしろ	shrine *[arch]*
yashi-yu	椰子油；やしゆ	coconut oil *[mat]*
yashi-zu	矢視図	arrow view *{n} [graph]*
ya'shoku	夜食	supper; nighttime meal *[cook]*
yasō	野草	wild grass; wild flower *[bot]*
yasō'kyoku	夜想曲	nocturne *[music]*
yasui	安い	cheap; inexpensive *{adj} [econ]*
yasu-mono	安物	cheap goods *[ind-eng]*
yasumu	休む	rest; repose *{vb}*
yasuri	鑢；やすり	file; rasp *{n} [eng]*
yasuri-shi'age	やすり仕上	filing; file finishing *{n} [eng]*
yatai-bone	屋台骨	framework; foundation (of a house) *[arch]*
yato-mō	野兎毛	hare hair *[tex]*

yatsu'gashira	八つ頭	taro; yam (vegetable) *[bot]*
yatsu'gashira	載勝；やつがしら	hoopoe; hoopoo (bird) *[v-zoo]*
yatsu'me-unagi	八つ目鰻	lamprey; lamprey eel *[v-zoo]*
yatsu'ori-ban	八つ折り版	octavo (pl -vos) *{n} [pr]*
yattoko	鋏；やっとこ	flat pliers; nippers; tongs *[eng]*
ya'uo	矢魚；やうお	darter (fish) *[v-zoo]*
yawarakai	柔かい；軟かい	soft; pliant; tender *{adj}*
yawaraka-sa	柔かさ；軟かさ	softness *[mat]*
yo	夜	night; nighttime *[astron]*
yō	陽	positive *{n} [elec]*
"	陽	plus; plus side *{n}*
yō	葉	lobe *{n} [anat]*
-yō	-洋	-Ocean *{suffix}*
-yō	-葉	-sheet(s) *{suffix} [pr]*
yo'ake	夜明け	dawn *{n} [astron]*
yo'atsu	与圧；與圧；よあつ	pressurization *[eng]*
yo'atsu-shitsu	与圧室	pressurized cabin *[mech-eng]*
yō'bai	溶媒；ようばい	solvent *[ch]*
yōbai'ka	溶媒化	solvation *[ch]*
yōbai'wa	溶媒和	solvation *[ch]*
yōbai'wa-denshi	溶媒和電子	solvated electron *[phys]*
-yōbi	-曜日	-day of week *{suffix}*
yobi-chōsa	予備調査；豫備調査	pretest *{n} [math]*
yobi'dashi	呼出し；よびだし	call; calling; paging *{n} [comm]*
yobi'dashi-fugō	呼出符号	call sign *[comm]*
yobi'dashi-jikan	呼出時間	access time; latency time *[comput]*
yobi'dashi-kei'retsu	呼出系列	calling sequence *[comm]*
yobi'dashi-on	呼出音	ringing tone *[comm]*
yobi-dasu	呼び出す	call; call out *{vb} [comm]*
yobi'hin	予備品；よびひん	spare (article or part) *{n} [ind-eng]*
yobi-in'satsu	予備印刷	preprinting *{n} [pr]*
yobi-ka'netsu	予備加熱	preheating *{n} [plas]*
yobi-kansō	予備乾燥	predrying *{n} [plas]*
yobiko	呼子	call; pipe *{n} [comm]*
yobi-mizu	呼び水	priming water *[hyd]*
yobi-rin	呼鈴	call bell *[comm]*
yobi-seikei	予備成形	preforming *{n} [plas] [met]*
yobi-senro	予備線路	spare line *[elec]*
yobi-sunpō	呼び寸法	nominal dimension; nominal size *[mech]*
yobi-yōryō	予備容量	reserve capacity *[ind-eng]*
yobō	予防；よぼう	precaution; prevention *{n}*
yobō-hozen	予防保全	preventive maintenance *[eng]*

yobō-i'gaku	予防医学	preventive medicine *[med]*
yō'boku	幼木	young tree *[bot]*
yobō-sesshu	予防接種	vaccination *[immun]*
yobu	呼ぶ	call; summon *{vb t}*
yobun	余分；餘分；よぶん	extra; excess; surplus; superfluity *{n}*
yōbyō-ki	揚錨機	windlass *[nav-arch]*
yō'chaku	溶着；ようちゃく	melt-deposition *[met]*
yō'chaku-bu	溶着部	weld; welded part *{n} [met]*
yō'chaku-kinzoku	溶着金属	weld metal *[met]*
yochi	予知；豫知	prediction; foreknowledge *[psy]*
yō'chōseki	葉長石	petalite *[miner]*
yōchū	幼虫；幼蟲	grub; larva (pl -ae) *[i-zoo]*
yō-chū'i	要注意	Caution Necessary
yōdan-denryū	溶断電流	fusing current *[elec]*
yōdan-shiji-ki	溶断指示器	blown-fuse indicator *[elec]*
yō-denka	陽電荷；ようでんか	positive charge *[elec]*
yō-denshi	陽電子	positive electron; positron *[part-phys]*
yō'denshi-hōkai	陽電子崩壊	positron decay *[nuc-phys]*
yō'denshi-hō'shutsu-dansō-zōhō	陽電子放出断層=像法	positron emission tomography; PET *[med]*
yōdō-ban	揺動板；搖動板 ようどうばん	oscillating board; rocking board *[mech-eng]*
yodomi	淀み；よどみ	stagnation *[hyd]*
yodomi-ten	淀み点；澱み点	stagnation point *[fl-mech]*
yōeki	溶液；ようえき	solution *[ch]*
yōeki	葉腋	axil (of leaf) *[bot]*
yōeki-jūgō	溶液重合	solution polymerization *[poly-ch]*
yōen	葉縁；葉緣	leaf margin *[bot]*
yōga	葉芽	leaf bud *[bot]*
yōgan	溶岩	lava *[geol]*
yōgan-daichi	溶岩台地	lava plateau *[geol]*
yō'gasa	洋傘	umbrella *[eng]*
yōga-zō	陽画像	positive image *[photo]*
yogen	予言	forewarning; prediction *{n} [psy]*
yogen	余弦；よげん	cosine *[math]*
yogen-hōsoku	余弦法則	cosines, law of *[math]*
yōgen-kin	溶原菌	lysogen *[microbio]*
yogen-kyoku'sen	余弦曲線	cosine curve *[math]*
yōgin	洋銀	German silver; nickel silver *[mat]*
yōgo	用語；ようご	term; terminology; vocabulary; wording *{n} [comm] [ling]*
yōgo-jiten	用語辞典	glossary (pl -ries) *[pr]*

yogore	汚れ	soiling; soil; stain *{n}*
yogoreta	汚れた；よごれた	soiled; dirty *{adj}*
yōgo-saku'in	用語索引	concordance *[pr]*
yōgo'shū	用語集	glossary (pl -ries) *[pr]*
yogosu	汚す	soil *{vb}*
yōgu	用具	tool; implement *{n} [eng]*
yōgyo	幼魚	fry (young fish) *{n} [v-zoo]*
yōgyo	養魚	pisciculture *[ecol]*
yō'gyō	窯業；ようぎよう	ceramics (the industry) *[eng]*
yo'haku	余白	margin (of page) *[pr]*
yōhaku	洋白	nickel silver *[mat]*
yōhan	羊斑	flocculus (pl -li) *[astron]*
yōhei	葉柄	petiole *[bot]*
yōheki	擁壁；ようへき	retaining wall *[civ-eng]*
yōhen'jō no	葉片状の	laminar *{adj} [sci-t]*
yōhen-sei	揺変性；搖變性	thixotropy *[p-ch]*
yōhen'sei-jushi	揺変性樹脂	thixotropic resin *[resin]*
yōhi	羊皮	sheepskin *[v-zoo]*
yōhi'shi	羊皮紙	parchment; vellum *[mat]*
yohō	予報；豫報；よほう	forecast *{n} [meteor]*
yōhō	養蜂	beekeeping; bee culture *{n} [agr]*
yōhō'jo	養蜂所	apiary (pl -aries) *[agr]*
yoi	宵；よい	early evening *[astron]*
yo'ido	余緯度	colatitude *[geod]*
yo'in	余音	lingering sound; aftersound *[acous]*
yō'in	要因	factor *{n} [eng] [sci-t]*
yōi na	容易な	easy; simple *{adj}*
yo'insū	余因数	cofactor; minor *[math]*
yō-ion	陽イオン	cation; positive ion *[ch] [phys]*
yō'ion-kaimen-kassei-zai	陽イオン界面活性剤	cationic surfactant *[mat]*
yō'ion-kōkan-jushi	陽イオン交換樹脂	cation-exchange resin *[org-ch]*
yōi suru	用意する	ready; prepare *{vb}*
yōji	幼時	childhood *[psy]*
yōji	楊枝；楊子；ようじ	toothpick *[eng]*
yo'jigen	四次元	fourth dimension *[math]*
yōji-ki	幼時期	childhood *[psy]*
yoji-kōzō	四次構造	quaternary structure (of protein) *[bioch]*
yōjin	用心	caution; precaution; care *{n} [psy]*
yoji'nobori-shokubutsu	攀じ上り植物	climbing plant *[bot]*
yo-jishō	余事象	complementary event *[math]*
yōji-uo	楊枝魚	pipefish; Syngnathus schlegeli *[v-zoo]*
yojō	余剰；餘剰	surplus; excess; remainder; margin *{n}*

yōjo	葉序；ようじょ	phyllotaxy; leaf arrangement *[bot]*
yōjō	葉状	lobate; leaf-form *[bio]*
yōjō	養生	curing (cement) *{n} [civ-eng]*
yōjō-roka-ki	葉状沪過器	leaf filter *[ch]*
yōjō-shokubutsu	葉状植物	thallophyte *[bot]*
yōjō'tai no	葉状体の	frondose *{adj} [bot]*
yōka'butsu	沃化物；ようかぶつ	iodide *[ch]*
yōka-chisso	沃化窒素	nitrogen iodide
yōka-dō	沃化銅	copper iodide
yōka-gin	沃化銀	silver iodide
yōkai	溶解；ようかい	dissolution; melting *[ch]*
"	溶解	lysis (of cells) *[cyt]*
yōkai-do	溶解度	solubility *[p-ch]*
yōkai'do-keisū	溶解度係数	solubility parameter *[poly-ch]*
yōkai-netsu	溶解熱	heat of solution *[thermo]*
yōkai-sei	溶解性	dissolving property *[ch]*
yōkai'sei-denpun	溶解性澱粉	soluble starch *[mat]*
yōkai'so	溶解素	lysin *[immun]*
yo'kajū	予荷重	preloading *{n} [eng]*
yōka-keiso	沃化硅素	silicon iodide
yōka-kin	沃化金	gold iodide
yo-kaku	与格	dative case *[gram]*
yo-kaku	余角；餘角	complementary angles *[math]*
yōkan	羊羹；ようかん	bean-jelly confection *[cook]*
yōka-namari	沃化鉛	lead iodide
yo'kansū	余関数	cofunction *[math]*
"	余関数	complementary functions *[math]*
yōka-suigin	沃化水銀	mercury iodide
yōka-suiso	沃化水素	hydrogen iodide
yōka'suiso-san	沃化水素酸	hydroiodic acid
yōka-suzu	沃化錫	tin iodide
yo'katsu	余割	cosecant *[math]*
yōkei	養鶏；ようけい	poultry farming *{n} [agr]*
yōken	要件	requirement(s) *[comput]*
yōketsu	溶血	hemolysis *[physio]*
yōki	容器	container; receptacle; vessel *[eng]*
yōkin	溶菌；ようきん	bacteriolysis *[microbio]*
yōkin-kansen	溶菌感染	lytic infection *[microbio]*
yōkin-kōso	溶菌酵素	lytic enzyme *[bioch]*
yōki'seki	陽起石	actinolite *[miner]*
yoko	横；よこ	transverse; width *{n} [math] [phys]*
"	横	side; sidewise *{n}*

yoko	余弧	complementary arc *[math]*
yokō	余効	aftereffect *[ind-eng]*
yōkō	熔鋼；ようこう	ingot steel; molten steel *[steel]*
yoko-bai	横這い	sidewise moving (curve) *{n} [math]*
yoko-biki	横引き	crosscut sawing *{n} [eng]*
yoko'biki-nokogiri	横引き鋸	crosscut saw *[eng]*
yōkō-chū	陽光柱	positive column *[electr]*
yoko'dachi-sōchi	横裁ち装置	cross cutter *[eng]*
yoko-dansei-keisū	横弾性係数	modulus of transverse elasticity *[mech]*
yoko-enshin-ki	横延伸機	tenter *{n} [tex]*
yoko-giru	横切る	cross; traverse *{vb t}*
yoko'gitte-	横切って-	trans- *{prefix}*
Yokohama	横浜；横濱	Yokohama *[geogr]*
yōkō-hi	揚抗比	lift-to-drag ratio *[aero-eng]*
yoko-hippari-zuyo-sa	横引張強さ	transverse tensile strength *[mech]*
yoko-hōkō	横方向	cross(wise) direction; transverse direction *[plas] [sci-t]*
yoko-hōkō-jika	横方向磁化	transverse magnetization *[eng-acous]*
yoko'ho'sen	横帆船	square-rigger *[nav-arch]*
yoko-ito	緯糸；横糸	weft; filling; woof *{n} [tex]*
yoko-jiku	横軸；よこじく	abscissa; axis of abscissas *[math]*
yoko-kaze	横風	crosswind *[meteor]*
yoko-keisha	横傾斜	list; listing (of a ship) *{n} [nav-arch]*
yoko-kōka	横効果	transverse effect *{n}*
yoko-nami	横波	transverse wave *[phys]*
yo'kongō-ki	予混合機	premixer *[eng]*
yoko ni naru	横になる	lie down *{vb i}*
yoko ni suru	横にする	lay on side *{vb t}*
yōkō-ro	溶鉱炉；溶鑛爐	blast furnace *[met]*
yoko-ryoku	横力	lateral force *[mech]*
yoko-shindō	横振動	transverse vibration *[mech]*
yoko-suberi	横滑り；横辷り	sidewise skidding; skid *{n} [traffic]*
yoko-yure	横揺れ；横搖れ	roll(ing) *{n} [mech] [nav-arch]*
yoko'yure-shindō	横揺れ振動	rocking vibration *[spect]*
yoko-zahyō	横座標	abscissa; axis of abscissas *[math]*
yoku	浴	bath *[ch] [met]*
yoku	翼	airfoil *[aero-eng]*
yoku; tsubasa	翼；よく；つばさ	wing *{n} [aero-eng] [zoo]*
yoku	翼	blade; vane *[eng]*
yoku'gata	翼型	airfoil *[aero-eng]*
yoku-haba	翼幅	wing span *[aero-eng]*
yoku-hi	浴比	bath ratio *[tex]*

yoku-jitsu	翌日；よくじつ	next day *{n}*
yoku kiku	よく聞く	listen carefully *{vb}*
yoku kiku	よく効く	work well; be very effective *{adj}*
yoku'on-pu	抑音符	grave accent *[ling]*
yoku'sei	抑制；よくせい	suppression *[comput]*
"	抑制	inhibition *[psy] [sci-t]*
"	抑制	repression *[psy]*
yokusei-inshi	抑制因子	repressor *[bioch]*
yokusei-zai	抑制剤	inhibitor *[ch]*
yokushi	抑止	deterrence *{n} [mil]*
yoku-shitsu	浴室	bathroom *[arch]*
yoku-shōmei-tō	翼照明灯	wing-illuminating light *[aero-eng]*
yoku'shū	抑銹；よくしゅう	rust prevention *[met]*
yokushū-zai	抑銹剤	rust inhibitor *[met]*
yoku'tan-tō	翼端灯	wingtip light *[aero-eng]*
yoku-to'shutsu-ryū	翼吐出流	blade-expelled flow *[fl-mech]*
yoku'yō	抑揚	inflection *[ling]*
yoku-yoku-jitsu	翌翌日；翌々日	next next day; day after next *{n}*
yoku'yō-onpu	抑揚音符	circumflex accent *[ling]*
yō-kyaku	葉脚	leaf base *[bot]*
yō-kyoku	陽極；ようきょく	anode; positive electrode *[elec] [p-ch]*
yōkyoku-dasshi	陽極脱脂	anodic degreasing *{n} [met]*
yōkyoku-denryū	陽極電流	anode current *[electr]*
yōkyoku-eki	陽極液	anolyte *[ch]*
yōkyoku-kenma	陽極研摩	anodic polishing *{n} [met]*
yōkyoku-sen	陽極線	anode rays *[electr]*
yōkyoku-shori	陽極処理	anodic treatment; anodizing *{n} [met]*
yōkyū	要求；ようきゅう	request; requirement(s); demand *[comput]*
yōkyū-hannō	溶球反応	bead reaction *[an-ch]*
yōmaku	羊膜	amnion *[embryo]*
yōmaku-rui	羊膜類	amniota *[v-zoo]*
yome'na	嫁菜	aster; starwort; Aster yomena (vegetable)
yomi	読み；讀み；よみ	reading *{n} [eng]*
"	読み	Japanese pronunciation (for a Chinese character) *[ling]*
yomi'dashi-sen'yō-ki'oku	読み出し専用記憶	read-only memory; ROM *[comput]*
yomi'gaeru	蘇る；甦る	revive *{vb i}*
yomi'mono	読物	reading material *[comm] [pr]*
yomi'tori	読み取り	reading *{n} [comput]*
yomi'tori-bōen'kyō	読取り望遠鏡	reading telescope *[opt]*
yomi'tori-kenbi'kyō	読取り顕微鏡	reading microscope *[opt]*
yomi'tori-sen'yō-memorī	読取専用メモリー	read-only memory; ROM *[comput]*

yomi'tori-sōchi	読(み)取り装置	reader *[comput]*
yōmō	羊毛; ようもう	wool (of sheep) *[tex] [v-zoo]*
yomogi	蓬; 艾; 蒿; よもぎ	mugwort; wormwood; Artemisia princeps *[bot]*
yomogi'giku	蓬菊	tansy (pl tansies) *[bot]*
yōmō-hyō'haku	羊毛漂白	wool bleaching *{n} [tex]*
yōmō-rō	羊毛蠟	wool wax; wool fat; wool grease *[mat]*
yōmō-seiren	羊毛精練	wool scouring *{n} [tex]*
yōmō'shi	羊毛脂	lanolin *[mat]*
yomu	読む	read *{vb t} [comm] [comput]*
yō-myaku	葉脈	leaf vein; nervure *[bot]*
yō'nashi'gata no	洋梨形の	pyriform *{adj} [bot] [i-zoo]*
yonbai	四倍; よんばい	quadruple *{n}*
yonbai ni naru	四倍になる	quadruple *{vb i}*
yonbai ni suru	四倍にする	quadruple *{vb t}*
yonbai-seido	四倍精度	quadruple precision *[eng]*
yon'bun no ichi	四分の一	one-quarter; quarter *{n} [math]*
yonbun-no-ichi hachō'ban	四分の一波長板	quarter-wave plate *[opt]*
yon'bun no san	四分の三	three-quarters *{n} [math]*
yo'netsu	予熱; 豫熱; よねつ	preheating *{n} [thermo]*
yo'netsu	余熱; 餘熱	residual heat *[thermo]*
yongen-gōkin	四元合金	quaternary alloy *[met]*
yongen-kyōjūgō'tai	四元共重合体	quadripolymer *[poly-ch]*
yo'nin; yottari	四人	four persons *{n}*
yonji no	四次の	biquadratic *{adj} [math]*
yonka	四価; 四價	quadrivalence; tetravalence *[ch]*
yonka-arukōru	四価アルコール	tetrahydric alcohol *[org-ch]*
yonka no	四価の	quadrivalent; tetravalent *{adj} [ch]*
yon'kyoku-hō	四極法	four-electrode method *[electr]*
yon'kyoku-kan	四極管	tetrode *[electr]*
yon'kyū'ka	四級化	quaternization *[ch]*
yonrin(-jidō)'sha	四輪(自動)車	four-wheeled vehicle *[mech-eng]*
yonrin-kudō	四輪駆動	four-wheel drive *[mech-eng]*
yonryō'tai	四量体	tetramer *[poly-ch]*
yonsen'shiki-kaisen	四線式回線	four-wire circuit *[comm]*
yonsen'shiki-tsūshin-ro	四線式通信路	four-wire channel *[comm]*
yontan'tō	四炭糖	tetrose *[bioch]*
yontō-rui	四糖類	tetrasaccharides *[bioch]*
yon'yōso-moderu	四要素モデル	four-parameter model *[poly-ch]*
yō'on	拗音; ようおん	contracted sound *[ling]*
yō'on	揚音	accent *{n} [ling]*
yō'on-pu	揚音符	acute accent *[ling] [pr]*
yore	寄れ	slippage *{n} [tex]*

yorei-ki	予冷器	precooler *[mech-eng]*
yō'ren	溶錬; 溶錬	smelting *{n} [met]*
yori	撚り; より	twist *{n} [tex]*
yōri	溶離	elution *[ch]*
yori-chijimi	撚縮み	twist shrinkage *[tex]*
yori chīsai	より小さい	smaller than *{adj} {n} [math]*
yori-ito	撚糸	twisted thread *[tex]*
yori-keisū	撚り係数	twist coefficient *[tex]*
yori'modoshi	撚り戻し	untwisting *{n} [tex]*
yori-ni'sen	撚り二線	twisted pair *[elec]*
yori ōi	より多い	more than *[math]*
yori-sen	撚り線	twisted wire *[mat]*
yori sukunai	より少い	less than *[math]*
yōri-zai	溶離剤; ようりざい	eluent *[mat]*
yōro	窯炉; 窯爐; ようろ	furnace; kiln; oven *[eng]*
yoroi	鎧; よろい	armor *{n} [ord]*
yoroi'do	鎧戸	shutter *{n} [arch]*
yōrō-seki	葉蠟石	pyrophyllite *[miner]*
yoru; yo	夜; よる; よ	night *{n} [astron]*
yōryō	用量	dose; dosage *{n} [med] [pharm]*
yōryō	容量; ようりょう	volume; capacity *[math] [sci-t]*
yōryō-bunseki	容量分析	volumetric analysis *[an-ch]*
yōryō-hyaku'bun-ritsu	容量百分率	volume percent(age) *[math]*
yō-ryoku	揚力	dynamic lift; lift *{n} [aero-eng]*
yō'ryoku-keisū	揚力係数	lift coefficient *[aero-eng]*
yōryoku'so	葉緑素	chlorophyll *[bioch]*
yōryoku'tai	葉緑体	chloroplast *[bot]*
yōryō'sei-fuka	容量性負荷	capacitive load *[elecmg]*
yōryō'teki-ketsu'gō	容量的結合	capacitive coupling *{n} [elec]*
yosa	良さ	goodness; merit *{n} [ind-eng]*
yōsai-rui	葉菜類	leafy vegetables *[cook]*
yosan	予算; 豫算	budget; estimated cost *[econ]*
yōsan	葉酸	folic acid *[bioch]*
yōsan	養蚕; ようさん	sericulture; silkworm culture *[tex]*
yosa no shisū	良さの指数	figure of merit *[electr]*
yose	寄せ	justification *[comput] [graph]*
yose'gi	寄木; よせぎ	marquetry; parquetry *{n} [furn]*
yose'gi-zaiku	寄木細工	wooden mozaic *[furn]*
yose'gi-zaiku no yuka	寄木細工の床	parquet flooring *{n} [arch]*
yōsei	幼生	larva (pl larvae) *[i-zoo]*
yōsei-shokubutsu	陽生植物	sun plant *[bot]*
yōseki	容積; ようせき	volume *[math]*

yōseki-ritsu	容積率	volume fraction *[math]*
yōsen	溶銑；ようせん	molten pig iron *[met]*
yōsen	葉先	leaf apex *[bot]*
yose-nabe	寄せ鍋	chowder *[cook]*
yosete kaesu nami	寄せて返す波	backwash *{n} [ocean]*
yo'setsu	余接	cotangent *[math]*
yō'setsu	溶接；ようせつ	welding *{n} [met]*
yōsetsu-bu	溶接部	weld joint *[met]*
yōsetsu-chūzō	溶接鋳造	fusion casting *{n} [met]*
yose-tsugi	寄せ接ぎ	approach grafting *{n} [bot]*
yōsetsu-ki	溶接機	welder (machine) *[eng] [met]*
yōsetsu-mippei	溶接密閉	hermetic seal *[eng]*
yōsetsu-sei	溶接性	weldability *[met]*
yōsetsu'shi	溶接士	welder (a person) *[met]*
yose-zan	寄せ算	addition *[math]*
yōsha	溶射	flame spray; spray coating *{n} [eng]*
yoshi	葦；蘆；よし	reed *[bot]*
yōshi	用紙	form; sheet (of paper) *[paper] [pr]*
yōshi	羊脂	lanolin; mutton tallow; wool grease *[mat]*
yōshi	要旨	substance; content matter *[pr]*
yōshi	陽子；ようし	proton *[phys]*
yōshi-fusoku	用紙不足	Paper Insufficiency (copier) *[electr]*
yōshi-ken	陽子圏	protosphere *[meteor]*
yō'shiki	様式；様式	format; layout *{n} [comput] [pr]*
yoshi-kiri	葦切り	reed warbler (bird) *[v-zoo]*
yoshi'kiri-zame	葦切り鮫	blue shark; Glyphis glaucus (fish) *[v-zoo]*
yoshin	予震；豫震；よしん	foreshock *[geophys]*
yoshin	余震；餘震	aftershock *[geophys]*
yōshin	溶浸	infiltration *[geol] [hyd] [met]*
yō-shin	葉身	leaf blade; lamina *[bot]*
yōshi-okuri-kikō	用紙送り機構	sheet feeder *[comput]*
yō'shitsu	溶質	solute *[ch]*
yōshō	葉鞘；ようしょう	leaf sheath *[bot]*
yo-shoku	余色	complementary color *[opt]*
yō'shoku	洋食	Western (Occidental) food *[cook]*
yō'shu	洋酒	Western wines and liquor *[cook]*
yo'shuku saseta	予縮させた	preshrunk *{adj} [cl] [tex]*
yō-shutsu	溶出	elution *[ch]*
yō'shutsu-eki	溶出液	eluate *[ch]*
yō'shutsu-zō	溶出像	elution profile *[ch]*
yoso	他所；余所；餘所	elsewhere; strange parts *{adv} {n}*
yosō	予想	prediction *[psy]*

yōso	沃素；ようそ	iodine (element)
yōso	要素	element *[math] [sci-t]*
yōsō	沃曹	abbreviation for sodium iodide
yōsō	葉層	lamina (pl -nae, -nas) *[geol]*
yōso-ion	沃素イオン	iodide ion *[ch]*
yōso-ka	沃素価	iodine number; iodine value *[an-ch]*
yo'soku	予測	forecast; prediction *{n} [psy]*
yōson	溶損	melting loss *[met]*
yōso-san	沃素酸	iodic acid *[inorg-ch]*
yōso'san-en	沃素酸塩	iodate *[inorg-ch]*
yosō-seinō-kyoku'sen	予想性能曲線	predicted performance curve *[eng]*
yōso-taisha	沃素代謝	iodine metabolism *[physio]*
yōso-teki'tei	沃素滴定	iodometry; iodimetry *[an-ch]*
yosu	止す；よす	stop; desist; discontinue *{vb}*
yōsu	様子；樣子；ようす	aspect; state *{n}*
yosū	余数；餘數	complementary number *[math]*
yōsui	用水	service water *[hyd]*
yōsui	揚水；ようすい	pumping (of water) *{n} [hyd]*
yosui-ro	余水路	spillway *[civ-eng]*
yōsui-senshi	羊水穿刺	amniocentesis *[med]*
yōsui-sō	用水槽	rainwater barrel (or tank) *[eng]*
yo'sumi	四隅	four corners, the *{n}*
yōtai'ka-shori	溶体化処理	solution (heat treatment) *{n} [met]*
yo'taka	夜鷹	nightjar (bird) *[v-zoo]*
yotei	予定；よてい	schedule; program; plan *{n} [sci-t]*
yōtei	揚程	pump head *[fl-mech]*
yotei-bi	予定日	scheduled date *[ind-eng]*
yōteki	溶滴	droplet; globule (welding) *[met]*
yōten	要点	critical point; key point *{n}*
yō-tetsu	熔鉄；熔鐵	ingot iron *[met]*
yōto	用途；ようと	use; service *{n}*
yōtō	溶湯	liquid metal; molten metal; melt *{n} [met]*
yōton	養豚	hog raising *{n} [agr]*
yotsu'ashi-jū	四足獣	tetrapod; four-footed animal *[v-zoo]*
yotsu'ba	四つ葉	quatrefoil *{n} [arch] [lap]*
yotsu'gumi	四つ組	tetrad *[math]*
yotsu'ori-ban	四つ折り版	quarto (pl quartos) *[pr]*
yotsu'wari	四つ割り	quartering *{n}*
yotsu'zume-ikari	四つ爪錨	grapnel *[nav-arch]*
yotto	ヨット	yacht *[nav-arch]*
yowai	齢；齡；よわい	age *{n} [physio]*
yowai	弱い	weak; fragile *{adj} [mech]*

yowamari	弱まり	weakening *{n} [mech]*
yowa-sa	弱さ	weakness *{n} [mech]*
yōyaku	要約；ようやく	summary (pl -ries) *{n} [pr]*
yo'yaku suru	予約する	preengage; reserve *{vb} [ind-eng] [trans]*
yōyaku suru	要約する	summarize; condense; abstract *{vb} [comm]*
yōyō'tai	幼葉態	aestivation *[bot]*
yōyū	溶融；ようゆう	fusion; melting *{n} [p-ch]*
yōyū-aen-mekki	溶融亜鉛めつき	galvanized coating; zinc hot dipping *{n} [met]*
yōyū-bōshi	溶融紡糸	melt spinning *{n} [tex]*
yōyū-kyōdo	溶融強度	melt strength *[mech]*
yōyū-mekki	溶融めっき	hot dipping *{n} [met]*
yōyū-mekki-hō	溶融めっき法	hot-dipping method *[met]*
yōyū-mekki-yoku	溶融めっき浴	hot dipping bath *[met]*
yōyū-namari-mekki	溶融鉛めっき	lead hot dipping *{n} [met]*
yōyū-nendo	溶融粘度	melt viscosity *[fl-mech]*
yōyū-ryūdō-sei	溶融流動性	melt fluidity *[fl-mech]*
yōyū-seikei	溶融成形	melt molding *{n} [poly-ch]*
yōyū-seki'ei	溶融石英	fused quartz *[miner]*
yōyū-suzu-mekki	溶融錫めっき	tin hot-dipping *{n} [met]*
yōzai	溶剤	solvent *[ch]*
yōzai-dasshi	溶剤脱脂	solvent degreasing *{n} [ch-eng]*
yōzon-enrui	溶存塩類	dissolved salts *[ch]*
yōzon-sanso	溶存酸素	dissolved oxygen *[ch]*
yō'zu	要図；要圖	sketch map; key chart *[graph] [map]*
yu	湯	bath; molten metal *[mat]*
yū	優	excellent (grade) *[ind-eng]*
yu-aka	湯垢	dross (casting) *[met]*
yu'atsu	油圧；油壓	oil pressure *[mech]*
yu'atsu-kōgaku	油圧工学	oil hydraulics *[fl-mech]*
yu'atsu-kudō-ben	油圧駆動弁	hydraulically operated valve *[mech-eng]*
yu'atsu no	油圧の	hydraulic *{adj} [eng]*
yuba	湯葉；ゆば	dried bean curds *[cook]*
yūbe	夕べ	evening *{n} [astron]*
yūbe	昨夜；ゆうべ	yesterday evening *{n}*
yubi	指；ゆび	finger *{n} [anat]*
yūbin	郵便；ゆうびん	mail *{n} [comm]*
yūbi na	優美な	elegant; graceful *{adj}*
yūbin-bakari	郵便秤	letter scale; postal scale *[eng]*
yūbin-bangō	郵便番号	postal code; zip code (U.S.) *[comm]*
yūbin-ha'gaki	郵便葉書	postcard *[comm]*
yūbin-kikai	郵便機械	mailing machine *[eng]*

yūbin'kyoku	郵便局	post office *[arch] [comm]*
yūbin'sen	郵便船	mail boat *[nav-arch]*
yūbin'sha	郵便車	mail car *[rail]*
yūbi-rui	有尾類	Urodela *[v-zoo]*
yubi'saki	指先	fingertip *[anat]*
yubi-wa	指環	ring (to wear on finger) *{n} [lap]*
yu'chō	油長	oil length *[mat]*
yūchō-kasen	有潮河川	tidal river *[hyd]*
yu-chū-sui no	油中水の	water-in-oil *{adj} [org-ch]*
yūchū-tō	誘虫灯	light trap *[eng]*
yu-danbō	湯暖房；ゆだんぼう	hot-water heating *{n} [mech-eng]*
yu'dashi	湯出し	tapping *{n} [met]*
yuden	油田	oil field *[petrol]*
yūden-bun'kyoku	誘電分極	dielectric polarization *[elec]*
yūden-hizumi	誘電歪み	dielectric strain *[elec]*
yūden-isō-kaku	誘電位相角	dielectric phase angle *[elec]*
yūden-ritsu	誘電率	dielectric constant; permittivity *[elec]*
yūden-sei'setsu	誘電正接	dielectric dissipation factor; dielectric loss tangent *[elec]*
yūden-son'shitsu	誘電損失	dielectric loss *[elecmg]*
yūden'tai	誘電体	dielectric *{n} [mat]*
yūden'tai-son	誘電体損	dielectric loss *[elecmg]*
yūden'tai-zōfuku-ki	誘電体増幅器	dielectric amplifier *[electr]*
yūden-yokō	誘電余効	dielectric aftereffect *[elec]*
yuderu	茹でる；ゆでる	boil *{vb} [cook]*
yude-tamago	茹で卵	boiled egg *[cook]*
yūdo	尤度；ゆうど	likelihood *[math]*
yūdo	裕度	tolerance *[eng]*
yūdo	融度	fusibility *[thermo]*
yūdō	遊動	nomadism *[bio]*
yūdō	誘導；ゆうどう	induction *[elec] [elecmg]*
"	誘導	guidance *{n}*
yūdō-buki-kansei	誘導武器管制	guided-weapon control *[ord]*
yūdō-busshitsu	誘導物質	inducer *[bioch]*
yūdō-dan	誘導弾	guided missile *[ord]*
yūdō-den'atsu	誘導電圧	induced voltage *[elec]*
yūdō-dendō-ki	誘導電動機	induction motor *[elec]*
yūdō-denkai	誘導電界	induction field *[elecmg]*
yūdō-denryū	誘導電流	induced current *[elecmg]*
yūdō-hō'shutsu	誘導放出	stimulated emission *[electr]*
yūdō-kenbi'kyō	遊動顕微鏡	traveling microscope *[opt]*
yūdō-ki	誘導期	induction period *[p-ch]*

yūdō-ki	誘導器	inductor *[elecmg]*
yūdō-kiden-ryoku	誘導起電力	induced electromotive force *[elecmg]*
yū'doku no (-na)	有毒の；(-な)	venomous; poisonous *{adj} [physio]*
yū'doku-gyo	有毒魚	toxic fish *[v-zoo]*
yūdō'ro-tō	誘導路灯	taxiway light *[navig]*
yūdō'shi	誘導子	inductor *[elecmg]*
yūdō'tai	誘導体	derivative *[ch]*
yū'ei-seibutsu	遊泳生物	nekton *[i-zoo]*
yu'en	油煙；ゆえん	lampblack *[mat]*
yūen'chi	遊園地	recreation ground *[civ-eng]*
yue ni	故に	therefore *{conj}*
yū'etsu	優越；ゆうえつ	dominance *[zoo]*
yufu	油布	oil cloth *[tex]*
yūga-giku	柚香菊；ゆうがぎく	Kalimeris pinnatifida (herb) *[bot]*
yūgai-busshitsu	有害物質	noxious substance *[med]*
yūgai'butsu	有害物	deleterious material *[ind-eng]*
yūgai-kasha	有蓋貨車	box car *[rail]*
yūgai no (yūgai na)	有害の；(有害-な)	detrimental; harmful *{adj}*
yūgai'sei-	有害性-	harmful-; poisonous- *{prefix}*
yugameru	歪める	distort; warp *{vb t}*
yugami	歪(み)	distortion *[math] [opt] [sci-t]*
"	歪み；ゆがみ	skew; skewness *{n} [sci-t]*
"	歪み	warping *{n} [tex]*
yugamu	歪む	contort; be distorted *{vb i}*
yū'gao	夕顔	bottle gourd; calabash (flower) *[bot]*
yūgata	夕方	evening *[astron]*
yūga-tō	誘蛾灯；ゆうがとう	light trap *{n} [agr]*
yūgeki	遊隙	clearance *[mech-eng]*
yūgen	有限；ゆうげん	finite *{n} [math]*
yūgen-bo'shūdan	有限母集団	finite population *[math]*
yūgen-kaku	雄原核	generative nucleus *[bot]*
yūgen-shinpuku	有限振幅	finite amplitude *[math]*
yūgen-shōsū	有限小数	terminating decimal *[math]*
yūgen-yōso-hō	有限要素法	finite element method *[math] [mech]*
yūgō	融合；ゆうごう	fusion *[nuc-phys] [p-ch]*
yūgō-bun'retsu-konsei-ro	融合分裂混成炉	fusion-fission hybrid reactor *[nucleo]*
yu'guchi	湯口	sprue *[met]*
yūhai-shokubutsu-rui	有胚植物類	Embryophyta *[bot]*
yū'hatsu	誘発；誘發	induction *[bio] [elecmg] [nucleo]*
"	誘発；ゆうはつ	stimulation *[elecmg]*
yūhatsu-hakkō	誘発発光	induced (stimulated) emission *[elecmg]*
yūhatsu-hōsha	誘発放射	induced radiation *[nucleo]*

yūhatsu-sōkyoku'shi	誘発双極子	induced dipole *[elec]*
yūhatsu(-totsu'zen)-hen'i	誘発(突然)変異	induced mutation *[gen]*
yūho'jō	遊歩場	foyer; promenade *{n} [arch]*
yūhō-kō	有泡鋼; 有泡鋼	blister steel *[steel]*
yū'i	有意; ゆうい	significant *{n} [math]*
yū'i-kankaku	有意間隔	significant interval *[comput]*
yūin-busshitsu	誘引物質	attractant; decoy *[agr] [bioch] [mat]*
yūin-zai	誘引剤	attractrant *[agr] [bioch] [mat]*
yū'i-sei	有意性	significance *{n} [math]*
yū'i'sei-kentei	有意性検定	test of significance *[math]*
yū'i-suijun	有意水準	level of significance *[math]*
yūjin-jinkō-eisei	有人人工衛星	manned satellite *[aerosp]*
yūjin-shimei	有人使命	manned mission *[aerosp]*
yūjin-un'yō	有人運用	manned operation *[ind-eng]*
yujō-busshitsu	油状物質	oily matter *[mat]*
yuka	床; ゆか	floor *{n} [arch]*
yu-kagaku	油化学	oil chemistry *[ch]*
yu-kagen	湯加減	water temperature *[thermo]*
yuka-haichi-zu	床配置図	floor plan *[arch] [graph]*
yū'kahō-sei	優加法性	superadditivity *[math]*
yūkai	融灰	molten ash *[geol]*
yūkai	融解; ゆうかい	fusion; melting; dissolution *{n} [ch]*
yūkai-netsu	融解熱	heat of fusion *[thermo]*
yūkai-ro	融解炉	melting furnace; smelting furnace *[eng]*
yūkai-seki'ei	融解石英	fused silica; silica glass *[mat]*
yūkai-sennetsu	融解潜熱	latent heat of fusion *[thermo]*
yūkai-ten	融解点	melting point *[thermo]*
yuka'jiki	床敷き	floormat *[furn]*
yū-kaku	優角	major angle; superior angle *[math]*
yuka-menseki	床面積	floor area *[arch]*
yukan	油管	oil pipe *[eng]*
yūkan	夕刊; ゆうかん	evening (news)paper *[pr]*
yūkan-jishin	有感地震	felt earthquake *{n} [geophys]*
yuka no hone'gumi	床の骨組み	floor framing *{n} [arch]*
yūkari	有加利; ゆうかり	eucalyptus (pl -ti, -tuses) *[bot]*
yukata	浴衣; ゆかた	bathrobe kimono; informal kimono; Japanese bath gown *[cl]*
Yukawa'gata-sōgo-sayō	湯川型相互作用	Yukawa interaction *[nuc-phys]*
yu-ketsu	輸血	blood transfusion *[med]*
yu-ketsu'gan	油頁岩	oil shale *[geol]*
yuki	雪	snow *{n} [meteor]*
yūki	有機; ゆうき	organic(ness) *{n} [org-ch]*

yūki-bunseki	有機分析	organic analysis *[org-ch]*
yūki'butsu	有機物	organic substance *[mat]*
yūki-dō-kagō'butsu	有機銅化合物	organocopper compound *[org-ch]*
yuki'doke	雪解け；雪融け	snowthaw *{n}* *[climat]*
yuki-domari; iki-domari	行き止り	dead end *[traffic]*
yūki-eiyō-seibutsu	有機栄養生物	heterotroph *[bio]*
yuki'ge-mizu	雪消水；雪解水	meltwater *{n}* *[hyd]*
yūki-hiso-kagō'butsu	有機砒素化合物	organoarsenic compound *[org-ch]*
yuki-hyō	雪豹；ゆきひょう	snow leopard (animal) *[v-zoo]*
yūki-kagaku	有機化学	organic chemistry *[ch]*
yūki-kagō'butsu	有機化合物	organic compound *[org-ch]*
yuki'kaki-guruma	雪掻車	snowplow car *[mech-eng]*
yuki'kaki-ki	雪掻機；ゆきかきき	snowplow *{n}* *[mech-eng]*
yūki-ka'sanka'butsu	有機過酸化物	organic peroxide *[org-ch]*
yūki-kinzoku-kagō'butsu	有機金属化合物	organometal(lic) compound *[org-ch]*
yūki-kōka	誘起効果	inductive effect *[elec]*
yūki-namari-kagō'butsu	有機鉛化合物	organolead compound *[org-ch]*
yuki-no-shita	雪の下	strawberry geranium (flower) *[bot]*
yūki-rin-kagō'butsu	有機燐化合物	organophosphorus compound *[org-ch]*
yūki'rin-zai	有機燐剤	organophosphorus compound *[org-ch]*
yūki-san	有機酸	organic acid *[org-ch]*
yū'kiseki	黝輝石；ゆうきせき	spodumene *[miner]*
yūki-sen'i-kinzoku-kagō'butsu	有機遷移金属化合物	organotransition metal compound *[org-ch]*
yūki'shitsu-hiryō	有機質肥料	organic fertilizer *[agr]*
yūki-shi'yaku	有機試薬	organic reagent *[org-ch]*
yūki-shokubai	有機触媒	organic catalyst *[bio]*
yūki'tai	有機体	organism *[bio]*
yūki'teki-kankyō	有機的環境	organic environment *[ecol]*
yuki-zumari; iki-zumari	行き詰り	blind ally *[traffic]*
yukkuri to	ゆっくりと	slowly; gradually *{adv}*
yūko	優弧	major arc *[math]*
yūkō	有効；ゆうこう	effectiveness; validity; availability *{n}*
yūkō-ba'riki	有効馬力	effective horsepower *[mech]*
yūkō-bun	有効分	active component *[ch]* *[elec]*
yūkō-chi	有効値	effective value *[math]*
yūkō'chū-rui	有孔虫類	Foraminifera *[i-zoo]*
yūkō-den'ryoku	有効電力	effective power *[elec]*
yūkō-do	有孔度	porosity (pl -ties) *[phys]*
yūkō-enso	有効塩素	available chlorine *[ch]*
yūkō-gakki-rui	有簧楽器類	reed instruments *[music]*
yūkō-han'i	有効範囲	scope *[comput]*

yūkō-jumyō	有効寿命；有效壽命	service life; useful life *[ind-eng]*
yūkō-kaku-denka	有効核電荷	effective nuclear charge *[elec]*
yūkō-kasoku-do	有効加速度	effective acceleration *[mech]*
yūkō-keisū	有効係数	effectiveness factor *[p-ch]*
yūkō-kō'ritsu	有効効率	effective efficiency *[eng]*
yūkō-kōtei	有効行程	effective stroke *[mech-eng]*
yūkō-ondo	有効温度	effective temperature *[astrophys]*
yūkō-ritsu	有孔率	porosity *[phys]*
yūkō-sei	有効性	validity; effectiveness *{n}*
yūkō-seibun	有効成分	effective component *[ch]*
yūkō-shitsu'ryō	有効質量	effective mass *[sol-st]*
yūkō-sui	有効水	available water *[hyd]*
yūkō-suibun	有効水分	available moisture *[hyd]*
yūkō-sūji	有効数字	significant figure; significant digit *[math]*
yūkō-sūji-enzan	有効数字演算	significant digit arithmetic *[math]*
yuku'saki	行く先	destination *[trans]*
yūkyō'gaku	優境学	euthenics *[bio]*
yūkyoku'sei-bunshi	有極性分子	polar molecule *[p-ch]*
yūkyoku-tessen	有刺鉄線	barbed wire *[met]*
yu-maku	油膜	oil film; oil slick *[petrol]*
yume	夢；ゆめ	dream *{n}* *[psy]*
yume'gaku	夢学	oneirology *[psy]*
yumen-hyōji-ki	油面表示器	oil level indicator *[eng]*
yumen-kei	油面計	oil gage *[eng]*
yumen-sokutei-bō	油面測定棒	oil dipstick *[eng]*
yumi	弓；ゆみ	archery; bow *{n}* *[ord]*
yu-michi	湯道	runner channel *[resin]*
yumi-gata	弓形	bow; bow shape *{n}* *[arch]* *[math]*
"	弓形	segment *[math]*
yumi'gata-danmen	弓形断面	ogival section *[arch]*
yumi'noko	弓のこ	hacksaw *{n}* *[eng]*
yu'mitsu-shiken	油密試験	oiltightness test *[ind-eng]*
yumo	油母	kerogen *[geol]*
yumo-ketsu'gan	油母頁岩	oil shale *[geol]*
yū-nagi	夕凪；ゆうなぎ	evening calm *[meteor]*
yūnō-den'ryoku	有能電力	available power *[elec]*
yu'nomi	湯呑	teacup *[cer]*
yu'noshi	湯熨；ゆのし	steam ironing *{n}* *[tex]*
yu'nyō-kan	輸尿管	ureter *[anat]*
yu'nyū	輸入；ゆにゅう	importing; importation *{n}* *[econ]*
yu'nyū-zei	輸入税	import duty *[econ]*

yū'ō	雄黄；ゆうおう	orpiment *[miner]*
yuragi	揺らぎ；搖らぎ	fluctuation *[electr] [sci-t]*
"	揺らぎ	sway *{n} [ocean]*
yūraku'jō	遊楽場	recreation area *[civ-eng]*
yuran-kan	輸卵管	oviduct *[anat]*
yūran'sen	遊覧船	pleasure boat *[nav-arch]*
yure	揺れ；ゆれ	swinging *{n} [electr] [navig]*
"	揺れ	quake *{n} [geophys]*
"	揺れ	shaking; quivering *{n}*
yu-rei	油冷	oil quenching *{n} [met]*
yūrei-sen	幽霊線；幽靈線	ghost line *[met]*
yure'mo	揺藻；ゆれも	Oscillatoria *[bot]*
yū'renseki	黝簾石	zoisite *[miner]*
yureru	揺れる	shake; quake; sway *{vb}*
yū'retsu no hō'soku	優劣の法則	dominance, law of *[ecol]*
yure-ude	揺れ腕	rocker arm *[mech-eng]*
yuri	百合；ゆり	lily (pl lilies) (flower) *[bot]*
yūri	有利	advantageous; profitable *{n} [econ]*
yūri	遊離；ゆうり	separation; isolation *{n} [ch]*
yūri-iō	遊離硫黄	free sulfur *[ch]*
yuri-isu	揺り椅子	rocking chair *[furn]*
yūri'ka	有理化	rationalization *[math]*
yuri-kago	揺り籠	cradle (for a baby) *{n} [furn]*
yuri-kamome	百合鷗；ゆりかもめ	hooded gull (bird) *[v-zoo]*
yuri'kei-ka	百合形花	liliaceous flower *[bot]*
yūri-ki	遊離基	free radical *[ch] [org-ch]*
yūri'ki-hosoku-zai	遊離基捕捉剤	free radical scavenger *[bioch]*
yuri-ne	百合根	lily bulb *[bot] [cook]*
yūri no	遊離の	free (unsecured) *{adj} [ch]*
yūrin-rui	有鱗類	Squamata *[v-zoo]*
yūri-san	遊離酸	free acid *[ch]*
yūri-shibō-san	遊離脂肪酸	free fatty acid *[org-ch]*
yūri-shisū	有理指数	rational index *[math]*
yūri-shisū no hōsoku	有理指数の法則	rational indices, law of *[math]*
yūri-sū	有理数	rational number *[math]*
yūri suru	遊離する	isolate; separate; extricate; release; liberate; set free *{vb}*
yūri-tanso	遊離炭素	free carbon *[met]*
yu-ritsu	輸率	transference number; transport number *[p-ch]*
yūropyūmu	ユーロピウム	europium (element)
yurui	緩い；弛い；ゆるい	flaccid *[bot]*

yurui	緩い；弛い	loose *(adj)*
yurume	緩め；弛め	loosening; release *(n)*
yurumeru	緩める；弛める	loosen; relax *(vb t)*
yurumi	緩み；弛み	looseness; slack *(n) [mech]*
yūryō-dōro	有料道路	tollroad; turnpike *[civ-eng]*
yūryō-kyō	有料橋	toll bridge *[civ-eng]*
yusa	油砂；ゆさ	oil sand *[geol]*
yūsai-shoku	有彩色	chromatic color *[opt]*
yusei	油井	oil well *[petrol]*
yusei	油性	oiliness *(n) [eng]*
yūsei	優性；ゆうせい	dominance *[gen]*
yūsei	優勢	predominant; preponderant *[bio]*
yūsei'gaku	優性学	eugenics *[gen]*
yūsei-ha'guruma	遊星歯車	planet(ary) gear *[mech-eng]*
yūsei-ha'guruma-sōchi	遊星歯車装置	epicyclic train *[mech-eng]*
yusei-kōryō	油性香料	oily perfume *[mat]*
yusei-kussaku	油井掘削	oil-well excavation *[petrol]*
yusei no	油性の	oleaginous (or oleagenous) *(adj) [bio]*
"	油性の	oily *(adj)*
yūsei-seishoku	有性生殖	sexual reproduction *[bio]*
yūsei-so'shitsu	優性素質	dominant trait *[gen]*
yusei-toryō	油性塗料	oil paint; oil varnish *[mat]*
yūsei-totsu'zen-hen'i	優性突然変異	dominant mutation *[gen]*
yūsei-wa	遊星輪	planet wheel *[mech-eng]*
yusei-yagura	油井櫓	derrick; oil derrick *[petrol]*
yusen	油腺	oil gland; uropygial gland *[v-zoo]*
yu'sen	湯煎；ゆせん	water-bath heating *(n) [ch]*
yūsen-bi	優先日	priority date *[pat]*
yūsen-denshin	有線電信	wire telegraph(y) *[comm]*
yūsen-denwa	有線電話	wire telephone *(n) [comm]*
yūsen-do	優占度	dominance *[bio]*
yūsen-do	優先度；ゆうせんど	priority (pl -ties) *[comput]*
yūsen-fusen	優先浮選	differential flotation *[min-eng]*
yūsen-hō'i	優先方位	preferred orientation *[crys]*
yūsen-jun'i	優先順位	precedence; priority *[comput]*
yūsen-ken	優先権；優先權	priority right *[pat]*
yūsen'ken-shuchō	優先権主張	priority claim; declaration of priority *[pat]*
yūsen'ken-shuchō-koku	優先権主張国	country claiming priority right *[pat]*
yūsen-shinsa	優先審査	priority examination *[pat]*
yūsen suru	優先する	take precedence *(vb)*
yūsen'teki na	優先的な	preferential *(adj) [pat]*

yūsen-terebi-hōsō	有線テレビ放送	cable television service *[comm]*
yūsen-wari'komi	優先割り込み	priority interrupt *[comput]*
yū'setsu	融接	fusion welding *{n} [met]*
yushi	油脂	fats and oils *[bioch] [mat]*
yūshi	遊糸; ゆうし	gossamer (of spider) *{n} [i-zoo]*
yūshi-bun'retsu	有糸分裂	mitosis (pl -toses) *[cyt]*
yū'shikai-hikō	有視界飛行	contact flight (or flying) *{n} [navig]*
yu-shin	油浸	oil immersion *[elec] [opt]*
yū'shitsu-dōbutsu	有櫛動物	Ctenophora; comb jellies *[i-zoo]*
yu'shitsu no	油質の	oleaginous (oleagenous) *{adj} [bio]*
yūshō	湧昇	upswelling *{n} [ocean]*
yūshu-dōbutsu	有鬚動物	Pogonophora *[i-zoo]*
yu'shutsu	輸出; ゆしゅつ	export *{n} [econ]*
yu'shutsu-nyū	輸出入	export and import *[econ]*
yusō	輸送	transportation; transport *[bio] [eng]*
yusō'boku	癒瘡木; ゆそうぼく	guaiac wood; Guajacum officinale *[bot]*
yūsō-dōbutsu	有爪動物	Onychophora *[i-zoo]*
yusō'sen	油送船	tanker *[nav-arch]*
yusō'sen	油槽船	oiler *[nav-arch]*
yūsō'shi	遊走子	zoospore *[bio]*
yusō-yōki	輸送容器	shipping container *[ind-eng]*
yusuri-ka	揺り蚊	midge *[i-zoo]*
yusuru	揺する; 搖する	shake *{vb t}*
yūtai-moku	有袋目	Marsupialia *[v-zoo]*
yu'tanpo	湯湯婆; ゆたんぽ	hot water bottle *[pharm]*
yūten	遊転; 遊轉	idle; idling *{n} [mech-eng]*
yūten	融点; ゆうてん	fusing point *[met] [thermo]*
"	融点	melting point *[thermo]*
yuten-gomu	油展ゴム	oil-extended rubber *[rub]*
yutō-zai	油糖剤	oil-sugar *{n} [pharm]*
yutori	ゆとり	allowance; leeway; latitude *{n} [eng]*
yu'wakashi	湯沸し	kettle (for water boiling) *[cook]*
yu'wakashi'gata-genshi'ro	湯沸し型原子炉	water-boiler reactor *[nucleo]*
yu'wakashi-sōchi	湯沸し装置	water heater *[mech-eng]*
yūwaku-shoku	誘惑色	alluring coloration *[zoo]*
yū-yake	夕焼	sunset glow *[geophys]*
yūyaku	釉薬; ゆうやく	glaze *{n} [cer]*
yū'yami	夕闇	dusk *[astron]*
yu-yoku	油浴	oil bath *[eng] [met]*
yuyō-sei	油溶性	oil-solubleness *{n} [mat]*
yūyō-sei	有用性	usefulness *{n}*
yuyō'sei-jushi	油溶性樹脂	oil-soluble resin *[org-ch]*

yuyō-senryō	油溶染料	oil-soluble dye *[org-ch]*
yūzai	融剤；融劑	flux *{n} [mat]*
yūzen-nori	友禅糊	Yuzen paste; Yuzen thickener *[tex]*
yūzen-zome	友禅染め	Yuzen dyeing; Yuzen print *{n} [tex]*
yuzu	柚；柚子；ゆず	Chinese lemon; citron (fruit) *[bot]*
yū'zui; o-shibe	雄蕊	stamen *[bot]*
yuzure-hyō'shiki	譲れ標識	yield sign *[traffic]*
yuzuri'uke'nin	譲受人；讓受人	assignee *{n} [legal]*
yūzū-sei	融通性	adaptability; versatility *{n}*

Z

za	座；ざ	locus (pl -i) *[gen]*
"	座	place *{n}* *[math]*
"	座	boss (casting) *{n}* *[met]*
zabon	朱欒；ざぼん	zamboa; pomelo; shaddock (fruit) *[bot]*
za'gane	座金	washer *{n}* *[eng]*
za'guri	座繰り；生繰	countersinking *{n}* *[mech-eng]*
"	座繰り；ざぐり	hand-reeling (of silk) *{n}* *[tex]*
za'hyō	座標；ざひょう	coordinates *[math]*
zahyō-jiku	座標軸	axis of coordinates *[math]*
zahyō-kika'gaku	座標幾何学	coordinate geometry *[math]*
zahyō-kōshi	座標格子	coordinate grid *[math]*
zahyō-tenkan	座標転換	coordinate transformation *[math]*
zai	材	wood *[bot]*
zai	財	wealth *{n}* *[econ]*
zaidan(-hōjin)	財団(法人)	foundation *[econ]*
zaiko-kanri	在庫管理	inventory control; stock control *[ind-eng]*
zaiko'ryō	在庫量	inventory; stock *{n}* *[econ]* *[ind-eng]*
zai'moku	材木；ざいもく	lumber; timber; wood *[mat]*
zaimu-kaikei	財務会計	financial accounting *{n}* *[econ]* *[math]*
zai'ryō	材料；ざいりょう	materials *[eng]*
zairyō-hyō	材料表	materials list *[ind-eng]*
zairyō-kagaku	材料科学	materials science *[eng]*
zairyō-riki'gaku	材料力学	strength of materials *[mech]*
zakka-yōhin	雑貨用品	general merchandise; sundries *[econ]*
zakkin	雑菌；雜菌	miscellaneous germs *[microbio]*
zakkin-konnyū	雑菌混入	contamination (by germs) *[microbio]*
zakkin-o'sen	雑菌汚染	contamination (by germs) *[microbio]*
zakkin'sei no	雑菌性の	saprophytic *{adj}* *[microbio]*
zakkoku	雑穀	miscellaneous cereals; minor grains *[agr]*
zako	雑魚	small fish *[cook]*
zakuro	柘榴；榴；ざくろ	pomegranate (fruit) *[bot]*
zakuro-hi	柘榴皮	pomegranate bark *[bot]*
zakuro-ishi	柘榴石	garnet *[miner]*
za'kutsu	座屈	buckling *{n}* *[mech]*
zamen	座面	bearing surface *[mech-eng]*
zanbun	残分；殘分	residue *[ch-eng]*
zangō	塹壕；ざんごう	trench *{n}* *[civ-eng]*
zangyō-jikan	残業時間	overtime *{n}* *[ind-eng]*
zanji-sei	残磁性；殘磁性	retentivity *[elecmg]*

zankō	残光；ざんこう	afterglow *[atom-phys] [meteor]*
zankō	残効	aftereffect; residual effect *[pharm]*
zankō	残項	residual term *[math]*
zankyō	残響；殘響	reverberation *[acous]*
zankyō-zukei	残響図形	decay pattern *[acous]*
zankyū	残丘	monadnock *[geol]*
zan'on	残音	aftersound *[acous]*
zanryū	残留；ざんりゅう	residual; remaining portion *{n}*
zanryū-bun'kyoku	残留分極	remanence *[elecmg]*
zanryū'butsu	残留物	residue *[ch-eng]*
zanryū-jika	残留磁化	residual magnetization *[elecmg]*
zanryū-ō'ryoku	残留応力	residual stress *[mech]*
zanryū-sen	残留線	residual rays *[nucleo]*
zansa	残差	residue *[math]*
zansai	残菜	garbage *[civ-eng]*
zansa no	残差の	residual *{adj} [math]*
zanson-bōchō	残存膨張	residual expansion *[phys]*
zanson-kōbutsu	残存鉱物	relict mineral *[miner]*
zanson-kōzō	残存構造	relict structure *[geol]*
zan'tō	残糖	residual sugar (in wine) *[cook]*
zan'yo-chisso	残余窒素	residual nitrogen *[inorg-ch]*
zanzō	残像	lag *{n} [electr]*
"	残像	afterimage *[physio]*
zara-gami	更紙；ざら紙	groundwood paper *[mat]*
zara'me	双目；ざらめ	crystal sugar *[cook]*
zara'tsuki	ざらつき	rough surface; rough deposits *{n} [eng]*
zara'tsuki-kan	ざらつき感	roughness feel *{n} [physio]*
zari-gani	蝲蛄；ざりがに	crawfish; crayfish (crustacean) *[i-zoo]*
zaru	笊；ざる	bamboo colander *[cook] [food-eng]*
zaseki-beruto	座席ベルト	seat belt *[eng]*
zaseki-kairo	座席回路	position circuit *[elec]*
zaseki-sū	座席数	seating capacity *[aero-eng]*
zaseki-yoyaku	座席予約	seat reservation *{n} [trans]*
zashō	座礁；ざしょう	grounding; running aground *{n} [navig]*
zasshi	雑誌；雜誌；ざっし	magazine; periodical *{n} [pr]*
zasshi-yōshi	雑誌用紙	magazine paper *[mat]*
zasshoku-sei	雑食性	omnivorous(ness) *{n} [zoo]*
zasshoku'sei-dōbutsu	雑食性動物	omnivore *[zoo]*
zasshu	雑種	hybrid *{n} [bot] [gen]*
zasshu-keisei	雑種形成	hybridization *[bot]*
zassō	雑草；雜草	weed *{n} [bot]*
zassōchi	雑装置	miscellaneous equipment *[ind-eng]*

zatō-kujira	座頭鯨	humpback whale *[v-zoo]*
zatsu-jikan	雑時間	incidental time; miscellaneous time; downtime {n} *[comput] [ind-eng]*
zatsu'on	雑音；ざつおん	noise *[acous]*
zatsu'on-hassei-ki	雑音発生器	noise generator *[electr]*
zatsu'on-kei	雑音計	noise meter *[electr]*
zatsu'on-shisū	雑音指数	noise factor *[electr]*
za-yaku	座薬；坐薬	suppository (pl -ries) *{n} [pharm]*
za-yō	座葉	sessile leaf *[bot]*
za'zai	坐剤	suppository *{n} [pharm]*
zazen'sō	座禅草	skunk cabbage *[bot]*
zei	税；ぜい	tax *{n} [econ]*
zei'biki-go-ri'eki	税引後利益	net after taxes *[econ]*
zei'biki-mae-ri'eki	税引前利益	net before taxes *[econ]*
zei'ginkō	脆銀鉱；脆銀鑛	stephanite *[miner]*
zei'jaku	脆弱；ぜいじゃく	fragility; weakness; delicacy *{n} [sci-t]*
zei'jaku-sei	脆弱性	vulnerability *[mil]*
zei'ka	脆化	brittling; degradation; embrittling; embrittlement *{n} [mech]*
zeikan	税関；税關	customs office *[econ]*
zeikan-kanshi'sen	税関監視船	revenue cutter *[nav-arch]*
zei'ka-ondo	脆化温度	brittle temperature *{n} [thermo]*
zei'ka-ten	脆化点	brittle point *[thermo]*
zeikin	税金	tax *{n} [econ]*
zei-niku	贅肉	flab; superfluous flesh *{n} [physio]*
zeiri'shi	税理士	tax accountant (a person) *[econ]*
zeisei	脆性；ぜいせい	brittleness *[mech]*
zeisei-hakai	脆性破壊	brittle fracture *[met]*
zeisei-zairyō	脆性材料	brittle material *[mat]*
zekkan-jūryō	絶乾重量	absolute(ly) dry weight *[mech]*
zekkei-dōbutsu	舌形動物	Linguatulida; Pentastomida *[i-zoo]*
zen-	全-	omni-; pan-; whole- *{prefix}*
zen-bibun	全微分	total differential *[math]*
zenbu	全部	all; the entire lot *{n}*
zenbu	前部	front part; forward part *{n}*
zenbu	膳部；ぜんぶ	meal set (trayful) *{n} [cook]*
zenchi'shi	前置詞	preposition *[gram]*
zenchi-zō'fuku-ki	前置増幅器	preamplifier *[electr]*
zen-chō	全長	overall length *[mech]*
zenchō-genshō	前兆現象	precursory phenomenon *{n} [geophys]*
zenchū	蠕虫；ぜんちゅう	helminth; parasitic worm *[i-zoo]*
zenchū'gaku	蠕虫学	helminthology *[bio]*

zenchū-ki	前中期	prometaphase *[bio]*
zendō(-undō)	蠕動(運動)	peristalsis (pl -stalses) *[physio]*
zen'en	前縁	leading edge *[aero-eng] [comput]*
zen'en no	全縁の	entire *{adj} [bot]*
zen-genzan-ki	全減算器	full-subtracter *[electr]*
zengo	全語	full word *[comput]*
zengo	前後；ぜんご	order; sequence *{n} [math]*
"	前後	ahead and behind *{n}*
zengo-denkai-hi	前後電界比	front-to-back ratio *[elecmg]*
zen-hansha	全反射	total reflection *[opt]*
zen-hansō'ha	全搬送波	full carrier *[comm]*
zen-hizumi	全歪み	total distortion *[opt]*
zen-hōkō no	全方向の	omnidirectional *{adj} [electr]*
zen'hōkō-sei	全方向性	isotropicness *{n}*
zeni	钱；錢；ぜに	cash; money *{n} [econ]*
zen'i	前位	front position *{n}*
zen'i	善意	good faith *[psy]*
zen-iō	全硫黄；ぜんいおう	total sulfur *[inorg-ch]*
zen-jidō	全自動	fully automatic *[eng]*
zen-jishō	全事象	universal event *{n} [math]*
zenkaku	全角	em *[comput] [pr]*
zen-kasan-ki	全加算器	full adder *[electr]*
zenkei	前景	foreground *[graph]*
zenken	前件	antecedent *{n} [logic]*
zenki	前期	prophase *[bio]*
zenkin-kidō	漸近軌道	asymptotic orbit *[math]*
zenkin-sen	漸近線	asymptote *[math]*
zenkō	前項	antecedent *{n} [logic]*
zen-kokei'bun	全固形分	total solids *[an-ch]*
zenku-busshitsu	前駆物質	precursor *[bio] [ch]*
zenku'tai	前駆体；前驅體	precursor *[bio] [ch]*
zenkyū'shi	前臼歯	premolar *[anat] [dent]*
zen-kyūshū	全吸収；全吸收	total absorption *{n} [opt]*
zenmai	薇；ぜんまい	flowering fern *[bot]*
zenmai-bakari	発条秤；發條秤	spring scale *[eng]*
zenmen no	前面の	frontal *{adj}*
zen-nijū-tsū'shin	全二重通信	full-duplex communication *[comm]*
zennō-keiki	前納計器	prepayment meter *[trans]*
zennyū	全乳；ぜんにゅう	whole milk *[cook]*
zen'on	全音	whole tone *[acous] [music]*
zenpa	全波	full wave *[electr]*
zenpa-jushin-ki	全波受信機	all-wave receiver *[electr]*

zenpa-seiryū	全波整流	full-wave rectification *[electr]*
zenpa-seiryū-ki	全波整流器	full-wave rectifier *[electr]*
zenpō	前方；ぜんぽう	forward side; ahead *{n}*
zenpō-shiya-han'i	前方視野範囲	front visibility *[navig]*
zen-puku	全幅；ぜんぷく	overall width *[mech]*
zen-retsu no	全裂の	dissected *{adj} [bot]*
zenrin-kudō	前輪駆動	front-wheel drive *[mech-eng]*
zen'ritsu-sen	前立腺	prostate gland *[anat]*
zenryū-fun	全粒粉	whole-grain flour *[cook]*
zensai	前菜；ぜんさい	appetizer *[cook]*
zen-sanso-yōkyū'ryō	全酸素要求量	total oxygen demand *[bio] [microbio]*
zensei-setsu	前成説	preformation theory *[embryo]*
zen'sekai'teki na	全世界的な	cosmopolitan; world-wide *{adj}*
zensen	前線；ぜんせん	front *{n} [meteor]*
zensen-tai	前線帯；前線帶	frontal zone *[meteor]*
zensen-teiki'atsu	前線低気圧	frontal cyclone *[meteor]*
zen'senzō-kaku	前潜像核	latent preimage speck *[photo]*
zenshi-funnyū	全脂粉乳	whole milk powder *[cook]*
zenshi-katō-rennyū	全脂加糖煉乳	whole sweetened condensed milk *[cook]*
zenshin	全身；ぜんしん	whole body; entire body *[anat]*
zenshin	前進	advance(ment); ahead; go ahead *[navig]*
zenshin	前震；ぜんしん	foreshock (earthquake) *{n} [geophys]*
zenshin-ha	前進波；ぜんしんは	advancing wave *[phys]*
zenshin-kaku	前進角	angle of advance *[mech-eng]*
zenshin-keisū-sōchi	全身計数装置	whole-body counter *[nucleo]*
zen-shin'puku	全振幅	total amplitude *[phys]*
zenshin-shugi	漸進主義	gradualism *[bio]*
zenshin suru	前進する	advance *{vb} [navig]*
zen-shitsu	前室；ぜんしつ	anteroom *[arch]*
zenshō-kigō	全称記号	universal quantifier *[logic]*
zen'shoku('sei) no	全色(性)の	panchromatic *{adj} [opt]*
zen'shori; mae-shori	前処理；前處理	preprocessing *{n} [ind-eng]*
zenshō'tō	前照灯	headlight *[navig] [opt]*
zenshū	全集	complete works *[pr]*
zen-shuku	全縮；ぜんしゅく	full condensation *[ch]*
zen-shutsu'ryoku	全出力	full power *[mech-eng] [phys]*
zen-soku	全速	full speed *[mech]*
zen'soku	喘息；ぜんそく	asthma *[med]*
zensō'kyoku	前奏曲	overture; prelude *[music]*
zen-suibun	全水分	total moisture *[meteor] [p-ch]*
zensū-tansaku	全数探索	exhaustive search *{n} [comput]*
zentai	全体；全體	the whole; entirety *{n}*

zen'taiden	前帯電	precharging *{n} [elec]*
zentai-heikin	全体平均	overall mean *[math]*
zentai-zu	全体図；全體圖	full view; full drawing *{n} [graph]*
zen-tanso	全炭素；ぜんたんそ	total carbon *[inorg-ch]*
zentei	前庭	front yard *[arch] [civ-eng]*
zentei	前提；ぜんてい	antecedent; premise *{n} [logic]*
zentei-jōken	前提条件；前提條件	prerequisite *{n} [comput] [stat]*
zentei-ka	禅庭花；ぜんていか	orange-yellow day-lily *[bot]*
zen-tenkō-kōkō	全天候航行	all-weather navigation *[navig]*
zen'wachū	全話中	all busy (telephone) *[comm]*
zenwan	前腕；ぜんわん	forearm *{n} [anat]*
zen-yūki-tanso	全有機炭素	total organic carbon *[org-ch]*
zeppan	絶版	out of print *{n} [pr]*
zeppeki	絶壁；ぜっぺき	cliff; precipice *[geol]*
zeppen	舌片；ぜっぺん	tongue piece (shoe) *[cl]*
zerachin	ゼラチン	gelatin *[org-ch]*
zero-jōtai	零状態	zero state; zero condition *[comput]*
"	零状態	null state; nought state *[math]*
zetsu'en	絶縁；ぜつえん	insulation *{n} [elec]*
zetsuen'butsu	絶縁物	insulator; insulating material *[mat]*
zetsuen-hakai	絶縁破壊	dielectric breakdown *[electr]*
zetsuen-hi'maku	絶縁被膜	insulating covering film *[mat]*
zetsuen-kyori	絶縁距離	insulation distance *[elec]*
zetsuen-maku	絶縁膜	insulating (insulation) film *[elec]*
zetsuen-shi	絶縁紙；ぜつえんし	insulating paper *[mat]*
zetsuen-sō	絶縁層	insulation layer *[elec] [mat]*
zetsuen'tai	絶縁体；絶縁體	insulator; nonconductor *[elec] [mat]*
zetsuen-tai'ryoku	絶縁耐力	dielectric strength *[elec]*
zetsuen-teikō	絶縁抵抗	insulation resistance *[elec]*
zetsuen-toryō	絶縁塗料	insulating paint; insulating varnish *[mat]*
zetsuen-zai	絶縁剤；絶縁材	insulating material *[mat]*
zetsu'jō no	舌状の	lingulate *{adj} [bio]*
zetsu'metsu	絶滅；ぜつめつ	extinction; destruction *[bio]*
zetsu'metsu-sunzen no shu	絶滅寸前の種	endangered species *[bio]*
zettai	絶対；絶對	absolute *{n}*
zettai-atsu	絶対圧	absolute pressure *[phys]*
zettai-chi	絶対値；ぜったいち	absolute value; modulus *[math]*
"	絶対値	magnitude *[math]*
zettai-dansei-ritsu	絶対弾性率	absolute modulus of elasticity *[mech]*
zettai-kenki'sei-seibutsu	絶対嫌気性生物	obligate anaerobe *[bio]*
zettai no	絶対の	absolute *{adj}*
zettai-ondo	絶対温度	absolute temperature *[thermo]*

zettai-onkan	絶対音感	absolute pitch sense *[physio]*
zettai-reido	絶対零度	absolute zero (point) *[thermo]*
zettai-shitsu'do	絶対湿度; 絶對濕度	absolute humidity *[phys]*
zettai-tan'i	絶対単位	absolute unit(s) *[phys]*
zettai-tan'i-kei	絶対単位系	absolute system of units *[phys]*
zettai-tōkyū	絶対等級	absolute magnitude *[astron]*
zetto	ゼット	Z (the letter)
zetto-heimen	ゼット平面	z plane *[math]*
zetto-henkan	ゼット変換	z transformation *[math]*
zetto-jiku	ゼット軸	z axis *[math]*
zō	象; ぞう	elephant (animal) *[v-zoo]*
zō	像	statue *[furn]*
"	像	image *{n} [math] [opt]*
zō'atsu-denchi	増圧電池	booster battery *[elec]*
zō-azarashi	象海豹	elephant seal (mammal) *[v-zoo]*
zōbai-keisū	増倍係数	multiplication factor *{n}*
zōbai-ritsu	増倍率	multiplication factor *[nucleo]*
zōbi'chū	象鼻虫	weevil (insect) *[i-zoo]*
zōbun	増分; 增分	increment *[comput] [math] [sci-t]*
zōbun-sakuzu-ki	増分作図機	incremental plotter *[comput]*
zōbun-seki'bun-ki	増分積分機	incremental integrator *[comput]*
zōbun'shiki-keisan-ki	増分式計算機	incremental computer *[comput]*
zō'ei-zai	造影剤	contrast medium (x-ray) *[phys]*
zō'en	造園; ぞうえん	landscape architecture *[arch]*
zō'en-hasshoku	造塩発色; 造鹽發色	halochromism *[ch]*
zō'en'jutsu	造園術	landscape gardening *{n} [arch] [bot]*
zōfuku(-do)	増幅(度)	amplification *[electr]*
zōfuku-kairo	増幅回路	amplifier circuit *[elec]*
zōfuku-ki	増幅器; ぞうふくき	amplifier *[eng]*
zōfuku-ritsu	増幅率	amplification factor *[electr]*
zōfuku-teisū	増幅定数	amplification constant *[electr]*
zō-game	象亀; ぞうがめ	giant tortoise; elephant tortoise *[v-zoo]*
zōgan	象嵌; 象眼	inlay (ornament) *{n} [graph]*
zōgan-zaiku	象眼細工	marquetry; marqueterie *[furn]*
zōge	象牙; ぞうげ	ivory (pl -ries) *{n} [mat] [v-zoo]*
zōge-iro	象牙色	ivory (color) *{n} [opt]*
zōge'shitsu	象牙質	dentin; odontoblast *[hist]*
zōho	増補	augmentation *[pr]*
zōka	造花	artificial flower(-making) *[eng] [furn]*
zōka'daka	増加高	increment *{n} [econ]*
zō-kai	造塊	ingot-making *{n} [met]*
zōka-kansū	増加関数	increasing function *[math]*

zō'kaku-zai	造核剤	nucleating agent *[crys]*
zōkan	増感；ぞうかん	sensitization *[photo]*
zōkan-sei	増感性	sensitizing property *[photo]*
zōkan-zai	増感剤	sensitizer *[photo]*
zōka-ryō	増加量	increment; amount increased *[math]*
zōkei-ki	造型機	molding machine *[eng]*
zō'ketsu-zai	造血剤	hematinic *{n} [pharm]*
zōki	臓器；ぞうき	organ(s) *[anat]*
zōkin	雑布；雜布	mopping cloth *[tex]*
zōkin-baiyō	増菌培養	enrichment culture *[microbio]*
zokkaku	属格	genitive case *[gram]*
zōkō-ryō	増香料	fragrance-enhancing agent *[physio]*
zoku	族	group *{n} [ch]*
"	族；ぞく	tribe *[bio] [syst]*
"	族	family (pl -lies) *[math]*
zoku	属；屬；ぞく	genus (pl genera) *[bio] [syst]*
"	属	group *{n} [bio]*
zoku'hen	続編；續編；続篇	sequel *{n} [pr]*
zoku'in	属員	crew *{n}*
zoku-mei	属名	generic name *[bio] [sci-t]*
zoku-sei	属性	attribute(s) *{n} [bio] [comput]*
zoku'sei-sayō	続成作用	diagenesis *[geol]*
zōkyō-zai	増強剤	enhancer; potentiator *[cook] [pharm]*
zō'maku-hannō	造膜反応	tarnishing reaction *[miner]*
zō-men	像面；ぞうめん	image surface *[opt]*
zō-mushi	象虫	weevil (insect) *[i-zoo]*
zō'nen-sei	増粘性	thickening property *[ind-eng]*
zō'nen suru	増粘する	thicken *{vb t} [ch-eng]*
zō'nen-zai	増粘剤；増粘劑	binder; thickener *[mat]*
zōni	雑煮；ぞうに	rice cake soup *[cook]*
zōran-ki	造卵器；蔵卵器	archegonium (pl -nia) *[bot]*
zōri	草履；ぞうり	zori (Japanese footwear) *[cl]*
zō'riku-undō	造陸運動	epeirogeny (pl -nies) *[geol]*
zōri-mushi	草履虫	paramecium (pl -cia) *[i-zoo]*
zoru	ゾル	sol *[ch]*
zō-rui	象類；象類	Elephantidea (elephants, mammoths) *[v-zoo]*
zōryō-ginu	増量絹	weighted silk *[tex]*
zōryō-zai	増量剤	extending agent; extender; loading material; weighting agent *[mat]*
zōryū	造粒；ぞうりゅう	pelletization *[min-eng]*
"	造粒	granulation *[sci-t]*
zōryū-ki	造粒機	granulator *[cook] [eng]*

zōryū-ki	造粒機	pelletizer *[eng]*
zōryū-tan	造粒炭	pellet coal *[geol]*
zōryū-zai	造粒剤	granulation agent *[mat]*
zōsei-ki	造精器; 蔵精器	antheridium (pl -idia) *[bot]*
zōsen	造船; ぞうせん	shipbuilding *{n} [nav-arch]*
zōsen'gaku	造船学; 造船學	naval architecture *[eng]*
zōsen'jo	造船所	shipyard *[nav-arch]*
zōsen'ka	造船家	naval architect *[nav-arch]*
zō'setsu	増設; ぞうせつ	expansion (of facility) *[ind-eng]*
zō'setsu-denwa	増設電話	extension telephone *[comm]*
zōshin	増進	enhancement *[bio]*
zō-shingō	像信号	picture signal *[comm]*
zō'shitsu-sōchi	増湿装置	humidifier *[mech-eng]*
zōsho-hyō	蔵書標; 藏書標	book plate *[pr]*
zō'shoku	増殖; ぞうしょく	multiplication; proliferation; propagation *[bio]*
"	増殖	breeding *{n} [nucleo]*
zōshoku-baichi	増殖培地	growth medium (pl -s, media) *[bioch]*
zōshoku-hi	増殖比	breeding ratio *[nucleo]*
zōshoku-inshi	増殖因子	growth stimulator *[bio]*
zōshoku-jūgō	増殖重合	proliferous polymerization *[poly-ch]*
zōshoku-ro	増殖炉	breeder reactor *[nucleo]*
zō-shōten	像焦点	image focal point *[opt]*
zō-ten	像点; ぞうてん	image point *[opt]*
zōzan-ki	造山期	orogenic period *[geol]*
zōzan-tai	造山帯	orogenic belt *[geol]*
zōzan-undō	造山運動	orogenesis; orogeny *[geol]*
zu	図; 圖; ず (づ)	chart; diagram; figure *{n} [graph] [math]*
"	図	map; picture; plot; view *{n} [graph]*
-zu	-図	-figure; graph; picture *{suffix}*
zu'an	図案; ずあん	design; device; sketch; plan *{n} [graph]*
zubon	ズボン	trousers; pants; slacks *[cl]*
zu-dai	図題	illustration caption *[graph]*
zuga	図画; 圖畫; ずが	drawing; picture *{n} [graph]*
zugai-kotsu; tōgai-kotsu	頭蓋骨	skull *[anat]*
zu'gaku	図学	descriptive geometry *[math]*
"	図学	graphic science; graphics *[sci-t]*
zu-gara	図柄; ずがら	figure in relief; picture pattern *[graph]*
zuga-yōshi	図画用紙	drawing paper *[paper]*
zuhan	図版	plate; figure; illustration *{n} [graph]*
zuhō	図法	draftsmanship; drawing; projection *[graph]*
zuhyō	図表; ずひょう	chart; diagram; graph *{n} [graph] [math]*

zu-hyōji	図表示	graphical representation *[graph]*
zuhyō-sen	図表線	diagram line *[graph]*
zui	髄；髓；ずい	pith *[bot]*
"	髄	marrow *[hist]*
zuibun	随分；隨分	very; fairly; quite *{adv}*
zuihan-bibun-hōtei'shiki	随伴微分方程式	adjoint differential equation *[math]*
zuihan-gyō'retsu	随伴行列	adjoint matrix *[math]*
zui'i-kin	随意筋；ずいいきん	voluntary muscle *[physio]*
zui'i ni	随意に	voluntarily; optionally *{adv} [psy]*
zui'i no	随意の	volitional *{adj} [psy]*
zui'ji	随時；ずいじ	anytime *{adv}*
zui'ji (ni)	随時(に)	as called for; from time to time; on occasion *{adv}*
zuiki	芋茎；ずいき	taro stem *[cook]*
zui'maku'en-kin	髄膜炎菌	meningococcus (pl -cocci) *[microbio]*
zui'shitsu	髄質；髓質	medulla (pl -las, -lae) *[anat]*
zui-shō	髄鞘；ずいしょう	myelin sheath *[hist]*
zujō-yo'geki	頭上余隙	overhead clearance *[navig]*
zukai	図解；ずかい	explanatory diagram; diagrammatic chart *[graph]*
zukai-hō	図解法	iconography *[graph]*
zukai-jiten	図解辞典	illustrated dictionary *[pr]*
zukai-riki'gaku	図解力学	graphical statics *[mech]*
zukan	図鑑；ずかん	illustrated book; pictorial book *[pr]*
zukei	図形；ずけい	figure; graphic form; pattern *{n} [graph]*
zukei-gengo	図形言語	graphic language *[comput]*
zukei-hyōji	図形表示	graphic display *[comput]*
zukei-kigō	図形記号	graphic symbol *[comput]*
zukei-kika'gaku	図形幾何学	descriptive geometry *[math]*
zukei-hyōji-sōchi	図形表示装置	graphic display unit *[comput]*
zukei-moji	図形文字	graphic character *[comput]*
zukei-nin'shiki	図形認識	pattern recognition *[comput]*
zu-keisan	図計算；ずけいさん	graphical calculation *[math]*
zukei-shōhyō	図形商標	figure trademark; pictorial trademark *[graph] [pat]*
zukei-shori	図形処理	graphics processing *{n} [comput]*
zukei(-shori)-gengo	図形(処理)言語	graphic language *[comput]*
zukei-shutsu'ryoku	図形出力	graphic output *[comput]*
zu-kigō	図記号	graphic symbol *[comput]*
zukin	頭布；ずきん	hood *[cl]*
zukku	ズック	canvas *[tex]*
zukō	図工	draftsman (a person) *[eng]*

zu-men	図面	diagram; drawing; illustration; plan *{n} [graph]*
zumen-hyō	図面表	table of drawings *[pr]*
zurasu	ずらす	displace; offset; shift; stagger *{vb}*
zure	ずれ	shift (in casting) *{n} [plas]*
"	ずれ	slip(page); displacement *{n} [geol]*
zure'ba-ha'guruma	ずれ歯歯車	stepped gear *[mech-eng]*
zure-dansei-ritsu	ずれ弾性率	shear modulus *[mech]*
zure-ō'ryoku	ずれ応力	shearing stress *[mech]*
zure-ritsu	ずれ率	shear rate *[fl-mech]*
zureru	ずれる	slip out of place *{vb i}*
zure-sokudo	ずれ速度	shear rate *[fl-mech]*
zuri	硏；ずり	debris; refuse (of coal) *{n} [geol]*
zuri	ずり	shear (machine) *{n} [mech-eng]*
zushi	図紙；圖紙；ずし	chart; chart paper *{n} [mat] [paper]*
zushi-hō	図示法；ずしほう	graphic method *[graph]*
zu-shiki	図式；ずしき	diagram; chart; schema; graph *{n} [graph]*
zu'shiki-bibun	図式微分	graphical differentiation *[math]*
zu'shiki-ga	図式画；圖式畫	schematic drawing *{n} [graph]*
zu'shiki-hō	図式法	graphology *[math]*
zu'shiki-hyōji	図式表示	pictorial display *[comput]*
"	図式表示	schematic plot *{n} [graph]*
zu'shiki-jōhō	図式情報	pictorial information *[comm]*
zu'shiki'ka	図式化	diagramming; schematizing *{n} [graph]*
zu'shiki-kaihō	図式解法	graphical solution method *[math]*
zu'shiki-keisan-hō	図式計算法	graphical calculation method *[math]*
zu-shiki no	図式の	graphical; graphic *{adj} [graph] [math]*
zu'shiki-rikigaku	図式力学；圖式力學	graphical mechanics *[mech]*
zu'shiki-sekibun	図式積分	graphical integration *[math]*
zu-shin	図心	centroid *{n} [math]*
zushi no	図示の	illustrated; pictured *{adj} [graph]*
zushi-paneru	図示パネル	graphic panel *[cont-sys]*
zu'sho	図書	pictures and books *[pr]*
zu'tsū	頭痛；ずつう	headache *[med]*
zu'tsū-yaku	頭痛薬	headache medicine *{n} [pharm]*
zu'yō	図様；圖様	illustrated state *[graph]*
zu'zō	図像；ずぞう	icon *{n} [graph]*

PART II

English-into-Japanese Section

A

aardvark (animal) *[v-zoo]*	土豚；つちぶた	tsuchi-buta
ab- *{prefix}*	から離れて-	kara hanarete-
" *{prefix}*	遠くに-	tōku ni-
abacus (pl -es, abaci) *[arith]*	十露盤；そろばん	soroban
abaft *{prep} [nav-arch]*	船尾の方に	senbi no hō ni
abalone (shellfish) *[i-zoo]*	鮑；あわび	awabi
abandon; waive *{vb t} [legal]*	放棄する	hōki suru
abandonment *[legal]*	廃棄；廢棄；はいき	haiki
abatement *[legal]*	排除	haijo
" *[med]*	減退；げんたい	gentai
" *{n}*	減少	genshō
abaxial *{adj} [bio]*	背軸(の)	hai'jiku (no)
abbreviate; abridge; shorten *{vb}*	短縮する	tanshuku suru
abbreviation *[comput]*	略記法	ryakki-hō
abbreviation: abbreviated term *[pr]*	略語；りゃくご	ryaku'go
abbreviation: abbreviated name *{n}*	略名	ryakumei
abbreviation: abbreviated name *{n}*	略称；略稱	ryakushō
abdomen *[anat]*	腹；はら	hara
abdominal *{adj} [anat]*	腹の	hara no
abdominal area *[anat]*	腹部；ふくぶ	fuku-bu
abdominal cavity *[anat]*	腹腔	fukkō
abeam *{adv} [nav-arch]*	真横に	ma-yoko ni
aberrant *[bio]*	異常形	ijō-kei
aberration (of light) *[astron]*	光行差	kō'kōsa
" *[opt]*	収差；收差	shūsa
aberration ellipse *[astron]*	光行差楕円	kōkōsa-da'en
abherent *[mat]*	粘着防止剤	nenchaku-bōshi-zai
abhesive *{adj} [adhes]*	非接着性の	hi'setchaku'sei no
abiotic environment *[ecol]*	非生物的環境	hi'seibutsu'teki-kankyō
ablation *[geol]*	削摩；さくま	sakuma
abnormal *{adj}*	異常な	ijō na
abnormality *[math]*	異常	ijō
abort *{vb} [comput]*	打切る	uchi-kiru
abortion *[med]*	流産	ryūzan
about *[math]*	位；ぐらい；くらい	gurai; kurai
about; extent *{n}*	程；ほど	hodo
about center *[math]*	中頃	naka-goro
above *[math]*	上；うえ	ue
aboveground-laid pipe *[hyd]*	地上敷設管	chijō-shisetsu-kan

above sea level *[geol]*	海抜	kai'batsu
abrasion *[eng] [geol]*	摩耗	mamō
abrasion fastness *[mat] [tex]*	摩擦堅牢度	masatsu-kenrō-do
abrasion loss *[plas]*	摩耗減量	mamō-genryō
abrasion resistance *[mat]*	耐摩耗性	tai'mamō-sei
abrasion-resistant cast iron *[met]*	耐摩耗鋳鉄	tai'mamō-chū-tetsu
abrasive *{n} [mat]*	研摩剤	kenma-zai
abrasive grain *[mech-eng]*	砥粒; とりゅう	toryū
abrasiveness *[mat]*	研摩性	kenma-sei
abrupt *{adj}*	突然の; 突然の	totsu'zen no
abscissa *[an-geom]*	横軸	yoko-jiku
" *[an-geom]*	横座標	yoko-zahyō
abscission *[bot]*	切断; 切斷	setsu'dan
abscopal effect *[med]*	遠達作用	en'tatsu-sayō
absence *{n}*	不在	fuzai
absolute *{n}*	絶対; ぜったい	zettai
" *{adj}*	絶対の; 絶對の	zettai no
absolute humidity *[phys]*	絶対湿度	zettai-shitsu'do
absolute judgment	絶対判断	zettai-handan
absolute(ly) dry weight *[mech]*	絶乾重量	zekkan-jūryō
absolute magnitude *[astron]*	絶対等級	zettai-tōkyū
absolute modulus of elasticity *[mech]*	絶対弾性率	zettai-dansei-ritsu
absolute pitch sense *[physio]*	絶対音感	zettai-onkan
absolute pressure *[phys]*	絶対圧	zettai-atsu
absolute system of units *[phys]*	絶対単位系	zettai-tan'i-kei
absolute temperature *[thermo]*	絶対温度	zettai-ondo
absolute unit(s) *[phys]*	絶対単位	zettai-tan'i
absolute value *[alg]*	絶対値	zettai-chi
absolute zero (point) *[thermo]*	絶対零度	zettai-reido
absorbance *[p-ch]*	光学濃度	kōgaku-nōdo
" *[p-ch]*	吸光度	kyūkō-do
absorbent *[mat]*	吸収剤; 吸收劑	kyūshū-zai
absorbent cotton *[tex]*	脱脂綿	dasshi-men
absorptance *[phys]*	吸収率	kyūshū-ritsu
absorptiometer *[an-ch]*	吸光光度計	kyūkō-kōdo-kei
absorption *[ch]*	吸収; きゅうしゅう	kyūshū
absorption coefficient *[acous] [phys]*	吸収係数	kyūshū-keisū
absorption factor *[acous] [phys]*	吸収因子	kyūshū-inshi
" *[acous] [phys]*	吸収率	kyūshū-ritsu
absorption spectrophotometry *[spect]*	吸光光度分析	kyūkō-kōdo-bunseki
absorptive *{adj} [ch]*	吸収性の	kyūshū'sei no
absorptivity *[an-ch]*	吸光率	kyūkō-ritsu

absorptivity *[an-ch]*	吸収力	kyūshū-ryoku
abstract *{n} [pr]*	抜粋；ばっすい	bassui
" *{n} [pr]*	抄録	shōroku
" *{n} [pr]*	摘要	teki'yō
" *{adj}*	抽象的な	chūshō'teki na
abstract; summarize *{vb} [comm]*	要約する	yōyaku suru
abstract algebra *[math]*	抽象代数学	chūshō-daisū'gaku
abstraction *{n} [ch]*	抽出	chū'shutsu
" *{n}*	抽象性	chūshō-sei
abstraction reaction *[ch]*	引抜き反応	hiki'nuki-hannō
abstract science *[sci-t]*	抽象科学	chūshō-kagaku
abstract space *[math]*	抽象空間	chūshō-kūkan
abstract symbol *[graph]*	抽象記号	chūshō-kigō
abundant; much *[math]*	多い；おおい	ōi
abut *{vb}*	当接する	tō'setsu suru
abutment *[arch]*	迫受台	seri'uke-dai
" (of a bridge) *[civ-eng]*	橋台	hashi-dai
abyss *{n} [geol]*	深淵	shin'en
abyss; deep sea *[ocean]*	深海	shinkai
Acanthocephala *[i-zoo]*	鉤頭虫類	kōtō'chū-rui
acarology *[zoo]*	壁蝨学；蜱学	dani'gaku
accelerate *{vb}*	促進する	sokushin suru
" *{vb}*	早める	hayameru
accelerated motion *[phys]*	加速運動	kasoku-undō
acceleration *[mech]*	加速	ka'soku
" *[mech]*	加速度；かそくど	kasoku-do
acceleration of free fall *[mech]*	重力加速度	jūryoku-ka'sokudo
acceleration of gravity *[mech]*	重力加速度	jūryoku-ka'sokudo
acceleration voltage *[elec]*	加速電圧	kasoku-den'atsu
accelerator *[ch]*	促進剤	sokushin-zai
" *[mech-eng]*	加速装置	kasoku-sōchi
accelerometer *[eng]*	加速(度)計	kasoku(do)-kei
accent *{n} [ling]*	揚音	yō'on
acceptance *[ind-eng]*	合格	gōkaku
" *[ind-eng]*	受入れ	uke'ire
" *[pat]*	受理	juri
acceptance inspection *[ind-eng]*	受入検査	uke'ire-kensa
acceptance test *[ind-eng]*	受入試験	uke'ire-shiken
acceptor *[ch]*	受容体	juyō'tai
" *[sol-st]*	受体	jutai
" *{n}*	取手	tori'te
access; look up *{vb} [comput]*	参照する	sanshō suru

accessory (pl -ries) *[cl] [furn]*	装飾品	sōshoku'hin
accessory constituent *[ch]*	副成分	fuku-seibun
access time *[comput]*	呼出時間	yobi'dashi-jikan
accident; fortuity *[math]*	偶然；ぐうぜん	gūzen
accident *{n}*	災害	saigai
accidental *{adj} [math]*	偶発的な	gūhatsu'teki na
accidental error *[math]*	偶然誤差	gūzen-gosa
accident rate *[ind-eng]*	事故率；じこりつ	jiko-ritsu
acclimating cultivation *[microbio]*	順養培養	jun'yō-baiyō
acclimatization *[bio]*	馴化	junka
" *{n}*	順養；じゅんよう	jun'yō
acclivity *[geol]*	上り傾斜	nobori-keisha
accomodation *[physio]*	順応；じゅんのう	junnō
" *{n}*	収容；收容	shūyō
accomodation coefficient *[stat-mech]*	適応係数；適應係數	tekiō-keisū
accompanying document *[pr]*	添付書類	tenpu-shorui
accord *{n} [acous]*	和音；わおん	wa'on
account *{n} [econ]*	口座	kōza
accountant (a person) *[econ]*	会計士；會計士	kaikei'shi
accounting *{n} [econ]*	会計(学)	kaikei('gaku)
accounting machine *[eng] [comput]*	会計機；かいけいき	kaikei-ki
accretion *[astron]*	付着	fu'chaku
accumulate *{vb}*	蓄積する	chiku'seki suru
" *{vb}*	堆積する	taiseki suru
accumulate; save; amass *{vb t}*	溜める；貯める	tameru
accumulation *{n} [elec] [math]*	集積	shūseki
" *{n}*	蓄積	chiku'seki
accumulative cultivation *[microbio]*	集積培養	shūseki-baiyō
accumulator *[comput]*	累算器	ruisan-ki
accuracy *[math] [sci-t]*	精度	seido
" *[math] [sci-t]*	正確度	seikaku-do
" *[math] [sci-t]*	正確さ；せいかくさ	seikaku-sa
accurate; correct *{adj} [math]*	正確な	seikaku na
accusative case *[gram]*	対格；對格	taikaku
acetate *[ch]*	酢酸塩	sakusan-en
acetic acid	酢酸	sakusan
acetic anhydride	無水酢酸	musui-sakusan
achene *[bot]*	痩果；そうか	sōka
achromatic color *[opt]*	無彩色	musai-shoku
Achyranthes japonica *[bot]*	牛膝；いのこずち	inokozuchi
acicular *{adj}*	針状の	hari'jō no; shin'jō no
acid *[ch]*	酸；さん	san

acid anhydride *[ch]*	無水酸	musui-san
" *[ch]*	酸無水物	san-musui'butsu
acid chloride *[ch]*	酸塩化物	san-enka'butsu
acid clay *[geol]*	酸性白土	sansei-hakudo
acid content *[ch]*	酸分	sanbun
acid dianhydride *[ch]*	酸二無水物	san-ni'musui'butsu
acid dye *[org-ch]*	酸性染料	sansei-senryō
acid-fast *{adj} [mat]*	抗酸性の	kōsan-sei no
acidic *{adj} [ch]*	酸性の	sansei no
acidic lead arsenate	酸性砒酸鉛	sansei-hisan-namari
acidic oxide *[ch]*	酸性酸化物	sansei-sanka'butsu
acidity *[ch]*	酸度	sando
" *[ch]*	酸性度；さんせいど	sansei-do
acidophil *{n} [bio]*	好酸性物質	kōsan'sei-busshitsu
acidophilous milk *[cook]*	乳酸菌牛乳	nyūsan'kin-gyūnyū
acidproofness *[mat]*	耐酸性	tai'san-sei
acid rain *[meteor]*	酸性雨	sansei'u
acid resistance *[mat]*	耐酸性	taisan-sei
acid-resistant *{adj} [mat]*	耐酸性の	tai'san'sei no
acid salt *[ch]*	酸性塩	sansei-en
acid steel *[steel]*	酸性鋼	sansei-kō
acidulant *[cook] [mat]*	酸味付与剤	sanmi-fuyo'zai
acid value *[ch]*	酸価	sanka
acknowledgment *{n} [pat]*	承認	shōnin
" *{n}*	肯定	kōtei
" *{n}*	認定	nintei
acmite *[miner]*	錐輝石	kiri'kiseki
aconite *[bot]*	鳥兜；とりかぶと	tori-kabuto
aconite root *[bot] [pharm]*	烏頭；うず	uzu
acorn *[bot]*	団栗；どんぐり	donguri
acoustic absorptivity *[acous]*	吸音率	kyūon-ritsu
acoustical board *[acous]*	防音板	bōon-ban
acoustic coupler *[eng-acous]*	音響結合装置	onkyō-ketsugō-sōchi
acoustic delay line *[electr]*	超音波遅延線	chō'onpa-chi'en-sen
acoustic efficiency *[acous]*	音響効率	onkyō-kō'ritsu
acoustic fidelity *[acous]*	音響的忠実性	onkyō'teki-chūjitsu-sei
acoustic pressure *[acous]*	音圧；音壓	on-atsu
acoustics *[phys]*	音響学；音響學	onkyō'gaku
acoustic velocity *[acous]*	音速	onsoku
acoustic wave *[acous]*	音波	onpa
acoustoelectronics *[acous]*	音響電子工学	onkyō-denshi-kōgaku
acoustooptics *[opt]*	音響光学	onkyō-kōgaku

acquired character *[bio]*	獲得形質	kakutoku-keishitsu
acquired immune deficiency syndrome; **AIDS** *[med]*	獲得免疫不全症；（エイズ）	kakutoku-men'eki-fuzen'shō; (eizu)
acquired immunity *[immun]*	獲得免疫	kakutoku-men'eki
acrid substance *[cook]*	辛味物質	shinmi-busshitsu
acronym *[comm]*	頭(文)字語	kashira(-mo)jigo
" *[comm]*	頭辞語	tōji'go
act *{n}* *[legal]*	法令；ほうれい	hōrei
actinic focus *[opt]*	化学焦点	kagaku-shōten
actinic ray *[opt]*	活性線	kassei-sen
actinium (element)	アクチニウム	akuchinyūmu
actinolite *[miner]*	緑閃石	ryoku'senseki
" *[miner]*	陽起石；ようきせき	yōki'seki
actinology *[phys]*	化学線学	kagakusen'gaku
actinometer *[eng]*	日射計	nissha-kei
Actinomyces *[microbio]*	放線菌	hōsen-kin
action *{n}* *[ch]* *[phys]*	作用	sayō
" *{n}* *[psy]*	行動；こうどう	kōdō
" *{n}*	動作	dōsa
action and reaction, law of *[phys]*	作用反作用の法則	sayō-han'sayō no hōsoku
activate *{vb t}* *[bioch]*	賦活する	fukatsu suru
" *{vb t}* *[comput]*	起動させる	kidō saseru
" *{vb t}* *[eng]*	付勢する	fusei suru
activated charcoal *[mat]*	活性炭	kassei-tan
activated clay *[mat]*	活性白土	kassei-haku'do
activated sludge *[civ-eng]*	活性汚泥	kassei-o'dei
activated water *[hyd]*	活性水	kassei-sui
activation *[ch]*	賦活；ふかつ	fukatsu
" *[ch]*	活性化	kassei'ka
" *[comput]*	活用化	katsuyō'ka
activation energy *[p-ch]*	活性化エネルギー	kassei'ka-enerugī
activator *[ch]*	活性剤	kassei-zai
active *[gram]*	能動；のうどう	nōdō
active; aggressive *{adj}*	能動的な	nōdō'teki na
active center *[astron]*	活性点；活性點	kassei-ten
" *[bioch]*	活性中心	kassei-chūshin
active component *[ch]* *[elec]*	有効分	yūkō-bun
" *[elec]*	活性成分	kassei-seibun
active fault *[geol]*	活断層；活斷層	katsu'dan-sō
active group *[ch]*	活性基	kassei-ki
active site *[astron]*	活性点	kassei-ten
" *[bioch]*	活性部位	kassei-bu'i

active species *[ch]*	活性種	kassei-shu
active transport *[cyt] [physio]*	能動輸送	nōdō-yusō
active voice *[gram]*	能動態	nōdō-tai
active volcano *[geol]*	活火山	kakkazan
activity *[ch]*	活性	kassei
" *[p-ch]*	活動度	katsudō-do
" *[thermo]*	活量	katsu'ryō
activity coefficient *[p-ch]*	活動係数	katsudō-keisū
actual *{adj} [math]*	実際の	jissai no
actual article *{n}*	実物；實物	jitsu-butsu
actual decimal point *[arith]*	実際の小数点	jissai no shōsū-ten
actuality *{n}*	事実；事實；じじつ	ji'jitsu
actual measurement *[sci-t]*	実測	jissoku
actual profit *[econ]*	実益	jitsu'eki
actual proof *[legal]*	実証	jisshō
actual results; record of performance *[ind-eng]*	実績	jisseki
actual stress *[mech]*	真応力	shin-ō'ryoku
actuate; operate *{vb t} [ind-eng]*	作動させる	sadō saseru
acuate; pointed *{adj} [bio]*	尖った	togatta
acuity (pl -ities) *{n} [bio]*	鋭さ	surudo-sa
acupuncture *[med]*	鍼；はり	hari
acute accent *[ling] [pr]*	揚音符	yōon-pu
acute angle *[geom]*	鋭角	eikaku
acute triangle *[geom]*	鋭角三角形	eikaku-sankak(u)kei
acyclic *{adj} [org-ch]*	非環状の	hi'kanjō no
adaptability *{n} [physio]*	適応性	tekiō-sei
" *{n}*	融通性	yūzū-sei
adaptation *[gen] [physio]*	適応	teki'ō
adaptation; arrangement *[music]*	編曲	henkyoku
adaptation *[physio] [psy]*	順応；じゅんのう	junnō
adaptive enzyme *[bioch]*	適応酵素	tekiō-kōso
adaxial *{adj} [bio]*	向軸(の)	kō'jiku (no)
add *{vb} [arith]*	加える	kuwaeru
" *{vb} [arith]*	足す	tasu
" *{vb} [ch] [sci-t]*	添加する	tenka suru
add dropwise *{vb} [ch]*	滴下する	tekika suru
add for fragrance *{vb} [food-eng]*	加香する	kakō suru
add together *{vb t}*	合わせる	awaseru
addend *[arith]*	加数；加數	kasū
addendum *[pr]*	補遣	ho'i
adder *[comput]*	加算器	kasan-ki

addiction *[med]*	耽溺；たんでき	tan'deki
addictive drug *[pharm]*	耽溺薬	tan'deki-yaku
addition *[arith]*	加法	kahō
" *[arith]*	加え算	kuwae-zan
" (cumulative) *[arith]*	積算	sekisan
" *[arith]*	足し算	tashi-zan
" *[arith]*	寄せ算	yose-zan
" *[ch]*	付加	fuka
" *[ch]*	添加	tenka
" *[pat]*	追加	tsuika
additional *{adj}*	追加の	tsuika no
additional patent *[pat]*	追加特許	tsuika-tokkyo
addition complex *[ch]*	付加錯体	fuka-sakutai
addition metal *[met]*	差物；さしもの	sashi-mono
addition polymerization *[poly-ch]*	付加重合	fuka-jūgō
additive *{n} [mat]*	添加物（添加剤）	tenka'butsu (tenka-zai)
additive formula *[alg]*	加法公式	kahō-kōshiki
additive primary color *[opt]*	加色法の原色	kashoku-hō no genshoku
additivity *[ch]*	加成率	kasei-ritsu
" *[ch]*	加成性	kasei-sei
additivity rule *[ch]*	加成則	kasei-soku
address *{n} [comm]*	宛名；あてな	ate'na
" *{n} [comm]*	住所	jūsho
" *{n} [comput]*	番地	banchi
adequate; sufficient *{adj}*	十分な	jūbun na
adhere *{vb} [mech]*	着く	tsuku
adherence *[adhes]*	固着	ko'chaku
" *[phys] [eng]*	密着	mitchaku
adherend *[adhes]*	被着体	hichaku'tai
" *[adhes]*	接着面	setchaku-men
adhering quantity *[adhes]*	被着量	hichaku-ryō
adhesion; bonding; cementing; glu(e)ing *{n} [adhes] [mech]*	接着；せっちゃく	setchaku
adhesive *{n} [mat]*	粘着性物質	nenchaku'sei-busshitsu
" (adhesive agent) *[mat]*	接着剤	setchaku-zai
adhesiveness property *[adhes]*	接着性	setchaku-sei
adhesive plaster (tape) *[pharm]*	絆創膏	bansō-kō
adhesive strength *[eng]*	粘着力	nenchaku-ryoku
adhesive tape; tacky tape *[mat]*	粘着テープ	nenchaku-tēpu
adiabatic *{n} [thermo]*	断熱；斷熱	dannetsu
adiabatic change *[thermo]*	断熱変化	dannetsu-henka
adiabatic expansion *[thermo]*	断熱膨張	dannetsu-bōchō

adiabatic material *[mat]*	断熱材；斷熱材	dannetsu-zai
adiabatic process *[thermo]*	断熱過程	dannetsu-katei
adipose tissue *[hist]*	脂肪組織	shibō-so'shiki
adit *[civ-eng]*	入路	nyūro
" *[civ-eng]*	横坑；おうこう	ōkō
" *[min-eng]*	通洞	tsūdō
adjacency effect *[photo]*	隣接効果	rinsetsu-kōka
adjacent angles *[geom]*	隣接角	rinsetsu-kaku
" *[geom]*	接角	sekkaku
adjective *[gram]*	形容詞	keiyō'shi
adjoining seas *[geogr]*	近海	kinkai
adjoint differential equation *[math]*	随伴微分方程式	zuihan-bibun-hōtei'shiki
adjoint matrix *[math]*	随伴行列	zuihan-gyō'retsu
adjudication *[legal]*	裁判；さいばん	saiban
adjust; coordinate *{vb}*	整合する	seigō suru
adjustable *{adj}*	調節出來る	chōsetsu dekiru
adjustment *[eng]*	調節	chōsetsu
" *[eng]*	整合	seigō
" *[psy]*	適応	tekiō
adjustment; regulation *{n}*	調整	chōsei
adjuvant *[immun]*	抗原補強剤	kōgen-hokyō-zai
" *[pharm]*	添加剤	tenka-zai
adlay; adlai *[bot]*	鳩麦；はとむぎ	hato-mugi
administer dropwise *{vb t} [cl]*	点滴する	ten'teki suru
administration; control; management *[econ] [ind-eng]*	管理	kanri
administration; operation *{n} [econ]*	運営	un'ei
administration *{n}*	行政	gyōsei
administrative center *[civ-eng]*	行政中心地	gyōsei-chūshin'chi
administrative disposition	行政処分	gyōsei-shobun
administrator; director	長官	chōkan
admixture *[mat]*	混合物	kongō'butsu
" *[mat]*	添加剤	tenka-zai
adolescence *[psy]*	青年期	sei'nen-ki
adrenal gland *[anat]*	副腎	fuku'jin
adrenalin *[bioch]*	アドレナリン	adorenarin
adsorbate; adsorptive *{n} [ch]*	吸着質	kyūchaku-shitsu
adsorbent *[ch]*	吸着剤；(-媒)	kyūchaku-zai; (-bai)
adsorption *[ch]*	吸着；きゅうちゃく	kyū'chaku
adsorption isotherm *[p-ch]*	吸着等温式	kyūchaku-tō'on-shiki
adult (person) *[bio]*	成人	seijin
adulterant *[ch]*	混在物	konzai'butsu

advance *{vb i} [navig]*	前進する	zenshin suru
" (progress) *{vb i}*	進む	susumu
advance and retreat; move *{vb}*	進退する	shintai suru
advancement; Go Ahead *[navig]*	前進	zenshin
advancing wave *[phys]*	前進波	zenshin-ha
advantage *{n}*	長所	chōsho
advantageous *{n}*	有利；ゆうり	yūri
advection *[meteor] [ocean]*	移流	i'ryū
adverb *[gram]*	副詞	fuku'shi
adverse weather *[meteor]*	悪天候；惡天候	aku-tenkō
advertisement *[comm]*	広告；廣告	kōkoku
advertisement tower *[arch]*	広告塔	kōkoku-tō
advice *{n}*	意見	iken
" *{n}*	助言	jogen
advisory *[meteor]*	注意報	chūi-hō
" *[meteor]*	状況報告	jōkyō-hōkoku
adze *[eng]*	手斧	chōna; te'ono
adzuki bean (a legume) *[bot]*	小豆；あずき	azuki
adzuki bean weevil (insect) *[i-zoo]*	小豆象虫	azuki-zō'mushi
Aedes (insect: mosquito) *[i-zoo]*	藪蚊；やぶか	yabu-ka
Aedes aegypti (insect) *[i-zoo]*	熱帯縞蚊	nettai-shima-ka
Aedes albopictus (insect) *[i-zoo]*	一筋縞蚊	hito'suji-shima-ka
aegerine *[miner]*	錐輝石	kiri'kiseki
aeolian deposit *[geol]*	風成層	fūsei-sō
aeration *[eng]*	曝気；曝氣	bakki
" *[eng]*	気曝	ki'baku
aeration solution *[ch]*	曝気液	bakki-eki
aeration tank *[ch]*	曝気槽	bakki-sō
aerial lead-in *{n} [elec]*	架空引込	kakū-hiki'komi
aerobe *[bio]*	好気菌	kōki-kin
" *[bio]*	好気性生物	kōki'sei-seibutsu
aerobic *{adj} [bio]*	好気性の	kōki'sei no
aerobic bacteria *[microbio]*	好気性微生物	kōkisei-bi'seibutsu
aerobic respiration *[bio]*	好気呼吸	kōki-kokyū
aerobiology *[bio]*	空中生物学	kūchū-seibutsu'gaku
aerodynamics *[fl-mech]*	航空力学	kōkū-riki'gaku
" *[fl-mech]*	空気力学	kūki-riki'gaku
aerolite *[geol]*	石質隕石	seki'shitsu-inseki
aeronautical *{adj} [aero-eng]*	航空の；こうくうの	kōkū no
aeronautics *[fl-mech]*	航空学	kōkū'gaku
aeronomy *[geoch]*	超高層大気化学	chō'kōsō-taiki-kagaku
" *[geophys]*	超高層大気物理学	chō'kōsō-taiki-butsuri'gaku

aerosol *[ch]*	煙霧体；煙霧體	enmu'tai
aerospace; airspace *{adj} [meteor]*	航空宇宙の	kōkū-uchū no
aerospace electronics *[electr]*	航空宇宙電子工学	kōkū'uchū-denshi-kōgaku
aerospace engineering *[eng]*	航空宇宙工学	kōkū'uchū-kōgaku
aerostat *{n} [aero-eng]*	気球	kikyū
aestivation *[bot]*	幼葉態	yōyō'tai
" *[physio]*	夏眠	ka'min
affidavit *[legal]*	宣誓供述書	sensei-kyō'jutsu'sho
affiliation; position; post *{n}*	所属	shozoku
affiliation *{n}*	提携；ていけい	teikei
affine transformation *[math]*	擬似変換	giji-henkan
affinity (pl -ties) *[ch]*	親和力	shinwa-ryoku
" *[ch]*	親和性	shinwa-sei
afflicted part; affected part *[med]*	患部	kanbu
afocal *{adj} [opt]*	無限遠焦点の	mugen'en-shōten no
afterburning *{n} [aero-eng]*	後燃え；あともえ	ato-moe
aftereffect *[ind-eng]*	余効；餘効	yokō
" *[pharm]*	残効；殘効	zankō
afterglow *[atom-phys] [meteor]*	残光	zankō
afterimage *[physio]*	残像	zanzō
afternoon *[astron]*	午後	gogo
after perpendicular *[nav-arch]*	船尾垂線	senbi-suisen
afterprocessing *{n} [mech-eng]*	後加工	ato-kakō
afterripening *{n} [photo]*	後熟成	ato-jukusei
aftershock *[geophys]*	余震	yoshin
aftershrinkage *[poly-ch]*	後収縮；後收縮	ato-shūshuku
aftersound *[acous]*	残音；殘音	zan'on
aftertaste *[cook] [physio]*	後味；あとあじ	ato-aji
aftertreatment *[ind-eng]*	後処理	ato-shori
aftervulcanization (cure) *[rub]*	後加硫	ato-karyū
agalloch (an incense) *[bot]*	沈香；じんこう	jinkō
agalmatolite *[miner]*	蠟石	rōseki
agar; agar-agar *[mat]*	寒天	kanten
" *[mat]*	天草；石花菜	ten-gusa
agaric; Pleurotus ostreatus *[bot]*	平茸；ひらたけ	hira-take
agate *[miner]*	瑪瑙；めのう	me'nō
agave *[bot]*	竜舌蘭；龍舌蘭	ryūzetsu'ran
age (in years) *{n} [bio]*	年齢	nenrei
" (in weeks) *{n} [bio]*	週齢	shūrei
" *{n} [bio] [physio]*	齢；よわい	yowai
" *[geol]*	期	ki
age hardening *{n} [met]*	時効硬化	jikō-kōka

agent (a person)	代理人	dairi'nin
" ; agency (an establishment)	代理店	dairi'ten
agglomerate *{n} [bot] [geol]*	集塊	shūkai
" *{n} [sci-t]*	団塊; 團塊	dankai
agglomeration *[met]*	塊状集積	kaijō-shūseki
agglutinate; condense *{vb t}*	凝集させる	gyōshū saseru
agglutinate onto *{vb} [ch]*	膠着する	kōchaku suru
agglutination *[ch] [immun]*	凝集; ぎょうしゅう	gyōshū
aggregate *{n} [geol] [mat]*	集合体; 集合體	shūgō'tai
" *{n} [mat]*	骨材	kotsu-zai
aggregate fruit *[bot]*	分離複果	bunri-fukka
aggregation *[phys]*	集合	shūgō
" (of powder) *[phys]*	凝集	gyōshū
aggressively positive; vigorous *{adj}*	積極的な	sekkyoku'teki na
aging *[ch] [eng]*	熟成	jukusei
" *[eng]*	時効	jikō
aging degradation *[eng]*	時効劣化	jikō-rekka
aging effect *[poly-ch]*	時効効果	jikō-kōka
aging property *[mat]*	老化性	rōka-sei
agitation *[mech-eng]*	攪拌; かくはん	kaku'han
agitator (apparatus) *[ch-eng]*	攪拌装置	kakuhan-sōchi
agonic line *[geophys]*	無方位角線	mu'hōi'kaku-sen
agree; match *{vb}*	一致する	itchi suru
agreement *{n} [legal]*	合意	gōi
" *{n} [legal]*	条約; 條約	jōyaku
" *{n} [legal]*	契約	keiyaku
" *{n}*	協定	kyōtei
" (matching) *{n}*	合致; がっち	gatchi
agricultural chemical *[mat]*	農薬; 農藥	nō-yaku
agricultural chemistry *[agr] [ch]*	農芸化学; 農藝化学	nōgei-kagaku
agricultural experiment station *[agr]*	農業試験場	nōgyō-shiken'jō
agricultural machinery *[agr] [eng]*	農機	nōki
agriculture *[agr]*	農業	nō'gyō
" *[agr] [bio]*	農学	nō'gaku
agrochemical *[mat]*	農薬	nōyaku
agronomy *[agr]*	耕種学	kōshu'gaku
ahead *{adv} {n}*	前; まえ	mae
" *{adv} {n}*	先; さき	saki
ahead and behind	前後	zengo
AIDS *[med]*	エイズ	eizu
AIDS-related complexes; ARC *[med]*	エイズ関連症候群	eizu-kanren-shōkō'gun
aileron *[aero-eng]*	補助翼	hojo-yoku

aim *{n}*	目安；めやす	meyasu
air *{n} [ch]*	空気；空氣；くうき	kūki
" (ventilate) *{vb t}*	通風する	tsūfū suru
airbag *[mech-eng]*	空気袋	kūki-bukuro
air bath *[thermo]*	空気浴	kūki-yoku
air bladder (of a fish) *[v-zoo]*	鰾；浮き袋	uki-bukuro
air blowing *{n} [ch-eng]*	送風	sōfū
airborne *{adj} [aero-eng]*	空輸の	kūyu no
air cargo; air freight *[trans]*	航空貨物	kōkū-ka'motsu
air cleaner *[eng]*	空気清浄器	kūki-seijō-ki
air conditioning *[mech-eng]*	空気調和	kūki-chōwa
" *[mech-eng]*	冷房	reibō
air-cooling *{n} [mech-eng]*	空冷	kūrei
air corridor *[navig]*	空中回路	kūchū-kairo
aircraft; airplane *[aero-eng]*	航空機	kōkū'ki
aircraft carrier *[nav-arch]*	(航空)母艦	(kōkū-)bokan
aircraft control *[navig]*	航空機制御	kōkūki-seigyo
air drying *{n} [eng]*	気乾；氣乾；きかん	kikan
" *{n} [eng]*	空気乾燥	kūki-kansō
airfoil *[aero-eng]*	翼；よく	yoku
" *[aero-eng]*	翼型	yoku'gata
air force *[mil]*	空軍	kūgun
airframe *[aero-eng]*	機体；機體；きたい	kitai
air freighting *{n} [trans]*	航空貨物輸送	kōkū-kamotsu-yusō
airglow *[geophys]*	大気光	taiki'kō
air intake *[aero-eng]*	空気取入口	kūki-tori'ire-guchi
airline company *[econ]*	航空会社；航空會社	kōkū-kaisha
airmail *[comm]*	航空便	kōkū'bin
air mass *[meteor]*	気団；氣團	kidan
air oxidation *[ch]*	空気酸化	kūki-sanka
air permeability *[fl-mech]*	透気度	tōki-do
airplane *[aero-eng]*	飛行機；ひこうき	hikōki
air pollution *[ecol]*	空気汚染	kūki-o'sen
" *[ecol]*	大気汚染	taiki-o'sen
airport *[civ-eng]*	空港；空港	kūkō
air quality *[meteor]*	大気質	taiki-shitsu
air ratio *[eng]*	空気比	kūki-hi
air route; airway *[navig]*	空路	kūro
airship *[aero-eng]*	飛行船	hikō'sen
airspeed *[aero-eng]*	対気速度；對氣速度	taiki-sokudo
airstrip; landing strip *[aero-eng]*	滑走路	kassō-ro
airtight *{adj} [eng]*	気密の	kimitsu no

airtightness *{n} [eng]*	気密度	kimitsu-do
air traffic *[navig]*	航空交通	kōkū-kōtsu
air-traffic control *[navig]*	航空交通管制	kōkū-kōtsu-kansei
airway beacon *[navig]*	航空路灯台	kōkū'ro-tōdai
aisle; passageway *[arch]*	通路; つうろ	tsūro
Akita dog (animal) *[v-zoo]*	秋田犬	Akita-inu; Akita'ken
alabaster *[miner]*	雪花石膏	sekka'sekkō
alarm *{n} [electr]*	警報; けいほう	keihō
alarm clock *[horol]*	目覚まし時計	me'zamashi-dokei
albacore (pl -core, -cores) (tuna fish) *[v-zoo]*	鬢長; びんなが; = びんちょう	binnaga (binchō)
albatross (bird) *[v-zoo]*	信天翁; 阿房鳥	ahō-dori
albinism; albino *[bio]*	白子	shira'ko
albite *[miner]*	曹長石	sō'chōseki
albumen *[bioch]*	卵白	ranpaku
" *[bot]*	胚乳; はいにゅう	hainyū
alcohol *[ch]*	アルコール	arukōru
" *[cook]*	酒精; しゅせい	shusei
alcoholic beverage(s) *[cook]*	酒精飲料	shusei-inryō
alcoholic beverages *[cook]*	酒類	sake-rui
alcove (for ornament) *[arch]*	床の間; とこのま	toko-no-ma
alcove shelves, staggered *[arch]*	違い棚	chigai-dana
aldehyde *[org-ch]*	アルデヒド	arudehido
alder (tree) *[bot]*	榛の木; はんのき	han no ki
alert signal *[comm]*	注意報	chūi-hō
alfalfa *[bot]*	紫馬肥	murasaki-uma'goyashi
alga (pl -ae) *[bot]*	藻; も	mo
algae *[bot]*	藻類; そうるい	sō-rui
algebra *[math]*	代数学; 代數學	daisū'gaku
algebraic language *[comput]*	代数処理言語	daisū-shori-gengo
algorithm *[comput]*	互除法	gojo-hō
" *[comput]*	算法; さんぽう	sanpō
algorithmic language *[comput]*	算法言語	sanpō-gengo
alias *[comput] [legal]*	別名	betsu'mei
alicyclic compound *[org-ch]*	脂環式化合物	shikan'shiki-kagō'butsu
alien *{n}*	外国人; 外國人	gaikoku'jin
alienate; estrange *{vb}*	疎外する	sogai suru
alignment *[eng]*	整列	sei'retsu
alimentary canal *[anat]*	食道	shoku'dō
aliphatic compound *[org-ch]*	脂肪族化合物	shibō'zoku-kagō'butsu
alkali *[ch]*	アルカリ	arukari
alkaline *{adj} [ch]*	アルカリ性の	arukari'sei no

alkyl *[org-ch]*	アルキル	arukiru
all *{n}*	あらゆる	arayuru
" *{n}*	皆; みな; みんな	mina; minna
" *{n}*	総て; 凡て; 全て	subete
" (the whole lot) *{n}*	全部; ぜんぶ	zenbu
allanite *[miner]*	褐簾石	katsu'renseki
all busy (telephone) *[comm]*	全話中	zen'wachū
allele *[gen]*	対立遺伝子	tairitsu-iden'shi
" *[gen]*	対立因子; 對立因子	tairitsu-inshi
allergy *[med]*	アレルギー	arerugī
alliance *[bot]*	群団; 群團	gundan
" *{n}*	同盟; どうめい	dōmei
alligation *[math]*	複合則	fukugō-soku
alligator (a reptile) *[v-zoo]*	鰐; わに	wani
alligatoring *{n} [mat] [met]*	鰐皮割れ	wani'gawa-ware
alligator skin *[met]*	鰐肌	wani-hada
allo- *{prefix}*	異る-	kotonaru-
allocation *[comput]*	割振り	wari'furi
" *[comput]*	割付; わりつけ	wari'tsuke
" *{n}*	配分; はいぶん	haibun
" *{n}*	割当	wari'ate
allotrope *[ch]*	同素体; どうそたい	dōso'tai
allowable current density *[elec]*	許容電流密度	kyoyō-denryū-mitsudo
allowable error *[elec] [phys]*	許容誤差	kyoyō-gosa
allowance *[econ]*	給与; 給與	kyūyo
" (leeway) *[eng]*	ゆとり	yutori
" (margin) *[tex]*	代; しろ	shiro
allowed limit *[ind-eng]*	許容限度	kyoyō-gendo
alloy *[met]*	合金; ごうきん	gōkin
alloy-plated steel sheet *[steel]*	合金化めっき鋼板	gōkin'ka-mekki-kōban
alloy steel *[steel]*	合金鋼	gōkin-kō
alloy tool steel *[steel]*	合金工具鋼	gōkin-kōgu-kō
alluring coloration *[zoo]*	誘惑色	yūwaku-shoku
alluvial fan *[geol]*	扇状地; 扇狀地	senjō-chi
alluvial plain; floodplain *[geol]*	沖積平野	chūseki-heiya
alluvial soil *[geol]*	沖積土	chūseki-do
all-wave receiver *[electr]*	全波受信機	zenpa-jushin-ki
all-weather navigation *[navig]*	全天候航行	zen-tenkō-kōkō
allyl alcohol	アリルアルコール	ariru-arukōru
almanac; calendar *[astron]*	暦; 曆; こよみ	koyomi
almandine *[miner]*	鉄礬柘榴石	tetsuban'zakuro-ishi
almond (tree, flower) *[bot]*	扁桃; へんとう	hentō

aloeswood (incense) *[bot]*	伽羅; きゃら	kyara
alpestrine; subalpine *{adj} [ecol]*	亜高山帯の	a'kōzan-tai no
alphabetical character *[comput]*	英字; えいじ	eiji
alphabetic string *[comput]*	英字列	eiji-retsu
alphanumeric character set *[comput]*	英数字集合	ei'sūji-shūgō
alpha rays *[nucleo]*	アルファー線	arufā-sen
alpine plant *[bot]*	高山植物	kōzan-shokubutsu
altazimuth; theodolite *[eng]*	経緯儀	kei'i-gi
alteration *[math]*	変更; 變更	henkō
alternate *{n} [comput]*	代替; だいたい	daitai
" *{n} [math]*	交互	kōgo
" *{adj} [bot]*	互生の	gosei no
alternate interior angles *[geom]*	錯角	sakkaku
alternate key *[comput]*	代替キー	daitai-kī
alternate route *[comput] [navig]*	代替経路	daitai-keiro
alternating application *[elec]*	交互印加	kōgo-inka
alternating copolymer *[poly-ch]*	交互共重合体	kōgo-kyōjūgō'tai
alternating current *[elec]*	交流; こうりゅう	kōryū
alternating-current power *[elec]*	交流電力	kōryū-denryoku
alternating interior angles *[geom]*	錯角; さっかく	sakkaku
alternating magnetic pole *[elecmg]*	交番磁極	kōban-ji'kyoku
alternating polarity *[phys]*	交番極性	kōban-kyoku'sei
alternating series *[alg]*	交代級数	kōtai-kyūsū
alternation pseudomorphism *[miner]*	変質仮像; 變質假像	hen'shitsu-kazō
alternative *{adj}*	交互の	kōgo no
" ; substituting *{adj}*	交代する; 交替する	kōtai suru
alternative hypothesis *[stat]*	対立仮説; 對立假說	tai'ritsu-ka'setsu
alternator *[elec]*	同期発電機	dōki-hatsu'den-ki
altimeter *[eng]*	高度計	kōdo-kei
" *[eng]*	高度測量器	kōdo-soku'ryō-ki
altitude *[eng] [geom]*	高度; こうど	kōdo
" *[eng] [geogr] [geom]*	高さ; たかさ	taka-sa
altocumulus *[meteor]*	高積雲	kōseki'un
altostratus *[meteor]*	高層雲	kōsō'un
alum *[inorg-ch]*	礬土; ばんど	bando
" *[inorg-ch] [miner]*	明礬; みょうばん	myōban
aluminum (element)	アルミニウム	aruminyūmu
aluminum sulfate	硫酸礬土	ryūsan-bando
alveolus (pl -li) *[anat]*	肺胞; 肺胞	haihō
Amanita pantherina (poison mushroom, fly killer) *[bot]*	天狗茸; てんぐたけ	tengu-take
amanthophilous *{adj} [bot]*	砂丘棲植物の	sa'kyū'sei-shokubutsu no

amara *[pharm]*	苦味薬	kumi-yaku
amber (color) *[opt]*	琥珀色；こはくいろ	kohaku-iro
ambergris *[physio]*	竜涎香；龍涎香	ryūzen'kō
ambidextrality *[physio]*	両手利き	ryō'te-kiki
ambience; aura *[psy]*	雰囲気；雰圍氣	fun'iki
ambient conditions *[phys]*	大気条件；大氣條件	taiki-jōken
ambiguity; vagueness *{n}*	曖昧(性)	aimai(-sei)
ambiguous case *[trig]*	不確定な場合	fu'kakutei na ba'ai
amblygonite *[miner]*	燐礬石	rinban'seki
ambulance *[med]*	救急車	kyūkyū'sha
ameiosis *[gen]*	不減数分裂	fu'gensū-bun'retsu
amendment; correction *[pat]*	補正(書)	hosei('sho)
" *[pat]*	訂正；ていせい	teisei
America *[geogr]*	アメリカ	Amerika
" *[geogr]*	米国；米國	Beikoku
American *{adj} [geogr]*	米国の	Beikoku no
americium (element)	アメリシウム	amerishūmu
Amethysanthus japonicus *[bot] [pharm]*	引起；ひきおこし	hiki-okoshi
amethyst *[miner]*	紫石英	murasaki-seki'ei
amidship *[nav-arch]*	船体中央(部)	sentai-chūō(-bu)
amidships *{adv} [nav-arch]*	船体中央部に	sentai-chūō'bu ni
amine *[org-ch]*	アミン	amin
amino acid *[bioch]*	アミノ酸	amino-san
amitosis *[cyt]*	無糸分裂	mushi-bun'retsu
ammeter *[eng]*	電流計	denryū-kei
ammonite *[paleon]*	菊石；きくいし	kiku-ishi
ammonium carbonate	炭酸アンモニユウム	tansan-anmonyūmu
ammonium carbonate (abbreviation)	炭安	tan'an
ammonium dihydrogen phosphate	燐酸二水素アンモ=ニウム	rinsan-ni'suiso-anmonyūmu
ammonium nitrate (abbreviation)	硝安	shōan
ammonium primary phosphate	第一燐酸アンモ=ニウム	dai'ichi-rinsan-anmonyūmu
ammonium secondary phosphate	第二燐酸アンモ=ニウム	dai'ni-rinsan-anmonyūmu
ammonium sulfate (abbreviation)	硫安	ryū'an
amniocentesis *[med]*	羊水穿刺	yōsui-senshi
amnion *[embryo]*	羊膜；ようまく	yōmaku
amniota *[v-zoo]*	羊膜類	yōmaku-rui
amorph *{n} [bio]*	無定形態	mu'teikei-tai
amorphous *{adj} [phys]*	無定形の	mu'teikei no
amorphous alloy *[met]*	非晶質合金	hishō'shitsu-gōkin

amorphous matter *[phys]*	非晶質	hishō-shitsu
amorphous polymer *[poly-ch]*	無定形高分子	mu'teikei-kōbunshi
amorphous region *[ch]*	非結晶領域	hi'kesshō-ryō'iki
amount *[math]*	量；りょう	ryō
amount of money invested *[econ]*	投資額	tōshi'gaku
amount of precipitation *[meteor]*	雨量	u'ryō
amperometric titration *[p-ch]*	電流滴定	denryū-teki'tei
amphi- *{prefix}*	両側に-	ryō'gawa ni-
" *{prefix}*	周囲の-	shū'i no-
Amphibia *[v-zoo]*	両棲類；両生類	ryōsei-rui
amphibious *{adj}* *[bio]*	水陸両性の	sui'riku-ryōsei no
amphidromic point *[map]*	無潮点	muchō-ten
amphipatic property *[org-ch]*	両親媒性	ryō'shinbai-sei
amphiphilic property *[ch]*	両親媒性	ryō'shinbai-sei
ampho-ion *[ch]*	両性イオン	ryōsei-ion
" *[ch]*	双性イオン	sōsei-ion
ampholyte *[ch]*	両性電解質	ryōsei-denkai'shitsu
amphoteric *[ch]*	両性の；兩性の	ryōsei no
amphoteric substance *[ch]*	両性電解質	ryōsei-denkai'shitsu
amplification *[electr]*	増幅(度)	zōfuku(-do)
amplification; dilatation *[comm]*	敷衍；ふえん	fu'en
amplification constant *[electr]*	増幅定数	zōfuku-teisū
amplification factor *[electr]*	増幅率	zōfuku-ritsu
amplifier *[eng]*	増幅器；增幅器	zōfuku-ki
amplifier circuit *[elec]*	増幅回路	zōfuku-kairo
amplitude *[alg]*	偏角	henkaku
" *[trig]*	巾；幅；はば	haba
" *[phys]*	振幅	shinpuku
amplitude distortion *[electr]*	振幅歪	shinpuku-hizumi
amplitude modulation; AM *[electr]*	振幅変調	shinpuku-henchō
amplitude transmittance *[elecmg]*	振幅透過率	shinpuku-tōka-ritsu
amusement district *[civ-eng]*	盛り場；さかりば	sakari'ba
anabolism *[bioch]*	同化作用	dōka-sayō
anaconda (reptile) *[v-zoo]*	大蛇；だいじゃ	daija
anaerobe *[bio]*	嫌気(性)菌	kenki(-sei)-kin
anaerobic decomposition *[bioch]*	嫌気分解	kenki-bunkai
anaerobic respiration *[bio]*	無気呼及	muki-kokyū
analcite; analcime *[miner]*	方沸石	hōfutsu'seki; hōfusseki
analgesic *{n}* *[pharm]*	鎮痛薬（鎮痛剤）	chintsū-yaku (chintsū-zai)
analogue *{n}* *[bio]*	類似体；類似體	ruiji'tai
analogy (pl -gies) *[bio]*	相似	sōji
analysis (pl -yses) *[ch-eng]*	分析	bunseki

analysis *[math]*	解析	kaiseki
analysis of covariance *[stat]*	共分散分析	kyō'bunsan-bunseki
analysis of variance *[stat]*	分散分析	bunsan-bunseki
analyte *[an-ch]*	被分析体	hi-bunseki'tai
analytical chemistry *[ch]*	分析化学	bunseki-kagaku
analytical dynamics *[mech]*	解析力学	kaiseki-riki'gaku
analytic geometry *[math]*	解析幾何学	kaiseki-kika'gaku
analyzer *[opt]*	検光子	kenkō'shi
anaphase *[bio]*	後期；こうき	kōki
anastigmatic aberration *[opt]*	非点収差	hiten-shūsa
anatase *[miner]*	鋭錐石	eisui'seki
anatomy (the discipline) *[bio]*	解剖学	kaibō'gaku
ancestor *[bio]*	先祖；せんぞ	senzo
anchor *{n}* *[nav-arch]*	錨；碇；いかり	ikari
anchorage dependence *[microbio]*	足場依存性	ashi'ba-izon-sei
anchor effect *[adhes]*	投錨効果	tōbyō-kōka
anchor light *[opt]*	停泊灯；碇泊燈	tei'haku-tō
anchovy (fish) *[v-zoo]*	片口鰯	kata'kuchi-iwashi
ancient times *[archeo]*	太古；たいこ	taiko
AND *[comput]*	論理積	ronri-seki
and *{conj}* *[logic]*	かつ	katsu
" *{conj}* *[logic]*	及び；および	oyobi
andalusite *[miner]*	紅柱石	kōchū'seki
AND circuit *[electr]*	論理積回路	ronri'seki-kairo
andesine *[miner]*	中性長石	chūsei'chōseki
andesite *[miner]*	安山岩	anzan'gan
andradite *[miner]*	灰鉄柘榴石	kaitetsu'zakuro-ishi
andromeda *[bot]*	馬酔木；あせび	asebi
anechoic room *[acous]*	無響室；無響室	mukyō-shitsu
anelasticity *[mech]*	擬弾性	gi'dansei
" *[mech]*	非弾性	hi'dansei
anemometer *[eng]*	風速計	fūsoku-kei
anesthetic *{n}* *[pharm]*	麻酔薬	masui-yaku
angelfish (fish) *[v-zoo]*	粕鮫；かすざめ	kasu-zame
angelica (herb) *[bot]*	犬独活；いぬうど	inu-udo
" (herb) *[bot]*	猪独活；ししうど	shishi-udo
" (herb) *[bot]*	当換；とうき	tōki
angelshark (fish) *[v-zoo]*	粕鮫；かすざめ	kasu-zame
anger shell (shellfish) *[i-zoo]*	筍貝	takenoko-gai
angiography *[med]*	血管造影法	kekkan-zōei-hō
angiosperm *[bot]*	被子植物	hishi-shokubutsu
angle *{n}* *[geom]*	角(度)	kaku(-do)

angle bar *[arch]*	山形材	yama'gata-zai
angle edge *[math]*	角稜; かくりょう	kaku-ryō
angle of advance *[mech-eng]*	前進角	zenshin-kaku
angle of attack *[aero-eng]*	迎角; げいかく	gei-kaku
angle of depression *[trig]*	俯角; ふかく	fu'kaku
" *[trig]*	伏角	fukkaku
angle of deviation *[opt]*	偏角	henkaku
angle of elevation *[eng] [trig]*	仰角; ぎょうかく	gyōkaku
angle of incidence *[opt]*	入射角	nyūsha-kaku
angle of reception of light *[opt]*	受光角	jukō-kaku
angle of reflection *[phys]*	反射角	hansha-kaku
angle of repose *[mech]*	安息角	ansok(u)kaku
angle of twist *[mech]*	捩れ角	nejire-kaku
anglerfish (fish) *[v-zoo]*	鮟鱇; あんこう	ankō
anglesite *[miner]*	硫酸塩鉱; 硫酸鹽鑛	ryūsan'enkō
angle steel *[arch]*	山形鋼	yama'gata-kō
angular frequency *[math] [phys]*	角振動数	kaku-shindō'sū
angular momentum *[mech]*	角運動量	kaku-undō'ryō
angular motion *[mech]*	角運動	kaku-undō
angular velocity *[mech] [phys]*	角速度; かくそくど	kaku-sokudo
anharmonic oscillator *[electr]*	非調和振動子	hi'chōwa-shindō'shi
anhydride *[ch]*	無水物	musui'butsu
anhydrite *[miner]*	硬石膏	kō'sekkō
anhydrous *{n} [ch]*	無水; むすい	mu'sui
anhydrous gypsum *[miner]*	無水石膏	musui-sekkō
animal *[zoo]*	動物; どうぶつ	dō'butsu
" *[zoo]*	獣; 獸	kedamono; jū
animal fiber *[tex]*	動物繊維	dōbutsu-sen'i
animal husbandry *[agr]*	畜産学	chikusan'gaku
Animalia *[syst]*	動物界	dōbutsu-kai
animal wax *[mat]*	動物蠟	dōbutsu-rō
animation; life; vitality *[bio]*	生気	seiki
anion *[ch]*	陰イオン	in-ion
anisogamete *[bot]*	異形配偶子	ikei-haigū'shi
anisogamy *[bot]*	異形配偶	ikei-haigū
anisometric *{adj} [crys]*	非等軸の	hi'tōjiku no
anisotropy *[phys]*	異方性	i'hō-sei
ankle (of foot) *[anat]*	足首	ashi-kubi
" (of foot) *[anat]*	踝; くるぶし	kurubushi
annatto; Bixa orellana *[bot]*	紅木; べにのき	beninoki
anneal *{vb} [eng]*	焼鈍す; やきなます	yaki-namasu
annealing *{n} [eng]*	焼鈍(し)	yaki'namashi

annealing; baking; enameling *[eng]*	焼付；焼き付け	yaki'tsuke
annealing furnace *[met]*	焼鈍炉；焼鈍爐	shōdon-ro
Annelida (earthworms) *[i-zoo]*	環形動物	kankei-dōbutsu
annihilation *[part-phys]*	消滅；しょうめつ	shōmetsu
annotated catalog *[pr]*	解題目録	kaidai-moku'roku
annotation*[pr]*	注記	chūki
" *[pr]*	注釈；ちゅうしゃく	chūshaku
announcement *[comm]*	告示	koku'ji
annual *{n} [bot]*	一年生	ichi'nen-sei
annual *{adj}*	一年の	ichi'nen no
annually *{adv}*	毎年	mai-toshi; mai'nen
annual plant; therophyte *{n} [bot]*	一年生植物	ichi'nen'sei-shokubutsu
annual report *[econ] [pr]*	年報；ねんぽう	nenpō
annular *{adj} [anat] [astron] [trans]*	環状の	kanjō no
annular clearance *[eng] [mech-eng]*	環状隙間	kanjō-sukima
annular eclipse *[astron]*	金環食	kinkan-shoku
annulation *{n} [i-zoo]*	環状構造	kanjō-kōzō
anode; positive electrode *[elec] [electr] [p-ch]*	陽極；ようきょく	yō-kyoku
anode current *[electr]*	陽極電流	yōkyoku-denryū
anode rays *[electr]*	陽極線	yōkyoku-sen
anodic degreasing *[met]*	陽極脱脂	yōkyoku-dasshi
anodic polishing *{n} [met]*	陽極研摩	yōkyoku-kenma
anodic treatment; anodizing *[met]*	陽極処理	yōkyoku-shori
anolyte *[ch]*	陽極液	yōkyoku-eki
anomalistic year *[astron]*	近点年	kinten-nen
anomalous propagation *[acous]*	異常伝搬	ijō-denpan
anomaly *[fl-mech] [opt]*	異常；いじょう	ijō
anomaly *[sci-t]*	変則(性)	hensoku(-sei)
anonymity *{n}*	匿名；とくめい	toku'mei
anonymous *{adj}*	匿名の	tokumei no
anorthite *[miner]*	灰長石	kai'chōseki
anoxic *{adj} [ch]*	無酸性の	mu'sansei no
answer; reply; response *{n} [comm]*	返答；へんとう	hentō
" *{n} [comm]*	答え；こたえ	kotae
" *{n} [comput]*	応答；おうとう	ōtō
ant (insect) *[i-zoo]*	蟻；あり	ari
antacid *{n} [pharm]*	制酸剤	seisan-zai
antagonism *[bio]*	拮抗作用	kikkō-sayō
" *[bio]*	拮抗性	kikkō-sei
" *[bio]*	対抗；對抗	taikō
antarctic *{adj}*	南極の	nan'kyoku no
Antarctic Circle (Zone) *[geod]*	南極圏；南極圏	Nankyoku-ken

Antarctic Ocean *[geogr]*	南極海	Nankyoku-kai
ante- *{prefix}*	以前の-	izen no-
ante- *{prefix}*	前の-	mae no-
anteater (animal) *[v-zoo]*	食蟻獣；ありくい	ari'kui
antecedent *{n} [logic]*	前項	zenkō
" *{n} [logic]*	前提	zentei
" *{n}*	先行	senkō
antelope (animal) *[v-zoo]*	羚羊；れいよう	reiyō
ante meridian; a.m. *[astron]*	午前	gozen
antenna (pl -nae, -s) *[elec]*	空中線	kūchū-sen
" *[i-zoo]*	触手；觸手	shoku'shu
anteroom *[arch]*	前室	zen-shitsu
anthelion (pl anthelia, -s) *[astron]*	反対幻日	hantai-genjitsu
anthelminthic *{n} [pharm]*	駆虫剤；驅蟲劑	kuchū-zai
anther *[bot]*	葯；やく	yaku
anther culture *[bot]*	葯培養	yaku-baiyō
antheridium (pl -idia) *[bot]*	造精器；蔵精器	zōsei-ki
anthracite coal *[miner]*	硬質炭	kōshitsu-tan
" *[miner]*	無煙炭	mu'en-tan
anthropogeomorphology *[bio]*	人類地形学	jinrui-chikei'gaku
Anthropoidea *[v-zoo]*	真猿類	shin'en-rui
anthropology *[bio]*	人類学	jinrui'gaku
anti- *{prefix}*	反-	han-
" *{prefix}*	抗-	kō-
" *{prefix}*	対-；對-	tai-
antiantibody *[immun]*	抗解毒剤	kō'gedoku-zai
" *[immun] [pharm]*	抗抗体	kō'kōtai
antiaromaticity *[ch]*	反芳香族性	han'hōkō'zoku-sei
antibacterial agent *[microbio]*	抗菌剤	kōkin-zai
antibacterial property *[microbio]*	抗菌性	kōkin-sei
antibiotics *[microbio] [pharm]*	抗生物質	kōsei-busshitsu
antibody *[immun]*	抗体；抗體	kōtai
antibromic *{n} [pharm]*	防臭剤	bōshū-zai
anticatalyzer; anticatalyst *[ch]*	触媒毒；觸媒毒	shokubai-doku
anticathode *[electr]*	対陰極；對陰極	tai'inkyoku
antichlor *[ch-eng]*	除塩素剤；除鹽素劑	jo'enso-zai
anticline *{n} [geol] [min-en]*	背斜	haisha
anticorrosive agent *[met]*	防食剤；防蝕劑	bōshoku-zai
anticyclone *[meteor]*	逆旋風	gyaku-senpū
antidepressant *[pharm]*	抗鬱薬；こぅうつ薬	kō'utsu-yaku
antiderivative *[calc]*	不定積分	futei-sekibun
antidiarrheal *{n} [pharm]*	止瀉剤	shisha-zai

antidinic *{n} [pharm]*	鎮暈薬（鎮暈剤）	chin'un-yaku (chin'un-zai)
antidiuretic *{n} [pharm]*	抗利尿剤	kō'ri'nyō-zai
antidontalgic *{n} [pharm]*	歯痛止め	shitsū-dome
antidote *{n} [pharm]*	解毒薬；げどくやく	ge'doku-yaku
antiemetic *{n} [pharm]*	鎮吐薬（鎮吐剤）	chinto-yaku (chinto-zai)
" *{n} [pharm]*	制吐薬（制吐剤）	seito-yaku (seito-zai)
antienzyme *{n} [bioch]*	抗酵素	kō'kōso
antifebrile *{n} [pharm]*	解熱剤；げねつざい	ge'netsu-zai
antiferroelectric (pl -s) *[sol-st]*	反強誘電体	han'kyō'yūden-tai
antiferromagnetic material *[sol-st]*	反強磁性体	han'kyōji'sei-tai
antiferromagnetism *[sol-st]*	反強磁性	han'kyōji-sei
antifoamer; antifoaming agent *[mat]*	消泡剤；消泡劑	shōhō-zai
antifoggant *[photo]*	かぶり防止剤	kaburi-bōshi-zai
antifouling *{n} [met]*	防汚；ぼうお	bō'o
antifouling coating *[mat]*	防汚塗料	bō'o-toryō
antifreeze *[ch]*	不凍液	futō-eki
antifungal antibiotics *[microbio]*	抗黴抗生物質	kō'kabi-kōsei-busshitsu
antigen *[immun]*	抗原	kōgen
antigenicity *[immun]*	抗原性	kōgen-sei
antihypnotic *{n} [pharm]*	覚醒薬	kakusei-yaku
antijamming *{n} [electr]*	妨害対抗	bōgai-taikō
antilogarithm *[alg]*	真数；真數	shinsū
antiluetic *{n} [pharm]*	駆梅薬（駆梅剤）	kubai-yaku (kubai-zai)
antimagnetic *{adj} [mat]*	耐磁の	taiji no
antimatter *[phys]*	反物質	han'busshitsu
antimetabolite *[pharm]*	代謝拮抗物質	taisha-kikkō-busshitsu
antimicrobial activity *[microbio]*	抗菌力	kōkin-ryoku
antimonite *[miner]*	輝安鉱；輝安鑛	ki'ankō
antimony (element)	アンチモン	anchimon
antimycotic *{adj} [pharm]*	抗黴性の	kō'kabi'sei no
antineutron *[phys]*	反中性子	han'chūsei'shi
antinode *[phys]*	波腹	ha'fuku
antinomy	二律背反	ni'ritsu-haihan
antinucleon *[part-phys]*	反核子；はんかくし	han'kakushi
antioxidant *[ch]*	酸化防止剤	sanka-bōshi-zai
antiparallel *[bioch] [gen] [phys]*	逆平行の(な)	gyaku-heikō no (na)
antiparticle *[part-phys]*	反粒子	han'ryūshi
antiperspirant *{n} [pharm]*	汗止め	ase'dome
" *{n} [pharm]*	制汗薬（制汗剤）	seikan-yaku (seikan-zai)
antiphlogistic *{n} [pharm]*	消炎剤	shō'en-zai
antipodal point *[bot]*	対蹠点	taiseki-ten
antipode *[ch]*	対掌体	taishō'tai

antipodes *[geod]*	対蹠地；たいせきち	taiseki'chi
antiprism *[geom]*	捩れプリズム	nejire-purizumu
antiproton *[part-phys]*	反陽子；はんようし	han'yōshi
antiprotozoic property *[microbio]*	抗原虫性	kō'genchū-sei
antipruritic *{n} [pharm]*	鎮痒薬（鎮痒剤）	chin'yō-yaku (chin'yō-zai)
" *{n} [pharm]*	痒み止め	kayu'mi-dome
antipsychotic *{n} [pharm]*	抗精神病薬	kō-seishin'byō-yaku
antipyrotic; burn medicine *[pharm]*	火傷薬	kashō-yaku; yakedo-gusuri
antiques; curios *[furn]*	骨董品	kottō'hin
antireciprocity *[math]*	逆相反性	gyaku-sōhan-sei
antiresonance *[elec]*	反共振	han'kyōshin
antiresonance frequency *[elec]*	反共振周波数	han'kyōshin-shūha'sū
antisensitization *[photo]*	反色増感	hanshoku-zōkan
antiseptic *{n} [pharm]*	防腐剤；ぼうふざい	bōfu-zai
antiserum *{n} [immun]*	抗血清	kō'kessei
antispasmodic; spasmolytic *{n} [pahrm]*	鎮痙薬（鎮痙剤）	chinkei-yaku (chinkei-zai)
antistatic agent *[mat]*	帯電防止剤	taiden'bōshi-zai
antistatic finish *[eng]*	帯電防止加工	taiden'bōshi-kakō
antisubmarine warfare *[mil]*	対潜作戦；對潜作戰	tai'sen-saku'sen
antisymmetry *[math]*	逆対称；逆對稱	gyaku-taishō
antisyphilitic *{n} [pharm]*	駆梅薬（駆梅剤）	kubai-yaku (kubai-zai)
antitoxin *{n} [immun] [pharm]*	抗毒素；こうどくそ	kō'doku'so
antituberclous drug *[pharm]*	抗結核剤	kō'kekkaku-zai
antitussive *{n} [pharm]*	鎮咳薬（鎮咳剤）	chingai-yaku (chingai-zai)
antler(s); horn(s) *[zoo]*	角；つの	tsuno
antonym *[gram]*	反意語；はんいご	han'i'go
" *[gram]*	（反）対語	(han)tai'go
anus *[anat]*	肛門；こうもん	kōmon
anvil *[anat]*	砧骨	kinuta-bone; chin-kotsu
" *[met]*	鉄敷き；かなしき	kana'shiki
anvil block *[met]*	鉄敷台；鐵敷臺	kana'shiki-dai
anytime *{adv}*	何時でも；いつでも	itsudemo
" *{adv}*	随時；隨時	zui'ji
aorta *[anat]*	大動脈	dai-dōmyaku
aoudad; Barbary sheep *[v-zoo]*	鬣羊	tate'gami-hitsuji
apatite *[miner]*	燐灰石	rinkai'seki
aperiodic *{adj} [phys]*	非周期的な	hi'shūki'teki na
aperitif *[cook]*	食前酒	shokuzen'shu
aperitive *{n} [pharm]*	食欲亢進薬	shoku'yoku-kōshin-yaku
aperture *[opt]*	開度	kaido
" *[opt]*	開口	kaikō
" *[opt]*	口径；こうけい	kōkei

aperture (opening) *{n}*	隙；すき	suki
apex *[geom]*	頂点	chōten
aphelion (pl aphelia) *[astron]*	遠日点	en'jitsu-ten
aphid (insect) *[i-zoo]*	油虫；あぶらむし	abura-mushi
" *[i-zoo]*	蟻巻；ありまき	arimaki
aphrodisiac *{n} [pharm]*	催淫薬	sai'in-yaku
apiary (pl -aries) *[agr]*	養蜂所	yōhō'jo
apical dominance *[bot]*	頂芽優勢	chōga-yūsei
apogalacticon *[astron]*	遠銀河中心点	en'ginga-chūshin-ten
apogee *[astron]*	遠地点	enchi-ten
apophyllite *[miner]*	魚眼石	gyogan'seki
apothem *[geom]*	辺心距離；邊心距離	henshin-kyori
apparatus *[eng]*	装置；そうち	sōchi
apparent *{n} [math]*	見掛け	mi'kake
" *{adj} [math]*	見掛けの	mi'kake no
apparent noon *[astron]*	真正午	shin-shōgo
apparent specific gravity *[mech]*	見掛比重	mikake-hijū
apparent viscosity *[fl-mech]*	見掛粘度	mikake-nendo
appeal *{n} [legal]*	上告	jōkoku
" *{n} [legal]*	訴訟；そしょう	soshō
appearance *{n}*	外見；がいけん	gaiken
" *{n}*	恰好；格好	kakkō
appended table *[graph] [pr]*	付表	fuhyō
appendix (pl -es, -pendices) *[anat]*	虫垂	chūsui
" *[anat]*	盲腸	mōchō
" *[pr]*	補遺；ほい	ho'i
appetizer *[cook]*	前菜	zensai
apple (fruit, tree) *[bot]*	林檎；りんご	ringo
applicability	適用性	tekiyō-sei
applicant *[pat]*	出願人	shutsugan'nin
application *[pat]*	願書	gansho
" *[pat]*	申請	shinsei
" *[pat]*	出願	shutsu'gan
" *[sci-t]*	応用；おうよう	ōyō
" *[sci-t]*	利用	riyō
" *[sci-t]*	適用	teki'yō
application; execution (of work) *[civ-eng]*	施工；せこう	sekō
application for design registration *[pat]*	意匠登録願	ishō-tōroku-negai
application for utility model registration *[pat]*	実用新案登録願	jitsuyō-shin'an-tōroku-negai

application study *[sci-t]*	応用研究	ōyō-kenkyū
applied chemistry *[ch]*	応用化学	ōyō-kagaku
applied information technology *[comm]*	応用情報技術	ōyō-jōhō-gi'jutsu
applied mechanics *[mech]*	応用力学; 應用力學	ōyō-riki'gaku
applied physics *[phys]*	応用物理学	ōyō-butsu'rigaku
applied science *[sci-t]*	応用科学	ōyō-kagaku
applied voltage *[elec]*	印加電圧	inka-den'atsu
appointment (of a person) *{n}*	選任; 選任	sennin
appraisal; expert opinion *{n}*	鑑定; かんてい	kantei
approach *{vb}*	接近する	sekkin suru
approach graft *[bot]*	寄せ接ぎ	yose-tsugi
appropriate; adequate *{adj}*	適切な	teki'setsu na
appropriateness *{n}*	適正性	tekisei-sei
" *{n}*	適性	tekisei
appropriate quantity *[math]*	適量	teki-ryō
appropriation; assignment *{n}*	充当; 充當	jūtō
approval *{n} [pat]*	認可	ninka
" *{n}*	同意	dō'i
" *{n}*	承認	shōnin
approved (an abbreviation) *[pat]*	適; てき	teki
approximate *{adj} [math]*	概略の	gairyaku no
approximate calculation *[math]*	概算	gaisan
approximately *{adv} [math]*	約	yaku
approximate value *[math]*	近似値	kinji-chi
apricot (fruit, tree) *[bot]*	杏; あんず	anzu
apricot stone *[pharm]*	杏仁; きょうにん	kyō'nin
April (month)	四月	shi'gatsu
apron *[cl]*	前掛け	mae-kake
" *[geol]*	沖積平地	chūseki-heichi
apron (at an airport) *[trans]*	駐機場	chūki'jō
aprotic solvent *[ch]*	非プロトン性溶媒	hi'puroton'sei-yōbai
apsis *[astron]*	軌道極点	kidō-kyokuten
aptitude *[psy]*	適性	tekisei
aptitude test *[psy]*	適性検査	tekisei-kensa
aquamarine *[miner]*	藍玉; らんぎょく	rangyoku
aqua regia *[ch]*	王水; おうすい	ō-sui
aquarium (pl -iums, -ia) *[arch]*	水族館	suizok(u)kan
Aquarius; Water Bearer *[constel]*	水瓶座	mizu'game-za
aquatic *{adj} [bio]*	水生の	suisei no
aquatic animal *[bio]*	水生動物; 水棲動物	suisei-dōbutsu
aquatic garden *[bio]*	水生園	suisei'en
aquatic plant *[bot]*	水生植物; 水棲植物	suisei-shokubutsu

aquatic product *[ocean]*	水産物	suisan'butsu
aqueous; of water *{adj} [ch]*	水の	mizu no
aqueous dispersed system *[ch]*	水分散系	mizu-bunsan-kei
aqueous solution *[ch]*	水溶液	sui-yōeki
aquiculture *[bio]*	水中生物培養	suichū-seibutsu-baiyō
aquifer *[geol]*	帯水層; 帶水層	taisui'sō
arabesque *[graph]*	唐草; からくさ	kara-kusa
arable soil *[agr]*	耕土	kōdo
Arachnida *[i-zoo]*	蛛(形)類	kumo('gata)-rui
aragonite *[miner]*	霰石; あられいし	arare-ishi
Aralia cordata (vegetable) *[bot]*	独活; うど	udo
araucaria (tree) *[bot]*	南洋杉	Nan'yō-sugi
arbitrage *[legal]*	裁定取引	saitei-tori'hiki
arbitral decision *[legal]*	裁定	saitei
arbitrary *[comput] [math]*	任意	nin'i
arbitrary constant *[math]*	任意定数	nin'i-teisū
arbitration	調停	chōtei
arboreal *{adj} [zoo]*	樹上性の	jujō'sei no
arborescent *{adj} [bio]*	樹枝状の	jushi'jō no
arboretum *[bot]*	樹木園	jumoku'en
arc *{n} [astron] [geom]*	弧; こ	ko
arc-extinguishing chamber *[electr]*	消弧室	shōko-shitsu
arch (of foot) *{n} [anat]*	土踏まず	tsuchi-fumazu
archegonium (pl -nia) *[bot]*	造卵器; 蔵卵器	zōran-ki
archeology *[sci-t]*	考古学	kōko'gaku
archerfish (fish) *[v-zoo]*	鉄砲魚; 鐵砲魚	teppō-uo
archery; bow *{n} [ord]*	弓; ゆみ	yumi
archipelago *[geogr]*	群島	guntō
architect (a person) *[arch]*	建築家	kenchiku'ka
architecture (the discipline) *[eng]*	建築学	kenchiku'gaku
archival memory *[comput]*	保存記憶	hozon-ki'oku
archive *{n} [comput]*	記録保管所	kiroku-hokan'jo
" *{vb} [comput]*	集積する	shūseki suru
archiving *{n} [comput]*	記録(-保管)	ki'roku (-hokan)
Arctic Circle (Zone) *[geod]*	北極圏; 北極圈	Hokkyoku-ken
Arctic Ocean *[geogr]*	北極海	Hokkyoku-kai
arcuate islands *[geol]*	弧状列島	kojō-rettō
area *[geom]*	面積	menseki
" *[geom]*	領域	ryō'iki
area of use *[pat]*	実施部門	jisshi-bumon
area percentage *[math]*	面積率	menseki-ritsu
areography *[astron]*	火星表面学	kasei-hyōmen'gaku

English	Japanese	Romanization
arete *[geol]*	山稜；さんりょう	sanryō
argentite *[miner]*	輝銀鉱	ki'ginkō
argentometry *[an-ch]*	銀滴定	gin-teki'tei
argon (element)	アルゴン	arugon
argument *[alg] [comput]*	引き数；ひきすう	hiki'sū
" *[alg]*	偏角	henkaku
" *[legal]*	弁論；辯論	benron
" *[logic]*	独立変項	doku'ritsu-henkō
" *[logic]*	論証	ronshō
" *{n}*	論争；論爭	ronsō
arid belt *[climat]*	乾燥帯	kansō-tai
aridity *[climat]*	乾燥状態	kansō-jōtai
Aries; the Ram *[constel]*	雄羊座；おひつじざ	o'hitsuji-za
arithmetic *[math]*	算術；さんじゅつ	san'jutsu
" *[math]*	算数	sansū
arithmetic expression *[comput]*	算術式	sanjutsu-shiki
arithmetic instruction *[comput]*	算術命令	sanjutsu-meirei
arithmetic mean *[alg]*	算術平均	sanjutsu-heikin
" *[alg]*	相加平均	sōka-heikin
arithmetic overflow *[comput]*	算術桁溢れ	sanjutsu-keta-afure
arithmetic progression *[alg]*	等差数列	tōsa-sū'retsu
arithmetic sequence *[alg]*	等差数列	tōsa-sū'retsu
arithmetic series *[alg]*	算術級数	sanjutsu-kyūsū
arithmetic unit *[comput]*	演算器	enzan-ki
" *[comput]*	演算装置	enzan-sōchi
ark shell (shellfish) *[i-zoo]*	赤貝；あかがい	aka-gai
arm *{n} [anat]*	腕；うで	ude
armature *[elecmg]*	電機子	denki'shi
" *[elecmg]*	接極子	sekkyoku'shi
armband *[cl]*	腕章	wanshō
armor *{n} [ord]*	鎧；よろい	yoroi
armoring *{n} [rub]*	鎧装	gaisō
armpit *[anat]*	腋窩	ekika
" *[anat]*	脇の下；腋の下	waki-no-shita
armrest *[furn]*	肘掛	hiji'kake
" (Japanese) *[furn]*	脇息	kyōsoku
arm shaft *[mech-eng]*	上軸；うわじく	uwa-jiku
army (pl armies) *[mil]*	陸軍	riku-gun
aroma *[physio]*	香気	kōki
aromatic; aromatic agent *[org-ch]*	芳香剤	hōkō-zai
aromatic compound *[org-ch]*	芳香族化合物	hōkō'zoku-kagō'butsu
aromaticity *[org-ch]*	芳香族性	hōkō'zoku-sei

aromatic nucleus *[org-ch]*	芳香核	hōkō-kaku
arrange; dispose *{vb t} [mech-eng]*	配置する	haichi suru
arrange in order *{vb t}*	揃える	soroeru
arrangement *[ch]*	配座	haiza
" *[math]*	配置	haichi
arrangement; setup *{n}*	仕組み；しくみ	shi'kumi
array; rank *{n} [comput] [math]*	配列	hairetsu
array name *[comput]*	配列名	hairetsu-mei
arrival *[trans]*	到着	tōchaku
arrival angle *[navig]*	到来角；到來角	tōrai-kaku
arrival of goods *[ind-eng]*	入荷	nyūka
arrive at *{vb} [trans]*	着く	tsuku
arrow *[graph]*	矢線	ya-sen
" *[ord]*	矢	ya
arrow; arrow mark *[comput] [graph]*	矢印	ya-jirushi
arrow diagram *[graph]*	矢線図	yasen-zu
arrowhead *[bot]*	慈姑；くわい	kuwai
arrowroot *[bot]*	葛；くず	kuzu
arrow signal; arrow light *[traffic]*	矢印灯；矢印燈	ya'jirushi-tō
arrow view *[graph]*	矢視図；矢視圖	yashi-zu
arsenate *[ch]*	砒酸塩	hisan-en
arsenic (element)	砒素；ひそ	hiso
arsenic acid	砒酸	hisan
arsenic chloride	塩化砒素	enka-hiso
arsenic hydride	水素化砒素	suiso'ka-hiso
arsenic oxide	酸化砒素	sanka-hiso
arsenic sulfide	硫化砒素	ryūka-hiso
arsenic trichloride	三塩化砒素	san'enka-hiso
arsenic trifluoride	三弗化砒素	san'fukka-hiso
arsenic trioxide	三酸化砒素	san'sanka-hiso
arsenide *[ch]*	砒化物	hi'ka-butsu
arsenious acid	亜砒酸；あひさん	a'hisan
arsenite *[ch]*	亜砒酸塩；亞砒酸鹽	a'hisan-en
arsenopyrite *[miner]*	硫砒鉄鉱；硫砒鐵鑛	ryūhi'tekkō
art *{n}*	美術	bi'jutsu
art; arts *{n}*	芸術；藝術	gei'jutsu
art bronze *[met]*	美術青銅	bijutsu-seidō
Artemisia princeps (herb) *[bot]*	蓬；艾；蒿；よもぎ	yomogi
artery *[anat]*	動脈	dōmyaku
artesian well *[geol]*	噴き井戸	fuki-ido
art gallery; art museum *[arch]*	美術館	bijutsu'kan
Arthropoda *[i-zoo]*	節足動物	sessoku-dōbutsu

article *[gram]*	冠詞	kanshi
" *[pat] [pr]*	条；條；じょう	jō
article: item *{n}*	事項	jikō
article: object *{n}*	物品	buppin
article: thesis *{n} [comm] [pr]*	論文	ronbun
article name; name of item	品名	hinmei
articulation *[comm]*	明瞭度	meiryō-do
artifact *[archeo]*	人工遺物	jinkō-i'butsu
artificial *{adj} [comput] [ind-eng]*	人工の	jinkō no
" *{adj} [ind-eng]*	人造の	jinzō no
" *{adj}*	擬似の；ぎじの	giji no
" *{adj}*	人為の	jin'i no
artificial aging *{n} [met]*	人工時硬	jinkō-jikō
artificial flower *[eng]*	造花	zōka
artificial flower-making *[eng]*	造花	zōka
artificial intelligence; AI *[comput]*	人工知能	jinkō-chinō
artificial language *[comput]*	人工言語	jinkō-gengo
artificial limb *[eng]*	義肢	gishi
artificial load *[mech-eng]*	擬似負加	giji-fuka
artificial rubber *[poly-ch] [rub]*	人造ゴム	jinzō-gomu
artificial sunlight *[elec]*	人工太陽光線	jinkō-taiyō-kōsen
artificial tooth root *[mat]*	人工歯根	jinkō-shikon
artificial ventilation *[min-eng]*	人工通気	jinkō-tsūki
artistic; esthetic *{adj}*	美術的な	bijutsu'teki na
aryl group *[org-ch]*	アリール基	arīru-ki
asbestos *[miner]*	石綿	ishi-wata; sekimen
as called for	随時(に)	zui'ji (ni)
ascend *{vb} [navig]*	登る	noboru
ascension *[navig]*	登り；昇り；上り	nobori
ascension; ascent *[aerosp]*	上昇	jōshō
ascertain; confirm; verify *{vb}*	確かめる	tashikameru
ascertainment	探知	tanchi
ascocarp *[mycol]*	子嚢果；しのうか	shinō'ka
Ascomycetes *[mycol]*	子嚢菌類	shinō'kin-rui
ascus (pl asci) *[mycol]*	子嚢	shi'nō
asexual reproduction *[bio]*	無性生殖	musei-seishoku
ash (tree) *[bot]*	梣；秦皮；とねりこ	toneriko
ash *[ch]*	灰；はい	hai
" *[ch]*	灰分；かいぶん	kaibun
ash content *[mat]*	灰分	kaibun
ashless coal *[geol]*	無灰炭	mukai-tan
Asia *[geogr]*	東洋	tōyō

Asian tiger mosquito (insect) *[i-zoo]*	一筋縞蚊	hito'suji-shima-ka
Asiatic beetle (insect) *[i-zoo]*	背斑黄金虫	se'madara-kogane-mushi
Asiatic locust (insect) *[i-zoo]*	殿様蝗；殿樣蝗	tonosama-batta
aspect; state *{n}*	様子；ようす	yōsu
aspect ratio *[eng]*	縦横比	jūō-hi
aspen; poplar (tree) *[bot]*	山鳴らし	yama-narashi
Aspergillus *[mycol]*	麹黴；糀黴	kōji-kabi
Aspergillus niger *[bot]*	黒黴；くろかび	kuro-kabi
asphalt *[mat]*	瀝青	reki'sei
aspherical mirror *[opt]*	非球面鏡	hi'kyūmen-kyō
aspidistra *[bot]*	葉蘭；はらん	haran
assay (verification) *[an-ch]*	検定；檢定	kentei
assay *[an-ch]*	試金	shikin
" *[an-ch]*	定量法	teiryō-hō
assemble *{vb}*	組(み)立る	kumi-tateru
assembly *[mech-eng]*	組立	kumi'tate
assembly drawing *[graph]*	組立図；組立圖	kumi'tate-zu
assertion	肯定	kōtei
assessment *[sci-t]*	査定	satei
asset *[econ]*	資産	shisan
assign *{n}* *[pat]*	譲渡；じょうと	jōto
" (a person) *{n}* *[legal]*	譲受人；讓受人	yuzuri'uke'nin
assigned frequency band *[comm]*	割当周波数帯	wari'ate-shūha'sū-tai
assignee *[legal]*	譲受人	yuzuri'uke'nin
assignment *{n}*	割当；割當	wari'ate
" *{n}*	割付	wari'tsuke
assimilate *{vb}* *[microbio]*	資化する	shika suru
assimilating property *[physio]*	資化性	shika-sei
assimilation *[geol]* *[physio]*	同化	dōka
assistant *[ch-eng]*	助剤	jozai
associate number *[math]*	同伴数	dōhan-sū
association *[ch]*	会合；會合	kaigō
" *[ch]*	結合	ketsu'gō
" *[math]*	組合せ	kumi-awase
" *[psy]*	連想	rensō
" (a group) *{n}*	団体；團體	dantai
associative law *[alg]*	結合法則	ketsugō-hōsoku
associative reaction *[ch]*	会合性反応	kaigō'sei-hannō
assumption; estimation *[math]*	仮定；假定；かてい	katei
" *[math]*	推定	suitei
astatic *[phys]*	無定位	mu'tei'i
astatine (element)	アスタチン	asutachin

aster; starwort (vegetable) *[bot]*	嫁菜；よめな	yome'na
asterisk *[comput] [pr]*	星印	hoshi-jirushi
asterism *[astron]*	星宿	seishuku
" *[crys]*	星形図形	hoshi'gata-zukei
asteroid *[astron]*	小惑星	shō-wakusei
Asteroidea *[i-zoo]*	海星類	hitode-rui; kaisei-rui
asthenosphere *[geol]*	岩流圏；岩流圈	ganryū-ken
asthma *[med]*	喘息；ぜんそく	zen'soku
astigmatism *[opt]*	乱視；亂視	ranshi
astringency *[physio]*	収斂性；收斂性	shūren-sei
" *[physio]*	渋味；しぶみ	shibu'mi
astringent *{n} [pharm]*	収斂剤	shūren-zai
" *{adj} [physio]*	渋い；澁い	shibui
astrocyte *[cytol]*	星状細胞	seijō-saibō
astronaut *[aerosp]*	宇宙飛行士	uchū-hikō'shi
astronautics *[aero-eng]*	空間航行学	kūkan-kōkō'gaku
astronavigation *[navig]*	天文航法	tenmon-kōhō
astronomer *[astron]*	天文学者	tenmon-gakusha
astronomical telescope *[opt]*	天体望遠鏡	tentai-bōenkyō
astronomical triangle *[astron]*	天文三角形	tenmon-sankaku-kei
astronomical twilight *[astron]*	天文薄明	tenmon-hakumei
astronomical unit *[astron]*	天文単位	tenmon-tan'i
astronomy *[sci-t]*	天文学	tenmon'gaku
astrophysics *[astron]*	宇宙物理学	uchū-butsuri'gaku
asymmetrical wave *[phys]*	非対称波；非對稱波	hi'taishō-ha
asymmetric carbon atom *[ch]*	不斉炭素原子	fusei-tanso-genshi
asymmetric synthesis *[org-ch]*	不斉合成；不齊合成	fusei-gōsei
asymmetry *[ch]*	不均整	fukin-sei
" *[math]*	不整	fusei
" *[math]*	非対称；非對稱	hi'taishō
asymptote *[an-geom]*	漸近線	zenkin-sen
asymptotic orbit *[an-geom]*	漸近軌道	zenkin-kidō
asynchronous *{adj} [comput] [phys]*	非同期の	hi'dōki no
" *{adj} [phys]*	非同時性の	hi'dōji'sei no
asynchronous transmission *[comput]*	非同期伝送	hi'dōki-densō
at about (time)	頃；ごろ；ころ	goro; koro
atacamite *[miner]*	緑塩銅鉱；綠鹽銅鑛	ryoku'en'dōkō
atactic *{adj} [poly-ch]*	不規則性の	fu'kisoku'sei no
athermal solution *[ch]*	無熱溶液	mu'netsu-yōeki
Atlantic Ocean *[geogr]*	大西洋	Taisei-yō
atlas (book) *[pr]*	地図書；地圖書	chizu'sho
at least *[math]*	以上	ijō

at least *[math]*	少くとも	sukunaku'tomo
atmometer; evaporimeter *[eng]*	蒸発計；蒸發計	jōhatsu-kei
atmophile element *[meteor]*	親気元素	shinki-genso
atmosphere *[meteor]*	大気	taiki
atmosphere, the *[meteor]*	大気圏；大氣圏	taiki-ken
atmosphere *(n)*	雰囲気；雰圍氣	fun'iki
atmospheric *(adj) [meteor]*	大気の	taiki no
atmospheric nitrogen fixation *[ch-eng]*	空中窒素固定	kūchū-chisso-kotei
atmospheric pressure *[phys]*	気圧；氣壓；きあつ	ki'atsu
" *[phys]*	大気圧	taiki-atsu
atmospherics *[geophys]*	空電	kūden
atmospheric tide *[meteor]*	大気潮	taiki-chō
at most *[math]*	多くて	ōkute
" *[math]*	以下	ika
atoll *[geogr]*	環礁；かんしょう	kanshō
atom *[ch]*	原子；げんし	genshi
atom bomb *[ord]*	原子爆弾	genshi-bakudan
atomic beam *[phys]*	原子線	genshi-sen
atomic clock *[horol]*	原子時計	genshi-dokei
atomic energy *[nucleo]*	原子力	genshi-ryoku
atomic explosion *[ord]*	原子爆発	genshi-baku'hatsu
atomic fraction *[phys]*	原子分率	genshi-bunritsu
atomic mass *[phys]*	原子質量	genshi-shitsu'ryō
atomic mass number *[phys]*	原子質量数	genshi-shitsuryō-sū
atomic mass unit *[phys]*	原子質量単位	genshi-shitsuryō-tan'i
atomic nucleus *[nuc-phys]*	原子核	genshi-kaku
atomic number *[nuc-phys]*	原子番号	genshi-bangō
atomic physics *[phys]*	原子物理学	genshi-butsuri'gaku
atomic power *[nucleo]*	原子力	genshi-ryoku
atomic powered ship *[nav-arch]*	原子力船	genshi'ryoku'sen
atomic propulsion *[mech]*	原子力推進	genshi'ryoku-suishin
atomic radius *[p-ch]*	原子半径	genshi-hankei
atomic susceptibility *[elecmg]*	原子磁化率	genshi-jika-ritsu
atomic symbol *[ch]*	原子記号	genshi-kigō
atomic volume *[p-ch]*	原子容	genshi-yō
atomic weight *[ch]*	原子量	genshi'ryō
atomization *[mech-eng]*	微粉化	bifun'ka
" *[mech-eng]*	霧化；むか	muka
atomization method *[mech-eng]*	噴霧法	funmu-hō
atomized powder *[mat]*	噴霧粉	funmu-ko
atomizer *[eng]*	噴霧器；ふんむき	funmu-ki
atonal music *[acous]*	無調音楽	muchō-ongaku

atrioventricular node *[anat]*	房室結節	bō'shitsu-kessetsu
atrium (pl atria) *[anat]*	心房	shinbō
" *[arch]*	中庭；なかにわ	naka-niwa
atrophy *[med]*	萎縮	i'shuku
attached document(s) *[pat]*	添付書類	tenpu-shorui
attaching, snapped (or fitted) in	嵌着；かんちゃく	kanchaku
attachment; accessory device *[eng]*	付属装置	fuzoku-sōchi
attachment *{n}*	装着；裝着	sōchaku
attach pivotally *{vb} [mech-eng]*	枢着する	sū'chaku suru
attack angle *[aero-eng]*	迎え角	mukae-kaku
attention *{n} [psy]*	注意；ちゅうい	chū'i
attenuate *{vb t}*	細くする	hosoku suru
attenuation *[phys]*	減衰；げんすい	gensui
attenuator *[electr]*	減衰器	gensui-ki
at the same time	同時	dōji
attic *[arch]*	屋根部屋	yane-beya
attitude *[psy]*	態勢；たいせい	taisei
attorney-at-law *[legal]*	弁護士；辯護士	bengo'shi
attractant *[agr] [bioch] [mat]*	誘引物質（誘引剤）	yūin-busshitsu (yūin-zai)
attractive force *[mech] [phys]*	引力	in-ryoku
attribute *{n} [bio] [comput]*	属性；屬性	zoku-sei
attune; acclimatize *{vb i}*	馴染む；なじむ	najimu
auburn; brown (color) *[opt]*	鳶色；とび色	tobi-iro
audibility *[acous]*	聴度	chōdo
audible *{adj}*	聞こえる	kikoeru
audible alarm *[comput]*	可聴警報	kachō-keihō
audible range *[acous]*	可聴範囲	kachō-han'i
audio frequencies *[acous] [comm]*	可聴周波数	kachō-shūha'sū
audio frequency oscillator *[electr]*	可聴周波発振器	kachō-shūha-hasshin-ki
audiometer *[eng]*	聴力計	chōryoku-kei
audio response unit *[comm]*	音声応答装置	onsei-ōtō-sōchi
audiovisual *[comm]*	聴視覚	chō'shikaku
auditing *{n} [comput] [econ]*	監査	kansa
audition (sense of) *[physio]*	聴覚；聽覺	chō'kaku
auditorium (pl -s, -ria) *[arch]*	講堂	kōdō
augend *[math]*	被加数	hi-kasū
augite *[miner]*	輝石；きせき	kiseki
augmentation *[pr]*	増補	zōho
August (month)	八月	hachi'gatsu
aurate *[ch]*	金酸塩；金酸鹽	kinsan-en
auric acid	金酸；きんさん	kinsan
aurichalcite *[miner]*	水亜鉛銅鉱	sui'aen'dōkō

Auricularia auricula (mushroom) *[bot]*	木耳；きくらげ	ki-kurage
aurora australis *[geophys]*	南極光	nankyoku-kō
aurora borealis *[geophys]*	北極光；北極光	hokkyoku-kō
aurora polaris *[geophys]*	極光	kyokkō
authentication *[comm]*	認証	ninshō
authenticity *{n}*	真偽；しんぎ	shingi
authority *{n}*	権限；權限	kengen
" *{n}* (agency)	機関	kikan
authorization *[legal]*	委任	i'nin
auto- *{prefix}*	自動の-	jidō no-
auto- *{prefix}*	自身の-	jishin no-
autocatalytic *{adj} [ch]*	自触的な	ji'shoku'teki na
autoclave *[eng]*	高圧滅菌器	kōatsu-mekkin-ki
" *[eng]*	オートクレーブ	ōtokurēbu
" *[eng]*	耐(加)圧器	tai(-ka)'atsu-ki
autodegradation *[microbio]*	自己分解	jiko-bunkai
autohesion *[org-ch]*	自着；じちゃく	ji'chaku
" *[rub]*	自己粘着	jiko-nenchaku
autoignition *[phys]*	自己着火	jiko-chakka
autoimmunity *[immun]*	自己免疫	jiko-men'eki
autoimmunization *[immun]*	自己免疫化	jiko-men'eki'ka
autolysis *[bioch] [microbio]*	自己消化	jiko-shōka
" *[microbio]*	自己融解	jiko-yūkai
automatic *{adj} [eng]*	自動的な	jidō'teki na
automatic check *[comput]*	自動検査	jidō-kensa
automatic control *[cont-sys]*	自動制御	jidō-seigyo
automatic data processing *[comp]*	自動データ処理	jidō-dēta-shori
automatic dictionary *[comput]*	自動辞書	jidō-jisho
automatic drawing instrument *[graph]*	自動製図機	jidō-seizu-ki
automatic gain control *[electr]*	自動利得制御	jidō-ri'toku-seigyo
automatic restart *[comput]*	自動再始動	jidō-sai'shidō
automatic scale *[eng]*	自動秤	jidō-bakari
automatic stop *[comput]*	自動停止	jidō-teishi
automatic two-side copier *[opt]*	自動両面複写機	jidō-ryōmen-fukusha-ki
automation *[comput]*	自動化	jidō'ka
" *[eng]*	自動操作化	jidō-sōsa'ka
automaton *[comput]*	自動機械	jidō-kikai
" *[eng]*	自動人形	jidō-ningyō
automobile; car *[mech-eng]*	自動車；じどうしゃ	jidō'sha
" *[trans]*	車；くるま	kuruma
automobile type *[trans]*	車種	shashu
automorphic; idiomorphic *[petr]*	自形	jikei

automotive *{adj} [mech-eng]*	自動推進の	jidō-suishin no
automotive industry *[mech-eng]*	自動車産業	jidōsha-sangyō
automotive-related *{adj} [mech-eng]*	自動車関係の	jidōsha-kankei no
autonomic nervous system *[anat]*	自律神経系	ji'ritsu-shinkei-kei
autopilot (of a ship) *[navig]*	自動操舵装置	jidō-sōda-sōchi
" (of a plane) *[navig]*	自動操従装置	jidō-sōjū-sōchi
autotroph; lithotroph *[bio]*	独立栄養生物	doku'ritsu-eiyō-seibutsu
" *[bio]*	無機栄養生物	muki-eiyō-seibutsu
autumn; Fall (season) *[astron]*	秋；あき	aki
" *[astron]*	秋季；秋期	shūki
autumnal equinox *[astron]*	秋分；しゅうぶん	shūbun
autumn bellflower; gentian *[bot]*	竜胆；龍膽	rindō
auxiliary; assistant *{n} [ch-eng]*	助剤；助劑	jozai
" *{adj}*	補助の	hojo no
auxiliary electrode *[elec]*	補助電極	hojo-denkyoku
auxiliary valence *[ch]*	側原子価	soku-genshi'ka
auxiliary view *[graph]*	補助図；ほじょず	hojo-zu
auxochrome *[ch]*	助色団；助色圖	joshoku'dan
auxotroph *[gen]*	栄養要求株	eiyō-yōkyū-kabu
" *[gen]*	栄養要求性変異体	eiyō-yōkyū'sei-hen'i'tai
auxotrophic mutation *[gen]*	栄養要求性変異	eiyō-yōkyū'sei-hen'i
available chlorine *[ch]*	有効塩素；有効鹽素	yūkō-enso
available moisture *[hyd]*	有効水分	yūkō-suibun
available power *[elec]*	有能電力	yūnō-den'ryoku
available water *[hyd]*	有効水	yūkō-sui
avalanche *{n} [hyd]*	雪崩；なだれ	nadare
aventurine; matte *{n} [mat]*	梨地；なしじ	nashi'ji
avenue; boulevard *[civ-eng]*	広路；廣路；ひろじ	hiroji
average *{n} [arith] [stat]*	平均(値)	heikin(-chi)
average access time *[comput]*	平均呼出(し)時間	heikin-yobi'dashi-jikan
average-quality article *[ind-eng]*	並品；なみひん	nami'hin
Aves; birds *[v-zoo]*	鳥類；ちょうるい	chō-rui
aviation *{adj} [aero-eng]*	航空の；こうくうの	kōkū no
aviation electronics; avionics *[eng]*	航空電子工学	kōkū-denshi-kōgaku
aviation fuel *[mat]*	航空燃料	kōkū-nenryō
aviation oil *[mat]*	航空潤滑油	kōkū-junkatsu-yu
avidity *[immun]*	結合活性	ketsu'gō-kassei
avifauna *[v-zoo]*	鳥(類)相	chō(rui)-sō
avoidance; aversion *[psy]*	回避；かいひ	kaihi
awl; gimlet; drill (a tool) *{n} [eng]*	錐；きり	kiri
awning; tent; cloth pavilion *[arch]*	天幕；てんまく	tenmaku
axe *{n} [eng]*	斧；おの	ono

axial flow *[fl-mech]*	軸流	jiku'ryū
axial ratio *[crys] [electr]*	軸比	jiku-hi
" *[crys]*	軸率	jiku-ritsu
axial skeleton *[anat]*	中軸骨格	chūjiku-kokkaku
axil (of a leaf) *[bot]*	葉腋；ようえき	yōeki
axinite *[miner]*	斧石	ono-ishi
axiom *[logic]*	公理	kōri
axis (pl axes) *[an-geom] [mech]*	軸；じく	jiku
" *[bio] [i-zoo]*	軸柱	jiku'chū
" (bone) *[v-zoo]*	軸椎(骨)	jiku'tsui(-kotsu)
axis of abscissas *[an-geom]*	横軸	yoko-jiku
" *[an-geom]*	横座標	yoko-zahyō
axis of coordinates *[math]*	座標軸	zahyō-jiku
axis of easy magnetization *[elecmg]*	磁化容易軸	jika-yōi'jiku
axis of ordinates *[an-geom]*	縦軸；縱軸	tate-jiku
" *[an-geom]*	縦座標	tate-zahyō
axis of symmetry *[an-geom] [mech]*	対称軸；對稱軸	taishō-jiku
axle *[mech-eng]*	軸	jiku
" *[mech-eng]*	車軸	sha'jiku
axle box *[eng]*	軸箱	jiku-bako
axle oil *[mat]*	車軸油	sha'jiku-yu
axonometry *[graph]*	軸測投像法	jiku'soku-tōzō-hō
ayu (fish) *[v-zoo]*	鮎；あゆ	ayu
azalea (flower, bush) *[bot]*	躑躅；つつじ	tsutsuji
azeotrope *[ch]*	共沸混合物	kyōfutsu-kongō'butsu
azeotropic mixture *[ch]*	共沸混合物	kyōfutsu-kongō'butsu
azeotropic point *[ch]*	共沸点	kyō'futten
azeotropic distillation *[ch]*	共沸蒸留	kyōfutsu-jōryū
azeotropy *[ch]*	共沸	kyō'futsu
azimuth *[astron]*	方位(角)	hō'i(-kaku)
azimuth error *[astron]*	方位角誤差	hōi'kaku-gosa
azure sky *[astron]*	瑠璃空；るりぞら	ruri-zora
azurite *[miner]*	藍銅鉱	ran'dōkō

B

baboon (animal) *[v-zoo]*	狒狒；狒々；ひひ	hihi
baby powder *[pharm]*	天花粉；天瓜粉	tenka-fun
bacillus (pl -li) *[microbio]*	桿菌	kankin
Bacillus subtilis *[microbio]*	枯草菌	kosō-kin
back *{n} [anat]*	背(中)	se(-naka)
" *{n}*	後方	kōhō
" *{n}*	後；うしろ	ushiro
backbone (spine) *[anat]*	脊柱	seki'chū
back-feed(ing) *{n} [mech-eng]*	逆送り	gyaku-okuri
background *[comput] [graph]*	背景	haikei
background noise *[acous]*	暗騒音	an'sōon
background processing *{n} [comput]*	背景処理	haikei-shori
backing material *[tex]*	裏当て材；裏當て材	ura'ate-zai
backlight *{n} [graph] [opt]*	背面光	haimen-kō
back mixing *{n} [ch-eng]*	逆混合	gyaku-kongō
back mutation; reverse mutation *[gen]*	復帰(突然)変異	fuk(u)ki(-totsu'zen)-hen'i
back of hand *[anat]*	手の甲；てのこう	te no kō
back part; hind part; rear part *{n}*	後部	kōbu
backplate *{n} [arch]*	裏金；うらがね	ura-gane
" *{n} [arch]*	後板；うしろいた	ushiro-ita
back pressure *[mech]*	逆圧；逆壓	gyaku-atsu
" *[mech]*	背圧	hai-atsu
backrest *[furn]*	背凭れ；せよれ	se'yore
backscattering *{n} [phys]*	後方散乱	kōhō-sanran
back side; other side *[paper] [tex]*	裏；うら	ura
backspace *{n} [comput]*	後退	kōtai
backspace character *[comput]*	後退文字	kōtai-moji
backstreet *[civ-eng]*	裏通り	ura-dōri
back teeth; molars *[anat] [dent]*	奥歯；奥齒；おくば	oku'ba
back titration *[ch]*	逆滴定	gyaku-teki'tei
backup *{n} [comput]*	補助	hojo
backward-facing *{adj}*	後向きの	ushiro-muki no
backward printing *{n} [comput]*	逆方向印刷	gyaku'hōkō-in'satsu
backwash *{n} [ocean]*	寄せて返す波	yosete kaesu nami
backwater *[hyd]*	背水	haisui
" *[nav-arch]*	跳ね水；はねみず	hane-mizu
backyard *[arch] [civ-eng]*	後庭	kōtei
bacon beetle (insect) *[i-zoo]*	鰹節虫	katsuo'bushi-mushi
bacteria (sing bacterium) *[microbio]*	黴菌；ばいきん	baikin

bacterial *(adj) [microbio]* 細菌の saikin no
bacterial cell *[mycol]* 菌体；菌體 kintai
bacterial strain *[mycol]* 菌株；きんかぶ kin-kabu
bacterial type *[mycol]* 菌種 kinshu
bactericide; bacteriocide *[mat]* 殺菌剤 sakkin-zai
bacteriological test *[bact]* 細菌試験 saikin-shiken
bacteriology *[microbio]* 細菌学 saikin'gaku
bacteriolysis *[microbio]* 溶菌 yōkin
bacteriostasis *[microbio]* 細菌発育阻止 saikin-hatsu'iku-soshi
" *[microbio]* 静菌作用 seikin-sayō
bacteriostat *(n) [mat] [pharm]* 細菌発育阻止剤 saikin-hatsu'iku-soshi-zai
bacteriostatic *(adj) [microbio]* 静菌性の seikin'sei no
bacteriostatic (agent) *(n) [pharm]* 静菌剤 seikin-zai
bacterium (pl -ria) *[microbio]* 細菌 saikin
Bactrian camel (animal) *[v-zoo]* 二瘤駱駝 futa'kobu-rakuda
bad *(adj)* 悪い；惡い；わるい warui
bad flavor *[cook]* 嫌味；いやみ iya'mi
badger (animal) *[v-zoo]* 穴熊；あなぐま ana-guma
badlands *[geogr]* 悪地；惡地 aku'chi
badness *[ind-eng]* 悪さ waru-sa
bad-quality *(adj)* 悪質の aku'shitsu no
bad solvent *[ch]* 貧溶媒 hin-yōbai
baffle board *[eng]* 邪魔板 jama-ban
bag *(n) [eng]* 袋；ふくろ fukuro
bagasse *[cook]* 砂糖黍搾り殻 satō'kibi-shibori'gara
baggage *[trans]* (旅行)荷物 (ryokō-)ni'motsu
baggage car *[rail]* 荷物車 ni'motsu-guruma
baggage room *[transp]* 荷物予り所 ni'motsu-azukari'jo
bagworm; fagot worm *[i-zoo]* 簑虫；みのむし mino-mushi
bait *(n) [bio]* 餌；えさ esa
baker's yeast *[cook]* パン酵母 pan-kōbo
bakery (pl -eries) *[arch] [cook]* パン屋 pan'ya
baking *(n) [cer] [paint]* 焼付け；焼付け yaki'tsuke
" *(n) [food-eng]* パン焼き pan-yaki
" *(n) [eng] [met]* 焼成 shōsei
baking finish *[paint]* 焼付塗料 yaki'tsuke-toryō
baking hardening *(n) [paint]* 焼付硬化 yaki'tsuke-kōka
baking powder *[cook]* 膨らし粉 fukurashi-ko
baking soda *[cook]* 重炭酸曹達 jū'tansan-sōda
" (abbreviation) *[cook]* 重曹 jūsō
balance *(n) [ch] [physio]* 平衡 heikō
" (uniformness) *(n) [ch]* 均整；均斉 kinsei

balance (a pair of scales) *{n} [eng]*	天秤；てんびん	ten'bin
" *{n} [mech-eng]*	釣り合い	tsuri-ai
balancing *{n}*	釣り合わせ	tsuri-awase
bald eagle (bird) *[v-zoo]*	白頭鷲	haku'tō-washi
bale (straw bag) *{n} [ind-eng]*	俵；ひょう；たわら	hyō; tawara
" *{n} [ind-eng]*	梱；こり	kori
baleen *[v-zoo]*	鯨鬚	geishu; kujira-hige
ball (a toy)	毬；まり	mari
ball; sphere *[mech-eng]*	球；たま	tama
ball-and-socket joint *[anat]*	球(窩)関節	kyū('ka)-kansetsu
ball-and-socket joint *[mech-eng]*	玉継手；玉繼手	tama-tsugi'te
ballast *{n} [aero-eng] [nav-arch]*	底荷；そこに	soko-ni
" *{n} [elec]*	安定器；あんていき	antei-ki
" *{n} [elec]*	安定抵抗	antei-teikō
ball bearing *[mech-eng]*	玉軸受	tama-jiku'uke
ballistics *[mech]*	弾道学；彈道學	dandō'gaku
balloon; aerostat *{n} [aero-eng]*	気球；ききゅう	kikyū
" (a toy) *{n} [eng]*	風船；ふうせん	fūsen
balloonfish; blowfish; puffer *[v-zoo]*	河豚；ふぐ	fugu
balloonflower; Chinese bellflower *[bot]*	桔梗；ききょう	kikyō
ball-point pen *[graph]*	ボールペン	bōru-pen
ballroom *[arch]*	舞踏室	butō-shitsu
bamboo *[bot]*	竹；たけ	take
bamboo: Phyllostachys pubescens	孟宗竹	mōsō'chiku
bamboo: Phyllostachys bambusoides	真竹；まだけ	ma'dake
bamboo blind; reed blind *[furn]*	簾；すだれ	sudare
bamboo colander *[cook] [food-eng]*	笊；ざる	zaru
bamboo-crafted ware *[furn]*	竹細工	take-zaiku
bamboo flute *[music]*	篠笛；しのぶえ	shino-bue
bamboo grass *[bot]*	笹；篠；ささ	sasa
bamboo grove; thicket *[ecol]*	藪；やぶ	yabu
bamboo leaf *[bot]*	笹の葉；ささのは	sasa no ha
bamboo shoot *[bot]*	筍；竹の子	takenoko
band *{n} [crys]*	帯；帶；たい	tai
bandage *{n} [med] [pharm]*	繃帯；包帯；包帶	hōtai
banded coal *[geol]*	縞状石炭	shima'jō-sekitan
banded structure *[met] [petr]*	縞状構造	shima'jō-kōzō
bandicoot (animal) *[v-zoo]*	鬼鼠；おにねずみ	oni-nezumi
band steel *[steel]*	帯鋼；おびこう	obi-kō
bandwidth *[comm]*	帯域巾；帶域幅	tai'iki-haba
banister; handrail; railing *[arch]*	手摺り；てすり	te'suri
banked fire *[ch]*	埋め火	ume-bi; uzume-bi

barbed *{adj} [bot] [mat]*	刺のある	toge no aru
barbed wire *[met]*	有刺鉄線	yūkyoku-tessen
bare wire *[elec]*	裸線；はだかせん	hadaka-sen
barge *[nav-arch]*	艀；はしけ	hashike
bar graph; bar chart *[stat]*	棒図表；棒圖表	bō-zuhyō
barite *[miner]*	重晶石	jūshō'seki
barium (element)	バリウム	baryūmu
bark *{n} [bot]*	樹皮	juhi
barley *[bot]*	麦芽	bakuga
" *[bot]*	大麦；大麥	ō-mugi
barley bran *[agr]*	大麦糠	ōmugi-nuka
barnacle *[nav-arch]*	富士壺；ふじつぼ	fuji'tsubo
baroclinity *[phys]*	傾圧性	kei'atsu-sei
barometer *[eng]*	気圧計；氣壓計	ki'atsu-kei
barometric pressure *[phys]*	(大)気圧	(tai-)ki'atsu
barotropy *[phys]*	順圧	jun'atsu
barracuda (fish) *[v-zoo]*	梭魚；魳；かます	kamasu
barrel; cask *{n} [eng]*	樽；たる	taru
barrenwort; Epimedium Herba *[bot]*	碇草；いかりそう	ikari'sō
barrier beach; barrier bar *[geol]*	沿岸洲	engan-su
barrier island *[geol]*	堡島；保島	hotō
barrier reef *[geol]*	堡礁；保礁	hoshō
bartack *{n} [eng]*	閂縫い	kannuki-nui
barycenter (of triangle) *[geom]*	重心	jūshin
baryon *[part-phys]*	重粒子	jū'ryūshi
barysphere *[geol]*	重圏；重圈	jūken
basalt *[miner]*	玄武岩	genbu'gan
basaltic lava *[miner]*	玄武岩質溶岩	genbu'gan'shitsu-yōgan
base *{n} [alg]*	台	dai
" *{n} [alg] [ch]*	基	ki
" *{n} [alg]*	基数	kisū
" *{n} [alg]*	底；てい	tei
" *{n} [ch]*	塩基	enki
" (less noble) *{n} [ch]*	卑；ひ	hi
" *{n} [comput]*	基準	kijun
" *{n} [comput]*	基底	kitei
" (base stand) *{n} [eng]*	基台	kidai
" *{n} [eng] [pharm]*	基材	kizai
" *{n} [geom]*	底辺	teihen
" *{n}*	元；もと	moto
base angle *[geom]*	底角	tei-kaku
baseboard *[arch]*	幅木；はばき	haba'ki

base catalyst *[ch]*	塩基触媒；鹽基觸媒	enki-shokubai
base-centered lattice *[crys]*	底心格子	teishin-kōshi
base coat *[mat]*	下塗り	shita-nuri
base color *[opt]*	基本色	kihon-shoku
base composition *[ch]*	基剤；基劑	kizai
base fabric *[poly-ch]*	基布	kifu
base-leveled plain *[geol]*	準平原	jun-heigen
base line *[navig] [sci-t]*	基線	kisen
base material *[poly-ch]*	基材；きざい	kizai
basement (floor); cellar *[arch]*	地階	chi'kai
basement; basement room *[arch]*	地下室；ちかしつ	chika-shitsu
base metal *[met]*	卑金属	hi-kinzoku
base number *[ch]*	塩基価；鹽基價	enki-ka
base oil *[mat]*	基油	kiyu
base pair *[mol-bio]*	塩基対	enki-tsui
base paper *[mat] [paper]*	原紙	genshi
base resin (adhesive) *[org-ch]*	主剤	shuzai
base unit *[phys]*	基本単位	kihon-tan'i
basicity *[ch]*	塩基度	enki-do
basic lead acetate	塩基性酢酸鉛	enkisei-sakusan-namari
basic lead sulfate	塩基性硫酸鉛	enkisei-ryūsan-namari
basic lead sulfite	塩基性亜硫酸鉛	enkisei-a'ryūsan-namari
basic oxide *[ch]*	塩基性酸化物	enkisei-sanka'butsu
basic salt *[ch]*	塩基性塩	enki'sei-en
basic unit *[econ]*	原単位	gen-tan'i
basidiocarp *[mycol]*	担子器果；擔子器果	tanshi'ki'ka
Basidiomycetes *[mycol]*	担子菌類；擔子菌類	tanshi'kin-rui
basidiospore *[mycol]*	担子胞子	tanshi-hōshi
basidium (pl -ia) *[mycol]*	担子器；たんしき	tanshi-ki
basin *[geol]*	盆地	bonchi
basin; valley *[geol]*	流域	ryū'iki
basis *{n} [sci-t]*	基準	kijun
basis weight *[mech]*	坪量	tsubo-ryō
basket *[eng]*	籠；かご	kago
bast fiber *[tex]*	靭皮繊維	jinpi-sen'i
basting *{n} [cl]*	仮縫い；假縫い	kari-nui
bat (mammal) *[v-zoo]*	蝙蝠；こうもり	kōmori
batch culture *[microbio]*	回分培養	kaibun-baiyō
batch process *[eng]*	回分操作	kaibun-sōsa
batch processing *{n} [ind-eng]*	一括処理	ikkatsu-shori
batfish (fish) *[v-zoo]*	燕魚；つばめうお	tsubame-uo
bath *[ch] [met]*	浴	yoku

bath *[met]*	湯；ゆ	yu
bathochromic shift *[p-ch]*	深色移動	shin'shoku-idō
batholith *[geol]*	底盤	teiban
bath ratio *[tex]*	浴比	yoku-hi
bathrobe kimono; informal kimono *[cl]*	浴衣；ゆかた	yukata
bathroom *[arch]*	風呂場；ふろば	furo'ba
" *[arch]*	浴室	yoku-shitsu
bathroom scale *[eng]*	体重秤；體重秤	taijū-bakari
batik *[tex]*	臈纈染め；臘纈染め	rō'ketsu-zome
battery *[elec]*	蓄電池	chiku'denchi
" *[elec]*	電池	denchi
battleship; warship *[nav-arch]*	軍艦	gunkan
bauxite *[miner]*	水礬土鉱	suiban'dokō
bay *{n} [aero-eng]*	格納室	kaku'nō-shitsu
" ; gulf *{n} [geogr]*	湾；灣；わん	wan
bay tree (tree) *[bot]*	月桂樹	gekkei-ju
bay window; oriel window *[arch]*	出窓；でまど	de-mado
" *[arch]*	張出し窓	hari'dashi-mado
be; exist *{vb i}*	存在する	sonzai suru
be axially hung *{vb} [mech-eng]*	軸架される	jiku'ka sareru
be effective; work *{vb}*	利く；きく	kiku
be insufficient	足りない	tarinai
be on time	間に合う	ma ni au
be optimistic *{vb} [psy]*	楽観する；樂觀する	rakkan suru
be pessimistic *{vb} [psy]*	悲観する	hikan suru
be present *{vb}*	居る	iru
be restored *{vb} [mech]*	復帰する	fukki suru
be sticky *{vb}*	べたべたする	betabeta suru
" *{vb}*	べたつく	beta-tsuku
be surprised *{vb} [psy]*	驚く	odoroku
be twisted; twist *{vb i} [eng]*	捩じれる	nejireru
beach *{n} [geol]*	浜；濱；はま	hama
" *{n} [geol]*	海岸	kaigan
" *{n} [geol]*	渚；なぎさ	nagisa
beach plover (bird) *[v-zoo]*	磯千鳥；いそちどり	iso-chidori
bead; ball; gem *[mat]*	玉；珠	tama
beading (of a paint) *[poly-ch]*	弾き；はじき	hajiki
bead polymer *[poly-ch]*	粒状重合体	ryūjō-jūgō'tai
bead reaction *[an-ch]*	溶球反応	yōkyū-hannō
bead tree, Japanese *[bot]*	栴檀；せんだん	sendan
beak *[v-zoo]*	嘴；くちばし	kuchibashi
beam *{n} [arch]*	梁；はり	hari

beam of light *[opt]*	光芒	kōbō
" *[opt]*	光束	kō-soku
beam scale *[eng]*	棹秤；さおばかり	sao-bakari
bean; kidney bean (legume) *[bot]*	莢隠元	saya-ingen
bean curd *[cook]*	豆腐；とうふ	tōfu
bean curd lees (dregs) *[cook]*	雪花菜；おから	okara
bean curd refuse *[cook]*	雪花菜；きらず	kirazu
bean jam *[cook]*	餡；あん	an
bean-jelly confection *[cook]*	羊羹；ようかん	yōkan
bean sprouts *[bot] [cook]*	萌し；もやし	moyashi
bear (animal) *[v-zoo]*	熊；くま	kuma
bearing *{n} [mech-eng]*	軸受	jiku-uke
" *{n} [navig]*	方角	hōgaku
" *{n} [navig]*	方位	hō'i
bearing angle *[astron]*	方位角	hō'i-kaku
bearing stand *[mech-eng]*	軸受台；軸受臺	jiku'uke-dai
bearing steel *[met]*	軸受鋼	jiku'uke-kō
bearing surface *[mech-eng]*	座面	zamen
beat *{n} [music]*	拍子；ひょうし	hyōshi
beat; beats *{n} [phys]*	唸り；うなり	unari
beat frequency *[electr]*	唸り周波数	unari-shūha'sū
beating (of pulp) *[eng]*	叩解；こうかい	kōkai
beat thin; hammer thin *{vb t}*	打ち延ばす	uchi-nobasu
beautiful *{adj}*	綺麗な；奇麗な	kirei na
" *{adj}*	美しい	utsukushī
beauty and health instrument *[eng]*	美容健康器	biyō-kenkō-ki
becalm; calm down *{vb} [meteor]*	凪ぐ；なぐ	nagu
become *{vb i}*	(に)なる	(ni) naru
become attuned *{vb i}*	馴染む	najimu
become effective	発効	hakkō
become hoarse *{vb i} [physio]*	掠れる；かすれる	kasureru
become light; become bright *{vb i}*	明るくなる	akaruku naru
become numb *{vb i} [physio]*	痺れる	shibireru
bed *{n} [furn]*	寝台；寢臺	shindai
" *{n} [furn]*	床；とこ	toko
bedbug (insect) *[i-zoo]*	床蝨；とこじらみ	toko-jirami
bedding *{n} [furn]*	夜具；やぐ	yagu
bedding closet *[arch]*	押し入れ	oshi-ire
bedrock *[geol]*	岩盤	ganban
" *[geol]*	床岩	shōgan
bedrock; base *[geol]*	基盤	kiban
bedroom *[arch]*	寝室	shin-shitsu

bee (insect) *[i-zoo]*	蜂	hachi
beech(nut), Japanese (tree) *[bot]*	山毛欅；ぶな	buna
bee glue *[pharm]*	蜂蠟；はちろう	hachi-rō
bee nettle *[bot]*	仏の座；佛の座	hotoke-no-za
bee toxin *[physio]*	蜂毒	hachi-doku
beef tallow *[mat]*	牛脂	gyūshi
beefsteak plant *[bot] [cook]*	紫蘇；しそ	shiso
beehive-like *{adj}*	蜂の巣様の	hachi'no'su-yō no
beekeeping; bee culture *[agr]*	養蜂	yōhō
beer *[cook]*	麦酒；ビール	bīru
beeswax *[mat]*	蜜蠟	mitsu'rō
beet; red turnip (vegetable) *[bot]*	赤蕪	aka-kabu
beetle (insect) *[i-zoo]*	兜虫；かぶとむし	kabuto-mushi
" (insect) *[i-zoo]*	甲虫	kōchū
beet sugar *[cook]*	甜菜糖	tensai'tō
before *{adv} {prep}*	前	mae
beginner; novice *{n}*	新米	shinmai
begonia (flower) *[bot]*	秋海棠	shūkaidō
behavior; action; conduct *[psy]*	行動；こうどう	kōdō
behavioral science *[psy]*	行動科学	kōdō-kagaku
behind; back *{adv}*	後	ato; ushiro
bell *[music]*	鐘	kane
bell bronze *[met]*	鐘青銅	kane-seidō
bell tower *[arch]*	鐘楼；鐘樓	shōrō
belong to; affiliated with *{vb}*	所属する	shozoku suru
below *{adv} [math]*	下；した	shita
below zero *[thermo]*	零下	reika
belt; band; strap; strip *{n} [eng]*	帯；帶；おび	obi
beluga (fish) *[v-zoo]*	大蝶鮫	ō-chō'zame
" (mammal) *[v-zoo]*	白海豚	shiro-iruka
bench *{n} [furn]*	腰掛	koshi'kake
bench mark *[eng]*	標線	hyōsen
bend (in a stratum) *{n} [geol]*	褶曲；しゅうきょく	shū'kyoku
" *{n} [geol]*	湾曲；灣曲	wan'kyoku
bend; bending *{n}*	曲げ	mage
bend *{vb t} [mech-eng]*	折る	oru
" (curve, fold) *{vb t} [mech-eng]*	(折)曲げる	(ori-)mageru
" (train) *{vb t}*	撓める	tameru
bendability *[mech]*	曲げ性	mage-sei
bending fatigue strength *[mech]*	曲げ疲労強度	mage-hirō-kyōdo
bending processing *{n} [ind-eng]*	曲折加工	kyoku'setsu-kakō
bending strength *[mech]*	(折)曲げ強さ	(ori)mage-tsuyo-sa

beneficial bird *[v-zoo]*	益鳥; えきちょう	eki-chō
beneficial insect *[i-zoo]*	益虫	eki-chū
beneficiation *[met]*	選鉱; 選鑛	senkō
benthic *{adj} [bio] [ocean]*	深海の	shinkai no
benthonic *{adj} [bio] [ocean]*	水底に住む	suitei ni sumu
benthos *{n} [ecol]*	底生生物	teisei-seibutsu
benumb *{vb i} [physio]*	痺れる	shibireru
benzoate *[ch]*	安息香酸塩	ansok(u)kō'san-en
benzoic acid	安息香酸	ansok(u)kō-san
benzoic anhydride	無水安息香酸	musui-ansokkō-san
benzoin *[bot]*	安息香樹	ansok(u)kō'ju
berkelium (element)	バークリウム	bākuryūmu
berm *[civ-eng]*	細道	hoso-michi
" *[civ-eng]*	路縁	ro'en
berry (pl berries) *[bot]*	漿果; しょうか	shōka
berth *{n} [nav-arch]*	停泊余地	teihaku-yochi
berthing (for ship) *{n} [nav-arch]*	係船位置; 繋船位置	keisen-ichi
" (for sleeping) *[nav-arch]*	寝台設備	shindai-setsubi
beryl *[miner]*	緑柱石	ryoku'chū'seki
beryllium (element)	ベリリウム	beriryūmu
betel nut *[bot]*	檳榔子; びんろうじ	binrōji
betel palm tree (tree) *[bot]*	檳榔樹	binrōju
between *{adv} [geom]*	間; あいだ	aida
between layers; interlayer	層間	sōkan
betweenness *[math]*	中間(にあること)	chūkan (ni aru koto)
between threads of a fabric *[tex]*	織り目	ori'me
bevel *{n} [eng]*	斜角	sha-kaku
bevel gear *[mech]*	傘歯車	kasa-ha'guruma
beverage(s); drink(s) *[cook]*	飲料; いんりょう	inryō
beyond right *{n} [legal]*	以遠権	i'en-ken
bi- *{prefix}*	複-	fuku-
" *{prefex}*	二つの-	futatsu no-
" *{prefix}*	双-	sō-
bias *[electr]*	偏倚	hen'i
" *[eng] [stat]*	偏り; 片寄り	katayori
biaural hearing *{n} [acous]*	両耳聴	ryō'mimi-chō
biaxial crystal *[crys]*	二軸性結晶	nijiku'sei-kesshō
biaxial orientation *[poly-ch]*	二軸配向	ni'jiku-haikō
biaxial stress *[mech]*	二軸応力	ni'jiku-ō'ryoku
biaxial stretching *[mech] [poly-ch]*	二軸延伸	ni'jiku-enshin
bibliography *[print]*	文献	bunken
bicarbonate *[ch]*	重炭酸塩	jū'tansan-en

biconcave *{adj} [opt]* 両凹の ryō-ō no
biconditional *[logic]* 相互条件的 sōgo'jōken'teki
biconditional sentence *[logic]* 双条件文；雙條件文 sō'jōken-bun
biconvex *{adj} [opt]* 両凸の ryō-totsu no
bicycle *[mech-eng]* 自転車；自轉車 jiten'sha
bicycle parking area *[civ-eng]* 駐輪場 chūrin'jō
bicyclic compound *[org-ch]* 二環式化合物 ni'kan'shiki-kagō'butsu
bid *{vb} [civ-eng] [econ]* 入札する nyū'satsu suru
biennial *{adj} [bot]* 二年毎の ni'nen'goto no
biennial plant *[bot]* 越年生植物 etsu'nen'sei-shokubutsu
bifunctionality; bifunctional *[ch]* 二官能性 ni'kannō-sei
big *{adj}* 大きい ōkī
bigarade; bitter orange (fruit) [bot] 橙；だいだい daidai
Big Bang theory *[astron]* 大爆発説 dai-baku'hatsu-setsu
Big Dipper *[astron]* 北斗(七)星 hoku'to(-shichi)'sei
bigeye tuna (fish) *[v-zoo]* 眼撥；めばち me'bachi
bight *[geol]* 湾曲部；灣曲部 wan'kyoku'bu
big snake *[v-zoo]* 大蛇；だいじゃ daija
big toe *[anat]* 親指 oya-yubi
big tree *[bot]* 喬木；きょうぼく kyōboku
bilateral symmetry *[bio]* 両側相称；兩側相稱 ryō'gawa-sōshō; ryōsoku-sōshō
" *[bio]* 左右相称 sayū-sōshō
bile *[physio]* 胆汁；膽汁 tanjū
bile acid *[bioch]* 胆汁酸 tanjū-san
bilinear *[math]* 双線形；雙線形 sōsen-kei
bill *{n} [econ]* 請求書 seikyū'sho
" (of a bird) *{n} [v-zoo]* 嘴；くちばし kuchibashi
bill of lading (ship) *[econ]* 船荷証券 funa'ni-shōken
billet (of steel) *{n} [steel]* 鋼片 kōhen
billion (American system) *[math]* 十億；じゅうおく jū-oku
" (British system) *[math]* 一兆；いっちょう itchō
billow (of waves) *{n} [ocean]* 波浪；はろう harō
bimolecular reaction *[ch]* 二分子反応 ni'bunshi-hannō
binary *{adj} [sci-t]* 二成分から成る ni-seibun kara naru
binary alloy *[met]* 二元合金 nigen-gōkin
binary arithmetic operation *[comput]* 二進演算 nishin-enzan
binary digit; bit *[comput]* 二進数子 nishin-sūji
binary condition *[sci-t]* 二元状態 nigen-jōtai
binary element string *[comput]* 二進要素列 nishin-yōso-retsu
binary fission *[bio]* 二分裂；にぶんれつ ni-bun'retsu
binary notation *[math]* 二進法 nishin-hō
binary number *[comput]* 二進数 nishin-sū

binary star; couple *{n} [astron]*	連星；れんせい	rensei
binary system *[math]*	二進法；にしんほう	nishin-hō
bind; restrict *{vb t}*	束縛する	soku'baku suru
bind; bind tightly *{vb t}*	緊縛する	kinbaku suru
binder; bonding agent *[adhes]*	結合剤；結合劑	ketsugō-zai
" *[adhes]*	固着剤	kochaku-zai
" *[adhes]*	粘結剤	nenketsu-zai
binding *{n} [bio]*	結合；けつごう	ketsu'gō
" *{n} [ch-eng]*	結着；けっちゃく	ketchaku
binding agent *[mat]*	結着剤	ketchaku-zai
binding filler *[mat]*	粘結性補填剤	nenketsu'sei-hoten-zai
binoculars *[opt]*	双眼鏡；雙眼鏡	sōgan'kyō
binocular vision *[physio]*	両眼視；兩眼視	ryōgan-shi
binomial *{n} [alg]*	二項式；にこうしき	nikō-shiki
binomial distribution *[stat]*	二項分布	nikō-bunpu
binomial expansion *[alg]*	二項展開	nikō-tenkai
binomial nomenclature *[syst]*	二(命)名法	ni(mei)mei-hō
binomial series *[alg]*	二項級数	nikō-kyūsū
binomial theorem *[alg]*	二項定理	nikō-teiri
binormal *{n} [math]*	陪法線	bai-hōsen
binuclear compound *[ch]*	二核体；二核體	ni'kaku-tai
bio- *{prefix}*	生物の-	sei'butsu no-
bioacoustics *[bio]*	生物音響学	seibutsu-onkyō'gaku
bioassay *[an-ch]*	微生物分析	bi'seibutsu-bunseki
" *[an-ch]*	生物学的定量法	seibutsu'gaku'teki-teiryohō
" *[an-ch]*	生物検定	seibutsu-kentei
biocatalyst *[bioch]*	生(体)触媒	sei(tai)-shokubai
biochemical oxygen demand; BOD *[microbio]*	生化学的酸素要求量	seikagaku'teki-sanso-yōkyū-ryō
biochemistry *[ch]*	生化学；せいかがく	sei'kagaku
biochip *[bio]*	生物化学素子	seibutsu-kagaku-soshi
bioclimatology *[climat]*	生気候学	sei-kikō'gaku
biodegradability *[mat]*	生物分解性	seibutsu-bunkai-sei
" *[mat]*	生体内分解性	seitai'nai-bunkai-sei
biodegradable *{adj} [mat]*	生物分解可能な	seibutsu-bunkai-kanō na
biodegradation *[bioch]*	生分解	sei'bunkai
bioelastic (pl -s) *{n} [mech]*	生体弾性体	seitai-dansei'tai
bioelectricity *[physio]*	生体電気	seitai-denki
bioelement *[ch] [bio]*	生元素；せいげんそ	sei'genso
bioengineering; biotechnology *[eng]*	生物工学	seibutsu-kōgaku
biogeochemical cycles *[geoch]*	生物地球化学的循環	seibutsu-chikyū'kagaku'teki-junkan

biogeochemistry *[geoch]*	生物地球化学	seibutsu-chikyū-kagaku
biogeography *[ecol]*	生物地理学	seibutsu-chiri'gaku
bioindustry *[eng]*	生物技術産業	seibutsu-gi'jutsu-sangyō
biological agent *[ord]*	生物剤; 生物劑	seibutsu-zai
biological clock *[physio]*	生物時計	seibutsu-dokei
biological control *[ecol]*	生物的防除	seibutsu'teki-bōjo
biological control (of insects) *[ecol]*	病害虫防除	byōgai'chū-bōjo
biological engineering *[sci-t]*	生体工学	seitai-kōgaku
biological warfare *[mil]*	生物(作)戦	seibutsu-(saku)sen
biological weapon *[ord]*	生物兵器	seibutsu-heiki
biologist *[bio]*	生物学者	seibutsu-gakusha
biology *[sci-t]*	生物学	seibutsu'gaku
bioluminescence *[bio]*	生物発光	seibutsu-hakkō
biomass *[ecol]*	生物(現存)量	seibutsu(-genson)'ryō
" *[ecol]*	生物体量	seibutsu'tai-ryō
biomaterial *[mat]*	生体材料	seitai-zairyō
biome *[bio]*	生物群系	seibutsu-gunkei
" *[bio]*	生物帯; 生物帶	seibutsu-tai
biomedicine *[med]*	生物医学; 生物醫學	seibutsu-i'gaku
biomembrane *[cytol]*	生体膜	seitai-maku
biometeorology *[meteor]*	生気象学	sei-kishō'gaku
biometrics *[stat]*	生物測定学	seibutsu-sokutei'gaku
biophile element *[bioch]*	親生元素	shinsei-genso
biophysics *[sci-t]*	生物物理学	seibutsu-butsuri'gaku
biopolymer *[bioch]*	生体高分子	seitai-kōbunshi
biosensor *[eng]*	生物化学検知器	seibutsu-kagaku-kenchi-ki
biosphere *[ecol]*	生物圏; 生物圈	seibutsu-ken
" *[ecol]*	生存圏	seizon-ken
biostatistics *[stat]*	生物統計学	seibutsu-tōkei'gaku
biosynthesis (pl -theses) *[bioch]*	生合成	sei'gōsei
biota *[bio]*	生物相	seibutsu-sō
biotechnology *[eng]*	生物技術	seibutsu-gi'jutsu
" *[eng]*	生物工学	seibutsu-kōgaku
" *[eng]*	生命技術	seimei-gi'jutsu
" *[eng]*	生命工学	seimei-kōgaku
biotic environment *[ecol]*	生物的環境	seibutsu'teki-kankyō
biotite; black mica *[miner]*	黒雲母; くろうんも	kuro-unmo
bioturbation *[ocean]*	生物による混濁化	seibutsu ni yoru kondakuka
bipinnately compound leaf *[bot]*	二回羽状複葉	nikai-u'jō-fuku'yō
bipolar transmission *[comm]*	両極性	ryōkyoku-sei
bipyramid *[geom]*	両錐(形)	ryōsui(-kei)
biquadratic *{adj}* *[math]*	四次の	yonji no

biquinary code *[math]*	二-五進(法)符号	ni'goshin(-hō)-fugō
birch (tree) *[bot]*	樺	kaba; kanba
bird *[v-zoo]*	鳥	tori
birdlime *[mat]*	鳥黐；とりもち	tori-mochi
bird-of-paradise (bird) *[v-zoo]*	風鳥	fū-chō
" *[v-zoo]*	極楽鳥；極樂鳥	goku'raku-chō
bird-of-paradise flower (flower) *[bot]*	極楽鳥花	gokuraku'chō-ka
birds; Aves *[v-zoo]*	鳥類；鳥類	chō-rui
bird's eye view *[navig]*	鳥瞰図	chōkan-zu
birectangular triangle *[geom]*	両直角三角形	ryō'chokkaku-sankak(u)kei
birefringence *[opt]*	複屈折	fuku-kussetsu
birth *{n} [bio]*	出生	shussei
" *{n} [bio]*	誕生；たんじょう	tanjō
birth and death process *[prob]*	出生死亡過程	shussei-shibō-katei
birthmark *[med]*	痣；あざ	aza
birthrate *[bio]*	出生率	shusseiritsu; shusshōritsu
biscuit firing *{n} [cer]*	絞焼；しめやき	shime-yaki
" *{n} [cer]*	素焼	su'yaki
biscuit kiln *[eng]*	素焼窯	suyaki-gama
bisect *{vb} [geom]*	二等分する	ni'tōbun suru
bisecting method *[geom]*	二分法	nibun-hō
bisection *[geom]*	二等分	ni'tōbun
bisector *[geom]*	二等分線(面)	ni'tōbun-sen; (-men)
bisectrix *[crys]*	二等分線	ni'tōbun-sen
bismuth (element)	ビスマス	bisumasu
" (element)	蒼鉛；そうえん	sō'en
bistable *{adj} [sci-t]*	二安定(の)	ni-antei (no)
" *{adj} [sci-t]*	双安定(の)	sō-antei (no)
bistable element *[ch]*	双安定素子	sō'antei-soshi
bisulfate *[ch]*	重硫酸塩	jū'ryūsan-en
bisulfite *[ch]*	重亜硫酸塩	jū'aryūsan-en
bit *{n} [comput]*	ビット	bitto
bit; binary digit *[comput]*	二進数子	nishin-sūji
bitartrate *[ch]*	重酒石酸塩	jū'shuseki'san-en
bite; crunch; gnaw *{vb}*	嚙む；咬む；かむ	kamu
biting midge (insect) *[i-zoo]*	糠蚊	nuka-ka
bitonality *[music]*	複調	fukuchō
bitter almond oil *[mat]*	苦扁桃油	ku'hentō-yu
bitter flavor; bitterness *[physio]*	苦味	niga'mi
bitter orange; bigarade (fruit) *[bot]*	橙；だいだい	daidai
bitter orange oil *[cook] [mat]*	橙皮油	tōhi-yu
bittern *[ch-eng]*	苦汁；にがり	nigari

bitterness *[cook] [physio]*	苦さ	niga-sa
bitters *[cook]*	苦味酒	nigami-zake
" *[pharm]*	苦味薬	kumi-yaku
" *[pharm]*	苦味薬	nigami-yaku; nigami-gusuri
bitumen *[mat]*	瀝青; れきせい	rekisei
bituminous coal *[geol]*	瀝青炭	rekisei-tan
bivalence *{n} [ch]*	二価; 二價	nika
bivalent *{adj} [ch]*	二価の	nika no
bivalves *[i-zoo]*	斧足類; 斧足類	fusoku-rui
" *[i-zoo]*	双殻類; 雙殼類	sōkaku-rui
biweekly *{adj} [pr]*	週二回の	shū-ni'kai no
black (color) *{n} [opt]*	黒; 黑; くろ	kuro
" *{adj} [opt]*	黒い	kuroi
black-and-white movie *[opt]*	黒白映画	koku'haku-eiga
black-and-white (sumi) painting *[graph]*	墨絵; 墨繪	sumi-e
black ant (insect) *[i-zoo]*	黒蟻	kuro-ari
black bamboo; Phyllostachys nigra *[bot]*	真竹; まだけ	ma-dake
blackberry (fruit) *[bot]*	黒苺; 黒苺	kuro-ichigo
blackberry lily (flower) *[bot]*	檜扇; ひおうぎ	hi'ōgi
blackbird (bird) *[v-zoo]*	烏; 鴉; からす	karasu
blackboard *[comm] [mat]*	黒板	kokuban
blackbody *[thermo]*	黒体	koku'tai
blackbody-radiation heat *[thermo]*	黒体放射熱	kokutai-hōsha-netsu
black drop *[astron]*	黒滴	kokuteki
black-eared kite (bird) *[v-zoo]*	鳶; とび	tobi
blacken *{vb t} [opt]*	黒くする	kuroku suru
blackening *{n} [opt]*	黒化	kokka
black-eyed pea (legume) *[bot]*	豇豆; さゝげ	sasage
black-filled period *[pr]*	中黒	naka-guro
black film *[p-ch]*	黒膜	kuro-maku
black-ink balance *[econ]*	黒字	kuro'ji
black koji mold *[mycol]*	黒麹菌	kuro-kōji-kin
black leopard (animal) *[v-zoo]*	黒豹; くろひょう	kuro-hyō
black liquor *[mat] [paper]*	黒液	koku-eki
black marlin (fish) *[v-zoo]*	黒梶木; 黒旗魚	kuro-kajiki
blackness *[opt]*	黒度	koku-do
black ore deposit *[geol]*	黒鉱鉱床; 黒鑛鑛床	kuro'kō-kōshō
blackout (of TV)) *[electr]*	帰線消去	kisen-shōkyo
black porgy (fish) *[v-zoo]*	黒鯛; くろだい	kuro-dai
blacksmith (a person) *[met]*	鍛冶屋	kaji'ya
black soybean *[bot] [cook]*	黒豆	kuro-mame

English	Japanese	Romanization
black spot *[electr]*	暗影	an'ei
black stream *[ocean]*	黒潮	kuro-shio
black swan (bird) *[v-zoo]*	黒鳥	koku'chō
black-tailed gull (bird) *[v-zoo]*	海猫; うみねこ	umi-neko
black tea *[cook]*	紅茶	kōcha
black walnut (tree) *[bot]*	黒胡桃	kuro-kurumi
black widow (spider) *[i-zoo]*	黒後家蜘蛛	kuro'goke-gumo
bladder *[anat]*	膀胱; ぼうこう	bōkō
blade; fan *{n} [eng]*	羽; はね	hane
" (of knife) *{n} [eng]*	刃; 刃; やいば	yaiba
" *{n} [eng]*	翼; よく	yoku
blade channel (gear) *[mech]*	歯溝; 齒溝	ha-mizo
blade-expelled flow *[fl-mech]*	翼吐出流	yoku-to'shutsu-ryū
blank; null *{n}*	空白	kūhaku
blank *{vb} [eng]*	打ち抜く	uchi-nuku
blanket *{n} [tex]*	毛布; もうふ	mōfu
blanking *{n} [eng]*	打抜き加工	uchi'nuki-kakō
blanking force *[eng]*	打抜き力	uchi'nuki-ryoku
blank test *[ind-eng]*	空試験	kū-shiken; kara-shiken
blast *{n} [phys]*	送風	sōfū
blast furnace *[met]*	高炉	kōro
" *[met]*	溶鉱炉; 溶鑛爐	yōkō-ro
blasting *{n} [eng]*	爆破	baku'ha
bleaching *{n} [cook] [tex]*	標白	hyōhaku
bleaching powder *[mat]*	晒し粉; さらしこ	sarashi-ko
bleeding *{n} [tex]*	滲み	nijimi
blended weave *[tex]*	交ぜ織	maze-ori
blended yarn *[tex]*	混紡糸	konbō-shi
blending agent *[mat]*	混和剤	konwa-zai
blimp *[aero-eng]*	軟式小型飛行船	nan'shiki-kogata-hikō'sen
blind *{adj} [med]*	盲の; めくらの	mekura no
blind alley *[civ-eng]*	袋小路	fukuro-kōji
" *[traffic]*	行き詰り	yuki-zumari
blind hole *[eng] [met]*	盲穴; めくら穴	mekura-ana
blind spot *[physio]*	盲点	mō-ten
blind zone *[comm]*	不感地帯; 不感地帶	fukan-chitai
blink *{n} [physio]*	瞬き; まばたき	mabataki
blinking *{n} [comm]*	明滅	mei'metsu
blister *{n} [eng]*	ふくれ	fukure
" *{n} [eng]*	痘痕	tō'kon
blister steel *[steel]*	有泡鋼; 有泡鋼	yūhō-kō
blizzard *[meteor]*	大吹雪	ō-fubuki

block *{n} [eng]*	滑車；かっしゃ	kassha
" *{n} [pr]*	版	han
blockade; blocking *{n} [ch]*	封鎖	fūsa
blockage; blockade *{n} [rail]*	閉塞	heisoku
blockage; clogging *{n}*	目詰まり	me'zumari
block diagram *[eng]*	構成図；構成圖	kōsei-zu
block polymerization *[poly-ch]*	塊状重合	kaijō-jūgō
block section *[rail]*	閉塞空間	heisoku-kūkan
blood *[hist]*	血；ち	chi
" *[hist]*	血液；けつえき	ketsu'eki
blood coagulation *[med]*	凝血	gyōketsu
blood pigment *[bioch]*	血色素	kesshikiso
" *[bioch]*	血液色素	ketsu'eki-shiki'so
blood plasma *[hist]*	血漿；けっしょう	kesshō
blood pressure *[physio]*	血圧；けつあつ	ketsu'atsu
bloodstone *[miner]*	血玉髄；血玉髓	ketsu'gyokuzui
blood substitute *[pharm]*	血液代用剤	ketsu'eki-daiyō-zai
blood transfusion *[med]*	輸血	yu-ketsu
blood type *[immun]*	血液型	ketsu'eki-gata
blood vessel *[anat]*	血管	kekkan
bloom; blossom; flower *{vb i} [bot]*	咲く	saku
bloom *{n} [steel]*	鋼片；こうへん	kōhen
" (of a paint) *{n} [poly-ch]*	曇り	kumori
blooming *{n} [steel]*	分塊	bunkai
blossom *{n} [bot]*	花；はな	hana
blossoming *{n} [bot]*	開花	kaika
blowfish (fish) *[v zoo]*	河豚；ふぐ	fugu
blowing *{n} [steel]*	吹錬；吹鍊	suiren
blowing agent *[mat]*	膨張剤	bōchō-zai
blown-fuse indicator *[elec]*	溶断指示器	yōdan-shiji-ki
blowsand *[geol]*	吹き砂	fuki-zuna
blue (color) *{n} [opt]*	青；あお	ao
" *{adj} [opt]*	青い	aoi
blue crab (crustacean) *[i-zoo]*	蝤蛑；がざみ	gazami
bluefin tuna (fish) *[v-zoo]*	黒鮪；くろまぐろ	kuro-maguro
blue-green algae *[bot]*	藍草	ransō
blue heat *[thermo]*	青熱	sei-netsu
blue-heat working *{n} [met]*	青熱塑性加工法	sei'netsu-sosei-kako-hō
blue mud *[geol]*	青泥；せいでい	seidei
blueprint *[graph]*	青焼	ao'yaki
blueprint; cyanotype *[arch]*	青写真；青寫真	ao-shashin
blue shark (fish) *[v-zoo]*	葦切り鮫	yoshi'kiri-zame

blue shift *{n} [astrophys]*	青方偏移	seihō-hen'i
blue vitriol; cupric sulfate *[mat]*	胆礬；たんばん	tanban
blue whale *[v-zoo]*	白長須鯨	shiro-nagasu-kujira
bluff *{n} [geol]*	断崖；斷崖	dangai
bluing (for laundry) *{n} [met]*	青味付	aomi'zuke
bluish purple (color) *[opt]*	紫紺；しこん	shikon
blunt end *[bioch]*	二本鎖末端	nihon'sa-mattan
boarding (a plane) *{n} [trans]*	搭乗；搭乘	tōjō
boat *[nav-arch]*	舟	fune
boat conformation *[org-ch]*	舟形配座	fune'gata-haiza
bobbin thread *[tex]*	裏糸	ura-ito
bodily fluid *[physio]*	体液；體液	tai-eki
bodily injury *[med]*	身体傷害	shintai-shōgai
body (pl bodies) *[anat]*	体；體；からだ	karada
" *[anat]*	身	mi
" *[anat]*	身体	shintai
" *[cer]*	素地；きじ	kiji
" (a pigment) *[ch]*	体質顔料	tai'shitsu-ganryō
" *[pat]*	本文	honbun; hon'mon
body-centered cubic crystal *[crys]*	体心立方晶	taishin-rippō-shō
body-centered cubic lattice *[crys]*	体心立方格子	taishin-rippō-kōshi
body-centered lattice *[crys]*	体心格子	taishin-kōshi
body color *[zoo]*	体色	tai-shoku
body heat *[bio]*	体熱	tai-netsu
body shape *[bio]*	体形	taikei
body strength *[physio]*	体力	tai-ryoku
body type *[physio]*	体質	tai-shitsu
body-warmer stick *[eng]*	懐炉；かいろ	kairo
body weight *[physio]*	体重	taijū
bog moss *[bot]*	水苔	mizu-goke
boil *{vb i} [ch] [cook]*	沸騰する	futtō suru
" *{vb t} [ch] [cook]*	沸騰させる	futtō saseru
" (as water) *{vb t} [cook]*	沸かす	wakasu
" (cook) *{vb t} [cook]*	煮る	niru
" (cook) *{vb t} [cook]*	茹でる	yuderu
boiled fish pudding dish *[cook]*	蒲鉾；かまぼこ	kamaboko
boiled rice *[cook]*	御飯；ごはん	gohan
boiler *[mech-eng]*	汽缶；汽罐	kikan
" *[mech-eng]*	煮沸器	sha'futsu-ki
boiler scale *[ch]*	缶石；罐石	kanseki
boiling *{n} [cook] [p-ch]*	沸騰；ふっとう	futtō
" *{n} [p-ch]*	煮沸	sha'futsu

boiling off; degumming; scouring *[tex]*	精練；精練	seiren
boiling point *[p-ch]*	沸点；沸點	futten
boiling water *[ch]*	沸騰水	futtō-sui
boiling water; hot water *[ch]*	熱湯；ねっとう	nettō
bold face *[comput] [pr]*	太字；ふとじ	futo'ji
bolide; fireball *[astron]*	火球；かきゅう	kakyū
bollard *[nav-arch]*	繋船柱	keisen'chū
boll weevil (insect) *[i-zoo]*	綿実花象虫	wata'mi'hana-zō'mushi
bolometric magnitude *[astron]*	放射等級	hōsha-tōkyū
bolt (for a door); latch *{n} [eng]*	閂；かんぬき	kannuki
bolt *{n} [mech-eng]*	締め釘	shime-kugi
" *{n} [mech-eng]*	締めねじ	shime-neji
bomb *{n} [ord]*	爆弾	bakudan
bomber *[aero-eng]*	爆撃機	bakugeki-ki
bond; bonding; association *{n} [ch]*	結合；けつごう	ketsu'gō
bond energy *[p-ch]*	結合エネルギー	ketsugō-enerugī
bonding *{n} [ch]*	結合	ketsu'gō
" ; cementing *{n} [adhes] [mech]*	接着；せっちゃく	setchaku
bonding agent; binder *[adhes] [mat]*	結合剤	ketsugō-zai
" *[adhes] [mat]*	接着剤	setchaku-zai
bonding force *[mech] [poly-ch]*	接着力	setchaku-ryoku
bonding strength *[mech]*	結合力	ketsugō-ryoku
bonding water; bound water *[ch]*	結合水	ketsugō-sui
bone *{n} [anat] [hist]*	骨；ほね；こつ	hone; kotsu
bone china *[cer] [mat]*	骨灰磁器	koppai-jiki
bone marrow *[hist]*	骨髄；骨髓	kotsu-zui
bone oil *[mat]*	骨油	kotsu-yu
bonito; Sarda orientalis (fish) *[v-zoo]*	歯鰹；はがつお	ha'gatsuo
bonito, dried *[cook]*	鰹節	katsuo'bushi
bonito, oceanic (fish) *[v-zoo]*	鰹；かつお	katsuo
bonito shark; mako (fish) *[v-zoo]*	青鮫；あおざめ	ao-zame
bony fishes; Osteichthyes *[v-zoo]*	硬骨魚類	kōkotsu'gyo-rui
book; volume *{n} [pr]*	本；ほん	hon
" *{n} [pr]*	書物；しょもつ	sho'motsu
bookbinding *{n} [pr]*	製本	seihon
bookbinding design(ing) *{n} [pr]*	装丁；装訂；装幀	sōtei
bookcase *[furn]*	本箱	hon-bako
bookkeeping *{n} [econ]*	簿記；ぼき	boki
book plate *[pr]*	蔵書標；藏書標	zōsho-hyō
books *[pr]*	書籍；しょせき	sho'seki
bookshelf (pl -shelves) *[arch]*	本棚	hon-dana
bookstore *[arch] [pr]*	書店	shoten

book title *[pr]*	書名	shomei
bookworm (insect) *[i-zoo]*	紙魚；しみ	shimi
boom *[aero-eng]*	張出し支柱	hari'dashi-shichū
" *[eng]*	防護材	bōgo-zai
" *[eng]*	防材	bōzai
booster battery *[elec]*	増圧電池	zō'atsu-denchi
boracite *[miner]*	方硼石	hō'hōseki
borage (herb) *[bot]*	瑠璃萵苣；るりじさ	ruri-jisa
borate *[ch]*	硼酸塩	hōsan-en
borax *[miner]*	硼砂	hōsha
border *{n}* *[math]*	縁；縁；へり	heri
" *{n}* *[pr]*	縁；ふち	fuchi
border; hem; fringe *{vb t}*	縁取る	fuchi-doru
border line *[math]*	境界線	kyōkai-sen
bore *{n}* *[paleon]*	穿孔	senkō
borehole *[min-eng]*	試錐孔；しすいこう	shisui-kō
boric acid	硼酸	hōsan
boride *[ch]*	硼化物	hōka'butsu
boring machine *[eng]*	穿孔機	senkō-ki
borneol *[org-ch]*	竜脳；龍脳	ryūnō
bornite *[miner]*	斑銅鉱	han'dōkō
borofluoric acid	硼弗化水素酸	hō'fukka-suiso-san
boron (element)	硼素；ほうそ	hōso
boron arsenate	砒酸硼素	hisan-hōso
boron carbide	炭化硼素	tanka-hōso
boron chloride	塩化硼素	enka-hōso
boron hydride	水素化硼素	suisoka-hōso
boron nitride	窒化硼素	chikka-hōso
boron oxide	酸化硼素	sanka-hōso
boron tribromide	三臭化硼素	san'shūka-hōso
boron trichloride	三塩化硼素	san'enka-hōso
boron trifluoride	三弗化硼素	san'fukka-hōso
boron trioxide	三酸化硼素	san'sanka-hōso
borrow *{vb}* *[arith]*	借りる	kariru
" *{vb}* *[arith]*	降ろす	orosu
borrow digit *[arith]*	借り数	kari-sū
borrowing *{n}* *[arith]*	借り	kari
boss (casting) *{n}* *[met]*	座	za
botanical garden *[bot]*	植物園	shokubutsu'en
botany *[bio]*	植物学	shokubutsu'gaku
both; pair *{n}*	双；雙；そう	sō
both *{pron}*	両方；兩方	ryōhō

bottle; jar *{n} [eng]*	瓶；壜；びん	bin
bottled goods *[food-eng]*	壜詰品	bin'zume-hin
bottle gourd *[bot]*	夕顔；ゆうがお	yūgao
bottleneck *[traffic]*	隘路；あいろ	airo
bottle opener *[eng]*	栓抜き	sen-nuki
bottle with ground stopper *[ch]*	共栓瓶	tomo'sen-bin
bottom *{n}*	底	soko
bottom cargo *[nav-arch]*	底荷	soko-ni
bottom of the sea *[ocean]*	水底	suitei
bottoms *[petrol]*	缶出液；罐出液	kan'shutsu-eki
bottom view *[graph]*	下面図；下面圖	kamen-zu
" *[graph]*	底面図	teimen-zu
botulism *[med]*	腸詰め中毒	chō'zume-chūdoku
bouillon *[microbio]*	肉汁	niku'jū
boulder *[geol]*	巨礫；きょれき	kyo'reki
boulder flow *[geol]*	土石流	doseki-ryū
boule *[crys]*	宝石原石	hōseki-genseki
boulevard *[civ-eng]*	広路；廣路	hiro'ji
boundary *[ch] [phys]*	界面	kaimen
" *[comput [geol] [sci-t]*	境界	kyōkai
boundary; limit *[math]*	限界	genkai
boundary condition *[math]*	境界条件；境界條件	kyōkai-jōken
boundary layer *[fl-mech] [phys]*	境界層	kyōkai-sō
bound electron *[atom-phys] [phys]*	束縛電子	soku'baku-denshi
bound water *[ch]*	結合水	ketsu'gō-sui
bournonite *[miner]*	車骨鉱	shakotsu'kō
bovine *{adj} [v-zoo]*	牛(属)の	ushi(-zoku) no
bow; bow shape *{n} [arch] [geom]*	弓形	yumi-gata
bow; fogbow *{n} [meteor]*	霧虹；むこう	mukō
bow (of a ship) *{n} [nav-arch]*	船首	senshu
" (weapon) *{n} [ord]*	弓	yumi
bowl (of chinaware) *[cer] [cook]*	碗	wan
" (basin) *{n} [cook] [eng]*	鉢	hachi
" (of wood or lacquer) *[cook]*	椀	wan
" (often with cover, for rice dishes) *[cer]*	丼；どんぶり	donburi
-bowl(s) *{suffix}*	-椀	-wan
bow wave *[fl-mech]*	頭部波	tōbu-ha
" *[nav-arch]*	船首波	senshu-ha; senshu-nami
box *{n} [eng]*	箱	hako
" *{n} [eng]*	筐体	kyōtai
box annealing *{n} [met]*	箱焼鈍し	hako-yaki'namashi

box car *[rail]*	有蓋貨車	yūgai-kasha
boxfish; cofferfish (fish) *[v-zoo]*	箱河豚	hako-fugu
box tree (tree) *[bot]*	黄楊；つげ	tsuge
box turtle (reptile) *[v-zoo]*	箱亀；はこがめ	hako-game
brace *{n}* *[civ-eng]*	控え	hikae
" *{n}* *[eng]*	受金物	uke-kana'mono
brace(s) *[comput]* *[pr]*	中括弧	chū-kakko
bracelet; bangle *[lap]*	腕輪	ude'wa
Brachiopoda *[i-zoo]*	腕足動物	wansoku-dōbutsu
bracken *[bot]*	蕨；わらび	warabi
bracket(s) *[comput]*	大括弧	dai-kakko
brackish water corrosion *[met]*	混水腐食	konsui-fushoku
bract *[bot]*	保護葉	hogo'yō
" *[bot]*	包葉；包葉	hōyō
braid *{n}* *[tex]*	編組；へんそ	henso
Braille *[comm]*	点字；點字；てんじ	tenji
brain *{n}* *[anat]*	脳；腦；のう	nō
brain coral *[i-zoo]*	脳珊瑚	nō-sango
brain stem *[anat]*	脳幹	nō-kan
brain wave *[physio]*	脳波	nōha
braking *{n}* *[mech-eng]*	制動	seidō
braking force *[mech]*	制動力	seidō-ryoku
bran; rice bran *[bot]* *[cook]*	米糠	kome-nuka
bran; wheat bran *[bot]* *[cook]*	麩；麸；ふすま	fusuma
branch *{n}* *[bot]*	枝	eda
branch; branching *{n}* *[ch]* *[comput]*	分岐	bunki
branch; department; section *{n}*	部門	bumon
branched *{adj}* *[comput]*	分岐した	bunki shita
branchial cleft (gill slit) *[v-zoo]*	鰓裂；さいれつ	sai-retsu
branching *{n}* *[bot]*	枝分れ	eda-wakare
branch office *[econ]*	分局	bun'kyoku
" *[econ]*	出張所	shutchō'jo
branch point *[comput]*	分岐点；分岐點	bunki-ten
branch store; branch office	支店	shiten
brand; brand name; make *{n}* *[econ]*	銘柄；めいがら	meigara
brass *[met]*	黄銅	ō'dō
" *[met]*	真鍮；しんちゅう	shinchū
Brassica Rapa var. pervidis (vegetable) *[bot]*	小松菜	komatsu'na
brazier for charcoal *[furn]*	火鉢	hibachi
bread *[cook]*	パン；麵麭；麵包	pan
" ; loaf bread *[cook]*	食パン	shoku-pan

breadboarding *{n} [electr]*	回路板上の実験	kairoban'jō no jikken
breadcrumb *[cook]*	パン粉	pan-ko
bread dough *[cook]*	パン生地	pan-kiji
bread mold *[mycol]*	黒黴; くろかび	kuro-kabi
breadth *[mech]*	広さ; 廣さ	hiro-sa
break; crack location *{n} [eng]*	毀れ目	koware'me
break *{vb t} [mech-eng]*	折る	oru
" (be broken) *{vb i}*	砕ける	kudakeru
" (destroy) *{vb t}*	毀す; 壊す; こわす	kowasu
breakdown *{n} [elec]*	事故	jiko
" *{n} [mech]*	機能停止	kinō-teishi
breakdown voltage *[elec]*	破壊電圧; 破壞電壓	hakai-den'atsu
breaker(s) *[ocean]*	砕け波	kudake-nami
breakfast *{n} [cook]*	朝御飯; 朝ごはん	asa-gohan
" *{n} [cook]*	朝食	chōshoku
breaking *{n} [mech]*	破壊; 破壞; はかい	hakai
" *{n} [mech]*	破砕; 破碎	hasai
breaking elongation *[mech]*	破断伸び; 破斷伸び	hadan-nobi
" *[mech]*	破断伸度	hadan-shindo
breaking length *[paper]*	裂断長	retsu'dan-chō
breaking load *[mech]*	破壊荷重	hakai-kajū
breaking strength *[mech]*	破壊強さ	hakai-tsuyo-sa
breaking stress *[mech]*	破壊応力	hakai-ō'ryoku
breaking tensile strength *[mech]*	破断強度	hadan-kyōdo
breaking toughness *[mech]*	破断靭性	hadan-jinsei
breaking wave *[ocean]*	砕け波	kudake-nami
break line *[graph]*	破断線	hadan-sen
break money *{vb t} [econ]*	崩す; くずす	kuzusu
breakpoint *[comput]*	区切り点	kugiri-ten
breakthrough time *[ch-eng]*	破過時間	haka-jikan
breakwater *[civ-eng]*	防波堤	bōha'tei
breast; chest; thorax *{n} [anat]*	胸	mune
breastbone (sternum) *{n} [anat]*	胸骨	kyō-kotsu
breath *[physio]*	息	iki
Breathalyzer *[eng]*	酒気検知器	shuki-kenchi-ki
breathing *{n} [physio]*	呼吸; こきゅう	kokyū
breccia *[miner]*	角蛮岩; 角蠻岩	kaku'bangan
" *[miner]*	角礫岩	kaku'rekigan
breed *{n} [agr] [bio]*	品種	hinshu
breeder reactor *[nucleo]*	増殖炉; 増殖爐	zōshoku-ro
breeding *{n} [agr]*	飼育	shi'iku
" *{n} [nucleo]*	増殖	zō'shoku

breeding ratio *[nucleo]*	増殖比	zōshoku-hi
breeze *[meteor]*	微風	bifū
" *[meteor]*	そよ風	soyo-kaze
brew (wine) *{vb t} [cook]*	醸造する; 釀造する	jōzō suru
brewage *{n} [cook]*	醸造酒	jōzō'shu
brewing *{n} [cook]*	醸造; じょうぞう	jōzō
brewing yeast; brewers' yeast *[cook]*	麦酒酵母	bīru-kōbo
brew tea *{vb} [cook]*	茶を入れる	cha o ireru
brick *[arch] [mat]*	煉瓦	renga
bridge *{n} [civ-eng]*	橋	hashi
" *{n} [nav-arch]*	艦橋; かんきょう	kankyō
" *{n} [nav-arch]*	船橋	senkyō
bridged complex *[ch]*	架橋錯体	kakyō-saku'tai
bridged structure *[ch]*	橋架け構造	hashi'kake-kōzō
bridge floor; bridge deck *[civ-eng]*	橋床	kyōshō
bridgehead position *[ch]*	橋頭位	kyōtō'i
bridge pier *[civ-eng]*	橋脚	kyō'kyaku
bridges *{n} [civ-eng]*	橋梁	kyōryō
bridging *{n} [ch]*	橋架け	hashi'kake
bridging; mediation *{vb t}*	橋渡し	hashi-watashi
bridle; rein *{n} [trans]*	手綱; たづな	ta'zuna
bright; light *{adj} [opt]*	明るい	akarui
bright annealing *{n} [met]*	光輝焼鈍	kōki-yaki'namashi
brightener *[met]*	光沢剤; 光澤剤	kōtaku-zai
bright field *[opt]*	明視野	mei-shiya
bright moon *[astron]*	明月	meigetsu
bright nebula (pl -s, -lae) *[astron]*	輝星雲	kisei'un
brightness; lightness *{n} [opt]*	明るさ	akaru-sa
brilliance *[opt]*	輝き; かがやき	kagayaki
" *[opt]*	光輝	kōki
brine *[mat] [ocean]*	食塩水; 食鹽水	shoku'en-sui
briquet coal *[mat]*	錬炭	rentan
" *[mat]*	成型炭	seikei-tan
brisance *[ord]*	猛度; もうど	mōdo
bristle *{n} [bio]*	剛毛	gōmō
British thermal unit; Btu *[thermo]*	英国熱量単位	Eikoku-netsu'ryō-tan'i
brittle fracture *[met]*	脆性破壊; 脆性破壞	zeisei-hakai
brittle material *[mat]*	脆性材料	zeisei-zairyō
brittleness *[mech]*	脆さ	moro-sa
" *[mech]*	脆性	zeisei
brittle point *[thermo]*	脆化点	zei'ka-ten
brittle star *[i-zoo]*	蜘蛛人手	kumo-hitode

brittle temperature *[thermo]* 脆化温度 zei'ka-ondo
brittling *{n} [mech]* 脆化 zei'ka
broadband *{adj} [comm]* 広帯域の; 廣帯域の kō-tai'iki no
broad bean (large) *[bot]* お多福豆 o'tafuku-mame
" (legume) *[bot]* 空豆 sora-mame
broadbill (fish) *[v-zoo]* 目旗魚; めかじき me'kajiki
broad-billed roller (bird) *[v-zoo]* 仏法僧; 佛法僧 buppōsō
broadcast *{n} [comm]* 放送 hōsō
broadcast seeding *{n} [agr]* 播種 hanshu
broaden; expand; spread; widen *{vt t}* 広げる; 廣げる hirogeru
broadside *{n} [nav-arch]* 玄側 gensoku
brocade *[tex]* 錦織り; にしきおり nishiki-ori
brochantite *[miner]* 水胆礬; 水膽礬 sui'tanban
broiled chicken *[cook]* 焼(き)鶏 yaki-tori
broiled dishes *[cook]* 焼き物 yaki-mono
broke *{n} [paper]* 損紙; そんし sonshi
broken line *[graph]* 破線 hasen
broken-out section *[graph]* 破断面 hadan-men
broken pulp *[paper]* 解砕パルプ kaisai-parupu
bromate *[ch]* 臭素酸塩 shūso'san-en
bromic acid 臭素酸 shūso-san
bromide *[ch]* 臭化物 shūka'butsu
bromide ion *[ch]* 臭素イオン shūso-ion
bromination *[ch]* 臭素化 shūso'ka
bromine (element) 臭素; しゅうそ shūso
bromine chloride 塩化臭素 enka-shūso
bromine number *[an-ch]* 臭素価 shūso-ka
bromine oxide 酸化臭素 sanka-shūso
bromine water *[ch]* 臭素水 shūso-sui
bromite *[ch]* 亜臭素酸塩 a'shūso'san-en
bromous acid 亜臭素酸 a'shūso-san
bromyrite *[miner]* 臭銀鉱 shū'ginkō
bronchus (pl bronchi) *[anat]* 気管支 kikan'shi
brontides *[geophys]* 地鳴り ji'nari
bronze *{n} [met]* 青銅 seidō
Bronze Age *[archeo]* 青銅器時代 seidō'ki-jidai
broom *[eng]* 箒; ほうき hōki
broth; soup *[cook]* 汁; つゆ tsuyu
brown (color) *[opt]* 茶色 cha-iro
" (color) *[opt]* 褐色; 褐色 kasshoku
brown algae (sing alga) *[bot]* 褐草 kassō
brown coal *[geol]* 褐炭 kattan

brownish green (color) *[opt]*	鶯色；うぐいすいろ	uguisu-iro
browsing *{n} [bio]*	若葉を食う（こと）	waka'ba o kū (koto)
brucite *[miner]*	水滑石	sui'kasseki
bruise *{n} [met]*	打ち疵	uchi'kizu
brush *{n} [eng]*	刷子；ブラシ	burashi
" *{n} [eng]*	刷毛；はけ	hake
" *{n} [graph]*	筆；ふで	fude
" (writing, painting) *{n} [graph]*	毛筆；もうひつ	mō'hitsu
brushing *{n} [paint]*	刷毛塗り	hake'nuri
brushwood *[bot]*	柴	shiba
bryology *[bot]*	蘚苔学	sentai'gaku
Bryophyta *[bot]*	苔類；こけるい	koke-rui
bubble(s) *[ch] [phys]*	泡；泡；あわ	awa
" *[ch] [phys]*	泡沫	hōmatsu
" *[ch] [phys]*	気泡	kihō
bubble cap *[ch-eng]*	泡鐘	hōshō
bubble chamber *[nucleo]*	泡箱	awa-bako
bubble tower *[ch-eng]*	気泡塔；氣泡塔	kihō-tō
bucket *[eng]*	桶；おけ	oke
buckeye (tree) *[bot]*	栃；とちのき；とち	tochinoki; tochi
buck grass *[bot]*	日陰葛；ひかげの=かずら	hikage-no-kazura
buckling *{n} [mech]*	座屈	za'kutsu
buckthorne *[bot]*	黒梅擬	kuro-ume'modoku
buckwheat noodles *[cook]*	蕎麦；そば	soba
bud *{n} [bot]*	芽；め	me
" *{n} [bot]*	蕾；つぼみ	tsubomi
budding *{n} [bot]*	分芽	bunga
" *{n} [bot]*	出芽	shutsu'ga
budgerigar (bird) *[v-zoo]*	背黄青鸚哥	sekisei-inko
bud scale *[bot]*	芽鱗	ga'rin
buffer; buffering agent *{n} [ch]*	緩衝剤	kanshō-zai
buffer *[comput]*	緩衝；かんしょう	kanshō
" (instrument) *[eng]*	緩衝器	kanshō-ki
buffer solution *[ch]*	緩衝液	kanshō-eki
bug *{n} [electr]*	音声監視装置	onsei-kanshi-sōchi
" *{n} [i-zoo]*	虫；蟲；むし	mushi
build (of body) *{n} [med]*	体格	taikaku
build; erect *{vb} [arch]*	立てる；建てる	tateru
building *{n} [arch]*	ビル	biru
" *{n} [arch]*	建築；けんちく	kenchiku
" *{n} [arch]*	建物	tate'mono

building blocks (toy) *[eng]*	積み木	tsumi-ki
building material *[arch] [civ-eng]*	建築材料	kenchiku-zairyō
" *[arch]*	建材	kenzai
buildup roll	肉盛ロール	niku'mori-rōru
buildup time *[comm]*	立ち上り時間	tachi'agari-jikan
buildup welding *{n} [met]*	肉盛溶接	niku'mori-yō'setsu
built-in function *[comput]*	組込み機能	kumi'komi-kinō
bulb *[bot]*	球根	kyūkon
bulb thermometer *[eng]*	球寒暖計	kyū-kandan-kei
bulbul (bird) *[v-zoo]*	鵯；ひよどり	hiyodori
bulfil *[bot]*	零余子；むかご	mukago
bulge *{n}*	脹らみ	fukurami
bulging *{n}*	張出し	hari'dashi
bulk *[pr]*	嵩；かさ	kasa
bulk density *[eng] [poly-ch]*	嵩比重	kasa-hijū
" *[eng] [poly-ch]*	嵩密度	kasa-mitsudo
bulk factor *[eng] [poly-ch]*	嵩係数	kasa-keisū
bulkhead *[aero-eng] [nav-arch]*	隔壁；かくへき	kaku'heki
" *[nav-arch]*	防水壁	bōsui'heki
bulk modulus *[mech]*	体積弾性係数	taiseki-dansei-keisū
" *[mech]*	体積弾性率	taiseki-dansei-ritsu
bulk polymerization *[poly-ch]*	塊状重合	kaijō-jūgō
bulk specific gravity *[mech]*	嵩比重	kasa-hijū
bull (animal) *[v-zoo]*	雄牛	osu-ushi
bullet *[ord]*	弾丸	dangan
bullet; shot; shell *{n} [ord]*	弾；彈（弾丸）	tama
bulletin *[pr]*	速報	soku'hō
bullet speed *[mech]*	弾速	dansoku
Bullet Train Line *[rail]*	新幹線	Shinkan-sen
bull frog (amphibian) *[v-zoo]*	牛蛙；うしがえる	ushi-gaeru
bullhead (fish) *[v-zoo]*	鰍；かじか	kajika
bullion *[met]*	金(属)塊	kin(-zoku)kai
bumping *{n} [ch]*	突沸；突沸	topputsu
bumps and dips *[civ-eng]*	凸凹；でこぼこ	deko-boko
bunch *{vb}*	束ねる	tsukaneru
bund *[civ-eng]*	沿岸道路	engan-dōro
bundle; bunch; sheave *{vb t}*	束ねる	tabaneru; tsukaneru
-bundle(s) *{suffix}*	-束	-taba
-bundle(s) *{suffix}*	-把	-wa; -pa
bundled-needle state	束針状	sokushin'jō
bunk *[arch] [furn]*	寝棚	ne-dana
buoy *{n} [eng]*	浮標；ふひょう	fuhyō

buoyancy; buoyant force *[fl-mech]*	浮力	fu'ryoku
buprestid (beetle) *[i-zoo]*	玉虫；たまむし	tama-mushi
burdock; cocklebur (vegetable) *[bot]*	牛旁；ごぼう	gobō
burglar alarm *[eng]*	盗難警報器	tōnan-keihō-ki
burial mound figure *[archeo]*	埴輪；はにわ	haniwa
burin *[graph]*	彫刻刀	chōkoku-tō
burn; scald *{n} [med]*	火傷；やけど	yakedo
burn *[vb i]*	燃える	moeru
burn; bake; fire; roast *{vb t}*	焼く；焼く；やく	yaku
burn; torch *{vb t}*	燃やす	moyasu
burnability *[cer]*	易焼成性	i'shōsei-sei
burning *{n} [eng]*	燃焼	nenshō
" *{n} [eng]*	焼成；しょうせい	shōsei
" *{n} [met]*	焼け	yake
burning-on *{n}*	焼付け	yaki-tsuke
burning resistance *[mat]*	耐燃性	tai'nen-sei
burnished gold *[met]*	磨き金	migaki-kin
burn mark *[plas]*	焼け	yake
burn medicine *[pharm]*	火傷薬	yakedo-gusuri
burnout *{n} [aero-eng]*	燃焼完了	nenshō-kanryō
" *[aero-eng]*	燃焼終了	nenshō-shūryō
burnout life *[elec]*	断線寿命；断線壽命	dansen-jumyō
burnt alum *[inorg-ch] [pharm]*	焼(き)明礬	yaki-myōban
burr (as of chestnut) *[bot]*	毬；いが	iga
" *[met]*	逆鉤；かえり	kaeri
burrow *{n} [paleon]*	潜穴	senketsu
" *{vb} [zoo]*	潜る；もぐる	moguru
bursting *{n} [paper]*	破裂；はれつ	ha'retsu
bursting strength *[mech]*	破裂強さ	haretsu-tsuyo-sa
bury; embed *{vb t}*	埋没する	mai'botsu suru
bury *{vb t}*	埋める	uzumeru
bus *{n} [comput] [geom]*	母線	bosen
bush clover (herb) *[bot]*	萩；はぎ	hagi
bush warbler (bird) *[v-zoo]*	鶯；うぐいす	uguisu
business *[econ]*	営業；營業	eigyō
" *[econ]*	業務	gyōmu
" *[econ]*	事業；じぎょう	jigyō
" *[econ]*	商業	shōgyō
business; trade; commerce *[econ]*	商売；商賣	shōbai
business card *[comm]*	名刺	meishi
business firm *[econ]*	商社	shōsha
business hours *[econ]*	営業時間	eigyō-jikan

business machine *[eng]*	事務機械	jimu-kikai
business office *[econ]*	営業所；營業所	eigyō'sho
busy (telephone) *[comm]*	話中；わちゅう	wa'chū
busy signal *[comm]*	話中音	wa'chū-on
butt against (one another) *{vb}*	突合わせる	tsuki-awaseru
butterbur(r) (vegetable) *[bot]*	蕗；ふき	fuki
butterbur(r) flower, stalk *[bot]*	蕗の薹	fuki no tō
butterfat *[cook]*	乳脂肪	nyū-shibō
butterfly (insect) *[i-zoo]*	蝶(蝶)；蝶(々)	chō(-chō)
butterfly fish (fish) *[v-zoo]*	蝶蝶魚；蝶々魚	chōchō-uo
butterfly ray (fish) *[v-zoo]*	燕鱝；つばくろ鱝	tsubakuro-ei
butterfly valve *[eng]*	蝶形弁；蝶形瓣	chōgata-ben
butt joint *[eng]*	突合(わ)せ継(ぎ)手	tsuki'awase-tsugi'te
buttocks *{n}* *[anat]*	尻；しり	shiri
button *{n}* *[cl]*	釦；鈕；ボタン	botan
button closure *[cl]*	釦閉め	botan-jime
buttonhead (screw) *[eng]*	丸頭；まるあたま	maru-atama
buttress *{n}* *[arch]*	控え壁	hikae-kabe
butt weld *{n}* *[met]*	突合せ溶接	tsuki'awase-yō'setsu
butyrate *[ch]*	酪酸塩	rakusan-en
butyric acid	酪酸；らくさん	raku-san
butyric anhydride	無水酪酸	musui-rakusan
buy; purchase *{vb}* *[econ]*	買う	kau
buyer *[econ]*	買手	kai'te
buzzing; hum *{n}* *[i-zoo]*	唸り；うなり	unari
bypass *{n}* *[civ-eng]*	側路	soku-ro
" *{n}* *[civ-eng]*	迂回路	ukai-ro
bypass valve *[eng]*	副制水弁；副制水瓣	fuku'seisui-ben
by-product *[eng]*	副産物	fuku-sanbutsu
" *[eng]*	副生物	fuku-seibutsu
byte *[comput]*	バイト	baito
bytownite *[miner]*	亜灰長石	a'kai'chōseki

C

cabbage (vegetable) *[bot]*	キャベツ	kyabetsu
cabdriver *[trans]*	運転手	unten'shu
cabin (of aircraft, seacraft) *[arch]*	機室	ki-shitsu
" (in a ship) *[nav-arch]*	船室	sen-shitsu
cabinetmaker *[furn]*	家具師	kagu'shi
cable (of steel) *[eng]*	鋼索	kō'saku
cable code *[comm]*	現波符号	genpa-fugō
cable television service *[comm]*	有線テレビ放送	yūsen-terebi-hōsō
cableway *[mech-eng]*	索道	saku'dō
caboose *[rail]*	車掌車	shashō'sha
cache *[comput]*	隠退蔵物；隠退藏物	in'taizō'butsu
" *[comput]*	隠し場	kakushi'ba
cacomistle (animal) *[v-zoo]*	麝香猫	jakō-neko
cactus (pl cacti, cactus, -es) *[bot]*	仙人掌；サボテン	saboten
cadmium (element)	カドミウム	kadomyūmu
caffeine *[org-ch]*	茶精	chasei
" *[org-ch]*	茶素	chaso
cage *{n} [eng]*	檻；おり	ori
cage rotor *[elec]*	籠形回転子	kago'gata-kaiten'shi
caking *{n} [eng]*	固化	koka
caking coal *[geol]*	粘結炭	nen'ketsu-tan
calabash (flower) *[bot]*	夕顔；ゆうがお	yū'gao
calcareous *{adj} [sci-t]*	石灰質の	sekkai'shitsu no
calcification *[geoch] [physio]*	石灰化	sekkai'ka
calcination *[eng]*	煆焼	kashō
" *[min-eng]*	焼成	shōsei
calcite *[miner]*	方解石	hō'kaiseki
calcium (element)	カルシウム	karushūmu
calcium dihydrogen phosphate	燐酸二水素カル=シウム	rinsan-ni'suiso-karushūmu
calcium hydroxide (slaked lime)	水酸化カルシウム	suisan'ka-karushūmu
calcium primary phosphate	第一燐酸カルシウム	dai'ichi-rinsan-karushūmu
calcium secondary phosphate	第二燐酸カルシウム	dai'ni-rinsan-karushūmu
calcium superphosphate	過燐酸石灰	ka'rinsan-sekkai
calculate; compute *{vb} [math]*	計算する	keisan suru
calculate; compute *{vb} [math]*	算出する	san'shutsu suru
calculation *[alg] [comput]*	演算；えんざん	enzan
" *[math]*	計算	keisan
calculus *[math]*	微分積分学	bibun-sekibun'gaku

caldera *[geol]*	大窪地	ō-kubo'chi
calendar *[astron]*	暦; 曆; こよみ	koyomi
calf (pl calves; calfs) *[anat]*	脹ら脛	fukura-hagi
calf; calved ice *[geol]*	氷塊; ひょうかい	hyōkai
calf *[v-zoo]*	子牛; 犢	ko-ushi
calibration *[sci-t]*	較正; 校正	kōsei
calibration curve *[eng]*	検量線	kenryō-sen
" *[eng]*	較正曲線	kōsei-kyoku'sen
calico (pl -coes, -cos) *[tex]*	更紗; さらさ	sarasa
californium (element)	カリホルニウム	karihorunyūmu
call *{n} [comm]*	呼出し	yobi'dashi
call; pipe *{n} [comm]*	呼子	yobiko
call *{vb} [comm]*	呼び出す	yobi-dasu
" *{vb}*	呼ぶ; よぶ	yobu
call bell *[comm]*	呼鈴	yobi-rin
called station *[comm]*	被呼局	hiko-kyoku
calligraphy *[comm]*	書道	shodō
calling *{n} [comm]*	呼出し	yobi'dashi
calling card; business card *[pr]*	名刺	meishi
calling sequence *[comm]*	呼出し系列	yobi'dashi-kei'retsu
calliper *[eng]*	パス	pasu
call sign *[comm]*	呼出符号	yobi'dashi-fugō
callus *[bot]*	肉状体	niku'jō'tai
" *[med]*	胼胝; べんち; たこ	benchi; tako
" *[med]*	仮骨; 假骨; かこつ	ka'kotsu
calms *{n} [meteor]*	凪; なぎ	nagi
calomel *[miner]*	甘汞; かんこう	kankō
calorie *[thermo]*	カロリー	karorī
calorimeter *[eng]*	熱量計	netsu'ryō-kei
camber *[eng]*	反り	sori
cambium (pl -s, -bia) *[bot]*	形成層	keisei-sō
camel (animal) *[v-zoo]*	駱駝; らくだ	rakuda
camel hair *[tex]*	駱駝毛	rakuda-mō
camellia (flower) *[bot]*	椿; つばき	tsubaki
camellia oil *[mat]*	椿油	tsubaki-yu
cameo (pl -eos) *[lap]*	浮彫細工	uki'bori-zaiku
camera *[opt]*	写真機; 寫眞機	shashin-ki
camouflage *[bio] [mil]*	擬装; ぎそう	gisō
camphor *[org-ch]*	樟脳	shōnō
camphoric acid	樟脳酸	shōnō-san
camphor oil *[mat]*	樟脳油	shōnō-yu
camphor tree *[bot]*	楠; 樟; くす(のき)	kusu(-noki)

Canada *[geogr]* カナダ Kanada
canal *[civ-eng]* 堀；濠；壕；ほり hori
" *[civ-eng]* 疏水；そすい so'sui
" *[civ-eng]* 運河；うんが un'ga
canal bridge *[civ-eng]* 運河橋 unga-kyō
canal elevator *[civ-eng]* 運河昇降機 unga-shōkō-ki
canard *[aero-eng]* 先尾翼 senbi'yoku
cancel *{n}* *[comput]* 取消し tori'keshi
" *{vb}* 取消す tori-kesu
cancel character *[comput]* 取消(し)文字 tori'keshi-moji
cancellation *[comput]* 取消し tori'keshi
" *[math]* 抹殺 massatsu
" *[pr]* 削除 sakujo
" ; withdrawal *{n}* 撤回 tekkai
cancellation; breaking off *[comm]* 破談 hadan
cancellation mark *[comm]* 消印；けしいん keshi-in
cancellation rule *[alg]* 簡約法則 kan'yaku-hōsoku
Cancer; the Crab *[constel]* 蟹座；かにざ kani-za
cancer *[med]* 癌 gan
candle *[mat]* 蠟燭 rōsoku
candlepower *[opt]* 燭光；しょっこう shokkō
" *[opt]* 燭；しょく shoku
candlestand *[furn]* 燭台 shoku'dai
candock *[bot]* 河骨；川骨；こほね kohone
cane *{n}* *[furn]* *[mat]* 籐類 tō-rui
cane sugar *[bioch]* 庶糖 shotō
" *[org-ch]* 甘蔗糖 kansho'tō
canine (tooth) *[dent]* 犬歯；けんし kenshi
canned goods *[cook]* *[food-eng]* 缶詰品；罐詰品 kan'zume'hin
cannery *[food-eng]* 缶詰製造所 kan'zume-seizō'sho
cannon; artillery (en masse) *[ord]* (大)砲；(大)砲 (tai-)hō
canola oil; rapeseed oil *[mat]* 菜種油；なたね油 natane-abura
can opener *[eng]* 缶切り；罐切り kan-kiri
canopy (pl -pies) *[aero-eng]* 円蓋 engai
" *[tex]* 天蓋 tengai
cantilever *[eng]* 片持梁 kata'mochi-bari
canvas *[tex]* 帆布；ほぬの ho-nuno
" *[tex]* ズック zukku
canyon *[geogr]* 峡谷 kyōkoku
cap (explosive) *{n}* *[ord]* 雷管 raikan
capacitive coupling *[elec]* 容量的結合 yōryō'teki-ketsu'gō
capacitive load *[elecmg]* 容量性負荷 yoryō'sei-fuka

capacitor *[elec]*	コンデンサー	kondensā
capacity *[sci-t]*	容量	yōryō
" ; accomodation *{n}*	収容; 收容	shūyō
capacity: capability *{n}*	能力	nō'ryoku
" **: capability** *{n}*	性能	seinō
cape *[geol]*	岬; みさき	misaki
cape jasemine (flower) *[bot]*	山梔子; くちなし	kuchinashi
capillarity (pl -ties) *[fl-mech]*	毛管現象	mōkan-genshō
capillary (pl -laries) *{n}* *[anat]*	毛細血管	mōsai-kekkan
capillary tube *[eng]*	毛管	mōkan
capital (funds) *[econ]*	資本	shihon
" (funds) *[econ]*	資金	shikin
capital (city) *[geogr]*	首府	shufu
capitol (Japanese national) *[arch]*	国会議事堂	kokkai-giji'dō
" (state building) *[arch]*	州会議事堂	shūkai-giji'dō
Capricorn; the Goat *[constel]*	山羊座	yagi-za
cap screw *[eng]*	押えねじ	osae-neji
capsicum *[bot]*	唐辛子; 唐芥子	tō'garashi
capsid *[microbio]*	頭殼; 頭殻	tō'kaku
capsomere *[bot]*	蒴; 朔; さく	saku
" *[bot]*	朔果	saku-ka
capstan *[nav-arch]*	車地	shachi
capsule *[bot]*	蒴(果)	saku(-ka)
" *[microbio]*	莢膜	kyōmaku
" *[pharm]*	カプセル	kapuseru
caption *[pr]*	表題	hyōdai
car; automobile *[mech-eng]*	車	kuruma
caramel *[cook]*	焼砂糖	yaki-zatō
carapace *[i-zoo]*	甲皮	kōhi
" *[v-zoo]*	甲羅	kōra
" *[v-zoo]*	背甲	sekō
carat *[lap]*	カラット	karatto
carbide *[inorg-ch]*	炭化物	tanka'butsu
carbide tool *[eng]*	超硬工具	chōkō-kōgu
carbohydrate *[bioch]*	糖質	tō-shitsu
" *[bioch]*	炭水化物	tansui'ka'butsu
carbohydrate metabolism *[bioch]*	糖質代謝	tō'shitsu-taisha
carbolic acid	石炭酸	sekitan-san
carbomonocyclic compound *[ch]*	炭素単環化合物	tanso-tankan-kagō'butsu
carbon (element)	炭素; たんそ	tanso
carbonaceous *{adj}* *[sci-t]*	炭素質の	tanso'shitsu no
carbonate *[ch]*	炭酸塩	tansan-en

carbonated water *[cook]*	ソーダ水	sōda-sui
carbon balance *[geoch]*	炭素収支；炭素收支	tanso-shūshi
carbon brush *[elec]*	炭素刷子	tanso-sasshi
carbon-carbon composite *[mat]*	炭素炭素複合物	tanso-tanso-fukugō'butsu
carbon dating *[nucleo]*	炭素年代測定	tanso-nendai-sokutei
carbon dioxide	二酸化炭素	ni'sanka-tanso
carbon dioxide assimilation *[bio]*	炭酸同化	tansan-dōka
carbon disulfide	二硫化炭素	ni'ryūka-tanso
carbon family element *[ch]*	炭素族元素	tanso'zoku-genso
carbon fiber *[mat] [tex]*	炭素繊維；炭素纖維	tanso-sen'i
carbon fluoride	弗化炭素	fukka-tanso
carbonic acid	炭酸；たんさん	tansan
carbonitriding *{n} [met]*	浸炭浸窒	shintan-shin'chitsu
carbonitriding furnace *[met]*	浸炭窒化炉	shintan-chikka-ro
carbonization *[ch]*	乾留；乾溜	kanryū
" *[ch] [geoch]*	炭化	tanka
carbonization of wood *[ch]*	木材乾留	mokuzai-kanryū
carbon monochloride	一塩化炭素	ichi'enka-tanso
carbon star *[astron]*	炭素星	tanso-sei
carbon steel *[steel]*	炭素鋼	tanso-kō
carbon sulfide	硫化炭素	ryūka-tanso
carbon tetrabromide	四臭化炭素	shi'shūka-tanso
carbon tetrachloride	四塩化炭素	shi'enka-tanso
carbon tetrafluoride	四弗化炭素	shi'fukka-tanso
carbon tetraiodide	四沃化炭素	shi'yōka-tanso
carbopolycyclic compound *[ch]*	炭素多環化合物	tanso-takan-kagō'butsu
carburetor *[ch-eng]*	気化器；氣化器	kika-ki
carburization *[met]*	浸炭	shintan
carburize *{vb t} [met]*	加炭する	katan suru
carcinogen *{n} [med]*	発癌物質	hatsu'gan-busshitsu
carcinogenic *{adj} [med]*	発癌性の	hatsu'gan'sei no
cardamon; cardamum (herb) *[bot]*	小豆蔻；しょうずく	shōzuku
cardboard *[mat] [paper]*	ボール紙	bōru-gami
" *[mat] [paper]*	板紙；いたがみ	ita-gami
cardiac *{n} [pharm]*	強心薬	kyōshin-yaku
cardiac muscle *[hist]*	心筋	shin-kin
cardinal (bird) *[v-zoo]*	猩猩紅冠鳥	shōjō-kōkan-chō
cardinal number *[arith]*	基数	kisū
cardinal points *[geod]*	基点	kiten
cardioid *{n} [geom]*	心臓形	shinzō'kei
cardiovascular *{adj} [anat]*	心(臓)血管の	shin(zō)-kekkan no
caret *[comput] [pr]*	脱字記号	datsu'ji-kigō

cargo *[aerosp] [nav-arch]*	荷役；にやく	ni'yaku
" (pl cargoes, cargos) *[trans]*	積荷	tsumi'ni
cargo vessel *[nav-arch]*	貨物船	kamotsu'sen
carious tooth *[dent]*	虫歯；齲歯；むしば	mushi'ba
carnallite *[miner]*	光鹵石	kōro'seki
carnivorous animal *[zoo]*	肉食動物	niku'shoku-dōbutsu
carnivorous plant *[bot]*	食虫植物	shoku'chū-shokubutsu
carotid artery *[anat]*	頸動脈	kei-dōmyaku
carp (fish) *{n} [v-zoo]*	鯉；こい	koi
carpenter (a person) *[eng]*	大工	daiku
carpenter's square *[eng]*	差し金	sashi-gane
carpet moth (insect) *[i-zoo]*	毛氈蛾	mōsen-ga
carpology *[bot]*	果実学；果實学	kajitsu'gaku
carriage *[comput]*	紙送り機構	kami'okuri-kikō
carriage return *[comput]*	改行復帰	kaigyō-fukki
carrier *[ch] [phys] [sol-st]*	担体；擔體	tantai
" *[eng]*	支持体	shiji'tai
carrier current *[comm]*	搬送電流	hansō-denryū
carrier part *[comm]*	搬送部	hansō-bu
carrier wave *[comm]*	搬送波	hansō-ha
carrion *{n} [zoo]*	腐肉食性動物	funiku'shoku'sei-dōbutsu
carrot (vegetable) *[bot]*	人参；にんじん	ninjin
carry *{n} [arith]*	桁上げ	keta'age
carry; transport *{vb t}*	運ぶ	hakobu
carry on back; shoulder; bear *{vb}*	背負う	se'ou
carrying; transporting *[trans]*	運搬	unpan
" *[trans]*	積載	sekisai
carry(ing) down *[arith]*	繰下げ	kuri'sage
carrying out; performing	施行；せこう	sekō
carry(ing) up *[arith]*	繰上げ	kuri'age
car telephone *[comm]*	自動車電話	jidōsha-denwa
cartesian coordinates *[trig]*	斜交座標	shakō-zahyō
cartilage *[hist]*	軟骨	nan'kotsu
cartilaginous fish *[v-zoo]*	軟骨魚類	nankotsu'gyo-rui
cartography *[graph]*	地図学；地圖學	chizu'gaku
" *[graph]*	製図	seizu
cartridge *[ord]*	薬包；藥包	yaku'hō
cartridge brass *[met]*	藥莢黄銅	yak(u)kyō-ō'dō
car type *[trans]*	車種	shashu
cascade amplifier *[electr]*	多段増幅器	tadan-zōfuku-ki
cascara *{n} [pharm]*	樹皮	juhi
" *{n} [pharm]*	核皮	kaku'hi

case (box for an instrument) *[eng]*	筐体；きょうたい	kyōtai
" ; box; etui *{n} [furn]*	手箱；てばこ	te'bako
" *{n} [gram]*	格	kaku
" ; matter *{n} [legal]*	事件	jiken
" ; instance *{n}*	場合；ばあい	ba'ai
" ; example *{n}*	事例	jirei
case hardening *{n} [met]*	肌焼き(処理)	hada-yaki(-shori)
case-making clothes moth *[i-zoo]*	衣蛾；いが	i-ga
casement window *[arch]*	開き窓	hiraki-mado
cash; money *{n} [econ]*	現金	genkin
" *{n} [econ]*	銭；ぜに	zeni
casserole (steam) baking *[cook] [food-eng]*	蒸し焼き	mushi-yaki
cassiterite *[miner]*	錫石；すずいし	suzu-ishi
Cassius' purple (pigment) *[mat]*	紫金	shikin
cassowary (bird) *[v-zoo]*	火喰鳥；ひくいどり	hikui-dori
casting *{n} [eng]*	注型；ちゅうけい	chūkei
" ; slip casting *{n} [cer] [eng]*	铸込(成形)	i'komi(-seikei)
casting oil *[mat]*	铸型油；鑄型油	i'gata-yu
casting resin *[poly-ch]*	注型用樹脂	chūkei'yō-jushi
" *[poly-ch]*	铸造樹脂	chūzō-jushi
casting sand *[eng]*	铸砂；いずな	i-zuna
cast(ing) surface *[eng]*	铸肌；いはだ	i-hada
cast iron *[met]*	铸鉄；ちゅうてつ	chū-tetsu
castor bean *[bot]*	唐胡麻の実	tō'goma no mi
castor oil *[mat]*	蓖麻子油	himashi-yu
castor-oil plant *[bot]*	唐胡麻；とうごま	tō-goma
castor-oil plant *[bot]*	蓖麻；篦麻；ひま	hima
cat (animal) *[v-zoo]*	猫；ねこ	neko
catabolism *[bioch]*	異化(作用)	ika(-sayō)
cataclysm *[geol]*	激変；げきへん	geki'hen
cataclysmic variable star *[astron]*	激変型変光星	gekihen'gata-henkō-sei
catalog *{n} [pr]*	目録；もくろく	moku'roku
cataloging *{n} [comput]*	目録作業	moku'roku-sagyō
catalpa, Japanese *[bot]*	木豇豆；きささげ	kisasage
catalysant *[ch]*	被接触体	hi-sesshoku'tai
catalysate *[ch]*	接触(反応)生成物	sesshoku(hannō)-seisei'butsu
catalyst *[ch]*	触媒；しょくばい	shoku'bai
catalyst activator *[ch]*	触媒活性化剤	shokubai-kassei'ka-zai
catalytic poison; anticatalyst *[ch]*	触媒毒	shokubai-doku
catalytic polymerization *[poly-ch]*	接触重合	sesshoku-jūgō
catalytic reaction *[ch]*	接触反応	sesshoku-hannō
" *[ch]*	触媒反応；觸媒反應	shokubai-hannō

catamaran ship *[nav-arch]* 双胴船 sōdō'sen
cataplasm *[pharm]* 巴布剤; ぱっぷざい pappu-zai
cataract *[hyd]* 瀑布 bakufu
" *[hyd]* 大滝; 大瀧 ō-taki
catastrophic *{adj}* 破局的な ha'kyoku'teki na
catch; capture; snare *{vb t}* 捕える; 捉える toraeru
catch onto; be caught on *[vt]* 引っ掛かる hikkakaru
catechu *[bot] [pharm]* 阿仙薬 asen-yaku
category *[ling]* 範疇; はんちゅう hanchū
" *[math]* 項類 kōrui
catenary *{pl -naries} [an-geom]* 懸垂線 kensui-sen
caterpillar (green) *[i-zoo]* 芋虫 imo-mushi
" (black) *[i-zoo]* 毛虫 ke-mushi
catfish; Parasilurus asotus *[v-zoo]* 鯰; なまず namazu
cathedral *[arch]* 大聖堂 dai-seidō
cathode *[elec]* 陰極; いんきょく in-kyoku
cathode current *[elec]* 陰極電流 inkyoku-denryū
cathode-ray tube; CRT *[electr]* 陰極線管 inkyoku'sen-kan
cathode rays *[electr]* 陰極線 inkyoku-sen
catholyte *[ch]* 陰極液 inkyoku-eki
cation *[ch]* 陽イオン yō-ion
cation-exchange resin *[org-ch]* 陽イオン交換樹脂 yō'ion-kōkan-jushi
cationic surfactant *[mat]* 陽イオン界面活性剤 yō'ion-kaimen-kassei-zai
catkin *[bot]* 尾状花序 bijō-kajo
catnip; Nepata cataria *[bot]* 犬薄荷 inu-hakka
catoptric system *[opt]* 反射光学系 hansha-kōgaku-kei
cat's eye *[miner]* 猫眼石 neko(no)me-ishi
cattail *[bot]* 蒲; がま gama
cattle (animal, collective) *[v-zoo]* 牛; うし ushi
caudal fin *[v-zoo]* 尾鰭 o-bire; o-hire
cauldron *[cook]* 大釜 ō-gama
cauliflower (vegetable) *[bot]* 花野菜; はなやさい hana-yasai
caulk *{n} [eng]* 加締め; かしめ kashime
causality (causation), law of 因果律 inga-ritsu
cause *{n} [math]* 原因 gen'in
cause to conform *{vb}* 適合させる tekigō saseru
caustic *{adj} [ch]* 苛性の kasei no
" *{adj}* 腐食性の fushoku'sei no
causticity *[ch]* 苛性度 kasei-do
caustic line *[mat] [opt]* 焦線 shōsen
caustic soda; sodium hydroxide *[ch]* 苛性ソーダ kasei-sōda
caution *{n} [psy]* 注意 chū'i

caution; precaution; care *{n} [psy]* 用心 yōjin
Caution Necessary 要注意 yō-chūi
cave *[geol]* 洞窟; どうくつ dōkutsu
cavern *[geol]* 大洞窟 ō-dōkutsu
cavitation *[fl-mech]* 空洞(現象) kūdō(-genshō)
cavity (pl -ties) *[bio]* 空洞 kūdō
" (in casting) *[met]* 巣; 巣; す su
cavity resonator *[elecmg]* 空胴共振器 kūdō-kyōshin-ki
cayenne pepper (seasoning) *[bot]* 唐辛; とうがらし tō'garashi
cease; halt; come to a stop *{vb}* 途絶える todaeru
cedar (tree) *[bot]* 杉 sugi
cede *{vb t} [pat]* 譲渡する; 讓渡する jōto suru
ceiling *[arch]* 天井; てんじょう tenjō
ceiling fan *[eng]* 天井扇風機 tenjō-senpū-ki
ceiling height *[arch]* 天井高 tenjō-daka
celadon *[cer]* 青磁 seiji
celery cabbage (vegetable) *[bot]* 白菜 hakusai
celestial almanac *[astron] [pr]* 天体暦; 天體暦 tentai-reki
celestial body *[astron]* 天体 ten-tai
celestial equator *[astron]* 天球赤道 tenkyū-seki'dō
" *[astron]* 天の赤道 ten no seki'dō
celestial horizon *[astron]* 天文水平 tenmon-suihei
celestial map *[astron]* 星図; せいず seizu
celestial meridian *[astron]* 天の子午線 ten no shigo-sen
celestial sphere *[astron]* 天球 tenkyū
celestite *[miner]* 天青石 tensei'seki
cell *[bio]* 細胞; 細胞 saibō; saihō
" *[elec]* 電池 denchi
" *[meteor]* 空気塊 kūki-kai
cell bank *[cyt]* 細胞銀行 saibō-ginkō
cell differentiation *[cyt]* 細胞分化 saibō-bunka
cell division *[cyt]* 細胞分裂 saibō-bun'retsu
cell-less *[bio]* 無細胞 mu'saibō
cell membrane *[cyt]* 細胞膜 saibō-maku
cellophane noodles *[cook]* 春雨; はるさめ haru'same
cell sorter *[eng]* 細胞選別器 saibō-senbetsu-ki
cell specialization *[cyt]* 細胞分化 saibō-bunka
cellular respiration *[bio]* 細胞呼吸 saibō-kokyū
cellulose *[bioch]* 繊維素; せんいそ sen'i'so
" *[bioch]* セルロース serurōsu
cellulose nitrate *[org-ch]* 硝化綿 shōka-men
cell wall *[cyt]* 細胞壁; 細胞壁 saibō-heki

Celsius temperature scale *[thermo]*	摂氏温度目盛	sesshi-ondo-me'mori
cement *{n} [mat]*	接着剤	setchaku-zai
cementation *[met]*	浸炭	shintan
cemented carbide *[inorg-ch]*	超硬合金	chōkō-gōkin
cemented carbide tool *[eng]*	超硬工具	chōkō-kōgu
cementing *{n} [mech] [poly-ch]*	接着; せっちゃく	setchaku
cementum (pl -ta) *[dent]*	白亜質; 白亞質	haku'a'shitsu
censorship *{n}*	検閲; 檢閱	ken'etsu
center *{n} [geom]*	中央	chū'ō
" *{n} [geom]*	中心	chū'shin
centerfold; center spread *[pr]*	中央見開き頁	chūō-mi'hiraki-pēji
center of buoyancy *[fl-mech]*	浮心	fu'shin
center of gravity *[mech]*	重心	jū'shin
center of inclination *[math]*	傾心	kei-shin
centigram (unit of mass) *[mech]*	甅; センチグラム	senchi'guramu
centimeter (unit of length) *[mech]*	糎; センチメートル	senchi'mētoru
centipede (insect) *[i-zoo]*	百足; むかで	mukade
central *{adj}*	中央の	chū'ō no
" *{adj}*	中心の	chūshin no
central angle *[geom]*	中心角	chūshin-kaku
centralized control *[comput]*	集中制御	shūchū-seigyo
central limit theorem *[stat]*	中心極限定理	chūshin-kyoku'gen-teiri
central nervous system *[anat]*	中枢神経系	chūsū-shinkei-kei
central office *[comm]*	中央局	chū'ō-kyoku
central processing unit; CPU *[comput]*	中央処理装置	chū'ō-shori-sōchi
centrifugal *{adj} [mech]*	遠心性の	enshin'sei no
centrifugal casting *{n} [poly-ch]*	遠心注型	enshin-chūkei
centrifugal dehydrator *[ch-eng]*	遠心脱水機	enshin-dassui-ki
centrifugal force *[mech]*	遠心力	enshin-ryoku
centrifugation *[ch-eng]*	遠心分離	enshin-bunri
centrifuge *{n} [eng]*	遠心機	enshin-ki
" *{n} [eng]*	遠心沪過機	enshin-roka-ki
centripetal *{adj} [mech]*	求心性の	kyūshin'sei no
centripetal acceleration *[mech]*	向心加速度	kōshin-ka'sokudo
centripetal force *[mech]*	向心力	kōshin-ryoku
" *[mech]*	求心力	kyūshin-ryoku
centriole *[cyt]*	中心小体	chūshin-shōtai
centroid *[calc]*	中心軌跡	chūshin-kiseki
" *[geom]*	図心; ずしん	zu-shin
centrosphere *[geol]*	中心圏; 中心圈	chūshin-ken
century (one hundred years)	百年間	hyaku'nen-kan
century	世紀	seiki

English	Japanese	Romanization
cephalization *[zoo]*	頭化	tōka
cephalopod *{adj} [i-zoo]*	頭足綱の	tōsoku-kō no
ceramic capacitor *[elec]*	磁気コンデンサ	jiki-kondensa
ceramic printing *{n} [cer] [pr]*	陶器印刷	tōki-in'satsu
ceramics *[cer]*	陶器; とうき	tōki
" *[cer]*	焼き物	yaki'mono
" (the industry) *[eng]*	窯業; ようぎょう	yō'gyō
cerargyrite *[miner]*	角銀鉱	kaku'ginkō
cereal; grain *[bot]*	穀物; こくもつ	koku'motsu
cerebellum (pl -s, -bella) *[anat]*	小脳; 小腦	shōnō
cerebral cortex *[anat]*	大脳皮質	dainō-hi'shitsu
cerebrum (pl -s, -bra) *[anat]*	大脳; だいのう	dainō
cerium (element)	セリウム	seryūmu
cermet *[cer]*	サーメット	sāmetto
cerography *[graph]*	蠟彫刻(術)	rō-chōkoku('jutsu)
ceropith; Ceropithecus (animal: monkey) *[v-zoo]*	尾長猿; おながざる	o'naga-zaru
certain; definite; reliable *{n} {adj}*	確かな	tashika na
certificate *[legal]*	証明書	shōmei'sho
certification *[legal]*	証明; 證明	shōmei
" ; authentication *{n}*	認証; にんしょう	ninshō
certified public accountant *[econ]*	公認会計士	kōnin-kaikei'shi
certify *{vb} [legal]*	認証する	ninshō suru
" *{vb} [legal]*	証明する	shōmei suru
cerusite *[miner]*	白鉛鉱	haku'enkō
cesium (element)	セシウム	seshūmu
cesspool *[civ-eng]*	汚水溜	o'sui-dame
Cestoda *[i-zoo]*	条虫類; 條虫類	jōchū-rui
Cetacea; whales *[v-zoo]*	鯨類; くじらるい	kujira-rui
chabazite *[miner]*	菱弗石	ryō'fusseki
Chaetognatha *[i-zoo]*	毛顎動物	mō'gaku-dōbutsu
chaff *{n} [mat]*	籾殻; もみがら	momi-gara
chain *[ch]*	鎖; さ	sa
" *[comput]*	連鎖	rensa
" *[mat]*	鎖; くさり	kusari
chain carrier *[poly-ch]*	連鎖担体; 連鎖擔體	rensa-tantai
chain compound *[ch]*	鎖式化合物	sa'shiki-kagō'butsu
chain explosion *[ch]*	連鎖爆発	rensa-baku'hatsu
chaining *{n} [comput]*	連鎖; れんさ	rensa
chain length *[poly-ch]*	鎖長	sa-chō
chain line *[graph]*	鎖線	sa-sen
chain molecule *[p-ch]*	鎖状分子	sajō-bunshi

chain polymer *[poly-ch]*	鎖状重合体	sajō-jūgō'tai
chain polymerization *[poly-ch]*	連鎖重合	rensa-jūgō
chain propagation reaction *[poly-ch]*	連鎖成長反応	rensa-seichō-hannō
chain reaction *[nucleo] [poly-ch]*	連鎖反応	rensa-hannō
chain reflex *[psy]*	連鎖反射	rensa-hansha
chain rule *[calc]*	連鎖律; れんさりつ	rensa-ritsu
chain stitch *[tex]*	鎖編み	kusari-ami
chain stopper *[poly-ch]*	連鎖停止剤	rensa-teishi-zai
chain termination reaction *[poly-ch]*	連鎖停止反応	rensa-teishi-hannō
chain transfer agent *[poly-ch]*	連鎖移動剤	rensa-idō-zai
chair *{n} [furn]*	椅子; いす	isu
chalcanthite *[miner]*	胆礬; 膽礬	tanban
chalcedony *[miner]*	玉髄; 玉髓	gyoku'zui
chalcocite *[miner]*	輝銅鉱; きどうこう	ki'dōkō
chalcography *[graph]*	銅版彫刻	dōban-chōkoku
chalcophile element *[geol]*	親銅元素	shindō-genso
chalcopyrite *[miner]*	黄銅鉱; 黃銅鑛	ō'dō'kō
chalk *{n} [mat]*	胡粉; ごふん	gofun
" *{n} [mat] [petr]*	白亜; 白亞	haku'a
chalking *{n} [rub]*	白亜化	haku'a'ka
chamber (of heart); ventricle *[anat]*	心室	shin-shitsu
chamotte *[cer]*	焼粉; 燒粉	yaki'ko
champignon (mushroom) *[bot]*	湿地; 占地; しめじ	shimeji
chance; hapstance *{n} [math]*	偶然	gūzen
change *{n} [comput] [math]*	変換	henkan
" (cash, money) *{n} [econ]*	釣(銭)	tsuri(-sen)
" (small change) *{n} [econ]*	銭; ぜに	zeni
" ; variation *{n} [math]*	変化; 變化; へんか	henka
" ; alteration *{n} [math]*]	変更	henkō
change; alter *{vb t}*	変える	kaeru
change; substitute, convert *{vb t}*	替える; 換える; 代える	kaeru
channel (ocean) *{n} [civ-eng] [navig]*	海峡; 海峽	kaikyō
" *{n} [civ-eng]*	溝; みぞ	mizo
" *{n} [comm] [elec]*	通話路	tsū'wa-ro
chaos *{n}*	混乱状態	konran-jōtai
chapter *[pr]*	章	shō
char *{n} [mat]*	炭; たん	tan
character *[comput] [pr]*	文字; もじ	moji
" *[gram]*	字	ji
" ; nature *{n} [psy] [sci-t]*	性質	sei'shitsu
character check *[comput]*	文字検査	moji-kensa

character code *[comput]*	文字符号；文字符號	moji-fugō
character expression *[comput]*	文字式	moji-shiki
character generator *[comput]*	文字発生器	moji-hassei-ki
characteristic *{n} [alg]*	指標	shihyō
" *{n} [sci-t]*	特徴	tokuchō
" *{n} [sci-t]*	特性；とくせい	toku'sei
" *{n} [sci-t]*	特質	toku'shitsu
characteristic curve *[math]*	特性曲線	tokusei-kyoku'sen
characteristic function *[math]*	特性関数	tokusei-kansū
characteristic property *[p-ch]*	固有性	koyū-sei
characteristic temperature *[sol-st]*	特性温度	tokusei-ondo
characteristic value *[math]*	固有値	koyū-chi
" *[math]*	特性値	tokusei-chi
character recognition *[comput]*	文字認識	moji-nin'shiki
character set *[comput]*	文字の組	moji no kumi
character space *[comput]*	文字間隔	moji-kankaku
character string *[comput]*	文字列	moji-retsu
charcoal *[mat]*	木炭；もくたん	mokutan
" *[mat]*	炭；すみ	sumi
charge; rate; fare *{n} [econ]*	料金	ryōkin
charge *{n} [elec]*	電荷	denka
" ; burden; onus *{n} [legal]*	負担；負擔；ふたん	futan
" (charging material) *{n} [met]*	装入物	sōnyū'butsu
charged body *[ch] [elec]*	帯電体；帶電體	taiden-tai
charged particle *[part-phys]*	荷電粒子	kaden-ryūshi
charge transfer *[p-ch]*	電荷移動	denka-idō
charge-transfer complex *[ch]*	電荷移動錯体	denka-idō-sakutai
charge-transfer polymerization *[poly-ch]*	電荷移動重合	denka-idō-jūgō
charging; charge *{n} [elec]*	充電	jūden
charging *{n} [eng] [met]*	装入；裝入	sōnyū
charr (fish) *[v-zoo]*	岩魚；いわな	iwana
chart *{n} [graph]*	図；圖；ず（づ）	zu
" *{n} [graph]*	図表	zuhyō
" *{n} [sci-t]*	一覧表	ichi'ran-hyō
chart paper *[graph] [paper]*	図紙	zushi
" *[ocean] [paper]*	海図用紙	kaizu-yōshi
chassis (pl chassis) *[mech-eng]*	車台；車臺	shadai
chauffeur *{n} [trans]*	運転手；運轉手	unten'shu
chaulmoolgra oil *[mat]*	大風子油	daifūshi-yu
cheap; inexpensive *{adj} [econ]*	安い	yasui
cheap goods *[ind-eng]*	安物	yasu-mono

check {n} [econ] 小切手 ko'gitte
checkerboard pattern [graph] 市松模様；市松模樣 ichi'matsu-moyō
checking {n} [comput] 検査；檢查 kensa
cheek {n} [anat] 頬；ほお（ほほ） hō
chelating agent [org-ch] キレート試薬 kirēto-shiyaku
Chelonia [v-zoo] 亀類；龜類 kame-rui
chemical; chemicals [ch] (化学)薬品 (kagaku-)yaku'hin
chemical affinity [ch] 化学親和力 kagaku-shinwa'ryoku
chemical agent [ord] 化学剤 kagaku-zai
chemical analysis [ch] 化学分析 kagaku-bunseki
chemical bond [ch] 化学結合 kagaku-ketsu'gō
chemical change [ch] 化学変化；化学變化 kagaku-henka
chemical constant [ch] 化学定数 kagaku-teisū
chemical engineering [eng] 化学工学 kagaku-kōgaku
chemical equation [ch] 化学方程式 kagaku-hōtei'shiki
chemical equivalent [ch] 化学当量；化学當量 kagaku-tōryō
chemical fiber [tex] 化学繊維；化学纖維 kagaku-sen'i
" (abbrev) [tex] 化繊；かせん kasen
chemical flavorings [cook] 化学調味料 kagaku-chōmi'ryō
chemical formula [ch] 化学式 kagaku-shiki
chemical injection pump [eng] 薬注ポンプ yakuchū-pompu
chemical injury [agr] 薬害；藥害 yaku'gai
chemically-synthesized product [mat] 化成品；かせいひん kasei'hin
chemical modification [ch] 化学修飾 kagaku-shūshoku
chemical oxygen demand (COD) value [org-ch] 化学的酸素要求量 kagaku'teki-sanso-yōkyū'ryō
chemical polishing {n} [met] 化学研摩 kagaku-kenma
chemical pollution [ecol] 薬品公害 yakuhin-kōgai
chemical product [mat] 化成品 kasei'hin
chemical property [ch] 化学的性質 kagaku'teki-seishitsu
chemical reaction [ch] 化学反応；化学反應 kagaku-hannō
chemical relaxation [ch] 化学緩和 kagaku-kanwa
chemical seasonings [cook] 化学調味料 kagaku-chōmi'ryō
chemical solution [ch] 薬液 yaku-eki
chemicals resistance [mat] 耐薬品性 tai'yakuhin-sei
chemical synthesis [ch] 化成 kasei
chemical vapor deposition [met] 化学蒸着 kagaku-jō'chaku
" [met] 化学気相成長法 kagaku-kisō-seichō-hō
" [met] 化学気相析出法 kagaku-kisō-seki'shutsu-hō
chemical warfare [mil] 化学戦 kagaku-sen
chemical weapon [ord] 化学兵器；化學兵器 kagaku-heiki
chemical weeding [agr] 薬剤除草 yakuzai-josō

chemisorption *[p-ch]*	化学吸着	kagaku-kyūchaku
chemist *[ch]*	化学者	kagaku'sha
chemistry *[sci-t]*	化学; 化學; かがく	kagaku
chemoautotroph *[microbio]*	化学独立栄養生物	kagaku-doku'ritsu-eiyō-seibutsu
chemolithotroph *[microbio]*	化学無機栄養生物	kagaku-muki'eiyō-seibutsu
chemoreceptor *[physio]*	化学受容器	kagaku-juyō-ki
chemosynthetic bacteria *[microbio]*	化学合成バクテリア	kagaku-gōsei-bakuteria
cherry (pl cherries) (fruit) *[bot]*	桜ん坊	sakuranbō
" (tree, wood, blossom) *[bot]*	桜; 櫻; さくら	sakura
cherry blossom *[bot]*	桜花	sakura-bana; ōka
" *[bot]*	桜の花	sakura no hana
cherry-stone clam *[i-zoo]*	蛤	hamaguri
chert *[geol]*	燧岩; すいがん	suigan
chest; breast; thorax *{n}* *[anat]*	胸	mune
chest; box; coffer *[furn]*	櫃; ひつ	hitsu
" (of drawers) *[furn]*	箪笥	tansu
chestnut (tree, nut) *[bot]*	栗	kuri
chew *{vb t}* *[physio]*	噛む; 咬む	kamu
chewing *{n}* *[physio]*	咀嚼	so'shaku
chick; fledgling (bird) *[v-zoo]*	雛	hina
chicken; hen; rooster *[v-zoo]*	鶏; 雞; にわとり	niwa'tori
chicken; hen; rooster *[v-zoo]*	鶏; 雞; とり	tori
chicken meat *[cook]*	鶏肉	tori-niku
chicken pox; varicella *[med]*	水疱瘡	mizu-bōsō
" ; varicella *[med]*	水痘	suitō
chickweed; Stellaria media *[bot]*	繁縷; はこべ	hakobe
chicory; Cichorium intybus (herb) *[bot]*	菊苦菜; きくにがな	kiku-niga'na
child (pl children) *[bio]*	子 (子供)	ko (kodomo)
" *[bio]*	小児; 小兒	shōni
childhood *[psy]*	幼時(期)	yōji(-ki)
children's park *[civ-eng]*	児童公園	jidō-kōen
child's room; children's room *[arch]*	子供部屋	kodomo-beya
chili pepper *[bot]*	唐辛子; 唐芥子	tō'garashi
chill *{n}* *[med]*	悪寒; おかん	o'kan
" (of paint) *{n}* *[poly-ch]*	曇り	kumori
chilly; cold *{adj}* *[physio]*	寒い	samui
chime *{n}* *[music]*	鉦; かね	kane
chimney *[arch]*	煙筒	entō
" *[arch]*	煙突	en'totsu
chin *[anat]*	顎さき	ago'saki
" *[anat]*	頤; おとがい	otogai

china *[cer] [mat]*	磁器	jiki
china clay *[mat]*	陶土	tōdo
chinaware *[mat]*	瀬戸物類; 瀬戸物類	seto'mono-rui
China wood oil *[mat]*	桐油	tōyu
Chinese bellflower *[bot]*	桔梗; ききょう	kikyō
Chinese black pine *[bot]*	槙; まき	maki
Chinese cabbage (vegetable) *[bot]*	白菜	hakusai
Chinese character *[pr]*	漢字; かんじ	kanji
Chinese characters for daily use	当用漢字	tōyō-kanji
Chinese character radicals *[pr]*	偏旁冠脚	hen'bō-kan'kyaku
Crown radical (examples next) *[pr]*	冠; かんむり	kanmuri
Bamboo (Example: cylinder)	竹冠: ⺮（筒）	take-kanmuri (Ex. tō)
Grass (Ex. bud)	草冠: ⺾, 艹（芽）	kusa-kanmuri (Ex. me)
Lid (Ex. redundancy)	ワ冠: 冖（冗）	wa-kanmuri (Ex. jō)
Pig's head (Ex. collection)	ヨ頭: ヨ, ⺕, ⺔（彙）	kei-gashira (Ex. i)
Roof (Ex. universe)	ウ冠: 宀（宇宙）	u-kanmuri (Ex. uchū)
Left-hand-side radical *[pr]*	扁; 偏; へん	hen
Left-hand-side radical *(suffix)*	-扁; -偏	-hen; -ben
Clothing (Example: sleeve)	衣扁: 衤（袖）	koromo-hen (Ex. sode)
Cow; bull (Ex. pasture)	牛扁: 牜（牧）	ushi-hen (Ex. boku)
Earth (Ex. place)	土扁: 土（場）	tsuchi-hen (Ex. ba)
Fire (Ex. flame)	火扁: 火（焔）	hi-hen (Ex. hono'o)
Fish (Ex. salmon)	魚扁: 魚（鮭）	uo-hen (Ex. sake)
Flesh (Ex. lung)	肉月: ⺼（肺）	niku-zuki (Ex. hai)
Grain (Ex. autumn)	ノ木扁: 禾（秋）	nogi-hen (Ex. aki)
Hand (Ex. push)	手扁: 扌（押）	te-hen (Ex. osu)
Ice (Ex. freeze, freezing)	二水: 冫（冷凍）	ni-sui (Ex. reitō)
Long-tailed bird (Ex. ostrich)	鳥扁: 鳥（鴕鳥）	tori-hen (Ex. dachō)
Man; person (Ex. body)	人扁: 亻（体）	nin-ben (Ex. karada)
Metal (Ex. iron)	金扁: 釒（鉄）	kane-hen (Ex. tetsu)
Mind, the (Ex. nature)	立心扁: 忄（性）	risshin-ben (Ex. sei)
Moon (Ex. misty)	月扁: 月（朦）	tsuki-hen (Ex. oboro)
Religion (Ex. celebration)	示扁: 礻（祝）	shimesu-hen (Ex. iwau)
Small community (Ex. shore)	阜扁: 阝（陸）	kozato-hen (Ex. riku)
Small walk (Ex. gain)	行人扁: 彳（得）	gyōnin-ben (Ex. toku)
Speech (Ex. notation)	言扁: 訁（記）	gon-ben (Ex. ki)
Sun (Ex. time)	日扁: 日（時）	hi-hen (Ex. toki)
Water (Ex. liquid)	三水: 氵（液）	sanzui (Ex. eki)
Woman (Ex. good)	女扁: 女（好）	onna-hen (Ex. yoshi)
Right-hand-side radical *[pr]*	旁; つくり	tsukuri
Bird (Example: wild duck)	鳥旁: 鳥（鴨）	tori-tsukuri (Ex. kamo)

Chinese character radicals (continued)		
Color (Ex. coloring)	彡旁（彩）	san-tsukuri (Ex. irodori)
Large community (Ex. portion)	大邑：阝（部）	ō-zato (Ex. bu)
Sword (Ex. divide)	立刀：刂（割）	rittō (Ex. waru)
Bottom radical *[pr]*	脚；きゃく	kyaku
Bird (Example: hawk)	鳥脚：鳥（鷹）	tori-kyaku (Ex. taka)
Fire (Ex. heat)	連火：灬（熱）	renga (Ex. netsu)
Bottom-surrounding radical (a type of kyaku, which see) *[pr]*	繞；にょう	nyō
To go (Example: extension)	廴繞：廴（延）	innyō (Ex. nobe)
To go (Ex. circumscribe)	之繞：辶（廻）	shinnyō (Ex. meguru)
To run (Ex. go over)	走繞：走（越）	sō-nyō (Ex. koeru)
Encasing radical *[pr]*	構え；かまえ	kamae; -gamae
Country (Ex. sphere of action)	国構：口（圏）	kuni-gamae (Ex. ken)
Gate (Ex. open and shut)	門構：門（開閉）	kado-gamae (Ex. kaihei)
Some other radicals:		
Roof: top-, side-covered (Ex. shop)	麻垂：广（店）	ma-dare (Ex. mise)
Short-tailed bird (Ex. sparrow)	旧鳥：隹（雀）	furu-tori (Ex. suzume)
Wild goose (Ex. field)	雁垂：厂（原）	gan-dare (Ex. hara)
Chinese compound word *[comm]*	熟語	juku'go
Chinese cork tree *[bot]*	黄蘗；きはだ	ki'hada
Chinese date tree *[bot]*	棗；なつめ	natsume
Chinese lantern plant *[bot]*	酸漿；鬼灯	hōzuki
Chinese lemon *[bot]*	柚；ゆず	yuzu
Chinese medicinal drug *[med]*	漢方薬	kanpō-yaku
Chinese milk vetch *[bot]*	蓮華草；れんげそう	renge'sō
Chinese noodles *[cook]*	老麺；拉麺	rāmen
Chinese watermelon *[bot]*	冬瓜	tōgan
Chinese yam *[bot]*	長芋	naga-imo
Chinese yellow *[miner]*	黄土	ō'do
chinook salmon (fish) *[v-zoo]*	鱒の介；ますのすけ	masu-no-suke
chinquapin (tree) *[bot]*	椎；しい	shī
chip *{n}* *[cer]*	欠け	kake
" *{n}* *[electr]*	半導体片	handōtai'hen
chip; dice; mince *{vb t}* *[cook]* *[eng]*	刻む；きざむ	kizamu
chips *[mat]*	切り屑	kiri'kuzu
chirality *[ch]*	対掌性	taishō-sei
chisel *{n}* *[eng]*	鑿；のみ	nomi
" *{n}* *[eng]*	鏨；たがね	tagane
chitin *[bioch]*	キチン	kichin
chive *[bot]*	蝦夷葱	ezo-negi
chlamydospore *[mycol]*	原膜胞子；厚膜胞子	genmaku-hōshi

chloric acid	塩素酸	enso-san
chloride *[ch]*	塩化物	enka'butsu
chloride ion *[ch]*	塩素イオン	enso-ion
chlorination *[ch]*	塩素化	enso'ka
" *[ch]*	塩素処理	enso-shori
chlorine (element)	塩素；鹽素；えんそ	enso
chlorine dioxide	二酸化塩素	ni'sanka-enso
chlorine oxide	酸化塩素	sanka-enso
chlorine trifluoride	三弗化塩素	san'fukka-enso
chlorine water *[ch]*	塩素水	enso-sui
chlorinity *[ocean]*	塩素量	enso-ryō
chlorite *[ch]*	亜塩素酸塩	a'enso'san-en
" *[miner]*	緑泥石	ryokudei'seki
chloroauric acid	塩化金酸	enka-kinsan
chlorofluorocarbon *[mat]*	クロロフルオロ=カーボン	kurorofuruorokābon
chlorophyll *[bioch]*	葉緑素	yōryoku'so
Chlorophyta *[bot]*	緑色植物	ryoku'shoku-shokubutsu
chloroplast *[bot]*	葉緑体；葉緑體	yōryoku'tai
chloroplatinate *[ch]*	塩化白金酸塩	enka'hakkin'san-en
chloroplatinic acid	塩化白金酸	enka'hakkin-san
chlorous acid	亜塩素酸	a'enso-san
choking *{n} [fl-mech]*	閉塞	hei'soku
cholagogue *{n} [pharm]*	利胆薬（利胆剤）	ritan-yaku (ritan-zai)
cholera bacillus *[microbio]*	コレラ菌	korera-kin
Chondrichthyes *[v-zoo]*	軟骨魚類	nankotsu'gyo-rui
choose; select *{vb t}*	選ぶ；撰ぶ；択ぶ	erabu
chopping board *[cook] [eng]*	俎；まないた	mana-ita
chopsticks *[cook] [eng]*	箸；はし	hashi
chord *[acous]*	和弦	wa'gen
" *[geom]*	弦	gen
" *[music]*	和音	wa'on
Chordata; chordates *[zoo]*	脊索動物	sekisaku-dōbutsu
chorographic map *[graph]*	地勢図；地勢圖	chisei-zu
chorography *[map]*	地方地誌	chihō-chishi
choroid *[anat]*	脈絡膜	myaku'raku-maku
chorus *[music]*	合唱団；合唱團	gasshō'dan
chowder *[cook]*	寄せ鍋	yose-nabe
chroma *[opt]*	彩度	saido
chromatic aberration *[opt]*	色収差；色收差	iro-shūsa
chromatic color *[opt]*	有彩色	yūsai-shoku
chromaticity *[opt]*	色度；しきど	shiki-do

chromaticity diagram *[opt]*	色度図	shiki'do-zu
chromatid *[cyt]*	染色分体	senshoku-buntai
chromatin *[bioch]*	染色質	senshoku'shitsu
chromatophore *[hist]*	色素胞; 色素胞	shikiso'hō
chrominance *[opt]*	色光度	shiki'kōdo
chromium (element)	クロム	kuromu
chromogenic development *[photo]*	発色現像	hasshoku-genzō
chromoisomerism *[opt]*	呈色異性	teishoku-i'sei
chromophore *[ch]*	発色団; 發色團	hasshoku-dan
Chromophyta *[bot]*	褐色植物	kasshoku-shokubutsu
chromosome *[cyt]*	染色体	senshoku'tai
chromosome number *[cyt]*	染色体数	senshoku'tai-sū
chromosphere *[astron]*	彩層; さいそう	saisō
chromotropy *[opt]*	呈色異性	tei'shoku-i'sei
chronic *{n}* *[med]*	慢性	mansei
" *{adj}* *[med]*	慢性の	mansei no
chronobiology *[bio]*	時間生物学	jikan-seibutsu'gaku
chronology (pl -gies) *[sci-t]*	年代学	nendai'gaku
chronometer *[horol]*	高精度時計	kō'seido-tokei
chrysalis *[i-zoo]*	蛹; さなぎ	sanagi
chrysanthemic acid	菊酸	kiku-san
chrysanthemum (flower) *[bot]*	菊; きく	kiku
Chrysanthemum coronarium (vegetable) *[bot]*	春菊; しゅんぎく	shun'giku
chrysanthemum structure *[mech-eng]*	菊目組織	kiku'me-soshiki
chrysoberyl *[miner]*	金緑石	kin'ryokuseki
Chrysophyta *[bot]*	黄金色植物	ōgon'shoku-shokubutsu
chrysotile (asbestos) *[miner]*	温石綿	on'sekimen
chub (fish) *[v-zoo]*	諸子; もろこ	moroko
chuck *{n}* *[eng]*	掴み	tsukami
church *[arch]*	教会堂	kyōkai'dō
chute *[eng]*	落とし樋	otoshi-hi
" *[hyd]*	射水路	shasui-ro
chymosin *[bioch]*	キモシン	kimoshin
cicada (insect) *[i-zoo]*	蜩; ひぐらし	higurashi
" (insect) *[i-zoo]*	蟬; せみ	semi
cilia (sing cilium) *[cyt]*	繊毛; 纖毛	senmō
Ciliata *[i-zoo]*	繊毛虫綱	senmō'chū-kō
cinchona bark *{n}* *[pharm]*	規那皮	kina-hi
cinder cone *[geol]*	噴石丘	funseki-kyū
cinders *[geol]*	噴石	funseki
cinerous vulture (bird) *[v-zoo]*	禿鷲; はげわし	hage-washi

cinnabar *[miner]*	辰砂	shinsha
" *[miner]*	朱	shu
cinnamate *[ch]*	桂皮酸塩	keihi'san-en
cinnamic acid	桂皮酸；けいひさん	keihi-san
cinnamon (spice) *[bot]*	肉桂	nikkei
cinnamon oil *[mat]*	桂皮油	keihi-yu
cipher; code *{n} [comm]*	暗号；暗號	angō
circa; (abbreviated ca.) *{prep}*	凡そ；およそ	oyoso
circadian rhythm *[bio]*	概日リズム	gai'nichi-rizumu
circle *{n} [geom]*	円；圓；えん	en
circle arc *[geom]*	円弧	enko
circle of position *[navig]*	位置の円	ichi no en
circuit *[elec]*	回路；かいろ	kairo
" *[elec]*	回線	kaisen
circuit breaker *[elec]*	（回路）遮断器	(kairo-)shadan-ki
circuit diagram *[elec]*	回路図；回路圖	kairo-zu
" *[elec]*	回線図	kaisen-zu
circuit load; line load *[elec]*	回線負荷	kaisen-fuka
circuit switching *{n} [elec]*	回線交換	kaisen-kōkan
circular arc *[geom]*	円弧線	enko-sen
circular cone *[geom]*	円錐；えんすい	ensui
circular cylinder *[math]*	円柱	enchū
circular functions *[trig]*	円関数；圓關數	en-kansū
circular motion *[mech]*	円運動	en-undō
circulation *[physio]*	循環流動	junkan-ryūdō
circulatory system *[anat]*	循環系	junkan-kei
circumcenter (of a triangle) *[geom]*	外心	gai'shin
circumcircle *[geom]*	外接円	gaisetsu-en
circumference *[geom]*	円周	enshū
circumferential strain *[mech]*	円周歪	enshū-hizumi
circumferential stress *[mech]*	円周応力	enshū-ō'ryoku
circumflex accent *[comput] [ling]*	抑揚音符	yoku'yō-onpu
circumpacific volcanic zone *[geol]*	環太平洋火山帯	kan'taiheiyō-kazan-tai
circumpolar star *[astron]*	周極星	shūkyoku-sei
circumscribed (-ing) *{adj} [geom]*	外接した	gaisetsu shita
circumscribed polygon *[geom]*	外接多角形	gaisetsu-takak(u)kei
circumscription *[math]*	外接	gai'setsu
circumstances; situation *{n}*	事情；じじょう	jijō
cirque *[geol]*	圏谷；圏谷	kenkoku
cirrocumulus *[meteor]*	巻積雲；絹積雲	kenseki'un
cirrostratus *[meteor]*	巻層雲；絹層雲	kensō'un
cirrus *[meteor]*	巻雲；巻雲；絹雲	ken'un

citation; quotation *[pr]*	引用	in'yō
cite *{vb} [pr]*	引用する	in'yō suru
" *{vb}*	援用する	en'yō suru
cited reference *[pr]*	引用文献	in'yō-bunken
citrate *[ch]*	枸櫞酸塩	kuen'san-en
citric acid	枸櫞酸；くえんさん	kuen-san
citron (fruit) *[bot]*	酸橘；すだち	su'dachi
" (fruit) *[bot]*	柚；柚子；ゆず	yuzu
citrus *[bot]*	柑橘	kankitsu
city; town; street *[geogr]*	町；街；まち	machi
city (pl cities) *[geogr]*	市	shi
city gas *[mat]*	都市ガス	toshi-gasu
city hall *[arch]*	市役所	shi-yakusho
city road *[civ-eng]*	市道	shidō
city streets *[civ-eng]*	市街	shigai
civet (animal) *[v-zoo]*	麝香猫	jakō-neko
civic center *[civ-eng]*	都心	to'shin
civil *{adj} [legal]*	民事の	min'ji no
civil engineering *[eng]*	土木工学	doboku-kōgaku
civil engineering and construction *[eng]*	土木建築；どぼく=けんちく	doboku-kenchiku
civil law *[legal]*	民法	minpō
civil time *[astron]*	常用時	jōyō-ji
civil twilight *[astron]*	常用薄明	jōyō-hakumei
civil year *[astron]*	常用年	jōyō-nen
cladding *{n} [eng]*	被覆加工	hifuku-kakō
Claim(s) *{n} [pat]*	特許請求の範囲	tokkyo-seikyū no han'i
clam: cherry-stone (shellfish)	蛤；はまぐり	hamaguri
clam: short necked (shellfish)	浅蜊；あさり	asari
clamp *{n} [eng]*	鎹；かすがい	kasugai
" *{n} [eng]*	締め具	shime'gu
" *{n} [eng]*	摑み	tsukami
clamping *{n} [mech-eng]*	型締め	kata'jime
clamping force *[mech]*	押付け力	oshi'tsuke-ryoku
clamping pressure *[mech-eng]*	型締圧力	kata'jime-atsu'ryoku
clapper (of a bell) *[music]*	鐘の舌	kane no shita
clarificant *[mat]*	清澄剤	seichō-zai
clarification; explanation *[comm]*	明確化	meikaku'ka
" ; explanation *[comm]*	説明；せつめい	setsu'mei
" (liquid) *[mech-eng]*	清澄化	seichō'ka
clarifier *[eng]*	浄化器；淨化器	jōka-ki
clarify (as soup) *{vb t}*	澄ます	sumasu

clarifying agent; clarifier *[mat]*	清澄剤	seichō-zai
clarifying filtration *[eng]*	清澄沪過；清澄濾過	seichō-roka
clasp *{n} [eng]*	留め金	tome'gane
class *{n} [bio] [syst]*	綱；こう	kō
" *{n} [gram] [math]*	集合	shūgō
" *{n} [stat]*	階級	kaikyū
" (grade) *{n} [syst]*	級	kyū
" *{n} [syst]*	類；類	rui
classical *{adj}*	古典の	koten no
classification *[comput]*	分類	bunrui
" *[syst]*	類別；類別	rui'betsu
classification by sex *[bio]*	性別	sei'betsu
classification index *[syst]*	分類目録	bunrui-moku'roku
classification system *[syst]*	分類体系	bunrui-taikei
classify *{vb t}*	分級する	bunkyū suru
classifying; classification *{n}*	分級	bunkyū
classroom *[arch]*	教室	kyō-shitsu
clause *[gram]*	節；せつ	setsu
" *[legal]*	條項	jōkō
claustrophobia *[psy]*	閉所恐怖症	heisho-kyōfu'shō
clavicle (bone) *{n} [anat]*	鎖骨；さこつ	sa-kotsu
claw *{n} [anat]*	鉤爪	kagi-zume
" *{n} [anat]*	爪	tsume
clay; china clay *[geol]*	白土	hakudo
clay *[geol]*	粘土	nendo
clay figure *[archeo]*	土偶	dogū
clay lump *[geol]*	粘土塊；ねんどかい	nendo-kai
clay plate *[ch]*	素焼き板	suyaki'ban; suyaki-ita
clean; tidy; refined *{adj}*	綺麗な；きれいな	kirei na
" ; purify *{vb t}*	清める	kiyomeru
" ; sweep; dust *{vb t}*	掃除する	sōji suru
clean coal *{n} [geol]*	精炭	seitan
clean copy; fair copy *[pat] [pr]*	清書	seisho
cleaning (scavengery) *{n} [civ-eng]*	清掃	seisō
" (of ore) *{n} [miner]*	精選；精選	seisen
" *{n}*	掃除；そうじ	sōji
cleanliness *[met]*	清浄度	seijō-do
clear *{n} [comput]*	消去	shōkyo
" (weather) *{n} [meteor]*	快晴	kaisei
" *{vb t } [comput]*	破算する	hasan suru
clearance *{n} [eng]*	隙間	sukima
" *{n} [mech-eng]*	遊隙	yūgeki

clearance gage *[eng]*	隙間ゲージ	sukima-gēji
clear-cut; explicit *{adj}*	明確な	meikaku na
clear ice; glaze *{n} [hyd]*	雨氷；うひょう	u'hyō
clear soup *{n} [cook]*	吸物；すいもの	sui'mono
cleavage *[bio]*	分割	bun'katsu
" *[bio] [geol]*	開裂；かいれつ	kai'retsu
" *[crys]*	劈開；へきかい	heki'kai
" *[embryo]*	卵割；らんかつ	ran'katsu
cleavage reaction *[crys]*	開裂反応	kairetsu-hannō
click beetle (insect) *[i-zoo]*	米搗虫	kome'tsuki-mushi
client; requestor (a person) *{n}*	依頼人	irai'nin
" ; consignor (a person) *{n}*	委託者	itaku'sha
cliff; precipice; bluff *[geogr]*	崖；がけ	gake
" *[geol]*	絶壁；ぜっぺき	zeppeki
climate *[climat]*	気候；氣候；きこう	kikō
climatologist *[climat]*	気候学者	kikō-gakusha
climatology *[meteor]*	気候学	kikō'gaku
climb; ascend *{vb} [navig]*	登る	noboru
climbing plant *[bot]*	攀じ上り植物	yoji'nobori-shokubutsu
clinical thermometer *[med]*	体温計	tai'on-kei
clinograph *[geol]*	傾斜計	keisha-kei
clinometer *[geol]*	傾斜儀	keisha-gi
clinozoisite *[miner]*	(単)斜黝簾石	(tan)sha'yū'renseki
clip; cull; pluck *{vb t}*	剪む；つむ	tsumu
clitellum *[i-zoo]*	環帯；環帶	kantai
cloaca *[v-zoo]*	総排出腔	sō-hai'shutsu-kō
clock; timepiece *{n} [comput] [horol]*	時計；とけい	tokei
clockwise rotation *[mech]*	時計回り	tokei-mawari
clockwork; clock mechanism *[horol]*	時計仕掛	tokei-ji'kake
clogging; choking *{n} [fl-mech]*	閉塞；へいそく	heisoku
clogs (footwear) *[cl]*	下駄；げた	geta
cloisonne ware *[cer] [furn]*	七宝焼；七寶焼	shippō-yaki
close; nearby *{adj} {adv}*	近い	chikai
close; shut *{vb t}*	閉じる；とじる	tojiru
closed circuit *[elec]*	閉路；へいろ	heiro
closed circulatory system *[anat]*	閉鎖循環系	heisa-junkan-kei
closed curve *[geom]*	閉曲線	hei-kyoku'sen
closed interval *[an-geom]*	閉区間	hei-kukan
closed mold *[eng]*	密閉型	mippei-gata
closed system *[eng]*	閉鎖系；へいさけい	heisa-kei
closed universe *[astron]*	閉じた宇宙	tojita-uchū
close-packed; minute; fine *[mat]*	緻密；ちみつ	chi'mitsu

close-packedness *[mat] [mech]*	緻密性	chi'mitsu-sei
closest-packed lattice *[crys]*	最密格子	saimitsu-kōshi
closest-packed structure *[crys]*	最密構造	saimitsu-kōzō
closest packing *{n} [crys]*	最密充填	saimitsu-jūten
closet *{n} [arch]*	物置；ものおき	mono'oki
" *{n} [arch]*	納戸；なんど	nando
closing of mold *[mech-eng]*	型締め	kata-jime
closing valve *[mech-eng]*	開閉弁；開閉瓣	kaihei-ben
closure property *[alg]*	閉包性；閉包性	heihō-sei
cloth *[tex]*	生地；きじ	kiji
clothes; clothing *[cl]*	衣類	i'rui
clothes tray *[cl]*	乱れ箱；亂れ箱	midare-bako
clothing, Japanese *[cl]*	着物；きもの	ki'mono
" *[cl]*	和服	wa-fuku
clothing stand *[cl] [furn]*	衣桁；いこう	ikō
clotting of blood *[physio]*	凝血	gyō'ketsu
cloud *{n} [meteor]*	雲；くも	kumo
cloud amount *[meteor]*	雲量	un-ryō
cloud chamber *[nucleo]*	霧箱	kiri-bako
cloud cover *[meteor]*	雲量率	un'ryō-ritsu
cloud form *[meteor]*	雲形	un-kei
clouding *{n} [plas]*	白化	hakka
clouding preventive *[mat]*	防曇剤	bō'un-zai
cloud mirror *[meteor]*	雲鏡	un-kyō
cloud point *[ch-eng]*	曇点；どんてん	donten
" *[ch-eng]*	曇り点	kumori-ten
cloud top *[meteor]*	雲頂	un-chō
cloudy *[meteor]*	曇り	kumori
clove (spice) *[bot] [cook]*	丁子	chōji
cloverleaf (pl -leafs, -leaves) *{n} [traffic]*	立体交差十字路	rittai'kōsa-jūji'ro
club moss *[bot]*	日陰葛；日陰の蔓	hikage-no-kazura
cluster *[astron]*	星団；星團	seidan
Cnidaria *[i-zoo]*	刺胞動物；刺胞動物	shihō-dōbutsu
co- (deputy-) *{prefix}*	副-	fuku-
" (together-) *{prefix}*	共同の-	kyōdō no-
" (mutual-) *{prefix}*	相互の-	sōgo no-
coagulant *{n} [ch]*	凝結剤	gyōketsu-zai
coagulated blood *[med]*	凝血	gyō'ketsu
coagulating bath *[ch]*	凝固浴	gyōko-yoku
coagulation *[ch]*	凝結；ぎょうけつ	gyō'ketsu
" *[ch]*	凝固	gyōko

coal *[geol]*	石炭；せきたん	seki'tan
coal conveyor *[min-eng]*	運炭機	untan-ki
coalesce (join into one) *{vb} [phys]*	合体する	gattai suru
" (join with) *{vb} [phys]*	合着する	gō'chaku suru
coalfield *[min-eng]*	炭田	tanden
coal furnace *[eng]*	石炭炉；石炭爐	sekitan-ro
coal mine; coalpit *[min-eng]*	炭坑	tankō
coal property *[mat]*	炭質	tan-shitsu
Coalsack *[astron]*	石炭袋	sekitan-bukuro
coarse aggregate *[geol]*	粗骨材	so-kotsu'zai
coarseness *{n} [eng]*	荒さ	ara-sa
coarse sand *[petr]*	粗砂；そしゃ	so-sha
coast *{n} [geogr]*	沿岸	engan
" *{n} [geogr]*	海岸	kaigan
" *{vb i} [navig]*	惰力で進む	da'ryoku de susumu
coastal *{adj} [geogr]*	臨海の	rinkai no
coasting *{n} [navig]*	惰力運転	da'ryoku-unten
coasting flight *[navig]*	惰性飛行	dasei-hikō
coast vegetation *[bot]*	海岸植生	kaigan-shokusei
coat *{n} [cl]*	外套	gaitō
coated paper *[mat] [paper]*	塗工紙；とこうし	tokō-shi
" *[paper] [resin] [rub]*	塗装紙	tosō-shi
coati(-mundi) (animal) *[v-zoo]*	鼻熊；はなぐま	hana-guma
coating (of paint) *[mat] [poly-ch]*	塗装	tosō
coating; covering *{n}*	被覆	hi'fuku
coating film *[mat]*	塗膜	to-maku
coating material *[mat]*	塗料	toryō
coaxial *{adj} [elec] [electr]*	同軸の	dō'jiku no
coaxial line *[elecmg]*	同軸線	dōjiku-sen
cobalt (element)	コバルト	kobaruto
cobblerfish (fish) *[v-zoo]*	糸引鰺	ito'hiki-aji
cobwebbing (paint) *{n} [org-ch]*	糸引き	ito'hiki
cocatalyst *[ch]*	共触媒；共觸媒	kyō'shokubai
Cocculus trilobus (herb) *[bot]*	青葛藤	ao'tsuzura-fuji
coccus (pl -i) *[microbio]*	球菌	kyūkin
coccyx (pl coccyges, coccyxes) *[anat]*	尾骨	bi-kotsu
cochlea (pl cochleae, -s) *[anat]*	蝸牛	kagyū
cock; plug *{n} [ch]*	栓	sen
cockatoo (bird) *[v-zoo]*	鸚哥；いんこ	inko
cockle (shellfish) *[i-zoo]*	鳥貝	tori-gai
cocklebur(r) *[bot]*	苓；おなもみ	onamomi
cockpit *[aero-eng]*	操縦室；操縱室	sōjū-shitsu

cockroach (insect) *[i-zoo]*	蜚蠊；ごきぶり	gokiburi
coco grass *[bot]*	浜菅	hama-suge
cocondensation resin *[poly-ch]*	共縮合樹脂	kyō'shukugō-jushi
coconut (fruit) *[bot]*	椰子の実	yashi no mi
coconut oil *[mat]*	椰子油；やしゆ	yashi-yu
cocoon *[i-zoo]*	繭；まゆ	mayu
cod (fish) *[v-zoo]*	鱈；たら	tara
code *{n}* *[comput]*	略語	ryaku'go
" *{n}* *[comput]* *[math]*	符号	fugō
coded decimal notation *[comput]*	符号化十進法	fugō'ka-jusshin-hō
coder; encoder *[comm]* *[comput]*	符号器；符號器	fugō-ki
coding *{n}* *[comput]*	符号化	fugō'ka
cod-liver oil *[mat]*	鱈肝油；たらかんゆ	tara-kan'yu
codominance *[bio]*	相互優勢	sōgo-yūsei
codon *[gen]*	遺伝暗号；遺傳暗號	iden-angō
coefficient *[alg]*	係数；けいすう	keisū
" *[alg]*	率；りつ	ritsu
coefficient of coupling *[phys]*	結合係数	ketsu'gō-keisū
coefficient of cubical expansion *[thermo]*	体積膨張係数	taiseki-bōchō-keisū
coefficient of determination *[stat]*	決定係数	kettei-keisū
coefficient of earth pressure *[civ-eng]*	土圧係数	do'atsu-keisū
coefficient of kinematic viscosity *[fl-mech]*	動粘性係数	dō-nensei-keisū
coefficient of kinematic viscosity *[fl-mech]*	動粘性率	dō-nensei-ritsu
coefficient of linear expansion *[thermo]*	線膨張係数；(-率)	sen-bōchō-keisū; (-ritsu)
coefficient of linear expansion *[thermo]*	線熱膨張率	sen-netsu'bōcho-ritsu
coefficient of mass-transfer capacity *[phys]*	物質移動容量係数	busshitsu-idō-yōryō-keisū
coefficient of static friction *[mech]*	静摩擦係数	sei-masatsu-keisū
coefficient of thermal conductivity *[thermo]*	熱伝導率	netsu-dendō-ritsu
coefficient of thermal expansion *[thermo]*	熱膨張係数；(-率)	netsu-bōchō-keisū; (-ritsu)
coefficient of variation *[stat]*	変動係数	hendō-keisū
coefficient of viscosity *[fl-mech]*	粘性係数；(-率)	nensei-keisū; (-ritsu)
coefficient of volume compressibility *[mech]*	体積圧縮係数	taiseki-asshuku-keisū

coefficient of volume expansion *[thermo]*	体膨張係数；(-率)	tai-bōchō-keisū; (-ritsu)
Coelenterata (jellyfish, corals) *[i-zoo]*	腔腸動物	kōchō-dōbutsu
coentro *[bot]*	胡荽；こえんどろ	koendoro
coenzyme *[bioch]*	補酵素；ほこうそ	ho'kōso
" *[bioch]*	助酵素	jo'kōso
coercive force *[elecmg]*	保磁力；ほじりょく	hoji-ryoku
cofactor *[bioch]*	補因子	ho'inshi
cofactor; minor *[math]*	余因数；餘因數	yo'insū
coffee *[bot] [cook]*	珈琲；コーヒー	kōhī
coffer; chest (wooden) *[furn]*	櫃；ひつ	hitsu
cofunction *[trig]*	余関数；よかんすう	yo'kansū
cog *{n} [eng]*	嵌め歯；はめば	hame'ba
cognate *{n} [ling]*	同族言語	dōzoku-gengo
cognate xenolith *[petr]*	同源捕獲岩	dōgen-hokaku'gan
cogon-grasses; eularias *[bot]*	萱；茅；かや	kaya
coherence *[phys]*	(可)干渉性	(ka)kanshō-sei
coherent scattering *{n} [elecmg]*	干渉性散乱	kanshō'sei-sanran
cohesion; cohesive force *[phys]*	凝集(力)	gyōshū(-ryoku)
cohesive failure *[adhes]*	凝集破壊；凝集破壞	gyōshū-hakai
cohesive force *[mech]*	抱合力；抱合力	hōgō-ryoku
cohesive strength *[mech]*	凝集力	gyōshū-ryoku
coiled molecule *[ch]*	糸毬状分子	itomari'jō-bunshi
coiling *{n} [ch] [tex]*	糸まり	ito'mari
" ; taking up *{n} [tex]*	巻取り；巻取り	maki-tori
coin *{n} [econ]*	硬貨	kōka
coincidence; agreement *[math]*	一致；いっち	itchi
" ; matching *{n}*	合致；がっち	gatchi
coincidence circuit *[electr]*	同時計数回路	dōji-keisū-kairo
coinventor; inventorship *[pat]*	共同発明者	kyōdō-hatsumei'sha
coitus; sexual intercourst *[zoo]*	性交	seikō
coking coal *[geol]*	強粘結炭	kyō'nenketsu-tan
coking property *[ch-eng]*	粘結性	nen'ketsu-sei
colatitude *[geod]*	余緯度；よいど	yo'ido
cold (to the feel); chilly; cool *{adj} [physio]*	冷い；つめたい	tsumetai
cold-blooded animal *[zoo]*	冷血動物	rei'ketsu-dōbutsu
cold brittleness *[mech]*	低温脆性	tei'on-zeisei
cold-cathode emission *[electr]*	冷陰極放出	rei-in'kyoku-hō'shutsu
cold drawing; cold stretching *{n} [plast]*	冷延伸	rei-enshin
cold eddy *[geol]*	冷水塊	reisui-kai
cold-forged product *[met]*	冷間鍛造品	reikan-tanzō'hin
cold front *[meteor]*	寒冷前線	kanrei-zensen

cold fusion *[nuc-phys]*	常温核融合	jō'on-kaku-yūgō
cold-headability *[met]*	冷間圧造性	reikan-atsuzō-sei
cold insulator *[mat]*	保冷剤	horei-zai
cold light *[opt]*	冷光	reikō
cold medicine *[pharm]*	風邪薬；かぜぐすり	kaze-gusuri
cold molding *{n} [plast]*	冷成形	rei-seikei
coldness (of weather) *[meteor]*	寒さ	samu-sa
" (sensation) *[physio]*	冷たさ	tsumeta-sa
cold resistance *[agr] [mat]*	耐寒性	taikan-sei
cold setting *{n} [poly-ch]*	常温硬化	jō'on-kōka
cold stretching *{n} [plas]*	冷延伸	rei-enshin
cold-water solubility *[ch]*	冷水可溶性	reisui-kayō-sei
cold wave *[meteor]*	寒波；かんぱ	kanpa
cold weather; chill *[meteor]*	冷気	reiki
cold working *{n} [met]*	冷間加工	reikan-kakō
cole (vegetable) *[bot]*	油菜；あぶらな	abura'na
colemanite *[miner]*	灰硼鉱	kai'hōkō
coliforms *[microbio]*	大腸菌	daichō-kin
collapse *{n}*	凹み；へこみ	hekomi
" *{vb t} [mech-eng]*	押し潰す	oshi-tsubusu
" *{vb i}*	崩壊する；崩壞する	hōkai suru
collapsing deformation *[mech]*	潰し変形	tsubushi-henkei
collar *{n} [cl]*	襟；えり	eri
collar; yoke *[eng]*	枠	waku
collar bone; clavicle *{n} [anat]*	鎖骨；さこつ	sa-kotsu
collate *{vb} [comput]*	照合する	shōgō suru
collating *{n} [comput]*	照合	shōgō
" *{n} [pr]*	丁合い調べ	chō'ai-shirabe
collator *[comput]*	照合機	shōgō-ki
collect; gather *{vb t}*	集める	atsumeru
collect exhaustively *{vb t}*	網羅する	mōra suru
collecting; picking; gathering *{n}*	採集	saishū
collecting together	集約	shūyaku
collection *{n}*	回収；回收	kaishū
collector *[bioch]*	捕収剤	hoshū-zai
" *[electr]*	集電器	shūden-ki
collider *[phys]*	衝突器	shōtotsu-ki
colliding-beam accelerator *[electr]*	衝突型加速器	shōtotsu'gata-kasoku-ki
colliery (pl -lieries) *[min-eng]*	炭鉱；炭鑛	tankō
colligative property *[p-ch]*	束一性	soku'itsu-sei
collimation *[phys]*	視準；しじゅん	shijun
collimation line *[opt]*	視準線	shijun-sen

collimator *[opt]*	視準器	shijun-ki
collinear *{adj} [geom]*	共線の	kyōsen no
collinear points *[geom]*	共線点; 共線點	kyōsen-ten
collision *[comput] [phys]*	衝突	shō'totsu
colloid *[ch]*	コロイド	koroido
colloid equivalence *[p-ch]*	コロイド当量	koroido-tōryō
colloquialism *[ling]*	口語的表現	kōgo'teki-hyōgen
colloquial speech *[comm]*	口語	kōgo
colloquial style *[gram]*	口語体	kōgo'tai
colloquial-style writing *[comm]*	口語文	kōgo'bun
collunarium *{n} [pharm]*	点鼻剤	tenbi-zai
collutorium *{n} [pharm]*	含嗽薬; うがいやく	ugai-yaku
collyrium *{n} [pharm]*	洗眼薬	sengan-yaku
" *[pharm]*	点眼液	tengan-eki
colon *[anat]*	結腸	ketchō
colonial organism *[bio]*	群体生物	guntai-seibutsu
colony (pl -nies) *[bio] [microbio]*	集落	shūraku
colophon *[pr]*	奥付け	oku'zuke
color *{n} [opt]*	色; いろ	iro
" *{n} [opt]*	色彩	shiki'sai
color; paint; dye *{vb t} [opt]*	彩る; 色取る	irodoru
colorant; coloring agent *[opt]*	着色剤	chaku'shoku-zai
color blindness *[med]*	色盲; しきもう	shiki'mō
color-change *[opt]*	変色; 變色	hen'shoku
color chart *[opt]*	色票	shiki-hyō
color-comparison tube *[an-ch]*	比色管	hishoku-kan
color developer *[opt]*	顕色剤; 顯色劑	ken'shoku-zai
color development speed *[photo]*	発色速度	hasshoku-sokudo
color difference *[opt]*	色差; しきさ	shiki-sa
color-difference meter *[eng] [opt]*	色差計	shikisa-kei
colorfastness *[eng]*	色堅牢度	iro-kenrō-do
" *[tex]*	染色堅牢度	senshoku-kenrō-do
" *[tex]*	耐変色性	tai'henshoku-sei
colorfastness to light *[tex]*	耐光堅牢度	taikō-kenrō-do
color fringe *[opt]*	色縁; いろぶち	iro-buchi
colorimeter *[opt]*	色彩計	shikisai-kei
colorimetry *[opt]*	測色	soku'shoku
color index *[astron]*	色指数	iro-shisū
coloring *{n} [opt]*	着色	chaku'shoku
coloring agent *[mat]*	着色剤	chaku'shoku-zai
coloring material *[pharm]*	着色料	chaku'shoku-ryō
coloring matter; pigment *[mat]*	色素	shikiso

coloring property *[opt]*	着色性	chaku'shoku-sei
coloring strength *[opt]*	着色力	chaku'shoku-ryoku
color-matching *{n} [opt] [poly-ch]*	色合せ	iro-awase
color name *[opt]*	色名	iro-mei
color number *[opt]*	色数；しきすう	shiki-sū
color of hair; color of fur *[bio]*	毛色	ke-iro
color order system *[opt]*	色票系	shiki'hyō-kei
color perception *[physio]*	色知覚；色知覺	iro-chi'kaku
color photograph *[photo]*	色彩写真；色彩寫真	shikisai-shashin
color reaction *[ch]*	呈色反応	teishoku-hannō
color scheme *[opt]*	配色	hai'shoku
color sensation *[physio]*	色感覚	iro-kankaku
color solid *[opt] [poly-ch]*	色立体	iro-rittai
color space *[opt]*	色空間	iro-kūkan
color temperature *[stat-mech]*	色温度	iro-ondo
color trademark *[econ]*	色彩商標	shikisai-shōhyō
color vision *[physio]*	色覚	shiki'kaku
columbine (flower) *[bot]*	苧環；おだまき	odamaki
column *[arch]*	円柱	enchū
" ; post *[arch]*	柱；はしら	hashira
column *[comput] [pr]*	段	dan
" *[comput] [pr]*	縦列	jū'retsu
" *[comput] [pr]*	欄；欄	ran
column; digit; order of magnitude *[math]*	桁；けた	keta
column *[math]*	列	retsu
columnar *[geom]*	円柱状	enchū'jō
column setting *{n} [comput]*	段組み	dan'gumi
colure *{n} [astron]*	分至経線	bunshi'keisen
comb *{n} [eng]*	櫛；くし	kushi
comb (one's hair) *{vb}*	梳る；くしけずる	kushi'kezuru
combination *[ch]*	化合	kagō
" *[crys]*	集形	shūkei
combinational circuit *[electr]*	組合せ回路	kumi'awase-kairo
combinations *[prob]*	組合せ	kumi-awase
combined flow *[hyd]*	合流	gōryū
combustibility *[ch]*	可燃性	ka'nen-sei
combustible matter *[mat]*	可燃物；かねんぶつ	ka'nen'butsu
combustion *[ch]*	燃焼	nenshō
combustion chamber *[mech-eng]*	燃焼室	nenshō-shitsu
comet *[astron]*	彗星；すいせい	suisei
" *[astron]*	箒星；ほうきぼし	hōki-boshi

comforter; bedquilt *[furn]*	蒲団；ふとん	futon
comfrey *[bot]*	鰭玻璃草	hire'hari'sō
coming into force	発効；發効	hakkō
comma *[gram] [pr]*	読点；讀點	tō'ten
command; order *{n} [comput]*	命令；めいれい	meirei
" ; instruction *{n} [comput]*	指示；しじ	shiji
" ; order *{n} [comput]*	指令；しれい	shirei
command language *[comput]*	指令言語	shirei-gengo
commensalism *[ecol]*	片利共生	henri-kyōsei
comment; annotation *{n} [comm]*	注釈；ちゅうしゃく	chūshaku
commentary; critique *{n} [pr]*	批評；ひひょう	hihyō
commercial *{adj} [econ]*	商用の	shōyō no
commercial district *[civ-eng]*	商業地域	shōgyō-chi'iki
commercial frequency *[elecmg]*	商用周波数	shōyō-shūha'sū
commerciality *[econ]*	商業性	shōgyō-sei
commercialization *[econ]*	商業化	shōgyō'ka
commercial name *[ind-eng]*	商号；商號	shōgō
commercial potential *[econ]*	企業性	kigyō-sei
commercial product *[ind-eng]*	製品；せいひん	seihin
commercial ship *[nav-arch]*	商船；しょうせん	shōsen
comminution; pulverization *[ch-eng]*	細末化；さいまつか	saimatsu'ka
" *[mech-eng]*	粉砕；ふんさい	funsai
commission; order *{vb t}*	注文する	chūmon suru
common area *[comput]*	共通域	kyōtsū-iki
common cold *[med]*	風邪；かぜ	kaze
common language *[comput]*	共通言語	kyōtsū-gengo
common logarithm *[alg]*	常用対数；常用對數	jōyō-taisū
common ratio *[arith]*	公比；こうひ	kōhi
common salt; table salt *[ch]*	食塩；食鹽	shoku'en
common-usage name *[sci-t]*	慣用名	kan'yō-mei
common year *[astron]*	平年；へいねん	hei'nen
communicable disease *[med]*	伝染病；傳染病	densen'byō
communication *[comm]*	連絡；れんらく	renraku
" *[comm]*	通信；つうしん	tsū'shin
communication line *[comm]*	通信回線	tsūshin-kaisen
communication network *[comm]*	通信網	tsūshin-mō
communications *[comm]*	通信情報伝達	tsūshin-jōhō-den'tatsu
communications satellite *[aero-eng] [comm]*	通信衛星	tsūshin-eisei
communication theory *[comm]*	通信理論	tsūshin-riron
community *[bio] [ecol]*	群集；ぐんしゅう	gunshū
commutable area *[civ-eng]*	通勤圏；通勤圈	tsūkin-ken

commutation *[elecmg]*	整流；せいりゅう	seiryū
" *[electr]*	転流；轉流	tenryū
commutative law *[alg]*	交換法則	kōkan-hōsoku
commutative property *[alg]*	可換性	kakan-sei
commutator *[alg]*	交換子	kōkan'shi
" *[elecmg]*	整流子	seiryū'shi
compandor *[electr]*	圧伸器；壓伸器	asshin-ki
companion star *[astron]*	伴星	hansei
companionway *[nav-arch]*	昇降口	shōkō-guchi
company (pl -nies) *[econ]*	会社；會社	kaisha
comparator *[comput] [eng]*	比較器；ひかくき	hikaku-ki
comparator circuit *[electr]*	比較回路	hikaku-kairo
compare *{vb}*	比較する	hikaku suru
" *{vb}*	比べる；較べる	kuraberu
compared to; compared with; versus	対；對；たい	tai
comparison *[math]*	比較	hikaku
compartment *[nav-arch]*	区画室；區畫室	kukaku-shitsu
compass *[eng]*	コンパス	konpasu
" *[eng]*	羅針盤	rashin-ban
" *[eng]*	羅針儀；らしんぎ	rashin-gi
" *[geom]*	両脚規	ryō'kyaku-ki
compass bearing *{n} [navig]*	羅針方位	rashin-hō'i
compass direction *[navig]*	磁針方向	jishin-hōkō
compatibility *[bot]*	和合性	wagō-sei
" *[ch]*	相溶性；相容性	sōyō-sei
compatibility; conformity	適合性	tekigō-sei
compensation *[econ]*	代償	daishō
" *[legal]*	補償；ほしょう	hoshō
compensation method *[econ]*	補償法	hoshō-hō
compensator *[opt]*	補償板	hoshō-ban
compete *{vb}*	競合する	kyōgō suru
competence *[immun]*	受容能力	juyō-nō'ryoku
competition *[ecol]*	競争；競爭	kyōsō
" *{n}*	競合	kyōgō
competitive addition *[ch]*	競合的付加	kyōgō'teki-fuka
competitive inhibitor *[bioch]*	拮抗阻害剤	kikkō-sogai-zai
competitiveness *[bio]*	拮抗性	kikkō-sei
competitive reaction *[ch]*	競争反応	kyōsō-hannō
complaint *{n}*	苦情；くじょう	kujō
complement *[math]*	補数	hosū
complementary angles *[geom]*	余角；餘角	yo-kaku
complementary arc *[geom]*	余弧	yoko

complementary color *[opt]*	補色	ho-shoku
" *[opt]*	余色	yo-shoku
complementary event *[prob] [stat]*	余事象；よじしょう	yo-jishō
complementary functions *[trig]*	余関数；餘關數	yo-kansū
complementary number *[math]*	余数	yosū
complementary operator *[comput]*	補数演算子	hosū-enzan'shi
complementation *[gen]*	相補性	sōho-sei
complement base *[comput] [math]*	補数の底	hosū no tei
complete *{vb}*	完成する	kansei suru
complete combustion *[ch]*	完全燃焼	kanzen-nenshō
completion *{n}*	完了	kanryō
complex *{n} [ch]*	錯体；さくたい	sakutai
complex compound *[ch]*	錯化合物	saku-kagō'butsu
complex conjugate *[alg]*	複素共役	fuku'so-kyōyaku
" *[alg]*	共役複素数	kyōyaku-fukuso'sū
complex fraction *[alg]*	繁分数	han-bunsū
complex index of refraction *[opt]*	複素屈折率	fukuso-kussetsu-ritsu
complex ion *[ch]*	錯イオン	saku-ion
complexity; complicatedness *{n}*	複雑さ；複雑さ	fuku'zatsu-sa
complex modulus of elasticity *[mech]*	複素弾性率	fukuso-dansei-ritsu
complex number *[alg]*	複素数	fuku-sosū
complex radical *[ch]*	錯基；さくき	saku-ki
complex salt *[inorg-ch]*	錯塩；さくえん	saku'en
complex variable *[math]*	複素変数	fukuso-hensū
complicated *{adj}*	複雑な	fuku'zatsu na
" *{adj}*	ややこしい	yayakoshī
complicated molded product *[ind-eng]*	複雑成形品	fuku'zatsu-seikei'hin
component *[ch] [elec] [math]*	成分	seibun
" *[ind-eng]*	部品	buhin
" *[sci-t]*	構成要素	kōsei-yōso
component of force *[phys]*	分力	bun-ryoku
composer (a person) *[music]*	作曲家	sakkyoku'ka
composite *[mat]*	複合物	fukugō'butsu
" *[mat]*	複合材	fukugō-zai
composite fiber *[tex]*	複合繊維	fukugō-sen'i
composite function *[alg]*	合成関数	gōsei-kansū
composite mold *[eng]*	複合金型	fukugō-kana'gata
composite(ness) *{n}*	複合；ふくごう	fuku'gō
composite number *[arith]*	合成数	gōsei-sū
composite plating *{n} [met]*	複合めっき	fukugō-mekki
composite tone *[acous]*	合成音	gōsei-on
composite volcano *[geol]*	複成火山	fukusei-kazan

composite wave *[phys]*	合一波	gō'itsu-ha
composition *[ch]*	組成；そせい	sosei
composition; writing *[pr]*	文章	bunshō
compost *{n} [agr]*	堆肥；たいひ	taihi
compound *{n} [ch]*	化合物	kagō'butsu
" *{vb t} [ind-eng]*	配合する	haigō suru
compounded coal *[geol]*	配合炭	haigō-tan
compounded spicery *[cook]*	調合香料	chōgō-kōryō
compound eye *[i-zoo]*	複眼；ふくがん	fuku-gan
compound fraction *[alg]*	繁分数	han-bunsū
compound fruit *[bot]*	複果	fuku-ka
compounding ingredient *[mat]*	配合剤	haigō-zai
compound leaf *[bot]*	複葉；ふくよう	fuku-yō
compound lipid *[bioch]*	複合脂質	fukugō-shi'shitsu
compound nucleus *[nuc-phys]*	複合核	fukugō-kaku
compound sentence *[logic]*	複合文	fukugō'bun
compound statement *[comput]*	複合文	fukugō'bun
" *[logic]*	合成命題	gōsei-meidai
compound tone *[acous]*	複合音	fukugō-on
compregnated wood *[mat]*	樹脂含浸圧縮木材	jushi-ganshin-asshuku-mokuzai
" *[poly-ch]*	強化木	kyōka-boku
compress; poultice *{n} [pharm]*	罨法物	anpō'butsu
compressed air *[mech]*	圧縮空気	asshuku-kūki
compressibility *[mech]*	圧縮率；壓縮率	asshuku-ritsu
compressibility factor *[thermo]*	圧縮因子	asshuku-inshi
compression *[comput] [mech]*	圧縮；あっしゅく	asshuku
" *[mech]*	圧搾；あっさく	assaku
compression breaking strength *[mech]*	圧縮破壊強さ	asshuku-hakai-tsuyo-sa
compression member *[eng]*	圧縮材	asshuku-zai
compression molding *{n} [poly-ch]*	圧縮成形	asshuku-seikei
compression (permanent) set *[poly-ch]*	圧縮永久歪	asshuku-eikyū-hizumi
compression ratio *[poly-ch]*	圧縮比	asshuku-hi
compression recovery rate *[poly-ch]*	圧縮復元率	asshuku-fuku'gen-ritsu
compression stroke *[mech-eng]*	圧縮行程	asshuku-kōtei
compression residual stress *[mech]*	圧縮殘留応力	asshuku-zanryū-ō'ryoku
compressive joint *[eng]*	圧縮接続	asshuku-setsu'zoku
compressive strength *[poly-ch]*	圧縮強さ	asshuku-tsuyo-sa
compressive stress *[mech]*	圧縮応力	asshuku-ō'ryoku
compressor *[comput] [mech-eng]*	圧縮機	asshuku-ki
computational chemistry *[ch]*	計算化学	keisan-kagaku
compute *{vb} [math]*	算出する	san'shutsu suru

comput(eriz)ed tomography; CT *[med]*	計算断層像法	keisan-dansō-zōhō
computer *[comput]*	電算機	densan-ki
computer; calculator *[comput]*	計算機；けいさんき	keisan-ki
computer-aided design; CAD *[comput]*	計算機援用設計	keisanki-enyō-sekkei
computer-aided manufacturing; CAM *[comput]*	計算機援用製造	keisanki-enyō-seizō
computer control *[comput]*	計算機制御	keisanki-seigyo
computer design language; CDL *[comput]*	計算機設計用語	keisanki-sekkei-yōgo
computerized typesetting *{n} [pr]*	電算植字	densan-shoku'ji
computer language *[comput]*	計算機言語	keisanki-gengo
computer network *[comput]*	計算機網	keisanki-mō
computer science *[comput]*	計算機科学	keisanki-kagaku
computer word *[comput]*	語	go
concatenate *{vb} [comput]*	連結する	ren'ketsu suru
concatenation *[comput]*	連結	ren'ketsu
concave *{n} [geom]*	凹；おう	ō
concave mirror *[opt]*	凹面鏡	ō'men-kyō
concave part *[math]*	凹部	ō-bu
concavo-concave *{adj} [opt]*	両凹の	ryō-ō no
concavo-convex *{adj} [opt]*	凹凸の；おうとつの	ō-totsu no
conceal *{vb t} [comput]*	隠す	kakusu
conceal; hide *{vb t}*	隠蔽する	inpei suru
concentrated solution *[ch]*	凝縮液	gyō'shuku-eki
concentrating; concentration *[ch]*	濃縮	nō'shuku
concentration *[ch]*	濃度；のうど	nōdo
concentration gradient *[ch]*	濃度勾配	nōdo-kōbai
concentration polarization *[p-ch]*	濃淡分極	nōtan-bun'kyoku
concentration scale factor *[ch]*	濃縮倍率	nō'shuku-bai'ritsu
concentric circles *[geom]*	同心円	dōshin-en
concept *[math] [psy]*	概念；がいねん	gai'nen
" *[psy]*	観念；觀念	kannen
conception *[psy]*	着想	chakusō
concerned party *{n}*	当事者；當事者	tōji'sha
concerted reaction *[org-ch]*	協奏反応	kyōsō-hannō
concert hall *[arch]*	協奏室	ensō-shitsu
concerto *[music]*	協奏曲	kyōsō'kyoku
conch (shellfish) *[i-zoo]*	法螺貝；ほらがい	hora-gai
conchology *[i-zoo]*	貝類学；貝類學	kairui'gaku
conclusion *[logic]*	結論	ketsu'ron
concordance *[pr]*	用語索引	yōgo-saku'in
concretion *[geol]*	固結	koketsu
concurrence *{n} [psy]*	同意	dō'i

concurrent *{n} [geom]*	共点	kyōten
" *{adj} [comput]*	同時平行の	dōji-heikō no
condensate *[mat]*	凝縮物	gyōshuku'butsu
" *[mat]*	縮合物	shukugō'butsu
" *[mat]*	縮合体	shukugō'tai
condensation *[ch] [mech]*	凝縮; ぎょうしゅく	gyō'shuku
" (of gas) *[ch]*	冷縮	rei'shuku
" *[ch]*	縮合	shuku'gō
condensation plane *[phys]*	凝縮面	gyōshuku-men
condensation polymerization *[poly-ch]*	重縮合	jū-shukugō
" *[poly-ch]*	縮(合)重合	shuku(gō)-jūgō
condensation trail *[meteor]*	航跡雲	kōseki'un
condensed milk *[cook]*	練乳; れんにゅう	rennyū
condenser; cooler *[ch] [mech-eng]*	冷却器	reikyaku-ki
condenser *[ch-eng]*	凝縮器	gyōshuku-ki
condensing *{n} [opt]*	集光	shūkō
condiment *[cook]*	香辛料	kōshin-ryō
" *[cook]*	薬味; やくみ	yaku'mi
condition *{n} [math]*	条件; 條件	jōken
" *{n} [phys]*	状態; 狀態	jōtai
conditional *{n} [alg]*	条件付	jōken-tsuki; jōken-zuki
" *{adj}*	条件付の	joken'tsuki no
conditional expression *[comput]*	条件式	jōken-shiki
conditional inequality *[alg]*	条件付不等式	jōken'tsuki-futō-shiki
conditional probability *[prob]*	条件付確率	jōken'tsuki-kaku'ritsu
conditional statement *[logic]*	条件命題	jōken-meidai
conditioning *{n} [eng]*	前処理; 前處理	mae-shori
" *{n} [poly-ch]*	状態調節	jōtai-chōsetsu
condor (bird) *[v-zoo]*	禿鷹; はげたか	hage-taka
conduct; behavior *{n} [psy]*	行動; こうどう	kōdō
conduction *[elec] [phys]*	伝導; 傳導	dendō
conductivity *[elec]*	伝導度	dendō-do
" *[elec]*	導電率	dōden-ritsu
conductometry *[an-ch]*	導電測定法	dōden-sokutei-hō
conductor *[elec]*	導体	dōtai
conduit; tube; pipe; duct *[eng]*	導管	dōkan
conduit *[eng]*	管路	kanro
cone *[bot]*	球果; 毬果; 球花	kyūka
" *[geom]*	(円)錐	(en)sui
" *[geom] [hist]*	錐状体	suijō'tai
cone shell (shellfish) *[i-zoo]*	芋貝; いもがい	imo-gai
confectionery *[cook]*	菓子; かし	kashi

confidence interval *[stat]*	信頼区間；信頼區間	shinrai-kukan
confidence level *[ind-eng]*	信頼水準	shinrai-suijun
confidence limits *[stat]*	信頼限界	shinrai-genkai
confidentiality *[comput]*	秘密性	himitsu-sei
configuration *[ch] [comput] [mech]*	構成；こうせい	kōsei
" *[ch] [math] [phys]*	配置；はいち	haichi
" *[ch] [math]*	立体配置	rittai-haichi
confirmation *{n}*	確認	kaku'nin
conflagration; fire *{n} [ch]*	火事；かじ	kaji
" (spread of fire) *[ch]*	延焼；延燒	enshō
conflict; infringement *{n} [legal]*	抵触；ていしょく	teishoku
" ; incompatibility *{n}*	矛盾；むじゅん	mujun
confluence *[hyd]*	合流；ごうりゅう	gōryū
confluent hypergeometric function *[math]*	合流型超何幾関数	gōryū'gata-chō'kika-kansū
conform *{vb}*	即応する；即應する	soku'ō suru
conformation; constellation *[org-ch]*	立体配座	rittai-haiza
conform to; correspond to *{vb}*	準ずる	junzuru
conform with *{vb}*	適合する	tekigō suru
confounding *{n} [math]*	交絡；こうらく	kōraku
confusing; misleading *{adj}*	紛らわしい	magirawashī
confusion *[psy]*	錯乱；さくらん	sakuran
congealing point *[ch]*	凝結点	gyōketsu-ten
conger; sea eel (fish) *[v-zoo]*	穴子；海鰻；あなご	anago
conglomerate *[petr]*	礫岩；れきがん	rekigan
congruence *[math]*	双；雙；そう	sō
congruent *{n} [geom]*	合同	gōdō
congruent melting *{n} [p-ch]*	調和融解	chōwa-yūkai
congruent triangles *[geom]*	合同三角形	gōdō-sankak(u)kei
conical form *{n} [geom]*	円錐形	ensui-kei
conic sections *[an-geom]*	円錐曲線	ensui-kyoku'sen
conidiophore *[mycol]*	分生子柄	bunseishi-hei
conidium; conidiospore *[mycol]*	分生子；ぶんせいし	bunsei'shi
conifer; softwood *[bot]*	針葉樹	shin'yō-ju
conjugate *{n} [alg]*	共役；きょうやく	kyō'yaku
" *{n} [bioch]*	抱合体；抱合體	hōgō'tai
" *{adj} [geom]*	共役な	kyōyaku na
conjugate angle *[geom]*	共役角	kyōyaku-kaku
conjugated double bond *[p-ch]*	共役二重結合	kyōyaku-nijū-ketsu'gō
conjugated system *[org-ch]*	共役系	kyōyaku-kei
conjugation *[alg] [ch]*	共役	kyōyaku
" *[bot] [i-zoo]*	接合；せつごう	setsu'gō

English	Japanese	Romanization
conjugation *[pharm]*	抱合；抱合	hōgō
conjunction *[astron]*	合	gō
conjunction; connective *[gram]*	接続詞	setsuzoku'shi
conjunction *[logic]*	連言	rengen
conjunctiva (pl -s, -tivae) *[anat]*	結膜；けつまく	ketsu'maku
connect *{vb t} [elec]*	結線する	kessen suru
connect (couple) directly *{vb t}*	直結する	chokketsu suru
connecting-link portion *[mech-eng]*	連結部	renketsu-bu
connecting passageway *[arch]*	連通路	rentsū-ro
connecting plate *[mech-eng]*	連結板	renketsu-ban
connecting rod *[mech-eng]*	連結杆	renketsu-kan
" *[mech-eng]*	連接棒	rensetsu-bō
connection *[comput] [sci-t]*	接続；せつぞく	setsu'zoku
" *[elec]*	結線	kessen
" *[mech-eng]*	結合	ketsu'gō
connection box *[comput]*	接合盤	setsugō-ban
connectivity *[comput]*	連結性	renketsu-sei
connector *[electr]*	結合子	ketsugō'shi
" *[electr]*	接合器	setsugō-ki
consciousness *[psy]*	意識；いしき	i'shiki
consecutive *{n} [sci-t]*	連続	renzoku
consecutive addition polymerization *[poly-ch]*	逐次付加重合	chikuji-fuka-jūgō
consecutive numbers *[arith]*	連続番号；連續番號	renzoku-bangō
" *[math]*	通し番号	tōshi-bangō
consecutive reaction *[ch]*	逐次反応；逐次反應	chikuji-hannō
consent *{n} [legal]*	承諾；しょうだく	shōdaku
" *{n}*	同意	dō'i
consequent *{n} [logic]*	後件	kōken
" *{adj} [geol]*	必従の；必從の	hitsu'jū no
" *{adj}*	結果の	kekka no
conservation of angular momentum, law of *[mech]*	角運動量保存則	kaku-undō'ryō-hozon-soku
conservation of electric power *[elec]*	省電力化	shō-denryoku'ka
conservation of energy law *[phys]*	エネルギー保存則	enerugī-hozon-soku
conservation of mass, law of *[phys]*	質量保存の法則	shitsu'ryō-hozon no hōsoku
conservation of matter, law of *[phys]*	物質保存の法則	busshitsu-hozon no hōsoku
conservation of momentum, law of *[mech]*	運動量保存則	undō'ryō-hozon-soku
conservation of natural resources	天然資源保護	tennen-shigen-hogo
consideration; examination	考察	kōsatsu
consistency *[mat]*	稠度；ちゅうど	chūdo

consistency (of facts) *[logic]*	辻褄；つじつま	tsuji'tsuma
consistent estimator *[stat]*	一致推定量	itchi-suitei'ryō
consistometer *[eng]*	稠度計	chūdo-kei
console *[comput]*	操作卓	sōsa'taku
consolidation *[civ-eng] [geol]*	圧密；壓密	atsu'mitsu
" *{n}*	統合；とうごう	tōgō
consonant (sound) *[ling]*	子音	shi'in; shi'on
consortium (pl -tia) *[bot]*	共同体	kyōdō'tai
" *[bot]*	集合体	shūgō'tai
" *[econ]*	資本合同	shihon-gōdō
conspectus; comprehensive survey *[pr]*	総覧；總覽	sōran
constant *{n} [math]*	恒数	kōsū
" *{n} [comput] [math]*	定数；ていすう	teisū
constant function *[alg]*	定数関数	teisū-kansū
constant heat summation, law of *[p-ch]*	総熱量保存の法則	sō-netsuryō-hozon no hōsoku
constant interfacial angle, law of *[crys]*	面角不変の法則	men'kaku-fuhen no hōsoku
constant interfacial angle, law of *[crys]*	面角一定の法則	men'kaku-ittei no hōsoku
constantly; incessantly *{adv}*	絶えず；たえず	taezu
constant speed *[mech] [phys]*	定速	teisoku
constant weight *[mech]*	恒量	kōryō
constellation *[astron]*	星座；せいざ	seiza
" *[org-ch]*	立体配座	rittai-haiza
constipation *[med]*	便秘	benpi
constituent *[sci-t]*	構成要素	kōsei-yōso
constitution *[legal]*	憲法	kenpō
" (make-up) *[mech]*	構成；こうせい	kōsei
constrained motion *[mech]*	束縛運動	soku'baku-undō
constraint *[mech]*	拘束	kōsoku
construct *{vb t} [arch]*	立てる；建てる	tateru
construct; install *{vb t}*	架設する	kasetsu suru
construction *[arch]*	構造；こうぞう	kōzō
" *[pat]*	構成	kōsei
construction plan *{n} [graph]*	構造図；構造圖	kōzō-zu
consulate *[arch]*	領事館	ryōji'kan
consulate general *[arch]*	総領事館	sōryōji'kan
consultation (conference) *{n} [comm]*	協議	kyōgi
" (discussion) *{n} [comm]*	相談；そうだん	sōdan
consume; exhaust *{vb}*	消耗する	shōmō suru
consumer *[econ]*	需用家	juyō'ka
" *[econ]*	消費者	shōhi'sha

consumer goods *[econ]* 消費財 shōhi-zai
contact; contact point *{n} [math]* 接点; せってん setten
contact adhesive *[poly-ch]* 接触型接着剤 sesshoku'gata-setchaku-zai
contact angle *[fl-mech]* 接触角 sesshoku-kaku
contact catalysis *[ch-eng]* 接触反応; 接觸反應 sesshoku-hannō
contact flight *[navig]* 有視界飛行 yū'shikai-hikō
contact inhibition *[cyt]* 接触阻止 sesshoku-soshi
contactless switch *[elec]* 無接点スイッチ mu'setten-suitchi
contact plug *[eng]* 接栓; せっせん sessen
contact printing *{n} [photo]* 密着焼付け mitchaku-yaki'tsuke
contact resistance *[elec]* 接点抵抗 setten-teikō
contagion *[med]* 伝染; でんせん densen
contagious *{adj} [med]* 伝染性の densen'sei no
container; receptacle; vessel *[eng]* 入れ物; いれもの ire'mono
" *[eng]* 容器 yōki
" *[rail]* コンテナ kontena
containment *[bioch]* 封じ込め fūji'kome
" *{n}* 収容; 收容 shūyō
contaminant; pollutant *[ecol]* 汚染物質 o'sen-busshitsu
contamination; soiling *{n} [ecol]* 汚染; おせん o'sen
contamination (by germs) *[microbio]* 雑菌混入 zakkin-konnyū
" (by germs) *[microbio]* 雑菌汚染 zakkin-o'sen
content (quantity) *[ch]* 含有量 gan'yū-ryō
" (quantity) *[ch]* 含量; がんりょう ganryō
content(s) *{n} [comput]* 内容 naiyō
content volume *[ind-eng]* 受け容積 uke-yōseki
contest; contend; argue *{vb} [comm]* 争う; 爭う arasou
context *[comput] [pr]* 文脈 bun'myaku
continent *{n} [geogr]* 洲; しゅう shū
" *{n} [geogr]* 大陸; たいりく tai'riku
continental air mass *[meteor]* 大陸気団; 大陸氣團 tairiku-kidan
continental climate *[climat]* 大陸気候 tairiku-kikō
continental divide *[geol]* 大陸分水界 tairiku-bunsui'kai
continental drift *[geol]* 大陸漂移 tairiku-hyō'i
continental shelf *[geol]* 大陸棚 tairiku-dana
continental slope *[geol]* 大陸斜面 tairiku-shamen
contingency; eventuality *[math]* 偶然性 gūzen-sei
" ; emergency *{n}* 非常; ひじょう hijō
contingency table *[stat]* 分割表 bun'katsu-hyō
contingent *{adj}* 臨時の; りんじの rinji no
continuance; persistance *{n} [bio]* 存続 sonzoku
continuity *[civ-eng]* 連続性; 連續性 renzoku-sei

continuity, law of *[math]*	連続の法則	renzoku no hōsoku
continuous *{n} [calc]*	連続; れんぞく	renzoku
" ; sequence *{n} [comput]*	一連	ichi'ren
continuous casting *{n} [met]*	連鋳; れんちゅう	renchū
" *{n} [met]*	連続鋳造	renzoku-chūzō
continuous culture *[microbio]*	連続培養	renzoku-baiyō
continuous extrusion molding *[eng]*	連続押出成形	renzoku-oshi'dashi-seikei
continuous form *[comput]*	連続帳票	renzoku-chōhyō
continuous furnace *[met]*	連続炉	renzoku-ro
continuous polymerization *[poly-ch]*	連続重合	renzoku-jūgō
continuous random variable *[stat]*	連続確率変数	renzoku-kaku'ritsu-hensū
continuum *[math] [phys]*	連続(体)	renzoku('tai)
contort; be distorted *{vb i}*	歪む; ゆがむ	yugamu
contour; delineation *[math] [sci-t]*	輪郭; りんかく	rinkaku
" *[sci-t]*	外郭; がいかく	gaikaku
contour line *[map] [meteor]*	等高線	tōkō-sen
contour planting *{n} [agr]*	等高線植え	tōkō'sen-ue
contraceptive (device) *[pharm]*	避妊具	hi'nin-gu
" (medicine) *[pharm]*	避妊薬 (避妊剤)	hi'nin-yaku (hi'nin-zai)
contract; agreement *{n} [econ] [legal]*	契約; けいやく	keiyaku
contract; shorten; shrink *{vb t}*	縮める	chijimeru
contracted sound *[ling]*	拗音; ようおん	yō'on
contractile vacuole *[cyt]*	収縮胞; 收縮胞	shūshuku-hō
contraction; shrinkage *[mech]*	収縮; しゅうしゅく	shū'shuku
contraction coefficient *[fl-mech]*	収縮係数	shūshuku-keisū
contractor (a person) *[ind-eng]*	請負人	uke'oi'nin
contract work *[civ-eng]*	請負工事	uke'oi-kōji
contradiction; opposition *[legal]*	反対; 反對	hantai
" ; incompatibility *[logic]*	矛盾; むじゅん	mujun
contrail; condensation trail *[meteor]*	航跡雲	kōseki'un
contrapositive *[logic]*	対偶命題	taigū-meidai
contrast *{n} [math]*	対照; たいしょう	taishō
" *{n} [opt]*	対比	taihi
" *{n}*	対立; 對立	tairitsu
contrast medium (x-ray) *[phys]*	造影剤	zō'ei-zai
contrast (control) solution *[ch]*	対照液	taishō-eki
contrivance; devisement; device *[pat]*	考案; こうあん	kōan
control *{n} [bio] [cont-syst]*	対照	taishō
" *{n} [comput] [math] [sci-t]*	制御; せいぎょ	seigyo
" ; management *{n} [ind-eng]*	管理	kanri
" ; adjustment *{n} [sci-t]*	調節; ちょうせつ	chōsetsu
control; check; restraint *{n}*	制止	seishi

control; regulate *{vb t}*	規制する	kisei suru
control character *[comput]*	制御文字	seigyo-moji
control command *[comput]*	制御指令	seigyo-shirei
controller *[comput]*	制御装置	seigyo-sōchi
control lever *[aero-eng]*	操縦桿；操縦桿	sōjū-kan
control operation *[comput]*	制御操作	seigyo-sōsa
control panel *[comput]*	制御盤	seigyo-ban
control relay *[comput] [elec]*	制御継電器	seigyo-keiden-ki
control rod *[nucleo]*	制御棒	seigyo-bō
control statement *[comput]*	制御文	seigyo'bun
control system *[eng]*	制御系	seigyo-kei
control systems *[eng]*	制御系学	seigyo'kei'gaku
control tower *[navig]*	管制塔	kansei-tō
control unit *[comput]*	制御装置	seigyo-sōchi
control word *[comput]*	制御語；せいぎょご	seigyo'go
controversy *{n} [comm]*	論争；論爭	ronsō
convection *[ocean] [phys]*	対流	tairyū
convection cell *[meteor]*	対流細胞；對流細胞	tairyū-saibō
convenience *{n}*	簡便さ；かんべんさ	kanben-sa
convention *[legal]*	条約；條約	jōyaku
convention center *[arch]*	集会総合施設	shūkai-sōgō-shisetsu
convergence *[math]*	収束；收束	shū'soku
" *[math] [phys]*	収斂	shūren
convergent series *[alg]*	収束級数	shūsoku-kyūsū
converging lens *[opt]*	集光レンズ	shūkō-renzu
" *[opt]*	収束レンズ	shūsoku-renzu
conversation *[comm]*	会話	kaiwa
" *[comm]*	対話；對話；たいわ	taiwa
conversational mode; interactive mode *[comput]*	会話形；會話型	kaiwa-gata
conversational mode; interactive mode *[comput]*	会話方式	kaiwa-hō'shiki
converse *{n} [logic]*	逆；ぎゃく	gyaku
" *{n} [logic]*	換位命題	kan'i-meidai
conversion *[ch]*	変移；變移	hen'i
" *[ch] [math] [nuc-phys]*	転換；轉換	tenkan
" *[ch-eng]*	反応率；反應率	hannō-ritsu
" *[ch-eng]*	転化	tenka
" *[comput] [math]*	変換	henkan
" *[comput]*	移行	ikō
" *[org-ch]*	転化率	tenka-ritsu
" *[poly-ch]*	重合率	jūgō-ritsu

conversion table *[math]*	換算表	kansan-hyō
conversion to high polymer *[poly-ch]*	高分子化	kōbunshi'ka
convert; translate *{vb} [comput]*	変換する	henkan suru
" (mathematically) *{vb} [math]*	換算する	kansan suru
converted paper *[mat] [paper]*	加工紙; かこうし	kakō-shi
converter *[comput]*	変換器; 變換器	henkan-ki
" (a furnace) *[met]*	転炉; 轉爐; てんろ	tenro
converter reactor *[nucleo]*	転化炉	tenka-ro
converter steel *[steel]*	転炉鋼	tenro-kō
convex; convexity *[geom]*	凸; とつ	totsu
convex mirror *[opt]*	凸面鏡	totsu'men-kyō
convexo-concave *{adj} [opt]*	凸凹の	totsu-ō no
convexo-convex *{adj} [opt]*	両凸の	ryō-totsu no
convex part *[mech]*	凸部	totsu-bu
convex polygon *[geom]*	凸多角形	totsu-takak(u)kei
convolute *{adj} [bot]*	回旋状の	kaisen'jō no
convoy *{n} [ord]*	護衛隊; ごえいたい	goei'tai
cook's knife (pl knives) *[cook]*	庖丁	hōchō
cookery; cooking *{n} [eng]*	料理; りょうり	ryōri
cooking (of food) *{n} [food-eng]*	調理	chōri
cool; chill *{vb t}*	冷す	hiyasu
coolant *[mat]*	冷媒; れいばい	reibai
coolant liquid *[mat]*	冷却液	reikyaku-eki
cool color *[opt]*	寒色	kan-shoku
cooled liquid *[phys]*	冷却液	reikyaku-eki
cooler *[ch]*	冷却器	reikyaku-ki
cool flame *[ch]*	冷炎	rei'en
cooling *{n} [eng]*	冷却; れいきゃく	rei'kyaku
" *{n} [mech-eng]*	冷房	reibō
cooling power *[thermo]*	冷却能	reikyaku-nō
cooling water; cooled water *[ch]*	冷却水	reikyaku-sui
coolness (of weather) *[meteor]*	涼しさ	suzushi-sa
cooper (a person) *[eng]*	樽師; たるし	taru'shi
cooperation *{n}*	協働	kyōdō
" *{n}*	提携	teikei
coordinate bond *[ch]*	配位結合	hai'i-ketsugō
coordinated water *[ch]*	配位水	hai'i-sui
coordinate geometry *[math]*	座標幾何学	zahyō-kika'gaku
coordinate grid *[math]*	座標格子	zahyō-kōshi
coordinates *{n} [an-geom]*	座標; ざひょう	za'hyō
coordinate transformation *[an-geom]*	座標転換	zahyō-tenkan
coordinating ion *[ch]*	配位イオン	hai'i-ion

coordination compound *[ch]*	配位化合物	hai'i-kagō'butsu
coordination isomerism *[ch]*	配位異性	hai'i-i'sei
coordination number *[phys]*	配位数；はいいすう	hai'i-sū
coordination polymer *[poly-ch]*	配位高分子	hai'i-kōbunshi
co-ownership *[econ]*	共有	kyōyū
coping *{n} [arch]*	笠木；かさぎ	kasa'gi
coplanar *{n} [geom]*	共角	kyōkaku
" *{adj} [geom]*	共面の	kyōmen no
copolycondensate *[poly-ch]*	共縮合体	kyō'shukugō'tai
copolycondensation *[poly-ch]*	共縮合	kyō'shuku'gō
copolymer *[poly-ch]*	共重合体	kyōjūgō'tai
copolymerization *[poly-ch]*	共重合	kyō'jūgō
copper (element)	銅；どう	dō
copper acetate	酢酸銅	sakusan-dō
copper arsenate	砒酸銅；ひさんどう	hisan-dō
copper boride	硼化銅	hōka-dō
copper borofluoride	硼弗化銅	hō'fukka-dō
copper bromide	臭化銅	shūka-dō
copper carbonate	炭酸銅	tansan-dō
copper chloride	塩化銅；鹽化銅	enka-dō
copper fluoride	弗化銅	fukka-dō
copper hydroxide	水酸化銅	suisan'ka-dō
copper iodide	沃化銅	yōka-dō
copper matrix *[met]*	銅基	dō-ki
copper nitrate	硝酸銅	shōsan-dō
copper oxide	酸化銅	sanka-dō
copper phosphate	燐酸銅	rinsan-dō
copper silicofluoride	珪弗化銅	kei'fukka-dō
copper suboxide	亜酸化銅	a'sanka-dō
copper sulfate	硫酸銅	ryūsan-dō
copper sulfide	硫化銅	ryūka-dō
coprecipitation *[ch]*	共沈；きょうちん	kyō'chin
coprecipitator *[bioch]*	共沈剤	kyōchin-zai
copy (auxiliary copy) *{n} [graph]*	副本	fuku-hon
" (pl copies) *{n} [graph]*	複写；複寫	fukusha
" *{n} [graph]*	写し；寫し；うつし	utsushi
" *{n} [legal]*	謄本	tōhon
" (volume) *{n} [pr]*	部	bu
" (replica; reproduction) *{n}*	模写	mosha
" (reproduce) *{vb t} [comput]*	複写する	fukusha suru
" *{vb} [graph]*	写す	utsusu
copying machine *[graph]*	複写機	fukusha-ki

copying paper *[paper]*	複写(用)紙	fukusha-(yō)shi
copyright *[legal]*	著作権	chosak(u)ken
coral *[i-zoo]*	珊瑚；さんご	sango
coral beads *[lap]*	珊瑚珠	sango-ju
coral evergreen *[bot]*	日陰の葛	hikage-no-kazura
coral reef *[geol]*	珊瑚礁	sango-shō
coral snake (reptile) *[v-zoo]*	珊瑚蛇	sango-hebi
corbicula (shellfish) *[i-zoo]*	蜆；しじみ	shijimi
cord; string *{n} [mat]*	紐；ひも	himo
cord fabric *[tex]*	簾織；すだれおり	sudare-ori
cordial *{n} [pharm]*	強心薬	kyōshin-yaku
cordierite *[miner]*	菫青石	kinsei'seki
cordless telephone *[comm]*	無電話線電話器	mu'denwa'sen-denwa-ki
core *{n} [geol]*	中心核	chūshin-kaku
" *{n} [geol] [nucleo] [sci-t]*	核	kaku
" *{n} [sci-t]*	芯；心；しん	shin
core: core bar *[met]*	芯金	shin-gane
core: core wire *[elec]*	心線	shinsen
cored steel *[steel]*	中空鋼	chūkū-kō
core metal *[met]*	芯金	shin-gane
core removal	脱心；脱芯	dasshin
core storage *[comput]*	磁心記憶	ji'shin-ki'oku
corkscrew *[eng]*	栓抜き	sen-nuki
corkscrew staircase *[arch]*	螺旋階段	rasen-kaidan
corm *[bot]*	球茎	kyūkei
cormorant (bird) *[v-zoo]*	鵜；う	u
corn (grain) *[bot]*	玉蜀黍	tō'morokoshi
cornea *[anat]*	角膜	kakumaku
corneous *{adj}*	角状の	tsuno'jō no
corner *{n} [arch] [civ-eng]*	角；かど	kado
corn oil *[cook] [mat]*	玉蜀黍油	tō'morokoshi-yu
corn poppy (flower) *[bot]*	虞美人草	gubijin'sō
" (flower) *[bot]*	雛芥子；ひなげし	hina'geshi
corn starch *[cook] [mat]*	玉蜀黍澱粉	tō'morokoshi-denpun
corolla *[bot]*	花冠	kakan
corollary (pl -laries) *[logic]*	系	kei
corona *[astron]*	光冠；光環	kōkan
coronal *{adj} [astron]*	王冠の	ō'kan no
corpuscular cloud *[meteor]*	微粒子雲	bi'ryūshi-un
corpuscular theory *[opt]*	粒子説	ryūshi-setsu
correct *{adj} [math]*	正確な	seikaku na
" *{vb} [pr]*	補正する	hosei suru

correcting plate *[opt]*	補正板	hosei-ban
correction *[acous] [math]*	調整	chōsei
" *[legal]*	訂正; ていせい	teisei
" *[pat]*	補正	hosei
" *{n}*	手直し	te'naoshi
correction factor *[eng]*	補正率	hosei-ritsu
corrective maintenance *[eng]*	修理保守	shūri-hoshū
correctness; properness; legality *{n}*	正しさ	tadashi-sa
correlation *[math] [stat]*	相関(性)	sōkan(-sei)
correlation coefficient *[stat]*	相関係数	sōkan-keisū
correlation diagram *[math]*	相関図; 相關圖	sōkan-zu
correlation equation *[stat]*	相関式	sōkan-shiki
correlative property *[stat]*	相関性	sōkan-sei
correspondence *[math] [phys]*	対応; 對應	tai'ō
corresponding *{adj} [math]*	対応する	taiō suru
corresponding angles *[geom]*	同位角; どういかく	dōi-kaku
corresponding sides *[geom]*	対応辺; 對應邊	taiō-hen
corresponding state *{n}*	対応状態	taiō-jōtai
corridor *[arch]*	歩廊; ほろう	horō
" *[arch]*	廊下	rōka
corrosion; decay; rot *[met]*	腐食; 腐蝕	fu'shoku
" *[met]*	侵食; 浸蝕	shin'shoku
corrosion percent *[met]*	腐触率	fushoku-ritsu
corrosion-proofing *{n} [met]*	防食; 防蝕	bōshoku
corrosion resistance *[mat]*	耐食性	tai'shoku-sei
corrosive *{adj} [ch]*	腐触性の	fushoku'sei no
corrosive sublimate *[inorg-ch]*	昇汞; しょうこう	shōkō
corrugated fiberboard *[mat]*	段ボール	danbōru
corrugated iron sheet *[met]*	海鼠板; ナマコ板	namako-ita
corrugating medium *[paper]*	中芯; なかしん	naka-shin
corsican weed *[bot]*	海人草; 海仁草; まくり	makuri
cortex (pl cortices, cortexes) *[anat]*	皮質	hi'shitsu
cortex (pl cortex) *[bot]*	皮層	hisō
Cortinellus shiitake (mushroom) *[bot]*	椎茸; しいたけ	shītake
corundum *[miner]*	鋼玉	kōgyoku
cosecant *[trig]*	余割; 餘割	yo'katsu
cosine *[trig]*	余弦	yogen
cosine curve *[trig]*	余弦曲線	yogen-kyoku'sen
cosmetics *[mat]*	化粧品	keshō'hin
cosmic dust *[astron]*	宇宙塵	uchū-jin
cosmic rays *[nuc-phys]*	宇宙線	uchū-sen

cosmic-ray telescope *[opt]*	宇宙線望遠鏡	uchū'sen-bōen'kyō
cosmogony *[astrophys]*	宇宙成因論	uchū-sei'in-ron
" *[astrophys]*	宇宙進化論	uchū-shinka-ron
cosmological principle *[astron]*	宇宙原理	uchū-genri
cosmology *[astron]*	宇宙論	uchū-ron
cosmopolitan; world-wide *{adj}*	全世界的な	zen'sekai'teki na
cost *{n} [econ]*	原価	genka
cost; costs *{n} [econ]*	費用; ひよう	hiyō
cost of goods sold *[econ]*	売上原価; 賣上原價	uri'age-genka
cotangent *[trig]*	余接; 餘接	yo'setsu
coterminal angles *[trig]*	両辺共有の角	ryōhen-kyōyū no kaku
cotidal line *[map]*	同時潮線	dōji-chōsen
cotter pin *[eng]*	割ピン	wari-pin
cotton *{n} [bot] [tex]*	木綿; もめん	momen
" *{n} [bot] [tex]*	綿; わた	wata
" *{n} [tex]*	綿; めん	men
cotton aphid (insect) *[i-zoo]*	綿油虫	wata-abura-mushi
cotton candy *[cook]*	綿菓子	wata-gashi
cotton carpet; cotton rug *[furn]*	段通; だんつう	dantsū
cotton flock *[tex]*	落綿	raku-men
cotton gin *[eng]*	綿繰り機	wata-kuri-ki
cotton picker *[eng]*	綿摘み機	wata-tsumi-ki
cotton planter *[eng]*	綿蒔き機	wata-maki-ki
cotton rose (flower) *[bot]*	芙蓉; ふよう	fuyō
cottonseed oil *[mat]*	綿実油	men'jitsu-yu
cotton stripper *[eng]*	綿剥ぎ機	wata-hagi-ki
cotton wax *[mat]*	綿蠟	men-rō
cotton yarn *[tex]*	綿糸	menshi
cotyledon *[bot]*	子葉	shiyō
cough *{n} [physio]*	咳	seki
" *{vb} [physio]*	咳く; せく	seku
coulometer *[p-ch]*	電量計	denryō-kei
coulometric titration *[an-ch]*	電量滴定	denryō-teki'tei
count *{n} [tex]*	密度; みつど	mitsudo
" *{n} [tex]*	打込(本)数	uchi'komi(-hon)-sū
count; compute *{vb} [arith]*	勘定する	kanjō suru
" *{vb} [arith]*	数える	kazoeru
" *{vb} [arith]*	計算する	keisan suru
countdown *{n} [aero-eng]*	秒読み	byō-yomi
counter *[comput]*	計数器; 計數器	keisū-ki
" *[nucleo]*	計数管	keisū-kan
counter- *{prefix}*	反対の-	hantai no-

counteraction *[phys]*	反動	handō
" *[phys]*	反作用	han'sayō
counterbalance; offset *{vb}*	見合う	mi'au
counterclockwise rotation *[mech]*	逆時計回り	gyaku'tokei-mawari
" *[mech]*	反時計回り	han'tokei-mawari
countercurrent *[fl-mech] [sci-t]*	向流；こうりゅう	kōryū
countercurrent distribution method *[ch-eng]*	向流分配法	kōryū-bunpai-hō
counterdepressant *{n} [pharm]*	抗抑制薬	kō'yokusei-yaku
counterelectrode *[elec]*	対(向)電極	tai(kō)-denkyoku
counterelectromotive force *[elecmg]*	逆起電力	gyaku-kiden'ryoku
counterfeit *{adj}*	紛いの；擬いの	magai no
counterfeiting *{n} [legal]*	偽造；ぎぞう	gizō
counterglow *{n} [astron]*	対日照；對日照	tai'jitsu'shō
counterion *[ch]*	対イオン	tai-ion
counterpoint *[music]*	対位法	tai'i-hō
counterrotation *[mech]*	逆回転	gyaku-kaiten
countersinking *{n} [mech-eng]*	座繰り；生繰り	za'guri
countervail *{vb}*	対抗する	taikō suru
countervalue strategy *[mil]*	対価値戦略	tai'kachi-sen'ryaku
counterweight *[mech-eng]*	平衡錘	heikō'sui
" *[mech-eng]*	釣合い錘	tsuri'ai-omori
counting method *[math]*	計数法	keisū-hō
country *[geogr]*	国；國；くに	kuni
country claiming priority right *[pat]*	優先権主張国	yūsen'ken-shuchō-koku
country of manufacture *[ind-eng]*	製造国	seizō-koku
country of origin *[ind-eng]*	生産国名	seisan'koku-mei
couple *{n} [astron]*	連星	rensei
couple of forces *[phys]*	偶力；ぐうりょく	gū-ryoku
coupler *[eng]*	連結器	renketsu-ki
coupling *{n} [elec]*	結合	ketsu'gō
" *{n} [eng]*	継手；つぎて	tsugi'te
coupling agent *[ch]*	縮合剤	shukugō-zai
coupling constant *[phys]*	結合係数	ketsugō-keisū
course; process; stage *{n}*	過程	katei
court session *[legal]*	開廷	kaitei
courthouse *[arch] [legal]*	裁判所庁舎	saiban'sho-chōsha
courtroom *[legal]*	法廷	hōtei
courtyard *[arch]*	中庭；なかにわ	naka-niwa
covalence *[ch]*	共有原子価	kyōyū-genshi'ka
covalent bond *[ch]*	共有結合	kyōyū-ketsu'gō
covariance *[stat]*	共分散	kyō'bunsan

cover *{n} [eng]*	蓋；ふた	futa
" *{n} [eng]*	覆い；おおい	ōi
coverage *[graph]*	占有面積率	sen'yū-menseki-ritsu
covered-arc welding *{n} [met]*	被覆アーク溶接	hifuku-āku-yōsetsu
covering *{n} [mat]*	覆い物	ōi'mono
covering fabric *[tex]*	外装地；がいそうじ	gaisō'ji
covering film *[mat]*	被膜	himaku
covering fur (of an animal) *[mat]*	被毛	himō
covering power *[eng]*	隠蔽力	inpei-ryoku
" (by paint) *[eng]*	被覆力	hifuku-ryoku
covulcanization *[poly-ch] [rub]*	共加硫	kyō-karyū
cow (animal) *[v-zoo]*	雌牛	me-ushi
" (animal) *[v-zoo]*	牛；うし	ushi
cowberry *[bot]*	苔桃；こけもも	koke-momo
cowrie *[i-zoo]*	宝貝；寶貝	takara-gai
coxscomb; comb *[v-zoo]*	鶏冠；鳥冠；とさか	tosaka
cozymase *[bioch]*	発酵補酵素	hakkō-ho'kōso
crab (crustacean) *[i-zoo]*	蟹；かに	kani
crab-eating macaque; Macaca irus (animal: monkey) *[v-zoo]*	蟹喰い猿；かに=くいざる	kani'kui-zaru
crab nebula *[astron]*	蟹星雲	kani-sei'un
crab spider (spider) *[i-zoo]*	蟹蜘蛛；かにぐも	kani-gumo
crack *{n} [geol]*	割れ目	ware'me
" *{n} [sci-t]*	亀裂；きれつ	ki'retsu
" *{n} [sci-t]*	裂け目	sake'me
" *{n} [sci-t]*	割れ	ware
" *{vb t}*	割る	waru
cracking *{n} [org-ch] [petrol]*	分解	bunkai
cradle (for a baby) *{n} [furn]*	揺り籠；搖り籠	yuri-kago
craft; skill; art *{n}*	技；ぎ；わざ	gi; waza
cranberry (fruit) *[bot]*	蔓苔桃	tsuru-koke'momo
crane *{n} [mech-eng]*	起重機	kijū-ki
crane (bird) *[v-zoo]*	鶴；つる	tsuru
crane, Japanese (bird) *[v-zoo]*	丹頂鶴	tanchō-zuru
cranium *[anat]*	頭蓋	tōgai; zugai
crank *{n} [mech-eng]*	回転腕	kaiten-ude
craping *{n} [tex]*	皺；しぼ	shibo
crater *[geol]*	火口	kakō
" *[geol]*	漏斗孔	rōto-kō
crater basin (of a volcano) *[geol]*	火口原	kakō-gen
crater lake *[hyd]*	火口湖	kakō-ko
crawfish; crayfish *[i-zoo]*	蝲蛄；ざりがに	zari-gani

crawl; creep *{vb}*	這う；はう	hau
crazing *{n} [cer] [met]*	罅割れ；ひび割れ	hibi'ware
" *{n} [eng]*	罅；ひび	hibi
cream of tartar *[cook]*	酒石英	shuseki'ei
crease; pucker; wrinkle *{n} [tex]*	皺；しわ	shiwa
creaseproof *{adj} [mat]*	耐皺性の	tai'shiwa'sei no
creaseproofness *[mat]*	耐皺性	tai'shiwa-sei
crease resistance *[tex]*	防皺性	bō'shiwa-sei
create *{vb}*	創作する	sō'saku suru
creek; inlet *[geogr]*	浦；うら	ura
creel; fish basket *[eng]*	魚籃；魚籠；びく	biku
creep *{n} [plas]*	匍匐；ほふく	ho'fuku
crepe; crepe cloth *[tex]*	縮緬；ちりめん	chirimen
crepe myrtle; crape myrtle; Lagerstroemia indica *[bot]*	百日紅；猿滑；さるすべり	saru-suberi
crescent moon *[astron]*	三日月；みかづき	mika-zuki
crest *{n} [arch] [cl] [naval arch]*	冠	kanmuri
" (screw thread) *{n} [eng]*	山；やま	yama
(of a wave) *{n} [ocean]*	山	yama
crest factor *[phys]*	波高率	hakō-ritsu
crest line *[geol]*	峰線	mine-sen
crevasse *[geol]*	裂け目	sake'me
crevice *[sci-t]*	間隙；かんげき	kangeki
" *[sci-t]*	空隙	kūgeki
crevice; gap; aperture *{n}*	隙間；すきま	sukima
crew *{n} [trans]*	乗組員；乗組員	nori'kumi'in
" *{n} [trans]*	属員	zoku'in
cricket (insect) *[i-zoo]*	蟋；こおろぎ	kōrogi
crimp *{n} [eng] [tex]*	捲縮	kenshuku
" *{n} [eng] [tex]*	皺；しぼ	shibo
crimson (color) *[opt]*	茜色；あかねいろ	akane-iro
" (color) *[opt]*	真紅；深紅	shinku
crimson sea bream (fish) *[v-zoo]*	血鯛；ちだい	chi-dai
crisis (pl crises)	危機	kiki
cristabolite *[miner]*	方珪石	hō'keiseki
criterion (pl -ria) *[math]*	目安；めやす	me'yasu
criterion; norm; standard *[ind-eng]*	規準	kijun
critical angle *[math]*	臨界角	rinkai-kaku
critical density *[astron]*	臨界密度	rinkai-mitsudo
critical frequency *[electr]*	臨界振動数	rinkai-shindō'sū
critical humidity *[ch-eng]*	臨界湿度	rinkai-shitsu'do
critical index *[phys]*	臨界指数	rinkai-shisū

criticality *[nucleo]*	臨界；りんかい	rinkai
critical-limit test *[mech-eng]*	限界試験	genkai-shiken
critical mass *[nucleo]*	臨界質量	rinkai-shitsu'ryō
critical micelle concentration; CMC *[p-ch]*	臨界ミセル濃度	rinkai-miseru-nōdo
critical point *[math] [p-ch]*	臨界点	rinkai-ten
critical point; key point *[comm]*	要点；要點	yōten
critical pressure *[fl-mech]*	臨界圧	rinkai-atsu
critical region *[stat]*	棄却域	ki'kyaku-iki
critical temperature *[p-ch]*	臨界温度	rinkai-ondo
critique; commentary *{n} [pr]*	批評；ひひょう	hihyō
croaker (fish) *[v-zoo]*	ぐち	guchi
" (fish) *[v-zoo]*	石首魚；石持	ishimochi
crock; earthenware jug; pot; urn *[cer]*	瓶；甕；かめ	kame
crockery; earthenware *[eng]*	土器；どき	doki
Crocodilia *[v-zoo]*	鰐類；わに類	wani-rui
crocodiling; alligatoring *{n} [mat] [met]*	鰐皮割れ	wani'gawa-ware
crocoite *[miner]*	紅鉛鉱	kōen'kō
crop; crops; farm products *{n} [agr]*	作物；さくもつ	saku'motsu
crop *{n} [v-zoo]*	嗉嚢；そのう	so'nō
crop science *[agr]*	作物学	sakumotsu'gaku
cross *{vb} [mech] [traffic]*	横断する	ō'dan suru
cross; pass (over, across) *{vb}*	越す	kosu
cross; traverse *{vb t}*	横切る	yoko-giru
crossarm; crosspiece; semaphore *[rail]*	腕木；うでぎ	ude'gi
crossbeak; crossbill (bird) *[v-zoo]*	鶍；いすか	isuka
crossbow *[ord]*	弩；大弓	ō-yumi
crosscurrent *[fl-mech]*	横流；おうりゅう	ō'ryū
crosscut adhesion test *[eng]*	碁板目試験	goban'me-shiken
crosscut saw *[eng]*	横引き鋸	yoko'biki-nokogiri
crosscut sawing *{n} [eng]*	横引き；よこびき	yoko-biki
cross cutter *[eng]*	横裁ち装置	yoko'dachi-sōchi
crossed helical gear; screw gear *[mech-eng]*	ねじ歯車	neji-ha'guruma
cross hair; cross wire *[eng]*	十字線	jūji-sen
crossing; intersection *{n} [traffic]*	横断；おうだん	ō'dan
crossing-over; cross *{n} [math]*	交叉；こうさ	kōsa
crossing symmetry *[part-phys]*	交差対称性	kōsa-taishō-sei
cross-linking *{n} [org-ch] [poly-ch]*	橋架け；はしかけ	hashi-kake
" *{n} [org-ch] [poly-ch]*	架橋；かきょう	kakyō
cross-linking agent *[poly-ch]*	架橋剤	kakyō-zai
crosslinking density *[org-ch]*	架橋密度	kakyō-mitsudo
cross-link site *[poly-ch]*	架橋点	kakyō-ten

cross orthogonally *{vb} [math]*	直交する	chokkō suru
cross product *[math]*	交差積；こうさせき	kōsa-seki
cross-reference table *[pr]*	対照表	taishō-hyō
crossrelaxation *[phys]*	交差緩和	kōsa-kanwa
cross-sealing *{n}*	交叉封鎖	kōsa-fūsa
cross section *[geol]*	断面(図)	danmen(-zu)
cross section *[math]*	横断面；横斷面	ō'dan-men
cross-sectional area *[math]*	断面積	dan'menseki
crosstalk *{n} [comm]*	漏話；ろうわ	rōwa
crosswalk *{n} [civ-eng]*	横断歩道	ō'dan-hodō
crosswind *[meteor]*	横風	yoko-kaze
cross(wise) direction; transverse direction *[plas] [sci-t]*	横方向	yoko-hōkō
cross wire *[eng]*	十字線	jūji-sen
croton (plant) *[bot]*	巴豆；はず	hazu
croton oil *[mat]*	巴豆油	hazu-yu
crow (bird) *[v-zoo]*	烏；鴉；からす	karasu
crown (of a tree) *{n} [bot]*	樹冠	jukan
crown *{n} [dent]*	歯冠	shikan
" *{n} [lap]*	王冠	ōkan
" *{n} [mech]*	冠；かんむり	kanmuri
crowned crane (bird) *[v-zoo]*	冠鶴	kanmuri-zuru
crown gear *[mech]*	冠歯車	kanmuri-ha'guruma
crown radical (of a kanji) *[pr]*	冠	kanmuri
crown saw *[eng]*	冠鋸	kanmuri-nokogiri
crow's foot *[cl]*	閂止め	kannuki-dome
crucian carp; Prussian carp (fish) *[v-zoo]*	鮒；ふな	funa
crucible *[sci-t]*	坩堝；るつぼ	rutsubo
cruciform *{adj} [sci-t]*	十字形の	jūji'kei no
crude drug *[pharm]*	生薬；しょうやく	shō-yaku
crude oil *[geol]*	原油	gen-yu
crude-oil-resistant steel *[steel]*	耐原油性鋼	tai-gen'yu'sei-kō
crude protein *[bioch]*	粗蛋白質	so-tanpaku'shitsu
crude rubber *[org-ch]*	生ゴム	nama-gomu
cruise (by ship) *{n} [trans]*	巡洋航海	jun'yō-kōkai
" (by ship) *{n} [trans]*	巡航；じゅんこう	junkō
cruiser *[nav-arch]*	巡洋艦	jun'yō'kan
cruise ship *[nav-arch]*	巡洋船	jun'yō'sen
crumble; collapse *{vb i}*	崩れる	kuzureru
crush (shatter) *{vb t}*	砕く	kudaku
crush (smash; mash) *{vb t}*	潰す	tsubusu

crusher *[eng]*	粉砕機	funsai-ki
" *[eng]*	破砕機	hasai-ki
crushing *{n} [min-eng]*	破砕	hasai
crust *[geol]*	地殻；地殼；ちかく	chi'kaku
Crustacea *[i-zoo]*	甲殻類；甲殼類	kōkaku-rui
crustacean *{adj} [i-zoo]*	甲殻綱の	kōkaku-kō no
crustal movement *[geol]*	地殻運動	chi'kaku-undō
crustal strain map *[geol]*	歪地図；ひずみちず	hizumi-chizu
crutches *[eng] [med]*	松葉杖	matsuba'zue
crying *{n} [physio]*	泣くこと	naku koto
cryogen *[cryo]*	寒剤	kanzai
" *[cryo]*	冷凍剤	reitō-zai
cryogenics *[phys]*	低温(物理)学	teion(-butsuri)'gaku
cryogenic separation *[eng]*	深冷分離	shinrei-bunri
cryohydrate *[ch]*	(含)氷晶	(gan)hyōshō
cryolite *[miner]*	氷晶石	hyō'shōseki
cryometer *[eng] [thermo]*	低温計	teion-kei
cryophyte *[ecol]*	氷雪植物	hyō'setsu-shokubutsu
cryopreservation *[eng]*	凍結保存	tōketsu-hozon
cryoprotectant *[mat] [microbio]*	凍結保護剤	tōketsu-hogo-zai
cryoscopy *[an-ch]*	氷点法	hyōten-hō
cryptic *{adj} [zoo]*	潜伏した	sen'puku shita
" *{adj} [zoo]*	潜在した	senzai shita
cryptogram *[comm]*	暗号；あんごう	angō
cryptography *[comm]*	暗号法	angō-hō
cryptomeria (tree) *[bot]*	杉；すぎ	sugi
cryptomeria resin *[mat]*	杉脂；杉やに	sugi-yani
crystal *[crys]*	結晶；けっしょう	kesshō
crystal clock; quartz clock *[horol]*	水晶時計	suishō-dokei
crystal form *[crys]*	結晶型	kesshō-gata
" *[crys]*	結晶形(態)	kesshō-kei(tai)
crystal habit *[crys]*	晶癖；しょうへき	shō'heki
crystal lattice *[crys]*	結晶格子	kesshō-kōshi
crystalline *{adj} [crys]*	結晶性の	kesshō'sei no
crystalline polymer *[poly-ch]*	結晶性重合体	kesshō'sei-jūgō'tai
crystalline rock *[geol]*	結晶片岩	kesshō-hengan
crystallinity *[crys]*	結晶度	kesshō-do
crystallite *[geol]*	微結晶	bi'kesshō
" *[geol]*	結晶子	kesshō'shi
crystallization *[crys]*	結晶化	kesshō'ka
" *[crys]*	晶析	shōseki
" *[crys]*	晶出	shō'shutsu

crystallization nucleus *[ch-eng]*	晶出核	shō'shutsu-kaku
crystallize *{vb i} [ch]*	結晶化する	kesshō'ka suru
crystallizing out *[ch-eng]*	晶出	shō'shutsu
crystallographic axis *[crys]*	結晶軸	kesshō-jiku
crystallography *[phys]*	結晶学	kesshō'gaku
crystalloid *{n} [bot]*	晶質; しょうしつ	shō-shitsu
crystal orientation *[crys]*	結晶方位	kesshō-hō'i
crystal-pulling method *[crys]*	結晶引上げ法	kesshō-hiki'age-hō
crystal resonator *[electr]*	水晶振動子	suishō-shindō'shi
crystal structure *[crys]*	結晶構造	kesshō-kōzō
crystal sugar *[cook]*	双目; ざらめ	zara'me
" *[cook]*	氷砂糖	kōri-zatō
crystal system *[crys]*	晶系	shōkei
crystal whisker *[crys]*	鬚結晶; ひげ結晶	hige-kesshō
Ctenophora; comb jellies *[i-zoo]*	有櫛動物	yū'shitsu-dōbutsu
cube *{n} [alg]*	三乗; 三乘	sanjō
" *{n} [geom]*	立方(体)	rippō('tai)
" *{n} [geom]*	立体; りったい	rittai
" *{n} [geom]*	正六面体	sei-roku'men'tai
" *{vb} [math]*	三乗する	sanjō suru
cube root *[alg]*	三乗根	sanjō-kon
" *[arith]*	立方根	rippō-kon
cubic *{adj} [geom]*	立体の	rittai no
cubic closest-packed structure *[crys]*	立方最密構造	rippō-saimitsu-kōzō
" *[crys]*	立体最密構造	rittai-saimitsu-kōzō
cubic crystal *[crys]*	立方晶	rippō-shō
cubic curve *[math]*	三次曲線	sanji-kyoku'sen
cubic system *[crys]*	立方晶系	rippō-shōkei
cuboctahedron *[geom]*	立方八面体	rippō-hachi'men'tai
cuboidal *{adj} [geom]*	立方形の	rippō'kei no
cuckoo (bird) *[v-zoo]*	杜鵑; 時鳥; 不如帰	hototogisu
cuckoo, Japanese (bird) *[v-zoo]*	郭公; かっこう	kakkō
cucumber (vegetable) *[bot]*	胡瓜; きゅうり	kyūri
cul-de-sac (pl culs-de-sac) *[civ-eng]*	袋小路	fukuro-kōji
Culex pipens pallens (insect: red house mosquito) *[i-zoo]*	赤家蚊; あかいえか	aka-ie-ka
culmination *[astron]*	南中	nanchū
cultivar; cultigen *[bot]*	栽培変種植物	saibai-henshu-shokubutsu
cultivate *{vb} [microbio]*	培養する	baiyō suru
cultivation *[agr]*	栽培	saibai
culture *{n} [microbio]*	培養; ばいよう	baiyō
culture medium *[microbio]*	培養基	baiyō-ki

culture medium *[microbio]*	培地；ばいち	baichi
cumulative curve *[math]*	累積曲線	rui'seki-kyoku'sen
cumulative distribution function *[stat]*	累積分布関数	rui'seki-bunpu-kansū
cumulative sum *[arith]*	累和	rui-wa
cumulonimbus *[meteor]*	入道雲	nyudō-gumo
" *[meteor]*	積乱雲	sekiran'un
cumulus *[meteor]*	積雲	seki'un
cuneiform *{adj}* *[ling]*	楔状の	kusabi'jō no
cuneiform character *[comm]*	楔形文字	kusabi'gata-moji
-cup; -cups *{suffix}*	-杯	-hai; -pai; -bai
cupboard *[arch]*	戸棚	to'dana
" (pantry; sideboard) *[furn]*	食器棚	shokki-dana
cup for Japanese wine *[cer]* *[cook]*	坏；盃；さかずき	sakazuki
cupric chloride	塩化第二銅	enka-daini-dō
cupric sulfate	硫酸第二銅	ryūsan-dai'ni-dō
cupro-nickel *[met]*	白銅	haku'dō
cuprous oxide	亜酸化銅	a'sanka-dō
curbstone *[civ-eng]*	縁石；ふちいし	fuchi-ishi
curcuma *[bot]*	鬱金；うこん	u'kon
curd *[cook]*	凝乳	gyōnyū
curdling *{n}* *[bioch]*	凝結	gyōketsu
" *{n}* *[ch]*	凝固	gyōko
cure *{n}* *[ch-eng]* *[rub]*	加硫	karyū
curing (cement) *{n}* *[civ-eng]*	養生；ようじょう	yōjō
curing *{n}* *[org-ch]*	硬化	kōka
curing agent *[poly-ch]*	硬化剤	kōka-zai
curing catalyst *[ch]*	硬化觸媒	kōka-shokubai
curios *[furn]*	骨董品	kottō'hin
curium (element)	キユリウム	kyuryūmu
curl *{n}* *[tex]*	捲縮	kenshuku
currency; current money *[econ]*	通貨	tsūka
current *{n}* *[hyd]*	流れ	nagare
" *{n}* *[elec]*	電流；でんりゅう	denryū
current; present *{adj}*	現在の	genzai no
current-carrying capacity *[elec]*	通電容量	tsūden-yōryō
current density *[elec]*	電流密度	denryū-mitsudo
current efficiency *[elec]*	電流効率	denryū-kō'ritsu
currently in use *{adj}* *[pat]*	現用の	gen'yō no
current price *[econ]*	時価	jika
curry (condiment) *[cook]*	カレー	karē
cursive-style writing *[comm]*	行書体	gyōsho-tai

cursive-style writing; enhanced-style writing *[pr]*	草字体；そうじたい	sōji-tai
cursor *[comput]*	位置表示機構	ichi-hyōji-kikō
" *[comput]*	滑子；かっし	kasshi
curtain *[furn]*	幕；まく	maku
curvature *[math]*	曲率；きょくりつ	kyoku-ritsu
curve (curved line) *{n} [geom]*	曲線	kyoku-sen
" (curving) *{n} [geom]*	湾曲；彎曲	wan'kyoku
curved *{adj} [geom]*	曲った；まがった	magatta
curved crystal *[crys]*	湾曲結晶	wan'kyoku-kesshō
curved surface *[geom]*	曲面	kyoku-men
curvilinear coordinates *[math]*	曲線座標	kyoku'sen-zahyō
cusp *[astron]*	先点；せんてん	senten
custard apple (fruit) *[bot]*	牛心梨	gyūshin'ri
custodian; trustee *[legal]*	保管者	hokan'sha
custody; safekeeping *{n} [legal]*	保管；ほかん	hokan
customer *[econ]*	客；きゃく	kyaku
customs office *[econ]*	税関；税關	zeikan
cut *{vb t}*	切る	kiru
cut horizontally *{vb}*	横断する	ō'dan suru
cut vertically *{vb t}*	縦断する；縱斷する	jūdan suru
cute; charming *{adj}*	可愛らしい	kawairashī
cuticle; epidermis *[hist]*	表皮	hyōhi
cutlass fish; scabbard fish *[v-zoo]*	太刀魚；たちうお	tachi-uo
cutlery *[cook] [food-eng]*	刃物；刄物；はもの	ha'mono
cutoff frequency *[electr]*	遮断周波数	shadan-shūha'sū
cutting *{n} [bot]*	切穂；きりほ	kiri'ho
" *{n} [mech-eng]*	切断；せつだん	setsu'dan
" *{n} [met]*	切削	sessaku
cutting bed (nursery) *[bot]*	挿木床；さしきどこ	sashi'ki-doko
cutting blade *[eng]*	切り刃	kiri-ha
cutting edge *[eng]*	切れ刃	kire-ha
cutting oil *[mat]*	切削油	sessaku-yu
cutting plane *[graph]*	切断面	setsu'dan-men
cutting pliers; pincers (a tool) *[eng]*	ペンチ	penchi
cutting quality; sharpness *[eng]*	切れ味；きれあじ	kire-aji
cutting working *{n} [met]*	切削加工	sessaku-kakō
cuttlefish; inkfish; squid *[i-zoo]*	烏賊；いか	ika
cut-to-order sheet *[mat]*	切り板；きりばん	kiri-ban
cutworm (insect) *[i-zoo]*	根切り虫	ne'kiri-mushi
cyan (color) *[opt]*	青緑色	sei'ryoku-shoku
cyanide ion *[ch]*	シアン(化物)イオン	shian('ka'butsu)-ion

cyaniding process *[met]*	青化法	seika-hō
cyanite; kyanite *[miner]*	藍晶石	ran'shōseki
Cyanophyta *[bot]*	藍色植物	ran'shoku-shokubutsu
cybernetics *[sci-t]*	人工頭脳学	jinkō-zunō'gaku
cycad *[bot]*	蘇鉄; 蘇鐵; そてつ	so'tetsu
cycle *{n} [comput]*	段階	dankai
" *{n} [sci-t]*	サイクル	saikuru
" *{n} [sci-t]*	周期; しゅうき	shūki
cyclic *{adj} [elec] [org-ch]*	環状の	kanjō no
cyclic compound *[org-ch]*	環式化合物	kan'shiki-kagō'butsu
cyclic hydrocarbon *[org-ch]*	環状炭化水素	kanjō-tanka'suiso
cyclic photophosphorylation *[bioch]*	循環的光燐酸化	junkan'teki-kō'rinsan'ka
cyclic storage *[comput]*	循環記憶	junkan-ki'oku
cyclization *[org-ch]*	環化	kanka
cycloid *[an-geom]*	擺線; はいせん	haisen
cyclone *[meteor]*	旋風	senpū; tsumuji-kaze
cyclopedia *[pr]*	事典	jiten
cyclopolymerization *[poly-ch]*	環化重合	kanka-jūgō
cyclotron *[nucleo]*	サイクロトロン	saikurotoron
Cygnus; the Swan *[constel]*	白鳥座	hakuchō-za
cylinder *[geom]*	柱	chū
" *[geom]*	(円)筒	(en)tō
cylindrical *{adj} [geom]*	円柱状の	enchū'jō no
cylindrical coordinates *[geom]*	円筒座標	entō-zahyō
" *[geom]*	柱座標	chū-zahyō
cylindrical grinding *{n} [mech]*	円筒研削	entō-kensaku
cylindrical wave *[elecmg]*	円筒波	entō-ha
cypress (tree) *[bot]*	檜; ひのき	hinoki
cyprinodont; Oryzias latipes (fish) *[v-zoo]*	目高; めだか	medaka
cyst *[med]*	嚢子; のうし	nōshi
" *[hist]*	嚢胞; 囊胞	nōhō
cytochemistry *[ch]*	細胞化学; 細胞化學	saibō-kagaku
cytokinesis *[cyt]*	細胞質分裂	saibō'shitsu-bun'retsu
cytology *[bio]*	細胞学	saibō'gaku
cytoplasm *[cyt]*	細胞質	saibō-shitsu

D

dace (pl dace) (fish) *[v-zoo]*	鮠；はえ；はや	hae; haya
" (fish) *[v-zoo]*	鯎；石斑魚；うぐい	ugui
daddy longlegs (spider) *[i-zoo]*	足高蜘蛛	ashi'daka-gumo
daffodil (flower) *[bot]*	喇叭水仙	rappa'zuisen
dagger; short sword *[ord]*	短剣；短劍	tanken
daikon radish (vegetable) *[bot]*	大根；だいこん	daikon
daily mean temperature *[meteor]*	日平均気温	nichi-heikin-ki'on
dairy farming *{n} [agr]*	酪農	raku-nō
dairy product *[agr]*	乳製品	nyū-seihin
dam *{n} [civ-eng]*	堰；せき	seki
damage *{n} [sci-t]*	損害	songai
" *{n} [sci-t]*	損傷	sonshō
damage; hurt *{vb t}*	痛める；傷める	itameru
damage compensation *[legal]*	損害賠償	songai-baishō
damage threat *[mil]*	損害脅威	songai-kyō'i
damask; satin damask *[tex]*	緞子；どんす	donsu
damp; humid; moist *{adj} [hyd]*	湿気のある	shikki no aru
damped oscillation *[phys]*	減衰振動	gensui-shindō
damper *[mech-eng]*	制動子	seidō'shi
" *[music]*	止音器	shi'on-ki
damp ground *[geol]*	湿地；濕地；しっち	shitchi
damping *{n} [phys]*	減衰	gensui
" *{n} [phys]*	制動	seidō
damping ratio *[phys]*	減衰比	gensui-hi
dandelion (flower) *[bot]*	蒲公英；たんぽぽ	tanpopo
danger; hazard *[ind-eng]*	危険；危險；きけん	kiken
dangerous *{adj}*	危い	abunai
dangerous object *[mat]*	危険物	kiken'butsu
daphne; Daphne odora (bush) *[bot]*	沈丁花	jinchōge
dark *{adj}*	暗い	kurai
dark adaptation *[opt]*	暗順応	an-junnō
dark color *[opt]*	暗色	an-shoku
dark-colored soy sauce *[cook]*	濃口醤油	koi'kuchi-shōyu
dark current *[electr]*	暗電流	an-denryū
" *[electr]*	暗流	anryū
darken; become dark *{vb i}*	暗くなる	kuraku naru
darken; make dark *{vb t}*	暗くする	kuraku suru
dark-field image *[opt]*	暗視野像	an'shiya-zō
dark lane *[astron]*	暗帯；暗帶	antai

dark line *[spect]*	暗線	ansen
dark nebula *[astron]*	暗黒星雲；暗黑星雲	ankoku-sei'un
darkness *{n} [astron]*	(暗)闇	(kura-)yami
dark reaction *[bioch] [ch]*	暗反応；暗反應	an-hannō
darkroom *[photo]*	暗室	an-shitsu
dark star *[astron]*	暗黒星；暗黑星	ankoku-sei
darter (fish) *[v-zoo]*	矢魚；やうお	ya'uo
dash (Morse code) *[comm]*	長点；長點	chōten
dashpot *[mech-eng]*	弾き壺	hajiki-tsubo
dasyure (animal) *[v-zoo]*	袋猫；ふくろねこ	fukuro-neko
data (sing datum) *[math]*	資料	shiryō
data processing *[comput]*	データー処理	dētā-shori
date; dating *{n} [comput] [pat]*	日付	hi'zuke
date (year, month, day)	年月日	nen'gappi
date line, international *[astron]*	日付変更線	hi'zuke-henkō-sen
date of birth *[bio]*	生年月日	sei-nen'gappi
date of mailing *[pat]*	発送日	hassō-bi
date palm (tree) *[bot]*	棗椰子	natsume-yashi
dating; estimation of date *[archeo]*	年代測定	nendai-sokutei
dative case *[gram]*	与格；與格；よかく	yo-kaku
daughter cell *[cyt]*	娘細胞；孃細胞	jō-saibō
davit *{n} [nav-arch]*	鉤柱	kagi-bashira
dawn *{n} [astron]*	暁；曉	akatsuki
" *{n} [astron]*	夜明け	yo'ake
day *[astron]*	日；ひ	hi
day after tomorrow *{n}*	明後日；あさって	asatte
" *{n}*	明後日；みょうご日	myōgo-nichi
day before yesterday *{n}*	一昨日	issaku-jitsu
" *{n}*	一昨日；おととい	ototoi
-day of week *{suffix}*	-曜日	-yōbi
dayfly (insect) *[i-zoo]*	蜉蝣；かげろう	kagerō
daylight *[astron]*	昼光；晝光	chūkō
" *[astron]*	日光	nikkō
daylight factor *[opt]*	昼光率	chūkō-ritsu
daylight lamp *[elec]*	昼光電球	chūkō-denkyū
daylight saving time *[astron]*	夏時間	natsu-jikan
" *[astron]*	日光節約時間	nikkō-setsu'yaku-jikan
daytime *[astron]*	昼；晝	hiru
" *[astron]*	日中	nitchū
dazzle *{n} [opt] [physio]*	ギラツキ	giratsuki
dazzlement; bewilderment *{n} [psy]*	眩惑；げんわく	gen'waku
deacidification *[ch]*	脱酸	dassan

deactivation *[ch]*	失活; しっかつ	shikkatsu
dead ahead (ship) *{adv} [navig]*	正前に	seizen ni
dead angle *[eng]*	死角	shikaku
dead astern (ship) *{adv} {n} [navig]*	正後に	seigo ni
dead center; dead point *[mech-eng]*	死点	shi-ten
dead end *[traffic]*	行き止り	yuki-domari
dead time *[comput]*	不感時間	fukan-jikan
" *[eng]*	無駄時間	muda-jikan
dead weight *[mech]*	死重; しじゅう	shijū
deaeration *[mech-eng]*	脱気	dakki
deaerator *[mech-eng]*	脱気装置	dakki-sōchi
deafness *[med]*	聾; つんぼ	tsunbo
deallocation *[comput]*	割振解除	wari'furi-kaijo
deashing *{n} [ch]*	脱灰	dakkai
deasphalting *{n} [ch-eng]*	脱瀝	datsu'reki
death *[bio]*	死	shi
death cup (poison mushroom) *[bot]*	天狗茸	tengu-take
death phase *[microbio]*	死滅期	shi'metsu-ki
death's head moth (insect) *[i-zoo]*	髑髏面形天蛾	dokuro-mengata-suzume
debris *[ecol] [geol]*	がらくた	garakuta
" (coal) *{n} [geol]*	硏; ずり	zuri
debugging *{n} [comput]*	手直し; てなおし	te'naoshi
" (eliminating insects) *{n}*	虫取り	mushi'tori
decade (ten years)	十年間	jūnen'kan
decagon *[geom]*	十角形	jikkaku-kei; jukkaku-kei
" *[geom]*	十辺形	jippen-kei
decagram (unit of mass) *[mech]*	瓧; デカグラム	deka'guramu
decalcification *[ch]*	脱灰	dakkai
decalescence *[met]*	減輝	genki
decameter (unit of length) *[mech]*	籵; デカメートル	deka'mētoru
decantation *[eng]*	傾瀉; けいしゃ	keisha
decarboxylation *[org-ch]*	脱炭酸	dattansan
decarboxylation polymerization *[poly-ch]*	脱炭酸重合	dattansan-jūgō
decarburization *[met]*	脱炭	dattan
decarburization annealing *[met]*	脱炭焼鈍	dattan-yaki'namashi
decavalent *{adj} [ch]*	十価の	jikka no; jukka no
decay *{n} [bio] [mat]*	腐れ	kusare
" (radioactive) *{n} [nuc-phys]*	崩壊; 崩壞	hōkai
" (radioactive) *{n} [nuc-phys]*	放射崩解	hōsha-hōkai
" ; rot *{vb i}*	腐る; くさる	kusaru
decay constant *[nuc-phys]*	崩壊定数	hōkai-teisū

English	Japanese	Romanization
decay pattern *[acous]*	残響図形；殘響圖形	zankyō-zukei
deceleration *[mech]*	減速；げんそく	gensoku
deceleration rate *[mech]*	減速率	gensoku-ritsu
decelerator *[eng]*	減速機	gensoku-ki
December (month)	十二月	jūni'gatsu
deciduous *{adj} [bot]*	落葉性の	raku'yō'sei no
deciduous teeth and molars *[dent]*	乳臼歯	nyūkyū'shi
decigram (unit of mass) *[mech]*	瓰；デシグラム	deshi'guramu
decimal *{adj} [arith]*	十進の；じっしんの	jisshin no
decimal fraction *[math]*	十進小数	jisshin-shōsū
decimal number *[arith]*	十進数	jisshin-sū
decimal numeration *[arith]*	十進法	jisshin-hō
decimal point *[arith]*	小数点	shōsū-ten
decimals *[math]*	小数	shōsū
decimal system *[arith]*	十進法	jisshin-hō; jusshin-hō
" *[arith]*	十進式	jisshin-shiki; jusshin-shiki
decision *{n}*	決定	kettei
decision-making *{n} [psy]*	意思決定	ishi-kettei
decision to reject *[pat]*	拒絶査定	kyo'zetsu-satei
deck (of a ship) *[nav-arch]*	甲板	kanban; kōhan
declarative statement *[comput]*	宣言文	sengen'bun
declare; proclaim *{vb}*	宣言する	sengen suru
declination *[astron]*	偏角；へんかく	henkaku
" *[astron]*	赤緯；せきい	seki'i
declination axis *[eng]*	赤緯軸	seki'i-jiku
declination circle *[eng]*	赤緯環	seki'i-kan
declivity (pl -ties) *[geol]*	下り勾配	kudari-kōbai
" *[geol]*	下向き傾斜	shita'muki-keisha
decoct *{vb t} [food-eng]*	煎じる	senjiru
decoction *[pharm]*	煎薬	sen-yaku
decoder *[comm]*	解読器；解讀器	kaidoku-ki
decolorization speed *[opt]*	消色速度	shōshoku-sokudo
decompose; resolve *{vb t}*	分解する	bunkai suru
decomposer *[ecol]*	分解者	bunkai'sha
decomposition *[ch]*	分解；ぶんかい	bunkai
decomposition point *[ch]*	分解点	bunkai-ten
decompression chamber *[eng]*	減圧室；減壓室	gen'atsu-shitsu
decontamination *[eng]*	浄化；淨化	jōka
" *[eng]*	除染	josen
decontamination index *[nucleo]*	除染指数	josen-shisū
decor *[furn]*	装飾	sōshoku
decorate; adorn *{vb}*	飾る	kazaru

decorative sheet *[mat] [poly-ch]*	化粧板	keshō-ban
decoy; lure *{n} [bioch]*	誘引物質	yūin-busshitsu
" *{n} [mil]*	囮；おとり	otori
decrease; reduce *{vb t}*	減らす；へらす	herasu
decreasing function *[alg]*	減少関数	genshō-kansū
decree *[legal]*	布告	fu'koku
decrement *[comput] [math]*	減分	genbun
" *[math]*	減少量	genshō'ryō
dedicated line *[comput]*	専用回線	sen'yō-kaisen
deduction *[logic] [math]*	演繹法	en'eki-hō
" *[logic]*	推論	suiron
deep *{adj}*	深い；ふかい	fukai
deep-drawability *[steel]*	深絞り性	fuka'shibori-sei
deep-drawing *{n} [eng] [steel]*	深絞り	fuka-shibori
deepen; intensify *{vb t}*	深める	fukameru
deep red (color) *[opt]*	臙脂色；えんじいろ	enji-iro
deep sea *{n} [geol]*	遠洋	en'yō
deep-sea *{adj} [ocean]*	深海の	shinkai no
deep-sea cable *[elec]*	深海線	fukami-sen
deep-sea floor *[geol]*	深海床	shinkai-shō
deer (stag, doe) (pl deer) (animal) *[v-zoo]*	鹿；しか	shika
deerfly (insect) *[i-zoo]*	虻；あぶ	abu
" (insect) *[i-zoo]*	恙虫；つつがむし	tsutsuga-mushi
defatted soybean *[cook]*	脱脂大豆	dasshi-daizu
default *{n} [legal]*	不履行	fu'rikō
" *{n} [legal]*	欠席	kesseki
" *{n} [legal]*	懈怠	ke'tai
default assumption *[comput]*	省略時解釈	shōryaku'ji-kaishaku
defect (flaw, fault) *{n} [ind-eng]*	疵；瑕；きず	kizu
" (flaw; fault) *{n} [sci-t]*	欠陥	kekkan
" *{n} [sci-t]*	欠点	ketten
" (shortcoming; demerit) *{n}*	短所	tansho
defense *[legal]*	弁護；辯護；べんご	bengo
" *{n}*	防衛	bōei
" *{n}*	防御；防禦	bōgyo
defense mechanism *[psy]*	防衛機構	bōei-kikō
defense reaction *[bio] [psy]*	防御反応	bōgyo-hannō
deferment *{n}*	延期	enki
defervescence *[p-ch]*	沸騰の停止	futtō no teishi
defined medium *[microbio]*	特定培地	toku'tei-baichi
definite article *[gram]*	定冠詞	tei-kanshi

definite composition, law of *[ch]*	定比例の法則	tei-hirei no hōsoku
definite integral *[calc]*	定積分	tei-sekibun
definite proportions *[ch]*	定比例	tei-hirei
definition *[ling] [math]*	定義；ていぎ	teigi
" *[photo]*	鮮明度	senmei-do
deflagration; detonation *[ch]*	爆燃	baku'nen
" *[ch]*	突燃；突燃	totsu'nen
deflect *{vb} [psy]*	離反する	rihan suru
deflection *[electr] [eng]*	偏向；へんこう	henkō
" *[eng]*	振れ	fure
" *[eng]*	偏位	hen'i
" *[mech]*	撓み；たわみ	tawami
deflection angle *[geod]*	偏角	henkaku
deflection plate *[electr]*	偏向板	henkō-ban
deflectometer *[eng]*	撓み計	tawami-kei
deflector *[eng]*	拡散板；擴散板	kakusan-ban
deflocculant *[ch-eng]*	解膠剤	kaikō-zai
deflocculation *[ch-eng]*	解膠；かいこう	kaikō
defoaming *{n} [ch-eng]*	泡消し；泡消し	awa'keshi
defoaming agent *[mat]*	消泡剤；消泡劑	shōhō-zai
defoliant *[mat]*	落葉剤	raku'yō-zai
deforestation *[for]*	森林伐採	shinrin-bassai
deformable *[math]*	変形しうる	henkei shi'uru
deformation *[mech]*	変形；變形	henkei
deformation luminosity *[opt]*	変角光度	henkaku-kōdo
defrosting *{n} [eng]*	霜落し	shimo-otoshi
degeneracy *[phys]*	縮退；しゅくたい	shuku'tai
degenerate matter *[astron]*	退化物質	taika-busshitsu
degenerate star *[astron]*	縮退星	shukutai-sei
degeneration; degeneracy *[bio]*	退化	taika
degeneration *[mat] [med]*	変質；變質	henshitsu
degradation *[ch] [eng]*	劣化	rekka
" *[mech]*	脆化；ぜいか	zeika
" (of enzyme) *[microbio]*	分解	bunkai
degradation factor *[comput]*	性能低下因数	seinō-teika-insū
degreasing *{n} [ch-eng]*	脱脂	dasshi
degree *[astron] [geogr] [geom] [thermo]*	度；ど	do
degree *[math]*	次数	jisū
degree Celsius (centigrade) *[thermo]*	摂氏度；攝氏度	sesshi-do
degree Farenheit *[thermo]*	華氏度	kashi-do
degree of accumulation *{n}*	集積度	shūseki-do

English	Japanese	Romanization
degree of beating *[eng]*	叩解度；こうかいど	kōkai-do
degree of consanguinity *[gen]*	親等	shintō
degree of crystallinity *[crys]*	結晶化度	kesshō'ka-do
degree of dilution *[ch]*	希釈度；稀釋度	kishaku-do
degree of dispersion *[ch]*	分散度	bunsan-do
degree of dissociation *[p-ch]*	解離度；かいりど	kairi-do
degree of dyeing power *[tex]*	染着度	senchaku-do
degree of exhaustion *[tex]*	染着率	senchaku-ritsu
degree of freedom *[mech] [p-ch] [stat]*	自由度	jiyū-do
degree of freeness *[paper]*	沪水性；濾水性	rosui-sei
degree of molecular orientation *[p-ch]*	分子配向度	bunshi-haikō-do
degree of orientation *[crys]*	配向度；はいこうど	haikō-do
" *[org-ch]*	配列度	hairetsu-do
degree of plasticity *[mech]*	可塑度；かそど	kaso-do
degree of polarization *[opt]*	偏光度	henkō-do
degree of polymerization *[poly-ch]*	重合度	jūgō-do
degree of preferred orientation *[petr]*	配向度	haikō-do
degree of refraction *[phys]*	屈折度	kussetsu-do
degree of substitution *[ch]*	置換度	chikan-do
degree of superheating *[thermo]*	過熱度	ka'netsu-do
degree of vacuum *[phys]*	真空度	shinkū-do
degree of yellowing *[opt]*	黄変度；黄變度	kōhen-do
degumming *{n} [tex]*	精練；精練	seiren
dehiscent fruit *[bot]*	裂開果	rekkai-ka
dehumidification *[mech-eng]*	脱湿；脱濕	dasshitsu
" *[mech-eng]*	除湿	jo'shitsu
dehydrate *{vb t} [ch]*	脱水する	dassui suru
dehydrating agent *[ch]*	脱水剤	dassui-zai
dehydration *[ch] [physio]*	脱水；だっすい	dassui
dehydration degree *[ch]*	脱水度	dassui-do
dehydration rate *[ch]*	脱水率	dassui-ritsu
dehydrator *[ch-eng]*	脱水機	dassui-ki
dehydrogenase *[bioch]*	脱水素酵素	dassuiso-kōso
dehydrogenation *[ch]*	脱水素	dassuiso
deicing *{n} [eng]*	氷結防止	hyōketsu-bōshi
deionized water *[hyd]*	脱イオン水	datsu-ion-sui
delay *{n}*	遅延	chi'en
" *{n}*	遅滞；遲滯；ちたい	chitai
" *{n}*	停滞	teitai
delayed action; delaying action	遅延作用	chi'en-sayō
delayed-action accelerator *[rub]*	遅効性促進剤	chikō'sei-soku'shin-zai
delayed fluorescence *[atom-phys]*	遅延蛍光	chi'en-keikō

delayed response *[comput]*	遅延応答; 遅延應答	chi'en-ōtō
delay line *[electr]*	遅延線	chi'en-sen
delay time *[electr]*	遅延時間	chi'en-jikan
delete *{vb} [comput]*	抹消する	masshō suru
" *{vb} [comput]*	削除する	sakujo suru
deleterious material *[ind-eng]*	有害物	yūgai'butsu
deletion *[comput]*	削除; さくじょ	sakujo
deliberately; intentionally *{adv} [psy]*	故意に; こいに	ko'i ni
" *{adv} [psy]*	わざと	wazato
delicacy (pl -cies) *[cook]*	珍味	chin'mi
" *{n}*	おいしい物	oishī mono
delicious *[cook] [physio]*	おいしい	oishī
deliming *{n} [ch]*	脱灰	dakkai
delimiter *[comput]*	区切り記号	kugiri-kigō
delineation *[math]*	輪郭; りんかく	rinkaku
deliquescence *[p-ch]*	潮解	chōkai
deliquoring degree *[ch]*	脱水度	dassui-do
delta *[geogr]*	三角洲	sankaku-su
deltoid; triangular *{adj} [math]*	三角状の	sankaku'jō no
delustering *{n} [opt]*	艶消し; つやけし	tsuya'keshi
delustering agent *[mat] [tex]*	艶消剤	tsuya'keshi-zai
demagnetization *[elecmg]*	減磁	genji
" *[elecmg]*	消磁	shōji
demagnetizing field *[elecmg]*	反磁場; はんじば	hanji-ba
demagnetizing force *[elecmg]*	減磁力	genji-ryoku
demand *{n} [econ]*	需用	juyō
" (for merchandise) *{n} [econ]*	売れ行; 賣れ行	ure'yuki
" *{n} [pat]*	請求	seikyū
" ; requirement *{n}*	要求; ようきゅう	yōkyū
demand for examination *[pat]*	審査請求	shinsa-seikyū
demarcate *{vb}*	区画する; 區畫する	kukaku suru
demerit mark *{n}*	罰点	batten
demineralization *[ch-eng]*	脱塩; 脱鹽	datsu'en
" *[dent]*	脱灰	dakkai
demodulation *[comm]*	復調	fukuchō
demodulator *[electr]*	復調器	fukuchō-ki
demography *[ecol]*	人口学	jinkō'gaku
demoisturization *[mech-eng]*	除湿; 除濕	jo'shitsu
demolish; level *{vb t}*	崩す; くずす	kuzusu
demulsification *[ch-eng]*	解乳化	kai'nyū'ka
demulsifier *[ch-eng]*	解乳化剤	kainyū'ka-zai
denaturant *[ch]*	変性剤; 變性劑	hensei-zai

denaturation *[ch]*	変性；変成	hensei
denatured alcohol *[ch]*	変成アルコール	hensei-arukōru
dendrite *[anat]*	樹状突起；樹狀突起	jujyō-tokki
" *[crys]*	樹枝状晶	jushi'jō-shō
dendritic growth *[sci-t]*	樹枝状成長	jushi'jō-seichō
dendrochronology *[geol]*	年輪年代学	nenrin-nendai'gaku
dendrology *[for]*	樹木学	jumoku'gaku
denial; disavowal; negation *{n}*	否定；ひてい	hitei
denitrification *[ch]*	脱窒素；だっちっそ	datchisso
" *[microbio]*	脱窒	datchitsu
denitrifying bacteria *[microbio]*	脱窒細菌	datchitsu-saikin
denominator *[arith]*	分母；ぶんぼ	bunbo
de novo synthesis *[bioch]*	新規の合成	shinki no gōsei
dense fog *[meteor]*	濃霧；のうむ	nōmu
dense hexagonal lattice *[crys]*	稠密六方格子	chū'mitsu-roppō-kōshi
densest array *[crys]*	最密配列	sai'mitsu-hai'retsu
densest structure *[crys]*	最密構造	sai'mitsu-kōzō
densimeter *[eng]*	比重計	hijū-kei
" *[eng]*	密度計	mitsudo-kei
density (pl -ties) *[mech]*	密度；みつど	mitsu'do
" *[photo]*	濃度；のうむ	nōdo
density current *[ocean]*	密度流	mitsudo-ryū
density function *[stat]*	密度関数	mitsudo-kansū
density gradient tube *[an-ch]*	密度勾配管	mitsudo-kōbai-kan
density wave theory *[astrophys]*	密度波理論	mitsudo'ha-riron
dent *{n} [geom] [mech-eng]*	凹み；へこみ	hekomi
dental calculi *[dent]*	歯石；齒石	shiseki
dental caries *[dent]*	齲食；齲蝕	u'shoku
dental plaque *[dent]*	歯垢；しこう	shikō
dental procelain *[mat]*	陶歯	tōshi
dental prosthesis *[eng]*	歯科補綴	shika-hotetsu
dental pulp *[dent] [hist]*	歯髄；しずい	shi-zui
dentate *{adj}*	歯状の	shijō no
denticulate *{adj} [zoo]*	小歯状の	shōshi'jō no
dentifrice *{n} [pharm]*	練歯磨；練齒磨	neri-ha'migaki
dentin *[hist]*	歯質	shi'shitsu
" *[hist]*	象牙質	zōge'shitsu
dentistry *[med]*	歯科医学；齒科醫學	shika-i'gaku
dentition *[dent] [physio]*	歯牙発生	shiga-hassei
denudation *[geol]*	削剝	saku'haku
deodorant *[mat] [pharm]*	脱臭剤	dasshū-zai
" *[mat] [pharm]*	消臭剤	shōshū-zai

deoxidation *[ch]*	脱酸(素)	dassan(so)
deoxidizer; deoxidizing agent *[ch]*	脱酸素剤	dassanso-zai
depart from; miss *{vb}*	外れる; はずれる	hazureru
department store *[arch]*	百科店	hyakka'ten
departure *[trans]*	出発; 出發	shuppatsu
dependence *[stat]*	依存関係	i'son-kankei
dependency *[comput]*	従属; じゅうぞく	jūzoku
dependent patent *[pat]*	従属特許	jūzoku-tokkyo
dependent variable *[alg]*	従(属)変数	jū(zoku)-hensū
dephosphorization *[ch] [met]*	脱燐; だつりん	datsu'rin
dephosphorylation *[org-ch]*	脱燐酸化	datsu'rinsan'ka
depilation *[eng]*	抜毛; ばつもう	batsu'mō
depletion layer *[electr]*	空乏層	kūbō-sō
deployment *{n} [mil]*	配置; はいち	haichi
depolarization *[elec]*	減極; げんきょく	genkyoku
" *[elec]*	消極(作用)	shōkyoku(-sayō)
depolymerization *[poly-ch]*	解重合	kai'jūgō
deposit; sediment *{n} [geol]*	堆積; たいせき	taiseki
deposit; silting *{n} [civ-eng]*	滞積; 滯積	taiseki
deposit at *{vb t} [microbio]*	寄託する	kitaku suru
deposition *[ch]*	析出; せきしゅつ	seki'shutsu
" *[microbio]*	寄託; きたく	kitaku
depreciation of cost *[econ]*	原価消却	genka-shō'kyaku
depressant *[ch]*	降下剤	kōka-zai
depression; dent; indentation *[geom]*	凹み; へこみ	hekomi
deprotecting agent *[ch]*	脱保護剤	datsu'hogo-zai
deproteinization *[bioch]*	除蛋白	jo'tanpaku
depth *[arch]*	奥行き	oku'yuki
" *[ocean]*	深さ; ふかさ	fuka-sa
depth; back; inner part *{n}*	奥; 奧; おく	oku
depth charge *[ord]*	(水中)爆雷	(suichū-)baku'rai
depth contour; isobath *[ocean]*	等深線	tōshin-sen
depth finder *[eng]*	測深儀	sokushin-gi
depth of cut; cut *{n} [eng]*	切り込み	kiri'komi
depth of cut *[eng]*	切込み深さ	kiri'komi-fuka-sa
deputy; agent (a person)	代理人; だいりにん	dairi'nin
derelict ship *[navig]*	漂流船	hyōryū'sen
derivative *[calc]*	導関数; 導關數	dō'kansū
" *[ch]*	誘導体	yūdō'tai
dermatophyte *[mycol]*	皮膚糸状菌	hifu-shijō-kin
dermis *{n} [anat]*	真皮; しんぴ	shinpi
derrick; oil derrick *[petrol]*	油井櫓	yusei-yagura

derusting *{n} [ch-eng]*	脱錆；だっせい	dassei
desalted water *[ch-eng]*	脱塩水	datsu'en-sui
desalting *{n} [ch-eng]*	脱塩；脱鹽	datsu'en
descend *{vb} [navig]*	下る	kudaru; oriru
descendant *[bio]*	子孫	shison
" *[bio]*	末；すえ	sue
descending node *[astron]*	降交点	kōkō-ten
descending order *[math]*	降順	kōjun
descension *[math]*	下降	kakō
descent *[aero-eng]*	降下	kōka
descent; descension *[navig]*	下り；くだり	kudari
description *[comput]*	記述	ki'jutsu
" *[pat]*	公表	kōhyō
descriptive geometry *[math]*	図形幾何学	zukei-kika'gaku
descriptive statistics *[stat]*	記述統計学	kijutsu-tōkei'gaku
descriptor *[comput]*	記述子	kijutsu'shi
desensitization *[immun]*	脱感作；だっかんさ	dakkansa
desensitizer *[photo]*	減感剤	genkan-zai
desert *{n} [geogr]*	乾荒原	kan-kōgen
" *{n} [geogr]*	荒原	kōgen
" *{n} [geogr]*	砂漠；さばく	sabaku
desert plant *[bot]*	砂漠植物	sabaku-shokubutsu
desiccant; desiccating agent *[ch]*	乾燥剤	kansō-zai
desiccation *[ch-eng]*	乾燥(作用)	kansō(-sayō)
" *[hyd]*	脱水	dassui
desiccator *[ch-eng]*	乾燥器；かんそうき	kansō-ki
design *{n} [comput] [sci-t]*	設計	sekkei
" *{n} [graph] [pat]*	意匠	i'shō
" (sketch; plan *{n} [graph]*	図案；圖案；ずあん	zu'an
" (design idea) *{n} [sci-t]*	構想	kōsō
designation of spacing *[comput]*	間隔設定	kankaku-settei
design drawing *{n} [graph]*	設計図	sekkei-zu
design engineer *[eng]*	設計技師	sekkei-gishi
design engineering *{n} [eng]*	設計工学	sekkei-kōgaku
desizing *{n} [tex]*	糊抜き	nori'nuki
desk *[furn]*	机；つくえ	tsukue
desk checking *{n} [comput]*	卓上型検査	takujō'gata-kensa
desktop electronic calculator *[electr]*	電卓	dentaku
desktop publishing *{n} [comput]*	机上出版	kijō-shuppan
desorption *[p-ch]*	脱着	datchaku
de(s)oxyribonucleic acid; DNA *[bioch]*	デオキシリボ核酸	deokishiribo-kakusan
dessert(s) *[cook]*	菓子；かし	kashi

English	Japanese	Romanization
destination *[comput]*	宛先；あてさき	ate'saki
" *[comput]*	入力先	nyū'ryoku-saki
" *[navig]*	目的地	mokuteki'chi
" *[navig]*	行く先	yuku'saki
destroyer *[nav-arch]*	駆逐艦	kuchiku'kan
destruction; demolition *[civ-eng]*	破壊；破壞；はかい	hakai
destruction test *[mech]*	破壊試検	hakai-shiken
destructive distillation *[org-ch]*	分解蒸留	bunkai-jōryū
desulfurization *[ch-eng]*	脱硫；だつりゅう	datsu'ryū
desulfurizer *[ch-eng]*	脱硫剤	datsuryū-zai
deswelling *{n} [ch]*	解膨潤	kai'bōjun
detach; remove; unfasten *{vb t}*	外す；はずす	hazusu
detachable; releasable *{adj} [adhes]*	剝離性の	hakuri'sei no
detail drawing *{n} [graph]*	詳細図	shōsai-zu
detailed; be familiar *{adj}*	詳しい；くわしい	kuwashī
detailed bill *[econ]*	内訳書；内譯書	uchi'wake'sho
detailed classification *[syst]*	細分類	sai-bunrui
detailed explanation *[pat]*	詳細な説明	shōsai na setsu'mei
detailed explanation of invention *[pat]*	発明の詳細な説明	hatsumei no shōsai na setsumei
details *{n}*	経緯	kei'i
details; particulars *{n}*	仔細；しさい	shisai
" *{n}*	詳細	shōsai
detection *[elec]*	検波；檢波	kenpa
" *[eng]*	検出	ken'shutsu
" *{n}*	探知	tanchi
detergency *[mat]*	洗浄力	senjō-ryoku
detergent *[mat]*	洗浄剤；洗淨劑	senjō-zai
deterioration *[ch] [eng]*	劣化	rekka
determinant *[alg]*	行列式	gyō'retsu-shiki
determination *[an-ch]*	定量；ていりょう	teiryō
deterrence *[mil]*	抑止	yokushi
detonating fuse *[eng]*	導爆線	dōbaku-sen
detonating gas *[eng]*	爆鳴気；ばくめいき	bakumei-ki
detonating sound *[acous]*	爆音	baku'on
detonation *[ch]*	爆轟	bakugō
detonator *[ord]*	雷管	raikan
detour *{n} [civ-eng] [traffic]*	回り道	mawari-michi
" *{n} [trans]*	迂回	u'kai
detour road *[civ-eng]*	迂回路	ukai-ro
detoxification *[bioch]*	解毒；げどく	ge'doku
detrimental *{adj}*	有害の；有害な	yūgai no; yūgai na

detritus *[geol]* 岩屑；がんせつ gansetsu
deuterium; heavy hydrogen *[ch]* 重水素 jū'suiso
Deuteromycetes *[mycol]* 不完全菌類 fu'kanzen'kin-rui
deuteron *[ch]* 重陽子 jū'yōshi
develop; grow *{vb i}* *[physio]* 発育する hatsu'iku suru
developer *[opt]* 顕色剤；顯色劑 kenshoku-zai
developer; developing agent *[photo]* 現像液（現像剤） genzō-eki (gensō-zai)
developer; developing agent *[mat]* 展開剤 tenkai-zai
development *[bio]* 発生 hassei
" (drafting) *[graph]* 展開図 tenkai-zu
" *[math]* 展開；てんかい tenkai
development; developing *{n}* *[photo]* 現像；げんぞう genzō
development *[sci-t]* 開発 kai'hatsu
deviation *[bio]* *[eng]* *[stat]* 偏差 hensa
" *[eng]* 偏移 hen'i
" *[eng]* 偏り；片寄り katayori
device; apparatus *[eng]* 装置；そうち sōchi
device control character *[comput]* 装置制御文字 sōchi-seigyo-moji
device dependence *[comput]* 装置依存 sōchi-i'zon
device independence *[comput]* 装置独立；裝置獨立 sōchi-doku'ritsu
devil's tongue *[cook]* 蒟蒻；こんにゃく konnyaku
devise *{vb}* 考案する kōan suru
devisement; device *{n}* *[pat]* 考案 kōan
devitrification *[ch]* 失透 shittō
devulcanization *[ch-eng]* 脱硫 datsu'ryū
dew *{n}* *[meteor]* 露；つゆ tsuyu
dew condensation *[meteor]* 結露 ketsu'ro
dew point *[ch]* *[meteor]* 露点 roten
dextral *{adj}* *[med]* 右側の；みぎがわの migi'gawa no
" *{adj}* *[med]* 右向きの migi'muki no
dextrorotatory *[opt]* 右旋光性 u'senkō-sei
" *[opt]* 右旋性 u'sen-sei
dextrorse *[bot]* 右巻きの；右卷きの migi'maki no
dextrose *[bioch]* 右旋糖 u'sen'tō
dezincification *[met]* 脱亜鉛；脫亞鉛 datsu'aen
diabetes mellitus *[med]* 糖尿病 tōnyō'byō
diacid base *[ch]* 二酸塩基 nisan-enki
diacritical mark *[pr]* 分音符号 bun'on-fugō
" *[pr]* 音符 onpu
diagenesis *[geol]* 続成作用；續成作用 zoku'sei-sayō
diagnostics *[comput]* 診断 shindan
diagonal *{n}* *[math]* 斜；はす hasu

diagonal *{n} [math]*	斜；ななめ	naname
diagonal line *[geom]*	対角線；對角線	taikaku-sen
diagonal matrix *[math]*	対角線行列	taikaku'sen-gyō'retsu
diagram *{n} [graph]*	線図；せんず	sen-zu
" ; figure *{n} [graph] [math]*	図；圖；ず	zu
" ; graph; chart *{n} [graph]*	図表	zuhyō
" ; drawing *{n} [graph]*	図面；ずめん	zumen
diagram; chart; graph; schema *[graph]*	図式	zu-shiki
diagram line *[graph]*	図表線	zuhyō-sen
diagramming; schematizing *[graph]*	図式化	zu'shiki'ka
diagrammatic explanation *[graph]*	図解；圖解；ずかい	zukai
dialog; dialogue *[comm]*	対話；對話；たいわ	taiwa
dialysis (pl -yses) *[p-ch]*	透析；とうせき	tōseki
dialyzate *[p-ch]*	透析物	tōseki'butsu
diamagnetic *{adj} [elecmg]*	反磁性の	han'jisei no
diamagnetic substance *[elecmg]*	反磁性体	hanjisei-tai
diamagnetic susceptibility *[elecmg]*	反磁化率	han'jika-ritsu
diamagnetism *[elecmg]*	反磁性；はんじせい	hanji-sei
diameter *[geom]*	直径；ちょっけい	chokkei
" *[geom]*	径；徑；けい	kei
diammonium hydrogen phosphate	燐酸水素二アンモ=ニウム	rinsan-suiso-ni'anmonyūmu
diamond *[miner]*	金剛石	kongō'seki
diapause *[physio]*	休眠；きゅうみん	kyūmin
diaphragm *[anat] [phys]*	隔膜；かくまく	kaku'maku
" *[anat]*	横隔膜	ō'kaku-maku
" *[eng-acous] [phys]*	振動板	shindō-ban
diaphragm; iris (pl -es, irides) *[opt]*	光束絞り	kōsoku-shibori
diarrhea *[med]*	下痢；げり	geri
diastatic enzyme *[bioch]*	糖化酵素	tōka-kōso
diastole *[physio]*	弛緩期	chikan-ki; shikan-ki
diatom *[i-zoo]*	珪藻植物	keisō-shokubutsu
diatomaceous earth; kieselguhr *[geol]*	珪藻土；けいそうど	keisō'do
dibasic acid *[ch]*	二塩基酸	ni'enki-san
dibasic potassium phosphate	燐酸二カリウム	rinsan-ni-karyūmu
dice (plural of die) *{n} [eng]*	骰子；賽子	saikoro
" ; mince; chip; chop *{vb} [cook]*	刻む；きざむ	kizamu
dichloride *[ch]*	二塩化物；二鹽化物	ni'enka-butsu
dichotomizing search *[comput]*	二分探索	nibun-tansaku
dichotomy (pl -mies) *[bio] [comput]*	二分法；にぶんほう	nibun-hō
dichroic fog *[photo]*	二色カブリ	ni'shoku-kaburi
dichroic ratio *[opt]*	二色性比	ni'shoku'sei-hi

English	Japanese	Romanization
dichroism *[opt]*	二色性	ni'shoku-sei
diction *[ling]*	発声法；發聲法	hassei-hō
dictionary *[pr]*	字引	ji'biki
" *[pr]*	辞書；辭書；じしょ	jisho
" *[pr]*	辞典	jiten
did (past tense of do) *{vb}*	した	shita
die; mold; model *{n} [eng]*	型；形；かた	kata
die *{n} [eng]*	ダイス	daisu
" *{n} [eng]*	金型；かながた	kana-gata
" *{n}* (pl dice)	骰子；賽子	saikoro
dielectric *{n} [mat]*	誘電体	yūden'tai
dielectric aftereffect *[elec]*	誘電余効	yūden-yokō
dielectric amplifier *[electr]*	誘電体増幅器	yūden'tai-zōfuku-ki
dielectric breakdown *[electr]*	絶縁破壊；絶縁破壞	zetsu'en-hakai
dielectric constant *[elec]*	誘電率	yūden-ritsu
dielectric dissipation factor *[elec]*	誘電正接	yūden-sei'setsu
dielectric flux *[elec]*	電束；でんそく	densoku
dielectric loss *[elecmg]*	誘電損失	yūden-son'shitsu
" *[elecmg]*	誘電体損	yūden'tai-son
dielectric loss tangent *[elec]*	誘電正接	yūden-sei'setsu
dielectric phase angle *[elec]*	誘電位相角	yūden-isō-kaku
dielectric polarization *[elec]*	誘電分極	yūden-bun'kyoku
dielectric strain *[elec]*	誘電歪み	yūden-hizumi
dielectric strength *[elec]*	絶縁耐力	zetsu'en-tai'ryoku
dietary *{adj} [cook] [med]*	飲食の	in'shoku no
dietary fiber *[cook]*	食物繊維	shoku'motsu-sen'i
diet building *{n} [arch]*	国会議事堂	kokkai-giji'dō
dietetics *[med]*	食餌療法学	shokuji-ryōhō'gaku
dietician *[bio]*	栄養士；えいようし	eiyō'shi
difference *[arith]*	差；さ	sa
difference tone *[acous]*	差音	sa'on
differentiable *{adj} [calc]*	微分可能な（の）	bibun-kanō na (no)
differential *{n} [calc]*	微分；びぶん	bibun
differential ammeter *[eng]*	差動電流計	sadō-denryū-kei
differential amplifier *[electr]*	差動増幅器	sadō-zōfuku-ki
differential analyzer *[comput]*	微分解析器	bibun-kaiseki-ki
differential centrifugation *[ch]*	分画遠心	bunkaku-enshin
differential electrometer *[eng]*	差動電位計	sadō-den'i-kei
differential equation *[calc]*	微分方程式	bibun-hōtei'shiki
differential flotation *[min-eng]*	優先浮選	yūsen-fusen
differential gear *[mech-eng]*	差動装置	sadō-sōchi
differential geometry *[math]*	微分幾何学	bibun-kika'gaku

differential means *[mech-eng]*	差動手段	sadō-shudan; sadō-te'date
differential permeability *[elecmg]*	微分透磁率	bibun-tōji-ritsu
differential refractometer *[eng]*	示差屈折計	shisa-kussetsu-kei
differentiate *{vb t} [calc]*	微分する	bibun suru
differentiating circuit *[elec]*	微分回路	bibun-kairo
differentiation *[bio] [cyt] [ling]*	分化	bunka
" *[calc]*	微分; びぶん	bibun
differentiator *[electr]*	微分器	bibun-ki
different variety *[bot]*	異種	ishu
difficult to do	為難い; しにくい	shi'nikui
difficult to see	見難い; 見悪い	mi'nikui
diffraction *[phys]*	回折; かいせつ	kaisetsu
diffraction angle *[phys]*	回折角	kaisetsu-kaku
diffraction grating *[spect]*	回折格子	kaisetsu-gōshi
diffraction pattern *[crys]*	回折図形	kaisetsu-zukei
diffused light *[opt]*	散光; さんこう	sankō
diffuse reflection *[opt]*	拡散反射	kakusan-hansha
" *[phys]*	乱反射; 亂反射	ran-hansha
diffuser method *[eng]*	散気式暴気方法	sanki'shiki-bakki-hōhō
diffusion *[opt] [phys]*	拡散; かくさん	kaku'san
diffusion coefficient *[phys]*	拡散係数	kakusan-keisū
diffusion-controlled current *[elec]*	拡散律速電流	kakusan-rissoku-denryū
diffusion dialysis *[p-ch]*	拡散透析	kakusan-tōseki
diffusion layer *[phys]*	拡散層	kakusan-sō
diffusion of gas *[phys]*	気体拡散; 氣體擴散	kitai-kakusan
diffusivity *[phys]*	拡散係数	kakusan-keisū
" *[phys]*	拡散率	kakusan-ritsu
dig *{vb}*	掘る	horu
" (drill; excavate) *{vb t}*	穿つ; うがつ	ugatsu
digestion *[ch-eng] [paper]*	蒸解; じょうかい	jōkai
" *[physio]*	消化; しょうか	shōka
digestion assistant *[ch-eng]*	蒸解助剤	jōkai-jozai
digestive enzyme *[bioch]*	消化酵素	shōka-kōso
digestive fluid *[physio]*	消化液	shōka-eki
digestive organ *[anat]*	消化器官	shōka-kikan
digestive system *[anat]*	消化系	shōka-kei
digestive tract *[anat]*	消化管	shōka-kan
digging *{n} [min-eng]*	採掘	saikutsu
digit *[arith]*	桁; けた	keta
" *[arith]*	数字; 數字; すうじ	sūji
" *[comput] [math]*	位数	isū
digital *[comput]*	計数型	keisū'gata

digitigrade animal *[v-zoo]*	趾行動物	shikō-dōbutsu
digitization *[comput]*	計数化	keisū'ka
" *[math]*	数字化	sūji'ka
digitize *{vb} [comput]*	計数化する	keisū'ka suru
dihedral angle *[geom]*	二面角	nimen-kaku
" *[geom]*	稜角；りょうかく	ryō-kaku
dihydric alcohol *[org-ch]*	二価アルコール	nika-arukōru
dikaryon; dicaryon *[mycol]*	二核共存体	ni'kaku-kyōson'tai
dilatant *{adj} [fl-mech]*	膨張性の	bōchō'sei no
dilatation *[phys]*	膨張度	bōchō-do
dilatometer *[eng]*	膨張計	bōchō-kei
diligently *{adv}*	鋭意	ei'i
dill *[bot] [cook]*	蒔蘿；いのんど	inondo
diluent *[ch]*	希釈剤；稀釋劑	kishaku-zai
dilute *{adj} [ch]*	薄い	usui
" (thin) *{vb t} [ch]*	希釈する	kishaku suru
" (thin) *{vb t} [ch]*	薄める	usumeru
dilute acid *[ch]*	希酸	kisan
dilute solution *[ch]*	希薄溶液	ki'haku-yō'eki
dilution *[ch]*	希釈；きしゃく	ki'shaku
dilution, degree of *[ch]*	希釈度	kishaku-do
dilution ratio *[ch]*	希釈率	kishaku-ritsu
dimension *[math]*	大きさ；おおきさ	ōki-sa
dimension(s) *[geom] [phys]*	次元	jigen
" *[geom] [phys]*	寸法	sunpō
dimensional stability *[mat]*	寸法安定性	sunpō-antei-sei
dimensionless function *[math]*	無次元関数	mu'jigen-kansū
dimensionless quantity *[math]*	無次元量	mu'jigen-ryō
dimensionless term *[math]*	無次元項	mu'jigen-kō
dimer *[poly-ch]*	二量体	niryō'tai
dimerization *[poly-ch]*	二量化	niryō'ka
diminution; decrease *{n}*	減少；げんしょう	genshō
dimmer *[elec]*	減光器	genkō-ki
dimming (of light) *{n} [elec]*	減光	genkō
dinette *[arch]*	小食堂	shō-shokudō
dingbat *[comput] [pr]*	装飾活字	sōshoku-katsu'ji
" *[comput] [pr]*	特別記号活字	toku'betsu-kigō-katsu'ji
dining car *[rail]*	食堂車	shokudō'sha
dining room *[arch]*	食堂	shoku'dō
" *[arch]*	食事室	shokuji-shitsu
dinitrogen monoxide	一酸化二窒素	issanka-ni'chisso
dinitrogen pentoxide	五酸化二窒素	go'sanka-ni'chisso

dinitrogen tetroxide	四酸化二窒素	shi'sanka-ni'chisso
dinitrogen trioxide	三酸化二窒素	san'sanka-ni'chisso
dinner; banquet *[cook]*	晩餐; ばんさん	bansan
Dinoflagellata *[i-zoo]*	渦鞭毛虫類	uzu'maki-benmō'chū-rui
Dinophyta *[bot]*	渦鞭毛植物	uzu'maki-benmō-shokubutsu
dinosaur (reptile) *[v-zoo]*	恐竜; 恐龍	kyōryū
diode *[electr]*	二極管	ni'kyoku-kan
diopside *[miner]*	透輝石; とうきせき	tō'kiseki
dioptase *[miner]*	翠銅鉱; 翠銅鑛	sui'dōkō
dioptric system *[opt]*	屈折光学系	kussetsu-kōgaku-kei
diorite *[miner]*	閃緑岩	senryoku'gan
dip *{n} [eng]*	俯角; ふかく	fu'kaku
" *{n} [eng]*	伏角; ふっかく	fukkaku
dip coating *{n} [eng]*	浸漬塗装	shinseki-tosō
dip corrosion *[met]*	浸漬腐食	shinseki-fu'shoku
dip dyeing *{n} [tex]*	浸染; しんせん	shinsen
Diplococcus *[microbio]*	双球菌; 雙球菌	sōkyū-kin
diploid *[gen]*	二倍体	nibai'tai
diploid number *[gen]*	二倍数	nibai-sū
diploidy *[gen]*	二倍性	nibai-sei
dipole *[elecmg]*	双極子; 雙極子	sōkyoku'shi
dipotassium hydrogen phosphate	燐酸水素二カリウム	rinsan-suiso-ni'karyūmu
dipping *{n} [met]*	浸漬; しんせき	shinseki
dipping degreasing *[ch-eng] [met]*	浸漬脱脂	shinseki-dasshi
direct *{adj}*	直接の	choku'setsu no
direct-acting *{adj} [elec]*	直動の	choku'dō no
direct benefit *[econ]*	直接便益	chokusetsu-ben'eki
direct control *[comput]*	直接制御	chokusetsu-seigyo
direct-coupled *[elec]*	直結形	chokketsu-kei
direct coupling *{n} [mech-eng]*	直接接合	chokusetsu-setsugō
direct current *[elec]*	直流; ちょくりゅう	choku'ryū
direct-current power *[elec]*	直流電力	chuku'ryū-denryoku
direct dye *[mat]*	直接染料	chokusetsu-senryō
direct illumination *[opt]*	直接照明	chokusetsu-shōmei
direction *[eng]*	方角; ほうがく	hōgaku
" *[eng]*	方位	hō'i
" *[eng]*	方向	hōkō
" *[eng]*	指向	shikō
directionality *[comm]*	指向性	shikō-sei
" *{n}*	方向性	hōkō-sei
direction cosines *[an-geom]*	方向余弦	hōkō-yogen
directions (of compass) *[geod]*	東西南北	tō'zai-nan'boku

directivity *[comm]*	指向性	shikō-sei
direct line *[comm]*	直通；ちょくつう	choku'tsū
directly *{adv}*	直接に	choku'setsu ni
direct motion *[astron]*	順行運動	junkō-undō
director *{n}*	長官	chōkan
Director-General of the Patent Office *[pat]*	特許庁長官	Tokkyo'chō-Chōkan
directory (pl -ries) *[comm] [pr]*	人名簿	jinmei'bo
" *[comm] [pr]*	人名録	jinmei'roku
" *[comm] [pr]*	氏名録	shimei'roku
" *[comput] [pr]*	登録簿；とうろくぼ	tōroku'bo
direct printing *{n} [tex]*	写し染め	utsushi-zome
direct projection *[astron]*	直射	choku'sha
directrix (pl -trixes, -trices) *[an-geom]*	準線	junsen
directrix *[an-geom]*	導面	dōmen
direct-viewing tube *[electr]*	直視管	chokushi-kan
direct vision *[opt]*	直視；ちょくし	choku'shi
direct-vision spectroscope *[spect]*	直視分光器	chokushi-bunkō-ki
dirty; soiled; messy *{adj}*	汚ない	kitanai
disability *[med]*	身体傷害	shintai-shōgai
disabled person *[med]*	身体傷害者	shintai'shōgai'sha
disaccharide *[bioch]*	二糖	nitō
disagreeableness *[cook]*	嫌味；いやみ	iya'mi
disassemble *{vb t}*	分解する	bunkai suru
disassembly *[ind-eng]*	分解；ぶんかい	bunkai
disaster *{n}*	災害	saigai
" *{n}*	災難	sai'nan
disaster prevention *[ind-eng]*	災害防止	saigai-bōshi
disc; disk *[geom]*	円板；圓板	enban
discard; discarding *{n}*	棄却	ki'kyaku
discard; throw away *{vb t}*	捨てる；棄てる	suteru
discarding *{n} [ecol]*	廃棄；廢棄；はいき	haiki
discerning power *[psy]*	識別能	shiki'betsu-nō
discharge *{n} [elec]*	放電	hōden
" *{n} [fl-mech]*	放出	hō'shutsu
" *{n} [hyd]*	放流	hōryū
" *{vb}*	発する	hassuru
discharge coefficient *[fl-mech]*	流量係数	ryūryō-keisū
discharge flow *[fl-mech]*	吐出流	to'shutsu-ryū
discharge lamp *[electr]*	放電灯	hōden-tō
discharge pipe *[civ-eng]*	吐出し管	haki'dashi-kan

discharge printing *{n} [graph]*	抜染；ばっせん	bassen
discharge rate *[elec]*	放電率	hōden-ritsu
discharge tube *[elec]*	放電管	hōden-kan
discipline (of study) *{n} [sci-t]*	学問領域	gakumon-ryōiki
disclination *[crys]*	回位	kai'i
disclose *{vb} [pat]*	開示する	kaiji suru
" *{vb} [pat]*	公開する	kōkai suru
disclosure *[pat]*	開示；かいじ	kaiji
discoloration *[opt]*	変色；變色	henshoku
discomfort index *[meteor]*	不快指数	fukai-shisū
disconnect *{vb} [elec] [eng]*	切り離す	kiri-hanasu
" *{vb} [elec] [eng]*	切断する	setsu'dan suru
disconnection *[elec]*	断線；だんせん	dansen
disconnect signal *[comm] [elec]*	切断信号；切斷信號	setsu'dan-shingō
discontinuance *{n}*	中止	chūshi
discontinue; interrupt *{vb}*	中止する	chūshi suru
discontinuity *{n}*	不連続；不連續	fu'renzoku
discontinuous *{adj} [math]*	不連続の	fu'renzoku no
discontinuous flow *[fl-mech]*	不連続流	fu'renzoku-ryū
discount *{n} [econ] [ind-eng]*	割引き；わりびき	wari'biki
discount rate *[econ]*	割引率	wari'biki-ritsu
discover; find; detect *{vb} [pat]*	発見する	hakken suru
discover; find; sight *{vb t}*	見付ける	mitsukeru
discoverer *[pat]*	発見者	hakken'sha
discovery *[min-eng]*	発見；發見	hakken
discrete *{adj} [math]*	離散の	risan no
discrete random variable *[stat]*	離散確率変数	risan-kaku'ritsu-hensū
discrete representation *[comput]*	離散的表現	risan'teki-hyōgen
discrete variable *[stat]*	離散変数	risan-hensū
discretionary; arbitrary *{adj}*	任意の	nin'i no
discriminant *[alg]*	判別式	han'betsu-shiki
discriminating power *[psy]*	識別能	shiki'betsu-nō
discrimination *[comm]*	弁別；辨別	ben'betsu
discriminator *[electr]*	弁別器；辨別器	benbetsu-ki
discussion; debate *[comm]*	討論	tōron
discussion *[pr]*	考察	kōsatsu
disc valve *[mech-eng]*	円板弁；圓板瓣	enban-ben
disease; illness *[med]*	病気；びょうき	byōki
" *[med]*	疾患	shikkan
disentangle; untangle; loosen *{vb}*	解す；ほぐす	hogusu
dish *{n} [cer] [cook]*	皿	sara
dish antenna *[elecmg]*	椀形アンテナ	wan'gata-antena

dishcloth gourd *[bot]*	糸瓜；へちま	hechima
dished *{adj} [plas]*	上反りした	uwa'zori shita
dishwasher (electric applicance) *[eng]*	皿洗機	sara'arai-ki
disinfect; sanitize *{vb} [med]*	消毒する	shōdoku suru
disinfectant *[med]*	消毒薬	shōdoku-yaku
disinfection *[med]*	消毒；しょうどく	shōdoku
disintegrate *{vb} [nuc-phys]*	崩壊する；崩壞する	hōkai suru
disintegrating agent *[pharm]*	崩壊剤；崩壞劑	hōkai-zai
disjoint *{adj} [math]*	排反の	haihan no
disjunction *[logic]*	離接	ri'setsu
" *[logic]*	選言；選言	sengen
disk *[astron]*	面；めん	men
disk flower *[bot]*	中心花	chūshin-ka
dislocation *[crys] [photo]*	転位；轉位；てんい	ten'i
" *[geol]*	変位；變位；へんい	hen'i
dismissal *[pat]*	却下	kyakka
dismissal of revision *[pat]*	補正却下	hosei-kyakka
disodium hydrogen phosphate	二燐酸水素ナト=リウム	ni'rinsan-suiso-natoryūmu
disodium hydrogen phosphate	燐酸水素二ナト=リウム	rinsan-suiso-ni'natoryumu
disorder *[math]*	無秩序	mu'chitsu'jo
disperse phase *[ch]*	分散相	bunsan-sō
disperse system *[ch]*	分散系	bunsan-kei
dispersing agent; dispersant *[mat]*	分散剤	bunsan-zai
dispersion *[phys]*	分散；ぶんさん	bunsan
" *[stat]*	ばらつき	baratsuki
" *[stat]*	散らばり；ちらばり	chirabari
dispersion polymerization *[poly-ch]*	分散重合	bunsan-jūgō
dispersive medium *[elecmg]*	分散媒	bunsan-bai
dispersoid *[ch]*	分散質	bunsan'shitsu
displace *{vb}*	ずらす	zurasu
displacement *[ch]*	置換；ちかん	chikan
" *[comput] [elec] [phys]*	変位	hen'i
" (of air, gas) *[fl-mech]*	排気量	haiki-ryō
" (of water) *[fl-mech]*	排水量	haisui-ryō
" *[fl-mech]*	押除け量	oshi'noke-ryō
displacement detector *[eng]*	変位検出器	hen'i-ken'shutsu-ki
display (apparatus) *{n} [electr]*	表示装置	hyōji-sōchi
" *{n} [comput] [electr]*	表示	hyōji
" *{n}*	展示；てんじ	tenji
" *{vb}*	展示する	tenji suru

display console *[comput]*	表示操作卓	hyōji-sōsa'taku
displayed value; indicated value *[electr] [math]*	表示値；ひょうじち	hyōji-chi
display tube *[electr]*	表示管	hyōji-kan
disposal; dealing; disposition *{n}*	処分；處分	shobun
dispose; arrange *{vb} [mech-eng]*	配置する	haichi suru
disposition *[mech-eng]*	配置	haichi
disproportionated rosin; rosin acid *[rub]*	樹脂酸；じゅしさん	jushi-san
disproportionation *[ch]*	不均化	fukin'ka
" *[ch]*	不斉化	fusei'ka
disruption *[crys]*	乱れ；亂れ；みだれ	midare
dissatisfaction *{n}*	不服	fu'fuku
dissected *{adj} [bot]*	全裂の	zen-retsu no
dissent *[pat]*	異議	igi
dissimilarity *{n}*	異同	i'dō
dissipate *{vb}*	消耗する	shōmō suru
dissipation *[phys]*	発散；發散	hassan
" *[phys]*	逸散；いっさん	issan
" *[phys]*	散逸；さんいつ	san'itsu
dissipation; expiation *{n}*	消滅	shōmetsu
dissipation function *[p-ch]*	散逸関数	san'itsu-kansū
dissociation *[p-ch]*	解離；かいり	kairi
dissociation constant *[p-ch]*	解離定数	kairi-teisū
dissolution; melting *[ch]*	溶解；ようかい	yōkai
" *[ch]*	融解；ゆうかい	yūkai
dissolve *{vb i} [ch]*	溶ける	tokeru
" *{vb t} [ch]*	溶かす	tokasu
dissolved oxygen *[ch]*	溶存酸素	yōzon-sanso
dissolved salts *[ch]*	溶存塩類；溶存鹽類	yōzon-enrui
dissolving property *[ch]*	溶解性	yōkai-sei
dissymmetry; asymmetry *[ch]*	不整	fusei
distal *{adj} [med]*	遠位の	en'i no
distance *[geom] [mech]*	距離；きょり	kyori
" *[navig]*	行程	kōtei
distance modulus *[astron]*	距離指数	kyori-shisū
distant; far; remote *{adj} {adv}*	遠い	tōi
distillate *[ch]*	流出物	ryū'shutsu'butsu
distillation *[ch]*	蒸留；じょうりゅう	jōryū
distillation column *[ch]*	蒸留塔；蒸溜塔	jōryū-tō
distillation tower *[ch]*	蒸留塔	jōryū-tō
distilled liquor *[cook]*	蒸留酒	jōryū'shu

distilled water *[ch]*	蒸留水	jōryū-sui
distilling tube *[ch]*	蒸留管	jōryū-kan
distinction *[psy]*	弁別；辨別	ben'betsu
distinguishing power *[psy]*	識別能	shiki'betsu-nō
distort; warp *{vb t}*	歪める	yugameru
distortion *[math] [opt]*	歪み；ゆがみ	yugami
" *[mech]*	歪；いびつ	ibitsu
" *[photo]*	歪曲	wai'kyoku
" (strain) *[phys]*	歪；ひずみ	hizumi
distortion factor *[comm]*	歪み率	hizumi-ritsu
" *[comm]*	歪率；わいりつ	wai-ritsu
distributing valve *[mech-eng]*	分配弁；分配瓣	bunpai-ben
distribution *[stat] [math]*	分布	bunpu
distribution coefficient *[p-ch]*	分配係数	bunpai-keisū
distribution curve *[stat]*	分布曲線	bunpu-kyoku'sen
distribution function *[stat]*	分布関数	bunpu-kansū
distribution of power *[elec]*	配電	haiden
distribution ratio *[ch]*	分配比	bunpai-hi
distribution structure *[econ]*	流通機構	ryūtsū-kikō
distribution voltage *[elec]*	配電電圧	haiden-den'atsu
distributive law *[alg]*	分配法則	bunpai-hōsoku
distributive property *[alg]*	分配性	bunpai-sei
district *[geogr]*	地区；地區；ちく	chiku
" (e.g. Hokkaidō) *[geogr]*	道；どう	dō
disturbance *{n}*	擾乱	jōran
disulfur monoxide	一酸化二硫黄	issanka-ni'iō
disulfur trioxide	三酸化二硫黄	san'sanka-ni'iō
ditch (drain) *{n} [civ-eng]*	溝；どぶ	dobu
" *{n} [civ-eng]*	堀；濠；壕	hori
diuretic *{n} [pharm]*	利尿薬（利尿剤）	ri'nyō-yaku (ri'nyō-zai)
diurnal animal *[zoo]*	昼行動物	chūkō-dōbutsu
diurnal libration *[astron]*	日周秤動	nisshū-hyōdō
diurnal motion *[astron]*	日周運動	nisshū-undō
diurnal parallax *[astron]*	日周視差	nisshū-shisa
diurnal tide *[ocean]*	日周潮	nisshū-chō
diurnal variation *[astron]*	日変化	nichi-henka
divalence *[ch]*	二価	nika
divalent *{adj} [ch]*	二価の	nika no
dive *{n}*	潜水	sensui
diver (a person) *[ocean]*	潜水夫；せんすいふ	sensui'fu
divergence *[bio]*	分岐	bunki
" *[math] [opt]*	発散	hassan

divergence line *[math]*	発散線	hassan-sen
divergent series *[alg]*	発散級数	hassan-kyūsū
diverging lens *[opt]*	発散レンズ	hassan-renzu
diverging wave *[ocean]*	八字波；はちじなみ	hachi'ji-nami
diverse; diversity *{n} [comm]*	(多種)多様性	(tashu-)tayō-sei
diversity *[bio]*	分化的差異	bunka'teki-sa'i
divide; watershed *{n} [geogr]*	分水界	bunsui-kai
divide *{vb} [civ-eng]*	区画する；區畫する	kukaku suru
" *{vb} [comput]*	分割する	bun'katsu suru
divide; separate *{vb}*	分かつ；わかつ	wakatsu
divided application *[pat]*	分割出願	bunkatsu-shutsu'gan
divide into pieces *{vb t}*	分断する	bundan suru
dividend *[arith]*	被除数；ひじょすう	hijo-sū
divider *[eng] [math]*	両脚規	ryōkyaku-ki
dividing *{n} [ind-eng]*	分断；分斷	bundan
" *{n} [mech-eng]*	割出し	wari'dashi
dividing circuit *[electr]*	分周回路	bunshū-kairo
diving *{n} [ocean]*	潜水	sensui
diving rudder *[aero-eng]*	水平舵；すいへいだ	suihei-da
diving suit *[eng]*	潜水服	sensui-fuku
divisibility *[arith]*	割り切れること	wari-kireru koto
divisible *{adj} [arith]*	割り切れる	wari-kireru
division *[arith]*	除法	johō
" *[arith]*	割り算	wari-zan
" *[bot] [syst]*	門；もん	mon
" *[cytol]*	分裂	bunretsu
" *[pat]*	分割	bunkatsu
division; section; group *{n}*	部；ぶ	bu
division; distribution; sharing *{n}*	分配	bunpai
division of application *[pat]*	出願の分割	shutsu'gan no bun'katsu
division of labor *[bio]*	分業	bungyō
division symbol *[arith]*	割算記号	wari'zan-kigō
divisor *[arith]*	除数	josū
do *{vb}*	為る；する	suru
do (or does) not do *{vb}*	為ない	shi-nai
do not feel *{vb} [physio]*	感じない	kanjinai
do not understand *{vb} [comm]*	分からない；解らない；判らない	wakaranai
docket *[legal]*	記録帳	kiroku'chō
dock fender *[civ-eng]*	防舷材	bōgen-zai
doctor; physician (a person) *[med]*	医者；醫者；いしゃ	i'sha
" (a person) *[med]*	医師	i'shi

document *[comput]*	文書	bunsho
" *[print]*	文献	bunken
documentation *[comm]*	文書化	bunsho'ka
document giving reason *[legal]*	理由書	riyū'sho
documents *[comm] [pr]*	書類；しょるい	shorui
dodecahedron(pl -s, -hedra) *[geom]*	十二面体	jūni'men'tai
does not move; unmoving *{adj}*	動かない	ugokanai
dog (animal) *[v-zoo]*	犬；いぬ	inu
doll (toy) *{n} [eng]*	人形；にんぎょう	ningyō
dollar *[econ]*	弗；ドル	doru
dolly *[eng]*	当て盤；當て盤	ate-ban
dolomite *[miner]*	苦灰石	ku'kaiseki
dolphin (mammal) *[v-zoo]*	海豚；いるか	iruka
" (mammal) *[v-zoo]*	真海豚	ma'iruka
domain *[alg] [comput]*	領域	ryō'iki
dome (ceiling) *[arch]*	丸天井	maru-tenjō
" (roof) *[arch]*	丸屋根	maru-yane
domestication (of animals) *{n}*	順養；じゅんよう	jun'yō
domestic consumption *[econ]*	内需	naiju
domicile *[legal]*	住所；じゅうしょ	jūsho
dominance *[bio]*	優占度	yūsen-do
" *[gen]*	優性	yūsei
" *[zoo]*	優越；ゆうえつ	yū'etsu
dominance, law of *[ecol]*	優劣の法則	yū'retsu no hō'soku
dominant mutation *[gen]*	優性突然変異	yūsei-totsu'zen-hen'i
dominant trait *[gen]*	優性素質	yūsei-so'shitsu
dominant wavelength *[phys]*	主波長	shu-hachō
donkey; ass (animal) *[v-zoo]*	驢馬；ろば	roba
donor *[ch]*	供体	kyōtai
" *[ch]*	供与体；供與体	kyōyo'tai
" *[sol-st]*	給体	kyūtai
door *[arch]*	戶	to
" *[arch]*	扉；とびら	tobira
doorknob *[arch]*	把手	totte
doorsill *[arch]*	敷居	shiki'i
dope; stock solution; undiluted solution *[ind-eng]*	原液	gen'eki
dormancy *[bot]*	休眠；きゅうみん	kyūmin
dormant state *[comput]*	休止状態	kyūshi-jōtai
dormant volcano *[geol]*	休火山	kyū-kazan
dormitory (pl -ries) *[arch]*	寄宿舎	kishuku'sha
dormouse; marmot (animal) *[v-zoo]*	山鼠	yama'ne

dorsal fin *[v-zoo]*	背鰭；せびれ	se-bire
dory; Zeus japonicus (fish) *[v-zoo]*	的鯛；まとだい	mato-dai
dosage *[med]*	線量率	senryō-ritsu
dosage meter; dosimeter *[nucleo]*	放射線計	hōshasen-kei
dose *{n} [med]*	照射量	shōsha-ryō
" (of radiation) *{n} [med]*	線量	senryō
dose; dosage *{n} [med] [pharm]*	用量	yōryō
dose rate *[nucleo]*	線量率	senryō-ritsu
dosimeter *[eng]*	線量計	senryō-kei
" *[nucleo] [pharm]*	薬量計	yakuryō-kei
dosimetry *[nucleo] [pharm]*	薬量学；藥量學	yakuryō'gaku
dot (of fruit) *{n} [bot]*	果点	katen
" (Morse code) *{n} [comm]*	短点	tanten
" (speck; spot) *{n} [graph]*	点；點；てん	ten
" *{n} [music]*	付点	futen
" *{n} [pr]*	ぽつ	potsu
dot electron formula *[atom-phys]*	点電子式	ten-denshi-shiki
dot product; inner product *[math]*	点乗積；點乗積	ten-jō'seki
dotted line *[graph]*	点線	tensen
double *{n} [math]*	(二)倍	(ni)bai
" *{adj} [math]*	(二)倍の	(ni)bai no
double-acting piston *[mech-eng]*	複動ピストン	fukudō-pisuton
double-angle formula *[trig]*	二倍角の公式	nibai-kaku no kō'shiki
double-blind test method *[ind-eng]*	二重盲検法	nijū-mōken-hō
double bond *[p-ch]*	二重結合	nijū-ketsu'gō
double-column setting *[pr]*	二段組み	nidan-gumi
double consonant *[ling]*	促音；そくおん	soku'on
double consonant mark; tsu-sound symbol *[pr]*	促音符	soku'on-pu
double-current system *[fl-mech]*	複流式	fukuryū-shiki
double-deck bridge *[civ-eng]*	二層橋	nisō-kyō
double decomposition *[ch]*	複分解	fuku-bunkai
double exposure *[photo]*	二重露光	nijū-rokō
double fertilization *[bot]*	重複受精	chōfuku-jusei
double-flowered *{adj} [bot]*	八重咲(の)	yae'zaki (no)
double-flowered cherry tree *[bot]*	八重桜	yae-zakura
double helical gear *[mech-eng]*	山歯歯車	yama'ba-ha'guruma
double integral *[calc]*	二重積分	nijū-seki'bun
double layer *[phys]*	二重層	nijū-sō
double melting point *[thermo]*	複融点	fuku-yūten
double plate *[pr]*	見開き図版	mi'hiraki-zuban
double precision *[comput]*	二倍精度	nibai-seido

double refraction *[opt]*	複屈折	fuku-kussetsu
double salt *[inorg-ch]*	複塩；複鹽	fuku-en
double-spread title page *[pr]*	見開き扉	mi'hiraki-tobira
double star *[astron]*	二重星	nijū-sei
doublet *[atom-phys]*	二重項	nijū-kō
double tide *[ocean]*	双潮；雙潮	sōchō
double-twisted yarn *[tex]*	合撚糸	gōnen-shi
doubt *{n} [psy]*	疑義；ぎぎ	gigi
doubtful; questionable *{adv}*	疑わしい	utagawashī
douche *{n} [pharm]*	膣洗剤	chitsu-senzai
douche; irrigation *{n} [pharm]*	灌注液；浣注液	kanchū-eki
dough coating for deep frying *[cook]*	衣；ころも	koromo
dovetail joint *[eng]*	蟻継ぎ	ari-tsugi
down; downward *{n} [math]*	下；した	shita
down (of bird) *{n} [v-zoo]*	綿毛	wata'ge
downdraft *[phys]*	下降気流	kakō-kiryū
" *[phys]*	下向き通風	shita'muki-tsūfū
downgust *[phys]*	下降突風	kakō-toppū
downstream *[hyd]*	川下；かわしも	kawa-shimo
downstream-side *[hyd]*	下流側	karyū-gawa
downtime *[ind-eng]*	停止時間	teishi-jikan
" *[ind-eng]*	雑時間；雜時間	zatsu-jikan
downward flow *[fl-mech]*	下向流	kakō-ryū
downwardly concave *[opt]*	下に凹	shita ni ō
downwardly convex *[opt]*	下に凸	shita ni totsu
downwind *[navig]*	風下；かざしも	kaza-shimo
draft *{n} [eng]*	抜き勾配	nuki-kōbai
" *{n} [graph]*	下絵；下繪；したえ	shita-e
" *{n} [nav-arch]*	喫水	kissui
" *{n} [pr]*	草稿	sōkō
" *{n} [steel]*	圧下率	akka-ritsu
draft dryer *[eng]*	通風乾燥機	tsūfū-kansō-ki
draft gage *[eng]*	通風計	tsūfū-kei
drafting *{n} [graph]*	製図；製圖；せいず	seizu
draftsman *[eng]*	図工	zukō
draftsmanship; drawing *{n} [graph]*	図法	zuhō
drag *{n} [aero-eng] [fl-mech]*	抗力	kōryoku
" *{n} [fl-mech]*	抵抗	teikō
" *{n} [met]*	下型	shita-gata
" (dragging) *{n}*	引き摺り；ひきずり	hiki'zuri
" (drag along) *{vb t}*	引き摺る	hikizuru
drag coefficient *[fl-mech]*	抵抗係数	teikō-keisū

dragonet (fish) *[v-zoo]*	喉腐り	nodo'kusari
dragonfly (insect) *[i-zoo]*	蜉蝣；かげろう	kagerō
" (insect) *[i-zoo]*	蜻蛉；とんぼ	tonbo
drain *{n} [arch]*	樋；とい	toi
" (let flow, flush) *{vb t}*	流す	nagasu
drainage surface *[hyd]*	排水面	haisui-men
drainage water *[hyd]*	排水	haisui
drainboard (in kitchen) *[cook]*	水切り板	mizu'kiri-ban
" *[cook]*	水切り台	mizu'kiri-dai
" *[cook]*	流し板	nagashi-ita
drain pipe *[civ-eng]*	排水管	haisui-kan
draw *{vb} [comput]*	描画する	byōga suru
" (sketch, paint) *{vb} [graph]*	描く；画く；えがく	egaku
drawability *[met]*	伸線性	shinsen-sei
draw a conclusion	結論を導く	ketsu'ron o michibiku
drawbridge *[civ-eng]*	跳ね橋	hane-bashi
drawer (of chest) *[furn]*	引出し	hiki'dashi
draw forming *{n} [eng]*	絞り成形	shibori-seikei
drawing *{n} [graph]*	描画；描畫	byōga
" *{n} [graph]*	絵図；繪圖	ezu
" *{n} [graph]*	製図	seizu
" *{n} [graph]*	図画	zuga
" *{n} [graph]*	図面	zumen
" *{n} [poly-ch]*	延伸	enshin
drawing board *[graph]*	製図板	seizu-ban
drawing paper *[graph]*	図画用紙	zuga-yōshi
drawing room; reception room *[arch]*	応接間；おうせつま	ō'setsu-ma
drawing up; cramping *{n} [med]*	引き攣り；ひきつり	hiki'tsuri
draw ratio *[poly-ch] [tex]*	延伸倍率	enshin-bai'ritsu
" *[poly-ch] [tex]*	延伸比；えんしんひ	enshin-hi
" *[poly-ch] [tex]*	延伸率	enshin-ritsu
dream *{n} [psy]*	夢；ゆめ	yume
dredging machine *[eng]*	浚渫機	shun'setsu-ki
dregs *[met]*	滓；かす	kasu
dressing (of ore) *[min-eng]*	選鉱	senkō
dressing room *[arch]*	更衣室	kō'i-shitsu
dried bean curds *[cook]*	湯葉；ゆば	yuba
dried bonito *[cook]*	鰹節；かつおぶし	katsuo'bushi
dried cuttlefish *[cook]*	鯣；するめ	surume
dried fish *[cook]*	干物	hi'mono
dried gourd strips *[cook]*	干瓢	kanpyō
dried orange peel *[pharm]*	陳皮	chinpi

dried small sardines *[cook]*	鱓；田作；ごまめ	gomame
" *[cook]*	煮干し	ni-boshi
dried squid *[cook]*	鯣；するめ	surume
drift *{n} [fl-mech]*	流動	ryūdō
" *{n} [navig]*	漂流；ひょうりゅう	hyōryū
" *{vb i}*	漂流する	hyōryū suru
drift net *[eng]*	流し網	nagashi-ami
drift velocity *[sol-st]*	流動速度	ryūdō-sokudo
driftwood *{n}*	流れ木	nagare'gi
drill *{n} [eng]*	錐；きり	kiri
" *[tex]*	太綾；ふとあや	futo-aya
drink *{vb} [physio]*	飲む；呑む；のむ	nomu
drinkable *{adj} [cook]*	飲むのに適した	nomu-noni-tekishita
drinkables; drinks *[cook]*	飲物；のみもの	nomi'mono
drinking fountain *[arch]*	水呑場	mizu-nomi'ba
drinking water *[hyd]*	飲料水	inryō-sui
drip *{n} [hyd]*	滴り；したたり	shitatari
drip; trickle *{vb} [ch]*	滴下する	tekika suru
drive *{n} [comput]*	駆動機構；驅動機構	kudō-kikō
" *{n} [psy]*	衝動；しょうどう	shōdō
drive clock *[mech-eng]*	運転時計	unten-dokei
driver (a person) *[trans]*	運転者；運轉者	unten'sha
drive shaft *[eng]*	駆動軸；驅動軸	kudō-jiku
drive train *[mech]*	駆動列	kudō-retsu
driving *{n} {vb}*	運転；うんてん	unten
driving axle *[mech-eng]*	動軸	dō-jiku
driving license *[trans]*	運転免許(状)	unten-menkyo('jō)
driving shaft *[mech-eng]*	原動軸	gendō-jiku
driving source *[mech-eng]*	駆動源；くどうげん	kudō-gen
driving wheel *[mech-eng]*	動輪	dōrin
" *[mech-eng]*	駆動車	kudō-guruma
drizzle *[meteor]*	霧雨；きりさめ	kiri'same
dromedary (animal) *[v-zoo]*	一瘤駱駝	hito'kobu-rakuda
drone (insect) *[i-zoo]*	雄蜂	osu-bachi
droop; shrivel; wilt *{vb i}*	萎む；しぼむ	shibomu
droop; hang; dangle; drip *{vb i}*	垂れる	tareru
drop (of liquid) *{n} [phys]*	滴；てき	teki
" ; fall *{vb i}*	落ちる	ochiru
" ; let fall *{vb t}*	落とす	otosu
drop below (a value) *[math]*	割る	waru
droplet *[ch] [meteor]*	液滴	eki'teki
" (welding) *[met]*	溶滴	yōteki

droplet infection *[med]*	飛沫感染	hi'matsu-kansen
dropout; dropping out *{n}*	脱落	datsu'raku
dropping funnel; drop funnel *[ch]*	滴下漏斗	tekika-rōto
drops; drop medicine *{n} [pharm]*	点滴薬	tenteki-yaku
drop test *[mech]*	落下試検	rakka-shiken
dropwort (herb) *[bot]*	芹；せり	seri
dross (casting) *[met]*	湯垢；ゆあか	yu-aka
dross; scum *[met]*	浮滓	uki-kasu
drought *[agr]*	日照り；ひでり	hideri
" *[climat]*	旱魃；かんばつ	kanbatsu
drought resistance *[agr]*	耐乾性	taikan-sei
drug; medicine *{n} [pharm]*	薬剤；藥劑	yaku'zai
drug *{n} [pharm]*	薬物；やくぶつ	yaku'butsu
drug metabolism *[physio]*	薬物代謝	yaku'butsu-taisha
drug resistance *[med]*	薬剤耐性	yakuzai-taisei
drugstore *[arch] [pharm]*	薬品雑貨店	yakuhin-zakka'ten
" *[arch] [pharm]*	薬店	yaku'ten
drum *{n} [music]*	太鼓；たいこ	taiko
drupe *[bot]*	核果	kaku'ka
" *[bot]*	石果	seki'ka
druse *[geol]*	晶洞	shōdō
dry; dryness; drying *{n} [sci-t]*	乾燥；かんそう	kansō
dry; dried up *{adj} [bot]*	枯れた	kareta
dry *{adj}*	乾いた	kawaita
" *{vb i}*	乾く；かわく	kawaku
" *{vb t}*	乾かす	kawakasu
dry-bulb thermometer *[eng]*	乾球温度計	kankyū-ondo-kei
dry cell *[elec]*	乾電池	kan'denchi
dry distillation *[ch]*	乾留；乾溜	kanryū
dryer; drier *[eng]*	乾燥機	kansō-ki
dry ice *[inorg-ch]*	固体炭酸（ドライ＝アイス）	kotai-tansan (dorai-aisu)
drying agent *[ch]*	乾燥剤	kansō-zai
drying oil *[mat]*	乾性油	kansei-yu
dry-lacquered image (ornament)	乾漆像	kan'shitsu-zō
dry plate *[photo]*	乾板；かんぱん	kanpan
dry point *[an-ch]*	乾点	kanten
dry process *{n} [pr]*	乾式法	kanshiki-hō
dry season *[climat]*	渇水期	kassui-ki
dry spinning *{n} [tex]*	乾式紡糸	kanshiki-bōshi
dry weight *[mech]*	乾量	kanryō
" *[mech]*	乾燥重量	kansō-jūryō

dual; double *{adj}*	二重の	nijū no
dual brightness *[opt]*	二重光度	nijū-kōdo
dual operation *[comput]*	双対演算	sōtai-enzan
dub; dubbing *{n} [electr]*	追加録音	tsuika-roku'on
duck (bird, domesticated) *[v-zoo]*	家鴨; あひる	ahiru
" (bird, wild) *[v-zoo]*	鴨; かも	kamo
duckweed *[bot]*	浮草	uki-kusa
duct *[mech-eng]*	管路	kanro
ductile *{adj} [mat]*	延性のある	ensei no aru
ductile fracture *[mech]*	延性破壊	ensei-hakai
ductility *[mat]*	延性; えんせい	ensei
" *[mat]*	伸度	shindo
ductility transition temperature *[met]*	延性遷移温度	ensei-sen'i-ondo
duet *[music]*	二重奏	nijū'sō
dug well *{n} [eng]*	掘井戸; ほりいど	hori-ido
dull (not sharp) *{adj}*	鈍い; にぶい	nibui
dulling; clouding *{n}*	曇り; くもり	kumori
dumbwaiter *[eng]*	小型エレベーター	kogata-erebēta
dummy *{n} [math]*	擬装物	gisō'butsu
dummy level *[stat]*	擬水準	gi-suijun
dune *[geol]*	砂丘; さきゅう	sakyū
duodecimal numeration *[math]*	十二進法	jūni'shin-hō
duodenum (pl -dena, -denums) *[anat]*	十二指腸	jūnishi'chō
duplex transmission *[comm]*	二重伝送	nijū-densō
duplicate *{n} [graph]*	副本	fukuhon
" *{vb} [comput]*	複製する	fukusei suru
" *{vb} [graph]*	重複する	chōfuku suru
duplicate key *[eng]*	合い鍵; あいかぎ	ai-kagi
duplicating *{n} [comput]*	複製	fukusei
duplication *[graph]*	重複; ちょうふく	chō'fuku
durability *[eng]*	持続性	jizoku-sei
" *[mat]*	耐久性	taikyū-sei
durable goods *[econ]*	耐久(消費)財	taikyū(-shōhi)-zai
duration *[mech]*	時間	jikan
dusk *[astron]*	夕闇	yū'yami
dust *{n} [geol]*	塵; 芥; ごみ	gomi
" *{n} [geol]*	塵埃	jin'ai
dust collection *[eng]*	集塵; しゅうじん	shūjin
dust counter *[eng]*	細塵計	saijin-kei
duster *[eng]*	叩き; はたき	hataki
dusting powder *[mat] [pharm]*	散布剤	sanpu-zai
" *[mat]*	打粉	uchi-ko

dustproofness *[eng]*	防塵性	bōjin-sei
dust resistance *[mat]*	耐塵性	taijin-sei
dust storm *[meteor]*	砂塵嵐	sajin-arashi
duty exemption *[econ]*	免税; めんぜい	menzei
duty-free goods *{n} [econ]*	免税品	menzei'hin
dwarfness *[med]*	矮性	wai-sei
dwarf star *[astron]*	矮星; わいせい	wai-sei
dwarf tree culture *[bot]*	盆栽; ぼんさい	bonsai
dwelling *{n} [arch]*	住宅	jūtaku
dyadic operation *[comput]*	二項演算	nikō-enzan
dye; dyestuff *{n} [ch]*	染料; せんりょう	senryō
dye; tint *{vb t} [opt]*	染める; そめる	someru
dye affinity *[tex]*	染着性	senchaku-sei
dye bleeding *{n} [tex]*	色流れ	iro-nagare
dye exclusion test *[cyt]*	色素排除試験	shikiso-haijo-shiken
dyefastness *[ch]*	染色堅牢度	senshoku-kenrō-do
dyeing *{n} [ch]*	染色	sen'shoku
" *{n} [ch-eng]*	染着	sen'chaku
dyeing and weaving *{n} [tex]*	染織	sen'shoku
dyeing auxiliary *[ch-eng]*	助剤	jozai
dyers grape *[bot]*	山牛蒡; やまごぼう	yama-gobō
dye toning *{n} [photo]*	染料調色	senryō-chō'shoku
dyke break *[civ-eng]*	波堤	hatei
dynamical parallax *[astron]*	力学視差	riki'gaku-shisa
dynamical time *[astron]*	力学時; 力學時	riki'gaku-ji
dynamic characteristic *[electr]*	動特性	dō-tokusei
dynamic electricity *[elec]*	動電気	dō-denki
dynamic lift *[aero-eng]*	揚力; ようりょく	yō-ryoku
dynamic load *[civ-eng]*	動荷重	dō-kajū
dynamic modulus *[mech]*	動的弾性率	dō'teki-dansei-ritsu
dynamic pressure *[mech]*	動圧; 動壓	dō-atsu
dynamic programming *{n} [comput]*	動的計画法	dō'teki-keikaku-hō
dynamic properties *[electr]*	動的性質	dō'teki-sei'shitsu
dynamics *[mech]*	動力学	dō'rikigaku
" *[mech]*	力学	riki'gaku
dynamo *[elec]*	発電機	hatsuden-ki
dynamometer *[eng]*	動力計	dōryoku-kei
dysentery bacillus *[microbio]*	赤痢菌; せきりきん	sekiri-kin
dysprosium (element)	ジスプロシウム	jisupuroshūmu

E

eagle (bird) *[v-zoo]*	鷲；わし	washi
eagle owl (bird) *[v-zoo]*	鷲木莵；鷲みみずく	washi-mimizuku
eagle ray (fish) *[v-zoo]*	鳶鱏；とびえい	tobi-ei
ear *[anat]*	耳	mimi
eardrop *[pharm]*	点耳薬	tenji-yaku
eardrum; typanic membrane *[anat]*	鼓膜	ko'maku
eared seal *[v-zoo]*	膃肭獣；おっとせい	ottosei
earlier document *[pr]*	先行文献	senkō-bunken
earlobe *[anat]*	耳朶；みみたぶ	mimi-tabu
early dawn *[astron]*	未明	mimei
early disclosure *[pat]*	公開	kōkai
early evening *[astron]*	宵；よい	yoi
early-maturing rice plant *[bot]*	早生；わせ	wase
earnings; income *[econ]*	所得	shotoku
earplug *[eng]*	耳栓	mimi-sen
earth *[astron]*	地球；ちきゅう	chikyū
" ; ground (earthing) *{n} [elec]*	アース	āsu
" *{n} [geol]*	地；ち	chi
" *{n} [geol]*	大地	daichi
" ; soil *{n} [geol]*	土；つち	tsuchi
earth avalanche *[geol]*	山崩れ	yama-kuzure
earth axis *[geod]*	地軸	chi-jiku
earth crust *[geol]*	地殻	chikaku
earth current *[geophys]*	地電流	chi-denryū
earth drill *[civ-eng]*	穴掘(り)機	ana'hori-ki
earthenware *[eng] [cer]*	土器	doki
" *[eng] [cer]*	瀬戸物；せともの	seto'mono
" *[eng] [cer]*	陶器	tōki
earthenware casserole *[cer] [cook]*	土鍋	do-nabe
earthenware jar *[cer]*	瓶；甕；かめ	kame
earthenware pipe *[civ-eng]*	土管	do'kan
earthenware teapot *[cer] [cook]*	土瓶	do'bin
earth fall *{n} [geol]*	崖崩れ	gake-kuzure
earthflow *[geol]*	山崩れ	yama-kuzure
" *[geol]*	山津波；やまつなみ	yama-tsu'nami
earth plate *[elec]*	接地板	setchi-ban
earth pressure *[civ-eng]*	土圧；土壓；どあつ	do-atsu
earthquake *[geophys]*	地震	ji'shin
earthquake disaster *{n} [geophys]*	震災	shinsai

earthquake mechanism *[geophys]*	発振機構	hasshin-kikō
earthquake-resistant design *[arch]*	耐震設計	taishin-sekkei
earth resistance *[elec]*	大地抵抗	daichi-teikō
earthshine *[astron]*	地球照	chikyū'shō
earth stratum (pl strata) *[geol]*	地層; 地層	chisō
earthworm *[i-zoo]*	蚯蚓; みみず	mimizu
earwax; cerumen *[physio]*	耳垢	jikō; mimi-aka
easel *[graph]*	画架; 畫架; がか	gaka
easement *[legal]*	地役権	chi'eki-ken
east (direction) *[geod]*	東; ひがし; あずま	higashi; azuma
East; Orient *[geogr]*	東洋; とうよう	tōyō
east and west *[geogr]*	東西	tōzai
east coast *[geogr]*	東岸	tōgan
easterly wind; easterlies *[meteor]*	東風	higashi-kaze; kochi
Eastern Hemisphere *[geogr]*	東洋球	tōyō-kyū
easy *{adj}*	易しい; やさしい	yasashī
" *{adj}*	容易な	yōi na
easy-flowing property *[fl-mech]*	易流動性	i'ryūdō-sei
easy to do	為易い; しやすい	shi-yasui
easy to hold	持ち易い	mochi-yasui
easy to see	見易い	mi-yasui
easy to understand *[comm]*	分かり易い	wakari-yasui
easy to use *[eng]*	使い易い	tsukai-yasui
easy-washing *{n} [tex]*	糊落性	nori'ochi-sei
eat; dine *{vb} [physio]*	食べる; たべる	taberu
eatables and drinkables *[cook]*	飲食物	in'shoku'butsu
eating *{n} [physio]*	食べること	taberu koto
eaves *[arch]*	庇; ひさし	hisashi
" *[arch]*	軒; のき	noki
ebb tide *[ocean]*	干潮	kanchō
ebony (tree) *[bot]*	黒壇; こくたん	koku'tan
ebullioscopy *[p-ch]*	沸点上昇法	futten-jōshō-hō
eccentric *{adj} [sci-t]*	偏心の	henshin no
eccentric anomaly *[astron]*	離心近点角	rishin-kinten-kaku
eccentricity (pl -ties) *[an-geom]*	離心率	rishin-ritsu
" *[math] [mech]*	偏心; へんしん	henshin
" *[mech]*	偏心率	henshin-ritsu
eccentric wheel *{n} [eng]*	偏心輪	henshin-rin
Echinodermata (starfish) *[i-zoo]*	棘皮動物	kyoku'hi-dōbutsu
echo (pl echoes) *{n} [acous]*	反響	hankyō
" *[acous]*	木霊; こだま	kodama
" *[acous]*	山彦; やまびこ	yama-biko

echo chamber *[acous]*	反響室	hankyō-shitsu
echo check *[comput]*	反響検査; 反響檢查	hankyō-kensa
echogram *[eng]*	音響測深記録図	onkyō-sokushin-kiroku-zu
echo sounder *[eng]*	音響測深機	onkyō-sokushin-ki
echo sounding *{n} [eng]*	音響測深	onkyō-soku'shin
echo suppressor *[electr]*	反響抑制器	hankyō-yokusei-ki
eclipse *{n} [astron]*	食(分); 蝕(分)	shoku(-bun)
eclipse year *[astron]*	食年; しょくねん	shoku-nen
eclipsing binary *[astron]*	食連星	shoku-rensei
ecliptic *[astron]*	黄道; こうどう	kōdō
ecoclimate *[bio]*	生態気候	seitai-kikō
ecological niche *[ecol]*	生態的地位	seitai'teki-chi'i
ecological succession *[ecol]*	生態遷移	seitai-sen'i
ecology *[bio]*	生態学	seitai'gaku
economics (the discipline) [ind-eng]	経済学; 經濟學	keizai'gaku
ecosystem *[ecol]*	生態系	seitai-kei
ectoderm; ectoblast *[embryo]*	外胚葉	gai'haiyō
ectoparasite *[ecol]*	外部寄生虫	gaibu-kisei'chū
" *[ecol]*	外部寄生者	gaibu-kisei'sha
Ectoprocta *[i-zoo]*	外肛動物	gaikō-dōbutsu
ectype *[graph]*	複写; 複寫	fukusha
eczema *[med]*	湿疹; しっしん	shisshin
edaphology *[ecol]*	農業土壌学	nōgyō-dojō'gaku
eddy *{n} [fl-mech]*	渦; うず	uzu
eddy current *[elec]*	渦(巻き)電流	uzu(maki)-denryū
eddy diffusion *[fl-mech]*	乱流拡散; 亂流擴散	ranryū-kakusan
eddy viscosity *[fl-mech]*	渦巻粘性; 渦巻粘性	uzu'maki-nensei
" *[fl-mech]*	渦粘性	uzu-nensei
edge *{n} [geom]*	縁; ふち	fuchi
" *{n} [geom]*	端; はし; たん	hashi; tan
" *{n} [geom]*	辺; 邊; へん	hen
" *{n} [geom]*	稜; りょう	ryō
" *{n} [pr]*	小口	ko-guchi
edge cracking *{n} [plas]*	耳割れ	mimi-ware
edge-curl *{n} [mat]*	縁反り; ふちぞり	fuchi-zori
edge dislocation *[photo]*	刃状転位; 刃狀轉位	yaiba'jō-ten'i
edge-jointing *{n} [adhes]*	縁継ぎ接着	heri'tsugi-setchaku
edible *{adj} [cook]*	食べられる	taberareru
edible chrysanthemum leaves *[cook]*	菊菜; きくな	kiku'na
edible fat *[cook]*	食用脂	shokuyō'shi; shokuyō-abura
edible oil *[cook]*	食用油	shokuyō'yu; shokuyō-abura
edit *{vb t} [comput] [pr]*	編集(する)	henshū (suru)

editing *{n} [comput] [pr]*	編集	henshū
edition *[pr]*	版；はん	han
editor (a person) *[pr]*	編集者	henshū'sha
editorialship *[pr]*	監修；かんしゅう	kanshū
educational district *[civ-eng]*	文教地区	bunkyō-chiku
eel (fish) *[v-zoo]*	鰻；うなぎ	unagi
eel, broiled glazed with soy *[cook]*	鰻の蒲焼	unagi no kaba'yaki
effect *{n} [ch] [phys]*	作用；さよう	sayō
" *{n} [math] [sci-t]*	効果	kōka
effective *{n}*	有効；ゆうこう	yūkō
" *{adj}*	実効的な	jikkō'teki na
effective acceleration *[mech]*	有効加速度	yūkō-kasoku-do
effective area *[mech]*	実効面積；實効面積	jikkō-menseki
effective charge *[elec]*	実効電荷	jikkō-denka
effective component *[ch]*	有効成分	yūkō-seibun
effective current *[elec]*	実効值；じっこうち	jikkō-chi
effective efficiency *[eng]*	有効効率	yūkō-kō'ritsu
effective horsepower *[mech]*	有効馬力	yūkō-ba'riki
effective mass *[sol-st]*	有効質量	yūkō-shitsu'ryō
effectiveness factor *[p-ch]*	有効係数	yūkō-keisū
effective nuclear charge *[elec]*	有効核電荷	yūkō-kaku-denka
effective power *[elec]*	有効電力	yūkō-den'ryoku
effective stroke *[mech-eng]*	有効行程	yūkō-kōtei
effective temperature *[astrophys]*	有効温度	yūkō-ondo
effective value *[math]*	有効值	yūkō-chi
effective voltage *[elec]*	実効電圧	jikkō-den'atsu
effervescence *[ch]*	泡立ち；泡立ち	awa'tachi; awa'dachi
effervescent powder *[pharm]*	沸騰散(剤)	futtō-san(zai)
efficacy; effect *[pharm]*	効能；こうのう	kōnō
efficiency *[eng] [phys]*	能率；のうりつ	nō'ritsu
" *[ind-eng]*	効率	kō'ritsu
" *[ind-eng]*	性能	seinō
effloresced coal *[geol]*	風化炭	fūka-tan
efflorescence *[ch]*	風解；ふうかい	fūkai
" *[miner]*	白華	haku'ka
effluent *[ch]*	流出液	ryū'shutsu-eki
" *[fl-mech]*	流出物流	ryūshutsu'butsu-ryū
effort *[phys]*	作用力	sayō-ryoku
effusion *[p-ch]*	噴散	funsan
effusion of gas, law of *[phys]*	気体流出の法則	kitai-ryū'shutsu no hōsoku
eft *[v-zoo]*	井守；いもり	imori
egg *[cook] [cyt]*	卵；らん；たまご	ran; tamago

egg beater; whisk *[eng] [cook]*	泡立て器	awa'date-ki
egg mushroom *[bot]*	卵茸	tamago-dake
eggplant (vegetable) *[bot]*	茄子; なす	nasu
eggroll *[cook]*	卵焼き	tamago-yaki
eggwhite *[bioch]*	卵白; らんぱく	ranpaku
egret (bird) *[v-zoo]*	(白)鷺	(shira-)sagi
eigentone *[acous]*	固有音; こゆうおん	koyū'on
eigenvalue *[math]*	固有値	koyū-chi
eight *{n} [math]*	八	hachi
einsteinium (element)	アインスタイニウム	ainsutainyūmu
Eisenia bicyclis (seaweed) *[bot]*	荒布; あらめ	arame
eject; project *{vb} [sci-t]*	投射する	tōsha suru
ejecta *[geol]*	噴出物	funshutsu'butsu
ejection *[eng]*	射出; しゃしゅつ	sha'shutsu
elaborate; amplify *{vb} [comm]*	敷衍する	fu'en suru
elastic after-effect *[mech]*	弾性余効; 彈性餘效	dansei-yokō
elastic coefficient *[mech]*	弾性係数	dansei-keisū
elastic deformation *[mech]*	弾性変形; 彈性變形	dansei-henkei
elastic fiber *[tex]*	弾性糸	dansei-shi
elasticity *[mech]*	弾性; だんせい	dansei
elastic limit *[mech]*	弾性限度	dansei-gendo
" *[mech]*	弾性限界	dansei-genkai
elastic material *[mat]*	弾性体	dansei-tai
elastic oscillation *[phys]*	弾性振動	dansei-shindō
elastic recovery *[rub]*	弾性戻り	dansei-modori
elastic strain *[mech]*	弾性歪	dansei-hizumi
elastic stress *[mech]*	弾性応力; 彈性應力	dansei-ō'ryoku
elbow *{n} [anat]*	肘; ひじ	hiji
electric; electrical *{adj}*	電気の; でんきの	denki no
electrical engineer *[elec]*	電気技師	denki-gishi
electrical engineering *[eng]*	電気工学	denki-kōgaku
electrical equipment *[elec]*	電装品	densō'hin
" *[elec]*	電装器; 電裝器	densō-ki
electrical insulating oil *[mat]*	電気絶縁油	denki-zetsu'en-yu
electrical insulation *[elec]*	電気絶縁	denki-zetsu'en
electrical length *[elecmg]*	電気長; 電氣長	denki-chō
electrically conductive material; electroconductive material *[mat]*	導電材料	dōden-zairyō
electrically operated *[elec]*	電気式	denki'shiki
electrical property *[phys]*	電気的性質	denki'teki-seishitsu
electrical resistivity *[elec]*	電気抵抗率	denki-teikō-ritsu
" *[elec]*	固有抵抗	koyū-teikō

electrical steel *[met]*	電気鋼	denki-kō
electrical steel band *[met]*	電気鋼帯	denki-kōtai
electric appliance *[eng]*	電気器具; 電氣器具	denki-kigu
electric bell *[comm] [elec]*	電鈴	denrei
electric cable *[elec]*	電力ケーブル	denryoku-kēburu
electric car *[trans]*	電車; でんしゃ	densha
electric charge *[elec]*	電荷; でんか	denka
electric circuit *[elec]*	電気回路	denki-kairo
electric computer *[comput]*	電気計算機	denki-keisan-ki
electric conductivity *[elec]*	導電率	dōden-ritsu
" *[elec]*	電気伝導度	denki-dendō-do
electric current *[elec]*	電流; でんりゅう	denryū
electric desalting *(n) [ch-eng]*	電気脱塩; 電氣脱鹽	denki-datsu'en
electric dipole *[elec]*	電気双極子	denki-sōkyokushi
electric displacement *[elec]*	電気変位	denki-hen'i
electric eye *[electr]*	電気の目	denki no me
electric field *[elec]*	電場; でんば	denba
" *[elec]*	電界; でんかい	denkai
electric field effect *[electr]*	電界効果	denkai-kōka
electric flux *[elec]*	電束; でんそく	densoku
electric force *[elec]*	電気力	denki-ryoku
electric furnace *[eng]*	電気炉; 電気爐	denki-ro
electric generation *[elec]*	起電	kiden
electric heat *[thermo]*	電熱	dennetsu
electric hygrometer *[eng]*	電気式湿度計	denki'shiki-shitsudo-kei
electrician (a person) *[eng]*	電気工	denki'kō
electric industry *[eng]*	電気工業	denki-kōgyō
electric insulating property *[elec]*	電気絶縁性	denki-zetsu'en-sei
electricity *[phys]*	電気; でんき	denki
" (the discipline) *[phys]*	電気学	denki'gaku
electric light *[elec]*	電灯; 電燈	dentō
electric line of force *[elec]*	電気力線	denki-rikisen
electric locomotive *[mech-eng]*	電気機関車	denki-kikan'sha
electric machine *[eng]*	電気機械	denki-kikai
electric motor *[elec]*	電動機	dendō-ki
electric-motor system *[elec]*	電動式	dendō'shiki
electric outlet *[elec]*	コンセント	konsento
electric parts *[elec]*	電装品	densō'hin
electric potential *[elec]*	電位	den'i
electric power *[elec]*	電力; でんりょく	den'ryoku
electric power consumption rate *(n)*	電力原単位	denryoku-gentan'i
electric railway *[rail]*	電気鉄道	denki-tetsudō

electric ray; Narke japonica *[v-zoo]*	痺れ鱏；しびれえい	shibire-ei
electric resistance *[elec]*	電気抵抗	denki-teikō
electric shaver *[eng]*	電気剃刀	denki-kami'sori
electric shock *[elec]*	感電	kanden
" *[physio]*	電撃；でんげき	dengeki
electric susceptibility *[elec]*	電気感受率	denki-kanju-ritsu
electric welding *{n} [met]*	電気溶接	denki-yōsetsu
electric wire *[mat]*	電線	densen
electrification *[elec]*	電化	denka
" *[elec]*	感電	kanden
" *[elec]*	帯電；帯電	taiden
electroacoustics *[acous]*	電気音響学	denki-onkyō'gaku
electroacoustic transducer *[eng-acous]*	電気音響交換器	denki-onkyō-kōkan-ki
electroanalysis *[p-ch]*	電気分析	denki-bunseki
electroanalytical chemistry *[p-ch]*	電気分析化学	denki-bunseki-kagaku
electrobalance *[an-ch]*	電気天秤	denki-tenbin
electrocapillarity *[phys]*	電気毛管現象	denki-mōkan-genshō
electrocast product *[mat]*	電铸品	denchū'hin
electrochemical affinity *[ch]*	電気化学親和力	denki'kagaku-shinwa-ryoku
electrochemical equivalent *[p-ch]*	電気化学当量	denki'kagaku-tōryō
electrochemical gradient *[ch]*	電気化学的勾配	denki'kagaku'teki-kōbai
electrochemical series *[p-ch]*	電気化学列	denki'kagaku-retsu
electrochemiluminescence *[p-ch]*	電気化学発光	denki'kagaku-hakkō
electrochemistry *[p-ch]*	電気化学	denki-kagaku
electrocoating *[met]*	電着	den'chaku
electroconductive layer *[poly-ch]*	導電層	dōden-sō
electroconductive polymer *[poly-ch]*	導電性高分子	dōden'sei-kōbunshi
electroconductive property *[elec]*	導電性	dōden-sei
electrocrystallization *[p-ch]*	(結晶)電析	(kesshō-)denseki
electrocycylic reaction *[p-ch]*	電子環状反応	denshi-kanjō-hannō
electrode *[elec]*	電極；でんきょく	den'kyoku
electrodeposition *[met]*	電着	den'chaku
" *[met]*	電析	denseki
electrodeposition coating *[mat]*	電着塗装	denchaku-tosō
electrodeposition stress *[met]*	電着応力	denchaku-ō'ryoku
electrode potential *[p-ch]*	電極電位	denkyoku-den'i
electrodialysis *[p-ch]*	電気透析	denki-tōseki
electrodynamics *[elecmg]*	電気力学	denki-rikigaku
electroencephalography; EEG *[med]*	脳波記録法	nō'ha-kiroku-hō
electroendosmosis *[phys]*	電気浸透	denki-shintō
electroforming *{n} [met]*	電铸；電鑄	denchū
electrogalvanizing *{n} [met]*	電気亜鉛めっき	denki-aen'mekki

electrojet *[geophys]*	高層電流	kōsō-denryū
electrokinetic phenomenon *[p-ch]*	動電現象	dōden-genshō
electrokinetic potential *[phys]*	界面動電位	kaimen-dōden'i
electroless plating *{n} [met]*	無電解めっき	mu'denkai-mekki
electroluminescence *[electr]*	電場発光	denjō-hakkō; denba-hakkō
" *[electr]*	電界発光	denkai-hakkō
" *[electr]*	放電発光	hōden-hakkō
electrolysis *[p-ch]*	電解; でんかい	denkai
" *[p-ch]*	電気分解 (電解)	denki-bunkai (denkai)
electrolyte *[p-ch]*	電解液	denkai-eki
" *[p-ch]*	電解質	denkai-shitsu
electrolytic bath *[met]*	電解浴	denkai-yoku
electrolytic capacitor *[elec]*	電解コンデンサ	denkai-kondensa
electrolytic cell *[p-ch]*	電解槽	denkai-sō
electrolytic copper *[met]*	電解銅	denkai-dō
electrolytic corrosion *[met]*	電食	denshoku
electrolytic dissociation *[ch]*	電離; でんり	denri
electrolytic extraction *[p-ch]*	電解抽出	denkai-chū'shutsu
electrolytic iron *[met]*	電解鉄; 電解鐵	denkai-tetsu
electrolytic polymerization *[poly-ch]*	電解重合	denkai-jūgō
electrolytic potential *[p-ch]*	電解圧	denkai-atsu
electrolytic reduction *[ch]*	電解還元	denkai-kangen
electrolytic separation *[p-ch]*	電解分離	denkai-bunri
electromagnet *[elecmg]*	電磁石; 電じしゃく	den'jishaku
electromagnetic field *[elecmg]*	電磁場; でんじば	denji-ba
" *[elecmg]*	電磁界	denji-kai
electromagnetic flowmeter *[eng]*	電磁流量計	denji-ryūryō-kei
electromagnetic force *[phys]*	電磁力	denji-ryoku
electromagnetic induction *[elecmg]*	電磁誘導	denji-yūdō
electromagnetic radiation *[elecmg]*	電磁放射線	denji-hōsha'sen
electromagnetic wave *[elecmg]*	電磁波; でんじは	denji-ha
" *[elecmg]*	電波	denpa
electromagnetism *[phys]*	電磁気	den'jiki
" (the discipline) *[phys]*	電磁気学	denjiki'gaku
electromechanical coupling factor *[mech-eng]*	電気機械結合係数	denki-kikai-ketsu'gō-keisū
electrometer *[eng]*	電位計	den'i-kei
electromotive force; emf *[p-ch]*	起電力	kiden-ryoku
electromotive series *[p-ch]*	電気化学列	denki-kagaku-retsu
electron *[phys]*	電子; でんし	denshi
electron affinity *[atom-phys]*	電子親和力	denshi-shinwa-ryoku
electron avalanche *[electr]*	電子雪崩	denshi-nadare

electron beam; electron ray *[electr]*	電子線；でんしせん	denshi-sen
electron-beam welding *{n} [met]*	電子投射溶接	denshi-tōsha-yō'setsu
electron carrier *[bioch]*	電子伝達体	denshi-den'tatsu'tai
electron current *[elec]*	電子流	denshi-ryū
electron diffraction *[phys]*	電子回折	denshi-kaisetsu
electronegativity *[p-ch]*	電気陰性度	denki-insei-do
electron emission *[phys]*	電子放出	denshi-hō'shutsu
electron equivalent *[phys]*	電子当量	denshi-tōryō
electron gun *[electr]*	電子銃	denshi-jū
electron hole droplets *[sol-st]*	電子正孔液滴	denshi-seikō-eki'teki
electronic *{adj} [electr]*	電子の	denshi no
electronic ballast *[electr]*	電子安定器	denshi-antei-ki
electronic calculator *[electr]*	電子式卓上計算機	denshi'shiki-taku'jō-keisan-ki
electronic cash register *[electr]*	電子式金銭登録機	denshi'shiki-kinsen-tōroku-ki
electronic computer *[comput]*	電子計算機	denshi-keisan-ki
electronic configuration *[atom-phys]*	電子配置	denshi-haichi
electronic editing *{n} [comput]*	電子編集	denshi-henshū
electronic instrument *[eng]*	電子機器	denshi-kiki
electronic mail *[comm]*	電子郵便	denshi-yūbin
electronic navigation *[navig]*	電子航法	denshi-kōhō
electronic notebook *[electr]*	電子手帳；電子手帖	denshi-techō
electronic oscillation *[electr]*	電子振動	denshi-shindō
electronic publishing *{n} [comput]*	電子出版	denshi-shuppan
electronic range *[cook] [eng]*	電子レンジ	denshi-renji
electronics *[phys]*	電子工学	denshi-kōgaku
electronic scanning *{n} [electr]*	電子走査	denshi-sōsa
electron impact spectroscopy *[phys]*	電子衝撃分光法	denshi-shōgeki-bunkō-hō
electron micrograph *[graph]*	電子顕微鏡写真	denshi-kenbikyō-shashin
electron microscope *[electr]*	電子顕微鏡	denshi-kenbi'kyō
electron multiplier *[electr]*	電子増倍管	denshi-zōbai-kan
electronograph *[electr]*	電子写真装置	denshi-shashin-sōchi
electron optics *[electr]*	電子光学	denshi-kōgaku
electron pair bond *[ch]*	電子対結合	denshi'tsui-ketsugō
electron paramagnetic resonance *[phys]*	電子常磁性共鳴	denshi-jōji'sei-kyōmei
electron shell *[atom-phys]*	電子殻	denshi-kaku
electron transport system *[bioch]*	電子伝達系	denshi-den'tatsu-kei
electron tube *[electr]*	電子管；でんしかん	denshi-kan
electronvolt; eV *[phys]*	電子ボルト	denshi-boruto
electrooptic effect *[opt]*	電気光学効果	denki'kōgaku-kōka
electrooptics *[opt]*	電気光学	denki-kōgaku

electroosmosis *[p-ch]*	電気浸透	denki-shintō
electrophilic property *[p-ch]*	求電子性	kyū'denshi-sei
electrophilic reagent *[p-ch]*	求電子試薬	kyū'denshi-shi'yaku
electrophoresis *[p-ch]*	電気泳動	denki-eidō
electrophorus *{n} [elec]*	電気盆；でんきぼん	denki-bon
" *{n} [elec]*	起電盤	kiden-ban
electrophotography *[graph]*	電子写真；電子寫真	denshi-shashin
electrophysiology *[med]*	電気生理学	denki-seiri'gaku
electroplating *{n} [met]*	電着	denchaku
" *{n} [met]*	電気めつき（鍍金）	denki-mekki
" *{n} [met]*	電鍍；でんと	dento
electropolishing *{n} [met]*	電解研摩	denkai-kenma
electrorefining *{n} [met]*	電解精錬	denkai-seiren
electroscope *[eng]*	検電器；檢電器	kenden-ki
electrostatic capacity *[elec]*	静電容量	seiden-yōryō
electrostatic field *[elec]*	静電界；靜電界	sei-denkai
electrostatic force *[elec]*	静電力	seiden-ryoku
electrostatic induction *[elec]*	静電誘導	seiden-yūdō
electrostatic potential *[elec]*	静電位；せいでんい	sei-den'i
electrostatic printer *[electr] [pr]*	静電式印書装置	seiden'shiki-insho-sōchi
electrostatic repulsion *[elec]*	静電反発	seiden-han'patsu
electrostatic storage *[comput]*	静電式記憶	seiden'shiki-ki'oku
electrostriction *[mech]*	電歪；でんわい	den'wai
electrothermal metallurgy *[met]*	電熱冶金	dennetsu-yakin
electrotypy *[met]*	電気版術	denki'han'jutsu
electrovalence *[p-ch]*	電気原子価	denki-genshi'ka
electroviscous effect *[fl-mech]*	電気粘性効果	denki-nensei-kōka
elegant; graceful *{adj}*	優美な	yūbi na
element *[ch]*	元素；げんそ	genso
" *[elec]*	素子	soshi
" *[math]*	元；げん	gen
" *[math] [sci-t]*	要素	yōso
elemental copper *[ch]*	素銅；そどう	sodō
elemental material *[ind-eng]*	素材；そざい	sozai
elementary charge *[elec]*	電気素量	denki-soryō
elementary particle *[part-phys]*	素粒子	so'ryūshi
elementary reaction *[p-ch]*	素反応；素反應	so-hannō
elephant (animal) *[v-zoo]*	象；ぞう	zō
Elephantidea *[v-zoo]*	象類	zō-rui
elephant seal (mammal) *[v-zoo]*	象海豹；象あざらし	zō-azarashi
elevation *[eng]*	標高	hyōkō
elevation, angle of *[eng]*	仰角	gyōkaku

elevation view *[graph]*	立面図	ritsu'men-zu
elevator (of aircraft) *[aero-eng]*	昇降舵	shōkō'da
" *[arch]*	昇降機	shōkō-ki
eligible; sound *{n}*	適格；てきかく	tekikaku
eliminate *{vb}*	消去する	shōkyo suru
elimination *[math]*	消去	shōkyo
elimination reaction *[org-ch]*	脱離反応	datsu'ri-hannō
elk (animal) *[v-zoo]*	箆鹿；へらじか	hera-jika
ellipse *[an-geom]*	長円	chō'en
" *[an-geom]*	楕円；楕圓；だえん	da'en
ellipsis (pl ellipses) *[pr]*	省略記号	shōryaku-kigō
ellipsoid *[an-geom]*	長円面	chōen'men
" *[an-geom]*	楕円面	daen'men
" *[geom]*	楕円体	daen'tai
ellipsoidal coordinates *[math]*	楕円座標	daen-zahyō
ellipsometry *[opt]*	偏光解析	henkō-kaiseki
elliptically polarized light *[opt]*	楕円偏光	daen-henkō
elliptic cone *[an-geom]*	楕円錐	daen-sui
elliptic cylinder *[an-geom]*	楕円柱	daen-chū
ellipticity *[electr]*	楕円率	daen-ritsu
" *[electr]*	扁平率	henpei-ritsu
elm (tree) *[bot]*	楡；にれ	nire
elm bark beetle (insect) *[i-zoo]*	楡木食虫	nire-ki'kui-mushi
elongation *[mech]*	伸び；のび	nobi
" *[mech]*	伸長；しんちょう	shinchō
elongation percentage *[mech]*	伸び率	nobi-ritsu
elongation rate *[mech]*	伸び率	nobi-ritsu
elongation ratio *[mech]*	伸長率	shinchō-ritsu
elsewhere; outside; other *{adv} {n}*	外；他；ほか	hoka
elsewhere; strange parts *{adv} {n}*	他所；余所；よそ	yoso
eluate *[ch]*	溶出液	yō'shutsu-eki
eluent *[mat]*	溶離剤	yōri-zai
elution *[ch]*	溶離；ようり	yōri
" *[ch]*	溶出	yō-shutsu
elution profile *[ch]*	溶出像	yō'shutsu-zō
elutriation *[ch-eng]*	水簸；すいひ	suihi
" *[ch-eng] [min-eng]*	懸濁分離法	kendaku-bunri-hō
elutriation velocity constant *[eng]*	飛出し速度定数	tobi'dashi-sokudo-teisū
em *{n} [comput] [pr]*	全角	zenkaku
emanation *{n} [nuc-phys]*	発散物	hassan'butsu
embankment *[civ-eng]*	堤防	teibō
embargo (pl -goes) *{n} [econ]*	出入港禁止	shutsu'nyūkō-kinshi

English	Japanese	Romanization
embassy *[arch]*	大使館；たいしかん	taishi'kan
embayment *[geol]*	湾入；灣入	wan'nyū
embedding (casting) *{n} [eng]*	封入注型	fūnyū-chūkei
" *{n} [sci-t]*	埋蔵；埋藏	maizō
embossing *{n} [eng]*	皺付け；しぼつけ	shibo'tsuke
embossing press *[pr]*	浮出し印刷機	uki'dashi-in'satsu-ki
embrittling; embrittlement *{n} [mech]*	脆化	zeika
embrocation *{n} [pharm]*	塗擦剤	to'satsu-zai
embroidery (pl -deries) *{n} [tex]*	刺繡；ししゅう	shishū
embryo *[bot]*	胚	hai
embryo bud *[bot]*	胚芽；はいが	haiga
embryo culture *[microbio]*	胚培養	hai-baiyō
embryogenesis *[embryo]*	胚形成	hai-keisei
embryology *[bio]*	発生学	hassei'gaku
" *[bio]*	胎生学	taisei'gaku
Embryophyta *[bot]*	有胚植物類	yūhai-shokubutsu-rui
emerge (from) *{vb}*	出る	deru
emergence *[i-zoo]*	羽化；うか	u'ka
emergency; crisis time *{n}*	非常(時)	hijō(-ji)
emergency exit *[arch]*	非常口	hijō-guchi
emergency maintenance *[comput]*	緊急保守	kinkyū-hoshu
emersion *[astron]*	出現	shutsu'gen
emery *[miner]*	金剛砂	kongō'sha
emery cloth *[mat]*	布鑢；ぬのやすり	nuno-yasuri
emigrant *[legal]*	移民	i'min
emission *[elecmg]*	発射	hassha
" *[phys]*	放射；ほうしゃ	hōsha
emission line *[spect]*	輝線	kisen
emission nebula *[astron]*	発光星雲	hakkō-sei'un
emission spectral analysis *[spect]*	発光分光分析	hakkō-bunkō-bunseki
emissive power *[thermo]*	放射度	hōsha-do
" *[thermo]*	放射強度	hōsha-kyōdo
emissivity *[thermo]*	放射率	hōsha-ritsu
" *[thermo]*	発散率	hassan-ritsu
emit; emanate *{vb}*	発する；發する	hassuru
emittance *[thermo]*	放射強度	hōsha-kyōdo
emitter *[electr]*	放射体	hōsha'tai
emollient *[pharm]*	皮膚軟化薬	hifu-nanka-yaku
emotion *[psy]*	感情	kanjō
emphasis; important point *{n}*	重点	jūten
emphasis dot marks *[pr]*	傍点；ぼうてん	bōten
empirical *{adj} [sci-t]*	経験的な；經驗的な	keiken'teki na

empirical formula *[ch]*	実験式；實驗式	jikken-shiki
empiricism *[sci-t]*	経験論	keiken-ron
employee *[ind-eng]*	従業者	jūgyō'sha
empress tree (tree) *[bot]*	桐；きり	kiri
empty; null *{n}*	空；から	kara
empty band *[chem]*	空帯；空帶	kūtai
empty event *[prob] [stat]*	空事象	kū-jishō
empty medium *[comput]*	空媒体	kū-baitai
empty set(s) *[math]*	空集合	kū-shūgō
empty tank *[ch-eng]*	空槽；くうそう	kūsō
emulsification *[ch]*	乳化；にゅうか	nyūka
emulsifying agent; disperser *[mat]*	乳化剤	nyūka-zai
emulsion *[ch]*	乳濁液	nyūdaku-eki
emulsion polymerization *[poly-ch]*	乳化重合	nyūka-jūgō
emulsoid *{n} [ch]*	乳濁質	nyūdaku'shitsu
en *{n} [comput] [pr]*	半角	hankaku
enabled *{adj} [comput]*	使用可能な(の)	shiyō-kanō na (no)
enactment; institution *[legal]*	制定	seitei
enamel *{n} [mat]*	琺瑯質	hōrō'shitsu
enameling *{n} [eng]*	施釉；せゆう	seyū
enantiomer *[ch]*	光学的対称体	kōgaku'teki-taishō'tai
enantiomorph *[ch]*	光学異性体	kōgaku-i'sei'tai
" *[ch]*	鏡像(異性)体	kyōzō(-i'sei)'tai
enantiomorphism *[ch]*	鏡像異性	kyōzō-i'sei
enantiotropy *[ch]*	互変；互變；ごへん	gohen
enargite *[miner]*	硫砒銅鉱	ryūhi'dōkō
encapsulating agent *[mat]*	封止剤	fūshi-zai
enclosure: encased matter *{n}*	封包物；封包物	fūhō'butsu
enclosure: enclosed matter *{n}*	同封物	dōfū'butsu
enclosure wall *[ind-eng]*	囲壁；圍壁；いへき	i'heki
encode *{vb} [comm]*	符号化する	fugō'ka suru
encourage *{vb}*	奨励する	shōrei suru
encrypt *{vb} [comm]*	暗号化する	angō'ka suru
encyclopedia *[pr]*	百科事典	hyakka-jiten
end; termination *{n} [phys]*	終端；しゅうたん	shūtan
end; closing; tip *{n}*	末；すえ	sue
endangered species *[bio]*	絶滅寸前の種	zetsu'metsu-sunzen no shu
endemic disease *[med]*	風土病	fūdo'byō
end group *[ch]*	末端基	mattan-ki
endless *{adj} [mech-eng]*	無端状の	mutan'jō no
endless chain *[mech-eng]*	輪鎖；わぐさり	wa-gusari
endocarp *[bot]*	内果皮	nai'kahi

endocrine gland *[physio]*	内分泌腺	nai'bun'pitsu-sen
endocrine system *[physio]*	内分泌系	nai'bun'pitsu-kei
endoderm; endoblast *[embryo]*	内胚葉	nai'haiyō
endoenzyme *[bioch]*	(細胞)内酵素	(saibō-)nai'kōso
end of work *[ind-eng]*	終業; しゅうぎょう	shūgyō
endomorph *[miner]*	内包鉱物; 内包鑛物	naihō-kōbutsu
endoparasite *[ecol]*	内部寄生虫	naibu-kisei'chū
" *[ecol]*	内部寄生者	naibu-kisei'sha
endoplasmic reticulum; ER *[cyt]*	小胞体; 小胞體	shōhō'tai
endosmosis (pl -moses) *[physio]*	内浸透	nai'shintō
endospore *[bio]*	内生胞子	naisei-hōshi
endothermic *{adj}* *[p-ch]*	吸熱性の	kyū'netsu'sei no
endothermic reaction *[p-ch]*	吸熱反応	kyū'netsu-hannō
end point *[an-ch]*	終点; 終點	shūten
end product *[ind-eng]*	仕上り品	shi'agari'hin
endurance limit *[mech]*	耐久限度	taikyū-gendo
endurance test *[ind-eng]*	耐久試験	taikyū-shiken
" *[ind-eng]*	疲れ試験	tsukare-shiken
endure; tolerate *{vb}*	耐える	taeru
end user *[comput]*	最終使用者	saishū-shiyō'sha
end view *[graph]*	端面図; たんめんず	tanmen-zu
enema *[pharm]*	灌腸剤; 浣腸剤	kanchō-zai
energize *{vb}* *[elec]*	付勢する	fusei suru
energy *[phys]*	エネルギー	enerugī
energy; spirit; essence *{n}* *[bio]*	精気	seiki
energy; vigor; vitality *{n}* *[bio]*	精力	seiryoku
engaged line *[comm]*	話中線	wachū-sen
engagement *[eng]*	嵌合; かんごう	kangō
engine (generator) *[mech eng]*	発電機	hatsu'den-ki
" (machine) *[mech-eng]*	機関; 機關; きかん	kikan
" *[rail]*	機関車	kikan'sha
engine bed *[mech-eng]*	機関台; 機關臺	kikan-dai
engineer *[ind-eng]*	技師	gishi
engineering *{n}* *[sci-t]*	工学; こうがく	kōgaku
engineering acoustics *[acous]*	工学音響学	kōgaku-onkyō'gaku
engine room *[nav-arch]*	機関室	kikan-shitsu
England *[geogr]*	英国; 英國	Eikoku
" *[geogr]*	イギリス	Igirisu
engraving *{n}* *[graph]* *[pr]*	彫刻	chōkoku
enhancement *[bio]*	増進; ぞうしん	zōshin
" *[comput]*	強調	kyōchō
enhancer *[cook]* *[pharm]*	増強剤	zōkyō-zai

enlarge *{vb} [eng]*	拡大する	kaku'dai suru
enlarging; enlargement *{n} [photo]*	引伸し; ひきのばし	hiki-nobashi
enquiry; inquiry; query *{n} [comm] [comput]*	問合せ; といあわせ	toi-awase
enriched fuel *[mat]*	強化燃料	kyōka-nenryō
" *[mat]*	濃縮燃料	nōshuku-nenryō
enrichment *[bio] [geol]*	富化; ふか	fuka
enrichment agent *[cook]*	強化剤	kyōka-zai
enrichment culture *[microbio]*	強化培養	kyōka-baiyō
" *[microbio]*	増菌培養	zōkin-baiyō
enrichment scale factor *[ch]*	濃縮倍率	nōshuku-bairitsu
ensign; national flag *[nav-arch]*	国旗; こっき	kokki
enstatite *[miner]*	頑火輝石	ganka'kiseki
enter; input *{vb} [comput]*	入力(する)	nyū'ryoku (suru)
" ; come (or go) into *{vb}*	入る; はいる; いる	hairu; iru
enteric *{adj} [pharm]*	腸溶性の	chōyō'sei no
enterprise; business; undertaking *[econ]*	事業; じぎょう	jigyō
entertainment; amusement *{n} [psy]*	娯楽; ごらく	goraku
enthalpy *[thermo]*	エンタルピー	entarupī
entire *{adj} [bot]*	全縁の; ぜんえんの	zen'en no
entomology *[i-zoo]*	昆虫学	konchū'gaku
Entoprocta; Endoprocta *[i-zoo]*	内肛動物	naikō-dōbutsu
entrance *[arch]*	玄関; げんかん	genkan
entropy *[thermo]*	エントロピー	entoropī
entry *[arch] [comput]*	入口	iri'guchi; iri'kuchi
" *[comput]*	記述項	kijutsu'kō
" *[comput]*	記入; きにゅう	ki'nyū
" *[comput]*	項目	kōmoku
" ; input *[comput]*	入力; にゅうりょく	nyū'ryoku
entry condition *[comput]*	入口条件	iri'guchi-jōken
entry point; entrance *[comput]*	入口点	iri'guchi-ten
entry sequence *[comput]*	入力順	nyū'ryoku-jun
enumerate *{vb} [math]*	羅列する	ra'retsu suru
enumeration *[math]*	列挙; れっきょ	rekkyo
envelope (stationery) *{n} [comm]*	封筒; ふうとう	fūtō
" *{n} [comm] [math]*	包絡線; 包絡線	hōraku-sen
" ; wrap *{vb t}*	包む; つつむ	tsutsumu
environment *[comput] [ecol]*	環境; かんきょう	kankyō
environmental pollution *[ecol]*	公害; こうがい	kōgai
environmental scanning electron microscope; ESEM *[electr]*	環境走査電子顕微鏡	kankyō-sōsa-denshi-kenbikyō
enzyme *[bioch]*	酵素; こうそ	kōso
enzyme degradation; zymolysis *[bioch]*	酵素分解	kōso-bunkai

English	Japanese	Romanization
enzyme immunoassay *[immun]*	酵素免疫測定法	kōso-men'eki-sokutei-hō
enzyme-linked immunosorbent assay *[immun]*	酵素結合免疫=吸着剤検定(法)	kōso-ketsugō-men'eki-kyūchaku'zai-kentei(-hō)
enzyme-linked immunosorbent assay *[immun]*	酵素免疫測定法	kōso-men'eki-sokutei-hō
enzyme-substrate complex *[bioch]*	酵素基質複合体	kōso-ki'shitsu-fukugō'tai
enzymology *[bioch]*	酵素学；こうそがく	kōso'gaku
epeirogeny (pl -nies); epeirogenesis *[geol]*	造陸運動；ぞう=りくうんどう	zō'riku-undō
ephedra *[bot]*	麻黄	ma'ō
ephemeris (pl ephemerides) *[astron]*	位置推算表	ichi-suisan-hyō
" *[astron] [pr]*	天体位置表	tentai-ichi-hyō
" *[astron] [pr]*	天体暦；天體暦	tentai-reki
epibenthos *[bio]*	表在底生物	hyōzai-teisei-seibutsu
epibion *[bio]*	表在生物	hyōzai-seibutsu
epicarp; exocarp *[bot]*	外果皮	gai'kahi
epicenter *[geophys]*	震央；しんおう	shin'ō
epicyclic train *[mech-eng]*	遊星歯車装置	yūsei-ha'guruma-sōchi
epicycloid *[an-geom]*	外擺線	gaihai-sen
epidemic *{n} [med]*	流行病	ryūkō'byō
epidemic encephalitis *[med]*	流行性脳炎	ryūkō'sei-nō'en
epidermis *[bot] [hist]*	表皮；ひょうひ	hyōhi
epididymis (pl -mides) *[anat]*	副精巣；副精巣	fuku-seisō
" *[bio]*	精巣上体	seisō-jōtai
epifauna *[bio]*	表在動物	hyōzai-dōbutsu
epigenetic *{adj} [embryo]*	後成的な	kōsei'teki na
epigraphy *[comm]*	碑銘研究	himei-kenkyū
epiphyte *[bot]*	着生植物	chakusei-shokubutsu
epitaxy *[crys]*	配向成長	haikō-seichō
epithelial tissue *[hist]*	上皮組織	jōhi-soshiki
epoch *[astron]*	元期	genki
" *[geol]*	世；せい	sei
epochal *{adj}*	画期的な；畫期的な	kakki'teki na
erg *[phys]*	エルグ	erugu
equal *{adj} {n} [math]*	等しい；ひとしい	hitoshī
equal-interval sampling method *[stat]*	等間隔抽出法	tō-kankaku-chū'shutsu-hō
equality (pl -ties) *[math]*	平等；びょうどう	byōdō
" *[math]*	相等	sōtō
" *[math]*	等号；等號	tōgō
equalization *[electr]*	等化	tōka
equalize *{vb t}*	平等化する	byōdō'ka suru
equalizer *[eng]*	釣合い装置	tsuri'ai-sōchi

equal sign *[math]*	等号；とうごう	tōgō
equation *[alg]*	方程式	hōtei'shiki
" *[alg]*	式；しき	shiki
equation of light *[astron]*	光差	kōsa
equation of motion *[fl-mech]*	運動方程式	undō-hōtei'shiki
equation of state *[p-ch]*	状態(方程)式	jōtai(-hōtei)-shiki
equation of the center *[astron]*	中心差	chūshin-sa
equation of time *[astron]*	均時差	kin'jisa
equator *[geod]*	赤道；せきどう	seki'dō
equatorial *{adj}* *[geod]*	赤道の	sekidō no
equatorial coordinates *[astron]* *[geod]*	赤道座標	sekidō-zahyō
equatorial telescope *[eng]*	赤道儀	sekidō-gi
equiangular polygon *[geom]*	等角多角形	tōkaku-ta'kak(u)kei
equilateral *{adj}* *[geom]*	等辺の；等邊の	tō'hen no
equilateral triangle *[geom]*	正三角形	sei-sankak(u)kei
equilibrium (pl -riums, -ria) *[ch]*	平衡；へいこう	heikō
" (equalizing) *{n}* *[mech]*	釣り合い	tsuri'ai
equilibrium condition *[ind-eng]*	平衡状態	heikō-jōtai
equilibrium constant *[ch]*	平衡定数	heikō-teisū
equilibrium point *[ch]*	均衡点	kinkō-ten
equiluminous *{adj}* *[opt]*	等照度の	tō'shō'do no
equinoctial colure *[astron]*	二分経線	nibun-keisen
equinoctial point *[astron]*	分点；分點	bunten
equinoctial tide *[ocean]*	彼岸潮	higan-jio
equinoctial week *[astron]*	彼岸；ひがん	higan
equipartition *[ch]*	等分配	tō'bunpai
equipartition law *[stat-mech]*	等分配則	tō'bunpai-soku
equipartition of energy, law of *[stat-mech]*	エネルギー均分法則	enerugī-kinbun-hōsoku
equipment *[eng]*	装置	sōchi
" *[eng]*	施設	shi'setsu
" *[ind-eng]*	装備	sōbi
equipotential *{n}* *[elec]*	等電位	tō'den'i
equipotential surface *[elec]*	等電位面	to'den'i-men
equip with *{vb}* *[ind-eng]*	装置する	sōchi suru
equivalence *[ch]* *[math]*	等価；等價；とうか	tōka
" *[math]*	同価	dōka
equivalence ratio *[p-ch]*	当量比；當量比	tōryō-hi
equivalent (in area) *{n}* *[geom]*	等積	tōseki
equivalent; equivalence *{n}* *[alg]*	同値	dōchi
equivalent; equivalency *[ch]* *[mat]*	当量；當量	tōryō
equivalent *{adj}* *[logic]*	同値の	dōchi no

English	Japanese	Romanization
equivalent *{adj} [logic]*	等価の	tōka no
" *{adj} [logic]*	対等な	taitō na
equivalent conductivity *[ch]*	当量伝導度	tōryō-dendō-do
" *[ch]*	当量伝導率	tōryō-dendō-ritsu
equivalent neutral printing density *[photo]*	等価中性焼付け=濃度	tōka-chūsei-yaki'tsuke-nōdo
equivalent reaction *[ch]*	当量反応	tōryō-hannō
equivalent resistance *[elec]*	等価抵抗	tōka-teikō
equivalent stress *[mech]*	相当応力; 相當應力	sōtō-ō'ryoku
equivalent weight *[ch]*	当量	tōryō
Equoidea *[v-zoo]*	馬類; うまるい	uma-rui
era *[geol]*	代; だい	dai
eradicate *{vb t} [eng]*	根絶させる	kon'zetsu saseru
erasable storage *[comput]*	消去可能記憶	shōkyo-kanō-ki'oku
erase; erasure *{n} [comput]*	消去; しょうきょ	shōkyo
erase *{vb t} [comput]*	抹殺する	massatsu suru
" *{vb t} [comput]*	抹消する	masshō suru
erase; extinguish *{vb t}*	消す; けす	kesu
eraser *[eng]*	消しゴム	keshi-gomu
erasure *[math]*	抹殺	massatsu
" *[math]*	抹消	masshō
erbium (element)	エルビウム	erubyūmu
erect image *[opt]*	正立像	sei'ritsu-zō
ergonomics *[ind-eng]*	人間工学	ningen-kōgaku
" *[ind-eng]*	生物工学	seibutsu-kōgaku
ergot *[mycol] [org-ch]*	麦角; 麥角	bakkaku
ermine (animal) *[v-zoo]*	貂; てん	ten
erodibility *[geol]*	侵食可能性	shinshoku-kanō-sei
erosion *[geol]*	浸食; しんしょく	shin'shoku
erosion control *[civ-eng]*	砂防	sabō
err; make a mistake *{vb}*	間違う; まちがう	machigau
error *{n} [comput]*	誤り; あやまり	ayamari
" *{n} [math]*	誤差; ごさ	gosa
" (mistake) *{n} [sci-t]*	間違い	machigai
" (mistake) *{n} [sci-t]*	錯誤	sakugo
error analysis *[comput] [math]*	誤り解析	ayamari-kaiseki
error condition *[comput]*	誤り状態	ayamari-jōtai
error control *[comput]*	誤り制御	ayamari-seigyo
error indicator *[comput]*	誤り標識	ayamari-hyō'shiki
error range *[comput]*	誤り範囲	ayamari-han'i
error recovery *[comput]*	誤り回復	ayamari-kaifuku
error variance *[stat]*	誤差分散	gosa-bunsan

eruption (of volcano) *[geol]*	噴火	funka
" ; spouting; jetting *{n}*	噴出；ふんしゅつ	fun'shutsu
erythrocyte; red blood cell *[hist]*	赤血球	sekkekkyū
escape *{n} [comput]*	拡張；擴張	kakuchō
" *{n} [comput]*	免責	menseki
" *{n}*	逃避	tōhi
" *{vb}*	逃げる	nigeru
escapement *[horol]*	脱進機	dasshin-ki
escape velocity *[aero-eng]*	脱出速度	dasshutsu-sokudo
escarpment *[geol]*	急斜面	kyū-shamen
Escherichia coli *[microbio]*	大腸菌	daichō-kin
esophagus *[anat]*	食道；しょくどう	shoku'dō
essential fatty acid *[org-ch]*	必須脂肪酸	hissu-shibō-san
essential oil *[mat]*	精油	seiyu
establish *{vb}*	設定する	settei suru
established theory *[sci-t]*	定説	teisetsu
estate *[legal]*	遺産	i'san
ester *[org-ch]*	エステル	esuteru
esterification *[org-ch]*	エステル化	esuteru'ka
esthetic area *[civ-eng]*	美観地区；美觀地區	bikan-chiku
estimated cost *[econ]*	予算；豫算；よさん	yosan
estimation *[math]*	見積り；みつもり	mi'tsumori
" *[math]*	推定	suitei
estimator *[stat]*	推定量	suitei'ryō
estuary *[geogr]*	河口	kakō
" *[geogr]*	三角江	sankaku-kō
et cetera; (abbreviated etc)	云々；うんぬん	un'nun
eternity *{n}*	永久	eikyū
ether *[org-ch]*	エーテル	ēteru
ethical drug *[pharm]*	処方箋による薬品	shohō'sen ni yoru yaku'hin
ethnobiology *[bio]*	民族生物学	minzoku-seibutsu'gaku
etiolation *[bot]*	白化	hakka
" *[bot]*	黄化；おうか	ō'ka
etui *[furn]*	手箱	te-bako
Eubacteriales *[microbio]*	真正細菌目	shinsei-saikin-moku
eucalyptus (pl -ti, -tuses) *[bot]*	有加利；ゆうかり	yūkari
eudiometer *[eng]*	水電量計	mizu-denryō-kei
eugenics *[gen]*	優性学	yūsei'gaku
euglena *[i-zoo]*	緑虫；みどりむし	midori-mushi
Euglenophyta *[bot]*	緑虫植物	midori'mushi-shokubutsu
euhedral *[petr]*	自形	jikei
eukaryote; eucaryote *[bio]*	真核生物	shinkaku-seibutsu

eukaryotic cell *[bio]*	真核細胞	shinkaku-saibō
eulalia *[bot]*	薄；すすき	susuki
Eumycetes *[mycol]*	菌界	kinkai
euphonic change; euphony *[ling]*	音便；おんびん	onbin
European brown bear (animal) *[v-zoo]*	赤熊；あかぐま	aka-guma
europium (element)	ユーロピウム	yūropyūmu
eutectic *{n}* *[met]*	共晶；きょうしょう	kyōshō
eutectic mixture *[p-ch]*	共融混合物	kyōyū-kongō'butsu
eutectic point *[p-ch]*	共融点；共融點	kyōyū-ten
eutectoid *[p-ch]*	共析晶	kyōseki'shō
euthenics *[bio]*	優境学	yūkyō'gaku
Eutheria *[v-zoo]*	正獣類；正獸類	seijū-rui
eutrophication *[ecol]*	富栄養化	fu'eiyō'ka
evaluation *[math]*	評価；ひょうか	hyōka
evaporated residue *[ch-eng]*	蒸発殘留物	jōhatsu-zanryū'butsu
evaporating dish *[ch]*	蒸発皿	jōhatsu-zara
evaporating to dryness *{n}* *[phys]*	蒸発乾固	jōhatsu-kanko
evaporation *[phys]*	蒸発；じょうはつ	jō'hatsu
evaporation heat *[thermo]*	気化熱	kika-netsu
evaporator *[ch-eng]* *[mech-eng]*	蒸発器；蒸發器	jōhatsu-ki
evaporimeter *[eng]*	蒸発計	jōhatsu-kei
evaporite *[geol]*	蒸発岩	jōhatsu'gan
evapotranspiration *[hyd]*	蒸発散	jō-hassan
evection *[astrophys]*	出差；しゅっさ	shussa
even function *[alg]*	偶関数；偶關數	gū-kansū
evening *[astron]*	夕べ	yūbe
" *[astron]*	夕方	yūgata
evening calm *[meteor]*	夕凪；ゆうなぎ	yū-nagi
evening newspaper *[pr]*	夕刊	yūkan
evenness; even parts; a draw *{n}*	五分五分；ごぶごぶ	gobu-gobu
even number *[arith]*	偶数；偶數	gūsū
even parity check *[comput]*	奇偶検査	kigū-kensa
event *[comput]* *[stat]*	事象；じしょう	jishō
eventuality *[math]*	偶然性	gūzen-sei
evergreen tree *[bot]*	常緑樹	jōryoku-ju
every day *{adv}*	毎日	mai-nichi
everyday wear; everyday clothes *[cl]*	普段着；不断着	fudan-gi
every other week; fortnightly *{adv}*	隔週の	kakushū no
every year; annually *{adv}*	毎年	mai'nen; mai-toshi
evidence *[legal]*	証拠；證據	shōko
evident *{adj}*	明白な	meihaku na
evolution (theory) *[bio]*	進化(論)	shinka(-ron)

evolution *{n}*	展開	tenkai
evolution equation *[quant-mech]*	発展方程式	hatten-hōtei'shiki
ewe (animal) *[v-zoo]*	雌羊；めすひつじ	mesu-hitsuji
exactly *{adv}*	丁度；ちょうど	chōdo
" (precisely) *{adv}*	きちんと	kichinto
" *{adv}*	ぴったり	pittari
examine *{vb}*	調べる	shiraberu
examiner *[pat]*	審査官	shinsa'kan
examiner of patents *[pat]*	特許審査官	tokkyo-shinsa'kan
example *[pat]*	実施例	jisshi-rei
" *[pat]*	実例；實例	jitsu-rei
" *[sci-t]*	例	rei
" *{n}*	事例；じれい	jirei
example of application *[pat]*	実施例	jisshi-rei
excavation mud *[geol]*	掘鑿泥水	kussaku-deisui
excavator *[eng]*	掘鑿機；くっさくき	kussaku-ki
excellent (grade) *[ind-eng]*	優；ゆう	yū
excenter (of triangle) *[geom]*	傍心	bōshin
exception; exclusion *[legal]*	除外	jogai
exception (to the rule) *{n}*	例外	rei'gai
excess *{n}*	過剰；過剩	kajō
excess carburizing *{n} [met]*	過剰浸炭	kajō-shintan
exchange *{n} [comput]*	取替え	tori'kae
" *{n} [math]*	交換；こうかん	kōkan
" *{vb t}*	交換する	kōkan suru
exchange current *[elec]*	交換電流	kōkan-denryū
exchange operator *[comm]*	交換扱者	kōkan-atsukai'sha
exchange rate *[econ]*	為替レート	kawase-rēto
excipient; mass; vehicle *{n} [pharm]*	賦形剤；ふけいざい	fukei-zai
excitation *[elecmg]*	励起；勵起；れいき	reiki
" *[phys]*	励振	reishin
excitation voltage *[elec]*	励発電圧	rei'hatsu-den'atsu
excited light *[opt]*	励起光	reiki-kō
excited singlet *[atom-phys]*	励起一重項	reiki-ichi'jū-kō
excited state *[atom-phys]*	励起状態	reiki-jōtai
excitement; excitation *[psy]*	興奮；こうふん	kōfun
exciton *[sol-st]*	励起子	reiki'shi
exclamation mark *[comput] [pr]*	感嘆符	kantan'fu
exclamation point *[comput] [pr]*	感嘆符	kantan'fu
exclusion *[comput]*	排他	hai'ta
exclusion principle *[quant-mech]*	排他律	haita-ritsu
exclusive industrial district	工業専用地区	kōgyō-senyō-chiku

English	Japanese	Romanization
exclusive right *[legal]*	独占権；獨占權	dokusen'ken
exclusive reference *[comput]*	排他的参照	haita'teki-sanshō
excrement; excreta *[physio]*	排泄物	haisetsu'butsu
excretion *[physio]*	排泄；はいせつ	hai'setsu
" *[physio]*	排出	hai'shutsu
excretory system *[anat]*	排泄系	haisetsu-kei
excurrent *{adj} [bio]*	流出の	ryū'shutsu no
executable statement *[comput]*	実行文	jikkō'bun
execute *{vb t} [comput]*	実行する	jikkō suru
execution *[comput]*	実行；實行	jikkō
" *[comput]*	命令実行	meirei-jikkō
execution time *[comput]*	実行時間	jikkō-jikan
exercise *{n} [math]*	演習	enshū
" (physical fitness) *{n}*	体操；たいそう	taisō
exercise; locomotion; athletics *{n}*	運動；うんどう	undō
exfoliation *[geol]*	剝脱作用	haku'datsu-sayō
exhaust air; exhaust gas *[mech-eng]*	排気；排氣；はいき	haiki
exhaustive search	全数探索	zensū-tansaku
exhaust pipe *[mech-eng]*	排気管	haiki-kan
exhaust smoke *[eng]*	排気煙；排氣煙	haiki-en
exhaust valve *[mech-eng]*	排気弁；排氣瓣	haiki-ben
exhibition *{n}*	展示	tenji
exist *{vb i}*	居る	iru
" *{vb i}*	在る	aru
exist displaced *{vb i} [math]*	偏存する	henzai suru
existence *{n}*	存在	sonzai
existential quantifier *[logic]*	存在記号	sonzai-kigō
exit *{n} [arch] [comput]*	出口	de'guchi
exit pupil *[opt]*	射出瞳	sha'shutsu-hitomi
exobiology *[bio]*	宇宙生物学	uchū-seibutsu'gaku
exocrine gland *[anat]*	外分泌腺	gai'bunpitsu-sen
exocyclic *{adj} [org-ch]*	環外の	kangai no
exoenzyme *[bioch]*	(細胞)外酵素	(saibō)gai-kōso
exoskeleton *[i-zoo]*	外骨格	gai'kokkaku
exosmosis *[physio]*	浸出	shin'shutsu
exosphere *[meteor]*	外気圏；外氣圏	gaiki-ken
" *[meteor]*	逸脱圏	itsu'datsu-ken
exothermic *{adj} [phys] [p-ch]*	発熱性の	hatsu'netsu'sei no
exothermic reaction *[ch]*	発熱反応；發熱反應	hatsu'netsu-hannō
expand *{vb} [comput]*	拡大する；擴大する	kaku'dai suru
expanded expression *[math]*	展開式	tenkai-shiki
expander *[electr]*	伸長器	shinchō-ki

expanding universe *[astrophys]*	膨張宇宙	bōchō-uchū
expansion *[math]*	展開；てんかい	tenkai
" *[phys]*	膨張；ぼうちょう	bōchō
expansion and contraction *[mech]*	屈伸	kusshin
expansion molding *{n} [eng] [plas]*	発泡成形	happō-seikei
expansion plan *[graph]*	展開図；展開圖	tenkai-zu
expectation *[stat]*	期待値	kitai-chi
expected time of departure *[trans]*	出発予定時間	shuppatsu-yotei-jikan
expectorant *{n} [pharm]*	去痰薬（去痰剤）	kyotan-yaku (kyotan-zai)
expel; spew; vomit *{vb}*	吐く	haku
expendable article *[ecol] [ind-eng]*	消耗品	shōmō'hin
expenditure *[econ]*	経費；經費；けいひ	keihi
" *[econ]*	消費	shōhi
expense(s); expenditure *[econ]*	費用	hiyō
expensive *{adj} [econ]*	高価な	kōka na
" *{adj} [econ]*	高い	takai
experience *{n} [psy]*	経験；經驗	keiken
" *{n} [psy]*	体験	taiken
experiment *{n} [sci-t]*	実験；實驗	jikken
" *{vb} [sci-t]*	実験(を)する	jikken (o) suru
experiment result(s) *[ch]*	実験結果	jikken-kekka
expert (a person)	専門家；專門家	senmon'ka
expert opinion *[legal]*	鑑定；かんてい	kantei
explanation *[comm]*	解説	kaisetsu
" *[comm]*	説明；せつめい	setsu'mei
explanatory diagram (drawing) *[graph]*	説明図	setsumei-zu
" *[graph]*	図解；圖解；ずかい	zukai
explant *{n} [cyt]*	外植片	gai'shoku-hen
explicit *{adj}*	明示の	meiji no
" *{adj}*	明確な	meikaku na
explicit declaration *[comput]*	明示宣言	meiji-sengen
exploded view *[graph]*	分解図	bunkai-zu
exploration; expedition *{n}*	探検；探檢	tanken
" ; research *{n}*	探究	tankyū
explore; look for *{vb}*	探る	saguru
explosion *[ch]*	爆発；ばくはつ	baku'hatsu
explosion acceptor *[ord]*	受爆薬	ju'baku-yaku
explosion donor *[ord]*	励爆薬；勵爆薬	reibaku-yaku
explosive fission *[nuc-phys]*	爆発性核分裂	bakuhatsu'sei-kaku-bunretsu
explosives *[mat]*	爆薬	baku'yaku
" *[ord]*	火薬類；火藥類	ka'yaku-rui
exponent *[arith]*	(冪)指数	(beki-)shisū

English	Japanese	Romanization
exponential equation *[alg]*	指数方程式	shisū-hōtei'shiki
exponential function *[alg]*	指数関数	shisū-kansū
exponential growth *[microbio]*	対数増殖；對數增殖	taisū-zōshoku
exponentiation *[arith]*	指数化；しすうか	shisū'ka
exponents, law of *[math]*	指数法則	shisū-hōsoku
export *{n} [econ]*	輸出	yu'shutsu
export and import *{n} [econ]*	輸出入	yu'shutsu-nyū
exposition; fair *{n}*	博覧会	hakuran'kai
exposure *[civ-eng] [min-eng]*	露出	ro'shutsu
" *[graph] [opt]*	暴露；ばくろ	bakuro
" *[nucleo]*	被曝	hibaku
" *[nucleo]*	照射(線)量	shōsha-(sen)'ryō
" *[photo]*	露光；ろこう	rokō
exposure meter *[eng]*	露光計	rokō-kei
exposure rate *[nucleo]*	照射線量率	shōsha'senryō-ritsu
express; high-speed *{adj} [trans]*	急速の	kyū'soku no
expression *[comput]*	表現	hyōgen
" *[comput]*	形式	keishiki
" *[cook]*	圧搾；壓搾	assaku
" (usage) *[ling]*	語法	gohō
" *[math]*	式；しき	shiki
" *{n}*	表示	hyōji
expressivity *[gen]*	表現度	hyōgen-do
express train *[rail]*	急行列車	kyūkō-ressha
expressway *[civ-eng]*	高速道路	kōsoku-dōro
extended storage *[comput]*	拡張記憶	kakuchō-ki'oku
extender pigment *[ch]*	体質顔料	tai'shitsu-ganryō
extending *{n}*	延展；えんてん	enten
extending agent; extender *[mat]*	増量剤	zōryō-zai
extensibility *[mech]*	伸び性	nobi-sei
" *[mech]*	伸展性	shinten-sei
" *[mech]*	展延性	ten'en-sei
extensible language *[comput]*	拡張可能言語	kakuchō-kanō-gengo
extensimeter *[eng]*	伸び計	nobi-kei
extension (telephone) *[comm]*	内線	naisen
" (prolongation) *[math]*	延長	enchō
" (extended line) *[math]*	延長線	enchō-sen
" *[mech]*	伸び；のび	nobi
" *[mech]*	伸長	shinchō
" *[phys]*	填充性	tenjū-sei
extension telephone *[comm]*	増設電話	zō'setsu-denwa
extent *[math]*	範囲；範圍	han'i

exterior angle *[geom]*	外角	gaikaku
exterior derivative *[math]*	外微分；がいびぶん	gai-bibun
exterior view *[arch]*	外景	gaikei
external angle *[geom]*	外角	gaikaku
external force *[mech]*	外力	gai-ryoku
external function *[comput]*	外部関数	gaibu-kansū
external indicator *[ch]*	外部指示薬	gaibu-shiji-yaku
external interrupt *[comput]*	外部割込み	gaibu-wari'komi
externally add *{vb t} [ch]*	外添する	gaiten suru
external noise *[electr]*	外部雑音	gaibu-zatsu'on
external plasticization *[poly-ch]*	外部可塑化	gaibu-kaso'ka
external pressure *[phys]*	外圧；外壓	gai-atsu
external sorting *{n} [comput]*	外部分類	gaibu-bunrui
external thread *[eng]*	雄螺子；おねじ	o-neji
extinction *[astron] [opt]*	減光	genkō
" *[bio]*	絶滅；ぜつめつ	zetsu'metsu
" *[p-ch]*	吸光度	kyūkō-do
extinction of arc *[elec]*	消弧	shōko
extinction rule *[crys]*	消滅則	shōgen-soku
extinct volcano *[geol]*	死火山	shi-kazan
extol *{vb}*	謳う；うたう	utau
extra *{n}*	余分；餘分；よぶん	yobun
extra- *{prefix}*	以外の-	i'gai no-
extract *{n} [ch] [cook] [pharm]*	エキス	ekisu
" *{n} [ch]*	抽出物	chūshutsu'butsu
" (excerpt) *{n} [pr]*	抜粋；ばっすい	bassui
" *{n} [pr]*	抄本	shōhon
" *{vb t} [ch]*	抽出する	chū'shutsu suru
" *{vb t} [comput]*	抜き出す	nuki-dasu
" (pull out) *{vb t}*	抜く；抜く；ぬく	nuku
extractability *[ch]*	抽出性	chūshutsu-sei
extractant *[ch]*	抽出用溶剤	chūshutsu'yō-yōzai
extract instruction *[comput]*	抽出命令	chūshutsu-meirei
extraction; extract *[ch]*	抽出；ちゅうしゅつ	chū'shutsu
extraction rate *[ch]*	採取率	saishu-ritsu
extractive distillation *[ch-eng]*	抽出蒸留	chūshutsu-jōryū
extractor *[eng]*	抽出装置	chūshutsu-sōchi
extra edition *[pr]*	号外；號外	gōgai
extragalactic nebula *[astron]*	銀河系外星雲	gingakei'gai-sei'un
extra-high voltage *[elec]*	超高圧	chō-kō'atsu
extraneous root *[alg]*	無縁根	mu'en-kon
extraordinary (light) ray *[opt]*	異常光線	ijō-kōsen

extrapolation *[math]*	外挿; がいそう	gaisō
" *[math]*	補外	ho'gai
extraterrestrial *{adj} [astron]*	地球外の	chikyū'gai no
extraterrestrial radiation *[astron]*	大気外太陽放射	taiki'gai-taiyō-hōsha
Extremely High Frequency; EHF *[comm]*	超高周波	chō-kō-shūha
Extremely Low Frequency; ELF *[comm]*	超低周波	chō-tei-shūha
extreme-pressure additive *[mat]*	極圧添加剤	kyoku'atsu-tenka-zai
extreme ultraviolet *[phys]*	極紫外	kyoku-shigai
extrinsic *{adj}*	外因性の	gai'in'sei no
extrinsic property *{n}*	外因的性質	gai'in'teki-seishitsu
extrinsic semiconductor *[electr]*	外來半導体	gairai-han'dōtai
extruder *[eng] [poly-ch]*	押出機	oshi'dashi-ki
extrusion *[eng] [tex]*	吐出; としゅつ	to'shutsu
" *[poly-ch]*	押出し; おしだし	oshi'dashi
extrusion molding *{n} [met] [poly-ch]*	押出し成形	oshi'dashi-seikei
extrusion pressure *[mech]*	吐出圧	to'shutsu-atsu
" *[poly-ch]*	押出し圧力	oshi'dashi-atsu'ryoku
extrusion rate *[poly-ch]*	押出し速度	oshi'dashi-sokudo
exudate *{n} [geol]*	浸出液	shin'shutsu-eki
" *{n} [med]*	浸出物	shin'shutsu'butsu
eye *{n} [anat]*	目; 眼; め	me
eyebrow *[anat]*	眉; まゆ	mayu
eye estimation *[physio]*	目測	moku-soku
eyeglasses *[opt]*	眼鏡; めがね	me'gane
eyelash(es) *[anat]*	睫毛	matsu'ge
eyelet *[eng]*	鳩目	hato'me
eyelid *[anat]*	眼瞼; 目蓋; まぶた	ma'buta
eye measure *{n} [sci-t]*	目分量	me-bunryō
eye of a hurricane *[meteor]*	台風の目	taifū no me
eyepiece *[opt]*	接眼レンズ	setsu'gan-renzu
" *[opt]*	対眼レンズ	taigan-renzu
eyespot *[i-zoo]*	眼点; 眼點	ganten
" *[i-zoo]*	目玉模様	me'dama-moyō
eyewash (eye lotion) *{n} [pharm]*	眼薬; めぐすり	me-gusuri; gan'yaku
" *{n} [pharm]*	洗眼液	sengan-eki
" *{n} [pharm]*	点眼液	tengan-eki

F

English	Japanese	Romanization
fabric *[geol]*	石理	sekiri
" *[geol]*	組織；そしき	so'shiki
fabric; textile; woven cloth *[tex]*	織物；おりもの	ori'mono
fabrication *[ind-eng]*	作製	sakusei
fabric-base laminate *[mat]*	布基材積層品	nuno'kizai-sekisō'hin
fabric cloth *[tex]*	織布	shoku'fu; ori-nuno
facade *[arch]*	正面；しょうめん	shōmen
face *{n} [anat]*	顔；かお	kao
" *{n} [anat] [crys] [geom]*	面	men
" ; confront; look out on *{vb}*	臨む；のぞむ	nozomu
face-centered cubic lattice *[crys]*	面心立方格子	menshin-rippō-kōshi
face-centered lattice *[crys]*	面心格子	menshin-kōshi
face-centered rhombohedral lattice *[crys]*	面心三方格子	menshin-sanbō-kōshi; menshin-sanpō-kōshi
face-centered tetragonal lattice *[crys]*	面心正方格子	menshin-seihō-kōshi
face powder *[mat]*	粉白粉	kona-oshiroi
facet *[mat] [met]*	面；めん	men
" *[met]*	彫面	chōmen
face width *[mech]*	歯幅	ha-haba
facies (pl facies) *[geol]*	相	sō
facility (pl -ties) *[comput]*	機能；きのう	kinō
" *{n} [eng]*	設備	setsubi
" *{n} [eng]*	施設；しせつ	shisetsu
facing *{n} [eng]*	上張り	uwa'bari
facsimile; FAX *[comm]*	模写電送	mosha-densō
fact *{n}*	事実；事實；じじつ	ji'jitsu
factor *{n} [alg] [gen]*	因子	inshi
" *{n} [alg]*	因数；いんすう	insū
" *{n} [eng] [sci-t]*	要因	yō'in
factorial *[prob]*	階乗；階乘	kaijō
factoring; factorization *{n} [alg]*	因数分解	insū-bunkai
factor theorem *[alg]*	因数定理	insū-teiri
factory *[ind-eng]*	工場	kōjō
factual *{adj} [math]*	事実の	ji'jitsu no
facula (pl -ae) *[astron]*	白斑；はくはん	haku'han
facultative *{adj} [microbio]*	条件的な	jōken'teki na
facultative anaerobe *[microbio]*	条件的嫌気性生物	jōken'teki-kenki'sei-seibutsu
facultative anaerobe *[microbio]*	通性嫌気性菌	tsūsei-kenki'sei-kin

English	Japanese	Romanization
fade *{vb} [tex]*	褪せる；あせる	aseru
fading *{n} [tex]*	退色	tai'shoku
failing grade *[ind-eng]*	不合格	fu'gōkaku
fail to work; be no good	利かない	kikanai
failure *[adhes]*	破壊；破壞；はかい	hakai
" *[elec] [mech]*	事故	jiko
" *[elec]*	故障	koshō
" (fail to pass) *[ind-eng]*	落第	rakudai
" *[mech]*	機能停止	kinō-teishi
fair (a grade) *[ind-eng]*	可	ka
fairing *{n} [aero-eng]*	整形	seikei
fairway; sea route *[navig]*	航路	kōro
fair weather *[meteor]*	晴天	seiten
fairy pitta (bird) *[v-zoo]*	八色鳥	ya'iro-chō
fake; sham *{n}*	偽物	nise'mono
falcon (bird) *[v-zoo]*	隼；はやぶさ	hayabusa
Fall (season) *{n} [astron]*	秋；あき	aki
fall *{n} [comm]*	立下り	tachi'sagari
falling *{n} [mech]*	落下	rakka
falling-ball impact test *[mech]*	落球衝撃試験	rakkyū-shōgeki-shiken
falling-ball strength *[mech]*	落球強度	rakkyū-kyōdo
falling body *[mech]*	落体	raku-tai
falling, law of *[mech]*	落下の定律	rakka no tei'ritsu
falling-sphere viscometer *[eng]*	落球粘度計	rakkyū-nendo-kei
falling-weight impact test *[mech]*	落錘衝撃試験	raku'sui-shōgeki-shiken
Fallopian tube *[anat]*	喇叭管；らっぱかん	rappa-kan
fallout (atomic) *{n} [nucleo]*	降下物	kōka'butsu
fall wind *[meteor]*	颪；おろし	oroshi
false *{n}*	擬似	giji
false- *{prefix}*	偽-	gi-
false; fake *{adj}*	偽りの	itsuwari no
false equilibrium *[phys]*	偽平衡；ぎへいこう	gi'heikō
false floor *[arch]*	二重床	nijū-yuka
false fruit *[bot]*	偽果	gika
false image *[opt]*	偽像	gizō
false signal *[comm]*	擬似符号	giji-fugō
false twist *[tex]*	仮撚り；假撚り	kari-yori
falsity *[math]*	虚偽；きょぎ	kyogi
family *[bio] [syst]*	科	ka
" (pl -lies) *[bio]*	家族	ka'zoku
" *[math]*	族	zoku
family group *[bot] [syst]*	上科	jōka

fan (electric) *{n} [eng]*	扇風機	senpū-ki
" (folding) *{n} [eng]*	扇；おうぎ	ōgi
" (folding) *{n} [eng]*	扇子；せんす	sensu
" (round) *{n} [eng]*	団扇；うちわ	uchiwa
fan; agitate *{vb t}*	煽る；あおる	aoru
fanlight *[arch]*	欄間	ranma
fanning *{n} [phys]*	送風	sōfū
fan shell *[i-zoo]*	玉珧；たいらぎ	tairagi
far; distant *{adv}*	遠い；とおい	tōi
faraday *[phys]*	ファラデー	faradē
far end *{n} [elec]*	遠端	entan
fare table; tariff table *[trans]*	運賃表	un'chin-hyō
far-infrared rays *[elecmg]*	遠赤外線	en'sekigai-sen
farm *{n} [agr]*	農場	nōjō
farmer (a person) *[agr]*	農業家	nōgyō'ka
farming *{n} [agr]*	農業；のうぎょう	nōgyō
farm products *[agr]*	作物；さくもつ	saku'motsu
farsightedness *[med]*	遠視	enshi
farther than *[navig]*	以遠	i'en
fashioning *{n} [eng]*	成形	seikei
fast *{adj} [phys]*	早い；速い；はやい	hayai
fastening *{n} [adhes]*	固着	ko'chaku
fastening; fastener *{n} [eng]*	留め金	tome'gane
fastening; tightening *{n}*	締結	teiketsu
fast-feed advance *[mech-eng]*	早送り前進	haya'okuri-zenshin
fastness *[ind-eng]*	堅牢度	kenrō-do
fat *{n} [bioch]*	脂；あぶら（し）	abura (shi)
" *{n} [bioch] [physio]*	脂肪；しぼう	shibō
fatal error *[comput]*	致命的誤り	chimei'teki-ayamari
fathogram (by echo sounder) *[eng]*	深度記録	shindo-ki'roku
fathom *{n} [ocean]*	尋；ひろ	hiro
fathometer *[eng]*	音響測深機	onkyō-sokushin-ki
fatigue *{n} [mech-eng] [physio]*	疲れ；つかれ	tsukare
" ; tiring *{n} [physio]*	疲労；疲勞；ひろう	hirō
fatigue cracking *[met]*	疲労亀裂	hirō-kiretsu
fatigue fracture *[mech]*	疲れ破損	tsukare-hason
fatigue limit *[mech]*	耐久限度	taikyū-gendo
fatigue strength *[mat]*	疲れ強さ	tsukare-tsuyo-sa
fats and oils *[bioch] [mat]*	油脂；ゆし	yushi
fat-soluble(ness) *[ch]*	脂溶性	shiyō-sei
fat tissue *[hist]*	脂肪組織	shibō-soshiki
fatty acid *[org-ch]*	脂肪酸	shibō-san

faucet *[eng]*	蛇口; じゃぐち	ja'guchi
" *[eng]*	給水栓	kyūsui-sen
fault *{n} [elec]*	異常	ijō
" *{n} [elec]*	事故	jiko
" *{n} [elec]*	故障; こしょう	koshō
" *{n} [eng]*	障害	shōgai
" *{n} [geol]*	断層; 斷層	dansō
" (mistake; misstep) *{n}*	過ち; あやまち	ayamachi
" *{n}*	誤り; 謬り	ayamari
fault line *[geol]*	断層線; 斷層線	dansō-sen
fault movement *[geol]*	断層運動	dansō-undō
fauna *[zoo]*	動物相	dōbutsu-sō
fava (bean) *[bot] [cook]*	空豆; そらまめ	sora-mame
favorable current *[navig]*	順流	junryū
favorable wind; fair wind *[navig]*	順風	junpū
favoring customer *[econ]*	顧客; 贔負; ひいき	hi'iki
favorite dish *[cook]*	好物	kō'butsu
favorite foods *[cook]*	嗜好品	shikō'hin
fawn (color) *[opt]*	仔鹿色; 子鹿色	ko'jika-iro
fayalite *[miner]*	鉄橄榴石	tetsu'kanran'seki
feather *{n} [v-zoo]*	毛羽; 毳; けば	keba
" *{n} [v-zoo]*	羽毛; うもう	umō
feather-duster worm (insect) *[i-zoo]*	毛槍虫; けやりむし	ke'yari-mushi
feature; characteristic *{n} [sci-t]*	特徴; とくちょう	tokuchō
February (month)	二月	ni'gatsu
fecal contamination *[ecol]*	糞便汚染	funben-o'sen
feces *[physio]*	糞便	funben
fee; recompense *[econ]*	報酬; ほうしゅう	hōshū
" ; charge(s) *[econ]*	料金	ryōkin
" ; handling fee *[econ]*	手数料	tesū'ryō
feeble magnetism *[phys]*	弱磁性	jaku'ji-sei
feed; feedstuff *{n} [agr]*	飼料	shiryō
feedback *[comput] [electr] [sci-t]*	帰還; 歸還; きかん	kikan
feeder layer *[microbio]*	支持細胞層	shiji-saibō-sō
feeder line *[elec]*	給電線	kyūden-sen
feeder root *[bot]*	細根	saikon
feed in; feeding in *{n} [mech-eng]*	送り込み	okuri'komi
feeding (of animal) *{n} [agr]*	給餌	kyūji
feeding apparatus *[mech-eng]*	送り装置	okuri-sōchi
feeding speed *[tex]*	給送速度	kyūsō-sokudo
feed motion *[mech-eng]*	送り運動	okuri-undō
feed nut *[mech-eng]*	送りナット	okuri-natto

feed pipe *[mech-eng]*	送り管	okuri-kan
feed rod *[mech-eng]*	送り杆	okuri-kan
feedstock *[agr]*	供給原料	kyōkyū-genryō
feel; experience; suffer *{vb} [physio]*	感じる	kanjiru
feel in use	使用感	shiyō-kan
feel to the palate *[cook]*	舌触り; したざわり	shita-zawari
feel to touch *[physio]*	肌ざわり	hada-zawari
feeler *[i-zoo]*	触手; 觸手	shoku'shu
feeling; sense *{n} [physio]*	感じ	kanji
feeling; feelings; emotion *{n} [psy]*	感情	kanjō
feeling (to touch) *{n} [tex]*	風合; ふうあい	fū'ai
feldspar *[miner]*	長石	chōseki
felt earthquake *{n} [geophys]*	有感地震	yūkan-jishin
female (human); woman; women *[bio]*	女性	josei
female *[zoo]*	雌; めす	mesu
feminine *{adj} [bio]*	女性的な	josei'teki na
femur (bone) *[anat]*	大腿骨	daitai-kotsu
fence *{n} [arch]*	塀	hei
" *{n} [arch]*	垣	kaki
" *{n} [arch]*	柵	saku
fender *{n} [civ-eng]*	防(護)材	bō(go)-zai
fennel; Foeniculum vulgare (herb) *[bot]*	茴香; ういきょう	uikyō
fenugreek (herb) *[bot]*	胡盧巴; ころは	koroha
ferberite *[miner]*	鉄重石; 鐵重石	tetsu'jūseki
ferment *{vb i} [microbio]*	発酵する; 發酵する	hakkō suru
fermentable *{adj} [microbio]*	発酵性の	hakkō'sei no
fermentation; zymosis *[microbio]*	発酵; はっこう	hakkō
fermentation technology *[micrbio]*	発酵工学	hakkō-kōgaku
fermentative *{adj} [microbio]*	発酵性の	hakkō'sei no
fermented soybeans *[cook]*	納豆; なっとう	nattō
fermium (element)	フェルミウム	ferumyūmu
fern *[bot]*	羊歯; しだ	shida
ferric nitrate	硝酸第二鉄	shōsan-dai'ni-tetsu
ferric sulfate	硫酸第二鉄	ryūsan-dai'ni-tetsu
ferroelastic (pl -s) *{n} [mat]*	強弾性体	kyō'dansei-tai
ferroelectric crystal *[sol-st]*	強誘電性結晶	kyō'yūden'sei-kesshō
ferroelectricity *[sol-st]*	強誘電性	kyō'yūden-sei
ferroelectric (pl -s) *{n} [sol-st]*	強誘電体	kyō'yūden-tai
ferromagnetic crystal *[sol-st]*	強磁性結晶	kyōji'sei-kesshō
ferromagnetic material *[sol-st]*	強磁性体	kyō'jisei-tai
ferromagnetism *[sol-st]*	強磁性	kyōji-sei
ferrous nitrate	硝酸第一鉄	shōsan-dai'ichi-tetsu

ferrous sulfate	硫酸第一鉄	ryūsan-dai'ichi-tetsu
ferry *{n} [nav-arch]*	渡し船	watashi-bune
fertility (of soil) *[agr]*	地力	chi-ryoku
fertilization *[agr]*	肥沃化; ひよくか	hiyoku'ka
" *[physio]*	受精; 授精	jusei
fertilizer *[agr] [mat]*	肥料; ひりょう	hiryō
fetal calf serum *[microbio]*	牛胎児血清	ushi-taiji-kessei
fetch *{n} [comput]*	取出し	tori'dashi
" *{vb} [comput]*	取出す	tori-dasu
fetus *{n} [bio] [med]*	胎児; 胎兒; たいじ	taiji
fever *{n} [med]*	熱; ねつ	netsu
feverfew; Chrysanthemum parthenium *[bot]*	夏白菊	natsu-shiro'giku
feverishness *{n} [med]*	熱気	netsu'ke
few; scanty *{adj} [math]*	少い	sukunai
fiber *[tex]*	繊維; 纖維; せんい	sen'i
fiber bundle *[tex]*	繊維束	sen'i-soku
fiber-reinforced metal *[met]*	繊維強化金属	sen'i-kyōka-kinzoku
fiber size *[mat] [tex]*	集束剤	shūsoku-zai
fiber thickening agent *[mat]*	繊維糊剤	sen'i-nori-zai
fibril *[bio]*	原繊維	gen-sen'i
fibrillatable *[tex]*	フイブリル化可能	fiburiru'ka-kanō
fibrin *[bioch]*	繊維素; せんいそ	sen'i-so
fibrinogen *[bioch]*	繊維素原	sen'i'so-gen
fibroblast *[hist]*	繊維芽細胞	sen'i-ga'saibō
fibrous matter *[mat]*	繊維質	sen'i-shitsu
fibrous root *[bot]*	鬚根; ひげね	hige-ne
fibula (bone) *[anat]*	腓骨; ひこつ	hi-kotsu
fiddler crab *[i-zoo]*	潮招き	shio-maneki
fidelity *[comm]*	忠実性	chū'jitsu-sei
fiducial point *[opt]*	起点	kiten
field (cultivated); farm *{n} [agr]*	畑; はたけ	hatake
" (rice or water field) *{n} [agr]*	田	ta
" *{n} [agr]*	田圃	tanbo
" *{n} [alg]*	体; 體	tai
" *{n} [comput]*	欄	ran
" *{n} [ecol]*	原	hara
" *{n} [elec]*	界	kai
" ; plain *{n} [geogr]*	野; の; や	no; ya
" (on site) *{n} [ind-eng]*	現場	genba
" (area; subject matter) *[pat]*	分野	bun'ya
" *{n} [phys]*	場; ば	ba
field angle *[opt]*	画角; 畫角; がかく	ga-kaku

field coil *[elecmg]*	界磁線輪	kaiji-senrin
field effect *[electr]*	電界効果	denkai-kōka
field emission *[electr]*	電界放出	denkai-hō'shutsu
field-emission microscope *[electr]*	電界放出顕微鏡	denkai-hō'shutsu-kenbi'kyō
field evaporation *[elec]*	電界蒸発	denkai-jōhatsu
fieldfare (bird) *[v-zoo]*	野原鶫	nohara-tsugumi
field horsetail *[bot]*	土筆; つくし	tsukushi
field magnet *[elecmg]*	界磁	kaiji
field poppy (flower) *[bot]*	虞美人草	gubijin'sō
field strength *[phys]*	場の強さ	ba no tsuyo-sa
field strength attenuation quantity *[elec]*	電界強度減衰量	denkai-kyōdo-gensui'ryō
fifty percent; one-half *[math]*	五分; ごぶ	gobu
fig (fruit) *[bot]*	無花果; いちじく	ichijiku
figure *{n} [graph]*	図; 圖; ず	zu
" *{n} [graph]*	図形	zukei
figured cloth *[tex]*	紋織物	mon-ori'mono
figured satin *[tex]*	綸子; りんず	rinzu
figure eight *{n} [math]*	八字形	hachi'ji-gata
figurehead *[nav-arch]*	船首像	senshu-zō
figure of merit *[electr]*	示性数	shisei-sū
" *[math]*	良さの指数	yosa no shisū
figure pattern; design *{n} [graph]*	図柄; ずがら	zu'gara
figure trademark *[graph] [pat]*	図形商標	zukei-shōhyō
figuring *{n} [opt]*	鏡面修正	kyōmen-shūsei
filament *[astron] [elec]*	線条; 線條	senjō
filamentous fungus *[microbio]*	糸状菌	shijō-kin
file; rasp *{n} [eng]*	鑢; やすり	yasuri
" ; rub; chafe *{vb i}*	摩る; 磨る; する	suru
filefish; leatherfish (fish) *[v-zoo]*	河剝ぎ	kawa-hagi
filiform corrosion *[met]*	糸錆; 糸さび	ito-sabi
filing *{n} [eng]*	鑢仕上; やすり仕上	yasuri-shi'age
filing date *[pat]*	出願日	shutsu'gan-bi
filler *[cer]*	骨材	kotsu-zai
" *[ch-eng]*	充填剤	jūten-zai
" *[mat]*	補填剤	hoten-zai
" *[mat]*	填料	tenryō
fillet *{n} [eng]*	隅肉; すみにく	sumi-niku
filling *{n} [dent]*	充填(物)	jūten('butsu)
filly (pl fillies) (animal) *[v-zoo]*	雌子馬	mesu-ko'uma
film *[ch-eng] [mat]*	膜; まく	maku
" (coating film) *[mat]*	被膜	himaku

film resistor *[elec]*	(被)膜抵抗器	(hi)maku-teikō-ki
filter *{n} [eng]*	沪波器；濾波器	roha-ki
" *{n} [eng]*	沪過器	roka-ki
" (strain) *{vb t}*	漉す；こす	kosu
" *{vb t}*	沪過する；ろかする	roka suru
filterability *[eng]*	沪過性	roka-sei
filter aid *[mat]*	沪過助剤	roka-jozai
filter bed *[ch]*	沪床	ro-shō
filter chamber *[ch]*	沪室	ro-shitsu
filter cloth *[mat]*	沪布	rofu
filter cloth speed *[mech]*	沪布速度	rofu-sokudo
filter feeder *[zoo]*	沪過摂食者	roka-sesshoku'sha
filtering speed *[mech]*	沪過速度	roka-sokudo
filter leaf *[ch]*	沪葉	royō
filter paper *[mat]*	沪紙	roshi
filter press *[eng]*	圧沪器；壓濾器	atsu'ro-ki
filtrate *[sci-t]*	沪液	ro'eki
filtration *[sci-t]*	沪過；濾過	roka
filtration circuit *[elec]*	沪波回路	roha-kairo
fimbria (pl fimbriae); pilus (pl pili) *[microbio]*	繊毛；せんもう	senmō
fin *[aero-eng]*	垂直安定板	sui'choku-antei'ban
" *[eng] [met]*	鋳ばり；鑄	i'bari
" *[v-zoo]*	鰭；ひれ	hire
final coat; finishing coat *[mat]*	上塗り	uwa-nuri
final total *[math]*	総計；總計	sōkei
final twist *[tex]*	上撚り	uwa-yori
final velocity *[mech]*	終速(度)	shū-soku(do)
finance *{n} [econ]*	金融	kin'yū
" *{n} [econ]*	財務	zaimu
financial accounting *[econ] [math]*	財務会計	zaimu-kaikei
finding(s) *{n} [med]*	所見	shoken
fine (weather) *{n} [meteor]*	晴；はれ	hare
fine aggregate *[mat]*	細骨材	sai-kotsu'zai
fine chemical *[ch]*	精薬品	sei-yaku'hin
fine-dividing *{n}*	細分化	saibun'ka
fine earthenware *[cer]*	精陶器；せいとうき	sei-tōki
fine Japanese sword *[ord]*	業物；わざもの	waza-mono
finely divided *{adj} [mech-eng]*	細かい	komakai
fineness (thread) *[tex]*	繊度	sendo
" *[mech-eng]*	細かさ	komaka-sa
fineness of powder *[met]*	粉末度	funmatsu-do

English	Japanese	Romanization
fine powder *[mat]*	(微)細粉(末)	(bi-)saifun(matsu)
fine spinning *{n} [tex]*	精紡	seibō
fine structure *[atom-phys]*	微(細)構造	bi(sai)-kōzō
fine-tuning *{n} [electr] [eng]*	微調整	bi'chōsei
finger *{n} [anat]*	指; ゆび	yubi
fingerprint *[anat] [graph]*	指紋	shi'mon
fingerprinting *{n} [legal]*	指紋採取	shimon-saishu
fingertip *{n} [anat]*	指先	yubi'saki
finings *{n} [cook]*	清澄剤	seichō-zai
finish; finishing *{n} [ind-eng]*	仕上; 仕上げ	shi'age
finish; completion *{n}*	完了	kanryō
" ; conclude *{vb}*	済ます; すます	sumasu
finished yarn *[text]*	加工糸	kakō-shi
finishing *{n} [eng]*	加工; かこう	kakō
finishing agent *[mat]*	仕上剤	shi'age-zai
finishing steel *[steel]*	仕上用工具鋼	shi'age'yō-kōgu-kō
finish time *[ind-eng]*	終了時刻	shūryō-jikoku
finite *{adj} [math]*	有限の	yūgen no
finite amplitude *[trig]*	有限振幅	yūgen-shinpuku
finite difference *[math]*	差分	sa'bun
finite element method *[math] [mech]*	有限要素法	yūgen-yōso-hō
finite population *[stat]*	有限母集団	yūgen-bo'shūdan
fir (tree) *[bot]*	樅; もみ	momi
fire *{n} [ch]*	火	hi
" ; conflagration *{n} [ch]*	火事; かじ	kaji
fire alarm *[eng]*	火災報知機	kasai-hōchi-ki
fireant; Solenopsis invicta (insect) *[i-zoo]*	針蟻; はりあり	hari-ari
fireball *[astron] [nucleo]*	火球	kakyū
fireclay *[geol]*	耐火粘土	taika-nendo
firecracker *[eng]*	爆筒	baku'tō
fire engine *[mech-eng]*	消防(自動)車	shōbō(-jidō)sha
fire extinguisher *[eng]*	消火器	shōka-ki
fire-extinguishing agent *[mat]*	消火剤	shōka-zai
firefly (insect) *[i-zoo]*	蛍; 螢; ほたる	hotaru
fire grate *[eng]*	火格子	hi-gōshi
fire hook *[eng]*	鳶口	tobi-guchi
fire insurance *[econ]*	火災保険	kasai-hoken
fire point *[ch]*	火点	ka'ten
fire power *[phys]*	火力	ka-ryoku
fireproofness *[mat]*	耐火性	taika-sei
fire protection wall *[civ-eng]*	防火壁	bōka-heki

fire resistance *[mat]*	耐火性	taika-sei
fire resisting property *[mat]*	耐火性	taika-sei
fire-retarding material *[mat]*	難燃剤	nannen-zai
fire shovel *[eng]*	十能	jūnō
fire station *[arch]*	消防署	shōbō'sho
firewood *[bot]*	柴	shiba
" *[mat]*	薪；まき	maki
fireworks *{n} [eng]*	花火	hanabi
firing *{n} [eng]*	発射	hassha
" *{n} [eng]*	焼成	shōsei
firn *[geol] [hyd]*	万年雪；萬年雪	mannen'setsu
first; earliest *{n}*	最初	saisho
first aid *[med]*	応急手当；應急手當	ōkyū-te'ate
first day of autumn *[astron]*	立秋	risshū
first day of spring *[astron]*	立春	risshun
first frost *[meteor]*	初霜；はつしも	hatsu-shimo
first of the season *[cook]*	走り	hashiri
first-order *[p-ch]*	一次；いちじ	ichi'ji
first-order reaction *[p-ch]*	一次反応	ichiji-hannō
first quarter (of moon) *[astron]*	上弦	jō-gen
first snow *[meteor]*	初雪	hatsu-yuki
first-step *[ind-eng]*	一次	ichi'ji
first strike *[mil]*	第一撃	dai'ichi-geki
fiscal year *[econ]*	会計年度；會計年度	kaikei-nendo
fish (pl fish, -es) *{n} [v-zoo]*	魚；さかな；うお	sakana; uo
fish detection *[ocean]*	魚群探知	gyogun-tanchi
fisheries *[ecol]*	水産業	suisan'gyō
fisheries experiment station *[ecol]*	水産試験所	suisan-shiken'jo
fisheries patrol boat *[nav-arch]*	漁業取締船	gyogyō-tori'shimari'sen
fishery harbor *[ocean]*	漁港；漁港	gyokō
fisheye *[mat]*	魚眼	gyogan
fish hawk *[v-zoo]*	鶚；みさご	misago
fishhook *[eng]*	釣(り)針；釣鉤	tsuri-bari
fishing *{n}*	(魚)釣り	(sakana-)tsuri
fishing industry *[ocean]*	漁業；ぎょぎょう	gyo'gyō
fishing net *[eng] [tex]*	漁網	gyomō
fishing port *[geogr]*	漁港；漁港	gyokō
fishing rod *[eng]*	釣り竿	tsuri-zao
fishing twine *[tex]*	漁網糸	gyomō-shi
fishing village *[geogr]*	漁村	gyo'son
fish market *[econ]*	魚河岸；うおがし	uo-gashi
fish meal *[mat]*	魚粉	gyofun

fish oil *[mat]*	魚油; ぎょゆ	gyo-yu
fish preserve; fishery *[ecol]*	生け洲; 生け簀	ike'su
fish sauce *[cook]*	魚醤油; うお醤油	uo-jōyu
fish-scale form *[graph]*	鱗片状	rinpen'jō
fishway *[civ-eng]*	魚道	gyo'dō
fissility *[geol]*	剥離性; はくりせい	hakuri-sei
" *[geol]*	裂開性	rekkai-sei
fission *[nuc-phys]*	(核)分裂	(kaku-)bun'retsu
fissionable material *[nucleo]*	核分裂性物質	kaku-bunretsu'sei-busshitsu
fission chain reaction *[nuc-phys]*	核分裂連鎖反応	kaku-bunretsu-rensa-hannō
fission reactor *[nucleo]*	核分裂炉	kaku-bunretsu-ro
fissure eruption (volcano) *[geol]*	割れ目噴出	ware'me-fun'shutsu
fit *{n} [eng]*	嵌め合; はめあい	hame'ai
fit; fitting-together *{n} [eng]*	嵌合; かんごう	kangō
fit together *{vb t} [eng]*	嵌合する	kangō suru
fit with; fit onto *{vb} [ind-eng]*	装置する	sōchi suru
fittings *{n} [arch]*	建具; たてぐ	tate'gu
" *{n} [eng]*	取付(け)部品	tori'tsuke-buhin
" *{n} [furn]*	取付け物	tori'tsuke-mono
five *{n} [math]*	五	go
five centroids (of triangle) *[geom]*	五心	go-shin
fix *{vb t} [civ-eng] [photo]*	定着する	teichaku suru
" ; secure *{vb t} [eng]*	固定する	kotei suru
fix; correct; repair; mend *{vb} [eng]*	直す; なおす	naosu
fixation *[ch-eng] [geod]*	固定; こてい	kotei
fixation; fixing *{n} [photo]*	定着	tei'chaku
fixative *[mat]*	揮発保留剤	ki'hatsu-horyū-zai
fixed bed *[ch-eng]*	固定床	kotei-shō
" *[ch-eng]*	固定層	kotei-sō
fixed capacitor *[elec]*	固定コンデンサ	kotei-kondensa
fixed carbon *[ch]*	固定炭素	kotei-tanso
fixed expenses *[econ]*	固定費	kotei-hi
fixed phase *[an-ch] [ch-eng]*	固定相	kotei-sō
fixed point *[ch] [math] [meteor]*	定点; 定點	teiten
" *[comput]*	不動点	fudō-ten
fixed-point number *[comput]*	固定小数点数	kotei-shōsū'ten-sū
fixed star *[astron]*	恒星; こうせい	kōsei
fixed storage *[comput]*	固定記憶	kotei-ki'oku
fixing; securing *{n}*	固定; こてい	kotei
fixing solution; fixer *[photo]*	定着液	teichaku-eki
fixtures *[arch]*	建具	tate'gu
flab *{n} [physio]*	贅肉; ぜいにく	zei-niku

flaccid *{adj} [bot]* 緩い；弛い；ゆるい yurui
flag *{n} [eng]* 旗 hata
flagellum (pl -la) *[bio]* 鞭毛 benmō
flagstone *[geol]* 板石 ita-ishi
flake graphite *[miner]* 片状黒鉛 henjō-koku'en
flaky graphite *[geol]* 鱗状黒鉛 rinjō-koku'en
flame *{n} [ch]* 焔；炎；ほのお hono'o
" *{n} [ch]* 火炎 ka'en
flame photometer *[spect]* 炎光光度計 enkō-kōdo-kei
flameproofing *{n} [eng]* 耐炎化 tai'en'ka
flameproofness *[ind-eng]* 防炎性 bōen-sei
flame reaction *[ch]* 炎色反応 en'shoku-hannō
flame resistance *[ch]* 難燃性 nannen-sei
flame-resisting treatment *[eng]* 耐炎化処理 tai'enka-shori
flame retardancy agent *[ch]* 難燃加工剤 nannen-kakō-zai
flame retardant finish *[mat]* 難燃加工 nannen-kakō
flame spray *{n} [eng]* 溶射 yōsha
flamingo (pl -gos, -goes) *[v-zoo]* 紅鶴；べにづる beni-zuru
flammable *{adj} [ch]* 可燃の ka'nen no
flammability *[mat]* 燃焼性 nenshō-sei
flange *[eng]* 鍔 tsuba
flank; side (of body) *{n} [anat]* 脇腹 waki-bara
flank wear *{n} [mech-eng]* 逃げ面摩耗 nige'men-mamō
flap (aircraft) *{n} [aero-eng]* 下げ翼 sage-yoku
flare bomb *[ord]* 照明弾 shōmei-dan
flare star *[astron]* 閃光星 senkō-sei
flash *{n} [eng] [met]* 铸ばり；鑄；いばり i'bari
" *{n} [opt]* 閃き hirameki
" *{n} [opt] [meteor]* 閃光 senkō
flashing *{n} [arch]* 雨押え ame-osae
" *{n} [arch]* 雨避け板 ame'yoke-ita
flashing on and off *{n} [navig]* 点滅 ten'metsu
flashlight *[eng]* 懐中電灯；懷中電燈 kaichū-dentō
flash photolysis *[p-ch]* 閃光光分解 senkō-kō'bunkai
flash point *[ch]* 引火点 inka-ten
flat *{adj} [geom]* 平たい hiratai
" ; horizontal *{adj} [geom]* 水平な suihei na
" *{adj} [geom]* 平な taira na
flatcar *[rail]* 台車；臺車 daisha
flat file *[eng]* 平鑢；ひらやすり hira-yasuri
flatfish (fish) *[v-zoo]* 比目魚；ひらめ hirame
" (fish) *[v-zoo]* 鰈；かれい karei

flat ground *[geol]*	平地	heichi
flathead (fish) *[v-zoo]*	鯒；牛尾魚；こち	kochi
flat pliers *[eng]*	鋏；やっとこ	yattoko
flattening *{n} [astron]*	偏率	hen-ritsu
flatworm *[i-zoo]*	扁形動物	henkei-dōbutsu
flavor *{n} [cook] [physio]*	味；あじ	aji
" *{n} [cook] [physio]*	風味	fūmi
" *{n} [cook] [physio]*	香味	kōmi
" *{n} [physio]*	香り；かおり	kaori
flavor and taste *[physio]*	香味	kōmi
flavoring *{n} [cook]*	味付け	aji'tsuke
flavoring agent *[cook] [pharm]*	着香料	chak(u)kō-ryō
" *[cook]*	呈味剤	teimi-zai
flavor-generating power *[cook]*	呈味力	teimi-ryoku
flavor-manifesting property *[cook]*	呈味性	teimi-sei
flavor with body *{n} [cook]*	コク味	koku-aji
flaw *{n} [ind-eng]*	疵；瑕；傷；きず	kizu
" *{n} [sci-t]*	欠陥	kekkan
flaw detector *[eng]*	探傷機	tanshō-ki
flax *[bot]*	亜麻；あま	ama
flaxseed *[bot]*	亜麻仁	amani(n)
flea (insect) *[i-zoo]*	蚤；のみ	nomi
fleet *{n} [ord]*	艦隊	kantai
flesh *{n} [anat]*	肉；にく	niku
flesh color *[opt]*	肌色	hada-iro
flex cracking *{n} [mech]*	屈曲亀裂	kukkyoku-ki'retsu
" *{n} [mech]*	曲げ亀裂	mage-ki'retsu
flexibility *[mech]*	撓み性	tawami-sei
flexibilizer *[poly-ch]*	可塑剤；かそざい	kaso-zai
flexible coupler *{n} [mech-eng]*	撓み連結器	tawami-ren'ketsu-ki
flexibleness; flexibility *[mech]*	可撓性	katō-sei
flexible plywood *[mat]*	可撓合板	katō-gōban
flexible response *[mil]*	柔軟対応力	jū'nan-taiō'ryoku
flexural modulus *[mech]*	曲げ弾性率	mage-dansei-ritsu
flexural rigidity *[mech]*	曲げ剛性	mage-gōsei
" *[mech]*	曲げ強さ	mage-kowa-sa
flexural strength *[mech]*	抗折力	kō'setsu-ryoku
" *[mech]*	曲げ強さ	mage-tsuyo-sa
flexural stress; bending stress *[mech]*	曲げ応力	mage-ō'ryoku
flexure; flexing *{n} [mech]*	撓み；たわみ	tawami
flicker *{n} [opt]*	ちらつき	chira'tsuki
" *{n} [opt]*	明滅	mei'metsu

flight; flying *{n} [aero-eng]*	飛行；ひこう	hikō
flight control *[aero-eng]*	航空管制	kōkū-kansei
flight deck *[aero-eng]*	飛行甲板	hikō-kanban
flight path *[navig]*	飛行径路	hikō-keiro
flight recorder *[eng]*	飛行記録器	hikō-ki'roku-ki
flint *[miner]*	燧石	hi'uchi-ishi; sui'seki
flip away *{vb t}*	弾く；はじく	hajiku
flipper *[v-zoo]*	鰭足；ひれあし	hire-ashi
float *{n} [eng]*	浮き(袋)	uki(-bukuro)
" *{n} [geol]*	浮鉱；浮鑛	fukō
" *{vb i}*	浮かぶ	ukabu
" *{vb i}*	浮く；うく	uku
floating *{n} [math]*	浮動；ふどう	fudō
floating decimal point *[arith]*	浮動小数点	fudō-shōsū-ten
floating pier *[civ-eng]*	浮桟橋；浮棧橋	uki-sanbashi
floating-point number *[comput]*	浮動小数点数	fudō-shōsū'ten-sū
floating solids *[phys]*	浮遊固形物	fuyū-kokei'butsu
floating weed *[bot]*	浮草	uki-kusa
float stitch *[tex]*	浮き編；うきあみ	uki-ami
flocculating agent; flocculant *[ch]*	凝集剤	gyōshū-zai
flocculating force *[ch]*	凝集力	gyōshū-ryoku
flocculus (pl -li) *[astron]*	羊斑；ようはん	yōhan
flocking *{n} [tex]*	植毛	shokumō
flood *{n} [hyd]*	洪水；こうずい	kōzui
flood damage *{n} [ecol]*	水害	suigai
floodgate *[civ-eng]*	水門	suimon
floodlight *[elec]*	投光照明(器)	tōkō-shōmei(-ki)
floodplain *[geol]*	沖積平野	chūseki-heiya
" *[geol]*	氾濫原	hanran'gen
floodtide *[ocean]*	高潮	taka-shio
floor *{n} [arch]*	床；ゆか	yuka
floor area *[arch]*	床面積	yuka-menseki
floor covering *[arch] [mat]*	敷物	shiki-mono
flooring of straw mat *{n} [arch]*	畳；たたみ	tatami
floormat *[furn]*	床敷き；ゆかじき	yuka'jiki
floor plan *{n} [arch] [graph]*	床配置図	yuka-haichi-zu
flora *[bot]*	植物相	shokubutsu-sō
floral art *{n}*	華道；かどう	kadō
floriculture *[bot]*	花卉園芸	kaki-engei
floss silk; floss *{n} [tex]*	真綿；まわた	mawata
flotation *[eng]*	浮選；浮選	fusen
" *[eng]*	浮遊選鉱；浮遊選鑛	fuyū-senkō

flotation agent *[ch]*	浮遊剤	fuyū-zai
flotation reagent *[min-eng]*	浮遊選鉱試薬	fuyū-senkō-shi'yaku
flounder (fish) *[v-zoo]*	比目魚；ひらめ	hirame
flour *{n}* *[cook]*	粉；こな	kona
flour mill *[agr]*	製粉機	seifun-ki
flour sifter *[cook]* *[eng]*	粉篩；粉ふるい	kona-furui
flow *{n}* *[comput]* *[fl-mech]*	流れ；ながれ	nagare
" *{n}* *[fl-mech]*	流動；りゅうどう	ryūdō
flow analysis *[comput]*	流れ解析	nagare-kaiseki
flow characteristics *[mech]*	流れ特性	nagare-tokusei
flow chart *[comput]* *[eng]*	流れ図；流れ圖	nagare-zu
flow coating *{n}* *[eng]*	流し塗り	nagashi-nuri
flow cytometry *[cyt]*	流動細胞計測法	ryūdō-saibō-keisoku-hō
flow diagram *[graph]*	流れ図	nagare-zu
flow direction *[eng]*	流れの向き	nagare no muki
flower *{n}* *[bot]*	花	hana
flower arrangement *{n}*	生け花；活け花	ike-bana
flower bed *[bot]*	花壇	kadan
flower bud *[bot]*	花芽	kaga
flowering fern *[bot]*	薇；ぜんまい	zenmai
flowering plant *[bot]*	顕花植物	kenka-shokubutsu
flowers of sulfur *[pharm]*	硫黄華	iō'ka
" *[pharm]*	昇華硫黄	shōka-iō
flowline *[eng]*	流れ線	nagare-sen
flow meter *[eng]*	流量計	ryūryō-kei
flow potential *[elec]*	流動電位	ryūdō-den'i
flow rate *[fl-mech]*	流量；りゅうりょう	ryū'ryō
flow stress *[mech]*	流動応力；流動應力	ryūdō-ō'ryoku
flow velocity *[ch]* *[geol]*	流速	ryū-soku
flu; influenza *[med]*	流行性感冒	ryūkō'sei-kanbō
fluctuate; waver *{vb}*	振れる	fureru
fluctuation *[electr]* *[sci-t]*	揺らぎ；搖らぎ	yuragi
flue *[eng]*	煙道；煙道	endō
flue cinder *[met]*	煙灰	enbai
fluff *{n}* *[mat]*	毛羽；けば	keba
fluff formation *[tex]*	毛羽立ち	keba'dachi
fluid *{n}* *[phys]*	流体；流體	ryū'tai
fluid (hydraulic) **coupling** *[mech-eng]*	流体継手	ryūtai-tsugi'te
fluid friction *[fl-mech]*	流体摩擦	ryūtai-ma'satsu
fluidity *[fl-mech]*	流動度	ryūdō-do
" *[fl-mech]*	流動率	ryūdō-ritsu
" *[fl-mech]*	流動性	ryūdō-sei

fluidization *[ch-eng] [fl-mech]*	流動化	ryūdō'ka
fluidized bed *[eng]*	流動床	ryūdō-shō
" *[eng]*	流動層	ryūdō-sō
fluidizing agent *[pharm]*	流動化剤	ryūdō'ka-zai
fluid mechanics *[mech]*	流体力学；流體力學	ryūtai-rikigaku
fluorescence *[atom-phys]*	蛍光；螢光	keikō
fluorescence emission spectrum *[spect]*	蛍光スペクトル	keikō-supekutoru
fluorescence microscopy *[opt]*	蛍光顕微鏡法	keikō-kenbikyō-hō
fluorescent brightening *[ch]*	蛍光増白	keikō-zō'haku
fluorescent dyeing *{n} [tex]*	玉虫(染め)	tama'mushi(-zome)
fluorescent indicator *[ch]*	蛍光指示薬	keikō-shiji-yaku
fluorescent lamp *[electr]*	蛍光灯	keikō-tō
fluorescent paint *[ch] [mat]*	蛍光塗料	keikō-toryō
fluorescent pigment *[ch]*	蛍光顔料	keikō-ganryō
fluorescent substance *[phys]*	蛍光体	keikō-tai
fluoridation *[eng]*	弗化物添加	fukka'butsu-tenka
" (of teeth) *[dent]*	弗素処理（歯の）	fusso-shori (ha no)
fluoride *[ch]*	弗化物；ふっかぶつ	fukka'butsu
fluoride resin *[poly-ch]*	弗素樹脂	fusso-jushi
fluorimetry *[an-ch]*	蛍光定量法	keikō-teiryō-hō
fluorine (element)	弗素；ふっそ	fusso
fluorine oxide	酸化弗素	sanka-fusso
fluorocarbon resin *[resin]*	弗素樹脂	fusso-jushi
fluorography *[graph]*	間接撮影	kan'setsu-satsu'ei
fluoroscopy *[eng]*	エックス線透視法	ekkusu'sen-tōshi-hō
fluorspar; fluorite *[miner]*	蛍石；ほたるいし	hotaru-ishi
flush *{vb t}*	流す	nagasu
flush end *[bioch]*	二本鎖末端	nihon'sa-mattan
flushing *{n} [physio]*	紅潮	kōchō
flute; pipe; whistle *{n} [music]*	笛	fue
flutter *{n} [eng]*	ばたつき	bata'tsuki
fluvial deposit *[geol]*	河成堆積物	kasei-taiseki'butsu
fluvial tide *[hyd]*	河潮	kachō
flux *{n} [mat]*	融剤；融劑	yūzai
" *{n} [phys]*	流速	ryū-soku
flux density *[phys]*	流速密度	ryūsoku-mitsudo
flux of light *[phys]*	光流	kōryū
fly (insect) *[i-zoo]*	蠅；はえ	hae
" (for fishing) *{n} [eng]*	毛鉤；けばり	ke-bari
" ; soar; jump *{vb i}*	飛ぶ；とぶ	tobu
" (let fly) *{vb t}*	飛ばせる	tobaseru
fly agaric (mushroom, fly killer) *[bot]*	紅天狗茸	beni-tengu-take

flyby *{n} [navig]*	天体接近飛行	tentai-sekkin-hikō
flying fish *[v-zoo]*	飛び魚	tobi-uo
flying frog (amphibian) *[v-zoo]*	飛び蛙	tobi-gaeru
flying lizard *[v-zoo]*	飛び蜥蜴	tobi-tokage
flying squirrel (animal) *[v-zoo]*	鼯鼠; むささび	musasabi
flywheel *[mech-eng]*	勢車; 弾車	hazumi-guruma
foam *{n} [ch]*	泡; 泡; あわ	awa
" *{n} [ch]*	泡沫	hōmatsu
foaming *{n} [ch-eng]*	泡立ち	awa'dachi
" *{n} [eng]*	起泡	kihō
foaming agent *[mat]*	泡立て剤	awa'date-zai
" *[mat]*	膨張剤	bōchō-zai
" *[mat]*	起泡剤; 起泡劑	kihō-zai
foaming property *[eng]*	起泡性	kihō-sei
focal length *[opt]*	焦点距離	shōten-kyori
focal plane *[opt]*	焦点面	shōten-men
focal ratio *[opt]*	焦点比	shōten-hi
focus (pl foci; focuses) *{n} [opt]*	焦点; しょうてん	shōten
" *{n} [opt]*	ピント	pinto
focusing *{n} [opt]*	集束	shūsoku
fog *{n} [meteor]*	霧; きり	kiri
" *{n} [photo]*	カブリ	kaburi
foghorn *[navig]*	霧笛	mu'teki
foglight *[navig]*	霧灯	mutō
fog quenching *{n} [met]*	噴霧焼入れ	funmu-yaki'ire
foil *{n} [mat]*	箔	haku
foil decorating *{n} [poly-ch]*	絵付成形	e-tsuke-seikei
fold *{n} [cl] [med]*	襞; ひだ	hida
" (in a stratum) *{n} [geol]*	褶曲	shū'kyoku
" *{vb t} [mech-eng]*	折る	oru
" *{vb t}*	折り畳む	ori-tatamu
fold; furl *{vb t}*	畳む; たたむ	tatamu
fold and fell *{vb t}*	折り倒す	ori-taosu
fold in; turn in *{vb t}*	畳み込む	tatami-komu
folding chair *[furn]*	折り畳み椅子	ori'tatami-isu
folding endurance; folding strength *[mat] [paper]*	耐折強さ	tai'setsu-tsuyo-sa
folding fan *[eng]*	扇; おうぎ	ōgi
" *[eng]*	扇子	sensu
folding screen *[furn]*	屛風	byōbu
folding strength *[mat]*	耐折強さ	tai'setsu-tsuyo-sa
foliage *[graph]*	唐草; からくさ	kara-kusa

foliation *[bot]*	発葉; 發葉	hatsu'yō
folic acid *[bioch]*	葉酸	yōsan
folio *[pr]*	二つ折り(判)	futatsu-ori(-ban)
folk art *{n}*	民芸; 民藝	mingei
follicle *[bio]*	卵胞; らんほう	ranhō
" *[bio]*	沪胞; 濾胞; ろほう	rohō
" *[bot]*	袋果	taika
following *{n}*	従動; 從動	jūdō
following; pursuing *{n}*	付き廻り	tsuki'mawari
fomentation *[pharm]*	罨法剤	anpō-zai
font *[comput] [graph]*	字体	jitai
food *[bio] [cook]*	食品; しょくひん	shoku'hin
" *[bio] [cook]*	食物; しょくもつ	shoku'motsu
" *[bio] [cook]*	食べ物; たべもの	tabe'mono
food additive *[cook]*	食品添加物	shokuhin-tenka'butsu
food and drink *[cook]*	飲食物	in'shoku'butsu
food boxes (stacked) *[cook] [furn]*	重箱; じゅうばこ	jū-bako
food chain *[ecol]*	食物連鎖	shoku'motsu-rensa
food chemistry *[food-eng]*	食品化学	shokuhin-kagaku
food color *[cook] [mat]*	食品染料	shokuhin-senryō
" *[cook] [mat]*	食用色素	shokuyō-shikiso
food engineering *{n} [eng]*	食物工学	shoku'motsu-kōgaku
" *{n} [eng]*	食糧工学	shokuryō-kōgaku
food flavor *[cook] [mat]*	食用香料	shoku'yō-kōryō
food hygiene *[med]*	食品衛生学	shokuhin-eisei'gaku
food industry *[ind-eng]*	食品工業	shokuhin-kōgyō
" *[ind-eng]*	食糧工業	shokuryō-kōgyō
food industry world *[ind-eng]*	食品業界	shokuhin-gyōkai
food irradiation *[food-eng]*	食品照射	shokuhin-shōsha
food oil *[cook]*	食用油	shokuyō'yu; shokuyō-abura
food poisoning *{n} [med]*	食中毒	shoku-chūdoku
food sanitation law *[cook] [legal]*	食品衛生法	shokuhin-eisei-hō
food web *[ecol]*	食物網	shoku'motsu-mō
foot (pl feet); leg *[anat]*	足; 脚; あし	ashi
foot (pl feet) (unit of length) *[mech]*	フート; 呎	fūto
foot; ft (unit of length) *[mech]*	フット	futto
foot, Japanese (unit of length) *[mech]*	尺; しゃく	shaku
footbridge *[civ-eng]*	人道橋	jindō'kyō
foothills *[geogr]*	丘陵地帯	kyūryō-chitai
footing *{n} [civ-eng]*	足場	ashi'ba
" *{n}*	立脚点	rikkyaku-ten
footnote *[comput] [pr]*	脚注; 脚註	kyaku'chū

foot-pound; ft-lb (unit of energy or work) *[mech]*	フート ポンド；呎磅	fūto-pondo
footprints; spoor *{n}*	足跡	ashi-ato
footwear *[eng]*	履き物；はきもの	haki'mono
foraging *{n} [v-zoo]*	食糧漁り	shokuryō-asari
Foraminifera *[i-zoo]*	誘孔虫類	yūkō'chū-rui
forbidden band *[ch]*	禁止帯；禁止帶	kinshi-tai
forbidden combination *[comput]*	禁止組合せ	kinshi-kumi'awase
forbidden line *[atom-phys]*	禁制線	kinsei-sen
force *{n} [mech]*	力；ちから	chikara
forced circulation *[mech-eng]*	強制循環	kyōsei-junkan
forced-circulation evaporator *[eng]*	強制循環形蒸発罐	kyōsei-junkan'gata-jō'hatsu-kan
forced draft *[mech-eng]*	押込通風	oshi'komi-tsūfū
forced drying *{n} [mech-eng]*	強制乾燥	kyōsei-kansō
force diagram *[graph] [mech-eng]*	示力図	shi'ryoku-zu
forced oscillation *[mech]*	強制振動	kyōsei-shindō
force factor *[phys]*	力係数	chikara-keisū
force field *[mech]*	力場；りきじょう	riki-jō
forearm *{n} [anat]*	前腕	zenwan
forecast *{n} [meteor]*	予報	yohō
" (prediction) *{n} [psy]*	予測	yo'soku
forefinger *[anat]*	人差指；食指	hito'sashi-yubi
foreground *[graph]*	前景	zenkei
forehead *{n} [anat]*	額；ひたい	hitai
foreign *{adj}*	外国の；外國の	gai'koku no
foreign capital *[econ]*	外資	gaishi
foreign character *[pr]*	外字	gaiji
foreign country *[geogr]*	外国；がいこく	gai'koku
foreign currency *[econ]*	外貨	gaika
foreigner *{n}*	外国人	gaikoku'jin
foreign exchange *[econ]*	外国為替	gaikoku-kawase
foreign language *[pr]*	外字	gaiji
foreign matter *[sci-t]*	異物	i'butsu
" *[sci-t]*	夾雑物	kyōzatsu'butsu
foreign-origin word *[ling]*	外来語；外來語	gairai'go
foreign trade *[econ]*	外国貿易	gaikoku-bōeki
foreman *[eng]*	職長	shokuchō
forenoon *[astron]*	午前	gozen
forensic chemistry *[ch]*	法化学	hō-kagaku
forensic pathology *[med]*	法病理学	hō-byōri'gaku
forensic science *[sci-t]*	法科学	hō-kagaku

foreshock (earthquake) *[geophys]*	前震；ぜんしん	zenshin
" *[geophys]*	予震；豫震；よしん	yoshin
forest; woodland; timberland *[ecol]*	森	mori
" *[ecol]*	森林	shin'rin
forest fire *[for]*	山火事；やまかじ	yama-kaji
forest product(s) *[ind-eng]*	林産物	rin-sanbutsu
forestry *[ecol]*	林学	rin'gaku
forewarning; prediction *{n} [psy]*	予言；豫言；よげん	yogen
forfeiture *[legal]*	没収；没收	bosshū
forged steel *[steel]*	鍛鋼	tankō
forget *{vb} [psy]*	忘れる	wasureru
forget-me-not (flower) *[bot]*	勿忘草	wasure'na-gusa
forge welding *{n} [met]*	鍛接	tan'setsu
forging *{n} [legal]*	偽造	gizō
" *{n} [met]*	鍛錬；鍛鍊	tanren
" *{n} [met]*	鍛造；たんぞう	tanzō
forked *{adj}*	分岐した	bunki shita
form *{n} [bio]*	品型	hinkei
" *{n} [bio] [syst]*	品種	hinshu
" *{n} [bio] [syst]*	型；かた	kata
" *{n} [comput]*	書式	sho'shiki
" (shape) *{n} [math] [phys]*	形；かたち	katachi
form; sheet *{n} [paper]*	用紙	yōshi
formability *[mat]*	成形性	seikei-sei
formal *{n}*	正式	seishiki
formal documents *[pr]*	定型書類	teikei-shorui
formality *[ch]*	式濃度	shiki-nōdo
formal clothes *[cl]*	礼服；禮服	rei-fuku
format *{n} [comput]*	様式	yō'shiki
" *{n} [comput] [pr]*	書式	sho'shiki
formate *[ch]*	蟻酸塩	gisan-en
formation *[paper]*	地合い	ji'ai
" *[p-ch]*	化成	kasei
" *{n}*	形成	keisei
formation voltage *[elec]*	化成電圧	kasei-den'atsu
formatted display *[comput]*	定様式表示	tei'yōshiki-hyōji
formatting *{n} [comput]*	書式作成	sho'shiki-sakusei
former year(s); earlier years	先年	sen-nen
form feed *[comput]*	書式送り	sho'shiki-okuri
formic acid	蟻酸；ぎさん	gisan
forming *{n} [rub]*	成形	seikei
forming property *[eng]*	加工性	kakō-sei

forming treatability *[eng]*	化成処理性	kasei-shori-sei
formula *[ch] [math]*	式；しき	shiki
" *[ch] [pharm]*	処方；處方	shohō
" (pl -s, -mulae) *[math]*	公式	kō'shiki
formulary (pl -laries) *[pharm]*	処方集	shohō'shū
formulation *[math]*	公式化	kōshiki'ka
formula weight *[ch]*	(化学)式量	(kagaku-)shiki'ryō
forsterite *[miner]*	苦土橄欖石	kudo'kanran'seki
fortified rosin *[mat]*	強化ロジン	kyōka-rojin
fortified wine *[cook]*	甘味果実酒	kanmi-ka'jitsu'shu
fortnightly *{adv}*	隔週の	kaku'shū no
" *{adv}*	二週に一回の	nishū ni ikkai no
forward direction *[electr]*	順方向	jun-hōkō
forward draft *[nav-arch]*	船首喫水	senshu-kissui
forward-facing *{adj}*	前向きの	mae-muki no
forward perpendicular *[nav-arch]*	船首垂線	senshu-suisen
forward reaction *[ch]*	正反応	sei-hannō
forward side; ahead *{adv} {n}*	前方	zenpō
forward voltage *[electr]*	順方向電圧	jun'hōkō-den'atsu
fossil *[paleo]*	化石；かせき	ka'seki
fossil fuel *[geol]*	化石燃料	kaseki-nenryō
fossil resin *[poly-ch]*	化石樹脂	kaseki-jushi
fouling *{n} [ecol]*	汚損	o'son
foundation *{n}*	素地	soji
foundation; groundwork *{n}*	下地	shita'ji
foundation drawing *{n} [graph]*	基礎図；基礎圖	kiso-zu
foundry *[eng]*	鋳物場；いものば	imono'ba
foundry resin *[poly-ch]*	鋳物用樹脂	imono'yō-jushi
fountain *[arch]*	噴水泉	funsui'sen
fountain; spring *{n} [hyd]*	泉；いずみ	izumi
fountain pen *[graph]*	万年筆；萬年筆	mannen'hitsu
four *{n} [math]*	四	shi; yon
four-component system *[eng]*	四成分系	shi-seibun-kei
four corners, the *{n}*	四隅	yo'sumi
four-electrode method *[electr]*	四極法	yon'kyoku-hō
four-parameter model *[poly-ch]*	四要素モデル	yon'yōso-moderu
four persons *{n}*	四人	yo'nin; yottari
four-saddle puffer (fish) *[v-zoo]*	北枕	kita'makura
fourth dimension *[math]*	四次元	yojigen
fourth proportional *[alg]*	第四比例項	dai'yon-hirei-kō
four-way valve *[mech-eng]*	四方弁；四方瓣	shihō-ben
four-wheel drive *[mech-eng]*	四輪駆動	yonrin-kudō

four-wheeled vehicle *[mech-eng]*	四輪(自動)車	yonrin(-jidō)'sha
four-wire channel *[comm]*	四線式通信路	yonsen'shiki-tsūshin-ro
four-wire circuit *[comm]*	四線式回線	yonsen'shiki-kaisen
fox (animal) *[v-zoo]*	狐; きつね	kitsune
foyer *[arch]*	入口の間	iri'guchi no ma
" *[arch]*	遊歩場	yūho'jō
fractal *{n} [math]*	次元分裂図形	jigen-bun'retsu-zukei
fraction *[arith]*	分数; ぶんすう	bunsū
" *[sci-t]*	画分; 畫分	kaku'bun
fractional crystallization *[ch]*	分別結晶(作用)	bunbetsu-kesshō(-sayō)
fractional distillation *[ch]*	分(別蒸)留	bun(betsu-jō)ryū
fractional precipitation *[ch]*	分別沈殿	bunbetsu-chinden
fractional sublimation *[ch]*	分別昇華	bunbetsu-shōka
fractionating column *[ch]*	分留管; 分溜管	bunryū-kan
fractionation *[bioch]*	分画	bunkaku
" *[ch]*	分別	bun'betsu
" (of crude oil) *[petrol]*	分別蒸留	bunbetsu-jōryū
fraction defective *[ind-eng]*	不良率	furyō-ritsu
fractocumulus *[meteor]*	片積雲	henseki'un
fractostratus *[meteor]*	断片層雲	danpen-sō'un
fracture (of bone) *{n} [med]*	骨折; こっせつ	kossetsu
" *{n} [miner]*	割れ口	ware'kuchi; ware'guchi
fracture(d) surface *[mech-eng]*	破(断)面	ha(dan)-men
fractured zone *[geol]*	断層破砕帯	dansō-hasai-tai
fracture mechanics *[phys]*	破壊力学; 破壞力學	hakai-rikigaku
fracture point *[mech]*	破断点; 破斷點	hadan-ten
fragile; fragility *{n} [sci-t]*	虚弱	kyojaku
fragile; fragility *{n} [sci-t]*	脆弱; ぜいじゃく	zei'jaku
fragility *{n} [mech]*	脆さ	moro-sa
fragment *{n}*	断片	danpen
fragmentation *{n}*	断片化	danpen'ka
fragrance *[mat]*	香料	kōryō
" *[org-ch]*	芳香剤	hōkō-zai
" *[physio]*	香り; 薫り; かおり	kaori
" *[physio]*	香気	kōki
fragrance-enhancing agent *[physio]*	増香料	zōkō-ryō
fragrant; delicious (aroma) *[cook]*	芳しい	kōbashī; kanbashī
fragrant olive *[bot]*	金木犀	kin-mokusei
fragrant sumac *[bot]*	匂い漆; におい漆	ni'oi-urushi
frame (of movie) *{n} [opt]*	駒	koma
" *{n} [stat]*	枠; わく	waku
frame wire *[met]*	枠線	waku-sen

framework; foundation (house) *[arch]* 屋台骨；やたいぼね yatai-bone
francium (element) フランシウム furanshūmu
frangipani (tree) *[bot]* 印度素馨 indo-sokei
frankincense *[mat]* 乳香 nyūkō
free (unsecured) *{adj} [ch]* 遊離の；ゆうりの yūri no
" (gratuitous) *{adj}* 只の；ただの tada no
free acid *[ch]* 遊離酸 yūri-san
free carbon *[met]* 遊離炭素 yūri-tanso
free competition *[econ]* 自由競争 jiyū-kyōsō
free-cutting steel *[met]* 快削鋼 kaisaku-kō
freedom *{n}* 自由；じゆう jiyū
free electron *[phys]* 自由電子 jiyū-denshi
free-electron laser *[opt]* 自由電子レーザー jiyū-denshi-rēzā
free fall *[{n} mech]* 自由落下 jiyū-rakka
free fatty acid *[org-ch]* 遊離脂肪酸 yūri-shibō-san
freely attachable and detachable *[mech-eng]* 着脱自在 chaku'datsu-jizai
freely insertable and extractable *[mech-eng]* 挿抜自在；そう=ばつじざい sō'batsu-jizai
freely slidable *[mech]* 摺動自在 shūdō-jizai
free market *[econ]* 自由市場 jiyū-shijō
freeness *{n} [paper]* 沪水度；濾水度 rosui-do
free path *[phys]* 自由行路 jiyū-kōro
" *[phys]* 自由行程 jiyū-kōtei
free radical *[ch] [org-ch]* 遊離基；ゆうりき yūri-ki
free-radical scavenger *[bioch]* 遊離基捕捉剤 yūri'ki-hosoku-zai
free-space wavelength *[phys]* 自由空間波長 jiyū-kūkan-hachō
free sulfur *[ch]* 遊離硫黄 yūri-iō
free water *[ch]* 自由水 jiyū-sui
freeze *{vb}* 氷結する hyō'ketsu suru
freeze drying *{n} [cook] [food-eng]* 凍結乾燥 tōketsu-kansō
freezer (machine or instrument) *[eng]* 凍結機；凍結器 tōketsu-ki
" (a room) *[mech-eng]* 氷室；ひょうしつ hyō-shitsu
freezing damage *[meteor]* 凍害 tōgai
freezing hardiness *[agr]* 耐凍性 taitō-sei
freezing mixture *[p-ch]* 寒剤 kanzai
" *[p-ch]* 起寒剤；きかんざい kikan-zai
freezing point *[p-ch]* 氷点；ひょうてん hyō-ten
freezing-point depressant *[p-ch]* 氷点降下剤 hyōten-kōka-zai
freezing-point method *[p-ch]* 氷点法 hyōten-hō
freight (cost of) *{n} [econ]* 運賃；うんちん un'chin
freight car *[rail]* 貨車 kasha

English	Japanese	Romanization
freighter *[nav-arch]*	貨物船	kamotsu'sen
frequency (pl -cies) *[math]*	度数	dosū
" *[math]*	頻度；ひんど	hindo
" *[phys]*	回数	kaisū
" *[phys]*	周波数	shūha'sū
frequency band *[phys]*	周波数帯域	shūhasū-tai'iki
frequency conversion *[electr]*	周波数変換	shūhasū-henkan
frequency demultiplier *[electr]*	周波数逓降器	shūhasū-teikō-ki
frequency discriminator *[electr]*	周波数弁別器	shūhasū-benbetsu-ki
frequency divider *[electr]*	分周器	bunshū-ki
frequency modulation; FM *[comm]*	周波数変調	shūhasū-henchō
frequency multiplier *[electr]*	倍周器	baishū-ki
" *[electr]*	周波数逓倍器	shūhasū-teibai-ki
frequency response *[eng]*	周波数応答	shūhasū-ō'tō
frequent *{adj}* *[math]*	頻繁な	hinpan na
fresh *{adj}* *[cook]*	新鮮な；しんせんな	shinsen na
freshness *[cook]*	(新)鮮度	(shin)sendo
fresh water *[hyd]*	真水；淡水；まみず	ma-mizu
" *[hyd]*	淡水；たんすい	tansui
freshwater fish *[v-zoo]*	淡水魚	tansui'gyo
freshwater lake *[hyd]*	淡湖	tanko
friability *[mat]*	砕け易さ	kudake-yasu-sa
friction *[mech]*	摩擦；まさつ	ma'satsu
frictional electricity *[elec]*	摩擦電気	masatsu-denki
friction coefficient *[mech]*	摩擦係数	masatsu-keisū
friction, law of *[mech]*	摩擦の法則	masatsu no hōsoku
friction welding *{n}* *[plast]*	摩擦溶接	masatsu-yō'setsu
friction wheel *[mech]*	摩擦車	masatsu-guruma
Friday	金曜日	kin'yōbi
fried bean curd *[cook]*	油揚げ	aburāge; abura'age
fried dishes *[cook]*	揚げ物	age-mono
fried eggs *[cook]*	目玉焼；めだまやき	me'dama-yaki
frigate bird *[v-zoo]*	軍艦鳥	gunkan-dori
frigid zone *[climat]*	寒帯；寒帶	kantai
frigorific *{adj}* *[thermo]*	起寒性の	kikan'sei no
frilled lizard *[v-zoo]*	襟巻き蜥蜴	eri'maki-tokage
fringing reef *[geol]*	裾礁；きょしょう	kyoshō
frizzle; fry *{vb t}* *[cook]*	炒める；煠める	itameru
frog (amphibian) *[v-zoo]*	蛙；かえる	kaeru
" *[rail]*	轍叉；てっさ	tessa
frogfish (fish) *[v-zoo]*	鮟鱇；あんこう	ankō
" (fish) *[v-zoo]*	躄魚；いざりうお	izari-uo

frog flounder (fish) *[v-zoo]*	目板鰈	me'ita-garei
frondose *{adj} [bot]*	葉状体の	yōjō'tai no
front *{adv} {n} [arch]*	表; おもて	omote
" *{n} [meteor]*	前線	zensen
" *{adv} {n}*	正面; しょうめん	shōmen
frontal *{adj}*	正面の	shōmen no
" *{adj}*	前面の	zenmen no
frontal cyclone *[meteor]*	前線低気圧	zensen-teiki'atsu
frontal zone *[meteor]*	前線帯; 前線帶	zensen-tai
front elevation *[graph]*	正面図	shōmen-zu
front part *{n}*	前部	zenbu
front position *{n}*	前位	zen'i
front sectional view *[graph]*	断正面図	dan'shōmen-zu
" *[graph]*	正面断面図	shōmen-danmen-zu
front teeth; incisors *[anat] [dent]*	前歯; 前齒; まえば	mae'ba
front-to-back ratio *[elecmg]*	前後電界比	zengo-denkai-hi
front view *[graph]*	正面図	shōmen-zu
front view cross section *[graph]*	正面断面図	shōmen-danmen-zu
front visibility *[navig]*	前方視野範囲	zenpō-shiya-han'i
front-wheel drive *[mech-eng]*	前輪駆動	zenrin-kudō
front yard *[civ-eng]*	前庭	zentei
frost *{n} [hyd]*	霜	shimo
frostbite *[med]*	霜焼け; しもやけ	shimo'yake
" *[med]*	凍傷; とうしょう	tōshō
frost column *[hyd]*	霜柱	shimo-bashira
frost damage *[cer]*	凍害	tōgai
frosted glass *[mat]*	磨き硝子	migaki-garasu
" *[mat]*	艶消し硝子	tsuya'keshi-garasu
frost heaving *{n} [geol]*	凍上	tōjō
froth *{n} [ch]*	泡沫; 泡沫	hōmatsu
" *{n} [fl-mech]*	泡; 泡; あわ	awa
frother *[poly-ch]*	起泡剤	kihō-zai
frothiness *[mat]*	多泡性	tahō-sei
frozen food(s) *[cook]*	冷凍食品	reitō-shoku'hin
fructose; fruit sugar *[bioch]*	果糖	katō
Frugivora *[v-zoo]*	食果類	shokka-rui
frugivorous *{adj} [zoo]*	果食性の	kashoku'sei no
fruit *{n} [bot]*	果実; かじつ	ka'jitsu
" *{n} [bot]*	果物; くだもの	kuda-mono
" *{n} [bot]*	実; 實; み	mi
fruit fly (insect) *[i-zoo]*	実蠅; みばえ	mi-bae
fruit grader; fruit grader *[agr]*	果実選別機	ka'jitsu-senbetsu-ki

fruiting body *[bot]*	子実体；子實體	shi'jitsu-tai
fruit orchard *[agr]*	果樹園	kaju'en
fruit sugar *[bioch]*	果糖	katō
fruit thinning *{n} [agr]*	摘果	teki-ka
fruit tree *[bot]*	果樹；かじゅ	kaju
fruit wine *[cook]*	果実酒	kajitsu'shu
frustum (pl -s, frusta) *[geom]*	台形；臺形	daikei
" *[geom]*	切頭体	settō'tai
fry (young fish) *{n} [v-zoo]*	幼魚	yōgyo
" *{vb t} [cook]*	揚げる	ageru
frying oil *[cook] [mat]*	揚げ油	age-abura
fucoxanthin *[bioch]*	褐藻素	kassō'so
fuel *{n} [mat]*	燃料；ねんりょう	nenryō
fuel cell *[elec]*	燃料電池	nenryō-denchi
fuel gage *[eng]*	燃料計	nenryō-kei
fuel oil *[mat]*	重油	jūyu
" *[mat]*	燃料油	nenryō-yu
fugacity *[ch]*	逸散性	issan-sei
" *[thermo]*	逃散度	tōsan'do
fulcrum *[eng]*	支柱	shichū
" *[mech]*	支点；支點；してん	shiten
fulcrum base *[mech]*	支点台	shiten-dai
full adder *[electr]*	全加算器	zen-kasan-ki
full band *[electr]*	充満帯；充滿帶	jūman-tai
full carrier *[comm]*	全搬送波	zen-hansō'ha
full condensation *[ch]*	全縮	zen-shuku
full-duplex communication *[comm]*	全二重通信	zen-nijū-tsū'shin
fuller's earth *[geol]*	漂布土	hyōfu-do
" *[geol]*	酸性白土	sansei-hakudo
fulling machine *[mech-eng]*	縮充機	shuku'jū-ki
full moon *[astron]*	望；ぼう	bō
" *[astron]*	満月	mangetsu
full power *[mech-eng] [phys]*	全出力	zen-shutsu'ryoku
full scale *[graph]*	原尺	gen'shaku
full-scale (size) drawing *[graph]*	原尺図	genshaku-zu
full speed *[mech]*	全速	zen-soku
full stop *[comput] [pr]*	句点	kuten
full subtracter *[electr]*	全減算器	zen-genzan-ki
full view; full drawing *{n} [graph]*	全体図；全體圖	zentai-zu
full wave *[electr]*	全波	zenpa
full-wave rectification *[electr]*	全波整流	zenpa-seiryū
full-wave rectifier *[electr]*	全波整流器	zenpa-seiryū-ki

English	Japanese	Romanization
full word *[comput]*	全語	zengo
fully automatic *[eng]*	全自動	zen-jidō
fulminate *[org-ch]*	雷酸塩	raisan-en
" *{vb} [ch]*	爆鳴する	bakumei suru
fulminic acid	雷酸	raisan
fumarole *[geol]*	噴気孔	funki-kō
fumes *{n} [ch]*	煙霧	enmu
fumigation *[eng]*	燻蒸；くんじょう	kunjō
fumigation sterilization *[eng]*	燻蒸消毒	kunjō-shōdoku
fuming nitric acid	発煙硝酸	hatsu'en-shōsan
fuming sulfuric acid	発煙硫酸	hatsu'en-ryūsan
function *{n} [alg]*	関数；函数	kansū
function; capability; facility *{n}*	機能；きのう	kinō
" *{n}*	作用	sayō
" *{vb}*	作用する	sayō suru
functional derivative *[math]*	汎関数微分	han'kansū-bibun
functional group *[org-ch]*	官能基	kannō-ki
functionality *[comput]*	機能性	kinō-sei
" (the property) *{n}*	官能性	kannō-sei
" (the numeral) *{n}*	官能数	kannō'sū
functionally *{adv} [ind-eng]*	作動的に	sadō'teki ni
functional materials *[mat]*	機能材料	kinō-zairyō
functional polymer *[org-ch]*	機能性高分子	kinō'sei-kōbunshi
function generator *[electr]*	関数発生器	kansū-hassei-ki
functioning *{n} [ind-eng]*	作動	sadō
function table *[comput]*	関数表	kansū-hyō
fundamental frequency *[phys]*	基本周波数	kihon-shūha'sū
fundamental number *[arith]*	基数	kisū
fundamentals *[sci-t]*	基礎；きそ	kiso
fundamental solution *[math]*	基本解	kihon-kai
fundamental star *[astron]*	基本星	kihon-sei
fundamental tone *[acous]*	基本音；きほんおん	kihon-on
fundamental unit *[phys]*	基本単位	kihon-tan'i
fundamental vibration *[mech]*	基本振動	kihon-shindō
fungal cell *[mycol]*	菌体	kintai
fungal strain *[mycol]*	菌株；きんかぶ	kin-kabu
fungal toxin *[mycol]*	黴毒	kabi-doku
fungicide; antiseptic *{n} [bioch]*	防黴剤	bōbai-zai
Fungi imperfecti *[mycol]*	不完全菌類	fu'kanzen'kin-rui
fungus (pl fungi) *[mycol]*	菌（類）	kin(-rui)
funicular polygon *[mech]*	糸多角形	ito-takak(u)kei
funicular railway *[rail]*	鋼索鉄道	kōsaku-tetsudō

funicular water *[ch]*	索状水	saku'jō-sui
funnel *{n} [arch]*	煙筒	entō
" *{n} [eng]*	漏斗；ろうと	rōto
funnel cloud *[meteor]*	漏斗雲	rōto-gumo
funnel tube *[ch-eng]*	漏斗管	rōto-kan
funny; amusing; ludicrous *{adj}*	おかしい	okashī
fur *[mat] [v-zoo]*	毛	ke
" *[v-zoo]*	毛皮	kegawa
furnace *[cer] [eng]*	窯；かま	kama
" *[eng]*	炉；爐；ろ	ro
furnace; kiln; oven *[eng]*	窯炉	yōro
furnishings *[furn]*	調度	chōdo
" *[furn]*	建具；たてぐ	tate'gu
furniture *[arch]*	家具	kagu
fur seal (mammal) *[v-zoo]*	膃肭獣；おっとせい	ottosei
fuse *{n} [eng]*	導火線	dōka-sen
" *{n} [eng]*	起爆装置	kibaku-sōchi
fused quartz *[miner]*	溶融石英	yōyū-seki'ei
fused silica *[mat]*	融解石英	yūkai-seki'ei
fuselage *[aero-eng]*	機胴	kidō
" *[aerosp]*	胴体	dōtai
fusibility *[thermo]*	可融性	kayū-sei
" *[thermo]*	融度；ゆうど	yūdo
fusible alloy *[met]*	易融合金	iyū-gōkin
fusing current *[elec]*	溶断電流	yōdan-denryū
fusing point *[met] [thermo]*	融点；ゆうてん	yūten
fusion *[nuc-phys] [p-ch]*	融合	yūgō
" *[p-ch]*	溶融	yōyū
" *[p-ch]*	融解	yūkai
fusion casting *{n} [met]*	溶接鋳造	yōsetsu-chūzō
fusion-fission hybrid reactor *[nucleo]*	融合分裂混成炉	yūgō-bun'retsu-konsei-ro
fusion welding *{n} [met]*	融接	yū'setsu
future *{n}*	未来	mirai
" *{n}*	將来	shōrai
fuzz *{n} [mat]*	毛羽；けば	keba

G

English	Japanese	Romanization
gabbro (pl gabbros) *[miner]*	斑糲岩	hanrei'gan
gadfly; deerfly (insect) *[i-zoo]*	虻; あぶ	abu
" (insect, a mite) *[i-zoo]*	恙虫; つつがむし	tsutsuga-mushi
gadolinium (element)	ガドリニウム	gadorinyūmu
gage; gauge *{n} [eng]*	計器	keiki
gage board *[electr]*	計器板	keiki-ban
gage element *[eng]*	測定子	sokutei'shi
gage length *[mech]*	標点距離	hyōten-kyori
gain; profit *{n} [econ]*	利益	ri'eki
" *{n} [econ] [elec]*	利得; りとく	ritoku
" (gainful) *{n} [econ]*	得	toku
" (obtain) *{vb}*	得る	uru
galactic coordinates *[astron]*	銀河座標	ginga-zahyō
galactic poles *[astron]*	銀河極	ginga-kyoku
galaxies *[astron]*	銀河; ぎんが	ginga
gale; strong wind *[meteor]*	強風	kyōfū
galena *[miner]*	方鉛鉱	hō'enkō
gall; insect gall *[bot]*	虫癭; ちゅうえい	chū'ei
gall; cecidium (pl -dia) *[bot]*	癭瘤	eiryū
gall *{n} [bot]*	虫瘤	mushi-kobu
" ; bile *{n} [hysio]*	胆汁; 膽汁	tanjū
" *{vb} [met]*	齧る; かじる	kajiru
" *{vb} [met]*	焼付く	yaki-tsuku
gallate *[ch]*	没食子酸塩	bosshokushi'san-en
gallbladder *[anat]*	胆嚢; たんのう	tannō
gallery (pl -leries) *[arch]*	歩廊	horō
" *[min-eng]*	坑道	kōdō
galley (of ship) *[nav-arch]*	厨房	chūbō
gallic acid	没食子酸	bosshoku'shi-san
gallium (element)	ガリウム	garyūmu
gallstone *[med]*	胆石; 膽石	tan'seki
galvanized coating *{n} [met]*	溶融亜鉛めっき	yōyū-aen-mekki
galvanized sheet iron *[met]*	亜鉛めっき鋼板	aen-mekki-kōban
" *[met]*	トタン板	totan-ita
galvanized steel wire *[met]*	亜鉛めっき鋼線	aen-mekki-kōsen
galvanizing *{n} [met]*	鍍金; めっき	mekki
galvanomagnetic effect *[elecmg]*	電流磁気効果	denryū-jiki-kōka
galvanometer *[eng]*	検流計	kenryū-kei
galvanometry *[elec]*	電流測定法	denryū-sokutei-hō

English	Japanese	Romanization
galvanoplastics (method) *[met]*	電铸法; 電鑄法	denchū-hō
galvanoplastics method *[met]*	電型法	denkei-hō
galvanotropism *[bio]*	向電性	kōden-sei
gametangium (pl -gia) *[bio]*	配偶子嚢	haigūshi-nō
gamete *[bio]*	配偶子; はいぐうし	haigū'shi
gametophyte *[bot]*	配偶体	haigū'tai
gamma rays *[nuc-phys]*	ガンマ線	ganma-sen
ganglion *[anat]*	神経節; 神經節	shinkei'setsu
gangplank *[nav-arch]*	道板; みちいた	michi-ita
gangrene *[med]*	壊疽; 壞疽; えそ	eso
gangue *[geol]*	脈石	myaku'seki
gannet (bird) *[v-zoo]*	白鰹鳥	shiro-katsuo'dori
gantry *[aerosp]*	発射整備塔	hassha-seibi'tō
gap *{n} [bot] [elec]*	間隙	kangeki
garage (for automobile) *[arch]*	車庫	shako
garbage *[civ-eng]*	厨芥; ちゅうかい	chūkai
" *[civ-eng]*	残菜	zansai
garden; yard *[arch] [bot]*	庭; にわ	niwa
" *[bot]*	庭園	tei'en
gardenia (flower) *[bot]*	山梔子; くちなし	kuchinashi
garden lantern *[furn]*	灯籠; 燈籠	tōrō
garden pea (vegetable) *[bot]*	莢豌豆	saya-endō
garden warbler (bird) *[v-zoo]*	庭虫食い	niwa-mushi'kui
garfish (fish) *[v-zoo]*	啄長魚; だつ	datsu
garganey (bird) *[v-zoo]*	島味; しまあじ	shima-aji
gargle *{n} [pharm]*	含嗽(薬)	ugai(-yaku)
" *{vb} [physio]*	含嗽する	ugai suru
garlic (bulb) *[bot]*	大蒜; にんにく	ninniku
garnet *[miner]*	柘榴石	zakuro-ishi
garnett *[text]*	反毛機	hanmō-ki
gas (pl -es) *{n} [phys]*	瓦斯; ガス	gasu
" *{n} [phys]*	気体; 氣体; きたい	kitai
gas constant *[thermo]*	気体定数	kitai-teisū
gaseous reaction, law of *[ch]*	気体反応の法則	kitai-hannō no hō'soku
gasification *[ch-eng]*	気化	kika
gas injection *[mech-eng]*	気体噴射	kitai-funsha
gasket *[nav-arch]*	括帆索	kappan'saku
gas law *[thermo]*	気体の法則	kitai no hō'soku
gas oil *[mat]*	軽油; 輕油; けいゆ	kei-yu
gas permeability *[fl-mech]*	通気性	tsūki-sei
gas phase *[ch]*	気相	kisō
gas-phase polymerization *[poly-ch]*	気相重合	kisō-jūgō

gastric digestion *[physio]*	胃内消化	i'nai-shōka
Gastrotrichia *[i-zoo]*	腹毛動物	fukumō-dōbutsu
gastrovascular system *[i-zoo]*	胃水管系	i'suikan-kei
gastrula (pl -las, -lae) *[embryo]*	原腸胚	genchō'hai
gastrulation *[embryo]*	原腸形成	genchō-keisei
gas wiper *[eng]*	気体絞り; 氣體絞り	kitai-shibori
gate; door *[arch]*	門; かど	kado
gate *[arch] [electr]*	門; もん	mon
gauze *[mat]*	ガーゼ	gāze
gauze fabric *[tex]*	絡み織り	karami-ori
gazette *[pat]*	公報; こうほう	kōhō
gazetteer *[pr]*	地名辞典	chimei-jiten
gear; toothed gear *[mech-eng]*	歯車; はぐるま	ha-guruma
gearbox; transmission *[mech-eng]*	変速機; 變速機	hensoku-ki
gear cutting *{n} [mech-eng]*	歯切り; はぎり	ha-giri; ha-kiri
gear generator; gear cutter *[mech-eng]*	歯切り盤	ha'giri-ban; ha'kiri-ban
gear-mesh; tooth-mesh *{vb} [mech-eng]*	嚙合する	shigō suru
gear ratio *[mech-eng]*	歯車比	ha'guruma-hi
gear roughing *{n} [mech-eng]*	荒歯切り	araha-giri
gear tooth *[mech-eng]*	歯; 齒; は	ha
gear train *[mech-eng]*	歯車列	ha'guruma-retsu
gecko; wall lizard (reptile) *[v-zoo]*	守宮; 家守; やもり	ya'mori
gel *{n} [ch]*	ゲル	geru
gelatin *[org-ch]*	ゼラチン	zerachin
gelatinization *[eng]*	糊化; こか	koka
gelidium jelly *[cook]*	心天; ところてん	tokoroten
gem; jewel; precious stone *[lap] [miner]*	宝玉; 寶玉	hōgyoku
" *[lap] [miner]*	宝石; ほうせき	hōseki
Gemini; the Twins *[constel]*	双子座; 雙子座	futago-za
gemology *[miner]*	宝石学	hōseki'gaku
gender; sex *[bio]*	性; せい	sei
gene *[gen]*	遺伝子; いでんし	iden'shi
gene bank *[gen]*	遺伝子銀行	idenshi-ginkō
gene(tic) engineering *[gen]*	遺伝子工学	idenshi-kōgaku
gene flow *[gen]*	遺伝子拡散	idenshi-kakusan
gene manipulation *[gen]*	遺伝子操作	idenshi-sōsa
gene pool *[gen]*	遺伝子給源	idenshi-kyūgen
general account; outline; review *[pr]*	概説; がいせつ	gaisetsu
" *[pr]*	総説; 綜説	sōsetsu
general catalog *[pr]*	総目録; 總目錄	sō-mokuroku
general condition *{n}*	一般状態	ippan-jōtai
general drawing; general view *[graph]*	一般図; 一般圖	ippan-zu

general information manual *[pr]*	概説書	gaisetsu'sho
general merchandise *[econ]*	雑貨用品	zakka-yōhin
general purpose *[ind-eng]*	一般用	ippan-yō
general purpose fuse *[elec]*	広域ヒューズ	kō'iki-hyūzu
general rules; basic principles	大綱; たいこう	taikō
general solution *[math]*	一般解	ippan-kai
general state *{n}*	一般状態	ippan-jōtai
general statement; review *[pr]*	概説; がいせつ	gaisetsu
general trend *{n}*	大勢	taisei
general view *[graph]*	一般図	ippan-zu
generate *{vb} [comput]*	生成する	seisei suru
" *{vb} [electr]*	発生する	hassei suru
generating; generation *{n}*	創成	sōsei
generating line *[geom]*	母線; ぼせん	bosen
generating point *[math]*	母点; 母點; ぼてん	boten
generating station *[mech-eng]*	発電所	hatsuden'sho
generation *[bio] [comput]*	世代; せだい	sedai
" *[comput]*	生成	seisei
generative nucleus *[bot]*	雄原核	yūgen-kaku
generator *[elec]*	発電機	hatsuden-ki
" *[electr]*	発生器	hassei-ki
generic name *[bio] [sci-t]*	属名	zoku-mei
" (for medicine) *[pharm]*	一般名	ippan-mei
" *[sci-t]*	総称(名)	sōshō(-mei)
genetic code; codon *[mol-bio]*	遺伝暗号; 遺傳暗號	iden-angō
genetic drift *[gen]*	遺伝的浮動	iden'teki-fudō
genetic map *[gen]*	遺伝子地図	idenshi-chizu
genetic recombination *[gen]*	遺伝的組換え	iden'teki-kumi'kae
genetics *[bio]*	遺伝学; いでんがく	iden'gaku
genetic stock *[gen]*	遺伝子材料	idenshi-zairyō
genitive case *[gram]*	属格	zokkaku
genome *[gen]*	ゲノム	genomu
" *[gen]*	遺伝子群	idenshi-gun
genotype *[gen]*	遺伝子型	idenshi-gata
" *[syst]*	模式種; もしきしゅ	moshiki'shu
genotypic frequency *[gen]*	遺伝子型頻度	idenshi'gata-hindo
gentian *[bot]*	竜胆; 龍膽	rindō
genus (pl genera) *[bio] [syst]*	属; 屬; ぞく	zoku
" *[syst]*	類; 類; るい	rui
geocentric coordinates *[astron]*	地心座標	chishin-zahyō
geocentric zenith *[astron]*	地心天頂	chishin-tenchō
geochemistry *[geol]*	地球化学	chikyū-kagaku

geochronology *[geol]*	地質年代学	chi'shitsu-nendai'gaku
geode *[geol]*	晶洞石	shōdō'seki
geodesic line *[geod]*	測地線	sokuchi-sen
geodesist (a person) *[geophys]*	測地学者	sokuchi-gakusha
geodesy *[geophys]*	測地学；そくちがく	sokuchi'gaku
geodetic triangle *[geod]*	測地三角形	sokuchi-sankak(u)kei
geographical features *[geogr]*	地勢	chisei
geography *[sci-t]*	地理学；ちりがく	chiri'gaku
geological age *[geol]*	地質時代	chi'shitsu-jidai
geological map *[graph]*	地質図；地質圖	chi'shitsu-zu
geological time scale *[mech]*	地質的年代目盛り	chishitsu'teki-nendai-memori
geology *[sci-t]*	地質学；ちしつがく	chi'shitsu'gaku
geomagnetic field *[geophys]*	地球磁場	chikyū-jiba
geomagnetic storm *[geophys]*	地磁気嵐	chi'jiki-arashi
geomagnetism *[geophys]*	地磁気；ちじき	chi'jiki
geometrical isomerism *[p-ch]*	幾何異性	kika-i'sei
geometrical optics *[opt]*	幾何光学	kika-kōgaku
geometric mean *[alg]*	幾何平均	kika-heikin
" *[alg]*	相乗平均；相乗平均	sōjō-heikin
" *[alg]*	等比中項	tōhi-chūkō
geometric mean rule *[alg]*	幾何平均則	kika-heikin-soku
geometric progression *[alg]*	等比数列	tōhi-sū'retsu
geometric series *[alg]*	等比級数	tōhi-kyūsū
geometry *[math]*	幾何学；きかがく	kika'gaku
geomorphology *[geol]*	地形学	chikei'gaku
geophone *[electr]*	地中聴音機	chichū-chō'onki
" *[electr]*	受振機	jushin-ki
geophysical prospecting *{n}* *[eng]*	物理探鉱	butsuri-tankō
geophysics *[geol]*	地球物理学	chikyū-butsuri'gaku
geostationary orbit *[aerosp]*	地球静止軌道	chikyū-seishi-kidō
geostrophic wind *[meteor]*	地衡風	chikō-fū
geosyncline *[geol]*	地向斜；ちこうしゃ	chi'kōsha
geothermal *{adj}* *[geophys]*	地熱の	chi'netsu no
geothermal power plant *[mech-eng]*	地熱発電所	chi'netsu-hatsuden'sho
geothermometer *[eng]*	地質温度計	chi'shitsu-ondo-kei
geotropism *[bot]*	重力屈性	jūryoku-kussei
geranium (flower) *[bot]*	風露草；ふうろそう	fūro'sō
Geranium thunbergii (flower) *[bot]*	現の証拠	gen'no'shōko
germ *[bot]*	胚芽；はいが	haiga
German cockroach (insect) *[i-zoo]*	茶翅蜚蠊	cha'bane-gokiburi
germanium (element)	ゲルマニウム	gerumanyūmu
German measles; rubella *[med]*	風疹；ふうしん	fūshin

German silver *[mat]*	洋銀	yōgin
germination; sprouting *{n} [bot]*	発芽	hatsu'ga
germ layer *[bio]*	胚葉	haiyō
germs; bacteria *[microbio]*	細菌; さいきん	saikin
germ warfare *[mil]*	細菌戦(争)	saikin-sen(sō)
germ wheat *[bot]*	麦芽; 麥芽; ばくが	baku'ga
geyser *[hyd]*	間欠(噴)泉	kanketsu-(fun)sen
ghost image *[electr]*	多重像	tajū-zō
ghost line *[met]*	幽霊線	yūrei-sen
giant (elephant) tortoise *[v-zoo]*	象亀; ぞうがめ	zō-game
giant Japanese spider crab *[i-zoo]*	高脚蟹; たかあし蟹	taka'ashi-gani
giant star *[astron]*	巨星	kyosei
gill *[v-zoo]*	鰓; えら	era
gill cover *[v-zoo]*	鰓蓋	era-buta; saigai
gimlet *[eng]*	錐; きり	kiri
ginger (herb) *[bot] [cook]*	生薑; しょうが	shōga
ginger, Japanese *[bot] [cook]*	茗荷	myōga
ginger, red pickled *[cook]*	紅生姜	beni-shōga
gingiva (pl -vae); gum *[anat]*	歯茎; 齒莖; はぐき	ha'guki
ginkgo (tree, nut) *[bot]*	銀杏; 公孫樹	ichō
ginkgo nut *[bot] [cook]*	銀杏; ぎんなん	ginnan
ginseng (root) (herb) *[bot]*	人参	ninjin
giraffe (animal) *[v-zoo]*	麒麟; きりん	kirin
girdle *[astron]*	獣帯; 獸帶	jūtai
" *[astron]*	黄道	kōdō
girth *{n} [mech]*	回り寸法	mawari-sunpō
give to the public *{vb} [legal]*	公共する	kōkyō suru
gizzard *[v-zoo]*	砂嚢	sa'nō
gizzard shad (fish) *[v-zoo]*	鮗; 鰶; このしろ	konoshiro
" (medium) *[v-zoo]*	小鰭; こはだ	ko'hada
glacial acetic acid	氷酢酸	hyōsaku-san
glacial epoch; glacial period *[geol]*	氷河期	hyōga-ki
glacial tongue *[hyd]*	氷舌; ひょうぜつ	hyō'zetsu
glaciation *[geol]*	氷河作用	hyōga-sayō
glacier *[hyd]*	氷河	hyōga
glaciogeology *[geol]*	氷河地質学	hyōga-chi'shitsu'gaku
glaciologist (a person) *[geol]*	氷河学者	hyōga-gakusha
glaciology *[geol]*	氷河学	hyōga'gaku
gland *[anat]*	腺	sen
glare *{n} [opt] [physio]*	ギラツキ	gira-tsuki
" *{n} [opt] [physio]*	眩しさ	mabushi-sa
glare resistance *[opt]*	防眩性	bōgen-sei

glass *[mat]*	硝子；ガラス	garasu
glasses *[opt]*	眼鏡；めがね	me'gane
glaucoma *[med]*	緑内障	ryoku'nai'shō
glaze *{n} [cer]*	釉薬	uwa-gusuri; yū'yaku
" *{n} [hyd]*	雨氷	u'hyō
glazing *{n} [cer]*	施釉；せゆう	seyū
glazing; lustering *{n} [tex]*	艶出し	tsuya'dashi
glide *{n} [crys]*	滑り	suberi
glide reflection *[crys]*	映進	eishin
Gliopeltis furcata *[bot]*	布海苔；ふのり	fu'nori
globe; terrestrial globe *[map]*	地球儀	chikyū-gi
globe valve *[mech-eng]*	玉形弁；玉形瓣	tama'gata-ben
Globicephala whale *[v-zoo]*	巨頭鯨；ごんどう鯨	gondō-kujira
globular; spherical *{adj} [geom]*	球形の	kyūkei no
globular star cluster *[astron]*	球状星団；球狀星團	kyūjō-seidan
globule (welding) *{n} [met]*	溶滴	yōteki
glossary (pl -ries) *[pr]*	語彙；ごい	go'i
" *[pr]*	小辞典；小辭典	shō-jiten
" *[pr]*	用語辞典	yōgo-jiten
" *[pr]*	用語集	yōgo'shū
glossed silk *[tex]*	練り絹；ねりぎぬ	neri-ginu
glossiness *[opt]*	光沢度；光澤度	kōtaku-do
" *[opt]*	光沢性	kōtaku-sei
gloss meter; glossimeter *[eng]*	光沢(度)計	kōtaku(-do)-kei
gloss retention *[ind-eng]*	光沢保持率	kōtaku-hoji-ritsu
glost firing *{n} [cer]*	本焼；本燒	hon-yaki
glucide *[bioch]*	糖質	tō'shitsu
glucose; grape sugar; dextrose *[bioch]*	葡萄糖；ぶどうとう	budō'tō
glucoside *[bioch]*	配糖体	haitō'tai
glue *{n} [mat]*	膠；にかわ	nikawa
" *{n} [mat]*	接着剤	setchaku-zai
glue plant *[bot]*	布海苔；ふのり	fu'nori
gluing-on *[adhes]*	貼着	tenchaku
" *[mech] [poly-ch]*	接着	setchaku
glutinous rice (cereal) *[cook]*	糯米；もちごめ	mochi-gome
glutinous-rice cake *[cook]*	餅；もち	mochi
glutinous rice with red beans *[cook]*	赤飯	sekihan
" *[cook]*	強飯；こわめし	kowa-meshi
glycogen *[bioch]*	グリコーゲン	gurikōgen
" *[bioch]*	糖原	tōgen
glycolipid *[bioch]*	糖脂質；とうししつ	tō'shi'shitsu
glycolysis *[bioch]*	解糖	kaitō

English	Japanese	Romanization
glycoprotein *[bioch]*	糖蛋白質	tō-tanpaku'shitsu
glycoside *[bioch]*	配糖体	haitō-tai
glyph *[graph] [pr]*	絵文字; 繪文字	e-moji
gnat; midge; sandfly (insect) *[i-zoo]*	蚋; ぶよ; ぶゆ	buyo; buyu
gneiss *[petr]*	片麻岩	henma'gan
gnomon *[eng]*	晷針; きしん	kishin
go *{vb}*	行く; いく; ゆく	iku; yuku
go out of business *{vb} [econ]*	廃業する	haigyō suru
go upstream; trace back; retroact	溯る; さかのぼる	saka'noboru
goal *{n}*	目標	moku'hyō
" (prospect) *{n}*	目処; めど	medo
goat (animal) *[v-zoo]*	山羊; やぎ	yagi
goat hair *[tex]*	山羊毛	yagi-mō
goat's milk *[cook]*	山羊乳	yagi-nyū
goby (pl gobies, goby) (fish) *[v-zoo]*	沙魚; はぜ	haze
goethite *[miner]*	針鉄鉱; 針鐵鑛	hari'tekkō
goiter *[med]*	甲状腺腫	kōjōsen'shu
gold (element)	金; きん	kin
" *[met]*	黄金; おうごん	ō'gon
" (color) *[opt]*	金色	kin-iro; kon'jiki
gold chloride	塩化金	enka-kin
goldcrest (bird) *[v-zoo]*	菊戴; きくいただき	kiku'itadaki
gold dust *[met]*	砂金	sa-kin
golden (color) *[opt]*	金色	konjiki; kin-iro
golden cut *[math]*	黄金分割	ōgon-bunkatsu
golden ratio *[math]*	黄金比	ōgon-hi
goldfinch (bird) *[v-zoo]*	五色鶸; ごしきひわ	go'shiki-hiwa
goldfish (fish) *[v-zoo]*	金魚	kingyo
gold flux *[cer]*	金付溶融剤	kin'tsuke-yōyū'zai
gold hydroxide	水酸化金	suisan'ka-kin
gold iodide	沃化金	yōka-kin
gold leaf; gold foil *[mat]*	金箔	kinpaku
gold mining *{n} [min-eng]*	採金	saikin
gold number *[an-ch]*	金数	kin-sū
gold oxide	酸化金	sanka-kin
gold paint *[mat]*	金泥	kindei
gold point *[thermo]*	金点; 金點	kin-ten
gold standard *[econ]*	金本位制度	kin-hon'i-seido
gold sulfide	硫化金	ryūka-kin
gold thread *[mat]*	金糸	kinshi
goldthread *[bot]*	黄連; 黃蓮	ō'ren
gonad *[bio]*	生殖腺	seishoku-sen

gong *[music]*	銅羅；どら	dora
" *[music]*	鉦；かね	kane
goniometer *[eng]*	測角器	sok(u)kaku-ki
good (a grade) *{n} [ind-eng]*	良；りょう	ryō
" *{adj}*	良い；いい；よい	ī; ii; yoi
good conductor *[elec]*	良導体	ryō'dōtai
good harvest *[agr]*	豊作	hōsaku
goodness; merit *{n} [ind-eng]*	良さ	yosa
good-quality *{adj}*	良質の	ryō-shitsu no
goods; object; article *{n}*	物品	buppin
good solvent *[ch]*	良溶媒	ryō'yōbai
good will *{n}*	好意	kō'i
goose (pl geese) (bird) *[v-zoo]*	鵝鳥；がちょう	gachō
gooseflesh *[physio]*	鳥肌	tori'hada
goosefoot; pigweed (edible) *[bot]*	藜；あかざ	akaza
gopher (animal) *[v-zoo]*	掘り鼠	hori-nezumi
gopher tortoise *[v-zoo]*	穴掘り亀	ana'hori-game
gorge *{n} [geogr]*	峡谷；峽谷	kyōkoku
gossamer (of spider) *{n} [i-zoo]*	遊糸	yūshi
gouge; prize; wrench; pry *{vb} [mech]*	抉る；くじる	kujiru
gouge *{vb t}*	抉る；刳る；えぐる	eguru
" *{vb t}*	抉る；こじる	kojiru
gourd *[bot]*	瓢箪；ひょうたん	hyōtan
gout *[med]*	痛風	tsūfū
grab bar *[eng]*	摑み棒	tsukami-bō
gradation *[graph]*	階調	kaichō
grade *{n} [ind-eng]*	品位	hin'i
" *{n} [math]*	等級	tōkyū
grade crossing *{n} [civ-eng]*	踏切	fumi'kiri
" *{n} [traffic]*	平面交差	heimen-kōsa
gradient *[calc]*	傾き	katamuki
" *[geol] [math]*	勾配；こうばい	kōbai
" *[photo]*	階調度	kaichō-do
gradient wind *[meteor]*	傾度風	keido-fū
grading *{n} [ind-eng]*	格付	kaku'tsuke; kaku'zuke
" *{n} [ind-eng]*	採点	saiten
gradualism *[bio]*	漸進主義	zenshin-shugi
graduated *{adj} [ch]*	目盛付きの	memori'tsuki no
graduation; scale *{n} [phys]*	目盛(り)	me'mori
graduation line *[an-ch]*	目盛り線	memori-sen
grafting *{n} [bio] [bot]*	移植	i'shoku
" (of branch) *{n} [bot]*	接木；つぎき	tsugi-ki

grain; cereal *{n} [bot]*	穀物；こくもつ	koku'motsu
" (of wood or cloth) *{n} [mat]*	目	me
" (of leather) *{n} [mat]*	皺；しぼ	shibo
" ; particle *{n} [phys] [mat]*	粒子	ryūshi
grain boundary *[crys]*	粒界；りゅうかい	ryūkai
grain diameter *[crys] [geol]*	粒径；粒徑	ryūkei
grain effect *[rub]*	列理	retsu'ri
graininess *[mat]*	多粒性	taryū-sei
" *[photo]*	粒状性	ryūjō-sei
grain interface *[crys]*	粒界	ryūkai
grain particle *[bot]*	穀粒；こくりゅう	koku'ryū
grain side (of skin) *[v-zoo]*	銀面	ginmen
grain size *[mech]*	粒度	ryūdo
grain-size number *[steel]*	結晶粒度	kesshō-ryūdo
grain texture *[crys]*	結晶集合組織	kesshō-shūgō-soshiki
gram (unit of mass) *[mech]*	瓦；グラム	guramu
gram-formula weight *[ch]*	グラム式量	guramu-shiki'ryō
grammage *[mech]*	目付；めつけ	me'tsuke
" *[mech]*	坪量；つぼりょう	tsubo'ryō
grammar *[comm] [comput] [ling]*	文法	bunpō
granite *[petr]*	花崗岩；かこうがん	kakō'gan
grant; granted; granting *[pat]*	付与；付與；ふよ	fuyo
granted *[pat]*	許可	kyoka
granularity *[photo]*	粒状度	ryūjō-do
granular powder *[mat]*	粒状粉	ryūjō-fun
granular texture *[petr]*	粒状組織	ryūjō-so'shiki
granulation *[astron]*	粒状組織	ryūjō-so'shiki
" *[sci-t]*	整粒	seiryū
" *[sci-t]*	造粒；ぞうりゅう	zōryū
granulation agent *[mat]*	造粒剤	zōryū-zai
granulator *[cook] [eng]*	造粒機	zōryū-ki
" *[eng]*	整粒機	seiryū-ki
granule *[bio]*	顆粒；果粒	karyū
" *[pharm]*	顆粒剤	karyū-zai
grape (fruit) *[bot]*	葡萄；ぶどう	budō
graph *[an-geom]*	図表；ずひょう	zuhyō
" *[an-geom]*	グラフ	gurafu
graphic *{adj} [comput]*	図形の；圖形の	zukei no
graphical *{adj} [math]*	図形の	zukei no
graphical calculation *[math]*	図計算；ずけいさん	zu-keisan
graphical determination method *[graph]*	作図法	sakuzu-hō
graphical differentiation *[calc]*	図式微分	zu'shiki-bibun

graphical integration *[calc]* 図式積分 zu'shiki-sekibun
graphical mechanics *[mech]* 図式力学；圖式力學 zu'shiki-rikigaku
graphical representation *[math]* 図式表示 zu'shiki-hyōji
graphical solution method *[math]* 図式解法 zu'shiki-kaihō
graphical statics *[mech]* 図解力学 zukai-rikigaku
graphic character *[comput]* 図形文字 zukei-moji
graphic display *[comput]* 図形表示 zukei-hyōji
graphic display unit *[comput]* 図形表示装置 zukei-hyōji-sōchi
graphic form *[graph]* 図形；ずけい zukei
graphic language *[comput]* 図形(処理)言語 zukei(-shori)-gengo
graphic method *[graph]* 図示法 zushi-hō
graphic output *[comput]* 図形出力 zukei-shutsu'ryoku
graphic panel *[cont-sys]* 図示パネル zushi-paneru
graphic science; graphics *[sci-t]* 図学 zu'gaku
graphics processing *[comput]* 図形処理 zukei-shori
graphic symbol *[comput]* 図(形)記号 zu(kei)-kigō
graphite *[miner]* 黒鉛；黑鉛 koku'en
" *[miner]* 石墨；石墨 seki'boku
graphitization *[met]* 黒鉛化 koku'en'ka
graphology *[math]* 図式法 zu'shiki-hō
graph paper *[graph]* 方眼紙 hōgan-shi
grapnel *[nav-arch]* 四つ爪錨 yotsu'zume-ikari
grasp *{vb t}* 握る nigiru
grass *[bot]* 草 kusa
grass cricket (insect) *[i-zoo]* 草雲雀；くさひばり kusa-hibari
grasshopper (long-horned) (insect) *[i-zoo]* 螽蟖；きりぎりす kirigirisu
grasshopper (short-horned) *[i-zoo]* 飛蝗；蝗；ばった batta
grassy plain *[geogr]* 草原 sōgen
grassy smell *[physio]* 青臭さ ao'kusa-sa
grate *{vb}* *[cook]* 下ろす orosu
grated daikon (q.v.) *[cook]* 下し；おろし oroshi
grated horseradish *[cook]* 山葵下ろし wasabi-oroshi
grated Japanese yam *[cook]* 薯蕷；とろろ tororo
graticule *[map]* *[opt]* 方眼区画図面 hōgan-kukaku-zumen
grating; grid *{n}* *[spect]* 格子；こうし kōshi
grating spectrometer *[spect]* 格子分光計 kōshi-bunkō-kei
grave accent *[ling]* 抑音符 yoku'on-pu
gravel *[geol]* 砂利；じゃり jari
gravimeter *[eng]* 重力計 jūryoku-kei
gravimetric analysis *[an-ch]* 重量分析 jūryō-bunseki
gravimetric titration *[an-ch]* 重量滴定 jūryō-teki'tei

gravimetry *[eng]*	重量測定	jūryō-sokutei
gravitation *[phys]*	重力；じゅうりょく	jū'ryoku
gravitational acceleration *[phys]*	重力加速度	jūryoku-kasoku-do
gravitational constant *[mech]*	重力定数	jūryoku-teisū
gravitational field *[mech]*	重力場	jūryoku'ba
gravitational mass *[phys]*	重力質量	jūryoku-shitsu'ryō
gravitational method *[phys]*	重力法	jūryoku-hō
gravitational shift *[astron]*	重力偏移	jūryoku-hen'i
gravitational wave *[rela]*	重力波	jūryoku-ha
graviton *[phys]*	重力量子	jūryoku-ryōshi
gravity *[mech] [phys]*	引力；いんりょく	in-ryoku
" *[mech] [phys]*	重力	jū-ryoku
gravity wave *[fl-mech]*	重力波	jūryoku-ha
gray (color) *[opt]*	灰色；はいいろ	hai-iro
graybody *[thermo]*	灰色体	kaishoku'tai
gray mullet (fish) *[v-zoo]*	鯔；ぼら	bora
gray starling (bird) *[v-zoo]*	椋鳥；むくどり	muku-dori
gray whale *[v-zoo]*	克鯨；こくくじら	koku-kujira
gray yarn *[tex]*	なま糸	nama-ito
grazing *{n} [v-zoo]*	剝ぎ取り	hagi'tori
" *{adj} [v-zoo]*	草を食う	kusa o kū
grazing occultation *[astron]*	接線食	sessen-shoku
great blue heron (bird) *[v-zoo]*	大青鷺	ō-ao'sagi
great circle *[geod]*	大円；大圓	dai'en
" *[geod]*	大圏；大圈	taiken
great circle route *[navig]*	大圏航路	taiken-kōro
great-circle sailing *{n} [navig]*	大圏航法	taiken-kōhō
Great Dipper *[astron]*	北斗星；ほくとせい	hoku'to-sei
greater than *[math]*	より大きい	yori ōkī
greatest; largest *[math]*	最大	saidai
greatest value *[math]*	最大値	saidai-chi
great excess *{n}*	大過剰；大過剩	dai-kajō
great spotted woodpecker *[v-zoo]*	赤啄木鳥；あかげら	aka'gera
great tit (Japanese bird) *[v-zoo]*	四十雀	shijū'kara
great trumpet flower; Campsis grandiflora *[bot]*	凌霄花；のうぜん=かずら	nōzen'kazura
grebe (bird) *[v-zoo]*	鸊鷉；かいつぶり	kai'tsuburi
green (color) *[opt]*	草色	kusa-iro
" (color) *[opt]*	緑；綠；みどり	midori
green algae *[bot]*	緑藻	ryoku-sō
green belt *[civ-eng]*	緑地帯；綠地帶	ryoku'chi-tai
green compact *[mat]*	圧粉体；壓粉體	appun-tai

green flash *[astron]*	緑色閃光	ryoku'shoku-senkō
greenhouse *[bot]*	温室；おんしつ	on-shitsu
greenhouse effect *[ecol] [meteor]*	温室効果	on'shitsu-kōka
green liquor *[paper]*	緑液	ryoku-eki
green lumber *[bot]*	生材木	nama-zaimoku
green mold *[mycol]*	青黴；あおかび	ao-kabi
green mustard; Japanese horseradish (a condiment) *[bot]*	山葵；わさび	wasabi
green onion (vegetable) *[bot]*	葱；ねぎ	negi
green peas (vegetable) *[bot]*	青豆	ao-mame
green plants *[bot]*	緑色植物	ryoku'shoku-shokubutsu
green rouge *[poly-ch]*	青棒	ao-bō
greens (vegetables) *{n} [bot]*	菜；な	na
green soy bean; young soy bean *[bot]*	枝豆	eda-mame
green tea *[cook]*	緑茶；りょくちゃ	ryoku-cha
green tea, middle grade *[cook]*	煎茶；せんちゃ	sencha
green tea, superior grade *[cook]*	玉露；ぎょくろ	gyoku'ro
green turtle *[v-zoo]*	青海亀；青海鼈	ao-umi'game
grid *[pr]*	母型盤	bokei-ban
grid detection *[electr]*	格子検波	kōshi-kenpa
griddle-roasted food *[cook]*	鉄板焼；鐵板燒	teppan-yaki
grilled chicken *[cook]*	焼き鳥	yaki-tori
grilled fish; broiled fish *[cook]*	焼き魚	yaki-zakana
grind (saw) *{vb t} [mech-eng]*	挽く；ひく	hiku
" (mash) *{vb t} [mech-eng]*	擂る；する	suru
" *{vb t} [mech-eng]*	研削する	kensaku suru
grindability *[mat]*	粉砕性	funsai-sei
" *[mat]*	研削性	kensaku-sei
grinding *{n} [mech-eng]*	粉砕；ふんさい	funsai
" *{n} [mech-eng]*	研削加工	kensaku-kakō
" *{n} [mech-eng]*	摩砕；まさい	masai
" *{n} [mech-eng]*	練肉；練肉	renniku
grinding burn *[mech-eng]*	研削焼	kensaku-yake
grinding cracks *[mat]*	研削割れ	kensaku-ware
grinding machine *[eng]*	研削盤	kensaku-ban
grindstone *[eng]*	砥石；といし	to'ishi
grip; gripper; pull *{n} [arch]*	把手；とって	totte
grip; gripper *{n} [eng]*	摑み(具)	tsukami('gu)
gripping *{n} [physio]*	摑み	tsukami
grip strength *[physio]*	握力；あくりょく	aku-ryoku
grit *{n} [mat]*	粗粒	soryū
grizzly bear (animal) *[v-zoo]*	灰色熊	hai'iro-guma

English	Japanese	Romanization
grocery store *[arch]*	八百屋；やおや	yao'ya
groin *{n} [anat]*	鼠径部	sokei'bu
groin; jetty (pl jetties) *[civ-eng]*	突堤	tottei
grommet *[eng]*	鳩目	hato'me
gromwell *[bot]*	紫	murasaki
groove *{n} [eng]*	溝；みぞ	mizo
gross section *[graph]*	総断面；總斷面	sō-danmen
grossular *[miner]*	緑柘榴石	midori'zakuro-ishi
grossularite *[miner]*	灰礬柘榴石	kaiban'zakuro-ishi
gross weight *[mech]*	総量	sōryō
ground *{n} [elec]*	アース	āsu
" *{n} [geol] [tex]*	地；じ	ji
" *{n} [mat]*	生地；きじ	kiji
" *{n}*	立脚点	rikkyaku-ten
ground base *[civ-eng]*	地盤	jiban
ground beetle (insect) *[v-zoo]*	歩行虫；おさむし	osa-mushi
ground-cherry *[bot]*	酸漿；鬼灯	hōzuki
ground crew *[aero-eng]*	整備員	seibi-in
ground fault *[elec]*	地絡	chi'raku
ground fault interrupter *[elec]*	漏電遮断器	rōden-shadan-ki
ground fog *[meteor]*	地霧；じぎり	ji-giri
grounding *{n} [elec]*	接地	setchi
" (of ship) *{n} [navig]*	座礁	zashō
groundmass *[petr]*	石基	sekki
ground metal *[met]*	地金；じがね	ji-gane
ground potential *[elec]*	大地電位	daichi-den'i
ground resistance *[elec]*	大地抵抗	daichi-teikō
grounds; reason *{n}*	原因	gen'in
groundsill *[arch] [civ-eng]*	土台；土臺；どだい	dodai
ground speed *[aero-eng]*	対地速度	taichi-sokudo
ground state *[quant-mech]*	基底状態	kitei-jōtai
ground surface *[geol]*	地肌面	ji'hada-men
" *[mech-eng]*	研削面	kensaku-men
ground transportation *[trans]*	地上輸送	chijō-yusō
groundwater *[hyd]*	地下水	chika'sui
ground wave *[comm]*	地上波	chijō-ha
groundwood *[paper]*	砕木；さいぼく	sai'boku
groundwood paper *[mat]*	更紙；ざらがみ	zara-gami
groundwork *{n}*	素地	soji
group (viruses) *{n} [bio] [syst]*	科	ka
" *{n} [bio] [gen]*	集団；集團	shūdan
" *{n} [ch]*	基	ki

group *{n} [ch]*	族；ぞく	zoku
" *{n} [syst] [zoo]*	群；ぐん	gun
" ; department; division; section *{n}*	部；ぶ	bu
" ; association; organization *{n}*	団体；團體	dantai
grouping *{n} [comm]*	群分け	gun'wake
group velocity *[phys]*	群速度	gun-sokudo
grow; develop *{vb i} [physio]*	発育する	hatsu'iku suru
growing; cultivation *{n} [agr]*	栽培；さいばい	saibai
grown junction *[electr]*	成長接合	seichō-setsu'gō
growth; development *[bio] [physio]*	生長；成長	seichō
" *[bio] [physio]*	生育；せいいく	sei'iku
growth factor *[physio]*	生長因子；成長因子	seichō-inshi
growth medium (pl -s, -a) *[bioch]*	増殖培地	zōshoku-baichi
growth reaction *[org-ch]*	生長反応	seichō-hannō
growth retardant *[agr]*	矮化剤；わいかざい	waika-zai
" *[bot]*	生長抑制剤	seichō-yokusei-zai
growth stimulator *[bio]*	増殖因子	zōshoku-inshi
grub; larva (pl -ae) *[i-zoo]*	幼虫；ようちゅう	yōchū
guaiac wood; Guajacum officinale *[bot]*	癒瘡木；ゆそうぼく	yusō'boku
gualou (fruit, seed) *[bot]*	括楼；括樓；栝樓	katsurō
guano *[mat]*	糞化石；ふんかせき	funkaseki
guarantee *{n} [econ]*	担保	tanpo
" *{n}*	保証；ほしょう	hoshō
guarantor (a person) *[legal]*	保証人	hoshō'nin
guard; hilt (of a sword) *[ord]*	鍔；つば	tsuba
guard (a person) *{n}*	番(人)	ban('nin)
" *{vb}*	守る；護る	mamoru
guidance *{n}*	誘導；ゆうどう	yūdō
guide; conduct *{vb t}*	案内する	annai suru
guide; conduct *{vb t}*	導く；みちびく	michibiku
guide in *{vb}*	導入する	dōnyū suru
guide bar *[mech-eng]*	案内棒	annai-bō
guide bearing *{n} [mech-eng]*	案内軸受	annai-jiku'uke
guided missile *[ord]*	誘導弾	yūdō-dan
guided wave *[elecmg]*	導波；どうは	dōha
guided-weapon control *[ord]*	誘導武器管制	yūdō-buki-kansei
guide plate *[mech-eng]*	案内板	annai-ban
guide pulley *[eng]*	案内車	annai-guruma
guideway *[civ-eng]*	案内面	annai-men
guiding center *[elecmg]*	案内中心	annai-chūshin
guiding-in *{n}*	導入	dōnyū
guiding telescope *[astron]*	案内望遠鏡	annai-bōenkyō

guinea fowl (bird) *[v-zoo]*	ほろほろ鳥	horo'horo-chō
guinea pig (animal) *[v-zoo]*	天竺鼠;てんじく鼠	tenjiku-nezumi
gulch *[geogr]*	峡谷; 峽谷	kyōkoku
gulf *[geogr]*	湾; 灣; わん	wan
Gulf Stream *[ocean]*	湾流	wan-ryū
gull (bird) *[v-zoo]*	鷗; かもめ	kamome
gullet (esophagus) *[anat]*	食道	shokudō
" (of a gear) *[mech]*	歯溝; 齒溝; はみぞ	ha-mizo
gum *[anat]*	歯茎; はぐき	ha'guki
" *[anat]*	歯肉	shi'niku
gun; firearm *[ord]*	鉄砲; 鐵砲	teppō
gunboat *[nav-arch]*	砲艦	hōkan
gunmetal *[met]*	砲金	hōkin
gunnel; Enedrias nebulosus (fish)	銀宝; ぎんぽ	ginpo
gunpowder; black powder *[ord]*	黒色火薬; 黒色火藥	koku'shoku-ka'yaku
guppy (pl guppies) (fish) *[v-zoo]*	虹目高	niji-medaka
gurnard (fish) *[v-zoo]*	魴鮄; ほうぼう	hōbō
gushing *{n}*	噴出	fun'shutsu
gusset; gore *{n}* *[cl]*	襠; まち	machi
gust (of wind) *{n}* *[meteor]*	突風	toppū
gustation *[physio]*	味覚	mikaku
guttation *[bot]*	排水	haisui
gutter *[arch]*	樋; とい	toi
" *[pr]*	喉開き; のどあき	nodo-aki
guy; guy wire *[eng]*	支え線	sasae-sen
guyot *[geol]*	平頂海山	heichō-kaizan
" *[geol]*	卓状海山	takujō-kaizan
gymnasium (pl -s, -nasia) *[arch]*	運動室	undō-shitsu
Gymnospermae *[bot]*	裸子植物	rashi-shokubutsu
gynecologist *[med]*	婦人科医	fujinka'i
gynecology *[med]*	婦人科	fujin'ka
gypsum *[miner]*	石膏	sekkō
gypsy moth (insect) *[i-zoo]*	舞舞蛾; まいまいが	maimai-ga
gyre *[ocean]*	環流	kanryū
gyrocompass *[navig]*	転輪羅針儀	tenrin-rashin-gi
gyromagnetic effect *[elecmg]*	磁気回転効果	jiki-kaiten-kōka
gyrostabilizer *[eng]*	回転安定器	kaiten-antei-ki

H

habit *[psy]*	癖；くせ	kuse
" ; custom *[psy]*	習慣；しゅうかん	shūkan
habitat *[ecol]*	生育地	sei'iku'chi
" *[ecol]*	生息地；棲息地	seisoku'chi
habit plane *[crys]*	晶癖面	shōheki-men
habituation *[bio]*	順化；馴化	junka
" *[bio]*	慣れ	nare
hackberry *[bot]*	榎；えのき	enoki
hacksaw *{n} [eng]*	弓のこ	yumi'noko
hafnium (element)	ハフニウム	hafunyūmu
hail; snow pellets; graupel *{n} [meteor]*	霰；あられ	arare
hail; hailstone *{n} [meteor]*	雹；ひょう	hyō
hail damage *[ecol]*	雹害	hyōgai
hair (of head) *{n} [anat]*	髪(の毛)	kami (no ke)
" (of head) *{n} [anat]*	頭髪；とうはつ	tō'hatsu
" *{n} [anat]*	毛	ke
" *{n} [anat]*	毛髪	mō'hatsu
hairline cracks *[met]*	毛割れ	ke-ware
hair ornament *[lap]*	髪飾り	kami-kazari
hairspring *[horol]*	髭薇；ひげ発条	hige-zenmai
hairtail (fish) *[v-zoo]*	太刀魚；たちうお	tachi-uo
halberd; halbert *[ord]*	薙刀；長刀	nagi'nata
half *{n} [math]*	半分	hanbun
" *{n} [math]*	二分の一	ni-bun no ichi
half-adder *[electr]*	半加算器	han'kasan-ki
halfbeak (fish) *[v-zoo]*	針魚；鱵；さより	sayori
half-cell *[p-ch]*	半電池	han-denchi
half-cell potential *[p-ch]*	半電池電位	han'denchi-den'i
half-duplex communication *[comm]*	半二重通信	han-nijū-tsūshin
half-life *[ch] [nucleo]*	半減期	hangen-ki
half moon *[astron]*	半月；はんげつ	han'getsu
half plane *[geom]*	半平面	han-heimen
half-subtracter *[electr]*	半減算器	han'genzan-ki
half tide *[ocean]*	半潮	hanchō
half-value layer *[phys]*	半価層；半價層	hanka-sō
half-value period *[ch]*	半減期；はんげんき	hangen-ki
half-value thickness *[phys]*	半減層	hangen-sō
half-value width *[math]*	半価幅；半價幅	hanka-haba
half-wavelength *[elecmg]*	半波長	han-hachō

half-wave potential *[an-ch]*	半波電位	hanpa-den'i
half-wave rectification *[electr]*	半波整流	hanpa-seiryū
half-width; half-value width *[math]*	半値巾；半値幅	hanchi-haba
" *[math]*	半価巾；半價幅	hanka-haba
halftone *[graph]*	中間調	chūkan'chō
" *[pr]*	網版；あみはん	ami'han
halfwave rectification *[electr]*	半波整流	hanpa-seiryū
halfwave reduction potential *[p-ch]*	半波還元電位	hanpa-kangen-den'i
halibut; flatfish; flounder *[v-zoo]*	比目魚；ひらめ	hirame
" (fish) *[v-zoo]*	烏鰈；からすがれい	karasu-garei
halite; rock salt *[miner]*	岩塩；がんえん	gan'en
hallucinogen; psychedelic drug *{n} [pharm]*	幻覚剤	genkaku-zai
halo (pl -s, -es) *{n} [astron]*	暈；かさ	kasa
halochromism *[ch]*	造塩発色	zō'en-hasshoku
halocline *[ocean]*	塩分躍層	enbun-yakusō
halogen(s) *[ch]*	ハロゲン	harogen
halophile *{n} [bio]*	好塩菌；好鹽菌	kōen-kin
halophilism *[bio]*	好塩性	kōen-sei
halophyte *[ecol]*	塩性植物	ensei-shokubutsu
halt *{n} [comput]*	停止；ていし	teishi
hamate *{adj} [bio]*	鍵状の	kagi'jō no
hammer; malleus (bone) *[anat]*	槌骨	tsuchi-bone; tsui-kotsu
hammer *{n} [eng]*	金槌；かなづち	kana'zuchi
" (made of iron) *{n} [eng]*	鉄槌；鐵槌	tettsui
" ; mallet *{n} [eng]*	槌；つち	tsuchi
hammerhead shark (fish) *[v-zoo]*	撞木鮫	shu'moku-zame
hammer mill *[eng]*	衝撃式製粉器	shōgeki'shiki-seifun-ki
hand *{n} [anat]*	手；て	te
handaxe *[eng] [paleon]*	握斧；あくふ	akufu
hand bellows *[met]*	手吹子；手ふいご	te-fuigo
hand-blowing (of glass) *{n} [cer]*	手吹き	te-fuki
handbow *[ord]*	手弓；てゆみ	te-yumi
handcuffs *[eng]*	手錠	tejō
hand-engraving *{n} [graph] [tex]*	手彫り；てぼり	te-bori
hand grenade *[ord]*	手榴弾	shuryū-dan
handgun *[ord]*	拳銃	kenjū
handicraft *{n} [eng]*	手芸；手藝	shugei
" *{n} [eng]*	手工；しゅこう	shukō
hand-knitting *{n} [tex]*	手編み	te-ami
handle; handhold *{n} [eng]*	把手；とって	totte
handle; deal with; sell *{vb t}*	捌く；さばく	sabaku
handling; feeling; hand feeling *{n} [tex]*	風合；ふうあい	fūai

hand loom *[tex]*	手機；てばた	te-bata
handmade *{adj} [cook] [food-eng]*	手打ちの	te-uchi no
handmade paper *[paper]*	手抄き紙；手漉き紙	te'suki-gami
hand-painting *{n} [graph] [tex]*	手描き	te-gaki
hand-printing *{n} [tex]*	手捺染	te-nassen; te-nasen
handrail *[arch]*	手摺り	tesuri
" *[eng]*	高欄；こうらん	kōran
hand-reeling (of silk) *{n} [tex]*	座繰り；ざぐり	za-guri
handsaw *[eng]*	片手鋸	kata'te-nokogiri
hand-shaped; handmade *{adj}*	手造りの	te-zukuri no
hand-shaping; hand-forming *{n} [cer]*	手造り	te-zukuri
handsheet *{n} [paper]*	手抄き紙；手漉き紙	te'suki-gami
hand tool *[eng]*	手工具	te-kōgu
hand truck *{n} [eng]*	手押し車	te'oshi-guruma
hand-washing *{n} [tex]*	手洗い	te-arai
" *{n} [tex]*	手洗濯；てせんたく	te-sentaku
hand-writing; handwritten *[comm]*	手書き	te-gaki
hand-written character *[comm]*	手書き文字	te'gaki-moji
hang (suspend) *{vb} [mech-eng]*	掛ける	kakeru
hang *{vb i}*	垂れる	tareru
" ; dangle; lower *{vb t}*	下げる	sageru
hang axially *{vb t} [mech-eng]*	軸架する	jik(u)ka suru
hang, mounted; mount hanging *{vb}*	装架する	sōka suru
hangar *[civ-eng]*	格納庫；かくのうこ	kakunō'ko
hanger *[eng]*	吊り手；釣手	tsuri'te
hanging picture (ornament) *[graph]*	掛け絵；掛け繪	kake-e
hanging scale *[eng]*	吊り秤	tsuri-bakari
hanging scroll (ornament) *[graph]*	掛け軸	kake-jiku
" (ornament) *[graph]*	掛け物	kake-mono
hank *[tex]*	綛；かせ	kase
haploid *[gen]*	半数体；半數體	hansū'tai
haploid number *[gen]*	半数	hansū
hapten *[immun]*	付着体	fuchaku'tai
harbor; port *{n} [geogr]*	港；湊；みなと	minato
harbor district *[civ-eng]*	港湾地区；港灣地區	kōwan-chiku
hard *{adj} [phys]*	固い；硬い；堅い	katai
hard acid *[ch]*	かたい酸	katai san
hard base *[ch]*	かたい塩基	katai enki
hardboard *[mat]*	硬質繊維板	kōshitsu-sen'iban
hard coal *[miner]*	硬質炭	kōshitsu-tan
hardenability *[met]*	焼入れ性	yaki'ire-sei
hardened oil *[mat]*	硬化油	kōka-yu

hardener *[poly-ch]*	硬化剤；こうかざい	kōka-zai
hardening *{n} [met]*	焼入；焼入	yaki'ire
" *{n} [phys]*	硬化	kōka
hardening agent *[mat]*	固化剤	koka-zai
hard human tissue *[hist]*	人体硬質組織	jintai-kō'shitsu-so'shiki
hardness *{n} [eng]*	硬さ；かたさ	kata-sa
hardness meter *[eng]*	硬度計	kōdo-kei
hardness number *[eng]*	硬さ数	katasa-sū
" *[eng]*	硬度数	kōdo-sū
hardpan; caliche *[geol]*	堅硬土層	kenkō'do-sō
hard porcelain *[mat]*	硬質磁器	kōshitsu-jiki
hard rubber *[mat]*	硬質ゴム	kōshitsu-gomu
hardtail (fish) *[v-zoo]*	縞鰺；しまあじ	shima-aji
hardware *[comput]*	硬物；かたもの	kata'mono
" *[eng]*	金具	kana'gu
" *[eng]*	金物；かなもの	kana'mono
hard water *[ch]*	硬水	kōsui
hardwood *[bot]*	広葉樹	kōyō'ju
" *[mat]*	堅木	kata'gi
" *{adj} [mat]*	堅木の	kenboku no
hare hair *[tex]*	野兎毛	yato-mō
harlequin snake; coral snake *[v-zoo]*	珊瑚蛇；さんごへび	sango-hebi
harmful *{adj}*	有害の(な)	yūgai no (na)
harmful- *{prefix}*	有害性-	yūgaisei-
harmful insect *[i-zoo]*	害虫	gaichū
harmonic; harmonics *{n} [acous]*	倍音	bai'on
harmonic oscillator *[electr]*	調和振動子	chōwa-shindō'shi
harmonic progression *[alg]*	調和数列	chōwa-sūretsu
harmonics *[acous]*	高調波	kōchō-ha
" *[math]*	調和関数	chōwa-kansū
harmonic sound *[acous]*	倍音	bai-on
harmonic wave *[phys]*	調波	chō-ha
harmony *[music]*	和声；和聲	wasei
harmotome *[miner]*	重十字沸石	jū'jūji'fusseki
harness *{n} [trans]*	馬具	bagu
harp *{n} [music]*	竪琴；たてこと	tate-koto
harpoon *{n} [eng]*	銛；もり	mori
harvest *[agr]*	収獲；收獲	shūkaku
harvest fish; Psenopsis anomala (fish) *[v-zoo]*	疣鯛；いぼだい	ibo-dai
harvest fish; Pampus argenteus (fish) *[v-zoo]*	真名鰹；まながつお	mana-gatsuo

harvesting *{n} [agr]*	刈り入れ	kari'ire
" *{n} [agr]*	採収	saishū
" *{n} [microbio]*	回収	kaishū
harvest moon *[astron]*	仲秋の満月	chūshū no mangetsu
" *[astron]*	仲秋の明月	chūshū no meigetsu
harvest mouse (animal) *[v-zoo]*	萱鼠; かやねずみ	kaya-nezumi
hat *[cl]*	帽子	bōshi
" (of bamboo or sedge) *[cl]*	笠	kasa
hatch; be hatched *{vb i} [agr]*	孵る; かえる	kaeru
hatchet *[eng]*	手斧	te-ono; chōna
hatchling of fish (or bird) *[zoo]*	孵化したばかりの=魚(鳥)	fuka shita bakari no sakana (tori)
haulage *[trans]*	運搬	unpan
hawk (bird) *{n} [v-zoo]*	鷹; たか	taka
hawk moth (insect) *[i-zoo]*	雀蛾; すずめが	suzume-ga
hawksbill turtle (reptile) *[v-zoo]*	玳瑁; たいまい	taimai
hawk's eye *[miner]*	鷹眼石	taka-no-me-ishi
hay fever *[med]*	枯草熱; こそうねつ	kosō-netsu
hazard *{n} [ind-eng]*	危険; きけん	kiken
hazardous material *[mat]*	危険物	kiken'butsu
haze *{n} [meteor]*	煙霧	enmu
" *{n} [meteor]*	霞; かすみ	kasumi
haze; hazing *{n} [poly-ch]*	曇り	kumori
hazel (tree) *[bot]*	榛; はしばみ	hashibami
hazel grouse (bird) *[v-zoo]*	蝦夷雷鳥	ezo-raichō
hazelnut *[bot]*	榛の実	hashibami no mi
haze value *{n}*	曇り価	kumori-ka
hazing *{n} [meteor]*	霞がかり	kasumi-gakari
hazy *{adj} [meteor]*	霞んだ	kasunda
head *{n} [anat]*	頭; あたま	atama
" *{n} [bot]*	頭花	tōka
" *{n} [fl-mech]*	落差	raku'sa
headache *[med]*	頭痛; ずつう	zu'tsū
headache medicine *{n} [pharm]*	頭痛薬	zu'tsū-yaku
headband *[cl]*	鉢巻; はちまき	hachi'maki
header word *[pr]*	見出し語	mi'dashi-go
headgear *[cl]*	帽子	bōshi
" *[eng]*	被り物	kaburi'mono
heading *{n} [pr]*	表題	hyōdai
heading; header *{n} [comput] [pr]*	見出し	mi'dashi
headlight *[navig] [opt]*	前照灯	zenshō'tō
headnote *[pat]*	頭書; とぅしょ	tōsho

headstock *[mech-eng]*	主軸台	shu'jiku-dai
head-to-head structure *[poly-ch]*	頭頭構造	tōtō-kōzō
head-to-tail structure *[poly-ch]*	頭尾構造	tōbi-kōzō
heal *{vb i} [med]*	治る; なおる	naoru
health *{n} [med]*	健康	kenkō
health center *[arch] [med]*	保健所	hoken'jo
health preservation *[med]*	保健	hoken
healthy *{adj} [bio]*	健康な	kenkō na
healthy and normal *[bio]*	健常	kenjō
heap *{vb}*	盛る	moru
" *{vb}*	堆積する	taiseki suru
hear *{vb} [physio]*	聞く; 聴く; きく	kiku
hearing *{n} [physio]*	聴覚	chō'kaku
" *{n} [physio]*	聴力; 聽力	chō'ryoku
hearing aid *[acous]*	補聴器; 補聽器	hochō-ki
hearsay *{n}*	伝聞; 傳聞	denbun
heart *{n} [anat]*	心臓; しんぞう	shinzō
heartwood *[bot]*	心材	shinzai
heat *{n} [thermo]*	熱; ねつ	netsu
" *{vb t} [thermo]*	加熱する	ka'netsu suru
heat balance *[thermo]*	熱収支; 熱收支	netsu-shūshi
heat capacity *[thermo]*	熱容量	netsu-yōryō
heat conduction coefficient *[thermo]*	熱伝導率	netsu-dendō-ritsu
heat content *[thermo]*	熱含量	netsu-ganryō
heat denaturation *[ch]*	熱変性; 熱變性	netsu-hensei
heat diffusivity *[thermo]*	熱拡散率	netsu-kakusan-ritsu
heat distortion *[phys]*	熱変形	netsu-henkei
heat-emitting body *[thermo]*	発熱体	hatsu'netsu-tai
heat engine *[mech-eng] [poly-ch]*	熱機関	netsu-kikan
heat exchanger *[eng]*	熱交換器	netsu-kōkan-ki
heat fatigue *[mech]*	熱疲労	netsu-hirō
heat fixing *{n} [eng]*	熱定着	netsu-teichaku
heat flow *[thermo]*	熱流	netsu-ryū
heat generation *[thermo]*	発熱	hatsu-netsu
heat history *[phys]*	熱履歴	netsu-ri'reki
heating (in a building) *{n} [eng]*	暖房; だんぼう	danbō
heating *{n} [thermo]*	加熱	ka'netsu
heat-insulating *{n} [civ-eng]*	断熱; 斷熱	dannetsu
heat-insulating material *[mat]*	断熱材	dannetsu-zai
" *[mat]*	保温剤	ho'on-zai
heat lamp *[elec]*	太陽灯	taiyō-tō
heat loss *[phys]*	熱損失	netsu-son'shitsu

heat medium *[phys]*	熱媒；ねつばい	netsu-bai
heat of absorption *[thermo]*	吸収熱；吸收熱	kyūshū-netsu
heat of activation *[p-ch]*	活性化熱	kassei'ka-netsu
heat of combustion *[p-ch]*	燃焼熱	nenshō-netsu
heat of crystallization *[thermo]*	結晶生成熱	kesshō-seisei-netsu
heat of decomposition *[p-ch]*	分解熱	bunkai-netsu
heat of dilution *[p-ch]*	希釈熱；稀釋熱	ki'shaku-netsu
heat of formation *[p-ch]*	生成熱	seisei-netsu
heat of fusion *[thermo]*	融解熱	yūkai-netsu
heat of hydration *[p-ch]*	水和熱；すいわねつ	suiwa-netsu
heat of neutralization *[p-ch]*	中和熱	chūwa-netsu
heat of reaction *[p-ch]*	反応熱	hannō-netsu
heat of solution *[thermo]*	溶解熱	yōkai-netsu
heat of vaporization *[thermo]*	気化熱	kika-netsu
heat plasticization *[eng]*	熱可塑化	netsu-kaso'ka
heatproof *{adj}* *[mat]*	耐熱(性)の	tai'netsu('sei) no
heat rays *[thermo]*	熱線；ねっせん	nessen
heat removal *[thermo]*	除熱	jo'netsu
heat resistance (degree) *[mat]*	耐熱度	tai'netsu-do
heat-resistant food utensil *[cook]*	耐熱食器	tai'netsu-shokki
heat resisting property *[mat]*	耐熱性	tai'netsu-sei
heat-sensitive adhesive agent *[mat]*	感熱接着剤	kannetsu-setchaku-zai
heat-setting property *[tex]*	熱セット性	netsu-setto-sei
heatsink (an instrument) *[elec]*	脱熱器	datsu'netsu-ki
" (a chemical) *[elec]*	脱熱剤	datsu'netsu-zai
heat-transfer coefficient *[thermo]*	熱伝達係数	netsu-den'tatsu-keisū
heat transmission *[thermo]*	伝熱；傳熱	dennetsu
heat treating *{n}* *[eng]*	焼成	shōsei
heat unit *[thermo]*	熱単位	netsu-tan'i
heat value; heat rate *[mech-eng]*	熱価；熱價	netsu-ka
heat wave *[meteor]*	熱波；ねっぱ	neppa
heat yellowing *{n}* *[opt]*	熱黄変	netsu-ō'hen
heavenly body *[astron]*	天体	tentai
heavily-loaded mix *[ch-eng]*	高充填配合	kō'jūten-haigō
heaviness *[mech]*	重さ	omo-sa
heavy *{adj}*	重い	omoi
heavy atom *[ch]*	重原子	jū-genshi
heavy chain *[ch]*	長鎖	chōsa
heavy hydrogen *[ch]*	重水素	jū-suiso
heavy industry *[ind-eng]*	重工業	jū-kōgyō
heavy metal *[met]*	重金属	jū-kinzoku
heavy oil *[mat]*	重油；じゅうゆ	jūyu

heavy-oil engine *[mech-eng]*	重油機関	jūyu-kikan
heavy oil stabilizer *[petrol]*	重油安定剤	jūyu-antei-zai
heavy seas *{n} [ocean]*	風浪；ふうろう	fūrō
heavy water *[inorg-ch]*	重水	jūsui
hectogram (unit of mass) *[mech]*	瓸；ヘクトグラム	hekuto'guramu
hedenbergite *[miner]*	灰鉄輝石	kaitetsu'kiseki
hedgehog (animal) *[v-zoo]*	蝟；はりねずみ	hari-nezumi
heel (of foot) *{n} [anat]*	踵；かかと	kakato
heeling (a shoe) *{n} [eng]*	踵付け	kakato'zuke
height (of body); tallness *[anat]*	丈；たけ	take
height *[geom]*	高さ	taka-sa
height; length; vertical *{n} [math]*	縦；縱；たて	tate
height above sea level *[eng]*	海抜高度	kai'batsu-kōdo
height equivalent to a theoretical plate *[ch-eng]*	一理論段に等価な	ichi-riron'dan ni tōka na
heir *[legal]*	相続人	sōzoku'nin
Heisei (era name) (1989-)	平成；へいせい	Heisei
helical gear *[mech]*	斜歯歯車	hasuba-ha'guruma
helical spring *[eng]*	蔓巻発条；蔓巻ばね	tsuru'maki-bane
helical tooth *[mech]*	斜歯；はすば	hasu-ba
heliocentric coordinates *[astron]*	日心座標	nisshin-zahyō
heliograph *[eng]*	日照計	nisshō-kei
heliographic coordinates *[astron]*	日面座標	nichi'men-zahyō
heliophyte *[ecol]*	好日性植物	kōnichi'sei-shokubutsu
helioscope *[opt]*	太陽鏡	taiyō-kyō
helioseismology *[astron]*	太陽地震学	taiyō-jishin'gaku
heliosphere *[geophys]*	太陽圏	taiyō-ken
heliotrope *[eng]*	日光反射計	nikkō-hansha-kei
" *[miner]*	血玉髄；血玉髓	ketsu'gyokuzui
heliotropism *[bio]*	向光性	kōkō-sei
" *[bio]*	向日性	kō'nichi-sei
helium (element)	ヘリウム	heryūmu
helix (pl helixes, helices) *[geom]*	螺旋；らせん	rasen
helix; spiral *[eng] [math]*	蔓巻線	tsuru'maki-sen
helix angle *[geom]*	蔓巻角	tsuru'maki-kaku
helminth; parasitic worm *[i-zoo]*	蠕虫；ぜんちゅう	zenchū
helminthology *[bio]*	蠕虫学	zenchū'gaku
hematinic *{n} [pharm]*	造血剤	zō'ketsu-zai
hematite *[miner]*	赤鉄鉱；赤鐵鑛	seki'tekkō
Hemichordata *[syst]*	半索動物	hansaku-dōbutsu
hemihedral *{adj} [crys]*	半完面体の	han'kanmen'tai no
hemihedral form *[crys]*	半面像	hanmen-zō

hemihydrate gypsum *[inorg-ch]*	半水石膏	hansui-sekkō
hemimorphic form *[crys]*	異極像	i'kyoku-zō
hemimorphite *[miner]*	異極鉱	i'kyoku'kō
hemiramph (fish) *[v-zoo]*	針魚; 鱵; さより	sayori
hemisphere *[geogr] [geom]*	半球	hankyū
hemlock *[bot]*	毒人参	doku-ninjin
hemoglobin *[bioch]*	ヘモグロビン	hemogurobin
" *[bioch]*	血色素	kesshikiso
hemolysis *[physio]*	溶血	yōketsu
hemorrhage *{n} [med]*	出血	shukketsu
hemorrhoid *{n} [med]*	痔; じ	ji
hemostatic *{n} [pharm]*	止血剤	shi'ketsu-zai
hemp *[bot]*	大麻	taima
" *[tex]*	麻; あさ	asa
hemp cloth *[tex]*	麻布	asa-nuno
hemp palm; Trachycarpus fortunei *[bot]*	棕櫚; しゅろ	shuro
hempseed *[bot] [cook]*	麻の実	asa no mi
hempseed oil *[mat]*	麻実油	asami-yu
hen; female bird *[v-zoo]*	雌鳥; めんどり	mendori
henbane; Hyoscyamus niger *[bot]*	菲沃斯; ひよす	hiyosu
hen's egg *[agr]*	鶏卵	keiran
hepatic portal system *[anat]*	肝門脈系	kan-mon'myaku-kei
hepatitis *[med]*	肝炎	kan'en
heptagon *[geom]*	七辺形	shichi'hen-kei
" *[geom]*	七角形	shichi'kak(u)kei
heptavalent *{adj} [ch]*	七価の	shichi-ka no
herb; herbage *[bot]*	草本; そうほん	sōhon
herb *[bot] [cook]*	風味用植物	fūmi'yō-shokubutsu
" *[bot] [cook]*	香味草	kōmi'sō
" *[bot] [pharm]*	薬草; やくそう	yaku'sō
" *[bot] [pharm]*	薬用植物	yaku'yō-shokubutsu
herbicide; weedkiller *[mat]*	除草剤	josō-zai
herbivore *[v-zoo]*	草食動物	sōshoku-dōbutsu
herbivorous *{adj} [v-zoo]*	草食性の	sō'shoku'sei so
here *{adv} {n} {pron}*	此処; 此所; ここ	koko
heredity; inheritance *[bio]*	遺伝; いでん	i'den
hermaphrodite *[bio]*	両性動物	ryōsei-dōbutsu
hermaphroditism *[bio]*	雌雄同体現象	shiyū-dōtai-genshō
hermetic container *[eng]*	密封容器	mippū-yōki
hermetic seal *{n} [eng]*	溶接密閉	yōsetsu-mippei
hermit crab (crustacean) *[i-zoo]*	宿借; 寄居虫	yado-kari
heron (bird) *[v-zoo]*	鷺; さぎ	sagi

herpetologist *[bio]*	爬虫類学者	hachū'rui-gakusha
herpetology *[bio]*	爬虫学	hachū'gaku
herring; Clupea pallasi (fish) *[v-zoo]*	鯡; 鰊; にしん	nishin
herringbone *{n} [cl] [graph]*	矢筈模様	ya'hazu-moyō
herring roe (fish egg) *[cook]*	数の子; かずのこ	kazu-no-ko
hesitate *{vb} [psy]*	躊躇する	chūcho suru
hetero-atom *[ch]*	異原子	i-genshi
heterobar *[ch]*	異重体	ijū'tai
heterochromatic photometry *[opt]*	異色測光	ishoku-shokkō
heterocyclic compound *[org-ch]*	複素環式化合物	fukusokan'shiki-kagō'butsu
" *[org-ch]*	異項環式化合物	ikōkan'shiki-kagō'butsu
heterocyclic ring *[org-ch]*	複素環	fukuso-kan
heteroduplex *[gen]*	異種二重螺旋	ishu-nijū-rasen
heterofermentative(ness) *[microbio]*	異種発酵(性)	ishu-hakkō(-sei)
heterogeneity *[bio] [gen]*	異質	i'shitsu
" *[ch]*	不均質性	fu'kinshitsu-sei
" *{n}*	不均一性	fu'kin'itsu-sei
heterogeneous *[ch]*	不均質	fu'kinshitsu
" *{adj}*	不均一な	fu'kin'itsu na
heterogeneous system *[sci-t]*	不均質系	fu'kinshitsu-kei
heterojunction *[electr]*	異質接合	i'shitsu-setsu'gō
heterokaryon *[mycol]*	異核接合体	i'kaku-setsugō'tai
" *[mycol]*	異種接合体	ishu-setsugō'tai
heterologous antigen *[immun]*	非対応抗原	hi'taiō-kōgen
heteroploidy *[gen]*	異数性; いすうせい	isū-sei
heteropolar bond *[ch]*	異極結合	i'kyoku-ketsu'gō
heteropolar compound *[ch]*	異極化合物	i'kyoku-kagō'butsu
heteropolarness *[elecmg]*	異極性	i'kyoku-sei
heteropoly acid *[inorg-ch]*	異種多重酸	ishu-tajū-san
heterosphere *[meteor]*	異質圏	i'shitsu-ken
heterothermy *[physio]*	異温性	i'on-sei
heterotroph *[bio]*	従属栄養生物	jūzoku-eiyō-seibutsu
" *[bio]*	有機栄養生物	yūki-eiyō-seibutsu
heulandite *[miner]*	輝沸石; きふっせき	ki'fusseki
heuristics *[psy]*	発見法	hakken-hō
hexadecimal notation *[comput]*	十六進法	jūroku'shin-hō
hexagon *[geom]*	六角形	rokkaku-kei
hexagonal array *[crys]*	最密配列	sai'mitsu-hairetsu
hexagonal closest-packed structure	六方最密構造	roppō-sai'mitsu-kōzō
hexagonal lattice *[crys]*	六方格子	roppō-kōshi
hexagonal structure *[crys]*	六面構造	rokumen-kōzō
hexagonal system *[crys]*	六方晶形	roppō-shōkei

hexagon socket screw *[eng]*	六角穴付きねじ	rokkaku'ana'tsuki-neji
hexahedron (pl -s, -hedra) *[geom]*	六面体	rokumen'tai
hexavalent *{adj} [ch]*	六価の; ろっかの	rokka no
hexode *[electr]*	六極管	rokkyoku-kan
hexose *[bioch]*	六炭糖	rokutan'tō
hiatus (pl -es) *[crys]*	隙間; すきま	suki'ma
" *[geol]*	割れ目	ware'me
hibernation *[physio]*	冬眠	tōmin
hibiscus *[bot]*	黄蜀葵	tororo'aoi
hiccup *{vb} [physio]*	しやっくりする	shakkuri suru
hiccups *{n} [physio]*	噦; 吃逆	shakkuri
hidden; invisible *{adj} [opt]*	隠れた; 隱れた	kakureta
hidden line *[comput]*	陰線	insen
hidden surface *[comput]*	陰面; 隠面	inmen
hide *{n} [cl]*	皮	kawa
" *{vb t} [comput]*	隠す; かくす	kakusu
hiding power *[eng]*	隠蔽力	inpei-ryoku
hiearchical structure *[comput]*	階層構造	kaisō-kōzō
hierarchy *[math]*	階層	kaisō
hieroglyph *[comm] [pr]*	象形文字	shōkei-moji
high (in rank) *{n} [ch]*	貴; き	ki
high *{n} [meteor]*	高気圧; 高氣壓	kō-ki'atsu
high; tall; elevated *{adj}*	高い; たかい	takai
high and low; undulations *{n}*	高低	kōtei
high-definition television *[electr]*	高精細度テレビ=(ジョン)	kō'seisaido-terebi(jon)
high degree of *{adj}*	高度の	kōdo no
higher *{adj} [math]*	上の	kami no
" (in grade) *{n}*	高級; こうきゅう	kōkyū
higher harmonics *[acous]*	高調波	kōchō-ha
higher-order structure *[mol-bio]*	高次構造	kōji-kōzō
higher oxide *[ch]*	高級酸化物	kōkyū-sanka'butsu
highest point; extreme point *{n}*	極点; 極點	kyoku-ten
highest priority *{n}*	最優先	sai-yūsen
high explosives *[ord]*	爆薬	baku'yaku
high fluidity (liquidity) *[fl-mech]*	高流動性	kō-ryūdō-sei
high frequency *[comm]*	高周波	kō-shūha
high-frequency induction heating *[eng]*	高周波誘導加熱	kō'shūha-yūdō-ka'netsu
high-grade (first-class) *{adj}*	上等な(の)	jōtō na (no)
" (high-level) *{adj}*	高位の	kō'i no
high-level language *[comput]*	高水準言語	kō-suijun-gengo
highly precise *{adj}*	高精度の	kō-seido no

high-molecular compound *[poly-ch]*	高分子；こうぶんし	kō'bunshi
" *[poly-ch]*	高分子化合物	kōbunshi-kagō'butsu
high-molecular surfactant *[poly-ch]*	高分子界面活性剤	kōbunshi-kaimen-kasseizai
high-order; high level *[math]*	高位	kō'i
high polymer *[poly-ch]*	高分子	kōbunshi
" *[poly-ch]*	高重合体	kōjūgō'tai
high polymerization *[poly-ch]*	高分子化	kōbunshi'ka
high pressure *{n} [meteor]*	高気圧	kō-ki'atsu
high pressure; high *{n} [meteor]*	逆旋風	gyaku-senpū
high pressure *{n} [phys]*	高圧；高壓	kō'atsu
" *{adj} [phys]*	高圧の	kō'atsu no
high-pressure chemistry *[p-ch]*	高圧化学	kōatsu-kagaku
high-pressure laminating *[poly-ch]*	高圧積層	kōatsu-sekisō
high-pressure molding *{n} [resin]*	高圧成形	kōatsu-seikei
high-pressure physics *[phys]*	高圧物理学	kōatsu-butsuri'gaku
high-pressure process *[ch-eng]*	高圧法	kōatsu-hō
high-pressure science *[phys]*	高圧科学	kōatsu-kagaku
high productivity *[ind-eng]*	高生産性	kō'seisan-sei
high-quality; fine-quality *{adj}*	上質の	jō-shitsu no
high-rise building *[arch]*	高層ビル	kōsō-biru
high seas *{n} [ocean]*	公海	kōkai
high-silicon steel *[met]*	高珪素鋼	kō'keiso-kō
high speed *{n} [mech] [phys]*	高速	kō-soku
high-speed carry *[comput]*	高速桁上げ	kōsoku-keta'age
high-temperature-oxidation resistance *[mat]*	耐高温酸化性	tai'kōon-sanka-sei
high-tenacity rayon *[tex]*	強力人絹	kyōryoku-jinken
high-tensile cast iron *[met]*	強靭鋳鉄	kyōjin-chū-tetsu
high-tensile steel *[met]*	高力鋼	kōryoku-kō
high-tensile-strength steel *[met]*	高張力鋼	kō-chō'ryoku-kō
high tide *[ocean]*	満潮	manchō
" *[ocean]*	高潮；たかしお	taka-shio
high-velocity star *[astron]*	高速度星	kō'sokudo-sei
high voltage *[elec]*	高圧；こうあつ	kō'atsu
high-voltage cable *[elec]*	高圧電線	kōatsu-densen
high-voltage insulator *[elec]*	高圧碍子	kōatsu-gaishi
high water *[ocean]*	高潮	kōchō
highway net *[civ-eng]*	道路網	dōro-mō
hijiki algae; Hizikia fusiforme *[bot]*	鹿尾菜；ひじき	hijiki
hill *[geogr]*	丘；岡	oka
hill (artificial) *[arch]*	築山	tsuki-yama
hills and valleys *{n}*	凹凸；おうとつ	ō-totsu

hillside *[geogr]*	山腹	san'puku
hilt (sword guard) *{n} [ord]*	鍔	tsuba
" (sword); haft (dagger) *[ord]*	柄；つか	tsuka
hinder; interfere with *{vb t}*	障る	sawaru
hindrance; interference *{n}*	防害；妨害	bōgai
hinge *{n} [mech-eng]*	蝶番	chō'tsugai
hinged joint *[mech-eng]*	滑節	kassetsu
hip *{n} [anat]*	股関節部	ko'kansetsu-bu
" *{n} [anat]*	腰；こし	koshi
hip joint *[anat]*	股関節	ko-kan'setsu
hippopotamus (animal) *[v-zoo]*	河馬；かば	kaba
hippurate *[ch]*	馬尿酸塩	ba'nyōsan-en
hippuric acid	馬尿酸	ba'nyō-san
histochemistry *[ch]*	組織化学	soshiki-kagaku
histogram *[stat]*	柱状図(表)	chūjō-zu(hyō)
" *[stat]*	度数分布図	dosū-bunpu-zu
histology *[anat]*	組織学；そしきがく	soshiki'gaku
historian (a person)	歴史家	rekishi'ka
history (pl -ries) *{n}*	歴史；歴史	rekishi
hit; strike *{vb t}*	打つ	utsu
Hizikia fusiforme (seaweed) *[cook]*	鹿尾菜；ひじき	hijiki
hobbing *{n} [mech]*	ホブ加工	hobu-kakō
" *{n} [mech]*	ホブ切	hobu-kiri
hobbing machine *[mech]*	ホブ盤	hobu-ban
hobby (pl hobbies)	趣味	shumi
hoe *{n} [agr] [eng]*	鍬；くわ	kuwa
hog raising *{n} [agr]*	養豚	yōton
hold (of a ship) *{n} [nav-arch]*	貨物室	kamotsu-shitsu
" (of a ship) *{n} [nav-arch]*	船倉	sensō
" *{vb t}*	把持する	haji suru
" (carry in hand) *{vb t}*	持つ	motsu
holdfast *{n} [bot]*	付着根	fuchaku-kon
holding power *{n} [adhes]*	保持力	hoji-ryoku
holding time *[comput]*	保留時間	horyū-jikan
hold over head or eyes; shade *{vb}*	翳す；かざす	kazasu
hold, sandwiched *{vb t} [mech-eng]*	挾持する	kyōji suru
hole *[comput]*	穿孔	senkō
" *[sol-st]*	空孔	kūkō
" *{n}*	穴；孔；あな	ana
hollow *{adj} [tex]*	中空の	chūkū no
" *{adj}*	空ろな；うつろな	utsuro na
hollow brick *[mat]*	空胴煉瓦	kūdō-renga

hollow fiber; hollow yarn *[tex]*	中空糸	chūkū-shi
hollow forgings *[met]*	中空鍛造品	chūkū-tanzō'hin
hollowness *{n}*	中空；ちゅうくう	chūkū
holly *[bot]*	柊；ひいらぎ	hīragi
hollyhock; mallow (flower) *[bot]*	葵；あおい	aoi
holmium (element)	ホルミウム	horumyūmu
holohedry; holohedral form *[crys]*	完面像	kanmen-zō
homeomorphism; homomorphy *[crys]*	異質同形	i'shitsu-dōkei
homeostasis *[bio]*	恒常性	kōjō-sei
homeotherm *[physio]*	常温動物	jō'on-dōbutsu
homespun; handwoven *{adj}* *[tex]*	手織りの	te'ori no
Hominidae *[bio]*	人科；ひとか	hito-ka
hominids *[v-zoo]*	人科動物	hito'ka-dōbutsu
Homo *[bio]*	人族；ひとぞく	hito-zoku
Homo sapiens; mankind *[bio]*	人類；じんるい	jinrui
" ; human being *[bio]*	人間；にんげん	ningen
homocyclic compound *[org-ch]*	単素環式化合物	tansokan'shiki-kagō'butsu
homofermentative *{adj}* *[microbio]*	同種発酵性の	dōshu-hakkō'sei no
homogeneity *[sci-t]*	同質性	dōshitsu-sei
homogeneous; homogeneity *{n}* *[sci-t]*	均質；きんしつ	kin'shitsu
homogeneous *{adj}* *[ch]* *[sci-t]*	均一な；きんいつな	kin'itsu na
homogeneous polymer *[poly-ch]*	均一重合物	kin'itsu-jūgō'butsu
homogeneous reaction *[ch]*	均一反応	kin'itsu-hannō
homogeneous transformation *[math]*	同次変換	dōji-henkan
homogenized milk *[cook]*	均質牛乳	kin'shitsu-gyūnyū
homoiothermy *[physio]*	恒温性	kō'on-sei
homolog *[ch]*	同属体	dōzoku'tai
homologous *{adj}* *[bio]*	同族の；どうぞくの	dōzoku no
" *{adj}* *[sci-t]*	相同の	sōdō no
homologous organ *[bio]*	相同器官	sōdō-kikan
homology *[bio]* *[ch]* *[org-ch]*	相同；そうどう	sōdō
" *[ch]*	同属関係	dōzoku-kankei
homomorphs *[ch]*	準同形	jun-dōkei
homomorphy *[bio]*	異質同形	i'shitsu-dōkei
homonym: same kun (q.v.) but of different characters *[ling]*	同訓異字語	dōkun-iji'go
homonym: same on (q.v.) but of different characters *[ling]*	同音異字語	dō'on-iji'go
homonym: same on (q.v.) but with different meaning *[ling]*	同音異義語	dō'on-igi'go
homophone *[ling]*	同音異字	dō'on-iji
homopolar bond *[p-ch]*	等極結合	tō'kyoku-ketsu'gō

homopolar crystal *[crys]* 等極結晶 tō'kyoku-kesshō
homopolar generator *[electr]* 単極発電機 tan'kyoku-hatsu'den-ki
homopolymer *[poly-ch]* 単独重合体 tandoku-jūgō'tai
homopolymerization *[poly-ch]* 単独重合 tandoku-jūgō
Homo sapiens; mankind *[bio]* 人類；じんるい jinrui
homoscedasticity *[stat]* 等分散性 tō'bunsan-sei
homosphere *[meteor]* 均質圏 kin'shitsu-ken
hondo jungle crow (bird) *[v-zoo]* 嘴太鴉；はしぶと鴉 hashi'buto-garasu
honewort; antirrhinum *[bot]* 金魚藻 kingyo'sō
" ; trefoil; wild chervil *[bot]* 三葉；みつば mitsuba
honey *[cook]* *[i-zoo]* (蜂)蜜 (hachi-)mitsu
honeybee (insect) *[i-zoo]* 蜜蜂；みつばち mitsu'bachi
honeycomb *[i-zoo]* 蜂の巣；蜂の巣 hachi no su
honey locust (tree) *[bot]* 皂莢；さいかち saikachi
honeysuckle, Japanese *[bot]* 忍冬；にんどう nindō (suikazura)
honing *{n}* *[mech-eng]* 研ぎ上げ togi'age
hood *[cl]* 頭布；ずきん zukin
hooded gull (bird) *[v-zoo]* 百合鴎 yuri-kamome
hoof (pl hooves) *[v-zoo]* 蹄；ひづめ hizume
hook *{n}* *[eng]* 鉤；かぎ kagi
" ; clasp; fastener; latch *{n}* *[eng]* 止め金；とめがね tome'gane
hooked punctuation mark (in Japanese) *[pr]* 鉤括弧：［ ］；「」 kagi-kakko
hook-tipped rope *[eng]* 鉤縄；鉤縄 kagi-nawa
hookworm *[i-zoo]* 鉤虫；こうちゅう kōchū
hoopoe; hoopoo (bird) *[v-zoo]* 戴勝；やつがしら yatsu'gashira
hop *{n}* *[bot]* 忽布；ほっぷ hoppu
horizon *[astron]* 地平線 chihei'sen
" *[astron]* 水平線 suihei'sen
horizontal; horizontality *{n}* *[sci-t]* 水平；すいへい suihei
" *{adj}* *[sci-t]* 水平な suihei na
horizontal deflection *[electr]* 水平偏向 suihei-henkō
horizontal parallax *[astron]* 地平視差 chihei-shisa
horizontal scanning *{n}* *[eng]* 水平走査 suihei-sōsa
horizontal supine position *[med]* 仰臥位；ぎょうがい gyōga'i
horizontal tail plane *[aero-eng]* 水平尾翼 suihei-bi'yoku
hormone *[bioch]* ホルモン horumon
horn; horns; antler(s) *[zoo]* 角；つの tsuno
hornbill (bird) *[v-zoo]* 犀鳥；さいちょう saichō
hornblende *[miner]* 角閃石 kaku'senseki
horned liverwort; hornwort *[bot]* 松藻；まつも matsu'mo
horned owl; eared owl (bird) *[v-zoo]* 木菟；みみずく mimizuku

horned shark; Port Jackson shark *[v-zoo]*	猫鮫；ねこざめ	neko-zame
horned toad (reptile) *[v-zoo]*	角蜥蜴；つのとかげ	tsuno-tokage
hornworm; green caterpillar *[i-zoo]*	芋虫；いもむし	imo-mushi
horology *[sci-t]*	測時学	sokuji'gaku
" *[sci-t]*	時計学；とけいがく	tokei'gaku
horse (animal) *[v-zoo]*	馬；うま	uma
horsebean; broad bean; fava *[bot]*	空豆；そらまめ	sora-mame
horse chestnut tree *[bot]*	栃；とちのき；とち	tochinoki; tochi
horsefly; deerfly (insect) *[i-zoo]*	虻；あぶ	abu
horsehead nebula *[astron]*	馬頭星雲	batō-sei'un
horse mackerel (fish) *[v-zoo]*	真鯵；まあじ	ma'aji
horsepower *[mech]*	馬力；ばりき	ba'riki
horseradish, Japanese *[bot] [cook]*	山葵；わさび	wasabi
horseshoe crab; helmet crab *[i-zoo]*	兜蟹；かぶとがに	kabuto-gani
horseshoeing *{n} [trans]*	装蹄	sōtei
horseshoe magnet *[elecmg]*	馬蹄形磁石	batei'kei-ji'shaku
horseshoe-shaped *{adj}*	馬蹄形の	batei'kei no
horticulture *[bot]*	園芸学；園藝學	engei'gaku
hormone *[bioch]*	ホルモン	horumon
hospital *[arch]*	病院；びょういん	byō'in
hospitalize *{vb t} [med]*	入院させる	nyūin saseru
host *{n} [bio]*	寄主；きしゅ	kishu
" *{n} [bio]*	宿主	shuku'shu; yado-nushi
" ; owner *{n} [comput]*	親；おや	oya
host cell *[gen]*	宿主細胞	shuku'shu-saibō
host language *[comput]*	親言語	oya-gengo; oya-kotoba
hot *{adj} [phys] [physio]*	熱い；あつい	atsui
hot-air heating *{n} [mech-eng]*	温風暖房	onpū-danbō
hotbed; frame *[bot]*	温床	onshō; on'doko
hot bending *{n} [met]*	熱間曲げ加工	nekkan-mage-kakō
hot-dip plating *{n} [met]*	熱浸鍍金	nesshin-mekki
hot dipping *[met]*	溶融めっき	yōyū-mekki
hot-dipping bath *[met]*	溶融めっき浴	yōyū-mekki-yoku
hot-dipping method *[met]*	溶融めっき法	yōyū-mekki-hō
hot electron *[electr]*	熱い電子	atsui denshi
hothouse; greenhouse *[bot]*	温室；おんしつ	on-shitsu
hot-melt adhesive *[poly-ch]*	熱溶融型接着剤	netsu-yōyū'gata-setchakuzai
hotness *{n} [phys] [physio]*	熱さ；あつさ	atsu-sa
hot rolling *{n} [met]*	熱間圧延	nekkan-atsu'en
hot-setting adhesive *[adhes]*	熱間硬化接着剤	nekkan-kōka-setchaku-zai
hot shortness *[met]*	熱間脆性	nekkan-zeisei

hot springs *[hyd]*	温泉; おんせん	onsen
hot-vulcanization *[rub]*	熱加硫	netsu-karyū
hot water *[ch]*	熱水	nessui
hot-water bottle *[pharm]*	湯湯婆; ゆたんぽ	yu'tanpo
hot-water heating *{n} [mech-eng]*	湯暖房	yu-danbō
hot-water solubility *[p-ch]*	熱水可溶性	nessui-kayō-sei
hot weather; warmth; heat *{n}*	暑さ	atsu-sa
hot working *{n} [met]*	熱間加工	nekkan-kakō
hour (unit of time) *[mech]*	時; じ	ji
" (time) *[mech]*	時間; じかん	jikan
hour angle *[astron]*	時角	ji-kaku
hour circle *[astron]*	時圏; 時圏; じけん	jiken
hourglass *[horol]*	砂時計	suna-dokei
hour hand *[horol]*	短針	tan'shin
house *{n} [arch]*	家; いえ	ie
household electric appliance *[elec]*	家庭用電気器具	katei'yō-denki-kigu
household utensil(s) *[eng]*	家庭用品	katei'yō'hin
house martin (bird) *[v-zoo]*	岩燕; いわつばめ	iwa-tsubame
house mosquito (insect) *[i-zoo]*	赤家蚊	aka-ie-ka
houseplant *[bot]*	鉢植	hachi'ue
house sparrow (bird) *[v-zoo]*	家雀	ie-suzume
house tick (insect) *[i-zoo]*	家蜱; いえだに	ie-dani
house wiring *{n} [elec]*	屋内配線	oku'nai-haisen
how *{adv}*	如何; どう	dō
" *{adv}*	如何に	ika ni
howitzer *[ord]*	曲射砲; 曲射砲	kyokusha-hō
hue *{n} [opt]*	色相	shiki'sō
hull (ship) *{n} [nav-arch]*	船体	sentai
human being *[bio]*	人間; にんげん	nin'gen
human ecology *[ecol]*	人間生態学	ningen-seitai'gaku
human growth hormone *[bioch]*	ヒト成長ホルモン	hito-seichō-horumon
human immunodeficiency virus; HIV *[microbio]*	ヒト免疫不全ウイ=ルス	hito-men'eki-fuzen-uirusu
humectant *[ch]*	保湿剤	ho'shitsu-zai
" *[ch]*	湿潤剤	shitsu'jun-zai
humerus (bone) *[anat]*	上腕骨	jō'wan-kotsu
humid *{adj} [hyd]*	湿気のある	shikki no aru
humidifier *[mech-eng]*	加湿器; 加濕器	ka'shitsu-ki
" *[mech-eng]*	増湿装置	zō'shitsu-sōchi
humidity (pl -ties) *[meteor]*	湿度; 濕度	shitsu-do
humidity index *[climat]*	湿度指数	shitsu'do-shisū
humidor *[ch-eng]*	水蒸気飽和室	sui'jōki-hōwa-shitsu

hummingbird (bird) *[v-zoo]*	蜂鳥；はちどり	hachi-dori
humpback whale *[v-zoo]*	座頭鯨	zatō-kujira
humus *[geol]*	腐植質	fushoku'shitsu
hunch *[psy]*	勘；かん	kan
hundred *{n} [math]*	百；ひゃく	hyaku
hunger; starvation; famine *[med]*	飢餓；きが	kiga
hungry, state of being *{n} [psy]*	空腹	kūfuku
hungry *{adj} [psy]*	空腹な	kūfuku na
hunter's moon *[astron]*	狩猟月	shuryō-zuki
hunting *{n} [cont-sys] [elec]*	乱調	ranchō
hurry *{vb}*	急ぐ	isogu
hurt *{vb i} [physio]*	痛む	itamu
" *{vb t} [ind-eng]*	傷める	itameru
husk; shell *{n} [mat]*	殻；殼；から	kara
" ; sheath *{n} [mat]*	鞘；さや	saya
hyaline; glassy *{adj} [geol] [i-zoo]*	硝子状の	garasu'jō no
hyalography *[graph]*	硝子彫刻技法	garasu-chōkoku-gihō
" *[graph]*	硝子画技法	garasu'ga-gihō
hybrid *{n} [bot] [gen]*	雑種；雜種	zasshu
" *{n} [gen]*	混種物	konshu'butsu
hybrid computer *[comput]*	混成型コンピュータ	konsei'gata-konpūtā
hybridization *[bot]*	雑種形成	zasshu-keisei
" *[gen]*	異種交配	ishu-kōhai
" *[gen]*	混成	konsei
" *[gen]*	交雑	kō'zatsu
hybridization breeding *{n} [gen]*	交雑育種	kōzatsu-iku'shu
hydnocarpic acid	大風子酸	daifūshi-san
hydrangea (flower) *[bot]*	紫陽花；あじさい	ajisai
hydrant *[civ-eng]*	消火栓	shōka-sen
hydrate *[ch]*	水化物	suika'butsu
" *[ch]*	水和物	suiwa'butsu
hydrated water *[ch]*	水和水	suiwa-sui
hydration *[ch]*	水化；すいか	suika
" *[ch]*	水和；すいわ	suiwa
hydration number *[ch]*	水和数	suiwa-sū
hydraulic *{adj} [eng]*	水圧の；水壓の	sui'atsu no
" *{adj} [eng]*	水力の	sui'ryoku no
" *{adj} [eng]*	油圧の	yu'atsu no
hydraulically operated valve *[mech-eng]*	油圧駆動弁；油壓=驅動瓣	yu'atsu-kudō-ben
hydraulic car jack *[mech-eng]*	液圧自動車押上装置	eki'atsu-jidōsha-oshi'age-sōchi

English	Japanese	Romanization
hydraulic dynamometer *[eng]*	水動力計	mizu-dōryoku-kei
hydraulic efficiency *[mech]*	水力効率	sui'ryoku-kō'ritsu
hydraulic engineering *[civ-eng]*	水圧工学；水壓工學	sui'atsu-kōgaku
hydraulic fluid *[mat]*	作動液	sadō-eki
" *[mat]*	水圧液体	sui'atsu-ekitai
hydraulic gear *[mech-eng]*	流体変速装置	ryūtai-hensoku-sōchi
hydraulic governor *[mech-eng]*	液圧式調速機	eki'atsu'shiki-chōsoku-ki
hydraulic gradient *[fl-mech]*	動水勾配	dōsui-kōbai
hydraulic mining *[min-eng]*	水力採鉱	suiryoku-saikō
hydraulic oil *[mat]*	作動油；さどうゆ	sadō-yu
hydraulic power *[mech]*	水力	sui'ryoku
hydraulic pressure *[mech]*	水圧；水壓	sui'atsu
hydraulic radius *[fl-mech]*	動水半径；動水半徑	dōsui-hankei
hydraulics *[fl-mech]*	水理学	suiri'gaku
hydride *[inorg-ch]*	水素化物	suiso'ka'butsu
hydro- *{prefix}* *[hyd]*	水の-	mizu no-
hydroacid *[ch]*	水素酸	suiso-san
hydrobiology *[bio]*	水生生物学	suisei-seibutsu'gaku
hydrocarbon *[org-ch]*	炭化水素	tanka'suiso
hydrocarbon-assimilating bacteria *[microbio]*	炭化水素資化菌	tanka'suiso-shika-kin
hydrocarbon radical *[org-ch]*	炭化水素基	tanka'suiso-ki
hydrochloric acid	塩酸；えんさん	ensan
hydrocracking *{n}* *[petrol]*	水素化分解	suiso'ka-bunkai
" *[petrol]*	水素添加分解	suiso-tenka-bunkai
hydrocyanic acid	シアン化水素酸	shian'ka-suiso-san
hydrodynamics *[fl-mech]*	流体力学	ryūtai-rikigaku
hydroelectric power generation	水力発電	sui'ryoku-hatsu'den
hydrofluoric acid	弗化水素酸	fukka-suiso-san
hydrofoil *[nav-arch]*	水中翼	suichū'yoku
hydrogen (element)	水素；すいそ	suiso
hydrogenated oil *[mat]*	硬化油	kōka-yu
hydrogenation *[org-ch]*	水素化	suiso'ka
" *[org-ch]*	水素添加（水添）	suiso-tenka (suiten)
hydrogen bomb *[ord]*	水素爆弾	suiso-bakudan
hydrogen bond *[p-ch]*	水素結合	suiso-ketsu'gō
hydrogen bromide	臭化水素	shūka-suiso
hydrogencarbonate *[ch]*	炭酸水素塩	tansan'suiso-en
hydrogen chloride	塩化水素	enka-suiso
hydrogen chloroaurate	金塩化水素酸	kin'enka'suiso-san
hydrogen cyanide	シアン化水素	shian'ka-suiso
hydrogen equivalent *[ch]*	水素当量；水素當量	suiso-tōryō

hydrogen exponent *[ch]*	水素指数	suiso-shisū
hydrogen fluoride	弗化水素	fukka-suiso
hydrogen iodide	沃化水素	yōka-suiso
hydrogen ion concentration *[ch]*	水素イオン濃度	suiso-ion-nōdo
hydrogen peroxide	過酸化水素	ka'sanka-suiso
hydrogen sulfide	硫化水素	ryūka-suiso
hydrogen-transfer polymerization *[poly-ch]*	水素転移重合	suiso-ten'i-jūgō
hydrogeology *[hyd]*	水文地質学	suimon-chishitsu'gaku
hydrographic chart *[map]*	海図；かいず	kaizu
hydrographic surveying *[ocean]*	河海測量	ka'kai-sokuryō
hydrography *[geogr] [navig]*	水路測量術	suiro-sokuryō'jutsu
hydroiodic acid	沃化水素酸	yōka'suiso-san
hydrolase *[bioch]*	加水分解酵素	kasui'bunkai-kōso
hydrologic cycle *[hyd]*	水分循環	suibun-junkan
" *[hyd]*	水の循環	mizu no junkan
hydrology *[geophys]*	陸水学	rikusui'gaku
" *[geophys]*	水文学	suimon'gaku
hydrolysate; hydrolyzate *[ch] [geol]*	水解物	suikai'butsu
hydrolysis *[ch]*	加水分解	kasui'bunkai
" *[ch]*	水解（加水分解）	suikai (kasui-bunkai)
hydromagnetics *[phys]*	流体磁気学	ryūtai-jiki'gaku
hydromechanics *[fl-mech]*	流体力学	ryūtai-riki'gaku
hydrometeorology *[meteor]*	水文気象学	suimon-kishō'gaku
hydrometer *[eng]*	比重計	hijū-kei
" *[eng]*	浮秤；うきばかり	uki-bakari
hydronium ion *[inorg-ch]*	ヒドロニゥムイオン	hidoronyūmu-ion
hydrophilic group *[ch]*	親水基	shinsui-ki
hydrophilicity *[ch]*	親水度	shinsui-do
hydrophilic property *[ch]*	親水性	shinsui-sei
hydrophobic group *[ch]*	疎水基；そすいき	sosui-ki
hydrophobicity *[ch]*	疎水度	sosui-do
hydrophobic property *[ch]*	疎水性	sosui-sei
hydrophone *[eng-acous]*	水中聴音器	suichū-chō'on-ki
hydroplane *[nav-arch]*	水上飛行機	suijō-hikōki
hydroponic farm *[agr] [bot]*	水耕農場	suikō-nōjō
hydroponics *[bot]*	水耕法	suikō-hō
hydrosilicofluoric acid	珪弗化水素酸	kei'fukka-suiso-san
hydrosphere *[hyd]*	水圏；水圏	suiken
hydrostatic equilibrium *[fl-mech]*	流体静力学平衡	ryūtai-sei'rikigaku-heikō
hydrostatic pressure *[fl-mech]*	静水圧	seisui-atsu
hydrostatics *[fl-mech]*	流体静力学	ryūtai-sei'rikigaku

hydrotaxis *[bio]*	走水性	sōsui-sei
hydrothermal deposit *[p-ch]*	熱水鉱床; 熱水鑛床	nessui-kōshō
hydrotropism *[bio]*	向水性	kōsui-sei
hydrousness *[ch]*	含水性	gansui-sei
hydroxide *[ch]*	水酸化物	suisan'ka'butsu
hydroxide salt *[ch]*	水酸化物塩	suisan'ka'butsu-en
hydroxyl *[ch]*	ヒドロキシル	hidorokishiru
hydroxylation *[org-ch]*	水酸化	suisan'ka
hydroxyl group *[ch]*	水酸基; すいさんき	suisan-ki
Hydrozoa *[i-zoo]*	ヒドロ虫類	hidoro'chū-rui
hygiene *[med]*	衛生; 衞生	eisei
hygienically harmful insect *[i-zoo]*	衛生害虫	eisei-gaichū
hygienic porcelain *[mat]*	衛生陶器	eisei-tōki
hygrograph *[eng]*	自記湿度計	jiki-shitsu'do-kei
hygrometer *[eng]*	湿度計; 濕度計	shitsu'do-kei
hygrophyte *[bot]*	湿性植物	shissei-shokubutsu
hygroscopicity *[bot] [ch]*	吸湿性	kyū'shitsu-sei
Hymenoptera *[i-zoo]*	膜翅目	maku'shi-moku
hyperbola (pl -las, -lae) *[an-geom]*	双曲線; 雙曲線	sōkyoku-sen
hyperbolic function *[calc]*	双曲線関数	sōkyoku'sen-kansū
hyperboloid *{n} [math]*	双曲面	sōkyoku-men
hyperchromic effect *[p-ch]*	濃色効果	nōshoku-kōka
hypercomplex number *[math]*	多元数	tagen-sū
hyperconjugation *[p-ch]*	超共役	chō'kyōyaku
hypereutectic *[met]*	過共晶	ka'kyōshō
hypereutectoid *[met]*	過共析晶	ka'kyōseki-shō
hyperfine processing *{n} [ind-eng]*	超微細加工	chō'bisai-kakō
hyperfine structure *[spect]*	超微細構造	chō'bisai-kōzō
hypergeometric distribution *[stat]*	超幾何分布	chō'kika-bunpu
hypergeometric function *[math]*	超幾何関数	chō'kika-kansū
hyperresolution *[comput]*	超分解	chō'bunkai
hypersensitivity *[immun]*	過敏性; かびんせい	kabin-sei
hypersensitization *[photo]*	超増感	chō'zōkan
hypersonic *{adj} [acous]*	極超音速の	kyokuchō-onsoku no
hypersonic flow *[fl-mech]*	極超音速流	kyokuchō-onsoku-ryū
hypersonic speed *[fl-mech]*	極超音速	kyokuchō-onsoku
hypersonic transport; HST *[trans]*	極超音速航空機	kyokuchō-onsoku-kōkūki
hyperspace *[math]*	超(三次元)空間	chō(-sanji'gen)-kūkan
hyper square *[math]*	超方格	chō-hōkaku
hyperstereoscopy (map) *[graph]*	超立体映像	chō'rittai-eizō
hypersthene *[miner]*	紫蘇輝石	shiso'kiseki
hypertension *[med]*	高血圧	kō-ketsu'atsu

hypertonic solution *[ch]*	高張液	kōchō-eki
hypervelocity physics *[phys]*	超高速(度)物理	chō'kōsoku(do)-butsuri
hypervelocity projectile *[ord]*	超高速発射体	chō'kōsoku-hassha'tai
hypha (pl hyphae) *[mycol]*	菌糸; きんし	kinshi
hyphen *[comput] [pr]*	連字符	renji'fu
hypnotic *(n) [pharm]*	催眠剤	saimin-zai
hypobromite *[ch]*	次亜臭素酸塩	jia'shūso'san-en
hypobromous acid	次亜臭素酸	jia'shūso-san
hypocenter (of earthquake) *[geophys]*	震源; しんげん	shingen
hypochlorite *[ch]*	次亜塩素酸塩	jia'enso'san-en
hypochlorous acid	次亜塩素酸	jia'enso-san
hypochromic effect *[p-ch]*	淡色効果	tanshoku-kōka
hypochromism *[opt]*	淡色効果	tanshoku-kōka
hypohalogenous acid salt *[ch]*	次亜ハロゲン酸塩	jia'harogen'san-en
hypoiodous acid	次亜沃素酸	jia'yōso-san
hyponitrite *[ch]*	次亜硝酸塩	jia'shōsan-en
hyponitrous acid	次亜硝酸	jia'shōsan
hypophosphate *[ch]*	次燐酸塩	ji'rinsan-en
hypophosphite *[ch]*	次亜燐酸塩	jia'rinsan-en
hypophosphoric acid	次燐酸	ji'rinsan
hypophosphorous acid	次亜燐酸	jia'rinsan
hypophysis (gland) *[anat]*	脳下垂体	nō'kasui'tai
hyposulfite *[ch]*	次亜硫酸塩	jia'ryūsan-en
hyposulfurous acid	次亜硫酸	jia'ryūsan
hypotensor *[pharm]*	降圧剤; 降壓劑	kōatsu-zai
hypotenuse (of a right triangle); hypothenuse *[geom]*	斜辺; しゃへん	shahen
hypothalamus *[anat]*	視床下部	shishō'ka-bu
hypothermia *[physio]*	体温低下	tai'on-teika
hypothesis *[logic] [sci-t]*	仮説; 假說; かせつ	ka'setsu
" *[sci-t]*	仮定	katei
hypothesis testing *(n) [stat]*	仮説検定; 假說檢定	kasetsu-kentei
hypothetical *(adj) [sci-t]*	仮説の	kasetsu no
hypotonic solution *[physio]*	低張液	teichō-eki
hypsochromic shift *[p-ch]*	浅色移動; 淺色移動	senshoku-idō
hypsography *[geogr]*	測高測深(術)	sokkō-soku'shin('jutsu)
hypsometric tinting *(n) [graph]*	測高彩色	sokkō-saishoku
hyssop (herb) *[bot]*	柳薄荷	yanagi-hakka
hysteresis (pl -eses) *[phys]*	履歴(現象)	ri'reki(-genshō)
hysteresis loss *[phys]*	履歴損	ri'reki-son

I

ibis (bird) *[v-zoo]*	鴇; 朱鷺; 鴾; とき	toki
ice *{n}* *[p-ch]*	氷; こおり	kōri
Ice Age *[geol]*	氷期	hyōki
ice bag (ice pack) *[med]*	氷嚢; ひょうのう	hyōnō
" (ice pillow) *[med]*	氷枕	kōri-makura
ice barrier *[hyd]*	防氷堤	bōhyō'tei
iceberg *[ocean]*	氷山	hyōzan
ice breaker *[eng]*	砕氷機	saihyō-ki
ice cap *[hyd]*	氷冠	hyōkan
ice compartment *[mech-eng]*	氷室	hyō-shitsu
ice crystal *[p-ch]*	氷晶	hyōshō
ice crystal rain *[meteor]*	氷晶雨	hyōshō'u
icefish; Silanx microdon *[v-zoo]*	白魚; しらうお	shira-uo
ice floe *[ocean]*	浮氷塊	fuhyō-kai
ice flow *[hyd]*	流氷	ryūhyō
ice fog *[meteor]*	氷霧; ひょうむ	hyōmu
ice island *[ocean]*	氷島	hyōtō
iceland spar *[miner]*	方解石	hōkai-seki
ice load *[eng]*	氷雪荷重	hyōsetsu-kajū
ice(-making) machine *[eng]*	製氷機	seihyō-ki
ice sheet *[hyd]*	氷床	hyōshō
ice shelf *[ocean]*	棚氷	hōhyō; tana-gōri
ice storm *[meteor]*	着氷性嵐	chaku'hyō'sei-arashi
ice water *[ch]* *[cook]*	氷水; こおりみず	kōri-mizu
ichnograph *[graph]*	平面図; 平面圖	heimen-zu
ichthyology *[zoo]*	魚類学; 魚類學	gyorui'gaku
icicle *[hyd]*	氷柱; つらら	tsurara
icing; ice coating *{n}* *[hyd]*	着氷	chaku'hyō
icon *[graph]*	図像; ずぞう	zu'zō
iconography *[graph]*	図解法	zukai-hō
icosahedron *[geom]*	二十面体	nijū'men'tai
" *[geom]*	正二十面体	shō-nijū'men'tai
icositetrahedron *[crys]* *[geom]*	二十四面体	nijūshi'men'tai
-ic salt *[ch]*	第二塩	dai'ni-en
idea; thought *[psy]*	考え; かんがえ	kangae
" ; conception *[psy]*	着想; ちゃくそく	chakusō
" ; general idea; concept *[psy]*	概念; がいねん	gai'nen
" ; notion; intention *[psy]*	観念; 觀念	kannen
ideal *{n}* *[psy]*	理想	risō

ideal *{adj} [math]*	理想的な	risō'teki na
ideal fluid *[fl-mech]*	理想流体	risō-ryūtai
ideal gas *[thermo]*	完全気体	kanzen-kitai
" *[thermo]*	理想気体	risō-kitai
ideal gas equation *[thermo]*	理想気体の式	risō-kitai no shiki
ideal gas law *[thermo]*	理想気体の法則	risō-kitai no hōsoku
ideal solution *[ch]*	理想溶液	risō-yōeki
identification *{n} [bio] [ch]*	同定；どうてい	dōtei
" *{n}*	同一性	dō'itsu-sei
" (of ship, etc.) *{n} [eng]*	確認；かくにん	kaku'nin
identification number *[math] [comput]*	識別番号	shiki'betsu-bangō
identify *[bio]*	同定する	dōtei suru
identify (establish identicalness between substances} *{vb} [ch]*	同定する	dōtei suru
identify *{vb} [legal]*	検証する；検證する	kenshō suru
" *{vb} [psy]*	識別する	shiki'betsu suru
identity (pl -ties) *{n} [alg]*	恒等式	kōtō-shiki
identity element *[alg]*	恒等元	kōtō-gen
" *[alg]*	単位元；たんいげん	tan'i-gen
identity matrix *[alg]*	恒等行列	kōtō-gyō'retsu
ideogram; ideograph *[pr]*	表意文字	hyō'i-moji
idiom *[ling]*	慣用語；かんようご	kan'yō'go
idiomorphic crystal *[crys]*	特異形結晶	toku'i'kei-kesshō
idiomorphic surface *[math]*	自形面	jikei-men
idle; idling *{n} [mech-eng]*	遊転；遊轉	yūten
idle; run idle *{vb} [mech-eng]*	空転する	kūten suru
idler gear *[mech-eng]*	遊び歯車	asobi-ha'guruma
if *{conj}*	若しも	moshimo
IF statement *[comput]*	部分条件文	bubun-jōken'bun
igneous rocks *[petr]*	火成岩；かせいがん	kasei-gan
ignitability *[petr]*	点火性；點火性	tenka-sei
ignite *{vb} [ch]*	点火する	tenka suru
ignition *[ch]*	灼熱；しゃくねつ	shaku-netsu
" *[ch]*	点火	tenka
ignition of arc; firing *[elec]*	点弧	tenko
ignition loss *[ch]*	強熱減量	kyō'netsu-genryō
ignition point *[ch]*	発火点；發火點	hakka-ten
ignition residue *[ch]*	強熱残分	kyō'netsu-zanbun
ignition temperature; kindling temperature *[ch] [thermo]*	発火温度；はっか=おんど	hakka-ondo
ilex (tree) *[bot]*	黐；もち	mochi
illegal *[legal]*	違法	ihō

illegal character *[comput]*	違法文字	ihō-moji
ill function *{n} [ind-eng]*	不具合；ふぐあい	fu'gu'ai
illness *[med]*	病気；びょうき	byōki
" *[med]*	病；やまい	yamai
illuminance; illumination *[opt]*	照度	shōdo
illuminance meter *[eng]*	照度計	shōdo-kei
illuminate *{vb} [opt]*	照明する	shōmei suru
illumination *[opt]*	照明	shōmei
illusion *[psy]*	錯覚；さっかく	sakkaku
illustrated; indicated *{adj} [graph]*	図示の	zushi no
illustrated book *[pr]*	図鑑	zukan
illustrated state *[graph]*	図様；圖樣	zu'yō
illustration *[graph]*	絵図；繪圖；えず	ezu
" (cut-in; cut) *[graph]*	挿絵；挿繪；さしえ	sashi-e
" *[graph]*	図面	zu-men
" (actual example) *[pat]*	実例；實例	jitsu'rei
illustration caption *[graph]*	図題	zu-dai
ilsemannite *[miner]*	藍水鉛華	ransui'enka
image *{n} [elec] [phys]*	影像	eizō
" *{n} [geom] [opt]*	像；ぞう	zō
image focal point *[opt]*	像焦点	zō-shōten
image formation *[graph]*	結像；けつぞう	ketsu-zō
image point *[opt]*	像点	zō-ten
image processing *{n} [comput]*	映像処理	eizō-shori
" *{n} [comput]*	画像処理	gazō-shori
image surface *[opt]*	像面	zō-men
imaginary; fanciful *{adj} [psy]*	架空の	kakū no
imaginary line *[math]*	仮想線；假想線	kasō-sen
imaginary number *[alg]*	虚数；きょすう	kyosū
imagination *[psy]*	想像(力)	sōzō(-ryoku)
imaging *{n} [graph]*	印写；印寫	insha
" *{n} [graph]*	結像	ketsu'zō
imaging power *[phys]*	影像力	eizō-ryoku
imbrication *[graph]*	鱗形；うろこがた	uroko'gata
imitate; copy *{vb}*	真似る；まねる	maneru
imitation; counterfeit *[legal]*	模造品	mozō'hin
imitation *{n} [eng]*	模造	mozō
" *{n}*	模倣；もほう	mohō
imitation sound *[acous]*	擬音	gi'on
immature *[bio]*	未熟	mijuku
immediate; immediately *{adv} {n}*	即時	sokuji
immediate command *[comput]*	即時指令	sokuji-shirei

immediately *{adv}*	即時に	soku'ji ni
immerse; soak; steep *{vb t}*	浸す；ひたす	hitasu
immersion *[astron]*	潜入	sennyū
" *[sci-t]*	浸漬	shinseki
immersion heater *[eng]*	投込電熱器	nage'komi-dennetsu-ki
immersion objective *[opt]*	液浸対物レンズ	eki'shin-tai'butsu-renzu
immigrant *[legal]*	移民	i'min
immiscible *{adj} [ch]*	混合しない	kongō shinai
immobile *{adj}*	動かない	ugokanai
immobile layer *[ch]*	不動層；ふどうそう	fudō-sō
immobilization *[bioch]*	運動抑制	undō-yokusei
" *{n} [bot]*	不動化	fudō'ka
immobilized enzyme *[bioch]*	固定化酵素	kotei'ka-kōso
immune globulin *[immun]*	免疫グロブリン	men'eki-guroburin
immune suppression *[immun]*	免疫抑制	men'eki-yokusei
immunity (pl -ties) *[immun]*	免疫(性)	men'eki(-sei)
immunization *[immun]*	免疫化；めんえきか	men'eki'ka
immunoadsorbent *[immun]*	免疫吸着剤	men'eki-kyūchaku-zai
immunoassay *[immun]*	免疫検定法	men'eki-kentei-hō
immunochemistry *[ch]*	免疫化学	men'eki-kagaku
immunogen *[immun]*	免疫原	men'eki-gen
immunogenetics *[med]*	免疫遺伝学	men'eki-iden'gaku
immunology *[bio]*	免疫学	men'eki'gaku
impact *{n} [mech]*	撃力	geki-ryoku
" *{n} [mech]*	衝動	shōdō
" (shock) *{n} [mech]*	衝撃；しょうげき	shō'geki
" (collision) *{n} [mech]*	衝突；しょうとつ	shō'totsu
impact fracture *[mech]*	衝撃破壊	shōgeki-hakai
impact printer *[graph]*	衝撃式印書装置	shōgeki'shiki-insho-sōchi
impact resistance *[mat]*	耐衝撃性	tai'shōgeki-sei
impact strength *[mech]*	衝撃強さ	shōgeki-tsuyo-sa
impact stress *[mech]*	衝撃応力	shōgeki-ō'ryoku
impartial; impartiality *{n}*	公平	kōhei
impartiality *[math]*	不偏；ふへん	fuhen
" ; fairness *{n}*	公平さ	kōhei-sa
impedance-matching box *[elec]*	整合器	seigō-ki
impediment *{n}*	妨害物	bōgai'butsu
impeller *[mech-eng]*	羽根車；はねぐるま	hane-guruma
" *[mech-eng]*	攪拌翼	kaku'han-yoku
impenetrability *{n} [mat]*	不可入性	fu'kanyū-sei
imperative statement *[comput]*	無条件命令	mu'jōken-meirei
imperfect crystal *[crys]*	不完全結晶	fu'kanzen-kesshō

imperial jade *[miner]*	翡翠；ひすい	hisui
impermeable layer *[geol]*	不透水層	fu'tōsui-sō
impersonal pronoun *[gram]*	不定代名詞	futei-daimei'shi
implant *{n} [med]*	移植片	i'shoku-hen
" *{n} [med]*	埋没物	maibotsu'butsu
implement *{n} [eng]*	道具；どうぐ	dōgu
" *{n} [eng]*	工具；こうぐ	kōgu
implication *[logic]*	含意	gan'i
implicit *{adj}*	黙示の；默示の	moku'ji no
implicit differentiation *[calc]*	陰関数微分法	in'kansū-bibun-hō
implicit identification *[psy]*	暗黙識別	anmoku-shiki'betsu
implicitly; tacitly *{adv}*	暗黙裡に	anmoku'ri ni
implosion *[phys]*	内破；ないは	naiha
implosion synthesis *[ch]*	爆縮合成	baku'shuku-gōsei
imply: suggest *{vb} [psy]*	暗示する	anji suru
imply: mean *{vb} [psy]*	含意する	gan'i suru
" *{vb} [psy]*	意味する	imi suru
importance; significance; weight *{n}*	重要性	jūyō-sei
important *{adj}*	重要な	jūyō na
" *{adj}*	大切な	taisetsu na
import duty *[econ]*	輸入税；ゆにゅう税	yunyū-zei
importing; importation *{n} [econ]*	輸入	yu'nyū
impregnant *[poly-ch] [rub]*	含浸剤	ganshin-zai
impregnated paper *[poly-ch]*	樹脂含浸紙	jushi-ganshin-shi
impregnating agent *[poly-ch] [rub]*	含浸剤	ganshin-zai
impregnation *[ch-eng] [rub]*	含浸	gan'shin
impressed voltage *[elec]*	印加電圧	inka-den'atsu
impression *[dent] [graph]*	印象	inshō
" *[geol]*	圧痕	akkon
" *[geol] [graph]*	痕；あと	ato
impression; issue *[pr]*	刷；さつ	satsu
impression agent *[graph]*	印象剤	inshō-zai
imprint *{n} [pr]*	刊記	kanki
imprinting *{n} [bio]*	刷り込み	suri'komi
improper fraction *[arith]*	仮分数；假分數	ka'bunsū; kari-bunsū
improvement *{n}*	改良	kairyō
improver (a chemical) *[mat]*	向上剤	kōjō-zai
impulse *[mech]*	力積；りきせき	riki-seki
" *[phys] [psy]*	衝動	shōdō
" *[phys] [zoo]*	衝撃；しょうげき	shō'geki
impulse wave *[phys]*	衝撃波	shōgeki-ha
impulsive force *[mech]*	撃力	geki-ryoku

impurity *[sci-t] [sol-st]*	不純物	fujun'butsu
in *{adv}*	中に	naka ni
in detail	逐一；ちくいち	chiku'ichi
in phase *[phys]*	同相	dōsō
in the name of *[legal]*	名義で	meigi de
in this connection; by the way	因みに；ちなみに	chinami ni
inaccurate *{adj}*	不正確な	fu'seikaku na
inactivation *{n}*	非働化	hidō'ka
inadequacy *{n}*	不足；ふそく	fu'soku
inanimate *{adj} [sci-t]*	無生物の	mu'seibutsu no
incandescence *[opt]*	白熱	haku-netsu
incandescent light bulb *[elec]*	電球；でんきゅう	denkyū
incense *{n} [mat]*	香；こう	kō
incense burner *[eng]*	香炉；香爐	kōro
incense powder *[mat]*	抹香	makkō
incense stick *[mat]*	線香	senkō
incenter *{n} [geom]*	内心	nai'shin
incentive *[ind-eng]*	奨励；奬勵	shōrei
incessant *{n}*	不断；不斷；ふだん	fudan
inch (unit of length) *[mech]*	吋；インチ	inchi
inching device *[eng]*	寸動装置	sundō-sōchi
inchworm (insect larva) *[i-zoo]*	尺取り虫	shaku'tori-mushi
incidental facility; supplemental facility *[ind-eng]*	付帯設備；ふたいせつび	futai-setsu'bi
incidental time *[comput]*	雑時間；雜時間	zatsu-jikan
incident angle *[opt]*	入射角	nyūsha-kaku
incident wave *[electr]*	入射波	nyūsha-ha
incineration *[ch]*	灰化；はいか	hai'ka
incinerator *[eng]*	焼却炉；燒却爐	shō'kyaku-ro
incipient *{adj} [med]*	初期の	shoki no
incircle *{n} [geom]*	内接円；内接圓	nai'setsu-en
incisor *{n} [anat]*	前歯；まえば	mae'ba
" *{n} [dent]*	門歯	monshi
inclination *[eng]*	伏角；ふっかく	fukkaku
" *[geol] [math]*	傾き；かたむき	katamuki
" *[geol] [math] [sci-t]*	傾角	keikaku
" *[math] [sci-t]*	傾斜角	keisha-kaku
" (of mind) *[psy]*	志向	shikō
incline (inclined shaft) *{n} [min-eng]*	斜坑	shakō
" *{n} [sci-t]*	勾配；こうばい	kōbai
inclined letter *[comput]*	斜体	shatai
inclined plane *[math]*	斜面	shamen

included angle *[geom]*	夾角；きょうかく	kyō-kaku
inclusion *[crys] [met] [petr]*	包有物；包有物	hōyū'butsu
" *[crys] [met] [petr]*	介在物	kaizai'butsu
incoherent scattering *[phys]*	非干渉性散乱	hi'kanshō'sei-sanran
incombustibility *[ch]*	不燃性	fu'nen-sei
income *[econ]*	所得	sho'toku
income tax *[econ]*	所得税；しょとく税	shotoku-zei
incoming wave *[phys]*	到来波	tōrai-ha
incoming wire *[elec]*	入り線；いりせん	iri-sen
incommensurate structure *{n}*	不整合構造	fu'seigō-kōzō
incompatibility (of facts) *[logic]*	矛盾	mujun
" *[math]*	不適合性	fu'tekigō-sei
incomplete *{adj} [math]*	不完備な	fu'kanbi na
incomplete combustion *[ch]*	不完全燃焼	fu'kanzen-nenshō
incompressibility *[mech]*	非圧縮率	hi'asshuku-ritsu
incompressible fluid *[fl-mech]*	縮まない流体	chijima'nai ryūtai
incongruent melting *{n} [p-ch]*	分解融解	bunkai-yūkai
inconsistency *{n} [logic]*	矛盾；むじゅん	mujun
inconsistent equations *[alg]*	不能な方程式	funō na hōtei'shiki
incorporate *{vb}*	組込む	kumi-komu
increase; add to *{vb t}*	殖やす；増やす	fuyasu
increasing function *[alg]*	増加関数；増加關數	zōka-kansū
increment *[econ]*	増加高	zōka'daka
" *[comput] [math] [sci-t]*	増分；増分	zō-bun
" (amount increased) *[math]*	増加量	zōka-ryō
incremental computer *[comput]*	増分式計算機	zōbun'shiki-keisan-ki
incremental integrator *[comput]*	増分積分機	zōbun-seki'bun-ki
incremental plotter *[comput]*	増分作図機	zōbun-sakuzu-ki
incremental representation *[comput]*	増分表示法	zōbun-hyōji-hō
incrustation *[geol]*	皮殻；皮殻	hi'kaku
incubation *[microbio]*	培養；ばいよう	baiyō
" *[microbio]*	温置	onchi
" *[zoo]*	孵化；ふか	fuka
incubation period *[med]*	潜伏期	sen'puku-ki
incubator *[agr]*	孵卵器	furan-ki
incurrent *{adj} [i-zoo]*	流入の	ryūnyū no
incurvate *{vb t}*	湾曲させる	wan'kyoku saseru
incus (pl incudes) (bone) *[anat]*	砧骨	chin-kotsu; kinuta-bone
incus; anvil cloud; thunderhead *[meteor]*	鉄床雲；かなとこ雲	kana'toko-gumo
indefinite article *[gram]*	不定冠詞	futei-kanshi
indefinite integral *[calc]*	不定積分	futei-sekibun

English	Japanese	Romanization
indefinite pronoun *[gram]*	不定代名詞	futei-daimei'shi
indehiscent fruit *[bot]*	閉果	heika
indemnification method *[econ]*	補償法	hoshō-hō
indemnity *[legal]*	損害賠償	songai-baishō
indentation *[comput]*	字下げ	ji'sage
" *[geom] [mech-eng]*	凹み；へこみ	hekomi
" *{n}*	窪み；くぼみ	kubomi
indentation hardness *[met]*	押込み硬度	oshi'komi-kōdo
independence *[math]*	独立性	doku'ritsu-sei
independent *[math]*	独立；獨立	doku'ritsu
independent assortment, law of *[gen]*	独立の法則	doku'ritsu no hōsoku
independent events *[stat]*	独立事象	doku'ritsu-jishō
independent ionic mobilities, law of *[p-ch] [phys]*	イオン独立移動の=法則	ion-dokuritsu-idō no hōsoku
independent variable *[alg]*	独立変数	doku'ritsu-hensū
indestructibility of matter, law of *[phys]*	物質不滅の法則	busshitsu-fu'metsu no hōsoku
indeterminancy *[math]*	不確定性	fu'kakutei-sei
index (pl -es, indices) *{n} [alg] [crys]*	指数	shisū
" *{n} [comput] [math]*	指標；しひょう	shihyō
" *{n} [pr] [comput]*	索引	saku'in
" *{vb} [mech-eng]*	割出す	wari-dasu
index finger *[anat]*	人差指；食指	hito'sashi-yubi
index gear mechanism *[mech-eng]*	割出し歯車装置	wari'dashi-ha'guruma-sōchi
index head; dividing head *[mech-eng]*	割出し台	wari'dashi-dai
indexing *{n} [comput] [pr]*	索引付け	saku'in-zuke
" *{n} [mech-eng]*	割出し	wari'dashi
index map *[comput]*	索引地図	saku'in-chizu
index of refraction *[phys]*	屈折率	kussetsu-ritsu
india ink *[mat]*	墨；墨；すみ	sumi
Indian hemp *[bot]*	大麻	taima
Indian Ocean *[geogr]*	印度洋；インド洋	Indo-yō
indicate; show *{vb}*	示す；しめす	shimesu
" (point out) *{vb}*	指摘する	shiteki suru
indicated value *[math]*	表示値	hyōji-chi
indication *[comput]*	表示；ひょうじ	hyōji
indicator *[ch]*	指示薬	shiji-yaku
" *[eng]*	指針	shishin
" *{n}*	指標	shihyō
indicator paper *[an-ch]*	指示薬試験紙	shiji'yaku-shiken-shi
indictment *[legal]*	起訴	kiso
indigo *[org-ch]*	藍；あい	ai

indigo blue (color) *[opt] [org-ch]*	藍色	ai-iro
indigo dyeing *{n} [tex]*	藍染め；あいぞめ	ai-zome
indigo (leaf) ball *[org-ch]*	藍玉	ai-dama
indirect *{adj}*	間接の；かんせつの	kan'setsu no
indirect analysis *[an-ch]*	間接分析	kansetsu-bunseki
indirect instruction *[comput]*	間接命令	kansetsu-meirei
indirectly *{adv}*	間接に	kan'setsu ni
indirect measurement *[math]*	間接測定	kansetsu-sokutei
indirect proof *[logic]*	間接証明	kansetsu-shōmei
indispensable; essential *{adj}*	不可欠な	fu'kaketsu na
indium (element)	インジウム	injūmu
individual; person *{n}*	個人；こじん	kojin
individual difference *[stat]*	個人差	kojin-sa
individual error *[stat]*	個人誤差	kojin-gosa
indoor shopping mall *[arch]*	屋内商店街	oku'nai-shōten'gai
induced current *[elecmg]*	誘導電流	yūdō-denryū
induced dipole *[elec]*	誘導双極子	yūdō-sōkyoku'shi
induced emission *[elecmg]*	誘発発光	yūhatsu-hakkō
induced mutation *[gen]*	誘発(突然)変異	yūhatsu(-totsu'zen)-hen'i
induced radiation *[nucleo]*	誘発放射	yūhatsu-hōsha
induced voltage *[elec]*	誘導電圧	yūdō-den'atsu
inducer *[bioch]*	誘導物質	yūdō-busshitsu
induction *[bio] [elecmg] [nucleo]*	誘発；ゆうはつ	yū'hatsu
" *[elec] [elecmg]*	誘導	yūdō
" *[logic]*	帰納；歸納；きのう	kinō
induction field *[elecmg]*	誘導電界	yūdō-denkai
induction motor *[elec]*	誘導電動機	yūdō-dendō-ki
induction period *[p-ch]*	励起期間	reiki-kikan
" *[p-ch]*	誘導期；ゆうどうき	yūdō-ki
inductive effect *[elec]*	誘起効果	yūki-kōka
inductor *[elecmg]*	誘導器	yūdō-ki
" *[elecmg]*	誘導子；ゆうどうし	yūdō'shi
industrial chemistry *[ch]*	工業化学	kōgyō-kagaku
industrial engineering *[eng]*	管理工学	kanri-kōgaku
" ; production control *[eng]*	生産管理	seisan-kanri
industrial equipment *[eng]*	産業機械	sangyō-kikai
industrialization *[ind-eng]*	工業化	kōgyō'ka
industrial machinery *[eng]*	産業機械	sangyō-kikai
industrial poisoning *{n} [med]*	産業中毒	sangyō-chūdoku
industrial pollution *[ecol]*	産業公害	sangyō-kōgai
industrial property right *[legal]*	工業所有権	kōgyō-shoyū-ken
industrial secret *[ind-eng]*	企業機密	kigyō-kimitsu

industrial waste water *[ecol]*	工場排水	kōjō-haisui
industrial water *[ch]*	工業用水	kōgyō-yōsui
industry (pl -tries) *[econ]*	工業; こうぎょう	kōgyō
" *[econ]*	産業; さんぎょう	sangyō
inelastic collision *[mech]*	非弾性衝突	hi'dansei-shōtotsu
inequality *[alg]*	不等(式)	futō(-shiki)
" *[astron]*	差; さ	sa
inert *{adj} [ch]*	不活性な	fu'kassei na
inert gas structure *[ch]*	希ガス構造	ki'gasu-kōzō
inertia *[mech]*	慣性; かんせい	kansei
inertia, law of *[mech]*	慣性の法則	kansei no hōsoku
inertial force *[mech]*	慣性力	kansei-ryoku
inertial guidance *[navig]*	慣性誘導	kansei-yūdō
inertial mass *[mech]*	慣性質量	kansei-shitsu'ryō
inexpensive *{adj} [econ]*	安い; やすい	yasui
infants and toddlers *[bio]*	乳幼児	nyūyō'ji
infauna *[zoo]*	内生動物	naisei-dōbutsu
infection *[med]*	感染; かんせん	kansen
infectious disease *[med]*	伝染病; 傳染病	densen'byō
" *[med]*	感染症	kansen'shō
infectious printing *{n} [photo]*	感染現像	kansen-genzō
inferior angle *[geom]*	劣角; れっかく	rekkaku
inferior arc *[geom]*	劣弧	rekko
inferior conjunction *[astron]*	内合	naigō
inferior planet *[astron]*	内惑星	naiwaku-sei
infertile land; barren soil *[agr]*	痩せ地; やせち	yase-chi
infiltration *[geol]*	浸入	shinnyū
" *[geol] [hyd] [met]*	溶浸	yōshin
infinite *[math]*	無限(な)	mugen (na)
infinite series *[alg]*	無限級数	mugen-kyūsū
infinite set *[math]*	無限集合	mugen-shūgō
infinitesimal *[calc]*	無限小	mugen-shō
infinitesimal cross section *[math]*	微小断面積	bishō-dan'menseki
infinitive *[gram]*	不定詞	futei'shi
infinitive mood *[gram]*	不定法	futei-hō
infinity *[arith]*	無限大	mugen-dai
inflection *[ling]*	抑揚; よくよう	yoku'yō
inflection point *[calc]*	変曲点; 變曲點	henkyoku-ten
" *[calc]*	屈折点	kussetsu-ten
inflorescence *[bot]*	花序	kajo
influence *{n}*	影響; えいきょう	eikyō
influence; power; strength *{n}*	勢力	seiryoku

influent *{n} [sci-t]*	流入液	ryūnyū-eki
influenza; flu *[med]*	流行性感冒	ryūkō'sei-kanbō
information; bulletin; message *[comm]*	通報；つうほう	tsūhō
" *[comput]*	情報；じょうほう	jōhō
information capacity *[photo]*	情報容量	jōhō-yōryō
information processing *{n} [comput]*	情報処理；情報處理	jōhō-shori
information retrieval *[comput]*	情報検索；情報檢索	jōhō-kensaku
information source *[comput]*	情報源	jōhō-gen
information technology *[comm]*	情報技術学	jōhō-gijutsu'gaku
infra- *{prefix}*	以下の-	ika no-
" ; sub-; under- *{prefix}*	下の-	shita no-
infraclass *[bio] [syst]*	下綱；かこう	kakō
infraction; breach *{n} [legal]*	違反；いはん	i'han
infrared absorption *[elecmg]*	赤外吸収	sekigai-kyūshū
infrared astronomy *[astron]*	赤外線天文学	sekigai'sen-tenmon'gaku
infrared homing *{n} [eng]*	赤外線自動追尾	sekigai'sen-jidō-tsui'bi
infrared radiation (rays)*[elecmg]*	赤外線；せきがい線	sekigai-sen
infrared spectrometer *[spect]*	赤外分光器	sekigai-bunkō-ki
infrared spectrophotometer *[spect]*	赤外分光光度計	sekigai-bunkō-kōdo-kei
infrared spectrophotometry *[spect]*	赤外線分光光度法	sekigai'sen-bunkō-kōdo-hō
infrared spectroscopy *[spect]*	赤外分光学	sekigai-bunkō'gaku
infrasound; inaudible sound *[acous]*	不可聴音	fu'kachō-on
infrastructure *[civ-eng]*	基盤構造	kiban-kōzō
infrasubspecies *[bio]*	亜種以下	a'shu-ika
infringement *[legal]*	侵害；しんがい	shingai
" *[legal]*	低触	teishoku
infructescence *[bot]*	果実序；果實序	ka'jitsu'jo
infusion; decoction *{n} [pharm]*	煎薬；せんやく	sen'yaku
" *{n} [ch]*	浸剤	shinzai
ingestion; intake *{n} [bio]*	接取	sesshu
ingot iron *[met]*	熔鉄；よぅてつ	yō-tetsu
ingot-making *{n} [met]*	造塊	zō-kai
ingot mold; mold *{n} [cer]*	鋳型；鑄型；いがた	i'gata
ingot steel; molten steel *[steel]*	熔鋼	yōkō
ingredient; ingredients *[cook]*	具；ぐ	gu
ingredient *[pharm]*	成分	seibun
inhalant *{n} [pharm]*	吸入剤	kyūnyū-zai
inheritance; heredity *[bio]*	遺伝；遺傳；いでん	i'den
" ; estate *[legal]*	遺産；いさん	i'san
inhibit *{n} [comput]*	禁止	kinshi
" ; prohibit *{vb} [sci-t]*	禁止する	kinshi suru
inhibition *[psy] [sci-t]*	抑制；よくせい	yokusei

English	Japanese	Romanization
inhibition *[sci-t]*	阻害；そがい	sogai
inhibitor *[bioch] [ch]*	阻害物質	sogai-busshitsu
" *[ch]*	抑制剤	yokusei-zai
" (polymerization) *[poly-ch]*	重合防止剤	jūgō-bōshi-zai
" (polymerization) *[poly-ch]*	重合禁止剤	jūgō-kinshi-zai
initial; initials *[pr]*	頭文字	kashira-moji
initial period *[sci-t]*	初期；しょき	shoki
initial-period condition *[ch]*	初期条件	shoki-jōken
initial value *[math]*	初期値	shoki-chi
initial velocity *[mech]*	初速（度）	sho'soku(do)
initiating explosive *[ord]*	起爆薬；きばくやく	kibaku-yaku
initiation reaction *[poly-ch]*	開始反応；開始反應	kaishi-hannō
initiator *[ord]*	起爆薬	kibaku-yaku
" (polymerization) *[poly-ch]*	重合開始剤	jūgō-kaishi-zai
injection *[mech-eng]*	注入	chūnyū
" *[med]*	注射；ちゅうしゃ	chūsha
injection blow molding *[eng]*	射出吹込成形	sha'shutsu-fuki'komi-seikei
injection molding *[eng] [poly-ch]*	射出成形	sha'shutsu-seikei
injection pressure *[eng]*	射出圧力	sha'shutsu-atsu'ryoku
injection rate *[eng]*	射出率	sha'shutsu-ritsu
injurious insect *[i-zoo]*	害虫	gaichū
injury surface *[bio]*	創面	sōmen
ink-attachment property *[pr]*	インク着肉性	inku-chaku'niku-sei
ink-brush painting *[graph]*	墨絵；すみえ	sumi-e
inkfish (mollusk) *[i-zoo]*	烏賊；いか	ika
inking *{n} [civ-eng] [graph]*	墨入れ	sumi-ire
" *{n} [pr]*	着肉	chaku'niku
inkstone *[comm]*	硯；すずり	suzuri
inland *[geogr]*	内陸	nai'riku
Inland Sea (of Japan) *[geogr]*	瀬戸内海；瀨戸內海	Seto-Naikai
inland sea *[ocean]*	内海	naikai
inland water *[geogr]*	陸水	riku-sui
inlay (ornament) *{n} [graph]*	象嵌；象眼	zōgan
inlet *[geogr]*	入江	iri'e
" ; *{n} [met] [nav-arch]*	入口；いりぐち	iri'guchi
inn *[arch]*	旅館；りょかん	ryokan
inner core *[geol]*	内核	nai-kaku
inner cylinder *[eng]*	内筒	naitō
inner layer *[geol]*	内層；ないそう	naisō
inner-lined steel pipe *[met]*	内張鋼管	uchi'bari-kōkan
inner lining *{n} [mat]*	内張；うちばり	uchi-bari
inner planet *[astron]*	内惑星	naiwaku-sei

inner product *[math]*	点乗積; 點乗積	ten-jōseki
inner side *[math]*	内側	uchi-gawa
inner tube *[eng]*	内筒	naitō
innovation; renovation *{n}*	革新	kakushin
inoculate *{vb} [bio]*	接種する	sesshu suru
inoculation *[bio]*	接種; せっしゅ	sesshu
inoculum (pl -la) *[microbio]*	接種物	sesshu'butsu
" *[microbio]*	接種材料	sesshu-zairyō
inorganic *{adj} [ch] [mat]*	無機の; むきの	muki no
inorganic acid *[ch]*	鉱酸; 鑛酸	kōsan
inorganic chemistry *[ch]*	無機化学	muki-kagaku
inorganic compound *[ch]*	無機化合物	muki-kagō'butsu
inorganic matter *[ch] [geol]*	無機質	muki-shitsu
inorganic peroxide *[ch]*	無機過酸化物	muki-ka'sanka'butsu
inorganic pigment *[mat]*	無機顔料	muki-ganryō
inorganic polymer *[poly-ch]*	無機高分子	muki-kōbunshi
in-plane anisotropy *[crys]*	面内異方性	mennai-ihō'sei
in-plane vibration *[crys]*	面内振動	mennai-shindō
input *{n} [comput] [electr]*	入力; にゅうりょく	nyū'ryoku
" *{vb} [comput]*	入力する	nyū'ryoku suru
input device *[eng]*	入力装置	nyū'ryoku-sōchi
input/output *{n} [comput]*	入出力	nyū'shutsu-ryoku
input port (or orifice) *[eng]*	投入口	tōnyū-kō
input unit; input device *[comput]*	入力装置	nyū'ryoku-sōchi
inquilinism *[zoo]*	住み込み共生	sumi'komi-kyōsei
inquiry (pl -ries) *[comput] [legal]*	照会	shōkai
" *[comput]*	問合せ; といあわせ	toi-awase
inscribed *[geom]*	内接する	naisetsu suru
inscribed polygon *[geom]*	内接多角形	naisetsu-takak(u)kei
inscription *[geom]*	内接	nai'setsu
" *[graph]*	銘刻	meikoku
insect *[i-zoo]*	昆虫; こんちゅう	konchū
" *[i-zoo]*	虫; 蟲; むし	mushi
Insecta *[i-zoo]*	昆虫類	konchū-rui
insecticide *[mat]*	殺虫剤	satchū-zai
Insectivora *[v-zoo]*	食虫類	shoku'chū-rui
insectivorous *{adj} [v-zoo]*	虫食性の	chū'shoku'sei no
insectivorous plant *[bot]*	食虫植物	shoku'chū-shokubutsu
insect-proofing preparation *[pharm]*	防虫製剤	bōchū-seizai
insect resistance *[bot]*	耐虫性	taichū-sei
insect wax *[mat]*	虫白蠟	chūhaku-rō
" *[mat]*	昆虫蠟	konchū-rō

insensitiveness *{n}*	不感度	fukan-do
insert *{vb} [comput]*	挿入する	sōnyū suru
" (set in; snap in) *{vb t}*	嵌める；はめる	hameru
" (place into) *{vb t}*	入れる；いれる	ireru
insert, fitting *{vb} [mech-eng]*	嵌挿する	kansō suru
inserted tool *[eng]*	植刃工具	ue'ba-kōgu
insertion *[comput] [mol-bio]*	挿入；そうにゅう	sōnyū
insertional inactivation *[microbio]*	挿入失活	sōnyū-shikkatsu
insertion sort *[comput]*	挿入法分類	sōnyū'hō-bunrui
inset; gore *[cl]*	襠；まち	machi
inside *{n} [math]*	中；なか	naka
inside; interior; one's home *{n}*	内；うち	uchi
inside appearance *{n}*	内相	naisō
inside diameter *[math]*	内径；ないけい	naikei
inside measure *[math]*	内法；うちのり	uchi-nori
inside (or indoor) wiring *{n} [elec]*	屋内配線	oku'nai-haisen
insolation *[astron]*	日射	nissha
inspection *[ind-eng]*	検査；檢查；けんさ	kensa
" *{n}*	検閲	ken'etsu
inspissation *[ch] [geoch]*	濃厚化	nōkō'ka
instability *[phys]*	不安定性	fu'antei-sei
install *{vb} [ind-eng]*	設置する	setchi suru
installation *[comput] [ind-eng]*	設置	setchi
" *[ind-eng]*	据え付け	sue-tsuke
instant *{n} [phys]*	瞬間；しゅんかん	shunkan
instantaneous value *[phys]*	瞬時値	shun'ji-chi
instantaneous velocity *[mech]*	瞬間速度	shunkan-sokudo
instantaneous voltage *[elec]*	瞬時電圧	shunji-den'atsu
instant noodles *[cook]*	即席麺	soku'seki-men
instar *[i-zoo]*	令；齢；れい	rei
instep (of foot) *[anat]*	足の甲	ashi no kō
instinct *[psy]*	本能	honnō
institute *{vb}*	設定する	settei suru
instruct; teach *{vb t} [comm]*	教示する	kyōji suru
instruction *[comput]*	命令；めいれい	meirei
" *[comput]*	指示	shiji
" *[comput]*	指令	shirei
instruction fetch *[comput]*	命令取出し	meirei-tori'dashi
instruction manual *[pr]*	取扱説明書	tori'atsukai-setsumei'sho
instrument *[eng]*	器具；きぐ	kigu
" *[eng]*	機器；きき	kiki
instrumental analysis *[eng]*	計装分析	keisō-bunseki

instrumental analysis *[eng]*	機器分析	kiki-bunseki
instrumental error *[eng]*	器差；きさ	kisa
instrument approach *[navig]*	計器着陸	keiki-chaku'riku
instrumentation *[eng]*	計装；計裝	keisō
" *[eng]*	計測；けいそく	kei'soku
instrument drawing *{n} [graph]*	器具製図；器具製圖	kigu-seizu
instrument flight *[navig]*	計器飛行	keiki-hikō
instrument landing system *[navig]*	計器着陸方式	keiki-chaku'riku-hōshiki
insufficiency; inadequacy *{n}*	不足；ふそく	fu'soku
insulating covering film *[mat]*	絶縁被膜	zetsuen-hi'maku
insulating (insulation) film *[elec]*	絶縁膜	zetsuen-maku
insulating material *[mat]*	絶縁剤；絶縁材	zetsuen-zai
insulating paint *[mat]*	絶縁塗料	zetsuen-toryō
insulating paper *[mat]*	絶縁紙；ぜつえんし	zetsuen-shi
insulation; nonconductor *[elec]*	絶縁(体)	zetsu'en(-tai)
insulation distance *[elec]*	絶縁距離	zetsuen-kyori
insulation layer *[elec] [mat]*	絶縁層	zetsuen-sō
insulation resistance *[elec]*	絶縁抵抗	zetsuen-teikō
insulator *[elec]*	碍子；がいし	gaishi
" *[elec] [mat]*	絶縁体；絶縁體	zetsuen'tai
" *[mat]*	絶縁物	zetsuen'butsu
insulin *[bioch]*	インシュリン	inshurin
insurance *[econ]*	保険；ほけん	hoken
insurance money *[econ]*	保険金	hoken-kin
intaglio (pl -glios) *[lap]*	凹板；おうばん	ō-ban
" *[lap]*	沈め彫(細工)	shizume-bori(-zaiku)
intaglio printing *{n} [pr]*	凹版印刷	ō'han-in'satsu
intake; ingestion *{n} [bio]*	接取；せっしゅ	sesshu
intake valve *[mech-eng]*	吸気弁；吸氣瓣	kyūki-ben
integer(s); whole number(s) *[arith]*	整数；せいすう	seisū
integer constant *[math]*	整定数	sei-teisū
integer control *[comput]*	整数制御	seisū-seigyo
integer programming *{n} [comput]*	整数計画法	seisū-keikaku-hō
integral; solution *{n} [calc]*	解；かい	kai
integral calculus *{n} [calc]*	積分；せきぶん	seki'bun
integral density *[photo]*	積算濃度	sekisan-nōdo
integrally; as a single body	一体的に	ittai'teki ni
integral number *[arith]*	整数；せいすう	seisū
integral scale *[calc]*	積分尺度	sekibun-shaku'do
integrand *[calc]*	被積分関数	hi-sekibun-kansū
integrate *{vb t} [calc]*	積分する	sekibun suru
integrated circuit; IC *[electr]*	集積回路	shūseki-kairo

integrated magnitude *[astron]*	積分等級	sekibun-tōkyū
integration *[calc]*	積分法	sekibun-hō
" *[comput]*	統合	tōgō
" *[electr]*	集積化	shūseki'ka
" *[eng] [math]*	積算；せきさん	sekisan
integration by parts *[calc]*	部分積分	bubun-sekibun
integration (-ing) circuit *[elec]*	積分回路	sekibun-kairo
integrator *[electr]*	積分器	sekibun-ki
integrity *[comput]*	保全(性)	hozen(-sei)
" *[comput]*	完全性	kanzen-sei
integument *[anat] [zoo]*	外皮；外被	gaihi
" *[anat]*	皮膚；ひふ	hifu
" *[bot]*	珠皮	shuhi
intellectual property right *[legal]*	知的所有権	chi'teki-shoyū'ken
intelligence *[comm] [psy]*	情報；じょうほう	jōhō
" *[psy]*	知能；ちのう	chi'nō
intelligence learning *{n} [comput]*	知能学習	chinō-gakushū
intelligence quotient *[psy]*	知能指数	chinō-shisū
intelligibility *[psy]*	了解度	ryōkai-do
intensification *[photo]*	補力	ho'ryoku
intensity (pl -ties) *[phys]*	強度；きょうど	kyōdo
" (of light) *[phys]*	強さ(光の)	tsuyo-sa (hikari no)
intensity of field *[phys]*	場の強さ	ba no tsuyo-sa
intensive *{adj} [agr]*	集約的な	shūyaku'teki na
intention *[legal] [psy]*	意思；いし	ishi
" *[psy]*	意図；意圖；いと	ito
intentionally *{adv} [psy]*	故意に	ko'i ni
inter- *{prefix}*	間の-	aida no-
interaction *[comput] [sci-t]*	相互関係	sōgo-kankei
" *[comput]*	対話；對話；たいわ	taiwa
" *[phys] [stat]*	交互作用	kōgo-sayō
" *[phys]*	相互作用	sōgo-sayō
interactive graphics *[comput]*	対話型図形	taiwa'gata-zukei
interactive mode *[comput]*	会話形；會話形	kaiwa-gata
interactive processing *{n} [comput]*	会話形処理	kaiwa'gata-shori
interbreeding *{n} [gen]*	異種交配	ishu-kōhai
intercalary day *[astron]*	閏日；うるうび	urū-bi
intercalation *[astron]*	置閏；ちじゅん	chijun
" *[geol]*	中間質	chūkan'shitsu
intercalation compound *[ch]*	挿入化合物	sōnyū-kagō'butsu
intercept *{n} [an-geom]*	切片	seppen
intercession; (inter)mediation *{n}*	仲介	chūkai

English	Japanese	Romanization
interchange *{n} [comput]*	交換；こうかん	kōkan
" *{n} [traffic]*	インターチェンジ	intāchenji
" *{vb}*	取り交わす	tori-kawasu
interchangeability *[comput]*	互換性	gokan-sei
intercom (instrument) *{n} [comm]*	相互通信器	sōgo-tsūshin-ki
intercommunicating system *[comm]*	相互通信方式	sōgo-tsū'shin-hō'shiki
intercontinental *{adj} [geogr]*	大陸間の	tairiku'kan no
intercontinental ballistic missile *[ord]*	大陸間弾道弾	tairiku'kan-dandō-dan
interdiffusion *[p-ch]*	相互拡散	sōgo-kakusan
interest *[econ]*	金利	kinri
" *[econ]*	利子；りし	rishi
" *[econ]*	利息	ri'soku
" *[psy]*	興味；きょうみ	kyōmi
interest; amusement *[psy]*	面白さ；おもしろさ	omoshiro-sa
interesting; amusing *{adj}*	面白い	omoshiroi
interface *[ch] [p-ch] [sci-t]*	界面；かいめん	kaimen
interfacial angle *[crys]*	面角	men-kaku
interfacial bond *[adhes]*	界面接着	kaimen-setchaku
interfacial electrochemistry *[p-ch]*	界面電気化学	kaimen-denki'kagaku
interfacial electrokinetic phenomenon *[p-ch]*	界面動電現象	kaimen-dōden-genshō
interfacial failure *[adhes]*	界面破壊；界面破壞	kaimen-hakai
interfacial polycondensation *[poly-ch]*	界面重縮合	kaimen-jū'shukugō
interfacial polymerization *[poly-ch]*	界面重合	kaimen-jūgō
interfacial tension *[phys]*	界面張力	kaimen-chō'ryoku
interference *[comm] [phys]*	防害	bōgai
" *[legal]*	紛争；紛爭	funsō
" *[phys]*	干渉；かんしょう	kanshō
interference fringes *[opt]*	干渉縞	kanshō-jima
interference spectroscopy *[spect]*	干渉分光方法	kanshō-bunkō-hōhō
interferometer *[opt]*	干渉計	kanshō-kei
interglacial period *{adj} [geol]*	間氷期	kanpyō-ki
intergranular corrosion resistance *[met]*	耐粒界腐食性	tai'ryūkai-fushoku-sei
interior (inside region) *{n}*	内部	nai'bu
" (inside) *{n}*	中；なか	naka
" (in room or house) *{n}*	室内	shitsu'nai
interior angle *[math]*	内角	nai-kaku
interjection *[gram]*	間投詞	kantō'shi
interlayer effect *[photo]*	重層効果	jūsō-kōka
interlocking *{n} [mech-eng]*	連動	rendō
interlocking; intertwining *{n} [tex]*	絡み合；からみあい	karami'ai

English	Japanese	Romanization
interlocking device *[mech-eng]*	連動装置	rendō-sōchi
interlude *{n}*	合間；あいま	aima
intermediate (a product) *[ch]*	中間体	chūkan'tai
intermediate (a substance) *{n}*	中間物	chūkan'butsu
" *{adj}*	中間の	chūkan no
intermediate frequencies *[elec]*	中間周波数	chūkan-shūha'sū
intermediate state *[quant-mech]*	中間状態	chūkan-jōtai
intermediate wave *[elecmg]*	中短波	chū-tanpa
intermeshing *{n} [mech-eng]*	噛み合い；かみあい	kami'ai
intermeshing precision *[mech-eng]*	噛合精度	kami'ai-seido
intermetallic compound *[ch]*	金属間化合物	kinzoku'kan-kagō'butsu
intermittent *{adj} [eng] [mech-eng]*	間欠の；かんけつの	kanketsu no
intermittent ringing *{n} [comm]*	断続信号；斷續信號	dan'zoku-shingō
intermittent signals *[comm]*	断続信号	dan'zoku-shingō
intermodulation crosstalk *[electr]*	相互変調漏話	sōgo-henchō-rōwa
intermolecular force *[ch]*	分子間力	bunshi'kan-ryoku
internal angle *[math]*	内角	nai-kaku
internal combustion engine *[mech-eng]*	内燃機関	nai'nen-kikan
internal diffusion *[ch-eng]*	内部拡散	naibu-kakusan
internal fertilization *[physio]*	体内受精	tai'nai-jusei
internal force *[mech]*	内力；ないりょく	nai-ryoku
internal gear *[mech-eng]*	内(歯)歯車	uchi(ba)-ha'guruma
internal gear drive *[mech-eng]*	内歯車駆動	uchi-ha'guruma-kudō
internal indicator *[ch]*	内部指示薬	naibu-shiji'yaku
internally add *{vb}*	内添する	naiten suru
internal medicine specialist *[med]*	内科医；内科醫	naika'i
internal memory *[comput]*	内部記憶	naibu-ki'oku
internal pressure *[phys]*	内圧；ないあつ	nai-atsu
internal procedure *[comput]*	内部手続	naibu-te'tsuzuki
internal resistance *[elec]*	内部抵抗	naibu-teikō
internal shrinkage *[met]*	内引け；うちびけ	uchi-bike
internal standard *[an-ch] [spect]*	内(部)標準法	nai(bu)-hyōjun-hō
" *[spect]*	内標準	nai-hyōjun
internal thread *[mech]*	雌螺子；めねじ	me-neji
international *{adj}*	国際の；國際の	koku'sai no
international date line *[astron]*	国際日付変更線	kokusai-hi'zuke-henkō-sen
international filing date *[pat]*	国際出願日	kokusai-shutsu'gan-bi
international filing number *[pat]*	国際出願番号	kokusai-shutsu'gan-bangō
international patent classification	国際特許分類	kokusai-tokkyo-bunrui
international search *[pat]*	国際調査	kokusai-chōsa
international searching authority *[pat]*	国際調査機関	kokusai-chōsa-kikan

international standard *[phys]*	国際規格	kokusai-kikaku
international unit *[bio]*	国際単位; 國際單位	kokusai-tan'i
international unit of enzyme activity *[bioch]*	酵素の国際単位	kōso no kokusai-tan'i
international waters *[ocean]*	公海; こうかい	kōkai
internuclear distance *[p-ch]*	核間距離	kaku'kan-kyori
interpenetrating network *{n}*	相互貫通網目	sōgo-kantsū-ami'me
interphase *{n} [cyt]*	静止期	seishi-ki
interplanar spacing *{n} [crys]*	面間隔	men-kankaku
interplanetary material *[astron]*	惑星間物質	wakusei'kan-busshitsu
interplanetary space *[astron]*	惑星間空間	wakusei'kan-kūkan
interpolation *[alg]*	補間	hokan
" *[alg]*	内挿; ないそう	naisō
interpret *{vb} [comm]*	解釈する	kaishaku suru
" *{vb} [comm]*	通訳する	tsū'yaku suru
interpretation *[comm]*	解釈; 解釋	kaishaku
" *[comm] [pr]*	解說	kaisetsu
interpreter (a person); interpreting *{n} [comm]*	通訳; つうやく	tsū'yaku
interrupt *{n} [comput]*	割込み	wari'komi
" *{vb}*	中断する	chūdan suru
interruption *{n} [comput]*	割込み	wari'komi
interruption *{n}*	中断; 中斷	chūdan
interruption control circuit *[elec]*	割込制御回路	wari'komi-seigyo-kairo
interrupt request *[comput]*	割込み要求	wari'komi-yōkyū
intersection *[comput]*	交差; こうさ	kōsa
" *[traffic]*	交差点	kōsa-ten
" *[math]*	交わり; まじわり	majiwari
intersection angle *[civ-eng]*	交角	kō-kaku
intersection number *[math]*	交わり数	majiwari-sū
interstellar space *[astron]*	星間空間	seikan-kūkan
interstitial matter *[sci-t]*	間隙物質	kangeki-busshitsu
interstitial position *[sci-t]*	割込み位置	wari'komi-ichi
intertidal zone *[geol]*	潮間帯; 潮間帶	chōkan-tai
interurban railway *[rail]*	市間鉄道	shikan-tetsu'dō
interval (music) *{n} [acous]*	音程	ontei
" (of time) *{n}*	間; ま; あいだ	ma; aida
interval timer *[eng]*	間隔計時機構	kankaku-keiji-kikō
intestine *{n} [anat]*	腸; ちょう	chō
intestine medicine *[pharm]*	整腸剤	seichō-zai
intoxication *[med]*	中毒; ちゅうどく	chūdoku
intra- *{prefix}*	内(側)に-	uchi(gawa) ni-

English	Japanese	Romanization
intraannular *{adj} [ch]*	環内の	kannai no
intramolecular rearrangement *[ch]*	分子内転位	bunshi'nai-ten'i
intransitive verb *[gram]*	自動詞；じどうし	ji-dōshi
intrinsic; inborn *{adj}*	本質的な	honshitsu'teki na
" *{adj}*	真性の；しんせいの	shinsei no
intrinsic property *[mat]*	真性的性質	shinsei'teki-seishitsu
intrinsic variable star *[astron]*	本質的変光星	honshitsu'teki-henkō-sei
intrinsic viscosity *[p-ch]*	固有粘度	koyū-nendo
" *[p-ch]*	極限粘度	kyoku'gen-nendo
introduction; preface *[pr]*	序論；じょろん	joron
introduction; foreword *[pr]*	緒言	shogen
introduction (into) *{n}*	導入；どうにゅう	dōnyū
intrusion; infiltration *[geol]*	浸入；しんにゅう	shinnyū
intrusion rock *[geol]*	迸入岩	hei'nyū-gan
intuition *[psy]*	直観；ちょっかん	chokkan
in vacuo *{adv} [phys]*	真空内で	shinkū-nai de
invalid; invalidity *{n} [legal]*	無効；むこう	mukō
invalid (a person) *{n} [med]*	病弱者	byōjaku'sha
invariable plane *[astron] [mech]*	不変面；ふへんめん	fuhen-men
invasion *[med] [mil]*	浸入	shinnyū
invention *[pat]*	発明；はつめい	hatsu'mei
inventiveness *[pat]*	進歩性；しんぽせい	shinpo-sei
inventor *[pat]*	発明者；發明者	hatsumei'sha
inventorship; coinventor *[pat]*	共同発明者	kyōdō-hatsumei'sha
inventory; stock *{n} [econ] [ind-eng]*	在庫量	zaiko-ryō
inventory control *[econ] [ind-eng]*	在庫管理	zaiko-kanri
inverse *{n} [alg]*	逆；ぎゃく	gyaku
" ; reverse *{adj} [math]*	逆(の)	gyaku no
inverse function *[alg]*	逆関数；逆闘数	gyaku-kansū
inverse hour; inhour *[nucleo]*	逆時間	gyaku-jikan
inversely proportional *{adj} [alg]*	反比例の	han'pirei no
inversely trigonometric functions *[trig]*	逆三角関数	gyaku-sankaku-kansū
inverse matrix *[alg]*	逆行列	gyaku-gyō'retsu
inverse ratio *[alg]*	逆比	gyaku-hi
inverse square law *[phys]*	逆二乗の法則	gyaku-nijō no hōsoku
inversion *[alg] [ch]*	反転；反轉	hanten
" *[meteor]*	逆転；ぎゃくてん	gyaku'ten
invert *{vb} [comput]*	反転する	hanten suru
" (turn upside down) *{vb t}*	逆さ(ま)にする	sakasa(ma) ni suru
invertebrates *[i-zoo]*	無脊椎動物	mu'sekitsui-dōbutsu
invertebrate zoology *[bio]*	無脊椎動物学	mu'sekitsui-dōbutsu'gaku

inverted *{n} [math]*	逆さ(ま)	sakasa(ma)
inverted image *[opt]*	倒(立)像	tō(ritsu)-zō
inverted layer *[meteor]*	逆転層; 逆轉層	gyaku'ten-sō
invertose; invert sugar *[cook]*	転化糖; てんかとう	tenka'tō
investigate *{vb}*	調べる	shiraberu
inviscid fluid *[fl-mech]*	非粘性流体	hi'nensei-ryūtai
" *[phys]*	完全流体	kanzen-ryūtai
invisibility; invisibleness *{n} [opt]*	不可視	fu'kashi
invisible *{adj} [opt] [graph]*	隠れた; かくれた	kakureta
" *{adj} [opt]*	見えない	mie'nai
invisible glass *[mat]*	不可視ガラス	fukashi-garasu
in vitro *{adj} [bio]*	生体外での	seitai-gai de no
in vitro *{adv} [bio]*	生体外で	seitai-gai de
in vivo *{adv} ({adj}) [bio]*	生体内で(の)	seitai-nai de (no)
invoice *{n} [econ]*	明細記入請求書	meisai-ki'nyū-seikyū'sho
" *{n} [econ]*	送り状	okuri'jō
involucre *[bot]*	総包; 總包	sōhō
involuntary muscle *[physio]*	不随意筋; 不隨意筋	fu'zui'i-kin
involute *{adj} [bot]*	内巻きの	uchi'maki no
inwardly directed *{adj}*	内向きの	uchi'muki no
inward membrane current *[physio]*	内向き膜電流	uchi'muki-maku-denryū
iodate *[inorg-ch]*	沃素酸塩; 沃素酸鹽	yōso'san-en
iodic acid *[inorg-ch]*	沃素酸	yōso-san
iodide *[ch]*	沃化物	yōka'butsu
iodide ion *[ch]*	沃素イオン	yōso-ion
iodine (element)	沃素; ようそ	yōso
iodine bromide	臭化沃素	shūka-yōso
iodine chloride	塩化沃素	enka-yōso
iodine fluoride	弗化沃素	fukka-yōso
iodine metabolism *[physio]*	沃素代謝	yōso-taisha
iodine monochloride	一塩化沃素	ichi'enka-yōso
iodine number; iodine value *[an-ch]*	沃素価	yōso-ka
iodine oxide	酸化沃素	sanka-yōso
iodine trichloride	三塩化沃素	san'enka-yōso
iodometry; iodimetry *[an-ch]*	沃素滴定	yōso-teki'tei
ion *[ch]*	イオン	ion
ion-exchange resin *[poly-ch]*	イオン交換樹脂	ion-kōkan-jushi
ionic bond(ing) *{n} [p-ch]*	イオン結合	ion-ketsu'gō
ionic radius *[p-ch]*	イオン半径	ion-hankei
ionic strength *[p-ch]*	イオン強度	ion-kyōdo
ionization *[ch]*	電離; でんり	denri
ionization chamber *[nucleo]*	電離箱	denri-bako

ionizing radiation *[nucleo]*	電離性放射線	denri'sei-hōsha'sen
ionosphere *[geophys]*	電離層	denri-sō
ion pair *[nucleo]*	イオン対	ion-tsui
ipecac; Cephaelis ipecacuanha *[bot] [pharm]*	吐根；とこん	tokon
iridescence *{n} [miner] [opt]*	暈色；うんしょく	un-shoku
iridescent *{adj} [opt]*	虹色の	niji-iro no
iridescent color *[opt]*	玉虫色	tama'mushi-iro
iris (pl -es, irides) *[anat] [opt]*	虹彩	kōsai
" (flower) *[bot]*	燕子花；杜若	kaki'tsubata
" (flower) *[bot]*	菖蒲；しょうぶ	shōbu
iris; stop *{n} [opt] [photo]*	(光束)絞り	(kōsoku-)shibori
iris diaphragm *[opt]*	虹彩絞り	kōsai-shibori
Irish moss *[bot]*	トチャカ	tochaka
iron (element)	鉄；鐵；てつ	tetsu
" (for ironing clothes) *{n} [eng]*	熨；のし	noshi
Iron Age *[archeo]*	鉄器時代；鐵器時代	tekki-jidai
iron arsenide	砒化鉄	hika-tetsu
iron black *[ch]*	鉄黒；鐵黑	tetsu-guro
iron boride	硼化鉄；ほうかてつ	hōka-tetsu
iron borofluoride	硼弗化鉄	hō'fukka-tetsu
iron carbide	炭化鉄	tanka-tetsu
iron carbonate	炭酸鉄	tansan-tetsu
iron castings *[met]*	銑ダライ粉	sendarai-ko
iron chloride	塩化鉄	enka-tetsu
iron chloride oxide	塩化酸化鉄	enka-sanka-tetsu
iron citrate	枸櫞酸鉄	ku'ensan-tetsu
iron complex *[ch]*	鉄錯体	tetsu-saku'tai
iron core *[elecmg]*	鉄心	tesshin
iron fluoride	弗化鉄	fukka-tetsu
iron gallate	没食子酸鉄	bosshokushi'san-tetsu
iron hydroxide	水酸化鉄	suisan'ka-tetsu
iron loss *[elecmg]*	鉄損	tesson
iron matrix *[met]*	地鉄；じてつ	ji-tetsu
iron modulus *[ch]*	鉄率	tetsu-ritsu
iron nitrate	硝酸鉄	shōsan-tetsu
iron nitride	窒化鉄	chikka-tetsu
iron ore *[geol]*	鉄鉱石；鐵鑛石	tetsu-kōseki
iron oxalate	蓚酸鉄	shūsan-tetsu
iron oxide	酸化鉄	sanka-tetsu
iron phosphate	燐酸鉄	rinsan-tetsu
iron phosphide	燐化鉄；りんかてつ	rinka-tetsu
iron pyrite *[miner]*	黄鉄鉱；黃鐵鑛	ō'tekkō

iron sand *[geol]*	砂鉄; さてつ	sa'tetsu
iron sesquioxide	三二酸化鉄	san-ni'sanka-tetsu
iron shop *[met]*	鉄工場	tekkō'jō
iron silicate	珪酸鉄	keisan-tetsu
iron silicofluoride	珪弗化鉄	kei'fukka-tetsu
iron succinate	琥珀酸鉄	kohaku'san-tetsu
iron sulfate	硫酸鉄	ryūsan-tetsu
iron sulfide	硫化鉄	ryūka-tetsu
iron works *[met]*	鉄工所	tekkō'sho
irradiance *[elecmg]*	輻射照度	fukusha-shōdo
irradiated light *[opt]*	放射光	hōsha-kō
irradiation *[eng]*	照射; しょうしゃ	shōsha
irrational numbers *[arith]*	無理数	muri-sū
irregularity *{n}*	不規則; ふきそく	fu'kisoku
irregular variable star *[astron]*	不規則変光星	fu'kisoku-henkō-sei
irreversibility *[bio] [math]*	不可逆性	fu'kagyaku-sei
" *[math]*	非可逆性	hi'kagyaku-sei
irreversible reaction *[ch]*	不可逆反応	fu'kagyaku-hannō
irrigation *[agr]*	灌漑; かんがい	kangai
" *[civ-eng]*	灌水	kansui
irritability *[bio] [physio]*	被刺激性	hi-shigeki-sei
irritant *[mat]*	刺激剤	shigeki-zai
irrotational flow *[fl-mech]*	非回転流	hi'kaiten-ryū
is (exists) *{vb i}*	在る; ある	aru
is eaten *{vb}*	食べられる	tabe'rareru
is edible *{adj}*	食べられる	tabe'rareru
is patterned after *{vb}*	準ずる	junzuru
isinglass; fish glue *[mat]*	鰾膠; にべ	nibe
island *[geogr]*	島	shima
island country *[geogr]*	島国; しまぐに	shima-guni
isoanthesic line *[meteor]*	等開花日線	tō'kaika'bi-sen
isobar *[meteor]*	等圧線; とうあつ線	tō'atsu-sen
" *[nuc-phys]*	同重核	dōjū-kaku
" *[nuc-phys]*	同重体	dōjū'tai
isobath *[ocean]*	等深線	tōshin-sen
isocenter *[graph]*	等中心	tō'chūshin
isoceraunic line *[meteor]*	等雷雨線	tō'rai'u-sen
isochromatic lines *[opt]*	等色線	tōshoku-sen
isochrone *[phys]*	等時線	tōji-sen
isochronism *[mech]*	等時性; とうじせい	tōji-sei
isoclinal *{n} [geophys]*	等傾斜線	tō'keisha-sen
isoclinic line *[geophys]*	等伏角線	tō'fukkaku-sen

isoclinic line *[geophys]*	等傾線	tōkei-sen
isodiapheres *[nuc-phys]*	同余体	dōyo'tai
isodose chart *[nucleo]*	等線量図	tō'senryō-zu
isoelectric *{adj} [elec]*	等電の	tōden no
isoelectric point *[p-ch]*	等電点	tōden-ten
isogamy *[bio]*	同形配偶	dōkei-haigū
isogonal map *[min-eng]*	等重力線図	tō'jū'ryoku-senzu
isogonic line *[geophys]*	等方位角線	tō-hō'i'kaku-sen
isohel *[meteor]*	等日照線	tō'nisshō-sen
isohyet *[meteor]*	等雨線; とううせん	tō'u-sen
isokinetic temperature *[meteor]*	等速温度	tōsoku-ondo
isolate *{vb} [ch]*	単離する	tanri suru
isolate; separate; extricate *[vb]*	遊離する	yūri suru
isolation *[ch]*	単離; 單離; たんり	tanri
" *[microbio]*	隔離	kaku'ri
isolux *{n} [opt]*	等照度	tōshō-do
isomer *[ch]*	異性体; いせいたい	isei'tai
isomerism *[ch]*	異性	i'sei
isomerization *[ch]*	異性化	isei'ka
isomerization polymerization *[poly-ch]*	異性化重合	isei'ka-jūgō
isomerized sugar *[bioch]*	異性化糖	isei'ka'tō
isometric contraction *[physio]*	等尺性収縮	tō'shaku'sei-shū'shuku
isometric crystal *[crys]*	等軸晶	tō'jiku-shō
isometric crystal system *[crys]*	立方晶系	rippō-shōkei
isometric drawing *[graph]*	等角投影図	tōkaku-tō'ei-zu
isometric matrix *[math]*	等長行列	tōchō-gyō'retsu
isometric system *[crys]*	等軸晶系	tō'jiku-shōkei
isometry *[an-geom]*	等長写像; 等長寫像	tōchō-shazō
isomorphism *[math] [p-ch]*	同形	dōkei
isomorphism, law of *[math]*	同形律	dōkei-ritsu
isophot *[opt]*	等照線	tōshō-sen
isophotal contour *[opt]*	等光度線	tō'kōdo-sen
isopiestic surface *[phys]*	等圧面; 等壓面	tō'atsu-men
isopycnic line *[meteor]*	等密度線	tō'mitsudo-sen
isosbestic point *[ch]*	等吸収点; 等吸收點	tō'kyūshū-ten
isosceles triangle *[geom]*	二等辺三角形	ni'tōhen-sankak(u)kei
isoseismal line *[geophys]*	等震線	tōshin-sen
isotach *[meteor]*	等速線	tōsoku-sen
isotherm *[geophys]*	等温線	tō'on-sen
isothermal *[thermo]*	等温; とうおん	tō'on
isothermal change *[ch]*	等温変化	tō'on-henka
isothermal distillation *[ch]*	等温蒸留	tō'on-jōryū

isothermal magnetic susceptibility *[elecmg]*	等温磁化率	tō'on-jika-ritsu
isotone *[nuc-phys]*	同中性子核	dō'chūsei'shi-kaku
" *[nuc-phys]*	同中性子体	dō'chūsei'shi-tai
isotonic solution *[p-ch] [physio]*	等張液	tōchō-eki
isotope *[nuc-phys]*	同位元素	dōi-genso
" *[nuc-phys]*	同位体；同位體	dōi'tai
isotopic exchange reaction *[ch]*	同位体交換反応	dōitai-kōkan-hannō
isotopic labeling *{n} [ch]*	同位体標識	dōitai-hyō'shiki
isotopic tracer *[ch]*	同位体追跡子	dōitai-tsui'seki'shi
isotropic *{adj} [phys]*	等方性の	tōhō'sei no
isotropicness *{n}*	全方向性	zen'hōkō-sei
isotropy; isotropism *[phys]*	等方性	tōhō-sei
isotypic *{adj} [crys]*	同一構造型の	dō'itsu-kōzō'gata no
isoviscous state *[fl-mech]*	等粘性状態	tō'nensei-jōtai
issuance date *[pat]*	発行日	hakkō-bi
issue *{n} [legal]*	下付；かふ	kafu
issue; issued *[pat] [pr]*	発行；はっこう	hakkō
issue; number *[pr]*	号；號；ごう	gō
isthmus *[geog]*	地峡；地峽	chikyō
itchiness *{n} [physio]*	痒み；かゆみ	kayu'mi
" *{n} [physio]*	痒さ	kayu-sa
item; entry *[math]*	項目；こうもく	kōmoku
" (a list of articles) *{n} [pr]*	品目	hinmoku
" *[pr]*	項	kō
" ; clause; article *{n} [pr]*	箇条；かじょう	kajō
" ; details; particulars *{n}*	細目；さいもく	saimoku
iteration *[math]*	反復	hanpuku
iterative *{adj} [math]*	反復の	hanpuku no
itinerant electron *[phys]*	遍歴電子	henreki-denshi
itinerary (pl -aries) *[trans]*	旅行日程	ryokō-nittei
" *[trans]*	旅程；りょてい	ryotei
-itis; inflammation *{suffix} [med]*	-炎	-en
ivory (pl -ries) *[mat] [v-zoo]*	象牙；ぞうげ	zōge
" (color) *[opt]*	象牙色	zōge-iro
ivy (pl ivies) *[bot]*	木蔦；きづた	ki'zuta
" *[bot]*	蔦；つた	tsuta

J

jack *[mech-eng]*	起重機	kijū-ki
" *[mech-eng]*	押上げ万力	oshi'age-manriki
jackdaw (bird) *[v-zoo]*	小鴉；こがらす	ko-garasu
jackfruit *[bot]*	波羅蜜；ぱらみつ	para'mitsu
jackhammer *{n}* *[mech-eng]*	手持ち鑿岩機	te'mochi-sakugan-ki
jackrabbit (animal) *[v-zoo]*	野兎	no'usagi
jade *{n}* *[miner]*	翡翠；ひすい	hisui
jadeite *[miner]*	翡翠輝石	hisui'kiseki
" *[miner]*	硬玉	kōgyoku
jalapa (root); Ipomoea purga *[bot]*	葯刺巴(根)；やらっぱこん	yarappa(-kon)
jamb *{n}* *[arch]*	竪枠；たてわく	tate-waku
jamesonite *[miner]*	毛鉱；毛鑛	mōkō
jamming; hindrance *{n}* *[electr]*	防害	bōgai
January (month)	一月	ichi'gatsu
Japan *[geogr]*	日本；にっぽん；にほん	Nippon; Nihon
Japanese (the language) *{n}* *[comm]*	日本語	Nippon'go; Nihon'go
" *{adj}* *[geogr]*	日本の	Nippon no; Nihon no
Japanese alphabet *[comm]* *[pr]*	仮名；假名；かな	kana
Japanese and Chinese *{adj}*	和漢の	wa'kan no
Japanese and Chinese medicines *[pharm]*	和漢薬	wa'kan-yaku
Japanese and English *{adj}*	和英の；わえいの	wa'ei no
Japanese angelica (tree) *[bot]*	楤の木；たらのき	tara-no-ki
Japanese apricot (fruit, tree) *[bot]*	梅；うめ	ume
Japanese apricot brandy *[cook]*	梅酒	ume'shu; ume-zake
Japanese bath gown *[cl]*	浴衣；ゆかた	yukata
Japanese bead tree *[bot]*	栴檀；せんだん	sendan
Japanese bear (animal) *[v-zoo]*	月の輪熊	tsuki-no-wa-guma
Japanese beetle (insect) *[i-zoo]*	豆黄金虫	mame-kogane-mushi
Japanese binding *[pr]*	和本	wa-hon
Japanese bitterling (fish) *[v-zoo]*	鱮；たなご	tanago
Japanese black pine (tree) *[bot]*	黒松；黑松	kuro-matsu
Japanese catalpa *[bot]*	木豇豆；きささげ	kisasage
Japanese character *[comm]* *[pr]*	国字；國字	koku'ji
" *[comm]* *[pr]*	和字	wa'ji
Japanese citron (fruit) *[bot]*	酸橘；すだち	su'dachi
Japanese clothing *[cl]*	和服	wa-fuku
Japanese colors *[opt]*	和絵具；和繪具	wa-enogu

Japanese crane (bird) *[v-zoo]*	丹頂鶴	tanchō-zuru
Japanese crested ibis (bird) *[v-zoo]*	鴇；朱鷺；とき	toki
Japanese cuckoo (bird) *[v-zoo]*	郭公；かっこう	kakkō
Japanese deer; sika (animal) *[v-zoo]*	日本鹿	Nihon-jika
Japanese ditto symbol *[pr]*	同前符：々	dōzen-pu
Japanese earth *[bot] [pharm]*	阿仙薬	a'sen-yaku
Japanese encephalitis *[med]*	日本脳炎	Nihon-nōen
Japanese garden *[bot]*	日本庭園	nihon-tei'en
Japanese genre painting *[graph]*	浮世絵；浮世繪	uki'yo-e
Japanese ginger *[bot]*	茗荷；みょうが	myōga
Japanese green gentian *[bot]*	千振；せんぶり	senburi
Japanese half-coat *[cl]*	羽織	ha'ori
Japanese harp *[music]*	琴	koto
Japanese honeysuckle *[bot]*	忍冬；にんどう	nindō
Japanese horseradish *[bot]*	山葵；わさび	wasabi
Japanese ink (black) *[mat]*	墨；墨；すみ	sumi
Japanese inn *[arch]*	旅館	ryokan
Japanese iron *[met]*	和鉄；和鐵	wa-tetsu
Japanese lacquer; japan *[mat]*	漆；うるし	urushi
Japanese larch (tree) *[bot]*	唐松；落葉松	kara-matsu
Japanese linden (tree) *[bot]*	科木；しなのき	shinanoki
Japanese livery half-coat *[cl]*	法被；半被	happi
Japanese lute *[music]*	琵琶	biwa
Japanese maple (tree) *[bot]*	紅葉；もみじ	momiji
Japanese mathematics *[math]*	和算	wasan; wazan
Japanese medicinal drug *[med]*	和薬；和藥	wa-yaku
Japanese medicine (the field) *[med]*	和方	wa-hō
Japanese monkey; Macaca fuscata (animal) *[v-zoo]*	日本猿；にほんざる	Nihon-zaru
Japanese mouth organ *[music]*	笙	shō
Japanese mullet (fish) *[v-zoo]*	赤目魚；めなだ	menada
Japanese music *[music]*	邦楽	hōgaku
Japanese name *[n]*	和名	wa-mei
Japanese noodles *[cook]*	饂飩；うどん	udon
Japanese nutmeg seed *[bot]*	榧子；ひし	hishi
Japanese oak (tree) *[bot]*	楢；なら	nara
Japanese pagoda tree (tree) *[bot]*	槐；えんじゆ	enju
Japanese painting *[graph]*	日本画；日本畫	Nihon-ga
Japanese pancake *[cook]*	お好み焼き	o'konomi-yaki
Japanese paper *[mat] [paper]*	和紙	washi
Japanese parsley *[bot] [cook]*	芹；せり	seri
Japanese pepper *[bot] [cook]*	山椒	sanshō

Japanese Pharmacopoeia *[pharm]*	日本薬局方	Nippon Yakkyoku-hō
Japanese phonetics of meaning for Chinese characters *[ling]*	訓；くん	kun
Japanese phonetics of sound for Chinese characters *[ling]*	音；おん	on
Japanese plum *[bot]*	李；すもも	sumomo
Japanese poem of 17 syllables *[pr]*	俳句	haiku
Japanese poem of 31 syllables *[pr]*	和歌	waka
Japanese product *[ind-eng]*	和製品	wasei'hin
Japanese pronunciation (of Chinese characters) *[comm]*	読み；よみ	yomi
Japanese quotation marks *[gram] [pr]*	引用符：「」	in'yō-fu
Japanese rice cake *[cook]*	餅；もち	mochi
Japanese rice cracker *[cook]*	煎餅；せんべい	senbei
Japanese rice wine *[cook]*	酒；さけ	sake
" *[cook]*	清酒	seishu
Japanese robin (bird) *[v-zoo]*	駒鳥	koma-dori
Japanese room *[arch]*	和室	wa-shitsu
Japanese rose; globe flower *[bot]*	山吹	yamabuki
Japanese sago palm *[bot]*	蘇鉄；蘇鐵；そてつ	so'tetsu
Japanese scale *[music]*	和音階	wa'on-kai
Japanese scops owl (bird) *[v-zoo]*	木の葉木菟	konoha-zuku
Japanese sewing *[cl]*	和裁	wa'sai
Japanese socks *[cl]*	足袋；たび	tabi
Japanese spaniel (dog) *[v-zoo]*	狆	chin
Japanese standard time *[astron]*	日本標準時	Nippon-hyōjun-ji
Japanese star anise (herb) *[bot]*	樒；しきみ	shikimi
Japanese steel *[met]*	和鋼	wa-hagane
Japanese stork (bird) *[v-zoo]*	鸛；こうのとり	kōnotori
Japanese-style *{adj}*	和様の	wa'yō no
Japanese-style food *[cook]*	和食	wa-shoku
Japanese-style room *[arch]*	日本間	Nihon-ma
Japanese swift moth (insect) *[i-zoo]*	蝙蝠蛾；こうもりが	kōmori-ga
Japanese sword *[ord]*	刀；かたな	katana
Japanese text *[pr]*	和文	wa'bun
Japanese trousers *[cl]*	袴	hakama
Japanese vellum *[paper]*	局紙	kyoku-shi
Japanese wine; rice wine *[cook]*	日本酒	Nihon'shu
Japanese writing *[comm] [pr]*	和文	wabun
Japanese yam; Dioscorea japonica *[bot]*	山の芋；薯蕷	yama-no-imo
Japanic acid *[ch]*	日本酸	Nihon-san
Japan Industrial Standards *[ind-eng]*	日本工業規格	Nippon Kōgyō Kikaku

japanning *{n} [met]*	漆塗り	urushi-nuri
Japan wax *[mat]*	木臘	mokurō
jar; pot; crock; urn *[eng]*	壺; つぼ	tsubo
jasmine *[bot]*	茉莉花; まつりか	matsuri'ka
" *[bot]*	素馨; そけい	sokei
jasmine tea *[cook]*	茉莉花茶	matsuri'ka-cha
" *[cook]*	香片茶	shanpen-cha
jasper *[petr]*	碧玉	heki'gyoku
jaundice *[med]*	黄疸; おうだん	ō'dan
jaw *{n} [anat]*	顎; あご	ago
jaw crusher *[eng]*	砕鉱機; 碎鑛機	saikō-ki
jawless fishes *[v-zoo]*	無顎類	mu'gaku-rui
jay (bird) *[v-zoo]*	樫鳥	kashi-dori
jelly *[ch] [geol]*	凝膠体	gyōkō'tai
jellyfish; medusa (pl -dusae) *[i-zoo]*	水母; 雲月; くらげ	kurage
jerboa (animal) *[v-zoo]*	跳び鼠	tobi-nezumi
jet *{n} [fl-mech]*	液柱	eki'chū
" *{n} [fl-mech]*	噴流; ふんりゅう	funryū
" *{n} [miner]*	黒玉炭; 黑玉炭	kuro'tama-tan
jet molding *{n} [eng]*	噴射成形	funsha-seikei
jet propulsion *[aero-eng]*	噴流推進	funryū-suishin
jet stream *[meteor]*	ジエット気流	jetto-kiryū
jettison *{vb t} [eng]*	投荷する	tōka suru
jetty *[civ-eng]*	突堤	tottei
jewel; gem *[miner]*	宝石; 寶石	hōseki
jewel bearing *{n} [horol]*	宝石軸受	hōseki-jiku'uke
jig *[mech-eng]*	治具; じぐ	jigu
jobsite *[ind-eng]*	現場	genba
jog *{n} [photo]*	食違い点	kui'chigai-ten
join; graft *{vb t} [bot]*	接ぐ; つぐ	tsugu
joining; jointing *{n} [civ-eng]*	接合	setsu'gō
joining surface *[adhes]*	接合面	setsugō-men
joint *{n} [anat]*	関節; 關節	kan'setsu
" *{n} [eng]*	目地	meji
" (joining part) *{n} [eng]*	接合部	setsugō-bu
" ; juncture; seam *{n} [eng]*	継ぎ目	tsugi'me
" ; splice; coupling *{n} [eng]*	継手; 繼手	tsugi'te
joint compound *[civ-eng]*	目地剤	meji-zai
joint displacement *[civ-eng]*	節点移動	setten-idō
jointed appendage *[anat]*	関節肢	kansetsu-shi
joint mixture *[civ-eng]*	目地剤	meji-zai
jointweed; Polygonum hydropiper *[bot]*	蓼; たで	tade

joist *[arch]*	梁; はり	hari
" *[arch]*	根太; ねだ	neda
jonquil (flower) *[bot]*	黄水仙; きずいせん	ki'zuisen
journal; diary *[comm]*	日誌	nisshi
-journal; -magazine *{suffix}* *[pr]*	-誌	-shi
journal box *[eng]*	軸箱	jiku-bako
journey *{n}* *[navig]*	行程	kōtei
judge; pass judgment *{vb t}*	裁く; さばく	sabaku
judgment *[legal]*	判定	hantei
juice *{n}* *[bot]* *[ch]*	液; えき	eki
" *{n}* *[bot]* *[cook]*	汁	shiru
jujube tree *[bot]*	棗; なつめ	natsume
July (month)	七月	shichi'gatsu
jump *{n}* *[comput]*	飛越し	tobi-koshi
jump fire *{n}*	飛び火	tobi-hi
jumping spider (spider) *[i-zoo]*	蠅取り蜘蛛	hae'tori-gumo
junction *[elec]*	接合(部)	setsu'gō(-bu)
" *[elec]*	接合点	setsu'gō-ten
junction box *[eng]*	接続箱; 接續箱	setsu'zoku-bako
junction circuit *[elec]*	中継回路	chūkei-kairo
junction port; junction harbor *[geogr]*	中継港; 中繼港	chūkei-kō
June (month)	六月	roku'gatsu
jungle *[ecol]*	密林	mitsu'rin
juniper; Juniperus (tree) *[bot]*	柏槇; びゃくしん	byaku'shin
Juniperus chinensis (tree) *[bot]*	伊吹; いぶき	ibuki
juniper berry *[bot]*	杜松実	toshō'jitsu
Jupiter (planet) *[astron]*	木星	mokusei
juridical person *[legal]*	法人	hōjin
jurisprudence *[legal]*	法学	hō'gaku
juristic person *[legal]*	法人	hōjin
justification *[comput]* *[graph]* *[pr]*	(位置)調整	(ichi-)chōsei
" *[graph]* *[pr]*	揃え; そろえ	soroe
" *[comput]* *[graph]* *[pr]*	詰め	tsume
" *[comput]* *[graph]* *[pr]*	寄せ	yose
jute *[bot]*	黄麻; こうま	kōma
juxtapose; put side by side *{vb}*	並べる	naraberu
juxtaposition *[math]*	並置	heichi

K

kakapo (bird) *[v-zoo]*	梟鸚鵡	fukurō-ōmu
kaleidoscope *[opt]*	万華鏡；まんげ鏡	mange'kyō
kana (Japanese alphabet) *[comm]*	仮名；假名；かな	kana
kana orthography *[comm]*	仮名遣い	kana-zukai
kana, cursive style *[pr]*	平仮名；ひらがな	hira-gana
kana, script style *[pr]*	片仮名；カタカナ	kata-kana
kana phonetics beside Chinese characters *[ling]*	送りがな	okuri-gana
kaoliang; koaliang; sorghum *[bot]*	高梁；コーリヤン；カォリヤン	kōryan; kaoryan
kaoliang oil *[mat]*	高梁油	kōryan-yu
kaolin; kaolinite *[miner]*	陶土	tōdo
karyoplasm; nucleoplasm *[cyt]*	核質	kaku'shitsu
karyotype *[cyt]*	核型	kaku'gata
katabatic wind *[meteor]*	颪；おろし	oroshi
katsura (tree) *[bot]*	桂	katsura
katydid (insect) *[i-zoo]*	蟋蟀；きりぎりす	kirigirisu
keel *[nav-arch]*	竜骨；龍骨	ryū'kotsu
keep; raise (animal, bird, fish) *{vb}*	飼う	kau
keepable foods *[food-eng]*	保存食	hozon-shoku
Keep to the Left *[traffic]*	左側通行	sa'soku-tsūkō
Keep to the Right *[traffic]*	右側通行	u'soku-tsūkō
kelp (seaweed) *[bot]*	昆布；こんぶ	konbu
Kelvin temperature scale *[thermo]*	ケルビン温度目盛	kerubin-ondo-memori
keratin *[bioch]*	角質	kaku'shitsu
keratinization *[hist]*	角質化	kaku'shitsu'ka
kerf (point of cut) *[eng]*	切り目	kiri'me
" (channel of cut) *[eng]*	切り溝	kiri-mizo
" *[eng]*	切断カーフ	setsu'dan-kāfu
kermesite *[miner]*	紅安鉱；紅安鑛	kōankō
kern *[pr]*	飾り鬚；かざりひげ	kazari-hige
kernel *[bot]*	仁	jin
" *[bot]*	核	kaku
" *[atom-phys]*	(原子)核	(genshi-)kaku
" (core part) *{n}*	中心部	chūshin'bu
kerogen *[geol]*	油母；ゆも	yumo
kerosine; kerosene *[mat]*	灯油；燈油	tōyu
kerosine engine *[mech-eng]*	石油機関	sekiyu-kikan
kestrel (bird) *[v-zoo]*	長元坊	chōgenbō

ketone *[org-ch]*	ケトン	keton
kettle *[cook]*	釜；かま	kama
" (of copper, brass) *[cook]*	薬缶；藥罐；やかん	yakan
" (for water boiling) *[cook]*	湯沸し	yu'wakashi
key *{n}* *[elec]*	電鍵；でんけん	denken
" *{n}* *[comput]* *[eng]*	鍵；かぎ	kagi
" (key to symbols) *{n}* *[graph]*	記号解；きごうかい	kigō-kai
key atom *[ch]*	鍵原子	kagi-genshi
keyboard *[comput]* *[music]*	鍵盤	kenban
keyhole *[eng]*	鍵穴	kagi-ana
key signature *[music]*	調号；調號	chōgō
kiaki (tree) *[bot]*	欅；けやき	keyaki
kidney *{n}* *[anat]*	腎臓；じんぞう	jinzō
kidney bean (legume) *[bot]*	隠元豆	ingen-mame
kieselguhr *[geol]*	珪藻土；けいそうど	keisō'do
killed ingot *[met]*	鎮静鋼塊	chinsei-kōkai
killer whale *[v-zoo]*	逆叉；さかまた	sakamata
" *[v-zoo]*	鯱；しゃち	shachi
killing frost *[meteor]*	枯らし霜	karashi-jimo
kiln *[eng]*	窯；かま	kama
" *[eng]*	炉；爐；ろ	ro
kilocalorie (unit of heat energy) *[thermo]*	キロカロリー	kiro'karorī
kilocalorie *[thermo]*	大カロリー	dai-karorī
kilogram (unit of mass) *[mech]*	瓩；キログラム	kiro'guramu
kiloliter (unit of volume) *[mech]*	竏；キロリットル	kiro'rittoru
kilometer (unit of length) *[mech]*	粁；キロメートル	kiro'mētoru
kilowatt-hour; kWh *[elec]*	キロワット時	kiro'watto-ji
kimono, Japanese *[cl]*	着物；きもの	ki'mono
kimono for visiting *[cl]*	訪問着	hōmon'gi
kimono overgarment *[cl]*	打(ち)掛け；裲襠	uchi'kake
kimono with long sleeves *[cl]*	振り袖	furi-sode
kimono with skirt design *[cl]*	裾模様	suso-moyō
kind; sort; variety; class *{n}*	類；類；たぐい	tagui
kindling temperature *[thermo]*	点火温度；點火溫度	tenka-ondo
kinematic coefficient of elasticity *[mech]*	動弾性係数；どう=だんせいけいすう	dō'dansei-keisū
kinematic friction *[mech]*	動摩擦	dō-masatsu
kinematics *[mech]*	運動学	undō'gaku
kinematic viscosity *[fl-mech]*	(運)動粘度	(un)dō-nendo
kinematic viscosity, coefficient of *[fl-mech]*	動粘性率；どう=ねんせいりつ	dō-nensei'ritsu

kinematic viscosity, coefficient of *[fl mech]*	動粘性係数	dō-nensei-keisū
kinetic energy *[mech]*	運動エネルギー	undō-enerugī
kinetic equation *[mech]*	運動論的方程式	undōron'teki-hōtei'shiki
kinetic mass action *[ch]*	動的質量作用	dō'teki-shitsu'ryō-sayō
kinetics *[mech]*	動力学	dō'rikigaku
" *[mech]*	速度論	sokudo-ron
kinetic temperature *[phys]*	運動温度	undō-ondo
king crab (crustacean) *[i-zoo]*	兜蟹; かぶとがに	kabuto-gani
" *[i-zoo]*	鱈場蟹; たらばかに	taraba-kani
kingdom *[bio] [syst]*	界	kai
kingfish; moonfish (fish) *[v-zoo]*	万鯛; まんだい	man-dai
kingfisher (bird) *[v-zoo]*	翡翠; かわせみ	kawasemi
king salmon (fish) *[v-zoo]*	鱒の介	masu-no-suke
kinnase *[bioch]*	燐酸化酵素	rinsan'ka-kōso
Kinorhyncha *[i-zoo]*	動吻動物	dōfun-dōbutsu
kitchen *[arch] [cook]*	台所; だいどころ	dai'dokoro
kitchen utensils; kitchenware *[eng]*	台所用品	dai'dokoro-yōhin
kite (bird) *[v-zoo]*	鳶; とび	tobi
" (toy) *[eng]*	凧; たこ	tako
kitten (animal) *[v-zoo]*	子猫	ko-neko
knead *{vb} [sci-t]*	捏る; こねる	koneru
" *{vb} [sci-t]*	混錬する; 混練する	konren suru
" *{vb} [sci-t]*	錬る; 練る	neru
kneader *[eng]*	捏和機; ねっかき	nekka-ki
kneading *{n} [eng]*	捏ね混ぜ	kone-maze
" *{n} [eng] [poly-ch]*	混錬	konren
" *{n} [eng]*	捏和	nekka
knee *{n} [anat]*	膝頭	hiza'gashira
knee brace *[arch]*	方杖; ほうづえ	hō'zue
kneecap; patella (pl patellae) *[anat]*	膝蓋骨	shitsu'gai-kotsu
knife (pl knives) (small) *{n} [eng]*	小刀	ko-gatana
knife-edge bearing *[mech-eng]*	刃受け; 刃受け	yaiba-uke
knife file *[eng]*	刃形やすり(鑢)	ha'gata-yasuri
knife tool *[eng]*	片刃バイト	kata'ha-baito
knit (knitting stitch) *{n} [tex]*	表目	omote'me
knitted work *[tex]*	編み物; あみもの	ami-mono
knitting machine *[eng]*	編み機	ami-ki
" *[eng]*	メリヤス機	meriyasu-ki
knitting needle *[eng]*	編み針	ami-bari
knob *[eng]*	握り	nigiri
knoll *[geol]*	小山	ko-yama

knot *{n} [mat]*	節；ふし	fushi
" *{n} [tex]*	結び目	musubi'me
knot hole *[civ-eng]*	節穴	fushi-ana
knotweed; knotgrass; Polygonum aviculare L (herb) *[bot]*	扁蓄；扁竹；へんちく	hen'chiku
knotweed; knotgrass (herb) *[bot]*	路柳；道柳	michi-yanagi
" (herb) *[bot]*	庭柳	niwa-yanagi
know *{vb} [psy]*	知る；識る；しる	shiru
knowledge *{n} [psy]*	知識；ちしき	chi'shiki
knowledge; awareness *{n} [psy]*	認識	nin'shiki
knowledge engineering *{n} [comput]*	知識工学	chishiki-kōgaku
known quantity *[math]*	既知数；既知數	kichi-sū
knuckle thread *[mech-eng]*	丸(山)ねじ	maru(yama)-neji
knurl *{n} [eng]*	刻み	kizami
knurling *{n} [met] [eng]*	彫刻	chōkoku
koala (animal) *[v-zoo]*	子守り熊	komori-guma
Kobe *[geogr]*	神戸；こうべ	Kōbe
kōji *[cook] [mycol]*	麹；こうじ	kōji
kōjic acid	麹酸	kōji-san
kōji mold *[mycol]*	麹黴；こうじきん	kōji-kabi
" *[mycol]*	麹菌	kōji-kin
kōji process *[microbio]*	麹法	kōji-hō
konimeter *[eng]*	塵埃測定器	jin'ai-sokutei-ki
konnyaku noodles *[cook]*	白滝；しらたき	shira-taki
koodoo; kudu (animal) *[v-zoo]*	捩じ角羚羊	neji'zuno-reiyō
krait (reptile) *[v-zoo]*	雨傘蛇	ama'gasa-hebi
krypton (element)	クリプトン	kuriputon
kudzu vine *[bot]*	葛；くず	kuzu
kudzu vine root *[pharm]*	葛根	kakkon
kumquat (fruit) *[bot]*	金柑	kinkan
kyanite; cyanite *[miner]*	藍晶石	ranshō-seki
kymograph *[med]*	動態記録器	dōtai-kiroku-ki
Kyoto *[geogr]*	京都；きょうと	Kyōto

L

label *{n} [comput]*	名札	na'fuda
" *{n} [paper]*	レッテル	retteru
labeling *{n} [nucleo]*	標識；ひょうしき	hyōshiki
labile; unstable *{adj} [sci-t]*	不安定な	fu'antei na
labor (giving birth) *{n} [med]*	分娩	bunben
labor; work *{n}*	労働；勞働；労仂	rōdō
laboratory (pl -ries) *[ch] [sci-t]*	実験室；實驗室	jikken-shitsu
" *[sci-t]*	研究所	kenkyū'jo
" *[sci-t]*	研究室	kenkyū-shitsu
labor cost *[econ]*	労働費	rōdō-hi
labor-saving *{adj} [ind-eng]*	省力の	shō'ryoku no
labor-saving investment *[econ]*	省力投資	shōryoku-tōshi
labradorite *[miner]*	曹灰長石	sōkai'chōseki
labyrinth *{n}*	迷宮	meikyū
laccolith; laccolite *[geol]*	餅盤；へいばん	heiban
laccolith *[geol]*	餅状岩体	heijō-gantai
lack; deficiency *{n}*	欠除；欠如	ketsu'jo
lacquer; lacquering *{n} [furn]*	蒔き絵；まきえ	maki-e
lacquer, Japanese *[mat]*	漆；うるし	urushi
lacquerware *[furn]*	漆器；しっき	shikki
lacrimal gland *{n} [anat]*	涙腺	rui-sen
Lactarius hatsudake (mushroom)	初茸	hatsu-take
lactate *[ch]*	乳酸塩	nyūsan-en
lactic acid	乳酸；にゅうさん	nyūsan
lactic acid bacteria *[microbio]*	乳酸菌	nyūsan-kin
lacto(butyro)meter *[eng]*	乳脂計	nyūshi-kei
lactose *[bioch]*	乳糖	nyūtō
ladder *[eng]*	梯子；はしご	hashi'go
ladder structure *[poly-ch]*	梯子構造	hashigo-kōzō
ladle (soup ladle) *{n} [cook] [eng]*	お玉杓子	o'tama'jakushi
" *{n} [met]*	杓；しゃく	shaku
" *{n} [met]*	取瓶；とりべ	tori'be
" *{n} [met]*	取り鍋	tori-nabe
ladybug (insect) *[i-zoo]*	天道虫；瓢虫	tentō-mushi
lag *{n} [comput] [phys]*	遅れ；遲れ；おくれ	okure
" *{n} [electr]*	残像	zanzō
Lagomorpha (rabbits, hares) *[v-zoo]*	兎類；兔類	usagi-rui
lagoon *[geogr]*	潟；かた	kata
" *[geogr]*	潟湖；せきこ	sekiko

lagoon *[geogr]*	礁湖；しょうこ	shō'ko
lag phase *[microbio]*	遅滞期；遲滯期	chitai-ki
laid open to public inspection *[pat]*	公開された	kōkai sareta
laid paper *[mat]*	簀の目紙	sunome-shi
lake *[hyd]*	湖；みずうみ	mizu'umi
" (color) *[opt]*	深紅色	shinkō-shoku
lakebed *[geol]*	湖床	koshō
lakes and marshes *[hyd]*	湖沼；しょうこ	koshō
lamb's wool *[tex]*	子羊の毛	ko-hitsuji no ke
lamella *[bot]*	薄板	hakuban
" *[mycol]*	襞；ひだ	hida
lamellar compound *[crys]*	成層化合物	seisō-kagō'butsu
lamina (pl -nae, -nas) *[geol]*	葉層	yōsō
lamina of mother of pearl *[i-zoo]*	螺鈿；らでん	raden
laminar *{adj} [sci-t]*	葉片状の	yōhen'jō no
laminar flow *[fl-mech]*	層流；そうりゅう	sōryū
laminate (laminated sheet) *{n} [mat]*	積層板	sekisō-ban
" (laminated article) *{n} [mat]*	積層物	sekisō'butsu
laminated glass *[mat]*	合せガラス	awase-garasu
laminated (iron) core *[elecmg]*	成層鉄心；成層鐵芯	seisō-tesshin
laminated spring *[eng]*	重ね板ばね	kasane-ita'bane
laminated wood *[mat]*	集成材	shūsei-zai
laminating; lamination *{n} [sci-t]*	積層；せきそう	sekisō
lampblack *[mat]*	油煙	yu'en
lamp oil *[mat]*	灯油；燈油；とうゆ	tōyu
lamp post *[civ-eng]*	灯柱	tōchū
lamprey; lamprey eel *[v-zoo]*	八つ目鰻	yatsu'me-unagi
lanceolate *{adj} [bot]*	皮針形の	hishin'kei no
land *{n} [geogr]*	陸(地)	riku(chi)
" *{vb} [aero-eng]*	着陸する	chaku'riku suru
land animal *[zoo]*	陸生動物；陸棲動物	rikusei-dōbutsu
land breeze *[meteor]*	陸風	riku'fū; riku-kaze
" *[meteor]*	陸軟風	riku-nanpū
land bridge *[geogr]*	陸橋	rik(u)kyō
land development *{n} [civ-eng]*	土地開発	tochi-kaihatsu
landfall *{n} [navig]*	陸地初認	rikuchi-sho'nin
landfall mark *[navig]*	陸地進入標識	rikuchi-shinnyū-hyo'shiki
landing (stairway) *{n} [arch]*	中休み段	naka'yasumi-dan
" *{n}* (stairway) *[arch]*	踊り場	odori'ba
" *{n} [navig]*	着地	chaku'chi
" ; touchdown *{n} [navig]*	着陸；ちゃくりく	chaku'riku
" ; disembarkation *{n} [navig]*	上陸	jō'riku

landing gear *[aero-eng]*	降着装置	kōchaku-sōchi
landing-gear door *[aerosp]*	脚扉; きゃくとびら	kyaku-tobira
landing-gear well *[aero-eng]*	脚室	kyaku-shitsu
landing light *[aero-eng]*	着陸灯	chaku'riku-tō
landing strip *[aero-eng]*	滑走路; かっそうろ	kassō-ro
landlocked country *[geogr]*	内陸国	nai'riku-koku
landmark *[eng] [navig]*	著明地形	chomei-chikei
" *[eng] [navig]*	陸標	riku-hyō
land mine *[ord]*	地雷	jirai
land plant *[bot]*	陸生植物; 陸棲植物	rikusei-shokubutsu
landscape *{n}*	風景	fūkei
landscape architecture *[arch]*	造園; ぞうえん	zō'en
landscape gardening *{n} [arch] [bot]*	造園術	zō'en'jutsu
landscape picture *[graph]*	風景画	fūkei-ga
landslide *[geol]*	崖崩れ; がけくずれ	gake-kuzure
" *[geol]*	地滑べり	ji-suberi
landslip *[geol]*	崖崩れ	gake-kuzure
landspout *[meteor]*	竜巻; 龍巻	tatsu'maki
land surveying *[eng]*	陸地測量	rikuchi-sokuryō
" *[eng]*	土地測量	tochi-sokuryō
language *[comm] [ling]*	言語; げんご	gengo
lanolin *[mat]*	羊(毛)脂	yō(mō)'shi
lantern *[furn] [opt]*	灯籠; 燈籠	tōrō
" (made of paper) *[opt]*	提灯; ちょうちん	chōchin
lanthanide series *[ch]*	ランタニド列	rantanido-retsu
lanthanum (element)	ランタン	rantan
lap *{n} [anat]*	膝; ひざ	hiza
lapidarist *[lap]*	宝石学者; 寶石學者	hōseki-gakusha
lapidary *[sci-t]*	宝石細工術	hōseki-saiku'jutsu
lapidist *[lap]*	宝石細工人	hōseki-saiku'nin
lapis lazuli *[miner]*	青金石	seikin'seki
" *[petr]*	瑠璃; るり	ruri
lap joint *[arch]*	重ね継ぎ; 重ね繼ぎ	kasane-tsugi
lapse rate *[meteor]*	逓減率	teigen-ritsu
lapwing (bird) *[v-zoo]*	田鳧; たげり	tageri
larch, Japanese (tree) *[bot]*	唐松; 落葉松	kara-matsu
lard; hog fat *[cook] [mat]*	豚脂	tonshi
large *{adj} {n} [math]*	大きい; おおきい	ōkī
large intestine *[anat]*	大腸	daichō
large numbers *[stat]*	大数; 大數	taisū
large-scale integrated circuit; LSI *[electr]*	大規摸集積回路	dai'kibo-shūseki-kairo

large-scale integration *[electr]*	大規摸集積化	dai'kibo-shūseki'ka
large ship *[nav-arch]*	大型船舶	ō-gata-senpaku
lark; skylark (bird) *[v-zoo]*	雲雀；ひばり	hibari
larva (pl larvae) *[i-zoo]*	幼虫	yōchū
" *[i-zoo]*	幼生	yōsei
larynx *[anat]*	喉頭；こうとう	kōtō
laser *[opt]*	レーザー	rēzā
last (of shoe) *{n}* *[cl]*	靴型	kutsu-gata
last; the end *{n}*	最後	saigo
last month	先月	sen-getsu
last name *{n}*	苗字；みょうじ	myōji
last night; yesterday evening	昨夜	saku'ya
last quarter (of moon) *[astron]*	下弦	kagen
last stop; terminal *[trans]*	終点	shūten
last term *[math]*	末項	makkō
latch *{n}* *[eng]*	掛け金；かけがね	kake'gane
" *{n}* *[eng]*	留め金	tome'gane
late *{adj}* *[mech]* *[phys]*	遅い；遲い；おそい	osoi
latency *[comput]*	待ち時間	machi-jikan
" *[med]* *[physio]*	潜在性	senzai-sei
latency time *[comput]*	呼出時間	yobi'dashi-jikan
latensification *[photo]*	潜像補力	senzō-ho'ryoku
latent force *[mech]*	潜力	sen-ryoku
latent heat *[thermo]*	潜熱；せんねつ	sennetsu
latent heat of fusion *[thermo]*	融解潜熱	yūkai-sennetsu
latent heat of vaporization *[thermo]*	蒸発潜熱	jō'hatsu-sennetsu
latent image *[graph]*	潜像；せんぞう	senzō
latent-image nuclei *[photo]*	潜像核	senzō-kaku
latent period *[med]*	潜伏期	sen'puku-ki
latent (pre)image speck *[photo]*	前潜像核	zen'senzō-kaku
latent subimage *[graph]*	亜潜像；亞潜像	a'senzō
lateral *{adj}* *[geom]*	側面の	soku'men no
lateral adaptation *[photo]*	側方順応	soku'hō-junnō
lateral edge *[geom]*	側縁	soku'en
lateral force *[mech]*	横力	ō'ryoku; yoko-ryoku
lateral line *[v-zoo]*	側線	soku-sen
lateral strain *[mech]*	横歪み；よこひずみ	yoko-hizumi
lateral view *[graph]*	側面図	sokumen-zu
lathe scraps *[met]*	銑ダライ粉	sendarai-ko
latitude *[geod]*	緯度	ido
lattice *[crys]*	格子；こうし	kōshi
lattice constant *[crys]*	格子定数	kōshi-teisū

lattice defect *[crys]*	格子欠陥	kōshi-kekkan
lattice point *[an-geom]*	格子点	kōshi-ten
lattice scattering *[sol-st]*	格子散乱	kōshi-sanran
lattice spacing *[crys]*	面間隔	men-kankaku
lattice water *[ch]*	格子水	kōshi-sui
latus rectum *[an-geom]*	通径; 通徑	tsūkei
laugh *{vb}* *[physio]*	笑う; わらう	warau
laughter *[physio]*	笑い	warai
laumontite *[miner]*	濁沸石	daku'fusseki
launch *{n}* *[aero-eng]*	発射; はっしゃ	hassha
" *{n}* *[aero-eng]*	打上げ	uchi'age
launching (of a ship) *{n}* *[nav-arch]*	進水	shinsui
launching *{n}* *[ord]*	発射	hassha
launch pad *[aerosp]*	発射台; 發射臺	hassha-dai
launch window *[aerosp]*	打上げ可能時間帯	uchi'age'kanō-jikan-tai
laundry; laundering *{n}*	洗濯; せんたく	sen'taku
laundry machine *[eng]*	洗濯機	sentaku-ki
laurel (tree) *[bot]*	月桂樹	gekkei'ju
lava *[geol]*	溶岩	yōgan
lava plateau *[geol]*	溶岩台地	yōgan-daichi
lavatory *[arch]*	洗面所	senmen'jo
lavatory basin *{n}*	洗面器	senmen-ki
lavender (color) *[opt]*	薄紫; うすむらさき	usu-murasaki
laver (sea weed) *[bot]*	海苔; のり	nori
law (the discipline) *[legal]*	法学	hō'gaku
law *[legal]*	法(律)	hō(ritsu)
" *[legal]* *[sci-t]*	法則; ほうそく	hōsoku
law of: action and reaction *[phys]*	作用反作用の法則	sayō-hansayō no hōsoku
" **averages** *[stat]*	平均の法則	heikin no hōsoku
" **causality** *[phys]*	因果律; いんがりつ	inga-ritsu
" **conservation of angular momentum** *[mech]*	角運動量保存則	kaku'undōryō-hozon-soku
" **conservation of energy** *[phys]*	エネルギー保存則	enerugī-hozon-soku
" **conservation of mass** *[phys]*	質量保存の法則	shitsuryō-hozon no hōsoku
" **conservation of momentum** *[mech]*	運動量保存則	undōryō-hozon-soku
" **constancy of facial angle** *[crys]*	面角一定の法則	men'kaku-ittei no hōsoku
" **constant heat summation** *[p-ch]*	総熱量保存の法則	sō-netsuryō-hozon no hōsoku
" **constant interfacial angle** *[crys]*	面角不変の法則	men'kaku-fuhen no hōsoku
" **continuity** *[math]*	連続の法則	renzoku no hōsoku

law of: contradiction *[logic]*	矛盾律	mujun-ritsu
" **corresponding states** *[ch]*	対応状態の法則	taiō-jōtai no hōsoku
" **cosines** *[trig]*	余弦法則；餘弦法則	yogen-hōsoku
" **diffusion of gas** *[phys]*	気体拡散の法則	kitai-kakusan no hōsoku
" **diminishing return** *[bio]*	報酬漸減の法則	hōshū-zengen no hōsoku
" **distribution of molecular velocities** *[stat-mech]*	分子速度分布則	bunshi-sokudo-bunpu-soku
" **dominance** *[bio]*	優劣の法則	yūretsu no hōsoku
" **effect** *[psy]*	効果の法則	kōka no hōsoku
" **effusion of gas** *[phys]*	気体流出の法則	kitai-ryū'shutsu no hōsoku
" **equipartition of energy** *[stat-mech]*	エネルギー均分の法則	energī-kinbun no hōsoku
" **excluded middle** *[logic]*	排中律	haichū-ritsu
" **exponents** *[math]*	指数法則	shisū-hōsoku
" **falling** *[mech]*	落下の法則	rakka no hōsoku
" **flotation** *[phys]*	浮遊の法則	fuyū no hōsoku
" **gaseous reaction** *[ch]*	気体反応の法則	kitai-hannō no hōsoku
" **gases**; gas law *[thermo]*	気体の法則	kitai no hōsoku
" **geometric progression** *[bio]*	幾何級数法則	kika-kyūsū-hōsoku
" **geometric series** *[bio]*	等比級数則	tōhi-kyūsū-soku
" **gravitation** *[mech]*	引力の法則	in'ryoku no hōsoku
" **grinding** *[mech-eng]*	粉砕法則	funsai-hōsoku
" **homology** *[bio]*	相同の法則	sōdō no hōsoko
" **identity** *[logic]*	同一律	dōitsu-ritsu
" **identity** *[logic]*	自同律；じどうりつ	jidō-ritsu
" **increase in size** *[bio]*	体大化の法則	tai-daika no hōsoku
" **independence** *[bio]*	独立の法則	doku'ritsu no hōsoku
" **independent assortment** *[bio]*	独立(遺伝)の法則	doku'ritsu(-iden) no hōsoku
" **independent ionic mobilities** *[phys]*	イオン独立移動=の法則	ion-doku'ritsu-idō no hōsoku
" **indestructibility of matter** *[phys]*	物質不滅の法則	busshitsu-fumetsu no hōsoku
" **inertia** *[mech]*	慣性の法則	kansei no hōsoku
" **isomorphism** *[crys]*	同形律	dōkei-ritsu
" **iterated logarithms** *[prob]*	重複対数の法則	jūfuku-taisū no hōsoku
" **kinetic mass action** *[ch]*	動的質量作用の=法則	dōteki-shitsuryō-sayō no hōsoku
" **large numbers** *[stat]*	大数の法則	taisū no hōsoku
" **least action** *[mech]*	最小作用の法則	saishō-sayō no hōsoku
" **legacy** *[math]*	遺産法則	i'san-hōsoku
" **logarithmic series** *[bio]*	対数級数則	taisū-kyūsū-soku
" **log normal distribution** *[bio]*	対数正規則	taisū-seiki-soku

law of: mass action *[p-ch]*	質量作用の法則	shitsuryō-sayō no hōsoku
" **minimum** *[bio]*	最少律	saishō-ritsu
" **mixture** *[sci-t]*	複合則	fukugō-soku
" **multiple proportions** *[ch]*	倍数比例の法則	baisū-hirei no hōsoku
" **nature** *[bio]*	自然法則	shizen-hōsoku
" **naught or one** *[math]*	悉無律; しつむりつ	shitsu'mu-ritsu
" **negative binomial distribution** *[bio]*	負の二項分布則	fu no nikō-bunpu-soku
" **octaves** *[ch]*	音程律	ontei-ritsu
" **parallelogram** *[math]*	平行四辺形の法則	heikō-shihen'kei no hōsoku
" **parsimony** *[sci-t]*	倹約律	ken'yaku-ritsu
" **partial pressure** *[ch-eng]*	分圧の法則	bun'atsu no hōsoku
" **partition** *[phys]*	分配の法則	bunpai no hōsoku
" **periodicity** *[ch]*	周期律	shūki-ritsu
" **photochemical absorption** *[p-ch]*	光化学吸収の法則	kō'kagaku-kyūshū no hōsoku
" **photochemical equivalent** *[p-ch]*	光化学当量の法則	kō'kagaku-tōryō no hōsoku
" **polar excitation** *[bio]*	極興奮の法則	kyoku-kōfun no hōsoku
" **progression** *[bio]*	人口法則	jinkō-hōsoku
" **quotient** *[math]*	商法則	shō-hōsoku
" **reciprocal proportions** *[ch]*	相互比例の法則	sōgo-hirei no hōsoku
" **reciprocity** *[math]*	相互法則	sōgo-hōsoku
" **rectilinear diameter** *[p-ch]*	直径線の法則	chokkei-sen no hōsoku
" **reflection** *[phys]*	反射の法則	hansha no hōsoku
" **refraction** *[phys]*	屈折の法則	kussetsu no hōsoku
" **segregation** *[bio]*	分離の法則	bunri no hōsoku
" **similitude** *[phys]*	相似法則	sōji-hōsoku
" **sines** *[trig]*	正弦法則	seigen-hōsoku
" **size distribution** *[phys]*	粒度分布則	ryūdo-bunpu-soku
" **small numbers** *[stat]*	少数の法則	shōsū no hōsoku
" **succession** *[bio]*	連続の法則	renzoku no hōsoku
" **superposition** *[geol]*	地層累重の法則	chisō-ruijū no hōsoku
" **supply and demand** *[econ]*	需要供給の法則	juyō-kyōkyū no hōsoku
" **tangents** *[trig]*	正接法則	seisetsu-hōsoku
" **the least action** *[mech]*	最小作用の法則	saishō-sayō no hōsoku
" **toleration** *[bio]*	耐忍の法則	tai'nin no hōsoku
" **trichotomy** *[math]*	三分法則	sanbun-hōsoku
" **velocity distribution** *[stat-mech]*	速度分布則	sokudo-bunpu-soku
lawn *[arch] [bot]*	芝生; しばふ	shiba'fu
lawn mower *[eng]*	芝刈機	shiba'kari-ki

lawrencium (element)	ローレンシウム	rōrenshūmu
laws of motion *[mech]*	運動の法則	undō no hōsoku
" **planetary motion** *[astron]*	惑星運動の法則	wakusei-undō no hō'soku
" **thought** *[logic]*	思考法則	shikō-hōsoku
lawyer *[legal]*	弁護士; 辯護士	bengo'shi
laxative *[pharm]*	緩下剤; かんげざい	kan-gezai
layer *{n}* *[geol]*	地層	chisō
" *{n}* *[geol]* *[geophys]* *[meteor]*	層; 層; そう	sō
lay on side *{vb t}*	横にする	yoko ni suru
layout *[graph]* *[math]*	配置; はいち	haichi
" (diagram) *[graph]*	配置図; 配置圖	haichi-zu
" *[pr]*	様式	yōshiki
lazulite *[miner]*	青燐鉱; 青燐鑛	seirin'kō
" *[miner]*	天藍石	tenran'seki
lazurite *[miner]*	青金石	seikin'seki
leachate *{n}* *[ch]*	浸出物	shin'shutsu'butsu
leached solution *[ch]*	浸出液	shin'shutsu-eki
leaching *{n}* *[ch-eng]* *[geoch]*	浸出; しんしゅつ	shin'shutsu
lead *{n}* (element)	鉛; なまり	namari
lead acetate	酢酸鉛	sakusan-namari
lead angle *[phys]*	進み角	susumi-kaku
lead borate	硼酸鉛	hōsan-namari
lead borofluoride	硼弗化鉛	hō'fukka-namari
lead carbonate	炭酸鉛	tansan-namari
lead chloride	塩化鉛; 鹽化鉛	enka-namari
lead dihydrogen phosphate	燐酸二水素鉛	rinsan-ni'suiso-namari
lead dioxide	二酸化鉛	ni'sanka-namari
leader stroke *[meteor]*	先駆放電	senku-hōden
lead fluoride	弗化鉛	fukka-namari
lead foil *[mat]*	鉛箔	en'paku
lead glaze *[cer]*	鉛釉; なまりぐすり	namari-gusuri
lead hot dipping *{n}* *[met]*	溶融鉛めっき	yōyū-namari-mekki
lead hydride	水素化鉛	suiso'ka-namari
lead hydroxide	水酸化鉛	suisan'ka-namari
leading edge *[aero-eng]* *[comput]*	前縁	zen'en
leading graphics *[comput]*	先行文字	senkō-moji
leading-in wire *[elec]*	導入線	dōnyū-sen
lead iodide	沃化鉛	yōka-namari
lead monohydrogen phosphate	燐酸一水素鉛	rinsan-ichi'suiso-namari
lead monoxide; litharge *[ch]*	一酸化鉛	issanka-namari
lead nitrate	硝酸鉛	shōsan-namari
lead oxide	酸化鉛	sanka-namari

lead phosphate	燐酸鉛	rinsan-namari
lead phosphite	亜燐酸鉛	a'rinsan-namari
lead screw *[eng]*	親ねじ	oya-neji
lead sheath *[mat]*	鉛被；えんぴ	enpi
lead silicate	珪酸鉛	keisan-namari
lead silicofluoride	珪弗化鉛	kei'fukka-namari
lead sulfate	硫酸鉛	ryūsan-namari
lead sulfide	硫化鉛	ryūka-namari
lead sulfite	亜硫酸鉛	a'ryūsan-namari
lead wire *[eng]*	口出線	kuchi'dashi-sen
leaf *{n} [bot]*	葉；は	ha
leaf apex *[bot]*	葉先；ようせん	yōsen
leaf base *[bot]*	葉脚	yō-kyaku
leaf blade; lamina *[bot]*	葉身	yō-shin
leaf bud *[bot]*	葉芽	yōga
leaf filter *[ch]*	葉状沪過器	yōjō-roka-ki
leaflet *[bot]*	小葉	shōyō
leaf margin *[bot]*	葉縁	yōen
leaf sheath *[bot]*	葉鞘；ようしょう	yōshō
leaf spring *[eng]*	板ばね	ita-bane
leaf vein; nervure *[bot]*	葉脈	yō-myaku
leafy vegetables *[cook]*	菜類	sai-rui; na-rui
" *[cook]*	葉菜類	yōsai-rui
leak *{n} [elec]*	漏電；ろうでん	rōden
leak; leakage *{n} [eng]*	漏れ	more
leak *{vb i}*	漏る；洩る；もる	moru
leakage current *[elec]*	漏れ電流	more-denryū
leakage flux *[elecmg]*	漏洩磁束	rō'ei-ji'soku
leakage transformer *[elecmg]*	磁気漏れ変圧器	jiki'more-hen'atsu-ki
lean gas *[ch]*	貧ガス	hin-gasu
leap second *[astron]*	閏秒；うるうびょう	urū-byō
leap year *[astron]*	閏年	urū-doshi
learn *{vb} [psy]*	習う	narau
" ; commit to memory *{vb} [psy]*	覚える；覺える	oboeru
learning; scholarship; science *{n} [sci-t]*	学術；學術；がくじゅつ	gaku'jutsu
learning *{n} [psy]*	学習	gakushū
learning process *[psy]*	学習過程	gakushū-katei
least action, principle of *[mech]*	最小作用の原理	saishō-sayō no genri
least common denominator *[arith]*	最小公分母	saishō-kō'bunbo
least common multiple *[arith]*	最小公倍数	saishō-kōbai'sū
least curvature *[math]*	最小曲率	saishō-kyoku'ritsu

least distance of distinct vision *[navig]*	明視距離；めいし=きょり	meishi-kyori
least squares estimator *[stat]*	最小二乗推定量	saishō-nijō-suitei'ryō
least time, principle of *[opt]*	最小時間の原理	saishō-jikan no genri
least value *[math]*	最小値	saishō-chi
leather *[cl]*	革；かわ	kawa
" *[mat]*	鞣し皮	nameshi-gawa
leatherback (turtle) *[v-zoo]*	長亀；おさがめ	osa-game
leaven *{n}* *[cook]*	パン種	pan-dane
leaving a space; with space left	間隔をへだてた	kankaku o hedateta
lee; leeward *{n}* *[sci-t]*	風下	kaza'shimo
leech (annelid) *[i-zoo]*	蛭；ひる	hiru
leek (vegetable) *[bot]*	韮；韭；にら	nira
lee(ward) *{adj}* *[ocean]*	風下の	kaza'shimo no
lee(ward) side *[sci-t]*	風下；かざしも	kaza'shimo
leeway; latitude *[eng]*	ゆとり	yutori
left; the left *[math]*	左；ひだり	hidari
left and right *[math]*	左右；さゆう	sayū
left-handed *[math]* *[physio]*	左利き	hidari-giki; hidari-kiki
left-hand polarized wave *[elecmg]*	左旋楕円偏波	sasen-daen-henpa
left-hand radical of a Chinese character *[pr]*	扁；偏；へん	hen
-left-hand radical *{suffix}*	-扁；偏	-hen; -ben
left-hand rule *[elecmg]*	左手の法則	hidari'te no hōsoku
left-hand side *{adv}*	左辺；左邊；さへん	sa'hen
left-justify *[comput]* *[graph]*	左詰め	hidari-tsume; hidari-zume
" *[comput]* *[graph]*	左寄せ	hidari-yose
left side view *[graph]*	左側面図	hidari-sokumen-zu
left turn *{n}* *[navig]*	左折	sa-setsu
left-turn lane *[traffic]*	左折車線	sa'setsu-shasen
leg *{n}* *[anat]*	脛；はぎ	hagi
" (of an article) *{n}*	脚；きゃく	kyaku; ashi
legal tender; lawful money *[econ]*	法貨	hōka
legend *[map]*	凡例	hanrei; bonrei
legibility *[pr]*	可読度；可讀度	ka'doku-do
legume *[bot]*	莢果；きょうか	kyōka
" *[bot]*	豆類	tō-rui; mame-rui
Leguminosae *[bot]*	豆科	mame-ka
leisure *{n}*	レジャー	rejā
lemma (pl lemmas, lemmata) *[logic]*	補題；ほだい	hodai
" *[logic]*	補助定理	hojo-teiri
lemur (animal) *[v-zoo]*	狐猿；きつねざる	kitsune-zaru

lend *{vb t}*	借す	kasu
length *[math]*	長さ；ながさ	naga-sa
" *[mech]*	縦；縱；たて	tate
lengthen; extend *{vb t}*	延ばす；伸ばす	nobasu
lengthen; stretch *{vb i}*	延びる；伸びる	nobiru
lengthwise direction *[eng]*	長手方向	naga'te-hōkō
lens *[opt]*	レンズ	renzu
lentil (legume) *[bot]*	平豆	hira-mame
Leo; the Lion *[constel]*	獅子座	shishi-za
leopard (animal) *[v-zoo]*	豹；ひょう	hyō
leopard flower; Belamcanda *[bot]*	檜扇；ひおうぎ	hi-ōgi
lepidolite *[miner]*	鱗雲母	uroko-unmo
lepidomelane *[miner]*	鉄雲母；てつうんも	tetsu-unmo
lepidopterology *[bio]*	蝶蛾学	chōga'gaku
" *[zoo]*	鱗翅学；りんしがく	rinshi'gaku
leptology *[atom-phys]*	微細構造学	bisai-kōzō'gaku
lepton *[part-phys]*	軽粒子	kei'ryūshi
less than *[math]*	より少い	yori sukunai
less than; fewer than *[math]*	以下	ika
lethal dose *[med]*	致死量	chishi-ryō
lethal mutation *[gen]*	致死変異	chishi-hen'i
let slip; let slide *{vb} [mech-eng]*	滑らす	suberasu
let stand; leave as is *{vb t}*	放置する	hōchi suru
letter *{n} [comm]*	手紙	tegami
" *{n} [comput] [pr]*	英字	eiji
" *{n} [graph] [ling] [pr]*	文字；もじ	moji
" *{n} [graph] [ling] [pr]*	欧字；歐字；おうじ	ōji
letter of attorney *[pat]*	委任状	i'nin'jō
letter paper *[comm]*	便箋；びんせん	binsen
letterpress printing *[pr]*	活版印刷	kappan-in'satsu
" *[pr]*	凸版印刷	toppan-in'satsu
letter scale *[eng]*	郵便秤	yūbin-bakari
lettuce (vegetable) *[bot]*	萵苣；ちしゃ	chisha
leucite *[miner]*	白榴石	hakuryū'seki
leukocyte; white blood cell *[hist]*	白血球	hakkekkyū
level (a tool) *{n} [eng]*	水準器	suijun-ki
" *{n} [comput] [math]*	水準	suijun
" ; stage *{n}*	段；だん	dan
level difference *{n}*	段差	dansa
leveling *{n}*	平滑化	heikatsu'ka
leveling property *[ch]*	均染性	kinsen-sei
leveling screw *[eng]*	整準ねじ	seijun-neji

level of significance *[stat]*	危険率；きけんりつ	kiken-ritsu
" *[stat]*	有意水準	yū'i-suijun
lever *[eng]*	梃子；てこ	teko
leverage *[mech]*	梃子比	teko-hi
levigation *[ch-eng]*	水簸；すいひ	suihi
levitation *[phys]*	空中浮揚	kūchū-fuyō
levorotatory *[opt]*	左旋(光)性	sa'sen(kō)-sei
levulose *[bioch]*	左旋糖；させんとう	sasen'tō
lexicon *[pr]*	辞典；辭典；じてん	jiten
liability; responsibility *[legal]*	責任；せきにん	seki'nin
liaison *[comm]*	連絡	renraku
libethenite *[miner]*	燐銅鉱	rin'dōkō
Libra; the Balance *[constel]*	天秤座；てんびんざ	tenbin-za
library (pl -braries) *[arch]*	資料館	shiryō'kan
" *[arch]*	図書館；圖書館	tosho'kan
libration *[astron]* *[phys]*	秤動；ひょうどう	hyōdō
libration of moon *[astron]*	月の秤動	tsuki no hyōdō
license *[legal]*	鑑札；かんさつ	kan'satsu
" *[legal]*	免許	menkyo
" *[pat]*	実施権；實施權	jisshi'ken
license plate *[legal]*	認可番号札	ninka-bangō'satsu
license plate light *[elec]*	番号灯；番號燈	bangō-tō
Lichenes; lichens *[bot]*	地衣界	chi'i-kai
" *[bot]*	地衣植物	chi'i-shokubutsu
lichi (tree, fruit) *[bot]*	茘枝；れいし	reishi
licorice (herb) *[bot]*	甘草；かんぞう	kanzō
lid; cover *[eng]*	蓋；ふた	futa
lie down *{vb}*	横になる	yoko ni naru
life (pl lives) *[bio]*	命；いのち	inochi
" *[bio]*	生命；せいめい	seimei
life and death *[bio]*	生死	seishi
lifeboat *[nav-arch]*	救命艇	kyūmei'tei
life cycle *[bio]*	生活環	seikatsu-kan
life expectancy *[bio]*	期待寿命；期待壽命	kitai-jumyō
life history *[bio]*	生活史	seikatsu-shi
life jacket *[nav-arch]*	救命胴衣	kyūmei-dō'i
life preserver *[nav-arch]*	救命具	kyūmei'gu
life sciences *[bio]*	生命科学	seimei-kagaku
life support system *[eng]*	生命維持系統	seimei-iji-keitō
lifetime *[eng]*	寿命；壽命	jumyō
lift *{n}* *[aero-eng]*	揚力	yō-ryoku
lift; lift up *{vb t}*	持上げる	mochi-ageru

lift coefficient *[aero-eng]*	揚力係数	yō'ryoku-keisū
lifting *{n} [adhes]*	浮き；うき	uki
lifting gear *[mech-eng]*	吊上げ装置	tsuri'age-sōchi
lift-off *[aero-eng]*	離昇	rishō
lift-to-drag ratio *[aero-eng]*	揚抗比；ようこうひ	yōkō-hi
ligament *[hist]*	靭帯；靭帯	jintai
ligand *[ch]*	配位子	hai'i'shi
ligand field *[ch]*	配位子場	hai'ishi-ba
ligase *[bioch]*	合成酵素	gōsei-kōso
ligation *[mol-bio]*	連結反応；連結反應	renketsu-hannō
light (lamp) *{n} [opt]*	明かり	akari
" *{n} [opt]*	光；ひかり	hikari
" (lightweight) *{adj}*	軽い；輕い	karui
" (turn the light on) *{vb}*	点す；ともす	tomosu
light-; photo- *{prefix}*	光-	kō-; hikari-
light adaptation *[physio]*	明順応	mei-junnō
light bulb *[elec]*	電球；でんきゅう	denkyū
light buoy *[eng]*	灯浮標	tō-fu'hyō
light chain *[immun]*	短鎖	tansa
light color; pale color *[opt]*	淡色；たんしょく	tan-shoku
light-colored soy sauce *[cook]*	薄口醤油	usu'kuchi-shōyu
light curve *[astrophys]*	光度曲線	kōdo-kyoku'sen
light detection *[opt]*	光検知；光檢知	kō-kenchi
light distribution curve *[opt]*	配光曲線	haikō-kyoku'sen
light electrical appliance *[elec]*	弱電機器	jaku'den-kiki
light-emitting diode *[electr]*	発光ダイオード	hakkō-daiōdo
light flavor *{n} [cook]*	薄味；うすあじ	usu-aji
light-gathering power *[opt]*	集光力	shūkō-ryoku
lighthouse *[arch] [navig]*	灯台；燈臺	tōdai
light industry *[ind-eng]*	軽工業；輕工業	kei-kōgyō
lighting *{n} [opt]*	照明；しょうめい	shōmei
lighting fixture *[opt]*	照明器具	shōmei-kigu
light metal *[met]*	軽金属	kei-kinzoku
lightness *{n} [opt]*	明度	meido
lightning *{n} [geophys]*	電光	denkō
" *{n} [geophys]*	稲光	ina'bikari
" *{n} [geophys]*	稲妻；いなずま	ina'zuma
lightning arrester *[elec]*	避雷器	hirai-ki
lightning conductor *[elec]*	避雷針；ひらいしん	hirai-shin
lightning discharge *[geophys]*	雷放電	kaminari-hōden; rai-hōden
lightning rod *[elec]*	避雷針	hirai-shin
light oil *[mat]*	軽(質)油	kei('shitsu)-yu

light quantity *[opt]*	光量；こうりょう	kōryō
light quantum *[quant-mech]*	光量子	hikari-ryōshi; kō'ryōshi
" *[quant-mech]*	光子	kōshi
light rays *[opt]*	光線	kōsen
light resistance *[electr] [mat]*	耐光性	taikō-sei
light-resisting stability *[opt]*	耐光安定性	taikō-antei-sei
light scattering *{n} [opt]*	光散乱	kō-sanran
light second *[mech]*	光秒；こうびょう	kō-byō
light-sensitive material *[photo]*	感光体	kankō-tai
light source *[opt]*	光源	kōgen
light spot *[phys]*	光点；光點	kōten
light structure *[poly-ch]*	軽構造	kei-kōzō
light time *[astron]*	光差	kōsa
light trap *[agr] [eng]*	誘虫灯；誘蟲燈	yūchū-tō
" *[agr] [eng]*	誘蛾灯	yūga-tō
light water *[nucleo]*	軽水；けいすい	keisui
light-water reactor *[nucleo]*	軽水炉；輕水爐	keisui-ro
light wave *[opt]*	光波	kōha
lightweight fabric *[tex]*	薄物；うすもの	usu-mono
light-year *[astrophys]*	光年	kō-nen
lignaloe (incense) *[bot]*	沈香；じんこう	jinkō
lignification *[bot]*	木化	mok(u)ka
lignin *[bioch]*	木質素；もくしつそ	moku'shitsu'so
lignite *[geol]*	亜炭；亞炭；あたん	a'tan
likelihood *[math]*	尤度；ゆうど	yūdo
like terms *[alg]*	同類項	dōrui-kō
lilac (color); lavender *[opt]*	藤色	fuji-iro
liliaceous flower *[bot]*	百合形花	yuri'kei-ka
lily (pl lilies) (flower) *[bot]*	百合；ゆり	yuri
lily bulb *[bot] [cook]*	百合根	yuri-ne
lily of the valley (flower) *[bot]*	鈴蘭	suzu'ran
limb *[astron]*	周辺；しゅうへん	shūhen
" *[bio]*	肢；し	shi
limb brightening *[astrophys]*	周辺増光；周邊增光	shūhen-zōkō
limb darkening *[astrophys]*	周辺減光	shūhen-genkō
limbs *[anat]*	四肢	shishi
lime *{n} [inorg-ch]*	石灰；せっかい	sekkai
lime emulsion *[mat]*	石灰乳液	sekkai-nyū'eki
lime grotto *[geol]*	石灰洞	sekkai-dō
limelight *[eng]*	石灰光	sekkai-kō
" *[opt]*	灰光灯	kaikō-tō
limen *[psy]*	閾(値)；いき(ち)	iki(-chi)

lime nitrogen *[inorg-ch]*	石灰窒素	sekkai-chisso
limestone *[petr]*	石灰岩	sekkai'gan
" *[petr]*	石灰石	sekkai'seki
limit *[calc]*	極限；きょくげん	kyoku'gen
" *[math]*	限度	gendo
limitation *[legal]*	限定	gentei
" *{n}*	限界	genkai
limiter *[electr]*	制限器；せいげんき	seigen-ki
limiting factor *[bio]*	制限因子	seigen-inshi
limiting viscosity number *[p-ch]*	極限粘度	kyoku'gen-nendo
limit of elasticity *[mech]*	弾性限度；彈性限度	dansei-gendo
limit of proportionality *[mech]*	比例限度	hirei-gendo
limnology *[ecol]*	陸水学	rikusui'gaku
limonite *[miner]*	褐鉄鉱；褐鐵鑛	kattekkō
limpet; mother of pearl; Notoacmea schrenckii (shellfish) *[i-zoo]*	青貝；あおがい	ao-gai
linarite *[miner]*	青鉛鉱	sei'enkō
linden, Japanese (tree) *[bot]*	科木；しなのき	shinanoki
lindera *[bot]*	黒文字；くろもじ	kuro-moji
line *[comput] [pr]*	行；ぎょう	gyō
" ; straight line *[geom]*	直線；ちょくせん	choku'sen
" *[elec]*	回線	kaisen
" *[electr] [geom]*	線	sen
" *[gen]*	系統	keitō
linear *{n} [math]*	一次；いちじ	ichi'ji
" *{n} [math]*	線状；線狀	senjō
" *{n} [math]*	線形	senkei
linear accelerator (linac) *[electr]*	直線加速装置	chokusen-kasoku-sōchi
linear amplification *[electr]*	直線増幅	chokusen-zōfuku
linear equation *[alg]*	一次方程式	ichi'ji-hōtei'shiki
linear function *[alg]*	一次関数；一次關數	ichi'ji-kansū
linear molecule *[ch]*	直線形分子	chokusen'gata-bunshi
" *[p-ch]*	線状分子	senjō-bunshi
linear polymer *[poly-ch]*	鎖状重合体	sajō-jūgō'tai
" *[poly-ch]*	線状重合体	senjō-jūgō'tai
" *[poly-ch]*	線状高分子	senjō-kōbunshi
linear programming *{n} [alg]*	線形計画(法)	senkei-keikaku(-hō)
linear regression equation *[math]*	一次回帰式	ichi'ji-kaiki-shiki
linear velocity *[mech]*	線速度；せんそくど	sen-sokudo
line drop *[elec]*	線路電圧効果	senro-den'atsu-kōka
line feed *[comput]*	改行	kaigyō
line loss *[elec]*	線路損	senro-son

linen *[tex]*	亜麻布	ama-nuno
" *[tex]*	麻；あさ	asa
line number *[comput]*	行番号；行番號	gyō-bangō
line of action *[mech-eng]*	作用線	sayō-sen
line of flow *[fl-mech]*	流線；りゅうせん	ryūsen
line of force *[phys]*	力線	riki-sen
line of sight *[sci-t]*	視線	shisen
line printer *[comput]*	行印字機	gyō-inji-ki
liner (a ship) *[nav-arch]*	定期船	teiki'sen
line-ruled paper *[mat]*	罫紙；けいし	keishi
line-ruling *{n} [eng]*	罫書き	ke'gaki
line segment *[geom]*	線分	senbun
line speed *[comm]*	回線速度	kaisen-sokudo
line up; be in a row; rank with *{vb}*	並ぶ	narabu
lineup *{n} [ch]*	整列	sei'retsu
lingering sound; aftersound *[acous]*	余音；餘音；よいん	yo'in
linguistics *[comm]*	言語学	gengo'gaku
linguistic sense *[ling]*	語感	gokan
lingulate *{adj} [bio]*	舌状の	zetsu'jō no
liniment *{n} [pharm]*	糊膏；ここう	kokō
" *{n} [pharm]*	擦剤；擦劑	satsu-zai
" *{n} [pharm]*	塗擦剤；とさつざい	to'satsu-zai
lining *{n} [cl]*	裏地；うらじ	ura'ji
" *{n} [mat]*	裏打ち	ura'uchi
" *{n} [tex]*	裏張	ura'bari
lining brick *[mat]*	内張り煉瓦	uchi'bari-renga
link; linkage *{n} [comput]*	連結	renketsu
link *{n} [mech-eng]*	節；せつ	setsu
linkage isomerism *[ch]*	結合異性	ketsu'gō-i'sei
" *[ch]*	連結異性	renketsu-i'sei
linnet (bird) *[v-zoo]*	胸赤鶸	mune'aka-hiwa
linseed *[bot]*	亜麻仁；あまに(ん)	amani(n)
linseed oil *[mat]*	亜麻仁油	amani(n)-yu
lint *[tex]*	糸屑；いとくず	ito-kuzu
" *[tex]*	綿屑	wata-kuzu
lion (animal) *[v-zoo]*	獅子；しし	shishi
lionfish; Pterolis lunulata *[v-zoo]*	簑笠子	mino-kasago
lip *{n} [anat]*	唇；くちびる	kuchibiru
lipid *[bioch]*	脂質	shi'shitsu
lipophilic *{adj} [ch]*	親脂(性)の	shinshi('sei) no
" *{adj} [ch]*	親油(性)の	shin'yu('sei) no
lipophilicity *[ch]*	脂溶性	shiyō-sei

lipoprotein *[bioch]*	リポ蛋白質	ripo-tanpaku'shitsu
liquefaction *[phys]*	液化; えきか	eki'ka
liquefying power *[ch]*	液化力	eki'ka-ryoku
liquescent *{adj} [ch]*	液化性の	eki'ka'sei no
liqueur *[cook]*	混成酒	konsei'shu
liquid *{n} [ch] [phys]*	液(体); えき(たい)	eki(tai)
" *{adj} [ch] [phys]*	液状の; 液狀の	eki'jō no
liquid air *[phys]*	液体空気; 液體空氣	ekitai-kūki
liquid carbon dioxide *[mat]*	液体炭酸	ekitai-tansan
liquid column *[phys]*	液柱	ekichū
liquid crystal *[p-ch]*	液晶; えきしょう	ekishō
liquid crystal display *[electr]*	液晶表示	ekishō-hyōji
liquid-evacuation vessel *[ch]*	排液槽	hai'eki-sō
liquid film *[p-ch]*	液膜	eki-maku
liquid fuel *[mat]*	液体燃料	ekitai-nenryō
liquid insulator *[mat]*	液体絶縁物	ekitai-zetsuen'butsu
liquid-jet cutting *{n} [mech-eng]*	流体噴流切	ryūtai-funryū'giri
liquid junction potential *[p-ch]*	液界電位	ekikai-den'i
liquid-liquid extraction *[ch-eng]*	液液抽出	eki-eki-chū'shutsu
liquid lubrication *[mech-eng]*	液体潤滑	ekitai-junkatsu
liquid medicine *[pharm]*	水薬	sui-yaku
" *[pharm]*	薬液; 藥液	yaku-eki
liquid metal *[mat]*	溶湯; ようとう	yōtō
liquid nitrogen *[mat]*	液体窒素	ekitai-chisso
liquid nutrient medium *[microbio]*	液体栄養培地	ekitai-eiyō-baichi
liquid paraffin *[mat]*	流動パラフィン	ryūdō-parafin
liquid phase *[ch]*	液相	ekisō
liquid rubber *[poly-ch] [rub]*	液状ゴム	eki'jō-gomu
liquid soap *[mat]*	液状石鹼	eki'jō-sekken
liquor *[ch-eng] [pharm]*	液; えき	eki
list; listing (of a ship) *{n} [eng]*	横傾斜	yoko-keisha
list *{n} [comput] [math]*	表; ひょう	hyō
listed company *[econ]*	上場会社; 上場會社	jōjō-kaisha
listen *{vb} [physio]*	聞く; 聴く; きく	kiku
listen carefully *{vb}*	よく聞く	yoku kiku
litchi (tree, fruit) *[bot]*	茘枝; れいし	reishi
liter (unit of volume) *[mech]*	立; リットル	rittoru
literal equation *[alg]*	文字方程式	moji-hōtei'shiki
literary language *[pr]*	文語	bungo
literary property *[legal]*	著作権; 著作權	chosak(u)ken
" *[pr]*	文芸; 文藝	bungei
literary-style writing *[pr]*	文語文	bungo'bun

literature (the list) *[pr]*	文献；ぶんけん	bunken
litharge *[miner]*	一酸化鉛	issanka-namari
" *[miner]*	密陀僧；みつだそう	mitsuda-sō
lithium (element)	リチウム	richūmu
lithogeochemistry *[geoch]*	岩石地球化学	ganseki-chikyū-kagaku
lithographic press *[pr]*	平板印刷機	heiban-in'satsu-ki
lithography *[pr]*	石版印刷	sekiban-in'satsu
lithology *[geol]*	岩石学	ganseki'gaku
lithophile *{n} [geoch]*	含酸素鉱生成元素	gan'sanso'kō-seisei-genso
lithophile element *[geoch]*	親石元素	shinseki-genso
lithophyte *[ecol]*	岩生植物	gansei-shokubutsu
Lithospermi Radix *[bot] [pharm]*	紫根；しこん	shikon
lithosphere *[geol]*	岩石圏；岩石圈	ganseki-ken
lithotroph *[bio]*	無機栄養生物	muki-eiyō-seibutsu
litigation *[legal]*	訴訟；そしょう	soshō
litmus *[mat]*	リトマス	ritomasu
little; limited in amount *{adj}*	少い	sukunai
little (fifth) finger *{n} [anat]*	小指；こゆび	ko-yubi
littoral current *[ocean]*	沿岸潮流	engan-chōryū
live ammunition *[ord]*	実弾薬；實彈藥	jitsu-dan'yaku
live fish carrier (a ship) *[nav-arch]*	活魚運搬船	katsu'gyo-unpan'sen
liver *{n} [anat]*	肝臓；かんぞう	kanzō
" *{n} [anat]*	肝；きも	kimo
liver oil *[mat]*	肝油	kan-yu
liverwort; Hepaticae *[bot]*	苔類	tai-rui
live steam *[phys]*	生蒸気；生蒸氣	nama-jōki
livestock *[agr]*	畜産物	chiku'san'butsu
" *[agr]*	家畜	ka'chiku
livestock farming *[agr]*	牧畜；ぼくちく	boku'chiku
live wire *[elec]*	活線	kassen
live yeast *[mycol]*	生酵母；なまこうぼ	nama-kōbo
living body *[bio]*	生体	seitai
living language *[ling]*	現用語	gen'yō'go
living organism *[bio]*	生物	seibutsu
living room *[arch]*	居間	i'ma
lizard (reptile) *[v-zoo]*	蜥蜴；とかげ	tokage
loach (fish) *[v-zoo]*	泥鰌；土鰌	dojō
load *{n} [elec] [mech]*	負荷	fuka
" *{n} [mech]*	荷重；かじゅう	kajū
load-bearing wall *[arch]*	耐力壁	tai'ryoku-heki
load-carrying capacity *[trans]*	耐荷重能	tai'kajū-nō
loaded circuit *[elec]*	装荷回線	sōka-kaisen

load factor *[civ-eng]*	負荷率；ふかりつ	fuka-ritsu
loading (machining) *{n} [met]*	目詰まり；めづまり	me'zumari
" (of weapon) *{n} [ord]*	填薬	ten-yaku
" ; carrying *{n} [transp]*	積載；せきさい	sekisai
loading material; filler *[mat]*	填料；てんりょう	tenryō
" ; extender *[mat]*	増量剤	zōryō-zai
load loss *[elec]*	負荷損	fuka-son
loam; kieselguhr *[geol]*	珪藻土；けいそうど	keisō'do
" ; earth; soil *{n} [geol]*	土；つち	tsuchi
lobate; leaf-form *{adj} [bio]*	葉状の	yōjō no
lobe *[anat]*	葉；よう	yō
" *[bot]*	裂片；れっぺん	reppen
" ; leaflet *[bot]*	小葉；しょうよう	shōyō
lobster, spiny (crustacean) *[i-zoo]*	伊勢蝦；いせえび	ise-ebi
lobster trap *[eng]*	笯；ど	do
" *[eng]*	筌；うえ	ue
local *{adj} [comput]*	局所的な	kyokusho'teki na
" (of the locality) *{adj} [geogr]*	土地の；とちの	tochi no
" (not express) *{adj} [trans]*	各駅停車の	kaku'eki-teisha no
local anesthetic *{n} [pharm]*	局所麻酔薬	kyoku'sho-masui-yaku
local current *[elec]*	局部電流	kyokubu-denryū
locally severe rainstorm *[meteor]*	集中豪雨	shūchū-gō'u
local purchase *[econ]*	現地購入	genchi-kōnyū
local terminal *[comput]*	構内端末	kōnai-tanmatsu
local time *[horol]*	地方時；ちほうじ	chihō-ji
location *[comput]*	位置；いち	ichi
" *[math]*	位；くらい	kurai
" *{n}*	所在；しょざい	shozai
location parameter *[math]*	位置母数	ichi-bosū
lock *{n} [eng]*	錠；じょう	jō
lock nut *[eng]*	止めナット	tome-natto
locking of mold; clamping *{n} [mech-eng]*	型締め	kata-jime
locking pin; locking device *[eng]*	割ピン	wari-pin
locking pressure *[mech-eng]*	型締圧力	kata'jime-atsu'ryoku
locksmith (a person) *[arch] [eng]*	錠前屋	jō'mae'ya
locomotion; walking *[physio]*	歩行；ほこう	hokō
" *[sci-t]*	移動運動	idō-undō
locomotive; engine *[rail]*	機関車；きかんしゃ	kikan'sha
locus (pl loci) *[gen] [geom]*	座；ざ	za
" *[geom]*	軌道；きどう	kidō
" *[geom] [geophys]*	軌跡；きせき	kiseki
locust (insect) *[i-zoo]*	飛蝗；蝗；ばった	batta

locust (insect) *[i-zoo]*	蝗；いなご	inago
lodestone *[miner]*	天然磁石	tennen-jishaku
loellingite *[miner]*	砒鉄鉱；砒鐵鑛	hi'tekkō
loess *[geol]*	黄土；おうど	ō'do
log *{n} [comm] [comput] [navig]*	経過記録；徑過記錄	keika-kiroku
" (of wood) *{n} [mat]*	丸太；まるた	maruta
logarithm *[alg]*	対数；對數	taisū
logarithmic decrement *[math] [phys]*	対数減衰率	taisū-gensui-ritsu
logarithmic growth *[microbio]*	対数増殖	taisū-zō'shoku
logarithmic mean *[math]*	対数平均	taisū-heikin
logarithmic paper *[math]*	両対数目盛	ryō'taisū-memori
logarithmic scale *[math]*	対数目盛	taisū-me'mori
logarithmic viscosity *[fl-mech]*	対数粘度	taisū-nendo
log book *[navig]*	航海日誌	kōkai-nisshi
loggerhead (a turtle) *[v-zoo]*	赤海亀	aka-umi'game
log hauling *{n} [agr]*	運材；うんざい	unzai
logic *[electr]*	論理；ろんり	ronri
" *[math]*	論理学；論理學	ronri'gaku
logical design *[comput]*	論理設計	ronri-sekkei
logical expression *[comput]*	論理式	ronri-shiki
logic(al) operation *[comput]*	論理演算	ronri-enzan
logical operator *[comput]*	論理演算子	ronri-enzan'shi
logical sum *[comput]*	論理和；ろんりわ	ronri-wa
logical unit *[comput]*	論理単位	ronri-tan'i
logic element *[comput]*	論理素子	ronri-soshi
logogram *[graph]*	略符；りゃくふ	ryaku'fu
logotype *[graph]*	標識図案；標識圖案	hyō'shiki-zu'an
lone; isolated; solitary *{adj}*	孤立した	ko'ritsu shita
lone electron pair *[p-ch]*	孤立電子対	ko'ritsu-denshi-tsui
long *[math]*	長い；ながい	nagai
longan (tree, fruit) *[bot]*	竜眼；りゅうがん	ryūgan
long and narrow *[math]*	細長い	hoso-nagai
long axis; major axis *[an-geom]*	長軸；ちょうじく	chō-jiku
longbill; snipe (bird) *[v-zoo]*	鴫；鷸；しぎ	shigi
long chain; heavy chain *[ch]*	長鎖	chōsa
long-distance call *[comm]*	長距離呼出	chō-kyori-yobi'dashi
long-distance line *[comm]*	長距離線路	chō-kyori-senro
long-distance navigation *[navig]*	長距離航法	chō-kyori-kōhō
long-distance telephone *[comm]*	長距離電話	chō-kyori-denwa
long-distance transmission *[comm]*	長距離送電	chō-kyori-sōden
long hand; minute hand (of clock) *[horol]*	長針；ちょうしん	chōshin
longish direction *[eng]*	長手方向	nagate-hōkō

longitude *[geod]*	経度；經度；けいど	keido
longitudinal direction *[eng]*	長手方向	nagate-hōkō
" *[eng]*	縦方向；縱方向	tate-hōkō
longitudinal division *[bio]*	縦分裂	tate-bun'retsu
longitudinal magnetization *[eng-acous]*	長さ方向磁化	nagasa-hōkō-jika
longitudinal oscillation *[phys]*	縦振動	tate-shindō
longitudinal shear *[mech]*	縦剪断	tate-sendan
longitudinal strain *[mech]*	縦歪み；たてひずみ	tate-hizumi
longitudinal vibration *[mech]*	縦振動	tate-shindō
longitudinal wave *[phys]*	縦波；縱波	tate-nami
long-lasting; wears well *[ind-eng]*	持ちが良い	mochi ga ī
long-length *[mech]*	長尺；ちょうしゃく	chō-shaku
long period *[phys]*	長周期	chō-shūki
long-range *{adj}*	長距離の	chō-kyori no
long-range force *[nuc-phys]*	遠達力	en'tatsu-ryoku
long-range forecast *[meteor]*	長期予報	chōki-yohō
long-range plan *{n}*	大計	taikei
long sea route *[navig]*	八潮路；やしおじ	ya'shio-ji
long shunt *[elec]*	外分巻；外分卷	soto-bun'maki
long sound; long vowel *[ling]*	長音	chō-on
long-tailed cock (bird) *[v-zoo]*	尾長鶏；おながどり	o'naga-dori
long-tailed tit (bird) *[v-zoo]*	尾長雀；おながが ら	o'naga-gara
long wave *[comm]*	長波	chō-ha
loom *{n} [tex]*	機；はた	hata
loon (bird) *[v-zoo]*	阿比；あび	abi
loop *{n} [geom]*	輪；環；わ	wa
loop checking *{n} [comput]*	返送照合	hensō-shōgō
loop pile *[tex]*	輪奈；わな	wana
loop strength *[tex]*	引掛強さ	hikkake-tsuyo-sa
loose *{adj}*	緩い；弛い；ゆるい	yurui
loosen; relax *{vb t}*	緩める；弛める	yurumeru
looseness *{n}*	緩み；弛み	yurumi
loosening; release *{n}*	緩め；弛め	yurume
loquat (tree, fruit) *[bot]*	枇杷；びわ	biwa
loran *[navig]*	自位置測定装置	ji'ichi-sokutei-sōchi
lose *{vb t}*	失う	ushinau
loss (of weight) *[mech]*	減量；げんりょう	genryō
" *{n}*	損失	son'shitsu
loss by drying *[mech]*	乾燥減量	kansō-genryō
loss factor *[elec]*	損率；そんりつ	son-ritsu
loss modulus *[mech]*	損失弾性係数	son'shitsu-dansei-keisū
" *[mech]*	損失弾性率	son'shitsu-dansei-ritsu

lot; plot *{n} [civ-eng]*	画地	kaku'chi
lotus (flower) *[bot]*	芙蓉；ふよう	fuyō
" *[bot]*	蓮；はす	hasu
lotus root *[bot] [cook]*	蓮根；れんこん	renkon
lotus seed *[bot] [cook]*	蓮実；蓮實	ren'jitsu; hasu no mi
loudness *[acous]*	音量	onryō
" *[acous]*	音の大きさ	oto no ōki-sa
loudspeaker *[acous]*	拡声器；擴聲器	kakusei-ki
louse (pl lice) *[i-zoo]*	虱；しらみ	shirami
loveable; sweet *{adj}*	可愛い；かわいい	kawaī; kawaii
lovebird (bird) *[v-zoo]*	牡丹鸚哥	botan-inko
low *{adj} [math]*	低い	hikui
low; low pressure *{n} [meteor]*	低気圧；低氣壓	tei-ki'atsu
low clouds *[meteor]*	下層雲	kasō'un
lower *{n}*	下；した	shita
" *{adj}*	下の	shita no; shimo no
" (lessen the height) *{vb t}*	低める	hikumeru
" (bring or let down) *{vb t}*	降ろす；下ろす	orosu
" *{vb t}*	下げる	sageru
lowercase letter *[comput] [pr]*	小文字	shō-moji
lower limit *[math]*	下限	kagen
lower side *[math]*	下方	kahō
lower-part radical of a Chinese character *[pr]*	脚；きゃく	kyaku
lowest *{adj} [math]*	最下位の	saika'i no
low-fat *{adj} [cook]*	低脂肪分の	tei-shibō'bun no
lowfat milk *[cook]*	低脂肪乳	tei'shibō-nyū
low frequencies *[acous]*	低音	tei'on
low-grade distilled spirit *[cook]*	焼酎；しょうちゅう	shōchū
low-grade polymer *[poly-ch]*	低重合体	tei-jūgō'tai
low land *[geogr]*	低地	teichi
low-molecular-weight compound *[ch]*	低分子量化合物	tei-bunshi'ryō-kagō'butsu
low-order *{n} [comput] [math]*	低次	teiji
low-pass filtration circuit *[elec]*	低域沪波回路	tei'iki-roha-kairo
low polymer *[poly-ch]*	低重合体	tei-jūgō'tai
low pressure *{n} [meteor] [phys]*	低圧；ていあつ	tei-atsu
" *{adj} [phys]*	低圧の；低壓の	tei-atsu no
low-pressure molding *{n} [eng]*	低圧成形	tei'atsu-seikei
low temperature *[thermo]*	低温；ていおん	tei'on
low-temperature-hardening property *[poly-ch]*	低温硬化性	teion-kōka-sei
low-temperature thermometry *[phys]*	低温度測定法	tei'ondo-sokutei-hō

low tension *[elec]*	低圧；低壓	tei-atsu
low tide *[ocean]*	干潮；かんちょう	kanchō
low voltage *{n} [elec]*	低圧	tei-atsu
" *{adj} [elec]*	低圧の	tei-atsu no
low-voltage cable *[elec]*	低圧電線	tei'atsu-densen
lubricant *[mat]*	潤滑剤	jun'katsu-zai
" *[mat]*	滑剤；かつざい	katsu-zai
lubricating oil *[mat]*	潤滑油	jun'katsu-yu
lubricity *[mat]*	滑らかさ	nameraka-sa
luffa; loofah *[bot]*	糸瓜；へちま	hechima
lug *{n} [eng]*	耳；みみ	mimi
luggage *[trans]*	荷物；にもつ	ni'motsu
" *[trans]*	旅行荷物	ryokō-ni'motsu
lumber *[mat]*	材木	zai'moku
lumber industry *[eng]*	木材工業	mokuzai-kōgyō
lumen *[sci-t]*	内腔	naikō
luminance *[opt]*	輝度	kido
luminescence *[phys]*	発光	hakkō
luminophor *[phys]*	蛍光体；螢光體	keikō'tai
" *[phys]*	燐光体	rinkō'tai
luminophore group *[phys]*	発光団；發光團	hakkō-dan
luminosity *[opt]*	光度	kōdo
luminous cloud *[meteor]*	夜光雲；やこううん	yakō'un
luminous density *[photo]*	明度濃度	meido-nōdo
luminous efficiency *[opt]*	発光効率	hakkō-kō'ritsu
luminous emittance *[opt]*	光束発散度	kōsoku-hassan-do
luminous flame *[ch]*	輝炎；きえん	ki'en
luminous flux *[opt]*	光束	kō-soku
luminous intensity *[opt]*	光の強度	hikari no kyōdo
luminous intensity *[opt]*	光度	kōdo
luminous paint *[mat]*	発光塗料	hakkō-toryō
" *[mat]*	夜光塗料	yakō-toryō
luminous radiance *[opt]*	光束発散度	kōsoku-hassan-do
lump *{n} [cook]*	継粉；ままこ	mamako
lump; mass; clod *{n}*	固まり；塊	katamari
lump coal *[geol]*	塊状炭	kaijō-tan
lumped-constant equivalent circuit *[elec]*	集中定数等価回路	shūchū-teisū-tōka-kairo
lunar crater *[astron]*	月面火口状窪み	getsu'men-kakō'jō-kubomi
lunar cusps *[astron]*	月の尖端	tsuki no sentan
lunar day *[astron]*	太陰日	tai'in-jitsu
lunar eclipse *[astron]*	月食；月蝕	gesshoku

lunar halo *[meteor]*	月の暈；つきのかさ	tsuki no kasa
lunar map *[astron]*	月面図；月面圖	getsu'men-zu
lunar month; lunation *[astron]*	太陰月	tai'in-getsu
lunar occultation *[astron]*	星食；星蝕	seishoku
lunar orbiter *[aero-eng]*	月周回宇宙船	tsuki-shūkai-uchū'sen
lunar probe *[aero-eng]*	月探査機	tsuki-tansa-ki
lunar tide *[ocean]*	太陰潮	tai'in-chō
lunar year *[astron]*	太陰年	tai'in-nen
lunch, boxed or picnic *{n} [cook]*	弁当；辨當	bentō
lung *{n} [anat]*	肺(臓)	hai(zō)
lungfish (fish) *[v-zoo]*	肺魚；はいぎょ	haigyo
lunisolar calendar *[astron]*	太陰太陽暦	tai'in-taiyō-reki
lunisolar precessions *[astrophys]*	日月歳差	nichi'getsu-saisa
luster *[opt]*	光沢；光澤	kō'taku
" *[opt]*	艶；つや	tsuya
lustering agent *[mat] [tex]*	艶出し剤	tsuya'dashi-zai
lute, Japanese *[music]*	琵琶；びわ	biwa
luxury ship *[nav-arch]*	豪華船	gōka'sen
Lycopsida *[bot]*	ひかげのかずら類	hikage'no'kazura-rui
lye; ash; harshness (of taste)	灰汁；あく	aku
Lyme disease *[med]*	ライム病	raimu'byō
lymph *[hist]*	リンパ	rinpa
lymphatic system *[anat]*	リンパ系	rinpa-kei
lynx (animal) *[v-zoo]*	山猫；やまねこ	yama-neko
lyophilicness *[ch]*	親液性	shin'eki-sei
lyophilization *[ch-eng]*	凍結乾燥	tōketsu-kansō
lyophobicness *[ch]*	疎液性；そえきせい	so'eki-sei
lyotropic series *[ch]*	離液順列	ri'eki-jun'retsu
lyrebird (bird) *[v-zoo]*	琴鳥；ことどり	koto-dori
lysimeter *[eng]*	浸漏計	shinrō-kei
lysin *[immun]*	溶解素；ようかいそ	yōkai'so
lysis (of cells) *[cyt]*	溶解（細胞の）	yōkai (saibō no)
lysogen *[microbio]*	溶原菌	yōgen-kin
lytic enzyme *[bioch]*	溶菌酵素	yōkin-kōso
lytic infection *[microbio]*	溶菌感染	yōkin-kansen

M

macaw (bird) *[v-zoo]*	鸚哥；いんこ	inko
maceration *[paper] [pharm] [zoo]*	離解	rikai
machinability *[mech]*	被削性	hi'saku-sei
" *[met]*	切削性	sessaku-sei
machine; machinery *[mech-eng]*	機械；きかい	kikai
-machine(s) *{suffix} [mech-eng]*	-機	-ki
machine direction *[eng]*	縦方向；縱方向	tate-hōkō
machine element *[eng]*	機素	kiso
machine gun *[ord]*	機関銃	kikan-jū
machine language *[comput]*	機械(言)語	kikai-(gen)go
" *[comput]*	機械言葉	kikai-kotoba
machine material *[mech-eng]*	機材	kizai
machine operation *[comput]*	機械操作	kikai-sōsa
machinery and tools *[eng]*	機器；きき	kiki
machine-shave *{vb t} [met]*	機削する	ki'saku suru
machine tool *[eng]*	工作機械	kōsaku-kikai
machine type *[mech-eng]*	機種	kishu
machining *{n} [met]*	加工；かこう	kakō
" *{n} [met]*	機械加工	kikai-kakō
mackerel (fish) *[v-zoo]*	鯖；さば	saba
mackerel pike (fish) *[v-zoo]*	秋刀魚；さんま	sanma
mackerel scad (fish) *[v-zoo]*	室鰺；むろあじ	muro-aji
macro- (abnormal-) *{prefix}*	異常な-	ijō na-
" (long-) *{prefix}*	長い-	nagai-
" (large-) *{prefix}*	大きい-	ōki-
macroanalysis *[an-ch]*	常量分析	jōryō-bunseki
macroclimate *[climat]*	大気候；大氣候	dai-kikō
macrocrystal *[crys] [petr]*	巨大結晶	kyodai-kesshō
macromolecular *{adj} [poly-ch]*	高分子の	kō'bunshi no
" *{adj} [poly-ch]*	巨大分子の	kyodai-bunshi no
macromolecular chemistry *[ch]*	高分子化学	kōbunshi-kagaku
macromolecule *[org-ch] [poly-ch]*	高分子；こうぶんし	kō'bunshi
" *[org-ch] [poly-ch]*	巨大分子	kyodai-bunshi
macron *[pr]*	長音符	chō'on-pu
macronucleus *[i-zoo]*	大核	dai-kaku
macrophage; histiocyte *[histol]*	大食細胞；大食細胞	daishoku-saibō
macropolymerization *[poly-ch]*	巨大重合	kyodai-jūgō
macroscopic *{adj} [sci-t]*	巨視的な	kyoshi'teki na
macrostructure *[met]*	肉眼組織	niku'gan-so'shiki

madder (dye) *{n} [bot] [tex]*	茜；あかね	akane
madder red (color) *[opt]*	茜色	akane-iro
magazine *[ord]*	倉庫；そうこ	sōko
" *[pr]*	雑誌；雜誌；ざっし	zasshi
magazine paper *[mat]*	雑誌用紙	zasshi-yōshi
magenta (color) *[opt]*	深紅色	shinkō-shoku
maggot (insect larva) *[i-zoo]*	蛆；うじ	uji
magic number *[math]*	魔術数	ma'jutsu-sū
magic square *[math]*	魔法陣；まほうじん	mahō-jin
magma (pl -s, -ta) *[geol]*	岩漿	ganshō
" *[geol]*	海底体；海底體	kaitei'tai
magnesite *[miner]*	菱苦土鉱	ryō'kudo'kō
" *[miner]*	菱苦土石	ryō'kudo'seki
magnesium (element)	マグネシウム	maguneshūmu
magnet *[elecmg]*	磁石；じしゃく	ji'shaku
magnetic *{adj} [phys]*	磁気の；磁氣の	jiki no
magnetic core *[electr]*	磁心；磁芯；じしん	ji'shin
magnetic declination *[geophys]*	磁気偏角	jiki-henkaku
magnetic delay line *[electr]*	磁気遅延線	jiki-chi'en-sen
magnetic domain *[sol-st]*	磁区；磁區	jiku
magnetic equator *[geophys]*	磁気赤道	jiki-sekidō
magnetic excitation *[elecmg]*	励磁；勵磁；れいじ	rei'ji
magnetic field *[elecmg]*	磁場；じば	jiba
" *[elecmg]*	磁界	ji-kai
magnetic fluid *[mat]*	磁性流体	jisei-ryūtai
magnetic flux *[elecmg]*	磁束	ji-soku
magnetic flux density *[elecmg]*	磁束密度	ji'soku-mitsudo
magnetic force *[elecmg]*	磁力	ji-ryoku
magnetic heading *{n} [navig]*	機首磁方位	kishu-ji'hō'i
magnetic induction *[elecmg]*	磁気誘導	jiki-yūdō
magnetic-ink printing *{n} [pr]*	磁気印刷	jiki-in'satsu
magnetic lines of force *[elecmg]*	磁力線	ji'ryoku-sen
magnetic meridian *[geophys]*	磁気子午線	jiki-shigo'sen
magnetic monopole *[elecmg]*	磁気単極	jiki-tan'kyoku
magnetic needle *[elecmg]*	磁針	ji'shin
magnetic north *[geophys]*	磁北；磁北；じほく	ji'hoku
" *[geophys]*	磁気北極	jiki-hokkyoku
magnetic permeability *[elecmg]*	透磁率	tōji-ritsu
" *[elecmg]*	透磁性	tōji-sei
magnetic permeating force *[elecmg]*	透磁力	tōji-ryōku
magnetic pole *[elecmg] [geophys]*	磁極	ji-kyoku
magnetic potential *[elecmg]*	磁位	ji'i

magnetic property; magnetism *[phys]*	磁性；じせい	ji-sei
magnetic resonance *[phys]*	磁気共鳴	jiki-kyōmei
magnetic-resonance imaging; MRI *[phys]*	磁気共鳴映像	jiki-kyōmei-eizō
magnetics (the discipline) *[elecmg]*	磁気学；磁氣學	jiki'gaku
magnetic semiconductor *[sol-st]*	磁性半導体	jisei-handōtai
magnetic storm *[geophys]*	磁気嵐	jiki-arashi
magnetic strain *[elecmg]*	磁気歪；じきひずみ	jiki-hizumi
magnetic susceptibility *[elecmg]*	磁化率	jika-ritsu
magnetic thin film *[sol-st]*	磁気薄膜	jiki-haku'maku
magnetic transformation *[elecmg]*	磁気変態	jiki-hentai
magnetic wall *[phys]*	磁壁	ji-heki
magnetism *[phys]*	磁気；磁氣；じき	jiki
magnetite *[miner]*	磁鉄鉱；磁鐵鑛	ji'tekkō
magnetite sand; magnetic sand *[miner]*	砂鉄；さてつ	sa-tetsu
magnetization *[elecmg]*	磁化	jika
magnetize *{vb}* *[elecmg]*	励磁する；勵磁する	reiji suru
magnetized pole *[elecmg]*	着磁極	chaku'ji-kyoku
magnetoelectric effect *[elecmg]*	磁気電気効果	jiki'denki-kōka
magnetoelectricity *[elecmg]*	磁気電気	jiki-denki
magnetoencephalography *[med]*	脳磁気図記録法	nō-jiki'zu-kiroku-hō
magnetofluid dynamics *[phys]*	電磁流体力学	denji-ryūtai-riki'gaku
magnetogram *[elecmg]*	磁力記録	jiryoku-kiroku
magnetograph *[elecmg]*	磁力記録機	jiryoku-kiroku-ki
magnetohydrodynamics; MHD *[phys]*	電磁流体力学	denji-ryūtai-rikigaku
magnetomechanical ratio *[phys]*	磁気角運動量比	jiki-kaku'undōryō-hi
magnetometer *[eng]*	磁力計	jiryoku-kei
magnetomotive force *[elecmg]*	起磁力；きじりょく	kiji-ryoku
magneton *[phys]*	磁子	jishi
magnetooptical *{adj}* *[opt]*	磁気光学の	jiki'kōgaku no
magnetooptics *[opt]*	磁気光学	jiki-kōgaku
magnetopause *{n}* *[geophys]*	磁気圏界面(-線)	jiki'ken-kaimen (-sen)
magnetosphere *[geophys]*	磁気圏；磁氣圏	jiki-ken
magnetostriction *[elecmg]*	磁気歪	jiki-hizumi
" *[elecmg]*	磁歪；じわい	ji'wai
magnetotail *[geophys]*	磁気圏尾	jiki'ken-bi
magnification *[opt]*	倍率；ばいりつ	bai'ritsu
magnifying glass *[opt]*	拡大鏡；擴大鏡	kakudai-kyō
" *[opt]*	虫眼鏡；むしめがね	mushi-megane
magnifying power *[opt]*	拡大率	kakudai-ritsu
magnitude *[astron]* *[math]*	等級	tōkyū
" *[math]*	大きさ；おおきさ	ōki-sa
" *[math]*	絶対値；絶對値	zettai-chi

magnolia (tree, flower) *[bot]*	木蓮；もくれん	moku'ren
magnon *[sol-st]*	スピン波量子	supin'ha-ryōshi
magpie, Korean (bird) *[v-zoo]*	鵲；かささぎ	kasasagi
mahuang *[bot]*	麻黄；まおう	ma'ō
mail *{n}* *[comm]*	郵便；ゆうびん	yūbin
mail boat *[nav-arch]*	郵便船	yūbin'sen
mailbox *[comm]*	ポスト	posuto
mail car *[rail]*	郵便車	yūbin'sha
mailing machine *[eng]*	郵便機械	yūbin-kikai
main artery *{n}* *[anat]*	主動脈	shu-dōmyaku
main bedroom *[arch]*	主寝室；主寢室	shu-shin'shitsu
main component *[ch]* *[math]*	主成分	shu-seibun
main dish *{n}* *[cook]*	主食	shu'shoku
main effect *[stat]*	主効果；しゅこうか	shu-kōka
main entrance *[arch]*	正面玄関	shōmen-genkan
mainframe *[comput]*	本体；本體	hontai
main line *[rail]*	本線	honsen
main memory *[comput]*	主記憶	shu-ki'oku
main rope *[civ-eng]*	親綱；おやづな	oya-zuna
main season crop *[agr]*	表作；おもてさく	omote-saku
main sequence *[astron]*	主系列	shu-kei'retsu
main sequence stars *[astron]*	主系列星	shu'keiretsu-sei
main shock (earthquake) *[geophys]*	本震	honshin
main storage *[comput]*	主記憶(機構)	shu-ki'oku(-kikō)
main store *[arch]* *[econ]*	本店	honten
main street; highway *[traffic]*	大通り；おおどおり	ō-dōri
maintainability *[eng]*	保守性	hoshu-sei
maintaining cooled *{vb}*	保冷(する)	horei (suru)
maintenance *[comput]* *[ind-eng]*	保守；ほしゅ	hoshu
" *[comput]* *[ind-eng]*	維持	iji
" *[ind-eng]*	保存	hozon
" *[ind-eng]*	整備	seibi
maintenance time *[ind-eng]*	保持時間	hoji-jikan
" *[ind-eng]*	保守時間	hoshu-jikan
major angle *[geom]*	優角	yū-kaku
major arc *[geom]*	優弧；ゆうこ	yūko
major axis *[an-geom]*	長軸；ちょうじく	chō-jiku
major industry *[econ]* *[ind-eng]*	重要産業	jūyō-sangyō
majority *{n}* *[comput]*	多数決；たすうけつ	tasū'ketsu
majority operation *[comput]*	多数決演算	tasū'ketsu-enzan
major port; major harbor *[geogr]*	主港；主港	shukō
major premise *[logic]*	大前提	dai-zentei

Major Section (of village) *[geogr]*	大字；おおあざ	Ō-aza
make; create; construct *{vb t}*	作る；造る	tsukuru
" ; manufacture; fashion *{vb t}*	拵える；こしらえる	koshiraeru
make a sound *{vb}*	音を立てる	oto o tateru
make bright(er); brighten *{vb t}*	明るくする	akaruku suru
make deep; deepen *{vb t}*	深くする	fukaku suru
make do; get by with *{vb}*	間に合わせる	ma ni awaseru
make higher; heighten *{vb}*	高める	takameru
make large; enlarge *{vb t}*	大きくする	ōkiku suru
make light(er) *{vb t}*	明るくする	akaruku suru
make narrow; narrow down *{vb t}*	狭くする	semaku suru
make narrower; pucker *{vb}*	窄める；すぼめる	subomeru
make shallow *{vb t}*	浅くする；淺くする	asaku suru
make slow; slow down *{vb}*	遅くする	osoku suru
make slow(er); slow down *{vb}*	遅める；遲める	osomeru
make small; reduce in size *{vb}*	小さくする	chīsaku suru
make straight; straighten *{vb t}*	真っ直ぐにする	massugu ni suru
make supine; face upward *{vb t}*	仰むきにする	ao'muki ni suru
make tight; tighten *{vb t}*	きつくする	kitsuku suru
make uniform; unify *{vb t}*	揃える；そろえる	soroeru
make wide; widen *{vb t}*	広くする；廣くする	hiroku suru
maker; manufacturer *[ind-eng]*	製造会社；製造會社	seizō-gaisha
makeshift *{n} [ind-eng]*	間に合わせ	ma-ni-awase
mako; Isurus glaucus (fish) *[v-zoo]*	青鮫	ao-zame
makuwa melon (fruit) *[bot]*	真桑瓜；まくわうり	makuwa-uri
malachite *[miner]*	孔雀石	kujaku'seki
malachite green (color) *[opt]*	常磐色；ときわいろ	tokiwa-iro
malate *[ch]*	林檎酸塩	ringo'san-en
male (human); man; men *[bio]*	男性	dansei
" (animal) *[bio]*	雄；牡；おす	osu
male fern *[bot]*	雄羊歯；おしだ	o'shida
maleic anhydride	無水マレイン酸	musui-marein-san
malfunction *{n} [ind-eng]*	誤作動	go'sadō
malic acid	林檎酸；りんごさん	ringo-san
mall *[arch]*	屋内商店街	oku'nai-shōten'gai
mallard (bird) *[v-zoo]*	真鴨	ma'gamo
malleability *[met]*	可鍛性；かたんせい	katan-sei
" *[met]*	展性	tensei
malleable cast iron *[met]*	可鍛铸鉄	katan-chū-tetsu
mallet *[eng]*	槌；つち	tsuchi
malleus (pl -lei) (bone) *{n} [anat]*	槌骨	tsuchi-bone
malnutrition *[med]*	栄養失調；榮養失調	eiyō-shitchō

malodorous; ill-smelling *{adj}*	臭い；くさい	kusai
malt *{n} [cook]*	麦芽；麥芽；ばくが	baku'ga
malted rice; malt *[cook]*	麹；糀；こうじ	kōji
malt juice *[cook]*	麦芽汁	bakuga-jū
maltose; malt sugar *[bioch]*	麦芽糖	bakuga'tō
mammal *[v-zoo]*	哺乳動物	ho'nyū-dōbutsu
Mammalia; mammals *[v-zoo]*	哺乳類	ho'nyū-rui
mammary gland *[anat]*	乳腺	nyū-sen
man (pl men: people) *[bio]*	人；ひと	hito
man (pl men) *[bio]*	男；おとこ	otoko
management *{n} [ind-eng]*	管理	kanri
" *{n} [ind-eng]*	経営；經營	kei'ei
" *{n} [ind-eng]*	運営；うんえい	un'ei
management organization *[econ]*	経営組織	kei'ei-soshiki
manager (a person) *[ind-eng]*	管理者	kanri'sha
manatee; sea cow; cowfish *[v-zoo]*	海牛	kaigyū
mandarin duck (bird) *[v-zoo]*	鴛鴦；おしどり	oshi-dori
mandarin orange; tangerine (fruit) *[bot]*	蜜柑；みかん	mikan
mandarin orange (fruit) *[bot]*	橘；たちばな	tachibana
mandatory clause *[legal]*	必須条件；必須條件	hissu-jōken
mandible *[anat]*	大顎	ō-ago; dai'gaku
mane *[zoo]*	鬛；鬣；たてがみ	tategami
maneuverability *[navig]*	操縦性；操縱性	sōjū-sei
manganese (element)	マンガン	mangan
manifest; reveal *{vb}*	現わす	arawasu
manifold *{n} [math]*	多様体；多樣體	tayō'tai
manipulate; maneuver *{vt t}*	操る	ayatsuru
mankind *[bio]*	人	hito
man-made *{adj} [ind-eng]*	人工の	jinkō no
" *{adj} [ind-eng]*	人造の	jinzō no
manmade fiber *[tex]*	化学繊維；化學纖維	kagaku-sen'i
manmade rubber *[poly-ch] [rub]*	人造ゴム	jinzō-gomu
manmade seawater *[hyd]*	人工海水	jinkō-kai'sui
man-month *[econ]*	人月；にんげつ	nin-getsu
manned mission *[aerosp]*	有人使命	yūjin-shimei
manned operation *[ind-eng]*	有人運用	yūjin-un'yō
manned satellite *[aerosp]*	有人人工衛星	yūjin-jinkō-eisei
manometer *[eng]*	圧力計；壓力計	atsu'ryoku-kei
manometry *[eng]*	検圧法；檢壓法	ken'atsu-hō
manpower *[ind-eng]*	人手	hito'de
mantis crab; squilla *[i-zoo]*	蝦蛄；しゃこ	shako

mantissa *[alg]*	仮数; 假數; かすう	kasū
mantle *[cl]*	外套	gaitō
manual (a document) *{n} [pr]*	手引書	tebiki'sho
manual arts *[eng]*	手工	shukō
manually operated *[ind-eng]*	手動式	shudō'shiki
manual operation *[comput]*	手動(操)作	shudō(-sō)sa
" *[ind-eng]*	手作業; てさぎょう	te-sagyō
manual system *[comput]*	手動系	shudō-kei
manual welding *{n} [met]*	手溶接	te-yō'setsu
manufacture *{n} [ind-eng]*	製造; せいぞう	seizō
" *{vb} [ind-eng]*	生産する	seisan suru
manufacture; prepare *{vb} [ind-eng]*	調製する	chōsei suru
manufactured product *[ind-eng]*	製品	seihin
manufacture paper *{vb} [paper]*	漉く; すく	suku
manufacturer *[ind-eng]*	製造業者	seizō-gyōsha
" (maker) *[ind-eng]*	製造元	seizō-moto
manure *[agr]*	下肥	shimo'goe
" *[agr]*	厩肥; うまやごえ	umaya-goe
man-week *[econ]*	人週	nin-shū
man-year *[econ]*	人年	nin-nen
many-membered ring *[org-ch]*	多員環	ta'inkan
map *{n} [graph]*	地図; 地圖; ちず	chizu
" *{n} [graph]*	図; ず	zu
" *{vb t} [comput]*	写像する	shazō suru
" *{vb t} [graph]*	写す; 寫す	utsusu
maple (tree) *[bot]*	楓; かえで	kaede
" (tree, leaf) *[bot]*	紅葉; もみじ	momiji
mapping *{n} [alg] [comput]*	写像	shazō
" *[graph]*	地図作成	chizu-sakusei
marble *{n} [petr]*	大理石; だいりせき	dairi'seki
marcasite *[miner]*	白鉄鉱; 白鐵鑛	haku'tekkō
March (month)	三月	san'gatsu
mare (pl maria) *[astron]*	月の海	tsuki no umi
margin *[sci-t]*	縁; ふち	fuchi
margin (of page) *[pr]*	余白; 餘白	yo'haku
margin *[pr] [sci-t]*	限界	genkai
" *[tex]*	代; しろ	shiro
marginal checking *{n} [electr]*	限界検査; 限界檢查	genkai-kensa
marginal note *[pr]*	傍注	bōchū
marginal probability *[prob]*	周辺確率	shūhen-kaku'ritsu
marginal ray *[opt]*	周縁光線	shū'en-kōsen
marginal sea *[geogr]*	沿岸海	engan-kai

mariculture *[agr]*	海中栽培	kaichū-saibai
marina *[civ-eng]*	繋船池；係船池	keisen'chi
marinade *[cook]*	垂；たれ	tare
marine *{adj} [nav-arch]*	船舶(用)	sen'paku('yō)
" (of the sea) *{adj} [ocean]*	海の；うみの	umi no
marine algae *[bot]*	海草	kaisō
marine biology *[bio]*	海洋生物学	kaiyō-seibutsu'gaku
marine blue (color) *[opt]*	海碧；かいへき	kai'heki
marine climate *[climat]*	海洋気候	kaiyō-kikō
marine product *[ocean]*	海産物	kaisan'butsu
marine swamp *[ecol]*	海岸沼沢地	kaigan-shōtaku'chi
marine terrace *[geol]*	海岸段丘	kaigan-dankyū
maritime law *[legal]*	海事法	kaiji-hō
maritime mobile radiotelephone *[comm]*	船舶電話	sen'paku-denwa
maritime satellite *[aero-eng]*	海事衛星	kaiji-eisei
mark *{n} [comm]*	印；しるし	shirushi
" *{n} [navig] [traffic]*	標識；ひょうしき	hyō'shiki
" *{n} [navig]*	標点	hyō-ten
" *{vb} [graph] [pr]*	印す	shirusu
marked line *[eng]*	標線	hyōsen
marker enzyme *[bioch]*	指標酵素	shi'hyō-kōso
market; marketplace *{n} [econ]*	市場	ichiba; shijō
market *{vb} [econ]*	市販する	shihan suru
marketability *[econ]*	市場性	shijō-sei
market value *[econ]*	時価；時價；じか	jika
marking-off *{n} [eng]*	罫書き	ke'gaki
marks; merit marks *{n} [ind-eng]*	点数；點數	tensū
marlin; spearfish (fish) *[v-zoo]*	真旗魚；まかじき	ma'kajiki
marmot (animal) *[v-zoo]*	山鼠；やまね	yama'ne
maroon (color) *[opt]*	蝦茶色	ebi'cha-iro
marquetry; marqueterie *[furn]*	象眼細工	zōgan-zaiku
marquetry; parquetry *[furn]*	寄木	yose'gi
marrow *{n} [hist]*	髄；髓；ずい	zui
Mars *[astron]*	火星；かせい	kasei
marsh *[ecol]*	沼地	numa'chi
marsh tit (bird) *[v-zoo]*	嘴太雀	hashi'buto-gara
Marsupialia *[v-zoo]*	有袋目	yūtai-moku
marten (animal) *[v-zoo]*	(黒)貂；(くろ)てん	(kuro-)ten
masculine (of humans) *{adj} [bio]*	男性的な	dansei'teki na
mash *{vb} [mech-eng]*	擂る；する	suru
mask; face mask *{n} [cl]*	面	men
masking *{n} [electr]*	遮蔽	shahei

mass *{n} [mech]*	量；りょう	ryō
" *{n} [mech]*	質量；しつりょう	shitsu'ryō
mass; vehicle; excipient *[pharm]*	賦形剤；ふけいざい	fukei-zai
mass absorption coefficient *[phys]*	質量吸収係数	shitsuryō-kyūshū-keisū
mass action, law of *[p-ch]*	質量作用の法則	shitsuryō-sayō no hōsoku
massage *{n} [med]*	按摩；あんま	anma
mass balance *[phys]*	物質収支；物質收支	busshitsu-shūshi
massicot *[miner]*	金密陀；きんみつだ	kin'mitsuda
massive coal *[geol]*	塊状炭	kaijō-tan
mass number *[nuc-phys]*	質量数	shitsuryō-sū
mass polymerization *[poly-ch]*	塊状重合	kaijō-jūgō
mass production *[ind-eng]*	量産	ryōsan
" *[ind-eng]*	大量生産	tairyō-seisan
mass ratio *[astrophys]*	質量比	shitsuryō-hi
mass spectrograph *[eng]*	質量分析器	shitsuryō-bunseki-ki
mass spectrometer *[eng]*	質量分析計	shitsuryō-bunseki-kei
mass spectrometry *[an-ch]*	質量分析学	shitsuryō-bunseki'gaku
mass storage *[comput]*	大(容量)記憶	dai(-yōryō)-ki'oku
mass-transfer rate *[phys]*	物質移動速度	busshitsu-idō-sokudo
master *{n} [comput]*	基本	kihon
" *{n} [pr]*	主；しゆ	shu
master batch *[poly-ch] [rub]*	親練り；おやねり	oya-neri
master bedroom *[arch]*	夫婦寝室	fūfu-shin'shitsu
master controller *[comput]*	主幹制御器	shukan-seigyo-ki
master gear *[mech-eng]*	種歯車	tane-ha'guruma
master mold *[eng]*	元型；もとがた	moto-gata
mastic *[mat]*	乳香	nyūkō
mastication *[ch-eng] [rub]*	素練り；素練り	su'neri
Mastigomycotina *[mycol]*	鞭毛菌類	benmō'kin-rui
Mastigophora *[i-zoo]*	鞭毛虫類	benmō'chū-rui
match; equal *{n}*	類；類；たぐい	tagui
match *{vb t} [comput]*	突き合わせる	tsuki-awaseru
" *{vb t}*	合わせる	awaseru
matching *{n} [comput]*	整合；せいごう	seigō
" *{n} [math]*	一致	itchi
matching box *[elec]*	整合器	seigō-ki
material *[ch] [phys]*	物質；ぶっしつ	busshitsu
" *[tex]*	代；しろ	shiro
" *[sci-t]*	資材	shizai
material balance *[ch-eng]*	物質収支	busshitsu-shūshi
material cloth *[tex]*	原反；げんたん	gentan
material for tools *[mech-eng]*	器材；きざい	kizai

materials *[eng]*	材料；ざいりょう	zai'ryō
" *{n}*	資材	shizai
materials list *[ind-eng]*	材料表	zairyō-hyō
materials science *[eng]*	材料科学	zairyō-kagaku
mathematical check *[comput]*	数学的検査	sūgaku'teki-kensa
mathematical induction *[logic]*	数学的帰納法	sūgaku'teki-kinō-hō
mathematical programming *[comput]*	数理計画法	sūri-keikaku-hō
mathematics *[sci-t]*	数学；數學	sū'gaku
" (Japanese) *[math]*	和算；わさん	wasan
matrix (pl -trices, -trixes) *[alg]*	行列；ぎょうれつ	gyō'retsu
" *[bio] [geol]*	基質	ki'shitsu
" *[bot]*	母体	botai
" *[met]*	地(鉄)	ji(-tetsu)
" *[petr]*	石基	sekki
matsudake mushroom *[bot]*	松茸；まつたけ	matsu-take
matte *{n} [mat]*	梨地	nashi'ji
" *{n} [met]*	鈹；かわ	kawa
matte coating film *[mat]*	艶消(し)被覆膜	tsuya'keshi-hifuku-maku
matted *{adj} [mat]*	艶消しの	tsuya'keshi no
matte enamel *[mat]*	艶消し琺瑯	tsuya'keshi-hōrō
matter (substance) *{n} [phys]*	物質；ぶっしつ	busshitsu
" (facts) *{n}*	事項	jikō
" (essence) *{n}*	実質；實質	jisshitsu
matter being copied *[opt]*	被複写原稿	hi-fukusha-genkō
matting agent *[mat]*	艶消剤	tsuya-keshi-zai
maturation *[bio]*	熟成；じゅくせい	jukusei
" *[resin] [rub]*	寝かし	nekashi
maturing *{n} [bio] [ch] [cook]*	熟成	jukusei
mauve (color) *[opt]*	藤色	fuji-iro
maxilla (pl -illae) *[anat]*	小顎	ko-ago; shō'gaku
maximum(pl -ma) *{n} [an-geom]*	極大；きょくだい	kyoku'dai
" *{n} [an-geom]*	最大	saidai
maximum likelihood estimator *[stat]*	最尤推定量	saiyū-suitei'ryō
maximum-minimum thermometer *[thermo]*	最高最低温度計	saikō'saitei-ondo-kei
maximum permissible amount *[ind-eng]*	最大許容量	saidai-kyoyō-ryō
maximum value *[math]*	極大値	kyokudai-chi
maximum work *[mech] [thermo]*	最大仕事	saidai-shigoto
May (month)	五月	go'gatsu
mayfly (insect) *[i-zoo]*	蜉蝣；かげろう	kagerō
maze *{n} [psy]*	迷路	mei'ro
meadow *[ecol]*	草地	kusa'chi
meadow grass *[bot]*	牧草	boku'sō

meal *[cook]*	食事	shoku'ji
" *[cook]*	御飯；ごはん	gohan
meal served a priest *[cook]*	斎；齋；とき	toki
mealy-bug (insect) *[i-zoo]*	粉貝殻虫	kona-kaigara-mushi
mean *{n} [arith]*	平均(値)	heikin(-chi)
mean density *[mech]*	平均密度	heikin-mitsudo
mean error *[math]*	平均誤差	heikin-gosa
mean free path *[acous] [phys]*	平均自由行程	heikin-jiyū-kōtei
mean free time *[acous] [phys]*	平均自由時間	heikin-jiyū-jikan
meaning (significance) *{n} [comm]*	意義	igi
" (sense) *{n} [comm] [ling]*	意味；いみ	imi
" (opinion; aim) *{n} [comm]*	趣旨	shushi
" (sense) *{n} [comm]*	訳；譯；わけ	wake
mean life *[phys]*	平均寿命；平均壽命	heikin-jumyō
mean number of moles added *[ch]*	平均付加モル数	heikin-fuka-moru-sū
mean proportional *[geom]*	比例中項	hirei-chūkō
mean residence time *[nucleo]*	平均滞留時間	heikin-tairyū-jikan
mean sea level *[ocean]*	平均海面	heikin-kaimen
mean square error *[cont-sys]*	平均二乗誤差	heikin-nijō-gosa
mean time *[astron]*	平均時；へいきんじ	heikin-ji
mean value *[arith]*	平均(値)	heikin(-chi)
mean value theorem *[calc]*	平均値の定理	heikin-chi no teiri
measles *[med]*	疹；麻疹；はしか	hashika
measure *{n} [math]*	測度	soku'do
" *{n} [music]*	拍子；ひょうし	hyōshi
measure; gage; survey *{vb t}*	計る；測る	hakaru
measurement *[sci-t]*	測定	sokutei
measuring *{n} [sci-t]*	計量；けいりょう	keiryō
measuring cup *[eng]*	計量カップ	keiryō-kappu
measuring instrument *[eng]*	計測機器	keisoku-kiki
measuring spoon *[eng]*	計量匙	keiryō-saji
meat *[cook]*	肉；にく	niku
meat broth *[microbio]*	肉汁	niku'jū
mechanic (a person) *[mech-eng]*	機械修理工	kikai(-shūri)'kō
mechanical advantage *[mech-eng]*	力の拡大率	chikara no kakudai-ritsu
mechanical engineering *[eng]*	機械工学	kikai-kōgaku
mechanical loss *[eng]*	機械損；きかいそん	kikai-son
mechanical quality factor *[ind-eng]*	機械的品質係数	kikai'teki-hin'shitsu-keisū
mechanical ventilation *[eng]*	機械換気	kikai-kanki
mechanics *[phys]*	力学；りきがく	riki'gaku
mechanism; machinery *[mech-eng]*	機構；きこう	kikō
mechanization *[eng]*	機械化	kikai'ka

English	Japanese	Romanization
mechatronics *[electr]*	機電学	kiden'gaku
median *{n} [geom] [stat]*	中央値	chūō-chi
" *{n} [stat]*	中位数	chū'i-sū
" *{adj} [geom]*	中央の	chū'ō no
median fin *[v-zoo]*	正中鰭	seichū-bire
median line *[geom]*	中線	chū-sen
mediation *{n}*	調停	chōtei
medical electronics *[electr]*	電子医学；電子醫學	denshi-i'gaku
medical treatment *[med]*	医療；いりょう	i'ryō
medicament *{n} [med]*	医薬；醫藥；いやく	i'yaku
medicated soap *{n} [mat] [pharm]*	薬用石鹼	yaku'yō-sekken
medicated wine *[cook]*	薬味酒	yakumi'shu
medication *{n} [pharm]*	薬物；藥物	yaku'butsu
medicinal chemistry *[ch]*	医薬品化学	i'yaku'hin-kagaku
medicinal crop *[agr]*	薬用作物	yaku'yō-saku'motsu
medicinal herb *[pharm]*	薬草	yaku'sō
medicinal pill *[pharm]*	丸薬	gan'yaku
medicinal plant *[bot] [pharm]*	薬草(植物)	yaku'sō(-shokubutsu)
medicinal tea *[cook]*	茶剤	cha-zai
medicinal wafer *[pharm]*	オブラート	oburāto
medicine (the discipline) *[med]*	医学；醫学；いがく	i'gaku
" (medical drug) *[pharm]*	薬；藥；くすり	kusuri
" *[pharm]*	薬物	yaku'butsu
" *[pharm]*	薬品	yaku'hin
medicine chest *[furn]*	薬箪笥	kusuri-dansu
medicine for external application *[pharm]*	外用薬；がいよう=やく	gaiyō-yaku
medicine for internal use *[pharm]*	内服薬	nai'fuku-yaku
medium *{n} [ch-eng] [phys]*	媒質；ばいしつ	bai'shitsu
" *{n} [ch-eng] [comput] [phys]*	媒体；媒體	baitai
" *{n} [microbio]*	培地；ばいち	baichi
medium; middle *{n} [math]*	中	chū
medium-degree *{adj}*	中位いの	chū-gurai no
medium-scale integrated circuit; MSI *[electr]*	中規模集積回路	chū'kibo-shūseki-kairo
medium-scale map *[map]*	中縮尺図	chū-shuku'shaku-zu
medulla (pl -las, -lae) *{n} [anat]*	髄質；髓質	zui'shitsu
medulla oblongata *[anat]*	延髄；えんずい	en'zui
medusa *[i-zoo]*	水母；海月；くらげ	kurage
megaphone *[acous]*	拡声器；擴聲器	kakusei-ki
meeting; assembly *{n}*	会合；會合	kaigō
Meiji (era name) (1868-1912)	明治；めいじ	Meiji

meionite *[miner]*	灰柱石	kai'chūseki
meiosis *[cyt]*	減数分裂	gensū-bun'retsu
melanin *[bioch]*	黒色素；黑色素	koku-shiki'so
melissa (herb) *[bot]*	西洋山薄荷	seiyō-yama'hakka
mellitic anhydride	無水メリト酸	musui-merito-san
mellowness (flavor) *[cook]*	丸味	maru'mi
melody *[acous] [music]*	旋律；せんりつ	sen'ritsu
melon (fruit) *[bot]*	瓜；うり	uri
melt *{n} [met]*	溶湯	yōtō
" (dissolve) *{vb i} [ch]*	溶ける	tokeru
" (dissolve) *{vb t} [ch]*	溶かす	tokasu
melt-deposition *[met]*	溶着	yō'chaku
meltdown *{n} [nucleo]*	炉心溶融	roshin-yōyū
melt fluidity *[fl-mech]*	溶融流動性	yōyū-ryūdō-sei
melting *{n} [p-ch]*	溶融；ようゆう	yōyū
" *{n} [p-ch]*	融解；ゆうかい	yūkai
melting furnace *[eng]*	融解炉；融解爐	yūkai-ro
melting loss *[met]*	溶損	yōson
melting point *[thermo]*	融解点	yūkai-ten
" *[thermo]*	融点；融點	yūten
melt molding *{n} [poly-ch]*	溶融成形	yōyū-seikei
melt spinning *{n} [tex]*	溶融紡糸	yōyū-bōshi
melt strength *[mech]*	溶融強度	yōyū-kyōdo
melt viscosity *[fl-mech]*	溶融粘度	yōyū-nendo
meltwater *[hyd]*	雪消水；雪解水	yuki'ge-mizu
member *{n} [comput]*	子；こ	ko
" (member part) *{n} [ind-eng]*	部材	buzai
membrane *[ch-eng]*	膜；まく	maku
membrane capacitance *[elec]*	膜容量	maku-yōryō
membrane current *[physio]*	膜電流	maku-denryū
membrane equilibrium *[ch-eng]*	膜平衡	maku-heikō
membrane filter *[sci-t]*	膜沪過器；膜濾過器	maku-roka-ki
membrane potential *[physio]*	膜電位	maku-den'i
membrane resistance *[ch-eng]*	膜抵抗	maku-teikō
memo (note; entry) *{n} [comm]*	控え	hikae
" *{n} [comm]*	覚書；覺書	oboe-gaki
memorization *[psy]*	暗記；あんき	anki
memorize *{vb} [psy]*	暗記する	anki suru
memory *[comput] [psy]*	記憶；きおく	ki'oku
" (device) *[comput]*	記憶素子	ki'oku-soshi
memory; storage *[comput]*	記憶装置	ki'oku-sōchi
memory protection *[comput]*	記憶保護	ki'oku-hogo

mendelevium (element)	メンデレビウム	menderebyūmu
meningococcus (pl -cocci) *[microbio]*	髄膜炎菌	zui'maku'en-kin
menstruation *[physio]*	月経；月經	gekkei
mensuration *[sci-t]*	求積法	kyūseki-hō
mental calculation *[math]*	暗算；あんざん	anzan
mental disease *[med]*	精神病	seishin'byō
menthol *[org-ch]*	薄荷脳；はっかのう	hakka'nō
menu (pl menus) *[comput]*	目録	moku'roku
" *[cook]*	献立(表)	kon'date(-hyō)
merchandise *[ind-eng]*	商品	shōhin
merchant ship *[nav-arch]*	商船	shōsen
mercuric oxide	酸化第二水銀	sanka-dai'ni-suigin
mercuric sulfide	硫化第二水銀	ryūka-dai'ni-suigin
mercurous sulfide	硫化第一水銀	ryūka-dai'ichi-suigin
mercury (element)	水銀；すいぎん	suigin
Mercury (planet) *[astron]*	水星	suisei
mercury barometer *[eng]*	水銀気圧計	suigin-ki'atsu-kei
mercury chloride	塩化水銀	enka-suigin
mercury column *[phys]*	水銀柱	suigin-chū
mercury fulminate	雷酸水銀	raisan-suigin
mercury iodide	沃化水銀	yōka-suigin
mercury nitrate	硝酸水銀	shōsan-suigin
mercury oxide	酸化水銀	sanka-suigin
mercury storage *[comput]*	水銀記憶	suigin-ki'oku
mercury sulfate	硫酸水銀	ryūsan-suigin
mercury sulfide	硫化水銀	ryūka-suigin
mercury vapor lamp *[electr]*	水銀灯	suigin-tō
mercury-wetted contact relay *[electr]*	水銀接点継電器	suigin-setten-keiden-ki
mere; simple; sheer *{adj}*	単なる	tan naru
merely; simply; solely; only *{adv}*	単に	tan ni
merge; merging *{n} [comput]*	併合	heigō
merge; merging; combination *[comput]*	組合せ	kumi'awase
merge *{vb} [comput]*	併合する	heigō suru
meridian *[astron] [geod]*	子午線	shigo-sen
meridional circulation *[meteor]*	南北循環	nan'boku-junkan
meristem; meristematic tissue *[bot]*	分裂組織	bun'retsu-so'shiki
merit (strong point; virtue) *{n}*	長所	chōsho
" ; worth; value *{n}*	価値；價値；かち	kachi
" (goodness) *{n}*	良さ；よさ	yosa
merohedrism *[crys]*	欠面像	ketsu'men-zō
mesa *[geol]*	地卓；ちたく	chi'taku
mesh *[tex]*	網目	ami'me

mesh-clogging *{n} [tex]*	目詰まり；めづまり	me-zumari
mesh grid *[elec]*	網状格子	ami'jō-kōshi
meshing *{n} [eng]*	係合	keigō
mesh(ing) together *{vb} [eng]*	係合する	keigō suru
meso- *{prefix} [ch]*	中央の-	chūō no-
" *{prefix}*	中位の-	chū'i no-
mesocarp *[bot]*	中果皮；ちゅうかひ	chū-kahi
mesolite *[miner]*	中沸石	chū'fusseki
Mesolithic Age *[archeo]*	中石器時代	chū-sekki-jidai
mesomorphic state *[zoo]*	中間形態	chūkan-keitai
meson *[phys]*	中間子	chūkan'shi
mesophyte *[bot]*	中生植物	chūsei-shokubutsu
mesosphere *[meteor]*	中間圏	chūkan-ken
mesotherm *[bot]*	中温植物	chū'on-shokubutsu
message *[comm] [comput]*	電文	denbun
" *[comm]*	通信文	tsūshin'bun
meta- *{prefix}*	変化-	henka-
" *{prefix}*	二次的-	niji'teki-
" *{prefix}*	の後-	no ato-
" *{prefix}*	を越えた-	o koeta-
metabolic product *[bioch]*	代謝産物	taisha-san'butsu
metabolism *[physio]*	(新陳)代謝	(shin'chin-)taisha
metabolite *[bioch]*	代謝物質	taisha-busshitsu
metacenter *[fl-mech]*	傾きの中心	katamuki no chūshin
metachromasia *[ch]*	異調染色性	i'chō-senshoku-sei
" *[ch]*	異染性；いせんせい	i'sen-sei
metagalaxy *[astron]*	超銀河	chō-ginga
metal *[mat]*	金；かね	kane
" *[mat]*	金属；きんぞく	kin'zoku
metalanguage *[comput]*	超言語	chō'gengo
metal fatigue *[met]*	金属の疲労	kinzoku no hirō
metallic bond *[p-ch]*	金属結合	kinzoku-ketsu'gō
metallic luster *[opt]*	金属光沢；金屬光澤	kinzoku-kōtaku
metallic soap *[org-ch]*	金属石鹸	kinzoku-sekken
metallizing *{n} [eng]*	真空蒸着	shinkū-jōchaku
metallography *[met]*	金属(組織)学	kinzoku(-soshiki)'gaku
metalloid *{n} [ch]*	半金属	han'kinzoku
metalloid element *[ch]*	類金属元素	rui'kinzoku-genso
metallurgical microscope *[opt]*	金属顕微鏡	kinzoku-kenbi'kyō
metallurgy *[sci-t]*	冶金学	yakin'gaku
metalworker (a person) *[met]*	鍛工；たんこう	tankō
metamathematics *[math]*	超数学	chō'sūgaku

English	Japanese	Romanization
metamorphic rock *[geol]*	変成岩	hensei-gan
metamorphism *[geol]*	変成作用	hensei-sayō
metamorphosis (pl -phoses) *[bio]*	変形；變形	henkei
metaphase *[bio]*	中期	chūki
metastable *[sci-t]*	準安定	jun-antei
metastable phase *[p-ch]*	準安定相	jun'antei-sō
Metatheria *[v-zoo]*	後獣類；後獸類	kōjū-rui
Metazoa *[zoo]*	後生動物	kōsei-dōbutsu
meteor *[astron]*	流星；りゅうせい	ryūsei; nagare-boshi
meteor burst communications *[comm]*	流星宛突発通信	ryūsei'ate-toppatsu-tsūshin
meteoric iron *[astron]*	隕鉄；いんてつ	in'tetsu
meteorite *[astron]*	隕石	inseki
meteorite crater *[astron]*	隕石孔	inseki-kō
meteorology *[sci-t]*	気象学；氣象學	kishō'gaku
meteor shower *[astron]*	流星雨	ryūsei'u
meter (unit of length) *[mech]*	米；メートル	mētoru
methanol *[org-ch]*	木精（メタノール）	mokusei (metanōru)
method *[sci-t]*	方法	hōhō
-method *(suffix) [sci-t]*	-法	-hō
method of least squares *[stat]*	最小二乗法	saishō-nijō-hō
methyl alcohol *[org-ch]*	木精；もくせい	moku'sei
metrics *[math]*	計量(学)	keiryō('gaku)
metrology *[phys]*	計測学	keisoku'gaku
metropolis *[geogr]*	都；と	to
metropolis and districts; urban and rural prefectures (Japan) *[geogr]*	都道府県；とどうふけん	to'dō-fu'ken
metropolitan area *[civ-eng]*	大都市圏	dai-toshi-ken
mezzanine *[arch]*	中二階	chū'nikai
mica *[miner]*	雲母；うんも	un'mo (un'bo)
mica flake *[miner]*	雲母片	unmo-hen
micro- *(prefix)*	微小-	bishō-
microanalysis *[an-ch]*	微量分析	biryō-bunseki
microbalance *[eng]*	微量天秤	biryō-tenbin
microballoon *[mat]*	微小中空球	bishō-chūkū-kyū
microbe *[microbio]*	黴菌；ばいきん	baikin
" *[microbio]*	微生物；びせいぶつ	bi'seibutsu
" *[microbio]*	細菌；さいきん	saikin
microbioassay *[bio]*	微生物学的定量法	bi'seibutsu'gaku'teki-teiryō-hō
microbiology *[bio]*	微生物学	bi'seibutsu'gaku
microcarrier *[microbio]*	微粒子担体	bi'ryūshi-tantai
microclimate *[climat]*	小気候	shō-kikō

microcline *[miner]*	微斜長石	bisha'chōseki
microcomponent *[elec] [sci-t]*	超小型構成部分	chō'kogata-kōsei-bubun
microcomputer *[comput]*	マイクロコン=ピュータ	maikuro-konpūta
microcrystal; microlite *[crys]*	微結晶	bi'kesshō
microcrystalline texture *[crys]*	微小質組織	bishō'shitsu-so'shiki
microdose *{n} [med]*	微量；びりょう	biryō
microearthquake *[geophys]*	微小地震	bishō-jishin
microelectronics *[electr]*	超小形電子工学	chō'kogata-denshi-kōgaku
microelement; trace element *[geoch]*	微量元素	biryō-genso
microfauna (pl -s, -faunae) *[zoo]*	微小動物類	bishō-dōbutsu-rui
microfiber; fine fiber *[tex]*	微繊維；微纖維	bi'sen'i
microflora (pl -s, -florae) *[bot]*	微小植物類	bishō-shokubutsu-rui
microinjection *[cyt]*	微量注射法	biryō-chūsha-hō
microlite *[crys]*	微結晶	bi'kesshō
" *[crys]*	微晶；びしょう	bishō
" *[miner]*	微晶石	bishō'seki
micrometeorite *[astron]*	流星塵	ryūsei'jin
microminiaturization *[electr]*	超小形化	chō'kogata'ka
micron (pl -s, micra) *[mech]*	ミクロン	mikuron
micronucleus *[i-zoo]*	小核；しょうかく	shōkaku
micronutrient *[bioch]*	微量栄養	biryō-eiyō
" *[bioch]*	微量養素	biryō-yōso
microorganism *[microbio]*	微生物；びせいぶつ	bi'seibutsu
microparticle; fine particle *[poly-ch]*	微粒子；びりゅうし	bi'ryūshi
microphotometer *[eng]*	微小物測光器	bishō'butsu-sokkō-ki
micropore *[geol]*	細孔；さいこう	saikō
microradiograph *[an-ch]*	微細エクス線写真	bisai-ekusu'sen-shashin
microscope *[opt]*	顕微鏡；顯微鏡	kenbi'kyō
microscopy *[opt]*	顕微鏡検査法	kenbikyō-kensa-hō
microspectrophotometry *[spect]*	顕微分光分析	kenbi-bunkō-bunseki
microstructure *[sci-t]*	微構造；びこうぞう	bi'kōzō
" *[sci-t]*	超小形構造	chō'kogata-kōzō
microwavable *[cook] [electr]*	電子レンジ調理可能	denshi'renji-chōri-kanō
microwave *[elecmg]*	極超短波	kyoku'chō-tanpa
" *[elecmg]*	マイクロ波	maikuro-ha
" *{vb t} [cook]*	電子レンジ調理する	denshi'renji-chōri suru
microwave oven *[cook] [eng]*	電子オーブン	denshi-ōbun
mid- *{prefix}*	中間の-	chūkan no-
midair *[navig]*	中空；ちゅうくう	chūkū
midday; noon; daytime *[astron]*	昼；晝；ひる	hiru
midday meal; lunch *[cook]*	昼食	chūshoku; chūjiki

middle; midway *[math]*	中間	chūkan
middle finger *[anat]*	中指；なかゆび	naka-yubi
middle frequencies *[acous]*	中音	chū-on
middle oil *[mat]*	中油	chū-yu
midge (insect) *[i-zoo]*	蚋；ぶよ；ぶゆ	buyo; buyu
" (insect) *[i-zoo]*	揺り蚊；搖り蚊	yusuri-ka
midline; median line *[geom]*	中線	chū-sen
midnight; dead of night *[astron]*	夜半；やはん	ya'han
midpoint *[geom]*	中点；中點	chū-ten
midship *{n} [nav-arch]*	船の中央部	fune no chūō'bu
midway, more or less *[math]*	中程；なかほど	naka-hodo
migrating bird *[v-zoo]*	渡り鳥	watari-dori
migrating property *[tex]*	移染性	i'sen-sei
migration *[ch]*	泳動；えいどう	eidō
" *[ch] [petrol]*	移動	idō
" *[poly-ch]*	移行	ikō
" *[tex]*	移染；いせん	i'sen
" *[v-zoo]*	渡り	watari
" *{n}*	移送	isō
mildew *{n} [mycol]*	べと病	beto'byō
" *{n} [mycol]*	黴；かび	kabi
" *{n} [mycol]*	白黴	shiro-kabi
mildewproofing agent *[bioch]*	防黴剤	bō'kabi-zai
mild steel *[met]*	軟鋼；なんこう	nankō
mile (unit of length) *[mech]*	哩；マイル	mairu
mileage *[pr]*	延び	nobi
military-industrial complex *[econ]*	軍産共同体	gun'san-kyōdō'tai
military tank *[ord]*	戦車；戰車	sensha
milk *{n} [bioch]*	乳汁	nyūjū
" (of cow) *{n} [cook] [physio]*	牛乳；ぎゅうにゅう	gyū'nyū
" *{n} [physio]*	乳；にゅう；ちち	nyū; chichi
milk fat *[bioch]*	乳脂	nyūshi
milk gland (mammary gland) *[physio]*	乳腺	nyūsen
milk sugar *[bioch]*	乳糖	nyū'tō
milkweed bug (insect) *[i-zoo]*	長椿象；ながかめ虫	naga'kame-mushi
milky quartz *[miner]*	乳石英	nyū-seki'ei
Milky Way *[astron]*	天の川；あまのがわ	ama-no-gawa
" *[astron]*	銀河；ぎんが	ginga
millenium (a thousand years)	千年間	sennen'kan
millet (cereal, grain) *[bot]*	粟；あわ	awa
" (cereal, grain) *[bot]*	黍；きび	kibi
millet jelly *[cook]*	水飴	mizu-ame

millibar (unit of pressure) *[mech]*	ミリバール	miri'bāru
milligram (unit of mass) *[mech]*	瓱；ミリグラム	miri'guramu
millimeter (unit of length) *[mech]*	粍；ミリメートル	miri'mētoru
millimicron (unit of length) *[mech]*	ミリミクロン	miri'mikuron
milling; grinding *{n} [mech-eng]*	摩砕	masai
" *{n} [mech-eng] [pr] [rub]*	錬り；練り；ねり	neri
" *{n} [mech-eng]*	錬肉	renniku
milling machine *[eng]*	フライス盤	furaisu-ban
million *[math]*	百万；ひゃくまん	hyaku'man
millipede (insect) *[i-zoo]*	蚰蜒；げじげじ	geji-geji
mimicry *[bio]*	真似；まね	mane
mind; mentality *{n} [psy]*	心；こころ	kokoro
" ; spirit *{n} [psy]*	精神；せいしん	seishin
mine *{n} [min-eng]*	鉱坑	kōkō
" *{n} [min-eng]*	鉱山；鑛山	kōzan
mineral *[geol]*	原鉱	genkō
" *[geol]*	鉱物；こうぶつ	kō'butsu
" *[geol]*	鉱石	kōseki
" ; inorganic matter *[geol]*	無機質；むきしつ	muki'shitsu
mineral acid; inorganic acid *[ch]*	鉱酸	kōsan
mineral deposit; ore deposit *[geol]*	鉱床；こうしょう	kōshō
mineral dressing; beneficiation *{n} [met]*	選鉱；選鑛	senkō
mineral fiber *[tex]*	鉱物繊維	kō'butsu-sen'i
mineralization *[geol]*	鉱化(作用)	kōka(-sayō)
mineralizer *[geol]*	鉱化剤；こうかざい	kōka-zai
mineral nutrient *[bot]*	無機養素	muki-yōso
mineralogy *[inorg-ch]*	鉱物学；鑛物學	kō'butsu'gaku
mineral oil *[mat]*	鉱油	kō'yu
mineral spring *[hyd]*	鉱泉	kōsen
mineral vein; streak *[min-eng]*	鉱脈；こうみゃく	kō-myaku
mineral water *[hyd]*	鉱水	kōsui
mineral wax; ozocerite *[geol]*	鉱物蠟	kō'butsu-rō
mineral wool; rock wool *[mat]*	岩綿；がんめん	ganmen
minesweeper *[nav-arch]*	掃海艇	sōkai'tei
mini- *{prefix}*	小規模の-	shō'kibo no-
miniature stone landscape *[arch]*	盆石；ぼんせき	bon'seki
miniaturization *[electr]*	小型化；こがたか	kogata'ka
minimal: least *{adj}*	最少の	saishō no
minimal: smallest *{adj}*	最小の	saishō no
minimal medium *[microbio]*	最小培地	saishō-baichi
minimum (pl -ma) *[an-geom]*	極小；きょくしょう	kyoku'shō
" (pl -ma) *[an-geom]*	最小；さいしょう	sai'shō

minimum documentation *[pr]*	最小限資料	saishō'gen-shiryō
minimum, law of *[bio]*	最少律	saishō-ritsu
minimum lethal dose *[med]*	最小致死量	saishō-chishi-ryō
minimum theoretical number of plates *[ch-eng]*	最小理論段数	saishō-riron-dansū
minimum value *[math]*	極小値	kyoku'shō-chi
minimum work *[mech] [thermo]*	最小仕事	saishō-shigoto
mining *{n} [min-eng]*	採鉱; 採鑛	saikō
" *{n} [min-eng]*	採掘	sai'kutsu
mining engineering *[eng]*	鉱山工学; 鑛山工學	kōzan-kōgaku
mining industry *[eng]*	鉱業; こうぎょう	kōgyō
mining right *[legal]*	鉱業権; 鑛業權	kōgyō'ken
-Ministry *{suffix}*	-省	-shō
minium *[miner]*	鉛丹; えんたん	entan
" *[miner]*	光明丹	kōmyō'tan
minor; minor determinant *[alg]*	小行列式	shō-gyō'retsu-shiki
minor angle *[geom]*	劣角; れっかく	rekkaku
minor arc *[geom]*	劣弧; れつこ	rekko
minor axis *[math]*	短軸	tan-jiku
minor premise *[logic]*	小前提	shō-zentei
Minor Section (of village) *[geogr]*	字; あざ	aza
mint *[bot] [cook]*	薄荷; はっか	hakka
minuend *[arith]*	被減数	hi-gensū
minus *{n} [math]*	負	fu
minute *{n} [geogr] [geom] [mech]*	分; ふん	fun
" *{adj} [math]*	微小な	bishō na
-minute; -minutes *{suffix}*	-分; -ふん; -ぷん	-fun; -pun
minute hand *[horol]*	長針	chō'shin
minuteness *[phys]*	微小	bishō
mirabilite *[ch]*	芒硝; ぼうしょう	bōshō
mirage *[opt]*	蜃気楼; 蜃氣樓	shinki'rō
mirror *{n} [opt]*	鏡; かがみ	kagami
mirror bronze *[met]*	鏡青銅	kagami-seidō
mirror-finished surface *[eng]*	鏡面	kyōmen
mirror image *[opt]*	鏡像; きょうぞう	kyōzō
mirror-image isomerism *[ch]*	鏡像異性	kyōzō-i'sei
mirror reflection *[opt]*	正反射	sei-hansha
mirror stand *[furn]*	鏡台	kyōdai
mirror surface *[opt]*	鏡面	kyōmen
miscellaneous cereals *[cook]*	雑穀; 雜穀	zakkoku
miscellaneous equipment *[ind-eng]*	雑装置	zassōchi
miscellaneous germs *[microbio]*	雑菌	zakkin

miscellaneous time *[comput]* 雑時間；雜時間 zatsu-jikan
miscibility *[ch] [sci-t]* 混和性；こんわせい konwa-sei
miscible *(adj) [ch]* 相溶性ある sōyō-sei aru
misfire *(n) [ch]* 不発；不發；ふはつ fu'hatsu
" *(n) [ch]* 不点火；不點火 fu'tenka
" *(n) [ch] [mech-eng]* 失火 shikka
mismatching *(n) [elec]* 不整合 fu'seigō
mispickel; arsenopyrite *[miner]* 硫砒鉄鉱；硫砒鐵鑛 ryūhi'tekkō
miss; be not on time 間に合わない ma ni awanai
missile *[ord]* ミサイル misairu
missing value *[math]* 欠測値 kessoku-chi
mist *(n) [meteor]* 霞；かすみ kasumi
" *(n) [meteor]* 靄；もや moya
mistake (in notation) *(n) [comm]* 誤記 goki
mistake; error *(n) [sci-t]* 間違い；まちがい ma'chigai
" *(n) [sci-t]* 錯誤 sakugo
Mister; Mr. *(suffix)* -氏 -shi
mistle thrush (bird) *[v-zoo]* 寄生木鶇 yadori'gi-tsugumi
mistranslate *(vb) [comm]* 誤訳する go'yaku suru
mite (insect) *[i-zoo]* 壁蝨；蜱；だに dani
mitochondrion (pl -dria) *[cyt]* 糸粒体 shiryū'tai
mitosis (pl -toses) *[cyt]* 核分裂 kaku-bun'retsu
" *[cyt]* 有糸分裂 yūshi-bun'retsu
mitotic index *[cyt]* 分裂指数 bun'retsu-shisū
mix; knead; work *(vb t) [sci-t]* 捏る；こねる koneru
mixed crystal *[crys]* 混晶 konshō
mixed examination *[an-ch]* 混融試験 kon'yū-shiken
mixed indicator *[ch-eng]* 混合指示薬 kongō-shiji-yaku
mixed melting *(n) [thermo]* 混融；こんゆう kon'yū
mixed melting point *[thermo]* 混融点 kon'yū-ten
mixed melting point test *[an-ch]* 混融試験 kon'yū-shiken
mixed number *[arith]* 混数 konsū
mixed-present matter *[ch]* 混在物 konzai'butsu
mixed valence *[ch]* 混合原子価 kongō-genshi'ka
mixed weave *[tex]* 交ぜ織 maze-ori
mixer tube *[electr]* 混合管 kongō-kan
mixing *(n) [eng] [poly-ch]* 混練；混練 konren
" *(n) [eng]* 混成 konsei
" *(n) [mech-eng]* 配合 haigō
" *(n) [pharm]* 調合 chōgō
" *(n) [rub]* 粗練り；あらねり ara-neri
" *(n) [sci-t]* 混合 kongō

mixture *[sci-t]*	混合物	kongō'butsu
mnemonic symbol *[comput]*	簡略記憶記号	kanryaku-ki'oku-kigō
mnemonic system *[psy]*	記憶法	ki'oku-hō
moat *[civ-eng]*	掘; 堀; 壕	hori
mobile phase *[ch-eng]*	移動相	idō-sō
mobility *[phys]*	移動度; 易動度	idō-do
mock gold *[met]*	模造金; もぞうきん	mozō-kin
mockingbird (bird) *[v-zoo]*	真似鶫; まねし鶫	maneshi-tsugumi
mock silver *[met]*	模造銀	mozō-gin
mode; method; system *[comput]*	方式	hōshiki
mode *[stat]*	並数; なみすう	nami'sū
" *[stat]*	最頻値	saihin-chi
model *[sci-t]*	模型; もけい	mokei
model(-wise) schematic *[graph]*	模式的概略図	moshiki'teki-gai'ryaku-zu
modem (modulator-demodulator) *[electr]*	変復調装置	hen'fukuchō-sōchi
mode of working *{n}* *[pat]*	実施態様	jisshi-taiyō
moderate (in degree); medium-degree	中等度	chūtō-do
moderator (instrument) *[eng]*	調節器	chōsetsu-ki
moderator *[nucleo]*	減速材	gensoku-zai
modern mathematics *[math]*	現代数学	gendai-sūgaku
modern times; the present	現代	gendai
modification (transformation) *[bio]*	変態; 變態	hentai
" *[mech-eng]*	変成	hensei
" *[plas]* *[rub]*	変性	hensei
" *[sci-t]*	修飾; しゅうしょく	shūshoku
modified starch *[bioch]*	化工澱粉	kakō-denpun
modifier; qualifier *[gram]*	修飾語	shūshoku'go
modifier *[mat]*	調整剤	chōsei-zai
" *[mat]*	変性剤	hensei-zai
" *[mat]*	条件剤; 條件劑	jōken-zai
" *[poly-ch]*	改質剤	kai'shitsu-zai
modifier gene *[gen]*	変更遺伝子	henkō-idenshi
modify *{vb t}*	変性する	hensei suru
modular *{adj}* *[arch]*	基本単位組合せ型の	kihon-tan'i-kumi'awase-gata no
" *{adj}* *[arch]*	積み木式の	tsumi'ki'shiki no
modular ratio *[mech]*	弾性係数	dansei-keisū
modulation *[comm]*	変調	henchō
modulator-demodulator; modem *[electr]*	変復調装置	hen'fukuchō-sōchi
modulus *[alg]*	係数; けいすう	keisū
" *[alg]*	絶対値; 絶對值	zettai-chi
modulus of elasticity *[mech]*	弾性率	dansei-ritsu

modulus of longitudinal elasticity *[mech]*	縦弾性係数	tate-dansei-keisū
modulus of longitudinal elasticity *[mech]*	縦弾性率	tate-dansei-ritsu
modulus of rigidity *[mech]*	剛性率	gōsei-ritsu
modulus of rupture *[mech]*	破壊係数；破壞係數	hakai-keisū
modulus of transverse elasticity *[mech]*	横弾性係数	yoko-dansei-keisū
moire (wood) *[bot]*	木目；もくめ	moku'me
moist *{adj} [hyd]*	湿気のある	shikki no aru
moisten *{vb t}*	湿らす；濕らす	shimerasu
moistening agent *[mat]*	湿潤剤	shitsu'jun-zai
moisture *[hyd]*	水分	suibun
moisture; humidity *[hyd] [meteor]*	湿気	shikki; shikke
moisture absorption *[ch-eng]*	吸湿；きゅうしつ	kyū'shitsu
moisture-adjusted coal *[geol]*	調湿炭	chō'shitsu-tan
moisture barrier *[mat]*	防湿層	bō'shitsu-sō
moisture content *[ch-eng]*	含水量	gansui-ryō
moisture permeability *[fl-mech]*	透湿性	tō'shitsu-sei
molality *[ch]*	重量モル濃度	jūryō-moru-nōdo
molal solution *[ch]*	モラル溶液	moraru-yōeki
molar *{n} [anat]*	(大)臼歯	(dai-)kyū'shi
" *{n} [anat]*	奥歯；奧齒；おくば	oku'ba
molar mass *[p-ch]*	モル質量	moru-shitsu'ryō
molar solution *[ch]*	モル液	moru-eki
molasses *[cook]*	糖蜜	tō-mitsu
mold *{n} [eng]*	鋳型；鑄型；いがた	i'gata
" *{n} [eng] [poly-ch]*	金型	kana-gata
" *{n} [eng]*	型；かた	kata
" *{n} [mycol]*	黴；かび	kabi
moldability *[mat]*	成形性	seikei-sei
mold curing *{n} [rub]*	型加硫	kata-karyū
mold fungus *[microbio]*	糸状菌	shijō-kin
molding *{n} [eng]*	成形；せいけい	seikei
molding machine *[eng]*	造型機	zōkei-ki
mold lubricant *[mat] [rub]*	離型剤	rikei-zai
mold releasability *[mech-eng]*	離型性	rikei-sei
mold releasing agent *[mat] [rub]*	離型剤	rikei-zai
mold shrinkage *[rub]*	型縮み	kata-chijimi
mole; mol *{n} [ch]*	モル	moru
mole (animal) *{n} [v-zoo]*	土竜；もぐら	mogura
mole cricket (insect) *[i-zoo]*	螻蛄；(お)けら	(o)kera

molecular absorptiometric analysis *[an-ch]*	吸光光度分析	kyūkō-kōdo-bunseki
molecular astronomy *[astron]*	分子天文学	bunshi-tenmon'gaku
molecular beam *[phys]*	分子線；ぶんしせん	bunshi-sen
molecular biology *[bio]*	分子生物学	bunshi-seibutsu'gaku
molecular breeding *{n} [ch]*	分子育種	bunshi-iku'shu
molecular cluster *[ch]*	分子团；分子團	bunshi-dan
molecular diagram *[ch]*	分子図；分子圖	bunshi-zu
molecular distillation *[ch]*	分子蒸留；分子蒸溜	bunshi-jōryū
molecular (electric) field *[p-ch]*	分子電場	bunshi-denba
molecular extinction coefficient *[an-ch]*	分子吸光係数	bunshi-kyūkō-keisū
molecular formula *[ch]*	分子式	bunshi-shiki
molecular orientation *[ch]*	分子配向	bunshi-haikō
molecular polarization *[p-ch]*	分子分極	bunshi-bun'kyoku
molecular physiology *[physio]*	分子生理学	bunshi-seiri'gaku
molecular sieve column *[an-ch]*	分子篩いカラム	bunshi-furui-karamu
molecular still *[ch]*	分子蒸留機	bunshi-jōryū-ki
molecular volume *[ch]*	分子体積	bunshi-taiseki
" *[ch]*	分子容	bunshi-yō
molecular weight *[ch]*	分子量	bunshi'ryō
molecule *[ch]*	分子；ぶんし	bunshi
Molluska; mollusks *[i-zoo]*	軟体動物	nantai-dōbutsu
molten ash *[geol]*	融灰	yūkai
molten metal *[mat]*	溶湯	yōtō
" *[mat]*	湯；ゆ	yu
molten pig iron *[met]*	溶銑	yōsen
molten steel *[met]*	溶鋼	yōkō
molybdenite *[miner]*	輝水鉛鉱	kisui'enkō
molybdenum (element)	モリブデン	moribuden
momentum (pl -s, momenta) *[mech]*	弾み；彈み；はずみ	hazumi
" *[mech]*	運動量	undō'ryō
monad *[ch]*	一価元素	ikka-genso
monadic *{adj} [comput]*	単項の；單項の	tankō no
monadic operation *[comput]*	単項演算	tankō-enzan
monadic operator *[comput]*	単項演算子	tankō-enzan'shi
monadnock *[geol]*	残丘；ざんきゅう	zankyū
monazite *[miner]*	モナズ石	monazu-ishi
Monday	月曜日	getsu'yōbi
money *[econ]*	貨幣；かへい	kahei
" *[econ]*	金銭；きんせん	kinsen
money invested *[econ]*	投資額	tōshi-gaku

monitor(ing) *{n} [ind-eng]*	監視；かんし	kanshi
monitoring device *[comm]*	聴話装置	chōwa-sōchi
monitor station *[comput]*	監視端末	kanshi-tan'matsu
monkey (animal) (pl -s) *[v-zoo]*	猿；さる	saru
monk fish; shark ray (fish) *[v-zoo]*	粕鮫；かすざめ	kasu-zame
monoacidic base *[ch]*	一酸塩基	ichi'san-enki
monoatomic *{adj} [ch]*	単原子の	tan'genshi no
monoatomic molecule *[ch]*	単原子分子	tan'genshi-bunshi
monoaural hearing *{n} [acous]*	片耳聴	kata'mimi-chō
monoaxial stretching *{n} [poly-ch]*	一軸延伸	ichi'jiku-enshin
monobasic acid *[ch]*	一塩基酸	ichi'enki-san
monobasic ammonium phosphate	燐酸一アンモニウム	rinsan-ichi-anmonyūmu
monobasic potassium phosphate	燐酸一カリウム	rinsan-ichi-karyūmu
monochromatic light *[opt]*	単色光	tanshokkō
monochromaticity *[opt]*	単色性	tanshoku-sei
monochromator *[spect]*	単色計	tanshoku-kei
monochrome *[opt]*	単色；たんしょく	tan-shoku
monocline *[geol]*	単傾斜(層)	tan-keisha(-sō)
" *[min-eng]*	単斜；たんしゃ	tansha
monoclinic crystal *[crys]*	単斜晶	tansha-shō
monoclinic system *[crys]*	単斜晶系	tansha-shōkei; tansha'shō-kei
monococcus (pl -cocci) *[microbio]*	単球菌	tankyū-kin
monocyclic compound *[org-ch]*	単環式化合物	tankan'shiki-kagō'butsu
monodentate ligand *[ch]*	単座配位子	tanza-hai'i'shi
monodisperse system *[poly-ch]*	単分散系	tan'bunsan-kei
monodispersing property *[poly-ch]*	単分散性	tan'bunsan-sei
monofilament; single yarn *[tex]*	単糸；たんし	tanshi
monofilament breakage *[tex]*	単糸切れ	tanshi-gire
monofunctional *{adj} [ch]*	単官能の	tan'kannō no
monofunctional compound *[ch]*	単官能化合物	tan'kannō-kagō'butsu
monograph *[pr]*	専攻論文	senkō-ronbun
monohydric alcohol *[org-ch]*	一価アルコール	ikka-arukōru
monolayer; monomolecular film *[p-ch]*	単分子膜	tan'bunshi-maku
" ; monomolecular layer *[p-ch]*	単分子層	tan'bunshi-sō
" ; single layer *[p-ch]*	単層；たんそう	tansō
monolayer culture *[microbio]*	単層培養	tansō-baiyō
monolith *[mat]*	一本石	ippon-seki
monomer *[poly-ch]*	単量体	tanryō'tai
monomer reactivity ratio *[poly-ch]*	単量体反応性比	tanryō'tai-hannō'sei-hi
monomial *{n} [alg]*	単項式	tankō'shiki
monomolecular chain *[ch]*	単分子鎖	tan'bunshi-sa
monomolecular film *[p-ch]*	単分子膜	tan'bunshi-maku

mononuclear compound *[org-ch]*	単環式化合物	tankan'shiki-kagō'butsu
monopoly (exclusive right) *[econ]*	独占権; 獨占權	dokusen'ken
" *[econ]*	専売(権); 專賣(權)	senbai('ken)
monosaccharide *[bioch]*	単糖; たんとう	tantō
monosaccharides *[bioch]*	単糖類; 單糖類	tantō-rui
monostable circuit *[electr]*	単安定回路	tan'antei-kairo
monotonic property *[sci-t]*	単調性	tanchō-sei
Monotremata *[v-zoo]*	単孔目	tankō-moku
monotropy *[fl-mech]*	単変; 單變	tanhen
monounsaturated fatty acid *[bioch]*	単不飽和脂肪酸	tan'fu'hōwa-shibō-san
monovalence *[ch]*	一価; 一價; いっか	ikka
monovalent *{adj} [ch]*	一価の	ikka no
monoxide *[ch]*	一酸化物	issan'ka'butsu
monsoon *[meteor]*	貿易風	bōeki'fū
" *[meteor]*	季節風	kisetsu'fū
month *[astron]*	月; がつ; げつ	gatsu; getsu
monthly mean temperature *[meteor]*	月平均気温	tsuki-heikin-ki'on
monthly periodical *[pr]*	月刊誌	gekkan'shi
monthly report *[pr]*	月報	geppō
moon *{n} [astron]*	月; つき	tsuki
moonquake *[astron]*	月面地震	getsu'men-jishin
moonrise *[astron]*	月の出	tsuki no de
moonset *[astron]*	月の入り	tsuki no iri
moon's path *[astron]*	白道	haku'dō
moonstone *[miner]*	月長石	getchō'seki
moor (a boat) *{vb t}*	舫う; もやう	moyau
mooring buoy *[eng]*	繋船ブイ	keisen-bui
mooring rope *[nav-arch]*	舫い綱	moyai-zuna
moot point; problem point	問題点	mondai-ten
mopping cloth *[tex]*	雑布; 雜布	zōkin
moraine *[geol]*	氷堆石	hyō'taiseki
" *[geol]*	堆石	taiseki
moray (fish) *[v-zoo]*	鱓; うつぼ	utsubo
mordant *[ch]*	媒染剤	baisen-zai
mordant dye; mordant color *[opt]*	媒染染料	baisen-senryō
mordanting *{n} [ch]*	媒染; ばいせん	baisen
more *{adv} [math]*	より多い	yori ōi
" *{adv}*	もっと	motto
more distant *[navig]*	以遠	i'en
more than *[math]*	以上	ijō
morning *[astron]*	朝	asa
" (forenoon) *[astron]*	午前	gozen

morning glory (flower) *[bot]*	朝顔; あさがお	asa-gao
morning newspaper *[pr]*	朝刊	chōkan
morphogenesis *[embryo]*	形態形成	keitai-keisei
morphology *[bio]*	形態学	keitai'gaku
mortality (rate) *[med]*	死亡率	shibō-ritsu
mortar *[mat]*	漆喰; しっくい	shikkui
" *[sci-t]*	乳鉢; にゅうばち	nyū'bachi
" *[sci-t]*	臼; うす	usu
mortar of earthenware *[cer] [cook]*	擂り鉢	suri-bachi
mortise joint *[eng]*	枘穴接合	hozo'ana-setsu'gō
Mosla japonica *[bot]*	山紫蘇; やまじそ	yama-jiso
mosquito (pl -es, -s) (insect) *[i-zoo]*	蚊; か	ka
mosquito net *[mat]*	蚊帳; かや	kaya
mosquito-repellent incense *[mat]*	蚊取り線香	ka'tori-senkō
moss *[bot]*	苔; こけ	koke
mosses; bryophyte *[bot]*	蘚苔	sentai
mossy *{adj}*	苔状の	tai'jō no
most abundant; most numerous *[math]*	最多	saita
most probable value *[math]*	最確値	saikaku-chi
moth (insect); Heterocera *[i-zoo]*	蛾; が	ga
mother (of records) *{n} [acous]*	母盤	boban
mother alloy *[met]*	母合金	bo-gōkin
mother board *[comput]*	主ボード	shu-bōdo
mother liquor *[ch-eng]*	母液	bo'eki
mother of pearl *[i-zoo]*	青貝	ao-gai
" *[i-zoo]*	真珠層	shinju-sō
moth-proofing *{n} [tex]*	防虫加工	bōchū-kakō
motif *[graph]*	画因; 畫因; がいん	ga'in
motile; movable *{adj} [bio]*	動き得る	ugoki-uru
motility *[bio]*	運動性	undō-sei
motion *{n} [mech]*	動作	dōsa
" *{n} [mech]*	運動; うんどう	undō
motion, laws of *[mech]*	運動の法則	undō no hōsoku
motion picture camera *[photo]*	撮影機	satsu'ei-ki
motive power *[phys]*	動力; どうりょく	dō'ryoku
motive-power load *[mech-eng]*	動力負荷	dōryoku-fuka
motor *{n} [elec]*	電動機	dendō-ki
motor cell *[physio]*	運動細胞	undō-saibō
motor gasoline *[mat]*	自動車ガソリン	jidōsha-gasorin
motor nerve *[physio]*	運動神経	undō-shinkei
mottle; mottled; mottling *[bio] [plas]*	斑; まだら	madara
mottled kidney beans *[bot] [cook]*	鶉豆; うずらまめ	uzura-mame

mottling *{n} [opt]*	色斑；いろむら	iro-mura
mount; mount upon *{vb}*	乗せる；乗せる	noseru
mount, fitted *{vb} [mech-eng]*	嵌装する	kansō suru
mount, hanging *{vb}*	装架する	sōka suru
mountain *[geogr]*	山岳；さんがく	san'gaku
" ; peak; hill *[geogr]*	山；やま	yama
mountain breeze *[meteor]*	山風	yama-kaze
mountain cranberry *[bot]*	苔桃；こけもも	koke-momo
mountaineering; mountain climbing *{n}*	登山	tozan
mountain fog *[meteor]*	山霧	yama-giri
mountainous land *[geogr]*	山地	sanchi
mountain range *[geogr]*	山脈	sanmyaku
mountain ridge *[geol]*	峠；とうげ	tōge
mountain rumble *[geol]*	山鳴り	yama-nari
mountainside *[geogr]*	山腹	sanpuku
Mount Fuji *[geogr]*	富士山；ふじさん	Fuji-san
mounting *{n} [elec]*	実装；實装	jissō
" *{n} [eng]*	取付け	tori'tsuke
mounting bracket *[civ-eng]*	取付け金具	tori'tsuke-kana'gu
mourning cloak (butterfly) *[i-zoo]*	黄縁立羽	ki'beri-tate'ha
mourning clothes *[cl]*	喪服	mo-fuku
mouse (pl mice) (animal) *[v-zoo]*	鼠；ねずみ	nezumi
mousetrap *[eng]*	鼠取り	nezumi-tori
mouth *{n} [anat]*	口；くち	kuchi
mouthwash *{n} [pharm]*	含嗽薬；うがいやく	ugai-yaku
movable core *[elecmg]*	可動コア	kadō-koa
movable property *[econ]*	動産	dōsan
move *{vb t} [comput]*	移動する	idō suru
" *{vb t} [comput]*	転送する；轉送する	tensō suru
" *{vb t} [trans]*	移送する	isō suru
" (budge) *{vb i}*	動く；うごく	ugoku
" *{vb t}*	動かす	ugokasu
move interlocked (interlinked) *{vb}*	連動する	rendō suru
movement *[mech]*	動作；どうさ	dōsa
" *{n}*	移動	idō
movie theater *[arch]*	映画館	eiga'kan
moving; mobile; working *{adj}*	動く	ugoku
moving average *[comput]*	移動平均	idō-heikin
moving bed *[ch-eng]*	移動床	idō-shō
" *[ch-eng]*	移動層	idō-sō
moving-bed catalytic cracking process *[ch-eng]*	移動床式接触分解法	idōshō'shiki-sesshoku-bunkai-hō

moving blade; moving vane *[mech-eng]*	回転羽根	kaiten-bane
moving sidewalk *[civ-eng]*	動く歩道	ugoku-hodō
moving sidewise (curve) *[math]*	横這い；よこばい	yoko-bai
moving star cluster *[astron]*	運動星団	undō-seidan
moving vane *[mech-eng]*	回転羽根	kaiten-bane
moving walkway *[civ-eng]*	動く歩道	ugoku-hodō
mow; cut down *{vb t}*	薙ぐ；なぐ	nagu
mowing machine *[eng]*	草刈機	kusa'kari-ki
moxa *{n} [bot]*	艾；もぐさ	mo'gusa
moxibustion *[med]*	灸；きゅう	kyū
mozaic gold *[met]*	彩色金	sai'shiki-kin
mozuku (seaweed) *[bot]*	水雲；もずく	mozuku
Mrs. (pl Mesdames) *{suffix}*	-夫人	-fujin
mucic acid	粘液酸	nen'eki-san
mucilage *[mat]*	粘質；ねんしつ	nen'shitsu
mucin *[bioch]*	粘素；ねんそ	nenso
muck soil *[geol]*	黒泥土；黒泥土	koku-deido
mucor *[mycol]*	毛黴；毛かび	ke-kabi
mucous *{adj} [physio]*	粘液性の	nen'eki'sei no
mucous membrane *[hist]*	粘膜；ねんまく	nenmaku
mucus *[physio]*	粘液	nen'eki
mud (muddy place) *[ecol]*	泥濘；ぬかるみ	nukarumi
" (mire) *{n} [geol]*	泥土；でいど	dei'do
" *{n} [geol]*	泥；どろ	doro
muddying; slurrying *{n} [ch-eng]*	泥漿化	deishō'ka
mudflat *[geol]*	干潟；ひがた	hi'gata
mudflow *[geol]*	泥流	deiryū
mud snail; pond snail *[i-zoo]*	田螺；たにし	ta'nishi
muffler; silencer *[eng]*	消音器	shō'on-ki
mugwort; wormwood *[bot]*	蓬；艾；よもぎ	yomogi
mulberry *[bot]*	桑；くわ	kuwa
mulberry bark *[bot] [pharm]*	桑白皮	sōhaku-hi
mule (animal) *[v-zoo]*	騾馬；らば	raba
mulling *{n} [eng]*	混練；混練	konren
multi- *{prefix} [sci-t]*	複-	fuku-
" *{prefix} [sci-t]*	多い-	ōi-
" *{prefix} [sci-t]*	多重-	tajū-
multicellular *{adj} [bio]*	多細胞の	ta'saibō no
multicolor *[opt]*	多彩；たさい	tasai
multicomponent catalyst *[ch]*	多成分触媒	ta'seibun-shokubai
multidimensional *{adj} [math]*	多次元の	ta'jigen no
multidimensional language *[comput]*	多次元言語	ta'jigen-gengo

multielectrode tube *[electr]*	多極管	ta'kyoku-kan
multielement model *[eng]*	多要素模型	ta'yōso-mokei
multifid leaf *[bot]*	多裂葉	ta'retsu-yō
multifunctional *{adj} [org-ch]*	多官能の	ta'kannō no
multilane road *[civ-eng] [trans]*	多車線道路	ta'shasen-dōro
multilayered circuit board *[elec]*	多層回路基板	tasō-kairo-kiban
multilayer extrusion *[eng]*	多層押出し	tasō-oshi'dashi
multilayer sheet *[mat]*	多層板	tasō-ban
multimer *[poly-ch]*	多量体	taryō'tai
multinational company *[econ]*	多国籍会社	ta'kokuseki-kaisha
multinomial *[alg]*	多項式; たこうしき	takō-shiki
multinucleate cell *[cyt]*	多核細胞	ta'kaku-saibō
multiple *{n} [math]*	倍数; ばいすう	baisū
multiple bond *[ch]*	多重結合	tajū-ketsu'gō
multiple correlation *[stat]*	重相関	jū-sōkan
multiple-cylinder engine *[mech-eng]*	多気筒機関	ta'kitō-kikan
multiple fluidized bed *[eng]*	多段流動層	tadan-ryūdō-sō
multiple fracture *[mech] [sci-t]*	多数破断	tasū-hadan
multiple image *[phys]*	複像	fuku'zō
multiple independently targetable reentry vehicle; MIRV *[ord]*	多弾頭各個目標再=突入弾	ta'dantō-kakko-mokuhyō-sai'totsunyū-dan
multiple intersection *[traffic]*	多枝交差	tashi-kōsa
multiple-layer plating *{n} [met]*	複層めっき	fukusō-mekki
multiple-precision *[comput]*	多倍精度	tabai-seido
multiple proportions, law of *[ch]*	倍数比例の法則	baisū-hirei no hōsoku
multiple regression *[stat]*	重回帰	jū-kaiki
multiple regression analysis *[stat]*	重回帰分析	jū'kaiki-bunseki
multiple scattering *{n} [phys]*	多重散乱	tajū-sanran
multiple star *[astron]*	多重星	tajū-sei
multiplet (term) *[quant-mech]*	多重項	tajū-kō
" (line) *[spect]*	多重線	tajū-sen
multiple warhead *[ord]*	多核弾頭	ta'kaku-dantō
multiplex operation *[comput]*	多重操作	tajū-sōsa
multiplicand *[arith]*	被乗数; 被乘數	hi-jō'sū
multiplication *[arith]*	乗法	jōhō
" *[arith]*	掛け算; かけざん	kake-zan
" *[bio]*	増殖	zō'shoku
multiplication factor *[nucleo]*	増倍率	zōbai-ritsu
multiplicity *[math]*	多重度	tajū-do
multiplier *[arith]*	乗数; じょうすう	jō'sū
" *[elec]*	倍率器	bai'ritsu-ki
multiplier phototube *[electr]*	光電子増倍管	kō'denshi-zōbai-kan

multiply *{vb} [arith]*	掛ける	kakeru
multiplying factor; scale factor *[eng] [math]*	倍率；ばいりつ	bairitsu
multiplying factor *[math]*	乗率	jō-ritsu
multiplying ratio *[math]*	倍率	bai-ritsu
multipoint line *[comm]*	分岐回線	bunki-kaisen
multipole radiation *[phys]*	多重極放射	ta'jūkyoku-hōsha
" *[phys]*	多極放射	ta'kyoku-hōsha
multipurpose satellite *[aero-eng]*	多目的衛星	ta'mokuteki-eisei
multi-screw extruder *[eng]*	多軸押出機	ta'jiku-oshi'dashi-ki
multistage *{adj} [eng]*	多段式の	tadan'shiki no
multistage amplifier *[electr]*	多段増幅器	tadan-zōfuku-ki
multivalent antibody *[bio]*	多価抗体	taka-kōtai
multiway layout *[stat]*	多元配置	tagen-haichi
mumps; parotitis *[med]*	お多福風	o'tafuku-kaze
munitions *[ord]*	軍需品	gunju'hin
mural *[graph]*	壁画；壁畫；へきが	heki-ga
murine *{adj} [v-zoo]*	鼠科の	nezumi-ka no
muscle *{n} [anat]*	筋肉	kinniku
" *{n} [anat]*	筋；すじ	suji
muscle fiber *[hist]*	筋(肉)繊維	kin(-niku)-sen'i
muscle relaxant *[pharm]*	筋弛緩薬	kin-shikan-yaku
muscology *[bot]*	苔学；こけがく	koke'gaku
muscovite; white mica *[miner]*	白雲母；しろうんも	shiro-unmo
muscular system *[anat]*	筋肉組織	kinniku-so'shiki
museum *[arch]*	博物館	haku'butsu'kan
museum beetle (insect) *[i-zoo]*	鰹節虫	katsuo'bushi-mushi
mushroom *{n} [bot]*	茸；たけ	take
" *{n} [mycol]*	茸；きのこ	kinoko
mushrooms *[bot]*	菌蕈；きんじん	kinjin
music (sheet) *[music] [pr]*	楽譜	gakufu
" (Japanese) *[music]*	邦楽；ほうがく	hōgaku
" *{n} [music]*	音楽；音樂	on'gaku
musical acoustics *[acous]*	音楽音響学	ongaku-onkyō'gaku
musical instrument *[music]*	楽器；樂器；がっき	gakki
musical tone; musical note *[acous]*	楽音	gaku'on
music hall *[arch]*	音楽会場	ongak(u)kai'jō
musician *[music]*	音楽家	ongak(u)'ka
music tape for sing-along *[music]*	空オケ	kara'oke
musk *[physio]*	麝香；じゃこう	jakō
musk deer (animal) *[v-zoo]*	麝香鹿	jakō-jika
musk ox (animal) *[v-zoo]*	麝香牛	jakō-ushi

mussel (shellfish) *[i-zoo]*	紫貽貝	murasaki-i'gai
mustache; beard; whiskers *[anat]*	鬚; 髭; 髯; ひげ	hige
mustard (condiment) *[bot]*	芥子; からし	karashi
mustard plant (vegetable) *[bot]*	芥菜; からしな	karashi'na
mustard (-seed) oil *[org-ch]*	芥子油	karashi-yu
mutagen *[gen]*	突然変異誘発物質	totsu'zen-hen'i-yūhatsu-busshitsu
" *[microbio]*	変異誘導源	hen'i-yūdō'gen
mutant *[gen]*	変異種; へんいしゅ	hen'i-shu
mutant strain *[microbio]*	変異株; へんいかぶ	hen'i-kabu
mutarotation *[ch]*	変旋光	hen'senkō
mutation *[gen]*	突然変異; 突然變異	totsu'zen-hen'i
mute (a person) *{n} [physio]*	唖; 啞; おし	oshi
mute letter *[ling]*	黙字; 默字	moku'ji
mutton tallow; lanolin *[mat]*	羊脂; ようし	yōshi
mutual cancellation *[math]*	相殺; そうさつ	sō'satsu
mutual division *[math]*	互除法; ごじょほう	gojo-hō
mutual induction *[elecmg]*	相互誘導	sōgo-yūdō
mycelial form *[mycol]*	菌糸形; きんしけい	kinshi-kei
mycelium (pl -a) *[mycol]*	菌糸体	kinshi'tai
mycetome *[i-zoo]*	菌器	kinki
mycology *[bot]*	菌(類)学; 菌(類)學	kin(rui)'gaku
mycorrhiza (pl -zae, -s) *[mycol]*	菌根; きんこん	kinkon
Mycota; Eumycetes *[mycol]*	菌界	kinkai
mycotoxin; fungal toxin *[mycol]*	黴毒; かびどく	kabi-doku
" *[mycol]*	真菌毒素	shinkin-doku'so
mydriatic *{n} [pharm]*	散瞳薬	sandō-yaku
myelin sheath *[hist]*	髄鞘; 髓鞘	zui-shō
myeloma cell *[cyt] [med]*	骨髄腫細胞	kotsuzui'shu-saibō
myopia; nearsightedness *[med]*	近視; きんし	kinshi
myrrh *[mat]*	没薬; もつやく	motsu-yaku
mysid; opossum shrimp *[i-zoo]*	糠蝦; あみ	ami
Myxobacterales *[microbio]*	粘液細菌目	nen'eki-saikin-moku
Myxomycota; slime molds *[microbio]*	変形菌類; 變形菌類	henkei'kin-rui

N

nacre *[i-zoo]*	真珠層	shinju-sō
nacreous cloud *[meteor]*	真珠雲	shinju'un
nadir *[astron]*	天底	tentei
Nagoya *[geogr]*	名古屋；なごや	Nagoya
nail *{n} [anat]*	爪；つめ	tsume
" *{n} [mat]*	釘；くぎ	kugi
nail clipper *[eng]*	爪切り	tsume-kiri
naked; nude; bare *{n}*	裸	hadaka
name *{n} [comput]*	名前	na'mae
" ; title; designation *{n}*	名称；名稱	meishō
" (of a person) *{n}*	名(乗り)	na('nori)
name card; business card *[comm]*	名刺；めいし	meishi
name change *[pat]*	名義変更	meigi-henkō
name of invention *[pat]*	発明の名称	hatsu'mei no meishō
nameplate *[comm]*	銘盤	meiban
nandin *[bot]*	南天；なんてん	nanten
nap *{n} [tex]*	毛羽；けば	keba
nape (of neck) *[anat]*	項；うなじ	unaji
Nara *[geogr]*	奈良；なら	Nara
narcissus (pl -s, -cissi, narcissus) (flower) *[bot]*	水仙	suisen
narcotic; narcotics *{n} [pharm]*	麻薬	ma'yaku
narcotic intoxication; narcotism *[med]*	麻薬中毒	mayaku-chūdoku
narcotics control law *[legal]*	麻薬取締法	mayaku-tori'shimari-hō
narrow *{adj} [math]*	細い；ほそい	hosoi
" *{adj}*	狭い	semai
narrow; become narrower *{vb i}*	細まる；ほそまる	hosomaru
" *{vb i}*	狭まる；せばまる	sebamaru
narrow; narrow down *{vb t}*	細める	hosomeru
" *{vb t}*	狭める；狹める	sebameru
narrowband *[comm]*	狭帯域；狹帯域	kyō-tai'iki
narrow-mouth bottle *[an-ch]*	細口瓶	hoso'kuchi-bin
narrows (ocean) *[geogr]*	海峡；海峽	kaikyō
narwhal(e); Monodon *[v-zoo]*	一角	ikkaku
nascency *[ch]*	発生；はっせい	hassei
nascent hydrogen *[ch]*	発生期水素	hassei'ki-suiso
nascent state *{n} [ch]*	発生期状態	hassei'ki-jōtai
national *{adj}*	国家の；國家の	kokka no
national highway *[civ-eng]*	国道	koku-dō

national park *[civ-eng]*	国立公園	koku'ritsu-kōen
native compound *[mat]*	天然化合物	tennen-kagō'butsu
native element(s) *[geol]*	元素鉱物；元素鑛物	genso-kōbutsu
native gold *[geol]*	自然金；しぜんきん	shizen-kin
native protein *[bioch]*	天然蛋白質	tennen-tanpaku'shitsu
natrolite *[miner]*	曹達沸石	sōda'fusseki
natural; nature *{n}*	天然；てんねん	ten'nen
natural *{adj}*	天然の	ten'nen no
natural aging *{n} [poly-ch]*	自然老化	shizen-rōka
natural binary *[math]*	自然二進	shizen-ni'shin
natural coloring material *[cook]*	天然着色料	tennen-chaku'shoku-ryō
natural convection *[thermo]*	自然対流；自然對流	shizen-tairyū
natural disaster; act of God	天災	tensai
natural dye *[mat]*	天然染料	tennen-senryō
natural enemy *[bio]*	天敵；てんてき	ten-teki
natural fiber *[tex]*	天然繊維	tennen-sen'i
natural gas *[mat]*	天然ガス	tennen-gasu
natural grade *[civ-eng]*	自然勾配	shizen-kōbai
natural history *[sci-t]*	博物学	haku'butsu'gaku
" *[sci-t]*	自然史	shizen-shi
natural illumination *[opt]*	採光	saikō
natural language *[comput]*	自然言語	shizen-gengo
natural logarithm *[calc]*	自然対数	shizen-taisū
naturally growing; wild *{adj} [bot]*	野生の；やせいの	yasei no
natural monument *{n}*	天然記念物	tennen-kinen'butsu
natural numbers *[arith]*	自然数	shizen-sū
natural pigment *[mat]*	天然色素	tennen-shiki'so
natural product *[mat]*	天然物	tennen'butsu
natural resin *[mat]*	天然樹脂	tennen-jushi
natural resource *[mat]*	天然資源	tennen-shigen
natural science *[sci-t]*	自然科学	shizen-kagaku
natural selection *[bio]*	自然選択；自然選擇	shizen-sentaku
" *[bio]*	自然淘汰	shizen-tōta
nature *{n}*	性質；せいしつ	seishitsu
" *{n}*	天然	ten'nen
nature; mother nature *{n}*	自然；しぜん	shizen
nature, law of *[bio]*	自然法則	shizen-hōsoku
nausea *[med]*	嘔気；おうき	ōki
nauseant *{n} [pharm]*	催吐薬；(-剤)	saito-yaku; (-zai)
nautical *{adj} [ocean]*	海事の	kaiji no
nautical almanac *[navig]*	航海暦	kōkai-reki
nautical mile *[navig]*	海里	kairi

nautical table *[navig]*	航海表	kōkai-hyō
nautilus (shellfish) *[i-zoo]*	鸚鵡貝；おうむがい	ōmu-gai
naval architect *[nav-arch]*	造船家	zōsen'ka
naval architecture *[eng]*	造船学；造船學	zōsen'gaku
naval brass *[met]*	海軍真鍮	kaigun-shinchū
navel *{n}* *[anat]*	臍；へそ	heso
navigation *[eng]*	航法	kōhō
" *[eng]*	航行	kōkō
navigation light *[nav-arch]*	航海灯；航海燈	kōkai-tō
navigation mark *[navig]*	航路標識	kōro-hyō'shiki
navigation pass *[navig]*	通船路；つうせんろ	tsūsen-ro
navigation satellite *[aero-eng]*	航行衛星	kōkō-eisei
navy *[mil]*	海軍	kaigun
navy blue (color) *[opt]*	紺；こん	kon
neap tide *[ocean]*	小潮	ko-shio
near-field pattern *[opt]*	近視野像	kin-shiya-zō
near-infrared radiation *[elecmg]*	近赤外線	kin-sekigai-sen
near-infrared region *[elecmg]*	近赤外領域	kin-sekigai-ryō'iki
nearly; nearly all *{adv}*	殆ど；ほとんど	hotondo
near seas; neighboring waters *[ocean]*	近海	kinkai
nearsightedness *[med]*	近視	kinshi
near-ultraviolet radiation *[elecmg]*	近紫外線	kin-shigai'sen
near-ultraviolet rays *[elecmg]*	近紫外線	kin-shigai'sen
neatsfoot oil *[mat]*	牛脚油	gyū'kyaku-yu
nebula (pl -las, -lae) *[astron]*	星雲	sei'un
nebular lines *[astrophys]*	星雲線	sei'un-sen
nebulization *[mech-eng]*	噴霧化	funmu'ka
necessarily; always *{adv}*	必ず；かならず	kanarazu
necessary *{n}* *[logic]*	必然的	hitsuzen'teki
" *{n}* *[math]*	必要；ひつよう	hitsu'yō
" *{adj}*	所要の	shoyō no
necessary condition *[math]*	必須要件	hissu-yōken
" *[math]*	必要条件	hitsu'yō-jōken
neck *{n}* *[anat]*	首；頸；くび	kubi
necklace *[lap]*	首飾り	kubi-kazari
nectar *[bot]*	果蜜	ka'mitsu
needle *{n}* *[eng]*	針；はり	hari
needle crystal *[crys]*	針状結晶	shin'jō-kesshō
needlefish (fish) *[v-zoo]*	啄長魚；だつ	datsu
needle-nose pliers *[eng]*	先尖りペンチ	saki'togari-penchi
needle thread *[cl]* *[tex]*	表糸	omote-ito
" *[cl]* *[tex]*	上糸；うわいと	uwa-ito

negate *{vb} [comput]*	否定する	hitei suru
negation *[logic]*	否定；ひてい	hitei
negative *{n} [elec]*	負(性)	fu(-sei)
" *{n} [comput]*	否定	hitei
" *{adj} [arith]*	負の	fu no
" *{adj} [elec]*	負性の；ふせいの	fusei no
negative acknowledge character *[comput]*	否定応答文字	hitei-ōtō-moji
negative afterimage *[physio]*	陰性残像	insei-zanzō
negative angle *[trig]*	負(の)角	fu(no)-kaku
negative catalysis *[ch]*	負触媒作用	fu-shokubai-sayō
negative charge *[elec]*	陰電荷	in-denka
negative crystal *[crys]*	負結晶	fu-kesshō
negative electrode *[elec]*	負極	fu-kyoku
" *[elec]*	陰極；いんきょく	in-kyoku
negative feedback *[electr]*	負帰還	fu-kikan
negative image *[photo]*	陰画像；陰畫像	inga-zō
negative ion *[ch] [phys]*	陰イオン	in-ion
negative modulation *[electr]*	負変調；負變調	fu-henchō
negative number *[alg]*	負(の)数	fu(no)sū
negative picture *[photo]*	陰画像	inga-zō
negative pressure *[phys]*	負圧；負壓；ふあつ	fu-atsu
negative valence *[ch]*	負原子価	fu-genshi'ka
negatron *[phys]*	陰電子	in'denshi
neighborhood park *[civ-eng]*	近隣公園	kinrin-kōen
nekton *[i-zoo]*	遊泳生物	yū'ei-seibutsu
Nemacystus decipiens (seaweed) *[bot]*	水雲；海蘊；もずく	mozuku
nematicide; nematocide *[mat]*	線虫駆除剤	senchū-kujo-zai
nematocide *{n} [pharm]*	殺線虫剤	sassenchū-zai
Nematoda (roundworms) *[i-zoo]*	線形動物	senkei-dōbutsu
nematode *[i-zoo]*	線虫；せんちゅう	senchū
Nematomorpha *[i-zoo]*	類線形動物	rui'senkei-dōbutsu
Nemertinea *[i-zoo]*	紐形動物	himo'gata-dōbutsu
neodymium (element)	ネオジム	neojimu
Neolithic Age *[archeo]*	新石器時代	shin-sekki-jidai
neon (element)	ネオン	neon
nepheline *[miner]*	霞石；かすみいし	kasumi-ishi
nephelometer *[meteor]*	雲量計	un'ryō-kei
" *[opt]*	比濁計	hidaku-kei
nephelometry *[opt]*	比濁法	hidaku-hō
nephridium (pl -ia) *[i-zoo]*	腎管	jinkan
nephrite *[miner]*	翡翠；ひすい	hisui

Neptune (planet) *[astron]*	海王星	kai'ō-sei
neptunium (element)	ネプツニウム	neputsunyūmu
neroli oil *[mat]*	橙花油；とうかゆ	tōka-yu
nerula *[embryo]*	神経胚	shinkei-hai
nerulation *[embryo]*	神経管形成	shinkei'kan-keisei
nerve *{n} [anat]*	神経；しんけい	shinkei
" *{n} [poly-ch] [rub]*	腰の強さ	koshi no tsuyo-sa
nerve agent *[ord]*	神経剤；神經劑	shinkei-zai
nerve cell *[hist]*	神経細胞；神經細胞	shinkei-saibō
nerve ending *{n} [anat]*	神経終末	shinkei-shūmatsu
nerve fiber *[cyt]*	神経繊維	shinkei-sen'i
nerve net *[i-zoo]*	神経網	shinkei-mō
nervous system *[anat]*	神経系	shinkei-kei
nervure *[bot]*	葉脈	yō'myaku
" *[i-zoo]*	翅脈；しみゃく	shi'myaku
nest; nesting *{n} [comput]*	入れ子	ire'ko
nest *{n} [met] [v-zoo]*	巣；巢；す	su
net *{n} [econ]*	正味；しょうみ	shōmi
" *{n} [tex]*	網；あみ	ami
net after taxes *[econ]*	税引後利益	zei'biki-go-ri'eki
net before taxes *[econ]*	税引前利益	zei'biki-mae-ri'eki
net charge *[elec]*	実効電荷；實効電荷	jikkō-denka
net plane *[crys]*	網平面	ami-heimen
net section *[graph]*	純断面；純斷面	jun'danmen
nettle tree (tree) *[bot]*	榎；えのき	enoki
net weight *[mech]*	正味重量	shōmi-jūryō
network *[comm]*	回路；かいろ	kairo
" *[comm] [elec]*	回路網	kairo-mō
" *[comput]*	網	ami; mō
" *[poly-ch]*	網目	ami'me
network chain *[poly-ch]*	網目鎖	ami'me-sa
network polymer *[poly-ch]*	網状重合体	amijō-jūgō'tai
" *[poly-ch]*	網状高分子	amijō-kōbunshi
network structure *[met]*	網状構造	amijō-kōzō
" *[met] [poly-ch]*	網目構造	ami'me-kōzō
network vein *[min-eng]*	網状鉱脈；網狀鑛脈	mōjō-kōmyaku
neuralgia *[med]*	神経痛	shinkei'tsū
neurobiologist *[bio]*	神経生物学者	shinkei-seibutsu-gakusha
neurobiology *[bio]*	神経生物学	shinkei-seibutsu'gaku
neuron *[hist]*	神経単位；神經單位	shinkei-tan'i
neuropharmacology *[pharm]*	神経薬理学	shinkei-yakuri'gaku
neuroreceptor *[bioch]*	神経性受容体	shinkei'sei-juyō'tai

neurotransmitter *[bioch]*	神経伝達物質	shinkei-den'tatsu-busshitsu
neutral; neutrality *{n}*	中性	chūsei
neutrality *{n}*	中立	chūritsu
neutralization *[ch]*	中和；ちゅうわ	chūwa
neutralization number *[an-ch]*	中和価；中和價	chūwa-ka
neutralize *{vb} [ch]*	中和する	chūwa suru
neutralized salt *[ch]*	中和塩；中和鹽	chūwa-en
neutralizing agent *[ch]*	中和剤	chūwa-zai
neutral line *[meteor]*	中性線	chūsei-sen
neutral salt *[ch]*	中性塩	chūsei-en
neutral transmission *[comm]*	単流式伝送	tanryū'shiki-densō
neutrino *[phys]*	中性微子	chūsei-bishi
neutron *[phys]*	中性子	chūsei'shi
neutron bomb *[ord]*	中性子爆弾	chūseishi-bakudan
neutron capture *[nuc-phys]*	中性子捕獲	chūseishi-hokaku
neutron counter *[nucleo]*	中性子計数器	chōseishi-keisū-ki
neutron cross section *[nuc-phys]*	中性子断面積	chūseishi-dan'menseki
neutron detector *[nucleo]*	中性子検出器	chūseishi-ken'shutsu-ki
neutron diffractometer *[phys]*	中性子回折計	chūseishi-kai'setsu-kei
neutron flux *[nucleo]*	中性子線束	chūseishi'sen-soku
neutron howitzer *[nucleo]*	中性子射出器	chūseishi-sha'shutsu-ki
neutron optics *[phys]*	中性子光学	chūseishi-kōgaku
neutron radiography *[nucleo]*	中性子線構造解析学	chūseishi'sen-kōzō-kaiseki'gaku
neutron scattering *{n} [nucleo]*	中性子散乱	chūseishi-sanran
neutron star *[astron]*	中性子星	chūseishi-boshi
neve *[geogr]*	粉状氷雪	funjō-hyō'setsu
" *[geogr]*	万年雪原；萬年雪原	mannen-setsu'gen
" *[hyd]*	粒状氷雪	ryūjō-hyō'setsu
new *{adj}*	新しい；あたらしい	atarashī
" (fresh) *{adj}*	新規の	shinki no
new-and-old contrast	新旧対比；新舊對比	shin-kyū-taihi
newel post *[civ-eng]*	親柱；おやばしら	oya-bashira
new-line character *[comput]*	改行復帰文字	kaigyō-fukki-moji
new moon *[astron]*	朔；さく	saku
" *[astron]*	新月；しんげつ	shin-getsu
news media *[comm]*	報道機関	hōdō-kikan
newspaper *[paper] [pr]*	新聞；しんぶん	shinbun
newsprint *[mat] [paper]*	新聞用紙	shinbun-yōshi
newt; water lizard *[v-zoo]*	井守；いもり	imori
next (the following) *{adj}*	次の	tsugi no
next-; following- *{prefix}*	次-	ji-

next day	翌日；よくじつ	yoku-jitsu
next month	来月；來月	rai-getsu
next next- *{prefix}*	来来-	rai-rai-
" *{prefix}*	翌翌-	yoku-yoku-
next next day; day after next	翌翌日	yoku-yoku-jitsu
next week	来週	rai-shū
next year	来年	rai-nen
nexus (pl nexuses, nexus) *[comm]*	繋ぎ；つなぎ	tsunagi
" *[geol]*	壁龕；へきがん	heki'gan
niche *[ecol]*	生態的地位	seitai'teki-chi'i
nickel (element)	ニッケル	nikkeru
nickel silver *[mat]*	洋銀	yōgin
" *[mat]*	洋白	yōhaku
nictitating membrane *[v-zoo]*	瞬膜；しゅんまく	shun-maku
night; nighttime *{n} [astron]*	夜；よ；よる	yo; yoru
night-blindness; nyctalopia *[med]*	夜盲症	yamō'shō
nightclothes *[cl]*	寝巻き；寝間着	ne'maki
night glass(es) *[opt]*	夜間(用)双眼鏡	yakan('yō)-sōgan'kyō
nightglow *[geophys]*	夜光	yakō
nightjar (bird) *[v-zoo]*	夜鷹	yo'taka
night sky *[astron]*	夜空；よぞら	yo-zora
night temperature *[thermo]*	夜温	ya'on
nighttime *[astron]*	晩；ばん	ban
" *[astron]*	夜間；やかん	yakan
night transmission *[comm]*	夜間通信	yakan-tsū'shin
nimbostratus *[meteor]*	乱層雲	ransō'un
nimbus (pl nimbi, -es) *[meteor]*	乱雲；亂雲	ran'un
nine *{n} [math]*	九	ku; kyū
nine's complement *[comput]*	九の補数	kyō no hosū
niobium (element)	ニオブ	niobu
nippers *[eng]*	鋏；やっとこ	yattoko
niter *[inorg-ch]*	硝石	shōseki
nitrate *[ch]*	硝酸塩	shōsan-en
nitration *[org-ch]*	硝化	shōka
nitric acid	硝酸；しょうさん	shōsan
nitric oxide; nitrogen monoxide	一酸化窒素	issanka-chisso
nitride *[ch]*	窒化物；ちっかぶつ	chikka'butsu
nitriding *{n} [ch]*	窒化	chikka
nitrification *[microbio]*	硝化(作用)	shōka(-sayō)
nitrifier (bacteria) *[microbio]*	硝化細菌	shōka-saikin
nitrite *[ch]*	亜硝酸塩	a'shōsan-en
Nitrobacter *[microbio]*	硝酸菌	shōsan-kin

nitrocellulose; nitrocotton *[org-ch]*	硝化綿	shōka-men
nitrogen (element)	窒素；ちっそ	chisso
nitrogen bromide	臭化窒素	shūka-chisso
nitrogen chloride	塩化窒素	enka-chisso
nitrogen cycle *[bio]*	窒素循環	chisso-junkan
nitrogen dioxide	二酸化窒素	ni'sanka-chisso
nitrogen iodide	沃化窒素	yōka-chisso
nitrogen monoxide	一酸化窒素	issanka-chisso
nitrogenous fertilizer *[agr]*	窒素肥料	chisso-hiryō
nitrogen oxide	酸化窒素	sanka-chisso
nitrogen sulfide	硫化窒素	ryūka-chisso
nitrolime; lime nitrogen *[inorg-ch]*	石灰窒素	sekkai-chisso
nitrous acid	亜硝酸	a'shōsan
nitrous oxide	亜酸化窒素	a'sanka-chisso
nivation *[geol]*	雪食；せっしょく	sesshoku
nobelium (element)	ノーベリウム	nōberyūmu
noble *{n}* *[ch]*	貴；き	ki
noble gas *[ch]*	希有気体	ke'u-kitai
noble metal *[met]*	貴金属	ki-kinzoku
Noctiluca *[i-zoo]*	夜光虫	yakō'chū
noctilucent cloud *[meteor]*	夜光雲	yakō'un
nocturnal animal *[v-zoo]*	夜行動物	yakō-dōbutsu
nocturnal radiation *[geophys]*	夜間放射	yakan-hōsha
nocturne *[music]*	夜想曲	yasō'kyoku
nodal point *[opt]*	節点；せってん	setten
node *[astron] [electr] [eng]*	節点	setten
" *[bot]*	節；ふし	fushi
" *[phys]*	波節	ha'setsu
nodical month *[astron]*	交点月；交點月	kōten-getsu
nodular graphite cast iron *[met]*	球状黒鉛铸鉄	kyūjō-koku'en-chū'tetsu
nodule *[anat]*	小結節	shō-kessetsu
" *[geol]*	団塊；團塊	dankai
no good *{n}*	駄目；だめ	dame
noise *[acous]*	騒音；そうおん	sō'on
" *[acous]*	雑音；雜音	zatsu'on
noise factor; noise index *[electr]*	雑音指数	zatsu'on-shisū
noise generator *[electr]*	雑音発生器	zatsu'on-hassei-ki
noise meter *[electr]*	雑音計	zatsu'on-kei
noise pollution *[ecol]*	騒音公害	sō'on-kōgai
noisy *{adj} [acous]*	騒々しい；騷騷しい	sōzōshī
" *{adj} [acous]*	喧しい；やかましい	yakamashī
no left turn *[traffic]*	左折禁止	sa'setsu-kinshi

nomadism *[bio]*	遊動；ゆうどう	yūdō
nomenclature *[sci-t]*	(学名-)命名法	(gakumei-)meimei-hō
" *[sci-t]*	名称；名稱	meishō
" *[sci-t]*	組織的命名法	soshiki'teki-meimei-hō
nominal *{n}* *[sci-t]*	公称；こうしょう	kōshō
nominal dimension; nominal size *[mech]*	呼び寸法	yobi-sunpō
nominal speed; rated speed *[comput]*	名目速度	meimoku-sokudo
nominal stress *[met]*	公称応力	kōshō-ō'ryoku
nominative case *[gram]*	主格；しゅかく	shukaku
nomograph *[math]*	計算図表；計算圖表	keisan-zuhyō
non- *{prefix}*	非-；否-	hi-
" *{prefix}*	無-	mu-
nonadhesive *{adj}* *[adhes]*	非接着性の	hi'setchaku'sei no
nonaging property *[eng]*	非時効性	hi'jikō-sei
nonagon *[geom]*	九角形	ku'kak(u)kei
nonaqueous *{adj}* *[ch]*	非水的な	hisui'teki na
nonaqueous solvent *[ch]*	非水溶媒	hisui-yōbai
nonavalent *{adj}* *[ch]*	九価の	kyūka no
noncaking coal *[geol]*	不粘結炭	fu'nenketsu-tan
noncentral force *[phys]*	無中心力	mu'chūshin-ryoku
noncentrality degree *[math]*	非心度	hi'shin-do
noncompetetive inhibition *[bioch]*	否拮抗的阻害	hi'kikkō'teki-sogai
nonconductor *[elec]* *[mat]*	不導体；ふどうたい	fudō'tai
" *[elec]* *[mat]*	絶縁体	zetsu'en'tai
nonconjugated *[org-ch]*	非共役	hi'kyōyaku
noncyclic photophosphorylation *[bioch]*	否循環的光燐酸化	hi'junkan'teki-kō'rinsan'ka
nondestructive *{adj}*	非破壊の；非破壞の	hi'hakai no
nondestructive readout *[comput]*	非破壊読出し	hi'hakai-yomi'dashi
nondestructive testing *{n}* *[eng]*	非破壊検査	hi'hakai-kensa
nondirectional *{adj}* *[steel]*	無方向性の	mu'hōkō'sei no
nondispersive-type *{adj}* *[math]*	非分散型の	hi'bunsan'gata no
nonelectrolyte *{n}* *[mat]*	非電解質	hi'denkai'shitsu
nonequilibrium thermodynamics *[thermo]*	非平衡熱力学	hi'heikō-netsu-rikigaku
nonequivalence element *[comput]*	不等価素子	fu'tōka-soshi
nonexecutable statement *[comput]*	非実行文	hi'jikkō'bun
nonfeasance; omission *[legal]*	不作爲；ふさくい	fu'saku'i
nonferrous metal *[met]*	非鉄金属；非鐵金屬	hi'tetsu-kinzoku
nonfinely-coking low-grade coal *[miner]*	非微粘結性劣質炭	hi-bi'nenketsu'sei-resshitsu-tan
nonglutinous rice *[bot]* *[cook]*	粳米；うるちまい	uruchi-mai
nonideal gas *[stat-mech]*	不完全気体	fu'kanzen-kitai
nonimpact printer *[pr]*	非衝撃式印字装置	hi'shōgeki'shiki-inji-sōchi

nonionic surfactant; nonionic surface-active agent *[mat]*	非イオン界面活性剤	hi'ion-kaimen-kassei-zai
nonlinear equation *[math]*	非線形方程式	hi'senkei-hōtei'shiki
nonlinearity *[math]*	非直線性	hi'chokusen-sei
" *[phys]*	非線形性	hi'senkei-sei
nonlinear programming *{n} [comput]*	非線形計画法	hi'senkei-keikaku-hō
nonlinear vibration *[mech]*	非線形振動	hi'senkei-shindō
nonlinear wave *[phys]*	非線形波動	hi'senkei-hadō
nonmagnetic *{adj} [elecmg]*	非磁性の	hi'jisei no
nonmagnetic steel *[met]*	非磁性鋼	hi'jisei-kō
nonmetal *{n} [mat]*	非金属	hi'kinzoku
nonmetallic inclusion *[mech-eng]*	非金属介在物	hi'kinzoku-kaizai'butsu
nonmetallic mineral *[miner]*	非金属鉱物	hi'kinzoku-kōbutsu
nonoriented *{adj} [steel]*	無方向性の	mu'hōkō'sei no
nonpetroleum oil *[mat]*	非鉱油; 非鑛油	hi'kōyu
nonpolar compound *[ch]*	非極性化合物	hi'kyokusei-kagō'butsu
nonpolar solvent *[mat]*	無極性溶媒	mu'kyoku'sei-yōbai
nonradiant *{adj} [phys]*	無輻射の	mu'fukusha no
nonrelative approximation *[phys]*	非相対論的近似	hi'sōtairon'teki-kinji
nonrenewable resources *[sci-t]*	更新しえない資源	kōshin-shi'enai shigen
nonreusable *{n} [comput]*	再使用不能	sai'shiyō-funō
nonrotatory(ness) *{n} [opt]*	比旋光性	hi'senkō-sei
nonslip finish *[rub]*	滑り止め加工	suberi'dome-kakō
nonstoichiometric compound *[ch]*	不定比化合物	fu'teihi-kagō'butsu
nontoxic *{adj} [med]*	無毒性の	mu'doku'sei no
nonvolatile matter *[mat]*	不揮発物	fu'kihatsu'butsu
nonwoven fabric *[tex]*	不織布	fushoku-fu
noodles; noodle-type food *[cook]*	麵類; めんるい	menrui
" (Japanese) *[cook]*	素麵; 索麵	sōmen
" (Japanese) *[cook]*	饂飩; うどん	udon
noon *[astron]*	正午	shōgo
no parking *[traffic]*	駐車禁止	chūsha-kinshi
no passing *[traffic]*	追越し禁止	oi'koshi-kinshi
no right turn *[traffic]*	右折禁止	u'setsu-kinshi
normal *{n} [geom]*	法線; ほうせん	hōsen
" *{adj} [sci-t]*	正常の	seijō no
normal chain *[ch]*	直鎖	choku'sa
normal coordinates *[mech]*	基準座標	kijun-zahyō
normal curve *[stat]*	正規曲線	seiki-kyoku'sen
normalcy; normalness *{n}*	正常; せいじょう	seijō
normal dielectric *[mat]*	常誘電体	jō-yūden'tai
normal dispersion *[phys]*	(正)常分散	(sei)jō-bunsan

normal distribution *[stat]*	正規分布	seiki-bunpu
normal force *[phys]*	垂直力	suichoku-ryoku
normal grain *[phys]*	正常粒	seijō-ryū
normal image *[opt]*	正像; せいぞう	seizō
normality *{n} [ch]*	規定度	kitei-do
" *{n} [ch]*	正規	seiki
" (normal state) *{n}*	常態; じょうたい	jōtai
normalization *[comput]*	規格化	kikaku'ka
" *[sci-t]*	正規化	seiki'ka
normalization factor *[photo]*	規格化係数	kikaku'ka-keisū
normalizing *{n} [met]*	焼準(し)	yaki'narashi
normal saline *[physio]*	生理食塩水	seiri-shoku'en-sui
normal salt *[ch]*	正塩; 正鹽	sei-en
normal segregtion *[met]*	正常偏析	seijō-henseki
normal solution *[ch]*	規定液; きていえき	kitei-eki
normal stress *[mech]*	法線応力	hōsen-ō'ryoku
normal structure *[sci-t]*	標準組織	hyōjun-so'shiki
normal temperature *[thermo]*	常温; じょうおん	jō'on
normal vibration *[phys]*	基準振動	kijun-shindō
north *[geod]*	北; 北; きた	kita
northerly wind; northerlies *[meteor]*	北風	kita-kaze
Northern Cross *[constel]*	北十字座	kita-jūji-za
North Pole *[geogr]*	北極; ほっきょく	hokkyoku
nose *{n} [anat]*	鼻; はな	hana
nosedrop; nasal drop *{n} [pharm]*	点鼻剤	tenbi-zai
nosology *[med]*	疾病分類学	shippei-bunrui'gaku
nostril *{n} [anat]*	鼻孔	bikō
NOT *[comput]*	否定; ひてい	hitei
not *{adv}*	でない	de'nai
NOT-AND *[comput]*	否定論理積	hitei-ronri-seki
NOT-AND element *[comput]*	否定論理積素子	hitei-ronri'seki-soshi
NOT-AND operation *[comput]*	否定論理積演算	hitei-ronri'seki-enzan
notation *[comput]*	記示法	kiji-hō
" *[math]*	記法	kihō
notch *{n} [eng]*	切欠き; きりかき	kiri'kaki
notched part *[eng]*	切欠(け)部	kiri'kake-bu
note *{n} [comm]*	覚書; おぼえがき	oboe'gaki
" (musical notation) *{n} [music]*	音符	onpu
" *{n} [pr]*	註; 注; ちゅう	chū
notebook; account book *[pr]*	帳面	chōmen
notebook; memo pad *[pr]*	手帳; 手帖	te'chō
not greater than *[math]*	以下; いか	ika

notice; notification *{n} [comm]*	通告	tsūkoku
notification; notice *{n} [comm]*	通知；つうち	tsūchi
" *{n} [comm]*	告示	koku'ji
notochord *[v-zoo]*	脊索；せきさく	seki'saku
NOT-OR *[comput]*	否定論理和	hitei-ronri-wa
noun *[gram]*	名詞	meishi
nova (pl novas, novae) *[astron]*	新星	shinsei
novelness *{n} [pat]*	新規性；しんきせい	shinki-sei
November (month)	十一月	jūichi'gatsu
now *{adv}*	今；いま	ima
noxious insect *[i-zoo]*	病害虫	byōgai'chū
noxious substance *[med]*	有害物質	yūgai-busshitsu
nozzle plate *[eng]*	紡糸口金	bōshi-kuchi'gane
nuclear adiabatic demagnetization *[phys]*	核断熱消磁	kaku-dan'netsu-shōji
nuclear astrophysics *[astron]*	天体核物理学	tentai-kaku-butsuri'gaku
nuclear breeder reactor *[nucleo]*	核増殖炉；核増殖爐	kaku-zōshoku-ro
nuclear chemistry *[atom-phys]*	核化学；核化學	kaku-kagaku
nuclear collision *[nuc-phys]*	核の衝突	kaku no shōtotsu
nuclear emulsion *[nucleo]*	核乳剤；核乳劑	kaku-nyūzai
nuclear energy *[nucleo]*	核エネルギー	kaku-enerugī
nuclear fission *[nuc-phys]*	核分裂	kaku-bun'retsu
nuclear force *[nuc-phys]*	核力；かくりょく	kaku-ryoku
nuclear fuel *[nucleo]*]	核燃料	kaku-nenryō
nuclear fusion *[nuc-phys]*	核融合	kaku-yūgō
nuclear induction *[phys]*	核磁気誘導	kaku-jiki-yūdō
nuclear magnetic resonance; NMR *[phys]*	核磁気共鳴	kaku-jiki-kyōmei
nuclear magnetism *[phys]*	核磁気；核磁氣	kaku-jiki
nuclear membrane *[cyt]*	核膜	kaku-maku
nuclear physics *[phys]*	核物理学	kaku-butsuri'gaku
nuclear radius *[nuc-phys]*	核半径；核半徑	kaku-hankei
nuclear reaction *[nuc-phys]*	核反応；核反應	kaku-hannō
nuclear reactor *[nucleo]*	原子炉；原子爐	genshi-ro
nuclear relaxation *[phys]*	核緩和	kaku-kanwa
nuclear resonance *[nuc-phys]*	核共鳴	kaku-kyōmei
nuclear warhead *[ord]*	原子(核)弾頭	genshi(-kaku)-dantō
nuclear yield *[nucleo]*	核出力	kaku-shutsu'ryoku
nuclease *[bioch]*	核酸分解酵素	kakusan-bunkai-kōso
nucleate *[bioch]*	核酸塩；核酸鹽	kakusan-en
nucleating agent *[crys]*	造核剤	zō'kaku-zai
nucleation *[ch]*	核形成	kaku-keisei
" *[ch]*	核生成	kaku-seisei

nucleic acid *[bioch]*	核酸；かくさん	kaku-san
nucleolytic enzyme *[bioch]*	核酸分解酵素	kakusan-bunkai-kōso
nucleon *[phys]*	核子	kaku'shi
nucleonics *[eng]*	原子核工学	genshi'kaku-kōgaku
" *[eng]*	核工学	kaku-kōgaku
nucleophilic reagent *[an-ch]*	求核試薬	kyūkaku-shiyaku
nucleus (pl nuclei, -es) *[astron]* *[cyt]* *[nuc-phys]* *[sci-t]*	核；かく	kaku
nucleus *[comput]*	中核	chūkaku
nuclide *[nuc-phys]*	核種	kakushu
Nudibranchia *[i-zoo]*	裸鰓類；らさいるい	rasai-rui
nugget *[geol]*	塊金	kaikin
null *{n}* *[math]*	空；から	kara
" *{n}* *[math]*	空白	kūhaku
" *{adj}* *[legal]*	無効の；むこうの	mukō no
null hypothesis *[stat]*	帰無仮説；歸無假說	kimu-kasetsu
null sets *[math]*	零集合	rei-shūgō
null state; nought state *[math]*	零状態	zero-jōtai
null string *[comput]*	空列	kūretsu
null system *[math]*	零系；れいけい	reikei
number *{n}* *[arith]* *[comput]*	番；ばん	ban
" *{n}* *[arith]* *[comput]*	番号；番號	bangō
" *{n}* *[arith]* *[comput]*	数；數；かず	kazu
" *{n}* *[arith]* *[comput]*	数；數；すう	sū
" *{n}* *[arith]* *[comput]*	数字	sūji
number-average degree of polymerization *[poly-ch]*	数平均重合度	sū-heikin-jūgō'do
number-average molecular weight *[ch]*	数平均分子量	sū-heikin-bunshi'ryō
number-average molecular weight *[ch]*	数平均分子量	kazu-heikin-bunshi'ryō
number line *[an-geom]*	数直線	sū-choku'sen
number of plates *[p-ch]*	段数；だんすう	dansū
number of stages *[p-ch]*	段数	dansū
number of steps *[an-ch]*	段数	dansū
number of teeth (of a gear) *[mech]*	歯数；齒數；はかず	ha-kazu
number of threads *[tex]*	条数；じょうすう	jōsū
number of tsubo; acreage *[math]*	坪数	tsubo-sū
number sign *[comput]*	番号記号	bangō-kigō
number system *[math]*	基数法	kisū-hō
numeral *[arith]*	数；數；すう	sū
" *[arith]*	数字	sūji
" *[gram]*	数詞	sūshi
numerator *[arith]*	分子；ぶんし	bunshi

English	Japanese	Romanization
numerical analysis *[math]*	数値解析	sūchi-kaiseki
numerical aperture *[opt]*	開口率	kaikō-ritsu
" *[opt]*	開口数	kaikō-sū
numerical control *[comput]*	数値制御	sūchi-seigyo
numerical control system *[cont-sys]*	数値制御系	sūchi-seigyo-kei
numerical equation *[math]*	数字方程式	sūji-hōtei'hiki
numerical expression *[math]*	数式；すうしき	sū-shiki
numerical formula *[math]*	数式	sū-shiki
numerical integration *[calc]*	数値積分	sūchi-sekibun
numeric character *[comput]*	数字	sūji
numeric word *[comput]*	数字語	sūji'go
numismatics *{n}*	貨幣学；かへいがく	kahei'gaku
" *{n}*	古銭学；古錢學	kosen'gaku
nurse (woman) *{n} [med]*	看護婦；かんごふ	kango'fu
" (man) *{n} [med]*	看護士	kango'shi
nursery powder *[pharm]*	天花粉；天瓜粉	tenka-fun
nut *[bot]*	堅果	kenka
nutation *[astron]*	章動；しょうどう	shōdō
nutcracker *[eng]*	胡桃割り(器)	kurumi-wari(-ki)
nut grass *[bot]*	浜菅；はますげ	hama-suge
nuthatch (bird) *[v-zoo]*	五十雀	gojū'kara
nutmeg (spice) *[bot]*	肉豆；にくずく	niku'zuku
nutrient *[bio]*	栄養素	eiyō'so
" *[bio]*	養分	yōbun
nutrient agar *[mat]*	普通寒天	futsū-kanten
nutrition *[bio]*	栄養；営養；營養	eiyō
nutritional value *[bio]*	栄養価；榮養價	eiyō-ka
nutrition disorder *[med]*	栄養障害	eiyō-shōgai
nutritionist *[bio]*	栄養学者	eiyō-gakusha
nutritive cell *[cyt]*	栄養細胞	eiyō-saibō
nux vomica *[bot]*	馬銭；まちん	machin
nymph (of insect) *[i-zoo]*	若虫	waka-mushi
nystagmus *[med]*	眼振症	ganshin'shō

O

oak (tree) *[bot]* 樫；かし kashi
" (tree) *[bot]* 柏；槲；かしわ kashiwa
" (Japanese tree) *[bot]* 楢；なら nara
oar; paddle *{n} [eng] [nav-arch]* 櫂 kai
oarfish (fish) *[v-zoo]* 竜宮の使い ryūgū no tsukai
oarlock *{n} [nav-arch]* 櫂受け kai-uke
oat *[bot]* 燕麦；燕麥 en'baku
object *{n} [gram]* 目的語 mokuteki'go
" *{n} [math] [phys]* 対象；對象 taishō
" ; article *{n} [phys]* 物体；物體 buttai
" ; contradict *{vb}* 反対する；反對する hantai suru
objective (lens) *{n} [opt]* 対物鏡；對物鏡 tai'butsu'kyō
" (lens) *{n} [opt]* 対物レンズ tai'butsu-renzu
" (aim) *{n}* 目的；もくてき mokuteki
" *{adj}* 客観的な kyakkan'teki na
objective case *[gram]* 目的格 mokuteki-kaku
objective function *[alg]* 目的関数 mokuteki-kansū
object language *[comput]* 目的言語 mokuteki-gengo
object point *[opt]* 物点；物點 butten
oblanceolate *{adj} [bot]* 倒皮針形の tō-hishin'kei no
oblateness *[astron]* 偏平率 henpei-ritsu
obligate aerobe *[bio]* 偏性好気性菌 hensei-kōki'sei-kin
obligate anaerobe *[bio]* 偏性嫌気性菌 hensei-kenki'sei-kin
" *[bio]* 絶対嫌気性生物 zettai-kenki'sei-seibutsu
oblique *{n} [math]* 斜；はす；ななめ hasu; naname
oblique angle *[geom]* 斜角；しゃかく sha-kaku
oblique coordinates *[trig]* 斜交座標 shakō-zahyō
oblique (light) ray *[opt]* 斜光線 sha-kōsen
oblique prism *[geom]* 斜角柱 sha-kakuchū
oblique triangle *[geom]* 非直角三角形 hi'chokkaku-sankaku-kei
obliteration *[math]* 抹殺；まっさつ massatsu
" *[math]* 抹消 masshō
oblong *{n} [geom]* 長方形 chōhō'kei
obovate *{adj} [bio] [geom]* 倒卵形の tōran'kei no
obsequent *{adj} [geol]* 逆従の；逆從の gyaku'jū no
observability *[cont-sys]* 可観測性 ka'kansoku-sei
observation *[math]* 観測；觀測 kan'soku
" *{n}* 観察 kan'satsu
observation car *[rail]* 展望車 tenbō'sha

observation tower *[arch]*	展望塔	tenbō-tō
observatory *[arch]*	観測所	kansoku-jo
" *[arch] [astron]*	天文台；てんもん台	tenmon-dai
observed value *[math]*	観察値	kan'satsu-chi
obsidian *[geol]*	黒曜石；黑曜石	kokuyō'seki
" *[geol]*	十勝石；とかちいし	to'kachi-ishi
obsolete *{adj}*	廃れた；すたれた	sutareta
obstacle; obstruction; impedance *{n}*	障害物	shōgai'butsu
obstruct; intercept; block *{vb t}*	遮る；さえぎる	saegiru
obstruction *[med]*	閉塞	heisoku
obtuse angle *[geom]*	鈍角；どんかく	don'kaku
obtuse triangle *[geom]*	鈍角三角形	donkaku-sankak(u)kei
obverse; reverse *{n} [math]*	裏；うら	ura
Occident; West *[geogr]*	西洋；せいよう	seiyō
Occidental cuisine *[cook]*	西洋料理	seiyō-ryōri
occlusion *[anat]*	咬合；こうごう	kōgō
" *[meteor]*	閉塞；へいそく	heisoku
" *[phys]*	吸蔵；吸藏	kyūzō
occultation *[astron]*	掩蔽；えんぺい	enpei
" (lunar) *[astron]*	星食；星蝕	seishoku
occupancy rate *{n}*	占積率	senseki-ritsu
occupation; title of office *{n}*	職名；しょくめい	shoku'mei
occupational disease *[med]*	職業病	shoku'gyō'byō
ocean; sea *[geogr]*	海洋；かいよう	kaiyō
" *[geogr]*	海；うみ	umi
ocean current *[ocean]*	海流；海流	kairyū
ocean floor; ocean bed *[ocean]*	海底	kaitei
oceanic noise *[acous]*	海鳴り；うみなり	umi-nari
ocean lane *[navig]*	遠洋航路	en'yō-kōro
ocean liner *[nav-arch]*	遠洋定期船	en'yō-teiki'sen
oceanography *[geophys]*	海洋学	kaiyō'gaku
ocean perch; rockfish (fish) *[v-zoo]*	目抜；めぬけ	me'nuke
ocean sound; ocean roar *[acous]*	潮騒；しおさい	shio'sai
ocean with high waves *[ocean]*	灘；なだ	nada
ocher; Chinese yellow *[miner]*	黄色土	kōshoku-do
" *[miner]*	黄土；おうど	ō'do
-o'clock *{suffix} [mech]*	-時	-ji
octagon *[geom]*	八辺形；八邊形	hachi'hen-kei
" *[geom]*	八角形	hakkaku-kei
octahedron (pl -s, -hedra) *[geom]*	八面体	hachi'men'tai
octal *{n} [comput]*	八進；はっしん	hasshin
octal notation *[math]*	八進表記	hasshin-hyōki

octal number *[math]*	八進数	hasshin-sū
octane number *[eng]*	オクタン価	okutan-ka
octavalent *[ch]*	八価の	hakka no; hachi'ka no
octavo (pl -vos) *{n}* *[pr]*	八つ折り版	yatsu'ori-ban
octet *[atom-phys]*	八隅子	hachigū'shi
" *[atom-phys]*	八重子	hachijū'shi
October (month)	十月	jū'gatsu
octode *[electr]*	八極管	hakkyoku-kan
octopus (pl -puses, -pi) *[i-zoo]*	蛸；章魚；たこ	tako
ocular *[opt]*	接眼レンズ	setsu'gan-renzu
" *[opt]*	対眼レンズ	taigan-renzu
odd; abnormal; illogical *{adj}* *{n}*	おかしい	okashī
odd-even check *[comput]*	奇偶検査	kigū-kensa
odd function *[alg]*	奇関数；奇關數	ki-kansū
odd number *[arith]*	奇数；きすう	kisū
odds-on-favorite *{n}*	本命	honmei; honmyō
odometer *[eng]*	走行距離計	sōkō-kyori-kei
odontoblast *[hist]*	象牙質；ぞうげしつ	zōge'shitsu
odor *[physio]*	臭い；におい	nioi
" *[physio]*	臭気	shūki
odorant *[mat]*	発臭剤；發臭劑	hasshū-zai
odor-correcting agent *[pharm]*	矯臭剤	kyōshū-zai
odoriferous *{adj}* *[physio]*	香い(臭い)を放つ	nioi o hanatsu
odorimeter *[eng]*	嗅覚計	kyūkaku-kei
odorizing *{n}* *[ch-eng]*	臭い付け	nioi'zuke
odorometer *[eng]*	臭度計	shūdo-kei
oenology *[cook]*	葡萄酒学	budōshu'gaku
of *{prep}*	の	no
off *[elec]*	開；かい	kai
office automation *[comput]*	事務処理の機械化	jimu-shori no kikai'ka
office hours *[econ]*	営業時間	eigyō-jikan
office instrument *[eng]*	事務機器	jimu-kiki
officer (of company)	役員	yaku'in
office work *[econ]*	事務	jimu
official *{adj}*	公定の	kōtei no
" *{adj}*	公の	ōyake no
official gazette *[pat]*	官報；かんぽう	kanpō
offline *{n}* *[comput]*	非直結	hi'chokketsu
off odor; off flavor *[cook]*	異臭	i'shū
offset *{n}* *[pr]*	裏移り	ura-utsuri
" *{vb}*	見合う	mi'au
" *{vb}*	ずらす	zurasu

English	Japanese	Romanization
offshore {n} [geol]	沖合；おきあい	oki'ai
offshore fisheries [ecol]	沖合漁業	oki'ai-gyogyō
offshore island [geogr]	沖合島	oki'ai-tō
offstreet parking {n} [civ-eng]	路外駐車	rogai-chūsha
ogival section [arch]	弓形断面	yumi'gata-danmen
oil {n} [geol]	油；あぶら；ゆ	abura; yu
oil absorptiveness [mech-eng]	吸油度	kyūyu-do
oil bath [eng] [met]	油浴；ゆよく	yu-yoku
oil cake; oil meal [agr]	油粕；油糟	abura-kasu
oil chemistry [ch]	油化学；脂化學	yu-kagaku
oil cleaner; oil purifier [eng]	油清浄器；油清淨器	abura-seijō-ki
oil cloth [tex]	油布；ゆふ	yufu
oil consumption [econ]	油消費量	abura-shōhi'ryō
oil content [ch-eng]	含油率	gan'yu-ritsu
oil dipstick [eng]	油面測定棒	yumen-sokutei-bō
oiler (ship) [nav-arch]	油槽船	yusō'sen
oil-extended rubber [rub]	油展ゴム	yuten-gomu
oil feeder; oil can [eng]	油差し	abura-sashi
oil field [petrol]	油田；ゆでん	yuden
oil film; oil slick [petrol]	油膜	yu-maku
oil gage [eng]	油面計	yumen-kei
oil gland (uropygial gland) [v-zoo]	油腺	yusen
oil-hardened steel [steel]	油焼入鋼	abura-yaki'ire-kō
oil hole [eng]	油穴；あぶらあな	abura-ana
oil hydraulics [fl-mech]	油圧工学；油壓工學	yu'atsu-kōgaku
oil immersion [elec] [opt]	油浸	yu-shin
oiliness {n} [eng]	油性	yusei
oiling {n} [mech-eng]	給油	kyūyu
oil-in-water [ch]	水中油	suichū-yu
oil length [mat]	油長；ゆちょう	yu'chō
oil level indicator [eng]	油面表示器	yumen-hyōji-ki
oil-modified resin [poly-ch]	油変性樹脂	abura-hensei-jushi
oil paint [mat]	油性塗料	yusei-toryō
oil painting {n} [graph]	油絵；あぶらえ	abura-e
oil paints; oil colors [mat]	油絵具；油繪具	abura-enogu
oil-paper sash [agr]	油障子	abura-shōji
oil pipe [eng]	油管	abura-kuda; yukan
oil press [agr] [eng]	搾油機；さくゆき	saku'yu-ki
oil pressure [mech]	油圧	yu'atsu
oilproof; oil resistant {adj} [mat]	耐油の	tai'yu no
oil quenching {n} [met]	油焼入れ	abura-yaki'ire
" {n} [met]	油冷；ゆれい	yu-rei

oil refinery *[ch-eng]*	製油所	seiyu'jo
oil repellency *[eng]*	撥油性	hatsu'yu-sei
oil sand *[geol]*	油砂；ゆさ	yusa
oil shale *[geol]*	油(母)頁岩	yu(mo)-ketsu'gan
oil soaking *{n} [met]*	浸油	shin-yu
oil-soluble *{n} [mat]*	油溶性	yuyō-sei
oil-soluble dye *[org-ch]*	油溶染料	yuyō-senryō
oil-soluble resin *[org-ch]*	油溶性樹脂	yu'yōsei-jushi
oilstone *[mat]*	油砥石	abura-to'ishi
oil-sugar *[pharm]*	油糖剤	yutō-zai
oil sump *[eng]*	油溜め；あぶらだめ	abura-dame
oiltightness test *[ind-eng]*	油密試験	yu'mitsu-shiken
oil varnish *[mat]*	油性塗料	yusei-toryō
oil well *[petrol]*	油井；ゆせい	yusei
oil-well digging (drilling) *[petrol]*	石油採掘	seki'yu-sai'kutsu
oil-well excavation *[petrol]*	油井掘削	yusei-kussaku
oily matter *[mat]*	油状物質	yujō-busshitsu
oily perfume *[mat]*	油性香料	yusei-kōryō
ointment *[pharm]*	軟膏	nankō
old *{adj}*	古い；旧い；ふるい	furui
oleaginous (oleagenous) *{adj} [bio]*	油性の	yusei no
" *{adj} [bio]*	油質の	yu'shitsu no
oleander (flower) *[bot]*	夾竹桃	kyōchiku'tō
oleaster *[bot]*	茱；胡頽子；ぐみ	gumi
oleophilicity *[ch]*	親油度	shin'yu-do
oleoresin *[mat]*	含油樹脂	gan'yu-jushi
olfaction *[physio]*	嗅覚；きゅうかく	kyūkaku
olfactometer *[eng]*	嗅覚計	kyūkaku-kei
olfactory cell *[physio]*	嗅細胞；嗅細胞	kyū-saibō
olibanum *[mat]*	乳香	nyūkō
oligo-; olig- *{prefix} [sci-t]*	小さい-	chīsai-
" *{prefix} [sci-t]*	少数の-	shōsū no-
Oligochaeta *[i-zoo]*	貧毛類	hinmō-rui
oligoclase *[miner]*	灰曹長石	kaisō'chōseki
oligosaccharide *[bioch]*	少糖	shōtō
olivenite *[miner]*	緑砒銅鉱；緑砒銅鑛	ryokuhi'dōkō
olive shell (shellfish) *[i-zoo]*	枕貝；まくらがい	makura-gai
olivine *{n} [miner]*	橄欖石	kanran'seki
ombre dyeing *{n} [tex]*	暈し染め	bokashi-zome
omission *[legal]*	不作爲；ふさくい	fu'saku'i
" ; falling away *{n}*	脱落	datsu'raku
" ; abridgment *{n}*	省略	shō'ryaku

omni- *{prefix}*	総ての-	subete no-
" *{prefix}*	全-	zen-
omnidirectional *{adj} [electr]*	全方向の	zen-hōkō no
omnivore *{n} [zoo]*	雑食性動物	zasshoku'sei-dōbutsu
omnivorous *{adj} [zoo]*	雑食性の	zasshoku'sei no
on *[elec]*	閉; へい	hei
once every two weeks *{adv}*	二週に一回の	nishū ni ikkai no
once more; again *{adv}*	再度	saido
oncogene *[gen]*	癌遺伝子	gan-idenshi
" *[gen]*	腫瘍遺伝子	shuyō-idenshi
oncogenesis *[med]*	腫瘍発生	shuyō-hassei
one *{n} [math]*	一; 壱	ichi
one "bu" (= 0.119 in.) *[mech]*	一分; いちぶ	ichi-bu
one by one; in detail *{adv}*	逐一; ちくいち	chiku-ichi
one century *{n}*	一世紀; いっせいき	isseiki
one-component adhesive *[poly-ch]*	一液性接着剤	ichi'eki'sei-setchaku-zai
one-eighth *[math]*	八分の一	hachi'bun no ichi
one fathom (= 6 feet) *[mech]*	一尋; ひとひろ	hito-hiro
one-fifth *[math]*	五分の一	go'bun no ichi
one-fourth *[arith]*	四分の一	yon'bun no ichi; shi'bun no ichi
one "gō" (= 0.381 pint) *[mech]*	一合	ichi-gō
one-half *[math]*	半分; はんぶん	hanbun
" *[math]*	二分の一	ni'bun no ichi
one hour	一時間	ichi-jikan
oneirology *[psy]*	夢学; 夢學	yume'gaku
one "jō" (= 3.314 yds.) *[mech]*	一丈; いちじょう	ichi-jō
one "ken" (= 5.965 ft.) *[mech]*	一間; いっけん	ikken
one layer *{n} [tex]*	一重	hito'e
one minute *[mech]*	一分; いっぷん	ippun
one-ninth *[math]*	九分の一	ku'bun no ichi
one o'clock (time)	一時	ichi-ji
one part; one section *[math]*	一部	ichi-bu
one person *{n}*	一人; ひとり	hitori
one-quarter *[arith]*	四分の一	yon'bun no ichi; shi'bun no ichi
one room *[arch]*	一間; ひとま	hito-ma
one's complement *[comput]*	一の補数	ichi no hosō
one-seventh *[math]*	七分の一	shichi'bun no ichi; nana'bun no ichi
one "shaku" (= 0.994 ft) *[mech]*	一尺; いっしゃく	isshaku
one "shō" (= 0.447 gallon) *[mech]*	一升; いっしょう	isshō

one-side-alloyed steel sheet *[steel]*	片面合金化鋼板	katamen-gōkin'ka-kōban
one-sixth *[math]*	六分の一	roku'bun no ichi
one skilled in the art *{n}*	当業者；當業者	tōgyō'sha
one's self *{n}*	自分	jibun
one-step cure *[rub]*	一段加硫	ichi'dan-karyū
one "sun" (= 1.193 in.) *[mech]*	一寸；いっすん	issun
one-tailed test *[stat]*	片側検定	kata'gawa-kentei
one-tenth *[math]*	十分の一	jū'bun no ichi
" *[math]*	割；わり	wari
one-third *[math]*	三分の一	san'bun no ich
one unit of house *[arch]*	一軒	ikken
one-way communication *[comm]*	単向通信	tankō-tsū'shin
one-way layout *[stat]*	一元配置	ichi'gen-haichi
one-way road *[traffic]*	一方通行路	ippō-tsūkō-ro
onion (vegetable) *[bot]*	玉葱；たまねぎ	tama-negi
online *{n}* *[comput]*	直結	chokketsu
only; merely *{adv}*	たった	tatta
on occasion; at any time *{adv}*	随時(に)；隨時(に)	zuiji (ni)
on one's side *[med]*	横臥位；おうがい	ōga'i
on site *{n}* *[ind-eng]*	現場；げんば	genba
ontogeny *[embryo]*	個体発生；個體發生	kotai-hassei
Onychophora *[i-zoo]*	有爪動物	yūsō-dōbutsu
onyx *[miner]*	縞瑪瑙；しまめのう	shima'menō
oolite; oolith *[petr]*	魚卵石	gyoran'seki
oolithic limestone *[petr]*	魚卵(状)石灰岩	gyoran('jō)-sekkaigan
oolong tea *[bot]* *[cook]*	烏龍茶	ūron-cha
oometer *[agr]* *[eng]*	卵選別機；卵選別機	tamago-senbetsu-ki
Oomycetes *[mycol]*	卵菌類	rankin-rui
opacifier *[mat]*	乳白剤	nyūhaku-zai
opacity *[opt]*	不透明度	fu'tōmei-do
" *[opt]*	乳白度	nyūhaku-do
opalescence *[opt]*	乳光；にゅうこう	nyūkō
" *[opt]*	蛋白光	tanpaku'kō
opaleye (fish) *[v-zoo]*	眼仁奈；めじな	mejina
opaque *{adj}*	不透明な	fu'tōmei na
opaque body *[opt]*	不透明体	fu'tōmei-tai
opaquefier; opalizer *[plas]*	乳白剤	nyūhaku-zai
opaque medium *[opt]*	不透明媒体	fu'tōmei-baitai
open *{n}* *[elec]*	開；かい	kai
" *{adj}*	開いた	aita; hiraita
" *{vb t}*	開ける	akeru
open air *{n}* *[meteor]*	野外；やがい	ya'gai

open circuit *[elec]*	開路	kairo
open circulatory system *[bio]*	開放循環系	kaihō-junkan-kei
open cluster *[astron]*	散開星団	sankai-seidan
open-ended *[computer]*	拡張可能	kakuchō-kanō
open for public inspection *[pat]*	公告; こうこく	kōkoku
open-hearth furnace *[steel]*	平炉; 平爐; へいろ	heiro
open-hearth process *[steel]*	平炉法	heiro-hō
opening (fibers) *{n} [tex]*	開繊	kaisen
" ; mouth *{n}*	口; くち	kuchi
open interval *[an-geom]*	開区間; 開區間	kai-kukan
open pit mining *{n} [min-eng]*	露天採鉱	roten-saikō
open sea *[ocean]*	公海; こうかい	kōkai
open sentence *[alg]*	開放文	kaihō'bun
open system *[eng]*	開放系	kaihō-kei
open universe *[astron]*	開いた宇宙	hiraita uchū
operability *[ind-eng]*	作業性	sagyō-sei
operand *[alg]*	被演算数	hi-enzan'sū
" *[comput]*	演算数	enzan'sū
operate *{vb} [ind-eng]*	操業する	sōgyō suru
operating; operation *{n} [ind-eng]*	運転; うんてん	unten
operating time *[comput]*	動作時間	dōsa-jikan
" *[comput]*	稼動時間	kadō-jikan
operating variable *[math]*	操作変数	sōsa-hensū
operation *[alg] [comput]*	演算; えんざん	enzan
" *[alg]*	運算	unzan
" *[comput]*	操作	sōsa
" *[comput]*	運転	un'ten
" (management) *[econ]*	経営; 經營	kei'ei
" (management) *[econ]*	運営	un'ei
" *[elec] [math]*	動作; どうさ	dōsa
" (work) *{n} [ind-eng]*	稼動	kadō
operational amplifier *[electr]*	演算増幅器	enzan-zōfuku-ki
operational effectiveness *[ind-eng]*	運用有効性	un'yō-yūkō-sei
operational reliability *[ind-eng]*	運用信頼度	un'yō-shinrai-do
operation time *[comput]*	演算時間	enzan-jikan
operator *[comput]*	演算子	enzan'shi
" (a person) *[comput]*	操作員	sōsa'in
operator console *[comput]*	制御卓	seigyo-taku
operator control panel *[comput]*	操作制御盤	sosa-seigyo-ban
operculum; gill cover *[v-zoo]*	鰓蓋	saigai; era-buta
opinion *{n} [psy]*	意見; いけん	iken
" ; thought; idea *{n} [psy]*	考え; かんがえ	kangae

opinion (speculation) *{n}*	思わく	omowaku
" (aim) *{n}*	趣旨; しゅし	shushi
opium *{n} [pharm]*	阿片	a'hen
opossum shrimp *[i-zoo]*	糠蝦; あみ	ami
opponent *{n}*	相手(方)	aite(-kata)
opportunity; chance; occasion *{n}*	機; き	ki
opposable thumb *[anat]*	向かい合わせ=られる親指	mukai-awase-rareru-oyayubi
opposing lane *[traffic]*	対向車線	taikō-shasen
opposing reaction *[ch]*	対向反応; 對向反應	taikō-hannō
opposite *{adj} [bot]*	対生の	taisei no
" *{adj} [math]*	反対の	hantai no
opposite angle *[geom]*	対頂角	taichō-kaku
opposite direction *[math]*	逆方向	gyaku-hōkō
opposite sex *[bio]*	異性	i'sei
opposite side *[geom]*	対辺; 對邊	taihen
opposition *{n} [astron]*	衝; しょう	shō
" *{n} [legal]*	反対; はんたい	hantai
" *{n}*	対向	taikō
" (contrast) *{n}*	対立	tai'ritsu
optical activity *[opt]*	光学活性	kōgaku-kassei
" *[opt]*	旋光性	senkō-sei
optical axis *[opt]*	光軸	kō-jiku
optical bistability *[opt]*	光学的双安定性	kōgaku'teki-sō'antei-sei
optical brightening agent *[ch]*	蛍光増白剤	keikō-zō'haku-zai
optical center *[opt]*	光心	kōshin
optical characteristic *[opt]*	光学特異性	kōgaku-tokui'sei
optical character reader *[comput]*	光学(式)文字読取=装置	kōgaku(shiki)-moji-yomi'tori-sōchi
optical character recognition *[comput]*	光学(式)文字認識	kōgaku(shiki)-moji-nin'shiki
optical communication *[comm]*	光通信	hikari-tsūshin
optical computer *[comput] [opt]*	光学計算機	kōgaku-keisan-ki
optical computing *{n} [comput]*	光学計算	kōgaku-keisan
optical depth *[opt]*	光学的深さ	kōgaku'teki-fuka-sa
optical fiber *[opt]*	光ファイバー	kō-faibā
" *[opt]*	光学繊維	kōgaku-sen'i
optical illusion *[physio]*	錯視; さくし	sakushi
optical instrument *[opt]*	光学機械	kōgaku-kikai
optical isomer *[ch]*	光学異性体	kōgaku-isei'tai
optical isomerism *[p-ch]*	光学異性	kōgaku-i'sei
optical lever *[opt]*	光挺子; ひかりてこ	hikari-teko
optically active *[opt]*	光学活性	kōgaku-kassei

optically active substance *[mat]*	光学活性体	kōgaku-kassei-tai
" *[mat]*	旋光性体	senkō'sei-tai
optical microscope *[opt]*	光学顕微鏡	kōgaku-kenbi'kyō
optical microstructure *[opt]*	光顕組織	kōken-so'shiki
optical oceanography *[ocean]*	光学海洋学	kōgaku-kaiyō'gaku
optical path *[opt]*	光路；こうろ	kōro
optical path difference *[opt]*	光路差	kōro-sa
optical path length *[opt]*	光路長	kōro-chō
optical (photo)micrograph *[opt]*	光学顕微鏡写真	kōgaku-kenbi'kyō-shashin
optical pyrometer *[eng]*	光高温計	hikari-kōon-kei
optical rotation *[opt]*	旋光度；せんこうど	senkō-do
optical rotatory power *[opt]*	旋光性	senkō-sei
optical system *[opt]*	光学系	kōgaku-kei
optical wedge *[opt]*	光学楔	kōgaku-kusabi
optics *[phys]*	光学；こうがく	kō'gaku
optimal color *[opt]*	最明色	saimei-shoku
optimism *[psy]*	楽観；らっかん	rakkan
optimization *[math] [eng]*	最適化	saiteki'ka
optimum (pl -tima) *{n} [math]*	至適	shiteki
" *{adj} [math]*	最適な	saiteki na
optimum temperature *[physio]*	至適温度	shiteki-ondo
option *[comput] [math]*	任意選択；任意選擇	nin'i-sentaku
optional *{adj} [comput] [math]*	任意の；にんいの	nin'i no
optional stop instruction *[comput]*	任意停止命令	nin'i-teishi-meirei
opto- *{prefix}*	視力の-	shi'ryoku no-
optoaccoustic effect *[phys]*	視覚効果	shi'kaku-kōka
optoelectronic integrated circuit *[electr]*	光電子集積回路	kō'denshi-shūseki-kairo
optoelectronics *[electr]*	光電子工学	kō'denshi-kōgaku
OR *[comput]*	論理和	ronri-wa
or *{conj}*	又は	matawa
oral; orally presented *[legal]*	口頭による	kōtō ni yoru
orange (color) *[opt]*	橙色	tō'shoku; daidai-iro
orange flower oil *[mat]*	橙花油	tōka-yu
orangutang (animal) *[v-zoo]*	猩猩；しょうじょう	shōjō
orbit *{n} [astron] [phys] [rail]*	軌道；きどう	kidō
orbital *{n} [atom-phys]*	軌道(関数)	kidō(-kansū)
orbital electron *[atom-phys]*	軌道電子	kidō-denshi
orbit determination *[astron]*	軌道決定	kidō-kettei
orbiter; space orbiter *[aero-eng]*	周回宇宙船	shūkai-uchū'sen
orchard; fruit orchard *[agr]*	果樹園	kaju'en
orchid (flower) *[bot]*	蘭；蘭；らん	ran

Orcinus orca *[v-zoo]*	鯱；しゃち	shachi
OR circuit *[electr]*	論理和回路	ronri'wa-kairo
order *{n} [bio] [syst]*	目；もく	moku
" *{n} [comput]*	命令；めいれい	meirei
" *{n} [comput]*	指令	shirei
" *{n} [gram]*	語順	gojun
" *{n} [math]*	次数；じすう	jisū
" (succession) *{n} [math]*	順位	jun'i
" (sequence) *{n} [math]*	前後	zengo
" *{n}*	注文；ちゅうもん	chūmon
" (place an order) *{vb t}*	注文する	chūmon suru
ordered *{adj} [an-geom]*	並べられた	naraberareta
ordered pair *[an-geom]*	順序対；順序對	junjo-tsui
ordering *{n} [alg]*	順序づけ	junjo'zuke
order of magnitude *[math]*	桁；けた	keta
order of reaction *[ch]*	反応次数；反應次數	hannō-jisū
ordinal number *[arith]*	序数	josū
" *[arith]*	順序数	junjo-sū
ordinance *[legal]*	法令；ほうれい	hōrei
" *[legal]*	令	rei
ordinary *{n}*	普通；ふつう	futsū
" (normal) *{adj}*	普通の	futsū no
ordinary differential equation *[calc]*	常微分方程式	jō'bibun-hōtei'shiki
ordinary grade *{n} [ind-eng]*	並級；なみきゅう	nami-kyū
ordinary ray *[opt]*	常光線	jō-kōsen
ordinary state *{n}*	常態	jōtai
ordinate *[an-geom]*	縦座標	tate-zahyō
ordnance *[eng]*	軍需品	gunju'hin
" *[ord]*	兵器	heiki
ore; paddle *{n} [eng]*	艪；艫；櫓；ろ	ro
ore *[geol]*	鉱石；鑛石	kōseki
ore deposit *[geol]*	鉱床；こうしょう	kōshō
organ *[anat]*	臓器；臟器	zōki
" *[bio]*	器官；器官	kikan
" *[pr]*	機関；機關；きかん	kikan
organ culture *[bot]*	器官培養	kikan-baiyō
organelle *[cyt]*	細胞器官；細胞器官	saibō-kikan
organic *{adj} [org-ch]*	有機の；ゆうきの	yūki no
organic acid *[org-ch]*	有機酸	yūki-san
organic analysis *[org-ch]*	有機分析	yūki-bunseki
organic catalyst *[bio]*	有機触媒；有機觸媒	yūki-shokubai
organic chemistry *[ch]*	有機化学	yūki-kagaku

organic environment *[ecol]*	有機的環境	yūki'teki-kankyō
organic fertilizer *[agr]*	有機質肥料	yūki'shitsu-hiryō
organic peroxide *[org-ch]*	有機過酸化物	yūki-ka'sanka'butsu
organic reagent *[org-ch]*	有機試薬	yūki-shi'yaku
organic substance *[mat]*	有機物	yūki'butsu
organism *[bio]*	生物; せいぶつ	sei'butsu
" *[bio]*	有機体	yūki'tai
organization *[bot]*	体制	taisei
" *[ind-eng]*	組織; そしき	so'shiki
" *{n}*	団体; 團體	dantai
organoarsenic compound *[org-ch]*	有機砒素化合物	yūki-hiso-kagō'butsu
organocopper compound *[org-ch]*	有機銅化合物	yūki-dō-kagō'butsu
organogenesis *[embryo]*	器官発生	kikan-hassei
organolead compound *[org-ch]*	有機鉛化合物	yūki-namari-kagō'butsu
organometallic compound *[org-ch]*	有機金属化合物	yūki-kinzoku-kagō'butsu
organophosphorus compound *[org-ch]*	有機燐化合物	yūki-rin-kagō'butsu
" *[org-ch]*	有機燐剤	yūki'rin-zai
organotransition metal compound *[org-ch]*	有機遷移金属化合物	yūki-sen'i-kinzoku-kagō'butsu
oriel *[arch]*	出窓	de-mado
Orient; The East *[geogr]*	東洋; とうよう	tōyō
Oriental beetle (insect) *[i-zoo]*	背斑黄金虫	se'madara-kogane-mushi
Oriental carpet *[furn]*	東洋絨毯	tōyō-jūtan
Oriental cockroach (insect) *[i-zoo]*	黄金蜚蠊	ko'gane-gokiburi
Oriental medicines *[pharm]*	和漢薬	wakan-yaku
Oriental poppy (flower) *[bot]*	鬼罌粟; おにげし	oni-geshi
Oriental rug *[furn]*	東洋段通	tōyō-dantsū
orientation *[crys] [p-ch]*	配向	haikō
" *[math]*	方位	hō'i
origin *[an-geom]*	原点; 原點	genten
" *[an-geom]*	起点	kiten
" *[comput]*	原; 源; げん	gen
" *[comput]*	基点	kiten
" *{n}*	元; もと	moto
original *[comput] [pr]*	原始	genshi
" *{adj}*	元の	moto no
original drawing *{n} [graph]*	元図; 元圖; もとず	moto-zu
original strain *[microbio]*	原株	gen-kabu
original text *[graph] [pr]*	原文	genbun
Orion; the Hunter *[constel]*	オリオン座	orion-za
ornament *[furn]*	飾り(物)	kazari(-mono)
" *[pr]*	装飾活字	sōshoku-katsu'ji

ornamental carving (antique) *{n}*	根付；ねつけ	ne'tsuke
ornamentation *[furn]*	装飾	sō'shoku
ornithology *[zoo]*	鳥類学；鳥類學	chōrui'gaku
orogenesis; orogeny *[geol]*	造山運動	zōzan-undō
orogenic belt *[geol]*	造山帯	zōzan-tai
orpiment *[miner]*	雄黃；ゆうおう	yū'ō
orthocenter (triangle) *[geom]*	垂心	sui'shin
orthoclase *[miner]*	正長石	sei'chōseki
orthodontics *[med]*	歯列矯正	shi'retsu-kyōsei
orthogonality *[math]*	直交性	chokkō-sei
Orthoptera *[i-zoo]*	直翅類	choku'shi-rui
orthorhombic system *[crys]*	斜方晶形	shahō-shōkei
Osaka *[geogr]*	大阪；おおさか	Ōsaka
oscillate; shake; vibrate *{vb}* *[phys]*	振動する	shindō suru
oscillating board *[mech-eng]*	揺動板；搖動板	yōdō-ban
oscillating radio wave *[elecmg]*	発振電波	hasshin-denpa
oscillation *[phys]*	発振；發振	hasshin
" *[phys]*	振動；しんどう	shindō
oscillator *[comm]* *[electr]*	発振器；發振器	hasshin-ki
Oscillatoria *[bot]*	揺藻；ゆれも	yure'mo
osculating orbit *[astron]*	接触軌道	sesshoku-kidō
osmium (element)	オスミウム	osumyūmu
osmometer *[an-ch]*	浸透圧計	shintō'atsu-kei
osmophore group *[phys]*	発香団；發香團	hakkō-dan
osmosis (pl osmoses) *{n}* *[p-ch]*	浸透；しんとう	shintō
osmotic coefficient *[p-ch]*	浸透係数	shintō-keisū
osmotic pressure *[p-ch]*	浸透圧	shintō-atsu
osprey (bird) *[v-zoo]*	鶚；みさご	misago
ossification *[physio]*	骨化	kokka
ossify *{vb}* *[physio]*	骨化する	kokka suru
Osteichthyes; bony fishes *[v-zoo]*	硬骨魚類	kōkotsu'gyo-ryu
ostrich (bird) *[v-zoo]*	駝鳥；だちょう	dachō
otter (animal) *[v-zoo]*	川獺；かわうそ	kawa'uso
ounce; oz (unit of mass) *[mech]*	オンス	onsu
-ous acid *[ch]*	亜酸；亞酸；あさん	a-san
-ous compound *[ch]*	第一化合物	dai'ichi-kagō'butsu
-ous salt *[ch]*	第一塩	dai'ichi-en
outboard motor *[eng]*	船外機	sengai-ki
outdoor; outdoors *[meteor]*	野外；やがい	yagai
outdoor exposure *[meteor]*	屋外暴露	oku'gai-bakuro
outdoors *[geogr]*	戸外	ko'gai
outer core *[geol]*	外核	gaikaku

outer cover *[bot]*	外皮	gaihi
outer harbor *[geogr]*	外港；外港	gaikō
outer-lined steel pipe *[mat]*	外張鋼管	soto'bari-kōkan
outer packaging *{n} [ind-eng]*	外装；外装	gaisō
outer perimeter *[geom]*	外郭；がいかく	gaikaku
outer planet *[astron]*	外惑星	gai-wakusei
outer side *{n}*	外側	soto-gawa; gai'soku
outer tube; outer cylinder *[mech-eng]*	外筒	gaitō
outgoing line *[comm]*	出線	de-sen
" *[elec]*	引出線	hiki'dashi-sen
outlet *{n}*	出口	de'guchi
outline *{n} [graph]*	輪郭；りんかく	rin'kaku
" (border line) *{n} [graph]*	輪郭線	rinkaku-sen
" *{n} [pr]*	概說；がいせつ	gaisetsu
" *{n} [pr]*	総說；綜說	sōsetsu
" (of history) *{n} [pr]*	大系	taikei
" *{n} [pr]*	大綱	taikō
out-of-plane deformation vibration *[p-ch]*	面外変角振動	mengai-hen'kaku-shindō
out-of-plane vibration *[p-ch]*	面外振動	mengai-shindō
out of print *[pr]*	絶版	zeppan
output *{n} [comput] [electr] [sci-t]*	出力；しゅつりょく	shutsu'ryoku
output device; output unit *[comput]*	出力装置	shutsu'ryoku-sōchi
outside; out front; outdoors *{n}*	表；おもて	omote
outside diameter *[math]*	外径；外徑	gaikei
outside measurement *[math]*	外法(寸法)	soto'nori(-sunpō)
outward flow *[fl-mech]*	外向き流れ	soto'muki-nagare
outward membrane current *[physio]*	外向き膜電流	soto'muki-maku-denryū
ovary (pl -ries) *[anat]*	卵巣；卵巢	ransō
" *[bot]*	子房	shibō
ovate *{adj} [geom]*	卵形の	rankei no
oven *[cook]*	窯；かま	kama
" *[cook] [eng]*	天火；てんぴ	tenbi
" *[eng]*	炉；爐；ろ	ro
over- *{prefix}*	越す-	kosu-
" *{prefix}*	上の-	ue no-
overaging treatment *[met]*	過時効	ka'jikō
overall efficiency *[ind-eng]*	総合効率	sōgō-kō'ritsu
overall length *[mech]*	全長	zen-chō
overall mean *[arith]*	全体平均	zentai-heikin
overall width *[mech]*	全幅	zen-puku
overcast (sky) *[meteor]*	曇り空	kumori-zora

overcoat *[cl]*	オーバー	ōbā
overcoating *{n} [mat]*	上塗り	uwa-nuri
overcure *{n} [poly-ch] [rub]*	過硬化	ka'kōka
overcuring (of resin) *{n} [ch-eng]*	押過ぎ	oshi'sugi
overflow *{n} [civ-eng] [comput]*	溢れ；あふれ	afure
" *{n} [civ-eng]*	溢流	itsu'ryū
" *{n} [sci-t]*	零；こぼれ	kobore
overflow indicator *[comput]*	溢れ標識	afure-hyō'shiki
overglaze color *[cer]*	上絵具；うわえのぐ	uwa-enogu
overhead clearance *[navig]*	頭上余隙	zujō-yo'geki
overhead valve *[mech-eng]*	頭上弁；頭上瓣	tōjō-ben; zujō-ben
overhead wire *[elec]*	架空電線	kakū-densen
overheating *{n} [met]*	焼き過ぎ	yaki'sugi
overlap *{n} [comput]*	平行	heikō
" *{n} [geom] [met]*	重なり	kasanari
overlap angle *[mech-eng]*	重なり角	kasanari-kaku
overlapping *{n} [math]*	重なり	kasanari
overline *{n} [pr]*	上線；うえせん	ue'sen
overload *{n} [electr]*	過負荷	ka'fuka
overlook; scenic overlook *[civ-eng]*	展望台；展望臺	tenbō-dai
overpacking *{n} [poly-ch]*	過剰充填；過剰充塡	kajō-jūten
overpass *{n} [civ-eng]*	高架道路橋	kōka-dōro'kyō
overpopulation *[bio]*	人口過剰	jinkō-kajō
overprint *{n} [pr]*	上重ね印刷	uwa'gasane-in'satsu
overproduction *[ind-eng]*	過剰生産	kajō-seisan
overreach *{n}*	飛越し	tobi'koshi
overshoot(ing) *{n} [elec]*	行過ぎ	yuki'sugi
overtime *[ind-eng]*	残業時間	zangyō-jikan
overtone *{n} [acous]*	倍音	bai'on
" *{n} [acous]*	上音	jō'on
" *{n}* (color) *[opt]*	上色；うわいろ	uwa-iro
overture *[music]*	前奏曲	zensō'kyoku
overturn *{vb i}*	引っくり返る	hikkuri-kaeru
overturn; upset; capsize *{vb t}*	引っくり返す	hikkuri-kaesu
overturned *[math]*	逆さ(ま)	sakasa(ma)
overturning; upsetting *{n}*	転倒；轉倒	tentō
overvoltage *[elec]*	過電圧	ka'den'atsu
over-vulcanization *[rub]*	過加硫	ka'karyū
overwhelming *{adj}*	圧倒的な	attō'teki na
oviduct *{n} [anat]*	輸卵管；ゆらんかん	yuran-kan
ovipositor *[i-zoo]*	産卵管	sanran-kan
ovoviviparism *[v-zoo]*	卵胎生	ran-taisei

ovule *[bot]*	胚珠	haishu
" *[cyt]*	卵細胞；卵細胞	ran-saibō
ovum *[bio]*	卵	ran
owl (bird) *[v-zoo]*	梟；ふくろう	fukurō
owner *[comput]*	親；おや	oya
" *[comput]*	所有者	shoyū'sha
oxalate *[org-ch]*	蓚酸塩	shūsan-en
oxalic acid	蓚酸；しゅうさん	shūsan
oxalic anhydride	無水蓚酸	musui-shūsan
oxalis *[bot]*	酢漿草；かたばみ	katabami
oxidase *[bioch]*	酸化酵素	sanka-kōso
oxidation *[ch]*	酸化；さんか	sanka
oxidation number *[ch]*	酸化数	sanka-sū
oxidation potential *[p-ch]*	酸化電位	sanka-den'i
oxidation-reduction enzyme *[bioch]*	酸化還元酵素	sanka'kangen-kōso
oxidation-reduction indicator *[an-ch]*	酸化還元指示薬	sanka'kangen-shiji-yaku
oxidation-reduction potential *[p-ch]*	酸化還元電位	sanka'kangen-den'i
oxidation resistance *[mat]*	耐酸化性	tai'sanka-sei
oxidative phosphorylation *[bioch]*	酸化的燐酸化	sanka'teki-rinsan'ka
oxidative polymerization *[poly-ch]*	酸化重合	sanka-jūgō
oxide *[ch]*	酸化物	sanka'butsu
oxide film *[ch]*	酸化皮膜	sanka-himaku
" *[ch]*	酸化膜	sanka-maku
oxidize *{vb}* *[ch]*	酸化する	sanka suru
oxidizer; oxidizing agent *[ch]*	酸化剤	sanka-zai
oxidizing flame *[ch]*	酸化炎	sanka-en
oxidoreductase *[bioch]*	酸化還元酵素	sanka-kangen-kōso
oxoacid; oxygen acid *[ch]*	酸素酸	sanso-san
oxygen (element)	酸素；さんそ	sanso
oxygen-free copper *[met]*	無酸素銅	mu'sanso-dō
oxygen point *[thermo]*	酸素点；酸素點	sanso-ten
oxyhydrogen flame *[ch]*	酸水素炎	san'suiso-en
oxyphil *{n}* *[bio]*	好酸性物質	kō'sansei-busshitsu
oxyphile *{n}* *[geoch]*	含酸素鉱生成元素	gan'sanso'kō-seisei-genso
oxy salt; oxide salt *[ch]*	酸化物塩	sanka'butsu-en
oyster (shellfish) *[i-zoo]*	牡蠣；かき	kaki
oyster crab (crustacean) *[i-zoo]*	隠れ蟹；かくれがに	kakure-gani
ozocerite *[geol]*	鉱物蠟；鑛物蠟	kōbutsu-rō
ozone *[ch]*	オゾン	ozon

P

pace; walk *{n}*	歩み	ayumi
Pacific halibut (fish) *[v-zoo]*	大鮃；おひょう	o'hyō
Pacific Ocean *[geogr]*	太平洋	Taihei-yō
pack; cram; stuff *{vb t}*	詰める	tsumeru
packaging; packing *{n} [ind-eng]*	包装；包裝	hōsō
packed column; packed tower *[ch-eng]*	充填塔	jūten-tō
packing *{n} [mat]*	充填物	jūten'butsu
packing machine; baling machine *[eng]*	荷造り機	ni'zukuri-ki
pad character *[comput]*	埋め込み文字	ume'komi-moji
padded welding *{n} [met]*	肉盛溶接	niku'mori-yō'setsu
paddle; blade *{n} [eng]*	羽；はね	hane
paddleboat *[nav-arch]*	外輪船	gairin'sen
page *{n} [pr]*	頁；ページ	pēji
pain *[physio]*	痛み；いたみ	itami
pain and agony *[physio]*	苦痛	kutsū
painfulness *[physio]*	痛さ	ita-sa
paint *{n} [mat]*	ペイント	peinto
" *{n} [mat]*	塗料；とりょう	toryō
paintbrush *[eng]*	塗料刷毛	toryō-bake
painted lady (butterfly) *[i-zoo]*	姫赤立羽	hime-aka-tate'ha
paint film *[mat]*	塗膜	to'maku
painting *{n} [arch]*	塗装	tosō
pair *{n} [math]*	対；對；つい	tsui
pair annihilation *[sci-t]*	対消滅	tsui-shōgen
pair creation *[sci-t]*	対生成	tsui-seisei
-pair(s) (of footwear) *{suffix}*	-足	-soku; -zoku
pale light *{n} [astron]*	薄明り	usu-akari
paleoanthropology *[paleon]*	古人類学	ko-jinrui'gaku
paleobotany *[paleon]*	古植物学	ko-shokubutsu'gaku
paleoclimatology *[geol]*	古気候学；古氣候學	ko-kikō'gaku
Paleolithic Age *[archeo]*	旧石器時代	kyū-sekki-jidai
paleomagnetics *[geophys]*	古地磁気学	ko-chi'jiki'gaku
paleontology *[bio]*	古生物学	ko-seibutsu'gaku
paleozoology *[paleon]*	古動物学	ko-dōbutsu'gaku
palimpsest *[geol]*	原岩形跡	gengan-keiseki
palisade *[geol]*	岩壁；がんへき	ganheki
palladium (element)	パラジウム	parajūmu
pallidness *[opt]*	淡色	tanshoku
palm (of hand) *{n} [anat]*	手の平；掌	te-no-hira

palm; palmetto (pl -tos, -toes) *[bot]*	椰子；やし	yashi
palmately compound leaf *[bot]*	掌状複葉	shōjō-fuku'yō
palm oil; palm butter *[mat]*	パーム油	pāmu-yu
palynology *[paleon]*	花粉学	kafun'gaku
pan; pot *{n} [cook] [eng]*	鍋；なべ	nabe
pan- *{prefix}*	汎-	han-
" *{prefix}*	全-	zen-
Panaeolus papilionaceus (poisonous mushroom) *[bot]*	笑い茸	warai-take
Panax japonicus *[bot]*	栃葉人参	tochi'ba-ninjin
Panax quinquefolium; American ginseng *[bot] [pharm]*	西洋人參	Seiyō-ninjin
Panax rhizome *[bot]*	竹節人參	chiku'setsu-ninjin
panchromatic *{n} [opt]*	全色(性)	zen'shoku(-sei)
" *{adj} [opt]*	全色(性)の	zen'shoku('sei) no
pancreas (pl pancreata) *[anat]*	膵臓；すいぞう	suizō
Pandorina *[bot]*	桑味藻；くわのみも	kuwanomi'mo
panel *[comput]*	盤；ばん	ban
pangolin (animal) *[v-zoo]*	穿山甲	senzan'kō
panicle *[bot]*	円錐花序	ensui-kajo
panning *{n} [min-eng]*	椀掛け；わんがけ	wan'gake
pansy (flower) *[bot]*	三色菫	san'shoku-sumire
panther (animal) *[v-zoo]*	豹；ひょう	hyō
pantograph *[graph]*	縮図器；縮圖器	shuku'zu-ki
paper *{n} [mat] [paper]*	紙；かみ	kami
" (Japanese) *{n} [mat] [paper]*	和紙	wa'shi
paperboard *[mat] [paper]*	ボール紙	bōru-gami
" *[mat] [paper]*	板紙	ita-gami
paper container *[mat]*	紙器；しき	shiki
papercrafted ware *[furn]*	紙細工	kami-zaiku
paper feed mechanism *[comput]*	紙送り機構	kami-okuri-kikō
paper folding (a craft) *{n} [paper]*	折紙；おりがみ	ori-gami
paper folding paper *[paper]*	折紙(用紙)	ori'gami(-yōshi)
paper industry *[ind-eng] [paper]*	紙業	shigyō
Paper Insufficiency *[electr]*	用紙不足	yōshi-fusoku
papermaking *{n} [paper]*	製紙	seishi
papermaking water *[eng] [paper]*	抄造水	shōzō-sui
paper mill *[ind-eng] [paper]*	製紙工場	seishi-kōjō
paper money *[econ]*	紙幣	shihei
paper mulberry *[bot]*	楮；こうぞ	kōzo
paper-passage property *[eng]*	通紙性	tsūshi-sei
paper rectangle (for poetry) *[pr]*	短册；たんざく	tan-zaku

paper space *[pr]*	紙面	shi'men
paper square, colored *[pr]*	色紙；しきし	shiki-shi
paper stock *[mat] [paper]*	原紙	genshi
" *[mat] [paper]*	紙料	shiryō
paper-strength enhancing agent *[mat]*	紙力増強剤	shi'ryoku-zōkyō-zai
paper technology *[paper]*	紙工学	kami-kōgaku
paperweight *[furn]*	文鎮；ぶんちん	bunchin
papilonaceous flower *[bot]*	蝶形花	chōgata-ka
parabola *[an-geom]*	放物線	hōbutsu-sen
parabolic *{adj} [an-geom]*	放物線状の	hōbutsu'sen'jō no
parabolic reflector *[elecmg]*	放物面反射器	hōbutsu'men-hansha-ki
paraboloid *{n} [an-geom]*	放物面	hō'butsu'men
parachute *{n} [aero-eng]*	落下傘；らっかさん	rakka-san
paraffin series *[org-ch]*	パラフィン列	parafin-retsu
paragraph *[comput]*	文節	bun'setsu
" *[comput] [gram]*	段落	dan'raku
" *[pat]*	項；こう	kō
parakeet (bird) *[v-zoo]*	鸚哥；いんこ	inko
parallactic angle *[astron]*	極頂対角	kyoku'chō-tai'kaku
parallactic inequality *[astron]*	月角差；げっかくさ	gekkaku-sa
parallax *[math] [opt]*	視差	shisa
parallel (in parallel) *{n} [elec]*	並列；へいれつ	hei'retsu
" *{n} [geom] [comput]*	平行；へいこう	heikō
" *{adj} [geom]*	平行な	heikō na
parallel circuit *[elec]*	並列回路	hei'retsu-kairo
parallel crosses *[graph]*	井桁；いげた	i'geta
parallelepiped *[geom]*	平行六面体	heikō-rokumen'tai
parallel flow *[elec]*	並流	heiryū
parallel lines *[geom]*	平行線	heikō-sen
parallelogram *[geom]*	平行四辺形	heikō-shihen'kei
parallel planes *[geom]*	平行面	heikō-men
parallel processing *{n} [comput]*	平行処理；平行處理	heikō-shori
parallel resonance *[elec]*	並列共振	hei'retsu-kyōshin
parallel transmission *[comput]*	並列伝送；並列傳送	hei'retsu-densō
paralyzant *{n} [pharm]*	麻痺薬	mahi-yaku
paramagnetic *{adj} [elecmg]*	常磁性の	jōji'sei no
" *{adj} [elecmg]*	正磁気の	sei'jiki no
paramagnetic crystal *[elecmg]*	常磁性結晶	jōjisei-kesshō
paramagnetic resonance *[phys]*	常磁性共鳴	jōjisei-kyōmei
paramagnetic substance *[mat]*	常磁性体	jōjisei-tai
paramagnetism *[elecmg]*	常磁性	jōji-sei
paramecium (pl -cia) *[i-zoo]*	草履虫；ぞうりむし	zōri-mushi

parameter *[comput]*	媒介；ばいかい	baikai
" *[crys]*	軸標；じくひょう	jiku'hyō
" *[stat]*	母数；ぼすう	bosū
" *[stat]*	助変数；助變數	jo'hensū
paramorphism *[miner]*	同質異像仮像	dō'shitsu-izō-kazō
para rubber tree *[bot]*	パラゴムの木	paragomu no ki
parasexual hybridization *[gen]*	擬似有性的交雑	giji-yūsei'teki-kōzatsu
parasite *[bio]*	寄生生物	kisei-seibutsu
" *[bio]*	寄生体；寄生體	kisei'tai
parasitic *{adj}* *[bio]*	寄生性の	kisei'sei no
parasiticide *{n}* *[pharm]*	殺寄生虫薬	sak'kiseichū-yaku
parasitism *[ecol]*	寄生；きせい	kisei
parasympathetic nervous system *[anat]*	副交感神経系	fuk(u)kōkan-shinkei-kei
parathyroid gland *[anat]*	副甲状腺	fuku'kōjō-sen
parcel; small baggage *{n}* *[trans]*	小荷物；こにもつ	ko-ni'motsu
parchment; vellum *[mat]*	羊皮紙；ようひし	yōhi-shi
parent *{n}* *[bio]* *[comput]*	親；おや	oya
parent cell *[cyt]*	親細胞；親細胞	oya-saibō
parent character; root character *[pr]*	親文字；おやもじ	oya-moji
parenthesis (pl -ses) *[arith]* *[pr]*	括弧；かっこ	kakko
" (round) *[arith]* *[pr]*	小括弧	shō-kakko
parent material *[ind-eng]*	母材	bozai
parent strain *[microbio]*	親株；おやかぶ	oya-kabu
parhelion; mock sun *[astron]*	幻日；げんじつ	gen'jitsu
parity *[comput]*	奇偶性	kigū-sei
" *[math]*	等価；等價；とうか	tōka
" *[phys]*	偶奇性	gūki-sei
park (for recreation) *{n}* *[civ-eng]*	公園；こうえん	kōen
parking (of vehicle) *{n}* *[traffic]*	停車；ていしゃ	teisha
parking area *[civ-eng]*	駐車場	chūsha'jō
parking light *[elec]*	駐車灯；駐車燈	chūsha-tō
parlor; reception room *[arch]*	応接間；應接間	ō'setsu-ma
parquet flooring *[arch]*	寄木細工の床	yose'gi-zaiku no yuka
parquetry; marquetry *{n}* *[arch]* *[furn]*	寄木；よせぎ	yose'gi
parrot (bird) *[v-zoo]*	鸚鵡；おうむ	ōmu
parsley (vegetable) *[bot]*	芹；せり	seri
part; parts; component(s) *{n}* *[ind-eng]*	部品；ぶひん	buhin
part; portion *{n}*	部(分)	bu(-bun)
part drawing *{n}* *[graph]*	部品図；部品圖	buhin-zu
parthenogenesis *[i-zoo]*	単為生殖；單爲生殖	tan'i-seishoku
partial carry *[comput]*	部分桁上げ	bubun-keta'age
partial condensation *[ch]*	分縮；ぶんしゅく	bunshuku

partial derivative *[calc]*	偏導関数；偏導關數	hen-dō'kansū
partial differentiation *[math]*	偏微分；へんびぶん	hen-bibun
partial dislocation *[met]*	部分転位；部分轉位	bubun-ten'i
partial eclipse *[astron]*	部分食；部分蝕	bubun-shoku
partial fractions *[calc]*	部分分数	bubun-bunsū
partially finished goods *[ind-eng]*	仕掛け品	shi'kake'hin
partially hydrogenated oil *[mat]*	半硬化油	han'kōka-yu
partial pressure *[phys]*	分圧；分壓	bun'atsu
partial pressure, law of *[ch-eng]*	分圧の法則	bun'atsu no hōsoku
partial tone *[acous]*	部分音；ぶぶんおん	bubun-on
partial view *[graph]*	部分図；部分圖	bubun-zu
participial adjective *[gram]*	連体形；連體形	rentai-kei
particle; grain *{n} [mat] [phys]*	粒子；りゅうし	ryūshi
particle accelerator *[nucleo]*	粒子加速器	ryūshi-ka'soku-ki
particle diameter *[crys] [geol]*	粒径；粒徑	ryūkei
particle physics *[phys]*	素粒子物理学	so'ryūshi-butsuri'gaku
particle size *[mech]*	粒度；りゅうど	ryūdo
particle size distribution *[eng]*	粒度分布	ryūdo-bunpu
particle-unifying effect *[phys]*	整粒効果	seiryū-kōka
particulars; details *{n} [comm]*	詳細；しょうさい	shōsai
particular solution *[math]*	特殊解	toku'shu-kai
particulate *{adj} [phys]*	粒子状の	ryūshi'jō no
parting compound; parting agent *[mat]*	離型剤	rikei-zai
parting sand *[met]*	別れ砂；わかれずな	wakare-zuna
partition *{n} [comput]*	区分；區分	kubun
" *{n} [comput]*	区画；區畫；くかく	ku'kaku
partition coefficient *[an-ch]*	分配係数	bunpai-keisū
partition function *[stat-mech]*	分配関数	bunpai-kansū
partition, law of *[phys]*	分配の法則	bunpai no hōsoku
partnership *[econ]*	合名会社；合名會社	gōmei-gaisha
part(s) of speech *[gram]*	品詞	hin'shi
parturition; childbirth *[med]*	分娩；ぶんべん	bunben
parure *{n} [lap]*	一揃いの宝石	hito'soroi no hōseki
passage orifice *[eng]*	通過口	tsūka-kō
passageway; hallway *[arch]*	廊下；ろうか	rōka
passageway *[traffic]*	通路	tsūro
passenger *[trans]*	乗客；じょうきゃく	jō'kyaku
" *[trans]*	旅客	ryo'kaku; ryo'kyaku
passenger airplane *[aero-eng]*	旅客機	ryokak(u)ki
passenger car *[mech-eng]*	乗用車；乘用車	jōyō'sha
" *[rail]*	客車	kyaku'sha
passenger facilities *[trans]*	旅客設備	ryokaku-setsubi

passenger ship *[nav-arch]*	旅客船	ryokaku'sen
passenger ticket *[trans]*	乗車券；乘車券	jōsha-ken
passionflower; Passiflora coerulea (flower) *[bot]*	時計草；とけいそう	tokei'sō
passivation *[met]*	不働態化	fudōtai'ka
passive *{n} [microbio] [psy]*	受身；うけみ	uke'mi
" *{adj} [math]*	受動の	judō no
" (negative) *{adj}*	消極的な	shōkyoku'teki na
passive immunity *[immun]*	受身免疫	uke'mi-men'eki
passive state *[met]*	不働態；不動態	fudō-tai
passive transport *[cyt]*	受動輸送	judō-yusō
passive voice *{n} [gram]*	受動態	judō-tai
passport *[trans]*	旅券；りょけん	ryoken
password *[comm]*	合言葉；あいことば	ai'kotoba
past; in the past *{adv}*	過去	kako
paste *{n} [mat]*	糊剤；こざい	kozai
paste skipping *{n} [tex]*	糊飛び；のりとび	nori-tobi
pasteurization *[food-eng]*	低温殺菌(法)	tei'on-sakkin(-hō)
pasturage; meadowland *[ecol]*	牧草；ぼくそう	boku'sō
patagium *[v-zoo]*	飛膜	himaku
patent *{n} [legal]*	特許；とっきょ	tokkyo
patentability *[pat]*	特許(可能)性	tokkyo(-kanō)'sei
patent application publication *[pat]*	特許出願公告	tokkyo-shutsu'gan-kōkoku
patent attorney *[pat]*	弁理士；辯理士	benri'shi
patent cooperation treaty *[pat]*	特許協力条約	tokkyo-kyō'ryoku-jōyaku
patent early disclosure *[pat]*	特開	tokkai
patentee *[pat]*	特許権者	tokkyo'ken'sha
patent gazette *[pat]*	特許公報	tokkyo-kōhō
patent interference *[pat]*	特許紛争	tokkyo-funsō
patent medicine *[pat] [pharm]*	特許薬	tokkyo-yaku
" *[pharm]*	売薬；賣薬	bai-yaku
patent number *[pat]*	特許番号	tokkyo-bangō
patent of addition *[pat]*	追加特許	tsuika-tokkyo
Patent Office of Japan *[pat]*	特許庁；特許廳	Tokkyo-chō
patent register *[pat]*	特許原簿	tokkyo-genbo
patent right *[pat]*	特許権；特許權	tokkyo'ken
path; road *[civ-eng]*	道；みち	michi
path; pathway *[navig]*	経路；經路；けいろ	keiro
pathogen *[med]*	病原体；病原體	byōgen'tai
pathogenic microbe; pathogen *[med]*	病原微生物	byōgen-bi'seibutsu
pathology *[med]*	病理学；病理學	byōri'gaku
patina (pl -s, -inae) *[met]*	緑青；ろくしょう	rokushō

patrol boat *[nav-arch]*	巡視船	junshi'sen
pattern *{n} [graph]*	柄；がら	gara
" *{n} [graph]*	模型	mokei
" *{n} [graph]*	模様；もよう	moyō
" *{n} [graph]*	図形；圖形；ずけい	zukei
pattern paper *[mat] [paper]*	型紙	kata-gami
" *[mat] [paper]*	模様紙	moyō-shi
pattern recognition *[comput]*	図形認識	zukei-nin'shiki
" *[comput]*	パターン認識	patān-nin'shiki
paulownia (tree) *[bot]*	桐；きり	kiri
pause *{n} [comput]*	一時停止	ichi'ji-teishi
" *{n} [music]*	休止	kyūshi
pause instruction *[comput]*	一時停止命令	ichi'ji-teishi-meirei
paved road *[civ-eng]*	舗装路；ほそうろ	hosō-ro
pavement; paving *{n} [civ-eng]*	舗道	hodō
" *{n} [civ-eng]*	舗装；鋪装；ほそう	hosō
pawl *[mech-eng]*	爪；つめ	tsume
pay; wages; salary *{n} [econ]*	給料	kyūryō
pay *{vb t} [econ]*	払う；拂う；はらう	harau
pay the bill; settle accounts *{vb}*	勘定をする	kanjō o suru
payment *{n} [econ]*	報奨金	hōshō-kin
" *{n} [econ]*	支払い；支拂い	shi'harai
pea (sometimes bean) *[bot]*	豆；まめ	mame
pea (vegetable) *[bot]*	莢豌豆	saya-endō
peace *[mil]*	平和	heiwa
peach (fruit, tree) *[bot]*	桃；もも	momo
" ; white flesh peach (fruit) *[bot]*	水蜜桃	sui'mitsu'tō
peach oil *[mat]*	桃油	tōyu
peacock (bird) *[v-zoo]*	孔雀；くじゃく	kujaku
pea crab (crustacean) *[i-zoo]*	隠れ蟹；かくれがに	kakure-gani
peak factor *[phys]*	波高率	hakō-ritsu
peak inverse voltage *[electr]*	逆耐電圧	gyaku-taiden'atsu
peanut (legume) *[bot]*	落花生；らっかせい	rakka'sei
peanut oil *[mat]*	落花生油	rakka'sei-yu
pear (fruit, tree) *[bot]*	梨；なし	nashi
pearl *[mat]*	真珠；しんじゅ	shinju
pearly luster *[miner]*	真珠光沢	shinju-kōtaku
pearlwort; Sagina japonica *[bot]*	爪草	tsume-kusa
pear-shaped *{adj} [math]*	梨状の	nashi'jō no
peat deposit *[geol]*	泥炭層	deitan-sō
peat moss *[bot]*	水苔；みずごけ	mizu-goke
pebble *[geol]*	(小)砂利	(ko-)jari

pectolite *[miner]*	曹灰針石	sōkai'shinseki
pectoral fin *[v-zoo]*	胸鰭；むなびれ	muna-bire
peculiar *{adj}*	特異な	toku'i na
pedestal *[arch] [civ-eng] [eng]*	台座；臺座；だいざ	daiza
" *[mech-eng]*	軸受台	jiku'uke-dai
pedestrian *{n} [traffic]*	歩行者	hokō'sha
pedestrian bridge *[civ-eng]*	人道橋	jindō-kyō
pedicel *[bot]*	小花柄	shō-kahei
" *[bot]*	薹；とう	tō
pedigree *[gen]*	系統	keitō
" ; blood line *[gen]*	血統；けっとう	kettō
pediment *[geol]*	山麓緩斜面	san'roku-kansha'men
pedology *[geol]*	土壌学；土壤學	dojō'gaku
" *[med]*	児童学	jidō'gaku
peek in; peep; look; snoop *{vb}*	覗く	nozoku
peel *{vb}*	剝ぐ；はぐ	hagu
" *{vb}*	剝く；むく	muku
peelable; strippable *{adj} [adhes]*	剝離性の	hakuri'sei no
peeling *{n} [mech]*	剝れ	hagare
peephole *[mech-eng]*	覗き穴	nozoki-ana
Pekinese (animal) *[v-zoo]*	狆；ちん	chin
pelagic *{adj} [geol]*	外洋性の	gaiyō'sei no
" *{adj} [ocean]*	遠洋の	en'yō no
pellet *{n} [sci-t]*	小球	shōkyū
" ; tablet *{n} [pharm]*	錠剤	jōzai
" *{n} [pharm]*	小丸薬	shō-gan'yaku
pellet coal *[geol]*	造粒炭	zōryū-tan
pelletization *[min-eng]*	造粒；ぞうりゅう	zōryū
pelletizer *[eng]*	造粒機	zōryū-ki
pellucid *{adj}*	透明な	tōmei na
pelvis (pl -es, pelves) *{n} [anat]*	骨盤；こつばん	kotsu'ban
penalty; fine (of money) *[legal]*	罰金；ばっきん	bakkin
pencil *{n} [eng]*	鉛筆；えんぴつ	en'pitsu
pencil lead *[mat]*	鉛筆芯	enpitsu-shin
pendant *{n} [lap]*	垂れ飾り	tare-kazari
pendulum *[phys]*	振(り)子	furiko
pendulum clock *[horol]*	振(り)子時計	furiko-dokei
peneplain *[geol]*	準平原	jun-heigen
penetrate; perforate *{vb t} [sci-t]*	貫通する	kantsū suru
" (pierce through) *{vb}*	貫く；つらぬく	tsuranuku
penetrating agent *[mat]*	浸透剤	shintō-zai
penetration *[tex]*	裏通し	ura'dōshi

penetration, degree of *[eng]*	浸入度	shinnyū-do
penetration ratio *[mech-eng]*	浸入度比	shinnyū'do-hi
penetration test *[civ-eng]*	貫入試験	kannyū-shiken
penetrator *[eng]*	圧子；壓子；あっし	asshi
penetrometer *[eng]*	針入度計	shinnyū'do-kei
penicillum (pl -lia) *[mycol]*	青黴；あおかび	ao-kabi
peninsula *[geogr]*	半島	hantō
penis (pl penes, penises) *{n}* *[anat]*	陰茎	inkei
pen shell; razor shell *[i-zoo]*	玉珧；たいらぎ	tairagi
pentad *[ch]*	五価元素	goka-genso
" *[climat]*	五個一組	goko-hito'kumi
pentagon *[geom]*	五角形	gokak(u)kei
pentagonal prism *[math]*	五角柱	gokaku-chū
pentagonal pyramid *[math]*	五角錐	gokaku-sui
pentatonic scale *[music]*	五音；ごいん	go'in
pentavalent *[ch]*	五価の	goka no
penumbra (pl -brae, -bras) *[astron]*	半影	han'ei
peony (pl -nies) (herbaceous flower) *[bot]*	芍薬；しゃくやく	shaku'yaku
peony (flower, tree) *[bot]*	牡丹；ぼたん	botan
pepper (condiment) *{n}* *[cook]*	胡椒	koshō
" (Japanese condiment) *[cook]*	山椒；さんしょう	sanshō
peppermint *[bot]* *[cook]*	薄荷	hakka
peppermint oil *[cook]* *[mat]*	薄荷油	hakka-yu
peptide bond *[org-ch]*	ペプチド結合	pepuchido-ketsu'gō
peptization *[ch]*	解膠；かいこう	kaikō
per *{prep}*	につき	ni tsuki
per hour *[mech]*	毎時；まいじ	mai-ji
per minute *[mech]*	毎分	mai-fun
per second *[mech]*	毎秒	mai-byō
peracetic acid	過酢酸；かさくさん	ka'sakusan
peracid *[ch]*	過酸	kasan
perbenzoic acid	過安息香酸	ka'ansokukō-san
perborate *[ch]*	過硼酸塩	ka'hōsan-en
perboric acid	過硼酸；かほうさん	ka'hōsan
percarbonate *[ch]*	過炭酸塩	ka'tansan-en
percarbonic acid	過炭酸	ka'tansan
percent *[arith]*	百分率	hyaku'bun-ritsu
" *[arith]*	パーセント	pāsento
percentage *[arith]*	百分率	hyaku'bun-ritsu
" *[math]*	部合；ぶあい	bu'ai
" *[math]*	比率；ひりつ	hi-ritsu

percentage (rate; ratio) *[math]*	部合高；ぶあいだか	bu'ai-daka
" *[math]*	百分比	hyaku'bun-hi
percentile *[stat]*	百分位数	hyakubun'i-sū
percent transmission *[opt] [phys]*	透過率	tōka-ritsu
perception *[physio]*	知覚；知覺；ちかく	chi'kaku
perchlorate *[ch]*	過塩素酸塩	ka'enso'san-en
perchloric acid	過塩素酸	ka'enso-san
percitric acid	過枸櫞酸	ka'kuen-san
percolate liquid	浸出液	shin'shutsu-eki
percussion *[mech-eng]*	衝撃；しょうげき	shōgeki
percussion instruments *[music]*	打楽器類；打樂器類	da'gakki-rui
perennial (plant) *{n} [bot]*	多年生植物	ta'nensei-shokubutsu
" *{adj} [bot]*	多年生の	ta'nen'sei no
perennial herb *{n} [bot]*	多年生草本	ta'nensei-sōhon
perfect; complete *{adj}*	完全な；かんぜんな	kanzen na
" ; ideal *{adj}*	理想的な	risō'teki na
perfect crystal *[crys]*	完全結晶	kanzen-kesshō
perfect fluid *[phys]*	完全流体	kanzen-ryūtai
perfectness; perfection *{n}*	完全性	kanzen-sei
perfluorohydrocarbon *[org-ch]*	過弗化炭化水素	ka'fukka-tanka'suiso
perforate *{vb} [sci-t]*	貫通する	kantsū suru
" *{vb} [eng]*	鑽孔する	sankō suru
perform; do *{vb}*	為る；する	suru
performance *{n} [ind-eng]*	能率；のうりつ	nō'ritsu
" (live performance) *{n}*	実演	jitsu'en
" (of an act) *{n}*	実行；實行	jikkō
" (capacity; power) *{n}*	性能	seinō
" (of an act) *{n}*	遂行；すいこう	suikō
performance coefficient *[thermo]*	成積係数	seiseki-keisū
performic acid	過蟻酸；かぎさん	ka'gisan
perfume; fragrance; aromatic agent *{n} [mat] [org-ch]*	芳香剤；芳香劑；ほうこうざい	hōkō-zai
perfume *{n} [mat]*	香料	kōryō
" *{n} [mat]*	香水	kōsui
perhaps; probably; maybe *{adv}*	確か	tashika
periastron *[astron]*	近星点；近星點	kinsei-ten
pericardial cavity *[anat]*	囲心腔；圍心腔	ishin-kō
pericarp *[bot]*	果皮	kahi
perigalacticon *[astron]*	近銀河中心点	kin'ginga-chūshin-ten
perigee *[astron]*	近地点	kinchi-ten
perihelion (pl -helia) *[astron]*	近日点	kin'jitsu-ten
perilla *[bot] [cook]*	紫蘇；しそ	shiso

Perilla japonica *[bot]*	斎墩果; えごのき	egonoki
Perilla ocimoides *[bot]*	荏胡麻; えごま	e'goma
perilla oil *[mat]*	荏の油	e-no-abura
perimeter *[geom]*	周辺; 周邊	shūhen
" *[geom]*	周囲; 周圍	shū'i
perimorph *[miner]*	外包鉱物; 外包鑛物	gaihō-kōbutsu
period *{n} [ch] [phys] [trig]*	周期	shūki
" *{n} [geol]*	紀; き	ki
" (era; time) *{n} [geol]*	期; き	ki
" *{n} [gram] [pr]*	句点; くてん	kuten
" (full stop) *{n} [gram] [pr]*	終止符	shūshi'fu
" (of time) *{n} [phys]*	期間; きかん	kikan
periodate *[ch]*	過沃素酸塩	ka'yōso'san-en
periodic *{adj} [trig]*	周期的な	shūki'teki na
periodic acid	過沃素酸	ka'yōso-san
periodical *{n} [pr]*	定期刊行物	teiki-kankō'butsu
periodic classification *[ch]*	周期律(表)分類	shūki'ritsu(hyō)-bunrui
periodic function *[alg]*	周期関数	shūki-kansū
periodic law; law of periodicity *[ch]*	周期律	shūki-ritsu
periodic maintenance *[ind-eng]*	定期保全	teiki-hozen
periodic motion *[mech]*	周期運動	shūki-undō
periodic table of the elements *[ch]*	元素周期表	genso-shūki-hyō
" *[ch]*	周期律表	shūki'ritsu-hyō
period of limitation *[pat]*	時効	jikō
periodontium (pl -dontia) *[med]*	歯周組織	shishū-so'shiki
periosteum (pl -tea) *[anat]*	骨膜; こつまく	kotsu'maku
peripheral device *[comput]*	周辺装置	shūhen-sōchi
peripheral length *[geom]*	周辺長	shūhen-chō
peripheral nervous system *[anat]*	末梢神経系	masshō-shinkei-kei
peripherals *[comput]*	周辺機器; 周邊機器	shūhen-kiki
" *[comput]*	周辺装置	shūhen-sōchi
peripheral sealing property *[mat]*	辺縁封鎖性	hen'en-fūsa-sei
peripheral surface *[geom]*	外周面	gaishū-men
peripheral velocity *[mech]*	周速度	shū-sokudo
periphery *[geom]*	外周	gaishū
" *[geom]*	周縁; しゅうえん	shū'en
periplasmic space *[microbio]*	細胞周辺腔	saibō-shūhen-kō
periscope *[nav-arch] [opt]*	潜望鏡	senbō'kyō
perishable foods *[cook]*	生鮮食料(品)	seisen-shoku'ryō('hin)
peristalsis (pl -stalses) *[physio]*	蠕動(運動)	zendō(-undō)
peritectic *{n} [p-ch]*	包晶; ほうしょう	hōshō
peritoneal cavity *[anat]*	腹膜腔	fuku'maku-kō

peritrichous *{adj} [microbio]*	周毛性の	shūmō'sei no
periwinkle (groundcover) *[bot]*	日日草；にちにち草	nichi'nichi'sō
perlactic acid	過乳酸	ka'nyūsan
permafrost *[geol]*	永久氷結土	eikyū-hyōketsu'do
permanence *[mat]*	耐久度	taikyū-do
" *{n}*	永久；えいきゅう	eikyū
permanent *{adj}*	永久な	eikyū na
permanent dipole *[elecmg]*	永久双極子	eikyū-sōkyoku'shi
permanently *{adv}*	永久に	eikyū ni
permanent magnet *[elecmg]*	永久磁石	eikyū-ji'shaku
permanent magnetic field *[elecmg]*	永久磁場	eikyū-jiba
permanent resident *[comput]*	常駐；じょうちゅう	jōchū
permanent set *[mech]*	永久歪み	eikyū-hizumi
permanent storage *[comput]*	固定記憶	kotei-ki'oku
permanganate *[ch]*	過マンガン酸塩	ka'mangan'san-en
permanganic acid	過マンガン酸	ka'mangan-san
permeability *[ch]*	透過率；とうかりつ	tōka-ritsu
" *[ch]*	透過性	tōka-sei
" *[elecmg]*	透磁率；とうじりつ	tōji-ritsu
" *[fl-mech]*	浸透性	shintō-sei
permeability coefficient *[fl-mech]*	透過係数	tōka-keisū
permeable layer *[geol]*	透水層	tōsui-sō
permeameter *[eng]*	透磁率計	tōji'ritsu-kei
permease *[bioch]*	透過酵素	tōka-kōso
permeating property *[ch]*	透過性	tōka-sei
permeation *[ch]*	浸透；しんとう	shintō
" *[ch]*	透過	tōka
permissible content *[ind-eng]*	許容含有量	kyoyō-gan'yū-ryō
permissible dose *[nucleo]*	許容線量	kyoyō-senryō
permission *{n}*	許諾	kyodaku
" *{n}*	許可	kyoka
permittivity *[elec]*	誘電率	yūden-ritsu
permutation *[prob]*	置換；ちかん	chikan
permutations *[prob]*	順列	jun'retsu
peroxidase *[bioch]*	過酸化酵素	ka'sanka-kōso
peroxide *[ch]*	過酸化物	ka'sanka'butsu
peroxy acid *[ch]*	過酸	kasan
perpendicular *{n} [geom]*	垂線	suisen
" *{adj} [geom]*	垂直な	sui'choku na
perpetual motion *[phys]*	永久運動	eikyū-undō
persimmon (fruit, tree) *[bot]*	柿；かき	kaki
persimmon tannin *[org-ch]*	渋；澁；しぶ	shibu

persistence *{n}*	存続	sonzoku
person *{n}*	人	hito
personal care *{n}*	身嗜み；みだしなみ	mi-dashinami
personal difference *[stat]*	個人差	kojin-sa
personal error *[stat]*	個人誤差	kojin-gosa
personnel expenses *[econ]*	人件費	jinken-hi
perspective *[graph]*	遠近画法	en'kin'gahō
perspective view *[graph]*	透視図	tōshi-zu
perspiration; sweat *[physio]*	汗；あせ	ase
perspiring *{n} [physio]*	汗かき	ase-kaki
pertinence *{n}*	妥当性；だとうせい	datō-sei
pertinent data *[math]*	該当事項	gaitō-jikō
pervasion *[med]*	瀰慢；びまん	biman
pessimism *[psy]*	悲観；悲觀；ひかん	hikan
pest *{n} [bio]*	有害生物	yūgai-seibutsu
" *{n} [i-zoo]*	害虫；害蟲	gaichū
pesticide *[mat]*	農薬	nō-yaku
pestle *[eng]*	乳棒；にゅうぼう	nyūbō
pestle of wood *[cook] [eng]*	擂り粉木；すりこぎ	suriko'gi
petal *[bot]*	花弁；花瓣；花びら	hana'bira
petalite *[miner]*	葉長石	yō'chōseki
petiole *[bot]*	葉柄	yōhei
petition *{n} [legal]*	申請(書)	shinsei('sho)
" *{n} [pat]*	上申書	jōshin'sho
petrel (bird) *[v-zoo]*	海燕；うみつばめ	umi-tsubame
petrifaction *[geol]*	化石作用	kaseki-sayō
" *[geol]*	石化(作用)	sekka(-sayō)
petrochemical industry *[org-ch]*	石油化学工業	sekiyu'kagaku-kōgyō
petrochemical product *[mat]*	石油化学製品	sekiyu'kagaku-seihin
petrogeny *[geol]*	岩石成因学	ganseki-sei'in'gaku
petrography *[geol]*	岩石記載学	ganseki-kisai'gaku
petroleum *[mat]*	石油；せきゆ	seki'yu
petroleum aromatics *[org-ch]*	石油系芳香族炭化=水素	sekiyu'kei-hōko'zoku-tanka'suiso
petroleum coke *[mat]*	石油コークス	sekiyu-kōkusu
petroleum engine *[mech-eng]*	石油機関	sekiyu-kikan
petroleum fermentation *[microbio]*	石油発酵	sekiyu-hakkō
petroleum microorganism *[microbio]*	石油微生物	sekiyu-bi'seibutsu
petroleum refining *{n} [ch-eng]*	石油精製	sekiyu-seisei
petroleum resin *[org-ch]*	石油樹脂	sekiyu-jushi
petrology *[geol]*	岩石学	ganseki'gaku
Phaeophyta *[bot]*	褐藻植物綱	kassō-shokubutsu-kō

phagocyte *[cyt]*	食細胞; 食細胞	shoku-saibō
phagocytosis (pl -toses) *[cyt]*	食作用	shoku-sayō
pharmaceutical affairs law *[legal]*	薬事法; やくじほう	yakuji-hō
pharmaceutical chemistry *[ch-eng]*	薬理化学	yakuri-kagaku
pharmaceutical engineering *[pharm]*	薬品製造工学	yakuhin-seizō-kōgaku
pharmaceuticals *[pharm]*	医薬品; 醫藥品	i'yaku'hin
pharmaceutical science *[pharm]*	薬学	yaku'gaku
pharmaceutics *[pharm]*	薬剤学; 藥劑學	yakuzai'gaku
pharmacist (a person) *[pharm]*	薬剤師	yakuzai'shi
pharmacodynamic effect *[pharm]*	薬力学的効果	yaku'rikigaku'teki-kōka
pharmacodynamics *[pharm]*	藥効学	yakkō'gaku
pharmacogenetics *[gen]*	薬理遺伝学	yakuri-iden'gaku
pharmacognosy *[pharm]*	生薬学	shōyaku'gaku
pharmacokinetics *[pharm]*	薬物動態学	yaku'butsu-dōtai'gaku
pharmacology *[ch]*	薬理学	yakuri'gaku
pharmacopoeia *[pharm] [pr]*	薬局方	yakkyoku-hō
pharmacy *[arch] [pharm]*	薬局; やっきょく	yakkyoku
pharyngeal *{adj} [anat]*	咽頭の	intō no
pharynx (pl -es, pharynges) *[anat]*	咽頭; いんとう	intō
phase *{n} [astron] [elec] [meth] [phys]*	位相	isō
" *{n} [ch] [math] [phys]*	相; そう	sō
" *{n} [comput]*	段階	dankai
phase angle *[phys]*	位相角	isō-kaku
phase-contrast microscope *[opt]*	位相差顕微鏡	isō'sa-kenbi'kyō
phase deviation *[comm]*	位相偏移	isō-hen'i
phase diagram *[p-ch]*	状態図; 狀態圖	jōtai-zu
phase discrimination *[electr]*	位相弁別; 位相辨別	isō-ben'betsu
phase distortion *[comput]*	位相歪み	isō-yugami
phase inversion *[ch]*	転相; 轉相	tensō
phase modulation *[comm]*	位相変調	isō-henchō
phase rule *[ch]*	相律	sō-ritsu
phase shift *[electr]*	移相	isō
phase shifter *[elec]*	移相器	isō-ki
phase-switching interferometer *[opt]*	位相切換干渉計	isō-kiri'kae-kanshō-kei
phase transition *[phys]*	相転移; そうてんい	sō-ten'i
phase velocity *[phys]*	位相速度	isō-sokudo
phasing *{n} [electr]*	整相	seisō
pheasant (bird) *[v-zoo]*	雉子; きじ	kiji
phenol *[org-ch]*	フェノール	fenōru
" *[org-ch]*	石炭酸	sekitan-san
phenolphthalein *[org-ch]*	フェノールフタ=レイン	fenōrufutarein

phenomenon *[math]*	現象；げんしょう	genshō
phenotype *[gen]*	表現型	hyōgen'kei
philately; stamp collecting *[comm]*	切手収集；切手收集	kitte-shūshū
philology *[comm]*	言語学；言語學	gengo'gaku
philosophy (pl -phies)	哲学	tetsu'gaku
phlegm *[physio]*	痰	tan
phloem *[bot]*	師部；篩部；しぶ	shibu
phlogopite; bronze, brown mica *[miner]*	金雲母	kin-un'mo
Pholiota nameko (mushroom) *[bot]*	滑子；なめこ	name'ko
phoneme *[ling]*	音素	on'so
phonetic alphabet *[ling]*	音声字母	onsei-jibo
phonetic pronunciation (of Chinese characters) *[comm]*	音；おん	on
phonics *[ling]*	音声学	onsei'gaku
phonogram *[pr]*	表音文字	hyō'on-moji
phonometer *[acous]*	音声計；音聲計	onsei-kei
phonon *[sol-st]*	音子	onshi
Phoronidae *[i-zoo]*	箒虫動物	hōki'mushi-dōbutsu
phosgenite *[miner]*	角鉛鉱	kaku'enkō
phosphate *[ch]*	燐酸塩	rinsan-en
phosphate ion *[ch]*	燐酸イオン	rinsan-ion
phosphite *[ch]*	亜燐酸塩	a'rinsan-en
phospholipid *[bioch]*	燐脂質；りんししつ	rin-shi'shitsu
phosphoprotein *[bioch]*	燐蛋白質	rin-tanpaku'shitsu
phosphor *[phys]*	蛍光体；螢光體	keikō'tai
" *[phys]*	燐光体	rinkō'tai
phosphor bronze *[met]*	燐青銅	rin'seidō
phosphorescence *[atom-phys]*	燐光	rinkō
phosphorescent substance *[phys]*	燐光体	rinkō'tai
phosphoric acid	燐酸；りんさん	rinsan
phosphorolysis *[bioch]*	加燐酸分解	ka'rinsan-bunkai
phosphorous acid	亜燐酸	a'rinsan
phosphorus (element)	燐；りん	rin
phosphorus bromide	臭化燐	shūka-rin
phosphorus chloride	塩化燐；鹽化燐	enka-rin
phosphorus hydride	水素化燐	suiso'ka-rin
phosphorus nitride	窒化燐；ちっかりん	chikka-rin
phosphorus oxide	酸化燐	sanka-rin
phosphorus oxide chloride	酸化塩化燐	sanka-enka-rin
phosphorus pentachloride	五塩化燐	go'enka-rin
phosphorus pentoxide	五酸化燐	go'sanka-rin
phosphorus sulfide	硫化燐	ryūka-rin

phosphorus trichloride	三塩化燐	san'enka-rin
phosphorus trioxide	三酸化燐	san'sanka-rin
phosphorus trisulfide	三硫化燐	san'ryūka-rin
phosphorylation *[org-ch]*	燐酸化；りんさんか	rinsan'ka
photoacoustic spectroscopy *[spect]*	光音響分光法	hikari-onkyō-bunkō-hō
" *[spect]*	光音響分光法	kō'onkyō-bunkō-hō
photoautotroph *[bio]*	光独立栄養生物	hikari-doku'ritsu-eiyō-seibutsu
photobleaching reagent *[opt]*	光漂白試薬	kō'hyōhaku-shiyaku
photocatalytic reaction *[ch]*	光触媒反応	kō'shokubai-hannō
photocell *[electr]*	光電池	kō'denchi; hikari-denchi
photochemical *{adj}* *[p-ch]*	光化学の	kō'kagaku no
photochemical absorption, law of *[p-ch]*	光化学吸収の法則	kō'kagaku-kyūshu no hōsoku
photochemical activation, principle of *[p-ch]*	光化学活性の原理	kō'kagaku-kassei no genri
photochemical cell *[electr]*	光化学電池	kō'kagaku-denchi
photochemical equivalent *[p-ch]*	光化学当量	kō'kagaku-tōryō
photochemistry *[p-ch]*	光化学；こうかがく	kō'kagaku
photochemistry, first law of *[p-ch]*	光化学第一法則	kō'kagaku-dai'ichi-hōsoku
photochemistry, second law of *[p-ch]*	光化学第二法則	kō'kagaku-dai'ni-hōsoku
photochromics *[ch]*	光色性物質	kōshoku'sei-busshitsu
photocomposing machine *[graph]*	写真植字機	shashin-shoku'ji-ki
photocurrent *[p-ch]*	光電流	kō'denryū; hikari-denryū
photodisintegration *[nuc-phys]*	光壊変；光壞變	hikari-kaihen
photodissociation *[p-ch]*	光解離	hikari-kairi
photoelasticity *[opt]*	光弾性	kō'dansei
photoelectric cell *[electr]*	光伝導セル	kō'dendō-seru
photoelectricity *[electr]*	光電気；光電氣	kō'denki; hikari-denki
" *[electr]*	光電効果	kōden-kōka
photoelectric photometer *[eng]*	光電光度計	kōden-kōdo-kei
photoelectric reflectivity *[electr]*	光電反射率	kōden-hansha-ritsu
photoelectric spectrophotometer *[spect]*	光電分光光度計	kōden-bunkō-kōdo-kei
photo(electric) tube *[electr]*	光電管	kōden-kan
photoemission *[electr]*	光電子放出	kō'denshi-hō'shutsu
photoengraving *{n}* *[graph]*	写真彫刻	shashin-chōkoku
photo finish (in competition)	写真判定決勝	shashin-hantei-kesshō
photofission *[nuc-phys]*	光分裂	kō'bunretsu
photogrammetry *[eng]*	写真測量法	shashin-soku'ryō-hō
photograph *{n}* *[graph]*	写真；寫真	sha'shin
" *{vb}* *[graph]*	写真を写す	shashin o utsusu

photographic *{adj} [photo]*	写真の；しゃしんの	shashin no
photographic density *[opt] [photo]*	写真濃度	shashin-nōdo
photographic dry plate *[photo]*	乾板；かんぱん	kanpan
photographic emulsion *[photo]*	写真乳剤；寫真乳劑	shashin-nyūzai
photographic intensity *[opt]*	写真濃度	shashin-nōdo
photographic paper *[paper] [photo]*	印画(用)紙	inga(-yō)shi
photographic wet plate *[photo]*	湿板；濕板	shitsu-ban
photographic zenith tube *[opt]*	写真天頂筒	shashin-tenchō-tō
photography *[graph]*	写真(撮影)術	shashin(-satsu'ei)'jutsu
photohardening property *[opt]*	光硬化性	kō'kōka-sei
photoionization *[p-ch]*	光電離	hikari-denri
photolithography *[pr]*	写真平版	shashin-heihan
photolysis *[p-ch]*	光分解	hikari-bunkai; kō'bunkai
photomacrograph *[opt]*	低倍率顕微鏡写真	tei'bairitsu-kenbikyō-shashin
photometer *[eng]*	光度計	kōdo-kei
" *[eng]*	測光器	sokkō-ki
photometric equivalent *[photo]*	写真測光当量	shashin-sokkō-tōryō
photometry *[opt]*	測光；そっこう	sokkō
photomicrograph *[graph]*	顕微鏡写真	kenbi'kyō-shashin
photomultiplier tube *[electr]*	光電子増倍管	kō'denshi-zōbai-kan
photon *[opt]*	光子；こうし	kōshi
" *[quant-mech]*	光量子	hikari-ryōshi; kō'ryōshi
photooxidation *[ch]*	光酸化	hikari-sanka
photoperiodism *[physio]*	光周(期)性	hikari-shū(ki)-sei; kō-shūsei
photo plate *[graph]*	写真版	shashin-ban
photopolymer; photosentive resin *[poly-ch]*	感光性樹脂	kankō'sei-jushi
photopolymerization *[poly-ch]*	光重合	hikari-jūgō; kō'jūgō
photorefractive crystal *[crys]*	光屈折性結晶	kō'kussetsu'sei-kesshō
photorefractive nonlinear optics *[opt]*	光屈折性非線形光学	kō'kussetsu'sei-hi'senkei-kōgaku
photorefractive polymer *[poly-ch]*	光屈折性高分子	kō'kussetsu'sei-kōbunshi
photorespiration *[bioch]*	光呼吸	kō'kokyū
photosensitiveness *[photo]*	感光性	kankō-sei
photosensitizer *[opt]*	光増感剤	kō'zōkan-zai
photosilver *[photo]*	光銀；こうぎん	kōgin
photosphere *[astron]*	光球	kōkyū
photosynthesis *[bioch]*	光合成	hikari-gōsei; kō'gōsei
photosynthetic bacteria *[microbio]*	光合成細菌	hikari-gōsei-saikin
photosystem *[bioch]*	光化学系	hikari-kagaku-kei
phototaxis (pl -taxes) *[bio]*	走光性	sōkō-sei
phototelegraphy *[comm]*	写真電送	shashin-densō

phototube *[electr]*	光電管	kōden-kan
photovoltatic cell *[electr]*	光電気化学電池	hikari-denki'kagaku-denchi
" *[electr]*	光起電力セル	hikari-ki'denryoku-seru
photovoltatic effect *[electr]*	光起電力効果	hikari-ki'denryoku-kōka
" *[electr]*	光起電効果	kō'kiden-kōka
phrase *{n} [ling]*	句；く	ku
phthalic anhydride	無水フタル酸	musui-futaru-san
phyllotaxy; leaf arrangement *[bot]*	葉序；ようじょ	yōjo
phylogeny (pl -nies) *[bio]*	系統発生	keitō-hassei
phylum (pl phyla) *[bio] [syst]*	門；もん	mon
" *[ling]*	語族	go'zoku
physical biochemistry *[bioch]*	物理生化学	butsuri-sei'kagaku
physical change *[phys]*	物理変化	butsuri-henka
physical chemistry *[ch]*	物理化学	butsuri-kagaku
physical fitness *[med]*	体力；體力	tai-ryoku
physical photometry *[opt]*	物理測光	butsuri-sokkō
physical properties *[ch] [phys]*	物性；ぶっせい	bussei
" *[ch] [phys]*	物理的性質	butsuri'teki-seishitsu
physicist (a person) *[phys]*	物理学者	butsuri-gakusha
physics *[sci-t]*	物理学	butsuri'gaku
physiological chemistry *[ch]*	生理化学	seiri-kagaku
physiological saline *[physio]*	生理食塩水	seiri-shoku'en-sui
physiology *[bio]*	生理学	seiri'gaku
physique *[med]*	体格	taikaku
phytochemistry *[bot]*	植物化学	shokubutsu-kagaku
phytoplankton *[ecol]*	浮遊植物	fuyū-shokubutsu
pick; pickaxe; mandrel *{n} [eng]*	鶴嘴；つるはし	tsuru'hashi
pick; pluck; cull *{vb}*	摘む	tsumu
pickerelweed *[bot]*	蛭蓆；ひるむしろ	hiru-mushiro
picking *{n} [v-zoo]*	撮み取り	tsumami'tori
pickled daikon radish (q.v.) *[cook]*	沢庵；たくあん	taku'an
pickled scallions *[cook]*	辣韮漬	rakkyō-zuke
pickled ume (q.v.) *[cook]*	梅干；うめぼし	ume-boshi
pickles *[cook]*	香の物	kō-no-mono
" *[cook]*	漬け物；つけもの	tsuke'mono
pickling *{n} [met]*	酸洗い	san-arai
pickup tube; camera tube *[electr]*	撮像管	satsu'zō-kan
pictogram *[graph] [pr]*	絵文字；えもじ	e-moji
pictorial art *[graph]*	絵画；繪畫	kaiga
pictorial book *[pr]*	図鑑；圖鑑；ずかん	zukan
pictorial display *[comput]*	図式表示	zu'shiki-hyōji
pictorial information *[comm]*	図式情報	zu'shiki-jōhō

pictorial map *[graph]*	絵地図; 繪地図	e-chizu
pictorial trademark *[graph] [pat]*	図形商標	zukei-shōhyō
picture *{n} [comput]*	画面; 畫面	gamen
" *{n} [comput] [graph]*	画像; がぞう	gazō
" (depiction) *{n} [graph]*	描写; びょうしゃ	byōsha
" *{n} [graph]*	絵; 繪; 画; 畫; え	e
" *{n} [graph]*	図; 圖; ず	zu
picture current *[elec]*	画電流	ga-denryū
picture element (pel, pixel) *[electr]*	画素; 畫素; がそ	gaso
picture image *[electr]*	画像	gazō
picture image engineering *[electr]*	画像工学	gazō-kōgaku
picture map *[graph]*	絵図; 繪圖; えず	e-zu
picture pattern *[graph]*	図柄	zu'gara
picture postcard *[comm]*	絵葉書	e-ha'gaki
pictures and books *[pr]*	図書	zu'sho
picture scroll *[graph]*	絵巻物; 繪巻物	e-maki'mono
picture signals *[comm]*	像信号	zō-shingō
" *[elec]*	映像信号	eizō-shingō
picture tube *[electr]*	受像管	juzō-kan
piece; unit *{n}*	箇; 個; こ	ko
" (chess or game) *{n}*	駒; こま	koma
piece dyeing *{n} [tex]*	反染め; たんぞめ	tan-zome
piedmont *[geol]*	山麓地帯	san'roku-chitai
pied wagtail (bird) *[v-zoo]*	白鶺鴒; 白せきれい	haku-sekirei
pier *[civ-eng]*	橋脚	kyō'kyaku
" *[civ-eng]*	桟橋; 棧橋	san'bashi
pierce *{vb t}*	刺す	sasu
piezoelectrical effect *[sol-st]*	圧電効果	atsuden-kōka
piezoelectric ceramics *[cer]*	圧電性セラミックス	atsuden'sei-seramikkusu
piezoelectric coefficient *[sol-st]*	圧電率; 壓電率	atsuden-ritsu
piezoelectric effect *[sol-st]*	圧電効果	atsuden-kōka
piezoelectricity *[elec]*	圧電気; あつでんき	atsu'denki
piezoelectric modulus *[sol-st]*	圧電率	atsuden-ritsu
piezoelectric oscillator *[electr]*	圧電発振器	atsuden-hasshin-ki
piezoelectric polymer *[poly-ch]*	圧電性高分子	atsuden'sei-kōbunshi
piezoelectric resonator *[electr]*	圧電共振子	atsuden-kyōshin'shi
piezoelectric vibrator *[sol-st]*	圧電振動子	atsuden-shindō'shi
piezoresistance effect *[elec]*	圧抵抗効果	atsu'teikō-kōka
pig; hog; swine (animal) *[v-zoo]*	豚; ぶた	buta
pigeon; dove (bird) *[v-zoo]*	鳩; はと	hato
pig iron *[mat]*	銑鉄; せんてつ	sen-tetsu
pigment *[biochem] [mat]*	顔料; がんりょう	ganryō

pigmentation *[physio]*	色素沈着	shiki'so-chin'chaku
pignolia (nut) *[bot]*	松の実; 松の實	matsu no mi
pike (fish) *[v-zoo]*	河梭魚; 河かます	kawa-kamasu
pike conger (fish) *[v-zoo]*	鱧; はも	hamo
pilaster *[arch]*	付柱; つけばしら	tsuke-bashira
pilchard; (fish) *[v-zoo]*	鯯; 拶双魚; さっぱ	sappa
pile *{n} [eng]*	杭; くい	kui
" *{n} [tex]*	毛羽; けば	keba
piled lumber *{n}*	積み木	tsumi-ki
pile driver *[mech-eng]*	杭打ち機	kui'uchi-ki
pilings *[civ-eng]*	杭; くい	kui
pill *{n} [tex]*	玉毛	gyoku'mō
" *{n} [pharm]*	丸薬	gan-yaku
pillbox (antique) *{n}*	印籠; いんろう	inrō
pill bug (insect) *[i-zoo]*	草鞋虫; わらじむし	waraji-mushi
pilling *{n} [tex]*	ももけ	momoke
pillow lava *[geol]*	枕状溶岩	chin'jō-yōgan
pilot (a person) *{n} [aero-eng]*	操縦者; 操縱者	sōjū'sha
" (a person) *{n} [aero-eng]*	操縦士; 操縱士	sōjū'shi
" (a person) *{n} [navig] [ocean]*	水先(案内)人	mizu'saki(-annai)'nin
pilot boat *[navig]*	水先船	mizu'saki'sen
pinacoid *{n} [crys]*	卓面	taku'men
pincers *[eng]*	ペンチ	penchi
pinch; one pinch *{n} [cook]*	一撮み; ひとつまみ	hito-tsumami
pinch; pick *{vb t}*	抓む; 摘む; 撮む	tsumamu
pine (tree) *{n} [bot]*	松; まつ	matsu
pinecone *[bot]*	松毬; 松笠	matsu-kasa
pinecone fish; (fish) *[v-zoo]*	松毬魚; まつかさ魚	matsu'kasa-uo
pine mushroom *[bot]*	松茸; まつたけ	matsu-take
pine needle *[bot]*	松葉	matsu-ba
pine-needle oil *[mat]*	松葉油	matsu'ba-yu
pine nut *[bot]*	松の実	matsu no mi
pine oil *[mat]*	松油	shōyu
pine resin; colophony *[mat]*	松脂; 松やに	matsu-yani
pin file *[eng]*	ピン鑢; ピンやすり	pin-yasuri
pinhole *[opt]*	ピン穴	pin-ana
pinion *{n} [mech-eng]*	小歯車	ko-ha'guruma; shō-ha'guruma
pink; wild pink (flower) *[bot]*	撫子; なでしこ	nadeshiko
pink (color) *[opt]*	桃色	momo-iro
pink coral; rose coral *[i-zoo]*	桃色珊瑚	momo'iro-sango
pinnate *{adj} [bot]*	羽状の; うじょうの	u'jō no
pinnately compound leaf *[bot]*	羽状複葉	u'jō-fuku'yō

pinocytosis (pl -toses) *[cyt]*	飲作用；いんさよう	in-sayō
pinwheel *[mech-eng]*	ピン歯車	pin-ha'guruma
pipe *[eng] [mat]*	円管；圓管	enkan
" ; tube *[eng] [mat]*	管；くだ	kuda
pipefish; Syngnathus schlegeli *[v-zoo]*	楊子魚；ようじ魚	yōji-uo
pipeline *[eng]*	配管系	haikan-kei
piping; plumbing *{n} [civ-eng]*	配管；はいかん	haikan
piracy *[comput]*	盗作	tōsaku
Pisces; fish *[v-zoo]*	魚類；ぎょるい	gyo'rui
Pisces; the Fishes *[constel]*	魚座；うおざ	uo-za
pisciculture *[ecol]*	養魚	yōgyo
pisolite *[petr]*	豆石；まめいし	mame-ishi
pistil *[bot]*	雌蕊	shi'zui; me-shibe
piston flow; plug flow *[fl-mech]*	押出し流れ	oshi'dashi-nagare
pit; seed; stone (of fruit) *[bot]*	種；たね	tane
pitch (of sound) *{n} [acous]*	音調の高さ	onchō no taka-sa
" (of sound) *{n} [acous]*	音高	onkō
" (as of roof) *{n} [arch] [sci-t]*	勾配；こうばい	kōbai
" ; spacing *{n} [comput]*	間隔	kankaku
" ; bitumen *{n} [mat]*	瀝青(物質)	rekisei(-busshitsu)
" ; gradient *{n} [math] [sci-t]*	傾斜度；けいしゃど	keisha-do
" *{n} [mech]*	刻み；きざみ	kizami
" *{n} [mech]*	縦揺れ；縱搖れ	tate'yure
pitch-dark night *[astron]*	闇夜；やみよ	yami-yo
pith *[bot]*	髄；髓；ずい	zui
pitting (by corrosion) *{n} [met]*	点食；點蝕	ten'shoku
pitting corrosion *[met]*	孔食；孔蝕	kōshoku
pituitary (gland); hypophysis *[anat]*	脳下垂体	nō-kasui'tai
pit viper (reptile) *[v-zoo]*	蝮；まむし	mamushi
pivot *{n} [mech]*	要；かなめ	kaname
pixel *[electr]*	絵画素	e-gaso
" *[electr]*	絵素；えそ	eso
place; square; plaza; *{n} [civ-eng]*	広場；廣場；ひろば	hiro'ba
place; location *{n} [geogr]*	場所	basho
place *{n} [math]*	位；い	i
" *{n} [math]*	座；ざ	za
" ; column *{n} [math]*	桁；けた	keta
" ; dispose; arrange *{vb t}*	配置する	haichi suru
placebo *{n} [pharm]*	偽薬；にせぐすり	nise-gusuri
placement; arrangement; layout *[math]*	配置	haichi
placenta (pl -centas, -centae) *[bot]*	胎座	taiza
placenta *[embryo]*	胎盤；たいばん	taiban

place of business *[econ]*	営業地；營業地	eigyō'chi
place upon *{vb}*	載置する	saichi suru
place value *[math]*	桁の値	keta no atai
plain *{n} [geol]*	平原	heigen
" *{n} [geol]*	平野；へいや	heiya
" (solid-colored; unfigured) *[tex]*	無地；むじ	muji
plain-paper copier *[graph]*	電子複写機	denshi-fukusha-ki
plain weave *[tex]*	平織；ひらおり	hira-ori
plan *{n} [graph]*	設計	sekkei
" *{n} [graph]*	図面；圖面；ずめん	zumen
" (scheme; project) *{n}*	計画；計畫	keikaku
planar *{adj} [math]*	平面的な	heimen'teki na
plane (a tool) *{n} [eng]*	鉋；かんな	kanna
" *{n} [geom]*	平面；へいめん	heimen
" *{n} [geom]*	面；めん	men
plane angle *[geom]*	平面角	heimen-kaku
plane geometry *[math]*	平面幾何学	heimen-kika'gaku
plane mirror *[opt]*	平面鏡	heimen-kyō
plane of incidence *[opt]*	入射面	nyūsha-men
plane of reflection *[phys]*	反射面	hansha-men
plane-polarized light *[opt]*	平面偏光	heimen-henkō
plane strain *[mech]*	平面歪	heimen-hizumi
planet *[astron]*	惑星；わくせい	waku'sei
planetarium (pl -s, -ia) *[astron]*	星座投影機	seiza-tōei-ki
planet(ary) gear *[mech-eng]*	遊星歯車	yūsei-ha'guruma
planetary nebula *[astron]*	惑星状星雲	wakusei'jō-sei'un
planetary physics *[astrophys]*	惑星物理学	wakusei-butsuri'gaku
planetoid *[astron]*	微惑星；びわくせい	bi'wakusei
planetology *[astron]*	惑星学	wakusei'gaku
planetophysics *[astrophys]*	惑星物理学	wakusei-butsuri'gaku
planet wheel *[mech-eng]*	遊星輪	yūsei-wa
plane wave *[phys]*	平面波	heimen-ha
planimetry *[math]*	求積法	kyūseki-hō
planing machine *[eng]*	平削り盤	hira'kezuri-ban
planitesimal *{n} [astron]*	微惑星	bi'wakusei
planking *{n} [civ-eng]*	敷板；しきいた	shiki-ita
" *{n} [nav-arch]*	外板	soto-ita
plankton *[ecol]*	浮遊生物	fuyū-seibutsu
planoconcave *{adj} [opt]*	平凹の；へいおうの	hei-ō no
planoconvex *{adj} [opt]*	平凸の；へいとつの	hei-totsu no
plant; vegetation; flora *{n} [bot]*	植物；しょくぶつ	shoku'butsu
" *{n} [bot]*	草木；そうもく	sōmoku

plant (workshop) *{n} [ind-eng]*	工場；こうじょう	kōjō
" ; set (a plant) *{vb t} [bot]*	植える	ueru
Plantae (plant kingdom) *[bot]*	植物界	shokubutsu-kai
plantain *[bot]*	芭蕉；ばしょう	bashō
" *[bot]*	大葉子；車前草	ō-bako
plant chemistry *[bot]*	植物化学	shokubutsu-kagaku
plant community *[bot]*	植物群落	shokubutsu-gunraku
plant-growth substance *[bot]*	植物生長調整物質	shokubutsu-seichō-chōsei-busshitsu
plantigrade animal *[v-zoo]*	蹠行動物	shokō-dōbutsu
planting *{n} [agr]*	植付け	ue'tsuke
plantlike protista *[bio]*	植物性原生生物	shokubutsu'sei-gensei-seibutsu
plant patent *[pat]*	植物特許	shokubutsu-tokkyo
plan view *[graph]*	平面図	heimen-zu
plaque *[furn]*	飾り板；かざりいた	kazari-ita
" (bacteria) *[microbio]*	斑；はん	han
plasma *[hist]*	血漿；けっしょう	kesshō
" *[phys]*	プラズマ	purazuma
plasma cell *[hist]*	形質細胞	kei'shitsu-saibō
plasma cosmology *[astron]*	プラズマ宇宙論	purazuma-uchū-ron
plasma etching *{n} [graph]*	プラズマ加工	purazuma-kakō
plasmalemma; plasma membrane *[cyt]*	原形質膜	genkei'shitsu-maku
plasma physics *[phys]*	プラズマ物理学	purazuma-butsuri'gaku
plasmodium *[microbio]*	変形体；變形體	henkei'tai
plasmolysis *[physio]*	原形質分離	genkei'shitsu-bunri
plaster *{n} [mat]*	漆喰；しっくい	shikkui
" *{n} [pharm]*	硬膏剤	kōkō-zai
plasterer's lime *[mat]*	左官用消石灰	sakan'yō-shō'sekkai
plastic *{n} [mat]*	プラスチック	purasuchikku
" *{adj} [mat]*	塑性の	sosei no
plastic deformation *[mech]*	塑性変形	sosei-henkei
plastic flow *[phys]*	塑性流れ	sosei-nagare
plastic forming *{n} [eng]*	塑性加工	sosei-kakō
plasticity *[mech]*	塑性；そせい	sosei
" *[poly-ch]*	可塑性	kaso-sei
plasticity number *[mech]*	可塑度	kaso-do
plasticization *[eng] [poly-ch]*	可塑化	kaso'ka
plasticizer *[poly-ch]*	可塑剤	kaso-zai
plastics (the field) *[poly-ch]*	合成樹脂化学	gōsei-jushi-kagaku
plastic strain *[mech]*	塑性歪	sosei-hizumi
plastic substance *[mat] [poly-ch]*	可塑物；かそぶつ	kaso'butsu

plastic viscosity *[fl-mech]*	塑性粘度	sosei-nendo
plate *{n} [cer] [cook]*	皿；さら	sara
" *{n} [ch-eng] [p-ch]*	段；だん	dan
" *{n} [eng]*	板；いた	ita
" *{n} [geol]*	岩板	ganban
" (illustration) *[graph]*	図版；圖版；ずはん	zu'han
" *{n} [pr]*	版	han
plateau *[geol]*	台地；臺地	daichi
plate column *[ch-eng]*	(棚)段塔	(tana)dan-tō
plate crystal *[crys]*	板状結晶	banjō-kesshō
plate culture *[microbio]*	平板培養	heiban-baiyō
plate efficiency *[ch-eng]*	段効率	dan-kō'ritsu
plate glass *[mat]*	板ガラス	ita-garasu
platelet *[hist]*	血小板	kesshōban
platform *[arch]*	壇；だん	dan
" *[civ-eng] [rail]*	ホーム	hōmu
" *[rail]*	乗降場；乘降場	jōkō'jō
platform car *[rail]*	台車；臺車	daisha
platform scale *[eng]*	台秤；だいばかり	dai-bakari
platinate *[ch]*	白金酸塩	hakkin'san-en
plating; galvanizing *{n} [met]*	鍍金；めっき	mekki
plating bath *[met]*	鍍金浴；めっきよく	mekki-yoku
platinic acid	白金酸	hakkin-san
platinum (element)	白金；はっきん	hakkin
platinum chloride	塩化白金；鹽化白金	enka-hakkin
platinum hydroxide	水酸化白金	suisan'ka-hakkin
platinum oxide	酸化白金	sanka-hakkin
platinum-wire loop *[ch]*	白金耳；はっきんじ	hakkin-ji
Platyhelminthes (flatworms) *[i-zoo]*	扁形動物	henkei-dōbutsu
platypus *[v-zoo]*	鴨嘴；かものはし	kamonohashi
playback *{n} [eng-acous]*	再生	saisei
playground *[civ-eng]*	運動場	undō'jō
playing of a variation *{n} [music]*	変奏；變奏	hensō
pleasure boat *[nav-arch]*	遊覧船	yūran'sen
pleat *{n} [cl]*	襞；ひだ	hida
pleiotropic effect *[gen]*	多面的効果	tamen'teki-kōka
pleiotropy *[gen]*	多面発現性	tamen-hatsu'gen-sei
pleochroic *{adj} [opt]*	多色性の	ta'shoku'sei no
pleochroism *[opt]*	多色性	ta'shoku-sei
pleomorphism *[bio]*	多態性；たたいせい	tatai-sei
" *[crys]*	多形態性	ta'keitai-sei
Plimsoll mark *[nav-arch]*	乾舷標	kangen-hyō

plinth *[arch]*	幅木；はばき	haba-ki
ploidy *[gen]*	倍数性	baisū-sei
plot *{n} [graph]*	図；圖；ず	zu
" *{vb} [graph]*	作図する	saku'zu suru
plotosid (fish) *[v-zoo]*	権瑞；ごんずい	gonzui
plotter *[graph]*	作図装置	sakuzu-sōchi
plotting *{n} [math]*	点綴；點綴	tentei
" *{n} [graph]*	方眼	hōgan
plough *{n} [agr]*	鋤；すき	suki
plover (bird) *[v-zoo]*	千鳥；ちどり	chidori
pluck off *{vb}*	毟り取る	mushiri-toru
plug *{n} [elec] [sci-t]*	プラグ	puragu
" ; stopper *{n} [sci-t]*	栓；せん	sen
plugboard *[comput]*	配線盤	haisen-ban
plug flow *[fl-mech]*	押出し流れ	oshi'dashi-nagare
" *[fl-mech]*	栓流	senryū
plum (fruit) *[bot]*	西洋李	seiyō-sumomo
plumage *[v-zoo]*	羽；はね	hane
plumbate *[ch]*	鉛酸塩；鉛酸鹽	namari'san-en
plumber (a person) *[eng]*	配管工	haikan'kō
plumbic acid	鉛酸；なまりさん	namari-san
plumbing *{n} [civ-eng]*	配管	haikan
plumb line *[eng]*	鉛直線	enchoku-sen
plumbojarosite *[miner]*	鉛鉄明礬	namari'tetsu'myōban
plumb the depth of *{vb}*	究める；極める	kiwameru
Plumeria rubra; frangipani *[bot]*	印度素馨	Indo-sokei
plumous *{adj} [bot]*	羽毛状の	umō'jō no
plural *{n} [gram]*	複数	fuku-sū
plural equation *[math]*	多元方程式	tagen-hōtei'shiki
plural number *[math]*	複数；ふくすう	fukusū
plus; plus side *{n}*	陽；よう	yō
Pluto (planet) *[astron]*	冥王星	mei'ō-sei
plutonium (element)	プルトニウム	purutonyūmu
ply (pl plies) *{n} [mat]*	層；層；そう	sō
plywood *[mat]*	ベニヤ板	beniya-ita
" *[mat]*	合板	gōban; gōhan
pocket gopher (animal) *[v-zoo]*	掘り鼠；ほりねずみ	hori-nezumi
pocket heater *[eng]*	懐炉；懷爐；かいろ	kairo
pod *[bot]*	莢；さや	saya
" *[v-zoo]*	小群	shōgun
podium *[arch]*	演壇	endan
" *[music]*	指揮台	shiki'dai

podocard (tree) *[bot]*	槙；まき	maki
poetry; verse *{n} [pr]*	詩	shi
Pogonophora *[i-zoo]*	有鬚動物	yūshu-dōbutsu
poikilotherm *[zoo]*	変温動物	hen'on-dōbutsu
poinciana (tree) *[bot]*	鳳凰木	hō'ō'boku
point *{n} [geom]*	点；點；てん	ten
point at; indicate *{vb}*	指す	sasu
point charge *[elec]*	点電荷	ten-denka
pointed *{adj}*	尖った	togatta
pointer *[eng]*	指針	shishin
point group *[crys]*	点群；點群	tengun
point mutation *[gen]*	点変異	ten-hen'i
point source *[acous]*	点音源	ten-on'gen
poison *{n} [mat]*	毒(物)	doku('butsu)
poisoning *{n} [med]*	中毒；ちゆぅどく	chūdoku
poison ivy *[bot]*	蔦漆；つたうるし	tsuta-urushi
poisonous- *{prefix}*	有害性-	yūgai'sei-
pokeweed *[bot]*	山牛蒡；やまごぼう	yama-gobō
polar axis *[eng]*	極軸	kyoku-jiku
polar bear (animal) *[v-zoo]*	北極熊	hokkyoku-guma
" (animal) *[v-zoo]*	白熊；しろくま	shiro-kuma
polar cap *[astron] [hyd]*	極冠	kyoku-kan
polar compound *[ch]*	極性化合物	kyoku'sei-kagō'butsu
polar coordinates *[an-geom]*	極座表	kyoku-zahyō
polar diagram *[phys]*	極線図；極線圖	kyoku-sen'zu
polar front *[meteor]*	極前線	kyoku-zensen
polar group *[org-ch]*	極性基	kyoku'sei-ki
polarimeter *[opt]*	旋光計	senkō-kei
Polaris *[astron]*	北極星；北極星	hokkyoku-sei
polariscope *[opt]*	偏光器	henkō-ki
polarity *[math] [phys]*	極性；きょくせい	kyoku-sei
polarizability *[elec]*	分極率	bunkyoku-ritsu
polarization *[elec] [phys]*	分極	bun'kyoku
" *[opt]*	偏光	henkō
polarization spectroscopy *[spect]*	偏光分光	henkō-bunkō
polarized charge *[elec]*	分極電荷	bunkyoku-denka
polarized light *[opt]*	偏光；へんこう	henkō
polarized wave *[electr]*	偏波	henpa
polarizer *[opt]*	偏光子	henkō'shi
polarizing microscope *[opt]*	偏光顕微鏡	henkō-kenbi'kyō
polarizing plate *[opt]*	偏光板	henkō-ban
polar molecule *[p-ch]*	極性分子	kyoku'sei-bunshi

polar zone *[climat]*	寒帯；寒帶	kantai
pole *{n} [elec] [geogr] [math]*	極；きょく	kyoku
pole figure *[graph]*	極点図	kyoku'ten-zu
police box *[arch]*	交番	kōban
policeman	警官	keikan
police station *[arch]*	警察所；けいさつ所	keisatsu'sho
polish; polishing *{n} [mech-eng]*	磨き；みがき	migaki
" *{vb} [eng]*	磨く	migaku
polished rice *[agr] [cook]*	精白米；せいはく米	seihaku-mai
polished surface *[eng]*	研摩面	kenma-men
polishing *{n} [ch-eng]*	研摩	kenma
" *{n} [ch-eng]*	艶出し；つやだし	tsuya'dashi
" *{n} [mech-eng]*	磨き	migaki
polishing powder *[mat]*	磨粉；みがきこ	migaki-ko
pollen *[bot]*	花粉；かふん	kafun
pollen count *[bot]*	花粉の数	kafun no kazu
pollen culture *[bot]*	花粉培養	kafun-baiyō
pollen grain *[bot]*	花粉粒	kafun-ryū
pollination (pollen-receiving) *[bot]*	受粉；じゅふん	jufun
" (pollen-giving) *[bot]*	授粉	jufun
pollutant *[ecol]*	汚染物；(-物質)	o'sen'butsu; (-busshitsu)
pollution *[ecol]*	汚濁	o'daku
" *[ecol]*	汚染；おせん	o'sen
pollution index *[ecol]*	汚濁指数	o'daku-shisū
polonium (element)	ポロニウム	poronyūmu
polyacidic base *[ch]*	多酸塩基；多酸鹽基	ta'san-enki
polyaddition *[poly-ch]*	重付加	jū'fuka
polyandry *[bio]*	一雌多雄	isshi-tayū
polybasic acid *[ch]*	多塩基酸	ta'enki-san
polybasite *[miner]*	硫安銅銀鉱	ryūan'dōgin'kō
polycondensation *[poly-ch]*	重縮合	jū'shukugō
" *[poly-ch]*	縮重合	shuku-jūgō
polycrystalline *{adj} [crys]*	多結晶質の	ta'kesshō'shitsu no
polycrystalline graphite *[miner]*	多結晶性黒鉛	ta'kesshō'sei-koku'en
polycrystalline substance *{n} [crys]*	多結晶質	ta'kesshō-shitsu
polycyclic *{adj} [org-ch]*	多環式の	takan'shiki no
polydentate ligand *[ch]*	多座配位子	taza-hai'i'shi
polydisperse system *[poly-ch]*	多分散系	ta'bunsan-kei
polydispersity *[poly-ch]*	多分散；たぶんさん	ta'bunsan
polyelectrolyte *[org-ch]*	多価電解質	taka-denkai'shitsu
" *[poly-ch]*	高分子電解質	kōbunshi-denkai'shitsu
polyfunctionality *[org-ch]*	多官能性	ta'kannō-sei

polygen; polyvalent element *[ch]*	多原子価元素	ta'genshi'ka-genso
polygon (multi-sides) *[geom]*	多辺形；多邊形	tahen'kei
" (multi-angles) *[geom]*	多角形	ta'kak(u)kei
Polygonaceae *[bot]*	蓼科	tade-ka
polygonal cone *[geom]*	多角錐；たかくすい	ta'kaku-sui
polygonal graph *[graph]*	多角図；多角圖	ta'kaku-zu
Polygonum hydropiper; jointwood *[bot]*	蓼；たで	tade
polygyny *[bio]*	多雌性	tashi-sei
" *[bot]*	多雌蕊；たしずい	tashi-zui
polyhedron *[geom]*	多面体；多面體	tamen'tai
polyhydric phenol; polyphenol *[org-ch]*	多価フェノール	taka-fenōru
polymer *[poly-ch]*	重合体	jūgō'tai
" *[poly-ch]*	高分子(化合物)	kōbunshi(-kagō'butsu)
" *[poly-ch]*	ポリマー	porimā
polymer alloy *[met] [poly-ch]*	高分子合金	kōbunshi-gōkin
polymerase chain reaction *[bio] [bioch]*	ポリメラーゼ連鎖=反応	porimerāze-rensa-hannō
polymer chemistry *[ch]*	高分子化学	kōbunshi-kagaku
polymeric flocculant *[poly-ch]*	高分子凝集剤	kōbunshi-gyōshū-zai
polymeric plasticizer *[poly-ch]*	高分子可塑剤	kōbunshi-kaso-zai
polymerization *[poly-ch]*	重合；じゅうごう	jūgō
polymerization degree *[poly-ch]*	重合度	jūgō-do
polymerization inhibitor *[poly-ch]*	重合防止剤	jūgō-bōshi-zai
" *[poly-ch]*	重合禁止剤	jūgō-kinshi-zai
polymerization initiator *[poly-ch]*	重合開始剤	jūgō-kaishi-zai
polymerization retarder *[poly-ch]*	重合抑制剤	jūgō-yokusei-zai
polymerized oil *[mat]*	重合油	jūgō-yu
polymolecularness *[org-ch]*	多分子性	ta'bunshi-sei
polymorphic change *[bio] [crys]*	多形変化	takei-henka
polymorphism *[bio] [crys]*	多形性；たけいせい	takei-sei
" *[crys]*	同質多像	dō'shitsu-tazō
polynomial *[alg]*	整式；せいしき	sei-shiki
" ; multinomial *[alg]*	多項式；たこうしき	takō-shiki
polynuclear complex *[org-ch]*	多核錯体	ta'kaku-sakutai
polynuclear hydrocarbon *[org-ch]*	多環式炭化水素	takan'shiki-tanka'suiso
polyphase; multiphase *[elec]*	多相	tasō
polyphase current *[elec]*	多相交流	tasō-kōryū
polyploidy; ploidy *[gen]*	倍数性	baisū-sei
polypore; shelf fungus *[mycol]*	猿の腰掛け	saru no koshi'kake
polysaccharides *[bioch]*	多糖類；たとうるい	tatō-rui
polysulfide *[org-ch]*	多硫化物	ta'ryūka'butsu
polysulfide rubber *[rub]*	多硫化ゴム	ta'ryūka-gomu

polytonality *[music]*	多調	ta'chō
polytropic atmosphere *[meteor]*	多方大気	tahō-taiki
polytypic species *[ch]*	多型種；たけいしゅ	takei-shu
polyunsaturated fatty acid *[org-ch]*	多不飽和脂肪酸	ta'fuhōwa-shibō-san
polyurea fiber *[tex]*	尿素繊維	nyōso-sen'i
polyvalent *{adj} [ch]*	多価の；たかの	taka no
polyvalent antibody *[bio]*	多価抗体	taka-kōtai
polyvalent element *[ch]*	多原子価元素	ta'genshi'ka-genso
pomace *{n} [agr]*	林檎の絞り渣	ringo no shibori-kasu
pomegranate (fruit) *[bot]*	柘榴；ざくろ	zakuro
pomegranate bark *[bot]*	柘榴皮	zakuro-hi
pomelo; zamboa (fruit) *[bot]*	朱欒；ざぼん	zabon
pomfret (fish) *[v-zoo]*	縞鰹；しまがつお	shima-gatsuo
pond *[geogr]*	池；いけ	ike
pond smelt (fish) *[v-zoo]*	鰙；若鷺；公魚	waka'sagi
pond snail *[i-zoo]*	田螺；たにし	ta'nishi
pongee *[tex]*	紬；つむぎ	tsumugi
pontoon bridge *[civ-eng]*	浮橋	uki-bashi; uki-hashi
poor; unacceptable *{n} [ind-eng]*	不可	fuka
poor harvest *[agr]*	不作；ふさく	fu'saku
poor solvent *[ch]*	貧溶媒	hin-yōbai
Popilla japonica (insect) *[i-zoo]*	豆黄金虫	mame-kogane-mushi
pop out; leap out *{vb}*	飛出す	tobi-dasu
poppy (pl poppies) (herb) *[bot]*	罌粟；芥子；けし	keshi
poppy seed *[bot]*	罌粟の実	keshi no mi
" *[bot] [cook]*	罌粟粒；芥子粒	keshi-tsubu
poppy seed oil *[mat]*	罌粟油	keshi-yu
popularization *{n}*	普及	fukyū
population *[bio]*	人口；じんこう	jinkō
" *[stat]*	母集団；母集團	bo'shūdan
population genetics *[gen]*	集団遺伝学	shūdan-iden'gaku
population inversion *[atom-phys]*	反転分布	hanten-bunpu
porbeagle; mack (fish) *[v-zoo]*	鼠鮫；ねずみざめ	nezumi-zame
porcelain *[cer] [mat]*	磁器	jiki
" *[cer] [mat]*	瀬戸物；瀨戸物	seto'mono
" *[cer] [mat]*	陶材	tō-zai
porcelain enamel *[mat]*	瀬戸引き	seto'biki
porcelain stone *[cer]*	陶石；とうせき	tōseki
porcupine (animal) *[v-zoo]*	山荒し	yama-arashi
pore (air hole) *{n} [bio]*	気孔；氣孔；きこう	kikō
" (fine or narrow pore) *{n} [bio] [met]*	細孔	saikō
" (small pore) *{n} [bio]*	小孔	shōkō

porgy (fish) *[v-zoo]*	鯛；たい	tai
Porifera (sponges) *[i-zoo]*	海綿動物	kaimen-dōbutsu
porosimeter *[eng]*	多孔度計	ta'kōdo-kei
porosity (pl -ties) *[bio]*	気孔率；きこうりつ	kikō-ritsu
" *[phys]*	含気率	ganki-ritsu
" *[phys]*	間隙率	kangeki-ritsu
" *[phys]*	多孔性	takō-sei
" *[phys]*	有孔度	yūkō-do
" *[phys]*	有孔率	yūkō-ritsu
porous matter *[mat]*	多孔質	takō-shitsu
porous sheet *[ch]*	素焼き板	suyaki'ban; suyaki-ita
porphyrite *[miner]*	玢岩；ひんがん	hingan
porphyry *[petr]*	斑岩；はんがん	hangan
porpoise (mammal) *[v-zoo]*	海豚；いるか	iruka
" (mammal) *[v-zoo]*	鼠海豚	nezumi-iruka
port (direction from ship) *{n}* *[navig]*	左玄	sagen
portability *[comput]*	移植性	i'shoku-sei
" *[comput]*	携帯性；携帶性	keitai-sei
portable *{adj}* *[eng]*	携帯用の	keitai'yō no
portable bridge *[civ-eng]*	可搬橋	kahan-kyō
portable lamp *[eng]*	手提げ灯	te'sage-tō
portal vein *[anat]*	門脈	mon-myaku
porthole *[nav-arch]*	舷窓；げんそう	gensō
portion; part *{n}*	部分	bubun
port of entry *[trans]*	通関港；通關港	tsūkan-kō
portrait *[graph]*	肖像画；肖像畫	shōzō-ga
Portugese man o'war *[i-zoo]*	鰹の烏帽子	katsuo no eboshi
position *{n}* *[comput]* *[math]* *[navig]*	位置；いち	ichi
position *{n}*	立脚点	rikkyaku-ten
" (post) *{n}*	所属	shozoku
" (put in place) *{vb t}*	位置付けする	ichi'zuke suru
position astronomy *[astron]*	位置天文学	ichi-tenmon'gaku
position circuit *[elec]*	座席回路	za'seki-kairo
positioning *{n}* *[navig]*	位置調節	ichi-chōsetsu
positioning time *[comput]*	位置決め時間	ichi'gime-jikan
position sense *[physio]*	位置感覚	ichi-kankaku
position specificity *[bio]*	位置特異性	ichi-toku'i-sei
positive *{n}* *[elec]*	陽；よう	yō
" *{n}* *[elec]* *[math]* *[sci-t]*	正；せい	sei
" *{n}* *[photo]*	ポジ	poji
" *{adj}*	肯定的な	kōtei'teki na
positive charge *[elec]*	陽電荷	yō-denka

positive column *[electr]*	陽光柱	yōkō-chū
positive electrode *[elec]*	正極	sei-kyoku
" *[elec]*	陽極；ようきょく	yō-kyoku
positive electron *[part-phys]*	陽電子	yō-denshi
positive feedback *[electr]*	正帰還；正歸還	sei-kikan
positive image *[photo]*	陽画像	yōga-zō
positive ion *[ch] [phys]*	陽イオン	yō-ion
positive number *[arith]*	正の数	sei no sū
positive sign *[math]*	正号；正號	seigō
positron *[part-phys]*	陽電子；ようでんし	yō-denshi
positron decay *[nuc-phys]*	陽電子崩壊	yō'denshi-hōkai
positron emission tomography; PET *[med]*	陽電子放出断層像法	yō'denshi-hō'shutsu-dansō-zōhō
posology *[med]*	薬量学；藥量學	yakuryō'gaku
possessive case *[gram]*	所有格	shoyū-kaku
possibility *[math]*	可能性	kanō-sei
post *{n} [arch]*	柱；はしら	hashira
post- *{prefix}*	以後の-	igo no-
postal code *[comm]*	郵便番号	yūbin-bangō
postal scale *[eng]*	郵便秤	yūbin-bakari
postcard *[comm]*	(郵便)葉書	(yūbin-)ha'gaki
post-categorization *{n}*	後分類	ato-bunrui
postcure; postcuring *{n} [plas]*	後硬化	ato-kōka
posterior *{adj} [zoo]*	後の；うしろの	ushiro no
postexposure *[photo]*	後露光	ato-rokō
postforming *{n} [poly-ch]*	後成形	ato-seikei
postmark *[comm]*	消印；けしいん	keshi-in
postmeridian *[astron]*	午後；ごご	gogo
postmortem *[comput]*	事後分析	jigo-bunseki
post office *[arch] [comm]*	郵便局	yūbin'kyoku
postpolymerization *[poly-ch]*	後重合	ato-jūgō
postponement *{n}*	延期	enki
post-sorting *{n}*	後分類	ato-bunrui
postulate *{n} [logic]*	公準	kōjun
posture *{n} [physio]*	姿勢；しせい	shisei
pot *{n} [cook]*	釜；かま	kama
" *{n} [eng]*	鉢	hachi
potable *{adj} [cook]*	飲むのに適した	nomu-noni-tekishita
potable water *[sci-t]*	飲料水	inryō-sui
potassium (element)	カリウム	karyūmu
potassium dihydrogen phosphate	燐酸二水素カリウム	rinsan-ni'suiso-karyūmu
potassium nitrate; saltpeter *[ch]*	硝石	shōseki

potassium primary phosphate	第一燐酸カリウム	dai'ichi-rinsan-karyūmu
potassium secondary phosphate	第二燐酸カリウム	dai'ni-rinsan-karyūmu
potato (pl -toes) (vegetable) *[bot]*	馬鈴薯；ばれいしょ	bareisho
" (vegetable) *[bot]*	芋；いも	imo
pot-cooked dishes *[cook]*	鍋類（鍋物）	nabe-rui (nabe-mono)
potency (pl -cies) *[pharm]*	力価；力價；りきか	riki-ka
" (efficacy) *[sci-t]*	効力	kōryoku
potential difference *[elec]*	電位差	den'i-sa
potential energy *[mech]*	位置エネルギー	ichi-enerugī
potentiator *[cook] [pharm]*	増強剤	zōkyō-zai
potentiometer *[elec] [eng]*	分圧計	bun'atsu-kei
" *[elec] [eng]*	電位差計	den'isa-kei
potentiometric titration *[an-ch]*	電位差滴定	den'isa-teki'tei
potherb mustard *[bot]*	蟒草；うわばみそう	uwabami'sō
pothole *[geol]*	甌穴	ō'ketsu; kame-ana
pothook *[eng]*	自在鉤；じざいかぎ	jizai-kagi
potted plant *[bot]*	植木；うえき	ue'ki
potter's wheel *[eng]*	轆轤；ろくろ	roku'ro
pottery (pl -teries) *[cer]*	陶磁器	tōji'ki
" *[cer]*	焼き物；焼き物	yaki'mono
pottery stone *[cer]*	陶石	tōseki
pottery tube *[mat]*	陶管	tōkan
poultice *{n} [pharm]*	罨法物	anpō'butsu
" *{n} [pharm]*	巴布剤；ぱっぷざい	pappu-zai
poultry farming *{n} [agr]*	養鶏	yōkei
pound (currency) *[econ]*	磅；ポンド	pondo
" (unit of mass) *[mech]*	封度；ポンド	pondo
pour (into, onto); sprinkle *{vb t}*	注ぐ；灌ぐ；そそぐ	sosogu
" (e. g. tea) *{vb t}*	注ぐ；つぐ	tsugu
pouring *{n} [eng] [met]*	鋳込み	i'komi
pour point *[fl-mech]*	流動点	ryūdō-ten
powder *[mat]*	粉末；ふんまつ	fun'matsu
" *[mat]*	粉；こな	kona
powder coal *[geol]*	粉炭	funtan
powder coating *{n} [mat]*	粉体塗装	funtai-tosō
powdered medicine *[pharm]*	散薬；さんやく	san-yaku
powdered resin *[resin]*	粉末樹脂	funmatsu-jushi
powdered tea *[cook]*	抹茶；まっちゃ	matcha
powder metallurgy *[met]*	粉末冶金	funmatsu-yakin
powder room *[arch]*	化粧室	keshō-shitsu
power *{n} [arith]*	冪；べき	beki
" *{n} [arith]*	累乗；累乗	ruijō

English	Japanese	Romanization
power *{n} [math]*	濃度	nōdo
" (strength) *{n} [mech]*	力；ちから	chikara
" *{n} [phys]*	仕事率	shigoto-ritsu
power cable *[elec]*	電力ケーブル	denryoku-kēburu
power conduction *[elec]*	動力伝導	dō'ryoku-dendō
power-distribution unit *[elec]*	配電器	haiden-ki
power factor *[elec]*	力率；りきりつ	riki-ritsu
power factor meter *[eng]*	力率計	riki'ritsu-kei
power failure *[elec]*	停電；ていでん	teiden
powerful medicine *[pharm]*	劇薬；げきやく	geki-yaku
power generation *[elec]*	発電	hatsu'den
power line *[elec]*	電力線	denryoku-sen
power loss *[electr]*	電力損	denryoku-son
power number *[fl-mech]*	動力数	dōryoku-sū
power plant *[mech-eng]*	原動所	gendō'sho
" *[mech-eng]*	発電所	hatsu'den'sho
power series *[alg]*	冪級数	beki-kyūsū
power source *[elec]*	電源；でんげん	dengen
power supply *[elec]*	電源装置	dengen-sōchi
" *[electr]*	給電装置	kyūden-sōchi
power supply (system) *[elec]*	給電(系統)	kyūden(-keitō)
power train *[mech-eng]*	動力列	dō'ryoku-retsu
power transmission *[elec]*	送電	sōden
practicability *[pat]*	実用性；實用性	jitsu'yō-sei
practical art *{n}*	実用的技芸	jitsuyō'teki-gigei
practical capacity *[sci-t]*	実用容量	jitsuyō-yōryō
practical utility *[pat]*	実用性	jitsu'yō-sei
practice *{n} [pat]*	実施；じっし	jisshi
" *{n} [math]*	演習	enshū
pragmatics *[ling]*	語用論	go'yō-ron
prairie *[geogr]*	草原	sōgen
prairie chicken (bird) *[v-zoo]*	草原雷鳥	sōgen-raichō
praise; commend *{vb t}*	褒める	homeru
praseodym (element)	プラセオジム	puraseojimu
praseodymium (element)	プラセオジミウム	puraseojimyūmu
prawn (crustacean) *[i-zoo]*	車海老；くるまえび	kuruma-ebi
prawn; Palaemon nipponensis *[i-zoo]*	手長蝦	te'naga-ebi
praying mantis (insect) *[i-zoo]*	蟷螂；かまきり	kamakiri
pre- *{prefix}*	以前の-	i'zen no-
preamplifier *[electr]*	前置増幅器	zenchi-zō'fuku-ki
precaution *[med]*	予防	yobō
precaution; caution *{n} [psy]*	用心；ようじん	yōjin

precedence *[alg] [comput]*	先行	senkō
" *[comput]*	順位；じゅんい	jun'i
" *[comput]*	優先順位	yūsen-jun'i
precession *[mech]*	歳差運動	saisa-undō
precharging *{n} [elec]*	前帯電	zen'taiden
precious metal *[met]*	貴金属；ききんぞく	ki-kinzoku
precious stone *[miner]*	宝石；寶石	hōseki
precipice *[geol]*	断崖	dangai
precipitant *[ch]*	沈殿剤	chinden-zai
precipitate *[ch]*	沈殿物	chinden'butsu
precipitation *[ch]*	沈殿；沈澱	chinden
" *[met]*	析出	seki'shutsu
" *[meteor]*	降水(量)	kōsui(-ryō)
" (rainfall) *[meteor]*	降雨(量)	kō'u(-ryō)
precipitation titration *[ch]*	沈殿滴定	chinden-tekitei
precipitin *[immun]*	沈降素；ちんこうそ	chinkō'so
precision *[math] [sci-t]*	精度	seido
" *[math]*	精密さ	sei'mitsu-sa
precooler *[mech-eng]*	予冷器；豫冷器	yorei-ki
preculture *[microbio]*	前培養	mae-baiyō
precure; precuring *{n} [plas]*	前硬化；まえこうか	mae-kōka
precursor *[ch]*	先駆物質	senku-busshitsu
" *[ch] [bio]*	前駆物質	zenku-busshitsu
" *[ch] [bio]*	前駆体；前驅體	zenku'tai
precursory phenomenon *[geophys]*	前兆現象	zenchō-genshō
predator *[zoo]*	捕食者	hoshoku'sha
predatory; predaceous *{adj} [zoo]*	捕食性の	hoshoku'sei no
" *{adj} [zoo]*	食肉の	shoku'niku no
predefined *[comput]*	定義済み	teigi-zumi
" *{n}*	既定	kitei
predicate *{n} [gram]*	述部	jutsu'bu
predicted performance curve *[eng]*	予想性能曲線	yosō-seinō-kyoku'sen
prediction; foreknowledge *{n} [psy]*	予知；豫知；よち	yochi
prediction *{n} [psy]*	予言	yogen
" *{n} [psy]*	予想	yosō
" *{n} [psy]*	予測	yo'soku
predominant; preponderant *[bio]*	優勢；ゆうせい	yūsei
predrying *{n} [plas]*	予備乾燥	yobi-kansō
preengage *{vb} [ind-eng]*	予約する	yo'yaku suru
pre-exposure *[photo]*	前露光	mae-rokō
prefecture (e.g. Osaka-fu) *[geogr]*	府；ふ	fu
" *[geogr]*	県；縣；けん	ken

preferential *{adj} [pat]*	優先的な	yūsen'teki na
preferred orientation *[crys]*	優先方位	yūsen-hō'i
" *[petr]*	方位配列	hō'i-hai'retsu
prefix *{n} [gram]*	接頭語；せっとうご	settō'go
" *{n} [gram]*	接頭辞	settō'ji
prefix part *[gram]*	接頭部	settō-bu
preformation theory *[embryo]*	前成説	zensei-setsu
preforming *{n} [met] [plas]*	予備成形	yobi-seikei
preheating *{n} [plas]*	予備加熱	yobi-ka'netsu
" *{n} [thermo]*	予熱；豫熱；よねつ	yo'netsu
prehensile *{adj} [v-zoo]*	把握力のある	ha'aku'ryoku no aru
prehistory *[archeo]*	太古；たいこ	taiko
prehnite *[miner]*	葡萄石	budō'seki
prejudice *{vb} [legal]*	侵害する	shingai suru
preloading *{n} [eng]*	予荷重	yo'kajū
prelude *[music]*	前奏曲	zensō'kyoku
premature scouring *{n} [tex]*	若練り；若練り	waka-neri
premise *[logic]*	前提	zentei
premixer *[eng]*	予混合機	yo'kongō-ki
premolar *[anat]*	前臼歯	zenkyū'shi
preparation *{n}*	準備；じゅんび	junbi
" *{n}*	整備	seibi
" (prearrangement) *{n}*	下拵え	shita-goshirae
preparation(s); arrangements *{n}*	支度；仕度；したく	shi'taku
preparatory plan *[civ-eng]*	段取り	dan'dori
prepayment *[econ]*	前払い；前拂い	mae-barai
prepayment meter *[trans]*	前納計器	zennō-keiki
preposition *[gram]*	前置詞；ぜんちし	zenchi'shi
preprinting *{n} [pr]*	予備印刷	yobi-in'satsu
preprocessing *{n} [ind-eng]*	前処理；前處離	zen'shori; mae-shori
prerequisite *{n}*	前提条件	zentei-jōken
presbyopia *[med]*	老視	rōshi
prescribe *{vb} [med]*	処方する；處方する	shohō suru
prescription *[med] [pharm]*	処方(箋)	shohō(-sen)
prescription medicine *[pharm]*	処方薬	shohō-yaku
preservation; conservation *[ecol]*	保全	hozen
" *[ecol]*	保存	hozon
preservative *[cook] [mat]*	防腐剤；ぼうふざい	bōfu-zai
" *[mat]*	保存剤	hozon-zai
preshrunk *{adj} [cl] [tex]*	予縮させた	yo'shuku-saseta
press *{n} [mech-eng]*	圧搾器；壓搾器	assaku-ki
" ; squeeze; wring; extract *{vb}*	絞る；しぼる	shiboru

English	Japanese	Romanization
pressed barley *[agr]*	押麦；おしむぎ	oshi-mugi
pressed oil *[mat]*	圧搾油；あっさくゆ	assaku-yu
pressed powder core *[met]*	圧粉鉄心；壓粉鐵芯	appun-tesshin
pressing *{n} [cook]*	圧搾	assaku
press mark; indentation *{n} [mech-eng]*	圧痕；あっこん	akkon
pressure *[mech]*	圧力；あつりょく	atsu'ryoku
pressure, sense of *[physio]*	圧迫感	appaku-kan
pressure gage *[eng]*	圧力計；壓力計	atsu'ryoku-kei
pressure gradient *[fl-mech]*	圧力勾配	atsu'ryoku-kōbai
pressure medium *[mat]*	圧媒体	atsu'bai-tai
pressure plate *[mech-eng]*	押え板	osae-ita
" *[mech-eng]*	押し板	oshi-ita
pressure resistance *[fl-mech]*	圧力抵抗	atsu'ryoku-teikō
pressure-sensitive adhesive *[mat]*	圧感接着剤	akkan-setchaku-zai
" *[mat]*	粘着剤	nenchaku-zai
pressure-sensitive adhesiveness	粘着性	nenchaku-sei
pressure-sensitive paper *[paper]*	感圧紙；かんあつし	kan'atsu-shi
pressurization *[eng]*	与圧；與圧；よあつ	yo'atsu
pressurization method *[eng]*	気圧増大法	ki'atsu-zōdai-hō
pressurized cabin *[mech-eng]*	与圧室	yo'atsu-shitsu
pretest *{n} [arith]*	予備調査	yobi-chōsa
pretreating; pretreatment *{n} [eng]*	前処理；前處理	mae-shori
prevailing wind *[meteor]*	卓越風	taku'etsu-fū
prevention *{n}*	防止	bōshi
" *{n}*	予防；よぼう	yobō
preventive maintenance *[eng]*	予防保全	yobō-hozen
preventive medicine *[med]*	予防医学；豫防醫學	yobō-i'gaku
Priapuloidea; Priapulida *[i-zoo]*	鰓曳動物	era'hiki-dōbutsu
price *{n} [econ]*	値段；ねだん	ne'dan
price ceiling *[econ]*	限界価格	genkai-kakaku
price index *[econ]*	物価指数；物價指數	bukka-shisū
prickly ash leaves *[bot] [cook]*	山椒；さんしょう	sanshō
prickly ash sprouts *[cook]*	木の目；きのめ	ki-no-me
prill *{vb t} [ch-eng]*	金属粒にする	kinzoku'ryū ni suru
primarily *{adv}*	一義的；いちぎてき	ichigi'teki
primary *{n} [ch]*	第一級	dai'ikkyū
" *{n} [ch] [elec] [math]*	一次；いちじ	ichi'ji
" *{adj}*	一次の	ichi'ji no
primary carbon atom *[ch]*	第一炭素原子	dai'ichi-tanso-genshi
primary color(s) *[opt]*	原色	gen'shoku
primary compound *[ch]*	第一化合物	dai'ichi-kagō'butsu
primary crystal *[crys] [met]*	初晶	shoshō

primary culture *[microbio]*	一次培養	ichiji-baiyō
" *[microbio]*	初代培養	shodai-baiyō
primary electron *[phys]*	一次電子	ichiji-denshi
primary germ layer *[bot]*	一次胚葉	ichiji-haiyō
primary meaning *[ling]*	原義	gengi
primary metabolite *[bioch]*	一次代謝産物	ichiji-taisha-sanbutsu
primary mirror *[opt]*	主鏡; しゅきょう	shu-kyō
primary production *[bio]*	基礎生産	kiso-seisan
primary salt *[ch]*	第一塩; 第一鹽	dai'ichi-en
primary scale *[met]*	一次スケール	ichiji-sukēru
primary storage *[comput]*	一次記憶(装置)	ichiji-ki'oku(-sōchi)
primary structure *[aero-eng]*	一次構造	ichiji-kōzō
primary twist *[tex]*	下撚り; したより	shita-yori
primary winding *{n}* *[elecmg]*	一次巻線	ichiji-makisen
Primates *[bio]* *[v-zoo]*	霊長目; 靈長目	reichō-moku
prime cost *[econ]*	原価; 原價; げんか	genka
prime factors *[arith]*	素因子; そいんし	so'inshi
prime mover *[mech-eng]*	原動機	gendō-ki
prime number *[arith]*	素数; そすう	sosū
prime vertical *[astron]*	東西圏; 東西圈	tōzai-ken
priming water *[hyd]*	呼び水	yobi-mizu
primitive *{adj}* *[bio]*	原始の	genshi no
primitive lattice *[crys]*	単純格子	tanjun-kōshi
primordial *{n}* *[bio]*	原始的	genshi'teki
principal axis *[math]* *[mech]* *[opt]*	主軸; しゅじく	shu-jiku
principal focus (of lens) *[opt]*	主焦点	shu-shōten
principal plane *[opt]*	主平面	shu-heimen
principal point *[opt]*	主点; 主點	shuten
principal quantum number *[atom-phys]*	主量子数	shu-ryōshi-sū
principal ray *[opt]*	主光線	shu-kōsen
principal valence *[ch]*	主原子価	shu-genshi'ka
principal values *[trig]*	主値; しゅち	shuchi
principle *[ch]*	素; そ	so
" *[sci-t]*	原理	genri
" (rule) *[sci-t]*	原則	gensoku
" *[sci-t]*	法則; ほうそく	hōsoku
print; printing *{n}* *[comput]* *[graph]* *[pr]*	印字; いんじ	inji
print; printing *{n}* *[comput]* *[pr]*	印刷(術)	insatsu('jutsu)
" *{n}* *[comput]* *[pr]*	印書	insho
print (woodblock) *{n}* *[graph]*	版画; 版畫; はんが	hanga
" *{n}* *[photo]*	印画	inga

English	Japanese	Romanization
print *{vb t} [comput] [pr]*	印刷する	in'satsu suru
" *{vb t} [pr]*	刷る；する	suru
printability *[pr]*	印刷適性	insatsu-tekisei
print contrast ratio *[graph] [pr]*	印刷鮮明度	insatsu-senmei-do
printed circuit *[electr]*	印刷回路	insatsu-kairo
printed circuit board *[electr]*	プリント配線板	purinto-haisen-ban
printed wiring board *[electr]*	プリント基板	purinto-kiban
printer *[comput]*	印字装置	inji-sōchi
printing *{n} [comm] [graph]*	印刷術	insatsu'jutsu
" *{n} [photo]*	焼付け；やきつけ	yaki-tsuke
printing paper *[paper]*	印画(用)紙	inga(-yō)shi
printing paste *[tex]*	捺染糊	nasen-nori
printing press *[pr]*	印刷機；いんさつき	insatsu-ki
printing substrate *[pr]*	被印刷体	hi-in'satsu-tai
print line *[comput] [pr]*	印刷行	insatsu-gyō
" *[comput] [pr]*	印書行	insho-gyō
printout *[comput]*	印字出力	inji-shutsu'ryoku
" *[comput]*	印刷出力	insatsu-shutsu'ryoku
print station *[comput]*	印刷機構	insatsu-kikō
prior(-applied) invention *[pat]*	先願発明；先願發明	sengan-hatsu'mei
priority (pl -ties) *[comput]*	優先(度)	yūsen(-do)
" *[comput]*	優先順位	yūsen-jun'i
priority claim *[pat]*	優先権主張	yūsen'ken-shuchō
priority date *[pat]*	優先日；ゆうせんび	yūsen-bi
priority examination *[pat]*	優先審査	yūsen-shinsa
priority interrupt *[comput]*	優先割り込み	yūsen-wari'komi
priority right *[pat]*	優先権；優先權	yūsen-ken
prior technology *[pat]*	従来の技術	jūrai no gi'jutsu
prism *[crys] [geom] [opt]*	角柱	kaku'chū
" *[geom]*	多角柱	ta'kaku'chū
prison *[arch]*	刑務所	keimu'sho
privacy *[comput]*	機密；きみつ	kimitsu
private line *[comm]*	私設回路	shisetsu-kairo
private railway *[rail]*	民営鉄道	min'ei-tetsu'dō
private road *[civ-eng]*	私道	shidō
privilege *[legal]*	特権；とっけん	tokken
privileged instruction *[comput]*	特権命令	tokken-meirei
pro- *{prefix}*	副-	fuku-
" *{prefix}*	前の-；前へ-	mae no-; mae e-
probability *[math]*	確率；かくりつ	kaku'ritsu
probability density function *[prob]*	確率密度関数	kaku'ritsu-mitsudo-kansū
probability mass function *[prob]*	度数関数	dosū-kansū

English	Japanese	Romanization
probably *{adv}*	恐らく；おそらく	osoraku
" *{adv}*	多分	tabun
probe *{n} [aero-eng]*	探査機	tansa-ki
" *{n} [phys]*	探針	tanshin
problem *[math]*	問題；もんだい	mondai
proboscis *[i-zoo]*	吻；ふん	fun
" *[zoo]*	口先	kuchi'saki
proboscis monkey (animal) *[v-zoo]*	天狗猿；てんぐざる	tengu-zaru
procedure *[comput]*	手順	te'jun
" *[comput] [pat]*	手続き；手續き	te'tsuzuki
process *{n} [eng]*	工程；こうてい	kōtei
" *{n}*	過程	katei
process chart *[graph] [ind-eng]*	工程図；工程圖	kōtei-zu
process control *[eng]*	工程管理	kōtei-kanri
processed food *[cook] [food-eng]*	加工食品	kakō-shoku'hin
processibility *[eng]*	加工性	kakō-sei
processing *{n} [comput]*	処理；處理；しょり	shori
" *{n} [eng]*	加工；かこう	kakō
" *{n} [photo]*	現像処理	genzō-shori
processing aid *[poly-ch] [rub]*	加工助剤	kakō-jozai
procreation *[bio]*	生殖；せいしょく	seishoku
procure *{vb} [econ]*	調達する	chō'datsu suru
produce *{vb} [sci-t]*	生産する	seisan suru
producer (a furnace) *[ch]*	発生炉；發生爐	hassei-ro
" (person(s)) *[ecol]*	生産者	seisan'sha
" *[ind-eng]*	製造元	seizō-moto
producing device *[eng]*	製造装置	seizō-sōchi
product (of multiplication) *[arith]*	積；せき	seki
" (produced goods) *[ind-eng]*	産物	san'butsu
" (") *[ind-eng]*	製品；せいひん	seihin
" (") *[ind-eng]*	生成物	seisei'butsu
product engineer *[ind-eng]*	生産技師	seisan-gishi
production *[ind-eng]*	生産；せいさん	seisan
" *[ind-eng]*	製造；せいぞう	seizō
production control *[ind-eng]*	生産管理	seisan-kanri
production facilities *[ind-eng]*	生産設備	seisan-setsu'bi
production method *[ind-eng]*	製法	seihō
production process *[ind-eng]*	工程；こうてい	kōtei
production structure *[ind-eng]*	生産体制	seisan-taisei
production time *[comput]*	生産時間	seisan-jikan
productivity *[ind-eng]*	生産力	seisan-ryoku
" *[ind-eng]*	生産性	seisan-sei

English	Japanese	Romanization
profile; outline *{n} [graph]*	輪郭；りんかく	rinkaku
" *[math]*	断面；斷面	dan'men
profiling *{n} [eng]*	倣い；ならい	narai
profit; gain *{n} [econ]*	利益；りえき	ri'eki
profitable *[econ]*	採算が取れる	saisan ga toreru
" *[econ]*	有利	yūri
profit and loss *[econ]*	損益	son'eki
profit-making enterprise *[econ]*	営利企業；營利企業	eiri-kigyō
progeny *{n} [bio]*	子孫；しそん	shison
programming *{n} [comput]*	プログラミング	puroguramingu
progress; development; growth *{n}*	発展；發展	hatten
prohibit; inhibit *{vb}*	禁止する	kinshi suru
prohibition sign *[civ-eng]*	禁止標識	kinshi-hyō'shiki
projected image *[elec]*	映像；えいぞう	eizō
projectile *[ord]*	発射体	hassha'tai
" *[ord]*	放物体	hōbutsu'tai
projectile speed; bullet speed *[mech]*	弾速；だんそく	dansoku
project into; thrust into *{vb}*	突入する	totsu'nyū suru
projection *[geom] [graph] [math]*	射影；しゃえい	sha'ei
" *[map] [math]*	投影；とうえい	tō'ei
projection drawing *{n} [graph]*	投影図；投影圖	tō'ei-zu
projector *[eng] [opt]*	投影機	tō'ei-ki
prokaryolytic cell *[cyt]*	原核細胞	genkaku-saibō
prokaryote *[cyt]*	原核生物	genkaku-seibutsu
proliferation; multiplication *[bio]*	増殖；ぞうしょく	zō'shoku
proliferous polymerization *[poly-ch]*	増殖重合	zōshoku-jūgō
prolongation; extension *{n}*	延長；えんちょう	enchō
prolonged-sound symbol *[comput]*	長音符号	chō'on-fugō
prolong-sound mark *[pr]*	長音符	chō'on-pu
promenade; foyer *{n} [arch]*	遊歩場	yūho'jō
prometaphase *[biol]*	前中期	zenchū-ki
promethium (element)	プロメチウム	puromechūmu
prominence *[astron]*	紅炎；こうえん	kōen
promontory; cape *[geol]*	岬；みさき	misaki
promote; accelerate *{vb}*	促進する	soku'shin suru
promoter *[ch]*	助触媒；助觸媒	jo'shokubai
" ; accelerator *[ch]*	促進剤	sokushin-zai
prone; on one's face *{adj}*	俯伏せの；俯伏しの	utsubuse no; utsubushi no
pronghorn antelope (animal) *[v-zoo]*	枝角羚羊	eda'zuno-reiyō
pronoun *[gram]*	代名詞	daimei'shi
pronunciation *[ling]*	発音；はつおん	hatsu'on
" (Japanese) *[ling]*	訓；くん	kun

proof (alcohol) *[cook]*	標準強度	hyōjun-kyōdo
" *[legal]*	証拠; 證據	shōko
" *[logic]*	証明; しょうめい	shōmei
proof print *[photo]*	試し焼き	tameshi-yaki
proofreading *{n} [pr]*	校正	kōsei
proof stress *[mech] [met]*	耐力	tai-ryoku
propagate *{vb} [acous] [phys]*	伝搬する; 傳搬する	denpan suru
propagation *[bio] [microbio]*	繁殖; はんしょく	hanshoku
" *[phys]*	伝搬; でんぱん	denpan
propagation constant *[elecmg]*	伝搬定数	denpan-teisū
propagation delay *[electr]*	伝搬遅延	denpan-chi'en
propagation reaction *[org-ch]*	生長反応	seichō-hannō
propagule *[bot]*	零余子; むかご	mukago
propellant *[mat]*	噴射薬	funsha-yaku
" *[mat]*	推進薬	suishin-yaku
propeller *[mech-eng]*	プロペラ	puropera
propeller shaft *[mech-eng]*	推進軸	suishin-jiku
proper *{n}*	正式	seishiki
" *{n}*	適正; てきせい	tekisei
proper fraction *[arith]*	真分数	shin-bunsū
proper motion *[astron]*	固有運動	koyū-undō
proper time *[astron]*	固有時間	koyū-jikan
property *[econ]*	資産; しさん	shisan
" *[econ]*	財産	zaisan
property; properties *[ind-eng]*	性能	seinō
prophase *[bio]*	前期	zenki
propolis *{n} [pharm]*	蜂蠟; はちろう	hachi-rō
proportion; ratio *[alg]*	比例; ひれい	hirei
" *[alg]*	割合; わりあい	wari'ai
proportional limit *[mech]*	比例限界	hirei-genkai
proportionally *{adv} [math]*	比例して	hirei shite
proposal; suggestion *{n}*	提案	tei'an
proposition *[logic]*	命題; めいだい	meidai
proprietary article *[ind-eng]*	専売品; せんばい品	senbai'hin
proprietary name (medicine) *[pharm]*	商標名	shōhyō-mei
propulsion *[mech]*	推進力	suishin'ryoku
prosecution *[legal]*	起訴; きそ	kiso
prospecting *{n} [min-eng]*	探鉱; 探鑛	tankō
prostate gland *[anat]*	摂護腺; せつごせん	setsu'go-sen
" *[anat]*	前立腺	zen'ritsu-sen
prosthesis (pl -theses) *[med]*	人工装具	jinkō-sōgu
protease *[bioch]*	蛋白(質)分解酵素	tanpaku('shitsu)-bunkai-kōso

protect {vb t}	保護する	hogo suru
protection {n} [comput]	保護；ほご	hogo
protective agent [mat]	防護剤	bōgo-zai
" [mat]	保護剤	hogo-zai
protective clothing [cl]	保護服	hogo-fuku
protective coating [ind-eng]	保護塗装	hogo-tosō
" [photo]	保護層	hogo-sō
protective coloration [zoo]	保護色	hogo-shoku
protective gloves [cl]	保護手袋	hogo-te'bukuro
protein [bioch]	蛋白質；たんぱく質	tanpaku'shitsu
protein biosynthesis [bioch]	蛋白質の生合成	tanpaku'shitsu no sei'gōsei
protein denaturation [bioch]	蛋白質変性	tanpaku'shitsu-hensei
protein engineering [bioch]	蛋白工学	tanpaku-kōgaku
protein hydrolysate [bioch]	蛋白質加水分解物	tanpaku'shitsu-kasui-bunkai'butsu
protein resin [poly-ch]	蛋白質樹脂	tanpaku'shitsu-jushi
proteolytic enzyme [bioch]	分解酵素	bunkai-kōso
proteolytic power [bioch]	蛋白質分解力	tanpaku'shitsu-bunkai-ryoku
protic solvent [ch]	プロトン性溶媒	puroton'sei-yōbai
Protista [bio]	原生生物	gensei-seibutsu
proto- {prefix} [ch]	初級の-	shokyū no-
" {prefix}	原始の-	genshi no-
" {prefix}	最初の-	saisho no-
protoactinium; protactinium (element)	プロトアクチニウム	purotoakuchinyūmu
protocol [comput]	規約	ki'yaku
proton [phys]	陽子；ようし	yōshi
proton microscope [electr]	陽子顕微鏡	yōshi-kenbi'kyō
protoplanet [astron]	原始惑星	genshi-wakusei
protoplasm [cyt]	原形質	genkei'shitsu
protoplast [cyt]	原形質体	genkei'shitsu-tai
protosphere [meteor]	陽子圏；ようしけん	yōshi-ken
protostar [astron]	原始星	genshi-sei
prototroph [microbio]	原栄養株	gen-eiyō'kabu
prototype [eng]	原型；げんけい	genkei
Protozoa [i-zoo]	原生動物	gensei-dōbutsu
protractor [geom]	分度器；ぶんどき	bundo-ki
protruding strip [mech]	突条；突條	totsu-jō
protuberance [geol]	隆起；隆起	ryūki
proustite [miner]	淡紅銀鉱	tankō'ginkō
provide engraved {vb} [graph]	刻設する	koku'setsu suru
provision [legal]	条文；條文	jōbun
provision(s); definition {n} [legal]	規定	kitei

proviso (pl -sos, -soes); provision *[legal]*	但し書；ただしがき	tadashi'gaki
prow (of ship) *[nav-arch]*	船首	senshu
proximal *{adj} [med]*	近位の	kin'i no
proximity effect *[elec]*	近接効果	kin'setsu-kōka
pruning *{n} [bot]*	剪定；せんてい	sentei
Prussian blue (color) *[opt]*	紺青；こんじょう	konjō
prussic acid	青酸	seisan
pry open *{vb t}*	抉じ開ける	koji-akeru
pseudo- *{prefix}*	擬-	gi-
" *{prefix}*	擬似-	giji-
" *{prefix}*	仮の-；假の-	kari no-
pseudo acid *[ch]*	擬酸	gi-san
pseudoaromaticity *[org-ch]*	擬似芳香族性	giji-hōkō'zoku-sei
pseudoasymmetry *[p-ch]*	擬不斉；擬不齊	gi'fusei
pseudogene *[gen]*	偽遺伝子；僞遺傳子	gi-idenshi
Pseudomonas *[microbio]*	シュードモナス	shūdomonasu
pseudoperiod *{n} [geophys]*	擬似週期	giji-shūki
pseudoplasmodium *[i-zoo]*	偽変形体；僞變形體	gi'henkei'tai
pseudoplastic flow *[fl-mech]*	擬塑性流動	gi'sosei-ryūdō
pseudoplastic fluid *[fl-mech]*	擬塑性流動体	gi'sosei-ryūdō'tai
pseudopod(ium) (pl -podia) *[cyt]*	仮足；假足；かそく	ka'soku
pseudorandom number sequence *[comput]*	擬似乱数列	giji-ransū-retsu
pseudosymmetry *[crys]*	擬似対称；擬似對稱	giji-taishō
pseudotenacity; pseudotoughness *[mech]*	擬靭性；ぎじんせい	gi'jinsei
Psilocybe venetana (poisonous mushroom) *[bot]*	痺れ茸	shibire-take
Psilopsida *[bot]*	古生松葉蘭類	kosei-matsuba'ran-rui
p sound *[ling]*	半濁音	han-daku'on
p-sound mark *[pr]*	半濁音符	han'dakuon-pu
psychedelic drug *{n} [pharm]*	幻覚剤	genkaku-zai
psychedelic substance *[pharm]*	精神異常発現物質	seishin-ijō-hatsu'gen-busshitsu
psychology *[bio]*	心理；しんり	shinri
" (the discipline) *[bio]*	心理学	shinri'gaku
psychometrics *[psy]*	精神測定学	seishin-sokutei'gaku
psychoparmacology *[pharm] [psy]*	精神薬理学	seishin-yakuri'gaku
psychrometer *[eng]*	乾湿計；乾濕計	kan'shitsu-kei
Pteropsida *[bot]*	歯朶類；羊齒類	shida-rui
puberty *[physio]*	春機発動期	shunki-hatsu'dō-ki
public *{adj}*	公の；おおやけの	ōyake no
public accountant *[econ]*	計理士	keiri'shi

English	Japanese	Romanization
public address system *[acous]*	拡声装置；擴聲裝置	kakusei-sōchi
publication *[pat]*	公知	kōchi
" *[pat]*	公告；こうこく	kōkoku
" *[pat] [pr]*	発行；發行	hakkō
" *[pat] [pr]*	発表	happyō
" *[pr]*	出版；しゅっぱん	shuppan
publication date *[pat]*	発行日	hakkō-bi
publication of applicaton *[pat]*	出願公告	shutsu'gan-kōkoku
publications *[pr]*	書籍；しょせき	shoseki
public body *{n}*	公共団体；公共團體	kōkyō-dantai
public domain *[legal]*	公有	kōyū
public hazard *[ecol]*	公害；こうがい	kōgai
public health *[med]*	公衆衛生	kōshū-eisei
public loan *[econ]*	公債	kōsai
public opinion survey *[comm] [psy]*	世論調査	seron-chōsa
public road; highway *[civ-eng]*	公道	kōdō
public utility *[ind-eng]*	公用施設	kōyō-shisetsu
public water domain *[ind-eng]*	公共水域	kōkyō-sui'iki
publish *{vb t} [pr]*	発行する	hakkō suru
" *{vb} [pr]*	出版する	shuppan suru
published volume *[pr]*	成書	seisho
pucker *{n} [tex]*	皺；しわ	shiwa
puff; swelling *{n}*	脹らみ	fukurami
puffer; globefish; swellfish; blowfish; balloonfish (fish) *[v-zoo]*	河豚；ふぐ	fugu
pug dog (animal) *[v-zoo]*	狆；ちん	chin
pull *{n} [eng]*	引き	hiki
pull; tug; draw *{vb} [mech]*	引く	hiku
pulley (pl pulleys) *[eng]*	滑車；かっしゃ	kassha
" *[eng]*	轆轤；ろくろ	roku'ro
pull hand over hand *{vb t}*	手繰る；たぐる	taguru
pulmonary artery *[anat]*	肺動脈	hai-dōmyaku
pulmonary vein *[anat]*	肺静脈	hai-jōmyaku
pulp; dental pulp *[hist]*	歯髄；しずい	shizui
pulsating load *[mech-eng]*	片振(り)荷重	kata'buri-kajū
pulsating star *[astron]*	脈動星	myaku'dō-sei
pulsation *[elec]*	脈動；みゃくどう	myaku'dō
pulse *{n} [physio]*	脈拍	myaku'haku
" *{n} [physio]*	パルス	parusu
pulse height *[electr]*	波高	hakō
pulse modulation *[comm]*	位相変調；位相變調	isō-henchō
pulverization *[mech-eng]*	粉砕；ふんさい	funsai

pulverized coal *[geol]*	微粉炭	bifun-tan
pumice *[geol]*	軽石；かるいし	karu-ishi
pump *{n} [mech-eng]*	ポンプ	ponpu
pump head *[fl-mech]*	揚程；ようてい	yōtei
pumping (of water) *{n} [hyd]*	揚水	yōsui
" (of air) *{n} [mech-eng]*	排気；排氣；はいき	haiki
pumpkin (vegetable) *[bot]*	南瓜；かぼちゃ	kabocha
punch *{n} [eng]*	ポンチ	ponchi
punch *{n} [eng]*	打抜き；うちぬき	uchi'nuki
punch out *{vb} [eng]*	打ち抜く	uchi-nuku
punching *{n} [eng]*	穿孔；せんこう	senkō
" ; blanking *{n} [eng]*	打抜き	uchi'nuki
punching quality *[eng]*	打抜き加工性	uchi'nuki-kakō-sei
punctuation *[gram]*	句読法；句讀法	kutō-hō
punctuation character *[comput]*	句読文字	kutō-moji
punctuation mark *[gram] [pr]*	句読点；くとうてん	kutō-ten
pungency *[cook] [physio]*	辛味；からみ	kara'mi
punkie; biting midge (insect) *[i-zoo]*	糠蚊；ぬかか	nuka-ka
pupa (pl pupae) *[i-zoo]*	蛹；さなぎ	sanagi
pupil (of eye) *[anat]*	瞳孔	dōkō
" (of eye) *[anat]*	瞳；ひとみ	hitomi
" (school child) *{n}*	学童；學童	gaku'dō
puppy (animal) *[v-zoo]*	小犬	ko-inu
purchasing *{n} [econ]*	購買	kōbai
pure *{adj} [math]*	純粋の	junsui no
pure chance *[math]*	純偶然	jun-gūzen
pure color *[opt]*	単一色	tan'itsu-shoku
pure culture *[microbio]*	純(粋)培養	jun(sui)-baiyō
pure difference *[arith]*	生の差；きのさ	ki no sa
purely chemical method *[ch]*	純化学的方法	jun-kagaku'teki-hōhō
pure substance *[phys]*	純粋物質	junsui-busshitsu
pure tone; pure note *[acous]*	純音	jun'on
pure value *[math]*	生の値；きのあたい	ki no atai
purification *[eng]*	純化	junka
" *[eng]*	精製；せいせい	seisei
purified water *[hyd]*	精製水	seisei-sui
purl *{n} [tex]*	裏目	ura'me
purple (color) *[opt]*	紫；むらさき	murasaki
purple of the ancients (color) *[opt]*	古代紫	kodai-murasaki
purplish Washington clam *[i-zoo]*	ムラサキ貝	murasaki-gai
purpose; end; object *{n}*	目的	mokuteki
purslane; portulaca *[bot]*	滑り莧；すべりひゆ	suberi-hiyu

pursuit *[comput]*	追跡	tsui'seki
pus *[med]*	膿；うみ	umi
push; press *{vb} [mech]*	押す	osu
pushability *[mech]*	押し易さ	oshi-yasu-sa
push button *[electr]*	押釦；押ボタン	oshi-botan
pushdown list *[comput]*	後入れ先出し表	ato'ire-saki'dashi-hyō
push rod *[mech-eng]*	押し棒	oshi-bō
pussy willow *[bot]*	猫柳；ねこやなぎ	neko-yanagi
put away; finish *{vb}*	仕舞う	shimau
put in place *{vb t}*	据える；すえる	sueru
putrefaction *[bioch]*	腐敗	fuhai
putty *[mat]*	パテ	pate
pycnocline *[geophys]*	密度躍層	mitsudo-yakusō
pycnometer *[eng]*	比重瓶	hijū-bin
pylon *[civ-eng]*	目標塔	mokuhyō-tō
" *[civ-eng]*	鉄塔；鐵塔	tettō
pyramid *[geom]*	角錐；かくすい	kaku'sui
pyrargyrite *[miner]*	濃紅銀鉱	nōkō'ginkō
pyrgeometer *[eng]*	夜間放射計	yakan-hōsha-kei
pyriform *{adj} [bot]*	洋梨状の	yō'nashi'jō no
pyrite *[miner]*	黄鉄鉱；黄鐵鑛	ō'tekkō
pyritohedron *[crys]*	五角十二面体	go'kaku-jūni'men'tai
pyro- *{prefix} [ch]*	火の-	hi no-
" *{prefix} [thermo]*	熱の-	netsu no-
pyroelectricity *[sol-st]*	焦電気	shō'denki
pyroligneous *{adj} [ch-eng]*	焦木性の	shōmoku'sei no
pyroligneous acid *[org-ch]*	木酢液	moku'saku-eki
" *[org-ch]*	木酢酸	moku'saku-san
pyrolysis *[ch]*	熱分解	netsu-bunkai
pyrometer *[eng]*	高温計	kō'on-kei
pyromorphite *[miner]*	緑鉛鉱；緑鉛鑛	ryoku'enkō
pyrope *[miner]*	紅榴石	kōryū'seki
pyrophyllite *[miner]*	葉蠟石	yō'rōseki
pyrotechnics *[eng] [mat]*	火工術	kakō'jutsu
pyrrhotite *[miner]*	磁硫鉄鉱；磁硫鐵鑛	jiryū'tekkō
pyruvic acid *[bioch]*	ピルビン酸	pirubin-san
python (reptile) *[v-zoo]*	錦蛇；にしきへび	nishiki-hebi

Q

quadrangle *[civ-eng]*	方庭	hōtei
quadrant *[an-geom]*	四分円；四分圓	shi'bun-en
" *[an-geom]*	象限；しょうげん	shōgen
quadrantal *{adj} [navig]*	象限の	shōgen no
quadrate *{adj} [geom]*	四辺形の	shihen'kei no
quadratic *{adj} [math]*	二次の；にじの	ni'ji no
quadratic equation *[alg]*	二次方程式	niji-hōtei'shiki
quadratic form *[alg]*	二次形式	niji-kei'shiki
quadratic function *[alg]*	二次関数	niji-kansū
quadratic programming *{n} [comput]*	二次計画法	niji-keikaku-hō
quadratic surface *[math]*	二次曲面	niji-kyoku'men
quadrature *[astron]*	直角離角	chokkaku-ri'kaku
" *[math]*	求積	kyūseki
quadrature component *[elec]*	直角分	chokkaku-bun
quadric *{n} [math]*	二次曲線	niji-kyokusen
quadridentate ligand *[ch]*	四座配位子	shiza-hai'i'shi
quadrilateral *[geom]*	四辺形；四邊形	shihen-kei
quadrillion (U.S.) *[math]*	千兆；せんちょう	senchō
quadriplegia *[med]*	四肢麻痺	shishi-mahi
quadriplegic (a person) *[med]*	四肢麻痺患者	shishi-mahi-kanja
quadripolymer *[poly-ch]*	四元共重合体	shigen-kyōjūgō'tai
" *[poly-ch]*	四元共重合体	yongen-kyōjūgō-tai
quadrivalence *[ch]*	四価；よんか	yonka
quadrivalent *{adj} [ch]*	四価の；四價の	yonka no
quadrupeds *[v-zoo]*	四足類	shi'soku-rui
quadruple *{vb i}*	四倍になる	yonbai ni naru
" *{vb t}*	四倍にする	yonbai ni suru
quadruple point *[p-ch]*	四重点；四重點	shijū-ten
quadruple precision *[eng]*	四倍精度	yonbai-seido
quadruplicity *{n}*	四重性	shijū-sei
quadrupole *{n} [elecmg]*	四重極	shijū-kyoku
quadrupole resonance *[elecmg]*	四極子共鳴	shi'kyokushi-kyōmei
quail (bird) *{n} [v-zoo]*	鶉；うずら	uzura
quake *{n} [geophys]*	振動；しんどう	shindō
" *{n} [geophys]*	揺れ；搖れ；ゆれ	yure
qualifier *[comput] [gram]*	修飾子	shūshoku'shi
" *[gram]*	限定詞；げんていし	gentei'shi
qualitative *[ch]*	定性的	teisei'teki
qualitative analysis *[an-ch]*	定性分析	teisei-bunseki

quality *[ind-eng]*	品位；ひんい	hin'i
" *[ind-eng]*	品質；ひんしつ	hin'shitsu
quality assurance *[ind-eng]*	品質保証	hinshitsu-hoshō
quality change *{n}*	変質；變質	hen'shitsu
quality control *[stat]*	品質管理	hinshitsu-kanri
quality standard *[ind-eng]*	品質基準	hinshitsu-kijun
" *[ind-eng]*	品質規格	hinshitsu-kikaku
quantification *[math]*	数量化	sūryō'ka
" *[sci-t]*	量化	ryō'ka
quantifier *[gram]*	数量詞	sūryō'shi
quantitative *[ch]*	定量的	teiryō'teki
quantitative analysis *[an-ch]*	定量分析	teiryō-bunseki
quantity (pl -ties) *[math]*	量；りょう	ryō
" *[math]*	数量；數量	sūryō
quantity of electricity; charge *[elec]*	電気量	denki-ryō
quantity of heat *[thermo]*	熱量；ねつりょう	netsu-ryō
quantity of light *[opt]*	光量	kōryō
quantity of magnetism *[phys]*	磁気量；じきりょう	jiki-ryō
quantity penetrated *[civ-eng]*	貫入度	kannyū-do
quantity used *[sci-t]*	使用量	shiyō-ryō
quantization *[quant-mech] [sci-t]*	量子化	ryōshi'ka
quantum (pl quanta) *[quant-mech]*	量子；りょうし	ryōshi
quantum chemistry *[ch]*	量子化学	ryōshi-kagaku
quantum defect *[atom-phys]*	量子偏差	ryōshi-hensa
quantum efficiency *[electr]*	量子効率	ryōshi-kō'ritsu
quantum electrodynamics *[quant-mech]*	量子電気力学	ryōshi-denki'rikigaku
quantum leap; quantum jump *[phys]*	量子飛躍	ryōshi-hi'yaku
quantum mechanics *[phys]*	量子力学	ryōshi-rikigaku
quantum number *[quant-mech]*	量子数	ryōshi-sū
quantum statistics *[stat-mech]*	量子統計	ryōshi-tōkei
quantum yield *[p-ch]*	量子収量；量子收量	ryōshi-shūryō
quarantine *{n} [med]*	検疫；けんえき	ken'eki
quarantine anchorage *[civ-eng]*	検疫錨地	ken'eki-byōchi
quarantine ship *[nav-arch]*	検疫船	ken'eki'sen
quark *[part-phys]*	クォーク	kuōku
quarry *{n} [eng]*	石切場；いしきりば	ishi'kiri-ba
" *{n} [eng]*	採石場	saiseki-ba; saiseki'jō
quarrying *{n} [eng]*	採石	sai'seki
quarry stone *[mat]*	野面石；のづらいし	no'zura-ishi
quarter *{n} [arith]*	四分の一	shi'bun no ichi; yon'bun no ichi
quarter binding *{n} [pr]*	背革；せがわ	se'gawa

quarter deck *[nav-arch]*	船尾甲板	senbi-kōhan
quartering (lumber) *{n} [eng]*	四つ割り; よつわり	yotsu'wari
" *{n} [civ-eng]*	四分法	shibun-hō
quarterly (journal) *{n} [pr]*	年四回刊行物	nen'shikai-kankō'butsu
quartermaster (a person) *[navig]*	操舵手; そうだしゅ	sōda'shu
quarter-wave(length) plate *[opt]*	四分の一波長板	shibun-no-ichi hachō'ban; yonbun-no-ichi hachō'ban
quarter-wave(length) transformer *[elecmg]*	四分の一波長変成器	shibun-no-ichi (yonbun-no-ich) hachō hensei-ki
quartet *[music]*	四重奏	shijū'sō
" *[phys]*	四重項	shijū-kō
quartile *[stat]*	四分値; しぶんち	shi'bun-chi
" *[stat]*	四分位数	shibun'i-sū
quartile deviation *[stat]*	四分位偏差	shibun'i-hensa
quarto (pl quartos) *[pr]*	四つ折り版	yotsu'ori-ban
quartz *[miner]*	石英; せきえい	seki'ei
quartz clock *[horol]*	水晶時計	suishō-dokei
quartzite *[petr]*	石英岩	seki'ei'gan
" *[petr]*	珪岩	keigan
quartz lamp *[electr]*	石英水銀灯	seki'ei-suigin'tō
quartz porphry *[miner]*	石英斑岩	seki'ei-hangan
quartz sand *[geol]*	珪砂; けいしゃ	keisha
quartz schist *[miner]*	石英片岩	seki'ei-hengan
quartz spectrograph *[spect]*	水晶分光写真器	suishō-bunkō-shashin-ki
quartz vibrator *[electr]*	水晶振動子	suishō-shindō'shi
quartz wedge *[opt]*	水晶楔	suishō-kusabi
quasar *[astron]*	恒星状天体	kōsei'jō-tentai
quasi- *{prefix}*	擬似の-	giji no-
" *{prefix}*	準-	jun-
" *{prefix}*	類似の-	rui'ji no-
quasi-chemical method *[ch]*	準化学的方法	jun-kagaku'teki-hōhō
quasi-crystal *[crys]*	準結晶	jun-kesshō
quasi-crystalline phase *[crys]*	準結晶相	jun'kesshō-sō
quasi drug *[med]*	医薬部外品	i'yaku-bugai'hin
quasi-elastic light scattering *[opt]*	準弾性光散乱	jun'dansei-kō'sanran
quasi-elastic scattering *{n} [mech]*	準弾性散乱	jun'dansei-sanran
quasi-equilibrium *{n} [geol]*	準平衡	jun-heikō
quasi-isotropic(ness) *{adj} [phys]*	擬似等方	giji-tōhō
quasi-linear equation *[math]*	準線形方程式	jun'senkei-hōtei'shiki
quasi-particle *[phys]*	準粒子	jun-ryūshi
quasi-periodic *[sci-t]*	準周期的	jun-shūki'teki
quasi-stationary front *[meteor]*	準停滞前線	jun'teitai-zensen

quasi-stellar galaxy *[astron]*	恒星状銀河	kōsei'jō-ginga
quasi-stellar radio source; quasar *[astron]*	準(恒)星(状)電波源	jun(kō)sei('jō)-denpa'gen
quaternary alloy *[met]*	四元合金	yongen-gōkin; shigen-gōkin
quaternary amine *[org-ch]*	第四級アミン	dai'yonkyū-amin
quaternary carbon atom *[ch]*	第四炭素原子	dai'shi-tanso-genshi
quaternary compound *[ch]*	第四(級)化合物	dai'shi(kyū)-kagō'butsu
quaternary structure *[bioch]*	四次構造	yoji-kōzō
quaternion *[math]*	多元数	tagen'sū
quaternization *[ch]*	四級化	shikyū'ka; yonkyū'ka
quatrefoil *{n} [arch] [lap]*	四つ葉	yotsu'ba
quay *[civ-eng]*	埠頭; ふとう	futō
" *[civ-eng]*	波止場; はとば	hatoba
" *[civ-eng]*	桟橋; 棧橋	san'bashi
quaywall *[civ-eng]*	崖壁	ganpeki
queen bee (insect) *[i-zoo]*	王蜂; おうほう	ō-hō
queen cell *[i-zoo]*	王台; 王臺	ō-dai
queen substance *[i-zoo]*	女王物質	jo'ō-busshitsu
quench aging *{n} [met]*	焼入時効	yaki'ire-jikō
quenched thin strip *[met]*	急冷薄帯; 急冷薄帶	kyūrei-haku'tai
quencher (of light) *[opt]*	消光物質	shōkō-busshitsu
" *[opt]*	消光剤	shōkō-zai
quench hardening *{n} [met]*	焼(き)入れ	yaki'ire
quenching *{n} [eng] [met]*	焼入れ; やきいれ	yaki'ire
" *{n} [met]*	急冷	kyūrei
" *{n} [electr]*	消滅; しょうめつ	shōmetsu
" *{n} [sol-st]*	消光	shōkō
quench(ing) crack *[met]*	焼割れ	yaki-ware
quenching diagram *[met]*	焼入状態図	yaki'ire-jōtai-zu
quenching of polarized light *[opt]*	消偏光	shō-henkō
quenching oil *[met]*	焼(き)入れ油	yaki'ire-yu
quenching solution treatment *[met]*	急冷固溶処理	kyūrei-koyō-shori
quenching stress *[met]*	焼入応力; 焼入應力	yaki'ire-ō'ryoku
query *{n} [comm] [comput]*	問い合わせ	toi-awase
question (problem) *{n} [math]*	問題; もんだい	mondai
" (doubt) *{n}*	疑問; ぎもん	gimon
" (query) *{n}*	質問	shitsu'mon
" (query) *{n}*	問い	toi
question mark *[comput] [gram] [pr]*	疑問符	gimon-fu
queue; queueing *{n} [comput] [math]*	(待ち)行列	(machi-)gyō'retsu
quick *{adj} [phys]*	早い; 速い; はやい	hayai
quick-acting fertilizer *[agr]*	速効性肥料	sokkō'sei-hiryō

quick-acting relay *[elec]*	速動継電器	soku'dō-keiden-ki
quick-break switch *[elec]*	速切りスイッチ	haya'kiri-suitchi
quick freezing *{n} [eng]*	急速凍結	kyūsoku-tōketsu
quick-hardening admixture *[mat]*	急硬剤	kyūkō-zai
quicklime *[inorg-ch]*	生石灰	sei'sekkai
quick return mechanism *[mech-eng]*	早戻り機構	haya'modori-kikō
quicksand *[geol]*	流砂；りゅうさ	ryū'sa
quick-setting *{adj} [eng] [mat]*	急結の	kyū'ketsu no
quicksilver; mercury *[ch]*	水銀；すいぎん	suigin
quick start *[mech-eng]*	急速起動	kyūsoku-kidō
quiescence *{n} [zoo]*	休止	kyūshi
quiescent *{adj}*	静止した	seishi shita
quiescent point *[elec]*	静止点；靜止點	seishi-ten
quiescent prominence *[astrophys]*	静かな紅炎	shizuka na kōen
quiet *{adj} [acous]*	静かな；しずかな	shizuka na
" (calm down) *{vb t}*	静める	shizumeru
quiet sun *[astrophys]*	静かな太陽	shizuka na taiyō
quill *{n} [v-zoo]*	羽軸根；うじくこん	ujiku'kon
" *{n} [eng]*	管；くだ	kuda
quillwort; Isosetes *[bot]*	水韮；みずにら	mizu-nira
quilt weave *[tex]*	刺し子織り	sashi'ko-ori
quinary *{n} [comput]*	五進法	goshin-hō
quinhydrone electrode *[an-ch]*	キンヒドロン電極	kinhidoron-denkyoku
quinine; cinchona *[org-ch]*	規那；きな	kina
quintessence (extraction) *{n} [ch]*	高純抽出物	kōjun-chū'shutsu'butsu
quintet *[quant-mech]*	五重項	gojū-kō
" *[spect]*	五重線	gojū-sen
quintile *{n} [stat]*	五分位数	go-bun'i-sū
quintuplets *[bio]*	五つ子	itsutsu'go
quire (paper) *[mat]*	帖；じょう	jō
quod vide (abbrev q.v.); which see	そこを見よ	soko o miyo
quoin *{n} [arch]*	隅石；すみいし	sumi-ishi
quonset hut *[arch]*	半月型小屋	han'getsu'gata-koya
quota *[math]*	割当量；割當量	wari'ate-ryō
quota method *[math]*	割当法	wari'ate-hō
quotation *[pr]*	引用	in'yō
quotation mark *[gram] [comput] [pr]*	引用符；いんようふ	in'yō-fu
quotient *[arith]*	商；しょう	shō

R

rabbit (animal) *[v-zoo]*	兎；兔；うさぎ	usagi
rabbitfish; Siganus fuscescens (fish) *[v-zoo]*	藍子；あいご	aigo
rabbit hair *[mat] [tex]*	家兎毛	kato-mō
" (rabbit fiber) *[tex]*	兎毛；ともう	to-mō
rabies; hydrophobia *[med]*	恐犬病	kyōken'byō
raccoon (animal) *[v-zoo]*	洗い熊	arai-guma
raccoon dog (animal) *[v-zoo]*	狸；たぬき	tanuki
race *{n} [bio]*	人種	jinshu
" *{n} [bio]*	民族；みんぞく	min'zoku
" (tribe; family) *{n} [bio]*	種族	shu'zoku
" *{n} [bot]*	品種	hinshu
race horse (animal) *[v-zoo]*	競争馬	kyōsō-ba
raceme *[bot]*	総状花序	sōjō-kajo
racemic modification *[org-ch]*	ラセミ体	rasemi-tai
racing *{n} [cont-sys] [elec]*	乱調；亂調	ranchō
rack *{n} [mech-eng]*	歯桿；はざお	ha'zao
radial (direction) *{adj} [geom]*	半径方向の	hankei-hōkō no
radial symmetry *[bot] [sci-t]*	放射相称	hōsha-sōshō
radial velocity *[mech]*	半径方向速度	hankei-hōkō-sokudo
" *[mech]*	視線速度	shisen-sokudo
radian measure *[geom]*	弧度法；こどほう	kodo-hō
radiant *{n} [phys]*	光点；光點	kōten
radiant efficiency *[opt]*	放射効率	hōsha-kō'ritsu
radiant flux *[opt]*	放射束	hōsha-soku
radiant heat *[thermo]*	輻射熱	fukusha-netsu
radiation *[phys]*	輻射；ふくしゃ	fuku'sha
" *[phys]*	発散；發散	hassan
" *[phys]*	放射(線)	hōsha('sen)
radiation chemistry *[nucleo]*	放射線化学	hōshasen-kagaku
radiation damage *[nucleo]*	放射線損傷	hōshasen-sonshō
radiation dose *[nucleo]*	放射線量	hōshasen-ryō
radiation dose rate *[nucleo]*	放射線量率	hōshasen'ryō-ritsu
radiation effect *[bio]*	放射線効果	hōshasen-kōka
radiation hazard *[med]*	放射線危険性	hōshasen-kiken-sei
radiation-induced mutation *[gen]*	放射線誘発突然変異	hōshasen-yūhatsu-totsu'zen-hen'i
radiation polymerization *[poly-ch]*	放射線重合	hōshasen-jūgō
radiation sickness; radiotoxemia *[med]*	放射線宿酔	hōshasen-yado'yoi

radiation shield *[eng]*	放射線遮蔽	hōshasen-shahei
radiation therapy *[med]*	放射線治療	hōshasen-chiryō
radiation warning symbol *[nucleo]*	放射能標識	hōsha'nō-hyō'shiki
radiator *[eng]*	放熱器	hō'netsu-ki
radical *{n} [alg]*	根基；こんき	konki
" *{n} [alg]*	累乗；累乘	ruijō
" *{n} [org-ch]*	基；き	ki
radicand *[alg]*	被開平数	hi-kaihei'sū
radioactive *{adj} [nuc-phys]*	放射性の	hōsha'sei no
radioactive dating *{n} [nucleo]*	放射性年代測定	hōshasei-nendai-sokutei
radioactive decay *[nuc-phys]*	放射性崩壊	hōshasei-hōkai
radioactive dust *[nuc-phys]*	放射性塵	hōshasei-chiri
radioactive element *[ch] [nuc-phys]*	放射性元素	hōshasei-genso
radioactive fallout *[nucleo]*	放射性降下物	hōshasei-kōka'butsu
radioactive rain *[meteor]*	放射能雨	hōsha'nō-ame
radioactive rays *[nuc-phys]*	放射線	hōsha-sen
radioactive substance *[nuc-phys]*	放射性物質	hōshasei-busshitsu
radioactive tracer *[nucleo]*	放射性追跡子	hōshasei-tsuiseki'shi
radioactive waste disposal *[nucleo]*	放射性廃棄物処理	hōshasei-haiki'butsu-shori
radioactivity *[nuc-phys]*	放射能	hōsha'nō
radioassay *[an-ch]*	放射能分析	hōsha'nō-bunseki
radio astronomy *[astron]*	電波天文学	denpa-tenmon'gaku
radio beacon *[navig]*	航空標識	kōkū-hyō'shiki
radiobiology *[bio]*	放射線生物学	hōshasen-seibutsu'gaku
radiocarbon dating *{n} [nucleo]*	放射性炭素年代測定	hōshasei-tanso-nendai-sokutei
radiochemistry *[ch]*	放射化学	hōsha-kagaku
radio frequency *[elecmg]*	無線周波	musen-shūha
radio galaxy *[astrophys]*	電波銀河	denpa-ginga
radioimmunoassay *[immun]*	放射線免疫検定法	hōshasen-men'eki-kentei-hō
" *[immun]*	放射免疫測定法	hōsha-men'eki-sokutei-hō
radio interference *[comm] [phys]*	混信；こんしん	konshin
radioisotope *[nuc-phys]*	放射性同位元素	hōshasei-dō'i-genso
" *[nuc-phys]*	放射線同位体	hōshasen-dōi'tai
Radiolaria *[i-zoo]*	放射虫類	hōsha'chū-rui
radiologist (a person) *[med]*	放射線専門医	hōshasen-senmon'i
radiology *[med]*	放射線医学	hōshasen-i'gaku
radioluminescence *[phys]*	放射性発光	hōshasei-hakkō
radiometer *[electr]*	放射計	hōsha-kei
radiometric analysis *[an-ch]*	放射線測定分析	hōshasen-sokutei-bunseki
radiometric titration *[an-ch]*	放射滴定	hōsha-teki'tei
radiomimetic substance *[ch]*	放射線類似作用物質	hōshasen-ruiji'sayō-busshitsu

radio navigation *[navig]*	無線航行	musen-kōkō
radio objects *[astron]*	電波天体	denpa-tentai
radiophototelegraphy; radiophoto; radiophotography *[comm] [graph]*	無線写真電送	musen-shashin-densō
radio screening *{n} [elecmg]*	電波遮蔽	denpa-shahei
radio source *[elecmg]*	電波源	denpa-gen
radio station *[comm]*	無線局	musen-kyoku
radiotelegraph station *[comm]*	無線電信局	musen-denshin'kyoku
radiotelegraphy *[comm]*	無線電信	musen-denshin
radiotelephony *[comm]*	無線電話	musen-denwa
radiotelescope *[eng]*	電波望遠鏡	denpa-bōenkyō
radio wave *[elecmg]*	電波	denpa
radium (element)	ラジウム	rajūmu
radius (pl radii) (bone) *[anat]*	橈骨; とうこつ	tō-kotsu
" *[geom]*	半径; はんけい	hankei
radius of curvature *[geom]*	曲率半径	kyoku'ritsu-hankei
radius vector *[astron]*	動径; 動徑	dōkei
radix *[math]*	基数; きすう	kisū
radix complement *[math]*	基数の補数	kisū no hosū
radix point *[math]*	基点	kiten
" *[math]*	小数点	shōsū-ten
raffinate *[ch]*	抽残液	chūzan-eki
" *[ch-eng]*	精整油	seisei-yu
rail (bird) *[v-zoo]*	水鶏; くいな	kuina
rail gun *[ord]*	電磁砲; 電磁砲	denji-hō
railing *{n} [arch]*	手摺り; てすり	te'suri
" *{n} [eng]*	高欄	kōran
railroad; railway *[rail]*	鉄道; てつどう	tetsu'dō
railroad crossing *{n} [civ-eng]*	踏切	fumi-kiri
railroad track *[rail]*	線路	senro
railway ferry *[nav-arch]*	鉄道連絡船	tetsudō-renraku'sen
railway freight *[trans]*	貨物	ka'motsu
railway radiotelephone *[comm]*	列車電話	ressha-denwa
railway signal *[rail]*	鉄道信号	tetsudō-shingō
rain *{n} [meteor]*	雨; あめ	ame
rain upon; fall; descend *{vb}*	降る	furu
rainbow *[opt]*	虹	niji
rainbow trout (fish) *[v-zoo]*	虹鱒; にじます	niji-masu
rain damage *[ecol]*	雨害	u'gai
raindrop *[hyd]*	雨滴; うてき	u'teki
rainfall *[meteor]*	降雨量	kō'u'ryō
" *[meteor]*	雨量	u'ryō

rainforest *[ecol]*	雨林; うりん	u'rin
rain gage *[eng]*	雨量計	u'ryō-kei
rain green forest *[ecol]*	雨緑樹林	u'ryoku-jurin
rain shower (early winter) *[meteor]*	時雨; しぐれ	shigure
rainstorm *[meteor]*	暴風雨	bōfū'u
rainwash *[geol]*	雨食	u'shoku
rainwater *[hyd]*	雨水	u'sui
rainwater barrel (or tank) *[eng]*	用水槽	yōsui-sō
rainy season *[meteor]*	梅雨期	bai'u-ki; tsuyu-ki
" *[meteor]*	雨季	u-ki
rainy weather *[meteor]*	雨天	u-ten
raise (erect) *{vb t} [arch]*	立てる	tateru
" (lift) *{vb t} [eng]*	上げる; 揚げる	ageru
" *{vb t} [math]*	累乗する	ruijō suru
" (breed animal, etc.) *{vb}*	飼う; かう	kau
raisin *[cook]*	干葡萄; 乾し葡萄	hoshi-budō
raising (crops) *{n} [agr]*	栽培; さいばい	saibai
" (animals) *{n} [agr]*	飼育	shi'iku
rake *{n} [eng]*	熊手; くまで	kuma'de
rake angle *[eng]*	掬い角; すくいかく	sukui-kaku
rake ratio *[nav-arch]*	傾斜比	keisha-hi
ram (animal) *[v-zoo]*	雄羊; おすひつじ	osu-hitsuji
ramie *[tex]*	麻; あさ	asa
ramp *[civ-eng]*	斜面	sha'men
rancidification *[cook]*	酸敗	sanpai
rancidity *[cook]*	酸敗度	sanpai-do
random *{n} [math]*	無作為; むさくい	mu'saku'i
random access *[comput]*	即時呼び出し	soku'ji-yobi'dashi
random access memory; RAM *[comput]*	即時呼び出し記憶	sokuji-yobi'dashi-ki'oku
randomization *[math]*	無作為化	mu-saku'i'ka
randomness *[math]*	無秩序; むちつじょ	mu'chitsujo
random numbers *[math]*	乱数; 亂數	ransū
random number sequence *[math]*	乱数配列	ransū-hai'retsu
random variable *[stat]*	確率変数	kaku'ritsu-hensū
range *{n} [alg]*	値域; ちいき	chi'iki
" (of flight) *{n} [navig]*	飛程	hitei
" *{n} [stat]*	範囲; 範圍; はんい	han'i
range-finding *{n} [eng]*	距離測定	kyori-sokutei
ranid; true frog (amphibian) *[v-zoo]*	赤蛙; あかがえる	aka-gaeru
rank *{n} [alg]*	階数	kaisū
" *{n} [bot]*	階級; かいきゅう	kaikyū
" *{vb}*	位置付けする	ichi'zuke suru

ranking; order; succession {n} [math]	順位；じゅんい	jun'i
rank of coalification [geol]	石炭化度	sekitan'ka-do
rape; cole (vegetable) [bot]	油菜；あぶらな	abura'na
rape blossom [bot]	菜の花；なのはな	na-no-hana
rapeseed oil; canola oil [mat]	菜種油	na'tane-abura
" [mat]	種油；たねあぶら	tane-abura
rapid; fast; quick {adj} [phys]	早い；速い；はやい	hayai
rapid analysis [ch]	迅速分析	jin'soku-bunseki
rapidly {adv}	早く；速く	hayaku
rapids [hyd]	早瀬；早瀨；はやせ	haya'se
" [hyd]	急流；きゅうりゅう	kyūryū
" (a current) [hyd]	瀬；瀨；せ	se
Raptores (birds of prey) [v-zoo]	猛禽類	mōkin-rui
rare; rarefied {adj} [fl-mech]	希薄な；きはくな	ki'haku na
" ; infrequent {adj}	稀な；まれな	mare na
rare acids [ch]	希酸類	kisan-rui
rare earth element [ch]	希土類元素	kido'rui-genso
rare element [ch]	希有元素；けう元素	ke'u-genso
rarefaction; dilution [acous] [ch]	希薄化	kihak(u)-ka
rare gas; noble gas [ch]	希有気体	ke'u-kitai
" [ch]	希ガス	ki-gasu
rare gas compound [ch]	希ガス化合物	kigasu-kagō'butsu
rare gas element [ch]	希ガス類元素	kigasu'rui-genso
rash (skin) {n} [med]	気触；かぶれ	kabure
raspberry; (fruit) [bot]	木苺；きいちご	ki-ichigo
ratchet [eng]	爪車；つめぐるま	tsume-guruma
rate; ratio (pl ratios) {n} [arith]	部合；ぶあい	bu'ai
rate; ratio; percentage {n} [arith]	比率；ひりつ	hi-ritsu
" {n} [arith]	率；りつ	ritsu
" ; price {n} [econ]	値段；ねだん	nedan
" ; speed {n} [mech]	速さ；早さ；はやさ	haya-sa
" ; speed; velocity {n} [mech]	速度；そくど	soku'do
" (of pace) {n} [physio]	歩度；ほど	hodo
rate constant [p-ch]	速度定数	sokudo-teisū
rate-controlling {n} [ch]	律速；りつそく	rissoku
rate-determining process [ch]	律速過程	rissoku-katei
rate-determining step [ch]	律速段階	rissoku-dankai
rated output [mech-eng]	定格出力	teikaku-shutsu'ryoku
rated speed [comput]	定格速度	teikaku-sokudo
rated voltage [elec]	定格電圧	teikaku-den'atsu
rate equation [ch]	反応速度式	hannō-sokudo-shiki
" [math]	速度式；そくどしき	sokudo-shiki

rate of operation *[ind-eng]*	稼動率；かどうりつ	kadō-ritsu
rate of production *[ind-eng]*	工率	kō'ritsu
" *[ind-eng]*	仕事率	shigoto-ritsu
ratfish; chimera; Chimera phantasma (fish) *[v-zoo]*	銀鮫；ぎんざめ	gin-zame
rating *{n}* *[elec]*	定格	teikaku
ratio (pl ratios) *[arith]*	比率	hi-ritsu
" *[arith]*	率；りつ	ritsu
" *[arith]*	割合；わりあい	wari'ai
ratio detector *[electr]*	比検波器	hi-kenpa-ki
rational *{adj}* *[math]*	合理的な	gōri'teki na
rational formula *[ch]*	示性式；しせいしき	shisei-shiki
rational index *[math]*	有理指数	yūri-shisū
rational indices, law of *[math]*	有理指数の法則	yūri-shisū no hōsoku
rationality *{n}*	合理性	gōri-sei
rationalization *[alg]*	有理化	yūri'ka
" *[legal]*	合理化	gōri'ka
rational number *[arith]*	有理数；ゆうりすう	yūri-sū
rational root *[alg]*	有理根	yūri-kon
ratio of rigidity *[mech]*	比剛性	hi-gōsei
rattan *[mat]*	籐類；とうるい	tō-rui
rattlesnake (reptile) *[v-zoo]*	がらがら蛇	garagara-hebi
ravine; gorge; defile; narrow pass *[geogr]*	狭間；はざま	hazama
raw *{n}* *[cook]*	生；なま	nama
" (uncooked) *{adj}* *[cook]*	生の；なまの	nama no
" (pure) *{adj}*	生の；きの	ki no
raw cotton *[bot]* *[tex]*	原綿	genmen
raw data *[comput]*	生データ	nama-dēta
raw hair *[tex]*	原毛	genmō
rawhide *[mat]*	生皮；なまかわ	nama-kawa
raw material *[ind-eng]*	原料；げんりょう	genryō
" (copy) *{n}* *[pr]*	素材	sozai
raw rubber *[org-ch]*	生ゴム	nama-gomu
raw sewage *[civ-eng]*	生下水；なまげすい	nama-ge'sui
" (sludge) *[civ-eng]*	屎尿汚泥	shi'nyō-o'dei
raw sheet *[mat]*	生シート	nama-shīto
raw silk *[tex]*	生糸；きいと	ki-ito
raw stock *[tex]*	原綿	genmen
raw sugar *[bioch]*	粗糖；そとう	sotō
raw water *[civ-eng]* *[hyd]*	原水	gensui
raw wool *[tex]*	原毛	genmō

ray *[astron] [geom]*	射線；しゃせん	shasen
" *[geom]*	放射線	hōsha'sen
" *[math] [opt]*	線；せん	sen
" ; skate (fish) *{n} [v-zoo]*	鱝；えい	ei
ray flower *[bot]*	周辺花；周邊花	shūhen-ka
ray fungus; Actinomyces *[microbio]*	放線菌	hōsen-kin
re- *{prefix}*	再び-	futatabi-
reaching the top; topping out	頭打ち；あたまうち	atama-uchi
react *{vb} [ch]*	反応する	hannō suru
reactant(s) *[ch]*	反応体；反應體	hannō'tai
reacting period *[ch]*	反応期	hannō-ki
reaction *[ch]*	反応；はんのう	hannō
" ; counteraction *[phys]*	反動；はんどう	handō
" ; counteraction *[phys]*	反作用；はんさよう	han-sayō
reaction chamber *[ch-eng]*	反応室	hannō-shitsu
reaction-controlling agent *[ch]*	反応制御剤	hannō-seigyo-zai
reaction injection molding; RIM *{n} [eng]*	反応射出成形	hannō-sha'shutsu-seikei
reaction intermediate *[ch]*	反応中間体	hannō-chūkan'tai
reaction order *[ch]*	反応次数	hannō-jisū
reaction path *[ch]*	反応経路；反応径路	hannō-keiro
reaction principle *[miner]*	反応原理	hannō-genri
reaction product; resultant *[ch]*	反応生成物	hannō-seisei'butsu
reaction (-ed, -ing) solution *[ch]*	反応液	hannō-eki
reaction system *[ch]*	反応系	hannō-kei
reactive dye *[mat]*	反応性染料	hannō'sei-senryō
reactive force *[mech]*	反力；はんりょく	han-ryoku
reactive plasticizer *[mat]*	反応性可塑剤	hannō'sei-kaso-zai
reactivity *[ch]*	反応度	hannō-do
" *[ch]*	反応性	hannō-sei
reactor *[ch]*	反応器；反應器	hannō-ki
" ; nuclear reactor *[nucleo]*	原子炉；げんしろ	genshi-ro
reactor core *[nucleo]*	炉心；爐芯；ろしん	ro-shin
reactor vessel *[nucleo]*	原子炉容器	genshi'ro-yōki
read *{vb t} [comm] [comput]*	読む；讀む；よむ	yomu
readability *[comm]*	判読率	handoku-ritsu
reader *[comput]*	読(み)取り装置	yomi'tori-sōchi
reading *{n} [comput]*	読み取り	yomi'tori
" *{n} [eng]*	読み	yomi
reading material *[comm] [pr]*	読み物；よみもの	yomi'mono
reading microscope *[opt]*	読取り顕微鏡	yomi'tori-kenbi'kyō
reading stand *[furn]*	見台；けんだい	kendai
reading telescope *[opt]*	読取り望遠鏡	yomi'tori-bōen'kyō

readjustment; later adjustment *{n}*	手直し；てなおし	te'naoshi
read-only memory; ROM *[comput]*	読出し専用記憶	yomi'dashi-sen'yō-ki'oku
" *[comput]*	読取専用メモリー	yomi'tori-sen'yō-memorī
ready; prepare *{vb}*	用意する	yōi suru
ready-made *{adj} [ind-eng]*	既製の	kisei no
ready-to-wear clothing *[cl]*	既製服	kisei-fuku
reagent *[an-ch]*	試薬；しやく	shi'yaku
reagent bottle *[an-ch]*	試薬瓶	shiyaku-bin
real constant *[comput]*	実定数	jitsu-teisū
realgar *[miner]*	鶏冠石	keikan'seki
real image *[opt]*	実像；じつぞう	jitsu-zō
reallocation *{n}*	再配分	sai'haibun
real numbers *[arith]*	実数；實數	jissū
real time *[comput]*	実時間；じつじかん	jitsu-jikan
real-time calculation *[comput]*	実時間計算	jitsu-jikan-keisan
ream (paper) *{n} [mat] [paper]*	連；れん	ren
reamer (hole reamer) *[eng]*	穴刳り具	ana'guri'gu
" (for fruit juice) *[eng]*	果汁絞り具	kajū-shibori'gu
rear *{n}*	後方；こうほう	kōhō
" *{n}*	後；うしろ	ushiro
rearmost *{adj}*	最後尾の	sai'kōbi no
rear position *{n}*	後位	kō'i
rearrange *{vb} [math]*	整理する	seiri suru
rearrangement *[ch]*	転位；轉位；てんい	ten'i
rear side *{n}*	後方	kōhō
rear view *[graph]*	背面図；背面圖	haimen-zu
rear wheel drive *[mech-eng]*	後輪駆動	kōrin-kudō
reason; cause; grounds *{n}*	理由；りゆう	riyū
reason; right; justice; truth *{n}*	道理	dōri
reasonable *{adj}*	合理的な	gōri'teki na
reasoning *{n} [geom]*	推論	suiron
reasoning power *[psy]*	推理力	suiri-ryoku
rebate *{n} [econ]*	割戻し	wari'modoshi
reboiler *[eng]*	再沸器；さいふつき	sai'futsu-ki
rebonding *{n} [ch]*	再結合	sai'ketsugō
rebound *{n} [mech]*	跳ね返り	hane'kaeri
" *{vb}*	跳ね返る	hane-kaeru
rebound elasticity *[mech]*	反撥弾性；反発弾性	han'patsu-dansei
rebound reaction *[ch]*	反跳反応	hanchō-hannō
rebound resiliency *[mech]*	反撥弾性；反發彈性	han'patsu-dansei
rebuilding *{n} [arch]*	改造；かいぞう	kaizō
recalescence *[met]*	再輝；さいき	saiki

recarburization *[steel]*	復炭；ふくたん	fuku'tan
recede; retreat *{vb i}*	後退する	kōtai suru
receipt; acceptance *{n}*	受領；じゅりょう	juryō
receive *{vb}*	受ける；うける	ukeru
" *{vb}*	受取る	uke-toru
receive and support *{vb} [mech-eng]*	支承する	shishō suru
receiver (receptacle) *[ch]*	受け器	uke-ki
" (telephone) *[comm]*	受話機；じゅわき	juwa-ki
" *[electr]*	受信機	jushin-ki
receiving; reception *{n} [comm] [electr]*	受信；じゅしん	jushin
" *{n} [comput]*	受取り	uke'tori
receiving box *[traffic]*	収納箱；收納箱	shūnō-bako
receiving dish; saucer *[cer]*	受け皿	uke-zara
recently; lately *{adv}*	最近；さいきん	saikin
receptacle; container *[eng]*	容器；ようき	yōki
reception; receiving *[electr]*	受信	jushin
reception room; drawing room *[arch]*	応接間；應接間	ō'setsu-ma
receptor field stop *[opt]*	受光絞り	jukō-shibori
receptor unit *[opt]*	受光部；じゅこうぶ	jukō-bu
recessed *{adj} [arch]*	引っ込んだ	hikkonda
recessive *{n} [gen]*	劣性；れっせい	ressei
" *{adj} [gen]*	劣性の	ressei no
recessive trait *[gen]*	劣性素質	ressei-so'shitsu
recipe; formula *{n} [ch] [pharm]*	処方；處方	shohō
" ; cooking method *{n} [cook]*	調理法	chōri-hō
recipient; taker *{n}*	取手；とりて	tori'te
" ; acceptor *{n}*	受け手	uke'te
reciprocal *{n} [alg]*	逆数；ぎゃくすう	gyaku'sū
" *{adj} [alg]*	相反の；そうはんの	sōhan no
" *{adj} [math]*	逆の	gyaku no
" *{adj}*	相互的な	sōgo'teki na
reciprocal proportions, law of *[ch]*	相互比例の法則	sōgo-hirei no hōsoku
reciprocate *{vb} [mech-eng]*	往復動する	ō'fuku'dō suru
reciprocating motion *[mech-eng]*	往復運動	ō'fuku-undō
reciprocity (pl -ties) *[math]*	相反性	sōhan-sei
reciprocity law *[graph]*	相反則	sōhan-soku
reclaimed land *[civ-eng]*	埋立地；うめたてち	ume'tate-chi
reclaimed rubber *[rub]*	再生ゴム	saisei-gomu
reclaiming agent *[rub]*	再生剤；再生劑	saisei-zai
reclamation *[civ-eng]*	埋立て；うめたて	ume'tate
reclined twill *[tex]*	緩斜文織	kan'shamon-ori
recognition *[comput] [psy]*	認識；にんしき	nin'shiki

recoil *{n} [mech]*	反跳；はんちょう	hanchō
recombinant *{n} [gen]*	組換え体	kumi'kae'tai
recombinant DNA *[bioch]*	組換えDNA	kumi'kae-DNA
recombinant gene *[gen]*	組換え遺伝子	kumi'kae-iden'shi
recombination *[gen]*	組換え；くみかえ	kumi-kae
recombination gene *[gen]*	組換え遺伝子	kumi'kae-idenshi
recombination process *[astron]*	再結合過程	sai'ketsugō-katei
reconfiguration *[comput]*	再構成	sai'kōsei
reconnaissance *[mil]*	偵察	tei'satsu
" *[eng]*	踏査	tōsa
record *{n} [comput] [pr]*	記録；きろく	kiroku
" *{vb} [acous]*	録音する	roku'on suru
" (information) *{vb}*	記録する	kiroku suru
" (information) *{vb}*	収録する；収禄する	shūroku suru
recorder *[eng]*	記録計	kiroku-kei
recording *{n} [acous]*	録音(体)	roku'on('tai)
" *{n} [pr]*	記録	kiroku
recording density *[comput]*	記録密度	kiroku-mitsudo
recording mechanism *[eng]*	記録機構	kiroku-kikō
recording medium *[electr]*	記録媒体	kiroku-baitai
recording surface *[eng]*	記録面	kiroku-men
recording wattmeter *[eng]*	記録電力計	kiroku-den'ryoku-kei
recovery *[comput] [med] [met] [sci-t]*	回復；かいふく	kaifuku
recovery (salvage) *[ecol]*	回収；回收	kaishū
" *[met]*	採取率	saishu-ritsu
recovery exponent *[nuc-phys]*	回復指数	kaifuku-shisū
recovery rate *[nuc-phys]*	回復率	kaifuku-ritsu
recreation area *[civ-eng]*	遊楽場	yūraku'jō
recreation ground *[civ-eng]*	遊園地；ゆうえんち	yūen'chi
recrystallization *[crys]*	再結晶	sai'kesshō
rectangle *[geom]*	長方形	chōhō'kei
" *[geom]*	矩形；くけい	kukei
rectangular coordinates *[an-geom]*	直角座標	chokkaku-zahyō
" *[an-geom]*	直交座標	chokkō-zahyō
rectangular parallelepiped *[geom]*	直方体	chokuhō'tai
rectangular waveguide *[elecmg]*	矩形導波管	kukei-dōha'kan
rectification *[ch]*	精留；精溜	seiryū
" *[elec]*	整流；せいりゅう	seiryū
rectifier *[elec]*	整流器	seiryū-ki
rectifier circuit *[electr]*	整流回路	seiryū-kairo
rectify *{vb} [elec]*	精留する	seiryū suru
rectifying column (or tower) *[ch-eng]*	精留塔	seiryū-tō

rectilinear diameter, law of *[p-ch]* 直径線の法則 chokkei-sen no hōsoku
recto (pl -s) *{n} [pr]* 奇数頁 kisū-pēji
rectum (pl -s, recta) *[anat]* 直腸 choku'chō
recurrent nova *[astron]* 再帰新星；再歸新星 saiki-shinsei
recursion *[comput]* 反復；はんぷく hanpuku
" *[logic]* 回帰(法) kaiki(-hō)
recursive *{adj} [comput]* 再帰的な saiki'teki na
recursive function *[comput]* 再帰的関数 saiki'teki-kansū
recurved *{adj} [sci-t]* 後方へ曲った kōhō e magatta
" *{adj} [sci-t]* 内側に曲った uchi'gawa ni mattata
recycle *{vb} [ecol] [eng]* 再生利用する saisei-riyō suru
recycling *{n} [ecol] [ind-eng]* 廃物再生利用 hai'butsu-saisei-riyō
red (color) *{n} [opt]* 赤；あか aka
" (color) *{adj} [opt]* 赤い akai
red algae *[bot]* 紅色植物 kōshoku-shokubutsu
" *[bot]* 紅藻；こうそう kōsō
red blood cell (erythrocyte) *[hist]* 赤血球 sekkekkyū
red copper ore *[miner]* 赤銅鉱；赤銅鑛 seki'dōkō
red coral *[i-zoo]* 赤珊瑚 aka-sango
red flag *[comm]* 赤旗 aka-hata
red flare *[comm]* 信号紅炎 shingō-kō'en
red giant *[astron]* 赤色巨星 seki'shoku-kyosei
red heat *[thermo]* 赤熱；せきねつ seki-netsu
red iron ore *[miner]* 赤鉄鉱；赤鐵鑛 seki-tekkō
red relief-patterned lacquerware *[furn]* 堆朱；ついしゅ tsui'shu
red lead *[miner]* 鉛丹 entan
red meat *[cook]* 赤身；あかみ aka-mi
red mercuric sulfide 朱；しゅ shu
red mud *[geol]* 赤泥 seki'dei
redox potential *[p-ch]* 酸化還元電位 sanka-kangen-den'i
redox reaction *[ch]* 酸化還元反応 sanka-kangen-hannō
red pepper *[bot]* 唐辛子；唐芥子 tō'garashi
red phosphorus *[ch]* 赤燐 seki-rin
red poppy *[bot]* 雛芥子；ひなげし hina'geshi
red rust *[met]* 赤錆 aka-sabi
red sea bream; porgy (fish) *[v-zoo]* 真鯛；まだい ma'dai
red shift *[astrophys]* 赤方偏移 sekihō-hen'i
red silk cloth *[tex]* 紅絹；もみ momi
redstart (bird) *[v-zoo]* 尉鶲；上鶲 jō'bitaki
red tide *[bio]* 赤潮 aka-shio
reduce *{vb t} [ch]* 還元する kangen suru
" ; curtail; contract *{vb}* 縮小する shuku'shō suru

reduced pressure *[ch-eng]*	減圧; 減壓	gen'atsu
reduced scale *[graph]*	縮尺; しゅくしゃく	shuku-shaku
reduced viscosity *[poly-ch]*	還元粘度	kangen-nendo
reducing agent; reductant *[ch]*	還元剤	kangen-zai
reducing flame *[ch]*	還元炎	kangen'en
reducing sugar *[org-ch]*	還元糖	kangen-tō
reductant *[ch]*	還元体	kangen'tai
reduction *[bio]*	減退	gentai
" *[ch] [comput]*	還元; かんげん	kangen
" *[math]*	減少	genshō
reduction (of a fraction) *[arith]*	約分	yaku'bun
reduction gear *[mech-eng]*	減速歯車	gensoku-ha'guruma
" *[mech-eng]*	減速機	gensoku-ki
reduction potential *[p-ch]*	還元電位	kangen-den'i
reduction printer *[graph]*	縮写機; 縮寫機	shuku'sha-ki
reduction ratio *[eng]*	粉砕比	funsai-hi
redundancy *[comput] [math]*	冗長性	jōchō-sei
red wine *[cook]*	赤葡萄酒	aka-budō'shu
redwing (bird) *[v-zoo]*	脇赤鶇	waki'aka-tsugumi
reed *[bot]*	葦; あし	ashi
" *[bot]*	葦; 蘆; よし	yoshi
" (for weaving) *[tex]*	筬; おさ	osa
reed instruments *[music]*	有簧楽器類	yūkō-gakki-rui
reed warbler (bird) *[v-zoo]*	葦切り; よしきり	yoshi-kiri
reef *[geol]*	(浅瀬)礁; (淺瀬)礁	(asa'se-)shō
" *[geol]*	岩礁; がんしょう	ganshō
" *[geol]*	砂洲; さす	sa'su
reel *{n} [tex]*	桛; かせ	kase
reeling *{n} [tex]*	桛揚げ	kase'age
reference *{n} [comput] [opt] [sci-t]*	基準; きじゅん	kijun
" *{n} [comput]*	参照	sanshō
" *{n} [pr]*	引用	in'yō
" *{n} [sci-t]*	対照; 對照	taishō
" *{n}*	参考; さんこう	sankō
" *{n}*	参照	sanshō
reference electrode *[p-ch]*	基準電極	kijun-denkyoku
reference line *[graph]*	参考線	sankō-sen
" *[navig]*	基準線	kijun-sen
reference number *[sci-t]*	整理番号	seiri-bangō
reference point *[navig]*	基準点	kijun-ten
reference solution *[an-ch]*	参照液	sanshō-eki
" *[an-ch]*	対照液	taishō-eki

refined, concentrated soy sauce *[cook]*	溜まり醬油	tamari-jōyu
refined sugar *[bioch]*	精製糖	seisei'tō
refinery (pl -eries) *[ch-eng]*	精製所	seisei'jo; seisei'sho
" *[met]*	精錬所；精鍊所	seiren'jo
refining *{n} [eng]*	精錬；せいれん	seiren
reflectance *[opt] [phys]*	反射率	hansha-ritsu
reflecting microscope *[opt]*	反射顕微鏡	hansha-kenbi'kyō
reflecting telescope *[opt]*	反射望遠鏡	hansha-bōenkyō
reflection *[an-geom]*	反転；反轉	hanten
" *[phys]*	反射；はんしゃ	hansha
reflection factor *[opt] [phys]*	反射率	hansha-ritsu
reflection, law of *[phys]*	反射の法則	hansha no hōsoku
reflection nebula *[astron]*	反射星雲	hansha-sei'un
reflection plane *[opt]*	鏡映面	kyō'ei-men
reflection stereoscope *[opt]*	反射実体鏡	hansha-jittai'kyō
reflectivity *[opt] [phys]*	反射率	hansha-ritsu
reflex *[physio]*	反射	hansha
reflexive *{adj} [gram]*	再帰の；再歸の	saiki no
" *{adj}*	反射的な	hansha'teki na
reflexive property *[alg]*	反射律	hansha-ritsu
" *[alg]*	反射的性質	hansha'teki-seishitsu
reflux; refluxing *{n} [ch-eng]*	還流；かんりゅう	kanryū
reflux *{n}*	逆流	gyaku'ryū
reflux condenser *[ch-eng]*	還流冷却機	kanryū-rei'kyaku-ki
reflux ratio *[ch-eng]*	還流比	kanryū-hi
reforestation *[ecol] [for]*	再植林	sai'shokurin
refracted wave *[phys]*	屈折波	kussetsu-ha
refracting telescope *[opt]*	屈折望遠鏡	kussetsu-bōenkyō
refraction *[phys]*	屈折；くっせつ	kussetsu
refraction, law of *[phys]*	屈折の法則	kussetsu no hōsoku
refractive index *[opt]*	屈折率	kussetsu-ritsu
refractive power *[phys]*	屈折力	kussetsu-ryoku
refractoriness *[cer]*	耐火度；たいかど	taika-do
refractometer *[eng]*	屈折計	kussetsu-kei
refractory (pl refractories) *{n} [mat]*	耐火物	taika'butsu
refractory clay *[mat]*	耐火粘土	taika-nendo
refrainment *[legal]*	回避；かいひ	kaihi
refreshing feel *[physio]*	清涼感	seiryō-kan
refrigerant *[mat]*	冷媒	reibai
refrigerate *{vb} [eng]*	冷凍する	reitō suru
refrigerating machine *[eng]*	冷凍機	reitō-ki
refrigeration *[mech-eng]*	冷凍；れいとう	reitō

refrigerator *[mech-eng]*	冷蔵庫；れいぞうこ	reizō'ko
refrigerator car *[rail]*	冷蔵車；冷藏車	reizō'sha
refuge *[arch]*	避難所	hinan'jo
refugee *{n}*	避難民	hinan'min
refuse *{n} [ecol]*	廃物；廢物	hai'butsu
" (of coal) *{n} [geol]*	硑；ずり	zuri
" *{n} [met]*	滓；かす	kasu
" *{vb}*	拒絶する	kyo'zetsu suru
refutation; counterevidence *[legal]*	反証	hanshō
regain *{n} [tex]*	水分率	suibun-ritsu
regelation *[hyd]*	復氷；ふくひょう	fuku'hyō
regenerated cellulose *[tex]*	再生繊維素	saisei-sen'i'so
regenerated fiber *[tex]*	再生繊維	saisei-sen'i
regeneration *[bio] [comput] [sci-t]*	再生；さいせい	saisei
regenerative repeater *[comm]*	再生式中継器	saisei'shiki-chūkei-ki
regenerator *[eng]*	再生機	saisei-ki
region *[comput]*	領域；りょういき	ryō'iki
regional planning *{n} [civ-eng]*	地方計画	chihō-keikaku
" *{n} [civ-eng]*	地域計画	chi'iki-keikaku
register *{vb} [comput] [pat]*	登録する	tōroku suru
registered trademark *[pat]*	登録商標	tōroku-shōhyō
registration *[pat]*	登記	tōki
" *{n}*	登録；とうろく	tō'roku
regolith *[geol]*	表土	hyōdo
regression *[psy]*	退行	taikō
" *[stat]*	回帰；回歸；かいき	kaiki
regrouping *[math]*	再編成	sai'hensei
regularity; constancy; steadiness	定常性	teijō-sei
regular octahedron *[geom]*	正八面体	sei-hachi'men'tai
regular polygon *[geom]*	正多角形	sei-takak(u)kei
regular polyhedron *[geom]*	正多面体	sei-ta'men'tai
regular polymer *[poly-ch]*	規則性高分子	kisoku'sei-kōbunshi
regular prism *[geom]*	正角柱	sei-kakuchū
regular reflection *[opt]*	正反射	sei-hansha
regulate (readjust; control) *{vb t}*	規正する	kisei suru
regulate *{vb t}*	律する；りっする	rissuru
regulating agent *[ch]*	調整剤	chōsei-zai
regulating valve *[mech-eng]*	調整弁；調整瓣	chōsei-ben
regulation *[bio]*	調整；ちょうせい	chōsei
" *[bio]*	調節	chōsetsu
regulation(s) *{n}*	規則；きそく	kisoku
regulator *[mat]*	調整剤	chōsei-zai

reindeer (animal) *[v-zoo]*	馴鹿；となかい	tonakai
reinforce *{vb} [eng]*	補強する	hokyō suru
reinforced concrete *[mat]*	鉄筋コンクリート	tekkin-konkurīto
reinforced fiber sheet *[tex]*	強化繊維シート	kyōka-sen'i-shīto
reinforcing *{n} [civ-eng]*	強化	kyōka
reinforcing agent *[mat]*	補強剤	hokyō-zai
" *[mat]*	強化剤	kyōka-zai
reins *{n} [trans]*	手綱；たづな	ta'zuna
reinstatement *{n}*	復位	fuku'i
reiteration *[math]*	反復	hanpuku
rejection; rejected *{n} [ind-eng]*	不合格	fu'gōkaku
rejection reaction *[ch]*	拒絶反応	kyo'zetsu-hannō
rejuvenation; rejuvenescence *{n}*	若返り；わかがえり	waka'gaeri
related control system *[eng]*	関連制御系	kanren-seigyo-kei
relation (relationship) *[alg]*	関係；關係	kankei
" (relationship) *[alg]*	関連	kanren
" *[comput]*	比較；ひかく	hi'kaku
relational expression *[math]*	関係式	kankei-shiki
relationship *[bio]*	類縁関係；類緣關係	rui'en-kankei
relative; kinsman *{n} [bio]*	親類	shinrui
" *{n} [bio]*	親戚；しんせき	shinseki
relative aperture *[opt]*	口径比；口徑比	kōkei-hi
relative error *[math]*	誤差率；ごさりつ	gosa-ritsu
" *[math]*	相対誤差	sōtai-gosa
relative humidity *[meteor]*	相対湿度；相對濕度	sōtai-shitsu-do
relative quantum efficiency *[electr]*	相対量子効率	sōtai-ryōshi-kō'ritsu
relative spectral distribution *[opt]*	相対的分光分布	sōtai'teki-bunkō-bunpu
relative viscosity *[fl-mech]*	相対粘度	sōtai-nendo
relative volatility *[thermo]*	比揮発度	hi-ki'hatsu-do
relativity (pl -ties) *[math] [phys]*	相対性(原理)	sōtai-sei(-genri)
relaxation *[mech]*	緩和；かんわ	kanwa
relaxation oscillator *[electr]*	弛張発振器	shichō-hasshin-ki
relaxation ratio *[mech]*	緩和率	kanwa-ritsu
relaxation time *[phys]*	緩和時間	kanwa-jikan
relay *{n} [elec]*	継電器；繼電器	keiden-ki
relaying *{n} [comm]*	中継；中繼	chūkei
releasability *{n} [adhes]*	剝離性；はくりせい	hakuri-sei
release *[comput]*	解放	kaihō
" *{n}*	緩め；弛め；ゆるめ	yurume
" (set free; free) *{vb}*	遊離する	yūri suru
release paper *[adhes]*	剝離紙	haku'ri-shi
" *[plas]*	離型紙	rikei-shi

releasing agent *[mat]*	離型剤	rikei-zai
relevancy *[math]*	関連性	kanren-sei
relevant data *[math]*	該当事項	gaitō-jikō
reliability *[comput] [eng] [stat]*	信頼度；しんらいど	shinrai-do
" *[comput] [eng] [stat]*	信頼性	shinrai-sei
relict mineral *[miner]*	残存鉱物	zanson-kōbutsu
relict structure *[geol]*	残存構造	zanson-kōzō
relief; relief engraving *{n} [lap]*	浮彫；うきぼり	uki-bori
relief plate *[pr]*	凸版；とっぱん	toppan
relief printing *{n} [tex]*	凸版捺染	toppan-nasen
relief valve *[mech-eng]*	逃し弁；逃し瓣	nigashi-ben
relocate *{vb} [comput]*	再配置する	sai'haichi suru
relocation *[comput]*	再配置；さいはいち	sai'haichi
reluctance *{n} [elecmg]*	磁気抵抗	jiki-teikō
remainder *[arith]*	余り；餘り；あまり	amari
" *[arith]*	剰余(項)；剰餘(項)	jōyo(-kō)
remainder theorem *[alg]*	剰余の定理	jōyo no teiri
remanence *[elecmg]*	残留分極	zanryū-bun'kyoku
remark(s) (note) *[pr]*	備考；びこう	bikō
" (note) *[pr]*	註；注；ちゅう	chū
remember *{vb} [psy]*	記憶する	ki'oku suru
" (recall) *{vb} [psy]*	思い出す	omoi-dasu
remilling *{n} [rub]*	再錬り；再練り	sai'neri
remittance check *[econ]*	送金小切手	sōkin-ko'gitte
remodeling *{n} [arch]*	改造	kaizō
remote *{adj}*	遠い；とおい	tōi
remote control *[elecmg]*	遠隔制御	enkaku-seigyo
" *[navig]*	遠隔操縦	enkaku-sōjū
remote sensing *{n} [eng]*	遠隔感知	enkaku-kanchi
" *{n} [eng]*	遠隔探査	enkaku-tansa
removal of harsh taste *{n} [food eng]*	灰汁抜き；あくぬき	aku-nuki
remove (take out) *{vb t}*	出す	dasu
" (eliminate) *{vb t}*	除く	nozoku
remuneration *[econ]*	報謝	hōsha
render fire-resistant *{vb t} [civ-eng]*	難燃化する	nan'nen'ka suru
rendering anistropic *[phys]*	異方性化	ihō'sei'ka
render superfine *{vb t}*	超微細化する	chō'bisai'ka suru
renewable resources *[sci-t]*	更新しうる資源	kōshin shi'uru shigen
renewal *[comput]*	更新；こうしん	kōshin
rennen *[bioch]*	凝乳酵素	gyōnyū-kōso
rennin *[bioch]*	凝固素；ぎょうこそ	gyōko'so
reorganize *{vb} [comput]*	再編成する	sai'hensei suru

repair {n} [eng]	補修	hoshū
" {n} [eng]	修復	shūfuku
" {n} [eng]	修理	shūri
" {n} [eng]	修繕; しゅうぜん	shūzen
" (patch) {n}	繕い	tsukuroi
" (mend; darn) {vb t}	繕う; つくろう	tsukurou
repeatability [comput]	再現性	saigen-sei
repeated permutation [prob]	重複順列	chōfuku-jun'retsu
repeater [electr]	中継器; 中繼器	chūkei-ki
" [eng]	再生機; さいせいき	saisei-ki
repeating decimal [arith]	循環小数	junkan-shōsū
repeat-sound mark [pr]	反復音符	hanpuku'on-pu
repel {vb t}	弾く; はじく	hajiku
" {vb t}	反発する; 反撥する	han'patsu suru
repellency [mech]	反発性	han'patsu-sei
repellent [bioch] [mat]	忌避剤; きひざい	kihi-zai
repetition [arith]	繰返し; くりかえし	kuri'kaeshi
" [comput]	反復	hanpuku
repetitive movement [ind-eng]	反復動作	hanpuku-dōsa
repetitive operation [comput]	繰返し演算	kuri'kaeshi-enzan
replace {vb t} [comput]	置き換える	oki-kaeru
replacement [comput]	取替え	tori'kae
replenisher solution [mat]	増し液	mashi-eki
replenishment {n}	補充; ほじゅう	hojū
replica plating method [bioch]	平板反復法	heiban-hanpuku-hō
replication [bioch] [gen] [graph] [mol-bio]	複製; ふくせい	fukusei
replication [stat]	反復	hanpuku
replotting {n} [civ-eng]	換地	kanchi
reply {n} [comm]	答え; こたえ	kotae
" {n} [comm]	答弁; 答辯	tōben
report (news) {n} [comm]	報道	hōdō
" {n} [comm] [pr]	報告; ほうこく	hōkoku
" {vb} [comm]	報告する	hōkoku suru
report generation [comput]	報告書作成	hōkoku'sho-sakusei
reprecipitation [ch]	再沈	sai'chin
representation [comput]	表示	hyōji
" {n}	代理	dairi
representation right [legal]	代理権; 代理權	dairi'ken
representative {n}	代表; だいひょう	daihyō
repression [psy]	抑制; よくせい	yoku'sei
repressor [bioch]	抑制因子	yokusei-inshi

reprint; reproduce *{vb t} [pr]*	複製する	fukusei suru
reproduce *{vb t} [pr]*	複製する	fukusei suru
reproducibility *[math]*	再現性	saigen-sei
reproduction *[bio]*	生殖	seishoku
" *[graph]*	複写; 複寫	fuku'sha
reproduction factor *{n}*	増倍係数	zōbai-keisū
reproductive potential *[bio]*	生殖能力	seishoku-nō'ryoku
reproductive organ *[anat]*	生殖器	seishok(u)ki
reptile *[v-zoo]*	爬行動物	hakō-dōbutsu
Reptilia; reptiles *[v-zoo]*	爬虫類	hachū-rui
repulsion *[mech]*	斥力; せきりょく	seki-ryoku
repulsive force (repulsion) *[mech]*	反発力	han'patsu-ryoku
request *{n}*	請求	seikyū
" *{n}*	要求	yōkyū
" *{vb}*	所望する	shomō suru
requestor *{n}*	依頼人; いらいにん	irai'nin
" *{n}*	委託者	itaku'sha
requiem shark (fish) *[v-zoo]*	目白鮫; めじろざめ	mejiro-zame
requirements (demand) *{n}*	要件	yōken
" (request) *{n}*	要求; ようきゅう	yōkyū
requisition *[mil]*	徴発	chōhatsu
rerun (redoing) *{n} [comput]*	再実行; 再實行	sai'jikkō
" (redoing) *{n} [comput]*	遣り直し	yari'naoshi
rescue ship *{n} [nav-arch]*	救難船	kyūnan'sen
research *{n} [sci-t]*	研究; けんきゅう	kenkyū
" (study) *{vb}*	研究する	kenkyū suru
researcher (a person) *[sci-t]*	研究者	kenkyū'sha
research institute *[sci-t]*	研究所	kenkyū'jo
research study member *{n}*	研修員	kenshū'in
resemblance *{n}*	類似性; るいじせい	ruiji-sei
resemble *{vb}*	似る	niru
reserve(s) *{n} [eng]*	埋蔵量; 埋藏量	maizō'ryō
" *{vb} [trans]*	予約する; 豫約する	yo'yaku suru
reserve capacity *[ind-eng]*	予備容量	yobi-yōryō
reservoir *[civ-eng]*	貯水池; ちょすいち	chosui'chi
residence *[arch]*	住宅	jūtaku
residence telephone *[comm]*	住宅電話	jūtaku-denwa
residence time *[ch-eng] [nucleo]*	滞留時間; 滯留時間	tairyū-jikan
residential sector *[civ-eng]*	住居地(区)	jūkyo-chi(ku)
residual *{n}*	残留; ざんりゅう	zanryū
" *{adj} [math]*	残差の	zansa no
residual effect *{n}*	残効; 殘効	zankō

residual error rate *[comput]*	見逃し誤り率	mi'nogashi-ayamari-ritsu
residual expansion *[phys]*	残存膨張	zanson-bōchō
residual heat *[thermo]*	余熱；餘熱；よねつ	yo-netsu
residual magnetism *[elecmg]*	残留磁化	zanryū-jika
residual nitrogen *[inorg-ch]*	残余窒素；殘餘窒素	zan'yo-chisso
residual rays *[nucleo]*	残留線	zanryū-sen
residual stress *[mech]*	残留応力	zanryū-ō'ryoku
residual sugar (in wine) *[cook]*	残糖；ざんとう	zan'tō
residual term *[math]*	残項	zankō
residue *[ch-eng]*	残分	zanbun
" *[ch-eng]*	残留物	zanryū'butsu
" *[math]*	残差	zansa
residue check *[comput]*	冗長検査	jōchō-kensa
residue solution *[ch-eng]*	釜残液	kama'zan-eki
resin *[org-ch] [poly-ch]*	樹脂；じゅし	jushi
resin acid *[ch]*	樹脂酸	jushi-san
resin cure *[rub]*	樹脂加硫	jushi-karyū
resin-finishing agent *[mat]*	樹脂加工剤	jushi-kakō-zai
resinification *[org-ch] [poly-ch]*	樹脂化	jushi'ka
resinous exudate *[bot]*	脂；やに	yani
resin-treatment agent *[mat]*	樹脂加工剤	jushi-kakō-zai
resist; resistance *{n} [tex]*	防染剤	bōsen-zai
resistance *[bot]*	抵抗性	teikō-sei
" *[elec] [mech] [phys]*	抵抗；ていこう	teikō
resistance element *[elec]*	抵抗素子	teikō-soshi
resistivity; specific resistance *[elec]*	比抵抗	hi-teikō
resistivity *[elec]*	比電気抵抗	hi-denki-teikō
" *[elec]*	抵抗率	teikō-ritsu
resistor *[elec]*	抵抗器	teikō-ki
" *[elec]*	抵抗体	teikō'tai
resist printing *{n} [tex]*	防染；ぼうせん	bōsen
resolution *[opt] [spect]*	分解(能)	bunkai(-nō)
" *[opt]*	分割	bun'katsu
" *[opt]*	解像(度)	kaizō(-do)
" *[opt]*	解像力	kaizō-ryoku
" *[photo]*	鮮明度；せんめいど	senmei-do
resolution of forces *[math]*	力の分解	chikara no bunkai
resolving power *[opt]*	分解能	bunkai-nō
" *[opt]*	解像力	kaizō-ryoku
resonance *[acous]*	反響；反響	hankyō
" *[elec] [phys] [quant-mech]*	共鳴；きょうめい	kyōmei
" *[elec] [phys]*	共振；きょうしん	kyōshin

resonance circuit *[elec]*	共振回路	kyōshin-kairo
resonance fluorescence *[atom-phys]*	共鳴蛍光	kyōmei-keikō
resonance frequency *[phys]*	共振周波数	kyōshin-shūha'sū
resonance scattering *{n} [nuc-phys]*	共鳴散乱	kyōmei-sanran
resonator *[phys]*	共振子	kyōshin'shi
resource(s) *[sci-t]*	資源；しげん	shigen
resource-converting property *[microbio]*	資化性	shika-sei
resource management *[comput]*	資源管理	shigen-kanri
respiration *[physio]*	呼吸；こきゅう	kokyū
respiratory enzyme *[bioch]*	呼吸酵素	kokyū-kōso
respiratory quotient *[physio]*	呼吸商	kokyū-shō
respiratory system *[anat]*	呼吸器系	kokyū'ki-kei
respiratory tract *[anat]*	気道；氣道；きどう	kidō
respond *{vb}*	応ずる；應ずる	ōzuru
responding-light quantity *[opt]*	感応光量	kannō-kōryō
respond optically *{vb} [opt]*	感光する	kankō suru
response *[bio] [comm] [comput]*	応答；おうとう	ō'tō
" *[comm]*	答弁書；答辯書	tōben'sho
response time *[comput]*	応答時間	ō'tō-jikan
responsibility *[legal]*	責任；せきにん	seki'nin
rest *{n} [med]*	安静	ansei
" (pause) *{n} [music]*	休止	kyūshi
" (stopping) *{n}*	止まり	todomari
" ; repose *{vb i}*	休む；やすむ	yasumu
restart *{n} [comput]*	再開始	sai'kaishi
" *{n} [comput]*	再始動；さいしどう	sai'shidō
restaurant *[arch]*	料理店	ryōri'ten
resting stage *[cytol]*	静止期	seishi-ki
restitution *[legal]*	補償；ほしょう	hoshō
rest mass *[rela]*	静止質量	seishi-shitsu'ryō
restoration *{n}*	修復；しゅうふく	shūfuku
restore *{vb} [comput]*	復元する	fukugen suru
" *{vb} [eng]*	修復する	shūfuku suru
restoring force *[mech-eng]*	復原力；復元力	fukugen-ryoku
restraining condition *[math]*	束縛条件	soku'baku-jōken
restraining order *[legal]*	禁止命令	kinshi-meirei
restraint *[mech]*	拘束	kōsoku
restretching *{n} [plas]*	再延伸	sai'enshin
Restricted Area *[civ-eng]*	立入り禁止区域	tachi'iri-kinshi-ku'iki
restriction endonuclease; restriction enzyme *[bioch]*	制限酵素	seigen-kōso

restriction gene *[gen]*	制限遺伝子	seigen-iden'shi
rest room; toilet; lavatory *[arch]*	手洗い；てあらい	te-arai
" *[arch]*	便所；べんじょ	benjo
" *[arch]*	洗面所	senmen'jo
restructuring; reconfiguration *{n} [comput]*	再構成	sai-kōsei
result; showing; record *{n} [econ]*	成積；せいせき	seiseki
" *{n} [math]*	結果；けっか	kekka
resultant; reaction product *[ch]*	反応生成物	hannō-seisei'butsu
" (an equation) *[mech]*	終結式	shūketsu-shiki
resultant of forces *[mech]*	合力；ごうりょく	gō-ryoku
resultant wind *[meteor]*	合成風	gōsei-fū
resurfacing *{n} [civ-eng]*	再舗装；さいほそう	sai'hosō
resuscitation *[med]*	蘇生(法)	sosei(-hō)
retain; hold *{vb t}*	保持する；ほじする	hoji suru
" ; detain *{vb t}*	保留する	horyū suru
retaining member *[eng]*	保持部材	hoji-buzai
retaining wall *[civ-eng]*	擁壁；ようへき	yōheki
retard; delay *{vb t}*	遅延させる	chi'en saseru
retardation; delaying *{n}*	遅延；ちえん	chi'en
retardation agent *[ch]*	遅延剤；遅延劑	chi'en-zai
retardation time *[electr]*	遅延時間	chi'en-jikan
retarder (polymerization) *[poly-ch]*	重合抑制剤	jūgō-yokusei-zai
retention *[ch-eng]*	保持；ほじ	hoji
retention period *[comput]*	保持時間	hoji-jikan
retention time *[an-ch]*	保持時間	hoji-jikan
" *[ch-eng] [poly-ch]*	滞留時間	tairyū-jikan
retention volume *[an-ch]*	保持容量	hoji-yōryō
retentivity *[elecmg]*	残磁性；ざんじせい	zanji-sei
reticulate *{adj} [bio] [geol]*	網状構造の	amijō-kōzō no
retina (pl -s, -inae) *[anat]*	網膜；もうまく	mōmaku
retouching *{n} [photo]*	修正	shūsei
retract *{vb t}*	後退させる	kōtai saseru
retreat *{vb i}*	後退する	kōtai suru
" (recede) *{vb i}*	退く；しりぞく	shirizoku
retrial *[legal]*	再審；さいしん	saishin
retrieval *[comput]*	検索；けんさく	kensaku
retrieval function *[comput]*	検索機能	kensaku-kinō
retrieve *{vb} [comput]*	検索する	kensaku suru
retrograde motion *[astron]*	逆行運動	gyak(u)kō-undō
retrograde movement *[astron]*	後退；こうたい	kōtai
retrograding *{n} [med]*	退行	taikō
retrograding wave *[elec]*	後進波	kōshin-ha

English	Japanese	Romanization
retry *{n} [comput]*	再試行	sai'shikō
retting *{n} [ch-eng] [tex]*	浸水	shinsui
return (reversion) *{n} [comput]*	復帰；ふっき	fukki
" (reversion) *{n} [comput]*	戻り；もどり	modori
" (come or go back) *{vb i}*	帰る；歸る；かえる	kaeru
" (go back; revert) *{vb i}*	戻る	modoru
" (give back) *{vb t}*	返す	kaesu
" (give back; restore) *{vb t}*	戻す	modosu
reusable *[comput]*	再使用可能	sai'shiyō-kanō
revenue cutter *[nav-arch]*	税関監視船	zeikan-kanshi'sen
revenue stamp *[econ]*	収入印紙	shūnyū-inshi
reverberation *[acous]*	残響；殘響	zankyō
reverberatory furnace *[eng]*	反射炉；反射爐	hansha-ro
reversal processing *{n} [photo]*	反転現像処理	hanten-genzō-shori
reverse; reversal *[ch]*	反転；はんてん	hanten
reverse *{n} [math]*	逆；ぎゃく	gyaku
" (back side) *{n} [math]*	裏；うら	ura
" (driving) *{n} [mech-eng]*	逆転；逆轉	gyaku'ten
" (go into reverse) *{vb i}*	逆転する	gyakuten suru
" *{vb i}*	反転する	hanten suru
" (invert) *{vb t}*	逆転させる	gyakuten saseru
" (turn around) *{vb t}*	反転させる	hanten saseru
reverse current *[elec]*	逆電流	gyaku-denryū
" *[electr]*	逆流；ぎゃくりゅう	gyaku'ryū
reverse direction *[math]*	逆方向	gyaku-hōkō
reverse hobbing *{n} [mech]*	逆手ホブ加工	saka'te-hobu-kakō
reverse mutation *[microbio]*	復帰(突然)変異	fukki(-totsu'zen)-hen'i
reverse osmosis *[ch-eng] [p-ch]*	逆浸透	gyaku-shintō
reverse-osmosis membrane *[p-ch]*	逆浸透膜	gyaku-shintō'maku
reverse printing *{n} [pr]*	反転印刷	hanten-in'satsu
reverse reaction *[ch]*	逆反応	gyaku-hannō
reverse transcriptase *[microbio]*	逆転写酵素	gyaku-tensha-kōso
reversibility *[math]*	可逆性	ka'gyaku-sei
reversible *{adj} [ch]*	可逆の	ka'gyaku no
reversible polymerization *[ch]*	可逆重合	ka'gyaku-jūgō
reversible reaction *[ch]*	可逆反応	ka'gyaku-hannō
reversing layer *[astrophys]*	反彩層	hansai-sō
reversion *[microbio]*	復帰変異	fukki-hen'i
" *[steel]*	復元	fukugen
revert to the original *{vb} [mech]*	復帰する	fukki suru
review *{n} [legal] [pat]*	審査；しんさ	shinsa
revision *[pat]*	補正	hosei

revision (revised edition) *[pr]*	改訂版	kaitei-ban
revive *{vb i}*	蘇る；甦る	yomi'gaeru
revocation *{n}*	取消し；とりけし	tori'keshi
revolute *{adj} [bot]*	外巻きの	soto'maki no
revolution *[mech]*	回転；回轉	kaiten
revolution (around common axis) *{n} [astron]*	公転；こうてん	kōten
revolution (around own axis) *{n} [astron]*	自転；じてん	jiten
revolutionary *{adj}*	画期的な	kakki'teki na
revolve *{vb i} [math]*	回転する	kaiten suru
" (rotate) *{vb t}*	回す；まわす	mawasu
revolve around a common axis *{vb} [astron]*	公転する	kōten suru
revolve around own axis *{vb} [astron]*	自転する	jiten suru
revolved section *{n} [graph]*	回転断面図	kaiten-danmen-zu
revolving door *[arch]*	回転扉	kaiten-tobira
revolvingly contact *{vb} [mech-eng]*	転接する	ten'setsu suru
reward; recompense *{vb t}*	報いる；むくいる	mukuiru
rewind *{vb t}*	巻き戻す	maki-modosu
rewinding *{n} [mech-eng]*	巻き戻し	maki'modoshi
rework *{n} [ind-eng]*	再加工	sai'kakō
" *{n} [ind-eng]*	手直し	te'naoshi
rewriting *{n} [comput]*	書き直し	kaki'naoshi
rhaxa; raxa *[tex]*	羅紗；らしゃ	rasha
rhenium (element)	レニウム	renyūmu
rheology *[mech]*	流動学	ryūdō'gaku
rheomorphism *[petr]*	流動変成作用	ryūdō-hensei-sayō
rheooptics *[opt]*	流動光学	ryūdō-kōgaku
rheostat *[elec]*	加減抵抗器	kagen-teikō-ki
rhesus monkey; Macaca mulatta (animal) *[v-zoo]*	赤毛猿；あかげざる	aka'ge-zaru
rhinoceros (pl -es, -eri) (animal) *[v-zoo]*	犀；さい	sai
rhizoid *{n} [bot]*	仮根；假根；かこん	kakon
rhizome *{n} [bot]*	根茎；根莖	konkei
Rhizopus *[microbio]*	蜘蛛の巣かび	kumo'no'su-kabi
Rhodea japonica *[bot]*	万年青；おもと	omoto
rhodium (element)	ロジウム	rojūmu
rhododendron (flower, bush) *[bot]*	躑躅；つつじ	tsutsuji
rhodonite *[miner]*	薔薇輝石	bara'kiseki
Rhodophyta *[bot]*	紅色植物	kōshoku-shokubutsu

rhombic dodecahedron *[crys]*	斜方十二面体	shahō-jūni'men'tai
rhombohedral system *[crys]*	菱面体晶系	ryōmen'tai-shōkei
rhombohedron *[geom]*	菱面体	ryōmen'tai
rhomboid *{adj} [geom]*	菱形の	hishi'gata no
rhombus (pl -es, rhombi) *{n} [geom]*	菱形；ひしがた	hishi-gata
" *{n} [geom]*	斜方形	shahō-kei
rhubarb (vegetable) *[bot]*	蕗；ふき	fuki
rhumb line *[map]*	航程線	kōtei-sen
" *[map]*	等角航路	tōkaku-kōro
rhyolite *[miner]*	流紋岩	ryūmon'gan
rib *{n} [anat]*	肋；あばら	abara
" *{n} [arch]*	肋材	rokuzai
rib bone *[anat]*	肋骨；ろっこつ	abara-bone; rokkotsu
ribonucleic acid; RNA *[bioch]*	リボ核酸	ribo-kaku'san
rib weave *[tex]*	畝織り	une-ori
rice *[bot]*	米穀	beikoku
" (raw) *[cook] [bot]*	米；こめ	kome
rice and wheat; grains *[bot]*	米麦；べいばく	beibaku
rice bowl *[cook]*	茶碗	chawan
rice bran *[bot]*	(米)糠	(kome-)nuka
rice bran wax *[mat]*	米糠蠟	kome'nuka-rō
rice cake *[cook]*	餅；もち	mochi
rice cake soup *[cook]*	雑煮；ぞうに	zōni
rice chaff; chaff *[bot]*	籾殻	momi-gara
rice cooker *[cook] [eng]*	炊飯器	suihan-ki
rice cracker *[cook]*	煎餅；せんべい	senbei
rice gruel *[cook]*	粥；かゆ	kayu
" (thin) *[cook]*	重湯；おもゆ	omo'yu
rice oil *[mat]*	糠油	nuka-yu
rice paper *[mat] [paper]*	通草紙；つうそうし	tsūsō-shi
rice plant *[bot]*	稲；いね	ine
rice steamed with ingredients *[cook]*	炊込御飯	taki'komi-gohan
rice weevil (insect) *[i-zoo]*	穀象虫	kokuzō-mushi
rich mixture *[ch]*	濃厚混合気	nōkō-kongō-ki
rickets *[med]*	佝僂病	kuru'byō
ride; mount *{vb}*	乗る；乘る；のる	noru
rider *[ch-eng] [eng]*	馬乗り分銅	uma'nori-fundō
ridge *{n} [geol]*	稜；りょう	ryō
" ; furrow; groove *{n} [agr] [geol]*	畝；うね	une
" (of pressure) *{n} [meteor]*	尾根	o'ne
ridge line *[meteor]*	尾根線	o'ne-sen
rift valley *[geol]*	地溝	chikō

right (privilege) *{n} [legal]*	権利；權利；けんり	kenri
" (direction) *{n} [math]*	右；みぎ	migi
" (proper; correct) *{adj} {n}*	正しい	tadashī
right angle *[geom]*	直角；ちょっかく	chokkaku
right-angle cylinder *[geom]*	直角柱	chokkaku-chū
right ascension *[astron]*	赤経；赤經	sekkei
right bank *[geol]*	右岸	u-gan
right circular cone *[geom]*	直円錐	choku-ensui
right circular cylinder *[geom]*	直円柱	choku-enchū
right-handed *{n} [physio]*	右利き	migi-kiki
right-handed elliptically polarized wave *[elecmg]*	右旋楕円偏波	u'sen-da'en-henpa
right-handed screw *[mech-eng]*	右ねじ	migi-neji
right-hand rule *[elecmg]*	右手の法則	migi'te no hōsoku
right-hand side *[math]*	右辺；右邊；うへん	u-hen
right-justifying *{n} [comput] [pr]*	右詰め	migi-tsume
right of decision *[legal]*	決定権	kettei'ken
right of exploitation *[pat]*	実施権；じっしけん	jisshi'ken
right-of-way *[traffic]*	通行権	tsūkō'ken
right prism *[geom]*	直角柱	chokkaku-chū
right side view *[graph]*	右側面図	migi-sokumen-zu
right triangle *[geom]*	直角三角形	chokkaku-sankak(u)kei
right turn *[traffic]*	右折；うせつ	u-setsu
right-turn lane *[traffic]*	右折車線	u'setsu-shasen
rigid body *[mech]*	剛体	gō'tai
rigidity *[mech]*	剛性；ごうせい	gō-sei
" *[mech]*	剛さ	kowa-sa
rigor mortis *[path]*	死後硬直	shigo-kōchoku
rill *{n} [astron]*	谷；たに	tani
rim *{n} [eng]*	周縁	shū'en
" *{n}*	縁；ふち	fuchi
rime *[meteor]*	霧氷；むひょう	mu'hyō
rind (of fruit) *[bot]*	果皮	kahi
ring *{n} [astron]*	環；わ	wa
" *{n} [eng]*	輪環；りんかん	rinkan
" (circle) *[geom]*	輪；わ	wa
" (to wear on finger) *{n} [lap]*	指環	yubi'wa
" (resound) *{vb i}*	鳴る	naru
" (sound; toll) *{vb t}*	鳴らす	narasu
ring and ball method *[ch-eng]*	環球法	kankyū-hō
ring connection *[elec]*	輪形接続	wa'gata-setsu'zoku
ring finger; fourth finger *{n} [anat]*	薬指；くすりゆび	kusuri-yubi

ring formation *[org-ch]*	環形成	kan-keisei
ring gear *[mech-eng]*	輪歯車	wa-ha'guruma
ringing tone *[comm]*	呼出音	yobi'dashi-on
ring isomerism *[ch]*	環異性	kan-i'sei
ring kiln *[cer] [eng]*	輪窯；わがま	wa-gama
ring magnet *[elecmg]*	輪形磁石	wa'gata-ji'shaku
Ring Nebula *[astron]*	環状星雲	kanjō-sei'un
ring-opening polymerization *[poly-ch]*	開環重合	kaikan-jūgō
ring-opening reaction *[poly-ch]*	開環反応	kaikan-hannō
ring spring *{n} [mech-eng]*	輪ばね	wa-bane
ring-tailed lemur (animal) *[v-zoo]*	輪尾狐猿	wa'o-kitsune-zaru
ring valve *[mech-eng]*	輪形弁；輪形瓣	wa'gata-ben
rinse; wash *{vb t}*	濯ぐ；嗽ぐ；雪ぐ	susugu
rinsing (with water) *{n}*	水洗	suisen
ripe *{adj} [bot]*	熟した；じゅくした	jukushita
ripening *{n} [bio] [ch]*	熟成	jukusei
ripple *{n} [ocean]*	漣；小波；さざなみ	saza'nami
ripple voltage *[elec]*	リプル電圧	ripuru-den'atsu
ripsawing *{n} [mech-eng]*	縦引き	tate'biki
riptide *[ocean]*	潮衝；ちょうしょう	chōshō
rise *{n} [math] [rub]*	立ち上り	tachi'agari
rise part *[comm]*	立ち上り部分	tachi'agari-bubun
rise time *[comm] [cont-syst]*	立ち上り時間	tachi'agari-jikan
rising; rise *{n} [astron]*	出；で	de
risk *{n} [math]*	危険度	kiken-do
risk factor *{n}*	危険率	kiken-ritsu
river *[hyd]*	川；河；かわ	kawa
riverbed *[geol]*	川床；かわどこ	kawa'doko
river boat *[nav-arch]*	川船	kawa-bune
river crab *[i-zoo]*	沢蟹；さわがに	sawa-gani
river discharge *[hyd]*	河川の流量	kasen no ryūryō
river ice *[hyd]*	河川氷	kasen-hyō
riverside; waterfront *[geol]*	河岸；かし	kashi
rivet *{n} [mat]*	鋲；びょう	byō
roach (fish) *[v-zoo]*	諸子；もろこ	moroko
road *[civ-eng]*	道路	dōro
" *[civ-eng]*	道；みち	michi
road accident *[med]*	交通事故	kōtsū-jiko
roadbed material *[mat]*	路盤材；ろばんざい	roban-zai
road bridge *[civ-eng]*	道路橋	dōro-kyō
road lighting *{n} [civ-eng]*	道路照明	dōro-shōmei
road map *[map]*	道路地図	dōro-chizu

road shoulder *[civ-eng]*	路肩；ろかた	ro-kata
roadside *[civ-eng]*	路側	ro'soku
roast *{vb t} [cook]*	焙る；あぶる	aburu
" *{vb t} [cook]*	煎る；炒る；いる	iru
roasted dishes *[cook]*	焼き物；焼き物	yaki'mono
roaster *[eng]*	焙り焼き器具	aburi'yaki-kigu
roasting *{n} [met]*	焙焼；ばいしょう	baishō
" *{n} [min-eng]*	焼成	shōsei
robin (Japanese bird) *[v-zoo]*	駒鳥	koma-dori
" (bird) *[v-zoo]*	駒鶫；こまつぐみ	koma-tsugumi
robot *[cont-sys]*	ロボット	robotto
robustness *[ind-eng]*	頑健性	ganken-sei
" *[stat]*	安定性	antei-sei
rock *{n} [petr]*	石；いし	ishi
rock cod; rockfish *[v-zoo]*	眼張；めばる	mebaru
rock crystal *[miner]*	水晶；すいしょう	suishō
rock-cutter (ship) *[nav-arch]*	砕岩船	saigan'sen
rock drill *[eng]*	鑿岩機；さくがんき	sakugan-ki
rocker arm *[mech-eng]*	揺れ腕；搖れ腕	yure-ude
rockfish; rosefish (fish) *[v-zoo]*	目抜；めぬけ	me'nuke
rocking board *[mech-eng]*	揺動板；搖動板	yōdō-ban
rocking chair *[furn]*	揺り椅子；ゆりいす	yuri-isu
rocking vibration *[spect]*	横揺れ振動	yoko'yure-shindō
rock phosphate *[geol]*	燐鉱石	rin'kōseki
rock salt *[miner]*	岩塩；岩鹽	gan'en
rock trout; (fish) *[v-zoo]*	鮎並；あいなめ	ainame
rock wool *[mat]*	岩綿	ganmen
rod *{n} [anat] [microbio]*	桿状体；杆状体	kanjō'tai
" (pole) *{n} [eng]*	棒；ぼう	bō
" *{n} [eng]*	桿；かん	kan
rod crystal *[crys]*	柱状晶	chūjō-shō
rodent *[v-zoo]*	齧歯動物	kesshi-dōbutsu
rodenticide *[mat]*	殺鼠剤	sasso-zai
rodform *[microbio]*	桿状	kanjō
roll *{n} [aerosp]*	横転；おうてん	ō'ten
" *{n} [mech]*	横揺れ	yoko'yure
" *{vb i}*	転がる；ころがる	korogaru
" *{vb t}*	転がす	korogasu
rollability *[mech]*	押し易さ	oshi-yasu-sa
" *[mech]*	転がり易さ	korogari-yasu-sa
rolled core *[elecmg]*	巻鉄心；巻鐵芯	maki-tesshin
rolled steel *[met]*	圧延鋼材	atsu'en-kōzai

roller *[eng]*	転；ころ	koro
roller bottle *[cyt] [eng]*	回転瓶	kaiten-bin
rolling *{n} [mech]*	転がり；轉がり	korogari
" *{n} [mech]*	転動；てんどう	tendō
" *{n} [mech] [nav-arch]*	横揺れ	yoko-yure
" *{n} [steel]*	圧延；あつえん	atsu'en
rolling force *[met]*	圧延力；壓延力	atsu'en-ryoku
" *[steel]*	圧下力	akka-ryoku
rolling friction *[mech]*	転がり摩擦	korogari-ma'satsu
rolling pin *[cook] [eng]*	麺棒；めんぼう	menbō
roll magnetic anisotropy *[elecmg]*	圧延磁気異方性	atsu'en-jiki-ihō'sei
roll mill *[mech-eng]*	圧延機	atsu'en-ki
roll up; wind up *{vb t}*	巻き上げる	maki-ageru
roof *{n} [arch]*	屋根；やね	yane
roofing material *[arch]*	屋根葺き材(料)	yane'buki-zai(ryō)
roof tile; roofing tile *[arch]*	屋根瓦	yane-gawara
rooftop *[arch]*	屋上；おくじょう	okujō
rookery *[zoo]*	繁殖地	hanshoku'chi
" *[zoo]*	繁殖所	hanshoku'jo
room *[arch]*	部屋；室；へや	heya
" *[arch]*	間；ま	ma
room-arrangement plan *[graph]*	間取り	ma-dori
room temperature *[thermo]*	室温；しつおん	shitsu-on
room-temperature cure *[poly-ch]*	常温硬化	jō'on-kōka
rooster; male bird (bird) *[v-zoo]*	雄鳥；おんどり	ondori
root *{n} [alg]*	乗根；乘根	jōkon
" *{n} [alg] [dent]*	根；こん	kon
" *{n} [bot]*	根；ね	ne
" *{n} [dent]*	歯根	shikon
root canal *[dent]*	根管	konkan
rootlet *[bot]*	根小毛	kon-shōmō
root-mean-square (rms) value *[elec]*	実効値；實効値	jikkō-chi
root nodule *[bot]*	根粒	konryū
rootstock *[bot]*	根茎；根莖	konkei
root system *[bot]*	根系	konkei
rope *{n} [mat]*	縄；繩；なわ	nawa
" *{n} [mat]*	綱；つな	tsuna
rope ladder *[eng]*	縄梯子；なわばしご	nawa-bashigo
ropeway *[min-eng]*	索道	sakudō
rosasite *[miner]*	亜鉛孔雀石	aen'kujaku'seki
rose (flower) *[bot]*	薔薇；ばら	bara
rosemary (pl -maries) (herb) *[bot]*	迷迭香	mannenrō

rostrum *[arch]*	演壇	endan
rot; decay *{n} [mat]*	腐食；ふしょく	fushoku
" ; decay *{vb} [mat]*	腐る；くさる	kusaru
rotamer *[ch]*	回転異性体	kaiten-isei'tai
rotary; rotating *{adj} [math]*	回転する	kaiten suru
rotary intersection *[traffic]*	環状交差路	kanjō-kōsa-ro
rotary kiln *[eng]*	回転炉；回轉爐	kaiten-ro
rotary letterpress machine *[pr]*	凸版輪転機	toppan-rinten-ki
rotary machine *[mech-eng]*	回転機(器)	kaiten-ki(ki)
rotary shaking culture *[microbio]*	回転振盪培養	kaiten-shintō-baiyō
rotating axis *[mech]*	回転軸	kaiten-jiku
rotation *[an-geom] [mech]*	回転；かいてん	kaiten
rotational isomer *[ch]*	回転異性体	kaiten-isei'tai
rotational molding *{n} [poly-ch]*	回転成型	kaiten-seikei
rotational motion *[mech]*	回転運動	kaiten-undō
rotational vibration *[spect]*	回転振動	kaiten-shindō
rotatory inversion *[math]*	回転反転（回反）	kaiten-hanten (kaihan)
rotatory reflection *[opt]*	回転鏡映	kaiten-kyō'ei
Rotifera *[i-zoo]*	輪形動物	rinkei-dōbutsu
" *[i-zoo]*	輪虫類；わむしるい	wa'mushi-rui
rotogravure printing *[pr]*	グラビア印刷	gurabiya-in'satsu
rotor *[elec]*	回転子	kaiten'shi
roughage *{n} [cook]*	繊維質食品	sen'i'shitsu-shoku'hin
rough aggregate *[geol] [mat]*	疎凝集体	so-gyōshū'tai
roughening *{n} [plas]*	肌荒れ；はだあれ	hada-are
rough gear cutting *{n} [mech-eng]*	荒歯切加工	araha'giri-kakō
rough idling *{n} [mech-eng]*	乱調；亂調	ranchō
roughing *{n} [mech-eng]*	荒引き；あらびき	ara'biki
roughness *[eng]*	荒さ	ara-sa
roughness feel *[physio]*	ざらつき感	zara'tsuki-kan
round *{adj} [geom]*	丸い；円い；圓い	marui
round bar (of steel); rod *[civ-eng]*	丸鋼；まるこう	maru-kō
round down *{vb} [math]*	切り捨てる	kiri-suteru
rounding (rounding off) *{n} [arith]*	四捨五入	shi'sha-go'nyū
" *{n} [comput]*	丸め	marume
rounding error *[arith]*	丸めの誤差	marume no gosa
roundness *[geom]*	丸味；まるみ	maru'mi
" *[geom]*	真円度；真圓度	shin'en-do
round number *[arith]*	丸めた数	marumeta sū
round(ing) **off** *{n} [math]*	丸め	marume
roundworm *[i-zoo]*	回虫	kaichū
" *[i-zoo]*	線形動物	senkei-dōbutsu

route; course; channel *{n}* 经路；經路 keiro
routine *{n} [comput]* 手順；てじゅん te'jun
routine test *[ind-eng]* 定期試験 teiki-shiken
routing *{n} [comput]* 径路選択；經路選擇 keiro-sentaku
roving *{n} [tex]* 粗紡；そぼう sobō
roving frame *[tex]* 練紡機；練紡機 renbō-ki
row *{n} [comput] [math]* 行；ぎょう gyō
" (a boat) *{vb}* 漕ぐ kogu
rowboat *[nav-arch]* 漕ぎ舟 kogi-fune
royalty *[legal]* 印税 in'zei
rub (scour) *{vb t}* 擦る；こする kosuru
" (crumble) *{vb t}* 揉む；もむ momu
" (chafe) *{vb t}* 摩る；磨る；擦る suru
rubber *[org-ch]* ゴム；護謨 gomu
rubbish; scraps *[ecol]* 屑；くず kuzu
rubble *[geol]* 角礫；かくれき kaku'reki
rubidium (element) ルビジウム rubijūmu
rudite *[geol]* 礫岩；れきがん reki'gan
rudder *[aero-eng]* 方向舵 hōkō'da
" *[eng]* 舵；かじ kaji
ruffed grouse (bird) *[v-zoo]* 襟巻き雷鳥 eri'maki-raichō
rugosity *[bio]* 皺度 shiwa-do
ruins *{n} [archeo]* 址；あと ato
rule *{n} [math]* 規則；きそく kisoku
" *{n} [sci-t]* 法則；ほうそく hōsoku
ruler *[eng]* 定規；じょうぎ jōgi
rules and regulations *{n} [legal]* 法規 hōki
run *{n} [comput]* 実行；實行 jikkō
" *{n} [comput]* 運転 unten
" *{n} [ind-eng]* パス pasu
" *{vb t} [comput]* 実行させる jikkō saseru
" *{vb i}* 走る hashiru
" (speed; race) *{vb i}* 駆ける；驅ける kakeru
runaway *{n} [ch]* 暴走；ぼうそう bōsō
runaway reaction *[ch]* 反応暴走 hannō-bōsō
runner *[bot]* 匍匐枝；ほふくし ho'fuku-shi
" *[mech-eng]* 羽根車 hane-guruma
runner channel *[resin]* 湯道；ゆみち yu-michi
running *{n} [econ]* 経営；經營 kei'ei
" *{n} [ind-eng]* 運転 unten
" *{n} [physio]* 走ること hashiru koto
" *{n}* 走行 sōkō

running light (at sea) *[navig]*	航海灯; 航海燈	kōkai-tō
" (in air) *[navig]*	航空灯	kōkū-tō
running time *[comput]*	実行時間; 實行時間	jikkō-jikan
running water *[hyd]*	流水	ryūsui
runoff *{n} [hyd]*	流れ水	nagare-mizu
" *{n} [hyd]*	流出; りゅうしゅつ	ryū'shutsu
runway *[aero-eng]*	滑走路; かっそうろ	kassō-ro
" *[civ-eng]*	助走路	josō-ro
" *[civ-eng]*	走路	sōro
rupture *{n} [mech]*	破壊; 破壞; はかい	hakai
" ; explosion *{n}*	破裂; はれつ	ha'retsu
rural road *[civ-eng]*	田園道路	den'en-dōro
rush *{n} [bot]*	藺(草)	i(-gusa)
rush mat, bound at edges *[mat]*	茣蓙; 御座; ござ	goza
rust (wheat disease) *{n} [bot]*	銹病	sabi'byō
" *{n} [met]*	錆; さび; せい	sabi (sei)
" *{n} [met]*	銹; さび; しゅう	sabi (shū)
" *{n} [met]*	鏽; さび; しゅう	sabi (shū)
rust inhibitor *[met]*	抑銹剤	yokushū-zai
rust prevention *[met]*	防錆	bōsei
" *[met]*	抑銹	yoku'shū
rustproofing *{n} [met]*	錆止め	sabi'dome
rust resistance *[mat] [met]*	耐銹性	taishū-sei
ruthenium (element)	ルテニウム	rutenyūmu
rutile *[miner]*	金紅石	kin'kōseki
rye (cereal) *[bot]*	ライ麦	rai-mugi

S

sable (animal) *[v-zoo]*	黒貂；くろてん	kuro-ten
sablefish; Anoplopona fimbria (fish) *[v-zoo]*	銀鱈；ぎんだら	gin-dara
saccharic acid	糖酸	tōsan
saccharides *[bioch]*	糖類；とうるい	tō-rui
saccharification *[bioch]*	糖化	tōka
saccharimeter *[eng]*	検糖計	kentō-kei
saccharin *[org-ch]*	サッカリン	sakkarin
sacrum (bone) *{n} [anat]*	仙骨	sen-kotsu
saddle *{n} [transp]*	鞍；くら	kura
saddle point *[math]*	鞍点	anten
saddle reef *[geol]*	鞍状鉱床	kura'jō-kōshō
saddle stitching (in bookmaking) *[pr]*	中綴じ；なかとじ	naka-toji
safe *{adj}*	安全な	anzen na
safekeeping *{n}*	保管	hokan
safety; security *{n}*	安全；あんぜん	anzen
safety belt *[eng]*	安全ベルト	anzen-beruto
safety factor; margin of safety *{n}*	安全率	anzen-ritsu
safety glasses *[ind-eng] [opt]*	安全眼鏡	anzen-me'gane
safety valve *[eng]*	安全弁；安全瓣	anzen-ben
safety zone *[civ-eng]*	安全地帯	anzen-chitai
safflower (herb) *[bot] [cook]*	紅花	beni'bana
saffron (yellow color) *[opt]*	鬱金色；うこんいろ	u'kon-iro
sag; slack *{n} [eng]*	弛；たるみ	tarumi
" *{n} [rub]*	だれ	dare
" *{vb}*	だれる	dareru
sagger *{n} [cer]*	匣鉢；サヤ	saya
sagging *{n} [mech]*	へたり	hetari
" *{n} [poly-ch] [paint]*	垂れ下がり	tare'sagari
Saghalien (Sakhalin) dog (animal) *[v-zoo]*	樺太犬；からふと=いぬ	Karafuto-inu
Sagittarium; the Archer *[constel]*	射手座	ite-za
sagittated squid *[i-zoo]*	鯣烏賊：するめいか	surume-ika
sail *{n} [nav-arch]*	帆；ほ	ho
sailboat *[nav-arch]*	帆船	han'sen; ho-bune
sailfish (fish) *[v-zoo]*	芭蕉梶木；芭蕉旗魚	bashō-kajiki
Saint John's wort *[bot]*	弟切草；おとぎり草	otogiri'sō
sakaki (tree) *[bot]*	榊；賢木；さかき	sakaki
sake (Japanese rice wine) *[cook]*	日本酒；にほんしゅ	Nihon'shu

sake *[cook]*	酒；さけ	sake
" *[cook]*	清酒	seishu
sake (see above) lees *[cook]*	酒粕；酒糟	sake-kasu
sake server; wine pourer *[cer]*	銚子；ちょうし	chōshi
" *[cer]*	徳利	tokuri; tokkuri
sake-steamed dishes *[cook]*	酒蒸し	saka-mushi
salamander (amphibian) *[v-zoo]*	山椒魚	sanshō'uo
salary *[econ]*	俸給	hōkyū
" *[econ]*	給料；きゅうりょう	kyūryō
sale; sales; selling *{n}* *[econ]*	販売；販賣	hanbai
sales *{n}* *[econ]*	売れ行	ure'yuki
salinimeter *[eng]*	塩度計；鹽度計	endo-kei
salinity *[ch]*	塩度；えんど	endo
" *[ocean]*	塩分	enbun
salinometer *[eng]*	塩分計	enbun-kei
" *[eng]*	塩量計	enryō-kei
saliva *[physio]*	唾液；だえき	da'eki
" *[physio]*	唾；つば	tsuba
salivary gland *[physio]*	唾液腺	da'eki-sen
salmon; Oncorhynchus keta (fish)	鮭；さけ	sake
salmon roe (indvidually) *[cook]*	イクラ	ikura
" (en masse) *[cook]*	筋子	suji'ko
salt *{n}* *[ch]*	塩；鹽	shio; en
salt; table salt; common salt; halite *[ch]* *[cook]*	食塩；しょくえん	shoku'en
salt anticline *[geol]*	岩塩背斜	gan'en-haisha
salt-bath quenching *{n}* *[met]*	塩浴焼入れ	en'yoku-yaki'ire
salt bridge *[p-ch]*	塩橋	enkyō
salt damage *[ind-eng]*	塩害；えんがい	engai
salted fish guts *[cook]*	塩辛；しおから	shio-kara
saltiness *{n}* *[cook]* *[physio]*	辛味	kara'mi
" *{n}* *[cook]* *[physio]*	塩辛さ	shio-kara-sa
salting (preservation) *{n}* *[food-eng]*	塩蔵；鹽藏	enzō
" (salt pickling) *{n}* *[food-eng]*	塩漬	shio'zuke
salting in *[ch-eng]*	塩溶	en'yō
salting out *[ch-eng]*	塩析	en'seki
salt isomerism *[ch]*	塩異性；えんいせい	en-i'sei
salt lake; saline lake *[hyd]*	塩湖	enko
salt mine *[min-eng]*	岩塩坑	gan'en-kō
saltpeter (potassium nitrate)	硝石；しょうせき	shōseki
salt-spray test *[met]*	塩水噴霧試験	ensui-funmu-shiken
salt tolerance *[agr]*	耐塩性	tai'en-sei

salt water *[ch] [cook]*	食塩水	shoku'en-sui
salt-water resistance *[mat]*	耐塩水性	tai'ensui-sei
salt-water spraying *{n} [met]*	塩水噴霧	ensui-funmu
salvage synthesis *[bioch]*	再利用合成	sai'riyō-gōsei
samarium (element)	サマリウム	samaryūmu
same; sameness *{n}*	同じ；おなじ	onaji
sameness *{n}*	同一性	dō'itsu-sei
sampan *[nav-arch]*	舢板；三板	sanpan
sample *{n} [ch]*	試料；しりょう	shiryō
" *{n} [ind-eng]*	見本	mihon
" *{n} [sci-t] [stat]*	標本；ひょうほん	hyōhon
sampling *{n} [sci-t]*	標本化	hyōhon'ka
" *{n} [stat]*	標本抽出	hyōhon-chū'shutsu
sampling inspection *[ind-eng]*	抜取検査	nuki'tori-kensa
sanctuary *[arch]*	避難所	hinan'jo
sand *{n} [geol]*	砂；すな	suna
sandalwood (tree) *[bot]*	白檀；びゃくだん	byaku'dan
sandbank *[geol]*	砂嘴；さし	sa'shi
sand bar *[geol]*	砂洲	sasu
sand bath *[med]*	砂浴	sa-yoku
sandblasting *{n} [eng]*	砂吹き	suna-fuki
sand crab (crustacean) *[i-zoo]*	磯蟹；いそがに	iso-gani
sand dollar *[i-zoo]*	菓子パン	kashi-pan
sand dune *[geol]*	砂丘	sakyū
sand eel; lance (fish) *[v-zoo]*	玉筋魚；いかなご	ikanago
sandfish (fish) *[v-zoo]*	鰰；はたはた	hatahata
sandfly (insect) *[i-zoo]*	蚋；ぶよ	buyo; buyu
sandglass *[horol]*	砂時計	suna-dokei
sandpaper *{n} [mat]*	紙鑢；紙やすり	kami-yasuri
sandstone *[petr]*	砂岩	sagan
sandstorm *[meteor]*	砂嵐；すなあらし	suna-arashi
sandwich *{vb t}*	挟む；挾む；はさむ	hasamu
sanidine *[miner]*	玻璃長石	hari'chōseki
sanitary cotton *[tex]*	脱脂綿	dasshi-men
sanitize *{vb} [med]*	消毒する	shōdoku suru
sap *{n} [bot]*	樹液	ju'eki
sap fruit *[bot]*	液果	eki'ka
sapling; young tree *[bot]*	若木；わかぎ	waka-gi
saponification *[ch]*	鹼化；けんか	kenka
saponification number *[an-ch]*	鹼化価	kenka-ka
saprobe *[microbio]*	腐生菌	fusei-kin
saprolite *[geol]*	腐食岩石	fushoku-ganseki

saprophagous *{adj} [bio]*	腐食性の	fushoku'sei no
saprophyte *[bot]*	腐生植物	fusei-shokubutsu
saprophytic *{adj} [bio]*	腐生(性)の	fusei('sei) no
" *{adj} [bio]*	非病菌(性)の	hi'byōkin('sei) no
" *{adj} [microbio]*	雑菌性の	zakkin'sei no
Sarcodina *[i-zoo]*	肉質(虫)類	niku'shitsu(-chū)-rui
sard *[miner]*	紅玉髄; 紅玉髓	kō'gyokuzui
sardine (fish) *[v-zoo]*	鰯; いわし	iwashi
sargasso (pl -s); gulfweed *[bot]*	馬尾藻; ほんだわら	hondawara
sasanqua (flower, bush) *[bot]*	山茶花; さざんか	sazan'ka
sash (for kimono, q.v.) *[cl]*	帯; 帶; おび	obi
satellite *[aerosp] [astron]*	衛星; えいせい	eisei
satellite communication *[comm]*	衛星通信	eisei-tsūshin
satellite computer *[comput]*	衛星計算機	eisei-keisan-ki
satinet *[tex]*	毛繻子; けじゅす	ke'jusu
satin finish *[mat]*	梨地仕上	nashi'ji-shi'age
satin weave *[tex]*	朱子織り; 繻子織り	shusu-ori
satisfactorily; favorably *{adv}*	順調に	junchō ni
satisfy *{vb t}*	満足させる	manzoku saseru
saturated atmosphere *[meteor]*	飽和大気; 飽和大氣	hōwa-taiki
saturated solution *[ch]*	飽和溶液	hōwa-yōeki
saturated steam *[phys]*	飽和蒸気	hōwa-jōki
saturation *[opt]*	彩度; さいど	saido
" *[phys]*	飽和; 飽和; ほうわ	hōwa
saturation magnetic flux density *[elecmg]*	飽和磁束密度	hōwa-ji'soku-mitsudo
saturation vapor pressure *[thermo]*	飽和蒸気圧	hōwa-jōki-atsu
Saturday	土曜日; どようび	do'yōbi
Saturn *[astron]*	土星; どせい	dosei
satyr (butterfly) *[i-zoo]*	蛇の目蝶	ja'no'me-chō
sauce *[cook]*	垂れ; たれ	tare
sauce-dressed dishes *[cook]*	和え物; あえもの	ae'mono
saucepan *[cook]*	鍋; なべ	nabe
saucer *[cook]*	受け皿	uke-zara
saurel (fish) *[v-zoo]*	鰺; あじ	aji
saury-pike (fish) *[v-zoo]*	梭魚; 魳; かます	kamasu
sausage *[cook]*	腸詰め	chō'zume
savage; wild; feral; untamed *{adj}*	野性の	yasei no
savannah *[ecol]*	大草原	dai-sōgen
save *{n} [comput]*	保管	hokan
" *{vb t} [comput]*	保存する	hozon suru
saw (a tool) *{n} [eng]*	鋸; のこぎり	nokogiri

sawing *{n} [eng]*	鋸断；きょだん	kyodan
sawmill *[ind-eng]*	製材場	seizai'jō
saw shark (fish) *[v-zoo]*	鋸鮫；のこぎり鮫	nokogiri-zame
sawtooth wave *[electr]*	鋸歯状波	kyoshi'jō-ha
" *[electr]*	鋸波	nokogiri-ha
scabbard fish (fish) *[v-zoo]*	太刀魚；たちうお	tachi'uo
scaffold(ing) *{n} [civ-eng]*	足場	ashi'ba
scale *{n} [ch]*	缶石；罐石	kanseki
" (for weighing) *{n} [eng]*	秤；量り；はかり	hakari
" (measure) *{n} [eng]*	尺度	shaku-do
" (on a map) *{n} [graph]*	縮尺線	shuku'shaku-sen
" (insect) *{n} [i-zoo]*	貝殻虫；貝殻蝨	kaigara-mushi
" *{n} [music]*	音階	onkai
" (Japanese) *{n} [music]*	和音階	wa'on-kai
" *{n} [phys]*	目盛	me'mori
" (of fish) *{n} [v-zoo]*	鱗；うろこ	uroko
" (scope; plan) *{n}*	規模	kibo
scale factor *[eng]*	基準化因数	kijun'ka-insū
" *[math]*	倍率	bai'ritsu
scale mark; graduation line *[graph]*	目盛線；めもりせん	memori-sen
scalene triangle *[geom]*	不等辺三角形	futō'hen-sankak(u)kei
scalenohedron *[crys] [geom]*	偏三角面体	hen'sankaku'men-tai
scaling *{n} [comput] [eng]*	基準化	kijun'ka
" *{n} [comput]*	位取り	kurai'dori
" (of paint) *{n}*	剥れ；はがれ	hagare
scaling down (of size, shape) *{vb}*	小型化	kogata'ka
scallion (vegetable) *[bot]*	韮；にら	nira
" (vegetable) *[bot]*	辣韮；らっきょう	rakkyō
scallop (shellfish) *[i-zoo]*	帆立貝；ほたてがい	ho'tate-gai
scalp *{n} [anat]*	頭皮	tōhi
scaly anteater (animal) *[v-zoo]*	穿山甲	senzan'kō
scan *{n} [comm] [comput]*	走査；そうさ	sōsa
scandium (element)	スカンジウム	sukanjūmu
scanner *[comput] [comm] [electr]*	走査器	sōsa-ki
" *[comput] [comm] [electr]*	走査装置	sōsa-sōchi
scanning *{n} [electr]*	走査	sōsa
scanning acoustic microscope *[opt]*	超音波顕微鏡	chō'onpa-kenbi'kyō
scanning electron micrograph *[electr]*	走査電子顕微鏡写真	sōsa-denshi'kenbikyō-shashin
scanning electron microscope *[electr]*	走査電子顕微鏡	sōsa-denshi-kenbi'kyō
scanning rate (or speed) *[electr]*	走査速度	sōsa-sokudo
scanning tunneling microscope *[electr]*	走査型トンネル=顕微鏡	sōsa'gata-tonneru-kenbikyō

scant; a few; slight *{adj}*	僅かな；わずかな	wazuka na
scapula *{n} [anat]*	肩甲骨	kenkō-kotsu
scar *{n} [bio]*	瘢痕	hankon
" *{n} [med]*	傷跡；きずあと	kizu-ato
scarlet (color) *[opt]*	緋；ひ	hi
scarlet phosphorus *[ch]*	紅燐	kōrin
scarlet tanager (bird) *[v-zoo]*	赤風琴鳥	aka-fūkin-chō
scatter *{n}*	ばらつき	bara'tsuki
" *{n}*	飛散	hisan
" *{vb i}*	散る	chiru
" *{vb t}*	散らす	chirasu
scatter diagram *[stat]*	散布図；散布圖	sanpu-zu
scattered clouds *[meteor]*	千切れ雲；ちぎれ雲	chi'gire-gumo
scattered light *[opt]*	散乱光；散亂光	sanran-kō
scattered wave *[phys]*	散乱波	sanran-ha
scattering *{n} [phys]*	散乱；さんらん	sanran
scattering angle *[phys]*	散乱角	sanran-kaku
scattering intensity distribution *[opt]*	散乱強度分布	sanran-kyōdo-bunpu
scaup duck (bird) *[v-zoo]*	鈴鴨；すずがも	suzu-gamo
scavenger *[ch]*	掃鉛剤	sōen-zai
" *[nucleo]*	捕収剤；捕收劑	hoshū-zai
" (animal) *[zoo]*	清掃動物	seisō-dōbutsu
scend (of ship) *[eng]*	持上り（船の）	mochi'agari (fune no)
scenery *[graph]*	風景	fūkei
schedule *{n} [comput] [sci-t]*	計画；計畫	keikaku
" ; program; plan *{n} [sci-t]*	予定；よてい	yotei
scheduled date *[ind-eng]*	予定日	yotei-bi
scheduled maintenance *[ind-eng]*	計画保守	keikaku-hoshu
scheelite *[miner]*	灰重石	kai'jūseki
schema *{n} [comput]*	概略；がいりゃく	gai'ryaku
schematic diagram *[graph]*	概要図	gaiyō-zu
schematic drawing *{n} [graph]*	図式画	zu'shiki-ga
schematic plot *[graph]*	図式表示	zu'shiki-hyōji
schematic view *[graph]*	略図；略圖	ryaku-zu
scheme drawing *{n} [graph]*	計画図	keikaku-zu
schist *[geol]*	（結晶）片岩	(kesshō-)hengan
Schizomycetes *[microbio]*	分裂菌類	bunretsu'kin-rui
Scholastic Aptitude Test; SAT *[psy]*	学習能力適性試驗	gakushū-nō'ryoku-tekisei-shiken
school *{n} [arch]*	学校；がっこう	gakkō
schorlite; schorl *[miner]*	黒電気石；黒電氣石	kuro-denki'seki

science *[sci-t]*	学術；學術	gaku'jutsu
" *[sci-t]*	科学；科學；かがく	kagaku
scientific *{adj}*	学術的な	gakujutsu'teki na
scientific method *[sci-t]*	科学的方法	kagaku'teki-hōhō
scientific name *[comm]*	学名	gaku'mei
scientific notation *[alg] [sci-t]*	科学的記数法	kagaku'teki-kisū-hō
scientific programming *{n} [comput]*	科学的計画法	kagaku'teki-keikaku-hō
scientific (scholarly) society *[sci-t]*	学会；學會	gakkai
scientific symbol *[sci-t]*	科学記号	kagaku-kigō
scientist *[sci-t]*	科学者	kagaku'sha
scission *[ch]*	切断	setsu'dan
scission polymerization *[poly-ch]*	開環重合	kaikan-jūgō
scissors *[eng]*	鋏；はさみ	hasami
sclera (pl -s, sclerae) *[anat]*	強膜；鞏膜	kyōmaku
scolecite *[miner]*	灰沸石	kai'fusseki
scoop *{n} [eng]*	箆；へら	hera
scope *[comput]*	有効範囲	yūkō-han'i
" *[math]*	範囲；範圍；はんい	han'i
Scopolia japonica (herb) *[pharm]*	走野老；走りどころ	hashiri'dokoro
scorch; singe *{n} [cook]*	焦げ；こげ	koge
" *{n} [rub]*	焼け；焼け	yake
score *{n} [music]*	楽譜；樂譜；がくふ	gaku'fu
scoria (pl- riae) *[geol]*	岩滓	gan'sai; ganshi
" *[geol]*	火山岩滓	kazan-gan'sai; kazan-ganshi
scorification *[met]*	焼溶試金法	shōyō-shikin-hō
scoring *{n} [eng]*	罫書き	ke'gaki
Scorpio; the Scorpion *[constel]*	蠍座；さそり座	sasori-za
scorpion (insect) *[i-zoo]*	蝎；蠍；さそり	sasori
scorpionfish (fish) *[v-zoo]*	笠子；かさご	kasago
scotopic vision *[physio]*	暗所視	ansho'shi
scoured yarn *[tex]*	湅り糸；練り糸	neri-ito
scouring *{n} [tex]*	精湅；精練	seiren
scrap wool *[tex]*	屑毛；くずけ	kuzu-ke
scratch; claw; maul *{vb t}*	引っ掻く；ひっかく	hikkaku
" *{vb t}*	掻く	kaku
scratch hardness *[mech]*	引掻き硬さ	hikkaki-kata-sa
scratch resistance *[mat]*	耐引っ掻き性	tai'hikkaki-sei
scree *[geol]*	がれ(場)	gare('ba)
screen *{n} [comput] [graph]*	表示(画)面	hyōji-(ga)men
" *{n} [furn]*	屏風；びょうぶ	byōbu
" *{n} [furn]*	衝立；ついたて	tsui'tate
" (cover; shield) *{vb}*	遮蔽する	shahei suru

screening *{n} [elecmg] [sci-t]*	遮蔽；しゃへい	shahei
" *{n} [eng]*	整粒	seiryū
screw *{n} [eng]*	螺子；捻子；捩子	neji
screw axis *[crys]*	螺旋軸	rasen-jiku
screw dislocation *[photo]*	螺旋状転位	rasen'jō-ten'i
screw driver *[eng]*	ねじ回し	neji-mawashi
screw gear *[mech-eng]*	ねじ歯車；螺子歯車	neji-ha'guruma
screw-mesh *{vb} [mech]*	螺合する	ragō suru
screw thread *[eng]*	ねじ山	neji-yama
scribe-cutting *{n} [eng]*	罫書き切断	ke'gaki-setsu'dan
script (print) style writing (of Chinese characters) *[comm]*	楷書体	kaisho-tai
scroll (scrolling) *{n} [comput]*	画面移動	gamen-idō
" (rolled book) *{n} [graph] [pr]*	巻物；巻物	maki'mono
scrollwork *[graph]*	唐草；からくさ	kara-kusa
scrubbing bottle *[an-ch]*	ガス洗浄瓶	gasu-senjō-bin
scrub brush *[cook]*	束子；たわし	tawashi
sculpting *{n} [graph]*	彫刻	chōkoku
Scyphozoa *[i-zoo]*	鉢虫類	hachi'chū-rui
scythe *{n} [agr]*	大鎌	ō-gama
sea *[geogr]*	海；うみ	umi
sea anemone *[i-zoo]*	磯巾着	iso'ginchaku
sea bass; Lateolabrax japonicus (fish) *[v-zoo]*	鱸；すずき	suzuki
sea berth *[nav-arch]*	海上停泊施設	kaijō-tei'haku-shi'setsu
seabird (bird) *[v-zoo]*	海鳥	kaichō
sea bream (fish) *[v-zoo]*	鯛；たい	tai
sea breeze *[meteor]*	海風	umi-kaze; kaifū
sea breeze front *[meteor]*	海風前線	umi'kaze-zensen
sea cow *[v-zoo]*	海牛；かいぎゅう	kaigyū
sea cucumber (echinoderm) *[i-zoo]*	海鼠；生子；なまこ	namako
sea eel (fish) *[v-zoo]*	穴子；海鰻	anago
" (fish) *[v-zoo]*	鱧；はも	hamo
sea fan (coral) *[i-zoo]*	海団扇；うみうちわ	umi-uchiwa
sea floor spreading theory *[geol]*	海洋底拡大説	kaiyō'tei-kaku'dai-setsu
sea fog *[meteor]*	海霧	umi-giri
seafood; fish and shellfish *[cook]*	魚介類	gyokai-rui
sea gravel *[geol]*	海砂利；うみじゃり	umi-jari
sea ground *[elec]*	海地気	umi-chiki
seagull (bird) *[v-zoo]*	鴎；かもめ	kamome
sea hare; Aplysia kurodai *[i-zoo]*	雨降らし	ame-furashi
seahorse (fish) *[i-zoo]*	竜の落(と)し子	tatsu no otoshi'go

sea interferometer *[opt]*	海面干渉計	kaimen-kanshō-kei
sea-island form *[geol]*	海島状	kaitō'jō
seal (stamp) *{n} [graph]*	印章；いんしょう	inshō
" (leakage stopper) *{n} [mech-eng]*	漏れ止め	more'dome
" (marine mammal) *{n} [v-zoo]*	海豹；あざらし	azarashi
" (fur seal) *{n} [v-zoo]*	膃肭獣；おっとせい	ottosei
" *{n}*	封(印)	fū'in
" (seal air-tight) *{vb t}*	密封する	mippū suru
sea layer *[geol]*	海成層	kaisei-sō
sealed (sealable) container *{n}*	密閉器	mippei-ki
sea lettuce; laver *[bot]*	石蓴；あおさ	aosa
sea-level pressure *[meteor]*	海面気圧；海面氣壓	kaimen-ki'atsu
sealing *{n} [eng]*	密封	mippū
" (imprinting) *{n} [graph]*	捺印；なついん	natsu'in
sealing wax *[mat]*	封蠟	fūrō
sealwort; Polygonatum falcatum *[bot]*	鳴子百合	naruko-yuri
sea mail *[comm]*	船便；ふなびん	funa-bin
seamount *[geol]*	海山	kaizan
sea mud *[geol]*	海土	kaido
seam welding *{n} [met]*	嚙合せ溶接	kami'awase-yō'setsu
Sea of Japan *[geogr]*	日本海	Nihon-kai
sea onion *[bot]*	海葱；かいそう	kaisō
sea pen *[i-zoo]*	海鰓；うみえら	umi-era
seaport *[civ-eng]*	海港	kaikō
search *{n} [comput]*	探索；たんさく	tansaku
" *{n} [navig]*	調査	chōsa
search for *{vb}*	探す	sagasu
searchlight *[opt]*	探照灯；探照燈	tanshō'tō
sea robin (fish) *[v-zoo]*	魴鮄；ほうぼう	hōbō
sea salt *[cook] [geol] [ocean]*	海塩；海鹽	umi-jio; kai'en
sea sand *[geol]*	海砂；かいしゃ	kaisha
sea shipping *{n} [trans]*	海運	kai'un
sea shock; sea quake *[geophys]*	海震	kaishin
seashore; beach *[geogr]*	浦；うら	ura
seasickness *[med]*	船酔い	funa-yoi
seaside atmosphere *[meteor]*	海岸大気	kaigan-taiki
sea slug (gastropod) *[i-zoo]*	海鼠；なまこ	namako
season (time) *{n} [astron]*	時期	jiki
" *{n} [climat]*	季節；きせつ	ki'setsu
seasonal variation *[geophys]*	季節変動；季節變動	kisetsu-hendō
season crack *[met]*	置割れ	oki-ware
seasoning *{n} [cook]*	調味料	chōmi'ryō

seasoning; condiment(s) *{n} [cook]*	香辛料	kōshin'ryō
seasoning *{n} [steel]*	枯し; からし	karashi
seasons (of the year) *{n} [astron]*	春夏秋冬	shun'ka-shū'tō
sea squirt *[i-zoo]*	海鞘; 老海鼠; ほや	hoya
sea surface *[ocean]*	海面	kaimen
seat; seating; chair *{n} [furn]*	腰掛; こしかけ	koshi'kake
seat belt *[eng]*	座席ベルト	zaseki-beruto
seating area *[arch]*	桟敷; さじき	sa'jiki
seating capacity *[aero-eng]*	座席数	zaseki-sū
sea toad; anglerfish; frogfish *[v-zoo]*	鮟鱇; あんこう	ankō
seat reservation *[trans]*	座席予約	zaseki-yoyaku
sea turtle *[v-zoo]*	海亀; うみがめ	umi-game
sea urchin *[i-zoo]*	雲丹; うに	uni
seawall; tide embankment *[civ-eng]*	防潮提	bōchō'tei
seawater *[ocean]*	海水	kaisui
seaway; sea route *[navig]*	海路; かいろ	kairo
seaweed *[bot]*	海草	kaisō
seaweed bed *[bot] [ecol]*	藻場; もば	moba
seaworthiness *{n} [nav-arch]*	耐航性	taikō-sei
sebaceous gland *[physio]*	皮脂腺; ひしせん	hishi-sen
sebaceous grime *[physio]*	皮脂汚垢	hishi-okō
secant *[geom]*	割線	kassen
" *[trig]*	正割; せいかつ	seikatsu
second (unit of plane angle) *[math]*	秒; びょう	byō
" (unit of time) *[mech]*	秒	byō
" (in order) *{adj} [math]*	第二の	dai'ni no
secondarily *{adv}*	二義的に	nigi'teki ni
secondary *{adj} [ch]*	第二級の	dai'nikyū no
" *{adj} [math]*	二次の; にじの	ni'ji no
" ; auxiliary *{adj}*	補助の	hojo no
secondary-; -sub *{prefix}*	副-	fuku-
secondary carbon atom *[ch]*	第二炭素原子	dai'ni-tanso-genshi
secondary cell *[elec]*	二次電池	niji-denchi
secondary circuit *[elec]*	二次回路	niji-kairo
secondary color *[opt]*	等和色	tōwa-shoku
secondary compound *[ch]*	第二化合物	dai'ni-kagō'butsu
secondary ion mass spectroscopy; SIMS *[phys]*	二次イオン質量=分析法	niji-ion-shitsu'ryō-bunseki-hō
secondary metabolite *[bioch]*	二次代謝産物	niji-taisha-sanbutsu
secondary mirror *[opt]*	副鏡; ふくきょう	fuku-kyō
secondary rainbow *[opt]*	副虹; ふくこう	fuku'kō
secondary salt; -ic salt *[ch]*	第二塩; だいにえん	dai'ni-en

secondary scale *[met]*	二次スケール	niji-sukēru
secondary winding *{n} [elecmg]*	二次巻線；二次卷線	niji-maki'sen
secondhand; used *{adj}*	中古の	chū'buru no
second-order *{adj} [math]*	二次の	niji no
second-order reaction *[ch]*	二次反応；二次反應	niji-hannō
second-order transition point *[p-ch]*	二次転移点	niji-ten'i-ten
secret *[ind-eng]*	秘密；ひみつ	hi'mitsu
secretion *[physio]*	分泌；ぶんぴつ	bunpitsu
section *{n} [bio] [comput] [syst]*	節；せつ	setsu
" *{n} [math]*	区分；區分；くぶん	kubun
" ; signature *{n} [pr]*	折り丁；おりちょう	ori'chō
sectional area *[math]*	断面積；斷面積	dan'menseki
sectional radiology; tomography *[electr]*	断層撮影法	dansō-satsu'ei-hō
sector (of a circle) *{n} [geom]*	扇形；おうぎがた	ōgi-gata
secular acceleration *[astron]*	永年加速	ei'nen-kasoku
secular parallax *[astron]*	永年視差	ei'nen-shisa
secular perturbation *[astrophys]*	永年摂動	ei'nen-setsu'dō
secular variation *[astron]*	経年変化；經年變化	kei'nen-henka
" *[astron]*	永年変化	ei'nen-henka
secure *{vb}*	固定する	kotei suru
securing; fixing *{n}*	固定；こてい	kotei
security *[comput]*	秘密保護	hi'mitsu-hogo
" ; collateral *[econ]*	担保；擔保；たんぽ	tanpo
sedative; tranquilizer *{n} [pharm]*	鎮静薬（鎮静剤）	chinsei-yaku (chinsei-zai)
sediment; deposit *{n} [geol]*	堆積物	taiseki'butsu
sediment (precipitated matter) *[geol]*	沈殿物；沈澱物	chinden'butsu
" (settled matter) *[geol]*	沈降物	chinkō'butsu
" (clastics; detritus) *[geol]*	砕屑物	saisetsu'butsu
" (silting) *{n} [geol]*	堆砂；たいしゃ	taisha
sedimentary rock *[petr]*	水成岩	suisei-gan
" *[petr]*	堆積岩	taiseki-gan
sedimentation *[geol] [met] [sci-t]*	沈降；ちんこう	chin'kō
sedimentation accelerator *[ch-eng]*	沈降促進剤	chinkō-soku'shin-zai
sedimentation equilibrium *[an-ch]*	沈降平衡	chinkō-heikō
sedimentation potential *[elec]*	沈降電位	chinkō-den'i
sedimentology *[geol]*	堆積学	taiseki'gaku
sedum (orpine) *[bot]*	弁慶草；辨慶草	benkei'sō
" *[bot]*	万年草；萬年草	mannen-gusa
see; look at *{vb}*	見る；観る；みる	miru
seed; pit; stone (of fruit) *{n} [bot]*	種子；しゅし	shushi
" *{n} [bot]*	種；たね	tane
seedbed *[agr]*	苗床；なえどこ	nae'doko

seed crystal *[crys]*	種(結)晶	tane-(kes)shō
seedless grape (fruit) *[bot]*	種無し葡萄	tane'nashi-budō
seedling; sapling *[bot]*	苗；なえ	nae
seedling culture *[agr]*	育苗；いくびょう	iku'byō
seed malt *[cook]*	種麹；たねこうじ	tane-kōji
seed oil *[mat]*	種子油	shushi-yu; tane-abura
seed plant *[bot]*	種子植物	shushi-shokubutsu
seek *{vb}*	求める	motomeru
seeping out *[fl-mech]*	浸出；しんしゅつ	shin'shutsu
segment *{n} [comput]*	部分；ぶぶん	bubun
" ; partition *{n} [comput]*	区分；區分；くぶん	kubun
" ; line segment *{n} [geom]*	線分	senbun
" *{n} [geom]*	弓形；ゆみがた	yumi-gata
" *{n} [i-zoo]*	分節；ぶんせつ	bun'setsu
" *{n} [zoo]*	体節；體節	tai'setsu
segmented *{adj} [i-zoo]*	環節のある	kan'setsu no aru
segregation *[ch]*	凝離；ぎょうり	gyōri
" *[met]*	偏析	henseki
segregation, law of *[gen]*	分離の法則	bunri no hōsoku
seiche *[fl-mech]*	静振	seishin; seishi
seine net *[eng]*	引網；ひきあみ	hiki-ami
" *[eng]*	地引網；地曳網	ji'biki-ami
seismic *{adj} [geol] [geophys]*	地震(性)の	jishin('sei) no
seismic center; epicenter *[geophys]*	震央；しんおう	shin'ō
seismic focus; hypocenter *[geophys]*	震源	shingen
seismic intensity *[geophys]*	震度	shindo
seismic sea wave; tsunami *[ocean]*	津波；つなみ	tsu'nami
seismicity *[geophys]*	地震活動度	jishin-katsudō-do
seismic waves *[geophys]*	地震波；じしんは	jishin-ha
seismograph *[eng]*	地震計	jishin-kei
seismology *[geophys]*	地震学	jishin'gaku
seize; grasp; grab *{vb t}*	摑む；つかむ	tsukamu
seizure *{n} [met]*	焼(き)付き	yaki'tsuki
selachian; cartilaginous fish *[v-zoo]*	軟骨魚	nankotsu'gyo
select; selecting *{n} [comput]*	選択；選擇	sentaku
" ; choose *{vb t}*	選ぶ；択ぶ；えらぶ	erabu
selection *[comput]*	選別	sen'betsu
" *[comput] [sci-t]*	選択；せんたく	sentaku
" *[comput]*	選定	sentei
" *[gen]*	淘汰；とうた	tōta
selective absorption *[elecmg]*	選択吸収；選擇吸收	sentaku-kyūshū
selectivity *[an-ch]*	選択性	sentaku-sei

selectivity *[electr]*	選択率	sentaku-ritsu
selenite *[miner]*	透石膏	tō'sekkō
selenium (element)	セレン	seren
selenograph *[astron]*	月面図；げつめんず	getsu'men-zu
selenography *[astron]*	月面誌	getsu'men'shi
selenology *[astron]*	月学；つきがく	tsuki'gaku
selenomorphology *[astron]*	月形態学	tsuki-keitai'gaku
self- *{prefix}*	自-	ji-
self-consumed part *{n}*	自消分	jishō-bun
self-curing adhesive *[poly-ch]*	自己硬化接着剤	jiko'kōka-setchaku-zai
self-extinguishing property *[mat]*	自消性	jishō-sei
self-fire-extinguishing *{n} [mat]*	自己消火	jiko-shōka
self-hardening property *[steel]*	自硬性	jikō-sei
selfheal; Prunella vulgaris *[bot]*	靫草；うつぼぐさ	utsubo-gusa
self-induced vibration *[mech]*	自励振動；自勵振動	jirei-shindō
self-induction *[elecmg]*	自己誘導	jiko-yūdō
self-leveling property *[eng]*	自己水平性	jiko-suihei-sei
self-propelling *{adj} [mech-eng]*	自己推進の	jiko-suishin no
self-sustaining *{adj} [mech]*	自立の	ji'ritsu no
sell (on the market) *{vb} [econ]*	市販する	shihan suru
" *{vb} [econ]*	売る；賣る；うる	uru
seller *{n} [econ]*	売(り)手	uri'te
selvage *[tex]*	耳；みみ	mimi
selvage stretch *{n} [tex]*	耳延び	mimi-nobi
semantics *[comm] [comput] [ling]*	意味論	imi-ron
semaphore *[comput]*	腕木信号	ude'gi-shingō
" *[rail]*	腕木；うでぎ	ude'gi
semaphore signal *[comm]*	腕木信号機	ude'gi-shingō-ki
semen (pl -semina) *[physio]*	精液	sei-eki
semi- *{prefix}*	半-	han-
" *{prefix}*	片側の-	kata'gawa no-
semicircle *{n} [geom]*	半円；はんえん	han'en
semicircular canal *[anat]*	半規管	hanki-kan
semicircular shape *[geom]*	半円形；半圓形	han'en-kei
semiconductive polymer *[poly-ch]*	半導性高分子	handō'sei-kōbunshi
semiconductor *{n} [sol-st]*	半導体	han'dōtai
semiconductor chip *[electr]*	半導体片	handōtai-hen
semiconductor detector *[nucleo]*	半導体検出器	handōtai-ken'shutsu-ki
semiconductor laser *[opt]*	半導体レーザー	handōtai-rēzā
semiconductor wafer *[electr]*	半導体板	handōtai-ban
semidesert *[ecol]*	半砂漠；はんさばく	han'sabaku
semidiurnal tide *[ocean]*	半日周潮	han'nichi-shūchō

semidrying oil *[mat]*	半乾性油	han'kansei-yu
semifossil resin *[resin]*	半化石樹脂	han'kaseki-jushi
semigroup *{n} [math]*	半群；はんぐん	hangun
semihardboard *[mat]*	半硬質繊維板	han'kōshitsu-sen'i'ban
semiindirect lighting *{n} [eng]*	半間接照明	han'kansetsu-shōmei
semilogarithmic paper *[math]*	片対数目盛	kata'taisū-memori
semilog graph paper *[an-geom]*	半対数方眼紙	han'taisū-hōgan-shi
semimajor axis *[an-geom]*	半長径	han-chōkei
semimetal *{n} [ch]*	半金属	han-kinzoku
semimicro analysis *[an-ch]*	小量分析	shōryō-bunseki
semiminor axis *[an-geom]*	半短径	han-tankei
seminal fluid *[physio]*	精液	sei-eki
semiotics *[comm]*	記号論；きごうろん	kigō-ron
semipermeable membrane *[phys]*	半透膜	hantō-maku
semiprecious stone *[miner]*	準宝石；準寶石	jun-hōseki
semisolid *{n} [phys]*	半固体	han'kotai
semisynthetic fiber *[poly-ch] [tex]*	半合成繊維	han'gōsei-sen'i
semisynthetic polymer *[poly-ch]*	半合成高分子	han'gōsei-kōbunshi
sen (Japanese money) *[econ]*	銭；錢；せん	sen
senarmontite *[miner]*	方安鉱	hōan'kō
send *{vb t} [comm]*	発信する	hasshin suru
" *{vb t} [comm] [trans]*	移送する	isō suru
" *{vb t} [comm] [trans]*	送る；おくる	okuru
" *{vb t} [comm]*	送信する	sōshin suru
" *{vb t} [comput]*	送り出す	okuri-dasu
send back; dismiss *{vb t}*	帰す；歸す；かえす	kaesu
sending *{n} [comm]*	発信；はっしん	hasshin
" *{n} [comm]*	送信	sōshin
" (transporting) *{n} [trans]*	移送	isō
sensation *[physio]*	知覚；ちかく	chikaku
" *[physio]*	感じ	kanji
" *[physio]*	感覚；かんかく	kan'kaku
sense of hearing *[physio]*	聴覚；聽覺	chō'kaku
sense of incompatibility *[psy]*	違和感；いわかん	iwa-kan
sense of sight *[physio]*	視覚	shi'kaku
sense of smell *[physio]*	嗅覚；きゅうかく	kyūkaku
sense of taste *[physio]*	味覚	mi'kaku
sense of touch *[physio]*	触覚；しょっかく	shokkaku
sense organ *[physio]*	感覚器(官)	kankaku-ki(kan)
sense strand *[mol-bio]*	意味のある鎖	imino no aru sa
sensibility *[physio]*	感じ	kanji
sensible climate *[climat]*	体感気候	taikan-kikō

sensible heat *[thermo]*	顕熱；顯熱	kennetsu
sensing element; sensor *[eng]*	検出部	ken'shutsu-bu
sensing weight *[mech]*	感じ分銅	kanji-fundō
sensitive material *[photo]*	感光体；感光體	kankō-tai
sensitivity *[electr] [sci-t]*	感度；かんど	kando
" *[photo]*	感光度	kankō-do
" *[physio]*	敏感性	binkan-sei
sensitivity speck *[photo]*	感光核	kankō-kaku
sensitization *[med]*	鋭敏化	eibin'ka
" *[photo]*	増感；ぞうかん	zōkan
sensitized paper *[photo]*	感光紙	kankō-shi
sensitizer *[med]*	鋭感剤	eikan-zai
" *[photo]*	増感剤；増感劑	zōkan-zai
sensitizing paper *[paper]*	感光原紙	kankō-genshi
sensitizing property *[photo]*	増感性	zōkan-sei
sensitometer *[photo]*	感光計	kankō-kei
sensor *[eng]*	感知器	kanchi-ki
" *[eng]*	感出器	kan'shutsu-ki
" *[eng]*	探査器；たんさき	tansa-ki
sensory cell *[physio]*	感覚細胞；感覺細胞	kankaku-saibō
sensory nerve *[physio]*	感覚神経	kankaku-shinkei
sensory test *[physio]*	官能検査	kannō-kensa
sentence *[gram]*	(完結)文	(kanketsu')bun
sepal *[bot]*	萼片；がくへん	gaku'hen
separate; individual *{n}*	個別	ko'betsu
separate *{n}*	別個；べっこ	bekko
" (detach from) *{vb i}*	離れる	hanareru
" *{vb t}*	分離する	bunri suru
" (detach; disconnect) *{vb t}*	離す	hanasu
" (space apart) *{vb t}*	離間させる	rikan saseru
separate edition *[pr]*	別冊；べっさつ	bessatsu
separate-phase indicator *[an-ch]*	分相指示薬	bunsō-shiji-yaku
separate seat (of car) *[mech-eng]*	独立座席；獨立座席	doku'ritsu-za'seki
separate sheet *[pat]*	別紙	besshi
separation *[astron]*	分離角	bunri-kaku
" *[min-eng]*	選別；選別	senbetsu
" *{n}*	分離；ぶんり	bunri
separation factor *[sci-t]*	分離係数	bunri-keisū
separator *[comput]*	分離記号	bunri-kigō
separatory funnel *[ch]*	分液漏斗	bun'eki-rōto
septarium *{n} [geol]*	亀甲石；龜甲石	kikkō'seki
September (month)	九月	ku'gatsu

septet *[quant-mech]*	七重項	shichi'jū-kō
" *[spect]*	七重線	shichi'jū-sen
septic tank *[civ-eng]*	浄化槽	jōka-sō
sequel *{n} [pr]*	続編；続篇	zoku'hen
sequence *{n} [alg]*	数列；すうれつ	sū-retsu
" *{n} [comput]*	一連	ichi-ren
" *{n} [comput] [eng]*	順位	jun'i
" *{n} [comput]*	順序；じゅんじょ	junjo
" *{n} [eng]*	連続；連續	renzoku
" *{n}*	順	jun
sequence number *[math]*	通し番号	tōshi-bangō
sequence rule *[math]*	順位則	jun'i-soku
sequencing *{n} [bioch] [gen]*	配列決定	hai'retsu-kettei
" *{n} [ind-eng]*	順序付け	junjo-zuke; junjo-tsuke
sequential; serial *{adv} {n} [math]*	逐次；ちくじ	chiku'ji
sequential access *[comput]*	逐(次)呼出し	jun(ji)-yobi'dashi
sequential circuit *[elec]*	順序回路	junjo-kairo
sequential operation *[comput]*	順序動作	junjo-dōsa
sequential processing *{n} [comput]*	逐次処理；逐次處理	chikuji-shori
sequential replication *[mol-bio]*	順続的複製	junzoku'teki-fukusei
sequestering agent *[ch]*	金属イオン封鎖剤	kinzoku-ion-hōsa-zai
serac *{n} [hyd]*	氷塔	hyōtō
sergeant major (fish) *[v-zoo]*	おやびっちゃ	oyabitcha
serial *{n} [sci-t]*	直列；ちょくれつ	choku'retsu
serial arithmetic *[arith]*	直列演算	chokuretsu-enzan
serialization *[elec]*	直列化	chokuretsu'ka
serialize *{vb} [comput]*	直列化する	chokuretsu'ka suru
serial number *[math]*	一連番号	ichi'ren-bangō
" *[math]*	通し番号	tōshi-bangō
serial transmission *[comput]*	直列伝送	chokuretsu-densō
sericulture *[tex]*	養蚕；ようさん	yōsan
series (pl series) *[alg]*	級数	kyūsū
" *[bio] [syst]*	系；けい	kei
" *[elec]*	直列	choku'retsu
series resonance *[elec]*	直列共振	chokuretsu-kyōshin
series transformer *[elecmg]*	直列変圧器	chokuretsu-hen'atsu-ki
serology *[bio]*	血清学	kessei'gaku
serpentine *{n} [miner]*	蛇紋石	jamon'seki
serpentinite *{n} [petr]*	蛇紋岩	jamon'gan
serrate *{adj} [bot]*	鋸歯状の	kyoshi'jō no
serration *[bot]*	刻み；きざみ	kizami
" *[bot]*	鋸歯；きょし	kyoshi

serum (pl -s, sera) *[physio]*	血清；けっせい	kessei
serum-free medium *[microbio]*	無血清培地	mu'kessei-baichi
service entrance *[elec]*	引込口	hiki'komi-guchi
service interruption; outage *[elec]*	停電；ていでん	teiden
service life *[ind-eng]*	有効寿命；有効壽命	yūkō-jumyō
service voltage *[elec]*	供給電圧	kyōkyū-den'atsu
service water *[hyd]*	用水	yōsui
servicing *{n} [eng]*	修理；しゅうり	shūri
" *{n} [ind-eng]*	整備	seibi
serving; one serving *{n} [cook]*	一盛り	hito-mori
serving size *[cook]*	盛り分け量	mori'wake-ryō
sesame; sesame seed *[bot]*	胡麻；ごま	goma
seseli *[bot]*	伊吹防風	ibuki-bōfū
sessile *{adj} [bot]*	無柄の	muhei no
sessile leaf *[bot]*	座葉	za-yō
set *{n} [comput]*	親子集合	oyako-shūgō
" *{n} [comput] [math]*	集合；しゅうごう	shūgō
" *{n} [math]*	組；くみ	kumi
" *{vb t} [comput]*	設定する	settei suru
seta (pl setae) *{n} [bio]*	剛毛；ごうもう	gōmō
set at break *{n} [mech]*	破断歪；破斷歪	hadan-hizumi
set seal *{vb} [legal]*	捺印する	natsu'in suru
set strain *[mech]*	設定歪	settei-hizumi
setting; set *{n} [astron]*	没；ぼつ	botsu
" *{n} [astron]*	入；いり	iri
setting *{n} [civ-eng]*	凝結	gyō'ketsu
setting priorities *[sci-t]*	段取りを決める	dan'dori o kimeru
setting property *[eng] [mat]*	定着性	teichaku-sei
settlement *[legal]*	分与；分與；ぶんよ	bun'yo
" *[legal]*	調停	chōtei
settlement of accounts *[econ]*	決済；決濟	kessai
settling *{n} [eng] [met]*	沈降	chinkō
settling pond (or basin) *[civ-eng]*	沈殿池；沈澱池	chinden'chi
settling tank *[eng]*	沈降槽	chinkō-sō
seven *{n} [math]*	七；しち	shichi
seven-spice pepper (condiment) *[cook]*	七味唐辛子	shichi'mi-tō'garashi
sevum *[v-zoo]*	脂；あぶら	abura
sew; stitch *{vb t}*	縫う；ぬう	nū; nu'u
sewage *[civ-eng]*	下水；げすい	ge'sui
sewer *[civ-eng]*	溝；どぶ	dobu
" *[civ-eng]*	下水道	gesui'dō
sewer rat (animal) *[v-zoo]*	溝鼠；どぶねずみ	dobu-nezumi

sewing *{n} [cl] [tex]*	裁縫；さいほう	saihō
" (Japanese) *{n} [cl] [tex]*	和裁	wasai
sewing and stitching (bookbinding)	綴じ；とじ	toji
sex *[bio]*	性；せい	sei
sexagesimal scale *[geom]*	六十進法	roku'jū-shinpō
sex cells *[bio]*	性細胞；性細胞	sei-saibō
sex chromosome *[gen]*	性染色体	sei-senshoku'tai
sex determination *[gen]*	性決定	sei-kettei
sex factor *[microbio]*	性因子	sei-inshi
sex-linked inheritance *[gen]*	伴性遺伝	hansei-iden
sex pilus (pl pili) *[gen]*	性繊毛；性線毛	sei-senmō
sextant *[eng]*	六分儀；ろくぶんぎ	roku'bun-gi
sexual dimorphism *[bio]*	性的二形	sei'teki-nikei
sexual intercourse *[bio]*	性交	seikō
sexual reproduction *[bio]*	有性生殖	yūsei-seishoku
shaddock (fruit) *[bot]*	朱欒；ざぼん	zabon
shade *{n} [opt]*	陰；蔭；かげ	kage
shade off; gradate *{vb t}*	暈す；ぼかす	bokasu
shade plant *[bot]*	陰生植物	insei-shokubutsu
shade tree; street tree *[bot] [ecol]*	街路樹；がいろじゅ	gairo-ju
shading *{n} [graph]*	陰影	in'ei
" *{n} [graph]*	明暗法	mei'an-hō
shadow *{n} [opt]*	影；かげ	kage
shadowgraph *[graph]*	逆光線写真	gyak(u)kōsen-shashin
shadowing *{n} [graph]*	隈取り；くまどり	kuma'dori
" *{n} [opt]*	陰影	in'ei
shadow picture *[graph]*	影絵	kage-e
shaft *[mech-eng]*	軸	jiku
shaft core *[mech-eng]*	軸芯；じくしん	jiku-shin
shaft of light *[opt]*	光芒	kōbō
shake *{vb i}*	揺れる；搖れる	yureru
" *{vb t}*	振る	furu
" *{vb t}*	揺する；搖する	yusuru
shake culture *[microbio]*	振盪培養	shintō-baiyō
shake resistant *[mat]*	耐振	tai-shin
shaking; quivering *{n}*	揺れ；搖れ；ゆれ	yure
shakudo (copper-gold alloy) *[met]*	赤銅；しゃくどう	shaku'dō
shale *[geol]*	泥板岩	deiban'gan
" *[petr]*	頁岩；けつがん	ketsu'gan
shale oil *[mat]*	頁岩油	ketsugan-yu
shallot (vegetable) *[bot]*	韮；にら	nira
" (vegetable) *[bot]*	辣韮；辣韭	rakkyō

shallow *{adj}*	浅い；あさい	asai
shallowness *[ocean]*	浅さ；淺さ	asa-sa
shallow-water cable *[elec]*	淺海線；あさみせん	asami-sen
shape *{n} [math] [phys] [sci-t]*	形；かたち	katachi
shape control *[eng]*	形状制御	keijō-seigyo
shape factor *[fl-mech]*	形状係数	keijō-keisū
shape-freezability *[eng]*	形状凍結性	keijō-tōketsu-sei
shape-memory alloy *[met]*	形状記憶合金	keijō-ki'oku-gōkin
shaping (appearance) *{n} [arch]*	外見	gaiken
" (molding) *{n} [eng]*	成形；せいけい	seikei
shaping mold *[eng]*	成形金型	seikei-kana'gata
shard (of earthenware) *{n} [archeo]*	破片（土器の）	hahen (doki no)
" *{n} [archeo]*	欠片；かけら	kakera
shared logic *[comput]*	共用論理	kyōyō-ronri
shark (large) (fish) *[v-zoo]*	鱶；ふか	fuka
" (fish) *[v-zoo]*	鮫；さめ	same
shark liver oil *[mat]*	鮫肝油	same-kan'yu
shark ray (fish) *[v-zoo]*	粕鮫	kasu-zame
shark sucker (fish) *[v-zoo]*	小判鮫；こばんざめ	koban-zame
sharp *{adj} [eng]*	鋭い；するどい	surudoi
sharpness *[eng]*	切れ味	kire-aji
" *[opt]*	鋭さ	surudo-sa
" *[photo]*	鮮鋭度；せんえいど	sen'ei-do
sharp-tailed grouse (bird) *[v-zoo]*	細尾雷鳥	hoso'o-raichō
shave *{vb t} [eng]*	削る	kezuru
" (as a beard) *{vb t}*	剃る；そる	soru
shear *{n} [mech]*	剪断；せんだん	sendan
" *{n} [mech-eng]*	ずり	zuri
shearing *{n} [mech]*	剪断作用	sendan-sayō
shearing stress *[mech]*	剪断応力	sendan-ō'ryoku
" *[mech]*	ずれ応力；ずれ應力	zure-ō'ryoku
shear modulus *[mech]*	剛性率	gōsei-ritsu
" *[mech]*	ずれ弾性率	zure-dansei-ritsu
shear rate *[fl-mech]*	剪断速度	sendan-sokudo
" *[fl-mech]*	ずれ率	zure-ritsu
" *[fl-mech]*	ずれ速度	zure-sokudo
shearwater (bird) *[v-zoo]*	水凪鳥	mizu'nagi-dori
shear wave; distortional wave *[phys]*	捩れ波；ねじれなみ	nejire-nami
sheath; scabbard *{n} [ord]*	鞘	saya
sheathe; armor; wrap *{vb t} [eng]*	外装する	gaisō suru
sheen *[opt]*	煌；きらめき	kirameki
sheep (pl sheep) (animal) *[v-zoo]*	羊；ひつじ	hitsuji

sheepskin *[v-zoo]*	羊皮	yōhi
sheep wool *[tex] [v-zoo]*	羊毛；ようもう	yōmō
sheerness *[tex]*	薄さ	usu-sa
sheet (of board) *{n} [mat]*	薄板；うすばん	usu-ban
" (of paper) *{n} [mat]*	用紙	yōshi
" *{n} [pr]*	枚葉紙	maiyō-shi
-sheet(s) *{suffix} [pr]*	-枚	-mai
" *{suffix} [pr]*	-葉	-yō
sheet cracking *{n} [eng]*	板割れ	ita-ware
sheet feeder *[comput]*	用紙送り機構	yōshi-okuri-kikō
sheeting agent *[rub]*	圧延剤；壓延劑	atsu'en-zai
sheet iron *[elec] [met]*	薄鋼板	usu-kōhan
sheet ligntning *[geophys]*	幕電光	maku-denkō
sheet machine *[eng] [paper]*	手抄機；てすきき	te'suki-ki
sheet metal *[met]*	薄板金	usu-ita'gane
" *[met]*	薄金属板	usu-kinzoku'ban
sheet piling *{n} [civ-eng]*	矢板；やいた	ya-ita
sheet zinc *[met]*	トタン板	totan-ita
shelf life (storage stability) *[ind-eng]*	貯蔵安定性	chozō-antei-sei
" (storage life) *[ind-eng]*	貯蔵寿命；貯藏壽命	chozō-jumyō
" (keeping life) *[ind-eng]*	保管寿命	hokan-jumyō
" (storability) *[ind-eng]*	保存性	hozon-sei
shell; husk *{n} [eng] [mat]*	殻；かく；から	kaku; kara
" *{n} [zoo]*	貝(殻)	kai(-gara)
shellfish *[i-zoo]*	貝；かい	kai
" *[i-zoo]*	貝類	kai-rui
shell lime *[agr]*	貝灰	kai-bai
shell structure *[nuc-phys]*	核構造	kaku-kōzō
shelter *{n} [arch]*	避難所	hinan'jo
shepherd's purse *[bot]*	薺；なずな	nazuna
sherd *{n} [archeo]*	破片（土器の）	hahen (doki no)
shield *{n} [ord]*	盾；楯；たて	tate
shielding *{n} [elecmg]*	遮蔽	shahei
shield volcano *[geol]*	盾状火山；楯状火山	tate'jō-kazan
shift; shifting *{n} [comput]*	移動；いどう	idō
" *{n} [comput]*	桁移動	keta-idō
" *{n} [comput]*	(桁)送り	(keta-)okuri
shift (in casting) *{n} [plas]*	ずれ	zure
" *{vb t}*	移す	utsusu
" *{vb t}*	ずらす	zurasu
Shigella *[microbio]*	赤痢菌；せきりきん	sekiri-kin
shikimic acid *[ch]*	樒酸；しきみさん	shikimi-san

shim (of metal) *{n} [eng] [met]*	挟み金；挾み金	hasami-gane
" (of wood) *{n} [eng]*	挟み板；はさみいた	hasami-ita
shimmer; glitter *{vb i} [opt]*	ちらちら光る	chira-chira hikaru
shin (of leg) *{n} [anat]*	向こう脛	mukō'zune
" (of leg) *{n} [anat]*	脛；すね	sune
shine; sheen *{n} [opt]*	艶；つや	tsuya
shine; sparkle; glint *{vb} [opt]*	輝く	kagayaku
shingle *[mat]*	柿板；コケラ板	kokera-ita
shingle roof *[arch]*	柿屋根；コケラ屋根	kokera-yane
ship *{n} [nav-arch]*	船；ふね	fune
" *{n} [nav-arch]*	船舶	sen'paku
shipbottom *[nav-arch]*	船底	funa'zoko
shipbuilding *{n} [nav-arch]*	造船；ぞうせん	zōsen
shipping *{n} [ind-eng]*	出荷	shukka
" *{n} [nav-arch]*	船積み	funa'zumi
shipping charge; cartage *[econ]*	運賃；うんちん	un'chin
shipping container *[ind-eng]*	輸送用器	yusō-yōki
shipping lane *[navig]*	運航路	unkō-ro
ship's captain; commander; master	船長	senchō
shipyard *[nav-arch]*	造船所	zōsen'jo
shirt *[cl]*	シャツ	shatsu
shītake (or shiitake) mushroom *[bot]*	椎茸；しいたけ	shī-take
shivering *{n} [cer]*	剝離	haku'ri
" *{n} [physio]*	震え	furue
shoal; bar *{n} [geol]*	州；洲；す	su
shoaling beach *[ocean]*	遠浅；とおあさ	tō-asa
shock *{n} [mech]*	振動	shindō
" *{n} [mech]*	衝撃；しょうげき	shōgeki
shock absorber *[mech-eng]*	緩衝器	kanshō-ki
shock resistance *[mat]*	耐衝撃性	tai'shōgeki-sei
shock resistant *{adj} [eng]*	耐震の	tai'shin no
shock-resistant steel *[steel]*	耐震鋼	tai'shin-kō
shock wave *[phys]*	衝撃波	shōgeki'ha
shoe *{n} [cl]*	靴；くつ	kutsu
shoehorn *[cl] [eng]*	靴箆；くつべら	kutsu-bera
shoot *{n} [bot]*	苗条	byōjō
shooting star *[astron]*	流星；りゅうせい	ryūsei
shop; store *{n} [arch]*	販売店；販賣店	hanbai'ten
shopping center (or district) *[arch]*	商店街	shōten'gai
shopping district *[civ-eng]*	店舗地区	tenpo-chiku
shore *{n} [geol]*	沿岸	engan
" *{n} [geol]*	海岸；かいがん	kaigan

shorebird (bird) *[v-zoo]*	浜鳥；濱鳥	hama-dori
shoreline; coastline *[geol]*	海岸線	kaigan-sen
shore reef *[geol]*	岸礁；がんしょう	ganshō
short *{adj}* *[math]*	短い；みじかい	mijikai
shortage; out of stock; sold out *{n}*	品切れ	shina'gire
short axis *[math]*	短軸	tan-jiku
short chain *[ch]*	短鎖	tansa
short circuit *[civ-eng]* *[elec]*	短絡；たんらく	tan'raku
shortcoming *{n}* *[sci-t]*	欠点	ketten
" (demerit) *{n}*	短所	tansho
short-day plant *[bot]*	短日植物	tan'jitsu-shokubutsu
shorten *{vb t}*	短縮する	tanshuku suru
shorthand *[comm]*	速記	sokki
short hand (of clock) *[horol]*	短針	tanshin
shortish direction *[math]*	短手方向	mijika'de-hōkō
short-necked clam *[i-zoo]*	浅蜊；あさり	asari
short oil *[mat]*	短油	tan-yu
short-range force *[phys]*	近達力	kin'tatsu-ryoku
" *[phys]*	短距離力	tan'kyori-ryoku
short-range forecast *[meteor]*	短期予報	tanki-yohō
short shoot *[bot]*	短枝	tanshi
short shunt *[elec]*	内分巻；内分巻	uchi-bun'maki
shortstopping agent; shortstop *{n}* *[poly-ch]*	重合停止剤	jūgō-teishi-zai
short sword *[ord]*	短刀	tantō
" *[ord]*	脇差；わきざし	waki'zashi
short wave *[comm]*	短波	tanpa
shot *{n}* *[ord]*	散弾	sandan
shoulder *{n}* *[anat]*	肩；かた	kata
shoulder blade (scapula) *[anat]*	肩甲骨	kenkō-kotsu
shouldering pole *[eng]*	天秤棒	tenbin-bō
shovel; scoop *{n}* *[eng]*	スコップ	sukoppu
shoveler (bird) *[v-zoo]*	嘴広鴨；はしびろ鴨	hashi'biro-gamo
show; exhibit *{vb t}*	見せる	miseru
" *{vb t}*	示す	shimesu
Shōwa (era name) (1926-1989)	昭和；しょうわ	Shōwa
shower; sudden rain *{n}* *[meteor]*	驟雨；しゅうう	shū'u
showing; record *{n}* *[econ]*	成積	seiseki
showings; business showings *[econ]*	実積；實積	jisseki
showroom; display room *[arch]*	陳列室	chin'retsu-shitsu
" *[arch]*	展示室	tenji-shitsu
shredder *[eng]*	粉砕機	funsai-ki

shrew; shrewmouse *[v-zoo]*	地鼠；じねずみ	ji-nezumi
shrike (bird) *[v-zoo]*	百舌鳥；鵙；もず	mozu
shrimp (crustacean) *[i-zoo]*	蝦；海老；えび	ebi
shrine *[arch]*	社；やしろ	yashiro
shrink *{vb} [mech]*	収縮する；収縮する	shūshuku suru
" *{vb i} [tex]*	縮む；ちぢむ	chijimu
" *{vb t} [tex]*	縮める	chijimeru
shrinkage *[tex]*	縮み	chijimi
shrinkage allowance *[tex] [plas]*	縮み代	chijimi-shiro
shrinkage cavity *[met]*	収縮孔	shūshuku-kō
shrinkage fit *[mech-eng]*	焼嵌め；やきばめ	yaki-bame
shrinking; shrinkage *{n} [mech]*	収縮；收縮	shū'shuku
shrinking resistance *[tex]*	防縮性	bōshuku-sei
shrink mark *[met]*	引け	hike
shrub *[bot]*	低木	tei'boku
shrubbery *[arch] [bot]*	植込み；うえこみ	ue'komi
shunt *{n} [elec]*	分路	bunro
" *{n} [elec]*	分流器	bunryū-ki
shunt motor *[elec]*	分巻電動機	bun'maki-dendō-ki
shut; close *{vb}*	締める	shimeru
" *{vb}*	閉じる；とじる	tojiru
shutdown *{n} [comput]*	遮断	shadan
" *{n} [comput]*	運転停止	unten-teishi
shut-off valve *[mech-eng]*	閉止弁；閉止瓣	heishi-ben
shutter *{n} [arch]*	鎧戸	yoroi'do
shuttle *{n} [tex]*	杼；ひ	hi
Siberian yarrow *[bot]*	鋸草；のこぎりそう	nokogiri'sō
sickle *[agr] [eng]*	鎌	kama
sickness *[med]*	病気；びょうき	byōki
side *{n} [anat]*	脇；わき	waki
" *{n} [geom]*	辺；邊；へん	hen
" *{n}*	側；傍；そば	soba
" (sidewise direction) *{n}*	側方	soku'hō
" (sidewise) *{n}*	横	yoko
sideband wave *[elecmg]*	側帯波	sokutai-ha
side chain *[org-ch]*	側鎖	soku-sa
side dish *[cook]*	副食	fuku'shoku
side effect *[comput] [pharm]*	副作用；ふくさよう	fuku-sayō
side reaction *[ch]*	副反応；副反應	fuku-hannō
sidereal day *[astron]*	恒星日	kōsei-jitsu
sidereal period *[astron]*	恒星周期	kōsei-shūki
sidereal time *[astron]*	恒星時	kōsei-ji

siderite *[miner]*	隕鉄；隕鐵	in'tetsu
" *[miner]*	菱鉄鉱；菱鐵鑛	ryō'tekkō
side road *[civ-eng]*	脇道；わきみち	waki-michi
siderolite *[geol]*	石鉄隕石	seki'tetsu-inseki
siderophile element *[ch]*	親鉄元素	shin'tetsu-genso
siderosphere *[geophys]*	鉄圏；てっけん	tekken
siderostat *[opt]*	太陽鏡	taiyō'kyō
sidetone *[comm]*	側音	soku'on
side view *[graph]*	側面図	sokumen-zu
sidewalk *[civ-eng]*	歩道	hodō
sidewise skidding *{n} [traffic]*	横滑り；横辷り	yoko-suberi
side yard *[arch] [civ-eng]*	側庭	sokutei
sieve *{n} [eng]*	篩；ふるい	furui
sieve plate *[ch-eng]*	師板；篩板；しばん	shiban
sieve pore *[eng]*	篩孔	shikō
sieve residue *[eng]*	篩残分	furui-zanbun
sieve tube *[bot]*	師管	shikan
sifter *[cook] [eng]*	粉篩；粉ふるい	kona-furui
sight; vision *{n} [physio]*	視覚	shi'kaku
sight check *[comput]*	視覚検査	shikaku-kensa
sigmoid *{adj} [bio]*	エス(S)字形の	esuji'gata no
sign *{n} [arith]*	符号；符號；ふごう	fugō
" (mark) *{n} [navig] [traffic]*	標識	hyō'shiki
" (set signature) *{vb} [legal]*	署名する	shomei suru
signal *{n} [comm]*	信号	shingō
signal regeneration *[comm]*	信号再生	singō-saisei
signature *[legal]*	署名	shomei
" *[pr]*	折り丁	ori'chō
" *[pr]*	背丁；せちょう	se'chō
signboard; sign *{n} [econ]*	看板	kanban
signed integer *[comput]*	符号付き整数	fugō'tsuki-seisū
significance *[math]*	重み	omo'mi
" *[math]*	有意性；ゆういせい	yū'i-sei
significant *{adj} [artith]*	有意の	yū'i no
significant digit (or figure) *[math]*	有効数字	yūkō-sūji
significant digit arithmetic *[math]*	有効数字演算	yūkō-sūji-enzan
significant interval *[comput]*	有意間隔	yū'i-kankaku
silence *{n} [acous]*	静けさ	shizuke-sa
silencer *[eng]*	消音器	shō'on-ki
silent; quiet *{adj} [acous]*	静かな	shizuka na
silent discharge *[elec]*	無声放電	musei-hōden
silent gear *[mech-eng]*	無音歯車	mu'on-ha'guruma

silica *[miner]*	珪石；けいせき	keiseki
silica glass; fused silica *[mat]*	融解石英	yūkai-seki'ei
silica sand; quartz sand *[geol]*	珪砂；けいしゃ	keisha
silicate *[ch]*	珪酸塩	keisan-en
siliceous *{adj}* *[miner]*	石英質の	seki'ei'shitsu no
silicic acid	珪酸；けいさん	keisan
silicic anhydride	無水珪酸	musui-keisan
silicide *[ch]*	珪化物	keika'butsu
silicification *[geol]*	珪化；硅化；けいか	keika
silicofluoride *[ch]*	珪弗化物	kei'fukka'butsu
silicon (element)	珪素；硅素；ケイ素	keiso
silicon bromide	臭化珪素	shūka-keiso
silicon carbide	炭化珪素	tanka-keiso
silicon chloride	塩化珪素	enka-keiso
silicon dioxide; silica	二酸化珪素	ni'sanka-keiso
silicon disulfide	二硫化珪素	ni'ryūka-keiso
silicone resin *[poly-ch]*	珪素樹脂	keiso-jushi
silicon hydride	水素化珪素	suiso'ka-keiso
silicon iodide	沃化珪素	yōka-keiso
silicon monochloride	一塩化珪素	ichi'enka-keiso
silicon nitride	窒化珪素	chikka-keiso
silicon oxide	酸化珪素	sanka-keiso
silicon steel *[steel]*	珪素鋼；けいそこう	keiso-kō
silicon sulfide	硫化珪素	ryūka-keiso
silicon tetrabromide	四臭化珪素	shi'shūka-keiso
silicon tetrachloride	四塩化珪素	shi'enka-keiso
silicon tetrafluoride	四弗化珪素	shi'fukka-keiso
silicon tetraiodide	四沃化珪素	shi'yōka-keiso
silizue *{n}* *[bot]*	長角果	chōkaku-ka
silk *[tex]*	絹；きぬ	kinu
silk fabric *[tex]*	絹織物	kinu-ori'mono
silk gland *[i-zoo]*	絹糸腺；けんしせん	kenshi-sen
silk gut *[mat]* *[i-zoo]*	天蚕糸；てぐす	tegusu
silk reeling *{n}* *[tex]*	製糸	seishi
silk yarn *[tex]*	絹糸	kinu-ito; ken'shi
silkworm (a moth larva) *[i-zoo]*	蚕；かいこ	kaiko
silkworm culture; sericulture *[tex]*	養蚕；ようさん	yōsan
silkworm moth; Bombyx (insect) *[i-zoo]*	蚕蛾；かいこが	kaiko-ga
Sillago sihama (fish) *[v-zoo]*	鱚；きす	kisu
sillimanite *[miner]*	珪線石	keisen'seki
silt *{n}* *[geol]*	沈積土	chinseki'do
silver (element)	銀；ぎん	gin

silver borofluoride	硼弗化銀	hō'fukka-gin
silver bromide	臭化銀	shūka-gin
silver carbonate	炭酸銀	tansan-gin
silver chloride	塩化銀	enka-gin
silver fir (tree) *[bot]*	蝦夷松；えぞまつ	ezo-matsu
silverfish (insect) *[i-zoo]*	紙魚；しみ	shimi
silver fluoride	弗化銀	fukka-gin
silver gray (color) *[opt]*	銀ネズ	gin-nezu
silvering *{n} [met]*	銀引き	gin'biki
silver iodide	沃化銀；ようかぎん	yōka-gin
silver nitrate	硝酸銀	shōsan-gin
silver oxide	酸化銀	sanka-gin
silver oxide nitrate	硝酸酸化銀	shōsan-sanka-gin
silver paint *[mat]*	銀泥；ぎんでい	gindei
silver peroxide	過酸化銀	ka'sanka-gin
silver salmon (fish) *[v-zoo]*	銀鱒；ぎんます	gin-masu
silver salt *[ch]*	銀塩	gin'en
silversides (fish) *[v-zoo]*	藤五郎鰯	tō'gorō-iwashi
silver solder *[met]*	硬蠟；こうろう	kōrō
silver soldering *{n} [met]*	銀蠟付	gin-rō'zuke
silver sulfate	硫酸銀	ryūsan-gin
silver sulfide	硫化銀	ryūka-gin
silver thread *[tex]*	銀糸	ginshi
silvervine *[bot]*	木天蓼；またたび	matatabi
simian *{adj} [v-zoo]*	猿の	saru no
simian immunodeficiency virus; SIV *[immun]*	猿免疫不全ウィルス	saru-men'eki-fuzen-uirusu
similar *{adj} [geom]*	相似の；そうじの	sōji no
similarity *{n} [meteor]*	類似性	ruiji-sei
simmered dish *[cook]*	煮物；にもの	ni'mono
simple *{n} [pharm]*	薬草；藥草	yaku'sō
simple *{adj} [math]*	簡単な	kantan na
" *{adj} [math]*	単一の；たんいつの	tan'itsu no
simple beam *[eng]*	両持梁	ryō'mochi-bari
simple correlation *[stat]*	単相関；單相關	tan-sōkan
simple distillation *[ch]*	単蒸留	tan-jōryū
simple harmonic motion *[mech]*	単振動	tan-shindō
simple lattice *[crys]*	単純格子	tanjun-kōshi
simple leaf *[bot]*	単葉；たんよう	tan'yō
simple lipid *[bioch]*	単純脂質	tanjun-shi'shitsu
simple machine *[mech-eng]*	単純機械	tanjun-kikai
simple ointment *[pharm]*	単軟膏	tan-nankō

simple pendulum *[mech]*	単振子；たんふりこ	tan-furiko
simple sentence *[gram]*	単文	tanbun
simple substance *[ch]*	単体；單體	tantai
simple tone *[acous]*	純音；じゅんおん	jun'on
simplex circuit *[comm]*	単信回線	tanshin-kaisen
simplex communication *[comm]*	単向通信	tankō-tsū'shin
" *[comm]*	単信通信	tanshin-tsū'shin
simplex method *[alg]*	単体法	tantai-hō
simplex winding *{n}* *[elec]*	単重巻；單重巻	tanjū-maki
simplicity *{n}*	単純	tan'jun
Simulium (insect) *[i-zoo]*	蚋；ぶよ	buyo; buyu
simultaneous *{adj}*	同時の	dōji no
simultaneous broadcasting *{n}* *[comm]*	同時放送	dōji-hōsō
simultaneous equations *[alg]*	連立方程式	ren'ritsu-hōtei'shiki
simultaneously *{adv}*	同時に；どうじに	dōji ni
simultaneous reaction *[ch]*	同時反応	dōji-hannō
" *[ch]*	並発反応；並發反應	hei'hatsu-hannō
simultaneous stretching *{n}* *[poly-ch]*	同時延伸	dōji-enshin
simultaneous transmission *[comm]*	同時伝送	dōji-densō
sine *[trig]*	正弦	seigen
sine wave *[phys]*	正弦波；せいげんは	seigen'ha
sing; recite; chant *{vb}*	歌う；唄う；うたう	utau
singing *{n}* *[elec]*	鳴音	mei'on
single *{adj}*	単一の	tan'itsu no
single-bath process *[ch]*	一浴法	ichi-yoku-hō
single bond *[ch]*	単結合	tan-ketsu'gō
single-cell protein *[bioch]*	単細胞蛋白質	tan-saibō-tanpaku'shitsu
single crystal *[crys]*	単結晶	tan-kesshō
" *[crys]*	単晶	tanshō
single fiber *[tex]*	単繊維；たんせんい	tan-sen'i
single molecule *[ch]*	単分子	tan-bunshi
single-phase power *[elec]*	単相電力	tansō-den'ryoku
single-precision *[comput]*	単精度；たんせいど	tan-seido
single reaction *[ch]*	単一反応	tan'itsu-hannō
single-sideband transmission *[comm]*	単側波帯伝送	tan-sokuha'tai-densō
single speed *[phys]*	単速	tansoku
singlet *[quant-mech]*	一重項	ichi'jū-kō
single yarn *[tex]*	単糸	tanshi
single yarn breakage *[tex]*	単糸切れ	tanshi-gire
singular; singular number *[math]*	単数；單數	tansū
singular *{adj}*	特異的な	toku'i'teki na
singularity *{n}* *[math]*	特異点；特異點	toku'i-ten

singular perturbation *[math]*	特異摂動	toku'i-setsu'dō
singular solution *[math]*	特異解；とくいかい	toku'i-kai
sinistral (of the left side) *[med]*	左側の	hidari'gawa no
" (leftward facing) *{adj}*	左向きの	hidari'muki no
sinistrorse *{adj} [bot]*	左巻きの	hidari'maki no
sink *{n} [arch] [cook]*	流し	nagashi
" (be submerged) *{vb i}*	沈む	shizumu
" (let collapse) *{vb t}*	陥没させる	kan'botsu saseru
" (let sink) *{vb t}*	沈める	shizumeru
sinkhole *[geol]*	陥落孔；陷落孔	kanrak(u)kō
sink mark *[met]*	引け	hike
sinoatrial node *[anat]*	洞房結節	tōbō-kessetsu
sinterability *[met]*	焼結性	shōketsu-sei
sintered hard alloy *[met]*	超硬合金	chōkō-gōkin
sintered ore *[geol]*	焼結鉱；燒結鑛	shōketsu-kō
sintering *{n} [met]*	焼結；しょうけつ	shō'ketsu
" *{n} [met]*	焼成；しょうせい	shōsei
sinusoidal wave; sine wave *[phys]*	正弦波	seigen'ha
Sipunculoidea; Sipunculida *[i-zoo]*	星口動物	seikō-dōbutsu
siren *{n} [electr]*	警報	keihō
Sirius *[astron]*	天狼星	tenrō-sei
sit *{vb i}*	座る；坐る；すわる	suwaru
sit down *{vb i}*	腰を掛ける	koshi o kakeru
six *{n} [math]*	六	roku
six-wheel car *[mech-eng]*	六輪車	roku'rin'sha
size *{n} [math]*	大きさ	ōki-sa
" *{n} [math]*	寸法	sunpō
size distribution (of particles), law of *[eng]*	流動分布法則	ryūdo-bunpu-hōsoku
sizing; size; starching *{n} [tex]*	糊付け	nori'zuke
skate; thornback (fish) *[v-zoo]*	雁木鱏；がんぎえい	gangi'ei
skein *[tex]*	綛；かせ	kase
skeleton *[anat]*	骸骨；がいこつ	gai'kotsu
" *[anat]*	骨組み	hone'gumi
" ; physique *[anat]*	骨格	kokkaku
sketch *{n} [graph]*	見取図；みとりず	mi'tori-zu
" *{n} [graph]*	素描；そびょう	so'byō
sketch map; key chart *[graph] [map]*	要図；要圖	yō'zu
skew; skewness *{n} [sci-t]*	歪み；ゆがみ	yugami
skewer; spit *{n} [cook]*	串；くし	kushi
skewness *{n} [sci-t]*	非対称度；非對稱度	hi'taishō-do
" *{n} [stat]*	歪度；わいど	wai-do

skew-symmetric *{adj} [geom]*	歪対称の	hizumi-taishō no
skid *{n} [mech-eng]*	(横)滑り; 横辷り	(yoko-)suberi
skill *{n} [ind-eng]*	熟練; 熟練	juku'ren
skilled workman *[ind-eng]*	熟練工	jukuren'kō
skim milk; skimmed milk *[cook]*	脱脂乳	dasshi-nyū
skimming *{n} [steel]*	除滓; じょさい	josai
skin *{n} [anat]*	肌; 膚; はだ	hada
" *{n} [anat]*	皮膚; ひふ	hifu
" *{n} [anat]*	皮	kawa
" (of foam) *{n} [ch]*	表皮	hyōhi
" (leather) *{n} [cl]*	革; かわ	kawa
skin beautifier *[pharm]*	美肌剤	bi'hada-zai
skin lotion *[mat]*	化粧液; けしょう液	keshō-eki
skinning (of ink) *{n} [pr]*	皮張り	kawa'bari
skip (over); jump (over) *{vb}*	飛ばす	tobasu
skip(ping) *{n} [comput]*	飛越し; とびこし	tobi'koshi
skua (bird) *[v-zoo]*	盗賊鴎	tō'zoku-kamome
skull *[anat]*	頭蓋; とうがい; ずがい	tōgai; zugai
" *[anat]*	頭蓋骨	tōgai-kotsu; zugai-kotsu
skunk cabbage *[bot]*	座禅草; ざぜんそう	zazen'sō
sky (pl skies) *[astron]*	空	sora
" (heavens) *[astron]*	天	ten
sky blue (color) *[opt]*	空色	sora-iro
skylark (bird) *[v-zoo]*	雲雀; ひばり	hibari
skylight *[arch]*	天窓	ten-mado
sky wave *[elecmg]*	上空波	jōkū-ha
slab *[steel]*	鋼片	kōhen
slack *{n} [mech]*	撓み; たわみ	tawami
" *{n} [mech]*	緩み; ゆるみ	yurumi
slack variable *[alg]*	緩和変数	kanwa-hensū
slag *[met]*	鍰; からみ	karami
" *[met]*	鉱滓; 鑛滓	kōsai; kōshi
" *[met]*	鉄滓; 鐵滓	tessai; tesshi
slaked lime *[inorg-ch]*	消石灰	shō'sekkai
slaking *{n} [cer]*	消化; しょうか	shōka
slaking resistance *[cer]*	耐消化性	tai'shōka-sei
slant *{vb t}*	斜にする	hasu ni suru
" *{vb t}*	斜めにする	naname ni suru
slant culture *[microbio]*	斜面培養	shamen-baiyō
slash *{n} [comput]*	斜線; しゃせん	shasen
slate *{n} [petr]*	粘板岩	nenban'gan

sled *{n} [eng]*	橇；そり	sori
sledge hammer *[eng]*	向う槌	mukō-zuchi
sleep (slumber) *{n} [physio]*	眠り	nemuri
" (slumber) *{n} [physio]*	睡眠	suimin
sleep *{vb i} [physio]*	眠る；睡る；ねむる	nemuru
" *{vb i} [physio]*	寝る；ねる	neru
sleepiness *{n} [physio]*	眠気；ねむけ	nemu'ke
sleeping car *[rail]*	寝台車；寢臺車	shindai'sha
sleet *{n} [meteor]*	霙；みぞれ	mizore
sleeve *{n} [cl]*	袖	sode
" (barrel) *{n} [eng]*	筒；つつ	tsutsu
sliced raw fish *[cook]*	刺身；さしみ	sashi'mi
slide *{n} [crys]*	滑り	suberi
" (glide; slip) *{vb i} [mech]*	滑る；辷る；すべる	suberu
slide bar *[eng]*	滑り棒	suberi-bō
slide contacting against *{vb}*	摺接する	shū'setsu suru
slide rule *[math]*	計算尺	keisan-jaku
sliding door *[arch]*	引戸	hiki-do
sliding door; bamboo paper *{n}*	唐紙；からかみ	kara-kami
sliding friction *[mech]*	滑り摩擦	suberi-ma'satsu
sliding fusuma door *[arch]*	襖；ふすま	fusuma
sliding gear *[mech-eng]*	摺動歯車	shūdō-ha'guruma
sliding roof (of car) *[mech-eng]*	引戸式屋根	hikido'shiki-yane
sliding shōji door *[arch]*	障子；しょうじ	shōji
slight *{adj}*	軽微な	keibi na
slightly overcast *[meteor]*	薄曇り；うすぐもり	usu-gumori
slime bacteria *[microbio]*	粘液菌	nen'eki-kin
slime layer *[microbio]*	粘液層	nen'eki-sō
slime molds *[mycol]*	変形菌類	henkei'kin-rui
" *[mycol]*	粘菌類	nenkin-rui
sliminess *{n} [mat]*	ヌメリ	numeri
slip *{n} [crys]*	滑り；すべり	suberi
" *{n} [mat]*	泥漿；でいしょう	deishō
slip angle *[geophys]*	滑り角	suberi-kaku
slip casting *{n} [cer]*	鋳込成形；鑄込成形	i'komi-seikei
slip factor *{n} [eng]*	滑り率	suberi-ritsu
slip joint *[eng]*	滑り継手；滑り繼手	suberi-tsugi'te
slip line *[geophys]*	滑り線	suberi-sen
slip out of place *{vb i}*	ずれる	zureru
slippage *{n} [eng]*	滑り率	suberi-ritsu
" *{n} [tex]*	寄れ；よれ	yore
" ; slip *{n} [geol]*	ずれ	zure

slipperiness; sliminess *[mat]*	ヌメリ	numeri
slippery *{adj} [mat]*	滑り易い	suberi-yasui
slip plane *[crys]*	滑り面；すべりめん	suberi-men
slipproof finish *[rub]*	滑り止め加工	suberi'dome-kakō
slipstream (of propeller) *[aero-eng]*	後流；こうりゅう	kōryū
slit *{n} [eng]*	細隙	saigeki
slope *{n} [an-geom]*	傾き；かたむき	katamuki
" (ramp) *{n} [geol]*	斜面	shamen
sloping *{adj} [math]*	傾斜している	keisha shite iru
slot *{n} [eng]*	溝；みぞ	mizo
sloth (animal) *[v-zoo]*	樹懶；なまけもの	namake'mono
sloth bear (animal) *[v-zoo]*	懶け熊	namake-guma
slow *{adv} [mech] [phys]*	遅い；遅い；おそい	osoi
slow accelerator *[rub]*	緩加硫促進剤	kan'karyū-soku'shin-zai
slow combustion *[ch]*	緩慢燃焼	kanman-nenshō
slow cooling *{n} [thermo]*	徐冷	jorei
slow freezing *[p-ch]*	緩慢凍結	kanman-tōketsu
slowly; gradually *{adv}*	ゆっくり（と）	yukkuri (to)
sludge *[civ-eng]*	汚泥	o'dei
slug *[i-zoo]*	蛞蝓；なめくじ	namekuji
sluice *[civ-eng]*	水門	suimon
slurried explosive *[ord]*	含水爆薬	gansui-baku'yaku
slurry *[mat]*	泥漿；でいしょう	deishō
slush; melting snow *[meteor]*	水雪；みずゆき	mizu-yuki
small *{adj} [math] [mech]*	小さい	chīsai
small-angle scattering *{n} [phys]*	（微）小角散乱	(bi)shōkaku-sanran
small arm *[ord]*	小火器	shō-kaki
small change *[econ]*	小銭；小錢；こぜに	ko-zeni
small circle *[geod]*	小円；小圓	shō-en
smaller than *{adj} {n} [math]*	より小さい	yori chīsai
small fish *[cook]*	雑魚；雜魚；ざこ	zako
small intestine *[anat]*	小腸	shōchō
small island; key *{n} [geol]*	小島	ko-jima
small numbers, law of *[math]*	少数の法則	shōsū no hōsoku
small package; parcel *{n} [trans]*	小荷物	ko-ni'motsu
smallpox; variola *[med]*	疱瘡；ほうそう	hōsō
" *[med]*	天然痘	tennen'tō
" *[med]*	痘瘡	tōsō
small quantity; a bit *{n}*	少し；すこし	sukoshi
small-scale map *[map]*	小縮尺図	shō-shuku'shaku-zu
small ship *[nav-arch]*	小型船舶	kogata-senpaku
small-size *{adj}*	小形の	ko-gata no

smash (break) *{vb t}* 毀す；壊す；こわす kowasu
" (crush) *{vb t}* 砕く kudaku
smell (pleasant) *{n}* *[physio]* 匂い；香い；におい nioi
" (unpleasant) *{n}* *[physio]* 臭い；臭い nioi
" (pleasantly) *{vb}* *[physio]* 匂う；香う；におう niou
" (unpleasantly) *{vb}* *[physio]* 臭う niou
smelting *{n}* *[met]* 製錬；製錬 seiren
" *{n}* *[met]* 溶錬；ようれん yō'ren
smelting furnace *[eng]* 融解炉；融解爐 yūkai'ro
smith (a person) *[met]* 鍛工 tankō
smithsonite *[miner]* 菱亜鉛鉱 ryō'aen'kō
smoke *{n}* *[eng]* 煙；けむり kemuri
" (fumigate) *{vb t}* 燻す；いぶす ibusu
" (e.g. a cigarette) *{vb t}* 喫煙する kitsu'en suru
smoke consumer *[eng]* 消煙器 shō'en-ki
smoke damage *{n}* *[ind-eng]* 煙害；えんがい engai
smoked meat *[cook]* 燻製肉 kunsei-niku
smoked product *[cook]* 燻製品 kunsei'hin
smoke point *[eng]* 煙り点；けむりてん kemuri-ten
smoking *{n}* *[ch]* *[cook]* 燻煙 kun'en
" (barbequeing) *[cook]* 燻し ibushi
" *{n}* *[cook]* 燻製；くんせい kunsei
smoky quartz *[miner]* 煙石英 kemuri-seki'ei
smooth *{n}* *[mech]* 滑らか；なめらか nameraka
" *{n}* 平坦 heitan
" *{adj}* 滑らかな nameraka na
smoothing; flattening *{n}* 平滑化 heikatsu'ka
smoothness (slickness) *{n}* *[mat]* 滑らかさ nameraka-sa
" *{n}* *[sci-t]* 平滑度 heikatsu-do
smut spore *[mycol]* 焦胞子；焦胞子 shō'hōshi
snack (food between meals) *{n}* *[cook]* 間食 kanshoku
snail *[i-zoo]* 蝸牛；かたつむり kata'tsumuri
snake (reptile) *[v-zoo]* 蛇；へび hebi
snake-belly form *[graph]* 蛇腹状 da'fuku'jō
snake cucumber; snake gourd *[bot]* 烏瓜；からすうり karasu-uri
snake venom *[physio]* 蛇毒 da'doku; ja'doku
snapper; Lutianus rivulatus (fish) *[v-zoo]* 笛鯛；ふえだい fue-dai
snapping turtle *[i-zoo]* 鼈；すっぽん suppon
snap together (mutually) *{vb}* *[eng]* 嵌合する kangō suru
sneeze *{n}* *[physio]* 嚔 kushami; kusami
" *{vb}* *[physio]* 嚔する kushami suru; kusami suru

snipe (bird) *{n} [v-zoo]*	鴫；鷸；しぎ	shigi
snow *{n} [meteor]*	雪	yuki
snowfall *[meteor]*	降雪	kōsetsu
snow leopard (animal) *[v-zoo]*	雪豹；ゆきひょう	yuki-hyō
snowplow *[mech-eng]*	除雪機	jo'setsu-ki
" *[mech-eng]*	雪掻機	yuki'kaki-ki
snowplow car *[mech-eng]*	雪掻車	yuki'kaki-guruma
snowshoe hare (animal) *[v-zoo]*	橇兎；樏兎	kanjiki-usagi
snow storm *[meteor]*	吹雪；ふぶき	fubuki
snowthaw; snowmelt *{n} [climat]*	雪解け；雪融け	yuki'doke
snow valley; snow gorge *[geol]*	雪渓；雪溪	sekkei
soaking *{n} [steel]*	均熱	kinnetsu
" *{n} [tex]*	漬け込み	tsuke'komi
soap *{n} [mat]*	石鹼；せっけん	sekken
soapstone *[miner]*	凍石	tōseki
soapwater *[hyd]*	石鹼水	sekken-sui
socket *[elec]*	受口	uke'guchi
socket-and-spigot joint *[eng]*	印籠接手	inrō-tsugi'te
socks *{n} [cl]*	靴下	kutsu'shita
soda (sodium carbonate)	曹達；ソーダ	sōda
sodalite *[miner]*	方曹達石	hō'sōda'seki
sodium (element)	ナトリウム	natoryūmu
sodium bisulfite	重亜硫酸ソーダ	jū-a'ryūsan-sōda
sodium borohydride	ナトリウム硼水化物	natoryūmu-hō'suika'butsu
sodium chlorite	亜塩素酸ソーダ	a'enso'san-sōda
sodium dihydrogenphosphate	燐酸二水素=ナトリウム	rinsan-ni'suiso-natoryumu
sodium hydrogencarbonate	炭酸水素ナトリウム	tansan-suiso-natoryūmu
sodium hydrogensulfate	硫酸水素ナトリウム	ryūsan-suiso-natoryūmu
sodium hypophosphite	次亜燐酸ナトリウム	ji-a'rinsan-natoryūmu
sodium iodide	沃化ナトリウム	yōka-natoryūmu
sodium iodide (abbreviation)	沃曹	yōsō
sodium primary phosphate	第一燐酸ナトリウム	dai'ichi-rinsan-natoryūmu
sodium secondary phosphate	第二燐酸ナトリウム	dai'ni-rinsan-natoryūmu
soft; pliant; tender *{adj}*	柔かい；軟かい	yawarakai
soft coal *[geol]*	軟質炭	nan'shitsu-tan
soft drink *[cook]*	清涼飲料水	seiryō-inryō'sui
softener; softening agent *[mat]*	軟化剤	nanka-zai
softening *{n} [phys]*	軟化；なんか	nanka
softening agent *[mat]*	柔軟剤	jūnan-zai
softening point *[phys]*	軟化点	nanka-ten
soft magnetic property *[phys]*	軟磁性；なんじせい	nanji-sei

soft metal *[met]*	軟金属; 軟金屬	nan-kinzoku
softness *[mat]*	柔軟性	jū'nan-sei
" *[mat]*	柔かさ; 軟かさ	yawaraka-sa
soft rime *[meteor]*	樹氷; じゅひょう	juhyō
soft rubber *[rub]*	軟質ゴム	nan'shitsu-gomu
soft-shelled turtle (reptile) *[v-zoo]*	鼈; すっぽん	suppon
soft steel *[met]*	軟鋼; なんこう	nankō
soft twist *[tex]*	甘い撚り	amai yori
soft water *[ch]*	軟水	nansui
soft wood *[bot]*	針葉樹	shin'yō'ju
soil *{n} [agr] [geol]*	土; つち	tsuchi
" *{n} [geol]*	土壌; 土壤	dojō
" ; stain; spot *{n}*	汚れ; よごれ	yogore
" (dirty) *{vb t}*	汚す	yogosu
soil conservation *[ecol]*	土壤保全	dojō-hozen
soiled; dirty *{adj}*	汚れた	yogoreta
soiling *{n} [ecol]*	汚損	o'son
soiling prevention *[ind-eng]*	防汚; ぼうお	bō'o
soil map *[civ-eng]*	土性図; 土性圖	dosei-zu
soil mechanics *[eng]*	土質力学	do'shitsu-riki'gaku
soil-resistant finish *[mat]*	防汚加工	bō'o-kakō
soil profile *[geol]*	土壌断面	dojō-danmen
soil science *[geol]*	土壌学; 土壤學	dojō'gaku
soil test *[geol]*	土質試験	do'shitsu-shiken
sol *{n} [ch]*	ゾル	zoru
solar activity *[astron]*	太陽活動	taiyō-katsudō
solar apex *[astron]*	太陽向点	taiyō-kōten
solar cell *[electr]*	太陽電池	taiyō-denchi
solar constant *[meteor]*	太陽定数	taiyō-teisū
solar control *[meteor]*	日照調整	nisshō-chōsei
solar eclipse *[meteor]*	日食; にっしょく	nisshoku
solar flare *[astron] [meteor]*	太陽面爆発	taiyō'men-baku'hatsu
solar furnace *[eng]*	太陽炉; 太陽爐	taiyō-ro
solar parallax *[astron]*	太陽視差	taiyō-shisa
solar system *[astron]*	太陽系; たいよう系	taiyō-kei
solar tide *[ocean]*	太陽潮	taiyō-chō
solar thermal power *[phys]*	太陽熱力	taiyō-netsu-ryoku
solar time *[astron]*	太陽時	taiyō-ji
solar wind *[geophys]*	太陽風	taiyō-fū
solderability *[eng]*	半田付け性	handa'zuke-sei
soldering (soft) *{n} [eng]*	半田付け	handa'zuke
" (hard) *{n} [met]*	蠟付け; ろうづけ	rō'zuke

sole (of foot) {n} [anat]	足の裏	ashi no ura
sole (fish) [v-zoo]	舌平目; したびらめ	shita-birame
sole invention [pat]	単独発明; 單獨發明	tandoku-hatsu'mei
solid {n} [geom]	立体; りったい	rittai
" {n} [geom] [phys]	固体; 固體; こたい	kotai
solid angle [math]	立体角	rittai-kaku
solid dyeing {n} [tex]	無地染め	muji-zome
solid geometry [geom]	立体幾何学	rittai-kika'gaku
solidification [eng]	固化	koka
" [phys]	凝固; ぎょうこ	gyōko
" [phys]	硬化	kōka
solidifying agent [mat]	固化剤	koka-zai
solidifying point [phys]	凝固点; 凝固點	gyōko-ten
solid-liquid extraction [p-ch]	固液抽出	ko'eki-chū'shutsu
solid-logic technology [comput]	固体論理技術	kotai-ronri-gi'jutsu
solid matter [phys]	固形物	kokei'butsu
solid phase [phys]	固相; こそう	kosō
solid-phase reaction [ch]	固相反応	kosō-hannō
solid propellant [mat]	固体推進薬	kotai-suishin-yaku
solids content; solid components {n}	固形分	kokei-bun
solid solubility [phys]	固溶解度	ko'yōkai-do
solid solution [phys]	固溶体; こようたい	koyō-tai
solid-solution treatment [phys]	固溶化熱処理	koyō'ka-netsu-shori
solid-state component [comput]	固体素子	kotai-soshi
solid-state computer [comput]	固体回路計算機	kotai'kairo-keisan-ki
solid-state physics [phys]	固体物理学	kotai-butsuri'gaku
solid-state polymerization [poly-ch]	固相重合	kosō-jūgō
solitary {adj}	孤立した	ko'ritsu shita
solstices [astron]	至点; 至點; してん	shiten
solstitial colure [astron]	二至経線	nishi-keisen
solubility [p-ch]	溶解度; ようかいど	yōkai-do
solubility parameter [poly-ch]	溶解度係数	yōkai'do-keisū
solubilization [ch]	可溶化	kayō'ka
soluble {adj} [ch]	溶ける	tokeru
soluble starch [mat]	溶解性澱粉	yōkai'sei-denpun
solute [ch]	溶質	yō'shitsu
solution {n} [alg]	解	kai
" {n} [ch]	溶液; ようえき	yōeki
" (method) {n} [math]	解法	kaihō
" {n} [pharm]	水剤	suizai
solution (heat) treatment [met]	溶体化処理	yōtai'ka-shori
solution polymerization [poly-ch]	溶液重合	yōeki-jūgō

solution set *[alg]* 解の集合 kai no shūgō
solvated electron *[phys]* 溶媒和電子 yōbai'wa-denshi
solvation *[ch]* 溶媒和（溶媒化） yōbai'wa (yōbai'ka)
solve *{vb} [alg]* 解答する kaitō suru
" *{vb} [alg]* 解く；とく toku
solvent *[ch]* 溶媒；ようばい yō'bai
" *[ch]* 溶剤；ようざい yō'zai
solvent degreasing *{n} [ch-eng]* 溶剤脱脂 yōzai-dasshi
solventless adhesive *[adhes]* 無溶剤型接着剤 mu'yōzai'gata-setchaku-zai
solvolysis *[ch]* 加溶媒分解 ka'yōbai-bunkai
somatic *{adj} [bio]* 体の；からだの karada no
" *{adj} [bio]* 身体の shintai no
somatic cell *[bio] [cyt]* 体細胞；體細胞 tai-saibō
somatic hybridization *[gen]* 体細胞交雑 tai'saibō-kōzatsu
somatic nervous system *[physio]* 体性神経系 taisei-shinkei-kei
somatic sensation *[physio]* 体性感覚 taisei-kankaku
sonant *[ling]* 濁音 daku'on
sonar *[eng]* 水中音波探知機 suichū'onpa-tanchi-ki
song; poem *{n} [pr]* 歌；唄 uta
song thrush (bird) *[v-zoo]* 歌鶫；うたつぐみ uta-tsugumi
sonication *[acous] [cyt] [microbio]* 音波処理；音波處理 onpa-shori
sonic barrier *[aero-eng]* 音速障壁 on'soku-shōheki
sonic depth finder *[eng]* 音響測深機 onkyō-soku'shin-ki
sonic speed; sonic velocity *[acous]* 音速；おんそく on-soku
sonochemistry *[ch]* 音化学 on-kagaku
sonography *[eng]* 音波検査法 onpa-kensa-hō
soot *[mat]* 煤；すす susu
sooty smoke *[eng]* 焙煙；焙煙 bai'en
sorghum *[bot]* 高粱；カオリヤン kōryan; kaoryan
sorption *[p-ch]* 収着；收着 shūchaku
sort (sorting) *{n} [comput]* 分類；ぶんるい bunrui
" *{n} [comput]* 整列 seiretsu
" *{vb} [comput]* 分類する bunri suru
sorter *{n} [comput]* 分類機 bunrui-ki
sorting *{n} [met]* 選別；選別 senbetsu
sound *{n} [acous]* 音響；音響 onkyō
" *{n} [acous]* 音；おと oto
" *{n} [geogr]* 入江；いりえ iri'e
sound absorption coefficient *[acous]* 吸音率 kyū'on-ritsu
sound effect *[acous]* 擬音；ぎおん gi'on
sound field *[acous]* 音場 onba; onjō
sound intensity *[acous]* 音の強さ oto no tsuyo-sa

sound level meter *[eng]*	音量計	onryō-kei
sound pitch *[acous]*	音の高さ	oto no taka-sa
sound pressure *[acous]*	音圧; 音壓	on-atsu
soundproof chamber *[acous]*	防音室	bō'on-shitsu
sound recorder *[eng]*	録音機	roku'on-ki
sound recording *{n} [acous]*	録音; ろくおん	roku'on
sound volume *[acous]*	音量	on-ryō
sound wave *[acous]*	音波	onpa
soup *[cook]*	吸物; すいもの	sui'mono
soup stock *[cook]*	出し; だし	dashi
source *{n} [hyd]*	湧き出し; 涌き出し	waki'dashi
" *{n} [sci-t]*	源; みなもと	minamoto
source language *[comput]*	原始言語	genshi-gengo
" *[ling]*	起点言語	kiten-gengo
souring *{n} [tex]*	酸洗い	san-arai
" *{n} [tex]*	酸通し	san-tōshi
sour milk *[cook]*	酸乳; さんにゅう	sannyū
sourness *[cook] [physio]*	酸味	san'mi
south *[geod]*	南; みなみ	minami
south and north *[geod]*	南北	nan'boku
southerly wind; southerlies *[meteor]*	南風	minami-kaze
Southern Cross *[constel]*	南十字座	minami-jūji-za
South Pole *[geod]*	南極; なんきょく	Nankyoku
South Seas *[geogr]*	南洋	Nan'yō
sow (animal) *[v-zoo]*	雌豚; めすぶた	mesu-buta
" (plant seeds) *{vb t} [agr]*	蒔く; まく	maku
sow bug *[i-zoo]*	草鞋虫; わらじむし	waraji-mushi
soybean *[bot]*	大豆; だいず	daizu
soybean meal *[agr]*	大豆粕	daizu-kasu
soybean milk *[cook]*	豆乳; とうにゅう	tōnyū
soybean milk lees (dregs) *[cook]*	豆乳粕	tōnyū-kasu
soybean paste *[cook]*	味噌; みそ	miso
soybean paste soup *[cook]*	味噌汁	miso-shiru (omiotsuke)
soysauce *[cook]*	醤油; しょうゆ	shōyu
soysauce (dark-colored) *[cook]*	濃口醤油	koi'kuchi-shōyu
" (light-colored) *[cook]*	薄口醤油; 淡口醤油	usu'kuchi-shōyu
soysauce-broiled fish, chicken *[cook]*	照焼; 照り焼き	teri'yaki
soysauce-reduced food *[cook]*	佃煮; つくだに	tsukuda'ni
space *{n} [astron]*	宇宙空間	uchū-kūkan
" (gap) *{n} [comput] [pr]*	間隔	kan'kaku
" *{n} [comput]*	空間	kūkan
" (opening, distance, interval) *{n}*	間; あいだ	aida

space biology *[bio]*	宇宙生物学	uchū-seibutsu'gaku
space-charge effect *[electr]*	空間電荷効果	kūkan-denka-kōka
space-charge-limited current *[electr]*	空間電荷制限電流	kūkan-denka-seigen-denryū
spacecraft *[aero-eng]*	宇宙機；うちゅうき	uchū-ki
space factor *[elecmg]*	占積率	senseki-ritsu
space group *[crys]*	空間群	kūkan-gun
space lattice *[crys]*	空間格子	kūkan-gōshi; kūkan-kōshi
space polymer *[poly-ch]*	立体重合体	rittai-jūgō'tai
space shuttle *[aero-eng]*	宇宙連絡船	uchū-renraku'sen
spacesickness *[aerosp]* *[med]*	宇宙酔い	uchū-yoi
space station *[aero-eng]*	宇宙局	uchū-kyoku
space suit *[eng]*	宇宙服	uchū-fuku
space-time *[rela]*	時空；じくう	jikū
space-time continuum *[rela]*	時空連続体	ji'kū-renzoku'tai
space-time yield *[ch-eng]*	空時収率	kūji-shū'ritsu
space tracking *{n}* *[aero-eng]*	宇宙追尾	uchū-tsui'bi
space vehicle *[aero-eng]*	宇宙船	uchū'sen
space velocity *[astron]* *[ch-eng]*	空間速度	kūkan-sokudo
space walk *{n}* *[aero-eng]*	宇宙遊泳	uchū-yū'ei
spacing (between words) *[comp]* *[pr]*	語間	gokan
" (between lines) *[comp]* *[pr]*	行間	gyōkan
Spanish mackerel (fish) *[v-zoo]*	鰆；さわら	sawara
Spanish moss *[bot]*	猿麻桛擬	saru'ogase'modoki
spar *{n}* *[aero-eng]*	桁；けた	keta
spare (article or part) *{n}* *[eng]*	予備品	yobi'hin
spare line *[elec]*	予備線路	yobi-senro
spark *{n}* *[elec]*	火花；ひばな	hi-bana
spark chamber *[nucleo]*	放電箱	hōden-bako
spark discharge *[elec]*	火花放電	hi'bana-hōden
sparkle; glitter *{n}* *[opt]*	ピカピカ	pika-pika
sparkling wine *[cook]*	発泡酒；發泡酒	happō'shu
sparrow (bird) *[v-zoo]*	雀；すずめ	suzume
sparse *{n}* *[math]*	疎；そ	so
" (scattered; sporadic) *{n}*	疎ら；まばら	mabara
spasmolytic *{n}* *[pharm]*	鎮痙薬（鎮痙剤）	chinkei-yaku (chinkei-zai)
spatial *{adj}* *[phys]*	空間の	kūkan no
spatter *{n}*	飛散	hisan
spatterdock *[bot]*	河骨；川骨；こほね	ko'hone
spatula; scoop *{n}* *[cook]* *[eng]*	箆；へら	hera
speaker; talker *[comm]* *[comput]*	話者	wa'sha
speaking; to speak *[physio]*	話すこと	hanasu koto
spear; javelin; lance *{n}* *[ord]*	槍；やり	yari

special *{adj} [sci-t]*	特別な(の)	toku'betsu na (no)
" *{adj} [sci-t]*	特殊の; とくしゅの	toku'shu no
special alloy steel *[steel]*	特殊鋼	tokushu-kō
special character *[comput]*	特殊記号	tokushu-kigō
" *[comput]*	特殊文字	tokushu-moji
special feature; special function *[comput]*	特殊機能	tokushu-kinō
specialist; expert *{n}*	専門家; 專門家	senmon'ka
specialization *[bio]*	分化	bunka
specialized *{adj} [bio]*	分化した	bunka shita
specially designated *{n}*	特定	toku'tei
special-purpose computer *[comput]*	専用計算機	sen'yō-keisan-ki
special right(s) *[legal]*	特権; 特權	tokken
special symbol *[comput] [pr]*	特別記号	toku'betsu-kigō
specialty (of business) *[econ]*	専業; せんぎょう	sengyō
" (of knowledge, interest)	専門; 專門	senmon
specialty store *[arch] [econ]*	専門店	senmon'ten
species (pl species) *[bio] [syst]*	種; しゆ	shu
specific *{adj}*	特異的な	toku'i'teki na
" ; specified *{adj}*	特定の	tokutei no
specific activity *[microbio]*	比活生; ひかっせい	hi-kassei
specific adhesion *[adhes]*	比接着	hi-setchaku
specification; designation *{n}*	指定	shitei
specification(s) *[eng] [ind-eng]*	規格	ki'kaku
" *[pat]*	明細書; めいさい書	meisai'sho
specifications *[ind-eng]*	示方書	shihō'sho
" *[ind-eng]*	仕様書	shiyō'sho; shiyō-gaki
specific bursting strength *[mech]*	比破裂強度	hi-haretsu-kyōdo
specific compressive strength *[mech]*	比圧縮強度	hi-asshuku-kyōdo
specific conductivity *[elec]*	比電導度	hi-dendō'do
specific dispersion *[ch]*	比分散	hi-bunsan
specific gravity *[mech]*	比重; ひじゅう	hijū
specific heat *[thermo]*	比熱	hi-netsu
specific humidity *[meteor]*	比湿; 比濕; ひしつ	hi-shitsu
specific inductive capacity *[elec]*	比誘電率	hi-yūden-ritsu
specificity *[bio]*	特異性	toku'i-sei
specific modulus *[mech]*	比弾性率	hi-dansei-ritsu
specific multiplication rate *[bio]*	比増殖速度	hi-zō'shoku-sokudo
specific permeability *[mech]*	比透磁率	hi-tōji-ritsu
specific purpose *{n}*	特別目的	toku'betsu-moku'teki
specific radioactivity *[nuc-phys]*	比放射能	hi-hōsha'nō
specific reaction *[ch]*	特異反応	toku'i-hannō

specific refraction *[mech]*	比屈折	hi-kussetsu
specific resistance; resistivity *[elec]*	比抵抗；ひていこう	hi-teikō
" *[elec]*	固有抵抗	koyū-teikō
specific rotation *[opt]*	比旋光度	hi-senkō-do
specific strength *[mech]*	比強度；ひきょうど	hi-kyōdo
specific surface area *[eng]*	比表面積	hi-hyō'menseki
specific susceptibility *[p-ch]*	比磁化率	hi-jika-ritsu
specific viscosity *[fl-mech]*	比粘度；ひねんど	hi-nendo
specific volume *[mech]*	比体積	hi-taiseki
" *[mech]*	比容；ひよう	hiyō
specified chemical substance *[ch]*	特定化学物質	tokutei-kagaku-busshitsu
specified invention *[pat]*	特定発明	tokutei-hatsu'mei
specimen *[bioch]*	標品；ひょうひん	hyōhin
" ; sample *[ch]*	試料；しりょう	shiryō
" *[sci-t]*	標本	hyōhon
" ; testpiece *[sci-t]*	試験片；しけんへん	shiken'hen
specimen table; specimen stand *(n)*	試料台；試料臺	shiryō-dai
speck; speckle; spot *(n)* *[bio]* *[opt]*	斑点；はんてん	hanten
" ; spot; dot *(n)* *[graph]*	点；點；てん	ten
spectacled bear (animal) *[v-zoo]*	眼鏡熊；めがねぐま	megane-guma
spectral characteristic *[opt]*	分光特性	bunkō-tokusei
spectral concentration *[spect]*	分光密度	bunkō-mitsudo
spectral diffraction *[opt]* *[phys]*	分光；ぶんこう	bunkō
spectral distribution *[opt]* *[phys]*	分光分布	bunkō-bunpu
spectral reflectance *[an-ch]*	分光反射率	bunkō-hansha-ritsu
spectral sensitivity *[electr]*	分光感度	bunkō-kando
spectral sensitization *[photo]*	分光増感	bunkō-zōkan
spectral transmissivity *[opt]*	分光透過率	bunkō-tōka-ritsu
spectrocalorimeter *[spect]*	分光熱量計	bunkō-netsu'ryō-kei
spectrochemical analysis *[ch]*	分光分析	bunkō-bunseki
spectrochemistry *[ch]*	分光化学	bunkō-kagaku
spectrograph *[spect]*	分光写真器	bunkō-shashin-ki
spectroheliograph *[astron]*	分光太陽写真儀	bunkō-taiyō-shashin-gi
spectrometer *[spect]*	分光計	bunkō-kei
spectrophotometer *[spect]*	分光光度計	bunkō-kōdo-kei
spectrophotometry *[spect]*	分光測光学	bunkō-sokkō'gaku
spectropolarimeter *[opt]*	分光旋光計	bunkō-senkō-kei
spectroreflectance *[an-ch]*	分光反射率	bunkō-hansha-ritsu
spectroscope *[spect]*	分光器；ぶんこうき	bunkō-ki
spectroscopic binary star *[astron]*	分光連星	bunkō-rensei
spectroscopic characteristic *[spect]*	分光特性	bunkō-tokusei
spectroscopy *[phys]*	分光学	bunkō'gaku

spectrum *[math] [phys]*	スペクトル	supekutoru
spectrum analysis *[phys]*	分光分析	bunkō-bunseki
specular density *[photo]*	平行光濃度	heikō'kō-nōdo
specular gloss *[opt]*	鏡面光沢	kyōmen-kōtaku
specular reflection *[opt]*	正反射	sei-hansha
speculation *{n} [psy]*	思惑；おもわく	omo'waku
speculum (pl -la, -lums) *[opt]*	反射鏡	hansha'kyō
speech (oral presentation) *[comm]*	演説	enzetsu
" *[comm] [ling]*	言語；げんご	gengo
" *[physio]*	音声；音聲	onsei
speech recognition *[comput]*	音声認識	onsei-nin'shiki
speech sound *[physio]*	言語音	gengo'on
speech synthesis *[comput]*	音声合成	onsei-gōsei
speed (rapidity) *{n} [mech]*	速さ；早さ；はやさ	haya-sa
" *{n} [mech] [phys]*	速度	soku'do
" (hasten) *{vb t}*	早くする	hayaku suru
speed control apparatus *[eng]*	速度制御装置	sokudo-seigyo-sōchi
speed governor *[mech-eng]*	調速機	chōsoku-ki
speed limit *[traffic]*	速度制限	sokudo-seigen
speed of sound *[acous]*	音速；おんそく	on-soku
speedometer *[eng]*	速度計	sokudo-kei
speed reduction *[mech-eng]*	減速	gensoku
speleology *[geol]*	洞窟学	dōkutsu'gaku
spent fuel *[nucleo]*	使用済み燃料	shiyō'zumi-nenryō
spermaceti *[mat]*	鯨蠟；げいろう	gei'rō
spermatogenesis *[physio]*	精子形成	seishi-keisei
sperm cell *[bio]*	精子；せいし	seishi
sperm oil *[mat]*	抹香鯨油	makkō'kujira-yu
sphagnum *[bot]*	水苔；みずごけ	mizu'goke
sphalerite *[miner]*	閃亜鉛鉱	sen'aen'kō
sphene *[miner]*	榍石；せっせき	sesseki
sphenoid *{adj} [geom]*	楔状の	kusabi'jō no
Sphenosida *[bot]*	木賊類；砥草類	tokusa-rui
sphere *{n} [geom]*	球；きゅう	kyū
sphere of activity *[astron]*	勢力圏	sei'ryoku-ken
spherical *{adj} [geom]*	球形の	kyūkei no
spherical aberration *[opt]*	球面収差	kyūmen-shūsa
spherical astronomy *[astron]*	球面天文学	kyūmen-tenmon'gaku
spherical coordinates *[geom]*	球座標	kyū-zahyō
spherical interface *[math]*	球界面	kyū-kaimen
spherical mirror *[opt]*	球面鏡	kyūmen-kyō
spherical surface *[eng] [geom]*	球面；きゅうめん	kyūmen

spherical triangle *[geom]*	球面三角形	kyūmen-sankak(u)kei
sphericity *[math]*	真球度	shinkyū-do
spheroidal graphite *[geol]*	球状黒鉛	kyūjō-koku'en
spherulite *[geol]*	球晶	kyūshō
sphinx moth (insect) *[i-zoo]*	雀蛾; すずめが	suzume-ga
sphygmomanometer *[med]*	血圧計; 血壓計	ketsu'atsu-kei
spice *{n} [cook]*	香辛料	kōshin'ryō
" *{n} [cook]*	薬味; やくみ	yaku'mi
spicebush *[bot]*	黒文字; くろもじ	kuro-moji
spider (insect) *[i-zoo]*	蜘蛛; くも	kumo
spider monkey (animal) *[v-zoo]*	蜘蛛猿	kumo-zaru
spider web *[i-zoo]*	蜘蛛の巣	kumo-no-su
spike (ear) *{n} [bot]*	穂; 穗; ほ	ho
" *{n} [bot]*	花穂; かすい	ka'sui
" *{n} [bot]*	穂状花序	suijō-kajo
spill *{n} [eng]*	零; こぼれ	kobore
" *{vb t}*	零す; 溢す	kobosu
spillway *[civ-eng]*	余水路	yosui-ro
spin; make yarn *{vb} [tex]*	紡ぐ; つむぐ	tsumugu
spinach (vegetable) *[bot]*	菠薐草; 鳳蓮草	hōren'sō
spinal cord *[anat]*	脊髄; せきずい	seki-zui
spindle *[tex]*	紡錘	bōsui
" *[tex]*	錘	sui
spine (spinal column) *{n} [anat]*	脊柱; せきちゅう	seki'chū
spine *{n} [bot]*	針	hari
spinel *[miner]*	尖晶石	senshō'seki
spinnability *[poly-ch] [tex]*	曳糸性	eishi-sei
spinneret *[eng]*	紡糸口金	bōshi-kuchi'gane
spinning *{n} [mech]*	急回転	kyū-kaiten
" *{n} [tex]*	紡績; ぼうせき	bōseki
" *{n} [tex]*	紡糸	bōshi
" *{n} [tex]*	精紡	seibō
spinning solution *[tex]*	紡糸液	bōshi-eki
spinning wheel *[tex]*	糸車	ito-guruma
" *[tex]*	紡ぎ車	tsumugi-guruma
spinthariscope *[electr]*	スピンサリスコープ	supinsarisukōpu
spiny dogfish; Squalus acanthias (fish) *[v-zoo]*	油角鮫	abura-tsuno-zame
spiny dogfish; Squalus mitsukurii (fish) *[v-zoo]*	角鮫; つのざめ	tsuno-zame
spiral *{n} [math]*	螺旋; らせん	rasen
" *{n} [math]*	渦形; うずがた	uzu'gata

spiral bevel gear *[mech]*	曲り歯傘歯車	magari'ba-kasa-ha'guruma
spiral galaxy *[astron]*	渦状銀河	uzu'jō-ginga
spiral line *[math]*	渦巻線；渦巻線	uzu'maki-sen
spiral motion *[mech]*	螺旋運動	rasen-undō
spiral nebula *[astron]*	渦巻星雲	uzu'maki-sei'un
spirit *{n} [cook]*	蒸留酒	jōryū'shu
" *{n} [pharm]*	酒精剤	shusei-zai
spirits of wine *[cook]*	酒精；しゅせい	shusei
Spirochaetales *[microbio]*	スピロヘータ目	supirohētā-moku
splashed pattern *[tex]*	絣；かすり	kasuri
splashproof *{n} [ind-eng]*	防沫	bōmatsu
spleen *{n} [anat]*	脾臓；ひぞう	hizō
splice (rope) *{n} [eng]*	接着	setchaku
splint *{n} [med]*	副木	fuku'boku
split *{n}*	割れ目	ware'me
" *{vb} [comput]*	分割する	bun'katsu suru
" *{vb}*	割る；わる	waru
split bearing *{n} [eng]*	割り軸受	wari-jiku'uke
split gear *[mech-eng]*	割(り)歯車	wari-ha'guruma
split gene *[gen]*	分断遺伝子	bundan-idenshi
split-level house *[arch]*	段違いの家	dan'chigai no ie
split mold *[eng] [met]*	割り型	wari-gata
split-phase induction motor *[elec]*	分相誘導電動機	bunsō-yūdō-dendō-ki
split-phase-start motor *[elec]*	分相始動電動機	bunsō-shidō-dendō-ki
splitting *{n} [ch] [spect]*	分裂；ぶんれつ	bun'retsu
spodumene *[miner]*	黝輝石；ゆうきせき	yūki'seki
spoil; be damaged *{vb} [ind-eng]*	傷む；いたむ	itamu
spoilage *[bio] [mat]*	腐り	kusari
" *[bioch]*	腐敗	fuhai
spoken language *[comm]*	口語	kōgo
sponge *{n} [zoo]*	海綿；かいめん	kaimen
" (bread) *[cook]*	中種；なかだね	naka'dane
sponge iron *[met]*	海綿鉄	kaimen-tetsu
spontaneous coagulation *[rub]*	自然凝固	shizen-gyōko
spontaneous combustion *[ch]*	自然燃焼	shizen-nenshō
spontaneous emission *[electr]*	自然放出	shizen-hō'shutsu
spontaneous ignition *[ch]*	自然発火	shizen-hakka
spontaneous magnetization *[elecmg]*	自発磁化	ji'hatsu-jika
spontaneous mutation *[gen]*	自然突然変異	shizen-totsu'zen-hen'i
spontaneous polarization *[elec]*	自発分極	ji'hatsu-bun'kyoku
spoon *{n} [cook] [eng]*	匙；さじ	saji
sporangiophore *[bot]*	胞子嚢柄	hōshi'nō'hei

sporangium (pl -gia) *[bot]* 胞子嚢；胞子嚢 hōshi'nō
spore *[bio]* 胞子；胞子；ほうし hōshi
sporophyte *[bot]* 胞子体 hōshi'tai
Sporozoa *[i-zoo]* 胞子虫類 hōshi'chū-rui
sports clothes *[cl]* 運動服 undō-fuku
sports shoes *[cl]* 運動靴 undō-gutsu
spot *{n} [bio] [opt]* 斑点；斑點 hanten
spot-billed duck (bird) *[v-zoo]* 軽鴨；かるがも karu-gamo
spotted shark; (fish) *[v-zoo]* 星鮫；ほしざめ hoshi-zame
spotted shrimp (crustacean) *[v-zoo]* 桜蝦；櫻蝦 sakura-ebi
spotted triggerfish; Balistes niger (fish) *[v-zoo]* 紋柄皮剝；もん=がらかわはぎ mon'gara-kawa'hagi
spot welding *{n} [met]* 点溶接 ten-yō'setsu
spouse *[bio]* 配偶者 haigū'sha
spout *{n} [met]* 樋；とい toi
spray *{n} [eng]* 噴霧；ふんむ fun'mu
spray coating *{n} [eng]* 溶射 yōsha
spray drying *{n} [paint]* 噴霧乾燥 funmu-kansō
sprayer *[eng]* 噴霧器 funmu-ki
spraying *{n} [agr]* 散布 sanpu
" *{n} [eng]* 霧吹き；きりふき kiri'fuki
spray injury *[agr]* 薬害；藥害 yaku-gai
spray method *[mech-eng]* 噴霧法 funmu-hō
spray polymerization *[poly-ch]* 噴霧重合 funmu-jūgō
spray quenching *{n} [met]* 噴射焼入れ funsha-yaki'ire
spread *{n} [pr]* 見開き mi'hiraki
" *{vb t}* 広げる；廣げる hirogeru
" (lay down, as paper) *{vb t}* 敷く；布く；しく shiku
spread coating *{n} [paint]* 展延塗装 ten'en-tosō
spreading *{n} [adhes]* 貼着；てんちゃく tenchaku
" *{n} [tex]* 延展；えんてん enten
spreading agent *[mat]* 展着剤 tenchaku-zai
spreadsheet *[comput]* 簡易；かんい kan'i
sprig *[bot]* 若枝 waka-eda
Spring (season) *[astron]* 春；はる haru
" *[astron]* 春期 shunki
spring *{n} [hyd]* 泉；いずみ izumi
" *{n} [mech]* 弾み；彈み；はずみ hazumi
" *{n} [mech-eng]* 発条；ばね bane
spring balance *[eng]* ばね秤 bane-bakari
spring constant *[mech-eng]* ばね定数 bane-teisū
spring rain *[meteor]* 春雨；はるさめ haru'same

spring scale *[eng]* 発条秤；ぜんまい秤 zenmai-bakari
spring steel *[met]* 発条鋼；ばね鋼 bane-kō
spring tide *[ocean]* 大潮；おおしお ō-shio
spring water *[hyd]* 湧水 waki-mizu
sprinkler *[eng]* 散水装置 sansui-sōchi
sprocket *[eng]* 鎖歯車 kusari-ha'guruma
sprue *[met]* 湯口；ゆぐち yu'guchi
spun yarn *[tex]* 紡績糸 bōseki-shi
spur *{n} [transp]* 拍車 haku'sha
spur gear *[mech]* 平歯車 hira-ha'guruma
spurious; artificial *{adj}* 紛いの；擬いの magai no
sputum *{n} [physio]* 痰 tan
squall *{n} [meteor]* 突風；突風 toppū
" *{n} [meteor]* 疾風；早手；はやて haya'te
Squamata *[v-zoo]* 有鱗類 yūrin-rui
square *{n} [alg]* 二乗；二乗；自乗 ji'jō
" *{n} [alg]* 二乗；にじょう ni'jō
" (shape) *{n}[geom]* 角形；かくがた kaku'gata
" (shape) *{n} [geom]* 正方形 seihō-kei
" (shape) *{n} [geom]* 四角形 shikaku'kei
" *{adj} [geom]* 四角い；しかくい shikaku'i
square array *[math]* 正方配列 seihō-hai'retsu
square bar (of steel) *[civ-eng]* 角鋼 kaku-kō
square column *[geom]* 四角柱 shikaku-chū
square matrix *[alg]* 正方行列 seihō-gyō'retsu
square-rigger *[nav-arch]* 横帆船；よこほせん yoko'ho'sen
square root *[alg] [arith]* 平方根 heihō-kon
" *[alg] [arith]* 二乗根；二乗根 nijō-kon
square timber *[civ-eng]* 角材 kaku'zai
square wave *[elec]* 四角波 shikaku-ha
squeeze *{vb}* 絞る shiboru
squeeze (through hands); strip *{vb}* 扱く；しごく shigoku
squid (mollusk) *[i-zoo]* 烏賊；いか ika
squill *[bot]* 海葱；かいそう kaisō
squilla *[i-zoo]* 蝦蛄；しゃこ shako
squirrel (animal) *[v-zoo]* 栗鼠 risu
squirrel monkey (animal) *[v-zoo]* 栗鼠猿；りすざる risu-zaru
stab culture *[microbio]* 穿刺培養 senshi-baiyō
stability *[electr]* 安定性 antei-sei
" *[mech-eng]* 復元性 fuku'gen-sei
" *{n}* 安定；あんてい antei
stability constant *[ch]* 安定度定数 antei'do-teisū

stabilize *{vb t}* 安定化する antei'ka suru
stabilizer *[mat]* 安定(化)剤 antei('ka)-zai
stable (for horses) *{n} [arch]* 厩；馬屋；うまや uma'ya
" *{adj} [phys]* 安定な antei na
stable isotope *[nuc-phys]* 安定同位体 antei-dō'i'tai
stack (heap) *{vb t}* 重ねる kasaneru
" (heap; pile up) *{vb t}* 積む；つむ tsumu
stacking *{n}* 段積み；だんづみ dan'zumi
" *{n}* 積み重ね tsumi'kasane
stage (in a theater) *{n} [arch]* 舞台；ぶたい butai
" *{n}* [ch-eng] 段；だん dan
" (step; phase) *{n} [sci-t]* 段階 dankai
" *{n}* 過程 katei
" (process) *{n}* 経路；經路；けいろ keiro
stagger *{n} [mech-eng]* 食違い；くいちがい kui'chigai
" *{vb t}* ずらす zurasu
stagnant water *[hyd]* 停滞水；停滯水 teitai-sui
stagnation *[hyd]* 淀み；よどみ yodomi
stagnation point *[fl-mech]* 淀み点；澱み点 yodomi-ten
stain *{n} [ecol]* 汚点；汚點；おてん o'ten
" *{n}* 汚れ yogore
staining *{n} [bioch]* 染色；せんしょく senshoku
" *{n} [opt]* 着色 chaku'shoku
staining power *[opt]* 着色力 chaku'shoku-ryoku
stainless steel *[steel]* 不錆鋼；ふせいこう fusei-kō
stain spot *[tex]* 染み；しみ shimi
stairway *[arch]* 階段 kaidan
stalactite *[geol]* 鐘乳石 shōnyū'seki
" *[geol]* 氷柱石；つらら石 tsurara-ishi
stalagmite *[geol]* 石筍 seki'jun
stalagmometer *[eng]* 滴数計 tekisū-kei
stalk; stem *{n} [bot]* 茎；莖；くき kuki
stall; stalling *{n} [mech-eng]* 失速 shissoku
stallion (animal) *[v-zoo]* 雄馬 osu-uma
stamen *[bot]* 雄蕊 yū'zui; o-shibe
stamp *{n} [comm]* 切手；きって kitte
standard *{n} [comput] [phys]* 標準；ひょうじゅん hyōjun
" (norm) *{n}* 規準；きじゅん kijun
standard atmospheric pressure *[meteor]* 標準気圧 hyōjun-ki'atsu
standard cloth *[tex]* 規格布 kikaku-nuno
standard conditions *[phys]* 標準状態 hyōjun-jōtai
standard deviation *[stat]* 標準偏差 hyōjun-hensa

standard electrode *[elec]*	標準電極	hyōjun-denkyoku
standard example *[sci-t]*	常例	jōrei
standardization *[ch]*	標定；ひょうてい	hyōtei
" *[eng]*	標準化	hyōjun'ka
" *[eng]*	規格化；きかくか	kikaku'ka
standardize *{vb} [eng]*	規格する	kikaku suru
" *{vb} [sci-t]*	標準化する	hyōjun'ka suru
standard oxidation-reduction potential *[p-ch]*	標準酸化還元電位	hyōjun-sanka'kangen-den'i
standard position *[trig]*	標準の位置	hyōjun no ichi
standard solution *[ch]*	標準(溶)液	hyōjun-(yō)eki
standard state *[phys]*	標準状態	hyōjun-jōtai
standard time *[astron]*	標準時	hyōjun-ji
standard weight *[mech]*	基準分銅	kijun-fundō
standby time *[comput]*	待機時間	taiki-jikan
standing wave *[phys]*	定常波	teijō-ha
" *[phys]*	定在波；ていざいは	teizai-ha
standing wave ratio *[phys]*	定在波比	teizai'ha-hi
stand upright *{vb t}*	縦にする	tate ni suru
stannic acid	錫酸；すずさん	suzu-san
stannite *[miner]*	黄錫鉱	ō'shakkō
stannous chloride	第一塩化錫	dai'ichi-enka-suzu
stapes (bone) *{n} [anat]*	鐙骨	tō-kotsu; abumi-bone
staphylococcus (pl-i) *[microbio]*	葡萄(状)球菌	budō('jō)-kyūkin
staple food *[cook]*	主食	shu-shoku
star *{n} [astron]*	星；ほし；せい	hoshi; sei
star anise, Japanese *[bot]*	樒；しきみ	shikimi
star atlas; star chart (or map) *[astron]*	星図；星圖；せいず	seizu
starboard *[navig]*	右玄	u'gen
star catalog *[astron]*	星表	seihyō
starch *{n} [bioch]*	澱粉；でんぷん	denpun
starch; paste *{n} [bioch]*	糊；のり	nori
starching *{n} [tex]*	糊付け	nori'tsuke
starch paste *[bioch]*	澱粉糊	denpun-ko
starch-syrup state *[mech]*	水飴状	mizu'ame'jō
star cloud *[astron]*	恒星集団；恒星集團	kōsei-shūdan
star cluster *[astron]*	星団；せいだん	seidan
star connection *[elec]*	星形接続	hoshi'gata-setsu'zoku
star coral *[i-zoo]*	菊目石	kiku'me-ishi
starfish *[i-zoo]*	海星；ひとで	hitode
stargazer (fish) *[v-zoo]*	三島虎魚	mishima-okoze
star gear *[mech]*	星形歯車	hoshi'gata-ha'guruma

start *{n} [comput]*	始動；しどう	shidō
" (beginning; commencement) *{n}*	開始	kaishi
starter *[elec]*	起動機；きどうき	kidō-ki
starting; start *{n} [comput] [elec]*	起動	kidō
starting; starting-up *[mech-eng]*	始動	shidō
starting of work *{n}*	始業；しぎょう	shigyō
starting-up *{n} [ind-eng]*	開始	kaishi
start signal *[comput]*	開始信号	kaishi-shingō
start-stop system *[comput]*	調歩式	chōho-shiki
starvation; hunger *[physio]*	飢え；うえ	ue
starve *{vb} [physio]*	飢える；餓える	ueru
state; condition *{n} [comput] [p-ch] [sci-t]*	状態；狀態	jōtai
" *{n} [geogr]*	州	shū
statement *[comm]*	陳述	chin'jutsu
" *[comput]*	文；ぶん	bun
" *[comput]*	命令文	meirei'bun
statement of account *[econ]*	請求書	seikyū'sho
statement of opposition *[pat]*	異議申立	igi-mōshi'tate
state of the art *{n} [ind-eng]*	技術水準	gi'jutsu-suijun
" *{adj} [ind-eng]*	最新式の	saishin'shiki no
stateroom *[nav-arch]*	専用室（船の）	sen'yō-shitsu (fune no)
static *{n} [comm]*	空電；くうでん	kūden
" *{adj} [comput]*	静的の；靜的の	sei'teki no
static characteristic *[electr]*	静特性	sei-tokusei
static charge prevention *[poly-ch]*	帯電防止	taiden-bōshi
static culture *[microbio]*	静置培養	seichi-baiyō
static electricity *[elec]*	静電気；靜電氣	sei-denki
static force *[mech]*	静圧；せいあつ	sei-atsu
static head *[fl-mech]*	静落差；せいらくさ	sei-raku'sa
statics (the discipline) *[mech]*	静力学	sei-riki'gaku
static water pressure *[hyd]*	静水圧；靜水壓	sei-sui'atsu
station; depot *{n} [rail]*	駅；驛；えき	eki
stationary culture *[microbio]*	静置培養	seichi-baiyō
stationary current *[elec]*	定常電流	teijō-denryū
stationary flow *[fl-mech]*	定常流	teijō-ryū
stationary information source *[comput]*	定常情報源	teijō-jōhō'gen
stationary phase *[an-ch]*	固定相	kotei-sō
" *[microbio]*	定常期	teijō-ki
stationary point *[astron]*	留；りゅう	ryū
stationary state *[quant-mech]*	定常状態	teijō-jōtai
stationary wave *[phys]*	定常波	teijō-ha
stationery *{n} [mat]*	文房具；ぶんぼうぐ	bunbō'gu

station premise(s) *[civ-eng]*	駅構内	eki-kōnai
station-to-station call *[comm]*	番号通話	bangō-tsūwa
statistic *{n} [stat]*	統計量	tōkei'ryō
statistical inference *[stat]*	統計的推測	tōkei'teki-suisoku
statistical mechanics *[phys]*	統計力学	tōkei-riki'gaku
statistical weight *[stat]*	統計学的重率	tōkei'gaku'teki-jūritsu
statistics (the discipline) *[math]*	統計学	tōkei'gaku
" *[stat]*	統計；とうけい	tōkei
stator *[elec]*	固定子	kotei'shi
statuary bronze *[met]*	美術青銅	bi'jutsu-seidō
statue *[furn]*	彫像；ちょうぞう	chōzō
" *[furn]*	像	zō
status; state; condition *{n}*	状況	jōkyō
statutory units *[phys]*	法定計量単位	hōtei-keiryō-tan'i
staurolite *[miner]*	十字石	jūji'seki
stay (of aircraft) *{n} [aero-eng]*	支柱	shichū
" (shore) *{n} [civ-eng]*	控え；ひかえ	hikae
" *{vb}*	中止する	chūshi suru
steady reaction *[ch]*	定常反応；定常反應	teijō-hannō
steady state *[phys]*	定常状態	teijō-jōtai
steady-state vibration *[mech]*	定常振動	teijō-shindō
steady wind *[meteor]*	定常風	teijō-fū
stealth *{n}*	忍び；しのび	shinobi
steam *{n} [phys]*	蒸気；じょうき	jōki
" *{n} [phys]*	水蒸気；水蒸氣	sui'jōki
" *{vb t}*	蒸す	musu
steam cooking *[cook] [food-eng]*	蒸し煮；むしに	mushi-ni
steam distillation *[ch-eng]*	水蒸気蒸留	sui'jōki-jōryū
steamed dishes *[cook]*	蒸し物	mushi'mono
steamed egg custard *[cook]*	玉子豆腐	tamago-dōfu
steam engine *[mech-eng]*	蒸気機関	jōki-kikan
steamer *[cook] [eng]*	蒸し器	mushi-ki
steam ironing *{n} [tex]*	湯熨；ゆのし	yu'noshi
steam fog *[meteor]*	蒸気霧	jōki-mu
steam locomotive *[rail]*	蒸気機関車	jōki-kikan'sha
steam pressure *[mech]*	蒸気圧	jōki-atsu
steamship; steamer *[nav-arch]*	汽船	kisen
steam whistle *[eng]*	汽笛；きてき	kiteki
steatite *[miner]*	凍石	tōseki
steel *{n} [met]*	鋼；はがね；こう	hagane; kō
steel band *[mat]*	鋼帯；鋼帶	kōtai
steel castings *[mat]*	铸鋼品；鑄鋼品	chūkō'hin

steel for structures; structural steel *[arch] [mech-eng]*	構造用鋼(材)	kōzō'yō-kō(zai)
steel ingot *[met]*	鋼塊; こうかい	kōkai
steelmaking *{n} [met]*	製鋼	seikō
steel plate (thick) *[met]*	厚鋼板	atsu-kōban
steel sheet; steel plate *[mat]*	鋼板	kōban
steel wire *[mat]*	鋼線	kōsen
steel wool *[mat]*	鉄の綿	tetsu no wata
steepest descent, method of *[math]*	鞍点法	anten-hō
steering *{n} [navig]*	操縦; 操縱	sōjū
steering gear *[mech-eng]*	舵取り歯車	kaji'tori-ha'guruma
" *[mech-eng]*	操舵機	sōda-ki
steering system *[mech-eng]*	舵取り装置	kaji'tori-sōchi
Stegomyia fasciata (mosquito type) *[i-zoo]*	縞蚊; しまか	shima-ka
stellar evolution *[astrophys]*	恒星進化	kōsei-shinka
stellar population *[astron]*	種族; しゅぞく	shu'zoku
stellar statistics *[stat]*	恒星統計学	kōsei-tōkei'gaku
stellar wind *[astron]*	恒星風	kōsei'fū
stem; stalk *{n} [bot]*	茎; 莖; くき	kuki
" *{n} [ling]*	基幹	kikan
" *{n}* (of a ship) *[nav-arch]*	船首	senshu
stem cell *[cyt]*	幹細胞; 幹細胞	kan-saibō
stencil printing *{n} [tex]*	型紙捺染	kata'gami-nasen
stenography *[comm]*	速記	sokki
step *{n} [arch] [ch-eng]*	段	dan
" *{n} [arch]*	階段; かいだん	kaidan
" *{n}*	段階	dankai
step and advance *{vb}*	歩進する	hoshin suru
step difference *{n}*	段差	dansa
stephanite *[miner]*	脆銀鉱	zei'ginkō
stepped gear *[mech-eng]*	ずれ歯歯車	zure'ba-ha'guruma
stepping stone *[arch]*	飛石; とびいし	tobi-ishi
stereochemistry *[p-ch]*	立体化学; 立體化學	rittai-kagaku
stereoisomer *[org-ch]*	立体異性体	rittai-isei'tai
stereoisomerism *[org-ch]*	立体異性	rittai-i'sei
stereometry *[math]*	求積法	kyūseki-hō
stereophonic *{adj} [acous]*	立体音的な	rittai'on'teki na
stereophonic broadcast *[comm]*	立体放送	rittai-hōsō
stereophonic sound system *[acous]*	立体音響系	rittai-onkyō-kei
stereophony *[acous]*	立体音響効果	rittai-onkyō-kōka
stereoregular polymerization *[poly-ch]*	立体規則性重合	rittai-kisoku'sei-jūgō

stereoregularity; stereospecificity *[org-ch]*	立体規則度	rittai-kisoku-do
stereoregularity; stereospecificity	立体規則性	rittai-kisoku-sei
stereoscope *[opt]*	実体鏡；實體鏡	jittai'kyō
" *[opt]*	立体鏡	rittai'kyō
stereoscopic image *[opt]*	立体像；りったい像	rittai-zō
stereoscopic microscope *[opt]*	双眼実体顕微鏡	sōgan-jittai-kenbi'kyō
stereospecificity *[org-ch]*	立体規則性	rittai-kisoku-sei
stereospecific polymerization *[poly-ch]*	立体規則性重合	rittai-kisoku'sei-jūgō
stereospecific polymerization	立体特異性重合	rittai-tokui'sei-jūgō
steric factor *[p-ch]*	立体因子	rittai-inshi
steric hindrance *[ch]*	立体障害	rittai-shōgai
steric strain *[ch]*	立体歪	rittai-hizumi
sterile chamber *[microbio]*	無菌室	mukin-shitsu
sterilization *[biol]*	滅菌；めっきん	mekkin
" *[microbio]*	殺菌；さっきん	sakkin
sterilize *{vb} [microbio]*	殺菌する	sakkin suru
sterilizer *[eng] [med]*	滅菌機	mekkin-ki
" *[eng] [med]*	滅菌装置	mekkin-sōchi
sterilizing lamp *[opt]*	殺菌灯；殺菌燈	sakkin-tō
stern (of ship) *{n} [nav-arch]*	船尾	senbi
sternum (bone) *{n} [anat]*	胸骨；きょうこつ	kyō-kotsu
stethoscope *[med]*	聴診器；聽診器	chōshin-ki
stewed fish or vegetables *[cook]*	煮浸し；にびたし	ni-bitashi
stibnite *[miner]*	輝安鉱	ki'an'kō
stick; rod *{n} [eng]*	棒；ぼう	bō
stick together *{vb} [adhes]*	くっつく	kuttsuku
stickiness *{n} [mat]*	粘着性	nenchaku-sei
sticking *{n} [adhes]*	固着；こちゃく	ko'chaku
sticking agent *[mat]*	固着剤	kochaku-zai
sticky end *[bioch]*	付着端	fuchaku-tan
sticky tape *[mat]*	粘着テープ	nenchaku-tēpu
stiffness *[mech]*	腰；こし	koshi
" *[mech]*	剛さ；こわさ	kowa-sa
stigma (pl stigmata) *[bot]*	柱頭	chūtō
" *[i-zoo]*	眼点	ganten
stigmatic point *[opt]*	無収差点；無收差點	mu'shūsa-ten
stilbite *[miner]*	束沸石	soku'fusseki
still (an apparatus) *{n} [ch-eng]*	蒸留器；蒸溜器	jōryū-ki
still-air cooling *{n} [ch-eng]*	放冷	hōrei
still residue *[ch-eng]*	釜残；かまざん	kama'zan

stilt (bird) *[v-zoo]*	背高鴫	sei'taka-shigi
stilts *{n} [eng]*	竹馬; たけうま	take-uma
stimulant *[mat]*	刺激剤; しげきざい	shigeki-zai
" *[pharm]*	興奮薬	kōfun-yaku
stimulated emission *[elecmg]*	誘導放出	yūdō-hō'shutsu
" *[electr]*	誘発発光	yū'hatsu-hakkō
stimulation *[elecmg]*	誘発; ゆうはつ	yū'hatsu
stimulus (pl -uli) *[physio]*	刺激; 刺戟	shigeki
sting (by an insect) *{vb t}*	螫す; さす	sasu
stingray; Dasyatis akajei *[v-zoo]*	赤鱏; あかえい	aka-ei
stinkbug (insect) *[i-zoo]*	亀虫; 椿象	kame-mushi
stinking nightshade *[bot]*	非沃斯; ひよす	hiyosu
stipulate; ordain *{vb} [legal]*	規定する	kitei suru
stirrer; agitator *[ch-eng]*	攪拌機	kakuhan-ki
stirring *{n} [mech-eng]*	攪拌; かくはん	kaku'han
stirring rod *[an-ch]*	掻き混ぜ棒	kaki'maze-bō
" *[mat]*	攪拌棒	kakuhan-bō
stirrup; stapes (bone) *{n} [anat]*	鐙骨; あぶみ骨	tō-kotsu; abumi-bone
stirrup *{n} [trans]*	鐙	abumi
stitch *{n} [tex]*	(網み)目	(ami-)me
stitch extraction strength *[tex]*	抜糸強度	basshi-kyōdo
stitch number per inch *[tex]*	運針数	unshin'sū
stochastic process *[stat]*	確率過程	kaku'ritsu-katei
stock *{n} [econ]*	株; かぶ	kabu
stock control *[econ] [ind-eng]*	存庫管理	zaiko-kanri
stockfish (for soup) *[cook]*	干物; ひもの	hi'mono
stock paste *[mat]*	元糊; もとのり	moto-nori
stoichiometric *{adj} [p-ch]*	化学量論的な	kagaku'ryōron'teki na
stoichiometric equation *[p-ch]*	化学量論式	kagaku'ryōron-shiki
stoichiometric number *[p-ch]*	化学量数; 化學量數	kagaku'ryō-sū
stoichiometry *[p-ch]*	化学量論	kagaku'ryō-ron
stomach *{n} [anat]*	胃; い	i
stomach and intestines *[anat]*	胃腸; いちょう	i'chō
stomachic *{n} [pharm]*	健胃薬	ken'i-yaku
stone *{n} [bot]*	種子; しゅし	shushi
" *{n} [bot]*	種; たね	tane
" *{n} [geol]*	石; いし	ishi
Stone Age *[archeo]*	石器時代	sekki-jidai
stone bridge *[civ-eng]*	石橋	ishi-bashi
stonefish (fish) *[v-zoo]*	虎魚; おこぜ	okoze
stone fly; grannom (insect) *[i-zoo]*	川螻; かわげら	kawa-gera
stone lantern *[arch]*	石灯籠; 石燈籠	ishi-dōrō

stone quarry *[eng]*	石切場	ishi'kiri'ba
stonemason (a person) *[civ-eng]*	石工；いしく	ishi'ku
stoneware *[cer]*	炻器；石器；せっき	sekki
stop *{n} [opt]*	絞り	shibori
" ; stop dead *{vb i}*	止まる	tomaru
" ; halt; arrest *{vb t}*	止める	tomeru; todomeru
" ; cease; discontinue; end *{vb}*	止める	yameru
" ; desist; discontinue *{vb}*	止す	yosu
stop, linked *{vb} [mech]*	係止する	keishi suru
stop, wedged *{vb} [mech]*	楔止する	sesshi suru
stop-and-go driving *{n} [traffic]*	のろのろ運転	noro-noro-unten
stopcock *[eng]*	活栓	kassen
stoplight *[traffic] [opt]*	停止灯；停止燈	teishi'tō
stoppage; block *{n} [med]*	梗塞；こうそく	kōsoku
stopper *[sci-t]*	栓；せん	sen
stoppered container *[ind-eng]*	密閉器	mippei-ki
stopping (of vehicle) *{n} [traffic]*	停車	teisha
stopsign *[traffic]*	一時停止標識	ichiji-teishi-hyō'shiki
stop signal *[comput]*	停止信号；停止信號	teishi-shingō
storage *[comput]*	蓄積；ちくせき	chiku'seki
" *[comput]*	保存	hozon
" *[comput]*	記憶；きおく	ki'oku
" *[ind-eng]*	貯蔵	chozō
storage battery *[elec]*	蓄電池；ちくでんち	chiku'denchi
storage capacity *[comput]*	記憶容量	ki'oku-yōryō
storage element *[comput]*	記憶素子	ki'oku-soshi
storage life *[ind-eng]*	貯蔵寿命；貯藏壽命	chozō-jumyō
store *{n} [econ]*	店；みせ	mise
store of long standing *{n} [econ]*	老舗；しにせ	shi'nise
store *{vb} [comput]*	蓄積する	chiku'seki suru
" *{vb} [comput]*	記憶する	ki'oku suru
store (within); house (in) *{vb t}*	収める；納める	osameru
storehouse *[arch] [ind-eng]*	倉庫	sōko
store location *[comput]*	記憶位置	ki'oku-ichi
storm *{n} [meteor]*	嵐；あらし	arashi
" *{n} [meteor]*	暴風雨	bōfū'u
" *{n} [meteor]*	時化；しけ	shike
storm sewer *[civ-eng]*	雨水溝	u'sui'kō
storm window *[arch]*	雨戸；あまど	ama'do
storm(y) petrel (bird) *[v-zoo]*	海燕；うみつばめ	umi-tsubame
straight *{n} [geom]*	真っ直；まっすぐ	massugu
" *{adj} [geom]*	真っ直な	massugu na

straight angle *[geom]*	平角	hei'kaku
straight bevel gear *[mech-eng]*	直刃傘歯車	sugu'ba-kasa-ha'guruma
straight chain *[ch]*	直鎖；ちょくさ	choku-sa
straight-chain form *[ch]*	直鎖状	choku'sa'jō
straight-chain hydrocarbon *[org-ch]*	直鎖炭化水素	choku'sa-tanka'suiso
straight line; line *{n}* *[geom]*	直線；ちょくせん	choku'sen
straightness *[math]*	真直度	shin'choku-do
strain *{n}* *[microbio]*	株；かぶ	kabu
" *{n}* *[microbio]*	系統	keitō
" *{n}* *[phys]*	歪；ひずみ	hizumi
strain aging *{n}* *[met]*	歪時効	hizumi-jikō
strain at fracture *{n}* *[mech]*	破断歪	hadan-hizumi
strainer *[cook]* *[eng]*	裏漉し；うらごし	ura'goshi
strain gage *[eng]*	歪み計	hizumi-kei
strain hardening *{n}* *[met]*	歪硬化	hizumi-kōka
strain-induced *{adj}* *[mech]*	加工誘起の	kakō-yūki no
strain stress *[mech]*	歪応力；歪應力	hizumi-ō'ryoku
strait *{n}* *[geogr]*	海峡；海峽	kaikyō
strand *{n}* *[eng]*	素線	sosen
strange(ness); wonder; marvel *{n}*	不思議；ふしぎ	fushigi
strategy (pl -gies) *[mil]*	戦略	sen'ryaku
stratigraphy *[geol]*	層位学	sō'i'gaku
stratocumulus *[meteor]*	層積雲	sōseki'un
stratosphere *[meteor]*	成層圏；成層圈	seisō-ken
stratus (pl strati) *[meteor]*	層雲	sō'un
straw *[agr]* *[bot]*	藁；わら	wara
straw ashes *[agr]* *[ch]*	藁灰	wara-bai
strawberry (fruit) *[bot]*	苺；莓；いちご	ichigo
strawberry geranium (flower) *[bot]*	雪の下	yuki-no-shita
straw color *[opt]*	藁色	wara-iro
straw mat; rush mat *[mat]*	菰；薦；こも	komo
straw mat(ting) *[mat]*	蓆；筵；むしろ	mushiro
straw sack; straw bag *[agr]*	俵；たわら	tawara
stray capacitance *[electr]*	浮遊容量	fuyū-yōryō
" *[electr]*	漂遊容量	hyōyū-yōryō
stray current *[elec]*	迷走電流	meisō-denryū
stray electromagnetic field *[elecmg]*	漂遊電磁界	hyōyū-denji'kai
stray flux *[elecmg]*	漂遊磁束	hyōyū-ji'soku
stray in *{vb}*	紛れ込む	magire-komu
stray light *[opt]*	迷光；めいこう	meikō
streak *{n}* *[microbio]*	画線；かくせん	kaku'sen
" *{n}* *[min-eng]*	鉱脈；鑛脈	kōmyaku

streak *{n} [miner]*	条痕；條痕	jōkon
" *{n} [miner] [opt]*	縞；しま	shima
stream *{n} [hyd]*	流れ；ながれ	nagare
streamer *[geophys]*	流光	ryūkō
" *[geophys]*	射光	shakō
" *[phys]*	吹流し	fuki'nagashi
" *{n}*	幟；のぼり	nobori
streamline *[fl-mech]*	流線	ryūsen
streamlined *{n} [aero-eng]*	流線形	ryūsen-kei
street *[civ-eng] [traffic]*	街路；がいろ	gairo
" *[civ-eng] [traffic]*	道	michi
street light *[civ-eng]*	街灯	gaitō
street system *[civ-eng]*	街路系統	gairo-keitō
strength *[mech]*	力；ちから	chikara
" *[mech]*	強度	kyōdo
" *[mech]*	強さ；つよさ	tsuyo-sa
strengthen *{vb t} [eng]*	補強する	hokyō suru
strengthening *{n} [mech]*	強化	kyōka
strength of materials *[mech]*	材料力学	zairyō-riki'gaku
Streptococcus (pl -i) *[microbio]*	連鎖(状)球菌	rensa('jō)-kyūkin
stress *{n} [mech]*	応力；おうりょく	ō'ryoku
" (tension) *{n} [mech]*	歪力；わいりょく	wai-ryoku
stress birefringence *[poly-ch]*	応力複屈折	ōryoku-fuku'kussetsu
stress concentration *[mech]*	応力集中	ōryoku-shūchū
stress corrosion cracking *{n} [met]*	応力腐食割れ	ōryoku-fu'shoku-ware
stress cracking *{n} [poly-ch]*	応力亀裂	ōryoku-ki'retsu
stress diagram *[met] [graph]*	応力図；應力圖	ōryoku-zu
stress induction *[mech]*	応力誘起	ōryoku-yūki
stress relaxation *[met]*	応力緩和	ōryoku-kanwa
stress-relieving annealing *[met]*	歪取焼鈍	hizumi'tori-yaki'namashi
stretch *{n} [mech] [physio]*	伸び；のび	nobi
" *{n} [mech]*	伸長	shinchō
" *{vb}*	引っ張る	hipparu
stretcher *[med]*	担架；擔架；たんか	tanka
stretch forming *{n} [eng]*	絞り成形	shibori-seikei
stretching *{n} [poly-ch]*	延伸；えんしん	enshin
stretch ratio *[poly-ch] [tex]*	延伸(倍)率	enshin-(bai)ritsu
" *[poly-ch] [tex]*	延伸比	enshin-hi
stretch spinning *{n} [tex]*	緊張紡糸	kinchō-bōshi
striation *[geol]*	縞	shima
strike; beat; spank *{vb t}*	叩く；たたく	tataku
" ; thrust; pierce *{vb t}*	突く；つく	tsuku

strike-through *{n} [pr]* 裏抜け；裏脱け ura-nuke
string *{n}[comput]* 連糸 renshi
" *{n} [comput]* 列 retsu
" *{n} [mat]* 紐；ひも himo
" (chord) *{n} [music]* 絃 gen
" *{n} [tex]* 糸；絲；いと ito
" *{n}* 連 ren
stringed instruments *[music]* 絃楽器類；絃樂器類 gen'gakki-rui
stringiness *{n} [adhes] [poly-ch]* 糸引き ito'hiki
" *{n} [poly-ch] [tex]* 曳糸性；えいしせい eishi-sei
stripe *{n} [tex]* 縞 shima
striped mosquito (insect) *[i-zoo]* 縞蚊 shima-ka
" (insect) *[i-zoo]* 藪蚊；やぶか yabu-ka
strip mining *{n} [min-eng]* 露天採鉱 roten-saikō
strobilation *[i-zoo]* 横分体形成 ō'buntai-keisei
stroke *{n} [mech-eng]* 行程 kōtei
" (of a Chinese character) *[pr]* 画；かく kaku
strong *{adj}* 強い；つよい tsuyoi
strong acid *[ch]* 強酸；きょうさん kyō-san
strong base *[ch]* 強塩基 kyō-enki
strong electrolyte *[p-ch]* 強電解質 kyō-denkai'shitsu
strong flour *[cook]* 強力粉 kyō'riki-ko
strong heating *{n} [ch]* 灼熱；しゃくねつ shaku-netsu
strontium (element) ストロンチウム sutoronchūmu
structural adhesive agent *[adhes]* 構造用接着剤 kōzō'yō-setchaku-zai
structural drawing *{n} [graph]* 構造製図 kōzō-seizu
structural engineering *[civ-eng]* 構造工学 kōzō-kōgaku
structural formula *[ch]* 構造式 kōzō-shiki
structural geology *[geol]* 構造地質学 kōzō-chi'shitsu'gaku
structural integrity *[eng]* 構造完全性 kōzō-kanzen-sei
structural member *[mat]* 構造部材 kōzō-buzai
structural steel *[met]* 構造用鋼 kōzō'yō-kō
structural strength *[mech]* 構造強さ kōzō-tsuyo-sa
structural viscosity *[fl-mech]* 構造粘性 kōzō-nensei
structure *{n} [arch] [sci-t]* 構造；こうぞう kōzō
" *{n} [civ-eng]* 構造物 kōzō'butsu
" *{n} [sci-t]* 組織；そしき so'shiki
" (organization) *{n}* 機構；きこう kikō
strut *{n} [civ-eng]* 控え hikae
" *{n} [eng]* 支柱 shichū
" *{vb i}* 張り出す hari-dasu
stucco *[mat]* 漆喰；しっくい shikkui

study; library (a room) *{n} [arch]* 書斎 shosai
" ; research *{n} [sci-t]* 研究；けんきゅう kenkyū
sturdy *{adj}* 頑丈な ganjō na
" (healthy; robust) *{adj}* 丈夫な jōbu na
sturgeon (fish) *[v-zoo]* 蝶鮫；ちょうざめ chō-zame
style *{n} [bot]* 花柱 kachū
styptic *{n} [pharm]* 血液凝固促進薬 ketsueki-gyōko-sokushin-yaku
" *{n} [pharm]* 止血剤 shi'ketsu-zai
styrax oil *[mat]* 斎墩果油；えごゆ ego-yu
sub- *{prefix}* 亜；亞 a-
" *{prefix}* 副- fuku-
" *{prefix}* 次- ji-
" *{prefix}* 下の- shita no-
subagent (a person) *[legal]* 復代理人 fuku-dairi'nin
subalpine zone *[geogr]* 亜高山帯；亞高山帶 a'kōzan-tai
subarctic *{adj} [geogr]* 亜北極の a'hokkyoku no
subatomic *{adj} [phys]* 原子以下の genshi'ika no
subaudio; subaudible *{adj} [acous]* 可聴下の kachō'ka no
subbituminous coal *[geol]* 亜瀝青炭 a'rekisei-tan
subcellular *{adj} [bio]* 亜細胞性の a'saibō'sei no
subclass *{n} [bio] [syst]* 亜綱；あこう a'kō
subcontinent *[geogr]* 亜大陸；亞大陸 a'tairiku
subculture *{n} [microbio]* 継代培養 keidai-baiyō
subdivide *{vb} [math]* 分ける wakeru
subdivision *{n} [bot] [syst]* 亜門 a'mon
subdivision of lot *[civ-eng]* 分筆 bunpitsu
subdued refinement *{n}* 渋味；澁味；しぶみ shibu'mi
subdwarf star *[astron]* 順矮星 jun-waisei
subfamily *[bio] [syst]* 亜科 a'ka
subgenus (pl -genera) *[bio] [syst]* 亜属 a'zoku
subgiant *[astron]* 準巨星 jun-kyosei
subgravity *[mech]* 亜重力 a'jūryoku
subgroup *{n} [math]* 亜群 a'gun
" *{n} [math]* 部分群；ぶぶんぐん bubun'gun
subject *{n} [comput]* 主体 shutai
" *{n} [gram]* 主語 shugo
" *{n} [sci-t]* 対象；對象 taishō
subjective *{adj} [psy]* 主観的な shukan'teki na
subjective probability *[prob]* 主観確率 shukan-kaku'ritsu
subjective symptom *[med]* 自覚症状 ji'kaku-shōjō
subject matter *[pat]* 分野；ぶんや bun'ya
subkingdom *[bio] [syst]* 亜界 a'kai

sublanguage *[ling]*	準言語	jun'gengo
sublethal *[gen] [med]*	亜致死；あちし	a'chishi
sublimate *{n} [ch]*	昇華物	shōka'butsu
sublimation *[thermo]*	昇華；しょうか	shōka
sublime *{vb t} [thermo]*	昇華する	shōka suru
sublimed sulfur *[pharm]*	昇華硫黄	shōka-iō
subliminal *{adj} [psy]*	識閾下の	shiki'iki'ka no
submarine *{n} [nav-arch]*	潜水艦	sensui'kan
submarine eruption *[geophys]*	海底噴火	kaitei-funka
submarine volcano *[geol]*	海底火山	kaitei-kazan
submerge *{vb} [nav-arch]*	潜水する	sensui suru
submerged *{adj}*	沈下した	chinka shita
" *{adj}*	水中の	suichū no
submerged culture *[microbio]*	深部培養	shinbu-baiyō
suboptimal *{adj}*	亜最適の(な)	a'saiteki no (na)
suborder *[bio] [syst]*	亜目；あもく	a'moku
suboxide *[ch]*	亜酸化物	a'sanka'butsu
subphylum (pl -phyla) *[syst] [zoo]*	亜門	a'mon
subpolar zone *[geogr]*	亜寒帯；亞寒帶	a'kantai
subrogate *{vb} [legal]*	代位する	dai'i suru
subsatellite point *[astron]*	衛星直下点	eisei-chokka'ten
subscript *[alg]*	添字	tenji
" *[comput] [pr] [sci-t]*	添(え)字	soe'ji
subscript character *[pr]*	下付き文字	shita'tsuki-moji
subscript numeral *[pr]*	下付き数字	shita'tsuki-sūji
subscript symbol *[pr]*	下付き記号	shita'tsuki-kigō
subset *[math]*	部分集合	bubun-shūgō
subsidence *[min-eng]*	沈下	chinka
subsidiary company *[econ]*	子会社；子會社	ko-gaisha
subsoil water *[hyd]*	湧水；わきみず	waki-mizu
subsonic speed *[mech] [phys]*	亜音速	a'onsoku
subspecies *[bio] [syst]*	亜種	a'shu
substance *{n} [phys]*	物質；ぶっしつ	busshitsu
" (content matter) *{n}*	内容	naiyō
" *{n}*	要旨	yōshi
substance flow *[fl-mech]*	物流；ぶつりゅう	butsu'ryū
substantive *{adj}*	実体の；實體の	jittai no
substation *[elec]*	変電所；變電所	henden'sho
substituent *{n} [ch]*	置換基	chikan-ki
substitute *{n} [comput]*	置き換え	oki'kae
" *{n} [ind-eng]*	代替物	daitai'butsu
" *{n} [ind-eng]*	代用品	daiyō'hin

substitution *[alg] [ch] [mol-bio]*	置換；ちかん	chikan
" ; substituting-in *{n} [alg]*	代入；だいにゅう	dainyū
" ; replacement *{n}*	代替	daitai
substitution product *[ch]*	置換体；ちかんたい	chikan-tai
substrain *[microbio]*	亜系	a'kei
" *[microbio]*	亜株；あかぶ	a'kabu
substrate *[bioch] [ch]*	基質；きしつ	ki'shitsu
" *[electr]*	基板	kiban
" *[eng]*	基体	kitai
" ; base material *[eng]*	基材	kizai
substrate specificity *[bioch]*	基質特異性	ki'shitsu-toku'i-sei
substratum (pl -strata) *[geol]*	下層	kasō
" *[photo]*	下引き層	shita'biki-sō
substring *[comput]*	部分列；ぶぶんれつ	bubun-retsu
subterranean *{adj} [geogr]*	地下の	chika no
subtotal *{n} [math]*	小計；しょうけい	shōkei
subtract *{vb} [arith]*	引く	hiku
subtracter *[electr]*	減算器；げんざんき	genzan-ki
subtraction (method) *[arith]*	減法；げんぽう	genpō
" *[arith]*	引き算；ひきざん	hiki-zan
subtractive color mixing *{n} [opt]*	減法混色	genpō-kon'shoku
subtractive primary color *[opt]*	減色法の原色	genshoku-hō no genshoku
subtrahend *[arith]*	減数	gensū
subtribe *[bio] [syst]*	亜連；あれん	a'ren
" *[bio] [syst]*	亜族	a'zoku
subtropical zone *[geogr]*	亜熱帯；あねったい	a'nettai
subtype *{n} [microbio]*	亜型；あがた	a'gata
" *{n} [syst]*	亜類型	a'rui-kei
suburban industrial land *[civ-eng]*	郊外工業地	kōgai-kōgyō'chi
suburbs *[geogr]*	郊外；こうがい	kōgai
subway *[trans]*	地下鉄(道)	chika-tetsu(dō)
successive *{adj} [math]*	逐次の；ちくじの	chiku'ji no
successive addition polymerization	逐次付加重合	chikuji-fuka-jūgō
successive reaction *[ch]*	逐次反応	chikuji-hannō
succinate *[ch]*	琥珀酸塩	kohaku'san-en
succinic acid	琥珀酸；こはくさん	ko'haku-san
succinic acid peroxide	過酸化琥珀酸	ka'sanka-kohaku-san
succinic anhydride	無水琥珀酸	musui-kohaku-san
succulent fruit; sap fruit *[bot]*	液果；えきか	eki'ka
sucrose; cane sugar; table sugar *[bioch]*	蔗糖；しょとう	shotō
suction bottle *[an-ch]*	吸引瓶	kyū'in-bin
suction filtration *[sci-t]*	吸引沪過	kyū'in-roka

sudden {adj}	突然の；とつぜんの	totsu'zen no
sudden failure [ind-eng]	突発故障；突發故障	toppatsu-koshō
suffice; be adequate {vb}	足りる	tariru
sufficient {n} [logic]	十分；充分	jūbun
sufficient quantity {n} [math]	適量	teki'ryō
suffix [gram]	接尾語；せつびご	setsubi'go
" [gram]	接尾辞	setsubi'ji
" [math]	添数	tensū
suffix part [gram]	接尾部	setsubi-bu
sugar {n} [bioch] [cook]	砂糖；さとう	satō
" {n} [bioch]	糖(質)	tō(-shitsu)
sugarbeet [bot]	砂糖大根	satō-daikon
" [bot]	甜菜；てんさい	tensai
sugarcane [bot]	砂糖黍	satō-kibi
sugar-coated tablet {n} [pharm]	糖衣錠	tō'i-jō
sugar coating {n} [pharm]	糖衣	tō'i
sugar cube [cook]	角砂糖；かくざとう	kaku-zatō
sugariness [cook]	糖度	tōdo
sugar syrup [cook]	糖漿；とうしょう	tōshō
suicide gene [gen]	自殺遺伝子	ji'satsu-idenshi
suicide substrate [bioch]	自殺基質	ji'satsu-ki'shitsu
suit (man's suit) [cl]	背広；せびろ	se'biro
sukiyaki (Japanese dish) [cook]	鋤焼き；すきやき	suki'yaki
sulfate [ch]	硫酸塩；硫酸鹽	ryūsan-en
sulfated ash [ch]	硫酸灰分	ryūsan-haibun
sulfate group [ch]	硫酸基	ryūsan-ki
sulfate ion [ch]	硫酸イオン	ryūsan-ion
sulfate radical [ch]	硫酸根	ryūsan-kon
sulfation [ch]	硫酸化	ryūsan'ka
sulfide [ch]	硫化物	ryūka'butsu
sulfide staining {n} [met]	硫化黒変	ryūka-koku'hen
sulfite [ch]	亜硫酸塩	a'ryūsan-en
sulfolipid [bioch]	硫脂質	ryū-shi'shitsu
sulfur (element)	硫黄；いおう	iō
sulfur bridge [rub]	硫黄架橋	iō-kakyō
sulfur bromide	臭化硫黄	shūka-iō
sulfur chloride	塩化硫黄	enka-iō
sulfur-crested cockatoo [v-zoo]	黄巴旦；きばたん	ki'batan
sulfur crosslinking {n} [rub]	硫黄架橋	iō-kakyō
sulfur dichloride	二塩化硫黄	ni'enka-iō
sulfur dioxide	二酸化硫黄	ni'sanka-iō
sulfur dioxide gas	亜硫酸ガス	a'ryūsan-gasu

sulfur hexafluoride	六弗化硫黄	roku'fukka-iō
sulfuric acid	硫酸；りゅうさん	ryūsan
sulfuric anhydride	無水硫酸	musui-ryūsan
sulfurizing *{n} [ch]*	浸硫；しんりゅう	shinryū
sulfur monochloride	一塩化硫黄	ichi'enka-iō
sulfur monoxide	一酸化硫黄	issanka-iō
sulfurous acid	亜硫酸	a'ryūsan
sulfurous acid gas	亜硫酸ガス	a'ryūsan-gasu
"	二酸化硫黄	ni'sanka-iō
sulfur oxide	酸化硫黄	sanka-iō
sulfur trioxide	三酸化硫黄	san'sanka-iō
sum ; total *[arith]*	合計；ごうけい	gōkei
" *[arith]*	和；わ	wa
sumac *[bot]*	白膠木；ぬるで	nurude
sum frequency *[phys]*	和周波数	wa-shūha'sū
sumi (q.v.) **painting** *{n} [graph]*	墨絵；墨繪；すみえ	sumi-e
summary (pl -ries) *{n} [pr]*	要約	yōyaku
summation *[math]*	和分	wa'bun
summation notation *[alg]*	総和記号；總和記號	sōwa-kigō
summer *[astron]*	夏期；かき	kaki
" *[astron]*	夏；なつ	natsu
summer crop *[agr]*	表作；おもてさく	omote-saku
summer solstice *[astron]*	夏至；げし	geshi
sum of products *[math]*	積和	seki-wa
sum rule *[math]*	和の法則	wa no hōsoku
sun *{n} [astron]*	日；ひ	hi
" *{n} [astron]*	太陽；たいよう	tai'yō
sun bear (animal) *[v-zoo]*	椰子熊；やしぐま	yashi-guma
sunbird (bird) *[v-zoo]*	太陽鳥	taiyō-chō
sun bleaching *{n} [tex]*	天日晒し	tenbi-zarashi
Sunday	日曜日	nichi'yōbi
sundial *[horol]*	日時計；ひどけい	hi-dokei
sundries *[econ]*	雑貨用品	zakka-yōhin
sunfish; Mola mola (fish) *[v-zoo]*	翻車魚；まんぼう	manbō
sunflower (flower) *[bot]*	向日葵；ひまわり	hi'mawari
sunlamp *[elec]*	太陽灯	taiyō-tō
sun light; sun heat *[cook]*	天日	tenbi; tenpi
sunlight *[astron]*	日光；にっこう	nikkō
" *[astron]*	太陽光線	taiyō-kōsen
sun pillar *[meteor]*	太陽柱	taiyō-chū
sun plant *[bot]*	陽生植物	yōsei-shokubutsu
sunrise *[astron]*	日の出	hi-no-de

sunset *[astron]*	日没；にちぼつ	nichi'botsu
sunset glow *[geophys]*	夕焼	yū-yake
sunspot *[astron] [meteor]*	太陽黒点；太陽黑點	taiyō-koku'ten
sunspot cycle *[astron]*	黒点周期	koku'ten-shūki
sun visor *[mech-eng]*	日除け(板)	hi'yoke(-ita)
sun wheel *[eng]*	太陽歯車	taiyō-ha'guruma
super- *{prefix}*	極度に-	kyoku'do ni-
" *{prefix}*	上の-	ue no-
super-absorbent polymer *[mat]*	高吸水性高分子	kō-kyūsui'sei-kōbunshi
superacid *[ch]*	超酸；ちょうさん	chōsan
superadditivity *[math]*	優加法性	yū'kahō-sei
" *[photo]*	超加成性	chō'kasei-sei
superaerodynamics *[fl-mech]*	超空気力学	cho'kūki-rikigaku
superalloy *[met]*	超合金	chō'gōkin
super-bronze *[met]*	超青銅	chō-seidō
supercharger *[mech-eng]*	過給機	kakyū-ki
supercharging *{n} [mech-eng]*	過給	kakyū
superclass *[bio] [syst]*	上綱；じょうこう	jōkō
supercluster of galaxies *[astron]*	超銀河団	chō'ginga-dan
superconducting magnet *[elecmg]*	超伝導磁石	chō'dendō-ji'shaku
superconducting material *[sol-st]*	超伝導材料	chō'dendō-zairyō
superconducting quantum interference device; SQUID *[electr]*	超伝導量子干渉計	chō'dendō-ryōshi-kanshō-kei
superconductive state *[sol-st]*	超導電状態	chō'dōden-jōtai
superconductivity *[sol-st]*	超伝導；超傳導	chō'dendō
supercooling *{n} [thermo]*	過冷	karei
supercritical fluid *[phys]*	超臨界流体	chō'rinkai-ryūtai
superelasticity *[mech]*	超弾性	chō'dansei
super-express train (abbrev) *[trans]*	特急；とっきゅう	tokkyū
" *[trans]*	特別急行列車	toku'betsu-kyūkō-ressha
superfamily *[syst] [zoo]*	上科	jōka
superficial *{adj}*	外面の	gaimen no
" *{adj}*	皮相的な	hisō'teki na
superfine grain *[mech] [phys]*	超微粒子	chō'bi'ryūshi
superfluous flesh *[physio]*	贅肉；ぜいにく	zei-niku
supergiant star *[astron]*	超巨星	chō'kyosei
supergroup *[comm]*	超群	chō'gun
superheat *{vb t} [thermo]*	過熱する	ka'netsu suru
superheated vapor *[thermo]*	過熱蒸気	ka'netsu-jōki
superheavy atom *[phys]*	超重原子	chō'jū'genshi
superhighway *[civ-eng]*	超高速道路	chō'kōsoku-dōro
superimpose *{vb t}*	重ねる；かさねる	kasaneru

superimposition *[geom]*	重ね合せ	kasane-awase
superinfection *[microbio]*	重感染	jū'kansen
superior angle; major angle *[geom]*	優角; ゆうかく	yū-kaku
superior conjunction *[astron]*	外合; がいごう	gaigō
superior planet *[astron]*	外惑星	gai'wakusei
supermultiplet *[quant-mech]*	超多重項	chō'tajū-kō
supermultiplication *[quant-mech]*	超多重化	chō'tajū'ka
supernatant; supernatant liquid *[ch]*	上澄み液	uwa'zumi-eki
supernova (pl -novas, -novae) *[astron]*	超新星	chō'shinsei
superorder *[bio] [syst]*	上目; じょうもく	jōmoku
superplasticity *[met]*	超塑性	chō'sosei
superpolymer *[poly-ch]*	超高分子	chō'kōbunshi
superposition *{n} [geom]*	重畳; ちょうじょう	chōjō
" *{n} [geom]*	重ね合せ	kasane-awase
superradiance *[opt]*	超放射	chō'hōsha
superrefraction *[geophys]*	超屈折	chō'kussetsu
supersaturated solid solution *[p-ch]*	過飽和固溶体	ka'hōwa-koyō'tai
supersaturated solution *[ch]*	過飽和溶液	ka'hōwa-yōeki
supersaturation *[p-ch]*	過飽和; 過飽和	ka'hōwa
superscript *[comput] [pr] [sci-t]*	肩(文)字	kata-(mo)ji
supersensitization *[photo]*	超色増感	chō-iro'zōkan
supersonic *{adj} [phys]*	超音速の	chō'onsoku no
supersonic transport; SST *[aero-eng]*	超音速旅客機	chō'onsoku-ryokak(u)ki
superstring theory *[part-phys]*	超紘理論	chō'himo-riron
superstructure *[civ-eng] [nav-arch]*	上部構造	jōbu-kōzō
" *[sol-st]*	超格子構造	chō'kōshi-kōzō
supersymmetry *[part-phys]*	超対称性; 超對稱性	chō'taishō-sei
supersynchronous motor *[eng]*	超同期電動機	chō'dōki-dendō-ki
supervision *{n}*	監修	kanshū
supper; nighttime meal *{n} [cook]*	夜食; やしょく	ya'shoku
supple; pliant *{n} [mat]*	撓やか; 嫋やか	shinayaka
supplement *{n} [legal]*	補足; ほそく	ho'soku
" ; addition *{n} [pat]*	追加	tsuika
" ; separate publication *{n} [pr]*	別冊; べっさつ	bessatsu
" ; addendum *{n} [pr]*	補遺; ほい	ho'i
supplemental facility *{n} [ind-eng]*	付帯設備	futai-setsubi
supplementary (of an angle) *{adj} [geom]*	補角の	ho'kaku no
" (of an arc) *{adj} [geom]*	補弧の; ほこの	hoko no
supplementary angle *[geom]*	補角; ほかく	ho-kaku
supplementary note *[comm] [pr]*	付記	fuki
supplementary procedure *[ind-eng]*	補助的手段	hojo'teki-shudan
supply; procure *{vb}*	調達する	chōdatsu suru

supply and demand, law of *[econ]*	需要供給の法則	juyō-kyōkyū no hōsoku
supply orifice *[mech-eng]*	供給口	kyōkyū-kō
supply ship *[nav-arch]*	補給船	hokyū'sen
supply source *[ind-eng]*	給源	kyūgen
support *{n} [arch] [math]*	支持; しじ	shiji
" (bearing) *{n} [civ-eng]*	支承	shishō
" *{n} [comput]*	支援; しえん	shi'en
" *{n} [eng]*	支柱	shichū
" *{n} [eng]*	支持体	shiji'tai
" *{vb} [mech-eng]*	支承する	shishō suru
support frame *[eng]*	支持枠; しじわく	shiji-waku
supporting axle (or shaft) *[mech-eng]*	支軸	shi-jiku
supporting electrolyte *[p-ch]*	支持電解質	shiji-denkai'shitsu
supporting point *[mech]*	支点; 支點; してん	shi-ten
support stand *[eng]*	支持台; 支持臺	shiji-dai
supposition *[math]*	仮定; 假定; かてい	katei
suppository (pl -ries) *{n} [pharm]*	座薬; ざやく	za'yaku
" *{n} [pharm]*	坐剤	za'zai
suppression *[comput]*	抑制; よくせい	yokusei
supra- *{prefix}*	上の-	ue no-
surface (curved surface) *{n} [geom]*	曲面	kyoku'men
surface *{n} [math]*	表面; ひょうめん	hyō'men
" *{n} [math]*	面; めん	men
" (right side) *{n} [eng]*	表; おもて	omote
surface-active agent *[mat] [poly-ch]*	表面活性剤	hyōmen-kassei-zai
surface area *[geom]*	表面積	hyō'menseki
surface chemistry *[p-ch]*	界面化学	kaimen-kagaku
surface-covering agent *[mat]*	表面被覆剤	hyōmen-hifuku-zai
surface drainage *[civ-eng]*	路面排水	romen-haisui
surface electrical resistance *[elec]*	表面(電気)抵抗	hyōmen(-denki)-teikō
surface-hardened layer *[met]*	表面硬化層	hyōmen-kōka-sō
surface mail *[comm]*	船便; ふなびん	funa'bin
surface of curvature *[geom]*	曲率面	kyoku'ritsu-men
surface potential *[elec]*	表面電位	hyōmen-den'i
surface resistivity *[elec]*	表面抵抗率	hyōmen-teikō-ritsu
surface roughness *[met]*	肌荒れ; はだあれ	hada-are
surface-skin impression *[physio]*	表皮圧痕	hyōhi-akkon
surface tension *[fl-mech]*	表面張力	hyōmen-chō'ryoku
surface-to-air *{adj} [ord]*	地対空の	chi-tai-kū no
" *{adj} [ord]*	艦対空の	kan-tai-kū no
surface-to-surface *{adj} [ord]*	地対地の	chi-tai-chi no
" *{adj} [ord]*	艦対艦の	kan-tai-kan no

surface water *[hyd]*	地上水	chijō-sui
surfactant *[mat]*	表面活性剤	hyōmen-kassei-zai
" *[mat]*	界面活性剤	kaimen-kassei-zai
surf clam (shellfish) *[i-zoo]*	北寄貝；ほっきがい	hokki-gai
surge *{n}* *[ocean]*	波浪	harō
surgeon *[med]*	外科医	geka'i
surgeon fish *[v-zoo]*	仁座鯛；にざだい	niza-dai
surmise; estimate *{vb}*	察する	sassuru
surname; last name; family name *{n}*	苗字；名字	myōji
" ; last name; family name *{n}*	姓；せい	sei
surplus; excess; remainder; margin	余剰；餘剰	yojō
surprise *{n}*	驚き；おどろき	odoroki
" *{vb t}*	驚かす	odorokasu
surround; come round; gird *{vb}*	回る；繞る；めぐる	meguru
surroundings *{n}* *[ecol]* *[eng]* *[phys]*	周囲；周圍	shū'i
" *{n}*	周り	mawari
surveillance satellite *[aero-eng]*	監視衛星	kanshi-eisei
survey (pl -surveys) *{n}* *[eng]*	調査；ちょうさ	chōsa
" *{n}* *[eng]*	測量	soku'ryō
" (measure) *{vb t}*	測量する	sokuryō suru
survival of the fittest *[bio]*	適者生存	teki'sha-sei'zon
susceptibility *[physio]*	感受性	kanju-sei
suspendability *[ch-eng]*	懸濁性	kendaku-sei
suspended matter *[phys]*	浮遊物；ふゆうぶつ	fuyū'butsu
suspended solids *[phys]*	浮遊固形物	fuyū-kokei'butsu
suspending property *[ch-eng]*	懸垂性	kensui-sei
suspension *{n}* *[ch]*	懸濁(液)	ken'daku(-eki)
suspension bridge *[civ-eng]*	吊り橋；釣(り)橋	tsuri-bashi
suspension culture *[microbio]*	浮遊培養	fuyū-baiyō
suspension polymerization *[poly-ch]*	懸濁重合	kendaku-jūgō
suspension wire *[comm]*	吊り線	tsuri-sen
suspicious value *[math]*	疑わしい値	utagawashī atai
suture *{n}* *[med]*	縫合(線)	hōgō(-sen)
swaddling clothes; baby clothes *[cl]*	産衣；うぶぎ	ubu'gi
swallow (bird) *{n}* *[v-zoo]*	燕；つばめ	tsubame
" *{vb}* *[physio]*	飲む；呑む	nomu
swallowtail (butterfly) *[i-zoo]*	揚羽蝶	age'ha-chō
swamp *{n}* *[ecol]*	沼地	numa'chi
" *{n}* *[ecol]*	沼沢地	shōtaku'chi
swan (bird) *[v-zoo]*	白鳥；はくちょう	hakuchō
swash; uprush *[ocean]*	打上げ波	uchi'age-nami
sway *{n}* *[ocean]*	揺らぎ；搖らぎ	yuragi

sweat *{n} [physio]*	汗；あせ	ase
sweat gland *[physio]*	汗腺	kansen
sweating *{n} [physio]*	汗かき	ase'kaki
sweating-out *[plas]*	滲み出し	nijimi-dashi
sweep *{n} [electr]*	掃引；そういん	sō'in
" *{vb} [comput]*	掃引する	sō'in suru
" *{vb}*	掃く	haku
" (sweep away) *{vb}*	払う；拂う；はらう	harau
sweep circuit *[electr]*	掃引回路	sō'in-kairo
sweet *{adj} [physio]*	甘い；あまい	amai
sweet almond oil *[mat]*	甘扁桃油	kan'hentō-yu
sweet basil (herb) *[bot]*	目箒；めぼうき	me'bōki
sweet brier *[bot]*	浜茄子；はまなす	hama-nasu
sweetened condensed milk *[cook]*	加糖練乳	katō-rennyū
sweetening agent *[cook] [pharm]*	甘味料（甘味剤）	kanmi-ryō (kanmi-zai)
sweet fermented rice drink *[cook]*	甘酒；あまざけ	ama-zake
sweet flavor; sweetness *[cook]*	甘味	ama'mi; kan'mi
sweet hydrangea leaf *[cook]*	甘茶	ama-cha
sweetness *{n} [cook] [physio]*	甘さ	ama-sa
sweet potato (vegetable) *[bot]*	甘藷	kansho
sweet potato; Ipomoea batatas *[bot]*	薩摩芋；さつまいも	Satsuma-imo
sweets; confections *{n} [cook]*	菓子；かし	kashi
sweet sake *[cook]*	味醂；みりん	mirin
sweet sorghum *[bot]*	蘆票；ろぞく	rozoku
" *[bot]*	砂糖蜀黍	satō'morokoshi
swell *{n} [ocean]*	うねり	uneri
swell *{vb i}*	脹らむ	fukuramu
" (become sodden) *{vb i}*	ふやける	fuyakeru
swelling *{n} [med]*	腫；はれ	hare
" *{n} [phys]*	膨潤	bōjun
" *{n} [physio]*	脹らみ	fukurami
swift (bird) *[v-zoo]*	雨燕；あまつばめ	ama-tsubame
swim *{vb}*	泳ぐ	oyogu
swim bladder (of fish) *[v-zoo]*	鰾；浮き袋	uki-bukuro
swinging *{n} [electr] [navig]*	揺れ；搖れ；ゆれ	yure
swirl *{n} [fl-mech]*	渦；うず	uzu
" *{n} [mech-eng]*	渦巻；渦巻	uzu'maki
switch *{n} [elec]*	開閉器	kaihei-ki
switchboard *[comm]*	交換器	kōkan-ki
" *[elec]*	配電盤	haiden-ban
switching; switch *{n} [comput]*	切り替え	kiri'kae
" (interchange) *{n} [elec]*	交換；こうかん	kōkan

switching equipment *[eng]*	交換機	kōkan-ki
sword (Japanese) *[ord]*	刀；かたな	katana
" *[ord]*	剣；つるぎ	tsurugi
swordfish (fish) *[v-zoo]*	目旗魚；めかじき	me'kajiki
swords *[ord]*	刀剣	tōken
swordtail (fish) *[v-zoo]*	剣目高	tsurugi-medaka
syllable *[ling]*	音節；おんせつ	on'setsu
syllogism *[logic]*	三段論法	sandan-ronpō
sylvite *[miner]*	加里岩塩	kari-gan'en
symbiosis (pl -bioses) *[ecol]*	共生(-関係)	kyōsei(-kankei)
" *[ecol]*	相利共生	sōri-kyōsei
symbiotic objects *[astron]*	共存天体	kyōson-tentai
symbol *{n}* *[ch]* *[sci-t]*	記号；記號；きごう	kigō
" *{n}*	象徴；しょうちょう	shōchō
symmetric; symmetrical *{adj}* *[an-geom]*	対称的な	taishō'teki na
symmetric distribution *[stat]*	対称分布	taishō-bunpu
symmetric property *[alg]*	対称性；對稱性	taishō-sei
symmetry (pl -tries) *[an-geom]*	対称；たいしょう	taishō
" *[bio]* *[ch]* *[math]*	相称(性)	sōshō(-sei)
symmetry axis *[an-geom]* *[mech]*	対称軸	taishō-jiku
symmetry element *[crys]*	対称要素	taishō-yōso
symmetry plane *[opt]*	対称面	taishō-men
symmetry species *[spect]*	対称種	taishō-shu
sympathetic detonation *[ch]*	殉爆；じゅんばく	junbaku
sympathetic nervous system *[anat]*	交感神経系	kōkan-shinkei-kei
symphony *[music]*	交響曲	kōkyō'kyoku
symptom *[med]*	徴候；兆候	chōkō
syncarp; multiple fruit *[bot]*	多花果	taka'ka
synchronism *[phys]*	同期；どうき	dōki
synchronization *[eng]*	同期化	dōki'ka
synchronous *{adj}* *[phys]*	同期の	dōki no
synchronous motor *[elec]*	同期電動機	dōki-dendō-ki
synchronous operation *[comput]*	同期動作	dōki-dōsa
synchrony *[phys]*	同期状態	dōki-jōtai
syncline *{n}* *[geol]* *[min-eng]*	向斜；こうしゃ	kōsha
syndrome *[med]*	症候群	shōkō'gun
syneresis (pl -eses) *[ch]*	離液	ri'eki
" *[ch]*	離漿；りしょう	rishō
synergism *[ecol]*	協力作用	kyō'ryoku-sayō
" *[ecol]*	相剰作用；相剩作用	sōjō-sayō
synergist *[ch]*	相剰剤；相乗劑	sōjō-zai
synergistic effect *[bioch]*	相剰効果	sōjō-kōka

synodic period *[astron]*	会合周期；會合周期	kaigō-shūki
synonym *[gram]*	同意語	dōi'go
" *[gram]*	類語；るいご	ruigo
synopsis *[pr]*	摘要	teki'yō
synoptic climatology *[climat]*	総観気候学	sōkan-kikō'gaku
synoptic meteorology *[meteor]*	総観気象学	sōkan-kishō'gaku
syntactics *[ling]*	構文論	kōbun'ron
syntax *[comput] [ling]*	構文(法)	kōbun(-hō)
" *[ling]*	統語論(法)	tōgo-ron(pō)
synthesis *[ch] [poly-ch]*	合成；ごうせい	gōsei
synthetic chemist *[ch]*	合成化学者	gōsei-kagaku'sha
synthetic division *[alg]*	組立除法	kumi'tate-johō
synthetic fiber *[tex]*	合繊；合纖	gōsen
synthetic-fiber fabric *[tex]*	合繊織物	gōsen-ori'mono
synthetic leather *[poly-ch]*	合成皮革	gōsei-hi'kaku
synthetic paper *[poly-ch]*	合成紙；ごうせいし	gōsei-shi
synthetic resin *[poly-ch]*	合成樹脂	gōsei-jushi
synthetic seed *[bot]*	人工種子	jinkō-shushi
syntrophy; syntrophism *[bio]*	栄養共生	eiyō-kyōsei
sypher joint *[eng]*	殺ぎ接ぎ；そぎはぎ	sogi-hagi
syringe; injector *[med]*	注射器	chūsha-ki
syrup *[cook] [pharm]*	シロップ	shiroppu
system *[comput] [math]*	方式	hō'shiki
" *[math]*	系統；けいとう	keitō
" *[sci-t]*	系	kei
" *[sci-t]*	式	shiki
" *[sci-t]*	組織；そしき	so'shiki
" *[sci-t]*	体系；體系	taikei
systematic *{adj}*	組織的な	soshiki'teki na
systematics *[bio]*	分類学；分類學	bunrui'gaku
" *[bio]*	系統学	keitō'gaku
system of equations *[alg]*	方程式系	hōtei'shiki-kei
system of inequalities *[alg]*	連立不等式	ren'ritsu-futō-shiki
systems engineering *{n} [eng]*	システム工学	shisutemu-kōgaku
systole *{n} [physio]*	収縮期；收縮期	shū'shuku-ki
syzygy *{n} [astron]*	朔望；さくぼう	saku'bō

T

T (the letter)	丁字；ていじ	tei'ji
table (list) *{n} [comput] [math]*	表；ひょう	hyō
" *{n} [comput]*	一覧表	ichi'ran-hyō
" *{n} [furn]*	卓子；テーブル	tēburu
tableland; plateau *[geogr]*	台地；臺地；だいち	daichi
table lookup *{n} [comput]*	索表	saku'hyō
table of contents *[pr]*	目次；もくじ	moku'ji
table of drawings *[pr]*	図面表	zumen-hyō
table salt; common salt *[cook]*	食塩；しょくえん	shoku'en
tablespoon *[cook] [eng]*	大匙；おおさじ	ō-saji
tablet *{n} [pharm]*	錠；じょう	jō
tableting machine *[eng]*	打錠機	dajō-ki
tableware; dinnerware *[cer] [cook] [eng]*	食器；しょっき	shokki
tabular *{adj} [crys]*	平板状の	heiban'jō no
tabulate *{vb} [comput]*	製表する	seihyō suru
" *{vb} [math]*	作表する	saku'hyō suru
tabulating; tabulation *{n} [math]*	作表；さくひょう	saku'hyō
tabulator *{n} [eng]*	製表機	seihyō-ki
tachometer *[eng]*	回転速度計	kaiten-sokudo-kei
tack; tackiness; tenacity *{n} [mech]*	粘り強さ	nebari-zuyo-sa
tackifier *[mat]*	粘着(性)付与剤	nenchaku('sei)-fuyo-zai
tackifying property *[mech]*	粘着化性	nenchak'ka-sei
tackiness; stickiness *{n} [mat]*	粘着性	nenchaku-sei
tackle; compound pulley *{n} [mech-eng]*	複滑車	fuku'kassha
tactical aircraft *[aerosp]*	戦術用航空機	senjutsu'yō-kōkū-ki
tactics *[mil]*	戦術；戰術	sen'jutsu
tactile sense; sense of touch *[physio]*	触覚；觸覺	shokkaku
tadpole *[v-zoo]*	お玉杓子	o'tama'jakushi
tag *{n} [comm]*	札；ふだ	fuda
" *{n} [comput]*	標識；ひょうしき	hyō'shiki
" ; luggage tag *{n} [trans]*	荷札；にふだ	ni-fuda
tagged atom *[phys]*	標識原子	hyō'shiki-genshi
taiga; boreal coniferous forest *[ecol]*	亜寒帯林；亞寒帶林	a'kantai-rin
" *[ecol]*	北方針葉樹林	hoppō-shinyō'ju-rin
tail *{n} [v-zoo]*	尾；お	o
tailings *[min-eng]*	尾鉱；尾鑛；びこう	bikō
tailpipe *[mech-eng]*	尾管	bikan
tailstock *[mech-eng]*	心押台	shin'oshi-dai
tailwind *[meteor]*	追風；おいかぜ	oi-kaze

Taishō (era name) (1912-1926)	大正；たいしょう	Taishō
take *{vb t}*	取る	toru
take action; take measures *{vb}*	手を打つ	te o utsu
takedown *{n} [eng]*	分解；ぶんかい	bunkai
take out; extract; remove *{vb t}*	出す	dasu
taker; acceptor (a person) *{n}*	取手；とりて	tori'te
take root *{vb i} [bot]*	定着する	teichaku suru
take up; taking up; coiling *{n} [tex]*	巻取り；巻取り	maki'tori
take-up tension *[tex]*	巻取張力	maki'tori-chōryoku
talc *[miner]*	滑石；かっせき	kasseki
talk; speak *{vb} [comm] [physio]*	話す	hanasu
tallow *{n} [mat]*	獣脂；獸脂	jūshi
tally *{n} [math]*	勘定；かんじょう	kanjō
" *{n} [math]*	計算	keisan
talon *[zoo]*	爪	tsume
talus *[geol]*	崖錐；がいすい	gaisui
tan *{n} [org-ch]*	渋；澁；しぶ	shibu
tang *{n} [eng] [ord]*	中子	naka'go
" *{n} [eng] [ord]*	刀根	tō'kon
tangent (tangential line) *[geom]*	接線；せっせん	sessen
" *[trig]*	正接；せいせつ	sei'setsu
tangential section *[bot]*	接線縦断面	sessen-jūdan'men
tangential stress *[mech]*	接線応力；接線應力	sessen-ō'ryoku
tangential velocity *[mech]*	接線速度	sessen-sokudo
tangerine *[bot]*	蜜柑；みかん	mikan
tangerine; Citrus unshiu *[bot]*	温州蜜柑	Unshū-mikan
tangle (seaweed) *{n} [bot]*	昆布；こんぶ	konbu
" (snarl) *{vb i}*	縺れる；もつれる	motsureru
tank *[eng]*	槽	sō
tanker *[nav-arch]*	油送船	yusō'sen
tanned paper *[paper]*	渋紙；澁紙	shibu-gami
tannin *[org-ch]*	鞣質；じゅうしつ	jū'shitsu
tanning *{n} [eng]*	鞣し(法)	nameshi(-hō)
tansy (pl tansies) *[bot]*	蓬菊；よもぎぎく	yomogi'giku
tantalum (element)	タンタル	tantaru
tap *{vb} [steel]*	出銑する	shussen suru
tape grass; eel grass *[bot]*	石菖藻	sekishō'mo
tape measure *[eng]*	巻尺；巻尺	maki-jaku
taper (down) *{vb t}*	先細にする	saki'boso ni suru
tapered wing *[aero-eng]*	先細翼	saki'boso-yoku
tapeworm *[i-zoo]*	条虫類；條蟲類	jōchū-rui
taphole *[met]*	出銑口	shussen'kō

tapir (animal) *[v-zoo]*	貘；ばく	baku
tappet *[mech-eng]*	凸子；とっし	tosshi
tapping (listening in) *{n} [comm]*	盗聴；盜聽	tōchō
" (for screwing-in) *{n} [mech-eng]*	ねじ立て	neji'tate
" (for molten metal) *{n} [met]*	湯出し	yu'dashi
" (for molten steel) *{n} [steel]*	出鋼；しゅっこう	shukkō
taproot *[bot]*	直根	chokkon
" *[bot]*	主根	shukon
tap water *[hyd]*	水道水	suidō-sui
tar *{n} [mat]*	タール	tāru
tarantula (spider) *[i-zoo]*	舞踏蜘蛛	butō-gumo
Tardigrada *[i-zoo]*	緩歩動物	kanpo-dōbutsu
tare; packing *{n} [mech]*	風袋；ふうたい	fūtai
target *{n} [comput]*	目標	moku'hyō
" *{n} [comput]*	目的；もくてき	mokuteki
" *{n} [phys]*	的；まと	mato
" *{n}*	対象；對象	taishō
target drone *[aero-eng]*	無人標的機	mujin-hyō'teki-ki
target language *[ling]*	目標言語	moku'hyō-gengo
" *[ling]*	対象言語	taishō-gengo
tariff *[econ]*	関税；關稅	kanzei
" *[econ]*	運賃	unchin
tarlatan *[tex]*	薄紗；うすしゃ	usu-sha
tarnish (silver) *{n} [met] [miner]*	曇り	kumori
tarnish; tarnishing *{n} [met]*	変色；變色	henshoku
tarnishing reaction *[miner]*	造膜反応	zō'maku-hannō
taro (pl taros) (vegetable) *[bot]*	里芋；さといも	sato-imo
" (vegetable) *[bot]*	八つ頭；やつがしら	yatsu'gashira
taro stem *[cook]*	芋茎；芋莖；ずいき	zuiki
tar sand *[petr]*	瀝青砂岩	rekisei-sagan
tarsier (animal) *[v-zoo]*	眼鏡猿；めがねざる	me'gane-zaru
tartar *[dent]*	歯石；齒石	shi'seki
" *[org-ch]*	酒石；しゅせき	shu'seki
tartar emetic *[org-ch] [tex]*	吐酒石	toshu'seki
tartaric acid	酒石酸；しゅせき酸	shuseki-san
tartrate *[org-ch]*	酒石酸塩	shuseki'san-en
task fleet *[ord]*	任務艦隊	ninmu-kantai
task force *[ord]*	任務部隊	ninmu-butai
taste *{n} [physio]*	味；あじ	aji
tasteable component *[cook]*	呈味成分	teimi-seibun
tasteableness; flavor-manifesting property *{n} [cook] [physio]*	呈味性；ていみせい	teimi-sei

taste bud *[anat]*	味蕾；みらい	mirai
tastiness *[cook] [physio]*	旨味；うまみ	uma'mi
tasting *{n} [physio]*	味わうこと	ajiwau koto
tasty *[cook] [physio]*	おいしい	oishī
Taurus; the Bull *[constel]*	牡牛座；おうしざ	o'ushi-za
tautness *[mat]*	緊縮性	kin'shuku-sei
tautology *[logic]*	同語反復	dōgo-hanpuku
" *[logic]*	恒真式	kōshin-shiki
tautomer *[ch]*	互変異性体	gohen-isei'tai
tautomerism *[ch]*	互変異性	gohen-i'sei
tautonym *[bio]*	反復名	hanpuku'mei
tautophony *[ling]*	同音反復	dō'on-hanpuku
tawing *{n} [eng]*	明礬鞣し	myōban-nameshi
tax *{n} [econ]*	税(金)	zei(kin)
tax accountant (a person) *[econ]*	税理士；ぜいりし	zeiri'shi
tax-exempt goods *{n} [mat]*	免税品	menzei'hin
tax exemption *[econ]*	免税；めんぜい	menzei
taxis *[physio]*	走性	sōsei
taxiway light *[navig]*	誘導路灯	yūdō'ro-tō
taxon (pl taxa, taxons) *{n} [syst]*	分類単位；分類單位	bunrui-tan'i
taxonomy *[syst]*	分類学；分類學	bunrui'gaku
tea *[bot] [cook]*	茶；ちゃ	cha
" (inferior grade) *[cook]*	番茶	bancha
" (green, superior grade) *[cook]*	玉露；ぎょくろ	gyoku'ro
" (of well-known brand) *[cook]*	銘茶	mei-cha
tea: black tea *[cook]*	紅茶	kō-cha
tea: brick tea *[cook]*	磚茶；だんちゃ	dan-cha
tea: green tea *[cook]*	緑茶	ryoku-cha
tea: Uji tea *[cook]*	宇治茶	Uji tea
tea ceremony *{n}*	茶道	chadō; sadō
" *{n}*	茶の湯	cha-no-yu
tea ceremony room *[arch]*	茶室	cha-shitsu
teach (show) *{vb} [comm]*	教示する	kyōji suru
" (instruct) *{vb} [comm]*	教える	oshieru
teacup *[cer]*	湯呑；ゆのみ	yu'nomi
tea-doused cooked rice *[cook]*	茶漬け	cha'zuke
tea-grinding mortar *[eng]*	茶碾；茶臼	cha-usu
teal blue (color) *[opt]*	暗青緑色	an'seiryoku-shoku
teapot (small) *[cer]*	急須；きゅうす	kyūsu
tea powder container *[furn]*	棗；なつめ	natsume
tear (fluid) *{n} [physio]*	涙；淚；なみだ	namida
" (a rip) *{n}*	破れ目	yabure'me

tear (rip) *{vb t}*	破る；やぶる	yaburu
" (rip) *{vb t}*	(引き)裂く	(hiki-)saku
teardrop *[lap]*	涙滴	ruiteki
tearing; ripping*{n}* *[tex]*	引裂き	hiki'saki
tearing property *[physio]*	催涙性	sairui-sei
tear(ing) strength *[mech]*	引裂強さ	hiki'saki-tsuyo-sa
" *[mech]*	引裂強さ	in'retsu-zuyo-sa
teaspoonful *{n}* *[cook]*	茶匙量	cha'saji'ryō
technetium (element)	テクネチウム	tekunechūmu
technical level *[ind-eng]*	技術水準	gijutsu-suijun
technical person in charge; technical personnel *[ind-eng]*	技術担当者	gijutsu-tantō'sha
technical specification *[ind-eng]*	仕様；しよう	shi'yō
technical standard *[ind-eng]*	技術的標準	gijutsu'teki-hyōjun
" *[ind-eng]*	規格；きかく	ki'kaku
technical standard value *[ind-eng]*	規格値	kikaku-chi
technical term *[comm]*	技術用語	gijutsu-yōgo
technician *[ind-eng]*	技師	gishi
technology (technique) *[sci-t]*	技術；ぎじゅつ	gi'jutsu
technology *[sci-t]*	技術工学	gijutsu-kōgaku
" *[sci-t]*	科学技術	kagaku-gijutsu
" *[sci-t]*	工学；こうがく	kōgaku
" *[sci-t]*	工芸；工藝	kōgei
tectonics *[civ-eng]*	構造学	kōzō'gaku
" *[geol]*	(構造)地質学	(kōzō-)chi'shitsu'gaku
tele- *{prefix}*	遠距離-	en'kyori-
teleceptor *[physio]*	遠受容器	en'juyō-ki
telecommunication(s) *[comm]*	電気通信(学)	denki-tsūshin('gaku)
telecommunication facility *[comm]*	通信機能	tsūshin-kinō
telecommunication line *[comm]*	通信回線	tsūshin-kaisen
telecommunications network *[comm]*	遠距離通信網	en'kyori-tsūshin-mō
telecommunications satellite *[comm]*	通信衛星	tsūshin-eisei
teleconferencing *{n}* *[comm]*	遠隔会議	enkaku-kaigi
teleconnection *[meteor]*	遠隔相関	enkaku-sōkan
telegage *{n}* *[eng]*	遠隔式計器	enkaku'shiki-keiki
telegram *{n}* *[comm]*	電報；でんぽう	denpō
telegraph *{n}* *[comm]*	電信；でんしん	denshin
telegraph carrier (wave) *[comm]*	電信搬送波	denshin-hansō'ha
telegraph code *[comm]*	電信符号	denshin-fugō
telegraph-exchange; Telex *[comm]*	加入電信	ka'nyū-denshin
telegraph line *[comm]*	電信線路	denshin-senro
telegraphy *[comm]*	電信学	denshin'gaku

telemeteograph *[eng]*	遠隔自記気象計	enkaku-jiki'kishō-kei
telemeter *[eng]*	遠隔測定器	enkaku-sokutei-ki
" *[eng]*	測距儀	sok(u)kyo-gi
telemetering *{n} [eng]*	遠隔測定	enkaku-sokutei
teleology *[sci-t]*	目的論	mokuteki'ron
Teleostei *[v-zoo]*	硬骨類	kōkotsu-rui
telephone (telephoning) *{n} [comm]*	電話；でんわ	denwa
" (instrument) *{n} [comm]*	電話(機)	denwa(-ki)
" *{vb t} [comm]*	電話を掛ける	denwa o kakeru
" *{vb t} [comm]*	電話する	denwa suru
telephone answering machine *[comm]*	留守番電話	rusu'ban-denwa
telephone directory *[comm] [pr]*	電話帳	denwa'chō
telephone exchange *[comm]*	電話交換機	denwa-kōkan-ki
telephone line *[elec]*	電話線	denwa-sen
telephone number *[comm]*	電話番号	denwa-bangō
telephotography *[comm]*	望遠写真術	bōen-shashin'jutsu
" *[comm]*	写真電送	shashin-densō
telephoto lens *[opt]*	望遠レンズ	bōen-renzu
telephotometer *[eng]*	遠隔物光度測定器	enkaku'butsu-kōdo-sokutei-ki
teleprinter *[comm]*	印刷電信機	in'satsu-denshin-ki
telescope *{n} [opt] [eng]*	望遠鏡	bōen'kyō
telescopic antenna *[elecmg]*	伸縮式アンテナ	shin'shuku'shiki-antena
teleseismology *[geophys]*	遠隔地震学	enkaku-jishin'gaku
television *[electr]*	テレビ(ジョン)	terebi(jon)
television camera tube *[electr]*	撮像管	satsu'zō-kan
Telex; TEX *[comm]*	加入電信	ka'nyū-denshin
telluric line *[spect]*	地球大気線	chikyū-taiki-sen
tellurium (element)	テルル	teruru
telophase *[cytol]*	終期	shūki
temper *{n} [met]*	鍛錬；鍛鍊	tanren
temperament *[music]*	音律；おんりつ	on-ritsu
temperate climate *[climat]*	温帯気候；溫帶氣候	ontai-kikō
Temperate Zone *[climat]*	温帯	ontai
temperature *[thermo]*	温度；おんど	ondo
temperature coefficient *[phys]*	温度係数	ondo-keisū
temperature control *[[eng]*	温度制御	ondo-seigyo
temperature dependency *[sci-t]*	温度依存性	ondo-i'zon-sei
temperature gradient *[thermo]*	温度勾配	ondo-kōbai
temperature-humidity index *[meteor]*	温湿指数；溫濕指數	on-shitsu-shisū
temperature inversion *[meteor]*	温度逆転；溫度逆轉	ondo-gyaku'ten
temperature-rise coefficient *[mech-eng]*	温度上昇係数	ondo-jōshō-keisū

temperature scale *[thermo]*	温度標準	ondo-hyōjun
temperature-sensitive element *[electr]*	感温素子	kan'on-soshi
temperature-sensitive mutation *[gen]*	温度感受性変異	ondo-kanju'sei-hen'i
temper brittleness *[met]*	焼き戻し脆さ	yaki'modoshi-moro-sa
" *[met]*	焼き戻し脆性	yaki'modoshi-zeisei
temper color *[met]*	焼き戻し色	yaki'modoshi-iro
tempered glass *[mat]*	強化ガラス	kyō'ka-garasu
tempering *{n} [met]*	錬り; 錬り; ねり	neri
" *{n} [met]*	焼入れ; やきいれ	yaki'ire
" *{n} [met]*	焼き戻し	yaki'modoshi
temper rolling *{n} [met]*	調質圧延	chō'shitsu-atsu'en
template *[eng] [mol-bio]*	铸型; 鑄型	i'gata
" *[eng] [resin]*	型板	kata'ban; kata-ita
template reaction *[met]*	铸型反応	i'gata-hannō
temple *[arch]*	寺院	ji'in
temples and shrines *[arch]*	寺社	jisha
tempo (pl -s, tempi) *[music]*	緩急速度	kankyū-sokudo
temporal bone *[anat]*	側頭骨	soku'tō-kotsu
temporarily (for a while) *{adv}*	一時的に	ichiji'teki ni
temporary *{adj} [math]*	一時的な	ichiji'teki na
" *[math]*	一過性; いっかせい	ikka-sei
temporary; a time *{n}*	一時; いちじ	ichi'ji
temporary memory *[comput]*	一時(的)記憶	ichiji('teki)-ki'oku
temporary storage *[comput]*	一時(的)記憶	ichiji('teki)-ki'oku
tempura (Japanese dish) *[cook]*	天麩羅; てんぷら	tenpura
tenacity *[mech]*	粘り強さ	nebari-zuyo-sa
" *[tex]*	靭性; じんせい	jinsei
" *[tex]*	強靭性	kyōjin-sei
tendency; trend *{n} [math]*	傾向	keikō
tender (a ship) *{n} [nav-arch]*	補給船	hokyū'sen
tendon *[anat]*	腱; けん	ken
" *[anat]*	筋; すじ	suji
tendril *[bot]*	巻鬚; まきひげ	maki-hige
tenebrescence *[phys]*	変色蛍光; 變色螢光	henshoku-keikō
tennantite *[miner]*	砒四面銅鉱	hi'shimen'dōkō
" *[miner]*	砒黝銅鉱	hiyū'dōkō
tenon and mortise *[eng]*	枘と枘穴	hozo to hozo-ana
tenon joint *[adhes]*	枘継ぎ; ほぞつぎ	hozo-tsugi
tenorite *[miner]*	黒銅鉱; 黒銅鑛	koku'dōkō
ten *{n} [math]*	十; じゅう	jū
ten minutes	十分	juppun; jippun
ten's complement *[comput]*	十の補数	jū no hosū

tense; stretch *{vb}*	引張る；ひっぱる	hipparu
tensile breaking elongation; tensile fracture elongation *[mech]*	引張破断伸	hippari-hadan-nobi
tensile modulus *[mech]*	引張弾性率	hippari-dansei-ritsu
tensile product *[mech]*	抗張積	kōchō-seki
tensile residual stress *[mech]*	引張殘留応力	hippari-zanryū-ō'ryoku
tensile shear stress *[mech]*	引張剪断応力	hippari-sendan-ō'ryoku
tensile strength *[mech]*	引張強度	hippari-kyōdo
" *[mech]*	引張強さ	hippari-tsuyo-sa
" *[mech]*	抗張力	kōchō-ryoku
tensile stress *[mech]*	引張応力	hippari-ō'ryoku
tensile test; tension test *[mech]*	引張試験	hippari-shiken
tension *[mech]*	張力；ちょうりょく	chō'ryoku
" *[mech]*	引張力	hippari-ryoku
tension gage *[eng]*	張力計	chōryoku-kei
tension set *[mech] [poly-ch]*	永久伸び	eikyū-nobi
tension spring *[mech-eng]*	引張りばね	hippari-bane
tentacle; antenna; feeler *[i-zoo]*	触手；觸手	shoku'shu
" (of squid) *[i-zoo]*	触腕；しょくわん	shoku'wan
tentative standard *[ind-eng]*	仮標準；假標準	kari-hyōjun
" *[sci-t]*	仮規格；かりきかく	kari-kikaku
tenter *{n} [tex]*	横延伸機	yoko-enshin-ki
tentering *{n} [tex]*	幅出し；はばだし	haba'dashi
ten thousand *{n} [math]*	万；萬；まん	man
ten-thousandth of one *[math]*	糸；し	shi
ten thousand years	万年；萬年	mannen
tepal *[bot]*	花被片	kahi'hen
tephra *[geol]*	降下火山砕屑物	kōka-kazan-saisetsu'butsu
tepid; lukewarm *{adj} [thermo]*	生温い；なまぬるい	nama-nurui
teratogen *[med]*	催奇形性物質	sai'kikei'sei-busshitsu
teratology *[med]*	奇形学；きけいがく	kikei'gaku
terbium (element)	テルビウム	terubyūmu
terebra; anger shell *[i-zoo]*	筍貝；たけのこがい	takenoko-gai
term *{n} [alg] [comput] [spect]*	項；こう	kō
" (terminology) *{n} [comm]*	術語；じゅつご	jutsu'go
" (terminology) *{n} [comm]*	用語；ようご	yōgo
" (period; time limit) *{n}*	期限；きげん	kigen
terminal *{n} [comput]*	終端(装置)	shūtan(-sōchi)
" *{n} [comput] [elec]*	端末(装置)	tan'matsu(-sōchi)
" *{n} [comput] [elec]*	端子；たんし	tanshi
" *{n} [elec]*	終端；しゅうたん	shūtan
" *{n} [elec]*	端末；たんまつ	tan'matsu

terminal board *[elec]*	端子板	tanshi-ban
terminal box *[elec]*	終端箱	shūtan-bako
terminal bud *[bot]*	頂芽; ちょうが	chōga
terminal facilities *[civ-eng]*	臨港施設; 臨港施設	rinkō-shi'setsu
terminal group *[ch]*	末端基; まったんき	mattan-ki
terminal point *[math]*	終点; 終點	shūten
terminal side *[trig]*	終辺; 終邊	shūhen
terminal station *[rail]*	終着駅; 終着驛	shūchaku-eki
terminal strip *[elec]*	端子板	tanshi-ban
terminal transferase *[bioch]*	末端転移酵素	mattan-ten'i-kōso
terminal user *[comput]*	端末使用者	tan'matsu-shiyō'sha
terminal velocity *[phys]*	終端速度	shūtan-sokudo
terminal voltage *[elec]*	端子電圧	tanshi-den'atsu
terminating *{n} [elec]*	成端	seitan
terminating decimal *[arith]*	有限小数	yūgen-shōsū
termination *[comput]*	終結	shū'ketsu
" *[comput]*	終了; しゅうりょう	shūryō
termination reaction *[org-ch]*	停止反応	teishi-hannō
terminator *[astron] [geophys]*	明暗界線	mei'an-kaisen
" (terminating agent) *[mat]*	停止剤	teishi-zai
termite (insect) *[i-zoo]*	白蟻; しろあり	shiro-ari
tern; Sterna hirundo (bird) *[v-zoo]*	鯵刺; あじさし	aji'sashi
ternary *{adj} [logic] [math]*	三進の	san'shin no
" *{adj} [sci-t]*	三つ組の	mitsu'gumi no
ternary alloy *[met]*	三元合金	sangen-gōkin
ternary arrangement *[math]*	三元配置	sangen-haichi
ternary complex *[bioch]*	三重複合体	sanjū-fukugō'tai
ternary eutectic *[met]*	三元共晶	sangen-kyōshō
ternary notation *[math]*	三進法	sanshin-hō
ternate *{adj} [bot]*	三出の	san'shutsu no
terpenes *[ch]*	テルペン類	terupen-rui
terpolymer *[poly-ch]*	三元共重合体	sangen-kyōjūgō'tai
terra *[astron]*	月面陸地	getsu'men-riku'chi
terra alba *[mat]*	白土	haku'do
terrace *[geol]*	段丘; だんきゅう	dankyū
terra-cotta *[cer]*	土器; どき	do'ki
terrain *[agr] [eng]*	地形	chikei
" *[geol]*	地勢	chisei
terrain-following *{n} [navig]*	超低空飛行	chō'teikū-hikō
terrane *[geol]*	層群; 層群	sōgun
terrapin (tortoise) *[v-zoo]*	入江亀; いりえがめ	iri'e-game
terrazzo; artificial stone *[mat]*	人造石	jinzō'seki

terrestrial *{adj} [sci-t]*	地上の	chijō no
" *{adj} [zoo]*	陸生の	riku'sei no
terrestrial broadcasting *{n} [comm]*	地上放送	chijō-hōsō
terrestrial current *[geophys]*	地電流	chi-denryū
terrestrial gravitation *[geophys]*	地力; ちりょく	chi-ryoku
terrestrial heat *[geophys]*	地熱	chi-netsu; ji-netsu
terrestrial magnetism *[geophys]*	地磁気; 地磁氣	chi-jiki
terrestrial radiation *[geophys]*	地球放射	chikyū-hōsha
terrestrial stem *[bot]*	地上茎; 地上莖	chijō-kei
terrigenous sediment *[geol]*	陸源堆積物	riku'gen-taiseki'butsu
territoriality (a behavior) *[zoo]*	縄張(り)行動	nawa'bari-kōdō
territorial waters *[geogr]*	領海	ryōkai
territory *[math]*	領域; りょういき	ryō'iki
tertiary *{adj} [ch]*	第三級の	dai'san-kyū no
tertiary carbon atom *[ch]*	第三炭素原子	dai'san-tanso-genshi
tertiary compound *[ch]*	第三化合物	dai'san-kagō'butsu
test (testing) *{n} [ind-eng]*	試験; しけん	shiken
" ; try; sample *{vb t} [ind-eng]*	試す	tamesu
testa *[bot]*	種皮	shuhi
testicle (testis) *[anat]*	睾丸	kōgan
testing *{n} [ind-eng]*	検査; けんさ	kensa
" *{n} [ind-eng]*	検定; 檢定	kentei
testing apparatus *[eng]*	試験器具; 試驗器具	shiken-kigu
testis (pl testes) *[anat]*	睾丸; こうがん	kōgan
" *[anat]*	精巣; 精巢	seisō
test of significance *[stat]*	有意性検定	yū'i'sei-kentei
test paper *[ch]*	試験紙	shiken-shi
testpiece *[sci-t]*	試験片(試料片)	shiken'hen (shiryō'hen)
test statistic *[stat]*	検定統計量	kentei-tōkei'ryō
test tube *[an-ch]*	試験管	shiken-kan
test tube stand *[an-ch]*	試験管立	shiken'kan-tate
tetanus *[med]*	破傷風	hashō'fū
tetartohedrism *[crys]*	四半面性	shihan'men-sei
tetartohedry *[crys]*	四半面像	shihan'men-zō
tetrad *[bioch]*	四分子	shi'bunshi
" *[ch]*	四価元素	shika-genso
" *[math]*	四つ組; よつぐみ	yotsu'gumi
tetragonal crystal *[crys]*	正方晶	seihō-shō
tetragonal structure *[crys]*	四面構造	shimen-kōzō
tetragonal system *[crys]*	正方晶形	seihō'shō-kei
tetrahedrite *[miner]*	安四面銅鉱	an'shimen'dōkō
tetrahedron (pl -s, -hedra) *[geom]*	四面体; しめんたい	shimen'tai

tetrahydric alcohol *[org-ch]*	四価アルコール	yonka-arukōru
tetraiodine enneaoxide	九酸化四沃素	ku'sanka-shi'yōso
tetramer *[poly-ch]*	四量体	yonryō'tai; shiryō'tai
tetrapod; four-footed animal *[v-zoo]*	四足獣; 四足獸	yotsu'ashi-jū
Tetrapoda *[v-zoo]*	四肢動物	shishi-dōbutsu
tetrarsenic tetrasulfide	四硫化四砒素	shi'ryūka-shi'hiso
tetrasaccharides *[bioch]*	四糖類	yontō-rui
tetravalence *[ch]*	四価; 四價	yonka
tetravalent *{adj} [ch]*	四価の; よんかの	yonka no
tetrode *[electr]*	四極管	yon'kyoku-kan
tetrose *[bioch]*	四炭糖	shitan'tō; yontan'tō
text *[comput]*	文書	bunsho
textbook *[pr]*	教科書	kyōka'sho
textile *[mat]*	織物; おりもの	ori-mono
textile engineering; textiles *[tex]*	織物工学	ori'mono-kōgaku
textile finishing machine *[eng]*	織物仕上機	ori'mono-shi'age-ki
textile printing *{n} [graph]*	捺染	nasen; nassen
texture *[crys]*	集合組織	shūgō-so'shiki
" *[geol]*	石理; せきり	sekiri
" *[met]*	組織; そしき	so'shiki
" *[physio]*	感触	kanshoku
" (of skin) *[physio]*	肌理; きめ	kime
" (of cloth) *[tex]*	布目; ぬのめ	nuno'me
thalamus (pl -ami) *[anat]*	視床	shishō
thallium (element)	タリウム	taryūmu
thallophytes *[bot]*	葉状植物	yōjō-shokubutsu
that; those *{n}*	其れ; それ	sore
that; those *{adj}*	其の; その	sono
thaw *{vb} [climat]*	溶ける	tokeru
thawing *{n} [eng]*	解凍	kaitō
theater *[arch]*	劇場; げきじょう	geki'jō
thematic apperception test; TAT *[psy]*	課題統覚検査	kadai-tōkaku-kensa
thematic map *[graph]*	主題図	shudai-zu
then; in that case *{conj} [logic]*	それならば	sorenaraba
theodolite *[eng]*	経緯儀; 經緯儀	kei'i-gi
theorem *[logic]*	定理; ていり	teiri
theoretical mathematics *[math]*	理論数学	riron-sūgaku
theoretical number of plates *[ch-eng]*	理論段数	riron-dansū
theoretical physics *[phys]*	理論物理学	riron-butsuri'gaku
theory (pl -ries) *[math] [sci-t]*	理論; りろん	riron
" *[sci-t]*	説; せつ	setsu
theory of equations *[math]*	方程式論	hōtei'shiki-ron

theory of rate process *[ch] [phys]*	速度論；そくどろん	sokudo-ron
therapeutic index *[pharm]*	治療指数	chiryō-shisō
therefore *[conj]*	故に；ゆえに	yue ni
Theria; animals *[v-zoo]*	獣類；獸類	jū-rui
thermal aging *[n] [ch]*	熱老化	netsu-rōka
thermal analysis *[an-ch]*	熱分析	netsu-bunseki
thermal capacity *[thermo]*	熱容量	netsu-yōryō
thermal conductivity *[thermo]*	熱伝導度	netsu-dendō-do
" *[thermo]*	熱伝導率	netsu-dendō-ritsu
thermal convection *[meteor]*	熱対流；熱對流	netsu-tairyū
thermal converter *[electr]*	熱発電素子	netsu-hatsu'den-soshi
thermal cracking *[n] [ch]*	熱分解	netsu-bunkai
thermal cracking resistance *[mat]*	耐熱亀裂性	tai-netsu'kiretsu-sei
thermal decomposition *[ch]*	熱分解	netsu-bunkai
thermal degradation *[ch]*	熱劣化；ねつれっか	netsu-rekka
thermal denaturation *[ch]*	熱変性；熱變性	netsu-hensei
thermal derusting *[n] [met]*	熱脱錆	netsu-dassei
thermal diffusivity *[phys]*	熱拡散率	netsu-kakusan-ritsu
thermal efficiency *[ch-eng]*	熱効率	netsu-kō'ritsu
thermal emission *[therm]*	熱放射	netsu-hōsha
thermal equilibrium *[thermo]*	熱平衡	netsu-heikō
thermal equivalent *[thermo]*	熱当量；熱當量	netsu-tōryō
thermal expansion *[phys]*	熱膨張	netsu-bōchō
thermal neutron *[nucleo]*	熱中性子	netsu-chūsei'shi
thermal noise *[electr]*	熱雑音	netsu-zatsu'on
thermal plasticity *[poly-ch]*	熱可塑性	netsu-kaso-sei
thermal polymerization *[poly-ch]*	熱重合	netsu-jūgō
thermal power generation *[elec]*	火力発電	ka'ryoku-hatsu'den
thermal power plant *[mech-eng]*	火力発電所	ka'ryoku-hatsuden'sho
thermal resistivity *[thermo]*	熱抵抗率	netsu-teikō-ritsu
thermal stability *[phys]*	熱安定度	netsu-antei-do
" *[phys]*	熱安定性	netsu-antei-sei
thermal strain *[mech]*	熱歪；ねつひずみ	netsu-hizumi
thermal stress *[mech]*	熱応力；熱應力	netsu-ō'ryoku
thermal unit *[phys]*	熱単位	netsu-tan'i
thermic cumulus *[meteor]*	熱積雲	netsu-seki'un
thermion *[electr]*	熱電子	netsu-denshi
thermobalance *[eng]*	熱天秤	netsu-tenbin
thermochemical equation *[ch]*	熱化学方程式	netsu'kagaku-hōtei'shiki
thermochemistry *[p-ch]*	熱化学	netsu-kagaku
thermocline *[geophys]*	(水温)躍層	(sui'on-)yakusō
thermocouple *[eng]*	熱電対；熱でんつい	netsu-den'tsui

English	Japanese	Romanization
thermodynamic efficiency *[thermo]*	熱力学的効率	netsu'rikigaku'teki-kō'ritsu
thermodynamics *[phys]*	熱力学	netsu-riki'gaku
thermodynamics, first law of *[thermo]*	熱力学第一法則	netsu'rikigaku-dai'ichi-hōsoku
thermodynamics, second law of *[thermo]*	熱力学第二法則	netsu'rikigaku-dai'ni-hōsoku
thermodynamics, third law of *[thermo]*	熱力学第三法則	netsu'rikigaku-dai'san-hōsoku
thermoelasticity *[phys]*	熱弾性	netsu-dansei
thermoelectric *{adj} [phys]*	熱電の；ねつでんの	netsu'den no
thermoelectric generating element *[electr]*	熱発電素子	netsu-hatsu'den-soshi
thermoelectricity *[phys]*	熱電気；ねつでんき	netsu-denki
thermoelectric series *[met]*	熱電列	netsu'den-retsu
thermoelectron *[electr]*	熱電子	netsu-denshi
thermoelement *[electr]*	熱電対；熱電對	netsu-den'tsui
thermoforming *{n} [poly-ch]*	熱成形	netsu-seikei
thermogalvanometer *[eng]*	熱電検流計	netsu'den-kenryū-kei
thermogenesis *[thermo]*	熱発生	netsu-hassei
thermogram *[eng]*	温度記録図	ondo-kiroku-zu
thermograph *[eng]*	記録温度計	kiroku-ondo-kei
thermogravimetric analysis *[an-ch]*	熱重力分析	netsu-jū'ryoku-bunseki
thermohaline circulation *[ocean]*	熱塩循環	netsu'en-junkan
thermoluminescence *[atom-phys]*	温度冷光	ondo-reikō
thermomagnetism *[phys]*	熱磁気；ねつじき	netsu-jiki
thermomechanical effect *[fl-mech]*	熱力学効果	netsu-rikigaku-kōka
thermometer *[eng]*	寒暖計	kandan-kei
" *[eng]*	温度計	ondo-kei
thermometric titration *[an-ch]*	温度滴定	ondo-teki'tei
thermonuclear reaction *[nuc-phys]*	熱核反応	netsu'kaku-hannō
thermopile *[eng]*	熱電対列	netsu-den'tsui-retsu
thermoplastic *{adj} [poly-ch]*	熱可塑性の	netsu'kaso'sei no
thermoplastic resin *[mat]*	熱可塑性樹脂	netsu'kaso'sei-jushi
thermopolymerization inhibitor *[poly-ch]*	熱重合禁止剤	netsu'jūgō-kinshi-zai
thermoregulation *[physio]*	温調；おんちょう	onchō
thermorelay *[eng]*	熱電継電器	netsu'den-keiden-ki
thermos bottle *[eng]*	魔法瓶；まほうびん	mahō-bin
thermosensitive paper *[mat]*	感熱紙	kannetsu-shi
thermosetting property *[mat]*	熱硬化性	netsu'kōka-sei
thermosetting resin *[mat]*	熱硬化性樹脂	netsu'kōka'sei-jushi
thermosphere *[meteor]*	熱圏；熱圈	nekken; netsu-ken
thermostat *[eng]*	自動温度調節装置	jidō-ondo'chōsetsu-sōchi
" *[eng]*	恒温槽	kō'on-sō
" *[eng]*	整温器；整溫器	sei'on-ki

English	Japanese	Romanization
thermotropic liquid crystal *[p-ch]*	熱変性液晶	netsu'hensei-eki'shō
thesaurus (pl -sauri, -es) *[pr]*	類義語辞典	ruigi'go-jiten
" *[pr]*	類語集	ruigo'shū
thick *{adj} [mech]*	厚い；あつい	atsui
" *{adj} [mech]*	太い；ふとい	futoi
thicken *{vb t} [ch-eng]*	増粘する	zō'nen suru
thickener *[mat]*	糊料；こりょう	koryō
" *[mat]*	粘度付与剤	nendo-fuyo-zai
" *[mat]*	増粘剤；増粘劑	zōnen-zai
thickening *{n} [ch-eng]*	濃厚化	nōkō'ka
thickening property *[ind-eng]*	増粘性	zōnen-sei
thicket; bush; brush *{n} [ecol]*	藪；やぶ	yabu
thick film *[mat]*	厚膜	atsu-maku
thick-film conductor *[elec]*	厚巻導体；厚巻導體	atsu'maki-dōtai
thick-film wiring *{n} [elec]*	厚巻配線	atsu'maki-haisen
thick-haired codium *[bot]*	水松；海松；みる	miru
thickness *[math]*	厚み	atsu'mi
" *[math]*	厚さ	atsu-sa
thickness deviation *[eng]*	偏肉；へんにく	henniku
thickness gage *[eng]*	厚み計（厚さ計）	atsumi-kei (atsusa-kei)
" *[eng]*	隙間ゲージ	sukima-gēji
thigh *[anat]*	大腿	daitai
" *[anat]*	股；もも	momo
thin *{adj} {n} [math]*	細い；ほそい	hosoi
" (weak, dilute) *{adj}*	薄い；うすい	usui
" (as seedlings); cull *{vb t}*	間引く	ma-biku
" (dilute) *{vb t}*	薄める	usumeru
thin down; narrow *{vb t} [mech-eng]*	細める	hosomeru
thin film *[electr]*	薄膜；はくまく	haku-maku
thing; object; article *{n}*	物体；物體	buttai
" *{n}*	物	mono
thin-leaf shaped *{adj}*	薄片状の	haku'hen'jō no
think *{vb i} [psy]*	思う	omou
" *{vb i} [psy]*	考える；かんがえる	kangaeru
thinking *{n} [psy]*	考えること	kangaeru koto
thinner; reducer *[paint]*	薄め液	usume-eki
thinness *[tex]*	薄さ	usu-sa
thinning *{n} [agr]*	間引き；まびき	ma'biki
" (of trees) *{n} [ecol] [for]*	間伐	kanbatsu
thin paper *[mat]*	薄葉紙	usu-yōshi
thin section *[an-ch]*	薄片	haku-hen
thin sheet *[mat]*	薄板	haku-ban

English	Japanese	Romanization
Thiobacillus *[microbio]*	硫黄細菌	iō-saikin
third (one-third) *{n} [math]*	三分の一	san'bun no ichi
" (in order) *{adj} [math]*	第三の；だいさんの	dai'san no
third-order reaction *[p-ch]*	三次反応	sanji-hannō
third person; third party	第三者	dai'san'sha
thirst *{n} [physio]*	渇き（喉の）	kawaki (nodo no)
thirty *{n} [math]*	三十	sanjū
this *{n}*	此れ；是；之；これ	kore
" *{adj}*	此の；斯の；この	kono
thistle *[bot]*	薊；あざみ	azami
thixotropic resin *[resin]*	揺変性樹脂	yōhen'sei-jushi
thixotropy *[p-ch]*	チキソトロピー	chikisotropī
" *[p-ch]*	揺変性；搖變性	yōhen-sei
thoracic cavity *[anat]*	胸腔	kyōkō
thorax (pl thoraxes, thoraces) *[anat]*	胸；むね	mune
thorium (element)	トリウム	toryūmu
thorn *[bot]*	刺；棘；とげ	toge
thought (thinking; idea) *{n} [psy]*	考え	kangae
" *{n} [psy]*	思い	omoi
" *{n} [psy]*	思考；しこう	shikō
" *{n}* (idea) *[psy]*	思想	shisō
thousand *{n} [math]*	千	sen
thread (of screw) *{n} [eng]*	ねじ山	neji-yama
" *{n} [tex]*	糸；絲；いと	ito
thread-forming property *[tex]*	曳糸性	eishi-sei
thread guide *[tex]*	糸道	ito-michi
thread height (screw) *[eng]*	山の高さ	yama no taka-sa
thread strand *[tex]*	糸条；糸條	shijō
thread trailing property *[tex]*	曳糸性；えいしせい	eishi-sei
three *{n} [math]*	三；參	san
three-component system *[poly-ch]*	三液系	san'eki-kei
three days ago	一昨々日；さき＝おととい	saki-ototoi
three days from now	明々後日	myō'myōgo-nichi
three-dimensional *{adj} [math] [sci-t]*	立体（的）な	rittai('teki) na
" *{adj} [math] [sci-t]*	三次（元）の	sanji(-gen) no
three-dimensional display *[comput]*	三次元表示装置	sanji'gen-hyōji-sōchi
three-dimensional polymer *[poly-ch]*	三次元高分子	sanji'gen-kōbunshi
three dimensions *{n} [sci-t]*	三次元；さんじげん	sanji-gen
three-electron bond *[ch]*	三電子結合	san'denshi-ketsu'gō
three-fourths; three-quarters *{n} [arith]*	四分の三	yon'bun no san; shi'bun no san

three-phase circuit *[elec]*	三相回路	sansō-kairo
three-phase motor *[elec]*	三相電動機	sansō-dendō-ki
three-ply yarn *[tex]*	三子糸; みこいと	miko-ito
three-spined stickleback (fish) *[v-zoo]*	糸魚; 棘魚; いとよ	itoyo
three-toed sloth (animal) *[v-zoo]*	三つ指樹懶	mitsu'yubi-namake'mono
three-way valve; cross valve *[mech-eng]*	三方弁; 三方瓣	sanbō-ben
three-wire system *[elec]*	三線式	sansen-shiki
three years ago	一昨々年; さき=おととし	saki-ototoshi
thresher shark; thrasher; fox shark (fish) *[v-zoo]*	尾長鮫; おながざめ	o'naga-zame
threshold *[comput]*	敷居値; しきいち	shiki'i-chi
" *[math] [phys] [physio]*	閾(値)	iki(-chi); shiki'i(-chi)
" *[phys]*	臨界; りんかい	rinkai
threshold characteristic value	限界特性値	genkai-tokusei-chi
threshold of audibility *[acous]*	最小可聴値	saishō-kachō-chi
threshold value *[math]*	限界値; げんかいち	genkai-chi
thrip (insect) *[i-zoo]*	薊馬; あざみうま	azami'uma
throat *[anat]*	咽喉; いんこう	inkō
" *[anat]*	喉; のど	nodo
throttle *{n} [mech-eng]*	絞り; しぼり	shibori
throttle valve *[mech-eng]*	絞り弁; 絞り瓣	shibori-ben
throughput *[comput]*	処理能力	shori-nō'ryoku
" *[comput]*	処理量; 處理量	shori'ryō
throwing power *[met] [paint]*	付き廻り性	tsuki'mawari-sei
thrush (bird) *[v-zoo]*	鶫; つぐみ	tsugumi
thrust; propulsion *{n} [mech]*	推進力	suishin'ryoku
" ; punch; strike *{vb t}*	突く; つく	tsuku
thrust fault *[geol]*	突上げ断層	tsuki'age-dansō
thrust into; project into *{vb}*	突入する	totsu'nyū suru
thrust out; stretch *{vb}*	突っ張る	tsupparu
thulium (element)	ツリウム	tsuryūmu
thumb (of hand); big toe *{n} [anat]*	親指; おやゆび	oya-yubi
thumbtack; rivet *{n} [eng]*	鋲; びょう	byō
thunder *{n} [geophys]*	雷; かみなり	kaminari
" *{n} [meteor]*	雷鳴	raimei
thunderbolt *[geophys]*	雷電; らいでん	raiden
" *[meteor]*	落雷	rakurai
thundercloud *[meteor]*	雷雲; らいうん	rai'un
thunderstorm *[meteor]*	雷雨	rai'u
thunderstroke *[meteor]*	雷撃; らいげき	raigeki

English	Japanese	Romanization
Thursday	木曜日	moku'yōbi
thyme (herb) *[bot] [cook]*	立ち麝香草	tachi'jakō'sō
thymus (pl -es, thymy) *[anat]*	胸腺；きょうせん	kyō-sen
thyroid gland *[anat]*	甲状腺	kōjō-sen
tiara *[lap]*	頭飾り	atama-kazari
tibia (bone) *[anat]*	脛骨；けいこつ	kei-kotsu
-tic; -tical *{suffix}*	-的	-teki
tick (insect) *[i-zoo]*	壁蝨；蜱；だに	dani
tickle *{vb t} [physio]*	擽る；くすぐる	kusuguru
tidal action *[ocean]*	潮汐作用	chōseki-sayō
tidal bore *[ocean]*	潮津波；しおつなみ	shio-tsu'nami
tidal current *[ocean]*	潮流；ちょうりゅう	chōryū
tidal dissipation *[ocean]*	潮汐消散	chōseki-shōsan
tidal friction *[ocean]*	潮汐摩擦	chōseki-ma'satsu
tidal river *[hyd]*	有潮河川	yūchō-kasen
tidal table *[ocean]*	潮汐表	chōseki-hyō
tidal wave *[ocean]*	潮波；ちょうは	chō-ha
tidal wind *[meteor]*	潮汐風	chōseki-fū
tide *{n} [ocean]*	潮；うしお	ushio
tide embankment *[civ-eng]*	防潮提	bōchō'tei
tideland *[geogr]*	干潟；ひがた	hi'gata
tide level *[geol]*	潮位	chō-i
tide pool *[ocean]*	潮溜り；しおだまり	shio-damari
tide range *[ocean]*	潮差	chōsa
tide table *[ocean]*	潮汐表	chōseki-hyō
tidy; clear away *{vb t}*	片付ける	kata-zukeru
tie; sleeper *{n} [rail]*	枕木	makura'gi
tie-dyeing *{n} [tex]*	括り染め；絞り染め	kukuri-zome; shibori-zome
tiger (animal) *[v-zoo]*	虎；とら	tora
tiger lily (flower) *[bot]*	鬼百合；おにゆり	oni-yuri
tiger's eye; tigereye *[miner]*	虎眼石	tora-no-me-ishi
tight; strong; powerful *{adj}*	きつい	kitsui
tight-adhesion bending *{n} [eng]*	密着折曲げ	mitchaku-ori'mage
tight-adhesiveness *[phys]*	密着性	mitchaku-sei
tighten *{vb}*	締める；しめる	shimeru
tighten on *{vb t}*	締め付ける	shime-tsukeru
tight milling *{n} [rub]*	薄通し	usu-dōshi
tight rolling *{n} [met]*	密着圧延	mitchaku-atsu'en
tile (for roof) *{n} [mat]*	瓦；かわら	kawara
till (the soil) *{vb} [agr]*	耕す；たがやす	tagayasu
tillandsia *[bot]*	猿麻桛擬	saru'ogase'modoki
tiller *{n} [nav-arch]*	舵柄；だへい	dahei

English	Japanese	Romanization
tiltable *{adj} [eng]*	可傾式の	kakei'shiki no
tilt angle *[elecmg]*	傾斜角	keisha-kaku
timber; living tree(s) *[bot]*	立(ち)木；たちき	tachi'ki
" *[mat]*	材木；ざいもく	zaimoku
timberline *[ecol]*	樹木限界線	jumoku-genkai-sen
timber pond *[ecol]*	貯木池	cho'boku-chi
timber wolf (animal) *[v-zoo]*	森林狼	shinrin-ōkami
timbre *[acous]*	音色	ne'iro; onshoku
time *{n} [astron]*	時期	jiki
" *{n} [phys]*	時間；じかん	jikan
" *{n} [phys]*	時刻	jikoku
" *{n} [phys]*	時；とき	toki
" (interim; a while) *{n}*	間；あいだ	aida
time and labor *[ind-eng]*	手間；てま	tema
time constant *[phys]*	時定数	ji-teisū; toki-teisū
time contour map *[map]*	等時間線図	tō'jikan'sen-zu
time difference *[astron]*	時差；じさ	jisa
time dilation *[phys]*	時間膨張	jikan-bōchō
time division system *[comm]*	時分割方式	toki-bun'katsu-hō'shiki
timed relationship *[ind-eng]*	調時関係	chōji-kankei
time fuse *[eng]*	時限信管	jigen-shinkan
time instant; time; hour *{n} [phys]*	時刻	ji'koku
time lag *[phys]*	遅れ；遅れ；おくれ	okure
time-lapse photography *[photo]*	低速度撮影写真	tei'sokudo-satsu'ei-shashin
time limit; deadline *{n}*	期限	kigen
time occupancy *{n}*	時間占有率	jikan-sen'yū-ritsu
time point; point in time *[mech]*	時点；時點；じてん	ji-ten
timer *[electr]*	計時機構	keiji-kikō
time-release *{n}*	時間放出	jikan-hō'shutsu
times as much *{n} [math]*	倍；ばい	bai
time scale *[comput]*	時間目盛	jikan-me'mori
time series *[stat]*	時系列	toki-kei'retsu
time share *{vb} [comput]*	時分割する	ji-bun'katsu suru
time-sharing *{n} [comput]*	時分割；じぶんかつ	ji-bun'katsu
time signal *[comm]*	報時信号	hōji-shingō
time table *[trans]*	時刻表	ji'koku-hyō
time zone *[astron]*	時間帯；時間帯	jikan-tai
tin (element)	錫；すず	suzu
tin alloy *[met]*	錫合金	suzu-gōkin
tin arsenide	砒化錫	hika-suzu
tin borofluoride	硼弗化錫	hō'fukka-suzu
tin bromide	臭化錫	shūka-suzu

tin chloride	塩化錫；鹽化錫	enka-suzu
tin complex *[ch]*	錫錯体	suzu-sakutai
tincture *{n} [pharm]*	チンキ剤	chinki-zai
tine *[acous]*	叉；また	mata
tin fluoride	弗化錫；ふっかすず	fukka-suzu
tin fluorophosphate	弗化燐酸錫	fukka-rinsan-suzu
tin hot-dipping *{n} [met]*	溶融錫めっき	yōyū-suzu-mekki
tin hydride	水素化錫	suiso'ka-suzu
tinning *{n} [met]*	錫引き；すずびき	suzu'biki
tin oxide	酸化錫	sanka-suzu
tinplate *{n} [met]*	ブリキ	buriki
tin-plating *{n} [met]*	錫鍍金；錫めっき	suzu-mekki
tin silicofluoride	珪弗化錫	kei'fukka-suzu
tin sulfide	硫化錫	ryūka-suzu
tip (location) *{n}*	先；さき	saki
" (location) *{n}*	先端	sentan
tip part *{n}*	尖頭部；先頭部	sentō-bu
tire (of a vehicle) *{n} [eng]*	タイヤ	taiya
tire fabric *[tex]*	簾織；すだれおり	sudare-ori
tissue *[hist]*	組織；そしき	so'shiki
tissue culture *[cyt]*	組織培養	soshiki-baiyō
tissue paper *[mat] [paper]*	薄葉紙；うすようし	usu'yō-shi
tissue-specific antigen *[immunol]*	組織特異抗原	soshiki-toku'i-kōgen
titanium (element)	チタン	chitan
titer *[ch]*	力価；力價；りきか	riki-ka
title *[pr]*	題名	daimei
" *[pr]*	表題	hyōdai
title of invention *[pat]*	発明の名称	hatsu'mei no meishō
title page *[pr]*	扉；とびら	tobira
titmouse (bird) *[v-zoo]*	山雀；やまがら	yama'gara
titrant (solution) *[an-ch]*	滴定液	tekitei-eki
" (reagent) *[an-ch]*	滴定試薬	tekitei-shi'yaku
titration *[an-ch]*	滴定；てきてい	teki'tei
titration curve *[an-ch]*	滴定曲線	tekitei-kyoku'sen
titration ratio *[an-ch]*	滴定比	tekitei-hi
titrimetric analysis *[an-ch]*	滴定分析	tekitei-bunseki
toad (amphibian); bufo *[v-zoo]*	蟇蛙；ひきがえる	hiki'gaeru
toad; Buff vulgaris *[v-zoo]*	蝦蟇；がま	gama (hiki'gaeru)
toad lily (flower) *[bot]*	杜鵑草；ほととぎす	hototogisu'sō; hototogisu
toad poison *[pharm] [physio]*	蝦蟇毒	gama-doku
tobacco (pl -cos) *[bot]*	煙草；たばこ	tabako
toe *{n} [anat]*	足指	ashi-yubi

toffee (candy) *[cook]*	飴玉	ame-dama
toilet (commode) *[arch]*	便器	benki
" (rest room) *[arch]*	手洗い；てあらい	te'arai
toiletry *[mat]*	化粧品類	keshō'hin-rui
toilet soap *[mat]*	化粧石鹸	keshō-sekken
tokay (reptile) *[v-zoo]*	大守宮；おおやもり	ō-yamori
Tokyo *[geogr]*	東京；とうきょう	Tōkyō
tolerance (resistance) *[agr] [med]*	耐性	taisei
" *[eng]*	公差	kōsa
" *[eng]*	裕度；ゆうど	yūdo
" *[math]*	許容差；きょようさ	kyoyō-sa
tolerance limits *[eng] [stat]*	許容限界	kyoyō-genkai
tolerate *{vb i}*	我慢する	gaman suru
toll bridge *[civ-eng]*	有料橋	yūryō-kyō
tollgate *[civ-eng]*	通行料徴収門	tsūkō'ryō-chōshū-mon
" *[civ-eng]*	通行料料金所	tsūkō'ryō-ryōkin'jo
toll line (telephone) *[comm]*	市外線	shigai-sen
tollroad; turnpike *[civ-eng]*	有料道路	yūryō-dōro
tomato (pl -toes) (vegetable) *[bot]*	蕃茄；トマト	tomato
tomography *[electr]*	断層撮影法	dansō-satsu'ei-hō
tomorrow *{n}*	明日	ashita; myō'nichi
ton (unit of weight) *[mech]*	瓲；噸；トン	ton
tonality *[music]*	調性	chōsei
tone *[acous]*	音；おと	oto
tone color *[acous]*	音色；ねいろ	ne'iro
tone quality *[acous]*	音質	on-shitsu
" *[acous]*	音色；おんしょく	on-shoku
tongs *[eng]*	鋏；やっとこ	yattoko
tongue *{n} [anat]*	舌；した	shita
tongue and groove (joint) *[eng]*	実接ぎ；さねつぎ	sane'tsugi
tongue fabric (shoe) *[cl]*	舌布	shita-nuno
tongue piece (shoe) *[cl]*	舌片；ぜっぺん	zeppen
tongue sole (fish) *[v-zoo]*	牛の舌	ushi-no-shita
tonic *{n} [pharm]*	強壮剤	kyōsō-zai
toning *{n} [photo]*	調色	chōshoku
tonnage *[nav-arch]*	積量；せきりょう	seki'ryō
tonsils *{n} [anat]*	扁桃腺	hentō-sen
tool *{n} [eng]*	道具	dōgu
" *{n} [eng]*	工具；こうぐ	kōgu
" *{n} [eng]*	用具	yōgu
tool box *[eng]*	道具箱	dōgu-bako
tool steel *[met]*	工具鋼	kōgu-kō

tooth (pl teeth) *[anat] [dent]*	歯；齒；は	ha
" *[anat]*	歯牙；しが	shiga
toothache *[med]*	歯痛	shi'tsū; ha-ita
toothbrush *[dent]*	歯刷子；歯ブラシ	ha-burashi
tooth cavity *[dent]*	歯孔	shikō
tooth crown *[dent]*	歯冠；しかん	shikan
tooth decay *[dent]*	齲食；齲蝕	u'shoku
toothed *{adj} [bio]*	歯のある	ha no aru
toothed gear *[mech]*	歯車	ha-guruma
tooth flank (gear) *[mech]*	歯面	ha-men
tooth-mesh (with) *{vb} [mech]*	嚙合する	shigō suru
toothpaste *[pharm]*	練歯磨；練齒磨	neri-ha'migaki
toothpick *[eng]*	(爪)楊子；(爪)楊枝	(tsuma-)yōji
tooth profile *[mech]*	歯形	ha-gata
top (toy) *{n} [eng]*	独楽；獨樂；こま	koma
" (location) *{n} [math]*	上部	jōbu
topaz *[miner]*	琥珀；こはく	ko'haku
" *[miner]*	黄玉；おうぎょく	ō'gyoku
top coat *[paint]*	上塗り塗料	uwa'nuri-toryō
top color *[opt]*	上色；うわいろ	uwa-iro
topographic map *[map]*	地形図；地形圖	chikei-zu
topography *[geogr]*	地形	chikei
topology *[math]*	位相幾何学	isō-kika'gaku
" *[math]*	位相数学	isō-sūgaku
topsoil *[geol]*	表土	hyōdo
top speed *[nav-arch] [phys]*	最高速度	saikō-sokudo
top view *[graph]*	上面図	jōmen-zu
torch *{n} [opt]*	松明；炬火	taimatsu
torii (Shinto archway) *[arch]*	鳥居；とりい	tori'i
tornado *{n} [meteor]*	トルネード	torunēdo
toroid *[geom]*	円環面(体)	enkan'men(-tai)
" *[geom]*	円錐曲線回転面体	ensui-kyoku'sen-kaiten'men'tai
toroid; torus *[geom]*	環状面(体)	kanjō'men(-tai)
torpedo *{n} [ord]*	魚雷；ぎょらい	gyorai
" *{n} [ord]*	水雷	suirai
torrent *[hyd]*	激流；げきりゅう	geki'ryū
" *[hyd]*	渓流；溪流	keiryū
torrid zone *[climat]*	熱帯；熱帶	nettai
torsion *[mech]*	捩れ	nejire
" *[mech]*	捩り；ねじり	nejiri
" *[mech]*	捩率；れいりつ	rei-ritsu

torsional oscillation *[phys]*	捩り振動	nejiri-shindō
torsional rigidity *[mech]*	捩り剛性	nejiri-gōsei
" *[mech]*	捩り剛さ	nejiri-kowa-sa
torsional strength *[mech]*	捩り強さ	nejiri-tsuyo-sa
torsional vibration *[mech]*	捩り振動	nejiri-shindō
torsion balance *[eng]*	捩り秤	nejiri-bakari
torsion bar *[mech-eng]*	捩り棒; ねじりぼう	nejiri-bō
torso; trunk (of body) *{n} [anat]*	胴(体)	dō(-tai)
tortoiseshell *[v-zoo]*	鼈甲; べっこう	bekkō
tortoiseshell cat (animal) *[v-zoo]*	三毛猫; みけねこ	mike-neko
tortuosity; torsion; twist *{n} [mech]*	曲がり; まがり	magari
" *[mech]*	捩(じ)れ	nejire
torus; toroid *{n} [math]*	円環面(体)	enkan'men(-tai)
total; sum *{n} [arith]*	合計; ごうけい	gōkei
" *{n} [arith]*	計	kei
total absorption *{n} [opt]*	全吸収; 全吸收	zen-kyūshū
total amplitude *[phys]*	全振幅	zen-shin'puku
total carbon *[inorg-ch]*	全炭素; ぜんたんそ	zen-tanso
total differential *[calc]*	全微分; ぜんびぶん	zen-bibun
total distortion *[opt]*	全歪み; ぜんひずみ	zen-hizumi
total eclipse *[astron]*	皆既食; かいき食	kaiki-shoku
total moisture *[meteor] [p-ch]*	全水分	zen-suibun
total nitrogen *[inorg-ch]*	全窒素	zen-chisso
total organic carbon *[org-ch]*	全有機炭素	zen-yūki-tanso
total oxygen demand *[bio] [microbio]*	全酸素要求量	zen-sanso-yōkyū'ryō
total reflection *[opt]*	全反射	zen-hansha
total solids *[an-ch]*	全固形分	zen-kokei'bun
total sulfur *[inorg-ch]*	全硫黄; ぜんいおう	zen-iō
totipotency *[embryo]*	分化全能性	bunka-zennō-sei
toucan (bird) *[v-zoo]*	大嘴; おおはし	ō-hashi
touch; tactile sense *{n} [physio]*	触覚; 觸覺	shokkaku
" ; feel *{vb t}*	触れる; ふれる	fureru
" ; feel *{vb t}*	触る; さわる	sawaru
touching *{n} [physio]*	触れること	fureru koto
touchup (painting) *{n} [eng]*	繕い塗	tsukuroi-nuri
tough and hard steel *[steel]*	強靭鋼	kyōjin-kō
toughness *[mech]*	(強)靭性	(kyō)jin-sei
tourist *[trans]*	観光客	kankō-kyaku
" ; traveler; passenger *[trans]*	旅客	ryo'kaku; ryo'kyaku
tourmaline *[miner]*	電気石	denki'seki; denki-ishi
towboat; tugboat *[nav-arch]*	引船; ひきぶね	hiki-bune
towel; hand towel *{n} [tex]*	手拭い; てぬぐい	te'nugui

tower *{n} [arch]*	塔；とう	tō
" (turret; scaffold) *{n} [arch]*	櫓；矢倉	yagura
tower summit *[ch-eng]*	塔頂	tōchō
town *[geogr]*	町；街；まち	machi
tow rope *[eng]*	引き綱	hiki-zuna
toxic fish *[v-zoo]*	有毒魚	yū'doku-gyo
toxicity *{n} [pharm]*	毒性；どくせい	doku-sei
toxigenicity *[microbio]*	毒素産生性	dokuso-sansei-sei
toxin *[bioch]*	毒素	doku'so
toy *{n} [eng]*	玩具	omocha; gangu
trace *{n} [alg]*	跡；せき；あと	seki; ato
" *{n} [paleon]*	足痕	sokkon
" *{n} [sci-t]*	痕跡；こんせき	kon'seki
" *{vb} [comput]*	追跡する	tsui'seki suru
" (follow) *{vb}*	擦る；なぞる	nazoru
trace back to; retroact *{vb}*	溯る；さかのぼる	saka'noboru
trace component *[ch]*	微量成分	biryō-seibun
trace element *[geoch]*	微量元素	biryō-genso
tracer *[ch]*	追跡子	tsui'seki'shi
trachea (pl -cheae) *[anat]*	気管；きかん	kikan
Tracheophyta *[bot]*	維管束植物類	i'kansoku-shokubutsu-rui
trachyte *{n} [petr]*	粗面岩	so'men'gan
track (of flying object) *[aero-eng]*	飛跡；ひせき	hiseki
" *{n} [astron] [phys] [rail]*	軌道；きどう	kidō
" *{n} [paleon]*	足痕	sokkon
tracking profile mechanism *[eng]*	触針倣い機構	shoku'shin-narai-kikō
traction *[mech]*	牽引力	ken'in'ryoku
trade *{n} [econ]*	貿易；ぼうえき	bōeki
" *{n} [econ]*	営業；營業	eigyō
trademark *[ind-eng] [pat]*	商標	shōhyō
trademark registration *[pat]*	商標登録	shōhyō-tōroku
trade name *[pat]*	商号；しょうごう	shōgō
trade secret *[econ]*	営業秘密	eigyō-himitsu
trade wind *[meteor]*	貿易風	bōeki-fū
traffic *{n} [eng]*	交通；こうつう	kōtsū
traffic control *[traffic]*	交通制御	kōtsū-seigyo
traffic engineering *[civ-eng]*	交通工学	kōtsū-kōgaku
traffic light *[traffic]*	交通信号灯	kōtsū-shingō-tō
traffic signal *[traffic]*	交通信号	kōtsū-shingō
tragus *[anat]*	耳珠；じしゅ	jishu
trail *{n} [paleon]*	匐痕	fuk(u)kon
trailer *{n} [comput]*	後書き	ato'gaki

trailing edge *[aero-eng]*	後縁	kōen
train *{n} [astron]*	尾；お	o
" *{n} [mech-eng]*	列；れつ	retsu
" *{n} [phys]*	連続；連續	renzoku
" *{n} [rail] [transp]*	列車；れっしゃ	ressha
training *{n}*	訓練；訓練	kunren
" *{n}*	鍛錬；鍛錬	tanren
trajectory *[mech]*	弾道；彈道	dandō
" *[mech]*	軌線	kisen
" *[meteor]*	流跡線	ryūseki-sen
tram *[trans]*	電車	densha
tranquilizer *{n} [pharm]*	鎮静薬（鎮静剤）	chinsei-yaku; (chinsei-zai)
" *{n} [pharm]*	精神安定薬	seishin-antei-yaku
trans- *{prefix}*	を越えて-	o koete-
" *{prefix}*	横切って-	yoko'gitte-
transaction *[comput] [econ]*	取引き；とりひき	tori'hiki
transcendental number *[arith]*	超越数	chō'etsu-sū
transcribe *{vb} [comput]*	転記する	tenki suru
" *{vb} [comput]*	転写する	tensha suru
transcription *[comput]*	書き直し	kaki'naoshi
" *[graph] [mol-bio]*	転写；轉寫	tensha
transducer *[eng]*	変換器；變換器	henkan-ki
transduction *[microbio]*	形質導入	kei'shitsu-dō'nyū
transfer *{n} [bio]*	転移；てんい	ten'i
" *{n} [comm]*	伝達；傳達	den'tatsu
" *{n} [comput]*	転送	tensō
" *{n} [econ]*	振替；ふりかえ	furi'kae
" *{n} [graph]*	転写	tensha
" *{n} [trans]*	移送	isō
" *{n} [trans]*	移転	iten
" *{vb t} [comput]*	転送する	tensō suru
" *{vb t}*	移す；うつす	utsusu
transferable *{n} [pat]*	譲渡可能	jōto-kanō
transferase *[bioch]*	転位酵素	ten'i-kōso
transference *[math]*	移行	ikō
transference number *[p-ch]*	輸率；ゆりつ	yu-ritsu
transfer function *[cont-sys]*	伝達関数；傳達關數	den'tatsu-kansū
transfer paper *[graph]*	転写紙	tensha-shi
transfer printing *{n} [graph] [pr]*	転写；轉寫	tensha
transfer rate *[comput]*	転送率	tensō-ritsu
transfer rate (or speed) *[comput]*	転送速度	tensō-sokudo
transferred meaning *[comm]*	転義	tengi

transfer ribonucleic acid; tRNA *[mol-bio]*	転移RNA	ten'i-RNA
transform *{vb} [comput]*	変形する	henkei suru
transformant *[cyt]*	形質転換細胞	kei'shitsu-tenkan-saibō
transformation *[bio]*	変態；へんたい	hentai
" *[ch]*	化成	kasei
" *[comput] [math]*	変形；變形	henkei
" *[cyt]*	(形質)転換	(kei'shitsu-)tenkan
" *[mech-eng]*	変成	hensei
" *{n}*	転化；轉化；てんか	tenka
transformation efficiency *[cyt]*	転換効率	tenkan-kō'ritsu
transformation point *[met]*	変態点；變態點	hentai-ten
transformer *[elec]*	変成器；變成器	hensei-ki
" *[elecmg]*	変圧器；變壓器	hen'atsu-ki
transfusion (of blood) *[med]*	輸血；ゆけつ	yu'ketsu
transgenic animal *[gen] [v-zoo]*	遺伝子導入動物	idenshi-dō'nyū-dōbutsu
transient *{n}*	一時	ichi'ji
" *{adj}*	一時の	ichi'ji no
transient response *[phys]*	過渡応答	kato-ō'tō
transient solution *[math]*	一時解；いちじかい	ichi'ji-kai
transient state *[elec]*	過渡状態	kato-jōtai
transistor *[electr]*	トランジスター	toranjistā
transit (tool for surveying) *[eng]*	転鏡儀	tenkyō-gi
transit circle *[eng]*	子午環；しごかん	shigo-kan
transition *[comm]*	渡り	watari
" *[comput]*	遷移；せんい	sen'i
" *[met] [thermo]*	転移	ten'i
" *{n}*	過渡；かと	kato
transitional *{adj}*	推移的な	sui'i'teki na
transition element *[ch]*	遷移元素	sen'i-genso
transition metal *[ch]*	遷移金属；遷移金屬	sen'i-kinzoku
transition point *[thermo]*	転移点	ten'i-ten
transitive *{adj} [geom]*	推移的な	sui'i'teki na
transitive law *[math]*	推移律	sui'i-ritsu
transitive property *[alg]*	推移的な性質	sui'i'teki na seishitsu
transitive verb *[gram]*	他動詞；たどうし	ta-dōshi
transit passenger *[transp]*	通過旅客	tsūka-ryo'kyaku
transit time *[trans]*	走行時間	sōkō-jikan
translate for meaning; translate freely *{vb} [comm]*	意訳する；意譯する	i'yaku suru
translate for the words; translate strictly *{vb} [comm]*	語訳する；ごやく=する	go'yaku suru

translation *[an-geom]*	平行移動	heikō-idō
" *[an-geom] [mech]*	並進；へいしん	hei'shin
" *[comm] [mol-bio]*	翻訳；飜譯	hon'yaku
" *[comput] [math]*	変換；變換	henkan
translational motion *[mech]*	並進運動	heishin-undō
translator (a person) *[comm]*	翻訳者；ほんやく者	hon'yaku'sha
transliteration *[comput]*	字訳	ji'yaku
" *[comput] [ling]*	音訳	on'yaku
translocation *[bioch]*	転位；轉位；てんい	ten'i
" *[gen]*	転座	tenza
translucency *[opt]*	透光性	tōkō-sei
" *[opt]*	透視度；とうしど	tōshi-do
translucent body *[opt]*	半透明体	han'tōmei-tai
transmissibility *[mech]*	伝達率；傳達率	den'tatsu-ritsu
transmission *[acous] [opt]*	伝導；でんどう	dendō
" *[comm]*	伝信；でんしん	denshin
" *[comm]*	伝送；傳送；電送	densō
" *[electr]*	送受信	sō'jushin
" *[electr]*	送信	sōshin
" *[mech-eng]*	動力伝達装置	dō'ryoku-den'tatsu-sōchi
" *[mech-eng]*	変速機	hensoku-ki
" *[opt]*	透過；とうか	tōka
transmission coefficient *[opt]*	透過係数	tōka-keisū
transmission electron microscope *[electr]*	透過(型)電子顕微鏡	tōka('gata)-denshi-kenbi'kyō
transmission factor *[phys]*	透過率	tōka-ritsu
transmission line *[comm]*	伝送線(路)	densō-sen(ro)
" *[elec]*	送電線(路)	sō-densen(-ro)
transmission loss *[comm]*	伝送損失	densō-son'shitsu
transmission network *[comm]*	送電網	sōden-mō
transmissivity *[elecmg]*	透過率	tōka-ritsu
transmit *(vb t) [comm]*	伝送する	densō suru
transmittance *[an-ch] [opt] [phys]*	透過率	tōka-ritsu
transmitted light *[opt]*	透過光	tōka-kō
transmitted light microscopy *[opt]*	透過光顕微鏡法	tōka'kō-kenbikyō-hō
transmitter *[comm] [electr]*	伝送機；でんそうき	densō-ki
" *[comm] [electr]*	送信機	sōshin-ki
" *[comm] [electr]*	通信機	tsūshin-ki
transmutation *[ch]*	変移；へんい	hen'i
" *[ch]*	変素反応	henso-hannō
" *[ch]*	転換	tenkan
" *[plas] [rub]*	変性	hensei

transom *[arch]*	欄間；らんま	ran'ma
transonic speed *[fl-mech] [phys]*	遷音速	sen'on-soku
transparency *[graph]*	透明画；透明畫	tōmei-ga
" *[opt]*	透視度	tōshi-do
" (degree of) *[phys]*	透明度	tōmei-do
transparent *{n} [phys]*	透明；とうめい	tōmei
" *{adj} [phys]*	透明な	tōmei na
transparent body *[opt]*	透明体	tōmei-tai
transpassive region *[p-ch]*	過受動態域	ka'judōtai-iki
transpiration *[bio]*	発散；發散	hassan
" *[bio]*	蒸散；じょうさん	jōsan
transplant *{n} [med]*	移植体；移植體	i'shoku'tai
transplantation *[bio] [bot]*	移植	i'shoku
transponder *[comm]*	応答機；應答機	ō'tō-ki
transport; transportation *{n} [trans]*	移送	isō
" *{n} [trans]*	輸送；ゆそう	yusō
transporting; transfer *{n} [trans]*	運搬；うんぱん	unpan
transport number *[p-ch]*	輸率	yu-ritsu
transpose *{vb t} [alg] [stat]*	移項する	ikō suru
" *{vb t} [alg] [gram]*	転置する	tenchi suru
transposition *[comm]*	位置転換	ichi-tenkan
" *[math]*	互換	gokan
" *{n}*	転位；轉位；てんい	ten'i
transshipment *[trans]*	積替；つみかえ	tsumi'kae
transversal *{adj} [geom]*	横断線の	ōdan-sen (no)
transverse *{n} [math] [phys]*	横；よこ	yoko
transverse direction *[plas]*	横方向	yoko-hōkō
transverse effect *{n}*	横効果	yoko-kōka
transverse flow *[poly-ch]*	横断流	ōdan-ryū
transverse magnetization *[eng-acous]*	横方向磁化	yoko-hōkō-jika
transverse tensile strength *[mech]*	横引張強さ	yoko-hippari-zuyo-sa
transverse vibration *[mech]*	横振動	yoko-shindō
transverse wave *[phys]*	横波	yoko-nami
trap *{n} [aero-eng] [nav-arch]*	昇降用梯子	shōkō'yō-hashigo
" *{n} [comput]*	割(り)込み	wari'komi
" ; snare *{n} [eng]*	罠；わな	wana
trap circuit *[elec]*	罠回路	wana-kairo
trapdoor *[arch]*	跳ね上げ戸	hane'age'do
" *[arch]*	落とし戸	otoshi'do
trapdoor spider *[i-zoo]*	戸閉蜘蛛	to'tate-gumo
trapezohedron *[crys] [geom]*	偏四角面体	hen'shikaku'men-tai
trapezoid *{n} [geom]*	台形；だいけい	daikei

trapezoidal distortion *[electr]*	台形歪	daikei-hizumi
trapped light *[opt]*	捕獲された光	hokaku sareta hikari
trash *{n} [ecol]*	屑；くず（くづ）	kuzu
trauma *{n} [med]*	外傷	gaishō
travel; trip *{n} [trans]*	旅行；りょこう	ryokō
travel bureau *[trans]*	旅行案内所	ryokō-annai'jo
traveling microscope *[opt]*	運動顕微鏡	undō-kenbi'kyō
" *[opt]*	遊動顕微鏡	yūdō-kenbi'kyō
travel time *[geophys]*	走時	sōji
traverse *{n} [mech-eng]*	横行；おうこう	ō'kō
traverse sailing *[navig]*	連針路航法	renshin'ro-kōhō
travertine *{n} [geol]*	湖生白亜	kosei-haku'a
tray *[an-ch]*	棚段；たなだん	tana-dan
" *[ch-eng] [p-ch]*	段	dan
" *[cook]*	盆；ぼん	bon
treat (process) *{vb t} [ind-eng]*	処理する	shori suru
" (remedy) *{vb t} [med]*	治療する	chiryō suru
treatment *[comput] [ind-eng]*	処理；處理；しょり	shori
" *[med]*	治療；ちりょう	chiryō
treaty *{n} [legal]*	条約；條約	jōyaku
treble *{adj} [math]*	三倍の	sanbai no
tree *[bot]*	樹木	ju'moku
" *[bot]*	木；樹；き	ki
tree creeper (bird) *[v-zoo]*	木走り	ki'bashiri
treefrog (amphibian) Hyla arborea *[v-zoo]*	雨蛙；あまがえる	ama-gaeru
tree line *[ecol]*	高木限界線	kōboku-genkai-sen
tree resin *[bot] [resin]*	脂；やに	yani
trefoil (vegetable) *[bot]*	三葉	mitsu'ba
tremblor *[geophys]*	地震；じしん	ji'shin
tremolite *[miner]*	透（角）閃石	tō(kaku)'senseki
tremor *[geophys]*	振動	shindō
trench *{n} [civ-eng]*	塹壕；ざんごう	zangō
" *{n} [geogr]*	濠	gō
" ; trough (ocean) *{n} [geol]*	海溝	kaikō
trend *{n} [stat]*	傾向；けいこう	keikō
trepang *[i-zoo]*	海鼠；なまこ	namako
triacidic base *[ch]*	三酸塩基	san'san-enki
triad *[math]*	三つ組	mitsu'gumi
trial *[ind-eng]*	試験；しけん	shiken
" *[ind-eng]*	試行	shikō
" *[legal]*	裁判	saiban

trial *[legal]*	審判	shinpan
" ; attempt *(n)*	試み；こころみ	kokoromi
trial and error *[psy]*	試行錯誤	shikō-sakugo
trial calculation *[math]*	試算	shisan
trial decision; judgment *[pat]*	判決；はんけつ	han'ketsu
trial of invalidation *[legal]*	無効審判	mukō-shinpan
trial production *[ind-eng]*	試作	shi'saku
triangle (tool) *[eng]*	三角定規	sankaku-jōgi
" *[geom]*	三角(形)	sankaku(-kei)
triangle of forces *[mech]*	力の三角形	chikara no sankak(u)kei
triangle scale *[graph]*	三角尺	sankaku-jaku
triangular coordinates *[math]*	三角座標	sankaku-zahyō
triangular diagram *[ch-eng] [geol]*	三角図表；三角圖表	sankaku-zuhyō
triangular prism *[geom]*	三角柱	sankaku-chū
triangulation *[eng] [navig]*	三角測量	sankaku-soku'ryō
triaxial stress *[mech]*	三軸応力；三軸應力	san'jiku-ō'ryoku
tribasic acid *[ch]*	三塩基酸	san'enki-san
tribasic ammonium citrate	枸櫞酸三アンモニウム	ku'ensan-san'anmonyūmu
tribasic sodium phosphate	燐酸三ナトリウム	rinsan-san'natoryūmu
tribe *[bot] [syst]*	連；れん	ren
" *[bio] [syst]*	族；ぞく	zoku
triboelectricity *[elec]*	摩擦電気	masatsu-denki
Tribolodon (fish) *[v-zoo]*	鯎；うぐい	ugui
tribology *[phys]*	摩擦学	masatsu'gaku
tribophysics *[[phys]*	摩擦物理学	masatsu-butsuri'gaki
tributary (pl -taries) *[hyd]*	支流	shiryū
Trichosanthe cucumeroides *[bot]*	烏瓜；からすうり	karasu-uri
Trichosanthes kirilowii Maxim *[bot]*	括楼；括樓；栝樓	katsurō
Trichosanthes Radix (root) *[bot]*	括楼根；瓜呂根 か(つ)ろ(う)こん	karokon; karōkon; katsurō-kon
trichotomy *(n)*	三分法	sanbun-hō
trichromatic coefficient *[opt]*	三色係数	sanshoku-keisū
trichromatic expression *[opt]*	三色表示	sanshoku-hyōji
trichromatic system *[opt]*	三色系	sanshoku-kei
trickling filter bed *[civ-eng]*	点滴沪床；點滴濾床	tenteki-roshō
triclinic system *[crys]*	三斜晶形	sansha-shōkei
tricycle *[transp]*	三輪車	sanrin'sha
tricyclic compound *[org-ch]*	三環式化合物	sankan'shiki-kagō'butsu
trident *(n) [math]*	三叉曲線	sansa-kyoku'sen
tridymite *[miner]*	鱗珪石	rin'keiseki
trifoliate orange *[bot]*	枳殻；からたち	karatachi

trigger *{n} [eng]*	制動機	seidō-ki
" *{n} [ord]*	引金；ひきがね	hiki'gane
" *{vb t}*	付勢する	fusei suru
trigonal *{adj} [crys]*	三方晶形の	sanbō'shōkei no
" (triangular) *{adj} [geom]*	三角形の	sankak(u)kei no
trigonal bipyramid *[geom]*	三方両錐	sanbō-ryōsui
" *[geom]*	三角両錐	sankaku-ryōsui
trigonal (triangular) **prism** *[geom]*	三角柱	sankaku-chū
trigonal pyramid *[geom]*	三方錐	sanbō-sui
" *[geom]*	三角錐	sankaku-sui
trigonal scalenohedron *[geom]*	三方偏三角面体	sanbō-hen'sankaku'men-tai
trigonal system *[crys]*	三方晶形	sanbō-shōkei; sanpō-shōkei
trigonal trapezohedron *[geom]*	三方偏四角面体	sanbō-hen'shikaku'men-tai
trigonometric function *[trig]*	三角関数	sankaku-kansū
trigonometric identity *[trig]*	三角恒等式	sankaku-kōtō-shiki
trigonometric parallax *[astron]*	三角視差	sankaku-shisa
trigonometric ratio *[geom]*	三角比	sankaku-hi
trigonometry *[math]*	三角法	sankaku-hō
triiron tetroxide	四三酸化鉄	shi'san-sanka-tetsu
trilateration *[eng]*	三辺測量；三邊測量	sanpen-soku'ryō
trilead tetroxide	四酸化三鉛	shi'sanka-san'namari
trillion (U.S.) *[math]*	兆；ちょう	chō
" (England) *[math]*	百万兆	hyaku'man'chō
trim (a state) *[aero-eng]*	平衡状態	heikō-jōtai
" (of automobile) *{n} [mat]*	外装品	gaisō'hin
" (interior trim) *{n} [mat]*	内装；ないそう	naisō
trimer *[poly-ch]*	三量体	sanryō'tai
Trimeresurus flavoviridis; habu snake (reptile) *[v-zoo]*	波布；飯匙倩；はぶ	habu
trimmings *{n}*	耳；みみ	mimi
trimorphism *[bio]*	三形；さんけい	sankei
trinomial *[alg]*	三項式	sankō-shiki
triode *[electr]*	三極(真空)管	sankyoku(-shinkū)-kan
triose *{n} [bioch]*	三炭糖	santan'tō
triphylite *[miner]*	三燐石	sanrin'seki
triple *{adj} [math]*	三倍の	sanbai no
" *{adj}*	三重の	san'jū no
" *{vb i}*	三倍になる	sanbai ni naru
" *{vb t}*	三倍にする	sanbai ni suru
triple bond *[org-ch] [p-ch]*	三重結合	sanjū-ketsugō
triple point *[p-ch]*	三重点	sanjū-ten
triplet *[bioch]*	三塩基連鎖	san'enki-rensa

triplets *[bio]*	三生児	sansei'ji
" *[bio]*	三つ子；みつご	mitsu'go
tripod *[eng]*	三脚	san'kyaku
tripod kettle *[met]*	鼎；かなえ	kanae
trisaccharide *[bioch]*	三糖(類)	santō(-rui)
trisect *{vb} [math]*	三等分する	san'tōbun suru
trisecting method *[math]*	三分法	sanbun-hō
trisection *[math]*	三等分	san'tōbun
trisoctahedron *[geom]*	二十四面体	nijūshi'men'tai
tristimulus values *[opt]*	三刺激値；三刺戟値	san'shigeki-chi
tritium *[nuc-phys]*	三重水素	sanjū'suiso
triton (shellfish) *[i-zoo]*	法螺貝；ほらがい	hora-gai
" *[nuc-phys]*	三重陽子	sanjū'yōshi
trivalent *{adj} [ch]*	三価の；三價の	sanka no
trivial name *[bio]*	常用名	jōyō-mei
trochophore larva *[i-zoo]*	担輪子幼生	tanrin'shi-yōsei
trolley (pl -s, trollies) *[mech-eng]*	市街電車	shigai-densha
trolley car *[trans]*	電車；でんしゃ	densha
trophocyte *[zoo]*	栄養細胞；榮養細胞	eiyō-saibō
tropic *{n} [astron] [geod]*	回帰線；かいきせん	kaiki-sen
tropical *{adj} [climat]*	熱帯の；ねったいの	nettai no
tropical depression *[meteor]*	熱帯低気圧	nettai-tei'ki'atsu
tropical fish *[v-zoo]*	熱帯魚	nettai-gyo
tropical month *[astron]*	分点月	bunten-getsu
tropical oil *[mat]*	熱帯油；熱帶油	nettai-yu
tropical rainforest *[ecol]*	熱帯多雨林	nettai-ta'u-rin
tropical storm *[meteor]*	熱帯暴風	nettai-bōfū
tropical year *[astron]*	回帰年	kaiki-nen
tropical zone *[climat]*	熱帯；ねったい	nettai
tropic of Cancer *[geod]*	北回帰線	kita-kaiki-sen
tropic of Capricorn *[geod]*	南回帰線	minami-kaiki-sen
tropics *[climat]*	熱帯地方	nettai-chihō
tropism *[bio] [bot]*	向性	kōsei
" *[bio]*	屈性	kussei
tropopause *[meteor]*	圏界面	kenkai-men
troposphere *[meteor]*	対流圏；對流圏	tairyū-ken
trouble *{n} [ind-eng]*	事故；じこ	jiko
" *{n} [ind-eng]*	故障；こしょう	koshō
troubleshoot *{vb} [comput]*	障害探究する	shōgai-tankyū suru
troubleshooter (a person) *{n}*	故障検査人	koshō-kensa'nin
troubleshooting *{n} [comput]*	障害探究	shōgai-tankyū
" *{n} [comput]*	障害追究	shōgai-tsuikyū

English	Japanese	Romanization
trough *[geol]*	地溝	chikō
" *[meteor]*	気圧の谷	ki'atsu no tani
" (of pressure) *[meteor]*	谷；たに	tani
" *[ocean]*	海溝；かいこう	kaikō
trough line *[meteor]*	谷線	tani-sen
trough shell; Tresus keenae *[i-zoo]*	海松貝；水松貝	miru-gai
" (shellfish) *[i-zoo]*	海松食；水松食	miru'kui
trousers (Japanese) *[cl]*	袴；はかま	hakama
" *[cl]*	ズボン	zubon
trout; Oncorhynchus masou (fish) *[v-zoo]*	鱒；ます	masu
trowel *{n}* *[eng]*	鏝；こて	kote
trowel coating *[civ-eng]*	鏝塗り	kote-nuri
truck; pushcart; flatcar *{n}* *[mech-eng]*	台車；臺車	daisha
" *{n}* *[mech-eng]*	貨物自動車	ka'motsu-jidōsha
" *{n}* *[mech-eng]*	運搬車	unpan'sha
truck farm; vegetable farm *[agr]*	市場向け野菜農場	shijō'muke-yasai-nōjō
true *{adj}* *[logic]*	真の；しんの	shin no; makoto no
true complement *[math]*	真の補数	shin no hosū
true fruit *[bot]*	真正果実	shinsei-ka'jitsu
true density *[mech]*	真密度	shin-mitsudo
true heading *{n}* *[navig]*	機首直方位	kishu-choku'hō'i
trueing; truing *{n}* *[mech-eng]*	整形	seikei
tru(e)ing apparatus *[eng]*	定寸装置	teisun-sōchi
true measure *[sci-t]*	実尺；じっしゃく	jisshaku
true north *[navig]*	真北	ma-kita; shin'poku
true porcelain *[cer]*	硬磁器；こうじき	kō-jiki
true solar time *[astron]*	真太陽時	shin-taiyō-ji
true specific gravity *[mech]*	真比重	shin-hijū
true value *[math]*	真の値	shin no atai
truffle *[bot]* *[cook]*	松露；しょうろ	shōro
trumpet; bugle *{n}* *[music]*	喇叭；らっぱ	rappa
trumpet shell *[i-zoo]*	法螺貝；ほらがい	hora-gai
truncated cone *[geom]*	円錐台；圓錐臺	ensui'dai
truncated pyramid *[geom]*	角錐台	kakusui'dai
truncation *[math]*	切り捨て	kiri'sute
" *[comput]*	切ること	kiru koto
" *[comput]*	打切り	uchi'kiri
truncation error *[comput]*	切り捨て誤差	kiri'sute-gosa
" *[comput]*	打切り誤差	uchi'kiri-gosa
trunk (of body) *{n}* *[anat]*	胴(体)	dō(-tai)
" (of tree) *{n}* *[bot]*	幹；みき	miki

trunk circuit *[elec]*	中継(線)回路	chūkei(-sen)-kairo
trust (charge; commission) *{n}*	委託	i'taku
trustee *[legal]*	保管者	hokan'sha
" *[legal]*	管財人	kanzai'nin
truth; reality; fact *{n}*	真実; 真實	shin'jitsu
truth or falsehood *{n}*	真偽; 眞僞; しんぎ	shingi
truth table *[logic]*	真理値表	shinri'chi-hyō
truth value logic]	真理値	shinri-chi
Trypanosoma *[i-zoo]*	睡眠病病原虫	suimin'byō-byōgen'chū
try square *[math]*	直角定規	chokkaku-jōgi
tsu-sound mark *[pr]*	促音符; そくおんぷ	soku'on-pu
tube *[eng]*	管; くだ	kuda
" *[math]*	筒; つつ	tsutsu
tuber *[bot]*	塊茎; 塊莖	kaikei
tubercle *[met]*	錆瘤; さびこぶ	sabi-kobu
tuberculostatic drug *{n} [pharm]*	抗結核剤	kō'kekkaku-zai
tubular bowl centrifuge *[eng]*	円筒形遠心分離機	entō'gata-enshin-bunri-ki
Tuesday	火曜日	ka'yōbi
tufa *[geol]*	石灰華; せっかいか	sekkai'ka
tuff *[geol]*	凝灰岩	gyōkai'gan
tugboat; towboat *[nav-arch]*	引舟; ひきぶね	hiki-bune
tumbling granulator *[mech-eng]*	転動造粒装置	tendō-zōryū-sōchi
tumeric *[bot]*	鬱金; うこん	u'kon
tumor *[med]*	腫瘍	shuyō
tuna (fish) *[v-zoo]*	鮪; まぐろ	maguro
" (fish) *[v-zoo]*	鮪; しび	shibi
tuna abdomen flesh *[cook]*	とろ	toro
tuna liver oil *[mat]*	鮪肝油	maguro-kanyu
tundra *[ecol]*	凍土帯	tōdo-tai
tuned circuit *[electr]*	同調回路	dōchō-kairo
tung oil *[mat]*	桐油; とうゆ	tō-yu
tungsten (element)	タングステン	tangusuten
tunicates; Tunicata *[i-zoo]*	尾索類	bisaku-rui
tuning *{n} [acous]*	調律; ちょうりつ	chō'ritsu
" *{n} [comput] [electr]*	調整	chōsei
tuning fork *[acout]*	音叉; おんさ	on'sa
tunnel *{n} [eng]*	隧道; トンネル	tonneru
tunneling *{n} [eng]*	通洞	tsūdō
turbidimeter *[an-ch]*	比濁計	hidaku-kei
turbidimetry *[an-ch]*	比濁法	hidaku-hō
turbidity *[an-ch]*	濁度; だくど	daku-do
" *[an-ch]*	濁り度; にごりど	nigori-do

turbidization *[ch]*	濁化	dak(u)ka
turbo; turban shell *[i-zoo]*	栄螺；さざえ	sazae
turbopause *[meteor]*	乱流圏界面	ranryū'ken-kaimen
turbosphere *[meteor]*	乱流圏；亂流圏	ranryū-ken
turbot (fish) *[v-zoo]*	鰈；かれい	karei
turbulence *[fl-mech]*	乱れ；みだれ	midare
turbulent flow *[fl-mech]*	乱流；らんりゅう	ranryū
turbulent viscosity *[fl-mech]*	乱流粘性	ranryū-nensei
turf *[geol]*	芝土；しばつち	shiba-tsuchi
turkey (bird) *[v-zoo]*	七面鳥	shichi'men-chō
turn (revolve; spin) *{vb i}*	回る；廻る；まわる	mawaru
turn (revolve; spin) *{vb t}*	回す	mawasu
turn on own axis *{vb i} [astron] [mech]*	自転する	ji'ten suru
turn over (a page) *{vb t} [pr]*	捲る；めくる	mekuru
turn sour; spoil; go bad *{vb i}*	饐える；すえる	sueru
turn the light on *{vb} [elec]*	点灯する；點燈する	tentō suru
turn to face upward *{vb i}*	仰向けになる	ao'muke ni naru
turn toward; face *{vb}*	向く	muku
turning *{n} [mech-eng]*	旋回	senkai
" *{n} [mech-eng]*	旋削；せんさく	sensaku
" (revolution) *{n}*	回り；廻り；まわり	mawari
turning circle *[navig]*	旋回圏	senkai-ken
turning point; switching point *{n}*	転換点；轉換點	tenkan-ten
turning radius *[mech-eng]*	回転半径	kaiten-hankei
turning roadway (or lane) *[traffic]*	転向車道	tenkō-shadō
turnip (vegetable) *[bot]*	蕪；蕪青；かぶら	kabura; kabu
" (vegetable) *[bot]*	菘；すずな	suzu'na
turnover number *[bioch]*	代謝回転数	taisha'kaiten-sū
turn signal (automobile) *[mech-eng]*	方向指示器	hōkō-shiji-ki
turns ratio *[elec]*	巻数比；巻數比	maki'sū-hi
turntable *[eng-acous]*	回転盤	kaiten-ban
turpentine *[mat]*	生松脂	seishō'shi
turpentine oil *[mat]*	テレピン油	terepin-yu
turpentine soot *[mat]*	松煙；しょうえん	shōen
turquois; turquoise *[miner]*	トルコ石	toruko-ishi
turquoise blue (color) *[opt]*	トルコ青	toruko-ao
turret *[arch]*	小塔	shōtō
" *[ord]*	砲塔；砲塔	hōtō
turtle (reptile) *[v-zoo]*	亀；龜；かめ	kame
tusk; fang *[zoo]*	牙；きば	kiba
tussah silk *[tex]*	柞蚕絹；柞蠶絹	saku'san-kinu

tuyere *[steel]*	羽口; はぐち	ha'guchi
tweezers *[eng]*	毛抜き	ke'nuki
" *[eng]*	ピンセット	pinsetto
Twentieth century	二十世紀	nijisseiki; nijusseiki
twenty *{n} [math]*	二十	nijū
Twenty-first century	二十一世紀	nijū-isseiki
twenty-four hours (time: midnight)	二十四時	nijū-yoji
twenty-four hour system *[mech]*	二十四時間制	nijūyo'jikan-sei
twilight *[astron]*	薄明	hakumei
" *[astron]*	黄昏; たそがれ	tasogare
twill weave *[tex]*	斜紋織り	shamon-ori
twinkling (of stars) *{n} [astron]*	瞬き	matataki; mabataki
twin *{n} [crys]*	双晶; そうしょう	sōshō
twinning *{n} [crys]*	双晶形成	sōshō-keisei
twins *[bio]*	双生児; 雙生兒	sōsei'ji; futa'go
" *[bot]*	双生	sōsei
twist *{n} [mech]*	捩れ; ねじれ	nejire
" *{n} [tex]*	撚り; より	yori
" (go wrong) *{vb i}*	拗れる; こじれる	kojireru
" *{vb t}*	捻る; 撚る; 拈る	hineru
" (screw; wrench) *{vb t}*	捩じる; ねじる	nejiru
twist coefficient *[tex]*	撚り係数	yori-keisū
twist drill *[eng]*	捩れ錐	nejire-kiri
twisted pair *[elec]*	撚り二線	yori-ni'sen
twisted thread *[tex]*	捩り糸	nejiri-ito
" *[tex]*	撚糸	yori-ito
twisted wire *[mat]*	撚線; よりせん	yori-sen
twisting *{n} [eng]*	捩回	nenkai
" *{n} [tex]*	加撚	ka'nen
twisting vibration *[spect]*	捻り振動	hineri-shindō
twist shrinkage *[tex]*	撚縮み	yori-chijimi
two *{n} [math]*	二; 弐	ni
two-color pyrometer *[eng]*	二色高温計	ni'shoku-kō'on-kei
two-component adhesive *[adhes]*	二液性接着剤	ni'eki'sei-setchaku-zai
two-component relay *[elec]*	二元継電器	nigen-keiden-ki
two-component system *[ch]*	二成分系	ni'seibun-kei
two cylinder *[mech-eng]*	二気筒; 二氣筒	ni-kitō
two-dimensional *{adj} [math]*	平面的な	heimen'teki na
" *{adj} [math]*	二次元の	niji-gen no
two-dimensional deformation reflection *[phys]*	二次元変角反射	nijigen-hen'kaku-hansha
two dimensions *[sci-t]*	二次元; にじげん	niji-gen

two-element relay *[elec]*	二元継電器	nigen-keiden-ki
two-line ground *[elec]*	二線接地	nisen-setchi
two persons *{n}*	二人；ふたり	futari
two-ply yarn *[tex]*	双糸；雙絲；そうし	sōshi
two's complement *[comput]*	二の補数	ni no hosū
two-tailed test *[stat]*	両側検定；兩側檢定	ryō'gawa-kentei
two-wheeler *[mech-eng]*	二輪車	nirin'sha
two-wire circuit *[elec]*	二線式回線	nisen'shiki-kaisen
Tylopoda (camels, etc.) *[v-zoo]*	核脚類	kaku'kyaku-rui
tympanic membrane *[anat]*	鼓膜；こまく	komaku
type *{n} [comput] [pr]*	型；かた	kata
" *{n} [pr]*	活字；かつじ	katsu'ji
" (sort; variety) *{n} [sci-t]*	類；類；るい	rui
" (sort; variety) *{n} [sci-t]*	種類	shurui
typecasting machine *[pr]*	活字鋳造機	katsu'ji-chūzō-ki
type diagram *[graph]*	模式図；模式圖	moshiki-zu
typeface *[comput] [graph] [pr]*	字形	jikei
type font *[graph]*	活字形	katsu'ji-gata; katsu'ji-kei
type metal *[met]*	活字合金	katsu'ji-gōkin
type of packing *[ind-eng]*	荷姿；にすがた	ni-sugata
type-one error *[stat]*	第一種の誤り	dai'isshu no ayamari
typeset *{vb} [pr]*	植字する	shoku'ji suru
typesetting *{n} [pr]*	植字；しょくじ	shoku'ji
type specimen *[sci-t]*	模式標本	moshiki-hyōhon
type-two error *[stat]*	第二種の誤り	dai'nishu no ayamari
typhoid bacillus *[microbio]*	チフス菌	chifusu-kin
typhoid fever *[med]*	腸チフス	chō-chifusu
typhoon *[meteor]*	台風；颱風；大風	taifū
typhus *[med]*	発疹チフス	hasshin-chifusu
typical *{adj}*	典型的な	tenkei'teki na
typographic printing *{n} [pr]*	活版印刷	kappan-in'satsu
typography *[pr]*	活版(印刷)術	kappan(-in'satsu)'jutsu
Tyrian purple (color) *[opt]*	古代紫	kodai-murasaki

U

ubiquitous species *[bio]*	汎存種	hanson-shu
udometer; rain gage *[eng]*	雨量計	u'ryō-kei
ulcer *{n} [med]*	潰瘍；かいよう	kaiyō
ulcer preventive *{n} [pharm]*	抗潰瘍剤	kō'kaiyō-zai
ulexite *[miner]*	曹灰硼鉱	sōkai'hōkō
ulmin *[geol]*	腐植；ふしょく	fu'shoku
ulna (bone) *[anat]*	尺骨；しゃっこつ	shakkotsu
ultimate analysis *[an-ch]*	元素分析	genso-bunseki
ultimate carcinogen *{n} [pharm]*	究極発癌物質	kyūkyoku-hatsu'gan-busshitsu
ultimate mutagen *{n} [pharm]*	究極突然変異原	kyūkyoku-totsuzen-hen'i'gen
ultimate strength *[mech]*	極限強度	kyoku'gen-kyōdo
ultra- *{prefix}*	超-	chō-
" *{prefix}*	越えて-	koete-
ultra-accelerator *[poly-ch]*	超促進剤	chō'sokushin-zai
ultracentrifugal analysis *[ch-eng]*	超遠心分析	chō'enshin-bunseki
ultracentrifuge *{n} [eng]*	超遠心機	chō'enshin-ki
ultrafilter *{n} [eng]*	限外沪過器	gengai-roka-ki
ultrafiltration *[ch-eng]*	限外沪過；限外濾過	gengai-roka
ultrafiltration membrane *[ch-eng]*	限外沪過膜	gengai-roka-maku
ultrafine *{adj} [phys]*	超微細の	chō'bisai no
ultra-high-frequency; UHF *[comm]*	極超短波	kyoku'chō-tanpa
ultra-high strength *[phys]*	超高強度	chō'kō-kyōdo
ultra-high temperature sterilizing method *[microbio]*	超高温瞬間殺菌法	chō'kōon-shunkan-sakkin-hō
ultra-high vacuum *[phys]*	超高真空	chō'kō-shinkū
ultra-low temperature *[phys]*	超低温	chō-tei'on
ultramarine blue (color) *[inorg-ch]*	群青；ぐんじょう	gunjō
ultrametamorphism *[petr]*	過変成作用	ka'hensei-sayō
ultramicroanalysis *[an-ch]*	超微量分析	chō'biryō-bunseki
ultramicrochemistry *[ch]*	極微量化学	kyoku'biryō-kagaku
ultramicroscope *[opt]*	限外顕微鏡	gengai-kenbi'kyō
ultra-rapid cooling *{n} [p-ch]*	超急速冷却	chō'kyūsoku-rei'kyaku
ultrashort waves *[comm]*	極超短波	kyoku'chō-tanpa
ultrasonication *[microbio]*	超音波処理	chō'onpa-shori
ultrasonic cleaning *{n} [eng]*	超音波洗浄	chō'onpa-senjō
ultrasonic flaw detector *[eng-acous]*	超音波探傷器	chō'onpa-tanshō-ki
ultrasonic flowmeter *[eng]*	超音波流量計	chō'onpa-ryūryō-kei
ultrasonic wave *[acous]*	超音波	chō-onpa
ultrasonic welding *{n} [plas]*	超音波溶接	chō'onpa-yō'setsu

ultrasonogram *[eng]*	超音波記録	chō'onpa-kiroku
ultrasonography *[eng]*	超音波検査法	chō'onpa-kensa-hō
ultrasound *[acous]*	超可聴音	chō'kachō-on
" *[acous]*	超音；ちょうおん	chō-on
ultrastructure *[mol-bio]*	超微細構造	chō'bisai-kōzō
ultraviolet absorber *[mat]*	紫外線吸収剤	shigai'sen-kyūshū-zai
ultraviolet curing *{n} [ch]*	紫外線硬化	shigai'sen-kōka
ultraviolet light *[elecmg]*	紫外線；しがいせん	shigai-sen
ultraviolet microscope *[opt]*	紫外線顕微鏡	shigai'sen-kenbi'kyō
ultraviolet radiation (rays) *[elecmg]*	紫外線	shigai-sen
Ulvales *[bot]*	石蓴目；あおさ目	aosa-moku
umbel *[bot]*	散形花序	sankei-kajo
umber (color) *[opt]*	焦茶色	koge'cha-iro
umbilical cord *[aero-eng] [embryo]*	臍の緒	heso no o
" *[embryo]*	臍帯；臍帯	saitai; seitai
umbilicus; navel *[anat]*	臍；へそ	heso
umbo (pl umbones) *[i-zoo]*	殻頂；殻頂	kaku'chō
umbra (pl umbras, umbrae) *[astron]*	本影；ほんえい	hon'ei
umbrella *[eng]*	傘；かさ	kasa
" *[eng]*	洋傘；ようがさ	yō'gasa
ume (Japanese apricot) *[bot]*	梅；梅；うめ	ume
ume (q.v.) brandy *[cook]*	梅酒	ume-zake; ume'shu
ume (q.v.) vinegar; ume juice *[cook]*	梅酢；うめず	ume-zu
un-; non- *{prefix}*	否-；非-	hi-
U-nail; clamp *{n} [eng]*	鎹；かすがい	kasugai
unary operation *[math]*	単項演算	tankō-enzan
unary operator *[math]*	単項演算子	tankō-enzan'shi
unbalance *{n} [mech-eng]*	不平衡；ふへいこう	fu'heikō
unbiased *[math]*	不偏；ふへん	fuhen
unbiased estimator *[stat]*	不偏推定量	fuhen-suitei'ryō
unbiased variance *[stat]*	不偏分散	fuhen-bunsan
unbranched chain molecule *[ch]*	枝無し鎖状分子	eda'nashi-sajō-bunshi
uncertainty (inaccurateness) *{n} [ind-eng]*	不確かさ	fu'tashika-sa
" (indefiniteness) *{n}*	不定；ふてい	futei
uncertainty principle *[phys]*	不確定性原理	fu'kakutei'sei-genri
unchangeability *{n}*	不変；不變；ふへん	fuhen
uncompetetive inhibition *[bioch]*	不拮抗阻害	fu'kikkō-sogai
unconditional statement *[comput]*	無条件命令	mu'jōken-meirei
unconfined; unsealed *{adj}*	非密閉の	hi'mippei no
unconformity *[geol]*	不整合；ふせいごう	fu'seigō
undamped oscillation *[phys]*	不減衰振動	fu'gensui-shindō
undefined *{adj} [logic]*	不定の	futei no

undefined concept *[math]*	無定義概念	mu'teigi-gai'nen
undefined term *[logic]*	無定義用語	mu'teigi-yōgo
under- *{prefix}*	不完全-	fu'kanzen-
" *{prefix}*	劣る-	otoru-
" *{prefix}*	下の-	shita no-
undercarriage *[aero-eng]*	降着装置	kōchaku-sōchi
undercoat(er) *[mat]*	下塗り; したぬり	shita'nuri
undercurrent *[ocean]*	潜流; せんりゅう	senryū
underflow *{n} [comput]*	下位桁溢れ	ka'i-keta'afure
underflow water *[geol]*	伏流水	fuku'ryū-sui
underglaze *{n} [cer]*	下薬; 下ぐすり	shita-gusuri
underglaze color *[cer]*	下絵具; 下繪具	shita-enogu
underground water *[hyd]*	地下水	chika-sui
undergrowth *[ecol]*	下生え; したばえ	shita'bae
underline *{n} [comput] [pr]*	下線	ka'sen; shita-sen
undersea mining *{n} [min-eng]*	海中採鉱; 海中採鑛	kaichū-saikō
understand (acquiesce) *{vb} [psy]*	納得する	nattoku suru
" (comprehend) *{vb} [psy]*	分かる; 解る; 判る	wakaru
understanding *{n} [psy]*	理解; りかい	rikai
undersurface *{n}*	裏	ura
undertaking *{n} [econ]*	事業	jigyō
under thread *[tex]*	裏糸	ura-ito
undertow *[ocean]*	引き波; ひきなみ	hiki-nami
" *[ocean]*	波の底引き	nami no soko'biki
undervulcanization *[rub]*	加硫不足	karyū-fusoku
underwater gun *[ord]*	水中銃	suichū-jū
undetermined coefficient method *[math]*	未定係数法	mitei-keisū-hō
undo; unfasten; loosen *{vb t}*	解く; ほどく	hodoku
undulate; fluctuate *{vb}*	波動する	hadō suru
" *{vb}*	うねる	uneru
undulation *[geol]*	うねり	uneri
uneasiness *[psy]*	不安	fu'an
unequal *[math]*	等しくない	hitoshiku'nai
unevenness *[mech]*	凸凹	deko'boko; ō-totsu
" (highs and lows) *{n}*	高低	kōtei
" *{n}*	斑; むら	mura
" (irregularity) *{n}*	ばらつき	para'tsuki
unexploded *{n} [ch] [ord]*	不爆	fu'baku
unexpressed; implied *{adj} [comm]*	言外の	gengai no
unfavorable winds *[navig]*	逆風	gyaku-fū
unfinished *{adj}*	未完成の	mi'kansei no
unguent *{n} [pharm]*	軟膏; なんこう	nankō

ungulla (pl ungullae) *[geom]*	蹄状体	teijō'tai
unhardened concrete *[mat]*	生コンクリート	nama-konkurīto
unhulled rice (cereal) *[bot]*	籾；もみ	momi
uniaxial crystal *[crys]*	一軸性結晶	ichi'jiku'sei-kesshō
" *[crys]*	単軸結晶	tan'jiku-kesshō
uniaxial orientation *[crys]*	一軸配向	ichi'jiku-haikō
uniaxial stretching *{n} [poly-ch]*	一軸延伸	ichi'jiku-enshin
unicellular organism *[bio]*	単細胞生物	tan'saibō-seibutsu
Unicorn, the *[constel]*	一角獣座	ikkaku'jū-za
unidentate ligand *[ch]*	単座配位子	tanza-hai'i-shi
uniform *{n} [cl]*	制服；せいふく	seifuku
" *{adj}*	画一的な	kaku'itsu'teki na
uniform acceleration *[mech]*	等加速度	tō-ka'sokudo
uniform field *[phys]*	平等な場	byōdō na ba
uniform motion *[mech]*	等速運動	tō'soku-undō
uniform standard *[phys]*	一律基準	ichi'ritsu-kijun
unilateral network *[comm]*	単方向性回路網	tan'hōkō'sei-kairo-mō
unimolecular reaction *[ch]*	単分子反応	tan'bunshi-hannō
union *[math]*	合併；がっぺい	gappei
" *[math]*	和集合	wa-shūgō
union cloth *[text]*	交織織物	kōshoku-ori'mono
unipolar *{adj} [elec]*	単極性の	tan'kyoku'sei no
uniquely *{adv}*	一義的に	ichigi'teki ni
unit *[comput]*	装置；裝置；そうち	sōchi
" *[phys]*	単位；單位；たんい	tan'i
unit cell *[crys]*	単位格子	tan'i-kōshi
unit cost *[econ] [ind-eng]*	単価；單價	tanka
unite; fit together *{vb t}*	合わせる	awaseru
unit pricing *{n} [econ]*	単位価格表示	tan'i-ka'kaku-hyōji
unit strain *[mech]*	単位歪	tan'i-hizumi
unit stress *[mech]*	単位応力	tan'i-ō'ryoku
unity (pl -ties) *{n}*	単一性	tan'itsu-sei
" *{n}*	統一；とういつ	tō'itsu
univalence *[ch]*	一価；一價；いっか	ikka
univalve *[i-zoo]*	単殻類；單殼類	tankaku-rui
universal (of all nations) *{adj}*	万国の；萬國の	bankoku no
universal event *{n} [math]*	全事象	zen-jishō
universal gravitation constant *[phys]*	万有引力定数	ban'yū-in'ryoku-teisū
universality *[math]*	普遍；ふへん	fuhen
universal joint *[eng]*	自在継ぎ手	jizai-tsugi'te
universally-used *{adj}*	汎用の；はんようの	han'yō no
universal quantifier *[logic]*	全称記号；全稱記號	zenshō-kigō

universal set *[math]*	普遍集合	fuhen-shūgō
universal shunt *[elec]*	万能分流器	bannō-bunryū-ki
universal testing machine *[eng]*	万能試験機	bannō-shiken-ki
universal time *[astron]*	世界時; せかいじ	sekai-ji
universal-use property *[sci-t]*	汎用性	han'yō-sei
universe *[astron]*	宇宙; うちゅう	u'chū
unknown *{n}* *[math]*	未知	mi'chi
" (number; value) *{n}* *[math]*	未知数; みちすう	michi-sū
unlawfulness *[legal]*	不法性	fuhō-sei
unlimited *{adj}* *[meteor]*	無限の	mugen no
unloading port *[geogr]*	揚げ地; あげち	age-chi
unmanned aircraft *[aerosp]*	無人機	mujin-ki
unorganized carbon *[ch]*	未組織炭素	mi'soshiki-tanso
unoriented *{adj}* *[steel]*	無方向性の	mu'hōkō'sei no
unpaired electron *[phys]*	不対電子; 不對電子	fu'tsui-denshi
unpalatable *{adj}* *[cook]* *[physio]*	不味い; まずい	mazui
unplated *{adj}* *[met]*	不めっき; 不鍍金	fu'mekki
unpolished rice *[bot]* *[cook]*	玄米; げんまい	genmai
unpublished *{n}* *[pat]*	未公開; みこうかい	mi'kōkai
" *{adj}* *[pat]*	未公開の	mi'kōkai no
unquestionable *{adv}*	疑いのない	utagai no nai
unrestricted extrusion *[met]*	否拘束押出し	hi'kōsoku-oshi'dashi
unripe; green *{n}* *[bio]*	未熟; みじゅく	mi'juku
" *{adj}* *[bot]*	青い; あおい	aoi
unripe plum *[bot]*	青梅	ao-ume
unripe smell *[physio]*	青臭さ	ao-kusa-sa
unsaponifiable matter *[mat]*	不鹼化物	fu'kenka'butsu
unsaturated fatty acid *[org-ch]*	不飽和脂肪酸	fu'hōwa-shibō'san
unsaturated hydrocarbon *[org-ch]*	不飽和炭化水素	fu'hōwa-tanka'suiso
unsaturation *{n}* *[org-ch]*	不飽和; ふほうわ	fu'hōwa
unscheduled *{adj}* *[transp]*	不定期の	fu'teiki no
unshapely; mishappen *{adj}*	不恰好な	bu'kakkō na
unsized silk paper *[mat]*	雁皮紙; がんぴし	ganpi-shi
unstable *{adj}* *[sci-t]*	不安定な	fu'antei na
unsteady flow *[fl-mech]*	非定常流	hi'teijō-ryū
untoward event *{n}*	変異	hen'i
untrimmed paper *[mat]*	原紙	genshi
untwisting *{n}* *[tex]*	撚り戻し	yori'modoshi
ununiformity *{n}* *[ch]*	不均一性	fu'kin'itsu-sei
" *{n}*	斑; むら	mura
unusual example *{n}*	異例	i'rei
unwinding *{n}* *[mech-eng]*	巻き戻し	maki'modoshi

unwoven cloth *[tex]*	非製織布	hi'seishoku-nuno
up *{adv} {n}*	上	ue
up and down *[math]*	上下	jōge
update *{vb} [comput]*	更新する	kōshin suru
updating *{n} [comput]*	更新；こうしん	kōshin
updraft *[fl-mech]*	上昇気流	jōshō-kiryū
" *[fl-mech]*	上向き通風	uwa'muki-tsū'fū
upgust *[meteor]*	上向き突風	uwa'muki-toppū
upland *[geogr]*	高地；こうち	kōchi
" *[geol]*	台地；臺地；だいち	daichi
uplift; upheaval *{n} [geol]*	隆起；隆起	ryūki
upper (of shoe) *{n} [cl]*	甲布	kōfu
" (upward) *{n} [math]*	上；うえ	ue
" *{adj} [math]*	上の；かみの	kami no
upper arm; brachium *[anat]*	上膊；じょうはく	jō-haku
" *[anat]*	上腕	jō-wan
upper audible limit *[acous]*	最高可聴限	saikō-ka'chō-gen
uppercase letter *[comput] [pr]*	大文字；だいもじ	dai-moji
upper coating *{n} [rub]*	上引き	uwa-biki
upper limit *[math]*	上限	jō-gen
uppermost *{adj} [math]*	最上位の	saijō'i no
upper side *[math]*	上方	jōhō
upper thread *{n} [cl] [tex]*	表糸	omote-ito
upside down *{n} [math]*	逆さ(ま)	sakasa(-ma)
upstream *{n} [hyd]*	川上；かわかみ	kawa'kami
upstream side *[hyd]*	上流側	jōryū-gawa
upswelling *{n} [ocean]*	湧昇；ゆうしょう	yūshō
upward flow *[fl-mech]*	上向流	uwa'muki-ryū
upwardly concave *[math] [opt]*	上に凹	ue ni ō
upwardly convex *[math] [opt]*	上に凸	ue ni totsu
upwardly-directed *{adj}*	上向きの	uwa-muki no
uraninite *[miner]*	閃ウラン鉱	sen'uran'kō
uranium (element)	ウラン	uran
uranography *[astron]*	天体図学；天體圖學	tentai'zu'gaku
uranometry *[astron]*	恒星位置誌	kōsei-ichi'shi
Uranus (planet) *[astron]*	天王星；てんのう星	tennō-sei
urate *[ch]*	尿酸塩；尿酸鹽	nyōsan-en
urban and rural prefectures *[geogr]*	都道府県	to'dō-fu'ken
urban area *[civ-eng]*	市街地；しがいち	shigai-chi
urban center *[geogr]*	都心	to'shin
urbanology *[civ-eng]*	都市学	toshi'gaku
urban traffic *[traffic]*	都市交通	toshi-kōtsu

urea *[org-ch]*	尿素；にょうそ	nyōso
urea resin *[poly-ch]*	尿素樹脂	nyōso-jushi
urea resin coating *[met]*	尿素樹脂塗料	nyōso-jushi-toryō
ureter *[anat]*	輸尿管	yu'nyō-kan
urethra *[anat]*	尿道；にょうどう	nyōdō
urgency signal *[comm]*	緊急信号	kinkyū-shingō
urgent *[n]*	至急；しきゅう	shikyū
uric acid	尿酸；にょうさん	nyōsan
urinary system *[anat]*	泌尿器系	hi'nyō'ki-kei
urine *[physio]*	尿；にょう	nyō
urn; crock; jar; pot *[n] [cer] [furn]*	壺；つぼ	tsubo
Urodela *[v-zoo]*	有尾類	yūbi-rui
Ursa Major; Great Bear *[astron]*	大熊座；おおぐまざ	ō'guma-za
Ursa Minor; Little Bear *[astron]*	小熊座；こぐまざ	ko'guma-za
urushi yarn *[tex]*	漆糸；うるしいと	urushi-ito
usability *[comput]*	使用性	shiyō-sei
use; application; utilization *[n]*	利用；りよう	riyō
" ; working; employment *[n]*	運用	un'yō
" ; service *[n]*	用途；ようと	yōto
" ; employ *[vb t]*	使用する	shiyō suru
" ; employ *[vb t]*	使う；つかう	tsukau
used (not new, secondhand) *[adj]*	中古の	chū'buru no
useful life; service life *[ind-eng]*	有効寿命；有効壽命	yūkō-jumyō
usefulness *[n]*	有用性	yūyō-sei
use in combination *[vb]*	併用する	hei'yō suru
user (a person) *[comput]*	利用者	riyō'sha
" (a person) *[comput]*	使用者；しようしゃ	shiyō'sha
user's manual; instruction manual *[pr]*	取扱説明書	tori'atsukai-setsumei'sho
usnea *[bot]*	猿麻桛；さるおがせ	saru'ogase
utahite *[miner]*	藁鉄鉱；藁鐵鑛	wara'tekkō
uterus *[anat]*	子宮；しきゅう	shikyū
utility *[n] [comput]*	効用；こうよう	kōyō
utility model *[pat]*	実用新案	jitsu'yō-shin'an
Utility Model Early Disclosure *[pat]* (abbrev in Japanese)	実開；じっかい	jikkai
Utility Model Gazette (abbrev) *[pat]*	実公；じつこう	jikkō
" *[pat]*	実用新案公報	jitsu'yō-shin'an-kōhō
utility model registration *[pat]*	実用新案登録	jitsu'yō-shin'an-tōroku
utility room (in a home) *[arch]*	諸機設備室	shoki-setsubi-shitsu
utilization *[n]*	利用；りよう	riyō

V

vacancy *[ch]*	空孔；くうこう	kūkō
" *[crys]*	空位	kū'i
" *[crys]*	空格子点	kū'kōshi-ten
vacancy rate *[ch] [nucleo]*	空席率	kūseki-ritsu
vaccination *[immun]*	種痘；しゅとう	shutō
" *[immun]*	予防接種；豫防接種	yobō-sesshu
vaccine *[med]*	ワクチン	wakuchin
vacuole *[cyt]*	液胞；液胞	eki'hō
" *[cyt]*	空胞	kūhō
vacuum (pl -s, vacua) *{n} [phys]*	真空；しんくう	shinkū
vacuum bottle; thermos bottle *[eng]*	魔法瓶；まほうびん	mahō-bin
vacuum cleaner *[mech-eng]*	電気掃除機	denki-sōji-ki
" *[mech-eng]*	真空掃除機	shinkū-sōji-ki
vacuum deposition *[met]*	蒸着	jō'chaku
vacuum distillation *[ch-eng]*	減圧蒸留；減壓蒸溜	gen'atsu-jōryū
vacuum drying *{n} [eng]*	真空乾燥	shinkū-kansō
vacuum evaporation *[eng]*	蒸着；じょうちゃく	jō'chaku
" *[eng]*	真空蒸着	shinkū-jō'chaku
vacuum forming *{n} [eng]*	真空成形	shinkū-seikei
vacuum gage *[eng]*	真空計	shinkū-kei
vacuum melting furnace *[met]*	真空溶解炉	shinkū-yōkai-ro
vacuum metallizing *[met]*	真空めっき	shinkū-mekki
vacuum plating *{n} [met]*	真空蒸着	shinkū-jō'chaku
vacuum tube *[electr]*	真空管	shinkū-kan
vagina *[anat]*	膣；ちつ	chitsu
valence *[ch]*	原子価；げんしか	genshi'ka
" *[ch]*	イオン価	ion-ka
valence band *[sol-st]*	価電子帯；價電子帶	ka'denshi-tai
valence electron *[atom-phys]*	(原子)価電子	(genshi)ka'denshi
valence fluctuation *[ch]*	原子価揺動	genshi'ka-yōdō
valence force field *[ch]*	原子価力の場	genshi'ka-ryoku no ba
valerate *[ch]*	吉草酸塩	kissō'san-en
valerian; Valeriana fauriei *[bot]*	鹿の子草；かのこ草	kanoko'sō
" *[bot]*	吉草根	kissō'kon
valeric acid	吉草酸	kissō-san
validation *[comput]*	妥当性化；妥當性化	datō'sei'ka
" *[legal]*	確認	kaku'nin
" *{n}*	認証；認證	ninshō
validity *{n} [logic]*	妥当性；だとうせい	datō-sei

validity *{n}*	適正性	tekisei-sei
" ; effectiveness *{n}*	有効性	yūkō-sei
valley *[geogr] [geol]*	谷; たに	tani
valley breeze; valley wind *[meteor]*	谷風	tani-kaze
valley fog *[meteor]*	谷霧	tani-giri
valuation *[math]*	付値; 賦値; ふち	fuchi
value *{n} [math]*	値; あたい; ち	atai; chi
" *{n} [math]*	価; 價; か	ka
" *{n} [math] [sci-t]*	価値	kachi
" (color value) *[opt]*	明度	meido
" *{n} [sci-t]*	値打ち	ne'uchi
valve *[anat] [mech-eng]*	弁; 瓣; べん	ben
valve action *[mech-eng]*	弁作用	ben-sayō
valve guide *[mech-eng]*	弁案内	ben-annai
valve seat *[eng]*	弁座; べんざ	ben-za
valve spring *[mech-eng]*	弁ばね	ben-bane
vampire bat (mammal) *[v-zoo]*	血吸(い)蝙蝠	chi'sui-kōmori
vanadinite *[miner]*	褐鉛鉱; 褐鉛鑛	katsu'enkō
vanadium (element)	バナジウム	banajūmu
Van der Waals forces *[p-ch]*	ファデルワールス力	fanderuwārusu-ryoku
vane *[eng]*	翼; よく	yoku
vane anemometer *[eng]*	微風計	bifū-kei
vanishing point *[navig] [opt]*	消点; 消點	shōten
vapor *[thermo]*	蒸気; じょうき	jōki
vapor barrier *[civ-eng]*	防湿層; 防濕層	bō'shitsu-sō
vapor degreasing *[met]*	蒸気脱脂	jōki-dasshi
vapor density *[ch]*	蒸気密度	jōki-mitsudo
vaporimeter *[eng]*	蒸気圧計	jōki'atsu-kei
vaporization *[thermo]*	蒸着	jōchaku
" *[thermo]*	蒸発; じょうはつ	jōhatsu
vapor phase *[ch]*	気相	kisō
vapor pressure *[thermo]*	蒸気圧; 蒸氣壓	jōki-atsu
variable *{n} [alg] [comput]*	変数; へんすう	hensū
" *{n} [alg]*	数値変数; 數值變數	sūchi-hensū
" *{adj} [ch] [math]*	可変性の	kahen'sei no
variable capacitance *[elec]*	可変容量	kahen-yōryō
variable capacitor *[elec]*	可変コンデンサ	kahen-kondensa
variable costs; variable expenses *[econ] [ind-eng]*	変動費; へんどうひ	hendō-hi
variable power source *[elec]*	可変電源	kahen-dengen
variable speed gear *[mech-eng]*	変速装置	hensoku-sōchi
variable star *[astron]*	変光星	henkō-sei

variable symbol *[comput]*	可変記号；可變記號	kahen-kigō
variance *[stat]*	分散	bunsan
variation *[astron]*	偏差	hensa
" (of moon) *[astron]*	二均差；にきんさ	nikin'sa
" *[bio]*	変異；變異；へんい	hen'i
variety *[agr] [bio]*	品種；ひんしゅ	hinshu
" *[bio] [syst]*	変種	henshu
variole *[geol]*	球顆；きゅうか	kyūka
various dimensions *[math]*	諸元	shogen
various factors *[math]*	諸元	shogen
varix (pl varices) *[i-zoo]*	裸層；らそう	rasō
varnish *{n} [mat]*	ニス	nisu
" *{n} [mat]*	ワニス	wanisu
varnishing *{n} [eng]*	ニス引き	nisu'biki
varve *[geol]*	氷縞；ひょうこう	hyōkō
" *[geol]*	年層	nensō
vascular bundle *[bot]*	維管束；いかんそく	i'kan'soku
vascular plants *[bot]*	維管束植物類	i'kansoku-shokubutsu-rui
vascular system *[bot]*	維管束系	i'kansoku-kei
vascular tissue *[bot]*	導管組織	dōkan-so'shiki
vase; flower vase *[furn]*	花瓶；かびん	kabin
vaseline *[mat] [pharm]*	ワセリン	waserin
vasoconstrictor *[pharm]*	血管収縮薬	kekkan-shū'shuku-yaku
vasodilator *[pharm]*	血管拡張薬	kekkan-kaku'chō-yaku
vat *[ch]*	槽；そう	sō
vat dye *[text]*	建染め染料	tate'zome-senryō
vat dyeing *{n} [tex]*	建染め；縦染め	tate-zome
vat paper *[paper]*	手抄き紙；手漉紙	te'suki-gami
vault (ceiling) *{n} [arch]*	円筒形天井	entō'gata-tenjō
" (room in a bank) *{n} [arch]*	金庫室；きんこしつ	kinko-shitsu
" (ceiling) *{n} [arch]*	穹窿；きゆぅりゆぅ	kyūryū
vector *[math] [phys]*	ベクトル	bekutoru
veering *{n} [meteor]*	方向転換	hōkō-tenkan
vegetable *[agr] [cook]*	蔬菜；そさい	sosai
" *[bot] [cook]*	野菜；やさい	yasai
vegetable fiber *[tex]*	植物繊維；植物纖維	shokubutsu-sen'i
vegetable garden *[bot] [cook]*	菜園	sai'en
vegetable growing *{n} [agr]*	野菜栽培	yasai-saibai
vegetable oil *[mat]*	植物油	shokubutsu-yu
vegetable parchment *[mat]*	硫酸紙	ryūsan-shi
vegetable wax *[mat]*	木蠟；もくろう	moku-rō
" *[mat]*	植物蠟	shokubutsu-rō

vegetarian (a person)	菜食主義者	sai'shoku-shugi'sha
vegetation *[bot]*	植物；しょくぶつ	shoku'butsu
vegetative propagation *[bot]*	栄養繁殖	eiyō-hanshoku
" *[bot]*	栄養生殖	eiyō-seishoku
vegetative reproduction *[bot]*	栄養生殖	eiyō-seishoku
vehicle *[graph] [pr]*	展色剤	tenshoku-zai
" *[mech-eng]*	車；くるま	kuruma
" *[pharm]*	賦形剤；ふけいざい	fukei-zai
" *[trans]*	車両；しゃりょう	sharyō
vehicle body; car body *[mech-eng]*	車体；車體	shatai
vein *[anat]*	静脈；じょうみゃく	jō'myaku
" *[bot]*	葉脈	yō'myaku
" *[geol]*	鉱脈；鑛脈	kō'myaku
" *[geol] [min-eng]*	脈；みゃく	myaku
" ; nervure *[i-zoo]*	翅脈；しみゃく	shi'myaku
veinlet *[bot]*	細脈	sai'myaku
velamen *[bot]*	根被	konpi
vellum *[mat]*	羊皮紙	yōhi'shi
velocity (pl -ties) *[mech]*	速度；そくど	soku'do
velocity distribution, law of *[mech]*	速度分布則	sokudo-bunpu-soku
velocity gradient *[fl-mech]*	速度勾配	sokudo-kōbai
velocity head *[fl-mech]*	速度水頭	sokudo-suitō
velocity modulation *[electr]*	速度変調	sokudo-henchō
velocity of light *[opt]*	光速	kō-soku
velvet *[tex]*	ビロード	birōdo
velveteen *[tex]*	別珍；べっちん	betchin
venation *[bot]*	脈相	myaku-sō
veneer *[mat]*	単板	tanban
venereal disease *[med]*	性病	seibyō
venom *{n} [physio]*	毒(液)	doku(-eki)
venomous; poisonous *{adj} [physio]*	有毒の；有毒な	yū'doku no; yū'doku na
vent; vent hole *{n} [eng]*	通風孔	tsū'fū'kō
" (of a volcano) *{n} [geol]*	火道	kadō
ventifact *[geol]*	風触礫；風食礫	fūshoku'reki
ventilation *[eng]*	換気；換氣；かんき	kanki
ventilator *[eng]*	換気機	kanki-ki
ventral fin *[v-zoo]*	腹鰭；はらびれ	hara-bire
ventricle *[anat]*	心室	shin'shitsu
ventriloquism *[comm]*	腹話術	fuku'wa'jutsu
Venus (planet) *[astron]*	金星	kinsei
veranda *[arch]*	縁側；えんがわ	en'gawa
verb *[gram]*	動詞	dōshi

verb phrase *[gram]*	動詞句; どうしく	dōshi'ku
verdigris *[org-ch]*	緑青; ろくしょう	roku'shō
verification *[comput]*	検証; 檢證	kenshō
" *[math]*	検算	kenzan
" *{n}*	検査; けんさ	kensa
verify; identify *{vb} [legal]*	検証する	kenshō suru
vermicelli; fine noodles *[cook]*	素麺; そうめん	sōmen
vermicide *[pharm]*	駆虫剤; 驅蟲劑	kuchū-zai
vermiculite *[miner]*	蛭石; ひるいし	hiru-ishi
vermillion (color) *[opt]*	朱; しゅ	shu
" (mercuric sulfide) *[met]*	銀朱	ginshu
vernacular language *[ling]*	地方語; ちほうご	chihō'go
" *[ling]*	方言	hōgen
vernal equinox *[astron]*	春分; しゅんぶん	shunbun
vernier calliper *[eng]*	ノギス	nogisu
versatility; adaptability *{n}*	融通性	yūzū-sei
verso (pl versos) *[pr]*	偶数ページ	gūsū-pēji
versus *{prep} [math]*	対; 對; たい	tai
vertebra (pl -brae) *[anat]*	脊椎; せきつい	seki'tsui
" *[anat]*	椎骨	tsui-kotsu
vertebral column; spine *[anat]*	脊柱; せきちゅう	sekichū
Vertebrata; vertebrates *[v-zoo]*	脊椎動物	seki'tsui-dōbutsu
vertebrate zoology *[bio]*	脊椎動物学	seki'tsui-dōbutsu'gaku
vertex (pl -es, vertices) *[math]*	角頂; ちょうかく	kaku'chō
" *[geom]*	頂点; 頂點	chōten
" *[geom]*	最高点	saikō-ten
vertical *{n} [geom]*	鉛直; えんちょく	en'choku
" *{n} [geom]*	垂直; すいちょく	sui'choku
vertical angle *[geom]*	頂角; ちょうかく	chō-kaku
" *[geom]*	鉛直角	enchoku-kaku
vertical axis *[math]*	垂直軸	suichoku-jiku
vertical circle *[astron]*	垂直環	suichoku-kan
vertical displacement *[mech]*	鉛直変位	enchoku-hen'i
vertical force *[mech]*	鉛直力	enchoku-ryoku
vertical illumination *[opt]*	垂直照明	suichoku-shōmei
vertical injection molding machine *[eng]*	縦型射出成形機	tate'gata-sha'shutsu-seikei-ki
verticality *[math]*	垂直性	suichoku-sei
vertical line *[comput] [math]*	縦線; たてせん	tate-sen
" *[graph]*	鉛直線	enchoku-sen
vertical mill *[eng]*	立型粉砕機	tate'gata-funsai-ki
vertical motion *[mech]*	縦動; 縱動	jūdō

vertical plane *[math]*	垂直面	suichoku-men
vertical rudder *[aero-eng]*	方向舵；ほうこうだ	hōkō-da
vertical scanning *{n} [eng]*	垂直走査	suichoku-sōsa
vertical stabilizer *[aero-eng]*	垂直安定板	suichoku-antei'ban
vertical tail *[aero-eng]*	垂直尾翼	suichoku-bi'yoku
verticillation *[bot]*	輪生	rinsei
vertigo; dizziness *[physio]*	目眩い；眩暈	me'mai
very *{adv}*	甚だ；はなはだ	hanahada
" *{adv}*	非常に；ひじょうに	hijō ni
" *{adv}*	とても	totemo
" *{adv}*	随分；隨分	zuibun
very-high-frequency; VHF *[comm]*	超短波	chō'tanpa
very-low-frequency *[elecmg]*	超長波	chō'chōha
very small quantity *[math]*	微少	bishō
vessel *[ind-eng]*	器；器；うつわ	utsuwa
" *[ind-eng]*	容器	yōki
vest *[cl]*	チヨッキ	chokki
vestibule *[arch]*	入口の間	iri'guchi no ma
vestigial *{adj} [bio]*	痕跡の	konseki no
vestigial organ *[bio]*	痕跡器官	konseki-kikan
veterinary doctor *[med]*	獣医師；じゅういし	jū-i'shi
veterinary medicine *[med]*	獣医学；獸醫學	jūi'gaku
viability *[bio]*	生活力	seikatsu-ryoku
" *[bio]*	生存能力	seizon-nō'ryoku
" (ability to execute) *{n}*	実行可能性	jikkō-kanō-sei
viaduct *[civ-eng]*	陸橋	rik(u)kyō
vibrate; shake *{vb}*	振動する	shindō suru
vibration *[mech]*	振動；しんどう	shindō
vibrational level *[p-ch]*	振動準位	shindō-jun'i
vibrational quantum number *[p-ch]*	振動量子数	shindō-ryoshi-sū
vibration-sensitive *[geol]*	感震	kanshin
vibrator *[elec] [mech-eng]*	振動子	shindō'shi
vibrometer; vibrograph *[eng]*	振動計	shindō-kei
viceroy (butterfly) *[i-zoo]*	茶色一文字	cha'iro-ichi'monji
vicinal group *[org-ch]*	近接基	kin'setsu-ki
video *{n} [electr]*	ビデオ	bideo
view; figure *{n} [graph]*	図；圖；ず	zu
village *[geogr]*	村	mura; son
vinculum (pl -a) *[alg]*	括線	kassen
vine *[bot]*	蔓；つる	tsuru
vinegar; table vinegar *[cook]*	食酢；しょくず	shoku'zu
" *[cook]*	酢	su

vinegared dishes, salads *[cook]*	酢の物	su-no-mono
vinegared lotus root *[cook]*	酢蓮；すばす	su-basu
vinegared rice with fish *[cook]*	鮨；寿司；すし	sushi
vine snake (reptile) *[v-zoo]*	蔓蛇	tsuru-hebi
vineyard *[agr]*	葡萄園；ぶどうえん	budō'en
vintner *[cook]*	葡萄酒醸造業者	budō'shu-jōzō'gyō'sha
violet (flower) *[bot]*	菫；すみれ	sumire
" (color) *[opt]*	菫色	sumire-iro
violet quartz *[miner]*	紫石英	murasaki-seki'ei
viper; pit viper (reptile) *[v-zoo]*	鎖蛇；くさりへび	kusari-hebi
" *[v-zoo]*	蝮；まむし	mamushi
vireo (bird) *[v-zoo]*	鵙擬；百舌擬	mozu'modoki
virga *{n}* *[meteor]*	尾流雲	biryū'un
virgin medium *[comput]*	未使用媒体	mi'shiyō-baitai
Virgo; the Virgin *[constel]*	乙女座	otome-za
virology *[microbio]*	ウイルス学	uirusu'gaku
virtual circuit *[elec]*	仮想回路	kasō-kairo
virtual image *[opt]*	虚像；きょぞう	kyozō
virtual mass *[mech]*	仮想質量；假想質量	kasō-shitsu'ryō
virulence *[microbio]*	毒力	doku-ryoku
virus *[microbio]*	ウイルス	uirusu
visa *[trans]*	査証	sashō
viscid *{adj}* *[bot]*	粘質の	nen'shitsu no
viscoelasticity (pl -ties) *[mat]*	粘弾性	nendan-sei
viscometer *[eng]*	粘度計	nendo-kei
viscoplasticity (pl -ties) *[mech]*	粘塑性；ねんそせい	nenso-sei
viscosity (pl -ties) *[fl-mech]*	粘度；ねんど	nendo
" *[fl-mech]*	粘性	nensei
viscosity-average molecular weight *[ch]*	粘度平均分子量	nendo-heikin-bunshi'ryō
viscosity index *[ch-eng]*	粘度指数	nendo-shisū
viscosity number *[ch-eng] [poly-ch]*	粘度数	nendo-sū
viscous damping coefficient *[mech]*	粘性減衰係数	nensei-gensui-keisū
viscous drag *[fl-mech]*	粘性抵抗	nensei-teikō
viscous flow *[fl-mech]*	粘性流動	nensei-ryūdō
viscous fluid *[fl-mech]*	粘性流体	nensei-ryūtai
viscous fracture *[fl-mech]*	粘性破毀；粘性破壊	nensei-hakai
viscousness *{n}* *[fl-mech]*	粘稠性	nenchū-sei
vise; vice (a tool) *{n}* *[eng]*	万力；まんりき	man'riki
visibility (pl -ties) *[electr]*	鮮明度；せんめいど	senmei-do
" *[meteor]*	目視距離	moku'shi-kyori
" *[meteor]*	視程	shitei

visibility range *[navig]*	視程；してい	shitei
visible flame *[ch]*	可視炎；かしえん	kashi-en
visible light *[opt]*	可視光	kashi-kō
visible spectrum *[spect]*	可視スペクトル	kashi-supekutoru
vision *[physio]*	視覚；視覺；しかく	shi'kaku
visual acuity *[physio]*	視力	shi-ryoku
visual angle *[opt]*	視角	shi-kaku
visual binaries *[astron]*	実視連星；實視連星	jisshi-rensei
visual density *[photo]*	視覚濃度	shikaku-nōdo
visual display *[comput]*	表示装置	hyōji-sōchi
visual field *[physio]*	視野；しや	shi'ya
visual inspection *[ind-eng]*	目視検査	moku'shi-kensa
visualization; rendering visible *[opt]*	可視化	kashi'ka
visual magnitude *[astron]*	実視等級	jisshi-tōkyū
vitality; vigor; stamina *{n}*	元気；元氣；げんき	genki
vital phenomenon *[bio]*	生命現象	seimei-genshō
vitamin *[bioch]*	ビタミン	bitamin
vitreous enamel *[mat]*	瀬戸引き；せとびき	seto-biki
vitrescence *[mat]*	硝子化性質	garasu'ka-seishitsu
vitrification *[crys]*	透化	tōka
vitrified rock *[geol]*	硝子化石	garasu'ka-ishi
vitriol *[inorg-ch]*	礬；ばん	ban
vivianite *[miner]*	藍鉄鉱；藍鉄鑛	ran'tekkō
viviparity *[physio]*	胎生	taisei
viviparous seed *[bot]*	胎生種子	taisei-shushi
vivisection *[bio]*	生体解剖	seitai-kaibō
vocabulary *[comm] [ling]*	語彙；ごい	go'i
" *[comm] [ling]*	用語	yōgo
vocal cord *[anat]*	声帯；聲帶	seitai
vocation; calling *{n}*	業；ぎょう	gyō
voice *{n} [comm] [physio]*	声；聲；こえ	koe
" *{n} [physio]*	音声	onsei
voice box; larynx (pl larynges, -es) *[anat]*	喉頭；こうとう	kōtō
voiced sound *[ling]*	濁音	daku'on
voiced sound mark *[pr]*	濁音符；だくおんぷ	daku'on-pu
" *[pr]*	濁点	daku-ten
voiceless sound *[ling]*	清音	sei-on
voice-operated device *[electro]*	音声作動装置	onsei-sadō-sōchi
voice print *[eng-acous]*	声紋；聲紋	sei-mon
voice recorder *[eng]*	音声記録装置	onsei-kiroku-sōchi
voice unit *[acous]*	音声単位	onsei-tan'i

void {n} [mech-eng] 空隙；くうげき kūgeki
" {adj} [legal] 無効の mukō no
voidage [ch-eng] 空間率 kūkan-ritsu
void component {n} [mat] 空隙分 kūgeki-bun
void content [geol] [poly-ch] 空洞率 kūdō-ritsu
void ratio [geol] 空隙率 kūgeki-ritsu
void volume [poly-ch] 空隙率 kūgeki-ritsu
volatile {adj} [ch] 揮発性の kihatsu'sei no
volatile matter [mat] 揮発分；きはつぶん kihatsu'bun
volatile rust preventive oil [mat] 気化性錆止め油 kika'sei-sabi'dome-abura
volatile solvent [ch] 揮発性溶剤 kihatsu'sei-yōzai
volatility [comput] 非持久性 hi'jikyū-sei
" [thermo] 揮発度 kihatsu-do
volatilization [thermo] 揮発；きはつ ki'hatsu
volcanic ash [geol] 火山灰；かざんばい kazan-bai
volcanic island [geogr] 火山島 kazan-tō
volcanic smoke [geol] 噴煙 fun'en
volcano (pl -es, -s) [geol] 火山；かざん kazan
volcanology [geol] 火山学 kazan'gaku
vole (animal) [v-zoo] 畑鼠；はたねずみ hata-nezumi
volition; will {n} [psy] 意志 i'shi
volitional {adj} [psy] 随意の；隨意の zui'i no
volt; V [elec] ボルト boruto
voltage [elec] 電圧；でんあつ den'atsu
voltage amplifier [electr] 電圧増幅器 den'atsu-zōfuku-ki
voltage commutation [elec] 電圧整流 den'atsu-seiryū
voltage detector [eng] 検電器；檢電器 kenden-ki
voltage divider [elec] 分圧器；分壓器 bun'atsu-ki
voltage doubler tube [electr] 倍電圧整流管 bai-den'atsu-seiryū-kan
voltage drop [elec] 電圧降下 den'atsu-kōka
voltage gain [elec] 電圧利得 den'atsu-ritoku
voltage multiplier [electr] 電圧増倍器 den'atsu-zōbai-ki
voltage to ground {n} [elec] 対地電圧；對地電壓 taichi-den'atsu
volt ammeter [eng] 電圧電流計 den'atsu-denryū-kei
voltmeter [eng] 電圧計 den'atsu-kei
volume [geom] 量；りょう ryō
" [geom] 体積；體積 tai'seki
" [geom] 容量；ようりょう yōryō
" [geom] 容積；ようせき yōseki
" (a copy) [pr] 冊；さつ satsu
volume flow [fl-mech] 体積流れ taiseki-nagare
volume force [phys] 体積力 taiseki-ryoku

volume fraction *[math]*	体積率	taiseki-ritsu
" *[math]*	容積率	yōseki-ritsu
volume percent(age) *[math]*	容積百分率	yōseki-hyaku'bun-ritsu
volume percent(age) *[math]*	体積百分率	taiseki-hyaku'bun-ritsu
volume resistivity *[elec]*	体積固有抵抗	taiseki-koyū-teikō
volume specific resistance *[elec]*	体積固有抵抗	taiseki-koyū-teikō
volumeter *[eng]*	体積計	taiseki-kei
volumetric analysis *[an-ch]*	容量分析	yōryō-bunseki
volume viscosity *[fl-mech]*	体積粘性	taiseki-nensei
voluntarily; optionally *{adv}*	随意に; 隨意に	zui'i ni
voluntary muscle *[physio]*	随意筋	zui'i-kin
volute spring *[mech-eng]*	竹の子ばね	takenoko-bane
volvox *[bot]*	大鬚回	ō-hige'mawari
vortex (pl vortices, -es) *[astron]* *[fl-mech]*	渦; うず	uzu
vortex filament *[fl-mech]*	渦糸	uzu-ito
vortex line *[fl-mech]*	渦線	uzu-sen
vortex pair *[fl-mech]*	渦対; 渦對	uzu-tsui
vortex ring *[fl-mech]*	渦環	uzu-wa
vortex sheet *[fl-mech]*	渦層; うずそう	uzu-sō
vortex street *[fl-mech]*	渦列	uzu-retsu
vorticity *[fl-mech]*	渦度	uzu-do
vowel *[ling]*	母音	bo'in; bo'on
voyage *[navig]*	航海	kōkai
vug *[petr]*	がま	gama
" *[petr]*	晶洞; しょうどう	shōdō
vulcanization *[ch-eng]* *[rub]*	加硫; かりゅう	karyū
vulcanization accelerator *[ch-eng]*	加硫促進剤	karyū-soku'shin-zai
vulcanization pan *[rub]*	加硫缶; 加硫罐	karyū-kan
vulcanized oil *[mat]*	加硫油	karyū-yu
vulcanizing agent *[ch-eng]*	加硫剤	karyū-zai
vulnerability *[mil]*	脆弱性	zei'jaku-sei
vulture (bird) *[v-zoo]*	禿鷹; はげたか	hage'taka
vulva (pl vulvae) *{n}* *[anat]*	外陰	gai'in

W

wading bird *[v-zoo]*	渉禽；しょうきん	shōkin
wafer *[electr]*	半導体板	handōtai'ban
wages; pay *{n} [econ]*	給料	kyūryō
wait *{vb}*	待つ；まつ	matsu
waiting time; latency *[comput]*	待ち時間	machi-jikan
waive *{vb} [legal]*	放棄する	hōki suru
waiver; abandonment *{n} [legal]*	(任意)放棄	(nin'i-)hōki
wakame seaweed *[bot]*	若布；和布；わかめ	wakame
wake *{n} [fl-mech]*	後流；こうりゅう	kōryū
" (awake) *{vb i}*	目覚める；目覺める	me'zameru
wake up; awaken *{vb i} [physio]*	起きる；おきる	okiru
walk *{vb i}*	歩く；あるく	aruku
walking; locomotion *{n} [physio]*	歩行；ほこう	hokō
wall; fence *{n} [arch]*	塀；へい	hei
" *[arch]*	壁；かべ	kabe
wallboard *[mat]*	壁板	kabe-ita
wall clock *[horol]*	壁時計；かべどけい	kabe-dokei
wallet *[econ]*	財布；さいふ	saifu
wall framing *{n} [arch]*	軸組；じくぐみ	jiku-gumi
wall-mud color (color) *[opt]*	壁土色	kabe-tsuchi-iro
wall outlet *[elec]*	壁コンセント	kabe-konsento
wall painting; mural *{n} [graph]*	壁画；壁畫	heki-ga
wallpaper *[mat]*	壁紙	kabe-gami
walnut (tree, nut) *[bot]*	胡桃；くるみ	kurumi
walrus (marine mammal) *[v-zoo]*	海馬；海象	sei'uchi
waning moon *[astron]*	欠けてゆく月	kakete yuku tsuki
war *{n} [mil]*	戦争；戰爭	sensō
warble tone *[acous]*	震音	shin'on
ward (of a city) *{n} [geogr]*	区；區；く	ku
ward office *[arch]*	区役所	ku-yakusho
wardroom *[nav-arch]*	士官室	shikan-shitsu
warehouse; storehouse *[arch] [ind-eng]*	倉庫；そうこ	sōko
warhead *[ord]*	弾頭	dantō
warm *{adj} [thermo]*	暖かい；温かい	atatakai
" (make warm; heat) *{vb t}*	暖める；温める	atatameru
warm-blooded animal *[zoo]*	温血動物	on'ketsu-dōbutsu
warm color *[opt]*	暖色；だんしょく	dan-shoku
warm front *[meteor]*	温暖前線	ondan-zensen
warm rain *[meteor]*	暖い雨	atatakai ame

warm rolling *{n} [met]*	温間圧延	onkan-atsu'en
warm water *[hyd]*	温湯	ontō
warning; siren *{n} [electr]*	警報；けいほう	keihō
warning coloration *[bot]*	警告色	keikoku-shoku
warning light *[navig] [opt]*	警告灯	keikoku-tō
warp *{n} [cer]*	曲り	magari
" *{n} [geol]*	褶曲；しゅうきょく	shū'kyoku
" *{n} [tex]*	経糸；縦糸	tate-ito
" *{vb i} [mech]*	反る；かえる	kaeru
" (bend backward) *{vb i} [mech]*	反る；そる	soru
warpage *[mech]*	歪；いびつ	ibitsu
warping *{n} [tex]*	歪み	yugami
wart *{n} [med]*	疣	ibo
warthog (animal) *[v-zoo]*	疣猪；いぼいのしし	ibo-inoshishi
wash *{n} [aero-eng]*	後流	kōryū
" (rinse) *{vb t}*	洗う	arau
washcloth *[cook] [tex]*	布巾；ふきん	fukin
washer (washing machine) *[eng]*	洗浄機	senjō-ki
" (washing machine) *[eng]*	洗濯機	sentaku-ki
" *{n} [eng]*	座金；ざがね	za'gane
washing *{n} [ch]*	洗浄；せんじょう	senjō
" (with water) *{n}*	水洗	suisen
washing bottle *[an-ch]*	洗瓶	senbin
washing (fluid) *{n} [an-ch]*	洗液	sen'eki
" *{n} [an-ch]*	洗浄液	senjō-eki
wasp (insect) *[i-zoo]*	雀蜂；すずめばち	suzume-bachi
waste *{n} [ecol]*	廃物；廢物	hai'butsu
" *{vb t} [econ]*	浪費する	rōhi suru
wastebasket *[eng]*	屑籠；くずかご	kuzu'kago
waste cotton *[tex]*	落綿	raku-men
waste disposal *[ecol]*	廃棄物処理	haiki'butsu-shori
wastefulness *[ecol]*	浪費	rōhi
waste paper *[mat]*	故紙	ko-shi
waste product (of body) *[physio]*	老廃物	rōhai'butsu
wastes; discards *{n} [ecol]*	廃棄物；はいきぶつ	haiki'butsu
waste water *[ind-eng]*	廃水	hai'sui
watch *{n} [horol]*	腕時計；うでどけい	ude-dokei
watch glass *[ch]*	時計皿	tokei-zara
water *{n} [ch]*	水；みず	mizu
water absorption *[ch]*	給水率	kyūsui-ritsu
water absorption capacity *[mat]*	吸水容量	kyūsui-yōryō
water activity *[hyd]*	水分活性	suibun-kassei

water balance *[ind-eng]*	水収支；水收支	mizu-shūshi
water-base paint *[mat]*	水性塗料	suisei-toryō
water bath *[ch]*	水浴	sui-yoku
water-bath heating *{n}* *[ch]*	湯煎；ゆせん	yu'sen
water bed *[geol]*	水床	suishō
water-boiler reactor *[nucleo]*	湯沸かし型原子炉	yu'wakashi'gata-genshi-ro
water-borne infectious disease *[med]*	水系伝染病	suikei-densen'byō
water break *[paper]*	水切れ	mizu-gire
water buffalo (animal) *[v-zoo]*	水牛	suigyū
water caltrop *[bot]*	菱の実；菱の實	hishi no mi
water channel *[civ-eng]*	水路	suiro
water chestnut (fruit) *[bot]*	菱；ひし	hishi
" (fruit) *[bot]*	黒慈姑；くろぐわい	kuro-guwai
water clock *[horol]*	水時計	mizu-dokei
watercolor *[mat]*	水彩絵具	suisai-enogu
water column *[mech-eng]*	水柱	suichū
water conservation *[ecol]*	水の保全	mizu no hozen
water content *[ch] [hyd]*	水分；すいぶん	suibun
" *[ch-eng]*	含水率	gansui-ritsu
" *[ch-eng]*	含水量	gansui-ryō
water-cooled roll(er) *[ind-eng]*	水冷ロール	suirei-rōru
watercourse *[hyd]*	水路	suiro
water droplet; dewdrop *[hyd]*	水玉；みずたま	mizu-tama; mizu-dama
water equivalent *[meteor]*	水当量；水當量	mizu-tōryō
water erosion *[ecol]*	水食；水蝕	sui'shoku
waterfall *[hyd]*	瀑布	baku'fu
" *[hyd]*	滝；瀧；たき	taki
water flea; Daphnia pulex *[i-zoo]*	微塵子；みじんこ	mijin'ko
waterflooding *{n}* *[petr]*	水攻法	suikō-hō
waterfowl (bird) *[v-zoo]*	水鳥；水禽	mizu-tori
waterfront area *[civ-eng]*	水辺地区；水邊地區	suihen-chiku
water gage *[eng]*	水位計	sui'i-kei
water gas *[mat]*	水性ガス	suisei-gasu
water gate *[civ-eng]*	水門	sui-mon
water gel *[ord]*	含水爆薬	gansui-baku'yaku
water hammer *[fl-mech]*	水槌；みずつち	mizu-tsuchi
water hemlock *[bot]*	毒芹	doku-zeri
water-holding capacity *[hyd]*	保水力	ho'sui-ryoku
water horsepower *[hyd]*	水馬力；みずばりき	mizu-ba'riki
watering; irrigation *{n}* *[agr]*	灌漑；かんがい	kangai
water-in-oil *{adj}* *[org-ch]*	油中水の	yu-chū-sui no
water-jet pump; tap aspirator *[eng]*	水流ポンプ	suiryū-ponpu

waterleaf paper *[paper]* 水漉紙；みずすき紙 mizu'suki-gami
water level *[hyd]* 水位 sui'i
water lily (flower) *[bot]* 水蓮；すいれん suiren
waterline *[geol]* 水線 suisen
" *[nav-arch]* 喫水線 kissui-sen
water main *[civ-eng]* 給水本管 kyūsui-honkan
watermark *[paper]* 漉入；すき入れ suki'ire
water medium method *[ch]* 水媒法 suibai-hō
watermelon (fruit) *[bot]* 西瓜；すいか suika
water meter *[eng]* 水量計 suiryō-kei
water milfoil *[bot]* 房藻；ふさも fusa'mo
water moccasin (reptile) *[v-zoo]* 沼蝮；ぬままむし numa-mamushi
water of crystallization *[ch]* 結晶水 kesshō-sui
water of hydration *[ch]* 水和水；すいわすい suiwa-sui
water paint *[mat]* 水性塗料 suisei-toryō
water permeability *[fl-mech]* 透水性 tōsui-sei
water pollution *[ecol]* 水質汚濁 sui'shitsu-o'daku
waterpower *[mech]* 水力；すいりょく sui'ryoku
water pressure *[mech]* 水圧；水壓 sui'atsu
waterproofing agent *[mat]* 防水剤 bōsui-zai
waterproof(ness) *{n} [ind-eng]* 防水性 bōsui-sei
" *{n} [ind-eng]* 耐水性 taisui-sei
water-purifying plant *[civ-eng]* 浄水場；淨水場 jōsui'jō
water-purifying tank *[civ-eng]* 浄化槽 jōka-sō
water quality *[hyd]* 水質 sui-shitsu
water quenching *{n} [eng]* 水焼入れ mizu-yaki'ire
waterrail (bird) *[v-zoo]* 水鶏；くいな kuina
water repellency *[mat]* 撥水性 hassui-sei
water repellent *{n} [mat]* 撥水剤 hassui-zai
water-repellent finishing *{n} [mat]* 撥水加工 hassui-kakō
water resistance; waterproofness *[mat]* 耐水性 taisui-sei
water retentivity *[hyd]* 保水性 hosui-sei
water sanding (in painting) *[eng]* 水研ぎ；みずとぎ mizu-togi
watershed *{n} [hyd]* 流域 ryū'iki
water shield (plant) *[bot]* 蓴菜；じゅんさい junsai
water softener *[mat]* 硬水軟化剤 kōsui-nanka-zai
water-soluble resin *[poly-ch]* 水溶性樹脂 suiyō'sei-jushi
water solution *[ch]* 水溶液 suiyō-eki
waterspout *[meteor]* 竜巻；たつまき tatsu'maki
water supply *[hyd]* 配水 haisui
water table *[hyd]* 地下水面 chika-suimen
" *[hyd]* 水位 sui'i

water tank *[eng]*	水槽	suisō
water temperature *[thermo]*	湯加減；ゆかげん	yu-kagen
watertight *{adj} [nav-arch]*	水密の	sui'mitsu no
water tower *[civ-eng]*	貯水塔	chosui-tō
water transport *[trans]*	水運	sui'un
water turbine *[mech-eng]*	水車；すいしゃ	suisha
water vapor *[phys]*	水蒸気	sui'jōki
" *[thermo]*	蒸気；じょうき	jōki
water-vapor permeability *[fl-mech]*	透湿度；透濕度	tō'shitsu-do
waterwheel *[mech-eng]*	水車	suisha
waterworks; water supply *[hyd]*	上水道	jō'suidō
wattmeter *[eng]*	電力計	denryoku-kei
wave (in glass) *{n} [cer]*	うねり	uneri
" *{n} [fl-mech] [ocean] [phys]*	波；なみ	nami
" *{n} [phys]*	波動；はどう	hadō
wave crest *[phys]*	波高点	hakō-ten
" *[phys]*	波の峰；なみのみね	nami no mine
wave cycle *[elec]*	周波；しゅうは	shūha
wave equation *[phys]*	波動方程式	hadō-hōtei'shiki
waveform *{n} [phys]*	波形	hakei
wavefront *[ocean]*	波頭	hatō
" *[phys]*	波面	ha'men
waveguide *[elecmg]*	導波管；どうはかん	dōha'kan
waveguide path *[elecmg]*	導波路	dōha-ro
waveguide switch *[elecmg]*	導波管切換器	dōha'kan-kiri'kae-ki
wave height *[ocean]*	波高	hakō
wavelength *[phys]*	波長	hachō
wavellite *[miner]*	銀星石	ginsei-seki
wave mechanics *[quant-mech]*	波動力学	hadō-riki'gaku
wave motion *[phys]*	波動；はどう	hadō
wave node *[phys]*	波節	ha'setsu
wave number *[phys]*	波数；はすう	hasū
wave optics *[opt]*	波動光学	hadō-kōgaku
wave packet *[phys]*	波束	ha-soku
waves *{n} [ocean]*	波浪	harō
" (billows; surges) *{n} [ocean]*	波濤；はとう	hatō
" *{n} [ocean] [phys]*	波；なみ	nami
wave surface *[phys]*	波面	ha'men
wave trough *[phys]*	波窪；なみくぼ	nami-kubo
waviness *[plas]*	うねり	uneri
wavy surface *[cer]*	凸凹；でこぼこ	deko'boko
wax *{n} [mat]*	蠟；ろう	rō

wax gourd *[bot]*	冬瓜；とうがん	tōgan
wax impression *[eng]*	蠟型	rō-gata
waxing *{n} [ind-eng]*	蠟引き	rō'biki
waxing moon *[astron]*	満ちてゆく月	michite yuku tsuki
wax paper *[mat]*	蠟紙	rō-gami
wax pattern *[eng]*	蠟型	rō-gata
waxwing (bird) *[v-zoo]*	連雀；れんじゃく	renjaku
way; path *{n} [civ-eng]*	道	michi
weak (fragile) *{adj} [mech]*	弱い；よわい	yowai
" (feeble) *{adj} [physio]*	弱い	yowai
" (thin; dilute) *{adj}*	薄い	usui
weak acid *[ch]*	弱酸；じゃくさん	jaku-san
weak base *[ch]*	弱塩基	jaku-enki
weak electrolyte *[p-ch]*	弱電解質	jaku-denkai'shitsu
weakening *{n} [mech]*	弱まり	yowamari
weakness *[mech]*	弱さ	yowa-sa
wealth *{n} [econ]*	財；ざい	zai
weapon *[ord]*	武器	buki
wear (abrasion) *{n} [eng] [geol]*	摩耗；まもう	mamō
" (on feet) *{vb} [cl]*	穿く；履く；はく	haku
" (on body) *{vb} [cl]*	着る	kiru
weasel (animal) *[v-zoo]*	鼬；いたち	itachi
weather *{n} [meteor]*	気候	kikō
" *{n} [meteor]*	天気；てんき	tenki
weatherability; weather resistance *[mat]*	耐候性	taikō-sei
weather conditions *[meteor]*	気象；きしょう	kishō
weathered coal *[geol]*	風化炭	fūka-tan
weather deck *[nav-arch]*	露天甲板	roten-kōhan
weather forecast *[meteor]*	天気予報；天氣豫報	tenki-yohō
weathering *{n} [geol]*	風化	fūka
weathering resistance *[mat]*	耐風化性	tai'fūka-sei
weather map *[meteor]*	天気図；天氣圖	tenki-zu
weather report *[meteor]*	気象通報	kishō-tsū'hō
weather resistance *[mat]*	耐候性	taikō-sei
weather ship *[meteor] [nav-arch]*	気象観測船	kishō-kansoku'sen
weaverbird (bird) *[v-zoo]*	機織鳥；はたおり鳥	hata'ori-dori
weaving *{n} [tex]*	製織	seishoku
web *[bio]*	網；あみ	ami
" *[paper]*	巻取紙；巻取紙	maki'tori-gami
" (of bird's foot) *[v-zoo]*	水搔き	mizu'kaki
webbing clothes moth (insect) *[i-zoo]*	小衣蛾；こいが	ko-i'ga

wedge *{n} [eng] [meteor]*	楔；くさび	kusabi
wedge spectrograph *[spect]*	楔分光写真器	kusabi-bunkō-shashin-ki
Wednesday	水曜日	sui'yōbi
weed *{n} [bot]*	雑草；雜草	zassō
weedkiller *[mat]*	除草剤	josō'zai
week *{n} [astron]*	週；しゅう	shū
weekly *{adv} {n}*	毎週	mai-shū
" *{adv}*	週毎の	shū'goto no
weekly periodical *[pr]*	週刊誌	shūkan'shi
weekly report *[pr]*	週報；しゅうほう	shū-hō
weep hole *[civ-eng]*	排水孔	haisui-kō
weevil (insect) *[i-zoo]*	象鼻虫	zōbi'chū
" *[i-zoo]*	象虫；ぞうむし	zō-mushi
weft; filling *{n} [tex]*	緯糸；橫糸	yoko-ito
weigh *{vb} [mech]*	量る；秤る；はかる	hakaru
weighing *{n} [eng]*	秤量；ひょうりょう	hyō'ryō
" *{n} [sci-t]*	計量	keiryō
weighing bottle *[an-ch]*	量り瓶	hakari-bin
weight (of a balance) *{n} [mat]*	分銅；ふんどう	fundō
" *{n} [mech]*	重量；じゅうりょう	jūryō
" *{n} [mech]*	重さ；おもさ	omo-sa
" *{n} [mech]*	重し	omoshi
" *{n} [mech]*	押え	osae
" (of body) *{n} [physio]*	体重；體重	taijū
weight-average molecular weight *[ch]*	(重)量平均分子量	(jū)ryō-heikin-bunshi'ryō
weighted average *[stat]*	加重平均	kajū-heikin
weighted silk *[tex]*	増量絹	zōryō-ginu
weighting agent *[mat]*	増量剤	zōryō-zai
weighting instrument *[eng]*	重み付け器	omo'mi'zuke-ki
weightlessness *[aerosp] [mech]*	無重量状態	mu'jūryō-jōtai
weight-loss rate *[mech]*	減量率	genryō-ritsu
weight per area *[mech]*	目付；めつけ	me'tsuke
weight percentage *[mech]*	重量百分率	jūryō-hyaku'bun-ritsu
weight-reduction processing *[ind-eng]*	減量加工	genryō-kakō
weights and measures *[sci-t]*	度量衡	doryō'kō
weld; welded part *{n} [met]*	溶着部	yō'chaku-bu
weldability *[met]*	溶接性	yōsetsu-sei
welder (machine) *[eng] [met]*	溶接機	yōsetsu-ki
" (a person) *[met]*	溶接士	yōsetsu'shi
welding *{n} [met]*	溶接；ようせつ	yō'setsu
weld joint *[met]*	溶接部	yōsetsu-bu
weld metal *[met]*	溶着金属	yō'chaku-kinzoku

well (water-well) *{n} [eng]*	井戸	ido
well; healthy *{adj} [med]*	健康な；けんこうな	kenkō na
well-formed formula *[logic]*	正形論理式	seikei-ronri-shiki
well foundation *[civ-eng]*	井筒基礎	i'zutsu-kiso
welsh onion (vegetable) *[bot]*	分葱；わけぎ	wake'gi
west *[geod]*	西；にし	nishi
West *[geogr]*	西洋；せいよう	seiyō
west coast *[geogr]*	西岸	seigan
westerly wind; westerlies *[meteor]*	偏西風	hensei-fū
" *[meteor]*	西風	nishi-kaze
Western food; Western dishes *[cook]*	西洋料理	seiyō-ryōri
" *[cook]*	洋食；ようしょく	yō'shoku
Western hemisphere *[geogr]*	西洋球	seiyō-kyū
Western wines and liquor *[cook]*	洋酒	yō'shu
wet *{adj}*	濡れた；ぬれた	nureta
wet-and-dry-bulb hygrometer *[eng]*	乾湿球湿度計	kan'shikkyū-shitsudo-kei
wet-bulb thermometer *[eng]*	湿球温度計	shikkyū-ondo-kei
wet cell *[elec]*	湿電池；濕電池	shitsu-denchi
wet corrosion *[met]*	湿食；濕蝕	shisshoku
wet grinding *{n} [mech-eng]*	湿式粉砕	shisshiki-funsai
" *{n} [mech-eng]*	湿式研削	shisshiki-kensaku
wetlands *[ecol]*	湿地	shitchi
wet rubbing; wet sanding *{n} [eng]*	水研ぎ；みずとぎ	mizu-togi
wet strength *[met] [paper]*	湿潤強度	shitsu'jun-kyōdo
wettability *[ch]*	湿潤性	shitsu'jun-sei
wetting *{n} [ch]*	濡れ	nure
wetting agent *[mat]*	湿潤剤	shitsu'jun-zai
wetting property *[ch]*	濡れ性	nure-sei
whale (mammal) *[v-zoo]*	鯨；くじら	kujira
whale oil *[mat]*	鯨油	gei-yu
whale shark *[v-zoo]*	甚兵衛鮫	jinbē-zame
wharf; quay; pier *[civ-eng]*	埠頭	futō
what *{pron}*	何；なに	nani
what time *{adv}*	何時；なんじ	nan'ji
wheat *[bot]*	小麦	ko'mugi
" *[bot]*	麦；麥；むぎ	mugi
wheat germ *[bot]*	小麦の麦芽	ko'mugi no baku'ga
wheat gluten bread *[cook]*	麸；ふ	fu
wheat starch *[cook]*	小麦澱粉	ko'mugi-denpun
wheel *{n} [eng]*	車；くるま	kuruma
" *{n} [eng]*	車輪；しゃりん	sharin
" *{n} [eng]*	輪；わ	wa

wheelchair *[eng]*	車椅子；くるまいす	kuruma-isu
wheel spindle stock *[eng]*	砥石台；砥石臺	to'ishi-dai
when *{n} {adv}*	何時；いつ	itsu
where *{adv}*	何処；どこ	doko
whetstone *[mat]*	砥石；といし	to'ishi
whey *[cook]*	乳漿；にゅうしょう	nyūshō
whipgraft *[bot]*	舌継ぎ	shita'tsugi
whirl *{n} [phys]*	旋回	senkai
" *{n} [phys]*	渦巻；渦卷	uzu'maki
whirl chamber *[mech-eng]*	渦室；うずしつ	uzu-shitsu
whirligig beetle (insect) *[i-zoo]*	水澄まし	mizu-sumashi
whirlpool *[ocean]*	渦巻	uzu'maki
whirlwind *[meteor]*	旋風；つむじかぜ	tsumuji-kaze
whisper *{n} [acous] [physio]*	囁き；ささやき	sasayaki
" *{vb} [comm]*	囁く	sasayaku
whispering gallery *[acous] [arch]*	囁きの回廊	sasayaki no kairō
white (color) *{n} [opt]*	白色	haku-shoku
" (color) *{n} [opt]*	白	shiro
" *{adj} [opt]*	白い；しろい	shiroi
white admiral (butterfly) *[i-zoo]*	一文字蝶	ichi'monji-chō
whitebait (fish) *[v-zoo]*	白魚；しらうお	shira'uo
white blood cell (leukocyte) *[hist]*	白血球	hakkekkyū
whitecap *[ocean]*	波頭	hatō
white cedar (tree) *[bot]*	檜；ひのき	hinoki
white coral *[i-zoo]*	白珊瑚；しろさんご	shiro-sango
white dwarf *[astron]*	白色矮星	haku'shoku-waisei
white-eye (bird) *[v-zoo]*	目白；めじろ	me'jiro
white flag *[comm]*	白旗	shiro-hata
white-fly; whitefly (insect) *[i-zoo]*	(蜜柑)粉虱	(mikan-)kona'jirami
white gold *[met]*	白金	shiro-kin
white heat *[thermo]*	白熱；はくねつ	haku-netsu
white iron *[met]*	白銑	haku-sen
white Japan wax *[mat]*	白蠟；はくろう	haku-rō
white light *[opt]*	白色光	haku'shoku-kō
white liquor *[paper]*	白水	haku'sui
white muskmelon (fruit) *[bot]*	白瓜	shiro-uri
white mustard (condiment) *[bot]*	白辛子；しろがらし	shiro-garashi
whiteness *[opt]*	白(色)度	haku('shoku)-do
whitening *{n} [plas]*	白化	hakka
white noise *[phys]*	白色雑音	haku'shoku-zatsu'on
white pepper (condiment) *[bot]*	白胡椒	shiro-goshō
white phosphorus *[ch]*	黄燐；おうりん	ō'rin

white pig iron *[mat]*	白銑	haku-sen
white poplar (tree) *[bot]*	白楊	haku'yō
white rot *[bot]*	白色腐朽	haku'shoku-fukyū
white rouge *[mat]*	白棒	shiro-bō
white-rumped swift; Apus pacificus (bird) *[v-zoo]*	雨燕；あめつばめ	ama-tsubame
white sake (q.v.) *[cook]*	白酒	shiro-zake
white shirt *[cl]*	ワイシャツ	wai-shatsu
white soybean paste *[cook]*	白味噌；しろみそ	shiro-miso
whitethroat (bird) *[v-zoo]*	喉白虫食い	nodo'shiro-mushi'kui
white water *[ocean]*	白水	haku'sui
white wine *[cook]*	白葡萄酒	shiro-budō'shu
who *{pron}*	誰；だれ；たれ	dare; tare
whole body *[anat]*	全身	zenshin
whole-body counter *[nucleo]*	全身計数装置	zenshin-keisū-sōchi
whole-grain flour *[cook]*	全粒粉	zenryū-fun
whole human blood; preserved blood *[hist]*	保存血液	hozon-ketsu'eki
whole milk *[cook]*	全乳；ぜんにゅう	zennyū
whole milk powder *[cook]*	全脂粉乳	zenshi-funnyū
whole number(s) *[arith]*	整数；せいすう	seisū
wholesale *{n} [econ]*	卸売；卸賣	oroshi'uri
wholesaler *[econ]*	問屋；とんや	ton'ya
whole sweetened condensed milk *[cook]*	全脂加糖煉乳	zenshi-katō-rennyū
whole tone *[acous] [music]*	全音	zen'on
whooping cough; pertussis *[med]*	百日咳	hyaku'nichi-zeki
whorl (of a shell) *{n} [i-zoo]*	ひと巻き	hito'maki
whorl *{n} [math]*	渦形；うずがた	uzu'gata
whose *{pron}*	誰の	dare no
why *{adv}*	何故；なぜ	naze
why; how *{adv}*	如何して；どうして	dō'shite
wick *[tex]*	灯心；燈芯	tō'shin
wicker chest, trunk, basket *[furn]*	葛籠；つづら	tsuzura
wicker trunk; luggage *[transp]*	行李	kōri
wide; broad *{adj}*	広い；廣い；ひろい	hiroi
wideband *[comm]*	広帯域；廣帯域	kō-tai'iki
wideband antenna *[elecmg]*	広帯域空中線	kōtai'iki-kūchū-sen
wide-mouth bottle *[an-ch]*	広口瓶	hiro'kuchi-bin
widen (broaden; spread) *{vb i}*	広まる	hiromaru
" (broaden; spread) *{vb t}*	広める	hiromeru
" (expand) *{vb t} [mech-eng]*	広げる	hirogeru
" (spread) *{vb t} [mech-eng]*	押し広げる	oshi-hirogeru

English	Japanese	Romanization
width *[math]*	横	yoko
" *[mech]*	広さ	hiro-sa
wig *[cl]*	鬘; かつら; かずら	katsura; kazura
wiggler (mosquito larva) *[i-zoo]*	孑孒; ぼうふら	bōfura; bōfuri
wild animal *[v-zoo]*	野獣; 野獸	yajū
wild boar (animal) *[v-zoo]*	猪; いのしし	inoshishi
wildcat (animal) *[v-zoo]*	山猫; やまねこ	yama-neko
wild cocoon *[i-zoo]*	野蚕繭; やさんけん	yasan'ken
wildflower *[bot]*	野の花	no no hana
wild goose (bird) *[v-zoo]*	雁; がん; かり	gan; kari
wild grass; wildflower *[bot]*	野草	yasō
wildlife *[zoo]*	野生生物	yasei-seibutsu
wild silk *[tex]*	野蚕糸	yasan-shi
wild spinach (vegetable) *[bot]*	藜; あかざ	akaza
wild strain *[microbio]*	野生株	yasei-kabu
wild type *[bio] [gen]*	野生株	yasei-kabu
wild yeast *[mycol]*	野生酵母	yasei-kōbo
willemite *[miner]*	珪酸亜鉛鉱	keisan'aenkō
willow (tree) *[bot]*	柳; やなぎ	yanagi
willow tit (bird) *[v-zoo]*	小雀; こがら	ko-gara
willow warbler (bird) *[v-zoo]*	樺太虫食い	Karafuto-mushi'kui
willow wren (bird) *[v-zoo]*	虫食い	mushi'kui
wilt *{n} [bot]*	萎れ; しおれ	shiore
" (plant disease) *{n} [bot]*	立枯れ病	tachi'gare'byō
wilted *{adj} [bot]*	萎びた	shinabita
winch *{n} [mech-eng]*	巻揚げ機	maki'age-ki
wind *{n} [meteor]*	風; かぜ	kaze
" (wintry blast) *{n} [meteor]*	凩; 木枯らし	ko'garashi
wind; coil; roll *{vb t}*	巻く; 捲く	maku
wind and rain *[meteor]*	風雨; ふうう	fū'u
windbreak *{n} [eng]*	防風塀	bōfū-hei
wind-chill index *[meteor]*	風冷え指数	kaze'bie-shisū
wind direction *[meteor]*	風向	fūkō; kaza-muki
wind erosion *[geol]*	風食; 風蝕	fū-shoku
wind force *[meteor]*	風力	fū-ryoku
winding *{n} [elec]*	巻線; 巻線	maki'sen
wind instruments *[music]*	管楽器; 管樂器	kan-gakki
windlass *[nav-arch]*	揚錨機	yōbyō-ki
windmill *[mech-eng]*	風車	fūsha; kaza-guruma
window *[arch]*	窓; まど	mado
window frame *[arch]*	窓枠	mado-waku
window pane *[arch] [mat]*	窓硝子; 窓ガラス	mado-garasu

windpipe (trachea) *[anat]*	喉笛；のどぶえ	nodo'bue
wind power plant *[mech-eng]*	風力発電所	fū'ryoku-hatsu'den'sho
wind pressure *[mech]*	風圧；風壓	fū-atsu
windshield *[eng]*	防風ガラス	bōfū-garasu
windspout *[meteor]*	旋風	senpū
windstorm *[meteor]*	暴風	bōfū
wind tunnel *[eng]*	風胴	fūdō
wind velocity *[meteor]*	風速；ふうそく	fū-soku
windward *{n} [meteor]*	風上	kaza'kami
" *{adj} [ocean]*	風上の	kaza'kami no
wind wave *[meteor]*	風波；かざなみ	kaza-nami
wine *{n} [cook]*	葡萄酒；ぶどうしゅ	budō'shu
winecup-shaped *{adj} [geom]*	坏状の	sakazuki'jō no
wing *{n} [aero-eng] [v-zoo]*	羽	hane
" (aircraft) *{n}[aero-eng]*	主翼	shu'yoku
" *{n} [aero-eng] [zoo]*	翼；つばさ	tsubasa
" *{n} [aero-eng] [zoo]*	翼；よく	yoku
wing-illuminating light *[aero-eng]*	翼照明灯	yoku-shōmei-tō
wing nut *[mech-eng]*	蝶ねじ	chō-neji
wing span *[aero-eng]*	翼幅	yoku-haba
wingtip light *[aero-eng]*	翼端灯	yoku'tan-tō
winter *{n} [astron]*	冬；ふゆ	fuyu
" *{n} [astron]*	冬期	tōki
winter crop; off season crop *[agr]*	裏作；うらさく	ura-saku
wintergreen oil *[org-ch]*	冬緑油	tō'ryoku-yu
winter hardiness *[agr]*	耐冬性	taitō-sei
winter solstice *[astron]*	冬至；とうじ	tōji
wire *{n} [elec] [met]*	針金	hari'gane
" (electric wire) *{n} [mat]*	電線	densen
wiredrawing *{n} [met]*	伸線	shinsen
wire drawing property *[met]*	伸線性	shinsen-sei
wiregrating *{n} [elecmg]*	針金格子	hari'gane-kōshi
wire-net grilling *{n} [food-eng]*	網焼き	ami-yaki
wire netting; wire gauze *{n} [ch]*	金網；かなあみ	kana-ami
wire recording *{n} [elecmg]*	針金磁気録音	hari'gane-jiki-roku'on
wire rod *[mat]*	線材	senzai
wire strand; strand *[mat]*	素線	sosen
wire telegraphy *[comm]*	有線電信	yūsen-denshin
wire telephone *[comm]*	有線電話	yūsen-denwa
wire-wound resistor *[elec]*	巻線抵抗器	maki'sen-teikō-ki
wiring *{n} [elec]*	配線；はいせん	haisen
wiring diagram *[elec]*	配線図	haisen-zu

wisdom tooth *[dent]*	智歯	chishi
wish to do *{vb} [psy]*	為たい; したい	shitai
wish to know *{vb} [psy]*	しりたい	shiritai
wisp of rain *[meteor]*	雨足; あまあし	ama-ashi
wisteria (flower) *[bot]*	藤; ふじ	fuji
withdrawal *[pat]*	取下げ	tori'sage
" *{n}*	撤回	tekkai
" *{n}*	撤去	tekkyo
wither; dry up *{vb} [bot]*	枯れる	kareru
witherite *[miner]*	毒重石	doku'jūseki
within the law *{adj} [legal]*	合法の	gōhō no
withstand voltage *[elec]*	耐電圧	tai'den'atsu
witness *{n} [legal]*	証人	shō'nin
wolf (pl wolves) (animal) *[v-zoo]*	狼; おおかみ	ōkami
wolfsbane *[bot]*	鳥兜; とりかぶと	tori'kabuto
wollasynonite *[miner]*	珪灰石	keikai'seki
wolverine (animal) *[v-zoo]*	屈狸; くずり	kuzuri
woman (pl women) *[bio]*	女性	josei
" *[bio]*	女; おんな	onna
wood *[bot] [mat]*	木; き	ki
" *[bot]*	材; ざい	zai
" *[mat]*	材木	zaimoku
wood-block printing *{n} [graph]*	木版	moku-han
wood borer (insect) *[i-zoo]*	木食虫	ki'kui-mushi
woodcraft *[furn]*	木製工芸	moku'sei-kōgei
woodcutter; logger (a person) *[eng]*	樵夫; きこり	kikori
wood decay *[bot]*	木材腐朽	moku'zai-fukyū
wood-destroying fungus *[mycol]*	木材腐朽菌	moku'zai-fukyū-kin
wooden bridge *[civ-eng]*	木橋	mok(u)kyō
wooden mozaic *[furn]*	寄木細工	yose'gi-zaiku
wooden ship *[nav-arch]*	木船	moku'sen
wood fiber *[bot]*	木材繊維	mokuzai-sen'i
wood flour *[mat]*	木粉; もくふん	moku-fun
woodfree paper *[mat]*	上質紙	jō'shitsu-shi
wood grain *[mat]*	木地; きじ	kiji
wood graining *{n} [eng]*	木目付け	moku'me-zuke
wood meal *[mat]*	木粉	moku-fun
wood oil *[mat]*	桐油	tō'yu
woodpecker (bird) *[v-zoo]*	啄木鳥; きつつき	ki'tsutsuki
wood plug *[eng]*	木栓	moku-sen
wood-preservative oil *[mat]*	木材防腐油	mokuzai-bōfu-yu
wood product *[mat]*	木工製品	mokkō-seihin

wood-rotting fungus *[mycol]*	木材腐朽菌	mokuzai-fukyū-kin
wood screw *[mech-eng]*	木螺子；もくねじ	moku-neji
wood shavings *[mat]*	鉋屑；かんなくず	kanna-kuzu
wood sorrel *[bot]*	酢漿草；かたばみ	katabami
wood spirit *[org-ch]*	木精	moku'sei
wood sugar *[bioch]*	木糖	moku'tō
wood tin *[miner]*	木錫石	moku'shaku-seki
wood vinegar *[org-ch]*	木酢液	moku'saku-eki
woodworking tool *[eng]*	木工具；もっこうぐ	mokkōgu
woody plant *[bot]*	木本植物	moku'hon-shokubutsu
woof *{n} [tex]*	緯糸；横糸	yoko-ito
woofer *[eng-acous]*	低音拡声器	tei'on-kakusei-ki
wool (of sheep) *[tex] [v-zoo]*	羊毛；ようもう	yōmō
wool bleaching *{n} [tex]*	羊毛漂白	yōmō-hyō'haku
woolen cloth *[tex]*	毛織物；けおりもの	ke-ori'mono
wool grease; wool fat; wool wax *[mat]*	羊脂	yōshi
" *[mat]*	羊毛臘	yōmō-rō
wool hair *[bot]*	綿毛；めんもう	men'mō
wool scouring *{n} [tex]*	羊毛精練	yōmō-seiren
wool sorting *{n} [tex]*	選毛；選毛	senmō
wool wax; wool grease *[mat]*	羊毛蠟	yōmō-rō
wool yarn *[tex]*	毛糸	ke'ito
word *[gram]*	語；ご	go
" *[gram]*	単語	tango
" *[ling] [gram]*	言葉；ことば	kotoba
word feeling *{n} [ling]*	語感	gokan
wording *{n} [comm]*	用語	yōgo
word line *[comput]*	語線	go-sen
word meaning *{n} [comm]*	字義；じぎ	jigi
word origin *[comm]*	字源	jigen
word processor *[comput]*	ワープロ	wāpuro
word time *[comput]*	語時間	go-jikan
work *{n} [mech]*	仕事	shi'goto
" *{n}*	作業	sagyō
work (toil; labor) *{vb}*	働く；はたらく	hataraku
" *{vb}*	作業する	sagyō suru
(act upon) *{vb}*	作用する	sayō suru
" *{vb}*	仕事する	shi'goto suru
workability *[mat]*	加工性	kakō-sei
worked traces *[met]*	加工痕跡	kakō-konseki
worker ant (insect) *[i-zoo]*	働き蟻	hataraki-ari
worker bee (insect) *[i-zoo]*	働き蜂	hataraki-bachi

work function *[sol-st]*	仕事関数	shigoto-kansū
work hardening *{n} [met]*	加工硬化	kakō-kōka
working *{n} [ind-eng]*	作動	sadō
" *{n} [pat]*	実施	jisshi
" ; work; labor *{n}*	働き；はたらき	hataraki
working diameter *[math]*	加工径	kakō-kei
working load *[eng]*	使用荷重	shiyō-kajū
working property *[ind-eng]*	作業性	sagyō-sei
working stress *[met]*	加工応力	kakō-ō'ryoku
work input *[mech]*	入力；にゅうりょく	nyū-ryoku
work output *[mech]*	出力	shutsu-ryoku
workpiece *[eng]*	工作物	kōsaku'butsu
" (testpiece) *[ind-eng]*	試片	shihen
workpiece material *[mech-eng]*	被加工物	hi-kakō'butsu
work quality *[ind-eng]*	作業質	sagyō-shitsu
work queue entry *[comput]*	作業待行列項目	sagyō-machi'gyō'retsu-kōmoku
workroom *[arch]*	作業室	sagyō-shitsu
workshop *[ind-eng]*	工作場	kōsaku'ba
" *[mech-eng]*	仕事場；しごとば	shigoto'ba
worksite *[ind-eng]*	現場	genba
workstation *[comput]*	作業端末	sagyō-tan'matsu
work transformation *[met]*	加工変態	kakō-hentai
work well; be very effective *{adj}*	よく効く	yoku kiku
world *[geogr]*	世界；せかい	sekai
world intellectual property *[pat]*	世界知的所有権	sekai-chi'teki-shoyū'ken
world line *[rela]*	世界線	sekai-sen
worm *{n} [i-zoo]*	虫；蟲；むし	mushi
wormwood; Artemesia absinthium *[bot]*	苦艾；にがよもぎ	niga'yomogi
" *[bot]*	蓬；艾；蒿；よもぎ	yomogi
worry; concern *{n} [psy]*	心配	shinpai
worsted yarn *[tex]*	梳毛糸	somō-shi
wort *[cook]*	麦汁；麥汁	baku'jū
wound; injury *{n} [med]*	傷；創；きず	kizu
" *{n} [med]*	創傷	sōshō
wound iron core *[elecmg]*	巻鉄心；巻鉄芯	maki-tesshin
wound profile *[tex]*	巻形態	maki-keitai
woven cloth *[tex]*	織布	ori-nuno
" *[tex]*	製織布	seishoku-nuno
woven fabric *[tex]*	布帛；ふはく	fu'haku
" *[tex]*	織物；おりもの	ori'mono
wrap; encase *{vb t}*	包む；包む	tsutsumu
wraparound *{n} [comput]*	循環	junkan

wrapping cloth square *[mat]*	風呂敷; ふろしき	furo'shiki
wrasse; seawife (fish) *[v-zoo]*	遍羅; べら	bera
wren (bird) *[v-zoo]*	鷦鷯; みそさざい	misosazai
wrench *{n} [eng]*	スパナ	supana
wring *{vb}*	絞る; しぼる	shiboru
wringing rate *[tex]*	絞り率	shibori-ritsu
wrinkle *{n} [tex]*	皺; しわ	shiwa
wrist *[anat]*	手首	te-kubi
wristwatch *[horol]*	腕時計; うでどけい	ude-dokei
writability *[graph]*	筆記性	hikki-sei
write *{vb} [comput]*	書き出す	kaki-dasu
" *{vb} [comput]*	書き込む	kaki-komu
" *{vb} [comm]*	書く; かく	kaku
write protection *[comput]*	書き込み保護	kaki'komi-hogo
writing (to write) *{n} [comm]*	書く事	kaku koto
" (writing-out) *{n} [comput]*	書き出し	kaki'dashi
" (writing-in) *{n} [pr]*	書き込み	kaki'komi
writing pad *[comm]*	便箋; びんせん	binsen
written application *[legal]*	申請書	shinsei'sho
written language *[comm]*	文語	bungo
written notification *[legal]*	届け書; 届け書	todoke'sho
written response *[comm]*	答弁書; 答辯書	tōben'sho
written statement *[legal]*	陳述書	chin'jutsu'sho
wrong number *[comm]*	誤り番号	ayamari-bangō
wrong operation *[ind-eng]*	誤作動; ごさどう	go-sadō
wrought alloy *[met]*	錬合金	ren-gōkin
wrought iron *[met]*	錬鉄; 錬鐵	ren-tetsu
wrought steel *[met]*	錬鋼; れんこう	renkō
wrought tool *[eng]*	打物	uchi-mono
wrought tool steel *[mat]*	打刃物鋼	uchi-ha'mono-kō
wulfenite *[miner]*	黄鉛鉱	kōen'kō
" *[miner]*	水亜鉛鉱	sui'aen'kō
wurtzite *[miner]*	繊維亜鉛鉱	sen'i-aen'kō

X

xanthating machine; xanthator *[eng]*	硫化機；りゅうかき	ryūka-ki
xanthism *[bio]*	皮膚黄変	hifu-ō'hen
xanthophore *[cyt]*	黄色素胞	ō'shoku-sohō
xanthoxylum (tree) *[bot]*	山椒；さんしょう	sanshō
x axis *[an-geom]*	エックス軸	ekkusu-jiku
xenia *[bot]*	海薊；うみあざみ	umi-azami
xeniophyte *[bot]*	胚乳体	hainyū'tai
xenoantigen *[immun]*	異種抗原	i'shu-kōgen
xenocryst *[crys]*	外來結晶	gairai-kesshō
" *[crys]*	捕獲結晶	ho'kaku-kesshō
xenograft *[immun]*	異種移植片	ishu'ishoku-hen
xenolith *[petr]*	捕獲岩；ほかくがん	hokaku'gan
" *[petr]*	捕虜岩；ほりょがん	horyo'gan
xenon (element)	キセノン	kisenon
xenotransplantation *[immun]*	異種移植	ishu-i'shoku
xeric *{adj}* *[ecol]*	好乾性の	kōkan'sei no
Xeriscape *[bot]*	乾景；かんけい	kankei
xerogel *[ch]*	乾膠体；乾膠體	kankō'tai
" *[ch]*	キセロゲル	kiserogeru
xerographic printer *[pr]*	電子写真式印書装置	denshi-shashin'shiki-insho-sōchi
xerography *[graph]*	静電写真法	seiden-shashin-hō
xeromorphism *[bot]*	乾生形態	kansei-keitai
xerophilic *{adj}* *[bio]*	好乾性の	kōkan'sei no
xerophyte *[ecol]*	乾性植物	kansei-shokubutsu
xeroradiography *[graph]*	乾式エクス線撮影	kan'shiki-ekusu'sen-satsu'ei
" *[graph]*	乾燥放射線写真	kansō-hōsha'sen-shashin
xerosere *{n}* *[ecol]*	乾性(遷位)系列	kansei(-sen'i)-kei'retsu
x intercept *[an-geom]*	エックス切片	ekkusu-seppen
xiphoid *{adj}* *[anat]*	剣状の；劍狀の	kenjō no
x-rays *[phys]*	エ(ッ)クス線	ekkusu'sen (ekusu'sen)
x-ray small-angle scattering *[phys]*	エックス線小角散乱	ekkusu'sen-shōkaku-sanran
xylem *[bot]*	木部；もくぶ	moku'bu
xylem fiber *[bot]*	木部繊維；木部纖維	mokubu-sen'i
xylem ray *[bot]*	木部放射組織	mokubu-hōsha-so'shiki
xylem parenchyma *[bot]*	木部柔組織	mokubu-jū'soshiki
xylose; wood sugar *[bioch]*	木糖；もくとう	moku'tō

Y

yacht *{n} [nav-arch]*	ヨット	yotto
" *{n} [nav-arch]*	快走船	kaisō'sen
yak (animal) *[v-zoo]*	犛牛；ぼうぎゅう	bōgyū
yam (vegetable) *[bot]*	山の芋	yama-no-imo
" (vegetable) *[bot]*	八つ頭	yatsu'gashira
yamase *[meteor]*	山背；やませ	yama'se
yard *[arch] [bot]*	庭	niwa
" (unit of length) *[mech]*	碼；ヤード；ヤール	yādo; yāru
" (shipyard) *[mech-eng]*	造船所	zōsen'jo
yardage *{n} [mech]*	ヤード尺計算数	yādo'jaku-keisan'sū
yard lumber; cut lumber *[mat]*	挽材；ひきざい	hiki-zai
yarn *[tex]*	糸；絲；いと	ito
yarn breakage *[tex]*	糸切れ	ito-gire
yarn bundling machine *[eng]*	綛締め機	kase'jime-ki
yarn (count) number *[tex]*	番手；ばんて	ban-te
yarn-dyed fabric *[tex]*	糸染め織物	ito'zome-ori'mono
yarn-dyed silk satin *[tex]*	煉絹朱子	neri'ginu-shusu
yarn-feeding *{n} [tex]*	給糸	kyūshi
yarn scouring *[tex]*	糸煉り；糸練り	ito-neri
yarn steaming *[tex]*	糸蒸し	ito-mushi
yarn twisting machine *[eng]*	撚糸機	nenshi-ki
yarn unevenness *[poly-ch] [tex]*	糸斑；いとむら	ito-mura
yaw *{n} [mech]*	偏揺；偏搖	hen'yō
yaw(ing) *{n} [mech]*	船首揺れ	senshu-yure
yawmeter *[eng]*	片揺れ計	kata'yure-kei
" *[meteor]*	風向計	kaza'muki-kei
yawn *{n} [physio]*	欠伸；あくび	akubi
" *{vb} [physio]*	欠伸をする	akubi o suru
y axis *[an-geom]*	ワイ軸	wai-jiku
year *[astron]*	年；ねん；とし	nen; toshi
yearbook; almanac *[pr]*	年鑑；ねんかん	nenkan
year-end date *{n}*	年末日	nen'matsu-jitsu
yearly *{n} {adv}*	毎年	mai-toshi; mai'nen
" *{adj}*	一年の	ichi'nen no
yearly load curve *[ind-eng]*	年負荷曲線	nen-fuka-kyoku'sen
yearly load factor *[ind-eng]*	年負荷率	nen-fuka-ritsu
year name; name of era *{n}*	年号；年號	nengō
yeast *[mycol]*	酵母(菌)	kōbo(-kin)
yeast culture *[mycol]*	酒母；しゅぼ	shubo

yeast extract *[mycol]*	酵母エキス	kōbo-ekisu
yellow (color) *[opt]*	黄(色)；き(いろ)	ki(-iro)
yellow beeswax *{n} [pharm]*	蜜蠟；みつろう	mitsu-rō
yellow colorants *[mat]*	黄色系着色剤	kō'shoku'kei-chaku'shoku-zai
yellow earth *[agr] [geol]*	黄色土	ō'shoku-do
yellow fever *[med]*	黄熱；おうねつ	ō-netsu
yellow fever mosquito *[i-zoo]*	熱帯縞蚊	nettai-shima-ka
yellowfin tuna (fish) *[v-zoo]*	黄肌(鮪)	ki'hada(-maguro); kiwada
yellow flag *[med] [nav-arch]*	検疫旗；けんえきき	ken'eki-ki
yellow fog *[photo]*	黄色カブリ	kō'shoku-kaburi
yellow-green (color) *[opt]*	黄緑色	kō'ryoku-shoku
yellow-headed blackbird (bird) *[v-zoo]*	黄頭椋鳥擬；きがし =らむくどりもどき	ki'gashira-mukudori'modoki
yellowing; burning *{n} [met]*	焼け；焼け；やけ	yake
" *{n} [opt] [poly-ch] [tex]*	黄変	kōhen; ō'hen
yellow iris (flower) *[bot]*	黄菖蒲；きしょうぶ	ki-shōbu
yellow jacket (insect)*[i-zoo]*	頬長雀蜂；ほおなが =すずめばち	hō'naga-suzume-bachi
yellow phosphorus *[ch]*	黄燐；おうりん	ō-rin
yellow prussiate of potash *[ch]*	黄血塩	ō'ketsu-en
" *[ch]*	黄血カリ	ō'ketsu-kari
yellow prussiate of soda *[ch]*	黄血ソーダ	ō'ketsu-sōda
yellow sand *[geol]*	黄砂；こうさ	kōsa
yellowtail (fish) *[v-zoo]*	鰤；ぶり	buri
" (young fish) *[v-zoo]*	魬；はまち	hamachi
" (young fish) *[v-zoo]*	鰍；いなだ	inada
" ; hardtail (fish) *[v-zoo]*	縞鯵；しまあじ	shima-aji
yellow warbler (bird) *[v-zoo]*	黄色虫食い	ki'iro-mushi'kui
yellow zinc flowers *[mat]*	亜鉛華黄	aen'ka-ki
yen (Japanese currency) *[econ]*	円；圓；えん	en
yen exchange *[econ]*	円為替；圓為替	en-kawase
yesterday *{adv} {n}*	昨日	kinō; saku'jitsu
yesterday evening *{adv} {n}*	昨夜	saku'ya; yūbe
yew; Japanese yew (tree) *[bot]*	一位；水松；いちい	ichi'i
" , Western (tree) *[bot]*	西洋一位	seiyō-ichi'i
yew leaf *[bot]*	一位葉	ichi'i-yō
yield (yield percentage) *{n} [ind-eng]*	収率；收率	shū'ritsu
" (yield quantity) *{n} [ind-eng]*	収量；しゅうりょう	shūryō
" (yield rate) *{n} [ind-eng]*	歩留り；ぶどまり	bu'domari
yielding; yield *{n} [mech]*	降伏；こうふく	kō'fuku
yielding properties *[mech]*	降伏特性	kōfuku-tokusei
yield percentage *[ind-eng]*	歩留り率	bu'domari-ritsu

yield point *[mech]*	降伏点; 降伏點	kōfuku-ten
yield point in shear *[mech]*	剪断降伏点	sendan-kōfuku-ten
yield point phenomena *[phys]*	降伏点現象	kōfuku-ten-genshō
yield sign *[traffic]*	譲れ標識	yuzure-hyō'shiki
yield strain *[mech]*	降伏歪み	kōfuku-hizumi
yield strength *[mech]*	降伏強度	kōfuku-kyōdo
" *[mech] [met]*	耐力(強度)	tai'ryoku(-kyōdo)
yield stress *[mech]*	降伏応力	kōfuku-ō'ryoku
yield surface *[mech]*	降伏面	kōfuku-men
yield value *[mech]*	降伏値; こうふくち	kōfuku-chi
" *[mech]*	降伏価	kōfuku-ka
y intercept *[an-geom]*	ワイ切片	wai-seppen
yoghurt; yogurt *[cook]*	ヨーグルト	yōguruto
yoke *[arch]*	窓上枠	mado-ue'waku
" *[eng]*	枠	waku
" *[elecmg]*	継鉄; けいてつ	kei'tetsu
yoke joint *[mech]*	枠継手; わくつぎて	waku-tsugi'te
Yokohama *[geogr]*	横浜; 横濱	Yokohama
yolk (of egg) *[bioch]*	黄身; きみ	kimi
" (of egg) *[bioch]*	卵黄; らんおう	ran'ō
yolk sac *[embryo]*	卵黄嚢	ran'ō-nō
Yoshino cherry tree (tree) *[bot]*	吉野桜; 吉野櫻	Yoshino-zakura
Yoshino paper (tissue paper) *[paper]*	吉野紙	Yoshino-gami
young greens (vegetable) *[cook]*	若菜	waka'na
young ice *[hyd]*	新海氷	shin-kaihyō
young seedling (rice) *[bot]*	若苗; わかなえ	waka-nae
Young's modulus *[mech]*	縦弾性係数	tate-dansei-keisū
" *[mech]*	縦弾性率	tate-dansei-ritsu
Young's modulus in flexure *[mech]*	曲げ弾性率	mage-dansei-ritsu
young tree *[bot]*	幼木; ようぼく	yō'boku
" *[bot]*	若木; わかぎ	waka-gi
youth *[bio]*	青年期	sei'nen-ki
ytterbium (element)	イッテルビウム	itterubyūmu
yttrium (element)	イットリウム	ittoryūmu
yugawalite *[miner]*	湯河原沸石	yugawara-fusseki
Yukawa interaction *[nuc-phys]*	湯川型相互作用	Yukawa'gata-sōgo-sayō
Yukawa potential *[nuc-phys]*	湯川型ポテンシャル	Yukawa-gata-potensharu
Yuzen dyeing; Yuzen printing *[tex]*	友禅染め	yūzen-zome
Yuzen thickening *[n] [tex]*	友禅糊	yūzen-nori

Z

Z (the letter)	ゼット	zetto
z plane *[math]*	ゼット平面	zetto-heimen
z transformation *[math]*	ゼット変換	zetto-henkan
Zacco platypus (a type of carp) (fish) *[v-zoo]*	追河; おいかわ	oi'kawa
zamboa; pomelo (fruit) *[bot]*	朱欒; ざぼん	zabon
zealously; assiduously *{adv}*	鋭意	ei'i
zebra (animal) *[v-zoo]*	縞馬; しまうま	shima'uma
zebra crossing *[traffic]*	縞横断歩道	shima-ō'dan-hodō
zelkova (tree) *[bot]*	欅; けやき	keyaki
zenith *[astron]*	天頂; てんちょう	tenchō
zenith attraction *[astron]*	天頂引力	tenchō-in'ryoku
zenith distance *[astron]*	天頂距離	tenchō-kyori
zenith telescope *[opt]*	天頂儀	tenchō-gi
zeolite *[miner]*	沸石; ふっせき	fusseki
zero (pl zeros, zeroes) *[arith]*	零; れい; ぜろ	rei; zero
zero adjustment *[eng]*	零位調整	rei'i-chōsei
zero correction *[math]*	零位補正	rei'i-hosei
zero defects *[ind-eng]*	無欠点運動	mu'ketten-undō
zero gravity *[mech]*	無重力状態	mu-jūryoku-jōtai
zero line *[geophys]*	零線	reisen
" *[mech]*	基準線	kijun-sen
zero-order reaction *[p-ch]*	零次反応	rei'ji-hannō
zero point *[math]*	零点; 零點	rei-ten
zero potential *[phys]*	零電位; れいでんい	rei-den'i
zero-pressure molding *{n}* *[eng]*	無圧成形; 無壓成形	mu'atsu-seikei
zero state; zero condition *[comput]*	零状態	zero-jōtai
zeunerite *[miner]*	砒銅ウラン鉱	hidō'uran'kō
zigzag chain stitch *[cl]*	千鳥縫い	chidori-nui
zigzag planting *{n}* *[agr]*	千鳥植え	chidori-ue
zigzag rule (folding ruler) *[eng]*	折り畳み尺	ori'tatami-jaku
zigzag seaming *{n}* *[cl]*	継ぎ合せ縫い	tsugi'awase-nui
zigzag-striped leafhopper (insect) *[i-zoo]*	稲妻横這; いな=ずまよこばい	ina'zuma-yoko'bai
zinc (element)	亜鉛; 亞鉛; あえん	a'en
zinc arsenide	砒化亜鉛	hika-aen
zincate *[ch]*	亜鉛酸塩; 亞鉛酸鹽	aen'san-en
zinc black *[ch]*	亜鉛黒; あえんぐろ	aen-guro
zincblende *[miner]*	閃亜鉛鉱; 閃亞鉛鑛	sen'aen-kō

zinc borate	硼酸亜鉛	hōsan-aen
zinc borofluoride	硼弗化亜鉛	hō'fukka-aen
zinc bromide	臭化亜鉛	shūka-aen
zinc carbonate	炭酸亜鉛	tansan-aen
zinc chloride	塩化亜鉛	enka-aen
zinc dust *[met]*	亜鉛末; あえんまつ	aen-matsu
zinc fluoride	弗化亜鉛	fukka-aen
zinc hot dipping *(n) [met]*	溶融亜鉛めっき	yōyū-aen-mekki
zinc hydroxide	水酸化亜鉛	suisan'ka-aen
zincic acid	亜鉛酸; あえんさん	aen-san
zincite *[miner]*	紅亜鉛鉱	kō'aen'kō
zinc oxide	酸化亜鉛	sanka-aen
zinc peroxide	過酸化亜鉛	ka'sanka-aen
zinc phosphate	燐酸亜鉛	rinsan-aen
zinc phosphide	燐化亜鉛	rinka-aen
zinc point *[ch]*	亜鉛点	aen-ten
zinc silicate	珪酸亜鉛	keisan-aen
zinc silicofluoride	珪弗化亜鉛	kei'fukka-aen
zinc sulfate	硫酸亜鉛	ryūsan-aen
zinc sulfide	硫化亜鉛	ryūka-aen
zinc sulfite	亜硫酸亜鉛	a'ryūsan-aen
zinc yellow *[mat]*	亜鉛黄; あえんこう	aen-kō
" *[miner]*	紅柱石	kōchū-seki
zinc white; Chinese white *[ch]*	亜鉛華	aen-ka
zingiber (Japanese ginger) *[bot]*	茗荷; みょうが	myōga
zip code *[comm]*	郵便番号	yūbin-bangō
zircon *[miner]*	風信子鉱	fū'shinshi'kō
zirconium (element)	ジルコニウム	jirukonyūmu
zodiac *[astron]*	獣帯; 獸帶	jūtai
" *[astron]*	黄道十二宮	kōdō-jūni'kyū
zodiacal light *[astron] [geophys]*	黄道光	kōdō-kō
zoisite *[miner]*	黝簾石	yū'renseki
zonal harmonics *[math]*	帯球調和関数	taikyū-chōwa-kansū
zonal index *[meteor]*	東西指数	tōzai-shisū
zonal wind *[meteor]*	帯状風; 帶狀風	taijō-fū
zonation *[geol]*	帯状斑紋	obi'jō-hanmon
zone *[crys]*	晶帯; しょうたい	shōtai
" *[elec]*	帯域	tai'iki
" *[geogr]*	地帯	chitai
" *[geogr]*	帯; 帶; たい	tai
" *[geol]*	岩帯	gantai
" *[meteor]*	圏; 圈; けん	ken

zone axis *[crys]*	晶帯軸	shōtai-jiku
zone heating *[eng]*	区域暖房	ku'iki-danbō
" *[eng]*	帯域加熱	tai'iki-ka'netsu
zone, law of *[crys]*	晶帯の法則	shōtai no hōsoku
zone of avoidance *[astron]*	不透視域	fu'tōshi-iki
zone of silence *[acous]*	非音響範囲	hi'onkyō-han'i
zone of totality; zone of total eclipse *[astron]*	皆既地帯；かいき＝ちたい	kaiki-chitai
zone time *[astron]*	経帯時；けいたいじ	keitai-ji
zoning *{n}* *[civ-eng]*	地域制	chi'iki-sei
" (of ore) *{n}* *[met]*	帯状分布	taijō-bunpu
zoning law *[legal]*	土地使用制限法	tochi-shiyō-seigen-hō
zoo *[arch]* *[zoo]*	動物園	dōbutsu'en
zoochemistry *[ch]*	動物化学	dōbutsu-kagaku
zoodynamics *[bio]*	動物力学	dōbutsu-rikigaku
zoogeography *[bio]*	動物地理学	dōbutsu-chiri'gaku
zoographer *[pr]*	動物記載学者	dōbutsu-kisai-gakusha
zoography *[pr]*	動物誌；どうぶつし	dōbutsu'shi
zoology *[bio]*	動物学	dōbutsu'gaku
zoophagous *{adj}* *[zoo]*	動物食性の	dōbutsu'shoku'sei no
zooplankton *[ecol]*	浮遊動物	fuyū-dōbutsu
zooplasty; zoografting *[med]*	動物組織移植法	dōbutsu-soshiki-i'shoku-hō
zoosphere *[ecol]*	動物圏	dōbutsu-ken
zoospore *[bio]*	遊走子；ゆうそうし	yūsō'shi
zori (Japanese footwear) *[cl]*	草履；ぞうり	zōri
zygomorphy *[bot]*	左右相称；左右相稱	sayū-sōshō
Zygomycetes *[mycol]*	接合菌類	setsugō'kin-rui
zygosperm; zygospore *[embryo]*	接合胞子	setsugō-hōshi
zygote *[embryo]*	接合子；せつごうし	setsugō'shi
zygotic induction *[microbio]*	接合誘発	setsugō-yū'hatsu
zymochemistry *[ch]*	発酵化学	hakkō-kagaku
zymogen *[bioch]*	酵素前駆体	kōso-zenku'tai
zymolysis *[bioch]*	酵素分解	kōso-bunkai
zymometer *[microbio]*	発酵計	hakkō-kei
zymosis; fermentation *[microbio]*	発酵；はっこう	hakkō
zymurgy *[microbio]*	醸造学	jōzō'gaku